ISBN 978-0-265-65194-0
PIBN 10876200

LONDON, EDINBURGH, AND DUBLIN

PHILOSOPHICAL MAGAZINE

AND

JOURNAL OF SCIENCE.

CONDUCTED BY

SIR DAVID BREWSTER, K.H. LL.D. F.R.S.L. & E.

SIR ROBERT KANE, M.D., F.R.S., M.R.I.A.

WILLIAM FRANCIS, Ph.D. F.L.S. F.R.A.S. F.C.S.

JOHN TYNDALL, F.R.S. &c.

"Nec aranearum sane textus ideo melior quia ex se fila gignunt, nec noster vilior quia ex alienis libamus ut apes." Just. Lips. *Polit.* lib. i. cap. 1. Not.

VOL. XXI.—FOURTH SERIES.
JANUARY—JUNE, 1861.

LONDON.

TAYLOR AND FRANCIS, RED LION COURT, FLEET STREET,
Printers and Publishers to the University of London;

SOLD BY LONGMAN, GREEN, LONGMANS, AND ROBERTS; SIMPKIN, MARSHALL
AND CO.; WHITTAKER AND CO.; AND PIPER AND CO., LONDON:—
BY ADAM AND CHARLES BLACK, AND THOMAS CLARK,
EDINBURGH; SMITH AND SON, GLASGOW; HODGES
AND SMITH, DUBLIN; AND PUTNAM,
NEW YORK.

. "Meditationis est perscrutari occulta; contemplationis est admirari perspicua . . . . . Admiratio generat quæstionem, quæstio investigationem, investigatio inventionem."—*Hugo de S. Victore.*

—"Cur spirent venti, cur terra dehiscat,
Cur mare turgescat, pelago cur tantus amaror,
Cur caput obscura Phœbus ferrugine condat,
Quid toties diros cogat flagrare cometas ;
Quid pariat nubes, veniant cur fulmina cœlo,
Quo micet igne Iris, superos quis conciat orbes
Tam vario motu."

J. B. Pinelli ad Mazonium.

CONTENTS OF VOL. XXI.

The Rev. H. Mitchell on the position of the Beds of the

NUMBER CXLII.—JUNE.

the preceding eight aggregates become developed into thirty-two, which together constitute the exhaustive Table in question.

Page 376, line 2, *for* eight *read* thirty-two.

Fig. 1 ig: 3.

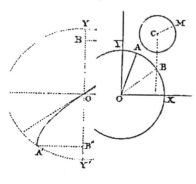

Fig: 4.

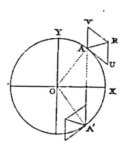

Fig: 7.

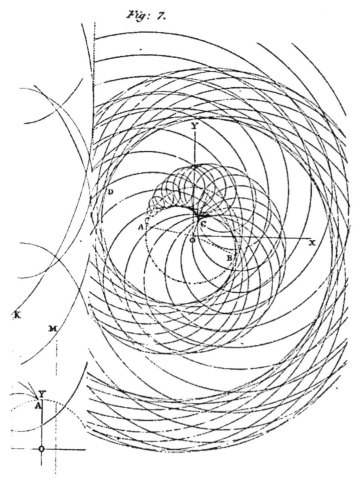

THE

LONDON, EDINBURGH AND DUBLIN

PHILOSOPHICAL MAGAZINE

AND

JOURNAL OF SCIENCE.

[FOURTH SERIES.]

JANUARY 1861.

I. *On Ripples, and their relation to the Velocities of Currents.* By T. ARCHER HIRST, *Mathematical Master at University College School, London*.*

[With a Plate.]

1. ALTHOUGH we are all familiar with the ripples which solid bodies produce upon the surface of a stream in which they are partially immersed, their precise nature and their relation to the velocity of the current appear to have received but little investigation. The reason of this is no doubt to be sought in the well-known difficulties presented by the hydrodynamical problem whose solution is here, strictly speaking, involved. Upon this problem Newton, Laplace, Lagrange, Poisson, Cauchy, and others have expended the greatest analytical power and mathematical skill, and in every case the inherent difficulties of the subject have compelled them to introduce hypotheses and restrictions which more or less vitiate the results at which they at length arrived. The brothers Weber, again, by their elaborate researches, have shown that, in the experimental investigation of the problem in question, difficulties of equal magnitude are encountered.

But instead of considering the phenomena of ripples as a particular case of this general and complicated hydrodynamical problem, the question arises, can we not in some more direct and simple manner arrive at the general relation which must exist between these beautifully symmetric ripples, and the velocities of the several parts of the current upon whose surface they are produced. Professor Tyndall, in his recent work 'On the Glaciers of the Alps' (p. 398), has, in fact, prepared the way for us, in his chapter 29, "On the ripple theory of the veined structure of

* Communicated by the Author.

glaciers,' by giving an exceedingly clear and simple explanation of the origin of ripples on the surfaces of streams. In the present paper I propose to pursue the subject somewhat further than he found it necessary to do, and to put his views into a mathematical form.

2. When a spherical body—or a drop of water—falls upon the surface of still water, a system of concentric and circular waves are formed around the point of impact. The foremost of these waves generally exceeds the rest in magnitude, and being on that account most visible, will be referred to as *the* wave : its height and breadth, as well as the velocity λ with which it recedes from the point of impact, all depend upon the magnitude of the body, and the height from which it fell. According to Weber[*], this velocity λ of propagation varies also with the time, or, more strictly, decreases as the radius of the circle formed by the wave increases : this variation, however, is admitted by Weber to be small[†], and according to Poisson's calculations, has no existence[‡]. In the present paper this possible variation of λ is not overlooked, although, to obtain definite results capable of being compared with those of experiment, λ is often treated as a constant; the error incurred by so doing being rendered less important by the circumstance that the waves with which we shall then be concerned cease, in reality, to be visible before their radii have reached any great magnitude.

3. If we suppose the spherical body to fall into a current of water whose velocity v is everywhere the same, the particles forming the surface of the current will still be relatively at rest, and the wave will again be circular in form, the centre of the circle being carried down the current whilst its radius increases with a velocity λ, which we may assume to be the same as before. If the velocity and direction of the current vary from point to point, the circular form of the wave will be destroyed as it floats downwards, and the variations of form through which it will pass will, as Weber remarks, indicate in some measure the variations in the direction and velocity of the current at its several points.

4. Let us next suppose a succession of drops to fall into the stream, the points of impact being fixed in space. Each drop will occasion a wave; and if there be no current, the several waves will form a system of concentric circles around the point of impact; if there be a current, however, and its velocity be not too small, the successive waves will intersect one another, and at the points of intersection the water will be raised to a height exceeding that of either of the intersecting waves. Lastly, if the

---

[*] *Wellenlehre,* p. 182. [†] Ibid. p. 210.
[‡] *Mémoires de l'Acad. Roy. des Sciences de l'Institut,* 1816, vol. i. p. 165. See also *Wellenlehre,* p. 423.

drops succeed each other with sufficient rapidity, in short, if they constitute a jet of water falling into the current, the successive waves will, by their intersection, give rise to a continuous series of points elevated above the general level of the liquid, and forming a *ripple* more visible than the waves which it envelopes. If we replace the jet by a solid cylinder, the effect will be essentially the same: the waves, it is true, may differ in form and even in the velocity with which they are propagated; but, as before, the ripple will be their envelope. It is in this manner that the pebbles and other partially immersed bodies on the banks of a stream give rise to ripples whose forms, as we shall see, indicate in every case the velocities of the adjacent parts of the current. It is scarcely necessary to add that bodies moving on the surface of still water produce precisely similar effects; the ripples caused by boats and water-fowl are examples familiar to all.

5. The relation between the form of the ripple and the velocity of the stream may be easily determined without even knowing the forms of the waves of which the ripple is the envelope. Deferring this determination, however, to art. 10, let us first, for the sake of completeness, consider the following question :—

The initial form of a wave being known, through what variations will it pass as it floats down a stream where the velocity and direction of the current vary from point to point according to a given law?

Let x and y be the coordinates of any point m on the surface of the stream, and let v and α be given functions of x and y, denoting respectively the velocity of the current at the point m, and the angle between its direction and the abscissa axis; the problem is to find the equation

$$y = f(x, t) \quad . \quad . \quad . \quad . \quad . \quad . \quad . \quad (1)$$

of the wave at the expiration of the time t, the equation

$$y = f(x, 0) \quad . \quad . \quad . \quad . \quad . \quad . \quad (2)$$

of the wave at the origin of that time being known.

At the time t in question, the point m of the wave bmc will have two velocities; one, λ, in the direction of the normal mn to the wave, and another, v, in the direction ma determined by the

angle $a\hat{m}X = \alpha$. At the expiration of the element of time dt, therefore, the point m will arrive at the opposite angle m' of a small parallelogram, whose sides $mn = \lambda dt$ and $ma = vdt$ have the directions above defined. If we call x' and y' the coordinates of m', and $\phi = n\hat{m}X$ the angle between the *external* normal

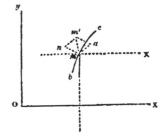

and the *positive* direction of the abscissa axis,

$$x' = x + (\lambda \cos \phi + v \cos \alpha)dt,$$
$$y' = y + (\lambda \sin \phi + v \sin \alpha)dt.$$

But on changing t into $t+dt$, the coordinates x', y' should satisfy the equation (1); hence

$$y + (\lambda \sin \phi + v \sin \alpha)dt = f[x + (\lambda \cos \phi + v \cos \alpha)dt, \quad t+dt].$$

Developing the function on the right, neglecting powers of dt higher than the first, and recalling the equation (1), we have

$$\lambda \sin \phi + v \sin \alpha = \frac{dy}{dx}(\lambda \cos \phi + v \cos \alpha) + \frac{dy}{dt};$$

or, since

$$\sin \phi = \frac{1}{\sqrt{1+\left(\frac{dy}{dx}\right)^2}}, \text{ and } \cos \phi = -\frac{\frac{dy}{dx}}{\sqrt{1+\left(\frac{dy}{dx}\right)^2}},$$

$$\lambda\sqrt{1+\left(\frac{dy}{dx}\right)^2} + v \sin \alpha = v \frac{dy}{dx}\cos \alpha + \frac{dy}{dt}. \quad . \quad . \quad (3)$$

This is the partial differential equation whose general integral will include the equations of all possible waves which can be formed under the given conditions. The arbitrary function which this integral involves will in each case be determined by the known equation (2) of the wave which corresponds to $t=0$. With respect to this equation, however, it must be remembered that λ varies with the nature of the displacement to which the waves owe their origin; in our case it depends upon the magnitude and velocity of the jet—a fact which Weber's experiments[*] establish beyond doubt. Apart from this variation, however, it follows from art. 2 that, even when the jet remains the same throughout, λ *may* vary from point to point of a wave; in other words, it may be a function of x and y. In assuming λ to be constant, therefore, approximate results can alone be expected.

6. If we assume the direction of the current to be everywhere the same, and parallel to the abscissa axis, then α is always zero, and the equation (3) becomes

$$\lambda\sqrt{1+\left(\frac{dy}{dx}\right)^2} = v\frac{dy}{dx} + \frac{dy}{dt}. \quad . \quad . \quad . \quad (4)$$

The velocity v still remains a function of both x and y; but without departing too much from the actual state of things in rectilinear streams, we may regard v as a function of y alone; that is to say, we may suppose the velocity of the current to

[*] *Wellenlehre*, p. 183.

remain the same at the same distance from its banks, and to vary only on *crossing* the stream. Under these conditions the equation (4) is integrable; and if λ be constant, one of its complete integrals will be found to be

$$\int dy \sqrt{(c-v)^2 - \lambda^2} = \lambda(x - ct) + \Phi(c); \quad \cdot \quad \cdot \quad (5)$$

where c is an arbitrary constant, and Φ an arbitrary function. The general integral will result from the elimination of c between this equation and its differential according to c, which is

$$\int \frac{(c-v)dy}{\sqrt{(c-v)^2 - \lambda^2}} + \lambda t = \Phi'(c). \quad \cdot \quad \cdot \quad \cdot \quad \cdot \quad (6)$$

7. The arbitrary function Φ may be determined from the known equation

$$F(x, y) = 0 \quad \cdot \quad \cdot \quad \cdot \quad \cdot \quad \cdot \quad \cdot \quad \cdot \quad (7)$$

of the initial wave in the following manner. Putting $t = 0$ in the equations (5) and (6), we know that the result of eliminating c from the equations

$$\int dy \sqrt{(c-v)^2 - \lambda^2} = \lambda x + \Phi(c), \quad \cdot \quad \cdot \quad \cdot \quad (5a)$$

$$\int \frac{(c-v)dy}{\sqrt{(c-v)^2 - \lambda^2}} = \Phi'(c) \quad \cdot \quad \cdot \quad \cdot \quad \cdot \quad \cdot \quad (6a)$$

must coincide with the equation (7). But eliminating c from these equations is equivalent to replacing c in the first by a function of x and y determined from the second. Now on differentiating (5a) on the hypothesis that c is a function of x and y thus determined, we have, in virtue of (6a),

$$dy \sqrt{(c-v)^2 - \lambda^2} = \lambda \, dx,$$

an equation which ought also to coincide with the result of differentiating (7), that is, with

$$\frac{dF}{dy} dy + \frac{dF}{dx} dx = 0.$$

We conclude then that

$$\frac{dF}{dx} \sqrt{(c-v)^2 - \lambda^2} + \lambda \frac{dF}{dy} = 0. \quad \cdot \quad \cdot \quad \cdot \quad \cdot \quad (8)$$

The result of eliminating x and y from the equations (7), (5a), and (8), therefore, will lead to the required relation between $\Phi(c)$ and c.

8. As an example, let the velocity v of the current be constant, or the same at all points. The complete integral (5) of the equation (4) will then become

$$\sqrt{(c-v)^2 - \lambda^2} = \lambda(x - ct) + c_1, \quad \cdot \quad \cdot \quad \cdot \quad (9)$$

where c_1 is a function of c to be determined from the initial form
of the wave. If we suppose this initial form to be that of a circle
with radius a around the origin, the required relation between c
and c_1 will result from eliminating x and y from the following
equations, to which (7), (5a), and (8) become respectively reduced:

$$\left.\begin{array}{c} x^2+y^2=a^2, \\ y\sqrt{(c-v)^2-\lambda^2}-\lambda x=c_1, \\ x\sqrt{(c-v)^2-\lambda^2}+\lambda y=0. \end{array}\right\} \quad \ldots \quad (10)$$

From these we easily deduce

$$a(c-v)=\pm c_1, \quad \ldots \ldots \ldots \quad (11)$$

by means of which (9) becomes

$$y\sqrt{c_1^2-a^2\lambda^2}=a\lambda(x-vt)+c_1(a\mp\lambda t).$$

Differentiating this according to c_1, we have, corresponding to
(6), the equation

$$\frac{c_1 y}{\sqrt{c_1^2-a^2\lambda^2}}=a\mp\lambda t.$$

The equation of the wave at the end of the time t results from
the elimination of c_1 from these two equations, or from the following two, to which they are equivalent:

$$c_1 y^2=a\lambda(x-vt)(a\mp\lambda t)+c_1(a\mp\lambda t)^2,$$

$$c_1=-a\lambda\frac{a\mp\lambda t}{x-vt}.$$

The result is clearly

$$y^2+(x-vt)^2=(a\mp\lambda t)^2, \quad \ldots \ldots \quad (12)$$

which, as might have been anticipated, is the equation of a circle
whose centre is on the abscissa axis at a distance from the origin
equal to vt—the space described in the time t by each point of
the current—and whose radius, from being a, has become $a+\lambda t$,
in consequence of the propagation of the wave with the constant
velocity λ. The upper sign in (12) is, of course, foreign to the
present inquiry; it refers to the propagation of the wave *inwards*,
a case which is included in the differential equation (4)*.

* It is worthy of notice that when $v=0$, the differential equation (3) or
(4) becomes

$$\left(\frac{dy}{dk}\right)=1+\left(\frac{dy}{dx}\right)^2.$$

where $k=\lambda t$, a relation which all *parallel curves* must satisfy, and which
may be at once obtained from the definition of these curves. If x, y, and
k be regarded as the coordinates of a point in space, the above partial differential equation represents a developable surface generated by a plane

9. Let us next consider the ripple which envelopes a system of waves having at their origin the same position and form. If, as in art. 5,

$$y = f(x, t) \quad . \quad . \quad (1)$$

represent the equation of any wave,

$$y = f(x, t + dt) = f(x, t) + \frac{df}{dt} dt + \&c...$$

will be that of the next preceding wave, and the values of x and y which satisfy both equations, or which satisfy (1) and the equation

$$\frac{dy}{dt} = 0, \quad . \quad . \quad . \quad . \quad . \quad . \quad . \quad . \quad (13)$$

will refer to a point on the required ripple. In short, the equation of this ripple will be the result of the elimination of t from the equations (1) and (13). If we differentiate (1), regarding t as a function of x determined by (13), and use brackets to distinguish partial from complete differential coefficients, we have

$$\frac{dy}{dx} = \left(\frac{dy}{dx}\right) + \left(\frac{dy}{dt}\right)\frac{dt}{dx};$$

hence by (13),

$$\left(\frac{dy}{dx}\right) = \frac{dy}{dx}; \quad . \quad . \quad . \quad . \quad . \quad . \quad (14)$$

an equation which merely expresses the well-known fact, that at their point of contact the wave and ripple have the same tangent. But it was shown in art. 5 that the equation (1) satisfies the partial differential equation (3); and the latter, by means of (13) and (14), becomes transformed into the ordinary differential equation

$$\lambda \sqrt{1 + \frac{dy^2}{dx^2}} + v \sin \alpha = v \frac{dy}{dx} \cos \alpha, \quad . \quad . \quad . \quad (15)$$

which is clearly that of the ripple. If the coordinate axes be turned around the origin until the abscissa axis is parallel to the direction of the current at any point M(xy) of the ripple, then, since $\alpha = 0$ at that point, (15) becomes

---

which is constantly inclined at an angle of 45° to the plane (xy) or axis k, and the sections of this surface made by planes parallel to (xy) will constitute a system of parallel curves. If the *base* of this system be a curve traced on the coordinate plane of (xy), the generating planes of the developable will always touch the same. The developable, in fact, has for its edge of regression a curve which cuts at an angle of 45° all the generators of a right cylinder whose base is the evolute of the curve traced on the plane (xy).

$$\frac{\frac{dy}{dx}}{\sqrt{1+\frac{dy^2}{dx^2}}} = \sin \theta = \frac{\lambda}{v}, \quad \ldots \ldots \quad (16)$$

where θ is the inclination of the ripple to the direction of the current at any point M of the former. We are thus led to the following simple and interesting result :—

At any point of a ripple, the sine of the angle between its direction and that of the current is inversely proportional to the velocity of the latter, and directly proportional to the velocity of propagation of the wave which touches the ripple in that point.

10. This result may be arrived at in a simpler manner. When the velocity and direction of the current remain the same at all points, the waves produced at a point A retain their circular form as they float down the current. If the radii of the several circular waves increase with the same velocity λ, then the ripple, their envelope, will clearly consist of two right lines diverging from A and touching all the circles. If from the centre B of any wave the radius B M be drawn to its point of contact with the ripple, then, since the wave has been propagated over the space B M in the same time that a point of the current has described the space A B, we have clearly

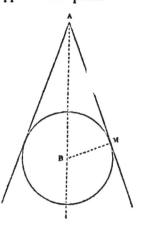

$$\frac{BM}{BA} = \sin BAM = \frac{\lambda}{v}. \quad . \quad (17)$$

If, as Weber asserts (see art. 2), the velocity λ with which the wave is propagated diminishes as its magnitude or radius BM increases, the ripple will no longer be rectilineal, but at the point M of the ripple the law of art. 9 will still hold. To prove this, it is only necessary to consider two immediately succeeding circular waves around B and B′, and from these points to let fall the perpendiculars B M, B′M′ upon their common tangent M M′, which will also be the tangent to the ripple at the point M ;

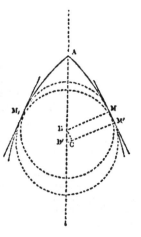

drawing B C parallel to M M', we have, as before,

$$\frac{B'C}{B'B} = \sin B'BC = \sin \theta = \frac{\lambda}{v},$$

where θ is the angle between the directions of the ripple and the current at the point M. The velocity v being constant, θ and λ will decrease simultaneously; so that, according to Weber, the ripple should consist of two curved lines, A M, A M$_1$, turning their concavities towards each other. This property of the wave suggests a crucial experiment as to the variation of λ; to apply it, however, a perfectly uniform current would be required, or what is equally difficult to realize, a jet must be made to describe a right line with perfectly uniform velocity over still water. In ordinary experiments, as will be hereafter seen, the ripple, as long as it remains visible, and as far as the eye can judge, is rectilinear.

Again, it may easily be shown that when the velocity and direction of the current vary from point to point, so as to destroy the circular form of the waves, the law of art. 9 is still fulfilled. In fact, in the immediate vicinity of the point M of the ripple we may regard this velocity and direction as constant; and in place of the non-circular wave, to whose intersection with the immediately preceding and succeeding waves the ripple at M is due, we may substitute a circular one osculating the real wave in M, increasing with the same velocity λ, and moving parallel to the current at M with the velocity v which exists at that point. This fictitious circular wave will clearly produce a ripple coincident with the actual one in the neighbourhood of the point M, and thus the relation (16) between λ, v, and θ will still exist at that point.

11. From the relation

$$\sin \theta = \frac{\lambda}{v},$$

it follows that when λ and v are equal, $\theta = \frac{\pi}{2}$; that is to say,

when the velocity of the current is equal to the velocity with which the wave is propagated, the several waves all touch a line at right angles to the direction of the current; they will consequently touch each other, and the ripple will become reduced to their point of contact. When λ exceeds v, θ is imaginary; in fact, in this case the waves will clearly be propagated up the stream, and will no longer intersect. Strictly speaking, however, this is the case onl when the waves are produced by a discontinuous series of drops; experiment shows that when a solid cylinder or jet is partially immersed in so slow a current, the water flows past it without its surface suffering any visible disturbance. From this it follows that a ripple which has been produced in a

stream, where the velocity diminishes from the centre towards
the sides, will end abruptly as soon as it has reached a point
where that velocity is less than λ; the pebbles on the banks of
such a stream will produce no ripples. In a similar manner, too,
λ may be regarded as a limit beyond which the velocity of a body
moving through still water cannot be increased without visibly
rippling its surface.

12. For the sake of further illustration, let us assume, as in
art. 6, that the direction of the current is everywhere parallel to
the abscissa axis, and that its velocity v varies only with the distance y from this axis. The differential equation (15) of the
ripple then becomes

$$dy \sqrt{v^2 - \lambda^2} = \lambda\,dx, \quad \ldots \ldots \quad (18)$$

whose integral, λ being considered as a constant, is

$$\int dy \sqrt{v^2 - \lambda^2} = \lambda x + C, \quad \ldots \ldots \quad (19)$$

where the constant C will be determined as soon as any point in
the ripple is known. By (18) we can determine v whenever the
form of the ripple is known, and by (19) we may find the equation of the ripple whenever the law in the variation of the current's velocity is given.

For instance, we may determine the nature of the current, the
ripples upon whose surface are parabolas. For in this case the
equation of any ripple being

$$y^2 = 2px,$$

we have at once

$$\frac{dy}{dx} = \frac{\lambda}{\sqrt{v^2 - \lambda^2}} = \frac{p}{y},$$

whence we deduce

$$\frac{v^2}{\lambda^2} - \frac{y^2}{p^2} = 1,$$

and conclude that, to produce parabolic ripples, the velocity of
the current at any distance y from the axis of the parabola will
be represented by the abscissa of a hyperbola having that axis
for transverse axis (2λ), the vertex of the parabola for centre, and
a conjugate axis equal to the parameter of the parabola.

Conversely, if the velocity of the current at any distance y
from the abscissa axis satisfy the relation

$$\frac{v^2}{a^2} + \frac{y^2}{b^2} = 1,$$

that is to say, if this velocity can be represented by the abscissa,
corresponding to the ordinate y, of an ellipse having its centre
at the origin, then the equation of the ripple, as given by (19),

would be

$$\int dy \sqrt{a^2-\lambda^2-c^2y^2}=\lambda x+C,$$

where $c=\dfrac{a}{b}$; on integrating, we find the equation of the ripple in this case to be

$$cy\sqrt{a^2-\lambda^2-c^2y^2}+(a^2-\lambda^2)\ \text{arc sin}\frac{cy}{\sqrt{a^2-\lambda^2}}=2c\lambda x,$$

where the arbitrary constant has vanished on assuming that the ripple passes through the origin. The tangent of the inclination of the ripple to the current at any point x, y has of course the value

$$\frac{dy}{dx}=\frac{\lambda}{\sqrt{a^2-\lambda^2-c^2y^2}},$$

which varies between the minimum limit $\dfrac{\lambda}{\sqrt{a^2-\lambda}}$ at the origin, and the maximum ∞ at the point A, whose coordinates are

$$x=\frac{\pi}{4}\frac{a^2-\lambda^2}{c\lambda}\ \text{and}\ y=\frac{\sqrt{a^2-\lambda^2}}{c}.$$

Further, since the equation is unchanged when x and y are replaced by $-x$ and $-y$, it is evident that the ripple consists of two similar branches in opposite quadrants, and has a point of inflexion at the origin, where it is least inclined to the current. Proceeding from this point, its inclination to the current increases until, at the points A and A', current and ripple are at right angles to each other. Here the latter ends abruptly, in consequence of the velocity of the current having become equal to that with which the wave is propagated (art. 11).

In Plate I. fig. 1, let O X be the direction of the current, and let its velocity at any point of a line parallel to O X be represented by half the chord which is intercepted upon that line by an ellipse, X Y X' Y'. Let the semi-chords A B and A' B' represent the velocity λ with which the wave is propagated. The curve A' O A will then represent the ripple produced by partially immersing a body at any one of its points, as A'. The tangent O C at the point of inflexion O is easily constructed, since it passes at the distance $\lambda=$XC$=$AB from the vertex X*.

13. In order to test by experiment the law enunciated in art. 9, two methods suggest themselves. *First,* to examine the ripples produced by a stationary jet falling into currents of different but known velocities; and *secondly,* to give to the jet or partially immersed body a definite motion, the water being motionless. This second method has many advantages, arising from the fact that the motion of the jet is more under our control than that of a current, which in general varies from point

* Compare the fig. in chapter 29 of ' Glaciers of the Alps.'

to point according to imperfectly known laws. Now of all conceivable motions which might be imparted to the jet, a circular one is beyond doubt most feasible; so that we are naturally led to inquire *what will be the form of the ripple produced by a jet which, as it falls into still water, describes a circle with a given constant velocity* u.

We shall throughout assume the velocity λ with which the waves are propagated to be independent of the magnitude of the latter, and, in accordance with art. 4, we shall seek the envelope of the several waves which the moving jet originates. There is one case where the nature of this envelope can be at once determined: it is when the jet moves with the same velocity λ as the waves. For at the moment when the jet arrives at a point A (fig. 2), the wave which it produced when at B will have acquired a radius equal to the arc A B; instead of intersecting, therefore, every two successive waves around B and B' will touch each other, the difference between their radii being equal to the distance B B' between their centres, and their point of contact A' will be in the tangent to the circle at the point B; in fact, the ripple in this case will be the *involute of the circle* A A' C, or the curve formed by unwrapping, under tension, a string originally wrapped round the circle as far as A. If the velocity of the jet were less than λ, the successive waves would precede the jet, and neither intersect nor touch each other; in other words, there would be no ripple. But if the velocity of the jet exceed λ, the ripple will separate into two branches (fig. 7), one of which will be outside the circle, whilst the other will enter it; but since the waves continue to increase in radius, this latter branch will necessarily leave the circle again, and never re-enter it. In art. 20 we shall find, in fact, that this branch, after approaching to within a certain distance of the centre, suddenly turns and recedes, the turning-point or cusp C being due to the intersection in that point of *three* successive waves. In experiments this cusp is tolerably well defined; and its position is the more important, since it bears a very simple relation to the velocities u and λ; in fact, we shall find that *the distance of the cusp from the centre of the circle described by the jet has to the radius of that circle the same ratio that the velocity* λ *of propagation has to the velocity* u *with which the jet moves.*

14. For the sake of future applications, it will be more convenient to deduce the form of the ripple above considered from the solution of the following more general problem.

A jet which describes, with uniform velocity u, a fixed circle with radius a, falls into a current whose velocity v and direction are the same at every point of its surface; required the form of the ripple.

Let the centre of the circle be taken as the origin of coordinate axes, one of which—the ordinate axis—is parallel to the direction of the current. To fix our ideas, let us suppose, too, that after having rotated for an indefinite period in the direction opposed to that of the hands of a clock, the jet has at length reached the position A (fig. 3) defined by the angle A O X=ψ. The centre of the circular wave which the jet originated when in any position B, will have been carried with the current along the line B C parallel to the ordinate axis, and its radius, from being zero, will have increased to a certain magnitude C M. The arc $\widehat{AB}$, the line B C, and the radius C M being described in the same time, we shall clearly have the proportions

$$\widehat{AB} : \dot{BC} : \overline{CM} = u : v : \lambda ;$$

so that, on representing the angle AOB by ϕ, and the ratios $\dfrac{\lambda}{u}$ and $\dfrac{v}{u}$ by α and β respectively, we shall have

$$\widehat{AB}=a\phi, \quad \overline{CM}=a\alpha\phi, \quad \overline{BC}=a\beta\phi ;$$

and the coordinates ξ, η of the centre C of the circular wave will be

$$\left.\begin{aligned}\xi &= a \cos (\psi-\phi), \\ \eta &= a \sin (\psi-\phi) + a\beta\phi ;\end{aligned}\right\} \quad \dots \quad (20)$$

whilst the equation of this wave will be

$$[x-a\cos(\psi-\phi)]^2 + [y-a\sin(\psi-\phi)-a\beta\phi]^2 - a^2\alpha^2\phi^2 = 0. \quad (21)$$

The equation of the immediately succeeding wave will be obtained from this by changing ϕ into $\phi+d\phi$; the intersections of the two waves will be two points on the ripple, and their coordinates will clearly satisfy the equation (21), as well as its differential according to ϕ. The equation of the ripple, therefore, at the moment the jet reaches the point A defined by the angle ψ, will result from the elimination of ϕ between (21) and the equation

$$[x-a\cos(\psi-\phi)]\sin(\psi-\phi)-[y-a\sin(\psi-\phi)-a\beta\phi]$$
$$[\cos(\psi-\phi)-\beta]+a\alpha^2\phi=0, \quad \dots \dots (22)$$

which is simply that of the chord of intersection of the two successive circular waves. In a similar manner, the equation of the chord of intersection of the second of the above waves, and a third, immediately following the same, will be found by putting $\phi+d\phi$ in place of ϕ in (22); and the coordinates of the point in which these two chords cut each other will satisfy both (22) and its differential according to ϕ, which is

$$[x-a\cos(\psi-\phi)]\cos(\psi-\phi)+[y-a\sin(\psi-\phi)-a\beta\phi]$$
$$\sin(\psi-\phi)+a[1-\alpha^2+\beta^2-2\beta\cos(\psi-\phi)]=0. \quad (23)$$

If the three successive waves intersect in a point, the two chords will also pass through that point, and its coordinates will consequently satisfy, simultaneously, the three equations (21), (22), (23). The position of the *cusps* of the ripple at the moment the jet reaches A will be found, therefore, from the two equations which result from eliminating ϕ from the last three equations. These two equations will of course contain the angle ψ which defines the position of the jet; if the latter be also eliminated, the resulting equation will be that of the curve described by the cusps as the jet rotates.

It will be at once observed that the elimination of ψ and ϕ from the above equations is equivalent to the elimination of the three variables $\sin(\psi-\phi)$, $\cos(\psi-\phi)$, and ϕ from those equations in conjunction with

$$\sin^2(\psi-\phi) + \cos^2(\psi-\phi) = 1 ;$$

so that in the result all circular functions will disappear; that is to say, as the jet rotates, the cusps of the ripple will describe an *algebraical* curve. Particular cases excepted, the order of this curve is high; for instance, when the three velocities λ, u, and v are equal, it reaches the eighth order; when v vanishes, however, it becomes a circle (art. 20).

If, lastly, we eliminate x and y from the equations (21), (22), (23), we shall obtain the relation between ϕ and ψ which corresponds to the cusps at the moment under consideration, *i. e.* when the jet reaches the position A.

15. Before proceeding further, however, it will be useful to examine the locus (20) of the centres of the circular waves at the moment under consideration. On eliminating ϕ from the equations (20), the equation of this locus will be found to be

$$\eta - a\beta\psi = \sqrt{a^2 - \xi^2} - a\beta \cos^{-1}\frac{\xi}{a} ;$$

from which we learn that the curve undulates between the two lines $\xi = \pm a$ parallel to the ordinate axis, and that the successive undulations are precisely similar, the length of each undulation being $2\pi a\beta$. The form of each undulation differs according as β is less than, equal to, or greater than 1, and in the following manner:—*First*, when $\beta < 1$, *i. e.* when the velocity of the current is less than that of the jet, the curve consists of a series of loops, A B C D E F (fig. 5), the distances between the points B, D, F, where it touches the line X'F, as well as between the points C, E, where it touches the line X E, being $2\pi a\beta$. The branches B C and C D, however, are not symmetrical. As β increases, the loops diminish until, *secondly*, when $\beta=1$, they become transformed into cusps B', D', at which the tangents are parallel to the abscissa axis. Lastly, when $\beta > 1$, the loops and cusps disappear, and are replaced by points of inflexion, H, K, which lie on the line $\xi = -\dfrac{a}{\beta}$. A somewhat clearer image of the ripple may be obtained by regarding it as the envelope of a circle whose centre M moves along one of the three curves of fig. 5, and whose radius increases proportionally to the distance M N, measured along a parallel to the ordinate axis, between its centre and that portion X Y' X' of the circumference of the circle whose concavity is turned in the same direction as that of the portion B' M C' of the curve along which the centre is moving.

16. To return to the equations of the ripple: let us, for brevity, put

$$\sin(\psi - \phi) = \mu, \quad \cos(\psi - \phi) = \nu;$$

the equations (21), (22), (23) will then become

$$\left.\begin{aligned}
& (x - \nu a)^2 + (y - \mu a - \beta\phi a)^2 - a^2\phi^2 a^2 = 0, \\
& \mu(x - \nu a) + (\beta - \nu)(y - \mu a - \beta\phi a) + a^2\phi a = 0,\\
& \nu(x - \nu a) + \mu(y - \mu a - \beta\phi a) + (1 - a^2 + \beta^2 - 2\nu\beta)a = 0.
\end{aligned}\right\} \quad (24)$$

The first two equations, when solved for x and y, give

$$\left.\begin{aligned}
\frac{x}{a} &= \nu - a\phi\,\frac{a\mu \pm (\beta - \nu)\sqrt{R}}{R + a^2}, \\
\frac{y}{a} &= \mu + \beta\phi - a\phi\,\frac{a(\beta - \nu) \mp \mu\sqrt{R}}{R + a^2},
\end{aligned}\right\} \quad \cdots \quad (25)$$

where

$$R = 1 - a^2 + \beta^2 - 2\beta\nu. \quad \cdots\cdots\cdots \quad (26)$$

In these values of the coordinates of any point of the ripple the upper and lower signs correspond, and refer to the two distinct branches into which the ripple divides itself.

On eliminating x and y from the three equations (24), the result will be found to be

$$[a^2 - (1 - \beta\nu)^2]a^2\phi^2 - 2a\beta\mu R a\phi + (1 + \beta^2 - 2\beta\nu)R^2 = 0. \quad (27)$$

This is the equation, mentioned at the end of art. 14, which is satisfied by the values of ϕ corresponding to the cusps.

17. When the value of ϕ is such as to make $R = 0$, the two branches of the ripple meet, and from (25) the points of junction lie in the curve represented by

$$\left.\begin{aligned}
x &= a\nu - a\mu\phi, \\
y &= a\mu + a\nu\phi,
\end{aligned}\right\} \quad \cdots\cdots \quad (28)$$

which may be easily shown to be the involute of the circle formed by unwrapping the circle *backwards* from the point A, where the jet has reached. Further, if $\hat{\lambda}$ be the angle, opposite to the side λ, in a triangle whose sides are respectively proportional to the velocities λ, u, and v, we shall have the relation

$$\lambda^2 = u^2 + v^2 - 2uv \cos \hat{\lambda}; \quad \ldots \quad (29)$$

or, dividing by u^2 and introducing the ratios α and β (art. 14),

$$\alpha^2 = 1 + \beta^2 - 2\beta \cos \hat{\lambda}.$$

This, compared with (26), shows that the condition $R = 0$ is satisfied when

$$\nu = \cos \hat{\lambda}, \text{ and } \mu = \pm \sin \hat{\lambda}; \quad \ldots \quad (30)$$

so that the result of eliminating μ, ν, and ϕ from (28) and the equation $R = 0$ will be

$$x \cos \hat{\lambda} \pm y \sin \hat{\lambda} = a.$$

This is the equation of the curve described by the junction points of the branches of the ripple as the jet describes its circle, which curve, as is at once seen, consists of two right lines touching that circle at points B and B' (fig. 6), whose angular distances B O X, B' O X from the abscissa axis are each equal to $\hat{\lambda}$. Now, since an angle $\hat{\lambda}$ fulfilling the condition (29) can always be found when the velocities λ, u, v form a triangle, that is to say, *when any two of these velocities together exceed the third*, we conclude that under these conditions the ripple will always break up into closed curves or loops, in consequence of the two branches of the ripple, which in other cases are always distinct, meeting each other. These meeting points describe the tangents (30) B$\hat{C}$, B'C' as the jet rotates, and their position at any moment is determined by the intersections C, C', &c. with these tangents, of the involute A C C' of the circle,—the latter being supposed to move with the jet.

The physical character of these meeting points, as will be immediately shown, is that the ripple is there least prominent, so that the tangents (30) B C, B' C' represent two lines along which the surface of the current is apparently *least disturbed*. If the angle between them, which is equal to $2\hat{\lambda}$, were determined by experiment, and the velocity u of rotation were also known, the equation (29) would serve to determine either of the velocities λ or v as soon as the other was given.

18. The above results will be further elucidated by considering the resultant relative velocity of the current and jet at any point A, fig. 4. This resultant will clearly be represented in direction and magnitude by the diagonal A R of a parallelogram whose sides A V parallel to O Y, and A U perpendicular to A O, respectively represent, in direction and magnitude, the velocities

of the current and jet. The angle at U being equal to the angle $AOX = \psi$, the magnitude of this resultant is expressed by

$$(R) = \sqrt{u^2 + v^2 - 2uv \cos \psi}, \quad \ldots \quad (31)$$

and becomes equal to λ, the velocity with which circular waves are propagated, when $\psi = \hat{\lambda}$, that is to say, when the jet reaches either of the points B, B' of the foregoing article (fig. 6). For all positions of the jet in the arc B X B' the resultant (R) will be less than λ, and for all points in the arc B X' B' greater; so that by art. 11, the jet will only commence producing a ripple when it reaches the point B, and will cease doing so as soon as it reaches the point B'. At these points, B and B', the immediately adjacent circular waves of the incipient ripple touch each other, and the point of contact is very slightly elevated above the surface of the current at adjacent points (art. 11); further, since the centres of these waves proceed with the same uniform velocity down the current, whilst their radii increase with the same velocity, they will clearly continue to touch each other along the line B C or B' C'; so that these lines, as above remarked, will be lines of least apparent disturbance. They do not in general touch the ripple; in fact, they do so only when the resultant velocity $(R) = \lambda$ coincides in direction with that of the radius O B (or O B'), in other words, when $v^2 = \lambda^2 + u^2$, as may be seen by a glance at fig. 4. This case of contact is represented in fig. 6, where A D C E represents the first closed loop of the ripple when the jet has reached the point A, and B' D, C, E, what this loop becomes when the jet reaches the point B'. The curve D' C' E' represents a portion of the second loop of the ripple corresponding to the position A; it is, in fact, a portion of what B' D, C, E, becomes as it floats down the current during the time that the jet is describing the arc B' B A.

19. The law enunciated in art. 9 leads to a simple method of determining the velocity of a current; to apply it, however, the velocity λ must be first determined, and the velocity of the current experimented upon must be greater than λ. The formula (31), however, suggests two other methods of determining the velocity of a current, which do not require a previous determination of λ, and may be applied to all currents. They are briefly the following:—

1st. The jet having a known velocity u, let the positions A and A' (fig. 4) be found at which the divergence between the branches of the ripple has a given magnitude, and call 2ψ the angle between the radii O A and O A'. The line bisecting this angle will be perpendicular to the direction of the current. Let the experiment be now repeated with a different velocity u_1, and let $2\psi_1$ represent the angle now subtended at the centre of the circle by the two positions of the jet at which the divergence between the branches of the ripple is the same as before. Then, since the divergences are the same in both experiments, it follows from art. 10 that the resultant velocities (R) and (R$_1$) will be equal; so that by (31) we have

$$u^2 + v^2 - 2uv \cos \psi = u_1{}^2 + v^2 - 2u_1 v \cos \psi_1,$$

whence we deduce

$$v = \frac{1}{2} \frac{u^2 - u_1{}^2}{u \cos \psi - u_1 \cos \psi_1}.$$

2nd. Let the positions A and A$'$ be found at which the angle between the branches of the ripple is bisected by the production of the radius. In this case the resultant velocity (R) will coincide in direction with the radius at each of the points A and A$'$, whose angular distance asunder is 2ψ, and from fig. 4 we deduce at once the relation

$$v = \frac{u}{\cos \psi}.$$

20. The case alluded to in art. 13, where the jet describes a circle in still water, may now be easily disposed of by making $v = 0$, and therefore $\beta = 0$, in the general case treated in art. 14. In this manner the equations (24) become

$$\left.\begin{array}{l} (x - av)^2 + (y - a\mu)^2 - \alpha^2 a^2 \phi^2 = 0, \\ (x - av)\mu - (y - a\mu)v + \alpha^2 a \phi = 0, \\ (x - av)v + (y - a\mu)\mu + (1 - \alpha^2)a = 0; \end{array}\right\} \quad . \quad . \quad (32)$$

or, by simplifying,

$$\left.\begin{array}{l} x^2 + y^2 - 2a(xv + y\mu) + a^2 - \alpha^2 a^2 \phi^2 = 0, \\ x\mu - yv + \alpha^2 a\phi = 0, \\ xv + y\mu - \alpha^2 a = 0; \end{array}\right\} \quad . \quad (33)$$

of which the first two represent the ripple, and all three are satisfied by the coordinates of its cusp. If ϕ_1 be the value of ϕ which corresponds to this cusp, we find immediately, by eliminating x and y from (32),

$$\phi_1 = \frac{\sqrt{1 - \alpha^2}}{\alpha} = \frac{\sqrt{u^2 - \lambda^2}}{\lambda}, \quad . \quad . \quad . \quad (34)$$

which will always be real and positive when u exceeds λ. This same value of ϕ might be obtained from (27) by putting $\beta = 0$; by means of it and the last of equations (33) the first becomes

$$x^2 + y^2 = \alpha^2 a^2,$$

which is the equation of the circle described by the cusp as the jet rotates. Putting r in place of $\sqrt{x^2+y^2}$, it leads to the relation

$$\frac{r}{a}=a=\frac{\lambda}{u} \quad \cdot \cdot \cdot \cdot \cdot \cdot \cdot \quad (35)$$

expressed in art. 13.

To obtain simple expressions for the coordinates x_1, y_1 of the cusp corresponding to a given position of the jet, it is merely necessary to observe that the form of the present ripple does not alter as the jet rotates; so that we may refer its equation to co-ordinate axes which rotate with the jet,—the abscissa axis being at the angular distance $\mathrm{AOX}=\phi_1$ behind the jet (fig. 7). By so doing μ and ν become respectively $\sin(\phi_1-\phi)$ and $\cos(\phi_1-\phi)$; and the two last of the equations (33) give, on substituting for ϕ its value (34),

$$\left.\begin{array}{r} y_1=aa\sqrt{1-a^2}, \\ x_1=a^2a\, ; \end{array}\right\} \quad \cdot \cdot \cdot \cdot \cdot \quad (36)$$

so that if we call θ_1 the angular distance AOC of the cusp C behind the jet, we have, in virtue of (35), and since $\mathrm{C\hat{O}X}=\phi_1-\theta_1$,

$$\left.\begin{array}{l} \sin(\phi-\theta_1)=\sqrt{1-a^2}, \quad \cos(\phi_1-\theta_1)=a, \\[2mm] \dfrac{y_1}{x_1}=\tan(\phi_1-\theta_1)=\phi_1, \\[2mm] \theta_1=\dfrac{\sqrt{1-a^2}}{a}-\tan^{-1}\dfrac{\sqrt{1-a^2}}{a}\, ; \end{array}\right\} \quad \cdot \cdot \quad (37)$$

i. e.

whence it follows that θ_1 increases as a diminishes, in other words, as u increases.

Again, since the tangent to the ripple at any point coincides with the tangent to the circular wave which there touches it, we have from the first of equations (32),

$$\frac{dy}{dx}=-\frac{x-a\nu}{y-a\mu}. \quad \cdot \cdot \cdot \cdot \cdot \cdot \cdot \quad (38)$$

With reference to our new coordinate axes, the values of ν and μ at the cusp are 1 and 0 respectively; and these, together with the values in (36), give, when substituted in (38),

$$\left(\frac{dy}{dx}\right)_1=\frac{a-x_1}{y_1}=\frac{\sqrt{1-a^2}}{a}=\frac{y_1}{x_1},$$

and show that the radius vector OC is the tangent at the cusp.

The polar equation of the ripple, with reference to an axis passing through the jet and moving with it, is easily seen to be

$$\left.\begin{array}{r} r^2-2ar\cos(\phi-\theta)=a^2a^2\phi^2-a^2, \\ r\sin(\phi-\theta)=a^2a\phi, \end{array}\right\} \quad \cdot \cdot \quad (39)$$

where θ and ϕ are estimated in the same direction.   To find the points A and B where the ripple cuts the circle, r must be replaced by a in these equations; on squaring and adding the results, it will be found that

$$\text{either}\quad \phi=0,\ \text{or}\ \phi=2\phi_1;$$

the former has reference to the point A, and the latter to the point B.   Substituting the latter value of ϕ in (39), we have in virtue of (37),

$$\tan(2\phi_1-\theta) = \frac{2\phi_1}{\frac{1}{a^2}-2\phi_1^2} = \frac{2\tan(\phi_1-\theta_1)}{\frac{1}{a^2}-\tan^2(\phi_1-\theta_1)} = \tan(2\phi_1-2\theta_1),$$

whence we conclude $\theta=2\theta_1$; that is to say, the radius OC to the cusp bisects the angle formed by the radii OA, OB to the two points where the ripple enters and leaves the circle.

Lastly, when $a=1$ or $\lambda=u$, the two first of equations (33) reduce themselves to the system (28), which, as we know, represents the involute of the circle.   This verification of the result contained in art. 13 may be most easily made by putting $a=1$, $\beta=0$, and hence $R=0$, in the equations (25), which are equivalent to the system (33).

II. *On certain Affections of the Retina.*
By Sir DAVID BREWSTER, *K.H., D.C.L., F.R.S.**

IN three articles published in the Philosophical Magazine for 1834[†], I have described several 'affections of the retina, which since that time I have frequently had occasion to study. Mrs. Mary Griffiths of New York had observed "a reticulated or network structure upon opening her eyes for the first time in the morning[‡]. "At one moment," she says, "the meshes of the network were of a dull brickdust colour, and the spaces between were of a pale dingy yellow; and in the next moment the case was reversed—the meshes and intersections were of this pale dingy yellow, and the spaces or interstices were of a dull brick-colour." She describes the meshes "as generally the *fifth* of an inch in diameter," but without telling us the distance of the surface upon which this measurement was made.

Having believed it possible to determine experimentally whether a beam of light was a continuous stream like a stream of water, or a stream in which the parts were at such a distance as to maintain a continued impression on the retina, I received a narrow beam of solar light upon a white ground, and made a strip of white card pass rapidly back and forwards across the beam,

---

* Communicated by the Author.
† Phil. Mag. 1834, pp. 115, 241, and 353.   ‡ Ibid. p. 43.

comparing the luminosity of the disc which it reflects with that of the fixed circular spot upon the paper. In these experiments I observed two interesting facts,—that the reflected disc of light exhibited different colours in different parts of it, and that it was more luminous than the beam received upon the white ground when no part of it was reflected by the moving card.

Before I had examined these phenomena more minutely, I found that M. Benedict Prevost had published an interesting paper in the *Mémoires de la Société de Physique et d'Histoire Naturelle de Genève*, and had anticipated me in the first of the above results. He agitated, to use his expression, a piece of white card across a sunbeam in a dark room, and he observed that the circular spot which it reflected, in place of being white, was *white* only in the centre. Around this white spot was a *violet* spot, growing deeper as it receded from the centre. The violet spot was surrounded with a zone of *deep indigo*, very well defined, and resembling the colour of the *Viola tricolor*. Round the indigo zone was one of *greenish yellow*, equally well defined, and round it a *red* shade. He considers the light as thus decomposed into seven principal colours!

"This phenomenon," he adds, "is produced by a single passage of the card across the sunbeam, which proves that it is independent of the fatigue of the eye. Nor does it depend immediately on the agitation or motion of the card, but only, without doubt, on some effect of this motion, and particularly on this, that the illuminated area strikes the eye only during a very short time; for if the card is sufficiently large, so that the illuminated space does not go out of it, and notwithstanding the agitation of it the eye continues always to see it, it appears *white* as if it was seen at rest, and there is no longer any appearance of the decomposition of the light."

Reckoning from the centre of the disc, the colours observed by M. Prevost are as follow:—

> White.
> Violet.
> Deep indigo.
> Greenish yellow.
> Red shade.

In the experiments which I made, I sometimes agitated a narrow slit, or a series of parallel slits, in a black card, across a circular aperture in another black card, or across an aperture in the window-shutter of a dark room; or I looked at a white surface, or a white disc through the slits of a revolving disc, as described in one of the papers already referred to. By these

* Vol. iii. part 2. p. 121.

methods I found that the white disc, or aperture, was *whitest* in the centre at certain velocities, and when the light was strong, but that it was bluish when the light was moderate.   The following was the order and character of the tints, reckoning from the centre:—

> White or bluish.
> Darker blue.
> White.
> A dark ring, pretty well defined.
> White.
> Greenish yellow.
> Reddish.

These colours vary, within certain limits, with the sensibility of the eye and the intensity of the light. I have sometimes observed the centre of the disc *darkish blue,* and sometimes *yellow.* When the velocity is great and the disc seen distinctly, there is not the slightest trace of colour.

Phenomena somewhat analogous to these may be seen in the flame of a spirit-lamp, and in other flames, where the effect is caused by the shooting up of the flame, which produces successive impulses on the retina. At the top of the flame are seen several curves, convex upwards, like elongated parabolas. They are generally of a *sap* or *olive-green* colour ; and sometimes the most brilliant *red* tints appear where the uppermost curve opens at its vertex, and leaves two lateral and almost parallel branches.

Colours similar to those discovered by M. Prevost have been recently observed by Mr. John Smith of the Perth Academy[*], who draws from them what he justly calls many "startling conclusions." M. Prevost and this writer have greatly misapprehended the nature of these phenomena. While the Swiss philosopher considers them as "independent of the fatigue of the eye," and as exhibiting a new decomposition of light by motion, Mr. Smith pronounces them to be "produced by alternate light and shade in various proportions ;" and he regards them " as *proving* the non-homogeneity of æther—as *proving* the undulatory hypothesis, but *opposing* the undulatory theory—as contrary to the idea of the waves of light having different lengths—as helping to explain many of the phenomena of polarization—and as giving a new explanation of prismatic refraction " different from that of Newton!

Although we cannot adopt these conclusions, yet the phenomena, when carefully studied, as being the effect of successive impulses on the retina, acting in the manner which I

---

[*] "On the Production of Colour and the Theory of Light," Report of the British Association at Aberdeen, 1859, Trans. p. 22.

have described in the Philosophical Magazine for 1834, have an interest of a different kind. When the luminous impressions succeed each other at a certain interval, the luminous disc, even when small, exhibits a reticulated structure like a mosaic pavement composed of distinct hexagons delineated in black lines (sometimes with a black spot in their centre), indicating that those portions of the retina are temporarily insensible to light.

In order to see this phenomenon distinctly, we must employ a large disc uniformly illuminated, such as a glass globe ground on both sides, or a plate of ground glass. When the revolving disc has a particular velocity, the illuminated surface will be seen covered with the hexagonal pattern already mentioned; and as the velocity diminishes, the pattern breaks up, exhibiting portions of circles and imperfect crosses, the exact forms of which it is difficult to describe. In daylight, when the light of the sky is used, the colours which accompany these phenomena are better seen, and also the variations which they undergo in reference to the *foramen centrale* of the retina, as described in a former paper.

When we maintain the revolving disc at that particular velocity which produces the hexagonal pattern, so that the retina may be greatly excited, a beautiful pattern of a very different kind makes its appearance, occasionally mixing itself with the first pattern, but most beautifully seen opposite the dark intervals between the slits of the disc when the velocity has become very slow, and the disc is nearly at rest. This pattern, which is too evanescent to permit it to be drawn, consists of a series of dark quadrangular spaces separated by a triple or multiple line of light. It has no relation whatever to the hexagonal pattern, and is never seen unless when the eye has been strongly impressed by the successive impulses of the luminous disc.

When we observe these phenomena at different distances from the illuminated disc, the hexagons always subtend the same angle, which I have found to correspond with a space on the retina equal to the 420th part of an inch; and there can be no doubt that they are produced by a structure of a hexagonal character. In so far as the human retina has yet been studied, no structures of a hexagonal character have been discovered. In the choroid coat, however, in front of the retina nucleated cells of a slightly hexagonal form have been seen in man and in almost all mammalia*; and it is not improbable that the parts of the retina immediately behind these hexagons may be so affected by them as to produce the hexagonal forms which we have described. With regard to the quadrangular pattern, in which the dark

* Nunneley, 'On the Organs of Vision,' p. 171, and plate 2. fig. 7.

quadrangles are separated by several parallel bright lines, it is obviously produced by some structure in the retina itself; and it is possible that it may arise from some regular arrangement of the rods in the columnar or bacillar layer which has not yet been detected by the microscope.

The hexagonal pattern is very distinctly seen in the flame of a coal or a wood fire, at that particular part of it where the jets of ignited gas succeed each other at the proper interval.

After making these experiments for some days in succession, I have found that the hexagonal pattern, and sometimes even the quadrangular one, is seen when the eye is accidentally directed to faintly illuminated surfaces. This fact seems to show that the pattern is rendered visible merely by the excitement of the retina with the action of light, and not by its successive impulses, the black lines of the hexagons and the dark spaces of the triangles requiring a longer time to exhibit the action of the faint light than the other parts of the retina—a property which I have found at the *foramen centrale,* and at various points of the retina at or near the *ora serrata,* its anterior margin.

As an inducement to optical observers to investigate any of the abnormal phenomena of vision which may come under their notice, I give the following list of properties or structures in the retina which have been discovered, or are manifested, by optical observations.

1. The polarizing structure which produces Haidinger's brushes.

2. The insensibility of the retina at the entrance of the optic nerve.

3. The exhibition of the *foramen centrale* by its inferior sensibility to feeble light.

4. The different sensibility to light possessed by different parts of the retina.

5. The inability of the retina beyond the foramen to maintain a sustained vision of objects.

6. The increased luminosity of objects seen indirectly, or by the parts of the retina beyond the foramen.

7. The hexagonal and quadrangular structure described in the preceding pages.

To these we may add the existence, in the vitreous humour, of the remains of vessels no longer required for its support, and the existence of cells in the same humour, as proved by the phenomena of Muscæ volitantes, formerly described in this Magazine.

Allerly by Melrose,
November 30, 1860.

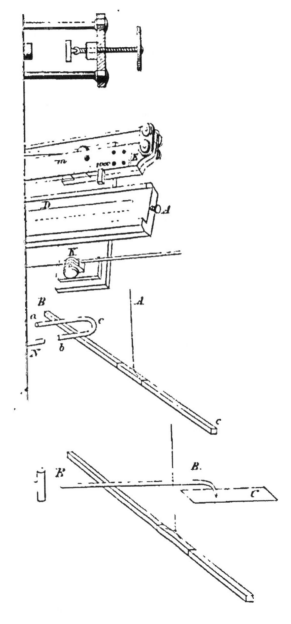

III. *Proposal for a new reproducible Standard Measure of Resistance to Galvanic Currents.* By M. WERNER SIEMENS*.

[With a Plate.]

THE want of a generally received standard measure of current resistance, and the great inconveniences thence arising, especially in technical physics, suggested to me some years ago the experiment I am about to describe.

My original object was to procure for Jacobi's standards a more general introduction into the arts. I soon found, however, that this could not be effected without inconvenience. On one occasion, several of Jacobi's standards that I had procured were so entirely at variance with each other, and their actual resistances agreed so little with what they ought to have been, that I should have been obliged to have had recourse to Jacobi's original measure, only that it was not then at my disposal. Independently of this, however, I became convinced that a standard measure of resistance is only adapted for general use when it is easily reproducible. Whether the resistance of a metal wire is altered by time, by the shaking of transport, by the passage of currents, or by any other cause, is not yet absolutely decided. It is, however, very probable that some such alteration takes place, and it is therefore altogether inadmissible to take the resistance of such a wire as the unit-measure of resistance. Moreover such standards being copies one of another, as must unavoidably be the case in the event of their general adoption, their deviation from accuracy would be continually increasing; and for the purpose of researches which are to be conducted with improved instruments and with great accuracy, mere copies which are themselves inaccurate are useless. Besides, it would be very desirable and convenient to be able to unite a definite geometrical conception with the standard measure of resistance, which could never be the case with a metallic wire, since the resistance of a solid body depends greatly on its molecular state, and on the almost unavoidable impurity of the metal.

The absolute standard measure appeared to me equally ill adapted to general use. It can only be reproduced by means of very perfect instruments, in places especially arranged for the purpose, and by those possessed of great manual dexterity; and moreover it is liable to the grave practical objection, that it does not exhibit itself in a physical form; and lastly, the numbers it involves are exceedingly inconvenient on account of their magnitude.

The only practicable method of establishing a standard measure of resistance which should satisfy all requirements, and

* Translated by Mr. F. Guthrie from Poggendorff's *Annalen*, No. 5, for 1860.

especially which any one could reproduce with ease and with sufficient accuracy, seemed to me to be to adopt the resistance of mercury as the unit. Mercury can easily be procured or rendered of sufficient, almost indeed of perfect purity. As long as it is fluid, it has no different molecular states affecting its conducting power; its resistance is more independent of temperature than that of any other simple metal; and finally, its specific resistance is very considerable, so that numerical comparisons founded on it as a standard are small and convenient.

I therefore determined to try if it were possible, by means of the ordinary glass tubes of commerce and purified mercury, to obtain by suitable methods fixed standard measures of resistance of sufficient accuracy. The great difficulty seemed to be the impossibility of obtaining glass tubes of a form exactly cylindrical. The diameter of the internal cavity of ordinary glass tubes generally varies irregularly. By gauging them, however, with a thread of mercury, some pieces of about 1 metre long may be selected out of a great number, the diameters of which vary almost uniformly. Such tubes as these may be regarded as truncated cones, the resistance of which can be calculated. The volume of the cone filled with mercury can be determined with great ease and accuracy by weighing the metal contained.

Let A B C D, Plate II. fig. 1, be such a truncated cone, r and R being the radii of its upper and lower circular extremities, and l its length. Let M N be a section parallel to the flame A B, and at the distance x from it; let dx be the thickness of this section, and z its radius. Then, if W be the resistance of the cone in the direction of its axis, and dW the resistance of the element M N in the same direction,

$$dW = \frac{dx}{\pi z^2}.$$

But

$$z = \frac{(R-r)x}{l} + r.$$

Therefore, differentiating with respect to x,

$$\frac{dz}{dx} = \frac{R-r}{l},$$

whence

$$dx = \frac{l}{R-r} dz;$$

and by substituting this value of dx in the first equation, we get

$$dW = \frac{l}{(R-r)\pi} \frac{dz}{z^2};$$

whence, integrating with respect to z,

$$W=\int_r^R \frac{l}{(R-r)\pi}\frac{dz}{z^2}=\frac{l}{(R-r)\pi}\left(\frac{1}{r}-\frac{1}{R}\right),$$

or

$$W=\frac{l}{Rr\pi}. \qquad \cdots \cdots \cdots \cdots \quad (1)$$

If, now, V be the volume of the truncated cone, G the weight of the mercury contained in it, and σ its specific gravity, then

$$V=(R^2+Rr+r^2)\frac{l\pi}{3};$$

and dividing both sides by Rr, we have

$$\frac{V}{Rr}=\left(\frac{R}{r}+1+\frac{r}{R}\right)\frac{l\pi}{3};$$

or calling $\dfrac{R^2}{r^2}=a$,

$$\frac{V}{Rr}=\left(\sqrt{a}+1+\frac{1}{\sqrt{a}}\right)\frac{l\pi}{3},$$

whence

$$Rr=\frac{V}{l\pi}\frac{3}{\left(\sqrt{a}+1+\frac{1}{\sqrt{a}}\right)};$$

and putting for V its value $\dfrac{G}{\sigma}$,

$$Rr=\frac{G}{l\pi\sigma}\frac{3}{1+\sqrt{a}+\frac{1}{\sqrt{a}}},$$

which, substituted in equation (I), gives

$$W=\frac{l^2\sigma}{G}\frac{1+\sqrt{a}+\frac{1}{\sqrt{a}}}{3}. \qquad \cdots \cdots \quad (2)$$

The value of W found from this equation is obviously correct for every conductor of pyramidical form, as long as a represents the ratio of the greatest and least sections. It is moreover equally true when, instead of a single truncated cone of length l, any number of equal cones be substituted whose collective length is l, provided only that in each the ratio of the greatest to the least section, or its reciprocal, is a.

For in this case, if

$$l=n\lambda,$$

where λ is the length of one of the cones,

$$W=n\frac{\lambda\sigma}{\frac{G}{n}}\cdot\frac{1+\sqrt{a}+\frac{1}{\sqrt{a}}}{3},$$

or

$$W = \frac{n^2 \lambda^2 \sigma}{G} \; \frac{1 + \sqrt{a} + \dfrac{1}{\sqrt{a}}}{3},$$

or

$$W = \frac{l^2 \sigma}{G} \; \frac{1 + \sqrt{a} + \dfrac{1}{\sqrt{a}}}{3}.$$

When R and r are nearly equal, the correction for the conical form of the conductor, viz. the factor

$$\frac{1 + \sqrt{a} + \dfrac{1}{\sqrt{a}}}{3} \quad \text{or} \quad \frac{1 + \dfrac{R}{r} + \dfrac{r}{R}}{3},$$

differs little from unity; whence every tube not accurately cylindrical may be regarded as a truncated pyramid without noticeable error, the ratio a being determined from the greatest and least lengths of the mercury thread employed in gauging its capacity.

By a series of experiments I now ascertained whether the calculated amount of the resistances of a number of tubes of very different mean sections agreed sufficiently well with the results of measurement. The method I pursued was as follows:—

Glass tubes, from about $\frac{3}{4}$ to 2 millims. internal diameter, were fastened to a long graduated scale, drops of mercury were introduced into them, and the length of the thread of mercury in the tube was measured. By inclining the tube under examination, the thread of mercury was made to pass gradually down its entire length; and thus every piece of tube of about 1 metre long which appeared most nearly cylindrical or uniformly conical, was detected. These pieces were then cut out of the tube, and, by means of a small apparatus constructed for that purpose by M. Halske, were reduced to the exact length of 1 metre. The tubes so prepared were then carefully cleaned. This was most conveniently done by twisting together two thin German silver or steel silk-covered wires, inserting them through the tube, twisting a pad of clean cotton to the projecting end, and drawing it slowly and cautiously through the tube. This operation required to be performed with some care to avoid, breaking the tube. The tube was then filled with clean mercury and its contents weighed. This operation was conducted as follows:—One end of the glass tube was fastened, by means of a piece of vulcanized caoutchouc, to the opening of a small retort receiver, such as is used in chemical laboratories, so that the end of the tube projected into the receiver. About the other end of the tube an iron collar was attached (fig. 2), by means of which a smooth iron plate could be screwed up against the mouth of the tube. The receiver was then properly secured and filled with clean mercury, which was suffered to run down the slightly inclined tube into a vessel placed below. When all the air-bubbles, which at first appeared adhering to the sides of the tube, seemed to be carried away by the descending stream of mercury, the lower orifice of the tube was tightly closed by means of the screw and plate; the tube was placed in a vertical position, and its upper end withdrawn from the caoutchouc covering. When this was done with care, the vertical tube was found entirely filled with a column of mercury which terminated in a projecting hemisphere of the metal. The upper orifice was next closed by pressing against it a glass plate ground flat, the superfluous metal being thus removed. The tube being then freed, by means of a brush, from any globules of mercury that might have adhered to its surface, was emptied into a glass vessel, and its contents weighed on an accurate chemical balance. When the precaution was taken of letting the mercury flow slowly from the tube by inclining it very slightly, and suffering the air to enter gradually at the upper orifice, it was found that no globules were left behind in the tube, as is generally the case under other circumstances. Warming the tube when full by contact with the naked hands was of course avoided. The temperature at the time the tube was filled was observed, and the weight of the contents corrected for the difference of this above 0° C. Of the following Tables, Table I. gives the lengths of the threads of mercury observed in gauging the capacity of the tubes employed, and the ratio *a* of the greatest and least section thence determined. Table II. the weight of the mercury at the actual temperature, and the same as corrected for 0° C.

Table I.

1.	2.	3.	4.	5.	6.
125·0	101·2	48·2	143·0	115	111
116·4	98·4	47·5	145·0	116	109
115·3	96·9	45·0	146·0	119	107
114·0	94·5	45·0	145·0	121	105
112·0	94·0	44·8	143·5	121	105
110·2	93·3	44·2	142·5	122	103
108·2	94·5	43·9	142·5	121	101
107·0	95·7	43·7	140·0	120	100
107·0	97·5	42·5	139·0	119	101
106·0	99·4	41·0	......	......	102
	101·1	40·1	......	......	100

Therefore $a = \dfrac{125}{106}$ $\dfrac{101·2}{93·3}$ $\dfrac{48·2}{40·1}$ $\dfrac{146}{139}$ $\dfrac{122}{115}$ $\dfrac{111}{100}$

And consequently the respective correction factors were—

1.	2.	3.	4.	5.	6.
1·00225	1·00055	1·00282	1·000201	1·000289	1·000906

Table II.

1.	2.	3.	4.	5.	6.
13·208	27·1915	24·3825	62·368	69·802	11·767
13·210	27·1900	24·3830	62·366	69·796	11·768
13·209	27·1915	24·3840	62·357	69·803	11·767
13·209	27·1920	24·3833			
at	at	at	at	at	at
13°·5 R.	14° R.	13°·5 R.	18° R.	14°·7 R.	15°·2 R.
			61·395	69·795	11·776
			62·398	69·795	11·777
			63·393	69·794	11·774
					11·774
			at	at	at
			14°·5 R.	18° R.	14°·7 R.

Weight in grammes at 0°.

1.	2.	3.	4.	!5.	6.
13·2491	27·277	24·457	62·774	70·057	11·808

If in the formula (2) for the resistance, found above, viz.

$$W = \frac{l^2\sigma}{G} \frac{1 + \sqrt{a} + \dfrac{1}{\sqrt{a}}}{3},$$

we substitute the values of G (in milligrammes) and the correction factor as determined by these Tables, and if we take for the specific gravity of mercury at 0°

$$\sigma = 13·557,$$

and for the mean length of all the tubes

$$l = 1000 \text{ millims.},$$

then we shall have the resistance of the tubes expressed in terms of the resistance of a cube of mercury 1 millim. each way. Table III. gives the values of W so calculated.

Table III.

1.	2.	3.	4.	5.	6.
1025·54	497·28	555·87	216·01	193·56	1148·9

The resistance of these tubes filled with mercury at 0° was next compared with one of Jacobi's standards (B), by means of a Wheatstone's bridge. As the bridge in question, in the form used by Halske and myself, is adapted to very accurate measurements, its more particular description will not be without interest.

Fig. 3 is a perspective view of the bridge. A A is a brass frame on which the slide B B moves. The button C on the slide B B is provided with a toothed wheel, which works into a toothed rack attached to the frame. The slide may therefore be moved either by direct pressure or by turning this button. Attached to the frame are the insulated pieces of metal E E, and the graduated scale $m\,m$ divided into millimetres. Between the insulated pieces of metal E E, whose inner surfaces are perpendicular to the scale $m\,m$, and are exactly 1000 millims. apart, a platinum wire of about 0·16 millim. in diameter is stretched. This wire, the ends of which correspond exactly with the division marks 0° and 1000 of the scale, is clasped by two small platinum rollers, whose axes are connected with the slide B by means of the springs G. The bodies whose resistances are to be compared are inserted between the metallic band H, which can be connected by means of the contact lever I with one pole of the battery, and the two thick copper rods, L and L, which move .freely in the eyes K and K. The other pole of the battery (a single-celled Daniell's battery was generally employed) is connected with the slide B and the platinum roller. The eyes K K, and the insulated pieces E E, which serve as points of attachment for the platinum wire, are placed in perfect connexion, by means of thick copper rods, with the four plates of the plug reverser S. By changing the plugs, the two resistances to be compared could be exchanged. The ends of the multiplying wire of the ga.vanometer employed are connected with the pieces of metal E E.1 In the following measurement I used a mirror galvanometer, with a round steel mirror 32 millims. in diameter, and 36,000 coils of a copper wire 0·15 millim. thick. The distance of the scale divided into millimetres from the mirror was about $6\frac{1}{2}$ metres.

The measurements obtained by means of this apparatus, which are collected together in the following Table, were for the most part determined by Dr. Esselbach. The method he pursued was as follows:—

Each end of the glass tube to be tested was inserted in a receiver, and retained there by means of a caoutchouc band. This receiver was so placed that the unused neck projected upwards, and in this position it was, together with the tube, plunged into a trough filled with lumps of ice. One of these receivers was then supplied with clean dry mercury, which filled thé tube and ran through it into the other. By the time the mercury in the two vessels was at the same level, the tube was generally found free from air-bubbles. Thick amalgamated copper wires were now inserted into the mercury through the top of·the receivers, and the resistance of the tube was compared with that of one of Jacobi's standards by means of the bridge above described*.

The resistance of the conducting rods was determined by plunging the amalgamated copper cylinder into a vessel filled with mercury. It was found to be quite evanescent in comparison with the resistance of the tubes.

The experiments whose results are collected in the following Table were conducted as follows:—In the first position of the commutator, the slide B B was moved to such a position that, on lowering the contact lever·I, the galvanometer exhibited no permanent deviation. The resistances to be compared were then changed by means of the commutator, and the slide was again adjusted. These two positions of the slide are given in the columns headed *a* and *b*. If the observations were free from error, the sum of these magnitudes would be exactly 1000, which was generally the case, or very nearly so. We must here remark that, after establishing the equality of the currents, a small temporary deviation of several divisions, indicating a greater· resistance on the part of the closely packed coils of Jacobi's standard, was observed immediately on completing the circuit. As on breaking the circuit an opposite deviation of the same magnitude resulted, this phenomenon was obviously to be attributed to the extra current of the wire coil of the Jacobi's standard. It appeared, moreover, that on a long continuance of the current, the mercury in the tube began to grow warm, even though only a single cell of a Daniell's battery was employed. On account of the slow oscillation of my mirror and the resistance encountered by its prolongations, the error arising from this cause was easily eliminated by allowing only short currents to traverse the instrument. The slide was always so placed that, immediately on completing the circuit, there was a slight movement towards the left, which, on the mercury becoming warm owing to the continuance of the current, gradually passed over

---

* At first, instead of amalgamated copper wires, iron cylinders were used as conductors; under these circumstances, however, we found that there was a very considerable resistance to the passage of the current from the iron to the mercury, even though the surface of the former was perfectly clean. This resistance, which was also considerable when unamalgamated copper wire was employed, was particularly strong when the cylinder had been some time exposed to the atmosphere after having been cleaned, so that it is probable that this phenomenon is to be attributed to the gaseous condensation on the surface of the metal.

into a deviation to the right. By again slightly altering the position of the slide, the first movement to the left could be rendered so small as to be imperceptible, and the effect of the change of temperature could thereby be entirely obviated.

Table IV.

Tubes	1.		2.		3.	
	a.	*b.*	*a.*	*b.*	*a.*	*b.*
Observed resist-ances .... {	605·7	394·3	429·1 429·0	570·9 571·1	456 456·3 456·2 456·2	543·7 543·6 543·3 543·6
Mean value	605·7	394·3	429·05	571·0	456·2	543·6
For $b=1$	1·536	......	0·7514	......	0·8392	
W_1	1016·52	......	497·28	......	555·38	
$\frac{W}{W_1}$	1·008	......	1·00	......	1·0008	
Tubes	4.		5.		6.	
	a.	*b.*	*a.*	*b.*	*a.*	*b.*
Observed resist-ances .... {	247·6	752·6	227·4 227·3	772·8 772·8	633·2 633·15 633·10	366·8 366·85 366·90
Mean value	247·6	752·6	227·35	772·8	633·15	366·85
For $b=1$	0·329	......	0·2942	......	1·726	
W_1	217·73	......	194·7	......	1142·3	
$\frac{W}{W_1}$	0·992	......	0·994	......	1·005	

The line distinguished by the letter W_1 is found from the preceding one by multiplication by 661·8, which number was furnished by comparing the calculated resistance of tube 2 with that of the Jacobi's standard employed. The numbers in this line ought therefore to agree with the resistances as derived from calculation contained in Table III. The numbers in the line headed $\frac{W}{W_1}$, which are the ratios of the calculated and observed resistances, show that these magnitudes do not differ more than was to be expected. The most considerable errors in our measurements arose from the fact that neither the temperature of the mercury nor that of the copper standard was constant. The temperature of the ice-water varied between 0° and 2° C., and that of the standard between 19° and 22° C.; and as the conductibility of copper is diminished about 0·4 per cent. for every degree Centigrade, the differences in the above measurements, which are all under 1 per cent., are thereby fully accounted for; and there can be no doubt that in this way standard measures of resistances can be reproduced of any degree of accuracy.

The observed resistances in Table IV. ought strictly to have been diminished by the resistance of the mercury in the glass vessel to the spreading of the current, that is, by the resistance encountered by the current in passing from the orifice of the tube to the amalgamated copper conductors. This resistance may, without any considerable error, be taken as the resistance of a hemispherical shell, whose lesser radius is equal to the inner radius of the tube, and whose greater radius, on account of its comparative magnitude, may be taken as infinite. Now the resistance of a hemispherical shell of radius x and thickness dx being called dW, we have

$$dW = \frac{dx}{2x^2\pi},$$

whence

$$W = \int_r^\infty \frac{dx}{2x^2\pi} = \frac{1}{2r\pi} = \frac{r}{2^{\prime 2}\pi}.$$

The resistance to the spreading of the current in the mercury at both ends of the tube is therefore equal to the resistance of a portion of the tube whose length is half its diameter. This resistance ought to be still further increased, because the surface of the mercury in the tube is flat and not hemispherical, as is assumed in the above calculation; but these two sources of error are each of them so trifling, that their joint effect may be neglected.

The straight tubes used in the experiments hitherto described cannot be very conveniently employed as standards; I therefore get M. Giessler of Berlin to make me some spiral glass tubes, having their extremities turned up and provided with small vessels in which to receive the conducting wires. These glass spirals were fastened, as shown in fig. 4, into the wooden cover of a broad glass vessel filled with water. The temperature of the water in the vessel was observed by means of a thermometer introduced through an opening in the cover. The glass spirals were easily filled with mercury, so as to be free from air-bubbles, in the following manner:—The orifice of the tube in one of the glass vessels was first stopped by means of a suitably shaped cork; the other vessel was then filled with mercury, the air being suffered to escape gradually by the cork at the other end, which was only.

removed when the mercury had slowly passed entirely through the windings of the tube.

As mercury is not to be found in the list of metals the alteration of whose specific resistance by heat has been determined by Arndtsen*, this deficiency had first to be remedied. This was effected by Dr. Esselbach, by means of the apparatus already described. The resistance of one of the spiral tubes was compared with that of the straight tube 2, first at the temperature of ice-water, and then when the temperature of the spiral was raised. If w represent the resistance of tube 2 (which according to Table III. is 498·7), and w_i the resistance of the spiral tube, then, since the resistance of the conducting wires was rendered equal for both tubes, and was equivalent to that of 11 cubes of mercury 1 millim. each way, we had

$$\frac{w_i+11}{w+11} = \frac{a}{b},$$

where a and b represent the pieces of platinum wire of the bridge when no current passed through the galvanometer. This was the case when

$$\frac{a}{b} = \frac{311\cdot3}{688\cdot7},$$

whence
$$w_i = 219\cdot4.$$

The temperature of the straight tube was now maintained at 0° by means of melting ice, while the temperature of the water surrounding the spiral was raised. In the following Table, t indicates the temperature of the straight, and t_i of the spiral tube, a and b the lengths of wire read off when the currents were equal, y the required coefficient calculated according to Arndtsen's formula,

$$\frac{w_i(1+yt_i)+11}{w(1+yt)+11} = \frac{a}{b}.$$

Table V.

t.	t_i.	a.	b.	y.
0°	47·0 C.	320·4	679·5	0·000964
0	34·5	318·0	682·0	0·000960
0	16·5	314·6	685·4	0·000981

From this it appears that, of all simple metals, mercury is that whose resistance is least increased by an increase of temperature.

By means of the coefficient y, the resistances of the two other spirals A and B were determined, which were afterwards used

* Poggendorff's *Annalen*, vol. cii. p. 1.

in constructing a standard of German silver wire. The resist-ance of spiral A at 0° was 514·45, that of B 678·9.

German silver wire is very well adapted for standards of resist-ance, because its conducting power is very small, and varies little with the temperature; according to Arndtsen, only about ·0004 per cent. for a degree Centigrade.

In the foregoing experiments, the resistance of a cube 1 millim. each way has been assumed as the unit. For small resistances, and especially in calculation, this has many conveniences. It appears, however, more advisable that the measure of resistance should be in entire agreement with the ordinary metrical scale. I therefore determined to take as the unit of resistance,—

The resistance of a prism of mercury 1 *metre long and* 1 *millim. in section at the temperature* 0° C.

If this proposal be generally acted on, all resistances can be at once reduced to the metrical scale.' Every experimenter will then be able to provide himself with a standard measure as accu-rate as his instruments permit or require, and to check the varia-tions of resistance of the more convenient metal standard. Of course also, in that case, the conducting power of mercury must be taken as the unit of conducting power, and not that of copper or silver. Unfortunately but few comparisons have been made between the conducting power of mercury and the solid metals, from which a table could be calculated in which the conducting power of mercury should be the unit; and in most of the com-parisons that have been made between the conducting power of the solid metals, it is not stated whether the wire employed was cold drawn or annealed. From the following Table, however, it appears that the conducting power of annealed wire is consider-ably greater than that of unannealed.

1.	2.	3.	4.	5.	6.
Species of wire.	Length in millims.	Weight in milli-grammes.	Specific gravity.	Resistance at 0° C.	Conducting power, that of mercury =1.
1. Silver, hard	4014·4	4884·9	10·479	614·55	56·252
„ „ annealed	4014·4	4889·1	10·492	537·2	64·38
2. „ hard	4014·4	3233·1	10·502	896·1	58·20
„ „ annealed	4014·4	3009·6	10·5132	889·08	63·31
3. Copper, hard............	4014·4	3099·5	8·925	890·5	52·109
4. „ hard............	4014·4	4409·1	8·916	622·7	52·382
„ „ annealed	4014·4	4355·2	8·903	599·05	52·013
5. „ hard............	2007·2	1260·4	8·916	545·8	52·217
„ „ annealed......	2007·2	1252·7	8·894	517	55·419
6. „ hard............	2007·2	1263·2	8 916	545·6	52·121
„ „ annealed	2007·2	1241·5	8 894	520·8	55·338
7. Platinum, hard	436·4	544·1	21·452	910·6	8 244
8. „ hard........	436·4	550·1	21·452	897·7	8 27
9. Brass, hard	1003·6	1406·1	8·473	530·6	11·439
„ annealed...........	1003·6	1397·8	8·464	451·7	13·502

It appears from this Table that the specific conducting power of annealed silver wire is 10 per cent. greater than that of the same wire unannealed; that of annealed copper wire on the average 6 per cent. greater. The difference of conducting power is especially great in the case of brass. As the density of wire depends on the amount it has been stretched since the last heating, that, as well as the conducting power, must vary much even when the metal is perfectly uniform. The temperature to which the wire was raised, the duration of the operation, and the rapidity of the cooling, all affect its spec fic conducting power. Column 5 of the above Table was calculated by means of the formula

$$W = \frac{P_\sigma}{Q} \frac{a + \sqrt{a} + \dfrac{1}{\sqrt{a}}}{3}.$$

The factor $\dfrac{a + \sqrt{a} + \dfrac{1}{\sqrt{a}}}{3}$, the correction for the conicality of the conductor, may, in the case of metal wires, be almost always neglected, since it never differs sensibly from unity. This method is obviously much more exact than that hitherto employed, in which the mean diameter of the wire had to be determined by direct measurement; and the square of the magnitude so obtained, which was never exact, entered into the calculation. By my method all the data may be determined with great accuracy, especially the length of the wire, which alone enters in the square.

If the above Table be compared with that of Arndtsen, it appears that the mean conducting power of unannealed platinum wire, viz. 8·257, and the smallest value found for unannealed silver wire, viz. 56·252, are exactly in the ratio given by Arndtsen; while the resistance of copper wire according to Arndtsen, agrees pretty well with that of annealed copper in the above Table. As the silver and platinum wires employed both by myself and Arndtsen were chemically pure, I have in the following Table taken the resistance of platinum and hard silver as a standard. The values taken from Arndtsen's Table are indicated by an A, those observed by myself by an S.

Table VI.

Conducting power of the metals at the temperature t compared with that of mercury at $0°$ C.

Mercury . . . . $\dfrac{1}{1+0.00095.t}$ (S).

Lead . . . . $\dfrac{5.1554}{1+0.00376.t}$ (A).

Platinum . . . . . . $\dfrac{8.257}{1+0.00376.t}$ (A. S).

Iron . . . . $\dfrac{8.3401}{1+0.00413.t+0.00000527.t^2}$ (A).

German silver . . . . $\dfrac{10.532}{1+0.000387.t-0.000000557.t^2}$ (A).

,,   annealed   4.137 (S).

Brass, hard . . . . . 11.439 (S).

,,   annealed . . . . 13.502 (S).

,, ... $\dfrac{14.249}{1+0.00166.t-8.00000203.t^2}$ (A).

Aluminium . . . . . $\dfrac{31.726}{1+0.003638.t}$

Copper . . . . . . $\dfrac{55.513}{1+0.00368.t}$ (A).

,,   hard . . . . . 52.207 (S).

,,   annealed . . . . 55.253 (S).

Silver, hard . . . . . $\dfrac{56.252}{1+0.0003414.t}$ (A. S).

,,   annealed . . 64.38 (S).

For convenience I have given Arndtsen's results, together with the coefficients for correcting the resistances for the temperature. Whether these are the same for annealed and unannealed wires I have not been able to determine. The brass that I experimented on was found on analysis to consist of 29.8 per cent. zinc, and 70.2 copper.

In conclusion I should mention, for the benefit of those who may wish to make standard measures of resistance in the manner here described, that it is necessary to warm the mercury to be employed for several hours under a covering of concentrated sulphuric acid mixed with a few drops of nitric acid, to get rid of all metallic impurities, as well as the free oxygen, which greatly *increase* its conducting power.

IV. *Remarks* on Mr. Harley's *paper on Quintics.*
By G. B. Jerrard.

[Concluded from vol. xix. p. 274.]

4. IN my former paper I contented myself with showing, from a comparison of certain results at which Mr. Harley had arrived, that there were decisive marks of the existence of an error in his processes, leaving to him and Mr. Cockle the task of tracing the error to its source. But after the lapse of many months they have as yet failed to perform the part thus fitly assigned to them. This is the more remarkable, as, in answer to an attempt made by the latter mathematician to infer the failure of my researches from the failure of theirs, I said at the time, in a postcript to m paper, what ought, I think, to have suggested to their minds the true origin of their error, and thence have shown them the irrelevancy of the objections urged against my method of solving equations of the fifth degree. Mr. Cockle, however, not seeing the meaning of my words, still adheres to his objections. I must now, therefore, abandoning suggestions the drift of which may escape observation, have recourse to statements of the plainest kind.

5. If I were asked in what respect the method first explained by me in the Philosophical Magazine for June 1845, and afterwards given at greater length in Chapter II. of my 'Essay on the Resolution of Equations*,' differed from all other methods constructed for the purpose of solving equations,—on what, while all these were seen to have failed, I grounded my hope of success, —I should answer by pointing to the theorem†

$$\Theta_{n, f(ab)(cd)..} =$$

$$\frac{\iota^\zeta}{5}(x_\alpha + \iota^{4n}x_\beta + \iota^{3n}x + \iota^{2n}x_\delta + \iota^n x_\epsilon)(ab)(cd).. \Big\} \quad \cdot \quad (v)$$

if $\quad \Theta_{n, f(ab)(cd)..} = \iota^\zeta(\Theta'_{n, f})(ab)(cd)..$

or

$$\Theta_{n, f(ab)(cd)..} =$$

$$\frac{\iota^\eta}{5}(x_\alpha + \iota^n x_\beta + \iota^{2n}x_\gamma + \iota^{3n}x_\delta + \iota^{4n}x_\epsilon)(ab)(cd).. \Big\} \quad \cdot \cdot \quad (w)$$

if $\quad \Theta_{n, f(ab)(cd)..} = \iota^n(\Theta''_{n, f})(ab)(cd)..$

From this theorem we learn that, in constructing equations

* Published by Taylor and Francis, Red Lion Court, Fleet Street, London.
† 'Essay,' p. 75.

between functions of the form

$$\Theta_{n, f(\underset{..}{a}\underset{..}{b})(\underset{..}{c}\underset{..}{d})..}$$

and those of the class

$$\frac{\iota^k}{5}(x_\alpha + \iota' x_\beta + \cdot\cdot)(\underset{..}{a}\underset{..}{b})(\underset{..}{c}\underset{..}{d})\cdot\cdot,$$

a complex process of a peculiar character must take the place of the simple method of substitutions.

Thus from (v$_2$) we see that

$$\Theta''_{n, f(\underset{..}{\beta}\epsilon)(\gamma\underset{..}{\delta})} = (\Theta'_{n, f})(\underset{..}{\beta}\epsilon)(\gamma\underset{..}{\delta}),$$

and analogously from (w$_2$), that

$$\Theta'_{n, f(\underset{..}{\beta}\epsilon)(\gamma\underset{..}{\delta})} = (\Theta''_{n, f})(\underset{..}{\beta}\epsilon)(\gamma\underset{..}{\delta});$$

in each of which equations the Θ of one member is accented differently from the Θ of the other in consequence of a transfer of Θ' and Θ'' from branch to branch for the affix $(\underset{..}{\beta}\epsilon)(\gamma\underset{..}{\delta})$.

This shows the necessity of carefully distinguishing between

$$\Theta_{n, f(\underset{..}{a}\underset{..}{b})(\underset{..}{c}\underset{..}{d})..} \text{ and } (\Theta_{n, f})(\underset{..}{a}\underset{..}{b})(\underset{..}{c}\underset{..}{d})\cdot\cdot$$

if we would avoid admitting relations among the roots through equations of the class

$$(\Theta''_{n, f})(\underset{..}{\beta}\epsilon)(\gamma\underset{..}{\delta}) = (\Theta'_{n, f})(\underset{..}{\beta}\epsilon)(\gamma\underset{..}{\delta}).$$

I might go on to elicit other properties of the theorem (v, w), but what I have already said is, I think, sufficient for my purpose.

6. Turning now to Mr. Harley's paper, I find that the very existence of the theorem (v, w) is ignored. No wonder then was it—where so much circumspection was needed, and such a safeguard against fallacies as the theorem (v, w) was flung aside —that he as well as Mr. Cockle fell into error.

7. But Mr. Cockle, not reflecting that the functions

$$\phi_n(x_1, x_2, .. x_5), \quad \chi_n(\dot{x}_1, \dot{x}_2, .. \dot{x}_5)$$

(wherein $\dot{x}_1, \dot{x}_2, .. \dot{x}_5$ are known to be such as not to involve any irreducible radical of the form $\sqrt[n]{z}$, while $x_1, x_2, .. x_5$ are not proved to be subject to any condition) do not come within the scope of Lagrange's theory for expressing one of two homogeneous functions of the roots of a *given equation* in rational terms of the other, attempts by means of that theory to apply objections drawn from the failure of the method given in Mr. Harley's paper to the method of solution in my 'Essay.' The nature of Mr. Cockle's second error is now manifest. He implicitly assumes as a direct consequence of Lagrange's theory, that he is permitted to express*

$$\chi_n(\dot{x}_1, \dot{x}_2, .. \dot{x}_5)$$

as a rational function of

$$\phi_n(x_1, x_2, .. x_5).$$

Such an assumption is clearly inadmissible. It is not a little remarkable that a most striking confirmation of the validity of my method may be derived from that very theory†.

8. Again, the objection which he urges against a statement of mine in reference to Taylor's theorem, is founded on a misapprehension of my meaning. The statement relates to equations of condition‡. But Mr. Cockle, ignoring their existence, speaks of a remainder. It would be easy, from what has been said in art. 7, to show the irrelevancy of the rest of his objections. It seems needless, however, to proceed further.

December 1860.

V. *On the Principles of the Science of Motion* (*Mechanics, Physics, Chemics*) §. *By* J. S. STUART GLENNIE, *M.A.*

1. IT is proposed in the following paper to give a brief introductory exposition of the conceptions by the analytical application of which it has been found that the laws and methods of mechanics can be more fully applied to the phenomena of physics and chemics.

2. In order that mechanical principles may be rigorously applied to physical and chemical phenomena, it seems clear that physical and chemical forces must be conceived in the same way as mechanical forces. Now a mechanical force, or the cause of a mechanical motion, we know to be, in general, the condition of a difference of pressure. Whether the motion is uniform, or accelerated, depends on whether the force, or difference, is instantaneous, or continuous. And whether that difference is between polar or extreme pressures, depends on whether the moved body transmits pressure with, or without, a tangential resistance. Hence the condition of the translation of a solid is

---

* See his paper in the Philosophical Magazine for March 1860. The same error runs through all his subsequent papers on the subject of equations, and thus renders nugatory every objection based on Mr. Harley's paper.

† Compare arts. 107, 112 of my 'Essay.'

‡ See note on art. 44 of my paper for June 1845.

§ Communicated by the Author, being a restatement of principles enounced in papers read at the two last Meetings of the British Association, and of which abstracts have been published in the 'Reports' for 1859 and 1860 (Trans. of the Math. and Phys. Section).

a difference of polar pressures; and of the motion of a fluid, a difference of maximum and minimum pressures; and similarly, it is evident that the causes of the other mechanical motions are, in fact, conditions of pressure. Thus, rotation is the effect of equal, a compound motion of translation and rotation of unequal, differences of polar pressures in opposite directions with respect to a line joining the points of application of each pair of polar pressures. It need hardly be said that it is not proposed that we should use the phrase "differences of polar pressures" instead of the term "forces." But, preparatory to a mechanical consideration of physical and chemical motions, it seems of importance clearly to see that, whatever force may be absolutely, whatever the unknowable ultimate cause of motion may be, or have been, the ordinary mechanical forces* called "a single force at the centre of gravity," "a couple," and "a single force not at the centre of gravity," are, in fact, such conditions of pressure as above stated.

3. Thus, as the cause of a mechanical motion is, in general, the condition of a resultant difference of the pressures on the moved body, and as the general law of the motion of the body is, that it is in the direction of the greater of the two opposed resultant pressures, if the great object of modern physical research is accomplishable, it must be attained by showing that physical forces are, like mechanical forces, conceivable as differences of pressure under certain characteristic but interchangeable conditions, and that, for instance, the movement of iron towards a magnet, or the movement of a paramagnetic body towards a place of stronger action, are motions in the direction of least resistance. And there is evidently this further condition of a general mechanical theory of physical and chemical phenomena —that pressure be, in physics and chemics, conceived, as in mechanics, as "a balanced force†," or "momentum virtually and not actually developed‡." Hence it appears that, if a general mechanical theory is possible, the ultimate property of matter must be conceived to be a mutual repulsion of its parts, and the indubitable Newtonian law of universal attraction be herefrom, under the actual conditions of the world, deduced.

4. The general experimental condition of the fitness of the mechanical conception of pressure as the basis of a general physical and chemical theory, evidently is that there be a plenum.

---

* "Toutefois, c'est une chose très-remarquable, qu'un même livre, écrit sur la science des forces, pourrait, sans cesser d'être exact et de traiter régulièrement la même science, être entendu de deux manières différentes, selon qu'on attacherait au mot de *force* l'idée d'une cause de translation, ou l'idée toute différente d'une cause de rotation."—Poinsot, *Théorie Nouvelle de la Rotation des Corps,* p. 13.

† Rankine, 'Applied Mechanics.'

‡ Price, 'Treatise on Infinitesimal Calculus,' vol. iii.

But such discoveries as that the tangential force of a resisting medium is given in the very formulæ of Encke's comet[*], and that of inductive action through contiguous particles, and not at a distance[†], will, it is believed, cause this first postulate of the theory to be generally granted[‡].

5. To give distinctness to this idea of the parts of matter as mutually repulsive, a molecule, or a body (an aggregate of molecules), is conceived as a centre of lines of pressure; the lengths and curves of these lines are determined by the relative pressure of the lines they meet; and lines from greater, are made up of lesser molecules and their lines, and so on *ad infinitum*.

In speaking of a molecule or body as such a centre of pressure, it will be convenient to have a technical name. Rather than coin a new term, it is proposed to use "atom" in this sense. In chemistry I shall use the term "equivalent" exclusively, and not, as at present, as more or less synonymous with "atom," which I have thus ventured to appropriate for a new conception. Atoms, or mutually determining centres of lines of pressure, may also be defined, and their relations analytically investigated, as mutually determining elastic systems with centres of resistance.

6. This is the fundamental conception (not hypothesis) of the theory. What can at present be called hypothetical in the theory, is this only—that the application of analysis to the conception of elastic systems with such conditions, will in all cases give results corresponding with phenomena. An atom, as above defined, is the postulate of a conception, not of an agent, of a relation, not of an entity. And this conception, it is believed, distinguishes the theory from the many others of which the general object has been the same[§]. But in this attempt to found a general theory cleared of æthers and fluids, of properties and virtues, the author owes whatever there may be of truth in its present imperfect form and incomplete application, chiefly to the discoveries of the immortal 'Experimental Researches;' and the constant endeavour has been to work out the theory in the

---

[*] *Ueber die Existenz eines widerstehenden Mittels im Weltraume,* von J. F. Encke, p. 52. Berlin, 1858.

[†] Faraday, 'Experimental Researches,' Series I., II., IV., IX., XI., XII., XIII.

[‡] See also on this point Humboldt, 'Cosmos,' iii. 33; and compare Newton's third letter to Bentley; the old aphorism, "Nature abhors a vacuum," and Grove's remark thereon, 'Correlation of Physical Forces.' And see Bacon, *Nov. Org.* ii. 8.

[§] But compare the conception of an atom by Boscovitch, in his *Theoria Philosophiæ Naturalis redacta ad unicam legem virium in Natura existentium.* Venetiis, 1763; and Faraday, in his 'Experimental Researches,' iii. 447. Also Seguin's memoir, *Sur l'origine et la propagation de la force,* and his *Considérations sur les causes de la cohésion.*

spirit of such remarks as these:—"What we really want is not a variety of different methods of representing the forces, but the one true physical signification of that which is rendered apparent to us by the phenomena, and the laws governing them. Of the two assumptions most usually entertained at present, magnetic fluids and electric currents, *one* must be wrong, perhaps *both* are. . . . . It is evident, therefore, that our physical views are very doubtful; and I think good would result from an endeavour to shake ourselves loose from such preconceptions as are contained in them, that we may contemplate for a time the force as much as possible in its purity*."

7. Now, in a system of atoms as above defined, let the centres be of equal mass, and at equal distances; there will be no difference of pressure on any one centre, no moving force will be developed, and the conditions of equilibrium will be satisfied. But it is clear that forces will be developed, or the general conditions of motion be fulfilled, either (a) by a difference in the masses of the centres, or (β) by a difference in the distances of the centres, in consequence of a displacement of any one of them, or (γ), supposing a state of dynamic equilibrium established in the system, by its being brought in contact with another system in a different state of such equilibrium.

8. Consider more particularly the first of these conditions of the development of a force, and with the postulate that the pressures of atomic lines are directly proportional to the mass of the atomic centre, and inversely to the square of the distance therefrom. But observe that this last property is conceived to arise, not from an absolute change in any individual line, whatever its distance from the centre, but because, from the lateral expansion of these lines, fewer will, when such expansion is unresisted, be cut at a greater than at a lesser distance by the same surface.

In a system of atoms in which there is a difference in the masses of any one or more centres, and in which the law of the pressure of the atomic lines is as above, any two masses, whether unequal or equal, will tend to approach.

And first, let the masses be unequal.

Then the pressure of the lines from the one is greater than that of the lines from the other; consequently these lines will be mutually deflected, and hence the centres approach.

Let the two masses be equal.

Then, if all the masses of the system were equal, and all at the same distance from each other, their mutual repulsions would be equal in all directions, and they would remain at rest. But if, though two masses may be equal, either has, on the other side of it, a mass of a greater size, or at a greater distance than

* Faraday, 'Experimental Researches,' vol. iii. p. 530.

the other, it is evident that the mutual pressures of these two equal masses will, under such conditions, be unequal, and hence, as in the first case, they will approach. It is also evident that a body may thus cause the approach to itself of another body, whatever the number of interposed bodies.

9. Thus, if the conception of atoms is applied to the unequal and unequally placed bodies of such a world as that presented to us, the law of universal attraction follows, and gravity is mechanically *explained*, that is, referred to a mechanical conception.

But it must be understood that the above proposition is given rather to show that, as an actual law, universal attraction may be deduced from the theoretical conception of universal repulsion, than with any pretension to its being the best attainable form of an explanation of the law. It may, however, be remarked that such an explanation is in accordance with the chief characteristics of the force of gravity: it is not polar; and it seems to be so far different in kind from other physical forces, that it is not interchangeable with them, as they are among each other; for the attraction of gravity is thus referred to difference of mass, either between the two attracted bodies, or in the system of which they are parts. And other attractions, as will presently appear, are referred to differences of tension, or of dynamic equilibrium; and such conceptions do, though that of difference of mass does not, involve polarity.

10. Consider now the second (β) of the above-stated (7) general conditions of the development of a force. And, to fix our thoughts, let the system consist of three, and not of an indefinite number of bodies. That it must consist of at least three bodies, will be evident from the general reasoning of this theory; for the fundamental difference between the established and the proposed conception of a force is this:—a cause of motion is ordinarily defined as "an action between *two* bodies;" it is here conceived as a differential action between *three* bodies,—a conception, it is submitted, equally defensible by metaphysical reasoning and by physical experiment.

11. The displacement of a molecule, conceived as an atomic centre, may be evidently under such physical and analytically expressible conditions as to cause such displacement to be permanent or alternating. The first gives the conception of tension, by which electric and magnetic, the second that of vibration and undulation, by which optic and acoustic, phenomena are explained.

Let the conditions of the displacement of the molecular centres of lines of pressure be such that they are (and consequently, the body they constitute) in a state of tension.

12. The theory of electricity here proposed is the development of the idea of the tension of atoms as above defined. Hence it is an immediate, not a mediate, mechanical conception of the phenomena; for the conceptions of atoms or of centres of lines of pressure are not hypotheses, but convenient forms of the general conception of the parts of matter as mutually repelling. And the ground of the theory is this, that if matter is so conceived, experimental facts may themselves at once, and without the aid of hypothetical virtues or æthers, be mechanically conceived. For the theory has arisen from finding that the *more simply* facts were worded the clearer expressions they became of the general laws of motion; and that hypotheses only obscured the meaning and necessity of the relations of phenomena. It should therefore seem to be here sufficient briefly but clearly to express the conception of tension in its general relations, without entering into any detailed explanation of how it applies to different phenomena; for if there is any truth in the theory, such application ought at once to appear to those who can look at phenomena, and not at their hypothetical representations only, at least admissible. The clear proof that phenomena can be thus immediately (from a general conception of matter, and without aid of hypotheses) mechanically conceived, rests, of course with analysis, on the data of recorded and further experiments.

13. Little more, therefore, will here be offered than *primâ facie* evidence of the truth of the theory. The above considerations suggest that this may perhaps best be given by a statement of those general mechanical conceptions of (I.) the nature, (II.) the states, and (III.) the effects of electricity, which appear to be rather inductive generalizations of phenomena than hypotheses by which they are to be explained. A few experimental facts will be recalled under each head, but rather as suggestive of others than as by any means exhaustive of those which might be cited in support of these generalizations.

14. As to (I.) the nature of electricity: it is conceived as a permanent (not alternating) displacement of molecular centres of lines of pressure. Hence duality and polarity. For by such displacement there is evidently a tendency to produce motion of opposite characters in opposite directions; the pressure of the lines being increased in the direction of displacement, and correspondingly diminished in the opposite direction. Hence also the identity of the various forms in which electricity may be developed; the various conditions of such displacement differing only in intensity and quantity; and these being inversely as each other. For if intensity is conceived as amplitude of displacement, it is clear that, as the condition of a great displacement of a centre of

lines of pressure is evidently the resistance or non-displacement of the centres of one or more of the limiting atoms in the direction of displacement, the more equally and quickly displacement of one atomic centre causes displacement of all neighbouring centres, the less will be the amplitude attained by the originally displaced molecule.

15. The fitness of the conceptions here offered of the *nature and states* of electricity will appear chiefly in the proof of the generalizations to which they lead of the *effects* of electricity. But the above view of the nature of electricity is also founded on such facts as these. The "sources de l'électricité*," mechanical, thermal, or chemical, are all motions, or conditions of displacement. These motions are of different momenta. Compare facts as to the heterogeneousness of the bodies in frictional and thermal electricity; and as to their compound character in chemical electricity, interpreted by the theory of the chemical constitution of bodies hereinafter enunciated. In reference to the above conception of the difference of electricities, compare facts proving "the identity of electricities derived from different sources†," and, more particularly as to intensity, the various facts defining the correlative conceptions of insulation and discharge.

16. (II.) The states of electricity are thus conceived. (1) In a statically electrified body, the tension, or displacement of the molecular centres of pressure, is in closed curves forming the surface of the body, and the direction of displacement is either outwards (positive) or inwards (negative); and it seems demonstrable that there will be such a displacement *only* on the surface of the body.

17. Compare such facts as :—it is the least resisting of two rubbed bodies that is found negatively electrified; if an electrified sphere is hollow, there will be little or no electricity on the inner surface, &c.

18 Two conditions have been well distinguished by Ampère‡ in (2) dynamically electrified bodies—that of "courant ouvert," and that of "courant fermé." The former is conceived as a condition of longitudinal, the latter of transverse or spiral tension. Hence a polarity corresponding to the duality of statically electrified bodies. For electric poles are conceived as extremities towards, and from, which molecular centres of lines of pressure have been moved, at which, therefore, there is not only a change in the mechanical relations to outward atoms of the lines of pressure from these extremities, but a change at the one extremity of increase, at the other of diminution, of pressure. And

* De la Rive, *Traité de l'Electricité*, cinquième partie, ii. 456—828.
† Faraday, 'Experimental Researches,' Series IV.
‡ Ampère, *Théorie des Phénomènes Electrodynamiques*. Paris, 1826.

thus a magnet is conceived as a body the molecular centres of pressure of which are transversely displaced about an axis joining its poles.

19. As.has been said, the fitness.of conceptions of the nature and states of electricity is to be proved chiefly by their explanations of the effects of electricity; but note also, in support of these conceptions of current electricity, such facts, as to the origin of open currents, as that mere difference in the motions of metals in contact develope in them electric currents; and, as to closed currents, note that the experimental conditions which define a closed current, and the facts as to the molecular structure of magnets, however much they may still require investigation, seem already to justify a conception of magnetism wholly referable to conditions of mechanical action and resistance—a conception the antithesis of that which "associates" with bodies "latent virtues" and "neutral fluids," and a conception which evidently differs also, though with respect, from the later theories in which mechanical principles are applied, not immediately to phenomena, but to, as it should seem, needless hypotheses of "æthers*."

20. (III). The effects of electricity may be generalized under these heads:—(A) induction; (a) insulation; (β) discharge; the latter distinguishable into (1) conduction; (2) electrolytic discharge; (3) disruptive discharge; (4) convection, or carrying discharge†: (B) motion (a) of bodies; conveniently classed as (1) ordinary attractions and repulsions; (2) paramagnetism and diamagnetism; (3) right- or left-handed deviations or rotations; (β) motion of the medium, including heat and light, &c. vibrations.

21. (A) Induction is conceived as the necessary mechanical consequence in a plenum of the displacement of centres of lines of pressure. For it is clear that, if the parts of matter are conceived as mutually repelling, and such repulsion is mechanically conceived in such forms as those above suggested—as the elasticity of atomic centres, or as lines of pressure from molecules, —the displacement of the molecules forming a line of centres of pressure implies a disturbance of the previously existing mechanical relations, not only at the extremities of, but all round such a line of displacement; and such a conception implies, further, that the character of such disturbance depends entirely on the relations between the direction and intensity of the original

---

* I would desire more especially to express the diffidence and respect with which I venture to differ from Professor Challis as to the necessity or advantage of the fundamental hypothesis of his theory. See Phil. Mag. February, October, and December 1860.

† Faraday, 'Experimental Researches,' Series VII. 1319.

displacement and the varying resistance offered by the diverse conditions of motion and aggregation of opposed centres of lines of pressure.

22. Now, as to the facts which justify the proposal of this conception as a true generalization by which mechanical principles can be immediately applied to the phenomena of induction, it may be remarked that, as the full proof of such a conception depends on an experimentally based, and analytically expressed, mechanical theory of the constitution of bodies, it is only surprising that, while so little has been done towards the establishment of such a theory, the general conclusions of researches on induction, independent of, or with wrong, theories of chemical constitution, should go so far to make induction mechanically conceivable. Among such conclusions, each of which will call up a vast number of experimental facts, may be mentioned:—Induction is the origin and effect of all electrical phenomena; and is an action, not "at a distance," but "through contiguous particles" in lines of any curve. Insulation and discharge, or conduction, are differences only of degree; and bodies have specific inductive capacities which are but degrees of resisting power. The degree to which particles are affected before discharge constitutes intensity; and, in order to discharge, intensity must be raised much higher for a solid than for a fluid, and higher for a fluid, than for a gaseous, dielectric. An electric current has not only polar, but lateral, inductive effects; and the "lines of force" about a magnet take the form of a "sphondyloid." As to such facts as that gases, having the same inductive capacity, differ in insulating power, and that the effect of a magnet on a body without it is not affected by great rarefication of the medium (or a vacuum), their explanation more particularly depends on a mechanical theory of the constitution of bodies, and the principle that the character of inductive effect depends on the conditions of molecular motion and aggregation of the body acted on.

23. The general principle by which this theory gives a common explanation of the above classified (B) (α) motions of bodies in presence of an electrified or magnetized body, is that the *mechanical motions of bodies* in such circumstances *are effects of differential molecular displacement*; or, as it may be otherwise expressed, if, of two bodies, one resists molecular displacement from a centre of disturbance less than the other, the former moves towards that centre as a direction of least resistance. For it is evident that if a force has its full effect in a molecular displacement, the body will, as far as the direct action of that force is concerned, remain at rest; and that if the molecules of a body resist displacement, the force will have its effect in the

repulsion of the body. Molecular and bodily motion, or resistance thereto, are inversely as each other. Hence, if a force has more effect in producing molecular displacement in one body than in another, the difference will be seen in a tendency to repel this second body, the reaction of which will evidently urge the first towards the centre of force.

24. A corollary of this theorem is, that electrified bodies of which either the molecular tension or the inductive lateral disturbance is in the same direction approach; or, as it may be otherwise more concretely expressed, opposite poles, and similar currents, attract. For it is evident that, when the directions of the molecular displacement of two bodies are in the same line, a point of increased, is opposite a point of diminished, molecular pressure. Hence, transmission of similar molecular displacement from the one is in this position less, in the opposite, more, resisted by the other than by the medium. And hence, as above, attraction in the former, and repulsion in the latter case.

Further, it is evident that, according as two parallel currents or lines of tension are in the same, or in opposite directions, will their lateral disturbance of equilibrium be in the same, or in opposite directions inwards; and hence, that the reverse lateral motions, or at least tendencies to motion, of the bodies, according as their currents are in the same, or opposite directions, are explicable in the same way as, above, the motions of bodies with the same, or opposite directions of molecular tension, that is, with opposite, or the same poles opposed.

25. The special facts which seem to justify the advancing of the above theorem and its corollary as a true generalization and mechanical explanation of electric and magnetic motions, may be summed up under the following experimental conclusions:—Paramagnetism and diamagnetism are not absolute, but relative conditions of bodies. Paramagnets tend to pass from weaker to stronger, and diamagnets from stronger to weaker, places of action. Two of either class repel, and one of each attract. These motions would be explicable as due to differences of conduction; but magnetic, is quite different from electric, conduction. As to these facts, if their mechanical meaning is not from the foregoing sufficiently clear, remark that, the tension of a magnetized body being spiral, while that of a body with an "open current" is longitudinal, the directions of the lateral inductive actions of a magnet and an ordinarily electrified body will be different, and hence the molecular conditions which permit of electric, will be different from those which favour magnetic, conduction. And if differences of conduction, that is, of molecular displacement, are thus admitted in the explanation of paramagnetic and diamagnetic phenomena, it is evident, from what has been already said of the

effect of differential molecular action, that a better conductor will, in moving in the direction of least resistance, pass to a stronger place of action.

The above conclusions are Faraday's[*]; but the theorem and corollary as generalizations rest also on the facts adduced by Ampère and his successors in support of his helix-theory of the magnet.

A third most important class of facts by which the view here given of the mechanical conditions of the electric motions of bodies may be supported, is the disposition of iron filings about one or more electrified wires or magnets in various positions, and the information given by a moving wire as to magnetic forces.

26. The fundamental importance of the conceptions of a force as, in general, a difference of pressures, and of polar attraction and repulsion as the effect of a differential molecular action, has induced me to give such disproportioned length to their illustration, that it will be impossible within the brief limits of this paper to do more than note the other chief points of this general theory.

What, therefore, has to be said on (B) (β) the effects of electricity as manifested in motions of the medium, must be referred to the paragraph on the correlation of forces.

27. There will not, it is hoped, be thought to be presumption in offering new views in theories which have been elaborated with such admirable genius as those of light and heat; for the most strenuous supporters of the present form of the undulatory theory candidly admit that "there undoubtedly are several classes of phenomena which the wave theory has not merely *failed to explain,* but which are apparently at *direct variance* with its principles[†]."

It will be evident that the chief new view necessitated by this general theory (and which alone can be here noticed) resolves itself into a theory of the connexion of the elastic medium with the vibrating molecules in it. Now, though according to the present theory "it is certain that light is produced by undulations propagated with transversal vibrations through a highly elastic æther, yet the constitution of this æther, and the laws of its connexion (if it has any connexion) with the particles of bodies, are utterly unknown[‡]." But the theory here proposed implies such "laws of connexion." For its practical result is, that the "æther" is conceived as the mutually determined lines of

[*] But compare Tyndall's Memoirs "On the Reverse Polarity of Bismuth" (Phil. Trans. 1855 and 1856).

[†] Baden Powell, 'Undulatory Theory,' Introduction, p. xxiv.

[‡] MacCullagh, "Laws of Crystalline Reflexion," &c., Mem. R. I. A., xviii. 38.

pressure from molecules of the size to give by their vibrations the sensations of light and heat.

It would seem that this conception of the "æther" leads to the explanation of more than one difficulty in the established theory; but nothing can, of course, be advanced on such a point except as the result of analysis.

It will be understood that I thus speak of the conception of atoms, as above defined, as a mode of conceiving the "æther" of the undulatory theory, only in order to make clear the application to that theory of the fundamental conception of the general theory here proposed; and that this in nowise contradicts what has been above said as to the seeming needlessness of hypotheses of *special* fluids, or æthers, acting on, or through, matter.

28. A general chemical theory is made up of two—a theory of the constitution, and a theory of the combination, of bodies. As to the constitution of bodies, the principal views here offered are:—Bodies are conceived as states of dynamic molecular equilibrium, that is, as states of molecular motion in which, while there is no decomposition, the intensity of motion is at all points equal. Hence, their differences are conceived as resulting from different conditions of molecular motion; and thus specific heat becomes one of the chief exponents of the nature of a body.

The distinguishing mechanical characteristic of the gaseous, fluid, and solid, states of matter is degree of tangential resistance. In an absolutely perfect gas the molecules would be of equal mass and at equal distances: hence perfect equality of resistance in all directions. But let there be inequality either in the masses or distances of the molecular centres of pressures, it is evident, from the same reasoning as that above applied to the mechanical explanation of gravity, that there would be a cohesive force developed, the consequence of which would be inequality of resistance. Fluids, therefore, as opposed to solids, are conceived as bodies the molecules of which are of sensibly equal masses and at equal distances; and gases, as opposed to liquids, as bodies in which the distances of the molecules, though equal, are greater than in liquids. Hence the greater amplitude of their motion, or specific heat.

The consideration of this theory of bodies with reference to that of Krönig[*] and Clausius[†], cannot be at present entered upon.

---

[*] " Grundzüge einer Theorie der Gase," Pogg. *Ann.* xcix. 315.

[†] " Ueber die bewegende Kraft der Wärme," Pogg. *Ann.* lxxix. 394; and " Ueber die Art der Bewegung welche wir Wärme nennen," ibid. c. 353. See also Prof. Maxwell's "Illustrations of the Dynamical Theory of Gases," Phil. Mag., January and July 1860. Dr. Tyndall's discoveries and researches " On the Transmission of Heat of different qualities through Gases of different kinds " (Proceedings of the Royal Institution, June 10, 1859), are of the greatest importance in such a dynamical theory of the constitution of bodies.

29. The combination of bodies, in whatever way it takes place, and not confining the meaning to the formation of salts*, is conceived as the establishment of a new state of dynamic equilibrium†. Hence there must necessarily be definite laws of chemical combination.

But as it is proposed to give the outlines of this mechanical theory of chemistry in a future paper or papers, it is unnecessary to enter more fully upon it at present. A theory so important has been introduced in a paper in which it must be so inadequately expressed, only to complete the general outline of the science of motion, and because a mechanical theory of chemistry is the necessary complement of the above-given mechanical theory of electricity.

30. In such a theory of forces as that here proposed, the great experimental truth of the " correlation of forces"‡ assumes an axiomatic clearness. Especially is it to be noted that it gives an account of the difference of the correlation between electricity and magnetism, and that of either of these with light, heat, &c. The former is a correlation of coexistence, the latter of change of conditions.

If I have been successful in making clear the conceptions offered in this theory of electricity and magnetism and of induction, it will be unnecessary here to say anything further of this correlation of coexistence; for it is implied in the conception of induction as a mechanical effect varying with resistance, and manifested at right angles to a longitudinal or transverse (spiral) line of tension, that is, of permanent (not alternating) molecular displacement. And if electricity is a permanent, light an alternating, molecular displacement, their correlation is evident, the conditions of their interchange assignable, and analytically expressible.

But not only is the correlation thus clear of physical motions among themselves, but also of these, as a class, with mechanical motions on one side, and chemical motions on the other—clear, that is, that the stopping of a mechanical, or the beginning of a chemical, motion implies more or less molecular displacement under such conditions as those above assigned for electricity, light, and heat. But what is the proof of such a correlation of forces but the sublime " persuasion that all the forces

---

* Berthelot's discoveries have definitively broken down the distinction between Inorganic and Organic Chemistry. See his *Chimie Organique.*

† Compare Williamson's " Theory of Etherification," Chem. Soc. Quart. Journ. iv. 110.

‡ Grove. See also Helmholtz in Taylor's Scientific Memoirs, 1853, p. 124; and compare Rendu's " *circulation* " of fire, light, electricity, and magnetism, *Théorie des Glaciers de la Savoie,* Memoirs of the Royal Academy of Sciences of Savoy, 1841, cited in Tyndall's ' Glaciers of the Alps,' p. 299.

of nature are mutually dependent, have one common origin, or rather are different manifestations of one fundamental power[*] " —become, at least for all the phenomena of Motion as distinguished from those of Growth, a scientific truth ?

31. For the sake of distinctly defining by its relations the Science of Motion, a few words may be added on the Classification of the Sciences proposed by the author. The general divisions of each of the two great classes—the Natural and Humane Sciences—are (A) the Systematic, (B) the Descriptive, and (C) the Historic Sciences.

The Systematic Natural Sciences are the sciences of (I.) Motion, (II.) Growth, (III.) Species ("the Classificatory Sciences" of Whewell)[†].

The twofold subdivision of each of the three sciences of motion will be evident from the foregoing.

Each of these rational sciences, as well as those making up the Science of Growth, have a corresponding applied science.

The Science of Motion, as a distinct science, and not as merely a general name for the sciences of mechanics, physics, and chemics, has a twofold division.

Under the first, the general relations, laws, or principles of motion, without regard to the particular conditions of its cause, would be considered: such general principles, for instance, as Galileo's laws of Inertia, and of the Composition of Motions; Newton's law of the Equality of Action and Reaction; its generalization in D'Alembert's principle; Jean Bernoulli's (?)[‡] principle of Virtual Velocities; Varignon's geometrical theorem on Moments; Poinsot's theory of Couples; Newton's principle of the Conservation of the movement of the Centre of Gravity; Kepler's principle of Areas; Laplace's theorem on the Invariable Plane; Euler's on Moments of Inertia, and Principal Axes; Huygens's principle of the Conservation of Vires vivæ; Daniel Bernoulli's theorem on the Coexistence of small Oscillations, &c.

The object of the second division of the General Science of Motion would be the generalization of the conditions giving rise to the various Forces of Motion, mechanical, physical, and chemical, and their correlations. This paper, therefore, has treated of the principles of but one division of the science.

The divisions of the special science of Mechanics, that on Solids, and that on Fluids, would be conveniently considered in kinematical [§], statical, and dynamical sections.

---

[*] Faraday, ' Experimental Researches,' 2702.
[†] History of the Inductive Sciences, iii. p. 211.
[‡] Poisson, *Traité de Mécanique*, i. p. 654.
[§] See Ampère's *Essai sur la Philosophie des Sciences*; Willi's to his ' Principles of Mechanism,' and the arrangement of I 'Manual of Applied Mechanics' (*Encyc. Metrop.*).

32. In conclusion, as every one of these theories is so dependent on every other, that it was necessary to give first a general outline of them all, it is hoped that allowance will be made for the imperfection, or even inaccuracy, unavoidable in so brief an exposition of each. And lest, from imperfect expression, the general principles themselves may be misunderstood, it may be added that the author's confidence in them arises only from their long-tested seeming accordance, not only with *experimental* generalizations, but with *scientific* metaphysics. For every physical theory must be implicitly, or explicitly founded on metaphysical views as to the nature of Knowledge, as to the mode in which Matter and Force are to be conceived, and as to the meaning of Law. And the metaphysical bases of this physical theory have been chiefly found in the Philosophy of Bacon, and of the Scottish School.

The necessary limitation of human knowledge, the central doctrine of Sir William Hamilton's system, and, indeed, the great result of the modern Critical School founded by Kant, is the citadel of those who would clear the Natural Sciences of absolute, and essentially distinct, force or forces acting on, associated with, or emanating from, matter. The fundamental conception of this theory—that of a force as a condition—is but a development of the doctrine of "Forms," "the investigation of which is the principal object of the Baconian method of induction *." The conception of matter is in accordance with that of substances, as "corpora individua edentia actus puros individuos ex lege † ;" and that of a Law, as not an imposed rule to be discovered, but a relation to be expressed, seems to agree no less with the principles of Bacon, with whom "the statement of the distinguishing character of the motion or arrangement, or of whatever else may be the form of a given phenomenon, takes the shape of a Law ‡," than with the principles of Mill §.

6 Stone Buildings, Lincoln's Inn,
  10th Dec. 1860.

VI. *On the Opacity of the Yellow Soda-Flame to Light of its own Colour.* By WILLIAM CROOKES ‖.

IN their remarkable investigations on the colours which certain substances impart to the flames in which they are heated ¶, Professors Kirchhoff and Bunsen describe certain experiments by

* Bacon's Works, by Ellis and Spedding, i. p. 31.
† Ibid.   ‡ Ibid. p. 29.
§ System of Logic, Ratiocinative and Inductive.
‖ Communicated by the Author.
¶ Chemical Analysis by Spectrum Observations.   By Professors Kirchhoff and Bunsen.   Phil. Mag. S. 4. vol. xx. p. 89. August, 1860.

which they prove that the luminous lines which are produced in the spectrum of a spirit- or gas-flame, when salts of the alkalies or alkaline earths are caused to volatilize therein, become *reversed* (*i. e.* that the bright lines become changed to dark ones) when a source of light of sufficient intensity, and giving a continuous spectrum, is placed behind the coloured flame.

During some researches on the spectra of artificial flames which I have been carrying on simultaneously with MM. Kirchhoff and Bunsen, one or two forms of experiment suggested themselves which, while they perfectly corroborate most of the facts mentioned by them, seem to merit attention, from the facility with which they may be performed, and the striking manner in which the phenomena can be exhibited to several persons at once without the necessity of employing any optical apparatus. The atmosphere of the room—or rather that part of it in which the illustration is performed—is to be first impregnated with soda-smoke, by igniting a piece of sodium, the size of a pea, on wet blotting-paper. Any flame, whether of gas, spirit, a candle, &c., which may now be burning anywhere in that part of the room, will exhibit in a marked manner the well-known yellow soda-flame; and if the full amount of gas in an ordinary wire-gauze air-burner is turned on and ignited, it will give a uniformly brilliant yellow flame, upwards of a foot high and 3 inches in diameter.

If a smaller flame be now moved in front of this large one, it will exhibit a curious phenomenon. Those parts of it which are ordinarily seen to be luminous will suffer no change, other than that slight diminution in intensity which might be anticipated from their projection in front of a broad but not very brilliant source of light; but beyond these there will appear a sharply-cut and intensely black narrow border, closely surrounding the visible flame, and presenting the curious appearance of the latter being set in an opake frame. A closer scrutiny will show that the position of this black rim is not, as I at first supposed, in that outer cone in which the yellow soda-flame is most distinctly seen, but that it lies in the dark space immediately outside the luminous part of the flame, affording proof of the existence of another invisible cone of vapour. The flame from a tallow candle shows this appearance better than that of wax or sperm, probably on account of its inferior luminosity. A small spirit- or gas-flame will also answer very well; but I think a tallow candle shows the phenomenon in a more striking manner.

The fact of the cone of yellow soda-flame being transparent, while the outer, non-luminous space is perfectly opake to the same kind of light placed behind it, appears worthy of attention. It seems to show that the yellow flame caused by the presence of incandescent solid particles of a sodium compound has no very marked absorptive action on light of its own colour; but that to give rise to this kind of opacity it is necessary that the sodium compound should be in the state of vapour. It appears, moreover, to prove that it is not necessary for this vapour to be in the metallic state; for it could hardly be supposed that so highly combustible a vapour as that of metallic sodium could be present in that part of the flame which is seen to possess this great opacity. That soda salts are easily volatile at the temperature of flame, is a fact abundantly proved by Bunsen*. The reason why the opacity is only exhibited by that part of the outer shell of vapour which is situated at the edge of the flame, and not by its entire extent, is owing to its thickness being insufficient to produce sensible absorption on rays which traverse it perpendicularly; an appreciable action taking place only when they pass as a tangent to the edge of the flame, and thus traverse a considerable extent of absorbing medium.

VII. *Experimental Researches on the Laws of Absorption of Liquids by Porous Substances.* By THOMAS TATE, *Esq.*

[Continued from vol. xx. p. 510.]

II. *On the Filtration of Liquids through different Porous Substances.*

FILTRATION is in general produced by the action of two forces, viz. by the force of absorption and that of pressure. Filters may be divided into two classes. The first class comprises those substances which are highly porous, and which undergo little or no change during the process of filtration. The second class comprises those substances with close pores, which under certain circumstances undergo a decided change during the process of filtration. The following laws (with certain limitations) apply to both kinds of filters:—

1. The rate of filtration, other things being the same, varies directly as the area of the surface of the filter in contact with the liquid, and inversely as the thickness.

2. The rate of filtration, other things being the same, increases in a high ratio with the increase of temperature.

3. The rate of filtration, other things being the same, varies as the depth of the column of liquid upon the filter.

And so on to other laws which will be hereafter illustrated.

These experiments were, for the most part, made with the apparatus represented in the annexed diagram.

* Phil. Mag. S. 4. vol. xviii. p. 513.

AB a wide glass tube, containing the liquid, about 2 feet in length, graduated into units of cubic inches, or into units of half cubic inches, as the case may be; C and D two equal plates of polished slate having equal orifices bored through their centres; the plate C is cemented (with a resinous cement) to the bottom of the tube, and the filter *e* is cemented to both plates, so that all lateral discharge from the filter is stopped, and at the same time the filter presents a surface, in contact with the liquid, equal to the section of the orifices of the plates. All liquids, before being used in the filtrometer, were twice filtered through ordinary filtering paper, and the top of the filter tube AB was covered during the experiment to prevent any dust from falling into the liquid. The tube AB was filled up, with the liquid, to a certain point of the graduation, and then the time at which the liquid, in its descent, arrived at the different points of the graduation was duly noted.

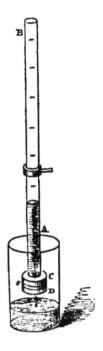

Experiment XVI.

The filter used in this experiment was common wood-charcoal half an inch in thickness. The liquid was distilled water. The diameter of the orifice of the plate was $\frac{4}{10}$ths of an inch. The filter-tube was graduated into 8 units, each containing half a cubic inch, and the 8 divisions measured 9·2 inches. The temperature was 56° throughout the experiment. The results recorded in the third column of the following Table are obtained by dividing the unit of space by the mean of the times taken in describing the two consecutive units of space; thus the velocity

at $4 = 1 \div \dfrac{46+55}{2} = \dfrac{1}{50}$.

Height of column of liquid, *h*,	Corresp. time in seconds, T.	Corresp. velocity of descent, *v*.	Value of *v* by formula $v = \dfrac{h}{2v_3}$	Value of T by formula (4).
8	0	...	$\frac{1}{35}$	0
7	27	$\frac{1}{30}$	$\frac{1}{36}$	26·5
6	58	$\frac{1}{42}$	$\frac{1}{43}$	57·5
5	94	$\frac{1}{47}$	$\frac{1}{50}$	94·0
4	140	$\frac{1}{50}$	$\frac{1}{60}$	139·0
3	195	$\frac{1}{78}$	$\frac{1}{77}$	196·0
2	272	$\frac{1}{103}$	$\frac{1}{101}$	277·0
1	400	...	...	415·0

The near coincidence of the results in the third and fourth columns shows that the *velocity of discharge varies directly as the height of the column of liquid upon the filter.*

Let S' = the whole depth of the liquid at the commencement of the experiment, in units of the divisions of the filtrometer; S = the descent of the liquid in the time T; v = the corresponding velocity of descent; α, γ, ρ = constants; then we have generally,

$$T = \alpha - \gamma \log (S' + \rho - S). \quad \cdots \quad (1)$$

By differentiation, we get

$$\frac{dS}{dT} = \frac{\log_e 10}{\gamma}(S' + \rho - S);$$

$$\therefore v = \frac{2 \cdot 30258}{\gamma}(h + \rho). \quad \cdots \cdots \quad (3)$$

In the foregoing experiment, $\rho = 0$, $\dfrac{2 \cdot 30258}{\gamma} = \dfrac{1}{203}$, and the relation of time and space is expressed by the formula

$$T = 415 \cdot 74 - 460 \cdot 4 \log h. \quad \cdots \cdots \quad (4)$$

This experiment, upon being repeated, gave a slight diminution in the velocity of descent of the liquid, showing that the filter had undergone only a very slight change during the process of filtration. This observation applies to the filters used in the three following experiments.

Experiment XVII.

The filter in this experiment was coke, $\frac{4}{10}$ths of an inch in thickness. The liquid was distilled water. The diameter of the orifice of the plate was $\frac{7}{10}$ths of an inch. The filter-tube was the same as in the last experiment, and the temperature was 57° throughout the experiment.

Height of column of liquid, h.	Corresp. time in seconds, T.	Corresp. velocity of descent, v.	Value of v by formula $v = \dfrac{h}{203}$.
8	0	...	$\frac{1}{25}$
7	27	$\frac{1}{29}$	$\frac{1}{29}$
6	59	$\frac{1}{34}$	$\frac{1}{34}$
5	95	$\frac{1}{40}$	$\frac{1}{40}$
4	139	$\frac{1}{50}$	$\frac{1}{51}$
3	195	$\frac{1}{68}$	$\frac{1}{68}$
2	275	$\frac{1}{101}$	$\frac{1}{101}$
1	400		

Here it will be observed that the liquid followed the same law of descent as that of the preceding experiment.

Reducing the coke filter to the same thickness and diameter of orifice as in the case of the charcoal filter, we find, under the same circumstances of pressure, &c., the filtering power of the charcoal to be 3¾ times that of the coke.

Experiment XVIII.

The filter in this experiment was stout woollen cloth. The liquid was distilled water. The diameter of the plate was $\frac{3}{10}$ths of an inch. Each half cubic inch graduation of the filter-tube measured on an average 1·55 inch, so that the height of the liquid column at the commencement of the experiment was 15·5 inches.

Height of column of liquid, h.	Corresp. time in seconds, T.	Corresp. velocity of descent, v.	Value of v by formula $v = \dfrac{A}{208}$.
10	0		
9·5	...		
9·0	22		
8·5	...		
8·0	46		
7·5	...		
7·0	74		
6·5	...		
6·0	106		
5·5	...		
5·0	143		
4·5	...		
4·0	190		
3·5	...		
3·0	250		
2·5	...		
2·0	330		
1·5	...		
1·0	480		

Here the results of the third and fourth columns show that the velocity of discharge varies according to the same law as in the two foregoing experiments.

Experiment XIX.

In this experiment the filter was sponge plugged tight into the bottom of the filter-tube, which was the same as that of the last experiment.

Height of column of liquid, h.	Corresp. time in seconds, T.	Corresp. velocity of descent, v.	Value of v by formula $v = \frac{h}{308}$.
11·5	0		
11·0	...		
10·5	28	$\frac{1}{34}$	$\frac{1}{34}$
10·0	...		
9·5	58	$\frac{1}{30}$	$\frac{1}{31}$
9·0	...		
8·5	92	$\frac{1}{34}$	$\frac{1}{34}$
8·0	...		
7·5	130	$\frac{1}{38}$	$\frac{1}{38}$
7·0	...		
6·5	174	$\frac{1}{44}$	$\frac{1}{44}$
6·0	...		
5·5	225	$\frac{1}{51}$	$\frac{1}{51}$
5·0	...		
4·5	286	$\frac{1}{61}$	$\frac{1}{61}$
4·0	...		
3·5	363	$\frac{1}{77}$	$\frac{1}{77}$
3·0	...		
2·5	460	$\frac{1}{97}$	$\frac{1}{102}$
2·0	...		
1·5	635	$\frac{1}{175}$	$\frac{1}{154}$

The results of this experiment show that, for the first nine units of descent, the velocity of discharge varies almost exactly as the height of the column of liquid. For the last two units, the rate of discharge considerably exceeded that which the formula $v = \dfrac{h}{308}$ would give, showing that the effect of the absorbent power of the filter greatly exceeded that which would be due to the pressure alone.

This experiment was repeated with little or no variation in the results.

Similar results were obtained from filters formed of plugs of different sorts of soft porous material; also from a filter formed of fine sea-sand laid upon a perforated plate.

The following experiment was made to determine the variation of the rate of filtration due to increase of temperature.

Experiment XX.

The liquid used in this experiment was distilled water, and the filters were those employed in Experiments XVI. and XIX.

With the charcoal filter under a constant pressure, the time required to discharge one cubic inch of liquid at 52° temperature was 114 seconds; whereas at the temperature of 90° it only required 65 seconds. In this case an increase of 38° of temperature increased the rate of discharge $1\frac{3}{4}$ times; that is, the rate of discharge was nearly doubled by the addition of 38° of temperature.

With the sponge-filter the following results were obtained :—

Time in seconds to discharge 11 cubic inches of water.

At 50°.	At 80°.	At 90°.	At 100°.
420″	286″	246″	204″

Here an increase of temperature from 50° to 90° caused the rate of discharge to be increased 1·707 times, a result nearly coinciding with that determined for the charcoal filter. At 100° temperature the rate of discharge is a little more than double the rate at 50°. These results further show that, *for equal volumes of discharge, the decrements of time are for the most part proportional to the increments of temperature.*

The following experiment was made to determine the rates of filtration of different liquids as compared with that of water.

Experiment XXI.

The filter used in this experiment was charcoal; and the liquids compared were distilled water, and three different solutions of carbonate of soda. Solution No. 1 contained 2 per cent. of carbonate of soda; No. 2 contained 4 per cent.; and No. 3 contained 8 per cent.; that is to say, the per-centage of the salt in these different solutions were in the geometrical progression 2, 4, 8. The discharge in each case was produced under the pressure of a column of 9·2 inches of the liquid.

Time in minutes to discharge one cubic inch of the liquids.

Water . . .	$1·73 = 1·73 \times 1·08^0$
No. 1. . . .	$1·87 = 1·73 \times 1·08^1$ nearly.
No. 2. . . .	$2·03 = 1·73 \times 1·08^2$,,
No. 3. . . .	$2·20 = 1·73 \times 1·08^3$,,

Here it will be observed that the rates of discharge are very nearly in geometrical progression. This property of filtration is analogous to that of absorption, as shown in connexion with experiments XI. and XII. Thus it appears that the chemical composition of a liquid affects its relative rate of filtration.

In general the rate of filtration or filtrativeness of a liquid seems to depend mainly upon its viscosity, and not so much upon its specific gravity. Alcohol, oils, &c., which have a less specific gravity than water, have a low rate of filtration. Solutions of sugar and starch, even when much diluted, have very low rates of filtration as compared with that of water; whilst diluted acids and weak solutions of alkaline salts, for the most part, have a rate of filtration nearly equal to that of water. In these respects the law of filtration is analogous to that of absorption; at the same time it must be observed that the relative absorbent power of a substance does not always correspond to its filtering power.

The filters used in the following experiment belong to the second class of filters.

Experiment XXII.

The filter used in this experiment was thick unsized paper, and the liquid was spring water. The diameter of the orifice of the plate was $\frac{4}{10}$ths of an inch. The filter-tube was the same as that of Experiment XVI. At the commencement of the experiment the liquid stood at the eighth division of the tube measured from the filter. The temperature was 64° throughout the experiment.

Descent of the liquid, S.	Corresp. time in seconds, T.	Velocity per second, v.	Value of v by formula $v = \frac{1}{80} \cdot 1.355^{-S}$
0	0	$\frac{1}{80}$	$\frac{1}{80}$
½	20		
1	97	$\frac{1}{108}$	$\frac{1}{108}$
1¼	125		
2	225	$\frac{1}{140}$	$\frac{1}{148}$
2¼	260		
3	400	$\frac{1}{200}$	$\frac{1}{198}$
3¼	450		
4	620	$\frac{1}{280}$	$\frac{1}{270}$
4¼	685		
5	870	$\frac{1}{380}$	$\frac{1}{365}$
5¼	960		
6	1360	$\frac{1}{500}$	$\frac{1}{497}$
6¼	1480		
7	1990	$\frac{1}{680}$	$\frac{1}{670}$
7¼	2160		

The formula expressing the relation between the time T and space S of descent is

$$T = 262.3 \, (1.355^{S} - 1). \quad \ldots \quad \ldots \quad (5)$$

It will be seen how very nearly the velocity of descent of the liquid is represented by the formula $v = \frac{1}{80} \cdot 1.355^{-S}$; showing that *if the spaces of descent be taken in arithmetical progression, the corresponding velocities of descent will be in geometrical progression.* The common ratio of the velocities in this experiment is $\frac{1}{1.355}$.

The relation of T and S for both kinds of filters may be represented by the general formula

$$T = \alpha(\beta^s - 1) - \gamma \log (S' + \rho - S),$$

where α, β, γ, and ρ are constants.

When $\beta = 1$, this formula becomes the same as equation (I); and when $\gamma = 0$, or very small, it becomes

$$T = \alpha(\beta^s - 1), \quad \cdots \cdots \cdots \quad (6)$$

which is the general formula applying to the second class of filters.

By differentiation, we find

$$v = \frac{1}{\alpha \log_e \beta} \cdot \beta^{-s}, \quad \cdots \cdots \cdots \quad (7)$$

which is a general formula for the rate of descent of the liquid.

In the foregoing experiment, $\alpha = 262 \cdot 3$, $\beta = 1 \cdot 355$, and

$$\therefore v = \frac{1}{80} \cdot 1 \cdot 355^{-s}. \quad \cdots \cdots \cdots \quad (8)$$

Towards the close of the experiment the velocity of descent became exceedingly small, showing that the adhesion of the liquid to the bottom of the filter-tube interfered with the law of velocity expressed by equation (8).

This experiment being immediately repeated, it was found that the rate of filtration had sensibly diminished, showing that the filtering power of the paper had undergone a decided change during the process of filtration. This change, as will be hereafter shown, is progressive, being in proportion (within certain limits) to the quantity of liquid filtered. But after the filter had been dried, it somewhat regained its original power.

In this manner various experiments were made, which gave precisely similar results.

Although the liquid in these experiments had been carefully filtered through ordinary filtering-paper, yet it is possible that certain minute particles may have passed through the filter, sufficient to deteriorate the filtering power of a small filtering surface such as that used in the foregoing experiment.

The following experiments were made on *upward* filtration.

Experiment XXIII.

In this experiment the filter was immersed in a jar of liquid to the depth of 7 units of the filter-tube. As the liquid rose in the tube through the filter, the liquid in the jar was maintained at a constant level. The filter was the unsized paper of the last experiment, the diameter of the plate being $\frac{4}{10}$ths of an inch. The liquid was a diluted solution of carbonate of soda.

Depth of the column of liquid, λ.	Corresp. time in minutes, T.	Corresp. velocity of ascent, v.	Value of v by formula $v = \frac{A}{3·09}$.
6	0	......	$\frac{1}{1·44}$
5½	......		
5	1·58	$\frac{1}{1·55}$	$\frac{1}{1·55}$
4½	......		
4	3·51	$\frac{1}{1·93}$	$\frac{1}{1·93}$
3½	......	$\frac{1}{2·4}$	$\frac{1}{2·48}$
3	5·91		
2½	......	$\frac{1}{2·85}$	$\frac{1}{3·49}$
2	9·26		
1½	......	$\frac{1}{3·65}$	$\frac{1}{5·79}$
1	14·91		
½	......	$\frac{1}{17·84}$	$\frac{1}{17·96}$
0	32·75		

The near coincidence of the results in the third and fourth columns shows that in this case the *rate of filtration varies directly as the pressure upon the filter.*

With a double filter the rate of discharge was found to be reduced to one-half very nearly; and with a filter of $\frac{2}{10}$ths of an inch diameter, that is, one-fourth of the surface of the filter in the above experiment, the rate of discharge was found to be reduced to one-fourth of that of the above experiment. Hence we conclude that *the rate of filtration varies directly as the areas of the surface of the filter in contact with the liquid, and inversely as the thickness.*

[To be continued.]

VIII. *A Theory of Magnetic Force.*
By Professor CHALLIS, *F.R.S.*[*]

IN my last communication, containing a theory of galvanism, no allusion was made to *thermo-electric* phenomena. This omission I propose to supply before proceeding to the theory of magnetism, the fact that galvanic currents may be produced by heat, being confirmatory of the explanation given by hydrodynamics of the generation of secondary currents generally, whether electric, or galvanic, or magnetic.

1. It is assumed in that explanation that in the neighbourhood of the earth's surface there are steady ætherial currents, which eventually may be found to be secondary with respect to other more general currents, but for considerable spaces may be considered to be uniform as to velocity and density. These currents are supposed to flow freely through the interior of bodies, with only such modifications as may result from the arrangement of

* Communicated by the Author.

the atoms, and the contraction of the channel by the atomic occupation of space. At the parts of the boundary of any substance where the stream *enters*, there will be a sudden *increment* of velocity, and at the parts where it *issues*, a sudden *decrement*; but not to a large amount, because there is reason to conclude that even in dense bodies the space occupied by atoms is very small compared to the vacant space. Under all circumstances the motion remains steady, if there be no extraneous force, and no variation of the primary current. If, throughout the interior of the body, the atoms within a given space (as one thousandth of a cubic inch) occupy a portion of it which has a constant ratio to the vacant space, they produce no acceleration or retardation of the velocity of the stream. But if from any cause there should be a gradation of internal density of the atoms, the atomic composition remaining the same throughout, secondary streams will be produced in the following manner:—Conceive the external primary current to be cut at right angles by a plane in a given position. Then since it is supposed to be steady and uniform, the velocity will be always the same at all points of the plane. Hence, tracing two contiguous and equal filaments of the stream into the interior of the body, it will be seen that if the density of the atoms be greater at certain portions of the course of one filament than at the adjacent portions of the course of the other, the velocity will be *greater* in any element of the *former* portions, than in the adjacent element of the *latter*, assuming, as may be done, that in other respects the channels of the two filaments are at adjacent parts of equal dimensions. Hence by the general hydrodynamical equation for steady motion, the density and pressure of the æther will be *less* in the *former* elements than in the *latter*. Consequently there will be an accelerative force of the fluid always tending from the *rarer* to the *denser* parts of the body, and taking effect, whatever be the direction of the original stream, in the directions of normals to surfaces of equal density, because in these directions the change of density of the atoms in a given space is greatest. These accelerative forces produce secondary currents, the velocities of which will depend on the magnitudes of the forces, and on the extent through which they act. Considering the vast elasticity of the æther, as shown by the rate of the propagation of light, a difference of its density equal to a ten-millionth part of the whole density would correspond to an enormous difference of pressure.

2. According to these views it might be expected that a lamina of metal, heated unequally at the two extremities, would become electro-dynamic, as Volta found to be the case with a lamina of silver. In this instance the heat produces local disturbances of the condition of *superficial* atoms, and from these disturbances

probably the internal gradation of density chiefly results, as in a body subject to the influence of electricity by induction. If a closed circuit of any metal be heated at any point, currents, according to this theory, will be generated in consequence of the different degrees of expansion of the metal at different points, and, flowing in opposite directions, will neutralize each other if they be equal. But if by any mechanical means they be made unequal, as by contortions of the metallic circuit, a galvanic current should result, as is found to be the fact by experiment. When the circuit is formed by two metals soldered together at their extremities, and heat is applied at one of the positions of junction, inequality of the opposite currents might be expected to arise from difference of the capacity of the two metals for generating currents, owing to difference of their atomic constitution; and accordingly it is found that under these circumstances a galvanic current is produced. I proceed now to the theory of magnetism.

3. In the preceding theories the generation of secondary ætherial currents has been ascribed to a disturbance of the atomic condition of bodies by external agency,—in electricity, by friction; in galvanism, by the mutual molecular actions of dissimilar substances in contact; and in thermo-electricity, by heat. In the theory of magnetism it must be assumed that there are substances in which a gradation of interior density exists independently and permanently,—that iron, for instance, is found in this state in nature; that the same state may be induced in steel by mechanical means, with different degrees of permanence; and that it may be momentarily induced in soft iron. Also it must be supposed that the direction of the gradation of density depends on the *form* of the magnetized body. These suppositions rest immediately on facts of experience, the explanations of which, since they relate to qualities of the bodies, and not to the agency by which magnetic phenomena are produced, are not now under consideration.

4. Let us take the case of a bar of magnetized steel of the form of a long rectangular parallelopiped, and let it be assumed that there exists permanently a uniform decrease of its atomic density from the end A to the end B. By the argument in art. 1, on the supposition of a steady and uniform primary current, there will be impressed on the æther within the bar a uniform accelerative force, acting throughout its length from B towards A. To the accelerative force in that half of the bar which lies towards B, may be ascribed the effect of overcoming the inertia of the æther in motion within and without the bar on the side of B, and to the accelerative force in the other half, the effect of overcoming the resistance opposed to the flow of the current by the

inertia of the æther on the side of A. Thus the motion will be maintained so as to be symmetrical with respect to a neutral position N, mid-way between A and B, the partial streams converging towards the parts about B, and diverging in like manner from the parts about A. The velocity of the æther will be greatest, and its density least, at N, and the velocity will decrease, and the density increase, in both directions from this position by the same gradations. Hence the atoms, assumed to be of finite dimensions, will be urged on both sides towards N, by reason of the excess of pressure on the halves of their surfaces turned from N, and the total moving forces in the opposite directions will be equal. Consequently the theory not only accounts for the well-known fact, that *the magnetism of a steel bar is equal and opposite on the opposite sides of a middle neutral line,* but explains also why terrestrial magnetism, being supposed, for reasons that will be hereafter adduced, to act as the primary current of the theory, produces no perceptible motion of translation of the bar. These inferences will not be altered when the dynamic effect of the *motion* in the secondary streams is taken into account, as will be shown in a subsequent part of the theory.

5. It also follows, conformably with experience, that if a magnetized bar be divided into two or more parts by being cut transversely, each part becomes a magnet, because it may be assumed that the gradation of density from end to end is the same, and in the same direction, in each, as when it formed a part of the whole bar.

6. The theory of the action between two magnets requires the investigation of the mutual influence of two steady streams, having separate origins and interfering courses. The following laws, which may be admitted on hydrodynamical principles, will suffice for my present purpose. The resultant of two interfering steady streams is steady. Where two streams meet, the velocity is in general less, and consequently the density greater, than that which would be due to either stream flowing separately. When two streams unite in the same course, the velocity is greater, and the density less, than in either of the component streams. When the courses of two streams cross at right angles, the gradations of density in the directions of the courses are very nearly the same as in the separate streams.

7. Conceive, now, that two bar-magnets, B N A, B′N′A′, are brought near each other, with their axes in the same straight line; and, for distinctness, let the axes be in the plane of the magnetic meridian, and let A, A′ be the ends from which the secondary streams always issue. The two currents under these circumstances will interfere with each other, so that the symmetrical distribution of density with reference to the neutral

positions N and N′ will be destroyed, and the disturbance will be in greater degree as the distance between the magnets is less. Also the consequences of the disturbance will be different according to the different directions of the currents. First, let B and A′ be adjacent ends. Then the streams, always flowing towards A and A′, will be in the *same* direction. Hence, by hydrodynamics, the density and pressure of the fluid are diminished by the junction of the streams, in greatest degree in the space intervening between B and A′, and in N B and N′A′ to a greater degree than in N A and N′B′. Thus the increment of density from the neutral lines towards the ends is less rapid in the adjacent halves, than in the remote halves. Hence the moving force of the æther urging the individual atoms towards the neutral lines is in excess in the remote parts, causing the magnets to move towards each other as if they were *attracted*. The same effect would be produced if the adjacent ends were B′ and A, because the two currents would still be in the same direction.

Next, let A and A′ be adjacent. In this case, the streams issue from the magnets in *contrary* directions towards the space between A and A′, and by their meeting the æther is condensed in such a manner that the decrements of density towards the neutral positions are more effective in the nearer halves of the magnets than in the more remote. The magnets consequently move from each other, or are apparently *repelled*. Lastly, let B and B′ be the adjacent ends. The streams now flowing from B towards A and from B′ towards A′, a diminution of velocity and consequent increment of density, result from their contrary tendencies between B and B′. The increase of density in this case may be conceived to be produced by the accelerations of the æther resolved in directions perpendicular to the common axis of the magnets, the resolved parts, in both streams and on both sides of the axis, conspiring to produce motion *towards* that line. Thus by the same reasoning as in the preceding case, the magnets will be *repelled*. Consequently the known law, that *like poles mutually repel and unlike poles mutually attract*, is accounted for on the principles of the hydrodynamical theory.

I take occasion to add that the above considerations respecting the mutual influence of opposing and conspiring magnetic streams, equally apply to the streams to which in the theory of electricity the attractions and repulsions of electrified bodies were attributed, and may be regarded as supplementary to the explanations given in arts. 18, 19, and 20 of the communication to the Number of the 'Philosophical Magazine' for last October.

8. In a similar manner the mutual action between magnets and galvanic currents may be explained. In the theory of galvanism, reasons were given for concluding that the movement of

galvanic currents relative to an electrode is composed of uniform
motions parallel to the electrode, and of uniform circular motions
about its axis. (See art. 10 of the communication to the 'Philosophical Magazine' for December.)   The theory I am about to
give of the mutual action between magnets and galvanic currents,
essentially depends on the existence of the circular motion : but
it is remarkable that the facts to be explained require that *the
direction of this motion should be always the same;* that is, if
the electrode be parallel to the earth's axis, and the current flow
from south to north, the circular motion must be in the same
direction as that of the earth's rotation about its axis. I do not
at present profess to account antecedently either for the circular
motion, or for its having a determinate direction; but I can
conceive that both these characteristics of the theoretical galvanic currents may be referable to the primary currents, which,
as I shall hereafter attempt to show, have their origin in the
earth's rotation.

9. Assuming galvanic currents to be such as are described
above, suppose a straight horizontal electrode to be placed in
the plane of the magnetic meridian, and the current to flow from
south to north; and let a horizontal magnetic needle be placed
directly underneath the electrode at a small distance from it.
Since the needle and electrode are parallel, if the galvanic current were wholly longitudinal, there would seem to be no cause
of disturbance of the needle, because the circumstances of the
æther would be alike on the opposite sides of the plane of the
magnetic meridian. But suppose the motion of the æther along
the electrode to be accompanied by a circular motion about its
axis in the direction from above towards the right hand of a
person looking northward; and calling the end of the needle
which points to the north A and the other end B, let A designate,
as heretofore, that end from which the magnetic streams *issue* in
curved diverging courses; also, abstract for the present from
the longitudinal motions both galvanic and magnetic.   Then
the circular motion produces a stream which crosses the magnet
from *east* to *west*. This stream *meets* the parts of the issuing
magnetic streams resolved in the horizontal direction, on the
east side of the *north* portion of the needle, and *flows with* them
on the *west* side.   There is consequently an increase of density
and pressure on the east side, and a diminution of density and
pressure on the west side, and the needle is consequently urged
towards the *west*. About the end B and the *south* portion of the
needle, where the entering streams converge in curved courses,
the parts resolved horizontally conspire with the circular streams
on the *east* side, and oppose them on the *west* side. The greater
density of the æther is therefore on the west side, and the south

end of the needle is consequently urged towards the *east.* As the actions on the opposite halves of the needle are equal and in contrary directions, the total effect is simply *a motion of rotation,* and the north end of the needle deviates towards the west, the direction of the galvanic current being from south to north.

If the direction of the current be reversed, the circular stream passes across the magnet from *west* to *east,* and consequently by the same reasoning as before, the *north* portion of the needle is urged *eastward,* and the *south* portion *westward.* It is clear that the directions of the deviations of a needle *above* the current are the opposite to those of a needle below, the direction of the effective parts of the circular streams being opposite.

This directive action of the galvanic current vanishes when the axis of the needle is transverse to the direction of the current. If the current acts simultaneously with the earth's magnetism, the needle must take a position intermediate to the transverse position and the plane of the magnetic meridian.

All these inferences from the theory are in accordance with well known results of experiment.

10. The curvilinear *paths of the ætherial streams* in the above theory of the magnet, correspond to Faraday's *lines of magnetic force,* and points of greater or less *velocity* correspond to points of greater or less *magnetic intensity.*

11. The explanation of the reciprocal action of magnets on galvanic currents on the same principles appears to be as follows. In the case first supposed, the magnet being *under* and *parallel* to the galvanic current, let the electrode be moveable about a vertical axis, and the magnet be fixed. Then the moving force which urges the magnet when moveable, reacts upon the circular current, and disturbs its uniformity. It has been previously argued that the circular movement is necessary for maintaining the galvanic current, by preventing its flowing towards the axis of the electrode. Hence under this disturbance there will be a tendency of the fluid to rush to the parts towards which the flow of the circular stream is impeded by the partial interruption, that is, to the *west* side of the *north* portion of the electrode. This impetus, being unopposed, will cause the electrode to deviate towards the *east.* Like considerations would show that the *south* portion of the electrode is made to deviate towards the *west.* Thus the magnet and electrode will have the same relative positions as when the former was moveable. Similar explanations may be given in the other cases. In general, it may be said that, as the galvanic current always tends to maintain itself by a uniform circular motion about the axis of the electrode, it tends also to impress on the electrode a movement by which the uniformity of the circular motion, when interrupted, may be most readily restored.

12. The directive action of *terrestrial magnetism* on galvanic currents admits of the following explanation. Let the current flow in a circular electrode moveable about a vertical axis, and let the electrode be so placed in the plane of the magnetic meridian, that the direction of the current at the lowest part may be from *north* to *south*. A diameter to the circle being drawn in the direction of the terrestrial current, that is, from *south* towards *north* in the line of the dipping needle, at and near the *south* extremity of the diameter this current will be confluent with the circular movement about the electrode on the *east* side, and opposed to it on the *west* side. Hence on the latter side there will be an excess of pressure which will tend to move the electrode *eastward*. At the other end of the diameter the terrestrial current flows with the circular stream on the *west* side and opposes it on the *east* side, and consequently tends to move the electrode *westward*. Hence on the whole the electrode will rotate in the direction which accords with observation, till it takes a position perpendicular to the plane of the meridian.

13. The directive action of terrestrial magnetism on the *dipping needle* is explained as follows. When the needle is in its normal position, the extremity A directed northward, the secondary streams flow longitudinally in the same direction as the primary terrestrial stream, and their transverse motions are perpendicular to the latter. There is consequently no tendency of the needle to move from this position. Now let the axis of the magnet be inclined to its normal position in any given direction. The secondary streams will not be altered by this change, because, as before explained, the accelerative forces to which they owe their origin are independent of the direction of the primary stream. But in the new position the two streams will influence each other so as to give rise to dynamic action on the needle. Conceive, for the sake of distinctness, the end A of the needle to deviate about 30° from the normal position towards the *west*. Then resolving both the primary and the secondary streams in the directions perpendicular and parallel to the needle, the respective perpendicular streams along the north portion of the needle (this being the portion from which the secondary streams *issue*) will be opposed to each other on the *west* side of the needle, and conspire on the *east* side. Thus the needle will be urged *eastward*. On the *south* portion of the needle, at which the secondary streams are confluent, the perpendicular components are opposed to each other on the *east* side and conspire on the *west* side, and that portion is consequently urged *westward*. As the two actions are equal and opposite, the needle has simply a *motion of rotation*, and the directive action is always *towards the normal position*.

14. It follows that the directive force of the earth's magnetism

is a measure of the *velocity* of the terrestrial current, that is of the *total intensity* of the magnetic force. According to the theory, this force may be supposed to be the resultant of horizontal and vertical components, and *the ratio of the latter to the former is the tangent of the Inclination.*

Cambridge Observatory,
December 22, 1860.

[To be continued.]

IX. *On a new Resistance Thermometer.*

To Professor John Tyndall, F.R.S., &c., Royal Institution.

3 Great George Street, Westminster, S.W.
December, 1860.

MY DEAR SIR,

YOU will probably be interested to hear about a very direct application of physical science to a purpose of considerable practical importance, which I had lately occasion to make. Having charge, for the British Government, of the Rangoon and Singapore telegraph cable, in so far as its electrical conditions are concerned, I was desirous to know the precise temperature of the coil of cable on board ship at different points throughout its mass, having been led by previous observations to apprehend spontaneous generation of heat. As it would have been impossible to introduce mercury thermometers into the interior of the mass, I thought of having recourse to an instrument based upon the well-ascertained fact that the conductivity of a copper wire increases in a simple ratio inversely with its temperature. The instrument consists of a rod or tube of metal about 18 inches long, upon which silk-covered copper wire is wound in several layers so as to produce a total resistance of, say 1000 (Siemens) units at the freezing temperature of water. The wire is covered for protection with sheet india-rubber, inserted into a tube and hermetically sealed. The two ends of the coil of wire are brought, by means of insulated conducting wires, into the observatory, where they are connected to measuring apparatus, consisting of a battery, galvanometer, and variable resistance coil. The galvanometer employed has two sets of coils, traversed in opposite directions by the current of the battery. One circuit is completed by the insulated thermometer coil, and the other by a variable resistance coil of German silver wire. Instead of the differential galvanometer, a regular Wheatstone's bridge arrangement may be employed.

You will readily perceive that if the thermometer coil before described were placed in snow and water, and the variable resistance coil were stoppered so as to present 1000 units of resistance, the currents passing through both coils of the differential galvanometer would equal one another, and produce, therefore, no deflection of the needle. If, however, the temperature of the water should rise, say 1° Fahr., its resistance would undergo an increase of $1000 \times \cdot 0021 = 2 \cdot 1$ units of resistance, necessitating an addition of $2 \cdot 1$ units to the variable resistance coil in order to re-establish the equilibrium of the needle.

The ratio of increase of resistance of copper wire with increase of temperature may be regarded as perfectly constant within the ordinary limits of temperature; and being able to appreciate the tenth part of a unit in the variable resistance coil employed, I have the means of determining with great accuracy the temperature of the locality where the thermometer resistance coil is placed. Such thermometer resistance coils I caused to be placed between the layers of the cable at regular intervals, connecting all of them with the same measuring apparatus in the cabin.

After the cable had been about ten days on board (having left a wet tank on the contractors' works), very marked effects of heat resulted from the indications of the thermometer coils inserted into the interior of the mass of the cable, although the coils nearer the top and bottom surfaces did not show yet any remarkable excess over the temperature of the ship's hold, which was at 60° Fahr. The increase of heat in the interior progressed steadily at the rate of about 3° Fahr. per day, and having reached 86° Fahr., the cable would have been inevitably destroyed in the course of a few days, if the generation of heat had been allowed to continue unchecked.

Considering the comparatively low temperature of the surface of the cable, much incredulity was expressed by lookers-on respecting the trustworthiness of these results; but all doubts speedily vanished when large quantities of cold water of 42° temperature were pumped upon the cable, and found to issue 72° Fahr. at the bottom.

. Resistance thermometers of this description might, I think, be used with advantage in a variety of scientific observations,— for instance, to determine the temperature of the ground at various depths throughout the year, or of the sea at various depths, &c. &c. In the construction of this instrument, care has to be taken that no sensible amount of heat is generated by the galvanic currents in any of the resistances employed.

By substituting an open coil of platinum wire for the insulated copper coil, this instrument would be found useful also as a pyrometer.

But, finding this letter already exceeds its intended limits, I shall not enlarge upon these applications, which, no doubt, are quite obvious to you.

I am, dear Sir,
Yours very truly,
C. Wm. Siemens.

X. *Proceedings of Learned Societies.*

March 22, 1860.—Sir Benjamin C. Brodie, Bart., Pres., in the Chair.

THE following communications were read:—

"On the Insulating Properties of Gutta Percha." By Fleeming Jenkin, Esq.

The experiments described in this paper were undertaken with the view of determining the resistance opposed by the gutta-percha coating of submarine cables at various temperatures to the passage of an electric current.

The experiments were made at the works of R. S. Newall and Co., Birkenhead. The relative resistance of the gutta percha at various temperatures was determined by measuring the loss on short lengths immersed in water. These experiments are described in the first part of the paper. The absolute resistance of gutta percha has been calculated from the loss on long submarine cables. These experiments and calculations are described in the second part of the paper.

PART I.

The loss of electricity was measured upon three different coils, each one knot in length. One was covered with pure gutta percha; the two remaining coils were covered with gutta percha and Chatterton's compound. The coils were kept at various temperatures by being covered with water in a felted tub; and the water was maintained at a constant temperature for twelve or fourteen hours before each experiment.

The loss or current flowing from the metal conductor to earth through the gutta-percha coating was measured on a very delicate sine-galvanometer. The loss from the connexions when the cable was disconnected, was measured in a similar manner. The electromotive force of the battery employed was on each occasion measured in the manner described by Pouillet. Corrections due to varying electromotive force and loss on connexions were made on the result of each experiment.

A remarkable and regular decrease in the loss was observed for some minutes after the first application of the battery to the cable; a phenomenon, which the author thinks may be due to the polarization of the molecules of gutta percha, or of the moisture contained in the pores of the gutta percha. The loss was therefore measured from minute to minute for five minutes, with each pole of the battery.

Nineteen tables containing the results, with the reductions and curves representing the results, accompany the paper. The following results were obtained from the first coil; this was prepared with Chatterton's patent compound. With a negative current between the limits of 50° and 80° Fahrenheit, the decrease of resistance is sensibly constant for equal increments of temperature; and the increase of resistance due to continued electrification is also nearly constant. At 60° the resistance increases about 20 per cent. in five minutes from this cause. With a positive current, similar results appear between the temperatures of 50° and 60°; but the resistance is somewhat greater than with the negative current. The extra resistance due to continued electrification is unchanged by a change in the sign of the current. Above the temperature of 63° great irregularities occur in the observations, which could not even be included in regular curves. The difference in the resistance of the gutta-percha coating when the copper is positively and negatively electrified, may be caused by the contact between the resinous compound and the copper: no such difference was observed when pure gutta percha was in contact with the copper.

The curves resulting from the experiments on the second coil, which was covered with pure gutta percha, present an entirely different character from those resulting from the first coil. The copper and gutta percha were of the same size in these two coils. The resistance of pure gutta percha at low temperatures is greater than that of the compound covering. At 65° the resistance of the two coverings is equal; at higher temperatures the resistance of pure gutta percha diminishes extremely rapidly. The curves obtained with positive and negative currents are identical up to about 75°; a slight difference occurs above this temperature, which may have been accidental. The extra resistance is less with pure gutta percha than with the compound; it increases slightly at high temperatures, and is not affected by a change in the sign of the current.

The curves derived from the experiments on the third coil, which contained a smaller proportion of Chatterton's compound than the first coil, appear in some respects intermediate between those derived from the first and second coils. The extra resistance due to continued electrification was still greater in this coil than in the others. 40 per cent. of the entire resistance is at 70° due to this cause. This increase is believed to be due to the greater mass of gutta percha used in covering this coil, which was of larger dimensions than the two others.

PART II.

Professor Thomson has supplied an equation expressing the law which connects the resistance of a cylindrical covering, such as that of a cable, with the resistance of the unit of the material forming the covering.

Let S be the specific resistance of the material, or the resistance of a bar one foot long, and one square foot in section; let G be the resistance of the cylindrical cover of a length of cable L; let $\frac{a}{b}$ be the ratio of the external to the internal diameter of the covering; then

$$S = \frac{2\pi LG}{\log \frac{a}{b}} \quad \cdots \cdots \cdots \quad (1)$$

The resistance G was calculated from cables of various lengths, lying in iron wells at the works of R. S. Newall and Co., Birkenhead.

The cables were not wet; but direct experiment proved that covering a sound iron-covered cable with water has no effect on the loss. The details of this experiment are given in the paper.

The resistance G was obtained in the following manner. The copper conductor of the cable to be tested was arranged so as to form a complete metallic arc with a battery of 72 cells and a tangent galvanometer: the deflection on this galvanometer was read and entered as the continuity test. Deflections were then read on the same galvanometer with the battery and several known resistances in circuit, for the purpose of measuring the resistance and electromotive force of the battery, in the manner described by Pouillet. The deflection caused by the loss was next read on a second tangent galvanometer: the same battery was used. This deflection was entered as the insulation test. The temperature of the tank containing the cable was observed by means of a thermometer inserted in a metal tube, extending from the circumference into the mass of the coil.

The relative delicacy of the galvanometers was ascertained by experiment, or, in other words, the coefficient was found by which the tangents of the deflections of the first were multiplied to render them directly comparable with the tangents of the deflections of the second galvanometer.

The resistances of the galvanometer coils, of the artificial resistance coils, and of the copper conductor of the cable were measured by Wheatstone's differential arrangement. Special experiments were made by means of this differential arrangement to determine the change of resistance of the copper conductor in the cable, produced by a change of temperature.

The equation (No. 2) $R = r(1 + 0.00192t)$ gives the value of the resistance R of the copper wire at any temperature $t + a$ in function of the resistance r at any temperature a (Fahrenheit). The length and temperature of any coil being known, the resistance of the copper wire was thus at once obtained from the resistance of one knot at 60°, which was very carefully determined.

Now let G = resistance of cylindrical coating.
  D = deflection called the continuity test.
  d = deflection called the insulation test.
  C = coefficient expressing the relative delicacy of the two galvanometers.
BR = resistance of the battery.
T_1 = resistance of the coil of first galvanometer.
T_2 = resistance of the coil of second galvanometer.

$$\text{Then } G = \frac{C \tan D \times (BR + T_1 + M)}{\tan d} - BR + T_2 + \frac{M}{2} \dots \dots (3)$$

G having been thus obtained in any desired units, S, the specific resistance of the material, can be at once obtained by equation No. 1, which appears from several experiments to give constant values for S when calculated from cables of different dimensions. In extreme cases, however, the influence of extra resistance would render the formula defective, especially after continued application of the cur-

rent : thus the resistance of a foot-cube would be very different to
that of an inch-cube.

The values of G for the covering of the Red Sea cable, after continued electrification for periods of one, two, three, four, and five
minutes, were calculated in Thomson's Absolute British Units, from
four sets of tests made specially for this purpose on four different
cables, each about 500 knots long. Tables containing the results of
these calculations accompany the paper.

A Table is also given of the resistance of the Red Sea covering
after one minute's electrification, and after five minutes' electrification,
at each degree of temperature, from 50° to 75° Fahrenheit. This
Table was formed by means of the temperature curves described in
the first part of the paper : this Table is here annexed (No. 1).

Similar Tables were given for the covering of the two experimental
coils mentioned in the first part of the paper. The coil composed
of pure gutta percha, gave very regular and complete results. An
abbreviation of the Table is annexed.

It was remarked that in the tests of the cable in the iron tanks,
the resistance after five minutes' electrification was invariably
greater with zinc than with copper to cable, whilst the reverse was
the case with the single knot covered by water. The length of the
cable, and the condition of immersion or non-immersion, have probably some influence on the phenomenon of extra-resistance. This
phenomenon appears to the author to be of much importance, and
to demand further investigation.

The values of G were also calculated from the daily tests of the
cables during manufacture at many temperatures. These values
agreed with those given in the Tables above described. The general
results of the experiments may be summed up as follows.

The relative loss at various temperatures through pure gutta
percha has been pretty accurately determined for all ordinary temperatures. To a less extent the same knowledge has been gained
concerning two other coatings containing Chatterton's compound.
The latter appears superior at high, and inferior at low temperatures.

Attention has been drawn to the considerably increased resistance
which follows the continued electrification of gutta percha and its
compounds. Some of the laws of this extra resistance have been
determined, and some suggestions made as to the cause of the
phenomenon.

The bounds have been pointed out within which formulæ may
be used, which consider gutta percha as a conductor of the same
nature as metals.

The resistance of gutta percha has been obtained in units, such as
are employed to measure the resistance of metals ; and by the use of
Professor Thomson's formula, the specific resistance of a unit of the
material has been fixed with some accuracy.

The resistance of other non-conductors, such as glass and the
resins, may probably, by comparison with gutta percha, be obtained
in the same units.

Incidentally, the increase of resistance in copper with increased
temperature has been given from new experiments ; and it has been

shown that the insulation of a sound wire-covered cable is little, if at all, affected by submersion.

Finally, tables and formulæ are given by which the resistance of, or the loss through any new cable coated with gutta percha, may be at least approximately estimated :—

TABLE I.

Specific Resistance in Thomson's Units of the Red Sea Covering at various Temperatures.

Tempera-ture.	Zinc to cable.		Copper to cable.	
	After electrification for one minute.	After electrification for five minutes.	After electrification for one minute.	After electrification for five minutes.
60°	2162×10^{17}	3330×10^{17}	2239×10^{17}	3405×10^{17}
65	$1810 \times$ „	$2947 \times$ „	$1720 \times$ „	$2770 \times$ „
70	$1460 \times$ „	$2378 \times$ „	$1318 \times$ „	$2239 \times$ „
75	$1160 \times$ „	$1753 \times$ „	$1000 \times$ „	$1739 \times$ „

TABLE II.

Specific Resistance in Thomson's Units of pure Gutta Percha at various Temperatures.

Tempera-ture.	Zinc to cable.		Copper to cable.	
	After electrification for one minute.	After electrification for five minutes.	After electrification for one minute.	After electrification for five minutes.
50	4113×10^{17}	5663×10^{17}	4113×10^{17}	5663×10^{17}
55	$2917 \times$ „	$3636 \times$ „	$2917 \times$ „	$3636 \times$ „
60	$2163 \times$ „	$2549 \times$ „	$2163 \times$ „	$2549 \times$ „
65	$1634 \times$ „	$1858 \times$ „	$1634 \times$ „	$1858 \times$ „
70	$1162 \times$ „	$1291 \times$ „	$1192 \times$ „	$1291 \times$ „
75	$805 \times$ „	$877 \times$ „	$796 \times$ „	$866 \times$ „
80	$566 \times$ „	$613 \times$ „	$548 \times$ „	$591 \times$ „

"On Scalar and Clinant Algebraical Coordinate Geometry, introducing a new and more general Theory of Analytical Geometry, including the received as a particular case, and explaining 'imaginary points,' 'intersections,' and 'Lines.'" By Alexander J. Ellis, Esq., B.A., F.C.P.S.

XI. *Intelligence and Miscellaneous Articles.*

THE LITHIUM SPECTRUM.

To the Editors of the Philosophical Magazine and Journal.

GENTLEMEN,

PERMIT me through you to ask MM. Kirchhoff and Bunsen, or other experimentalists in this country who may be in possession of the requisite apparatus, whether the "very weak yellow line Li β,"

described by them* as accompanying the brilliant red line Li α, requires any extraordinary precautions to render it visible. Two
specimens of lithium salts which I have examined in a very perfect
apparatus, somewhat similar to the one described by them†, have
failed to give the slightest evidence of its presence, although I have
repeatedly examined them with that object; and as I know that
other experimentalists in this country have been equally unsuccessful, it is possible that the presence of this line in the spectrum given
by MM. Kirchhoff and Bunsen's specimens of lithia may really be
due to the presence of another element hitherto unknown. The
spectrum of the new alkali metal Cæsium (which it may be of interest
to know I have detected in some highly concentrated mother-liquors
from sea-water) is in no respect similar to it. It will be a sufficient
proof of the delicacy of my apparatus, to say that it is constructed
principally of quartz, and that it easily separates the double line D.

I remain, Gentlemen,

Your obedient Servant,

WILLIAM CROOKES.

A CONSTANT COPPER-CARBON BATTERY.   BY JULIUS THOMSEN.

In the ordinary galvanic apparatus, zinc usually officiates as positive element. This metal is, however, readily attacked by acid if it is
not either chemically pure or well amalgamated. If the sulphuric
acid is not greatly diluted, the zinc cylinders are strongly attacked
by continuous use, in spite of the amalgamation, by which a great loss
of metal is caused; if, on the other hand, the acid is much diluted, it
is soon saturated, and the action of the apparatus is enfeebled.

In my investigations I use a galvanic apparatus consisting of *copper* in dilute sulphuric acid (1 part acid and 4 parts of water) as positive
element, and, as a negative element, *carbon* in the mixture of bichromate of potass, sulphuric acid, and water, recommended by Wöhler
and Buff. (Buff uses 100 parts water, 12 of bichromate, and 25 of
sulphuric acid.) The electromotive force of this combination is $\frac{8}{10}$ths
of that of a Daniell's battery.

Its advantages are as follows :—The copper is not at all attacked
by the acid when the circuit is open ; the resistance of the sulphuric
acid, from its being so little diluted, is a minimum : the sulphuric acid
is so strong that it can be used for months without becoming saturated. As, further, the mixture of chromate of potass and sulphuric
acid is inodorous, this combination is very convenient for working
with in closed spaces.

This combination is very interesting theoretically ; for as copper
cannot decompose dilute sulphuric acid, the copper-carbon element
is an example of a powerful apparatus in which chemical action and
the disengagement of electricity are quite inseparable.—Poggendorff's
Annalen, October 5, 1860.

* Phil. Mag. S. 4. vol. xx. p. 96. August, 1860.

† Ibid. p. 90.

THE

LONDON, EDINBURGH and DUBLIN

PHILOSOPHICAL MAGAZINE

AND

JOURNAL OF SCIENCE.

[FOURTH SERIES.]

FEBRUARY 1861.

XII. *Note respecting* Ampère's *Experiment on the Repulsion of a Rectilinear Electrical Current on itself.* By JAMES D. FORBES, *D.C.L., V.P.R.S.E., Principal of the United College, St. Andrews, late Professor of Natural Philosophy in the University of Edinburgh**.

[With a Plate.]

IN a communication to the Royal Society of Edinburgh, on the 3rd of January, 1859, I related some experiments on the vibrations of metallic bodies and of carbon, produced by the passage of electricity through them†. I learned soon after, from a paper by Dr. Tyndall‡, that these vibrations had been already described by Dr. Page of the United States§. The analogy of the experiment to that of Mr. Trevelyan, on the vibrations of metallic bodies by heat, led me to suspect that the resistance to the propagation of heat in the first case, and of electricity in the second, in its passage from the one to the other of the bodies in contact, was the cause of a sensible mechanical repulsion. So strongly was I persuaded of this, that twenty-seven years ago, when making my experiments on the "Trevelyan" bars, I attempted to set them in motion by electricity, though then ineffectually.

I am aware that the idea of the idio-repulsive quality of the heat-current has not been favourably received, and that the high authority of Mr. Faraday is still given for an explanation of the

---

* Communicated by the Author, having been read to the Royal Society of Edinburgh, January 7, 1861.

† Proceedings of the Royal Society of Edinburgh, vol. iv. p. 151; and Phil. Mag. vol. xvii. p. 358.

‡ Phil. Mag. vol. xvii. p. 358.

§ Silliman's Journal (1860), vol. ix. p. 105.

phenomena founded on the familiar laws of conduction and expansion alone. Dr. Tyndall thinks that the effects of electricity are also to be ascribed to the indirect action of the heat which it produces, in the way of expansion. I have never been satisfied with either of these explanations. The conviction to which I still hold is that the alleged expansion and contraction cannot sensibly operate in the almost infinitely short periods of successive contact and separation, and that the effect must be due to a molecular impulse of a far more sudden and instantaneous character*. We seem to be sufficiently conversant with such impulses in galvanic, and especially in secondary induced currents.

This inquiry, however, into the effects of an electrical current *upon itself*, and upon the different portions of one and the same conductor conveying it, is one of extraordinary difficulty. We cannot fail to be struck by the hesitation with which that great Master of this branch of science, Mr. Faraday, expresses himself in different parts of his writings respecting them †.

There is one experiment, quoted in modern treatises on electricity, which might seem to throw light on the subject; and that is Ampère's experiment on the repulsion of one portion of an electrical current upon another portion which is a continuation of it in the same right line. This repulsion appeared to be the necessary complement of Ampère's theory of the mutual action of currents placed in different positions relatively to one another. He considered it to be sufficiently demonstrated by the following experiment:—A and B (Plate II. fig. 5) are two little troughs filled with mercury, or rather a single trough divided lengthwise into two by the glass partition C D. Two straight pieces of copper wire *a, b,* united by the bridge *c,* float on the mercury, which, however, they are everywhere prevented from touching metallically by means of a coating of sealing-wax, except at the extreme ends nearest to A and B, where they are amalgamated, and of course touch the mercury. The two poles of a *powerful* battery being inserted in the mercury troughs near the letters A, B, the circuit is evidently completed through the wire *a c b.* It is stated that when this is done, the floating wire is repelled from the vicinity of the terminal wires of the battery. This repulsion is ascribed to the reaction of a rectilinear current from A to *a* through the mercury, upon the continuation of the current in the copper conductor *a,* and to a similar action on the side *b* B.

---

* That the electrical vibrations go on during the copious affusion of the apparatus with cold water, and that they take place with remarkable energy in an almost inexpansible substance like carbon (see my paper cited above), are direct arguments in favour of this conclusion.

† See particularly the Ninth and Thirteenth Series of his ' Researches.'

This remarkable experiment has been circumstantially described in more places than one by Ampère himself[*]; but the description is in all essentially alike, and the accounts, being nearly contemporaneous, may be assumed to refer to one and the same trial, which was made, it appears, at Geneva, and in the presence of M. Auguste De la Rive. This last is a circumstance of some importance; for it would perhaps be difficult to establish that it has ever been successfully repeated since. The authors whom I have consulted usually content themselves with the barest citation of Ampère's authority. The only exception which I know of is in Müller's *Lehrbuch der Physik* (vol. ii. p. 318), where the writer notes the difficulty he found in repeating the experiment: "Ferner muss noch angeführt werden dass dieser Versuch keineswegs zu den leicht gelingenden gerechnet werden kann." From which I infer that the writer had *not* succeeded in repeating it. He also raises doubts (which seem to be reasonable) as to whether, supposing it successful, the conclusion of Ampère could with certainty be legitimately drawn from it.

It appeared to me to be a matter of some interest in connexion with the experiment of electrical vibration, to repeat the observation of Ampère. I had an apparatus made for the purpose, but circumstances have hitherto prevented me from using it; and I have requested my successor in the Chair of Natural Philosophy, Professor Tait, to make trial of it when he happens to have a powerful voltaic battery in action. I did, however, attempt the experiment in a different form, and, as it appears to me, in one both more sensitive in its indications, and less ambiguous in its interpretation[†].

· I fitted up a Coulomb's torsion apparatus (one that had been used for experiments on diamagnetism) in the following way. It is represented in outline in fig. 6. A moderately fine platinum wire A, about 18 inches long, was used to suspend a wooden torsion-rod B C, near one end of which (laid on a piece of pasteboard fastened to the rod) was poised a copper wire bent in the form of a horseshoe, as shown in the figure at *a b*[‡]. Two strong copper wires P, N, the terminals of a Grove's battery of four moderate-sized pairs in good action, were brought into the position indicated in the figure, so as to be exactly opposite to, and in continuation of, the limbs of the horseshoe. The terminals P, N were kept firmly in their place, and the extremities *a, b,*

* *Recueil d'Observations Electro-dynamiques,* 8vo, pp. 285, 318. Also *Bibliothèque Universelle,* vol. xxi. p. 47; and Ampère's *Théorie,* p. 39 (this last cited by Dr. Roget in his 'Electro-Magnetism,' art. 187).

† The following experiments were made in December 1858 and January 1859.

‡ The distance *a b* might be half an inch, and the distance from the suspension wire was relatively greater than the figure indicates.

of the horseshoe were adjusted by moving it on the pasteboard shelf until simultaneous contact of the two limbs with the fixed wires was obtained on moving the graduated head of the torsion wire A from right to left. The completion of the current was effected without much difficulty by gently pressing the rough extremities of the wires into contact by the help of the torsion head.

I found it convenient to increase the tension of the current by introducing into the circuit an ordinary electrodynamic coil of stout wire, with or without a core of iron wires.

. So far from finding anything like *repulsion* of the moveable horseshoe on the continuation of the circuit, I soon observed that *attraction* took place, resisting the force of twist of the platinum wire exerted through a considerable arc, when it tended to produce separation. This attraction continued to subsist even after the current was withdrawn.

With a horseshoe composed of two straight limbs, *a* and *b*, of bismuth, the connexion *c* being of copper, a still more decided attraction was the result; and it continued to subsist after withdrawing the current of electricity, even if only one of the poles touched the horseshoe.

. When the terminals P, M were somewhat flatly rounded (as in fig. 6 *a*), and the extremities of the horseshoe also carefully rounded and polished, I was unable to obtain electrical contact by the pressure due to mere torsion. The current did not pass until the extremities were touched with nitrate of mercury and wiped dry, when it passed readily; and a marked adhesion took place while the current lasted, and for some time after it stopped. While the electricity passed, a fizzing noise was audible from the extremities of the wires. This *might* possibly be due to an insensible vibration. It thus appears that the irregularities of surface of the wires at the points of contact in the first form of the experiment was favourable to the electrical propagation, as it is in the case of ordinary frictional electricity.

A current from a very powerful Ritchie's induction coil, excited by four pairs of Grove's, was next passed through the apparatus last described. Distinct adhesion took place of the ends in contact, opposing a force amounting to 90° or more of torsion. It continued while the current passed, and for some time after. By and by they separated through the effect of torsion.

This experiment was very instructive. If it commenced when the terminations were a little apart, the high-tension electricity passed readily in the form of sparks from P and N to the horseshoe—the horseshoe steadily approaching—the sparks getting shorter and finally vanishing at both contacts, the attraction continuing and facilitating electrical contact. Apparently the very same process, of polarity preceding conduction, obtains in the ordinary galvanic current.

The form of the terminals P, N was again altered to a hammer-shape, as in fig. 6 *b*; and now electrical contact with the copper horseshoe was even more difficult to obtain than before. Though for intensity eight pairs of Grove's battery and a coil were used, the current did not pass without amalgamation, and also decided pressure being used; then the adhesion became strong, amounting to 140° of torsion of the suspending wire.

A similar result was obtained when the current was passed from one of the hammer-shaped poles A (fig. 7) through a straight wire B, laid on the torsion-rod and terminating in a small trough of mercury C, connected with the opposite pole of the battery. The adhesion and other phenomena were in all respects the same as before.

I think it will not be doubted that these experiments are far more delicate tests than Ampère's floating wire, of the mutual action of two portions of one and the same current, and also that the first form is free from the ambiguity which the introduction of a fluid conductor occasions. At the same time the result, with reference to Ampère's theory, would be more satisfactory if the two rectilinear currents whose mutual action is examined were independent of one another, and not parts of one and the same current; for the break in the conductor, necessary to the mobility of the parts, introduces a peculiarity due rather to the force required to propagate the current at all, than to the reciprocal action of two currents propagated independently.

It is needless to add that the result of these experiments is not favourable to the cause which I have assigned to the electrical vibrations; but I have not the less thought it desirable to record the result, though it leaves that experiment, in my opinion, more in need of explanation than ever.

I have not overlooked the probability of the maintaining power of electricity in the rocking-bar being due to instantaneous induction currents (Faraday's) excited at the two points of contact of the bar and block round which the rocking takes place; but some experiments made on this supposition led to no result.

I would, before concluding this paper, direct the attention of electricians to another of those isolated experiments connected with this subject which require confirmation: I mean the production of "Davy's cones," said to occur when two terminals of a very powerful battery, coated with sealing-wax and naked only at the ends, are introduced, parallel and vertically, through the bottom of a cup filled with mercury, and reach to within a little space of the surface of that fluid. On two occasions—once with 500 or 600 pairs of the Royal Institution battery, and again with

Mr. Pepys's gigantic single galvanic pair at the London Institution—cones of mercury rose over the general level of the fluid and above the polar wires[*]. Even Mr. Faraday has failed in repeating the experiment, which is remarkable since, though he only used 100 pairs, the electromotive power of modern batteries vastly exceeds that of those used in Davy's time, and the success with Pepys's apparatus shows that intensity is not indispensable[†]. Mr. Faraday thinks that Davy's cones and Ampère's repulsion experiment are due to one and the same cause. But I incline to believe that Davy's cones intimate an attraction, and not repulsion, and so far concur with the results of my experiments with the torsion-balance.

I have had Davy's apparatus constructed and placed in the Natural Philosophy Class Collection in Edinburgh, where I hope that it may one day be tested with an adequate power.

Pitlochry, N. B.,
  July 11, 1860.

Postscript.—Soon after writing the preceding notice, I requested Professor A. De la Rive to give me any confirmation which his memory could supply, of the success of Ampère's experiment; and this he has kindly done with all the precision which might be expected from him. He adds that the motion could not be due to the action of the earth's magnetism, as it was independent of the direction of the current.

United College, St. Andrews,
  November 26, 1860.

22 *January* 1861.—I have just learned from Professor Tait that he has succeeded in repeating Ampère's experiment.

XIII. *On the existence of a Fourth Member of the Calcium group of Metals. By* F. W. *and* A. Dupré[‡].

DURING our recent examination of London waters by the beautiful method of Kirchhoff and Bunsen, we several times noticed a faint blue line not due to strontium or potassium, or to the lately discovered cæsium. Having since worked with larger quantities of the deep well-water which had given this line most distinctly, we believe we are now justified in stating that the group of calcium, strontium, and barium, like that of the alkali-metals, contains a fourth member. This new metal gives but

---

  * Sir H. Davy in Philosophical Transactions for 1823. Compare Mr. Faraday's 'Researches in Electricity,' Ninth and Thirteenth series, arts. 1113 and 1609.
  † Mr. Faraday's 'Researches,' as above cited.
  ‡ Communicated by the Authors.

one blue line*, situated between the lines Sr δ† and K β, about twice as far from the former as from the latter. In brightness and sharpness of definition it is quite equal to the line Sr δ. The method which, after repeated trials, we found most advantageous for obtaining it in a state of comparative purity is the following. The deposit formed by boiling the water was dissolved in hydrochloric acid, and a small quantity of sulphuric acid added to the clear solution. The precipitate formed, consisting principally of sulphate of calcium, but containing some sulphate of barium, sulphate of strontium, and sulphate of the metal under consideration, was collected, washed, dried and fused with carbonate of sodium. The fused mass was then boiled with water, and the insoluble carbonates of the above-named metals collected and treated several times successively in the same manner as the original deposit. By these means the lime was gradually removed, the carbonates becoming proportionally richer in barium, &c. Owing to the small quantity of substance at our command, we did not attempt to remove the lime entirely. The carbonates finally obtained were once more dissolved in hydrochloric acid, and the solution, after considerable dilution, mixed with a few drops of sulphuric acid. After standing at rest for twenty-four hours, the slight quantity of deposit formed, consisting of almost pure sulphate of barium‡, was filtered off, and some alcohol added to the filtrate, by which a further deposit was obtained, composed chiefly of the sulphates of strontium and the new metal, though not quite free from sulphate of calcium. These sulphates were again converted into carbonates, which were then dissolved in hydrochloric acid, and the solution evaporated to dryness. When a portion of this dry residue was brought into the flame of the apparatus, the spectra of calcium and strontium appeared; and, in addition, beyond the line Sr δ, in the position indicated above, a blue line, rivalling the strontium-line in brilliancy and distinctness of outline.

As far as we have been at present able to ascertain, the carbonate, oxalate, and sulphate of the new metal are insoluble in water, the last-named possessing about the same insolubility as sulphate of strontium. The chloride of the metal does not seem to be hygroscopic, resembling in this respect that of barium rather than those of strontium and calcium.

* Having been unable to separate it completely from calcium and strontium, we are not quite positive whether or not it gives any lines at the red end of the spectrum.

† In our last communication there is a typographical error, Sr γ being put for Sr δ.

‡ We have detected the presence of barium in several London waters since the publication of our notice in the November Number of the Philosophical Magazine.

We hope soon to be in possession of a sufficient quantity of some salt of this metal to give a fuller account of its properties.

In reference to the letter of Mr. Crookes in the last Number of this Magazine, stating his inability to recognize the yellow line Li β, we may say that all the different specimens of lithium-salts we have experimented with have shown this line distinctly. Since the publication of his letter, we have examined some very pure specimens of lithium-salts, all of which gave the yellow line Li β with equal distinctness, though showing no other lines except Li α and β. The red line may certainly be recognized with quantities of lithium-salt much smaller than are required to exhibit the yellow line.

The apparatus we use is one similar to that described in Kirchhoff and Bunsen's paper, and we find no particular precautions necessary to exhibit the line in question, beyond using a comparatively large quantity of lithium-salt, say $\frac{1}{70}$ to $\frac{1}{50}$ of a grain, and a thin platinum wire of about $\frac{1}{2}$ millim. diameter.

We may here mention incidentally, that we have found it advisable to discontinue the use of a brass Bunsen's burner, to prevent the occasional appearance of a series of green and blue lines, due to copper. A Bunsen's burner made of steatite is quite free from any such defect. The flame of the burner should also be regulated in such a manner as to have the part placed opposite the slit free from any marked blue tint, as otherwise a number of lines, one green and three blue being the most conspicuous, make their appearance. The blue or lower part of the flame of a candle or of an oil-lamp shows precisely the same lines.

XIV. *On the Insulation of Antozone.   In a Letter to* Professor FARADAY. *By* Professor SCHŒNBEIN[*].

I HAVE been working very hard these many months to obtain the "antozone," or Θ, in its insulated state, and I flatter myself that I have succeeded, at least to a certain extent. You are aware that, from a number of facts, notably from the reciprocal deoxidizing influence exerted by many oxy-compounds upon each other, I drew the inference that there existed two series of oxides, one of which contains Θ, the other ·Θ,—the ozonides and antozonides. The mutual deoxidation of those compounds I made dependent upon the depolarization or neutralization of Θ and Θ into O. Now Θ and Θ being capable of being transformed into O, I thought it possible, even probable, that the contrary might be effected, *i. e.* the chemical polarization of O into Θ and Θ; and you know that in the course of the last and present year I

* Communicated by Professor Faraday.

have ascertained a great number of facts that speak, as far as I can see, highly in favour of that idea. As the typical or fundamental fact of this chemical polarization of O, I consider the simultaneous production of Θ and $HO + \Theta$ which takes place during the slow combustion of phosphorus. This simultaneity is such, that ozone never makes its appearance without its equivalent $HO + \Theta$. All the metals which oxidize slowly, HO being present, such as zinc, lead, iron, &c., give rise to the formation of $HO + \Theta$; and so do a great number of organic substances, such as ether, the tannic, gallic, and pyrogallic acids, hæmatoxidine, &c.; and even reduced indigo, associated with potash, &c., makes no exception to the rule. The same simultaneity takes place during the electrolysis of water; never ozone without peroxide of hydrogen. I admit therefore that O, on being brought into contact with an oxidable substance and water, undergoes that change of condition which I call "chemical polarization;" *i. e.* is converted into Θ and Θ, the latter of which combines with HO to form $HO + \Theta$, whilst Θ is associated to the oxidable matter, such as phosphorus, zinc, &c. In the preceding statements you have only a very rough outline of my late researches on the chemical polarization of neutral oxygen: the details on that subject are contained in a number of papers lately published, and of which your English periodicals have as yet not taken any notice. Having gone so far, I could not but be very curious to try whether it was not possible to obtain Θ in its insulated or free state. I directed of course my attention to that set of peroxides which I call "antozonides," and tried in different ways to eliminate from them that part of their oxygen which I consider to be Θ. Years ago I remarked, in accordance with an observation made by M. Houzeau, that the oxygen disengaged from the peroxide of barium by means of the monohydrate of SO^3, exhibits an ozone-like smell, and the power of turning my ozone test-paper blue. Not having at that time a notion of two opposite active conditions of oxygen, I was inclined to ascribe these properties to the presence of minute quantities of ozone in the said gas; but on examining it more closely, I found it to be neutral oxygen mixed up with a very small portion of antozone, or Θ. A most important and distinctive property of antozone is the readiness with which it unites with water to form peroxide of hydrogen, whilst ozone, like neutral oxygen, is entirely incapable of doing so. Hence it comes to pass that the oxygen disengaged from $BaO + \Theta$, under certain precautions, becomes inodorous on being shaken with water, and that this fluid contains HO^2. The simple cause of the minute quantities of Θ obtained from $BaO + \Theta$, is the heat disengaged during the action of SO^3 upon the peroxide, by which most of the Θ eliminated is transformed into O.

Now, what do you say to the extraordinary fact, that the antipode of ozone has these many thousand years been ready formed and incarcerated, only waiting for somebody to recognize and let it loose out of its prison ? A dark blue species of fluor-spar has for years been known by the German mineralogists, being distinguished by its property of producing a peculiarly disagreeable smell on being triturated. Many conjectures were made as to the chemical nature of the odorous matter emitted from the spar: chlorine, iodine, and even ozone were spoken of, but it turned out to be a different thing.

M. Schafhaeutl of Munich sent me, a month ago, some hundred grammes of the said fluor-spar (occurring within the veins of a granitic rock at Wulsendorf, a Bavarian village near Amberg), asking me to try my luck in ascertaining the nature of the smelling matter; and I think I have fully succeeded in making out what it is. Surprising as it may sound to you, and unique as the fact certainly is, that matter happens to be nothing but my insulated antozone. "But how do you prove that ?" you will ask me. In the first place, it smells exactly like Θ disengaged from BaO^2. "But smells are fallacious tests." They are. You shall have another proof that will irresistibly carry conviction with it: on triturating the fluor-spar with water, peroxide of hydrogen is formed, not in homœopathic, but very perceptible quantities. When I first found out this extraordinary fact (I think it was on the 17th of November last), I could not help laughing aloud, though I happened to be quite alone in my laboratory. I laughed because I strongly suspected my foe to be hidden in the spar, and I broke his mask under water with the view of catching him by that fluid. Indeed, it was to me as if I had caught a very cunning fox, long sought after, in a trap put up for him. To show you that in saying this I have neither been joking nor dreaming, I shall send you, as soon as I can, a sample of the wonderful spar, with which you may easily satisfy your curiosity and convince yourself of the correctness of my statements. I must not omit to tell you that, according to some previous experiments of mine, the fluor-spar of Wulsendorf contains $\frac{1}{3000}$th part of antozone—a quantity, as you see, not at all homœopathic. How that subtle matter got into the spar I cannot tell.

XV. *Note on the Remarks of* Mr. Jerrard.
By JAMES COCKLE, *Esq.*
[Concluded from vol. xix. p. 332.]

THE sequel to Mr. Jerrard's important 'Essay' may remove the new doubts and difficulties which arise upon his 'Remarks,' and which, numbered in conformity with his paragraphs, are here briefly noticed. Mr. Jerrard has to meet the additional objections that—

1, 2, 4. There is no error in Mr. Harley's processes. The error of inferring the sextic to be an Abelian I have already traced to its source (vol. xix. p. 197 *et seq.*).

3. If the functions $_1\Xi$, $_2\Xi$, $_3\Xi$, and $_4\Xi$ are not extraneous, let two quintic surds be expelled, as reducible, from them and from $_0\Xi$. Then, according to Abel's theorem, all the roots of (ab) are roots of (ac). Denote (ac) and (ab) by

$$\xi=0, \qquad \xi'=0$$

respectively. Then

$$\xi=\xi'(\Xi_\zeta-\mathrm{r}\{\mathrm{P}_{f(\beta_{..}\epsilon)}\}).$$

But what is the form of ζ? Certainly not f, for the root Ξ_f is common to (ab) and (ac). Unless Mr. Jerrard shows that ζ cannot be replaced by $f(\beta_{..}\epsilon)$, his conclusion is vitiated by an error in comparing (ab) and (ac).

5. The complex process need not supersede the simple method of substitutions when we are dealing with symmetric functions of Θ' and Θ''. Such functions are Mr. Jerrard's Ξ, or

$$\Theta_1{}^5+\Theta_2{}^5+\Theta_3{}^5+\Theta_4{}^5,$$

and my θ, or

$$5^4\Theta_1\,\Theta_2\,\Theta_3\,\Theta_4.$$

To take a simple example,

$$\Theta'_{n,f(\beta_{..}\epsilon)(\gamma_{..}\delta)}+\Theta''_{n,f(\beta_{..}\epsilon)(\gamma_{..}\delta)}=(\Theta''_{n,f}+\Theta'_{n,f})(\beta_{..}\epsilon)(\gamma_{..}\delta).$$

6. It can scarcely be said that Mr. Jerrard's theorem is ignored by Mr. Harley. That theorem is inapplicable to the processes dealt with in Mr. Harley's paper, into which no symbol corresponding to P enters.

7. Actual calculation shows that Ξ and θ are similar functions (*fonctions semblables*). And if in my function γ we substitute Ξ for θ, the result is an equation which, combined with (ac), shows that, if the latter be an Abelian, the root of a general quintic can be expressed without quintic surds.

8. If, in expanding a function of two independent quantities, we do not obtain an expression symmetric with respect to those quantities, and such as to admit of an interchange between them, I cannot consider the expansion as algebraically (rigorously) true.

In generalizing the theory of transcendental roots which I have sketched, and partially developed, in these pages, we may start from

$$x^n-nx+(n-1)a=0.$$

We next deduce an *n*-ic in *y*, in which

$$y = (1 - a^{n-1})\frac{dx}{da};$$

and, hence, we pass to an *n*-ic in x', in which

$$x' = P + Qy + Ry^2 + \ldots + Ty^m,$$

where *m* is not greater than $n-1$, and P, Q, R, .. T are so determined that the *n*-ic in x' is of the form

$$x'^n - nx' + (n-1)a' = 0.$$

From the equations in *x* and x' we obtain

$$x = \phi a, \qquad x' = \phi a', \qquad x'' = \phi a'', \ldots x^{(r)} = \phi a^{(r)},$$

whence the form of ϕ is to be sought.

Mr. Jerrard's trinomial transformation indicates that this process is applicable (*primâ facie*) to equations as high as the fifth degree inclusive. Beyond that degree a trinomial cannot be a general equation. But possibly the properties of trinomial equations, or of equations involving only one parameter, may enable us to extend the process further by obtaining the *n*-ic in x'.

4 Pump Court, Temple, London,
January 12, 1861.

XVI. *A Theory of Magnetic Force.*
By Professor CHALLIS, F.R.S.

[Continued from p. 73.]

IN art. 6 of the foregoing part of the Theory of Magnetic Force, certain laws of the motion of an elastic fluid, essential both to that theory and to the previous theories of electric and galvanic forces, were enunciated without being supported by mathematical evidence. Before proceeding further with the explanation of magnetic phenomena, I propose to adduce the mathematical investigation of those laws.

15. It was stated that the resultant of the composition of steady motions is steady motion. The proof of this theorem is as follows:—For cases of steady motion, the general hydrodynamical equation which expresses constancy of mass, viz.

$$\frac{d\rho}{dt} + \frac{d \cdot \rho u}{dx} + \frac{d \cdot \rho v}{dy} + \frac{d \cdot \rho w}{dz} = 0,$$

becomes, to the *second* order of approximation,

$$\frac{du}{dx} + \frac{dv}{dy} + \frac{dw}{dz} = 0,$$

because for that kind of motion $\frac{d\rho}{dt} = 0$, and $\frac{d\rho}{\rho dx}$, $\frac{d\rho}{\rho dy}$, $\frac{d\rho}{\rho dz}$ are each of the order of the *square* of the velocity, as is shown by the general differential equations involving the pressure, no extraneous force being supposed to act. Now if u_1, v_1, w_1; u_2, v_2, w_2, &c. be velocities due to different disturbances acting separately, and each set of values satisfy separately the second of the above equations, it is evident that if $u = u_1 + u_2 +$ &c., $v = v_1 + v_2 +$ &c., $w = w_1 + w_2 +$ &c., u, v, w will satisfy the same equation. From this reasoning it follows that when several causes of steady motion coexist, the velocity they produce conjointly at any point, is the resultant of the velocities which they would produce at the same point by acting separately. Hence since the component motions, being steady, are constant in magnitude and direction at each point, the resultant motions will be constant in magnitude and direction at each point; that is, the whole motion will be steady.

16. Hence in the general hydrodynamical equation applicable strictly to steady motion, when no extraneous force acts, viz.

$$a^2 \text{ Nap. } \log \rho = C - \frac{V^2}{2},$$

we may assume, since terms involving the square of the velocity have been taken into account, that V is the resultant of several velocities, v', v'', &c., given in magnitude and direction. Take, for instance, the two velocities v' and v'', and suppose their directions to make an angle a with each other. Then

$$V^2 = v'^2 + v''^2 + 2v'v'' \cos a.$$

From this equation it is seen that if the value of C be given, the velocity is greatest and the density and pressure least, where $a = 0$, that is, where the two streams coincide in direction; and that the velocity is least, and the density and pressure greatest, where $a = \pi$ and the two streams are in opposite directions.

In general the constant C will be different for different lines of motion. But in the cases of convergent streams flowing from an unlimited distance, and divergent streams flowing to an unlimited distance, at remote points $\rho = 1$ where $V = 0$, and consequently $C = 0$ for each of the lines of motion.

17. The hydrodynamical theorems above demonstrated are those which have been used in accounting for electric attractions and repulsions, the mutual action between two electrodes, that between an electrode and a magnet, and the mutual attractions and repulsions of magnets. It is to be remarked that in these explanations the moving forces of the æther acting on the ultimate atoms of bodies have been supposed to be due to variations of its density, producing statically differences of pressure at different points of each atom. But if motion resulted solely from variation of the density of the æther from point to point

of space, there would seem to be no reason why a magnet should not move various substances as an electrified body does. In the one case, as in the other, variation of pressure is produced, according to the theory, by ætherial streams. What, then, is the reason that an electrified body acts indifferently on many kinds of substances, whilst a magnet acts in a special manner on certain substances, which from this peculiarity have been named magnetic?

18. In answering this question, it is first to be considered that, according to the hydrodynamical theory of electricity, an electrified body has had its *superficial* atoms forcibly disturbed, and that this state of disturbance gives rise to *vibrations* of the æther, the dynamic effect of which on a neighbouring body is to put its superficial atoms also into a state of disturbance, and thus to *induce* electricity. It was explained in the theory referred to, that this state of induced electricity coexists with a gradation of internal density, which generates the secondary currents to which electric attractions and repulsions are attributable. But in a magnet there is nothing corresponding to that superficial disturbance, the internal gradation of density being an independent property of the body. Hence there are no *vibrations* to disturb the superficial atoms of adjacent bodies, the magnet acting dynamically wholly by *currents*, and consequently, as is known by experience, it is incapable of inducing electricity. At the same time, between a magnet and a body partially or inductively electrified, there must be interference of ætherial currents, and consequently mutual action, as is found experimentally to be the case.

Still it may be argued that magnetic streams, varying from point to point of space as to density and velocity, must act dynamically on the atoms of bodies of all kinds, independently of their electric or magnetic states. That such action does really take place to a sensible amount under the influence of *powerful* magnets, will be made apparent by the following considerations applied to the class of facts discussed experimentally by Faraday in the Twentieth and Twenty-first Series of his 'Researches in Electricity' (Phil. Trans. part 1. 1846).

19. It has been already stated that, according to hydrodynamics, a single spherical atom, subject to the action of a *uniform* steady current, suffers no change thereby of a condition either of rest or of uniform motion. But this is no longer true if, the stream being steady, the lines of motion are not parallel to each other, because in that case there will be inequality of the pressures on opposite hemispherical surfaces of the atom. Also when such a stream enters into any substance, the partial convergences or divergences of its course, which must result from its encountering an aggregation of atoms, may materially influence its

dynamic action on any given atom. We have not the data for inquiring what that action may be in any particular substance, because for that purpose we require to know its composition and the magnitudes and arrangement of its atoms. I shall therefore assume, as a result of the experiments above referred to, that divergent and convergent magnetic streams act dynamically in different manners and in different degrees on a great variety of substances which they permeate.

20. If the substance be *iron*, the moving force of the magnetic streams on its individual atoms, apart from any action resulting from interference with secondary streams, tends along the lines of motion *towards* the poles of the magnet. The following remarkable experiment by Faraday (2368) establishes this law. Cylindrical glass tubes containing ferruginous solutions of different degrees of strength were suspended with the axes vertical in a ferruginous solution of given strength between the poles of the magnet. If the solution in the tube was *stronger* than that of the surrounding fluid, the tube was *attracted* towards either pole; and if *weaker*, it was *repelled*. *Both* these effects may be explained on the assumption that the magnet *attracts* the solutions in proportion to the quantity of iron they contain. The tube was drawn towards the poles in the first case, because the contained fluid being more ferruginous, was more attracted than the surrounding fluid. In the other case, the hydrostatic pressure of the surrounding fluid, due to the assumed magnetic attraction towards the poles, is greater on the side of the tube nearest either pole, than on the opposite side, and this difference of pressure, not being wholly counteracted by the attraction on the contained weaker solution, causes a repulsion of the tube. The cases are exactly analogous to those of a body specifically heavier than water sinking in it, and a body specifically lighter rising in it, by the action of gravity. Now it is true that the magnetic forces here assumed have the same *directions* as those which the variation of the pressure of the æther would of itself produce. But if the action were either wholly or chiefly due to this cause, no reason appears why the observed motions should depend, not upon the difference of the specific gravities of the fluids within and without the tube, but upon one being more or less *ferruginous* than the other. The experiment, therefore, allows us to infer theoretically, that ætherial streams, flowing through *iron*, and affected by the number, size, and state of aggregation of its atoms, exert on each atom an accelerative force, which is in the direction of the current where it is convergent, and in the direction contrary to that of the current where it is divergent. The atoms being by hypothesis spherical, it may be inferred from

hydrodynamical considerations, that converging and diverging streams impress upon them the same moving forces, if the velocities be the same, and the degree of convergency be equal to that of divergency; but the forces will be in opposite directions with respect to the course of the streams. Hence, under those circumstances, if one magnetic pole be attractive, the other will be attractive also, and in the same degree. This result experience confirms.

21. If a bar of *bismuth* be suspended either horizontally or vertically between the poles of a magnet, it is found to be impelled in the directions of the lines of motion *from* the poles. The action is therefore opposed to that which results from variation of density of the æther. The same effect took place when the bismuth was broken into very small pieces. Also it appeared that two pieces of bismuth subject to the action of magnetic streams, exerted no influence on each other; that is, no secondary streams were generated. From these facts it must be concluded theoretically, that convergent and divergent magnetic streams, when they enter bismuth and substances of the same class, and are modified by the number, size, and arrangement of the constituent atoms, impress on the individual atoms accelerative forces, which impel them along the lines of motion from the magnetic poles. For the same reason as that above given in the case of iron, the action is alike from both poles, and in the same degree, although the streams diverge from one and converge towards the other.

22. It is found by experiment that *air* and *gases* are not sensibly acted upon by magnetic streams. This fact may be explained by the theory on the supposition that such substances, on account of their small density, do not perceptibly, by the state of aggregation of their atoms, modify the ætherial streams, which consequently act on each atom as if it were alone. Now in that case a divergent stream *impels* the atom in the direction of the course, and a convergent stream *draws* it in the direction contrary to the course, and it is possible that these actions may be just counteracted by the opposite effects of the variation of density of the æther.

23. Since the magnetic streams attract one class of substances (the magnetic) towards the poles, and repel another class (the diamagnetic) from the poles, we might expect to meet with other substances which are neither attracted nor repelled, or are only acted upon very feebly. This appears to be the case with *copper.* The peculiar phenomena (Faraday, 2309, &c.) which this metal exhibits when placed between the poles of a magnet, seem to admit of explanation on the principles of this theory, by supposing its atoms to be easily moveable from their normal

positions,—a property to which it may perhaps owe the facility with which it conducts galvanic currents.   Iron and bismuth, which are not good conductors of galvanism, present very different phenomena, in consequence, probably, of a great degree of fixity of their atoms, by reason of which, and of atomic arrangement and constitution, the magnetic streams permeating them may be so modified as to become respectively attractive and repulsive.   A bar of copper, when first acted upon by the magnetic streams, begins to move as if it were a magnetic body, this motion being probably due only to the variation of pressure from point to point of the streams.   But quickly its atoms, on account of their mobility, will be moved by that pressure into new positions, which may be such as to render it for a brief interval diamagnetic, so that the incipient motion is stopped, the atoms finally settling into positions for which the bar is neither magnetic nor diamagnetic.   These are positions of constraint, in which they are held by the dynamic action of the pressure of the magnetic streams.   When the streams are interrupted, the atoms, by the proper molecular action of the copper, return to their normal positions; and in the mean time the reaction of the æther on the bar produces the observed *revulsion*.   The magnetic streams continuing in action, if an impulse be given to the bar tending to move it into a position of *greater* magnetic action, for the same reason as above an increment of diamagnetic force is generated, and a momentary *excess* of diamagnetic action, arising from the atoms being carried by their momentum beyond the positions in which they finally settle, stops it; and if it be moved into a position of *less* magnetic action, a momentary *defect* of diamagnetic action, also due to the momentum of the atoms in taking up new positions, equally stops it.   The sluggish motion of the bar may be thus accounted for.   The movement which occurs when the streams commence, would not take place if the bar were placed in the axial or equatorial direction, because the initial magnetic action would be symmetrical with respect to each of these directions.   If the copper were of the form of a globe, with its centre on the axial direction, the sluggishness would still exist, because any movement of it about a vertical axis would bring each point into a position of either greater or less magnetic action.

24. Since it appears from the foregoing discussion that, independently of the dynamic effect of magnetic streams resulting from gradations of the density of the æther, they exert, when they enter a magnetic body, a particular moving force due to their divergence or convergence as affected by the atomic constitution of the body, this force ought to be taken into account in the theory of the magnet (art. 4), and in the explanation

of the mutual action between two magnets (art. 7). As the two kinds of force act along the same lines in the same directions, the consideration of this second force will not alter the conclusion arrived at in art. 4, as, in fact, is stated at the conclusion of that article. With respect to the manner in which the mutual action of two magnets is affected by the motion of the ætherial streams within them, it is to be remarked that divergence and convergence are increased by the opposition of two streams, and diminished by their confluence; and that the particular force under consideration is always *towards* positions from which the streams diverge, or towards which they converge. Hence this additional force, being some function of the degree of convergency or divergency, produces effects exactly like those of the pressures considered in art. 7, and the conclusion there drawn remains the same when both kinds of force are taken into account.

25. The following theory of *magnetic induction* rests on the foregoing views. It is found that when a bar of soft iron is placed in the plane of the magnetic meridian, it is magnetized inductively, the magnetism disappearing when its axis is placed perpendicular to that plane. The induced magnetism is greatest when the axis coincides with the direction of the dipping-needle; and the magnetism of its north end is the same as that of the north end of the needle. These facts admit of being explained on the supposition that the primary terrestrial stream in its passage through the bar exerts a force on the atoms in the direction from south to north, so as to produce a gradation of density of the bar towards the north end. This force may originate in the circumstance that when the stream enters the bar its velocity is increased by the contraction of the channel by the atoms, and, the density and pressure being consequently diminished, the surrounding fluid is drawn towards the axis of the stream. Thus there will be lines of motion without and within the bar converging northward, and consequently, from what is argued in art. 20, a force will be generated proper for producing the increment of density towards the north·end. The same effect would be produced even if the streams were *parallel*, because as *divergent* streams flowing through iron, draw the atoms towards the quarter from which they flow, à *fortiori*, parallel streams would do the same. Thus the bar is converted into a magnet. But it is to be observed that the atoms retain their positions in consequence of the counteraction of this disturbing force by the *molecular* repulsion of the iron due to the variation of its density. Hence when the bar, by being put into a position transverse to the magnetic meridian, ceases to be under the influence of the terrestrial current, the atoms return to their normal positions, and the bar ceases to be a magnet. It is clear that this

induction of magnetism will be greater, the more directly the primary stream flows through the bar, as is known to be the case by experiment.

26. When a *coil* of wire in the form of a *helix* surrounds a cylinder of soft iron, and a galvanic current is sent through the wire, magnetism is found to be induced in the iron. This fact is explained by the theory as follows. The turns of the helix being supposed to be very little apart, each turn will be nearly in a plane perpendicular to the axis of the cylinder. Hence the circular motions, which, according to the theory of galvanism, take place about the wire, will be very nearly in planes passing through that axis. Suppose the course of the current to be from the end B towards the end A of the cylinder, and the turns of the helix to proceed from the left hand, over the cylinder, to the right hand of a person looking from B towards A. Then, since the circular motion about any portion of an electrode in which the current is conceived to flow parallel to the earth's axis from south to north is always in the direction of the earth's rotation, it follows, in the case supposed, that if the circular motion within the cylinder be resolved into parts parallel and perpendicular to its axis, the former will all be in the direction from B towards A, and the latter will very nearly destroy each other. Thus the result will be a longitudinal stream from B towards A, the velocity of which, if the helix be pretty close to the cylinder, will be greatest along its axis. The explanation of the manner in which this stream induces magnetism in the iron core, is precisely the same as that applied in the preceding article to induction by terrestrial magnetism. The pole A of the magnetized cylinder corresponds to that pole of a magnetic needle which points northward. The contrary would plainly be the case if the helix, instead of being *dextrorsum*, as supposed above, had been *sinistrorsum*. All these results agree with well known experimental facts.

27. The theory I now proceed to give of *terrestrial magnetism* rests on no other hypotheses than those on which the general physical theory is based. It will be assumed that the earth and all the bodies of the solar system are composed of spherical inert atoms, of different magnitudes, and in different states of aggregation, and that the æther pervading their interiors is in the same state of density as in the external spaces. Any investigation of the motions impressed on the æther by the known motions of the bodies of the solar system, must, in order to be consistent with the whole preceding argument, set out from these hypotheses. The following considerations will, I think, show that they lead to conclusions which are in accordance with observed facts of terrestrial and cosmical magnetism.

We are to assume that the mass of the earth consists of discrete spherical atoms, and that the whole of its interior, excepting the space occupied by atoms, is filled with æther in the same state of density as in the external regions of space, with only such variations of density as are produced by its motion. The same assumption is to be extended to the sun, the moon, and all the planets. In the first place, let us endeavour to ascertain what kind of movement is impressed on the æther by the earth's *rotation*; and for the sake of distinctness, I shall, at first, suppose that the earth has no motion of translation, and that it is perfectly symmetrical with respect to its axis and its equatorial plane. Then as the terrestrial atoms will impress a part of their velocity on the æther, a rotatory motion of the latter will be produced, which by its centrifugal force will tend to draw the fluid from the earth's axis. This tendency cannot give rise to a stream from one pole to the other, because there is no reason why it should flow in one direction rather than the other. Neither can streams be generated setting from the equatorial parts towards the poles, because as the velocity will be greatest in and near the plane of the equator, the fluid would rather be urged *towards* this plane, and *circulate* by return currents along the axis towards the poles. The tendency of the centrifugal force to draw off the fluid from the axis of rotation, will be counteracted by accelerative forces due to increments of density of the æther which result from the resistance of its inertia to the centrifugal movement. These forces, acting always towards the axis, cause the circular motion immediately impressed by the atoms to extend into the æther beyond the earth's surface. But it is evident that so long as the terrestrial atoms have greater velocity than the æther contiguous to them, the centrifugal force is on the increase, and the permanent state is reached only when there is no motion of the atoms relative to the æther. The opposite forces then maintain a steady motion at all points, there being no reason why the motion should not be steady at one point rather than at another. Consequently in the earth's interior, and at the surface, and sensibly to considerable heights above, the æther, like the atmosphere, partakes of the earth's rotation. This gyratory motion must spread to remote distances, the velocity decreasing with the distance according to a law which eventually may be found to admit of mathematical investigation on hydrodynamical principles. As the motion of rotation must be combined with the circulating currents above mentioned, the total motion is *spiral*.

28. Now let the earth be supposed to have a motion of translation, either uniform or variable. Then on the principle that there can be neither gain nor loss of the momentum of the æther

due to the earth's rotation, it may be asserted that the gyratory motion is transferred without alteration through space with the earth. It may not be that the motion of a given particle of the fluid remains the same; but the motion is constantly the same in positions which in successive instants are the same relative to the earth's centre and equatorial plane. Coexistent with the gyrations of the æther, motions of translation must be impressed upon it, having constant relations to the earth's motion. With respect to that portion of æther which has sensibly the same rotation as the earth, the most probable supposition is that it has also very nearly the same motion of translation, and that consequently the relative motion of the earth's atoms and the æther in its interior is very small.

These conclusions apply to the other rotatory bodies of the solar system. All are consequently centres of gyratory motions, which, according to the reasoning in art. 15, may coexist in the cosmical spaces without interfering with each other.

29. But it is not true, as supposed, that the earth is symmetrical with respect to its axis and equatorial plane, the superficial distribution of land and water showing that this is not the case. On the supposition of symmetry, the centrifugal force of the earth's rotation would induce a variation of its internal density, proper, according to the theory, for generating secondary streams under the influence of a primary current. But as by reason of the symmetry equal and opposite effects would be produced on every diameter of the earth, no streams would become sensible. For the same reason no streams result from the increase of the earth's density towards the centre by the effect of gravity. In the actual case of deviation from symmetry, the motion of the æther cannot be wholly in symmetrical gyrations as above inferred, although it will be approximately such. There must be motion relative both to the equator and the axis, steady in its character, as depending on constant causes, and constituting a primary stream, which flowing through the earth, put into an unsymmetrical state of restraint by the centrifugal force, will give rise to secondary streams. The primary current may be of feeble intensity, and yet by originating accelerative forces which act through large spaces, may generate streams of great intensity. The earth's magnetic streams may be regarded as resulting from a combination of the primary and secondary streams. According to these views the deviations of the earth's form and matter from symmetry determine the directions of the magnetic streams, which appear from experiment to *enter* the earth on the *north* side of the magnetic equator, and to *issue* from it on the *south* side. *The earth is thus a vast magnet, the streams of which are of constant intensity, excepting so far as they may be disturbed by cosmical influences.*

The terrestrial streams probably *circulate*, those which issue
nearly perpendicularly to the earth's surface in the antarctic
regions, after rising to great heights, turning back in curved
courses distant from the earth, so as to enter again nearly perpendicularly in the arctic regions.   To this circulation must be
added that spoken of in art. 27, which conspires with the antarctic streams and opposes the arctic, and thus probably accounts
for the observed excess of antarctic magnetic intensity.

30. If the earth had been symmetrical with respect to its axis,
but not with respect to its equator, the magnetic streams would
also have been symmetrical with respect to the axis.   But the
actual irregular distribution of land and sea is opposed to this
law, and it is found in fact that there is an approximation to
two magnetic poles as well in the arctic as the antarctic regions.
Also it is established by observation that these poles have not
fixed positions on the earth's surface.   Taking account of this
circumstance, and of the character of the irregularities to which
the theory ascribes the generation of the currents, we might
infer that the magnetic poles would be unsymmetrically situated
with respect to the earth's poles, that the streams would be
disposed unsymmetrically about them, and that in their movement they would not retain the same relative positions.   These
inferences observation appears to countenance.

31. The *arches* and *streamers* of the *Aurora Borealis* and the
Aurora Australis are portions of terrestrial magnetic streams
made visible by particular disturbances.   According to my view
of the undulatory theory of light, I should have no difficulty in
admitting that ætherial streams that are steady, and on that
account emit no light, become luminous upon being disturbed.

32. We have now to consider the disturbances which the
steady terrestrial currents may undergo from cosmical influences.
And in the first place it may be stated, as a deduction from
hydrodynamical principles, that the steady streams to which
magnetic force has been attributed, and the vibrations which
were assumed to exist in the theory of the force of gravity, *may
coexist without mutual interference,* so that the two kinds of force
act independently of each other.

33. The effect which any motion of translation which the
earth's atoms may have relatively to the æther, may be presumed
to be the same, in respect to generation of secondary currents,
as that of a current within the earth at rest.   In the investigation of such effect, the motion of translation of the solar system
would have to be taken into account, together with the motion
of the earth in its orbit.   As observation has not detected a
variation of terrestrial magnetism corresponding to the law of
the variation of velocity which would result from the combina-

tion of these motions, it must be concluded, in conformity with what is said in art. 28, that there is no perceptible effect from motion of translation of the atoms relative to the æther.

34. The terrestrial streams will be disturbed by the streams due to the gyratory motion about the sun, and the direction and amount of the resultant, at a given place, of the combination of the solar and terrestrial streams, will vary with the time of day. Hence there will be a *diurnal variation* of magnetic intensity, as is also known from observation. The epoch of maximum intensity will depend in part on the position of the place relative to the magnetic poles.

35. The amount of the variation of intensity on a given day will depend on the sun's declination, and therefore will be subject to an *annual variation*, being in north latitudes greatest at the summer solstice, and least at the winter solstice. These results are confirmed by observation.

It may deserve to be considered whether the magnetic intensity is sensibly affected by currents which must be produced in the atmosphere by *solar heat*, setting in all directions *from* the parts of the atmosphere most heated by the sun. The effect of these currents would be subject to diurnal and annual variations, and if of considerable amount, would have an influence on the local hour of maximum intensity.

36. There will also be an *annual variation* of intensity independent of the sun's declination, being due solely to the change of the velocity of the solar gyrations, caused by change of the earth's distance from the sun. Such a variation has been detected; but its amount is very small.

37. Through the telescope the *sun* presents itself to our view as a globular body, surrounded by a thick envelope consisting of matter which accumulates and divides, floating and whirling most probably by the action of an atmosphere, and which is separated from the interior solid globe by an intervening space. We cannot be far wrong in calling such matter *cloud*. It is then reasonable to suppose that the observed changes are attributable to the action of galvanic or magnetic streams, generating and precipitating from time to time the cloudy matter. But *changes* can hardly be attributed to the proper magnetic streams of the sun, because these, as in the case of the earth, will most probably be of a constant character. The changes are rather due to *disturbances* of the solar streams by external influence, as by the gyratory streams of the *planets*. It is therefore quite in accordance with this theory to find, as has been done lately, that observations indicate periods of maxima and minima of *solar spots*. These periods will be determined by changes of the combined action of the gyratory motions about the planets, as

resulting from changes of their positions relative to the sun; and it is therefore not surprising to find that one of them (that of $11\frac{1}{9}$ years) differs little from the periodic time of Jupiter, the influence of this planet being likely to be predominant on account of his large size and rapid rotatory motion. The periodicity of the magnetic action of any planet must clearly depend in part on the position of the plane of its equator, the gyratory motion being at a maximum in this plane.

38. If the preceding account of the periodicity of the maxima of solar spots be true, it will be seen to be a necessary consequence that the planetary gyrations produce *disturbances of terrestrial magnetism,* having in the long run a regular and periodic character. The researches of General Sabine, applied to magnetic observations taken at various positions on the earth's surface, have in fact conducted him to the result that *magnetic storms* are periodic, and led him to assign to them a period of ten years. Considering that the same planetary magnetic streams must operate on the earth as on the sun, and that the sun's position is central with respect to the earth's orbit, it might have been anticipated, on theoretical grounds, that a like periodicity of effects would be detected, and that the periods in the two cases would not greatly differ. Possibly the period of the magnetic storms may eventually be found to be in some degree variable, and the mean period inferred from observations extended over a longer interval, to be somewhat different from ten years.

39. The theory, by giving to extraordinary magnetical disturbances a planetary origin, accounts for their occurring simultaneously at places on the earth's surface widely distant. But the *amounts* of disturbance at the same instant may differ greatly at different places, owing to difference of latitude, and difference of position relative to the magnetic poles. Considering, however, the changes of the configurations of the planets, the amounts of disturbance, regarded as functions of the time of day, may be expected in the long run to follow the same law at different places. This is found to be the case, although, as might have been anticipated, the local hours of the maximum values are different.

40. The fact of there being local hours of maximum disturbance, in the case of the planetary as in that of the solar disturbances, seems to be referable to the circumstance that the disturbances of the superior planets will be greatest when they are in opposition, or about midnight, they being then least distant from the earth, and the disturbances of the inferior planets will for the same reason be greatest when they are in inferior conjunction, or about mid-day. If this explanation be true, we should expect to find

a set of maxima prevailing in hours of the night, and another set in hours of the day, and that the directions of the deviations of the needle in the two cases would be opposite, the directions of the gyratory streams of the superior and inferior planets being opposite for their positions of maximum disturbance. These laws General Sabine has in fact detected by a discussion of numerous observations taken at Kew and Hobarton. (See 'Proceedings of the Royal Society,' vol. x. No. 41, p. 632.)

41. The facts relating to the movements of the earth's magnetic poles do not appear to be sufficiently well ascertained to allow of applying the theory to explain them. Conceive, for the sake of illustration, that there are two north magnetic poles at the same distance from the earth's pole, and 180° from each other, that the currents converging to them are of the same intensity, and that they have the same uniform movement from *east* to *west*. On these suppositions it is evident that when the poles are on the great circle passing through the pole of the earth and a given place, the Inclination is at a *maximum*, and the Declination is changing from *east* to *west*. When they have passed over 90° the Declination will have gone through a western maximum, and be on the point of changing from *west* to *east*, because the needle, being equally acted upon by the two poles in the new positions, must point exactly between them. At the same time the Inclination will be at a *minimum*. After the poles have advanced 90° further, the needle will have gone through a maximum eastern Declination, and returned to its original position, and the Inclination will again be at a maximum: and so on. These hypothetical circumstances are, perhaps, not so far different from the actual as to prevent our concluding from them that neither the direction nor the amount of the motion of the magnetic poles can be inferred with any certainty from mere observations of "points of convergence" of the horizontal needle—at any rate not with so much certainty as the direction of the movement may be inferred from the direction in which the Declination is changing when it is passing through zero at any place, as London or Paris, *if at the same time the Inclination is near its maximum at the same place.* Now early observations at Paris show that the Declination was zero there about the year 1663, and was changing from east to west, and observations both at Paris and London indicate a constant diminution of Inclination from that date, while, according to Hansteen's Isoclinal lines, the Inclination was *less* in 1600 than in 1700. It may therefore be inferred that a maximum of Inclination occurred between those two epochs, and consequently, from the above argument, that the motion of one of the north magnetic poles is from *east* to *west*. If this

law should eventually be found to be true of all the magnetic poles, the theory will account for it in this manner. The north magnetic streams may be supposed to descend into the earth from heights at which the gyratory motion of the æther is sensibly less than at the earth's surface, and the south magnetic poles to issue from the earth to heights at which the same circumstance prevails. The streams being supposed to *circulate* from south to north in courses distant from the earth's surface, that diminution of the velocity of gyration will cause them to lag behind the earth in rotatory motion, and thus to draw the magnetic poles *westward*.

I have now completed an outline of a general physical theory, of which the leading idea is that all quantitative physical laws may be mathematically deduced from a few fundamental facts, distinct conceptions of which may be formed from sensation and experience. The hypothetical facts on which the theory rests are, that all substances consist of minute spherical atoms, of different, but constant, magnitudes, and of the same intrinsic inertia, and that the dynamical relations and movements of different substances are determined by the motions and pressures of a uniform elastic medium pervading all space not occupied by atoms, and varying in pressure in proportion to variations of its density. I am well aware that many of the explanations I have given of physical phenomena on these hypotheses are expressed in general terms, and are too little supported by exact analytical or numerical calculation. Some of the explanations, requiring a knowledge of the interior constitution of bodies, could not be conducted in an exact manner in the present state of science. Still so comprehensive a theory, resting on so few hypotheses, could hardly fail of meeting with contradictions in the attempt to explain facts, unless the hypotheses were true. The number and the variety of the explanations of physical phenomena which have been drawn from them without the support of symbolical calculation, seem almost of themselves to justify the conclusion that the ultimate atoms of bodies are really such as they have been assumed to be, and that the physical forces are modes of action of a single elastic medium.

But no doubt the general theory and the explanations derived from it are not fully established till they have borne the test of numerical verification. And here I take occasion to add that every complete physical theory is necessarily *mathematical*. Observation and experiment are essential for furnishing facts both for the foundation and for the verification of a theory, and may also discover laws and relations of facts; but the theory is not complete till all facts not fundamental, and all laws, have been referred by mathematical calculation to the fewest possible fundamental facts. These principles have been admitted in physical astronomy, and belong equally to other departments of physical science. Hence if the hypothesis of the existence of the æther as the sole source of physical power be true, the mathematical investigation of the motions of an elastic fluid become essential; and the *new principles* of the application of partial differential equations to the determination of fluid motion, which I have proposed, will have the same relation to the future progress of theoretical physics, as the discovery by Newton of the principles of the application of differentials in dynamics had to the progress of physical astronomy. On account of the important applications those principles may eventually receive, I purpose, as soon as I shall be able, to bring them again under the notice of mathematicians.

Cambridge Observatory,
January 17, 1861.

XVII. *On an Alloy which may be used as a Standard of Electrical Resistance.* By A. Matthiessen, *Ph.D.**

THE expression of electrical resistances in the absolute measure proposed by Weber† is, and probably will always remain, the best; but its determination requires so much apparatus and room, as well as so much skill in manipulation, that it is placed beyond the means of most experimenters. I have therefore deemed it worth while to test some alloys, to see whether I should not be able to find one—

I. Whose resistance will remain the same, whether it be made of absolutely pure or commercially pure metals; in other words, that such an alloy may be made by any chemist or assayer, and its conducting power will always be the same.

II. That its conducting power will not be altered by the process of annealing.

III. That its conducting power will not vary much with an increase or decrease of temperature.

IV. That the alloy will not alter by exposure to the atmosphere.

The great difficulty in obtaining absolutely pure metals, together with the fact that the smallest traces of impurity materially increase the electrical resistance of most metals, pre-

* Communicated by the Author.
† *Pogg. Ann.* vol. lxxxii. p. 337.

clude the use of them as standards, and make it desirable that, instead of comparing, as heretofore, the conducting powers of metals, alloys, &c. with that of silver, copper, or, as has lately been proposed, with mercury*, they should in future be compared with an alloy having the properties mentioned above.

The alloy best adapted for a standard of resistance for galvanic currents, is that composed of

2 parts by weight of gold †,
1 ,, ,, ,,   silver :

for on looking at the curve which expresses the conducting powers of the gold-silver alloys‡, we find part of it almost at a straight line; that is to say, to the alloy containing 50 volumes per cent. of gold (which is the middle point of the straight line) one or two per cent. more or less gold may be added without altering its conducting power to any great extent.

In order to test the first condition, I have had the alloys made in different parts of the world; and my best thanks are due to those gentlemen who kindly undertook the making and procuring of them: The following order was given :—

Take 6 grammes proof, or purest gold, and 3 grammes proof, or purest silver; fuse and cast three times§, and then draw into wire of about 0·5 millim. diameter. The wire to be hard-drawn.

The alloys experimented with were :—

·No. 1. Made of pure gold and silver by Mr. R. Smith, in Prof. Percy's laboratory. Drawn by Messrs. Watts and Son, of Kirby Street. These wires were annealed in a red-hot crucible.

No. 2. Made and drawn in Brussels. Procured for me by my friend Mr. G. C. Foster, through the kindness of Prof. Stas. Annealed on wire-gauze over a four Bunsen burner.

No. 3. Made and drawn in New York. Procured through my friend Mr. C. M. Warren. Annealed in the same way as No. 2.

No. 4. Made and drawn in Paris. Procured through Mr. S. Reuter. This wire came to hand already annealed; therefore no determinations of the hard-drawn wire could be made.

---

* Siemens, Phil. Mag. (Jan. 1861); Schröder von der Kölk, Pogg. *Ann.* vol. cx. p. 452. It should be borne in mind that the use of this metal as a standard is open to the following grave objection : viz. that the copper wires or plates dipping in the mercury will after a time make it impure; and *as traces of foreign metals* (0·1 or 0·2 *per cent.*) *cause a decrement in the conducting power of pure mercury* (*not as stated by Siemens, an increment*), it would become necessary often to change it, thereby requiring a large supply of chemically pure metal.

† Corresponding nearly to equal volumes of gold and silver.

‡ Phil. Trans. 1860.

§ It would have been better to have added, *fuse with a little borax and saltpetre.*

No. 5. Made and drawn in Frankfort. Procured through my friend Dr. Dupré. These wires were annealed on a wire-gauze over a Bunsen's glass-blower's lamp.

No. 6. Made of proof-gold and silver, by Messrs. Johnson and Matthey, of Hatton Garden, and drawn by them. These wires were annealed in the same manner as No. 2.

No. 7. Made and drawn by myself, of proof-gold and silver, lent me by Professor Hofmann. Annealed in the same manner as No. 2.

No. 8. Made and drawn by myself, of pure gold and silver. Annealed as No. 2 was.

In Table I. are given the values found for conducting power of the above alloys. They have been compared with a hard-drawn silver wire, whose conducting power has been taken = 100 at 0° C.

Table I.

No. of alloy.	Hard-drawn.	Temp.	Annealed.	Temp.	Mean reduced to 0° C. Hard-drawn.	Annealed.
No. 1.						
1 wire	14·99	14·5 C.	15·02	15·0 C.		
2 wire	14·87	15·0	14·92	15·2	15·08	15·13
Mean	14·93	14·8	14·97	15·1		
No. 2.						
1 wire	14·98	9·8	15·02	10·0		
2 wire	14·95	9·6	15·04	10·4	15·06	15·14
Mean	14·96	9·7	15·03	10·2		
No. 3.						
1 wire	14·60	16·8	14·66	17·0		
2 wire	14·75	16·2	14·82	17·4	14·85	14·92
Mean	14·68	16·5	14·74	17·2		
No. 4.						
1 wire	......	...	14·94	6·0		
2 wire	......	...	15·02	9·8	......	15·06
Mean	......	...	14·98	7·9		
No. 5.						
1 wire	15·02	8·5	15·06	9·0		
2 wire	14·99	9·2	15·03	9·8	15·09	15·14
Mean	15·00	8·8	15·04	9·4		
No. 6.						
1 wire	14·87	14·8	14·82	16·0		
2 wire	14·79	17·2	14·86	15·0	14·99	15·00
Mean	14·83	16·0	14·84	15·5		
No. 7.						
1 wire	14·94	14·0	14·99	16·2		
2 wire	14·91	16·2	14·97	17·2	15·07	15·16
Mean	14·92	15·1	14·98	16·7		
No. 8.						
1 wire	14·86	18·4	14·92	18·2		
2 wire	14·88	16·6	14·90	17·0	15·05	15·10
Mean	14·87	17·5	14·91	17·6		
				Mean.........	15·03 = 100	15·08 = 100·3

Calling the mean of the values at 0° of the hard-drawn wires 100, we find that, as Table II. shows, the differences between the conducting powers of the alloys are very small, in fact, that the greatest is only 1·6 per cent.

Table II.

No. of alloy.	Hard-drawn.	Difference.		Annealed.	Difference.
1.	100·3	+ 0·3		100·6	+ 0·3
2.	100·2	+ 0·2		100·7	+ 0·3
3.	98·8	− 1·2		99·2	− 1·1
4.	......	...		100·2	− 0·1
5.	100·4	+ 0·4		100·7	+ 0·4
6.	99·7	− 0·3		99·8	− 0·8
7.	100·3	+ 0·3		100·8	+ 0·5
8.	100·1	+ 0·1		100·4	+ 0·1

The great concordance in the above results is partially due to the wire drawing so very well. It draws as well as, if not better than, any wire I have as yet tried.

For the sake of comparison, I will give in Table III. the values found by different experimenters for the metals with which the others are usually compared.

Table III.

	Becquerel [*].		Lenz[†].	Siemens[‡].		Matthiessen.	
	Hard-drawn.	Annealed.		Hard-drawn.	Annealed.	Hard-drawn.	Annealed.
Silver	100·0	107·0	100·0	100·0	108·8	100·0	110·0
Copper	95·3	97·8	73·3	89·7	95·2	99·5	102·0
Gold	68·9	70·0	58·5	......	......	78·0	80·0
Mercury		1·86	3·42 at 18°·7	1·72		1·63 at 22°·8	

When no temperature is given, the observations have been made at 0° C.

Part of these differences may be due to the silver, which in some cases may not have been chemically pure; but if any of the others be taken as unit, we do not arrive at any better result.

Table IV. shows the differences between annealed and hard-drawn wires,—the annealed being the better conductor, according to—

Table IV.

	Becquerel.	Siemens.	Matthiessen, Holzmann[§].
Silver	7·0 per cent.	8·8 per cent.	10·0 per cent.
Copper	2·6 per cent.	6·1 per cent.	2·5 per cent.
Gold	1·6 per cent.	....	2·6 per cent.

---

[*] *Ann. de Chim. et de Phys.* vol. xvii. (1846), p. 242.
[†] *Pogg. Ann.* vol. xxxiv. p. 418, and vol. xlv. 105.
[‡] *Loc. cit.*　　　　　　　　　§ Phil. Trans. 1860

The gold-silver alloy, however, only varies 0·3 per cent.,—another reason why the alloy may be drawn by anybody, and still have the same conducting power.

The following experiments show the effect of an increase of temperature on the conducting power of this alloy. The details of the experiments, together with the apparatus employed, will be published shortly in a paper by Dr. von Bose and myself, "On the Conducting Power of the Metalloids, Metals, and Alloys at different temperatures." Our results at present do not agree with those of Arndtsen and Siemens, who state that the resistances of most metals vary in direct ratio with the increase of temperature, but with those of Lenz, who calculates the conducting powers for different temperatures by the formula $\lambda = x + y\,t + z\,t^2$, where x is the conducting power at 0° C., y and z constants. I may also mention that the metalloids conduct better on being heated, being the reverse of that which the metals do.

An annealed wire of No. 1 alloy was heated in a glass tube placed in a water-bath (a), and afterwards heated in an oil-bath (b). Its conducting power was determined at different temperatures; and the values found are given in Table V. The values found for an annealed wire of No. 5 alloy (c), heated in an oil-bath, are also added.

Table V.

(a.)		(b.)		(c.)	
T.	Conducting power.	T.	Conducting power.	T.	Conducting power.
0·4 =	15·053	9·1 =	14·954	6·1 =	15·088
19·5 =	14·854	30·9 =	14·728	33·3 =	14·798
41·3 =	14·626	53·5 =	14·501	51·6 =	14·607
60·6 =	14·431	71·4 =	14·325	73·3 =	14·391
82·7 =	14·219	95·1 =	14·094	96·7 =	14·162
100·0 =	14·052	69·4 =	14·342	73·3 =	14·391
79·3 =	14·251	47·9 =	14·550	51·4 =	14·609
59·1 =	14·453	30·8 =	14·730	31·7 =	14·811
39·3 =	14·647	10·5 =	14·939	10·3 =	15·039
17·9 =	14·865				
0·7 =	15·049				

The diameter of (a) was 0·539 millim. and its length 683 millims., and that of (c) 0·915 millim. and its length 987 millims.

Taking the mean of the two temperatures and conducting powers, and calculating (by the method of least squares) the probable values for x, y, z for the formula $\lambda = x + yt + zt^2$, where λ = conducting power x at t degrees, x conducting power at 0° C., y and z constants, we find,

for   (a) $\lambda = 15\cdot059 - 0\cdot01077\,t + 0\cdot00000722\,t^2$,
           (b) $\lambda = 15\cdot052 - 0\cdot01074\,t + 0\cdot00000714\,t^2$.
           (c) $\lambda = 15\cdot152 - 0\cdot01098\,t + 0\cdot00000774\,t^2$.

Table VI. gives the mean of the observed conducting powers, those calculated from the above formulæ, and their differences.

Table VI.

(a.)			(b.)		
T.	Observed conducting power.	Calculated conducting power. Diff.	T.	Observed conducting power.	Calculated conducting power. Diff.
$0\cdot6 = 15\cdot051$		$15\cdot053 - 0\cdot002$	$9\cdot8 = 14\cdot946$		$14\cdot947 - 0\cdot001$
$18\cdot7 = 14\cdot860$		$14\cdot860\quad 0\cdot000$	$30\cdot8 = 14\cdot729$		$14\cdot728 + 0\cdot001$
$40\cdot3 = 14\cdot637$		$14\cdot638 - 0\cdot001$	$50\cdot7 = 14\cdot526$		$14\cdot526\quad 0\cdot000$
$59\cdot8 = 14\cdot442$		$14\cdot441 + 0\cdot001$	$70\cdot4 = 14\cdot333$		$14\cdot331 + 0\cdot002$
$81\cdot0 = 14\cdot235$		$14\cdot234 + 0\cdot001$	$95\cdot1 = 14\cdot094$		$14\cdot096 - 0\cdot002$
$100\cdot0 = 14\cdot052$		$14\cdot054 - 0\cdot002$			

(c.)		
T.	Observed conducting power.	Calculated conducting power. Diff.
$8\cdot2 = 15\cdot063$		$15\cdot063\quad 0\cdot000$
$32\cdot5 = 14\cdot800$		$14\cdot803 - 0\cdot003$
$51\cdot5 = 14\cdot608$		$14\cdot608\quad 0\cdot000$
$73\cdot3 = 14\cdot391$		$14\cdot389 + 0\cdot002$
$96\cdot7 = 14\cdot162$		$14\cdot162\quad 0\cdot000$

If we now take the conducting power at $0° = 100\cdot3$, being the mean of the observed conducting powers (see Table II.), we find

for   (a) $\lambda = 100\cdot3 - 0\cdot07216\,t + 0\cdot0000484\,t^2$,
           (b) $\lambda = 100\cdot3 - 0\cdot07196\,t + 0\cdot0000478\,t^2$,
           (c) $\lambda = 100\cdot3 - 0\cdot07247\,t + 0\cdot0000511\,t^2$;

and taking the mean of the mean of a and b (this being the same wire) and c, we find the conducting power of the annealed wire at different temperatures

$$= 100\cdot3 - 0\cdot07226\,t + 0\cdot0000496\,t^2.$$

The next step was to determine the conducting powers of hard-drawn wires for different temperatures: but here we had to contend with great difficulties; for when a hard-drawn wire is heated to 100°, a different conducting power is generally found on cooling; and to obtain concordant results, it is necessary to heat the wire several times; but when once obtained, the values found will remain the same, no matter how often the wire may be heated, showing that the apparatus and method are not faulty, whether by letting the wire remain for a length of time it will gradually assume its original conducting power or not is a question now under consideration. With annealed wires this is also the case, but in a much less degree. All the above wires, as well as the following, were heated several times before concordant results were obtained. I cannot at present state the cause of this behaviour; but experiments are now being made on the subject by Dr. von Bose and myself.

Hard-drawn wires of No. 5 (*a*) and No. 3 (*b*) alloys were heated, (*a*) in a glass tube in a water-bath, (*b*) in an oil-bath. The values found are given in Table VII.

Table VII.

(*a.*)		(*b.*)	
T.	Conducting power.	T.	Conducting power.
0·0 = 15·075		9·3 = 14·870	
20·9 = 14·858		35·6 = 14·608	
46·5 = 14·605		50·8 = 14·460	
76·4 = 14·318		67·6 = 14·300	
100·0 = 14·094		98·1 = 14·011	
71·6 = 14·370		70·4 = 14·278	
51·4 = 14·566		54·4 = 14·426	
23·8 = 14·838		35·2 = 14·613	
0·0 = 15·075		14·1 = 14·824	

The diameter of (*a*) was 0·616 millim., and its length 876 millims., and that of (*b*) 0·551 millim., and its length 348 millims. Taking, as before, the mean of the two temperatures and conducting powers, and calculating the values of x, y, z from them, we find

for   (*a*) $\lambda = 15\cdot074 - 0\cdot01012\, t + 0\cdot00000329\, t^2$,
         (*b*) $\lambda = 14\cdot964 - 0\cdot01011\, t + 0\cdot00000410\, t^2$.

Table VIII. gives the mean of the observed conducting powers, and those calculated from the above formulæ, with their differences.

Table VIII.

(*a.*)				(*b.*)			
T.	Observed conducting power.	Calculated conducting power.	Diff.	T.	Observed conducting power.	Calculated conducting power.	Diff.
0·0 = 15·075		15·074	+0·001	11·7 = 14·847		14·847	0·000
22·3 = 14·848		14·850	−0·002	35·4 = 14·610		14·611	−0·001
49·0 = 14·586		14·586	0·000	52·6 = 14·443		14·443	0·000
74·0 = 14·344		14·343	+0·001	69·0 = 14·287		14·286	+0·001
100·0 = 14·094		14·095	−0·001	98·1 = 14·011		14·011	0·000

Taking the conducting power at $0^\circ = 100$, we have

$$\text{for } (a), \quad \lambda = 100 - 0\cdot06714\,t + 0\cdot0000218\,t^2,$$
$$(b), \quad \lambda = 100 - 0\cdot06758\,t + 0\cdot0000274\,t^2,$$
$$\text{mean} \quad \lambda = 100 - 0\cdot06784\,t + 0\cdot0000246\,t^2,$$

which shows there is a difference in the conducting powers at different temperatures between annealed and hard-drawn wires of this alloy.

A similar difference we have already found between hard-drawn and annealed silver wires.

Although the values found for x and y do not agree very well with each other, yet for all purposes it will not make much difference which is used, as they will lead to the same results; thus, if we calculate the conducting power for the highest ordinary summer temperature (in a room), say 30° C, we find it in the one case to be

98·005, and in the other 97·999.

In Table IX., I have given the differences in the conducting powers of some metals between 0° and 100°, taking the conducting power at $0^\circ = 100$.

Table IX.

Silver......................	28·5 per cent. (annealed).
Copper	29·0 per cent. (annealed).
Gold	28·0 per cent. (annealed).
Mercury......................	8·7 per cent. (Siemens).
The gold-silver alloy......	6·5 per cent. (hard-drawn).
	6·7 per cent. (annealed).

From the foregoing Table it will be seen how well the alloy is adapted for use as a standard to compare the resistances of other metals.

Whilst making these experiments, I have found that as soon as most of the pure metals are alloyed with traces of any other, these differences rapidly decrease, in fact, almost in the same proportion as the conducting power of the metals themselves. This may explain why the copies of Weber's standard vary so much one from another. For instance, I have tested a commercial copper wire whose conducting power only varied between 0° and 100° 7 per cent (about), whilst pure copper varies 29 per cent. Now, suppose a wire of that copper whose conducting power only varies 7 per cent. between 0° and 100° be compared with one of Weber's standards at a certain temperature, and then with a pure copper wire at another temperature, say 20° difference, it is obvious that the pure copper wire will not have the same resistance as the original standard. It has as yet been generally assumed that the conducting power of all copper wire, whether pure or commercial, varies with an increase of temperature to the same degree, which, however, is far from true, and should be borne in mind when constructing a resistance thermometer as described by Siemens*. The fourth condition needs no comment. It is too well known how gold-silver alloys behave when exposed to the atmosphere.

With regard to the expense, the 9 grammes alloy cost, drawn into wire, about £1 4s., but the gold in it is always worth about 15s.; so that the real expense is very small. Care must, of course, be taken to prevent the alloy coming in contact with mercury, which amalgamates readily with all gold-silver alloys. The best way to prevent any such accident is to varnish the wires.

In having this alloy made, it would be advisable always to have two made by different parties, so as to be sure no mistake has occurred.

I therefore propose that all those who study the electrical resistance of metals, should compare one of their metals with this alloy, calling its conducting power 100 at 0° (hard-drawn wire of 1 millim. length and 1 millim. diam.); for then we should be able to compare the results obtained by different experimenters with one another.

I am sorry that I am not in a position to give the value of the absolute resistance of this alloy in terms of Weber's standard; for if this be once determined, we shall, of course, be able to reproduce an alloy of known resistance in absolute measure.

1 Torrington Street, January 1861.

XVIII. *Experimental Researches on the Laws of Absorption of Liquids by Porous Substances.* By THOMAS TATE, *Esq.*

[Concluded from p. 65.]

THE following experiments were made with filters of larger surfaces.

Experiment XXIV.

This experiment was made with upward filtration, as in the foregoing experiment. The filter was unsized paper, presenting a surface of $\frac{9}{10}$ths of an inch in contact with the water. The temperature was 43° throughout the experiment, other things being the same as in the last experiment.

The velocity of ascent of the liquid in the filter-tube for the first five units was found to be correctly represented by the formula

$$v = \frac{h}{11}.$$

Towards the close of the experiment, however, it was observed

* Phil. Mag. January 1861.

that the filtering power of the paper had sensibly decreased; but, owing to the rapidity of the process, the effect arising from this cause had not materially interfered with the normal law of filtration.

Upon repeating the process, it was found that the velocity of ascent followed, for the most part, the law expressed by the general equation (7).

Hence it would appear that the normal law is, that the rate of filtration varies directly as the pressure upon the filter, the deviations from this law being due to the change which takes place in the molecules of the filter during the process.

The following experiment shows that by alternating the *direction* of the current of filtration, the original power of the filter may be maintained nearly unimpaired.

Experiment XXV.

The filter in this experiment was coke, presenting an interior surface of half an inch in contact with the water, and an exterior surface of about two inches, other things being the same as in the last experiment.

By *upward* filtration, the law of ascent of the liquid was accurately expressed by the formula

$$v = \frac{h}{144},$$

where v is the velocity per second, h being the column of liquid pressure on the filter expressed in units of the graduation of the tube.

The water thus filtered into the tube was allowed to be discharged by *downward* filtration.   In this case the liquid followed nearly the same law of descent; but upon repeating this process of downward filtration, the velocity of descent was found to follow the law expressed by equation (7).   Upon reversing the *direction* of the current of filtration, the law of ascent was found to be accurately expressed by the formula $v = \frac{h}{180}$.   And so on to other alternations.

Similar results were obtained with wood-charcoal filters.

The following experiment was made to determine the rate of change which takes place in the filter during the process of filtration under a constant pressure.

Experiment XXVI.

This experiment was made with a close coke filter about an inch and a quarter in depth, and presenting three-fourths of an inch of surface in contact with the water, which had been carefully filtered through ordinary filtering-paper. The discharge was produced by downward filtration, the exterior portion of the filter being exposed to the air. The pressure on the filter, throughout the experiment, was produced by a column of 15 inches of the liquid; and the temperature was carefully maintained at 51° throughout the experiment. The successive intervals of time requisite for the discharge of one cubic inch of water were noted, and recorded in the following Table of results.

Succession of cubic inches discharged, n.	Corresp. time in minutes for each cubic inch, T.	Value of T by formula $T_{n-1} = 3{\cdot}16 \times 1{\cdot}42^{n-1}$.
1st cub. in.	3·16	3·16
2nd „	4·56	4·49
3rd „	6·36	6·37
4th „	9·00	9·05
5th „	13·50	12·85
6th „	18·00	18·24
7th „	26·00	25·91
8th „	33·00	36·80

Here the near coincidence of the results in the second and third columns shows that, *under a constant pressure on the filter* (within certain limits), *the times requisite to produce equal successive quantities of discharge are in geometrical progression.*

At or near to the eighth cubic inch of water discharged the progression seemed to have reached its limit; for the time required for the discharge of the succeeding cubic inch was found to be nearly the same as that of the eighth cubic inch.

The experiment, as above recorded, extended over a period of two hours nearly, and during that time the rate of filtration had changed from $\frac{1}{3 \cdot 16}$th of a cubic inch per minute to $\frac{1}{33}$rd of a cubic inch; that is, the rate of filtration had decreased eleven times nearly.

I offer no hypothesis on these remarkable results, beyond the mere statement of the fact that *during the process of filtration these filters undergo a progressive molecular change, causing the rate of filtration to decrease according to a general law expressed by the formula* $T_{n-1} = \alpha \beta^{n-1}$, *where α and β are constants for each particular filter, and T_{n-1} is the time required to filter the* nth *unit of water.*

At the close of the foregoing experiment, the water being discharged, the tube was filled with water by *upward* filtration, and then, this water being allowed to discharge itself by downward filtration, it was found that one cubic inch of water was discharged in 15·6 minutes, showing that by thus reversing the direction of the current of filtration, the filter had to a considerable extent regained its original power.

Precisely similar results were obtained with filters of wood-charcoal and unsized paper.

As it is scarcely possible to obtain filters, especially of charcoal and coke, of precisely the same internal structure, different filters, even of the same substance, vary very much with respect to the value of the ratio expressing the change in their filtering power. The rate of change of the filter in the foregoing experiment was unusually great.

The results recorded in Experiment XXI. were no doubt to some extent affected by the progressive deterioration of the filtering power of the coke; but it must be observed that this filter, having been previously tested, underwent an exceedingly small change during the process.

The following experiments were made to determine the analogy subsisting between the filtration of liquids through porous substances, and the discharge of liquids through small perforations made in thin plates.

The results of these experiments showed that the velocity of discharge through the orifices of [thin plates varies according to a certain power of the column of liquid pressure, that is, $v \propto h^n$, where the exponent n is constant for orifices of the same diameter, but for orifices of different diameters it varies between the limits $n = \frac{1}{2}$ and $n = 1$.

For orifices less than $\frac{1}{30}$th of an inch diameter, $n = \frac{1}{2}$; that is, the velocity of discharge, in this case, varies as the square root of the column of liquid pressure.

For orifices about $\frac{1}{100}$th of an inch diameter, $n = 1$, that is, the velocity, in this case, varies as the depth of the column of liquid pressure.

Moreover it was found that the velocity of discharge increases in a high ratio with the increase of temperature, especially for the smallest orifices, and also that it varies with the adhesiveness, and even, in some cases, with the chemical composition of the liquids.

Hence it would appear that the laws of filtration are, in some respects, analogous to the discharge of liquids through minute perforations not exceeding $\frac{1}{100}$th of an inch diameter.

Experiment XXVII.

In this experiment the liquid was discharged from a small orifice $\frac{1}{43}$th of an inch in diameter, made in a thin plate cemented to the bottom of the filter-tube of the foregoing experiments.

Depth of the column of liquid, h.	Corresp. time in seconds, T.	Value of T by formula (9).
6	0	0
5	33	32·6
4	68	68·0
3	108	108·2
2	173	179·0
1	218	218·0
0	378	368·0

The formula expressing the relation between T and h is

$$T = 368 - 150 h^{\frac{1}{2}}. \qquad \qquad (9)$$

Hence we find

$$v = \frac{h^{\frac{1}{2}}}{75}; \qquad \qquad (10)$$

that is to say, for orifices of this diameter, *the velocity of discharge varies directly as the square root of the depth of the column of liquid.*

In general it has been found that

$$v = \epsilon h^{n}. \qquad \qquad (11)$$

By integrating the equation $v = \dfrac{dh}{dT}$, we find

$$T = \alpha - \gamma h^{1-n}, \qquad \qquad (12)$$

where $\gamma = \dfrac{1}{1-n} \cdot \dfrac{1}{\epsilon}$.

In formulæ (9) and (10), $n = \frac{1}{2}$, $\epsilon = \frac{1}{75}$, and $\gamma = 150$. When $n = 1$, we find from equation (11),

$$T = \alpha - \gamma \log h, \qquad \qquad (13)$$

where $\gamma = \dfrac{2 \cdot 30218}{\epsilon}$.

This formula represents the law of descent of the liquid in the following experiment.

Experiment XXVIII.

In this experiment the diameter of the orifice was $\frac{1}{100}$th of an inch nearly.

Depth of the column of liquid, h.	Corresp. time in minutes, T.	Value of T by formula (14).
6	0	0
5	4·0	4·0
4	9·0	8·8
3	15·2	15·1
2	23·5	23·9
1	36·6	38·9

In this case the formula expressing the relation between T and h is

$$T = 38 \cdot 95 - 50 \log h, \quad \cdot \quad \cdot \quad \cdot \quad \cdot \quad (14)$$

and

$$\therefore v = \cdot 046h; \quad \cdot \quad \cdot \quad \cdot \quad \cdot \quad \cdot \quad \cdot \quad (15)$$

that is, *the velocity of discharge varies directly as the depth of the column of liquid.*

XIX. *Chemical Notices from Foreign Journals.* By E. ATKINSON, *Ph.D., F.C.S., Teacher of Physical Science in Cheltenham College.*

[Continued from vol. xx. p. 523.]

M. PASTEUR has for a considerable length of time been engaged in a research on the nature of alcoholic fermentation, a complete account of which he has pu lished in the *Annales de Chimie et de Physique.* Some of the results have already appeared in this Journal; but the importance of the subject induces us to lay before our readers the following brief summary of the whole investigation, taken from the *Répertoire de Chimie.*

According to Pasteur, *alcoholic fermentation* is the peculiar transformation which sugar experiences under the influence of beer yeast; the author shows that glycerine and succinic acid are products of the alcoholic fermentation, and treats of their estimation and separation.

When the fermentation is terminated, the fermented liquid is passed through a weighed filter; the increase in weight of the filter after being dried at 100°, gives the weight of the yeast which has collected at the bottom of the vessel. The filtered liquid is then gradually evaporated until it is reduced to a small bulk, and the evaporation is terminated *in vacuo.* This residue is exhausted with a mixture of alcohol and rectified ether, which dissolves out succinic acid and glycerine. This liquid is evaporated first on the water-bath, and then, water having been added, the evaporation is continued over a gentle flame in order to get rid of the ether. The liquid is next exactly neutralized by milk of lime, carefully evaporated, and exhausted by a mixture of alcohol and ether, which only dissolves the glycerine. The residue, which consists of impure succinate of lime, is digested with alcohol of 80°; this dissolves the foreign matters and leaves the succinate of lime, which is dried and weighed.

The alcoholic liquid, which contains glycerine, is also evaporated in the water-bath, and finally *in vacuo,* where it must not remain more than two or three days; for it loses weight *in vacuo,* even at the ordinary temperature, when free from water. The glycerine is then weighed.

Using a very small quantity of yeast to produce the fermentation, Pasteur finds that the weight of succinic acid obtained exceeds the total weight of the soluble matters contained in the yeast. The same is the case with the glycerine, as compared with that of the yeast.

Glycerine, succinic acid, alcohol, and carbonic acid are not the only products of fermentation. The yeast assimilates something from the sugar: in one experiment 100 grms. of sugar gave up 1½ grm. to the yeast; doubtless the cellulose of the new globules produced in the fermentation forms part of this increase.

The equation

$$C^{12} H^{12} O^{12} = 2 C^4 H^6 O^2 + 4 CO^2,$$

by which the alcoholic fermentation was formerly expressed, does not exactly represent the change. The quantity of carbonic acid formed is less than that required by the equation. Hence a certain quantity of sugar disappears without being accounted for.

Pasteur assumes that this portion of sugar is resolved into succinic acid, glycerine, and carbonic acid, and he represents the change by the equation

$$49(C^{12}H^{11}O^{11}) + 109HO = 12(C^8H^6O^8) + 72(C^6H^8O^6) + 60CO^2.$$

Succinic acid.   Glycerine.

Succinic acid and glycerine are constant and necessary products of alcoholic fermentation.

Lactic acid is an accidental production of the fermentation. Whenever it occurs (and it is very rare), the yeast must have contained some lactic ferment. Each of the ferments effects its usual transformation, and then the fermented liquid contains, besides succinic acid and glycerine, lactic acid and mannite, as well as a new acid.

In the second part of his research the author examines what becomes of the yeast in the fermentation. He shows that the nitrogen of the yeast is never changed into ammonia during alcoholic fermentation. Far from forming ammonia, that substance disappears; for yeast is formed in a mixture of sugar, an ammoniacal salt, and phosphates.

The globules of yeast are formed of small vesicles with elastic sides, full of a liquid containing a soft substance more or less granular and vascular. This is usually near the side; but in proportion as the globule becomes older, it tends towards the middle of the cell.

The globules are reproduced by means of gemmation, as M.

Cagniard Latour showed. The translucent globules without granular contents gemmate most rapidly; there are more granulations in proportion as the globule is older, less active, and less capable of gemmation.

Mitscherlich supposed these globules to burst and discharge their granules, dispersing seminules in the liquid, which increase and become globules of ordinary yeast. Pasteur does not agree with this.

It has been usually thought that, in the fermentation of solution of sugar, the ferment, far from being destroyed, is developed by gemmation; a close examination has shown the author that, in the fermentation of sugar in the presence of albuminous substances, *neither more nor less yeast is produced* than when the fermentation takes place with pure solution of sugar.

In all cases of the fermentation of pure solution of sugar, the weight of nitrogenous matter dissolved in the fermented liquid, added to the weight of the yeast, perceptibly exceeds the total weight of the original yeast. The increase amounts from 1·2 to 1·5 per cent. of the weight of the sugar.

The disappearance of the yeast in certain cases is merely apparent. Less yeast is obtained than was taken for fermentation because the quantity dissolved is greater than the weight of the new globules which are formed. In the fermentation of solutions of sugar containing albuminous matters, about 1 per cent. (of the sugar) of yeast and soluble products is formed, and therefore a little less than in working with yeast already formed, and with pure solution of sugar.

Hence the result is the same whether albuminous substances are present or not; the small difference observed arises doubtless from the fact that the globules, formed in a medium where the nitrogenized aliment is in excess, are more active, and for the same weight decompose more sugar than those formed in a medium poor in mineral or nitrogenized aliments.

Hence yeast, placed in solution of sugar, lives at the expense of the sugar and of its nitrogenous matter, which is dissolved or which becomes soluble from the changes taking place during fermentation between the principles which it contains. Fermentations which take place in the presence of excess of sugar are virtually of indefinite duration. This is readily conceived, for there is no destruction of nitrogenized matter; only displacements or modifications of this substance occur; and it remains in the complex state in which we are accustomed to meet with it in these products. The soluble part of the nitrogenous matter becomes partially fixed in the globules in an insoluble state. But the power of organization which these globules possess is such that the old globules can yield their nitrogenized matter in the soluble state to serve as food for the recent globules; and thus this fermentation continues for a long time.

The nitrogen of the yeast diminishes during fermentation from two reasons :—first, because the yeast increases in weight during fermentation by assimilating the elements of sugar, which contains no nitrogen; secondly, in consequence of the solubility of certain nitrogenized principles of the yeast.

In all alcoholic fermentation, part of the sugar becomes fixed on the yeast in the form of cellulose. When the yeast is formed in a medium consisting of pure sugar, of phosphates, and of an ammoniacal salt, it is clear that the cellulose is formed from the elements of the sugar, and that the ammonia combines with another part of the sugar to form the soluble and insoluble albuminous matters in the globules.

Are the phenomena analogous in the case in which sugar ferments in the presence of albuminous substances ? Experiment proves that there is more cellulose in the yeast after than before the fermentation ; so that it is very probable, if not certain, that all the cellulose of the yeast-globules is formed from the elements of sugar. But, besides the formation of cellulose, a perceptible quantity of sugar doubtless becomes assimilated by the yeast; for the weight of the yeast taken, added to the weight of the cellulose fixed during fermentation, does not equal the total weight of the yeast and of its soluble part, such as is found when the fermentation is terminated.

The weight of the cellulose increases considerably during fermentation, which furnishes a further proof of the vitality of the yeast during this act.

In every alcoholic fermentation, part of the sugar becomes assimilated to the yeast in the form of fatty matter. If solution of pure sugar be mixed with an aqueous extract of yeast which has been repeatedly extracted by alcohol and ether, and also with an imponderable weight of fresh globules, a few grammes of yeast are obtained containing one to two per cent. of its weight of fatty bodies, which are readily saponifiable, forming crystallizable fatty acids. This fat is formed from the elements of sugar ; for yeast prepared in a mixture of water, sugar, ammonia, and phosphates also forms fatty matter.

Permanent Vitality of Yeast.—When yeast is mixed with a proportionally small quantity of sugar, after the latter has been decomposed, the activity of the yeast continues, but is turned upon its own tissues with an extraordinary energy and rapidity ; a weight of alcohol and carbonic acid is thus obtained, exceeding that which the sugar could yield. Under these conditions the following facts are observed :—

1st. The action of the yeast is at first exerted on the sugar.

2nd. The yeast reacts on itself when the sugar has been completely destroyed.

3rd. The effect produced by the yeast on itself is not proportional to the weight of the yeast.

The author assumes that the globules formed by the fermentation of the sugar cannot attain their complete development for want of sufficient sugar, and that the young globules needing this nourishment live at the expense of the parent globules,—which produces a secondary fermentation, or the yeast destroys itself.

Lastly, M. Pasteur speaks of the application of some of the results to the composition of fermented liquors.

Since glycerine and succinic acid are constant products of the alcoholic fermentation, they ought to be found in wine, beer, cider, &c. This the author has already shown to be the case.

He terminates his memoir by the following passage, containing the fundamental conclusion which he draws from his important researches :—"As to the interpretation of the whole of the new facts which I have met with in my researches, I think that whoever considers them impartially will see that fermentation is a correlative act of life, and of the organization of globules, and not of death or the putrefaction of these globules; still less does it appear to be a phenomenon of contact, where the transformation of sugar proceeds in the presence of the ferment without yielding anything to it, or taking anything from it."

Kekulé has described* a mode of preparing the brominated derivatives of succinic acid, which is simpler than the methods hitherto employed.

To obtain bibromosuccinic acid, $C^4 H^4 Br^2 O^4$, he heats in sealed tubes (at 150°—180°) twelve parts of succinic acid with thirty-three parts of bromine and twelve parts of water. After the reaction is complete, the whole mass is changed into small greyish crystals, and on opening the tube a large quantity of hydrobromic acid escapes. The crystals are purified by washing with cold water, solution in boiling water, and treatment with animal charcoal. Bibromosuccinic acid is formed in all cases in which a small quantity of water is taken, even when the proportion of bromine is such as to form monobromosuccinic acid.

Monobromosuccinic acid, $C^4 H^5 Br O^4$, is obtained by heating succinic acid with bromine and with a *larger* quantity of water. It is purified in a similar manner, but, being more soluble, it crystallizes less easily.

* *Bulletin de la Société Chimique*, p. 208.

Butlerow[*] has investigated the action of ammoniacal gas on dioxymethylene[†]: these substances act on each other with great energy, and the mixture becomes ultimately converted into a magma of granular crystals, which, when purified, present the appearance of colourless, transparent, lustrous rhombohedra. This body can be sublimed when heated slowly. It has distinctly basic properties; it forms with hydrochloric acid a compound which crystallizes in long prismatic needles. The composition of the base is $C^6 H^{12} N^4$, and its hydrochlorate is $C^6 H^{12} N^4 HCl.$

He considers that it has the rational formula $\left. \begin{array}{l} 2(C H^2) N \\ 2(C H^2) N \\ 2(C H^2) N \end{array} \right\} N$, de-

rived from the type NH^3, in which three atoms of hydrogen are replaced by three atoms of dimethylenammonium. He names it *hexamethylenamine.* Its nearest congener is Debus's glycosine. Like that body, Butlerow's new base is formed with elimination of water, according to the equation

$$3 C^2 H^4 O^2 + 4 NH^3 = C^6 H^{12} N^4 + 6 H^2 O.$$

Butlerow has also found that diacetate of methylglycol, when heated with an excess of water in a sealed tube, is converted into dioxymethylene and free acetic acid.

The amalgam of sodium and mercury, obtained by adding sodium in small quantities to mercury, is very convenient for applying sodium in many reactions; in most cases it simply acts by dividing the sodium and increasing its surface.

Löwig and Scholz[‡] have investigated the action of this amalgam on a mixture of sulphide of carbon and iodide of ethyle. A very brisk reaction took place, and the vessel in which it was effected required cooling; the product was then treated with ether, which dissolved out a new body, as well as the excess of iodide of ethyle and bisulphide of carbon. The etherial solution, mixed with water, was distilled in the water-bath to drive off the ether, the excess of iodide of ethyle, and of the bisulphide of carbon; on cooling, the new body collected under water in the form of a yellow oil with a penetrating alliaceous odour. When this was fractionally distilled, it was found to consist principally of mercaptan, and of a new body which boiled at 188°, the analysis of which led to the composition $C^6 H^5 S^3$. It is formed thus:—

$$C^4 H^5 I + C^2 S^4 + 2 Na = C^6 H^5 S^3 + Na I + Na S.$$

It is a sulphur-yellow liquid, very fluid, and highly refracting.

[*] *Bulletin de la Société Chimique*, p. 221.
[†] Phil. Mag. vol. xviii. p. 287.
[‡] *Journal für Prakt. Chemie*, vol. lxxix. p. 441.

Its density is 1·012 at 15°. It has a very disagreeable odour; it is insoluble in water, but soluble in alcohol, ether, and sulphide of carbon. Nitric acid attacks it violently, as also do chlorine, bromine, and hypochlorite of lime. Its alcoholic solution, mixed with an alcoholic solution of bichloride of mercury, gives a white precipitate which dissolves in hot alcohol. It consists of the new body combined with 6 equivalents of bichloride of mercury.

Löwig[*] has examined the action of sodium on oxalic ether. In the presence of water, sodium acts on oxalic ether, liberating carbonic oxide and forming carbonic ether. The mass becomes coloured, and a new acid is formed which the author calls myrinic acid.

With a pulverized sodium amalgam the action is quite different; oxalate of soda is deposited, only a few bubbles of carbonic oxide are disengaged, and not a trace of carbonic ether is formed. Treated with water, a colourless solution is formed, which, agitated with ether, gives up a neutral body of a bitter taste, which, on evaporation of the ether, is left as a syrupy colourless liquid. In two or three days this deposits beautiful white crystals, which have the crude formula $C^4 H^3 O^3$.

Neither baryta water nor subacetate of lead act in the cold on the aqueous solution of these crystals; but on the application of heat, a white precipitate is formed which contains no oxalic acid.

Graham showed long ago that phosphuretted hydrogen, which is not spontaneously inflammable, is made so by admixture with a very small quantity of nitrous acid. Landolt describes[†] a convenient mode of performing this experiment. Phosphuretted hydrogen, which is not spontaneously inflammable, is generated by heating phosphorus with concentrated soda lye, to which about double its volume of alcohol has been added, and is passed through some nitric acid placed in a porcelain dish. If the nitric acid is about 1·34, and has been previously freed from hyponitric acid by boiling, the bubbles of gas burst without inflaming. But if now a few drops of fuming nitric acid be added, each bubble takes fire, forming the usual rings. The spontaneous inflammability is again destroyed by an excess of hyponitric acid, for then the phosphuretted hydrogen is destroyed in the liquid. The hyponitric acid doubtless causes the formation of a small quantity of the spontaneously inflammable liquid compound $\overline{P} H^2$. At the same time the nitric acid plays some part; for water to which hyponitric acid has been added produces no such effect.

[*] *Journal für Prakt. Chemie*, vol. lxxix. p. 453.
[†] Liebig's *Annalen*, November 1860.

XX. *Note on the Numbers of* Bernoulli *and* Euler, *and a new Theorem concerning Prime Numbers. By* J. J. Sylvester, *M.A., F.R.S., Professor of Mathematics at the Royal Academy, Woolwich*[*].

FOLLOWING the accepted *continental* notation, I denote by B_n[†] the positive value of the coefficient of t^{2n} in $\dfrac{t}{1-e^t}$ multiplied by the continual product $1.2.3\ldots n$.

The law which governs the fractional part of B_n was first given in Schumacher's *Nachrichten*, by Thomas Clausen in 1840; and almost immediately afterwards a demonstration was furnished by Professor Staudt in Crelle's Journal, with a reclamation of priority, supported by a statement of his having many years previously communicated the theorem to Gauss.

The law is this, that the positive or negative fractional residue of B_n (according as n is odd or even) is made up of the simple sum of the reciprocals of all the prime numbers which, respectively diminished by unity, are contained in $2n$. The proof, which is of an inductive kind, is virtually as follows: Suppose the law holds good up to $(n-1)$ inclusive; if we expand $\Sigma\,(x)^{2n}$ under the form $\dfrac{1}{e^{x}-1}-x^{2n}$, we shall evidently obtain $\dfrac{\Sigma(x)^{2n}}{x}\pm B_n$ under the form of a finite series, of which the terms are numerical multiples of the products of powers of x by the Bernoullian numbers of an order inferior to the nth. If, now, we make x equal to the product of all the primes which, diminished by unity, are contained in $2n$, it will at once be seen (on inspection of the series) that all its terms become integer numbers, and consequently $\dfrac{\Sigma x^{2n}}{x}\pm B_n$ becomes an integer; and therefore the law will hold good up to n, since it may easily be shown, by an application of Fermat's theorem and elementary arithmetical considerations, that if N be the product of any prime numbers whatever, and if p is the general name of such of them as diminished by unity are factors of μ, then $\dfrac{\Sigma N^{\mu}}{N}-\Sigma\dfrac{1}{p}$ is an integer.

Hence, since the law holds good for $n=1$, it is universally true.

* Communicated by the Author.

† Were it not for the general usage being as stated in the text, I certainly think it would be far more convenient to use a notation agreeing with the continental method as to sign, and nearly, but not quite, with Mr. DeMorgan's as to quantity, viz. to understand by B_n the coefficient of t^n in $\tfrac{1}{2}t\dfrac{e^t+1}{e^t-1}$ taken positively, so that B_n should be equal to zero for all the odd values of n, not excepting $n=1$.

This theorem, then, of Staudt and Clausen, *inter alia*, gives a rule for determining what primes alone enter into the denominators of the Bernoullian numbers when expressed as fractions in their lowest terms; it enables us to affirm that only simple powers of primes enter into those denominators, and to know *à priori* what those prime factors are.   This note is intended to supply a law concerning the *numerators* of the Bernoullian numbers, which I have not seen stated anywhere, and which admits of an instantaneous demonstration, *to wit*, that the whole of n will appear in the numerator of B_n, save and except such primes, or the powers of such primes, as we know by the Staudt-Clausen law must appear in the denominator.

I am inclined to believe that this law of mine was not known, at all events, in 1840, from the circumstance that in Rothe's Table, published by Ohm in Crelle's Journal in that year, which gives the values of B_n up to $n=31$, the numerators are, with one exception (about to be named), all exhibited in such a form as to show such low factors as readily offer themselves, but for B_{23} the fact of the divisibility of the numerator by 23 is not indicated.   This numerator is 5964511115939121163277961, which in fact $=23 \times 259326570258222267968607$. It is obvious, indeed, under my law, that whenever p is a prime number other than 2 and 3, the numerator of B_p must contain p, because in such case $p-1$ cannot be a factor of $2\,p$.   When $p=3$ or $p=2$, $2\,p$ always contains $(p-1)$, so that 2 and 3 are necessarily constant factors of the Bernoullian denominators, and can therefore never appear in the numerators.   In Schumacher the law of the denominator is given as "a passing" (or *chance*?) "specimen" of a promised memoir by Clausen on the Bernoullian numbers, which I shall feel obliged if any of the readers of this Magazine will inform me whether anywhere, and if so, where it has appeared.   Now for my demonstration of the law of the numerators.

By definition, $B_n = \Pi(2n) \times$ coefficient of t^{2n-1} in $\dfrac{1}{e^t-1}$.   Let μ be any integer number; then $\pm(\mu^{2n}-1)B_n = \Pi(2n) \times$ coefficient of t^{2n-1} in $\dfrac{\mu}{e^{\mu t}-1} - \dfrac{1}{e^t-1}$, or in

$$\frac{(\mu-1)-(e^{(\mu-1)t}+e^{(\mu-2)t}+ \ldots +e^t)}{e^{\mu t}-1},$$

or in

$$-\frac{e^{(\mu-2)t}+2e^{(\mu-3)t}+ \ldots +(\mu-2)e^t+(\mu-1)}{e^{(\mu-1)t}+e^{(\mu-2)t}+ \ldots +e^t+1}.$$

But obviously, by Maclaurin's theorem, the coefficient of t^{2n-1} in

the expansion of this last generating function will be of the form
$$\pm \frac{1}{\Pi(2n-1)} \cdot \frac{I}{\mu^{2n-1}},$$
where I is an integer, and therefore B_n will be of the form
$$\frac{2nI}{\mu^{2n-1}(\mu^{2n}-1)}.$$

Suppose now, when $\dfrac{2nI}{\mu^{2n-1} \cdot (\mu^{2n}-1)}$ is reduced to its lowest terms, that p (a prime contained in $2n$) does not appear in the numerator, this can only happen by virtue of p being contained in $\mu^{2n-1}(\mu^{2n}-1)$; let now μ be taken successively $2, 3, 4, \ldots (\mu-1)$, then $\mu^{2n}-1$ in all these cases is divisible by p; and therefore, by an obvious inverse of Fermat's theorem, $(p-1)$ must be contained in $2n$, i. e. p must be a factor of the denominator of B_n under the Staudt-Clausen law, which proves my theorem.

As a corollary to the foregoing, using Herschel's transformation, we see that if μ be taken any integer whatever,

$$\pm B_n = \frac{2n}{\mu^{2n}-1} \cdot \frac{(1+\Delta)^{\mu-2}+2(1+\Delta)^{\mu-3}+ \ldots +(\mu-1)}{\Delta^{\mu-1}+\mu\Delta^{\mu-2}+\mu\frac{\mu-1}{2}\Delta^{\mu-3}+ \ldots +\mu} 0^{2n}$$

$$= \frac{2n}{\mu^{2n}-1} \frac{\Delta^{\mu-2}+\mu\Delta^{\mu-1}+\mu\frac{\mu-1}{2}\Delta^{\mu-3}+ \ldots +\mu\frac{\mu-1}{2}}{\Delta^{\mu-1}+\mu\Delta^{\mu-2}+\mu\frac{\mu-1}{2}\Delta^{\mu-3}+ \ldots +\mu\frac{\mu-1}{2}\Delta+\mu} 0^{2n};$$

and if we write 0^{2n+1} instead of 0^{2n}, the result vanishes. For the case of $\mu=2$, this theorem accords with one well known. As this subject is so intimately related to that of the Herschelian differences of zero, I may take this occasion of stating a proposition concerning the latter, which (simple as it is) appears to have escaped observation, viz. that $\dfrac{\Delta^r 0^{n+r}}{\Pi(r)}$ is in fact the expression for the sum of the homogeneous products of the natural numbers from 1 to r, taken n and n together. For

$$\frac{1}{(x-n)(x-n+1)\ldots(x-1)x}$$

$$= \frac{1}{\Pi r}\left\{ \frac{1}{x-n} - \frac{r}{(x-n+1)} + \frac{r \cdot \dfrac{r-1}{2}}{(x-n+2)} \cdots \pm \frac{1}{r} \right\}$$

Hence obviously

$$\frac{1}{\Pi r}\left\{ r^n - r(r-1)^n + r \cdot \frac{r-1}{2}(r-2)^n \mp \&c. \right\},$$

i. e.

$$\frac{\Delta^r 0^n}{\Pi(r)} = \text{coefficient of } \frac{1}{x^{n-r}} \text{ in } \frac{1}{(x-n)(x-n+1)\dots(x-1)}$$

$$= \text{the sum of the } (n-r)\text{ary homogeneous products of}$$
$$1, 2, 3, \dots r.$$

Thus, then, we are able to affirm, from what is known concerning $\dfrac{\Delta^r 0^{r+n}}{\Pi r}$ (see Prof. De Morgan's Calculus), that the r-ary homogeneous product-sum of $1, 2, 3 \dots n$ (which is of the degree $2r$ in n) always contains the algebraic factor $n(n+1)\dots(n+r)$.

Addendum.—Since sending the above to press, I have given some further and successful thought to the Staudt-Clausen theorem. Staudt's demonstration labours under the twofold defect of indirectness and of presupposing a knowledge of the law to be established. In it the Bernoullian numbers are not made the subject of a direct contemplation, but are regarded through the medium of an alien function, one out of an infinite number, in which they are as it were latently embodied; and the proof, like all other inductive ones, whilst it convinces the judgment, leaves the philosophic faculty unsatisfied, inasmuch as it fails to disclose the reason (the title, so to say, to existence) of the truth which it establishes. I present below an immediate and a direct proof of this beautiful and important proposition, founded upon the same principle as gives the law of the necessary factor in the numerators (viz. the arbitrary decomposition of the generating function of Bernoulli's numbers into partial fractions), and resting upon a simple but important conception, that of *relative* as distinguished from absolute integers.

I generalize this notion, and define a quantity to be an integer relative to r (or, for brevity's sake, to be an rth integer) when it may be represented by a fraction of which the denominator does not contain r.

The lemma* upon which my demonstration rests is the fol-

* This lemma is the converse of a self-evident fact, and it virtually embodies a principle respecting an arithmetical fraction strikingly analogous to a familiar one respecting an algebraical one; viz. in the same way as a rational algebraical function of x can be expressed *in one, and only one, way* as an integral function augmented by a sum of negative powers of linear functions of x, so a rational arithmetical quantity can be expressed in one, and only one, way as an integer augmented by the sum of negative powers of simple prime numbers multiplied respectively by numbers less than such primes. In drawing this parallel, the arithmetical quantity $\dfrac{c}{p^i}$, where $c < p$, is regarded as the analogue of the algebraical one $\dfrac{1}{(ax+b)^i}$,

lowing, which is itself an immediate corollary from the arithmetical theorem that if $a, b, c, \ldots l$, with or without repetitions, are the distinct prime factors of the denominator of a fraction, the fraction itself may be resolved into the sum of simple fractions,

$$\frac{A}{a^{\alpha}} + \frac{B}{b^{\beta}} + \frac{C}{c^{\gamma}} + \&c. + \frac{L}{l^{\lambda}}$$

(itself a direct inference from the familiar theorem that if p, q be any two relative primes, the equation $p\,x - q\,y = c$ is soluble in integers for all values of c). The lemma in question is as follows: If the quantity above described is representable under the several forms,

$$\frac{a'}{a^{\prime}} + \text{ an } (a^{\text{th}}) \text{ integer } \frac{b'}{b^{\delta}} + \text{ a } (b^{\text{th}}) \text{ integer } \ldots \frac{l'}{l^{\lambda}} + \text{a } (k^{\text{th}}) \text{ integer,}$$

then it is equal to

$$\frac{a'}{a^{\prime}} + \frac{b'}{b^{\delta}} + \ldots + \frac{l'}{l^{\lambda}} + \text{ an } \textit{absolute} \text{ integer.}$$

From what has been already shown, it is obvious that μ being any prime number, the highest power of μ which *can* enter into the denominator of $(\mu^{2n} - 1)B_n$ is μ^{2n}, and consequently $\mu^{2n}B$ is an integer relative to μ. Also it is clear that only those values of μ can appear in the denominator of B_n which, diminished by unity, are factors of $2n$. We have, moreover,

$$(-)^{n-1}(\mu^{2n} - 1)B_n = \Pi(2n) \times \text{coefficient of } t^{2n-1} \text{ in } \frac{\mu}{e^{\mu t} - 1} - \frac{1}{e^t - 1},$$

i. e. coefficient of t^{2n-1} in $\dfrac{-N}{e^{\mu t} - 1}$, where

$$N = \Pi(2n)\left(e^{(\mu - 1)t} + e^{(\mu - 2)t} + \ldots + e^t - (\mu - 1)\right)$$
$$= \nu_1 t + \nu_2 t^2 + \quad \ldots \quad + \nu_{2n}t^{2n} + \&c.,$$

---

as is quite proper, for both of them are fractions in their simplest forms, which would not be the case for the former were c equal to or greater than p, since in such case $\dfrac{c}{p^i}$ could be more simply expressed under the form $\dfrac{\gamma}{p^{i-1}} + \dfrac{\gamma'}{p^i}$.

This principle amounts to an affirmation that the equation in positive integers,
$$(b \ldots kl)x + (ab \ldots l)y + \ldots + (ab \ldots k)t - (ab \ldots kl)u = N,$$
where $a, b, \ldots k, l$ are relative primes, and $N < (ab \ldots kl)$, always admits of a solution, which may be termed the primitive one, and which will be unique, that namely in which $x, y, \ldots z, t$ are respectively less than $a, b, \ldots k, l$.

K 2

where obviously $\nu_1, \nu_2, \ldots \nu_{2n}$ are all integers, and the last of them

$$= (\mu-1)^{2n} + (\mu-2)^{2n} + \ldots + 2^{2n} + 1^{2n}.$$

Suppose now that $2n$ contains $(\mu-1)$, then by Fermat's theorem

$$\nu_{2n} \equiv (\mu-1) \; [\mathrm{mod} \; \mu].$$

Again, a very slight consideration* will serve to show that when μ is any prime other than 2, $e^{\mu t}-1$ is of the form

$$\mu(t + \mu\delta_1 t^2 + \mu\delta_2 t^3 + \ldots + \mu\delta_{2n-1} t^{2n-1} + \&\mathrm{c.}),$$

where $\delta_1, \delta_2, \ldots \delta_{2n-1}$ are all integers *relative* to μ. Now suppose

$$\frac{\mu N}{e^{\mu t}-1} = q_0 + q_1 t + q_2 t^2 + \ldots + q_{2n-1} t^{2n-1};$$

then by multiplication and comparison of coefficients we obtain the identities following:

$$q_0 = \nu_1, \quad q_1 + \mu q_0 \delta_1 = \nu_2, \quad q_2 + \mu q_1 \delta_1 + \mu q_0 \delta_2 = \nu_3, \ldots$$
$$q_{2n-1} + \mu q_{2n-1} \delta_1 + \ldots \mu q_0 \delta_{2n-1} = \nu_{2n};$$

obviously therefore $q_{2n-1} = \mu \times$ (an integer relative to μ) $+ \nu_{2n}$. Hence

$$(-1)^n B_n = \text{(an integer relative to } \mu) - \frac{\nu_{2n}}{\mu}$$

$$= \text{(an integer relative to } \mu) + \frac{1}{\mu}.$$

---

* For μ being a prime number greater than 2, if we put $\frac{\mu^r}{\Pi(r)}$ (the coefficient of t^r in $e^{\mu t}-1$) under the form of (an integer *qud* μ) $\times \mu^i$, we have

$$i = r - E\frac{r}{\mu} - E\frac{r}{\mu^2} - E\frac{r}{\mu^3} - \&\mathrm{c.},$$

$$= \text{or} > r - \frac{r}{\mu-1} = \text{or} > \frac{r}{2} > 1 \text{ when } r > 2; \text{ also when } r=2, i = 2 - E\frac{2}{\mu} = 2.$$

When $\mu = 2$, this would be no longer true; and in fact it is easily seen that in this case, whenever r is a power of 2, i will be only equal to 1.

For the benefit of my younger readers, I may notice that the *direct* proof of the theorem that the product of any r consecutive numbers must contain the product of the natural numbers up to r, or, in other words, that the trinomial coefficient $\frac{\Pi n}{\Pi \nu \Pi \nu'}$, where $\nu + \nu' = n$ is an integer, is drawn from the fact that this fraction may be represented as an integer *qud* μ (any prime) multiplied by μ^i, where

$$i = \left(E\frac{\lambda}{\mu} - E\frac{\nu}{\mu} - E\frac{\nu'}{\mu}\right) \mp \left(E\frac{n}{\mu^2} - E\frac{\nu}{\mu^2} - E\frac{\nu'}{\mu^2}\right) + \&\mathrm{c.}$$

(Ex meaning the integer part of x), so that i is necessarily either zero or positive, because the value of each triad of terms within the same parenthesis is essentially zero or positive. This is the natural and only direct procedure for establishing the proposition in question.

And this relation obtains for any value of μ other than 2, which (or a power of which) *could* be contained in $2n$. When $\mu = 2$, the δ series will *not* all of them be the doubles of relative integers to 2; but the ν series, on account of the factor $\Pi(2n)$, will obviously, up to ν_{2n-1} inclusive, all contain 2 and $\nu_{2n} = 1$; consequently q_{2n} will be twice (an integer *qud* 2) $+1$, and B_n will still be (an integer relative to μ) $+ \dfrac{1}{\mu}$ as before. Hence it follows from the lemma that $(-1)^n B_n = $ an absolute integer $+ \Sigma \dfrac{1}{\mu}$, or

$$B_n = \text{an integer} + (-)^n \Sigma \frac{1}{\mu},$$

which is the equation expressed by the Staudt-Clausen theorem*.

My researches in the theory of partitions have naturally invested with a new and special interest (at least for myself) everything relating to the Bernoullian numbers. I am not aware whether the following expression for a Bernoullian of any order as a quadratic function of those of an inferior order happens to have been noticed or not. It may be obtained by a simple process of multiplication, and gives a means (not very expeditious, it is true) for calculating these numbers from one another without having recourse to the calculus of differences or Maclaurin's theorem, viz.

$$-\frac{B_n}{\Pi(2n)} = (2^2 - 1)\frac{B_1}{\Pi(2)} \cdot \frac{B_{n-1}}{\Pi(2n-2)} + (2^4 - 1)\frac{B_2}{\Pi(4)} \cdot \frac{B_{n-2}}{\Pi(2n-4)}$$

$$+ \&c. \ldots + (2^{2n-4} - 1)\frac{B_{n-2}}{\Pi(2n-4)} \cdot \frac{B_2}{\Pi(4)}$$

$$+ (2^{2n-2} - 1)\frac{B_{n-1}}{\Pi(2n-2)} \cdot \frac{B_1}{\Pi 2},$$

in which formula the terms admit of being coupled together from end to end, excepting (when n is even) one term in the middle.

To illustrate my law respecting the numerators of the numbers of Bernoulli, and its connexion with the known law for the denominators, suppose twice the index of any one of these

---

* I ought to observe that in all that has preceded I have used the word *integer* in the sense of positive or negative integer, and the demonstration I have given holds good without assuming B_n to be positive. That this is the case, or, in other words, that the signs of the successive powers of t^2 in $\dfrac{e^t - 1}{e^t + 1}$ are alternately positive and negative, may be seen at a glance by putting $t = 2\sqrt{-1}\theta$, and remembering that all the coefficients in the series for $\tan\theta$ in terms of θ are necessarily positive, because $\left(\dfrac{d}{d\theta}\right)^t \tan\theta$ obviously only involves positive multiples of powers of $(\tan\theta)$ and $(\sec\theta)$.

numbers to contain the factor $(p-1)p^i$, where p is any prime; then this number will contain the first power of p in its denominator; but if the factor p^i is contained in double the index in question, but $(p-1)$ *not*, then p^i will appear bodily as a factor of the numerator.

It has occurred to me that it might be desirable to adhere to the common definition of "*Bernoulli's numbers*," but at the same time to use the term Bernoulli's *coefficients* to denote the actual

coefficients in $\dfrac{e^t+1}{2(e^t-1)}$; so that if the former be denoted in general by B_n and the latter by β_n, we shall have

$$\beta_{2n}=(-)^{n-1}B_n,$$
$$\beta_{2n+1}=0.$$

In the absence of some such term as I propose, many theorems which are really single when affirmed of the *coefficients*, become duplex or even multifarious when we are restrained to the use of the *numbers* only.

Postscript.—The results obtained concerning Bernoulli's numbers in what precedes, admit of being deduced still more succinctly; and this simplification is by no means of small importance, as it leads the way to the discovery of analogous and unsuspected properties of Euler's numbers (namely the coefficients of $\dfrac{\theta^{2n}}{\Pi 2n}$ in the expansion of sec θ), and to some very remarkable theorems concerning prime numbers in general.

In fact, to obtain the laws which govern the denominators and numerators of Bernoulli's numbers, we need only to use the following principles:—(1) That μ being a prime, $\Sigma\mu^n=0$, or $=-1$ to the modulus μ, according as $\mu-1$ is, or is not, a factor of m, —the second part of this statement being a direct consequence of Fermat's theorem, the first part a simple inference from its inverse. (2) That $e^{\mu t}-1$ is of the form $\mu t+\mu^2 t^2 T$, where T is a series of powers of t, all of whose coefficients are integers relative to μ, except for the case of $\mu=2$, when $e^{\mu t}-1$ is of the form $2t+2t^2T$. We have then $(\mu^{2n}-1)(-)^{n-1}B_n=-\Pi(2n)\times$ coefficient of t^{2n-1} in $\dfrac{e^{(\mu-1)t}+e^{(\mu-3)t}+\ldots+e_t}{e^{\mu t}-1}$ by actual division (in virtue of principle (2)) $=I-\dfrac{R}{\mu}$, where I is an integer relative to μ, containing n, and $-R=-\dfrac{1}{\mu}(1^{2n-1}+2^{2n-1}+\ldots+(\mu-1)^{2n-1}$. Hence $(-)^n B_n=$ an integer relative to μ or to such integer $+\dfrac{1}{\mu}$, according as 2n does not or does contain $(\mu-1)$, which proves the law for the numerators; and so if μ^i is a factor of n, but $(\mu-1)$ not a factor of $2n$, $\frac{B}{\mu}$ will vanish, and $\mu^{2n}-1$ will not contain μ; hence $(\mu^{2n}-1)B_n$, and consequently B_n will be the product of μ^i by an integer relative to μ, which proves my numerator law.

So by extending the same method to the generating function $\frac{1}{e^t+\sqrt{-1}}$, it may very easily be proved that if we write

$$\sec\theta = E_0 + E_1\frac{\theta^2}{1.2}t + E_2\frac{\theta^4}{1.2.3.4} + \dots + E_n\frac{\theta^{2n}}{1.2.3\dots2n} + \&c.,$$

every prime number μ of the form $4n+1$, such that $(\mu-1)$ is a factor of $2n$, will be contained in E_n; and every such factor, when p is of the form $4m-1$, will be contained in $E_n+(-)^n2$.

I call the numbers $E_1, E_2, \dots E_n$ Euler's 1st, 2nd, ... nth numbers, as Euler was apparently the first to bring them into notice. In the *Institutiones Calculi Diff.* he has calculated their values up to E_9 inclusive: in this last there is an error, which is specified by Rothe in Ohm's paper above referred to; had Euler been possessed of my law this mistake could not have occurred, as we know that E_9+2 ought to contain the factors 19 and 7, neither of which will be found to be such factors if we adopt Euler's value of E_9, but both will be such if we accept Rothe's corrected value. But in still following out the same method, I have been led, through the study of Bernoulli's and the allied numbers, and with the express aid of the former, to a perfectly general theorem concerning prime numbers, in which Bernoulli's numbers no longer take any part. Fermat's theorem teaches us the residue of $q^{\mu-1}$ in respect to μ, viz. that it is unity; but I am not aware of any theorem being in existence which teaches anything concerning the relation of $\frac{q^{\mu-1}-1}{\mu}$ to μ (or, which is the same thing, of the relation of $q\mu-1$ to the modulus μ^2). I have obtained remarkable results relative to the above quotient, which I will state for the simplest case only, viz. that where q as well as μ is a prime number. I find that when q is any odd prime,

$$\frac{1-q^{\mu-1}}{\mu} \equiv \frac{c_1}{\mu-1} + \frac{c_2}{\mu-2} + \frac{c_3}{\mu-3} + \dots + \frac{c_{\mu-1}}{l},$$

where $c_1, c_2, c_3, \dots c_{\mu-1}$ are continually recurring cycles of the numbers $1, 2, 3, \dots r$, the cycle beginning with that number r'

which satisfies the congruence $pr' \equiv 1 \pmod{r}$. Since we kno

that $\dfrac{1}{\mu-1} + \dfrac{1}{\mu-2} + \dfrac{1}{\mu-3} + \ldots + \dfrac{1}{l} \equiv 0$ (to mod. μ) in pla

of the cycle $1, 2, 3, \ldots r$, we may obviously substitute the r

duced cycle

$$-\frac{r-1}{2}, \quad -\frac{r-3}{2}, \ldots -1, 0, 1, \ldots \frac{r-3}{2}, \frac{r-1}{2}.$$

Thus, *ex. gr.*, $\dfrac{3^{\mu-1}-1}{\mu}$, when μ is of the form $6n+1$,

$$\equiv \frac{1}{\mu-1} - \frac{1}{\mu-3} + \frac{1}{\mu-5} - \frac{1}{\mu-7} \ldots + 1, \text{to mod.} \mu,$$

and when μ is of the form $6n-1$,

$$\equiv \frac{-1}{\mu-2} + \frac{1}{\mu-3} - \frac{1}{\mu-4} + \frac{1}{\mu-5} \ldots -1, \text{to mod.} \mu.$$

When q is 2, the theorem which replaces the preceding is

follows: $\dfrac{2^{\mu-1}-1}{\mu}$, when μ is of the form $4n+1$,

$$\equiv \frac{1}{\mu-1} + \frac{1}{\mu-2} - \frac{1}{\mu-3} - \frac{1}{\mu-4} + \frac{1}{\mu-5} + \frac{1}{\mu-6}$$

$$- \frac{1}{\mu-7} - \frac{1}{\mu-8} + \frac{1}{\mu-9} \pm \&c.,$$

and when μ is of the form $4n-1$,

$$\equiv -\frac{1}{\mu-1} + \frac{1}{\mu-2} + \frac{1}{\mu-3} - \frac{1}{\mu-4} - \frac{1}{\mu-5}$$

$$+ \frac{1}{\mu-6} + \frac{1}{\mu-7} \mp \&c., \text{to mod. } \mu.$$

When q is not a prime, a similar theorem may be obtained
the very same method, but its expression will be less simp
The above theorems would, I think, be very noticeable were
only for the circumstance of their involving (as a condition) t
primeness as well of the base as of the augmented index of t
familiar Fermatian expression $q^{\mu-1}$,—a condition which h
makes its appearance (as I believe) for the first time in t
theory of numbers.

D.	Percentage of SO_4H	D_{15}	B_{15}°
	40.0		
101			34
1.32		1.31	
102	41.0	1.32	
1.33			35
103	42.0		
1.34		1.33	
104	43.0		36
105		1.34	
1.35	44.0		
106			37
1.36	45.0	1.35	
107			
			38
108	46.0	1.36	
109			
1.38	47.0	1.37	39
110			

Table giving the density at 0°C and at 15°C,
the degrees Beaumé and the corresponding Strength
of Aqueous Sulphuric Acid.

J. Basire Sc.

which satisfies the congruence $pr^l \equiv 1 \pmod{r}$. Since we know

that $\dfrac{1}{\mu-1} + \dfrac{1}{\mu-2} + \dfrac{1}{\mu-3} + \dots + \dfrac{1}{l} \equiv 0$ (to mod. μ) in place

of the cycle $1, 2, 3, \dots r$, we may obviously substitute the re duced cycle

$$-\frac{r-1}{2}, \quad -\frac{r-3}{2}, \dots -1, 0, 1, \dots \frac{r-3}{2}, \frac{r-1}{2}.$$

Thus, *ex. gr.*, $\dfrac{3^{\mu-1}-1}{\mu}$, when μ is of the form $6n+1$,

$$\equiv \frac{1}{\mu-1} - \frac{1}{\mu-3} + \frac{1}{\mu-5} - \frac{1}{\mu-7} \dots +1, \text{to mod.} \mu,$$

and when μ is of the form $6n-1$,

$$\equiv \frac{-1}{\mu-2} + \frac{1}{\mu-3} - \frac{1}{\mu-4} + \frac{1}{\mu-5} \dots -1, \text{to mod.} \mu.$$

When q is 2, the theorem which replaces the preceding is as

follows: $\dfrac{2^{\mu-1}-1}{\mu}$, when μ is of the form $4n+1$,

$$\equiv \frac{1}{\mu-1} + \frac{1}{\mu-2} - \frac{1}{\mu-3} - \frac{1}{\mu-4} + \frac{1}{\mu-5} + \frac{1}{\mu-6}$$

$$- \frac{1}{\mu-7} - \frac{1}{\mu-8} + \frac{1}{\mu-9} \pm \&c.,$$

and when μ is of the form $4n-1$,

$$\equiv -\frac{1}{\mu-1} + \frac{1}{\mu-2} + \frac{1}{\mu-3} - \frac{1}{\mu-4} - \frac{1}{\mu-5}$$

$$+ \frac{1}{\mu-6} + \frac{1}{\mu-7} \mp \&c., \text{to mod.} \mu.$$

When q is not a prime, a similar theorem may be obtained b the very same method, but its expression will be less simpl The above theorems would, I think, be very noticeable were only for the circumstance of their involving (as a condition) tl primeness as well of the base as of the augmented index of tl familiar Fermatian expression $q^{\mu-1}$,—a condition which he makes its appearance (as I believe) for the first time in tl theory of numbers.

D.	Percentage of SO_4H	D_{15}	B_{15}
	40.0		
			34
1.32		1.31	
	41.0		
1.33		1.32	35
	42.0		
1.34		1.33	
	43.0		36
1.35		1.34	
	44.0		37
1.36		1.35	
	45.0		
			38
1.37	46.0	1.36	
1.38	47.0	1.37	39

Table giving the density at 0°C and at 15°C, the degrees Beaumé and the corresponding Strength of Aqueous Sulphuric Acid.

J. Basire Sc.

XXI. *On a new Method of arranging Numerical Tables.* By
W. Dittmar, *Assistant in the Laboratory of Owens College,
Manchester*.*

[With a Plate.]

THE use of numerical tables constructed in the ordinary
manner is, for obvious reasons, inseparably connected with
interpolation calculations. Such calculations, although by no
means difficult, involve, as every practical mathematician knows,
much loss of time, and often give rise to mistakes. This is
especially the case when, from a given value of a dependent variable, the corresponding value of the independent, or of another
dependent variable has to be found. It is obvious that all interpolations could be avoided by giving all the values which can
possibly be required; this, however, is in most cases practically
impossible, as the tables would thus become inconveniently voluminous, and the chance of typographical error would be greatly
increased. I believe that the completeness thus attainable can
be arrived at by the use of the following graphical method,
while at the same time the size of the table will not extend
beyond the ordinary limits. Let it be required to construct
a table giving the values of several functions $y, z, w \ldots$ of a
variable x. Draw a system of vertical parallels, and call them
respectively the $x, y, z, w \ldots$ line. On each of the verticals
construct a scale, and let every point on each of the scales be the
symbol for a number equal either to the number of divisions (and
in general one fraction of a division) contained between the
origin and that point, or to a simple multiple of this number;
so that the marks on each scale represent the terms of an arithmetical series. These several scales must be so constructed that
the corresponding values of all the variables are found in one
and the same horizontal line. It is true that, strictly speaking,
this can only be the case when all the functions are linear ones; it
is, however, easy to show that, practically speaking, the problem
can always be solved with any degree of exactitude required,
provided that the functions are continuous. The mode of construction and use of such tables will perhaps be best explained
by an example.

Fig. 1, Plate III. represents the commencement of a Table of
logarithms and reciprocals, which is intended to afford about the
same degree of exactitude as a common 4-place logarithmic table.
Column II. contains the logarithm-scale; each of the divisions
is 4 millims. in length, and represents a logarithmic increment
of 0·001; each point in this scale has a twofold meaning; it
stands, namely, both for the mantissa $x = \lambda$, and for the (posi-

* Communicated by the Author.

tive) mantissa of $\frac{1}{x}$ which equals $1-\lambda$. When this scale had
been drawn, the integral numbers from 100 to 1000 were marked
on the other two columns by drawing horizontal lines opposite
to the corresponding logarithms; the numbers up to num.
(mant. $=\cdot5000$) were placed in column I. beyond this number
in column III. Lastly, the interval between every two such
marks was divided into ten equal parts. Since, now, within the
intervals $x = 100$ to $x = 101,$ $x = 101$ to $x = 102$, &c. the
quotients $\dfrac{\Delta (\log x)}{\Delta x}$ may in our case safely be taken as a constant, any $\Delta \log x$ within such an interval may, together with its
Δx, be graphically represented by one and the same straight
line, as is in fact done in the Table. It is now clear that any
horizontal line will cut the verticals in points which (whether
they coincide with marks or not) are symbols for respectively a
certain number x, $\log x$, and a number having the same succession of figures as $\dfrac{1}{x}$. The above will afford complete information for the practical use of the Table*. The division of the
spaces between the marks on the scales is to be made by the eye.
With some practice an error greater than one-tenth of an interval
in the logarithm-scale will rarely be made; the position of the
marks themselves may be by far more accurately determined†;
the error accompanying any logarithm taken out of the Table
will therefore scarcely ever exceed 0·0001, and any number
found by its help will be correct within about $\frac{1}{5000}$ of its value.
The reliability of the results is not so dependent upon the exactitude of the drawing as one might at first sight be disposed to
think, as only that portion of every number is really graphically
determined which in an ordinary 4-place table is found by compu-
tative interpolation. The Table is particularly handy for finding
the values of reciprocals, as these may be obtained directly with-

---

* Let us suppose the logarithm and the reciprocal of $1\cdot0653$ were to be
found. Divide the interval between the marks $106\cdot5$ and $106\cdot6$ on the
x-scale (in your mind) into ten equal parts, and through the third point
from $106\cdot5$ draw a horizontal line, which is best done by bringing a straight
line etched on one side of a piece of plate-glass into the right position. This
line will cut the log-scale in the point $\cdot0274 =$ mant. 10653, and the
$\frac{1}{x}$-scale in the point $938\cdot7 = 1000 \times \dfrac{1}{1\cdot0653}.$ At the same time $\log \dfrac{1}{1\cdot0653}$
may be read off directly, the point of intersection in the log-scale standing
also for mant. $\dfrac{1}{1\cdot0653} = \cdot9726.$

† When great exactitude is required, the figure may be drawn on a copper plate by help of a dividing engine, and copies printed from the copper.

out the intervention of logarithms. It will therefore be especially useful in chemical calculations, as, for instance, in converting specific gravities into specific volumes, for reducing per-centage compositions to the unit of weight of one constituent, &c. The general applicability of this method is evident. For the sake of illustration I append figs. 2 and 3, giving respectively the beginning of a general interpolation- and of a densimetric Table. The mode of construction and use of these Tables will be understood from the description given of the Table of logarithms.

XXII. *On Graphical Interpolation.* By W. Dittmar, *Assistant in the Laboratory of Owens College, Manchester* [*].

THE principles laid down in the preceding article for the construction of numerical tables may also be employed for the purpose of carrying out graphical interpolations. Let us suppose that the corresponding values $x_0 y_0$, $x_1 y_1$, $x_2 y_2 \ldots$ belonging to an unknown function $y = f(x)$ are given by observation, and that it is required to complete the series of variables. It is clear that the direct results of observation may be registered in a graphical table in the manner described above. For this purpose it is only necessary to draw a straight line, and to construct on one side of it a scale with a constant unit of length, the points of which are considered as representatives of the values of y, while on the other side of the line marks made opposite to the points y_0, y_1, y_2, $\ldots$ are taken as symbols of the respective values of x, i. e. x_0, x_1, x_2, $\ldots$ &c. The question now is, how can the gaps on the x-scale be filled up by graphic interpolation? This may be accomplished in the following way:—When there are reasons for supposing that $f(x)$ does not differ much from a linear function, all the divisions on the x-scale may be made equal to one another, and each so long that the points corresponding to x_0, x_1, &c. coincide as nearly as possible with those signifying respectively y_0, y_1, &c. on the y-scale. This is best done by dividing the distance between the two furthest points on the x-scale into the requisite number of equal parts, drawing lines from the points thus obtained to one point situated at some distance, and by moving the y-scale along in this system of lines parallel to the line divided[†], till a position is found in which the points of intersection of the radii with the y-scale yield an x-scale which agrees as closely as possible with the observed values.

---

* Communicated by the Author.

† A sharp-edged drawing measure is most conveniently employed for this purpose.

If a satisfactory result cannot be obtained in this way, it is best to try whether $f(x)$ can be practically represented by an expression of the form $A + Bx + Cx^2$, where A, B, and C signify constants. If this be the case, we have[*]

$$y = A + Bx + Cx^2. \quad \dots \dots \dots \quad (I.)$$

$$y - \Delta y = A + B(x - \Delta x) + C(x - \Delta x)^2.$$

$$\Delta y = B\Delta x - C(\Delta x)^2 + (2C\Delta x)x. \quad \dots \dots \quad (II.)$$

$$\Delta x \frac{y_p - y_n}{x_p - x_n} = B\Delta x - C(\Delta x)^2 + (2C\Delta x)\left(\frac{x_p + x_n + \Delta x}{2}\right). \quad (III.)$$

Comparing equation (III.) with (II.), we see that $\dfrac{y_p - y_n}{x_p - x_n}\Delta x$ is equal to *this* Δy, the graphical representative of which is contained between the two points in the x-scale corresponding to the numbers $\dfrac{x_p + x_n + \Delta x}{2}$ and $\dfrac{x_p + x_n + \Delta x}{2} - \Delta x$.

From the equations (II.) and (III.) the following method for constructing the x-scale may be found :—Combine the observed pairs of variables by twos, and find from every combination, with the help of equation (III.), a certain Δy, the graphical representative of which is contained between the two points which in the x-scale mean $x - \Delta x$ and x. Then construct a rectangular system of coordinates, and represent the values of x thus obtained (with an arbitrary unit of length) as abscissæ, the corresponding values of Δy as ordinates, using for the latter that length as unit which represents $\Delta y = 1$ in the y-scale. Next draw a straight line which passes as nearly as possible through the extreme points of the ordinates. The ordinates of this line corresponding respectively to Δx, $2\Delta x$, $3\Delta x$, ..., when put together in the right order, give the required x-scale. In order to obtain exact results, it is advisable to choose first such a large value for Δx that only a few points of the x-scale are obtained, and to determine the intermediate points by new constructions. As soon as so many points are determined that two successive intervals do not differ perceptibly in length, the subdivisions of each interval may be made equal to one another. Should the indications of a scale thus obtained not agree quite satisfactorily with the observed data, it may often be improved by slightly changing the unit of length used in the construction, and by altering its position with respect to the y-scale. A convenient method for doing this has been already described.

[*] x and y mean any values of the variables belonging together; x_p, y_p, and x_n, y_n mean particular pairs of variables; Δx stands for the constant numerical difference corresponding to one division in the x-scale; Δy for the *corresponding variable* increment of y.

If an interpolation of the second degree proves to be insufficient for representing the observations, the series of values given is divided into several intervals, and each of these is then treated in the manner described.

In some cases it will be advisable to represent, not y, but some function of y like y^n, $\log y$, &c., on a scale with equal divisions.

The advantages which the method of graphical interpolation described appears to me to possess, as compared with the usual one of drawing a curve in a rectangular system of coordinates, are the following :—

1. All the lines drawn are straight lines; the personal error in the drawing is therefore reduced to a minimum.

2. The drawing can be executed with less trouble and greater exactitude, and it takes up less space than in the ordinary way.

3. When the drawing is finished, the value of y belonging to any given x may be read off at once, and *vice versâ*.

XXIII. *Proceedings of Learned Societies.*

ROYAL SOCIETY.

[Continued from p. 79.]

March 22, 1860.—Sir Benjamin C. Brodie, Bart., Pres., in the Chair.

THE following communications were read :—

"On the Theory of Compound Colours, and the Relations of the Colours of the Spectrum." By J. Clerk Maxwell, Esq., Professor of Natural Philosophy, Marischal College and University, Aberdeen.

Newton (in his 'Optics,' Book I. part ii. prop. 6) has indicated a method of exhibiting the relations of colour, and of calculating the effects of any mixture of colours. He conceives the colours of the spectrum arranged in the circumference of a circle, and the circle so painted that every radius exhibits a gradation of colour, from some pure colour of the spectrum at the circumference, to neutral tint at the centre. The resultant of any mixture of colours is then found by placing at the points corresponding to these colours, weights proportional to their intensities; then the resultant colour will be found at the centre of gravity, and its intensity will be the sum of the intensities of the components.

From the mathematical development of the theory of Newton's diagram, it appears that if the positions of any three colours be assumed on the diagram, and certain intensities of these adopted as units, then the position of every other colour may be laid down from its observed relation to these three. Hence Newton's assumption that the colours of the spectrum are disposed in a certain manner in the circumference of a circle, unless confirmed by experiment, must be regarded as merely a rough conjecture, intended as an illustration of his method, but not asserted as mathematically exact. From the results of the present investigation, it appears that the colours of the spectrum, as laid down according to Newton's method from actual observation, lie, not in the circumference of a circle, but in the periphery of a triangle, showing that all the colours of the spectrum may be *chromatically* represented by three, which form the angles of this triangle.

Wave-length in millionths of Paris inch.

Scarlet........ 2328, about one-third from line C to D,
Green .... 1914, about one-quarter from E to F,
Blue .... 1717, about half-way from F to G.

The theory of three primary colours has been often proposed as an interpretation of the phenomena of compound colours, but the relation of these colours to the colours of the spectrum does not seem to have been distinctly understood till Dr. Young (Lectures on Natural Philosophy, Kelland's edition, p. 345) enunciated his theory of three primary sensations of colour which are excited in different proportions when different kinds of light enter the organ of vision. According to this theory, the threefold character of colour, as perceived by us, is due, not to a threefold composition of light, but to the constitution of the visual apparatus which renders it capable of being affected in three different ways, the relative amount of each sensation being determined by the nature of the incident light. If we could exhibit three colours corresponding to the three primary sensations, each colour exciting one and one only of these sensations, then since all other colours whatever must excite more than one primary sensation, they must find their places in Newton's diagram within the triangle of which the three primary colours are the angles.

Hence if Young's theory is true, the complete diagram of all colour, as perceived by the human eye, will have the form of a triangle.

The colours corresponding to the pure rays of the spectrum must all lie within this triangle, and all colours in nature, being mixtures of these, must lie within the line formed by the spectrum. If therefore any colours of the spectrum correspond to the three pure primary sensations, they will be found at the angles of the triangle, and all the other colours will lie within the triangle.

The other colours of the spectrum, though excited by uncompounded light, are compound colours; because the light, though simple, has the power of exciting two or more colour-sensations in different proportions, as, for instance, a blue-green ray, though not compounded of blue rays and green rays, produces a sensation compounded of those of blue and green.

The three colours found by experiment to form the three angles of the triangle formed by the spectrum on Newton's diagram, *may* correspond to the three primary sensations.

A different geometrical representation of the relations of colour may be thus described. Take any point not in the plane of Newton's diagram, draw a line from this point as origin through the point representing a given colour on the plane, and produce them so that the length of the line may be to the part cut off by the plane as the intensity of the given colour is to that of the corresponding point on

Newton's diagram. In this way any colour may be represented by a line drawn from the origin whose direction indicates the quality of the colour, and whose length depends upon its intensity. The resultant of two colours is represented by the diagonal of the parallelogram formed on the lines representing the colours (see Prof. Grassmann in Phil. Mag. April 1854).

Taking three lines drawn from the origin through the points of the diagram corresponding to the three primaries as the axes of coordinates, we may express any colour as the resultant of definite quantities of each of the three primaries, and the three elements of colour will then be represented by the three dimensions of space.

The experiments, the results of which are now before the Society, were undertaken in order to ascertain the exact relations of the colours of the spectrum as seen by a normal eye, and to lay down these relations on Newton's diagram. The method consisted in selecting three colours from the spectrum, and mixing these in such proportions as to be identical in colour and brightness with a constant white light. Having assumed three standard colours, and found the quantity of each required to produce the given white, we then find the quantities of two of these combined with a fourth colour which will produce the same white. We thus obtain a relation between the three standards and the fourth colour, which enables us to lay down its position in Newton's diagram with reference to the three standards.

Any three sufficiently different colours may be chosen as standards, and any three points may be assumed as their positions on the diagram. The resulting diagram of relations of colour will differ according to the way in which we begin; but as every colour-diagram is a perspective projection of any other, it is easy to compare diagrams obtained by two different methods.

The instrument employed in these experiments consisted of a dark chamber about 5 feet long, 9 inches broad, and 4 deep, joined to another 2 feet long at an angle of about 100°. If light is admitted at a narrow slit at the end of the shorter chamber, it falls on a lens and is refracted through two prisms in succession, so as to form a pure spectrum at the end of the long chamber. Here there is placed an apparatus consisting of three moveable slits, which can be altered in breadth and position, the position being read off on a graduated scale, and the breadth ascertained by inserting a fine graduated wedge into the slit till it touches both sides.

When white light is admitted at the shorter end, light of three different kinds is refracted to these three slits. When white light is admitted at the three slits, light of these three kinds in combination is seen by an eye placed at the slit in the shorter arm of the instrument. By altering the three slits, the colour of this compound light may be changed at pleasure.

The white light employed was that of a sheet of white paper, placed on a board, and illuminated by the sun's light in the open air; the instrument being in a room, and the light moderated where the observer sits.

Another portion of the same white light goes down a separate compartment of the instrument, and is reflected at a surface of blackened glass, so as to be seen by the observer in *immediate contact* with the compound light which enters the slits and is refracted by the prisms.

Each experiment consists in altering the breadth of the slits till the two lights seen by the observer agree both in colour and brightness, the eye being allowed time to rest before making any final decision. In this way the relative places of sixteen kinds of light were found by two observers. Both agree in finding the positions of the colours to lie very close to two sides of a triangle, the extreme colours of the spectrum forming doubtful fragments of the third side. They differ, however, in the intensity with which certain colours affect them, especially the greenish blue near the line F, which to one observer is remarkably feeble, both when seen singly, and when part of a mixture; while to the other, though less intense than the colours in the neighbourhood, it is still sufficiently powerful to act its part in combinations. One result of this is, that a combination of this colour with red may be made, which appears red to the first observer and green to the second, though both have normal eyes as far as ordinary colours are concerned; and this blindness of the first has reference only to rays of a definite refrangibility, other rays near them, though similar in colour, not being deficient in intensity. For an account of this peculiarity of the author's eye, see the Report of the British Association for 1856, p. 12.

By the operator attending to the proper illumination of the paper by the sun, and the observer taking care of his eyes, and completing an observation only when they are fresh, very good results can be obtained. The compound colour is then seen in contact with the white reflected light, and is not distinguishable from it, either in hue or brilliancy; and the average difference of the observed breadth of a slit from the mean of the observations does not exceed $\frac{1}{50}$ of the breadth of the slit if the observer is careful. It is found, however, that the errors in the value of the sum of the three slits are greater than they would have been by theory, if the errors of each were independent; and if the sums and differences of the breadth of two slits be taken, the errors of the sums are always found greater than those of the differences. This indicates that the human eye has a more accurate perception of differences of hue than of differences of illumination.

Having ascertained the chromatic relations between sixteen colours selected from the spectrum, the next step is to ascertain the positions of these colours with reference to Fraunhofer's lines. This is done by admitting light into the shorter arm of the instrument through the slit which forms the eyehole in the former experiments. A pure spectrum is then seen at the other end, and the position of the fixed lines read off on the graduated scale. In order to determine the wavelengths of each kind of light, the incident light was first reflected from a stratum of air too thick to exhibit the colours of Newton's rings. The spectrum then exhibited a series of dark bands, at intervals increasing from the red to the violet. The wave-lengths corresponding to these form a series of submultiples of the retardation ; and by counting the bands between two of the fixed lines, whose wave-lengths have been determined by Fraunhofer, the wave-lengths corresponding to all the bands may be calculated ; and as there are a great number of bands, the wave-lengths become known at a great many different points.

In this way the wave-lengths of the colours compared may be ascertained, and the results obtained by one observer rendered comparable with those obtained by another, with different apparatus. A portable apparatus, similar to one exhibited to the British Association in 1856, is now being constructed in order to obtain observations made by eyes of different qualities, especially those whose vision is dichromic.

POSTSCRIPT.

Account of Experiments on the Spectrum as seen by the Colour-blind.

The instrument used in these observations was similar to that already described. By reflecting the light back through the prisms by means of a concave·mirror, the instrument is rendered much shorter and more portable, while the definition of the spectrum is rather improved. The experiments were made by two colour-blind observers, one of whom, however, did not obtain sunlight at the time of observation. The other obtained results, both with cloud-light and sun-light, in the way already described. It appears from these observations—

I. That any two colours of the spectrum, on opposite sides of the line "F," may be combined in such proportions as to form white.

II. That all the colours on the more refrangible side of F appear to the colour-blind "blue," and all those on the less refrangible side appear to them of another colour, which they generally speak of as "yellow," though the green at E appears to them as good a representative of that colour as any other part of the spectrum.

III. That the parts of the spectrum from A to E differ only in intensity, and not in colour ; the light being too faint for good experiments between A and D, but not distinguishable in colour from E reduced to the same intensity. The *maximum* is about $\frac{2}{3}$ from D towards E.

IV. Between E and F the colour appears to vary from the pure "yellow" of E to a "neutral tint" near F, which cannot be distinguished from white when looked at steadily.

V. At F the blue and the "yellow" element of colour are in equilibrium, and at this part of the spectrum the same blindness of the central spot of the eye is found in the colour-blind that has been already observed in the normal eye, so that the brightness of the spectrum appears decidedly less at F than on either side of that line ; and when a large portion of the retina is illuminated with the light of this part of the spectrum, the *limbus luteus* appears as a dark spot, moving with the movements of the eye. The observer has not yet been able to distinguish Haidinger's "brushes" while observing polarized light of this colour, in which they are very conspicuous to the author.

VI. Between F and a point $\frac{1}{3}$ from F towards G, the colour appears to vary from the neutral tint to pure blue, while the brightness increases, and reaches a maximum at $\frac{2}{3}$ from F towards G, and then diminishes towards the more refrangible end of the spectrum, the purity of the colour being apparently the same throughout.

VII. The theory of colour-blind vision being "*dichromic,*" is confirmed by these experiments, the results of which agree with those obtained already by normal or "*trichromic*" eyes, if we suppose the "red" element of colour eliminated, and the "green" and "blue" elements left as they were, so that the "*red-making rays,*" though dimly visible to the dichromic eye, excite the sensation not of red but of green, or as they call it, "yellow."

VIII. The extreme red ray of the spectrum appears to be a sufficiently good representative of the defective element in the colour-blind. When the ordinary eye receives this ray, it experiences the sensation of which the dichromic eye is incapable; and when the dichromic eye receives it, the luminous effect is probably of the same kind as that observed by Helmholtz in the ultra-violet part of the spectrum—a sensibility to light, without much appreciation of colour.

A set of observations of coloured papers by the same dichromic observer was then compared with a set of observations of the same papers by the author, and it was found—

1. That the colour-blind observations were consistent among themselves, on the hypothesis of *two* elements of colour.

2. That the colour-blind observations were consistent with the author's observations, on the hypothesis that the two elements of colour in dichromic vision are identical with two of the three elements of colour in normal vision.

3. That the element of colour, by which the two types of vision differ, is a red, whose relations to vermilion, ultramarine, and emerald-green are expressed by the equation

$$D = 1 \cdot 198V + 0 \cdot 078U - 0 \cdot 276G,$$

where D is the defective element, and V, U and G the three colours named above.

April 26.—Sir Benjamin C. Brodie, Bart., President, in the Chair.

The following communication was read:—

"Note on Regelation." By Michael Faraday, D.C.L., F.R.S. &c.

The philosophy of the phenomenon now understood by the word Regelation is exceedingly interesting, not only because of its relation to glacial action under natural circumstances, as shown by Tyndall and others, but also, and as I think especially, in its bearings upon molecular action; and this is shown, not merely by the desire of different philosophers to assign the true physical principle of action, but also by the great differences between the views which they have taken.

Two pieces of thawing ice, if put together, adhere and become one; at a place where liquefaction was proceeding, congelation suddenly occurs. The effect will ke place in air, or in water, or in vacuo. It will occur at everypoint where the two pieces of ice touch; but not with ice below the freezing-point, *i. e.* with dry ice, or ice so cold as to be everywhere in the solid state.

Three different views are taken of the nature of this phenomenon. When first observed in 1850, I explained it by supposing that a particle of water, which could retain the liquid state whilst touching ice only on one side, could not retain the liquid state if it were touched by ice on both sides; but became solid, the general temperature remaining the same*. Professor J. Thomson, who discovered that pressure lowered the freezing-point of water†, attributed the regelation to the fact that two pieces of ice could not be made to bear on each other without pressure; and that the pressure, however slight, would cause fusion at the place where the particles touched, accompanied by relief of the pressure and resolidification of the water at the place of contact, in the manner that he has fully explained in a recent communication to the Royal Society‡. Professor Forbes assents to neither of these views; but admitting Person's idea of the gradual liquefaction of ice, and assuming that ice is essentially colder than ice-cold water, *i. e.* the water in contact with it, he concludes that two wet pieces of ice will have the water between them frozen at the place where they come into contact§.

Though some might think that Professor Thomson, in his last communication, was trusting to changes of pressure and temperature so inappreciably small as to be not merely imperceptible, but also ineffectual, still he carried his conditions with him into all the cases he referred to, even though some of his assumed pressures were due to capillary attraction, or to the consequent pressure of the atmosphere, only. It seemed to me that experiment might be so applied as to advance the investigation of this beautiful point in molecular philosophy to a further degree than has yet been done; even to the extent of exhausting the power of some of the principles assumed in one or more of the three views adopted, and so render our knowledge a little more defined and exact than it is at present.

In order to exclude all pressure of the particles of ice on each other due to capillary attraction or the atmosphere, I prepared to experiment altogether under water; and for this purpose arranged a bath of that fluid at 32° F. A pail, surrounded by dry flannel, was placed in a box; a glass jar, 10 inches deep and 7 inches wide, was placed on a low tripod in the pail; broken ice was packed between the jar and the pail; the jar was filled with ice-cold water to within an inch of the top; a glass dish filled with ice was employed as a cover to it, and the whole enveloped with dry flannel. In this way the central jar, with its contents, could be retained at the unchanging temperature of 32° F. for a week or more; for a small piece of ice floating in it for that time was not entirely melted away. All that was required to keep the arrangement at the fixed temperature, was to renew the packing ice in the pail from time to time, and also

---

* Researches in Chemistry and Physics, 8vo. pp. 373, 378.

† Mousson says that a pressure of 13,000 atmospheres lowers the temperature of freezing from 0° to −18° Cent.

‡ Phil. Mag. vol. xix. p. 391.

§ Proceedings of the Royal Society of Edinburgh, April 19, 1858.

that in the basin cover. A very slow thawing process was going on in the jar the whole time, as was evident by the state of the indicating piece of ice there present.

Pieces of good Wenham-lake ice were prepared, some being blocks three inches square, and nearly an inch thick, others square prisms four or five inches long: the blocks had each a hole made through them with a hot wire near one corner; woollen thread passed through these holes formed loops, which being attached to pieces of lead, enabled me to sink the ice entirely under the surface of the ice-cold water. Each piece was thus moored to a particular place, and, because of its buoyancy, assumed a position of stability. The threads were about $1\frac{1}{2}$ inch long, so that a piece of ice, when depressed sideways and then left to itself, rose in the water as far as it could, and into its stable position, with considerable force. When, also, a piece was turned round on its loop as a vertical axis, the torsion force tended to make it return in the reverse direction.

Two similar blocks of ice were placed in the water with their opposed faces about two inches apart; they could be moved into any desired position by the use of slender rods of wood, without any change of temperature in the water. If brought near to each other and then left unrestrained, they separated, returning to their first position with considerable force. If brought into the slightest contact, regelation ensued, the blocks adhered, and remained adherent notwithstanding the force tending to pull them apart. They would continue thus, even for twenty-four hours or more, until they were purposely separated, and would appear (by many trials) to have the adhesion increased at the points where they first touched, though at other parts of the contiguous surfaces a feeble thawing and dissecting action went on. In this case, except for the first moment and in a very minute degree, there was no pressure either from capillary action or any other cause. On the contrary, a tensile force of considerable amount was tending all the time to separate the pieces of ice at their points of adhesion; where still, I believe, the adhesion went on increasing—a belief that will be fully confirmed hereafter.

Being desirous of knowing whether anything like soft adhesion occurred, such as would allow slow change of position without separation during the action of the tensile force, I made the following arrangements. The blocks of ice being moored by the threads fastened to the lowest corners, stood in the water with one of the diagonals of the large surfaces vertical; before the faces were brought into contact, each block was rotated 45° about a horizontal axis, in opposite directions, so that when put together, they made a compound block, with horizontal upper edges, each half of which tended to be twisted upon, and torn from the other. Yet by placing indicators in holes previously made in the edges of the ice, I could not find that there was the slightest motion of the blocks in relation to each other in the thirty-six hours during which the experiment was continued. This result, as far as it goes, is against the necessity of pressure to regelation, or the existence of any condition like that of softness or a shifting contact; and yet I shall be able to show that there is either soft adhesion or an equivalent for it, and from that state draw still further cause against the necessity of pressure to regelation.

Torsion force was then employed as an antagonist to regelation. The ice-blocks, being separate, were adjusted in the water so as to be parallel to each other, and about 1¼ inch apart. If made to approach each other on one side, by revolution in opposite directions on vertical axes, a piece of paper being between to prevent ice contact, the torsion force set up caused them to separate when left to themselves; but if the paper were away and the ice pieces were brought into contact, by however slight a force, they became one, forming a rigid piece of ice, though the strength was, of course, very small, the point of adhesion and solidification being simply the contact of two convex surfaces of small radius. By giving a little motion to the pail, or by moving either piece of ice gently in the water with a slip of wood, it was easy to see that the two pieces were rigidly attached to each other; and it was also found that, allowing time, there was no more tendency to a changing shape here than in the case quoted above. If now the slip of wood were introduced between the adhering pieces of ice, and applied so as to aid the torsion force of one of the loops, *i. e.* to increase the separating force, but unequally as respects the two pieces, then the congelation at the point of contact would give way, and the pieces of ice would move in relation to each other. Yet they would not separate; the piece unrestrained by the stick would not move off by the torsion of its own thread, though, if the stick were withdrawn, it would move back into its first attached position, pulling the second piece with it; and the two would resume their first associated form, though all the while the torsion of both loops was tending to make the pieces separate.

If when the wood was applied to change the mutual position of the two pieces of ice, without separating them, it were retained for a second undisturbed, then the two pieces of ice became fixed rigidly to each other in their new position, and maintained it when the wood was removed, but under a state of restraint; and when sufficient force was applied, by a slight tap of the wood on the ice to break up the rigidity, the two pieces of ice would rearrange themselves under the torsion force of their respective threads, yet remain united; and, assuming a new position, would, in a second or less, again become rigid, and remain inflexibly conjoined as before.

By managing the continuous motion of one piece of ice, it could be kept associated with the other by a flexible point of attachment for any length of time, could be placed in various angular positions to it, could be made (by retaining it quiescent for a moment) to assume and hold permanently any of these positions when the external force was removed, could be changed from that position into a new one, and, within certain limits, could be made to possess at pleasure, and for any length of time, either a flexible or a rigid attachment to its associated block of ice.

So regelation includes a flexible adhesion of the particles of ice, and also a rigid adhesion. The transition between these two states takes place when there is no external force like .pressure tending to bring the particles of ice together, but, on the contrary, a force of torsion is tending to separate them ; and, if respect be had to the mere point of contact on the two rounded surfaces where the flexible adhesion is exercised, the force which tends to separate them may be esteemed very great. The act of regelation cannot be considered as complete until the junction has become rigid ; and therefore. I think that the necessity of pressure for it is altogether excluded. No external pressure can remain (under the circumstances) after the first rigid contact is broken. All the forces which remain tend to separate the pieces of ice ; yet the first flexible adhesions and all the successive rigid adhesions which are made to occur, are as much effects of regelation as those which occur under the greatest pressure.

The phenomenon of flexible adhesion under tension looks very much like sticking and tenacity ; and I think it probable that Professor Forbes will see in it evidence of the truth of his view. I cannot, however, consider the fact as bearing such an interpretation ; because I think it impossible to keep a mixture of snow and water for hours and days together without the temperature of the mixed mass becoming uniform ; which uniformity would be fatal to the explanation. My idea of the flexible and rigid adhesion is this :— Two convex surfaces of ice come together ; the particles of water nearest to the place of contact, and therefore within the efficient sphere of action of those particles of ice which are on both sides of them, solidify ; if the condition of things be left for a moment, that the heat evolved by the solidification may be conducted away and dispersed, more particles will solidify, and ultimately enough to form a fixed and rigid junction, which will remain until a force sufficiently great to break through it is applied. But if the direction of the force resorted to can be relieved by any hinge-like motion at the point of contact, then I think that the union is broken up among the particles on the opening side of the angle, whilst the particles on the closing side come within the effectual regelation distance ; regelation ensues there and the adhesion is maintained, though in an apparently flexible state. The flexibility appears to me to be due to a series of ruptures on one side of the centre of contact, and of adhesion on the other,—the regelation, which is dependent on the vicinity of the ice surfaces, being transferred as the place of efficient vicinity is changed. That the substance we are considering is as brittle as ice, does not make any difficulty to me in respect of the flexible adhesion ; for if we suppose that the point of contact exists only at one particle, still the angular motion at that point must bring a second particle into contact (to suffer regelation) before separation could occur at the first ; or if, as seems proved by the supervention of the rigid adhesion upon the flexible state, many particles are concerned at once, it is not possible that all these should be broken through by a force applied on one side of the place of adhesion, before particles on the opposite side should have the opportunity of regelation, and so of continuing the adhesion.

It is not necessary for the observation of these phenomena that a carefully-arranged water-vessel should be employed. The difference between the flexible and rigid adhesion may be examined very well in air. For this purpose, two of the bars of ice before spoken of, may be hung up horizontally by threads, which may be adjusted to give by torsion any separating force desired ; and when the ends of these bars are brought together, the adhesion of the ice, and the ability of placing these bars at any angle, and causing them to preserve that angle by the rigid adhesion due to regelation, will be rendered evident ; and though the flexible adhesion of the ice cannot in this way be examined alone, because of the capillary attraction due to the film of water on the ice, yet that is easily obviated by plunging the pieces into a dish of water at common temperatures, so that they are entirely under the surface, and repeating the observations there. All the important points regarding the flexible and rigid junction of ice due to regelation, can in this way be readily investigated.

It will be understood that, in observing the flexible and rigid state of union, convex surfaces of contact are necessary, so that the contact may be only at one point. If there be several places of contact, apparent rigidity is given to the united mass, though each of the places of contact might be in a flexible and, so to say, adhesive condition. It is not at all difficult to arrange a convex surface so that, bearing at two places only on the sides of a depression, it should form a flexible joint in one direction, and a rigid attachment in a direction transverse to the former.

It might seem at first sight as if the flexible adhesion of the ice gave us a point to start from in the further investigation of the principle of pressure. If the application of pressure causes ice to freeze together, the application of tension might be expected to produce the contrary effect, and so cause liquidity and separation at the flexible joint. This, however, does not necessarily follow ; nor do I intend to consider what might be supposed to take place whilst theoretically contemplating that case. I think the changes of temperature and pressure are too infinitesimal to go for anything; and in illustration of this, will describe the following experiment. Wool is known to adhere to ice in the manner, as I believe, of regelation. Some woollen thread was boiled in distilled water, so as thoroughly to wet it. Some clean ice was broken up small and mixed with water, so as to produce a soft mass, and, being put into a glass jar clothed in flannel that it might keep for some hours, had a linear depression made in the surface, so as to form a little ice-ditch filled with water ; in this depression some filaments of the wetted wool were placed, which, sinking to the bottom, rested on the ice only with the weight which they would have being immersed in water ; yet in the course of two hours these filaments were frozen to the ice. In another case, a small loose ball of the same boiled wool, about half an inch in diameter, was put on to a clean piece of ice ; that into a glass basin ; and the whole wrapped up in flannel and left for twelve hours. At the end of that time it was found that thawing had been going on, and that the wool had melted a hole in the ice, by the heat conducted through it to the ice from the air. The hole was filled with the water and wool, but at the bottom some fibres of the wool were frozen to the ice.

Is this remarkable property peculiar to water, or is it general to all bodies? In respect of water it certainly seems to offer us a glimpse into the joint physical action of many particles, and into the nature of cohesion in that body when it is changing between the solid and liquid state. I made some experiments on this point. Bismuth was melted and kept at a temperature at which both solid and liquid metal could be present; then rods of bismuth were introduced, but when they had acquired the temperature of the mixed mass, no adhesion could be observed between them. By stirring the metal with wood, it was easy to break up the solid part into small crystalline granules; but when these were pressed together by wood under the surface, there was not the slightest tendency to cohere, as hail or snow would cohere in water. The same negative result was obtained with the metals tin and lead. Melted nitre appeared at times to show traces of the power; but, on the whole, I incline to think the effects observed resulted from the circumstance that the solid rods experimented with had not acquired throughout the fusing temperature. Nitre is a body which, like water, expands in solidifying; and it may possess a certain degree of this peculiar power.

Glacial acetic acid is not merely without regelating force, but actually presents a contrast to it. A bottle containing five or six ounces, which had remained liquid for many months, was at such a temperature that being stirred briskly with a glass rod crystals began to form in it; these went on increasing in size and quantity for eight or ten hours. Yet all that time there was not the slightest trace of adhesion amongst them, even when they were pressed together; and as they came to the surface, the liquid portion tended to withdraw from the faces of the crystals; as if there were a disinclination of the liquid and solid parts to adhere together.

Many salts were tried (without much or any expectation),—crystals of them being brought to bear against each other by torsion force, in their saturated solutions at common temperatures. In this way the following bodies were experimented with:—Nitrates of lead, potassa, soda; sulphates of soda, magnesia, copper, zinc; alum; borax; chloride of ammonium; ferro-prussiate of potassa; carbonate of soda; acetate of lead; and tartrate of potassa and soda; but the results with all were negative.

My present conclusion therefore is that the property is special for water; and that the view I have taken of its physical cause does not appear to be less likely now than at the beginning of this short investigation, and therefore has not sunk in value among the three explanations given.

Dr. Tyndall added to one of his papers*, a note of mine "On ice of irregular fusibility" indicating a cause for the difference observed in this respect in different parts of the same piece of ice. The view there taken was strongly confirmed by the effects which occurred in the jar of water at constant temperature described in the beginning of the preceding pages, where, though a thawing process was set up, it was so slow as not to dissolve a cubic inch of ice in six or seven days. The blocks retained entirely under water for several days,

* Philosophical Transactions, 1858, p. 228.

became so dissected at the surfaces as to develope the mechanical composition of the masses, and to show that they were composed of parallel layers about the tenth of an inch thick, of greater and lesser fusibility, which layers appear, from other modes of examination, to have been horizontal in the ice whilst in the act of formation. They had no relation to the position of the blocks in the water of my experiments, or to the direction of gravity, but had a fixed position in relation to each piece of ice.

ADDENDUM.

The following method of examining the regelation phenomena above described may be acceptable. Take a rather large dish of water at common temperatures. Prepare some flat cakes or bars of ice, from half an inch to an inch thick ; render the edges round, and the upper surface of each piece convex, by holding it against the inside of a warm saucepan cover, or in any other way. When two of these pieces are put into the water they will float, having perfect freedom of motion, and yet only the central part of the upper surface will be above the fluid ; when, therefore, the pieces touch at their edges, the width of the water-surface above the place of contact may be two, three, or four inches, and thus the effect of capillary action be entirely removed. By placing a plate of clean dry wax or spermaceti upon the top of a plate of ice, the latter may be entirely submerged, and the tendency to approximation from capillary action converted into a force of separation. When two or more of such floating pieces of ice are brought together by contact at some point under the water, they adhere ; first with an apparently flexible, and then with a rigid adhesion. When five or six pieces are grouped in a contorted shape, as an S, and one end piece be moved carefully, all will move with i rigidly ; or, if the force be enough to break through the joint, the rupture will be with a crackling noise, but the pieces will still adhere, and in an instant become rigid again. As the adhesion is only by points, the force applied should not be either too powerful or in the manner of a blow. I find a piece of paper, a small feather, or a camel-hair brush applied under the water very convenient for the purpose. When the point of a floating wedge-shaped piece of ice is brought under water against the corner or side of another floating piece, it sticks to it like a leech ; if, after a moment, a paper edge be brought down upon the place, a very sensible resistance to the rupture at that place is felt. If the ice be replaced by like rounded pieces of wood or glass, touching under water, nothing of this kind occurs, nor any signs of an effect that could by possibility be referred to capillary action ; and finally, if two floating pieces of ice have separating forces attached to them, as by threads connecting them and two light pendulums, pulled more or less in opposite directions, then it will be seen with what power the ice is held together at the place of regelation, when the contact there is either in the flexible or rigid condition, by the velocity and force with which the two pieces will separate when the adhesion is properly and entirely overcome.

GEOLOGICAL SOCIETY.

[Continued from vol. xx. p. 486.]

November 21, 1860.—L. Horner, Esq., President, in the Chair.

The following communication was read :—

"On the Geology of Bolivia and Southern Peru." By D. Forbes, Esq., F.R.S., F.G.S. With Notes on the Fossils by Prof. Huxley, F.R.S., Sec. G.S., and J. W. Salter, Esq., F.G.S.

After some observations on the previous researches by others, and on the general features of the region, the author proceeded to describe the Post-tertiary formations of the maritime district. These beds, containing existing species of shells, occur at various heights up to 40 feet above the sea-level. Guano deposits are frequent along the coast, and deposits of salt also in raised beaches a little above the sea. The author could not verify Lieut. Freyer's statement of *Balani* and *Millepora* being attached high up the side of the Morro de Arica, a perpendicular cliff at the water's edge ; indeed, from the state of old Indian tumuli along the beach, and other circumstances, the author believes that no perceptible elevation has here taken place since the Spanish Conquest, although such an alteration of level has occurred in Chile. The sand-dunes of the coast, and their great mobility during the hot season, were noticed. From Mexillones to Arica the coast is steep and rugged, formed of a chain of mountains, 3000 feet high, consisting of rocks of the Upper Oolitic age. At Arica the high land recedes, leaving a wide plain formed of the débris of the neighbouring mountains; and in the middle of this area was observed stratified volcanic tuff contemporaneous with the formation of the gravel.

The saline formations were next treated of as three groups, according to their height above the sea-level, and were shown to be much more extensive than generally supposed, extending over the rainless regions of this coast for more than 550 miles. They are mostly developed, however, between latitudes 19° and 25° South. These salines are supposed to have originated in the evaporation of sea-water confined in them as lagoons by the longitudinal ranges of hills separating them from the ocean. The nitrate of soda had, in the author's opinion, resulted from the chemical reactions of sea-salt, carbonate of lime, and decomposing vegetable matter (both terrestrial and marine). The borate of lime, occurring with the nitrate, is connected with the volcanic conditions of the district, and was produced by fumaroles containing boracic acid. Where the highest range of salines extend beyond the rainless region, they are much modified in the rainy season, and generally take the form of salt plains encircling salt lakes or swamps.

The great Bolivian plateau, having an average elevation of 13,000 or 14,000 feet above the sea, consists of great gravel plains formed by the spaces between the longitudinal ranges of mountains being filled up by the débris of these mountains. The most western of these consists of Oolitic débris with volcanic tuff and scoriæ; it bears the salines above-mentioned, and is nearly destitute of water. The central range of plains, formed from the disintegration of red sandstones and marls, with some volcanic scoriæ, is well watered. The third range consists of plains made up of the débris of Silurian and granitic rocks, and is auriferous. The thickness of this accumulation of clays, gravel, shingle, and boulders is, at places, immense. At La Paz it is more than 1600 feet. Contemporaneous trachytic tuff was found also in these deposits. In freshwater ponds on this plateau, at a height of 14,000 feet (lat. 15° S.), Mr. Forbes found abundance of *Cyclas Chilensis,* formerly considered to be peculiar to the most southern and coldest part of Chile at the level of the sea (lat. 45° to 50° S.).

The volcanic formations were next noticed. Volcanic action has continued certainly from the pleistocene age to the present. The line of volcanic phenomena is nearly continuous N. and S. Cones are frequent, some of them 22,000 feet high and upwards; but craters are rare. Volcanic matter, both in ancient times and at present, has in a great part been erupted from lateral vents, often of great longitudinal extent; recent trachytic lavas from such orifices have covered in some cases more than 100 miles of country. Besides trachyte, there are great tracts of trachydoleritic and felspathic lavas. On the whole, in these South American lavas silex abounds, and it has been the first element in the rock to crystallize; whereas apparently in granite quartz is the last to crystallize and form the state of so-called "surfusion." Diorites (including the so-called "Andesite") occur in force along two parallel N. and S. lines of eruption in this region, reaching through Chile, Bolivia, and Peru, for more than 40 degrees of latitude. These diorites, and more especially the rocks which they traverse, are metalliferous; and the author looks upon the greater part of the copper, silver, iron, and other metallic veins of these countries as directly occasioned by the appearance of this rock.

Shales and argillaceous limestones, with clay-stones, porphyry-tuffs, and porphyries, form the mass of the Upper Oolite formation of Bolivia, equivalent to Darwin's Cretaceo-Oolitic Series of Chile. At Cobija these are traversed in all directions by metallic veins, chiefly copper, and which, as before mentioned, appear to emanate from the diorite.

Red and variegated marls and sandstones, with gypsum and cupriferous and yellow sandstones and conglomerates, come next in order; they have a thickness of 6000 feet, and are much folded and dislocated. These are considered by the author to resemble closely the Permian rocks of Russia. Fossil wood is not uncommon in some of these strata, which extend for at least 500 miles N. and S.

Carboniferous strata occur chiefly as a small, contorted, basin-shaped series of limestones, sandstones, and shales, with abundant characteristic fossils.

The quartzites which are generally supposed to represent the Devonian formation in Bolivia, but which the author is rather disposed to group as Upper Silurian, are really not of very great thickness, but are very much folded, and perhaps are about 5000 feet thick.

The Silurian rocks (perhaps 15,000 feet thick) are well developed over an area of from 80,000 to 100,000 miles of mountain country, including the highest mountains of South America, and giving rise to the great rivers Amazon, &c. These slates, shales, grauwackes, and quartzites yield abundant fossils even up to the highest point reached, 20,000 feet. The problematical fossils known as *Cruziana* or *Bilobites* occur not only in the lower beds, but (with many other fossils) in the higher part of the series.

Lastly, the differences between the sections made by M. D'Orbigny, M. Pissis, and the author were pointed out, though for the most part difficult of explanation. D'Orbigny makes the mountain Illimani to be granite; it is slate according to the author. M. Pissis describes as carboniferous the beds in which Mr. Forbes found Silurian fossils,—and so on.

" On a New Species of *Macrauchenia* (*M. Boliviensis*)." By Prof. T. H. Huxley, F.R.S., Sec. G.S. &c.

Some bones, fully impregnated with metallic copper, which had been brought up from the mines of Corocoro in Bolivia were submitted to Prof. Huxley for examination. The mines referred to are situated on a great fault; and the bones were probably part of a carcass that had fallen in from the surface,—the copper-bearing water of the mines having mineralized them. A cervical and a lumbar vertebra, an astragalus, a scapula, and a tibia show complete correspondence in essential characters with those bones of the great *Macrauchenia Patachonica* described by Prof. Owen in the Appendix to the ' Voyage of the Beagle;' but the relative size and proportions of the vertebra, the tibia, and the astragalus indicate a distinct species, much smaller and more slender; and in some points of structure this new form (*M. Boliviensis*) approaches more nearly to the recent *Auchenidæ* than to the larger and fossil species. The fragments of the cranium show some peculiarities of form, but, on the whole, it has many resemblances to that of the Vicugna.

Prof. Huxley pointed out that this slender and small-headed *Macrauchenia* may have been the highland-contemporary of the larger *M. Patachonica*; just as now-a-days the Vicugna prefers the mountains, whilst its larger congener the Guanaco roams over the Patagonian plains.

Lastly it was remarked that as *Macrauchenia* was an animal combining, to a much more marked degree than any other known recent or fossil mammal, the peculiarities of certain artiodactyles and perissodactyles, and yet was certainly but of postpleistocene age, it presents a striking exception to the commonly asserted doctrine that "more generalized" organisms were confined to the ancient periods of the earth's history. For similar reasons, the structure of the *Macrauchenia* is also inimical to the idea that an extinct animal can always be reconstructed from a single tooth or a single bone.

" On the Palæozoic Fossils brought by Mr. D. Forbes from Bolivia." By J. W. Salter, Esq., F.G.S.

The Fossils of Carboniferous age brought home by Mr. Forbes are the well-known species described by D'Orbigny. Several are identical with European forms (as *Productus Martini*, &c.), and are cosmopolitan; others are peculiar to the district (as *Spirifer Condor*, *Orthis Andii*, &c.).

Mr. Forbes has brought a "Devonian" trilobite (*Phacops latifrons* or *Ph. Bufo*), in a rolled pebble, from Oruro: it is a widely-spread species. Another allied form was found by Mr. Pentland, many years back, at Aygatchi. In other respects the "Devonian" evidence is scanty.

In Mr. Forbes's fine collection of Silurian fossils none of D'Orbigny's ten Silurian species occur; nearly all are such as are met with in Lower Devonian and in Upper Silurian rocks—*Homalonotus, Tentaculites, Orthis, Ctenodonta, Pileopsis* (?), *Strophomena, Bellerophon.* South Africa and the Falkland Isles yield a similar fossil fauna.

The *Bilobites* in this collection differ, some of them probably generically, from D'Orbigny's figured species. A little *Beyrichia* from the upper part of the Silurian series in Bolivia appears to be like a North American form figured by Emmons as Silurian.

XXIV. *Intelligence and Miscellaneous Articles.*

ON THE POLARIZATION OF LIGHT BY DIFFUSION. BY G. GOVI;

THE polarization of atmospheric light has long since proved that gases, as well as solid and liquid bodies, have the property of polarizing light; but I am not aware that any direct experiments have been made to prove the presence of polarizing power in the case of gases.

I was led to the consideration of this question by the polariscopic study of the light of comets; and the idea occurred to me to investigate how a pencil of light would be affected by being transmitted through a certain thickness of a gaseous medium in which it was reflected or diffused.

The experiment was made in the following manner:—A thick pencil of the sun's rays, reflected from a heliostat, was allowed to pass into a dark room, through a hole in the window-shutter. This light was principally reflected from metal, and showed very feeble traces of polarization. A large quantity of smoke was then produced by burning incense; and the pencil immediately expanded and formed a large cylinder, which diffused white light in all directions. This light, when investigated by a polariscope, was found to be polarized even when the cylinder was viewed at right angles to its axis; but the intensity of the polarization was truly extraordinary when the direction of the visual ray, on the side of the source of light, formed a somewhat small angle with the axis of the cylinder: one would have said that the phenomenon was caused by the action of a solid or liquid body on the molecules of the ether. Viewing the cylinder in this direction, the polarization perceptibly decreased on approaching the source of light or removing from it. The light proceeding from the column of smoke seen by reflexion upon the aperture was only feebly polarized.

Excepting in its intensity, the above phenomenon presents nothing extraordinary; but for a physicist the circumstance appears to me important, that the light polarized by diffusion does not seem to arise from a simple reflexion from gas molecules, for its plane of polarization is at right angles to the plane in which the reflexion ought to occur. For on examining the cylinder of light round its axis in the direction of the maxima of polarization, it was found that the light proceeding from it was polarized tangential to that point of the surface of the cylinder towards which the polariscope was directed. Whether the plane of polarization becomes removed by its repeated reflexions from the gas molecules, or whether the action of gases under certain circumstances is analogous to that of refracting bodies, are questions which hitherto I have not been able to decide by experiment.

I endeavoured to depolarize the light completely on its entry into the dark room, by allowing it to pass through a thin sheet of white paper; but the phenomena, with the exception of the intensity of the light, were quite the same.

Light polarized by reflexion from a black glass experienced no perceptible change by the action of the smoke, and its plane of polarization always retained its original direction.

It is possible, by suitably regulating the incident quantity of polarized light, to succeed in finding a limit to the action of the gas molecules, beyond which the original polarization of the pencil preponderates over the molecular forces of the medium which the light has to penetrate.

The relations which these facts may possibly bear to the phenomena of atmospheric polarization, and perhaps also to fluorescence and the peculiar colour of bodies, have induced me to publish these observations, spite of their incompleteness.

Some days after the preceding experiments had been laid before the Academy, I repeated them with more sensitive polariscopes, and I found exactly the same facts; I can further state that the plane of polarization of diffused light suddenly rotated 90°, on passing the direction in which I had seen all trace of polarization disappear in my previous experiments.

Thus on receiving in the polariscope the rays emanating from the luminous track produced by the passage of the sun's light, or of the electric light, through the smoke of incense, it is found that under a small inclination (the angles being measured from the luminous source) the polarization of diffused light is already very perceptible; that it increases up to a certain angle, which is the maximum; it then decreases, and at the normal it is almost nil. Up to this point the plane of polarization is perpendicular to the plane which passes through the source of light, the place observed, and the eye or the polariscope. Above 90°, the polarization, although very feeble, reappears, but its plane is then perpendicular to the first plane. Still further it diminishes very rapidly, and the diffused light soon shows no sensible traces of polarized rays.

I have investigated the smoke of tobacco in the same way; and the results were the same, though the angle at which I found the neutral point and the reversal of the plane of polarization was perhaps a little less than with the smoke of incense.

It is possible that the nature of the diffused particles has an appreciable influence on these phenomena, and that different gases (if gases do diffuse light), vapours, and powders may in this manner be distinguished. I propose to undertake a series of experiments from this point of view, the results of which I shall lay before the Academy.—*Comptes Rendus,* Sept. 3, and Oct. 29, 1860.

ON ELECTRIC ENDOSMOSE. BY M. C. MATTEUCCI.

Having recently had occasion to examine into the construction and operation of the galvanic batteries used in our telegraph offices, I have been led to make certain original experiments on the subject of electric endosmose, a short description of which, as they seem to throw some light on the true nature of the phenomenon in question, I beg to lay before the Academy. MM. Porret and Becquerel were the first who called attention to the fact that a liquid mass, separated into two compartments by a porous diaphragm, and traversed by an electric current, appears to be transported in the direction of that current; that is to say, the level of the liquid is lowered in the compartment that contains the positive pole, and raised in that which contains the negative pole. The determination of the law of this phenomenon is due to M. Wiedemann, who proved that the quantity of water transported is directly proportional to the intensity of the current and the electric resistance of the liquid. M. Wiedemann seems to have regarded this mechanical effect of the current as a different phenomenon from its electrolytic action; while other physicists have considered that the transportation of the liquid was only a secondary effect of electrolysis. I should mention also that MM. Van Breda and Logemann have in vain endeavoured to ascertain whether, in the absence of a diaphragm, there is any displacement of the electrolysed liquid, or whether a very light moveable diaphragm is itself displaced in the direction of the current. Theoretical considerations, which easily suggest themselves to the mind, and which I need not here specify, founded on the equality of the electrolytic effects, whether endosmose be produced or not, give probability to the conclusion that these phenomena are caused by some secondary action of electrolysis. The following experiments seem to show that this supposition is correct.

I divided a rectangular vessel of varnished wood into six compartments, by means of diaphragms of porous porcelain. All these compartments were filled to the same height with well-water, the level of which was indicated by a line of white varnish. A platinum plate of the same size as the diaphragms was placed in each of the end compartments. Through this apparatus I caused a current to pass, produced sometimes by 10, sometimes by 15, and sometimes by 20 cells of a Grove's battery. The endosmose became apparent after the current had lasted for some hours, and in every case the first effect produced was as follows :—The level of the liquid was raised in the compartment that contained the negative electrode, and was lowered in the compartment next to it, while it was lowered in the compartment that contained the positive electrode (though to a less degree than it was elevated in the compartment at the opposite extremity), and raised in the adjoining compartment. These effects were invariably produced, notwithstanding the change of the diaphragm and the reversal of the position of the vessel with respect to the electrodes. Floats may be put in all the compartments, except those containing the platinum plates, in which the liquid is too much agitated by gaseous bubbles due to electrolysation; and on viewing these floats with a glass, the displacements I have described become sensible much earlier. In the intermediate compartments the liquid remains stationary for several hours; but after a certain time the liquid begins to rise in the compartments towards the positive pole, and to fall in those towards the negative pole. I shall mention but one precaution which must not be neglected in these experiments, namely, that the diaphragms must be as equal as possible.

In a second series of experiments I closed one end of each of two glass tubes with a porcelain diaphragm fixed with mastic. Each of these tubes was then placed in a glass vessel, and both vessels and tubes were filled to the same height with well-water. The same current passed through both tubes, in each case passing from the water in the vessel to that in the tube, the only difference being in the position of the platinum electrodes, which in the one case were very near the diaphragm, while in the other they were placed at the greatest possible distance from it. Under these circumstances I invariably found that the electric endosmose made its appearance much sooner, and with much greater intensity in the first case than in the second.

I shall not stop to discuss the consequences of these experiments, since they appear to me to be obvious, and to prove that the phenomenon in question is no other than that mentioned above, that is to say, a case of endosmose produced by changes in the composition of the liquid in contact with the two electrodes. I should mention here that the liquid round the positive electrode always acquires an acid reaction, while that round the negative electrode becomes alkaline, and that these effects are produced even when distilled water is employed. I did not content myself with the ancient experiments of Dutrochet, which prove that there is a current of endosmose from an acid liquid to water, from water to an alkaline liquid, and from an acid to an alkaline liquid. I repeated the experiment with the two liquids which had been in contact with the electrodes as described above, sometimes making use of both of the liquids, sometimes testing each of them separately with pure water. I invariably found that there was endosmose from the liquid that had been in contact with the positive electrode to pure water, and from pure water to the liquid that had been in contact with the negative electrode. It appears therefore that the conditions for the production of ordinary endosmose are undoubtedly present in the phenomenon called electric endosmose. I should, however, observe that the amount of displacement by endosmose is much less when the liquids which have been in contact with the electrodes are experimented on simply without any electric current, and that it is hardly perceptible in the case of electrolysed distilled water. Without attempting to explain all the phenomena of electric endosmose, it seems natural to suppose that the presence of electricity, and the peculiar state in which the elements of electrolysation are produced, give to these products properties which influence the effect of endosmose, and which cease with the cessation of the current.—*Comptes Rendus*, Dec. 1860.

THE

LONDON, EDINBURGH AND DUBLIN

PHILOSOPHICAL MAGAZINE

AND

JOURNAL OF SCIENCE.

◆

[FOURTH SERIES.]

MARCH 1861.

XXV. *On Physical Lines of Force.* By J. C. Maxwell, *Professor of Natural Philosophy in King's College, London**.

PART I.—*The Theory of Molecular Vortices applied to Magnetic Phenomena.*

IN all phenomena involving attractions or repulsions, or any forces depending on the relative position of bodies, we have to determine the *magnitude* and *direction* of the force which would act on a given body, if placed in a given position.

In the case of a body acted on by the gravitation of a sphere, this force is inversely as the square of the distance, and in a straight line to the centre of the sphere. In the case of two attracting spheres, or of a body not spherical, the magnitude and direction of the force vary according to more complicated laws. In electric and magnetic phenomena, the magnitude and direction of the resultant force at any point is the main subject of investigation. Suppose that the direction of the force at any point is known, then, if we draw a line so that in every part of its course it coincides in direction with the force at that point, this line may be called a *line of force*, since it indicates the direction of the force in every part of its course.

By drawing a sufficient number of lines of force, we may indicate the direction of the force in every part of the space in which it acts.

Thus if we strew iron filings on paper near a magnet, each filing will be magnetized by induction, and the consecutive filings will unite by their opposite poles, so as to form fibres, and these fibres will *indicate* the direction of the lines of force. The beautiful illustration of the presence of magnetic force afforded by this experiment, naturally tends to make us think of

* Communicated by the Author.

Phil. Mag. S. 4. Vol. 21. No. 139. *March* 1861.　　　M

the lines of force as something real, and as indicating something more than the mere resultant of two forces, whose seat of action is at a distance, and which do not exist there at all until a magnet is placed in that part of the field. We are dissatisfied with the explanation founded on the hypothesis of attractive and repellent forces directed towards the magnetic poles, even though we may have satisfied ourselves that the phenomenon is in strict‧ accordance with that hypothesis, and we cannot help thinking that in every place where we find these lines of force, some physical state or action must exist in sufficient energy to produce the actual phenomena.

My object in this paper is to clear the way for speculation in this direction, by investigating the mechanical results of certain states of tension and motion in a medium, and comparing these with the observed phenomena of magnetism and electricity. By pointing out the mechanical consequences of such hypotheses, I hope to be of some use to those who consider the phenomena as due to the action of a medium, but are in doubt as to the relation of this hypothesis to the experimental laws already established, which have generally been expressed in the language of other hypotheses.

I have in a former paper* endeavoured to lay before the mind of the geometer a clear conception of the relation of the lines of force to the space in which they are traced. By making use of the conception of currents in a fluid, I showed how to draw lines of force, which should indicate by their number the amount of force, so that each line may be called a unit-line of force (see Faraday's 'Researches,' 3122); and I have investigated the path of the lines where they pass from one medium to another.

In the same paper I have found the geometrical significance of the "Electrotonic State," and have shown how to deduce the mathematical relations between the electrotonic state, magnetism, electric currents, and the electromotive force, using mechanical illustrations to assist the imagination, but not to account for the phenomena.

I propose now to examine magnetic phenomena from a mechanical point of view, and to determine what tensions in, or motions of, a medium are capable of producing the mechanical phenomena observed. If, by the same hypothesis, we can connect the phenomena of magnetic attraction with electromagnetic phenomena and with those of induced currents, we shall have found a theory which, if not true, can only be proved to be erroneous by experiments which will greatly enlarge our knowledge of this part of physics.

* See a paper "On Faraday's Lines of Force," Cambridge Philosophical Transactions, vol. x. part 1.

The mechanical conditions of a medium under magnetic influence have been variously conceived of, as currents, undulations, or states of displacement or strain, or of pressure or stress.

Currents, issuing from the north pole and entering the south pole of a magnet, or circulating round an electric current, have the advantage of representing correctly the geometrical arrangement of the lines of force, if we could account on mechanical principles for the phenomena of attraction, or for the currents themselves, or explain their continued existence.

Undulations issuing from a centre would, according to the calculations of Professor Challis, produce an effect similar to attraction in the direction of the centre; but admitting this to be true, we know that two series of undulations traversing the same space do not combine into one resultant as two attractions do, but produce an effect depending on relations of *phase* as well as intensity, and if allowed to proceed, they diverge from each other without any mutual action. In fact the mathematical laws of attractions are not analogous in any respect to those of undulations, while they have remarkable analogies with those of currents, of the conduction of heat and electricity, and of elastic bodies.

In the Cambridge and Dublin Mathematical Journal for January 1847, Professor William Thomson has given a "Mechanical Representation of Electric, Magnetic, and Galvanic Forces," by means of the displacements of the particles of an elastic solid in a state of strain. In this representation we must make the angular displacement at every point of the solid proportional to the magnetic force at the corresponding point of the magnetic field, the direction of the axis of rotation of the displacement corresponding to the direction of the magnetic force. The absolute displacement of any particle will then correspond in magnitude and direction to that which I have identified with the electrotonic state; and the relative displacement of any particle, considered with reference to the particle in its immediate neighbourhood, will correspond in magnitude and direction to the quantity of electric current passing through the corresponding point of the magneto-electric field. The author of this method of representation does not attempt to explain the origin of the observed forces by the effects due to these strains in the elastic solid, but makes use of the mathematical analogies of the two problems to assist the imagination in the study of both.

We come now to consider the magnetic influence as existing in the form of some kind of pressure or tension, or, more generally, of *stress* in the medium.

Stress is action and reaction between the consecutive parts of a body, and consists in general of pressures or tensions different in different directions at the same point of the medium.

The necessary relations among these forces have been investigated by mathematicians; and it has been shown that the most general type of a stress consists of a combination of three principal pressures or tensions, in directions at right angles to each other.

When two of the principal pressures are equal, the third becomes an axis of symmetry, either of greatest or least pressure, the pressures at right angles to this axis being all equal.

When the three principal pressures are equal, the pressure is equal in every direction, and there results a stress having no determinate axis of direction, of which we have an example in simple hydrostatic pressure.

The general type of a stress is not suitable as a representation of a magnetic force, because a line of magnetic force has direction and intensity, but has no third quality indicating any difference between the *sides* of the line, which would be analogous to that observed in the case of polarized light*.

We must therefore represent the magnetic force at a point by a stress having a single axis of greatest or least pressure, and all the pressures at right angles to this axis equal. It may be objected that it is inconsistent to represent a line of force, which is essentially dipolar, by an axis of stress, which is necessarily isotropic; but we know that *every* phenomenon of action and reaction is isotropic in its *results*, because the effects of the force on the bodies between which it acts are equal and opposite, while the nature and origin of the force may be dipolar, as in the attraction between a north and a south pole.

Let us next consider the mechanical effect of a state of stress symmetrical about an axis. We may resolve it, in all cases, into a simple hydrostatic pressure, combined with a simple pressure or tension along the axis. When the axis is that of greatest pressure, the force along the axis will be a pressure. When the axis is that of least pressure, the force along the axis will be a tension.

If we observe the lines of force between two magnets, as indicated by iron filings, we shall see that whenever the lines of force pass from one pole to another, there is *attraction* between those poles; and where the lines of force from the poles avoid each other and are dispersed into space, the poles *repel* each other, so that in both cases they are drawn in the direction of the resultant of the lines of force.

It appears therefore that the stress in the axis of a line of magnetic force is a *tension*, like that of a rope.

If we calculate the lines of force in the neighbourhood of two gravitating bodies, we shall find them the same in direction as

* See Faraday's 'Researches,' 3252.

those near two magnetic poles of the same name; but we know that the mechanical effect is that of attraction instead of repulsion. The lines of force in this case do not run between the bodies, but avoid each other, and are dispersed over space. In order to produce the effect of attraction, the stress along the lines of gravitating force must be a *pressure*.

Let us now suppose that the phenomena of magnetism depend on the existence of a tension in the direction of the lines of force, combined with a hydrostatic pressure; or in other words, a pressure greater in the equatorial than in the axial direction: the next question is, what mechanical explanation can we give of this inequality of pressures in a fluid or mobile medium? The explanation which most readily occurs to the mind is that the excess of pressure in the equatorial direction arises from the centrifugal force of vortices or eddies in the medium having their axes in directions parallel to the lines of force.

This explanation of the cause of the inequality of pressures at once suggests the means of representing the dipolar character of the line of force. Every vortex is essentially dipolar, the two extremities of its axis being distinguished by the direction of its revolution as observed from those points.

We also know that when electricity circulates in a conductor, it produces lines of magnetic force passing through the circuit, the direction of the lines depending on the direction of the circulation. Let us suppose that the direction of revolution of our vortices is that in which vitreous electricity must revolve in order to produce lines of force whose direction within the circuit is the same as that of the given lines of force.

We shall suppose at present that all the vortices in any one part of the field are revolving in the same direction about axes nearly parallel, but that in passing from one part of the field to another, the direction of the axes, the velocity of rotation, and the density of the substance of the vortices are subject to change. We shall investigate the resultant mechanical effect upon an element of the medium, and from the mathematical expression of this resultant we shall deduce the physical character of its different component parts.

Prop. I.—If in two fluid systems geometrically similar the velocities and densities at corresponding points are proportional, then the differences of pressure at corresponding points due to the motion will vary in the duplicate ratio of the velocities and the simple ratio of the densities.

Let l be the ratio of the linear dimensions, m that of the velocities, n that of the densities, and p that of the pressures due to the motion. Then the ratio of the *masses* of corresponding portions will be $l^3 n$, and the ratio of the velocities acquired in traversing similar parts of the systems will be m; so that $l^3 m\,n$ is the ratio of the momenta acquired by similar portions in traversing similar parts of their paths.

The ratio of the surfaces is l^2, that of the forces acting on them is $l^2 p$, and that of the times during which they act is $\frac{l}{m}$; so that the ratio of the impulse of the forces is $\frac{l^3 p}{m}$, and we have now

$$l^3 mn = \frac{l^3 p}{m},$$

or

$$m^2 n = p \,;$$

that is, the ratio of the pressures due to the motion (p) is compounded of the ratio of the densities (n) and the duplicate ratio of the velocities (m^2), and does not depend on the linear dimensions of the moving systems.

In a circular vortex, revolving with uniform angular velocity, if the pressure at the axis is p_0, that at the circumference will be $p_1 = p_0 + \frac{1}{2}\rho v^2$, where ρ is the density and v the velocity at the circumference. The *mean pressure* parallel to the axis will be

$$p_0 + \frac{1}{4}\rho v^2 = p_2.$$

If a number of such vortices were placed together side by side with their axes parallel, they would form a medium in which there would be a pressure p_2 parallel to the axes, and a pressure p_1 in any perpendicular direction. If the vortices are circular, and have uniform angular velocity and density throughout, then

$$p_1 - p_2 = \frac{1}{4}\rho v^2.$$

If the vortices are not circular, and if the angular velocity and the density are not uniform, but vary according to the same law for all the vortices,

$$p_1 - p_2 = C\rho v^2,$$

where ρ is the mean density, and C is a numerical quantity depending on the distribution of angular velocity and density in the vortex. In future we shall write $\frac{\mu}{4\pi}$ instead of $C\rho$, so that

$$p_1 - p_2 = \frac{1}{4\pi}\mu v^2, \quad \cdots \quad (1)$$

where μ is a quantity bearing a constant ratio to the density, and v is the linear velocity at the circumference of each vortex.

A medium of this kind, filled with molecular vortices having their axes parallel, differs from an ordinary fluid in having different pressures in different directions. If not prevented by properly arranged pressures, it would tend to expand laterally. In so doing, it would allow the diameter of each vortex to expand

and its velocity to diminish in the same proportion. In order that a medium having these inequalities of pressure in different directions should be in equilibrium, certain conditions must be fulfilled, which we must investigate.

Prop. II.—If the direction-cosines of the axes of the vortices with respect to the axes of x, y, and z be l, m, and n, to find the normal and tangential stresses on the coordinate planes.

The actual stress may be resolved into a simple hydrostatic pressure p_1 acting in all directions, and a simple tension $p_1 - p_2$, or $\frac{1}{4\pi}\mu v^2$, acting along the axis of stress.

Hence if p_{xx}, p_{yy}, and p_{zz} be the normal stresses parallel to the three axes, considered positive when they tend to increase those axes; and if p_{yz}, p_{zx}, and p_{xy} be the tangential stresses in the three coordinate planes, considered positive when they tend to increase simultaneously the symbols subscribed, then by the resolution of stresses[*],

$$p_{xx} = \frac{1}{4\pi}\mu v^2 l^2 - p_1$$

$$p_{yy} = \frac{1}{4\pi}\mu \, v^2 m^2 - p$$

$$p_{zz} = \frac{1}{4\pi}\mu v^2 n^2 - p_1$$

$$p_{yz} = \frac{1}{4\pi}\mu v^2 mn$$

$$p_{zx} = \frac{1}{4\pi}\mu v^2 nl$$

$$p_{xy} = \frac{1}{4\pi}\mu v^2 lm.$$

If we write

$$\alpha = vl, \quad \beta = vm, \text{ and } \gamma = vn,$$

then

$$
\left.
\begin{array}{ll}
p_{xx} = \dfrac{1}{4\pi}\mu\alpha^2 - p_1 & \quad p_{yz} = \dfrac{1}{4\pi}\mu\beta\gamma \\[2mm]
p_{yy} = \dfrac{1}{4\mu}\mu\beta^2 - p_1 & \quad p_{zx} = \dfrac{1}{4\pi}\mu\gamma\alpha \\[2mm]
p_{zz} = \dfrac{1}{4\pi}\mu\gamma^2 - p_1 & \quad p_{xy} = \dfrac{1}{4\pi}\mu\alpha\beta.
\end{array}
\right\} \quad (2)
$$

Prop. III.—To find the resultant force on an element of the medium, arising from the variation of internal stress.

[*] Rankine's 'Applied Mechanics,' art. 106.

We have in general, for the force in the direction of x per unit of volume by the law of equilibrium of stresses*,

$$X = \frac{d}{dx} p_{xx} + \frac{d}{dy} p_{xy} + \frac{d}{dz} p_{zx} \quad . \quad . \quad . \quad . \quad . \quad . \quad (3)$$

In this case the expression may be written

$$X = \frac{1}{4\pi} \left\{ \frac{d(\mu\alpha)}{dx} \alpha + \mu\alpha \frac{d\alpha}{dx} - 4\pi \frac{dp_1}{dx} + \frac{d(\mu\beta)}{dy} \alpha + \mu\beta \frac{d\alpha}{dy} \right.$$
$$\left. + \frac{d(\mu\gamma)}{dz} \alpha + \mu\gamma \frac{d\alpha}{dz} \right\}. \quad . \quad . \quad . \quad . \quad . \quad . \quad (4)$$

Remembering that $\alpha \dfrac{d\alpha}{dx} + \beta \dfrac{d\beta}{dx} + \gamma \dfrac{d\gamma}{dx} = \dfrac{1}{2} \dfrac{d}{dx} (\alpha^2 + \beta^2 + \gamma^2)$, this becomes

$$X = \alpha \frac{1}{4\pi} \left(\frac{d}{dx} (\mu\alpha) + \frac{d}{dy} (\mu\beta) + \frac{d}{dz} (\mu\gamma) \right) + \frac{1}{8\pi} \mu \frac{d}{dx} (\alpha^2 + \beta^2 + \gamma^2)$$
$$- \mu\beta \frac{1}{4\pi} \left(\frac{d\beta}{dx} - \frac{d\alpha}{dy} \right) + \mu\gamma \frac{1}{4\pi} \left(\frac{d\alpha}{dz} - \frac{d\gamma}{dx} \right) - \frac{dp_1}{dx}. \quad . \quad . \quad (5)$$

The expressions for the forces parallel to the axes of y and z may be written down from analogy.

We have now to interpret the meaning of each term of this expression.

We suppose α, β, γ to be the components of the force which would act upon that end of a unit magnetic bar which points to the north.

μ represents the magnetic inductive capacity of the medium at any point referred to air as a standard. $\mu\alpha$, $\mu\beta$, $\mu\gamma$ represent the quantity of magnetic induction through unit of area perpendicular to the three axes of x, y, z respectively.

The total amount of magnetic induction through a closed surface surrounding the pole of a magnet, depends entirely on the strength of that pole; so that if $dx \, dy \, dz$ be an element, then

$$\left(\frac{d}{dx} \mu\alpha + \frac{d}{dy} \mu\beta + \frac{d}{dz} \mu\gamma \right) dx \, dy \, dz = 4\pi m \, dx \, dy \, dz, \quad . \quad (6)$$

which represents the total amount of magnetic induction outwards through the surface of the element $dx \, dy \, dz$, represents the amount of "imaginary magnetic matter" within the element, of the kind which points north.

The *first term* of the value of X, therefore,

$$\alpha \frac{1}{4\pi} \left(\frac{d}{dx} \mu\alpha + \frac{d}{dy} \mu\beta + \frac{d}{dz} \mu\gamma \right), \quad . \quad . \quad . \quad (7)$$

may be written

$$\alpha m, \quad . \quad . \quad . \quad . \quad . \quad . \quad . \quad . \quad . \quad . \quad (8)$$

* Rankine's 'Applied Mechanics,' art. 116.

where α is the intensity of the magnetic force, and m is the amount of magnetic matter pointing north in unit of volume.

The physical interpretation of this term is, that the force urging a north pole in the positive direction of x is the product of the intensity of the magnetic force resolved in that direction, and the strength of the north pole of the magnet.

Let the parallel lines from left to right in fig. 1 represent a field of magnetic force such as that of the earth, *s n* being the direction from south to north. The vortices, according to our hypothesis, will be in the direction shown by the arrows in fig. 3, that is, in a plane perpendicular to the lines of force, and revolving in the direction of the hands of a watch when observed from *s* looking towards *n*. The parts of the vortices above the plane of the paper will be moving towards *e*, and the parts below that plane towards *w*.

We shall always mark by an arrow-head the direction in which we must look in order to see the vortices rotating in the direction of the hands of a watch. The arrow-head will then indicate the *northward* direction in the magnetic field, that is, the direction in which that end of a magnet which points to the north would set itself in the field.

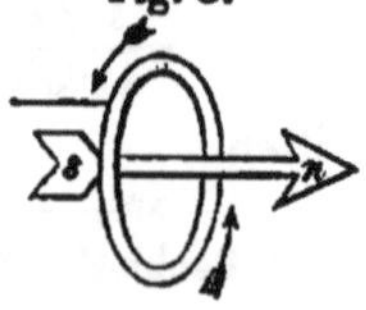

Now let A be the end of a magnet which points north. Since it repels the north ends of other magnets, the lines of force will be directed *from* A outwards in all directions. On the north side the line A D will be in the *same* direction with the lines of the magnetic field, and the velocity of the vortices will be *increased*. On the south side the line A C will be in the opposite direction, and the velocity of the vortices will be diminished, so that the lines of force are more powerful on the north side of A than on the south side.

We have seen that the mechanical effect of the vortices is to produce a tension along their axes, so that the resultant effect

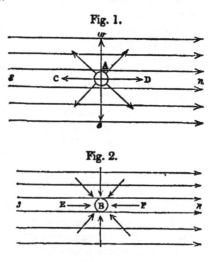

Fig. 1.

Fig. 2.

Fig. 3.

on A will be to pull it more powerfully towards D than towards C; that is, A will tend to move to the north.

Let B in fig. 2 represent a south pole. The lines of force belonging to B will tend *towards* B, and we shall find that the lines of force are rendered stronger towards E than towards F, so that the effect in this case is to urge B towards the south.

It appears therefore that, on the hypothesis of molecular vortices, our first term gives a mechanical explanation of the force acting on a north or south pole in the magnetic field.

We now proceed to examine the second term,

$$\frac{1}{8\pi}\mu\frac{d}{dx}(\alpha^2+\beta^2+\gamma^2).$$

Here $\alpha^2+\beta^2+\gamma^2$ is the square of the intensity at any part of the field, and μ is the magnetic inductive capacity at the same place. Any body therefore placed in the field will be urged *towards places of stronger magnetic intensity* with a force depending partly on its own capacity for magnetic induction, and partly on the rate at which the square of the intensity increases.

If the body be placed in a fluid medium, then the medium, as well as the body, will be urged towards places of greater intensity, so that its hydrostatic pressure will be increased in that direction. The resultant effect on a body placed in the medium will be the *difference* of the actions on the body and on the portion of the medium which it displaces, so that the body will tend to or from places of greatest magnetic intensity, according as it has a greater or less capacity for magnetic induction than the surrounding medium.

In fig. 4 the lines of force are represented as converging and becoming more powerful towards the right, so that the magnetic tension at B is stronger than at A, and the body A B will be urged to the right. If the capacity for magnetic induction is greater in the body than in the surrounding medium, it will move to the right, but if less it will move to the left.

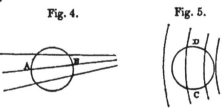

Fig. 4.   Fig. 5.

We may suppose in this case that the lines of force are converging to a magnetic pole, either north or south, on the right hand.

In fig. 5 the lines of force are represented as vertical, and becoming more numerous towards the right. It may be shown that if the force increases towards the right, the lines of force will be curved towards the right. The effect of the magnetic tensions will then be to draw any body towards the right with a force depending on the excess of its inductive capacity over that of the surrounding medium.

We may suppose that in this figure the lines of force are those surrounding an electric current perpendicular to the plane of the paper and on the right hand of the figure.

These two illustrations will show the mechanical effect on a paramagnetic or diamagnetic body placed in a field of varying magnetic force, whether the increase of force takes place along the lines or transverse to them. The form of the second term of our equation indicates the general law, which is quite independent of the direction of the lines of force, and depends solely on the manner in which the force *varies* from one part of the field to another.

We come now to the third term of the value of X,

$$-\mu\beta\frac{1}{4\pi}\left(\frac{d\beta}{dx}-\frac{d\alpha}{dy}\right).$$

Here $\mu\beta$ is, as before, the quantity of magnetic induction through unit of area perpendicular to the axis of y, and $\dfrac{d\beta}{dx}-\dfrac{d\alpha}{dy}$ is a quantity which would disappear if $\alpha dx+\beta dy+\gamma dz$ were a complete differential, that is, if the force acting on a unit north pole were subject to the condition that no work can be done upon the pole in passing round any closed curve. The quantity represents the work done on a north pole in travelling round unit of area in the direction from $+x$ to $+y$ parallel to the plane of xy. Now if an electric current whose strength is r is traversing the axis of z, which, we may suppose, points vertically upwards, then, if the axis of x is east and that of y north, a unit north pole will be urged round the axis of z in the direction from x to y, so that in one revolution the work done will be $=4\pi r$. Hence $\dfrac{1}{4\pi}\left(\dfrac{d\beta}{dx}-\dfrac{d\alpha}{dy}\right)$ represents the *strength of an electric current parallel to z* through unit of area; and if we write

$$\frac{1}{4\pi}\left(\frac{d\gamma}{dy}-\frac{d\beta}{dz}\right)=p,\quad \frac{1}{4\pi}\left(\frac{d\alpha}{dz}-\frac{d\gamma}{dx}\right)=q,\quad \frac{1}{4\pi}\left(\frac{d\beta}{dx}-\frac{d\alpha}{dy}\right)=r,\quad (9)$$

then p, q, r will be the quantity of electric current per unit of area perpendicular to the axes of x, y, and z respectively.

The physical interpretation of the third term of X, $-\mu\beta r$, is that if $\mu\beta$ is the quantity of magnetic induction parallel to y, and r the quantity of electricity flowing in the direction of z, the element will be urged in the direction of $-x$, transversely to the direction of the current and of the lines of force; that is, an *ascending* current in a field of force magnetized towards the *north* would tend to move *west*.

To illustrate the action of the molecular vor-
tices, let sn be the direction of magnetic force
in the field, and let C be the section of an
ascending magnetic current perpendicular to
the paper. The lines of force due to this current
will be circles drawn in the opposite direction
from that of the hands of a watch; that is, in the
direction $nwse$.   At e the lines of force will be
the sum of those of the field and of the current,
and at w they will be the difference of the two

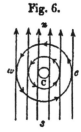

Fig. 6.

sets of lines; so that the vortices on the east side of the current
will be more powerful than those on the west side.   Both sets
of vortices have their equatorial parts turned towards C, so that
they tend to expand towards C, but those on the east side have
the greatest effect, so that the resultant effect on the current is to
urge it towards the *west*.

The fourth term,

$$+\mu\gamma\frac{1}{4\pi}\left(\frac{d\alpha}{dz}-\frac{d\gamma}{dx}\right),\text{ or }+\mu\gamma q,\quad \cdot\ \cdot\ \cdot\quad (10)$$

may be interpreted in the same way, and indicates that a current
q in the direction of y, that is, to the north, placed in a magnetic
field in which the lines are vertically upwards in the direction of
z, will be urged towards the *east*.

The fifth term,

$$-\frac{dp_1}{dx},\quad \cdot\ \cdot\ \cdot\ \cdot\ \cdot\ \cdot\ \cdot\quad (11)$$

merely implies that the element will be urged in the direction in
which the hydrostatic pressure p_1 diminishes.

We may now write down the expressions for the components
of the resultant force on an element of the medium per unit of
volume, thus:

$$X=\alpha m+\frac{1}{8\pi}\mu\frac{d}{dx}(v^2)-\mu\beta r+\mu\gamma q-\frac{dp_1}{dx},\quad \cdot\ \cdot\quad (12)$$

$$Y=\beta m+\frac{1}{8\pi}\mu\frac{d}{dy}(v^2)-\mu\gamma p+\mu\alpha r-\frac{dp_1}{dy},\quad \cdot\ \cdot\cdot\quad (13)$$

$$Z=\gamma m+\frac{1}{8\pi}\mu\frac{d}{dz}(v^2)-\mu\alpha q+\mu\beta p-\frac{dp_1}{dz}.\quad \cdot\ \cdot\quad (14)$$

The first term of each expression refers to the force acting on
magnetic poles.

The second term to the action on bodies capable of magnetism by induction.

The third and fourth terms to the force acting on electric currents.

And the fifth to the effect of simple pressure.

Before going further in the general investigation, we shall consider equations (12, 13, 14,) in particular cases, corresponding to those simplified cases of the actual phenomena which we seek to obtain in order to determine their laws by experiment.

We have found that the quantities p, q, and r represent the resolved parts of an electric current in the three coordinate directions. Let us suppose in the first instance that there is *no* electric current, or that p, q, and r vanish. We have then by (9),

$$\frac{d\gamma}{dy} - \frac{d\beta}{dz} = 0, \quad \frac{d\alpha}{dz} - \frac{d\gamma}{dx} = 0, \quad \frac{d\beta}{dx} - \frac{d\alpha}{dy} = 0, \quad . \quad (15)$$

whence we learn that

$$\alpha dx + \beta dy + \gamma dz = d\phi \quad . \quad . \quad . \quad . \quad . \quad . \quad (16)$$

is an exact differential of ϕ, so that

$$\alpha = \frac{d\phi}{dx}, \quad \beta = \frac{d\phi}{dy}, \quad \gamma = \frac{d\phi}{dz} : \quad . \quad . \quad . \quad . \quad (17)$$

μ is proportional to the density of the vortices, and represents the "capacity for magnetic induction" in the medium. It is equal to 1 in air, or in whatever medium the experiments were made which determined the powers of the magnets, the strengths of the electric currents, &c.

Let us suppose μ constant, then

$$m = \frac{1}{4\pi} \left(\frac{d}{dx} (\mu\alpha) + \frac{d}{dy} (\mu\beta) + \frac{d}{dz} (\mu\gamma) \right)$$

$$= \frac{1}{4\pi} \mu \left(\frac{d^2\phi}{dx^2} + \frac{d^2\phi}{dy^2} + \frac{d^2\phi}{dz^2} \right) \quad . \quad . \quad . \quad . \quad (18)$$

represents the amount of imaginary magnetic matter in unit of volume. That there may be no resultant force on that unit of volume arising from the action represented by the first term of equations (12, 13, 14), we must have $m = 0$, or

$$\frac{d^2\phi}{dx^2} + \frac{d^2\phi}{dy^2} + \frac{d^2\phi}{dz^2} = 0. \quad . \quad . \quad . \quad (19)$$

Now it may be shown that equation (19), if true within a given space, implies that the forces acting within that space are such as would result from a distribution of centres of force beyond that space, attracting or repelling inversely as the square of the distance.

Hence the lines of force in a part of space where μ is uniform, and where there are no electric currents, must be such as would result from the theory of "imaginary matter" acting at a distance. The assumptions of that theory are unlike those of ours, but the results are identical.

Let us first take the case of a single magnetic pole, that is, one end of a long magnet, so long that its other end is too far off to have a perceptible influence on the part of the field we are considering. The conditions then are, that equation (18) must be fulfilled at the magnetic pole, and (19) everywhere else. The only solution under these conditions is

$$\phi = -\frac{m}{\mu}\frac{1}{r}, \quad \cdots \quad \cdots \quad (20)$$

where r is the distance from the pole, and m the strength of the pole.

The repulsion at any point on a unit pole of the same kind is

$$\frac{d\phi}{dr} = \frac{m}{\mu}\frac{1}{r^2}. \quad \cdots \quad \cdots \quad (21)$$

In the standard medium $\mu = 1$; so that the repulsion is simply $\frac{m}{r^2}$ in that medium, as has been shown by Coulomb.

In a medium having a greater value of μ (such as oxygen, solutions of salts of iron, &c.) the attraction, on our theory, ought to be *less* than in air, and in diamagnetic media (such as water, melted bismuth, &c.) the attraction between the same magnetic poles ought to be *greater* than in air.

The experiments necessary to demonstrate the difference of attraction of two magnets according to the magnetic or diamagnetic character of the medium in which they are placed, would require great precision, on account of the limited range of magnetic capacity in the fluid media known to us, and the small amount of the difference sought for as compared with the whole attraction.

Let us next take the case of an electric current whose quantity is C, flowing through a cylindrical conductor whose radius is R, and whose length is infinite as compared with the size of the field of force considered.

Let the axis of the cylinder be that of z, and the direction of the current positive, then within the conductor the quantity of current per unit of area is

$$r = \frac{C}{\pi R^2} = \frac{1}{4\pi}\left(\frac{d\beta}{dx} - \frac{d\alpha}{dy}\right); \quad \cdots \quad \cdots \quad (22)$$

so that within the conductor

$$\alpha = -2\frac{C}{R^2}y, \quad \beta = 2\frac{C}{R^2}x, \quad \gamma = 0. \quad . \quad . \quad . \quad (23)$$

Beyond the conductor, in the space round it,

$$\phi = 2C \tan^{-1}\frac{y}{x}, \quad . \quad . \quad . \quad . \quad . \quad . \quad . \quad (24)$$

$$\alpha = \frac{d\phi}{dx} = -2C\frac{y}{x^2+y^2}, \quad \beta = \frac{d\phi}{dy} = 2C\frac{x}{x^2+y^2}, \quad \gamma = \frac{d\phi}{dz} = 0. \ (25)$$

If $\rho = \sqrt{x^2+y^2}$ is the perpendicular distance of any point from the axis of the conductor, a unit north pole will experience a force $= \dfrac{2C}{\rho}$, tending to move it round the conductor in the direction of the hands of a watch, if the observer view it in the direction of the current.

Let us now consider a current running parallel to the axis of z in the plane of xz at a distance ρ. Let the quantity of the current be c', and let the length of the part considered be l, and its section s, so that $\dfrac{c'}{s}$ is its strength per unit of section. Putting this quantity for ρ in equations (12, 13, 14), we find

$$X = -\mu\beta\frac{c'}{s}$$

per unit of volume; and multiplying by ls, the volume of the conductor considered, we find

$$X = -\mu\beta c'l$$

$$= -2\mu\frac{Cc'l}{\rho}, \quad . \quad . \quad . \quad . \quad . \quad (26)$$

showing that the second conductor will be attracted towards the first with a force inversely as the distance.

We find in this case also that the amount of attraction depends on the value of μ, but that it varies directly instead of inversely as μ; so that the attraction between two conducting wires will be greater in oxygen than in air, and greater in air than in water.

We shall next consider the nature of electric currents and electromotive forces in connexion with the theory of molecular vortices.

XXVI. On the Benzole Series.
By Arthur H. Church, B.A. Oxon., F.C.S.*

Part III. Note on the Oxidation of Nitrobenzole and its Homologues.

RECENT experiments by Hofmann, Berthelot, and others have shown, with reference to many important organic substances, how incorrect is any theory which does not permit them to be viewed in more than one aspect. Is nitrobenzole simply a hydrocarbon in which one equivalent of hydrogen is replaced by the group NO^4? or is it the hydride of nitrophenyle, $C^{12}\left(\begin{smallmatrix}H^4\\NO^4\end{smallmatrix}\right)$, H? or, again, is it the nitrite of phenyle, $C^{12} H^5, NO^4$? Some of the metamorphoses to which this interesting body is subject suggest one of these views, and some another. The ready production from nitrobenzole of phenylamine, in which the group phenyle may be reasonably supposed to exist, might induce us to regard nitrobenzole as a compound of phenyle, possibly the nitrite; while a reaction which I pointed out some time ago†, in which, by the action of sulphuric acid upon nitrobenzole, a compound acid was formed to which the empirical name of nitrosulphobenzolic acid was assigned, and which can hardly be regarded in any other way except as the sulphite of *nitrophenyle* and hydrogen, $C^{12}\left(\begin{smallmatrix}H^4\\NO^4\end{smallmatrix}\right), H, 2SO^3$, almost obliges us to view nitrobenzole as containing *nitrophenyle*. I propose in the present note to cite a few experiments which present some of the nitro-derivatives of the benzole series in a new light. I feel, however, that though an apology is due for the imperfections of the present account, yet a preliminary notice of my results (results which will require much time and labour to bring to a satisfactory conclusion) might not be unacceptable. But I should have deferred publishing any account of my inquiries for some time longer, had not an acid been lately discovered homologous with benzoic acid, and isomeric, but not, I think, identical with an acid mentioned in this paper: I refer to collinic acid, $C^{12} H^4 O^4$, obtained by the oxidation of gelatine. Then, too, several of the speculations and anticipations of Berthelot in his work on Organic Synthesis, trench somewhat closely upon one of the inquiries in which I have been lately engaged. I intend, when my inquiries are more advanced, to make the two new acids (phenoic and nitrophenoic) mentioned in the present notice the subject of another communication.

The oxidation of toluole, xylole, and cumole by means of bi-

---

* Communicated by the Author.
† Phil. Mag. April and June 1855.

chromate of potassium and sulphuric acid, has been found in each case to yield benzoic acid. Cymole, on the other hand, the last member of the series, when treated in the same way, yields the insolinic acid discovered by Dr. Hofmann, and not the toluic acid which Mr. Noad obtained by acting upon this hydrocarbon with dilute nitric acid. Benzole remains wholly unacted upon when boiled for a very long period with bichromate of potassium and sulphuric acid. Not so, however, with nitrobenzole, which is slowly converted by this powerful oxidizing mixture into a soft, white, crystalline mass of intensely acid reaction. In the first experiment, I made use of a very excellent sample of commercial nitrobenzole, a portion of which seemed to be acted on with greater facility than the rest. An analysis of the acid thus formed gave numbers which indicated a mixture of nitrobenzoic and *nitrophenoic* acids: I propose the latter name for the new acid, which I believe to be the next lower homologue of the benzoic. The view that the acid burnt was a mixture, is confirmed by another experiment, to be related further on. In order to secure the absence of nitrotoluole from the nitrobenzole operated upon, I converted some benzole from benzoic acid into nitrobenzole, and repeated my experiments with this. The action of a most concentrated oxidizing solution was now found to be very much slower than in the former case; but after long digestion the nitrobenzole solidified in great measure on cooling the mixture, while from the solution itself numerous white crystalline spangles separated. The liquid and solid parts were together poured into a funnel plugged with asbestos. To separate the unchanged nitrobenzole, the solid stopped by the filter was exhausted with boiling water, and the solution filtered twice through paper. When cold, the filtrate was full of large nacreous plates of a very pale straw colour, which by recrystallization became perfectly white. Having had but a few grains of this substance at my disposal, I have not yet made an accurate examination of its physical properties. I found, however, that its reaction is strongly acid, that it is fusible without decomposition, tolerably soluble in boiling water, and that it yields crystallizable salts. Not only does its origin and the method of its formation preclude the existence of more than twelve atoms of carbon in the new acid, but the following determination of silver in a carefully prepared specimen of its silver-salt points to the formula $C^{12}H^{2}Ag(NO^{4})O^{4}$ for this compound, and to $C^{12}H^{3}(NO^{4})O^{4}$ for the acid itself:—

·971 grm. of silver-salt gave ·537 grm. of chloride of silver = ·4041 grm. of silver, corresponding to a per-centage of—

	Experiment.	Theory, $C^{12}H^2Ag(NO^4)O^4$.
Silver . . .	41·61	41·54

So far as they have been yet examined, the properties of nitro-phenoic acid confirm the idea that it is a true homologue of nitrobenzoic acid.

I have made an attempt to prepare the original acid, the *phenoic*, of which I have supposed the above-noticed acid to be the nitro-derivative. Although a mixture of bichromate of potassium and sulphuric acid is without action on benzole, it acts most energetically on sulphobenzolic acid formed by dissolving benzole in Nordhäusen sulphuric acid. If to a slightly diluted solution of sulphobenzolic acid at about 70° C. minute fragments of bichromate of potassium be added, one at a time, and the action which ensues at each addition be moderated by cooling the apparatus, an acid distillate will be obtained, on the surface of which small brilliant crystals, generally accompanied by a few oily globules, will be found floating. It would seem that the oily and solid portions of the distillate are alike in composition, since the analysis of the silver-salts of the two bodies, separated mechanically as far as possible, gave almost the same numbers. Of the annexed determinations, I. was made with a silver-salt prepared from the oily part, and II. with one prepared from the crystalline part of the distillate.

I. ·789 grm. gave ·37 grm. of silver,

II. ·822 grm. gave ·168 grm. of silver;

corresponding to the following per-centages of silver:—

Experiment.		Theory, $C^{12}H^3AgO^4$.
I.	II.	
50·06	50·06	50·23

Sulphotoluolic and sulphocumolic acids, when oxidized as described above, yield benzoic acid in abundance. I have identified the product by all the usual tests. I have not yet experimented with the xylole series. Sulphocymolic acid yields a white powder, apparently identical with insolinic acid.

Nitrotoluole, when oxidized, yielded nitrobenzoic acid, which agreed in every respect with a pure specimen in my possession prepared from benzoic acid. I have before mentioned the difficulty with which nitrobenzole is acted on by the oxidizing mixture; and if this latter be somewhat diluted, it affords a means of separating the nitro-derivatives of toluole, &c. from nitrobenzole, which, when the action is complete, is siphoned off and washed with an alkaline solution to remove the nitrobenzoic acid formed.

Since sulphobenzolate of ammonium, when submitted to dry distillation, yields some quantity of benzole*, I imagined that the *nitro*sulphobenzolate would yield *nitro*benzole: experiment, however, has not corroborated this view.

In 1859 I showed† that nascent chlorine acts powerfully on toluole, xylole, and other homologues of benzole, yielding the chlorides of toluenyle, xylenyle, &c., from which the cyanides, and subsequently the acids (toluic and xyloic), are producible. I am pursuing some inquiries in this direction with benzole, upon which unfortunately nascent chlorine acts with more difficulty, and at the same time does not appear to yield such definite results.

There is a point of view from which some of the experiments which I have made acquire a fresh interest. If 1 vol. of light coal-naphtha containing, say 50 per cent. of benzole, be submitted to the action of 6 vols. of oil of vitriol previously diluted with 1 vol. of water, and the mixture heated for some time in a suitable condensing apparatus, the benzole will remain nearly, if not quite, unacted upon, while the other hydrocarbons will be dissolved by the sulphuric acid. If the acid be absorbed by small fragments of pumice and thus used, it exerts a much more rapid and effectual action on the naphtha. The benzole, after having been washed with water, is nearly pure. The other hydrocarbons which have been dissolved are now contained as sulphotoluolic and similar acids in the liquid, which is to be collected, diluted with half its bulk of water, and poured into a retort provided with a Liebig's condenser. Bichromate of potassium, about one-sixth part in weight of the acid present, is added gradually to the solution, and the mixture cautiously distilled. In this way a considerable proportion of benzoic acid may be obtained.

Postscript, February 8, 1861.

Since writing the above remarks, my attention has been directed to a short notice of some experiments by MM. Cloetz and Guignet, who also seem to have obtained a new acid by the oxidation of nitrobenzole. I think it right to say that I succeeded in producing the acid which I have termed nitrophenoic in June 1860.

* Phil. Mag. December 1859.
† Chemical News, December 10, 1859.

XXVII. *Note on the Theory of Determinants.* By A. CAYLEY, *Esq.*[*]

THE following mode of arrangement of the developed expression of a determinant had presented itself to me as a convenient one for the calculation of a rather complicated determinant of the fifth order; but I have since found that it is in effect given, although in a much less compendious form, in a paper by J. N. Stockwell, "On the Resolution of a System of Symmetrical Equations with Indeterminate Coefficients," Gould's 'Ast. Journal,' No. 139 (Cambridge, U. S., Sept. 10, 1860).

Suppose that the determinant

$$\begin{vmatrix} 11, & 12, & 13 \\ 21, & 22, & 23 \\ 31, & 32, & 33 \end{vmatrix}$$

is represented by $\{123\}$, and so for a determinant of any order $\{123\ldots n\}$.

Let $|1|$, $|2|$, $|12|$, $|123|$, &c. denote as follows: viz.

$$|1| = 11, \quad |2| = 22, \text{ &c.}$$
$$|12| = 12.21,$$
$$|123| = 12.23.31,$$
$$\text{&c.,}$$

where it is to be noticed that, with the same two symbols, *e. g.* 1 and 2, there is but one distinct expression $|12|$ (in fact $|21| = 21.12 = |12|$); with the same three symbols 1, 2, 3, there are two distinct expressions, $|123|$ $(=12.23.31)$ and $|132|$ $(=13.32.21)$; and generally with the same m symbols 1, 2, 3...m, there are $1.2.3\ldots m-1$ distinct expressions $|123\ldots m|$, which are obtained by permuting in every possible manner all but one of the m symbols.

This being so, and writing for greater simplicity $|1|2|$ to denote the product $|1| \times |2|$, and so in general, the values of the determinants $\{12\}$, $\{123\}$, $\{1234\}$, $\{12345\}$, &c. are as follows: viz.

---

[*] Communicated by the Author.

	No. of terms.	
	+	−
${12} = + \mid 1 \mid 2 \mid$	1	
$- \mid 1 \; 2 \mid$		1
	$1 + 1 = 2$	
${123} = + \mid 1 \mid 2 \mid 3 \mid$	1	
$- \mid 1 \; 2 \mid 3 \mid$ . . .		3
$+ \mid 1 \; 2 \; 3 \mid$	2	
	$3 + 3 = 6$	
${1234} = + \mid 1 \mid 2 \mid 3 \mid 4 \mid$ . .	1	
$- \mid 1 \; 2 \mid 3 \mid 4 \mid$ . .		6
$+ \mid 1 \; 2 \; 3 \mid 4 \mid$ . .	3	
$+ \mid 1 \; 2 \mid 3 \; 4 \mid$ . .	3	
$- \mid 1 \; 2 \; 3 \; 4 \mid$ . .		6
	$12 + 12 = 24$	
${12345} = + \mid 1 \mid 2 \mid 3 \mid 4 \mid 5 \mid$.	1	
$- \mid 1 \; 2 \mid 3 \mid 4 \mid 5 \mid$.		10
$+ \mid 1 \; 2 \; 3 \mid 4 \; 5 \mid$.	20	
$+ \mid 1 \; 2 \mid 3 \; 4 \mid 5 \mid$.	15	
$- \mid 1 \; 2 \; 3 \; 4 \mid 5 \mid$.		30
$- \mid 1 \; 2 \; 3 \mid 4 \; 5 \mid$.		20
$+ \mid 1 \; 2 \; 3 \; 4 \; 5 \mid$.	24	
	$60 + 60 = 120$	

where, as regards the signs, it is to be observed that there is a sign − for each compartment| |containing an even number of symbols; thus in the expression for ${1234}$, the terms $\mid 1 \; 2 \mid 3 \; 4 \mid$ have the sign $- - =: +$, and the terms $\mid 1 \; 2 \; 3 \; 4 \mid$ the sign −. Or, what comes to the same thing; when n is even, the sign is + or − according as the number of compartments is even or odd; and contrariwise when n is odd. As regards the remaining part of the expression, this merely exhibits the partitions of a set of n things; and the formulæ for the several determinants up to the determinant of a given order are all of them obtained by means of the form

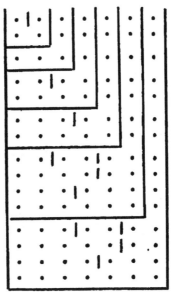

which is carried up to the order 7, but which can be further extended without any difficulty whatever.

It is perhaps hardly necessary; but I give at full length the expressions of the determinant of the third order: this is

$$
\{123\} = \quad \begin{aligned}
&|\,1\,|\,2\,|\,3\,| \\
&-|\,1\quad 2\,|\,3\,| \\
&-|\,2\quad 3\,|\,1\,| \\
&-|\,3\quad 1\,|\,2\,| \\
&+|\,1\quad 2\quad 3\,| \\
&+|\,1\quad 3\quad 2\,|
\end{aligned}
$$

And by writing down in like manner the expression for the twenty-four terms of the determinant of the fourth order, the notation will become perfectly clear.

The formula hardly requires a demonstration. The terms of a determinant $\{123\ldots n\}$, for example the determinant $\{1234\}$, are obtained by permuting in every possible manner the symbols in either column, say the second column, of the arrangement

$$
\begin{aligned}
&1\ 1 \\
&2\ 2 \\
&3\ 3 \\
&4\ 4
\end{aligned}
$$

and prefixing the sign $(+$ or $-)$ of the arrangement; and the resulting arrangements, for instance

$$+ \; 1\,1, \qquad - \; 1\,2, \qquad - \; 1\,2,$$
$$2\,2 \qquad\qquad 2\,1 \qquad\qquad 2\,3$$
$$3\,3 \qquad\qquad 3\,3 \qquad\qquad 3\,4$$
$$4\,4 \qquad\qquad 4\,4 \qquad\qquad 4\,1$$

are interpreted either into $+11.22.33.44$, $-12.21.33.44$, $-12.23.34.41$, or in the notation of the formula, into

$$+\,|1|2|3|4|, \qquad -\,|12|3|4|, \qquad -\,|1234|.$$

And so in general.

Suppose that any partition of n contains α compartments each of a symbols, β compartments each of b symbols ... $(a, b, \ldots$ being all of them different and greater than unity), and ρ compartments each of a single symbol, we have

$$n = \alpha a + \beta b + \ldots + \rho.$$

And writing, as usual, $\Pi a = 1.2.3\ldots a$, &c., the number of ways in which the symbols $1, 2,.3,\ldots n$, can be so arranged in compartments is

$$\frac{\Pi n}{(\Pi a)^{\alpha}(\Pi b)^{\beta}\ldots \Pi\alpha\, \Pi\beta\ldots\Pi\rho};$$

but each such arrangement gives $\big(\Pi(a-1)\big)^{\alpha} \cdot \big(\Pi(b-1)\big)^{\beta}$ terms of the determinant, and the corresponding number of terms therefore is

$$\frac{\Pi n}{a^{\alpha}\, b^{\beta}\ldots\Pi\alpha\,\Pi\beta\ldots\Pi\rho}.$$

The whole number of terms of the determinant is Πn, and we have thus the theorem

$$1 = \Sigma \frac{1}{a^{\alpha}\, b^{\beta}\ldots\Pi\alpha\,\Pi\beta\ldots\Pi\rho},$$

in which the summation corresponds to all the different partitions $n = \alpha a + \beta b \ldots + \rho$, where $a, b, \ldots$ are all of them different and greater than unity; a theorem given in Cauchy's *Mémoire sur les Arrangements,*' &c., 1844. But it is to be noticed also that, the number of the positive and negative terms being equal, we have besides

$$0 = \Sigma \frac{(-)^{\alpha(a-1)+\beta(b-1)+\ldots}}{a^{\alpha}\, b^{\beta}\ldots\Pi\alpha\,\Pi\beta\ldots\Pi\rho};$$

contained in the atmosphere, and from the presence of these lines should infer that of these various metals. The more intense luminosity of the sun's solid body, however, does not permit the spectrum of its atmosphere to appear; it *reverses* it, according to the proposition I have announced; so that instead of the bright lines which the spectrum of the atmosphere by itself would show, dark lines are produced. Thus we do not see the spectrum of the solar atmosphere, but we see a negative image of it. This, however, serves equally well to determine with certainty the presence of those metals which occur in the sun's atmosphere. For this purpose we only require to possess an accurate knowledge of the solar spectrum, and of the spectra of the various metals.

" I have been fortunate enough to obtain possession of an apparatus from the optical and astronomical manufactory of Steinheil in Munich, which enables me to examine these spectra with a degree of accuracy and purity which has certainly never before been reached. The main part of the instrument consists of four large flint-glass prisms, and two telescopes of the most consummate workmanship. By their aid the solar spectrum is seen to contain thousands of lines; but they differ so remarkably in breadth and tone, and the variety of their grouping is so great, that no difficulty is experienced in recognising and remembering the various details. I intend to make a map of the sun's spectrum as I see it in my instrument, and I have already accomplished this for the brightest portion of the spectrum—that portion, namely, included between Fraunhofer's lines F and D. By painting the lines of various degrees of shade and of breadth, I have succeeded in producing a drawing which represents the solar spectrum so closely, that, on comparison, one glance suffices to show the corresponding lines.

" The apparatus shows the spectrum of an artificial source of light, provided it possess sufficient intensity, with as great a degree of accuracy as the solar spectrum. A common colourless gas-flame in which a metallic salt volatilizes, is in general not sufficiently luminous; but the electric spark gives with great splendour the spectrum of the metal of which the electrodes are composed. A large Ruhmkorff's induction apparatus produces such a rapid succession of sparks, that the spectra of the metals may be thus examined with as great facility as the solar spectrum.

" By means of a very simple arrangement, the spectra of two sources of light may be compared. The rays from one source of light can be led through *one* half of the vertical slit, whilst those from another source are led through the *other* half. If this is done, the two spectra are seen directly under one another, separated only by an almost invisible dark line. By this arrangement it is extremely easy to see whether any coincident lines occur in the two spectra.

"I have in this way assured myself that all the bright lines characteristic of iron correspond to dark lines in the solar spectrum. In that portion of the solar spectrum which I have examined (between the lines D and F), I have had occasion to remark about seventy particularly brilliant lines as caused by the presence of iron in the solar atmosphere. Angström only observed three bright lines in this part of the spectrum of the electric spark; Masson noticed only a few more; Van der Willigen says that iron produces only a very few feeble lines in the spectrum of the electric spark. From the number of these lines which I have been able to observe with ease, and map with absolute certainty, some idea may be formed of the capabilities of the instrument which I am fortunate enough to possess.

"Iron is remarkable on account of the number of the lines which it causes in the solar spectrum; magnesium is interesting because it produces the group of Fraunhofer's lines which are most readily seen in the sun's spectrum, namely, the group in the green, consisting of three very intense lines to which Fraunhofer gave the name *b*. Less striking, but still quite distinctly visible, are the dark solar lines coincident with the bright lines of chromium and nickel. The occurrence of *these* substances in the sun may therefore be regarded as certain. Many metals, however, appear to be absent; for although silver, copper, zinc, aluminium, cobalt, and antimony possess very characteristic spectra, still these do not coincide with any (or at least with any distinct) dark lines of the solar spectrum. I hope before long to be in a position to publish more extended information on this point.

"The combination of Ruhmkorff's induction coil with the spectrum apparatus will doubtless also be of importance for the chemistry of terrestrial matter. Very many metallic compounds do not give the spectrum peculiar to the metal when placed in a flame, because they are not sufficiently volatile, but they give it at once when placed on the electrodes of an electric spark. These lines are then indeed seen, together with those of the metal of the electrode, and those of the air through which the spark passes; and owing to the great number of bright lines which compose the spectrum of every electric spark, it would be almost impossible, without a special arrangement, to distinguish the lines caused by the metal of the electrodes from those produced by the metallic salt added. The special arrangement which in this case removes all difficulty, consists in allowing the spark to pass at the same instant between two pairs of electrodes, in such a manner that the light of one spark passes through the upper half of the slit, whilst the light of the other spark passes through the lower half of the slit, so that the two spectra are seen one directly above the other. If both pair of electrodes are pure, both the spectra are alike; if a metallic salt is brought on to one of the electrodes, the lines peculiar to that metal appear in the one spectrum in addition to those present before. These are recognized at the first moment, because they are absent in the other spectrum. The lines which are common to the two spectra may serve, when they are once for all drawn, as the simplest mode by which to represent the position of the lines of the other metals employed.

"I have proved that in this way the metals of the rare earths, yttrium, erbium, terbium, &c., may be detected in the most certain and expeditious manner. Hence we may expect that, by help of Ruhmkorff's coil, the spectrum-analytical method may be extended to the detection of the presence of all the metals. I trust that this expectation may be borne out in the continuation of the research which Bunsen and I are jointly carrying on with the object of rendering this method practically applicable."

XXIX. *On Ripples, and their relation to the Velocities of Currents.* By T. Archer Hirst, *Mathematical Master at University College School, London.*

[Concluded from p. 20.]

[With a Plate.]

21. IN art. 20 it was shown that when a jet or partially immersed solid cylinder is made to describe a circle of radius a in still water, the ripple it produces has a cusp C (fig. 7, Plate I.), which describes another circle whose radius r has to a the same ratio that the velocity λ, with which the waves causing the ripple are propagated, has to the velocity u of the jet; and further, that the angle θ between the radii to the jet and to the cusp varies with r, in accordance with the relation

$$\tan(\phi_1 - \theta) = \phi_1 = \frac{\sqrt{a^2 - r^2}}{r}. \quad \dots \quad (40)$$

Hence if the axis of polar coordinates pass through, and rotate with the jet, this is the equation of a curve upon which the cusp of the ripple must lie, no matter with what velocity the jet may be rotating. The curve itself, as may be easily shown (Plate IV. fig. 2), lies entirely within the circle (a); it commences at the jet A, $r = a$, $\theta = 0$, where it touches the axis O A, and proceeds inwards, turning its convexity towards the centre O, until on arriving at

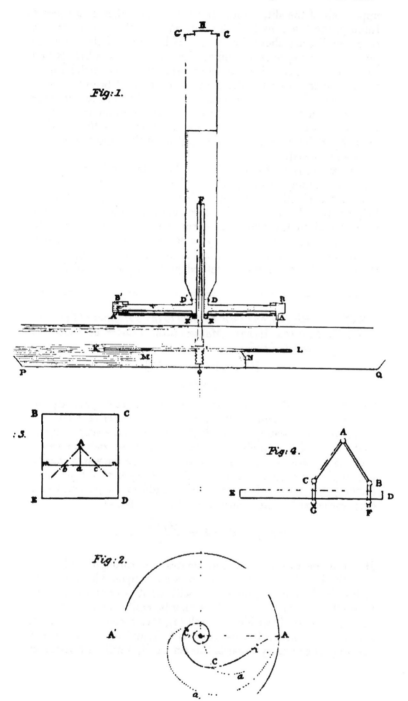

Fig: 1.

: 3.

Fig: 4.

Fig: 2.

a point of inflexion i, whose coordinates are $r = \dfrac{a}{\sqrt{2}}$, $\theta = 1 - \dfrac{\pi}{4}$ (equivalent to $12° 17' 44''$ nearly), it commences and continues ever after to turn its concavity towards the centre O, which point it approaches asymptotically.   In comparing the foregoing results with those of experiment, the above curve was found to be of service.

Experiments.

22. I proceed to describe a few experiments which will serve to verify the general principles upon which the foregoing results are based, though they are by no means sufficiently complete and accurate to decide, fully, the interesting question as to the velocities of differently shaped waves.

23. The experiments on the ripples produced by a jet or solid cylinder rotating in still water were made with the modification of Barker's mill represented in section by fig. 1, Pl. IV. D C C' D' is a tin tube 17 inches long and 2 inches in diameter, whose lower extremity, contracted to a diameter of 1 inch, is soldered into a hollow brass cap D D' E' E. Into this cap three brass tubes are fitted; one vertical, projecting 6 inches into the interior of the tin tube, and closed at its upper extremity by a plate F of hard steel; and two, D B and D' B', horizontal, in the same straight line, about $4\frac{1}{2}$ inches long, $\frac{3}{10}$ths of an inch in diameter, and terminated by two brass caps A B and A' B', which slide over the tubes and can be removed at pleasure.   In each cap a lateral aperture is made, into which can be fitted either a smaller brass tube suitable for the issue of a jet $\frac{1}{10}$th of an inch in diameter, or a solid brass cylinder of the same diameter, which completely prevents the efflux of water.   The whole weight of the portion of the instrument thus far described, and of the water it may contain, is supported by a steel pin O F about 8 inches long, whose hardened and pointed upper extremity enters a small cavity in the centre of the steel plate F, whilst its lower extremity is screwed firmly into a lead weight M N, placed at the bottom of a bath P Q R S.   By this arrangement, the water in the tin tube issuing at one or both of the orifices A, A' causes a rotation unaccompanied by sensible oscillations, and the issuing jets of water, by falling into the water in the bath, produce the ripples whose forms are to be examined.   When a constant velocity of rotation is required, it is merely necessary to retain the same depth of water in the tin tube; and this can be done either by regulating, with a cock, the supply of water entering at H, or by carrying a siphon through H to a reservoir kept at a constant level.   When necessary, too, the water in the bath can be kept at a constant level by similar means.

24. The moment the instrument rotates with sufficient velocity the ripples make their appearance, and the general resemblance between their forms and that of the one drawn according to theory in fig. 7, Plate I., is very striking. The external branch A X, and the portion of the internal one between the jet A and the cusp C are quite distinct and well defined; the return branch C B D is far less distinct, the height of this part of the ripple being small, and its breadth great; its existence, however, may be established by a closer inspection, or by a glance at the distorted image at the bottom of the bath caused by the irregular refraction of the light incident upon the rippled surface. Besides the principal ripple there are a number of secondary ones, similar in form but less in extent, which precede the former, and of course tend to diminish its prominence.

25. In order to compare the actual with the theoretical ripple more precisely, the following simple expedient was adopted:— A circular plate of glass K L a foot in diameter, and having a circular aperture at its centre to admit the screw O, was attached by means of this screw to the top of the lead weight. Its lower surface was silvered, and the silvering afterwards coated with shell-lac to protect it from the action of the water. The image of the ripple in this plane mirror was not only more distinct than the ripple itself, but the arrangement had the additional advantage of obviating errors due to parallax, since the image of the observer's eye could always be made to coincide with that of any part of the ripple under examination. Lastly, a second and transparent glass plate A A', 10 inches in diameter, was attached by means of a screw E E' to the cap at the bottom of the mill. This plate turned of course with the mill, and could be graduated in any required manner; it had a small indentation at a point in its circumference, in order that the axis of one, A, of the vertical jets (or cylinders) might be made to describe, precisely, the circumference of a circle 5 inches in radius.

26. By means of the equations (32) and (40), one or two ripples A a C and A $a_{\prime}$ C$_{\prime}$ (fig. 2), and the locus A C C$_{\prime}$ of their cusps, were carefully plotted on a sheet of paper; they corresponded to a value of $a = 5$ inches, and in the case of the ripples, to arbitrarily assumed values of the ratio α between the velocities λ and u of the wave and jet. These curves were next transferred, in white paint, to the glass plate A A', and one of the jets A made to coincide with their origin. The instrument being now filled with water, the jet at A' was made to issue horizontally in order to turn the mill, whilst the jet at A descended vertically into the bath and produced the ripple to be examined. As the velocity of rotation diminished, the cusp of the actual ripple described very accurately the curve C$_{\prime}$ C A traced on the glass plate A A'; and

with equal accuracy the portion of this ripple between the jet
and cusp coincided successively with the corresponding portions
A a, C$_{\prime}$ A a C of the theoretical ripples. To test the coincidence
of the portion of the ripple outside the circle described by the
jet (A X E in fig. 7; Plate I.), a hole was drilled through the glass
plate A A' at a distance of $2\frac{1}{2}$ inches from its centre, and a solid
cylinder thrust through the same so as to be partially immersed
in the water of the bath. The mill was turned as before by the
jet A' alone, the aperture at A being closed; and when the velocity of rotation was properly regulated, it was found that the
complete actual ripple produced by the immersed cylinder coincided very well with as much of the theoretical one as could be
traced upon the glass plate.

27. The slight divergence from perfect coincidence between
the theoretical and actual ripples being sufficiently accounted for
by the conditions under which the experiments were made—the
limited extent of the bath, the disturbing effect of the portion of
the axis O F immersed in the water, and the magnitude of the
jet (which in theory was disregarded),—the accuracy of the principles upon which the foregoing theory is based may be considered as sufficiently established, and we may proceed at once to
give an account of the measurements from which an approximate
value of the velocity λ with which the waves, producing the ripples in question, are propagated may be deduced.

For this purpose the curves on the glass plate A A' were obliterated, and replaced by concentric circles whose radii increased
by half inches from 1 to 5 inches. A constant rotation was
secured in the manner described in art. 23, and its velocity determined in each case by counting the number n of rotations
made in a given time t, the latter being measured by an accurate timepiece, the motion of whose index could be instantaneously
arrested or renewed by simply pressing or releasing a spring.

Since each jet issues at right angles to the horizontal arm B B',
it is clear that its effect, as far as the rotation of the mill is concerned, depends not only upon the height of the column of water
in the tin tube, but also upon the inclination of the jet to the
horizon; for upon the former depends the velocity of efflux, and
the uncompensated pressure against the part of the horizontal
tube opposite to the orifice through which the jet issues, whilst
upon the latter depends the magnitude of the horizontal component of this pressure, which component is alone effective in producing rotation. But every change in the velocity of efflux and
in the inclination of the jet to the horizon produces a corresponding change in the distance, from the axis, of the point at
which the jet falls into the water in the bath, and consequently
in the radius a of the circle described by the jet. This radius,

too, also depends upon the height of the water in the bath; so that at first sight it would appear, not only that the latter ought to be kept constant throughout one and the same experiment, but that a separate determination of a should be made in each case.

On setting both jets in action, however, at different inclinations, and comparing the two ripples which they produced, it was soon found that, however much the radii a and a' of the circles described by the jets might differ in magnitude, the radii r and r' of the circles described by the cusps of their ripples were in all cases *nearly* equal. This result is confirmed by theory, which also shows that the slight difference between the radii r and r' is due, solely, to the slight difference between the values of the velocities λ and λ' with which the waves, produced under these different circumstances, are propagated. In fact it follows at once from equation (35) of art. 20, that

$$\frac{\lambda}{u} : \frac{\lambda'}{u'} = \frac{r}{a} : \frac{r'}{a'} ;$$

and the angular velocity of rotation being the same for each jet, their actual velocities u and u' will clearly be proportional to the radii a and a' of the circles they describe; the above proportion, therefore, reduces itself to the simpler one,

$$\lambda : \lambda' = r : r', \quad \cdots \quad \cdots \quad (41)$$

and thus verifies the above remark. When one jet descended vertically and the other issued horizontally, the difference between r and r', although still very small, was at a maximum, and the cusp of the ripple produced by the vertical jet was always furthest distant from the axis; from which we may conclude that the velocity λ is greater when the jet descends vertically into the water of the bath, than when it strikes the surface of the latter obliquely. When the vertical jet was replaced by a solid cylinder, the difference between the positions of the cusps remained about the same as before, thus indicating that the velocity λ was appreciably the same for a jet and for a solid cylinder.

It is also worth noting here, that when the distance, from the axis, of the cusp of the ripple produced by the oblique jet exceeded the radius (5 inches) of the circle described by the vertical jet, the latter *produced no ripple whatever*. The reason of this has already been given in art. 13; and in accordance with the statement there made, it was found that in all such cases the velocity of the vertical jet was less than $7\frac{1}{2}$ inches per second, which, as we shall see, is the mean value of the velocity λ with which the waves produced by each jet are propagated. Attempts were made to regulate the velocity of rotation so that the cusp of the ripple produced by the oblique jet should be exactly 5 inches distant from the axis. Theoretically, the ripple corresponding to the vertical jet should then have been the *involute of a circle* (art. 13); in reality, however, the ripple was scarcely distinguishable in such cases, the disturbance which the vertical jet produced upon the surface of the water in the bath being still so small.

28. Although the apparatus was sufficiently delicate to establish the fact of slight variations in the velocity λ due to differences in the obliquity of the jet and the velocity of efflux, it did not appear suitable for the full investigation of the magnitude and conditions of these variations; I contented myself therefore with a few determinations of λ from a series of observations made under all possible conditions. From them the limits of the variation of λ may be to some extent ascertained and its mean value estimated. In order to calculate λ from the data supplied by experiment, the formula

$$\frac{\lambda}{u} = \frac{r}{a},$$

given in art. 20, requires a slight transformation. The circumference of the circle described by the jet being $2\pi a$, and n rotations being made in t seconds, the velocity u has the value

$$u = \frac{2\pi a n}{t},$$

so that the above equation giving the velocity λ per second becomes

$$\lambda = 2\pi \frac{nr}{t}. \quad \ldots \quad \ldots \quad (42)$$

According to this, and as already hinted in the last article, the velocity λ depends, solely, upon the number of rotations in a given time, and upon the distance r of the cusp of the ripple from the axis of the instrument; in other words, it is not dependent upon the radius of the circle described by the jet; so that it was not necessary to determine the distance, from the axis, of the point of impact between the jet and the water in the bath, and therefore not necessary to ascertain the height of the water in the bath, or to render the latter constant.

29. Some of the results of a great many accordant observations are shown in the following Table, wherein n represents the number of rotations per minute, r the distance (in inches) from the axis of the cusp of the ripple, and λ the velocity in inches per second with which the corresponding waves are propagated, calculated according to (42):—

n.	r.	λ.
78	0·5	4·1
63	0·8	5·5
59	0·9	5·6
56	1·0	5·9
53	1·2	6·5
50	1·3	6·6
43	1·5	6·6
41	1·6	6·9
32	2·1	7·0
31	2·3	7·3
28	2·5	7·4
26	2·7	7·4
25	2·8	7·4
24	2·9	7·3
22	3·2	7·3
21	3·5	7·7
20	3·7	7·7
18	4·0	7·5
15	4·4	7·8

Of these results the first five or six are less trustworthy than the rest,—and this for several reasons, amongst which the following may be cited:—The rapidity of the rotation renders the determination of the position of the cusp more difficult; the proximity of the immersed axis interferes with the clear definition of the cusp; and lastly, the consequences of a small error in estimating the distance of this cusp from the axis increase as this distance diminishes, since the cusp approaches the axis asymptotically (art. 21). On the other hand, the rapid diminution of λ shown in the Table, can in some measure be accounted for by the fact that, in order to avoid disturbing too much the surface of the water in the bath, it was found necessary, with rapid rotations, to use only one jet, and it of course issued at a small angle towards the horizon, so that the impact between the jet and the water in the bath was necessarily a very oblique one. But, as already mentioned in art. 20, the tendency of this obliquity is to diminish the velocity λ. There can be no doubt that the rapid diminution of λ in the Table is to be ascribed chiefly to these causes; though it is also worth mentioning that the property of the wave insisted upon by Weber, *i. e.* that its velocity diminishes as its radius increases (see art. 2), would also tend to produce the effect observed.

In the last twelve observations recorded in the Table, two jets (or a jet and a cylinder) were used, and the ripple of the vertical one was chiefly examined. As a consequence, the values of λ vary far less, and 7·5 inches per second may be taken as a mean value of the velocity with which waves thus produced are propagated.

80. As an interesting coincidence, it may be mentioned that Poisson, in his memoir before referred to[*], gives the results of four experiments on the velocity of waves made by Biot. In this case the waves were produced by suddenly withdrawing from the water a partially immersed solid of revolution. The velocity of the wave was found to vary with the form of the body, and with the radius of its section at the water's level. In one case, the body being an ellipsoid, and the radius in question equal to ⅔ of an inch, the velocity was about 5½ inches per second; in another case, where the body was a sphere and the radius of the section 1¼ inch, the velocity was 7·87 inches per second. The waves experimented upon by Weber had a far greater velocity: they were produced by allowing a column of liquid suspended in a tube to descend suddenly into the general mass; and their velocities varied from 17 to 34 inches per second.

31. In the foregoing experiments, the velocity λ was determined by causing an immersed cylinder to move with a given velocity in still water. A few experiments were next made with a view of ascertaining the value of λ when a cylinder is simply immersed in a current of known velocity. According to art. 10, the sine of half the angle 2θ between the branches of the ripple caused by immersing a cylinder in a current is inversely proportional to the velocity v of the current, and directly proportional to the velocity λ in question. In fact it was there shown that

$$\lambda = v \sin \theta = v \frac{\tan \theta}{\sqrt{1 + \tan^2 \theta}}. \quad \cdot \quad \cdot \quad \cdot \quad (43)$$

To determine the velocity v at any point of the surface of a current, a Wollaston's current-meter was used. As is known, this instrument consists of a screw which is made to rotate by the force of the current. The instrument must be immersed to the depth of two inches at least, in order that the screw may be completely covered by the water; and when so immersed, the rotation of the screw can be communicated at any instant to a divided wheel, and the communication as suddenly broken. The space described by the current during the interval between making and breaking this communication—an interval which can be measured by means of an ordinary watch with a seconds' hand—is at once read off on the wheel, the instrument having been previously carefully graduated.

The angle θ was determined by means of a simple instrument, to which we may give the name of *ripple-meter*. It consisted of a glass plate (BCDE, Pl. IV. fig. 3) 5 inches square, through which, at a point A, a hole 1/10th of an inch in diameter was drilled in

[*] *Mémoires de l'Acad. Roy. des Sciences de l'Institut.* Année 1816, vol. i. p. 173.

order to insert a solid cylinder of ivory about 1 inch long. Passing through the centre of A and parallel to the sides of the glass, a fine line A a, an inch long, was scratched on its surface with a diamond point; and through the extremity a of this line another, mn, was drawn perpendicular to the former, and graduated on each side from a into tenths of an inch. In using this ripple-meter, the plate of glass was held close and parallel to the surface of the current, so that the ivory pin, by becoming partially immersed, might cause a well-defined ripple, bAc, visible through the glass; the plate was then turned until A a bisected the angle between the branches of this ripple, when of course the ratio $\dfrac{ac}{aA}$ gave at once the tangent of the required angle θ. In this case, as in that of the rotating jet, a number of secondary ripples are also visible: after a little practice, however, and when the current was not too slow (the angle bAc too obtuse), it was not difficult to estimate, approximately, the value of tan θ corresponding to the principal ripple.

The following Table contains a few of the best results of many experiments made on streams. In it the first column gives the values of tan θ as read off on the ripple-meter; the second column shows the corresponding values of the velocity v as indicated in feet per minute by the current-meter; and the third column contains the respective values of λ, calculated according to the formula (43), in inches per second.

tan θ.	v.	λ.
·993	44	6·2
·775	55½	6·8
·705	59	6·8
·577	67	6·7
·545	71	6·8
·392	96	7·0
·388	105	7·6

The observations, of which the above are some of the most trustworthy results, were made at different periods on streams in Hampshire and Gloucestershire. When we take into consideration the fact that a stream could rarely be found where the velocity at any one point remained constant, that in all cases this velocity was determined with the current-meter at a point 2 or 3 inches below the surface, and lastly, that the method of estimating the angle formed by the branches of the ripple can only lead to approximate results, the general agreement of the values of λ with those obtained from the rotatory experiments is as close as could be expected. The results appear to indicate that, with one and the same immersed body, the value of λ varies somewhat with the velocity of the current; but there can be little doubt that, under more favourable circumstances, this variation would be found to be far less than that indicated by the Table.

In many cases no ripple whatever was produced by the immersion of the cylinder of ivory attached to the ripple-meter; and in all such cases the current-meter indicated a velocity less than 7 inches per second; the quickest of such currents in fact had only a velocity of 25 feet per minute, or 5 inches per second. This corroborates the explanation given in art. 11, where it was foreseen that a body immersed in a current whose velocity was less than that of the wave, would allow the water to flow past it without visibly rippling its surface.

The phenomenon represented by the second figure of art. 10, where the branches of the ripple turn their concavities towards each other, and which was shown to be a necessary consequence of Weber's statement, that the velocity of the wave diminishes as its radius increases, was never observed. A slight concavity, however, might easily have escaped detection.

32. The velocity λ being once determined for any ripple-meter, we can of course by its means determine, conversely, the velocity of a current, provided the latter exceeds the limit λ. For this purpose I made use of a more convenient, though perhaps less accurate ripple-meter, with a description of which I will conclude the present paper. A B and A C in fig. 4 represent two strips of brass, each 3 inches long, and made to turn round A. The ends B and C of these strips also turn on axes at the extremities of two brass stirrups B F and C G, through which passes a wooden scale D E divided into twentieths of an inch. The cylinder, of the same dimensions as before, by whose immersion in the current ripples are produced, is pushed through an aperture in the joint A. The stirrup B F being fixed at the zero of the scale, is held there by a clamping screw F, and the stirrup C G is made to slide along the scale until the strips A B, A C are parallel to the branches of the ripple. This adjustment once made, the distance B C, as read off from the scale, being directly proportional to $2 \sin \theta$, is clearly inversely proportional to the velocity of the current (arts. 10 and 31).

The decrease in velocity from the centre to the banks of a stream is clearly indicated by this little instrument. As an illustration, I give the results of a few observations made on a mill stream at Brimscombe near Stroud. A wooden plank was thrown across the stream, which was about 7 feet 6 inches wide, and upon it, commencing at one bank, marks were made 9 inches apart. By kneeling on the plank, the ripple-meter was immersed in the current exactly under each mark.   In explanation of the following Table, it is only necessary to add that the first column N shows the number of the mark on the plank; the second column D the distance, in inches, of that mark from the bank; the third column *d* the depth, in inches, of the stream at each mark; the fourth column the values proportional to $2 \sin \theta$ as read off on the ripple-meter; and the last column *v* the velocities in feet per minute, of the current as calculated, on the hypothesis of $\lambda = 7 \cdot 5$ inches per second, by formula (43) :

N.	D.	d.	$2 \sin \theta$.	v.
0	0	11	...	...
1	9	16	98	63
2	18	18	74	83
3	27	22	77	80
4	36	24	67	92
5	45	$23\frac{1}{2}$	60	103
6	54	18	57	106
7	63	14	59	104
8	72	13	68	91
9	81	13	92	65
10	90	12	...	...

In conclusion it may be added that, if desired, more accurate ripple-meters might easily be devised, and by means of such the velocities of currents might be determined from the forms of their ripples with a degree of precision little, if at all, inferior to that possessed by the methods now in use.   It is from a theoretical point of view, however, that the relation between waves and ripples, which we have endeavoured to establish, promises the greatest interest.   For there can be little doubt that a skilful experimenter, pursuing the subject in this direction, would greatly extend our present knowledge with respect to the changes in the velocity of a current at different points of its surface, and especially with respect to the velocities with which different kinds of waves are propagated on the surface of still water, and to the variation of this velocity during the propagation of one and the same wave.

January 15, 1861.

XXX. *On the Equilibrium of a Fluid Mass revolving freely within a Hollow Spheroid about an Axis which is not its Axis of symmetry.* By G. R. Dahlander[*].

IF we suppose a fluid ellipsoid to revolve alone about an axis which is not an axis of symmetry, we easily perceive that it cannot assume a position of equilibrium.   It is, however, dif-

* Communicated by the Author.

ferent if we suppose the fluid mass surrounded by a hollow spheroid, the two bounding surfaces of which are not concentric.
The fluid mass can then, under certain circumstances, actually
assume a position of equilibrium, even when its axis of rotation
is entirely external to the mass itself, and when it thus plays the
part of an inner satellite to the hollow spheroid, with which it
has a common motion of rotation. The existence of such singular states of equilibrium we will now demonstrate.

Suppose the hollow spheroid, together with the internal fluid
mass, to revolve about the axis of the outer bounding surface of
the spheroid, the axis of the inner bounding surface being parallel
to the axis of rotation. The fluid mass can then assume a position
of equilibrium in which its figure is that of an ellipsoid of rotation, whose axis is parallel to the above-mentioned axes.

Let the centre of the outer bounding surface be the origin of
a system of rectangular coordinates, the axis of z coinciding with
the axis of rotation. Let m, n, and p be the coordinates of the
centre of the inner bounding surface, and α, β, γ those of the
centre of the fluid mass. The equation of the surface of the fluid
will then be

$$\frac{(x-\alpha)^2+(y-\beta)^2}{a^2}+\frac{(z-\gamma)^2}{b^2}=1. \quad \ldots \quad (1)$$

If the component parts of the attraction parallel to the axes be
denoted by X, Y, Z, we have

$$X=-Mx+M'(x-m)-M''(x-\alpha),$$
$$Y=-My+M'(y-n)-M''(y-\beta),$$
$$Z=-Nz+N'(z-p)-N''(z-\gamma).$$

If therefore the angular velocity be denoted by w, we get for the
differential equation of the *surfaces de niveau*,

$$\left(-Mx+M'(x-m)-M''(x-\alpha)\right)dx+\left(-My+M'(y-n)\right.$$
$$\left.-M''(y-\beta)\right)dy+\left(-Nz+N'(z-p)-N''(z-\gamma)\right)dz$$
$$+w^2(xdx+ydy)=0, \quad \ldots \ldots \ldots \ldots \quad (2)$$

where M, M', M'', N, N', and N'' are independent of x, y, z.
Consequently by integration we get

$$(-M+M'-M''+w^2)(x^2+y^2)+(-N+N'-N'')z^2$$
$$+2(-M'm+M''\alpha)x+2(-M'n+M''\beta)y$$
$$+2(-N'p+N''\gamma)z=0. \quad \ldots \ldots \ldots \ldots \quad (3)$$

As equations (1) and (3) are to be identical, we have the following equations of condition:—

$$\left.\begin{aligned}
-\alpha &= \frac{-M'm + M''\alpha}{-M + M' - M'' + w^2}, \\
-\beta &= \frac{-M'n + M''\beta}{-M + M' - M'' + w^2}, \\
-\gamma &= \frac{-N'p + N''\gamma}{-M + M' - M'' + w^2} \frac{b^2}{a^2}, \\
\frac{a^2}{b^2} &= \frac{-N + N' - N''}{-M + M' - M'' + w^2}.
\end{aligned}\right\} \quad \dots \dots (4)$$

From which we obtain

$$\left.\begin{aligned}
\alpha &= \frac{-M'm}{M - M' - w^2}, \\
\beta &= \frac{-M'n}{M - M' - w^2}, \\
\gamma &= \frac{-N'p}{N - N'}.
\end{aligned}\right\} \quad \dots \dots \dots (5)$$

From the first two of equations (5) it follows that

$$\alpha : \beta = m : n.$$

The geometrical signification of this proportion is, that the centre of the fluid mass be in the plane which passes through the axis of rotation and the centre of the inner bounding surface of the spheroid. We find, moreover, that the position of this centre in a state of equilibrium with a fixed angular velocity is independent of the density of the fluid, supposing the form and density of the outer spheroid to be the same. Further, we find that if a fluid ellipsoid satisfy the conditions of equilibrium, all other similar ellipsoids of the same density will also satisfy these conditions. Generally, from equations (5), real finite positive or negative values of α, β, γ can be obtained if M, M', N, N', m, n, p, and w are given. But one or more of these values may become infinite under certain circumstances. This is the case when the two bounding surfaces of the solid spheroid are similar. Then $M = M'$ and $N = N'$, whence the value for γ would be infinite, unless at the same time $p = 0$, in which case γ can have any value whatever.

The last of equations (4) constitutes the real equation of condition, which determines the relation which must subsist between the density, form, and angular velocity of the fluid mass, in order that equilibrium may be possible for the given values of M, M', N, and N'. We shall separately consider the particular case when both the bounding surfaces of the surrounding spheroid are similar.

From what has before been stated, it is evident that p becomes $=0$.

We can take for the axis of y a line which lies in the plane passing through the centre of the fluid mass and the centres of the bounding surfaces. In this case $n=0$ and $\beta=0$, and also $M=M'$ and $N=N'$. We shall now examine if an oblate ellipsoid of rotation can satisfy the conditions of equilibrium. Supposing $\frac{a^2}{b^2}=1+\lambda^2$, and the density of the fluid $=\rho$, then the last of equations (4) will become

$$1+\lambda^2=\frac{-\frac{4}{3}\pi\rho f\frac{(1+\lambda^2)}{\lambda^3}(\lambda-\arctan\lambda)}{-\frac{2}{3}\pi\rho f\frac{1+\lambda^2}{\lambda^3}\left(\arctan\lambda-\frac{\lambda}{1+\lambda^2}\right)+w^2}.$$

If $\frac{w^2}{2\rho\pi f}$ be taken $=E$, the equation of condition becomes

$$E=\frac{\arctan\lambda}{\lambda^3}(\lambda^2+3)-\frac{3}{\lambda^2}. \quad \ldots \quad (6)$$

But this equation is just the same as that we obtained in determining the conditions of equilibrium of a freely revolving fluid mass whose particles attract each other. Thus we find that precisely the same conditions of equilibrium are involved when the fluid is revolving in a hollow ellipsoid with similar but eccentric bounding surfaces, and when it is perfectly free. To a given value for λ there is therefore always a corresponding angular velocity; and to a given angular velocity there corresponds either no ellipsoid, or one ellipsoid, or two different ellipsoids, according as E is $\gtrless 0\cdot2246$.

Between the rotation of a fluid mass confined in a hollow spheroid and a mass which revolves freely, there is, however, this important difference, that in the former case the rotation does not take place about the axis of symmetry of the fluid unless both the bounding surfaces of the spheroid are concentric, but about a parallel axis which is the axis of symmetry of the outer bounding surface of the spheroid,—the distance between the two axes being

$$a=\frac{M'm}{w^2}, \quad \ldots \quad \ldots \quad \ldots \quad (7)$$

whence

$$aw^2=M'm.$$

But M'm is the attraction which the hollow spheroid exerts on any point within it, and aw^2 is the centrifugal force at the axis of symmetry of the fluid mass. Whence we find that at the centre of the fluid mass the acting forces counterbalance each other, which might have been anticipated from a known theorem in mechanics.

Gothenburgh, January 7, 1861.

XXXI. *Remarks on* Sainte-Claire Deville's *Theory of Dissociation.* By THOMAS WOODS, *M.D.*

To the Editors of the Philosophical Magazine and Journal.

GENTLEMEN,   Parsonstown, February 1861.

IN an interesting paper by Sainte-Claire Deville published in this Magazine last December, that author gives his views on the decomposition of bodies by heat. His idea of the relation and behaviour of the constituents forming a compound, towards each other, is, physically considered, the same as that which I published in this Journal so long ago as January 1852; that is, that they act merely as the molecules of a simple body, differing in nothing from the latter, except that, being diverse, they are capable of attaining a greater proximity among themselves, and so of causing a greater opposite movement in the particles of other bodies. A reference to my paper will show a diagram I gave in order to explain this similarity of constitution. The paper was therefore the more interesting to me as it brings forward fresh ideas on a thought-of subject. As I do not, however, yet agree with what is new in it, I beg to offer a few remarks on some of its contents; and first with respect to his theory of *dissociation.*

Sainte-Claire Deville thinks that compound gases and vapours, when heated to a certain temperature, as steam at 1000°, undergo some such change as a solid body does when it liquefies; that the *constituent* particles being removed from each other, as well as the compound particles, the gas loses *stability*, and that heat is rendered latent thereby. This condition he calls the *dissociated* state. I do not find he offers any demonstration of this state, but that he only ascribes to its influence the production of some phenomena previously otherwise explained. For instance, to account for the heat of chemical combination, he takes for granted, as an example, that the molecules of chlorine and hydrogen are double, and, even at low temperatures, in the dissociated state; and then ascribes the heat produced by their union to the latent heat of this particular condition, which he imagines is given out when the gases by their combination get into a state of stability.

Now, if the heat produced in this instance is due to the change of state of the chlorine, how does it happen that the same amount is produced when the chlorine combines, not being in the gaseous state at all ? If hydrochloric acid and zinc are placed together, the chlorine unites with the zinc, and the same quantity of heat is evolved as when the zinc burns in the gas ; and the same amount is absorbed by the decomposition as if both the constituents again attained the gaseous state ; or if (to take an instance where no gas is present, either in combination or decomposition) zinc causes a deposition of copper from chloride of copper, exactly the same heat is produced by the combination of the chlorine and copper, and exactly the same quantity is absorbed by the decomposition, as if the chlorine and copper acted as gases, changing their state as they combined or decomposed. Unless, therefore, it is imagined that when the chlorine leaves the copper or hydrogen it becomes for a time a gas and enters into the *dissociated* state, absorbing heat, and again becomes solid, giving it up, I cannot see how the temperature is raised. But even granting that it does so the phenomenon could not be accounted for, because when zinc decomposes chloride of hydrogen or chloride of copper, more heat is produced by the combination than is lost by the decomposition : and such could not occur if it were due to the latent heat of dissociation ; for the heat would be taken in the first instance from the materials afterwards heated, and so an exchange only, and not an increase, would be effected. It might be said that the zinc influences the result ; that is, that metals have a certain amount of heat connected with them which is given out in combining, and that this being greater in some instances than in others, might account for the increase of temperature when the zinc displaces the hydrogen. But if all bodies have definite quantities of heat, as Sainte-Claire Deville seems to think, the same order ought to be observed in the amounts evolved by their combination with the gases. For instance, if an equivalent of chlorine, by uniting with zinc, copper, silver, &c., produces heat, the quantity of which varies in the order in which the metals are named, oxygen ought to do the same, if the heat evolved in combination was previously connected with the combining bodies : but it is known that such is not the case. Chlorine produces more heat with silver than it does with copper, and oxygen the reverse. Instances of this kind might, of course, be multiplied ; and they prove, I think, that no fixed amount of heat, resulting either from change of condition, or from latent heat becoming evolved as it does in the condensation of vapour, can be connected with matter as part of its constitution, independent of alteration of the relation of particles.

Deville, in fine, thinks that every body possesses a certain amount of heat, or condition in itself whereby heat can be produced; and therefore *one* of the combining bodies *might* evolve from itself the heat of combination; whereas the theory I published in 1852 divests particles of any influence except that of lessening the distance between themselves, or of destroying volume, as they come together, and so that at least *two* particles of matter are essential to its production. In his theory each body engaged in chemical action is said to give out heat by change of condition; in mine the volume only, or distance between the constituent particles, is supposed to be altered. I still think the latter theory is the more reasonable, and more in accordance with our present scientific knowledge.

M. Deville seems to reject this latter theory, because the contraction arising in chemical combination is not equivalent to the expansion or heat produced; and he calculates the contraction when oxygen and hydrogen unite, to show that it is not of the same value or extent as the increase of volume given to other bodies as the accompanying or opposite movement. He also shows how chlorine and hydrogen unite without contraction at all; yet that expansion in other bodies or rise of temperature is the result. This apparent argument against the theory, however, disappears when it is considered that the particles whose combination evolves the heat are not the same as those which determine the volume. When oxygen and hydrogen unite, these elemental gases themselves, by coming together, cause other bodies to expand, and so are said to give rise to heat; but the *volume* attained by the compound they produce is determined by the distance between, not the oxygen and hydrogen, but between the particles of the water that results. In order, therefore, to calculate the contraction which causes the heat, we should know what takes place between the constituents of the compound: the bulk or volume of the compound itself tells nothing.

An argument, therefore, for some necessary change of state in combining bodies as the cause of heat, drawn from the apparent want of coincidence between the contraction on the one hand and the heat or expansion on the other, is valueless. Besides, I have shown (Phil. Mag., January 1852) that the coefficient of expansion increasing with the dilatation, the nearer particles are to each other, the greater is the effect they produce by a given contraction in causing expansion in other bodies; so that it is not only necessary to know the amount of contraction amongst the constituents of a compound at the time they combine, but also the distance they ultimately arrive at with respect to each other, before we can calculate the amount of heat they ought to produce*.

* In the last edition of Grove's 'Correlation of the Physical Forces,' when speaking of the theory I brought forward in 1852, to account for the

But this state of *dissociation* is altogether founded on gratuitous assumptions. The ground from which it springs is this : that as compound bodies when heated expand, the constituents must recede from each other as well as the compound particles. But this proposition has yet to be proved. I believe many facts favour an opposite conclusion : for instance, not to speak of the manner in which solids and fluids, when *simple* bodies, expand, being somewhat similar to the same process in *compounds*, Gay-Lussac's law with respect to the equal expansion of all gases and vapours for equal increments of heat, would surely show that the constituents of a compound do not recede from each other in expanding. Hydrogen, or any other simple gas, and vapour of ether, or other compound gas, expand exactly according to the same law. Could this occur with the simple particles of hydrogen to the same extent precisely as with compound molecules of ether, where, instead of two, we have ten elementary atoms to divide the distance and moving force between them? The resistance to expansion of a gas by heat seems to be the weight of the atmosphere; and consequently in all gases, the same resistance being present, the same expansion is attained by a certain increase of temperature. Now in a simple gas the weight is the only resistance; whereas, if Deville's theory is correct, there is in compound gases not only the weight, but the affinity of the constituents to be partly overcome; and yet the same expansion is noticed for the same increase of temperature in both. But this would be impossible, except we imagine that the separation of these constituents does not absorb heat, which we know it does.

It seems to me that this fact alone, of the similar and equal expansion of simple and compound gases by heat, shows that no motion takes place in one which does not occur in the other; therefore that no expansion of the particles themselves, that is, that no separation of the simple constituents of the compound molecule is produced by raising its temperature, and consequently that this state of *dissociation* does not exist.

A consideration of other portions of the paper would lead me too far for the present, but I may recur to it if you think the subject sufficiently interesting for your Magazine.

Your obedient Servant,
THOMAS WOODS, M.D.

---

heat of chemical combination, he seems to think it an objection (page 177) that the whole expansion which would be an equivalent to the contraction of the combining particles is not seen in the compound produced; but surely, as this volume would be the *temperature* evolved by the combination, it cannot remain longer than a moment in the compound; it must be dispersed to surrounding bodies.

XXXII. *On the Temperature Correction of Siphon Barometers.* By
WILLIAM SWAN, *Professor of Natural Philosophy in the United
College of St. Salvator and St. Leonard, St. Andrews*[*].

IN the 'Athenæum' of the 5th of January, Admiral FitzRoy,
writing on the subject of the temperature correction of
siphon barometers, invites attention to an experiment recently
made by Mr. Negretti. A siphon barometer was heated to
about 110° from some lower temperature, when it was found
that, although the mercury rose in the long or vacuum leg of the
siphon, it did not rise, but seemed to be depressed, in the short,
or open leg. The late Mr. Robert Bryson of Edinburgh invented
a self-registering barometer, which is described in the Transactions of the Royal Society of Edinburgh for 1844[†]. In that
instrument, variations in the pressure of the atmosphere were
indicated by means of a float resting on the surface of the mercury in the open leg of a siphon tube, precisely as in the ordinary wheel barometer. Mr. Bryson was anxious to ascertain
whether his instrument required any notable correction for temperature; and to settle that point experimentally, a Bunten's
barometer was heated to a high temperature. In Bunten's barometer the effective height of the mercurial column is ascertained
by reading two verniers; one indicating the level of the upper,
and the other that of the lower surface of the mercury in a siphon
tube. Mr. Alexander Bryson, who made the experiment, found
that the reading of the upper vernier rapidly changed with increase
of temperature, while the reading of the lower vernier remained
sensibly constant,—proving that the level of the mercury in the
open leg of the siphon was very little affected by change of temperature. Mr. Bryson having communicated to me the result
of his experiment, I immediately gave him an investigation of
the temperature corrections of the two surfaces of the mercury
in the siphon barometer, of which the following is substantially
a reproduction. As Admiral FitzRoy has expressed some doubts
regarding the results of observations of siphon barometers "as
hitherto obtained," I have deemed it desirable to make the following investigation perfectly general, so as to include every form
of tube; and in the first instance I have avoided employing any
formula which is only approximately true.

Let h_1, h_2 be the vertical distances of the upper and lower surfaces of the mercury in a siphon barometer, reckoned from any

* Communicated by the Author; the results of the investigation having
been communicated, on the 2nd of February, to the Literary and Philosophical Society of St. Andrews.
† Vol. xv. p. 503.

horizontal plane below the instrument, and h the barometric pressure, all at a temperature of t degrees Centigrade. When the temperature rises to $t + \Delta t$ degrees, let the above quantities become

$$h_1 + \Delta h_1, \quad h_2 + \Delta h_2, \quad h + \Delta h;$$

then, if $m =$ cubic expansion of mercury for one degree Centigrade,

$$\Delta h = m h \Delta t.$$

And since

$$h = h_1 - h_2$$

and

$$h + \Delta h = (h_1 + \Delta h_1) - (h_2 + \Delta h_2),$$

we have

$$\Delta h_1 - \Delta h_2 = \Delta h = m h \Delta t.$$

Now if

    $c =$ volume of mercury in the barometer;
    $a =$ area of the bore of the tube at upper surface of mercury;
    $b =$ area of the bore of the tube at lower surface; all at t degrees;
    $g_1 =$ the superficial, and $g_2 =$ the cubic dilatation of glass;

it will be easily seen that, at the temperature $t + \Delta t$ degrees, the capacity of that part of the tube which was occupied by the mercury at t degrees will become $c(1 + g_2 \Delta t)$; while the capacities of the portions of the tube at the ends of the former mercurial column, which are now filled by the expanded mercury, and whose lengths are Δh_1, Δh_2, will be

$$a(1 + g_1 \Delta t) \Delta h_1, \quad b(1 + g_1 \Delta t) \Delta h_2.$$

The whole volume of the mercury will therefore be

$$(a \Delta h_1 + b \Delta h_2)(1 + g_1 \Delta t) + c(1 + g_2 \Delta t).$$

But the volume of the expanded mercury must also be

$$c(1 + m \Delta t);$$

whence

$$(a \Delta h_1 + b \Delta h_2)(1 + g_1 \Delta t) = c(m - g_2) \Delta t.$$

This equation, along with

$$\Delta h_1 - \Delta h_2 = m h \Delta t,$$

gives

$$\Delta h_1 = \frac{\{c(m - g_2) + b m h(1 + g_1 \Delta t)\} \Delta t}{(a + b)(1 + g_1 \Delta t)},$$

$$\Delta h_2 = \frac{\{c(m - g_2) - a m h(1 + g_1 \Delta t)\} \Delta t}{(a + b)(1 + g_1 \Delta t)}.$$

Now since m is greater than g_2, the coefficient of c in the above values of Δh_1, Δh_2 is positive; and the coefficient of h is also positive for all possible values of Δt—g_1 being a very small quantity. It is therefore obvious that Δh_1 can never vanish, but that Δh_2 may be positive, negative, or zero, according to the values which may be assigned to a, c, and h. The depression, by heat, of the mercury in the open leg of the siphon, or in other words, the negative value of Δh_2, observed by Mr. Negretti, and the value zero of the same quantity, observed by Mr. Alexander Bryson, are therefore both perfectly accounted for.

It also appears, and this seems to be of practical importance, that we can altogether get rid of the temperature correction for the lower surface of the mercury, for any *one* given atmospheric pressure, by properly adjusting the value of c, and that thus we shall be able to make the temperature corrections for *all* other pressures exceedingly small.

For this purpose it will be convenient to simplify the expressions for Δh_1, Δh_2 by rejecting small terms. We then obtain

$$\Delta h_1 = \frac{c(m-g_2)+bmh}{a+b} \cdot \Delta t,$$

$$\Delta h_2 = \frac{c(m-g_2)-amh}{a+b} \cdot \Delta t;$$

and Δh_2 will vanish when

$$c = \frac{amh}{m-g_2}.$$

A particular case will best illustrate this. Suppose that the siphon consists of two tubes of a uniform bore a, connected at the bottom by a narrow channel whose capacity may be neglected. We have then

$$c = a(h+2l);$$

$$\Delta h_1 = \{(m-g_2)l+(m-\tfrac{1}{2}g_2)h\}\,\Delta t;$$

$$\Delta h_2 = \{(m-g_2)l-\tfrac{1}{2}g_2 h\}\,\Delta t;$$

and when $\Delta h_2 = 0$,

$$l = \frac{g_2}{2(m-g_2)} \cdot h.$$

We must now select some particular value of h for which the temperature correction is to vanish; and we shall have, upon the whole, the smallest temperature corrections for *extreme* values of h if we make the temperature correction disappear for its *mean* value. Assuming then $h = 29 \cdot 5$ inches as sufficiently near the mean atmospheric pressure, and adopting for the coefficients of cubic expansion of mercury and glass for one degree Centigrade the values

$$m = \cdot0001803, \quad g_2 = \cdot0000258,$$

we obtain

$$l = 2 \cdot 463 \text{ inches.}$$

This indicates a perfectly practicable arrangement. To render the temperature correction insensible at mean atmospheric pressures when the siphon tube has a uniform bore, we must put so much mercury into the tube, that, when the pressure is 29·5 inches, there shall be a column of about 2·5 inches of mercury in the open leg. The temperature corrections throughout all ordinary fluctuations of atmospheric pressure for the *lower* surface of the mercury will then be extremely small, as will be seen by the following Table:—

Atmospheric pressure (*h*).	Column of mercury in open leg of siphon (*l*)	Displacement of upper surface for a difference of temperature Δt in Centigrade degrees.	Displacement of lower surface of mercury for difference of temp. Δt in Cent. degrees.
31 inches.	1·713 inch.	$+\cdot00545 \Delta t$	$-\cdot00014 \Delta t$
29·5 ,,	2·463 ,,	$+\cdot00531 \Delta t$	$\cdot00000 \Delta t$
28 ,,	3·213 ,,	$+\cdot00518 \Delta t$	$+\cdot00014 \Delta t$

It may be well to observe that the numbers in the last column of the Table show that if the barometer to which they refer were heated, as in the experiments already described, and if the atmospheric pressure were much greater than 29·5 inches, we should have the result obtained by M. Negretti—a depression of the mercury in the short leg of the siphon; while if the pressure were nearly 29·5, there would be no sensible change of level, as observed by Mr. Bryson.

I need scarcely remind the reader, in conclusion, that the formulæ I have investigated are intended to be employed when *only* the upper or *only* the lower surface of the mercury is observed. When *both* surfaces are observed, as in the Bunten barometer, we have simply to apply the ordinary and well-understood correction, due to the expansion of mercury by heat.

United College, St. Andrews,
February 16, 1861.

XXXIII. *Note on* Mr. Jerrard's *Researches on the Equation of the Fifth Order. By* A. CAYLEY, *Esq.*[*]

FUNCTIONS of the same set of quantities which are, by any substitution whatever, simultaneously altered or simultaneously unaltered, may be called *homotypical.* Thus all symmetric functions of the same set of quantities are homotypical: $(x+y-z-w)^2$ and $xy+zw$ are homotypical, &c.

It is one of the most beautiful of Lagrange's discoveries in the theory of equations, that, given the value of any function of the roots, the value of any homotypical function may be rationally determined[†]; in other words, that any homotypical function whatever is a rational function of the coefficients of the equation and of the given function of the roots.

The researches of Mr. Jerrard are contained in his work, "An Essay on the Resolution of Equations," London, Taylor and Francis, 1859. The solution of an equation of the fifth order is made to depend on an equation of the sixth order in W; and he conceives that he has shown that one of the roots of this equation is a rational function of another root: "The equation for W will therefore belong to a class of equations of the sixth degree, the resolution of which can, as Abel has shown, be effected by means of equations of the second and third degrees; whence I infer the possibility of solving any proposed equation of the fifth degree by a finite combination of radicals and rational functions."

The above property of rational expressibility, if true for W, will be true for any function homotypical with W; and conversely. I proceed to inquire into the form of the function W.

The function W is derived from the function P, which denotes any one of the quantities p_1, p_2, p_3. And if x_1, x_2, x_3, x_4, x_5 are the roots of the given equation of the fifth order, and if α, β, γ, δ, ϵ represent in an undetermined or arbitrary order of succession the five indices 1, 2, 3, 4, 5, and if ι denote an imaginary fifth root of unity (I conform myself to Mr. Jerrard's notation), then p_1, p_2, p_3 and the other auxiliary quantities t, u, are obtained from the system of equations—

---

* Communicated by the Author.

† The *à priori* demonstration shows the cases of failure. Suppose that the roots of a biquadratic equation are 1, 3, 5, 9; then, given $a+b=8$, we know that either $a=3$, $b=5$, or else $a=5$, $b=3$, and in either case $ab=15$; hence in the present case (which represents the general case), $a+b$ being known, the homotypical function ab is rationally determined. But if the roots are 1, 3, 5, 7 (where $1+7=3+5$), then, given $a+b=8$, this is satisfied by $\left(\begin{smallmatrix}a=3\\b=5\end{smallmatrix}\right)$ or by $\left(\begin{smallmatrix}a=1\\b=7\end{smallmatrix}\right)$, and the conclusion is $ab=15$ or 7; so that here ab is determined, not as before, rationally, but by a quadratic equation.

$$x_\alpha^3 + p_1 x_\alpha^2 + p_2 x_\alpha + p_3 = \quad t + \quad u,$$
$$x_\beta^3 + p_1 x_\beta^2 + p_2 x_\beta + p_3 = \iota t + \iota^4 u,$$
$$x_\gamma^3 + p_1 x_\gamma^2 + p_2 x_\gamma + p_3 = \iota^2 t + \iota^3 u,$$
$$x_\iota^3 + p_1 x_\iota^2 + p_2 x_\iota + p_3 = \iota^3 t + \iota^2 u,$$
$$x_\epsilon^3 + p_1 x_\epsilon + p_2 x_\epsilon + p_3 = \iota^4 t + \iota u.$$

If from these equations we seek for the values of $p_1, p_2, p_3, t, u,$ we have

$$1 : p_1 : p_2 : p_3 : -t : -u = \Pi_1 : \Pi_2 : \Pi_3 : \Pi_4 : \Pi_5 : \Pi_6,$$

where $\Pi_1, \Pi_2 \ldots$ denote the determinants formed out of the matrix

$$\begin{vmatrix} x_\alpha^3, & x_\alpha^2, & x_\alpha, & 1, & 1, & 1 \\ x_\beta^3, & x_\beta^2, & x_\beta, & 1, & \iota, & \iota^4 \\ x_\gamma^3, & x_\gamma^2, & x_\gamma, & 1, & \iota^2, & \iota^3 \\ x_\iota^3, & x_\iota^2, & x_\iota, & 1, & \iota^3, & \iota^2 \\ x_\epsilon^3, & x_\epsilon^2, & x_\epsilon, & 1, & \iota^4, & \iota \end{vmatrix}$$

i.e., denoting the columns of this matrix by 1, 2, 3, 4, 5, 6, we have $\Pi_1 = 23456$, $\Pi_2 = -34561$, $\Pi_3 = 45612$, &c. In particular, the value of Π_1 is

$$= \begin{vmatrix} x_\alpha^2, & x_\alpha, & 1, & 1, & 1 \\ x_\beta^2, & x_\beta, & 1, & \iota, & \iota^4 \\ x_\gamma^2, & x_\gamma, & 1, & \iota^2, & \iota^3 \\ x_\iota^2, & x_\iota, & 1, & \iota^3, & \iota^2 \\ x_\epsilon^2, & x_\epsilon, & 1, & \iota^4, & \iota \end{vmatrix}$$

And developing, and putting for shortness $\{\alpha\beta\} = x_\alpha x_\beta (x_\alpha - x_\beta)$, &c., we have

$$\Pi_1 = (\{\alpha\beta\} + \{\beta\gamma\} + \{\gamma\delta\} + \{\delta\epsilon\} + \{\epsilon\alpha\})(-2\iota + \iota^2 - \iota^3 + 2\iota^4)$$
$$+ (\{\alpha\gamma\} + \{\gamma\epsilon\} + \{\epsilon\beta\} + \{\beta\delta\} + \{\delta\alpha\})(+\iota + 2\iota^2 - 2\iota^3 - 2\iota^4).$$

And this is also the form of the other determinants, the only difference being as to the meaning of the symbol $\{\alpha\beta\}$, which, however, in each case denotes a function such that $\{\alpha\beta\} = -\{\beta\alpha\}$. Writing for greater shortness,

$$\{\alpha\beta\gamma\delta\epsilon\} = \{\alpha\beta\} + \{\beta\gamma\} + \{\gamma\delta\} + \{\delta\epsilon\} + \{\epsilon\alpha\},$$

and in like manner

$$\{\alpha\gamma\epsilon\beta\delta\} = \{\alpha\gamma\} + \{\gamma\epsilon\} + \{\epsilon\beta\} + \{\beta\delta\} + \{\delta\alpha\},$$

Π_1 is an unsymmetric linear function (without constant term) of

$\{\alpha\beta\gamma\delta_\epsilon\}$, $\{\alpha\gamma\epsilon\beta\delta\}$; or, what is all that is material, it is an unsymmetric function, containing only odd powers, of $\{\alpha\beta\gamma\delta\epsilon\}$, $\{\alpha\gamma\epsilon\beta\delta\}$.

If for
$$\alpha \quad \beta \quad \gamma \quad \delta \quad \epsilon$$
we substitute any one of the five arrangements
$$\alpha \ \beta \ \gamma \ \delta \ \epsilon,$$
$$\beta \ \gamma \ \delta \ \epsilon \ \alpha,$$
$$\gamma \ \delta \ \epsilon \ \alpha \ \beta,$$
$$\delta \ \epsilon \ \alpha \ \beta \ \gamma,$$
$$\epsilon \ \alpha \ \beta \ \gamma \ \delta,$$
then $\{\alpha\beta\gamma\delta\epsilon\}$ and $\{\alpha\gamma\epsilon\beta\delta\}$ will in each case remain unaltered.

But if we substitute any one of the five arrangements
$$\alpha \ \epsilon \ \delta \ \gamma \ \beta,$$
$$\epsilon \ \delta \ \gamma \ \beta \ \alpha,$$
$$\delta \ \gamma \ \beta \ \alpha \ \epsilon,$$
$$\gamma \ \beta \ \alpha \ \epsilon \ \delta,$$
$$\beta \ \alpha \ \epsilon \ \delta \ \gamma,$$
then in each case $\{\alpha\beta\gamma\delta\epsilon\}$ and $\{\alpha\gamma\epsilon\beta\delta\}$ will be changed into $-\{\alpha\beta\gamma\delta\epsilon\}$ and $-\{\alpha\gamma\epsilon\beta\delta\}$ respectively. Hence Π_1 remains unaltered by any one of the first five substitutions ; and it is changed into $-\Pi_1$ by any one of the second five substitutions. And the like being the case as regards Π_2, &c., it follows that the quotient $\dfrac{\Pi_1}{\Pi_2}$, or say P, remains unaltered by any one of the ten substitutions. Now the 120 permutations of α, β, γ, δ, ϵ can be obtained as follows, viz. by forming the 12 different pentagons which can be formed with α, β, γ, δ, ϵ (treated as five points), and reading each of them off in either direction from any angle. To each of the 12 pentagons there corresponds a distinct value of P, but such value is not altered by the different modes of reading off the pentagon; P is consequently a 12-valued function.

But there is a more simple form of the analytical expression of such a 12-valued function; in fact, if $[\alpha\beta\gamma\delta\epsilon]$ be any function which is not altered by any one of the above ten substitutions—if, for instance, $[\alpha\beta]$ is a symmetrical function of x_α, x_β, and
$$[\alpha\beta\gamma\delta\epsilon] = [\alpha\beta] + [\beta\gamma] + [\gamma\delta] + [\delta\epsilon] + [\epsilon\alpha],$$
and $\therefore$
$$[\alpha\gamma\epsilon\beta\delta] = [\alpha\gamma] + [\gamma\epsilon] + [\epsilon\beta] + [\beta\delta] + [\delta\alpha],$$
then any unsymmetrical function of $[\alpha\beta\gamma\delta\epsilon]$ and $[\alpha\gamma\epsilon\beta\delta]$ will be a 12-valued function homotypical with P.

Mr. Jerrard's function W is the sum of two values of his function P; the substitution by which the second is derived from the first can only be that which interchanges the two functions $[\alpha\beta\gamma\delta\epsilon]$ and $[\alpha\gamma\epsilon\beta\delta]$; and hence any symmetrical function of $[\alpha\beta\gamma\delta\epsilon]$ and $[\alpha\gamma\epsilon\beta\delta]$ is a function homotypical with Mr. Jerrard's W; such symmetric function is in fact a 6-valued function only. Indeed it is easy to see that the twelve pentagons correspond together in pairs, either pentagon of a pair being derived from the other one by *stellation*, and the six values of the function in question corresponding to the six pairs of pentagons respectively.

Writing with Mr. Cockle and Mr. Harley,

$$\tau = x_\alpha x_\beta + x_\beta x_\gamma + x_\gamma x_\delta + x_\delta x_\epsilon + x_\epsilon x_\alpha,$$

$$\tau' = x_\alpha x_\gamma + x_\gamma x_\epsilon + x_\epsilon x_\beta + x_\beta x_\delta + x_\delta x_\alpha,$$

then ($\tau + \tau'$ is a symmetrical function of all the roots, and it must be excluded; but) $(\tau - \tau')^2$ or $\tau \tau'$ are each of them 6-valued functions of the form in question, and either of these functions is linearly connected with the Resolvent Product. In Lagrange's general theory of the solution of equations, if

$$f\iota = x_1 + \iota x_2 + \iota^2 x_3 + \iota^3 x_4 + \iota^4 x_5,$$

then the coefficients of the equation the roots whereof are $(f\iota)^5$, $(f\iota^2)^5$, $(f\iota^3)^5$, $(f\iota^4)^5$, and in particular the last coefficient $(f\iota f\iota^2 f\iota^3 f\iota^4)^5$, are determined by an equation of the sixth degree; and this last coefficient is a perfect fifth power, and its fifth root, or $f\iota f\iota^2 f\iota^3 f\iota^4$, is the function just referred to as the Resolvent Product.

The conclusion from the foregoing remarks is that *if the equation for* W *has the above property of the rational expressibility of its roots*, the equation of the sixth order resulting from Lagrange's general theory has the same property.

I take the opportunity of adding a simple remark on cubic equations. The principle which furnishes what in a foregoing foot-note is called the *à priori* demonstration of Lagrange's theorem is that an equation need never contain extraneous roots; a quantity which has only one value will, if the investigation is properly conducted, be determined in the first instance by a linear equation; one which has two values by a quadratic equation, and so on; there is always enough, and not more than enough, to determine what is required.

Take Cardan's solution of the cubic equation $x^3 + qx - r = 0$, we have $x = a + b$, and thence $3ab = -q$, $a^3 + b^3 = r$; and to obtain the solution we write

$$a^3 b^3 = -\frac{q}{27}, \quad a^3 + b^3 = r.$$

But *these* two equations are not enough to precisely determine x, they lead to the 9-valued function

$$\sqrt[3]{\frac{r}{2}+\sqrt{\frac{r^2}{4}+\frac{q^3}{27}}}+\sqrt[3]{\frac{r}{2}-\sqrt{\frac{r^2}{4}+\frac{q^3}{27}}}:$$

in order to precisely determine x, it is (as everybody knows) necessary to use the original equation $ab=-\frac{q}{3}$. But seek for the solution as follows; viz. write $x=ab(a+b)$, which gives

$$8a^3b^3=-q, \quad a^3b^3(a^3+b^3)=r,$$

or what is the same thing,

$$a^3b^3=-\frac{q}{3}, \quad a^3+b^3=-\frac{3r}{q};$$

these equations give $x=ab(a+b)$, where

$$a=\sqrt[3]{-\frac{3r}{2q}+\sqrt{\frac{9r^2}{4q^2}+\frac{q}{3}}}, \quad b=\sqrt[3]{-\frac{3r}{2q}-\sqrt{\frac{9r^2}{4q^2}+\frac{q}{3}}},$$

which is a 3-valued function only, ab in this case being not given.

2 Stone Buildings, W.C.,
    January 28, 1861.

XXXIV. *On some further applications of the Ferrocyanide of Po-tassium in Chemical Analysis.* By EDMUND W. DAVY, *A.B., M.B., M.R.I.A., Professor of Agriculture and Agricultural Chemistry to the Royal Dublin Society*[*].

I HAVE recently been engaged in making some experiments on the ferrocyanide of potassium or yellow prussiate of potash, with a view to extend its applications in chemical analysis; for though this important salt has already been applied to a number of useful purposes in analytical research, still my experiments have shown me that its use might be advantageously extended, particularly as a reagent in volumetric analysis, a form of analysis which has of late come into very general adoption, especially for technical purposes, on account of the great quickness and at the same time accuracy with which different substances may by its means be determined. The principles upon which volumetric analysis depend are so well known, that I need not refer to them; and though it possesses so many advantages over the older gravimetrical method, in which the different substances are determined by weight instead of by volume, it yet has this drawback, that the preparation of the necessary standard solutions often takes considerable time, first, in order

[*] Part of a paper read before the Royal Dublin Society, December 17, 1860; and communicated by the Author.

to obtain the substance to be used for this purpose in a sufficiently pure and dry state, and secondly, to form a solution of it the exact strength of which may be known: for though it may appear a very simple operation to dissolve a known weight of a certain substance in a given bulk of water or other solvent, yet, when this has to be done with such great precision as is necessary in these cases, it is a tedious and troublesome operation, and any inaccuracy in the graduation of the standard solution will render all determinations made with it more or less inaccurate. It is obvious, therefore, that it would be most desirable that the substances which are intended to be used as reagents in volumetric analysis should be easily obtained in a pure state, and that where considerable time and trouble have been expended in graduating solutions of those substances, they should not be liable to undergo changes whereby their strength would be more or less altered, but that when standard solutions have once been made, they might be kept and used for a great number of determinations.

The ferrocyanide of potassium fulfils both those conditions; for it is in general met with in commerce almost chemically pure, and in a state in which it can at once be employed as a volumetric reagent; and if at any time it should happen to occur not quite so pure, it can readily be purified by recrystallization; and in addition to these important considerations, its solution is not prone to change, especially if it be not left exposed to the action of the light. In this latter respect it has a decided advantage over several of our most useful volumetric reagents, viz. the permanganate of potash, the protosalts of iron, sulphurous acid, &c., which, from their being so prone to undergo spontaneous decomposition, must be either freshly prepared, or the strength of their solutions accurately ascertained every time they are used, if a day or so has elapsed between each determination.

The employment of the ferrocyanide of potassium as a volumetric reagent depends on the following circumstances: viz., that it is readily converted into the ferridcyanide of potassium (red prussiate of potash) under different circumstances, and that the point where the whole of the former salt has been changed into the latter may easily be known, either by the use of a diluted solution of a persalt of iron (which gives with a drop of the mixture a blue or green coloration as long as any of the ferrocyanide remains unchanged) or by some other simple indication. Thus, for example, when chlorine is brought in contact with the ferrocyanide of potassium, this change, as is well known, takes place, which is expressed by the following symbols:—

$$2 (K^2, Fe\,Cy^3) + Cl = (K^3\,Fe^2\,Cy^6) + K\,Cl.$$

The same occurs, as far as the conversion of the ferrocyanide into ferridcyanide, when an acidified solution of the former salt is brought in contact with a solution of the permanganate of potash, which is instantly decolorized by the reducing action of the ferrocyanide of potassium, which is thereby converted into the ferridcyanide, and this decoloration of the permanganate continues as long as any of the ferrocyanide remains in the mixture.

Again, if a solution of the ferrocyanide of potassium, acidified strongly with either hydrochloric or sulphuric acid, be brought in contact with a solution of the bichromate of potash, the same change of the ferrocyanide into the ferridcyanide immediately takes place.

The first reaction has been long known, and ⠄ the means employed at present for obtaining the ferridcyaniq[3] or red prussiate of potash for manufacturing and other purposes; the second reaction has been more recently discovered; but I am not aware that the third, in the case of the bichromate, is generally known, or that the changes which occur in the reaction have been previously studied.

From experiments which I made, it would appear that when a solution of ferrocyanide of potassium, acidified with hydrochloric acid, was mixed with one of the bichromate of potash, the following reaction was produced, viz. $6(K^2 Fe Cy^3) + KO, 2 Cr O^3 + 7HCl = 8(K^3 Fe^2 Cy^6) + 4 KCl + Cr^2 Cl^3 + 7HO$; for, amongst other facts, I may observe that when I mixed together solutions of the two salts in the proportions corresponding to 6 equivalents of the ferrocyanide of potassium to 1 of the bichromate of potash (as indicated in the above formula), acidifying the mixture with hydrochloric acid, I found that the whole of the ferrocyanide was converted into the ferridcyanide, and that any quantity less than that proportion of the bichromate of potash left more or less of the ferrocyanide unchanged. The same results followed the use of sulphuric acid; and it appears that a similar reaction occurs with this acid as with hydrochloric acid, with the exception that in this case the 4 equivalents of chloride of potassium and the 1 equivalent of sesquichloride of chromium ⠤ are replaced by 4 equivalents of sulphate of potash and 1 of the sesquisulphate of chromium.

The proportion of either acid used, provided there is enough to strongly acidify the mixture, does not appear to affect the reaction; for I obtained precisely the same results where a very large amount of acid was employed as where the quantity necessary only to strongly acidify the mixture had been added.

On these three reactions which I have noticed, may be based the means of employing the ferrocyanide of potassium in several useful determinations, the first, and one of the most important, of which is *the ascertaining the amount of available chlorine in the chloride of lime or bleaching powder,* which is a matter of much importance in many of the chemical arts, but particularly in bleaching; for not only does the commercial value of this substance depend on the quantity of available chlorine that it contains, which is subject to great variation from exposure to the air and other causes, but likewise it is of the greatest importance that the bleacher should readily be able to determine from time to time the strength of the bleaching liquor which he employs: for if it be too strong, he knows that the fabric which he bleaches will be injured; and if too weak, it will not be sufficiently bleached, and the process must be repeated, which incurs much additional expenditure of time.

Various methods have from time to time been proposed for the determination of the value of chloride of lime; but the greater number of them, from the trouble required to make the test-solutions, and their not keeping when made, as well as the skill required in their use, render them inapplicable for general purposes.

I shall therefore merely refer to the two methods which are chiefly used at present to determine the value of this important substance. The first is Gay-Lussac's, in which the amount of chlorine is ascertained by seeing how much chloride of lime is necessary to convert a given quantity of arsenious into arsenic acid; the second is Otto's, in which protosulphate of iron is substituted for arsenious acid, and the determination of chlorine is made by seeing how much of the bleaching powder is required to change a given weight of the protosulphate of iron into a persalt of that metal: these processes are so well known that I need not describe them.

In both these methods I find that more or less chlorine is always lost, which, however, may be reduced to a minute quantity by very carefully adding the solution of chloride of lime· either to that of arsenious acid, or of protosalt of iron; but in ordinary hands they (especially the latter process) will yield results in which too small a proportion of chlorine will be indicated, from the loss of that substance which will invariably take place.

The ferrocyanide of potassium answers admirably for the estimation of available chlorine in the chloride of lime, when used in the manner I shall presently explain, and according to my experiments will give in ordinary hands far more accurate results than either Gay-Lussac's or Otto's method. I am aware, indeed, that this salt was proposed by Mr. Mercer some years ago for this. purpose; but the way which he recommended it to be used (which consisted in dissolving a certain weight of the ferrocyanide in water, acidifying it, and then adding the solution of bleaching powder from a burette till all the ferrocyanide was converted into ferridcyanide) is, I find, not a good manner of employing the ferrocyanide in this estimation, and, like the other methods, will lead to a loss of chlorine; for when the solution of chloride of lime is added to the acidified ferrocyanide, a portion of the chlorine is separated, especially if the bleaching liquor be added too quickly, or is not greatly diluted. But the way I propose of using the ferrocyanide of potassium in this important valuation, is to mix together a certain quantity of a standard solution of ferrocyanide with a given amount of a graduated solution of the chloride of lime, using more of the former salt than the latter can convert into ferridcyanide; then adding hydrochloric acid to dissolve the precipitate formed and render the mixture strongly acid, and finally ascertain, by means of a standard solution of bichromate of potash, how much of the ferrocyanide remained unconverted into the ferridcyanide by the action of the chlorine of the chloride of lime,—which is effected by adding slowly from a graduated burette the standard solution of bichromate till a minute drop taken from the well-stirred mixture by means of a glass rod, ceases to give, with a small drop of a very dilute solution of perchloride of iron placed on a white plate, a blue or greenish colour, but produces instead a yellowish brown*. When this latter effect is observed, it indicates that all the ferrocyanide has been converted into ferridcyanide; and as 147·59 (one equivalent) of bichromate of potash is capable of converting 1267·32 (six equivalents) of crystallized ferrocyanide of potassium into ferridcyanide, and as 422·44 (two equivalents) of the ferrocyanide are converted into the same substance by 35·5 (one equivalent) of chlorine, as is seen by the formulæ already given, knowing the amount of chloride of lime employed, we have all the data necessary to calculate the per-centage of chlorine.

Having made two standard solutions, the first containing 21·122 grammes of ferrocyanide of potassium in a litre of the solution, and the second 14·759 grammes of bichromate of potash in the same quantity of solution (weights which are to each other as their atomic equivalents), I made several estimations of chloride of lime with them, adopting the method I have just described, and found that it gave the most consistent results, and which agreed very closely with those obtained by Gay-Lussac's and Otto's methods when the latter were performed with the greatest care,—the only difference being that the results obtained by

* The yellowish-brown coloration which is at first produced when enough of the bichromate has been added, quickly changes to a greenish colour by some secondary reactions which take place when the persalt of iron is left in contact with the mixture. But this does not interfere with the test; for it is the first effect which is produced which indicates the completion of the reaction, and not the after changes which may result.

my method indicated a few hundredths of a part more of chlorine than either of those methods did, which may be accounted for by the unavoidable loss of a minute quantity of chlorine which takes place in those processes.

In order to simplify the process, and render the calculation as short as possible, I would recommend for commercial valuations the following way of carrying out this principle:— Having obtained a flat-bottom flask or bottle which will contain 10,000 grains of distilled water when filled up to a certain mark in the neck, make two standard solutions, the first by placing in the flask or bottle 1190 (or exactly 1189·97*) grains of the purest crystallized ferrocyanide of potassium (yellow prussiate of potash) reduced to powder, adding distilled water to dissolve the salt, and when this is effected, filling up with water to the mark; and having mixed the solution thoroughly, place it in a well-stoppered bottle. The second standard solution is made in the same manner, substituting for the ferrocyanide 138·6 (or exactly 138·58) grains of bichromate of potash which has been purified by recrystallization and fused in a crucible at as low a heat as possible. Both these solutions will keep unchanged, and will answer for a number of determinations if they are preserved in well-stoppered bottles, and the ferrocyanide solution be kept, when not in use, excluded from the light. Get a burette or alkalimeter capable of holding or delivering 1000 grains of distilled water, and divided into 100 equal divisions; also two small bottles, one capable of delivering 1000 grains, and the other 500 grains of distilled water when filled up to a certain mark on the neck of each†, which may both be readily made by filling them with water, emptying them, and after they have drained for a minute or two, weighing into each the above weights of distilled water; or, what will be sufficiently accurate for most purposes, pour from the burette into one 100 divisions of distilled water, and into the other 50, and mark with a file where the fluid stands in the neck of each bottle. Having these all ready, take an average specimen of chloride of lime, and weigh out 100 grains of it, and make in the usual way a solution of it by trituration in a mortar with some

---

* The above numbers are obtained as follows:—35·5 parts of chlorine are capable, as before stated, of converting 422·44 parts of the crystallized ferrocyanide of potassium into ferridcyanide; therefore 100 parts of the former will convert 1189·97 parts of the latter into the same compound. Again, as before observed, 1267·32 parts of the crystallized ferrocyanide require 147·59 parts of the bichromate of potash to convert them into the ferridcyanide; 1189·97 parts, therefore, will take 138·58 parts of that salt to produce the same effect.

† Two small pipettes capable of delivering the above quantities would be found still more convenient.

water; pour it into the flask which was used in preparing the two standard solutions, and having filled up with water to the mark in the neck, mix the solution thoroughly; and before each time that any of the chloride of lime is taken out, shake well the contents of the flask.

Measure out into a beaker-glass, by means of the two little bottles, 100 divisions of the chloride of lime solution, and 50 of the standard solution of ferrocyanide; and having mixed them well together, add some hydrochloric acid to dissolve the precipitate formed and acidify the mixture strongly; and having mixed the whole well, pour from the burette slowly the standard solution of bichromate (stirring well all the while) till a drop taken from the mixture and brought in contact with a drop of a very weak solution of perchloride of iron produces a yellowish-brown colour, as already noticed. Then read off the number of divisions of the standard solution of bichromate which was necessary to produce this effect; and this being deducted from 50, gives the per-centage by weight of chlorine.

For the standard solution of ferrocyanide having been made so that the 10000-grain measures should be equivalent to 100 grains of chlorine, and as every division of the burette equals 10 grains, each of these divisions of the ferrocyanide solution converted into ferridcyanide will indicate 0·1 grain of chlorine. Again, the 100 divisions of the solution of chloride of lime represent 10 grains of that substance, and we want to know how many divisions of the ferrocyanide solution its chlorine has converted into ferridcyanide. This is readily ascertained by the bichromate solution, which has been so graduated that each division represents a division of the ferrocyanide solution. So that to determine the per-centage of chlorine we have only to deduct, as before stated, the number of divisions of the bichromate solution employed from the 50 of the ferrocyanide solution, and the difference gives us the per-centage of chlorine by weight in the sample; thus in four experiments 50 divisions of the ferrocyanide solution, mixed with 100 divisions of the solution of chloride of lime, required 18·5 divisions of the bichromate solution to convert the whole of the ferrocyanide employed into ferridcyanide; this number, taken from 50, leaves 31·5 divisions of ferrocyanide, which were converted into ferridcyanide by the chlorine of the chloride of lime; and as each division represents 0·1 grain of chlorine, 31·5 will be equivalent to 3·15 grains of chlorine, which is the amount contained in 10 grains of the sample; consequently 100 grains will contain 31·5 grains of chlorine, which is the same amount as is obtained by simply deducting the number of divisions of bichromate solution employed from 50 of ferrocyanide used in the estimation.

Though this process appears a long one, from the details which are necessary to explain its principle, yet in practice it is very expeditious, and requires only a very few minutes for its performance, and is much quicker than either Gay-Lussac's or Otto's method.

Though I have as yet chiefly confined my attention to the use of the ferrocyanide of potassium in the estimation of chlorine in bleaching powder, I have no doubt that it may be advantageously employed in many other useful determinations by carrying out the principles already explained: thus, for example, it may be used as a means of determining the amount of bichromate of potash present in a sample of that salt, or the quantity of chromic acid that exists under different circumstances. Again, the same salt may be used in different determinations where a certain amount of chlorine is liberated, which represents a proportional quantity of some other substance: thus, for example, in the estimation of manganese ores for commercial purposes, where they are heated with hydrochloric acid, the quantity of chlorine disengaged will indicate a certain amount of peroxide of manganese in the ore, on the presence of which its commercial value almost entirely depends; and the chlorine evolved may be estimated by absorbing the gas in a dilute solution of caustic potash, and then determining the amount of chlorine in it by precisely the same process as that I have recommended in the valuation of chloride of lime. To test the accuracy of this method, I heated in a small flask a given quantity of pure bichromate of potash with an excess of strong hydrochloric acid, and collected the evolved chlorine by means of a dilute solution of caustic potash, employing the bulbed retort and curved dropping tube as recommended by Bunsen in the "Analysis of the Chromates" (see the last edition of Fresenius's 'Quantitative Analysis,' page 234), and ascertained afterwards, by the use of the ferrocyanide of potassium, the amount of chlorine evolved, which corresponded almost exactly with the calculated amount of that substance which should have been obtained by the action of the quantity of bichromate used on the hydrochloric acid. Again, a standard solution of ferrocyanide of potassium may be used, as E. de Haen has shown, to determine the strength of the permanganate of potash in the analyses of the ferrocyanide and ferridcyanide of potassium, as an acidified solution of the ferrocyanide, as before stated, rapidly decolorizes a solution of permanganate of potash, whereas the ferridcyanide has no action on that salt; and this reaction might be taken advantage of, in the valuation of chloride of lime, to determine the excess of ferrocyanide used in my process: but from my experiments I found that more precise and accurate results were obtained by the use of the bichromate of potash.

The reaction of the bichromate of potash on the ferrocyanide might be employed in the valuation of the ferrocyanide of potassium and other ferrocyanides—having previously, in the case of those which were insoluble, converted them into the ferrocyanide of potassium by boiling them with caustic potash, and separating the insoluble oxides by filtration.

It might also be employed for the valuation of the commercial red prussiate of potash, which is now to some extent employed as a bleaching agent in calico-printing, and which consists of varying quantities of ferro- and ferrid-cyanide of potassium together with chloride of potassium. By ascertaining first how much a given quantity of the sample requires of a standard solution of bichromate of potash to convert the ferrocyanide present into ferridcyanide, the per-centage of that substance would be known; and then by taking another portion of the sample and converting the ferridcyanide it contained, by reducing agents, such as the sulphites of soda and potash, &c., into the ferrocyanide, and finally determining the amount of bichromate necessary to bring the whole of the ferrocyanide then present into the state of ferridcyanide, the difference in the two results would indicate the proportion of ferridcyanide originally present in the sample.

The last application of ferrocyanide of potassium which I shall notice in the present communication, is its employment as a reducing agent. It has long been known that the cyanide of potassium possesses most powerful reducing properties, and has been very usefully employed for that purpose in the reduction of different metallic salts under various circumstances; but I am not aware that the ferrocyanide of potassium has been proposed or used for similar purposes: at least, I have referred to a great number of analytical and general chemical works, and in none of them is this salt recommended as a reducing agent, though the cyanide is so much extolled for that purpose. According to my experiments, the ferrocyanide is a far more convenient reducing agent than the cyanide, and may be substituted for it in many cases of reduction with the best results, as it possesses many unquestionable advantages over that salt for this purpose. Thus the ferrocyanide does not deliquesce and decompose when exposed to the air, whereas the cyanide rapidly absorbs moisture, and, unless kept in very well-stoppered bottles, becomes quite wet, and in this state quickly decomposes; and this deliquescence on the part of the cyanide is often a source of much inconvenience in its use as a reducing agent, owing to the almost unavoidable absorption of more or less moisture which takes place in mixing it with the substance to be reduced, and during the introduction of the mixture into the reducing tube. The ferrocyanide, on the other hand, in a thoroughly dried and finely powdered state, can be intimately mixed with the substance without any appreciable absorption of moisture. I made the following comparative experiment to ascertain the relative absorptive properties for moisture of the two salts under the same circumstances. Having thoroughly dried in a water oven, till it ceased to vary in weight, some finely powdered ferrocyanide, I placed 50 grains of it in a counterpoised watch-glass, and powdering in a warm mortar some fresh cyanide of potassium, I placed the same quantity of it in a similar counterpoised watch-glass, and left them both exposed to the air. On examining them after four hours' exposure, I found that the former had only gained $\frac{8}{100}$th parts of a grain of moisture, whereas the latter had taken up 3·6 grains, or sixty times as much moisture under the same circumstances. After two days' exposure I found that nearly all the cyanide had passed into the liquid condition, having taken up 46 grains of water; whereas the ferrocyanide appeared perfectly dry, and had only absorbed 1·4 grain.

The great fusibility of the cyanide is sometimes rather a disadvantage, which has to be lessened by mixing it with a certain proportion of dried carbonate of soda; but the ferrocyanide not fusing at so low a temperature, does not require in most cases this admixture to lessen its fusibility. Again, the ferrocyanide is not a poisonous salt, whereas the cyanide is highly so, and must be used with great caution; and lastly, the former salt is little more than half the price of the latter. Combined with the above advantages, I find that the ferrocyanide is equally effective in reducing metallic oxides and sulphurets, and is especially convenient for the reduction of different combinations of arsenic and mercury, which are reduced by it with the greatest ease.

I made several comparative experiments with the dried ferrocyanide and with the cyanide as reducing agents for the sulphuret of arsenic and arsenious acid, employing the same quantity of arsenical compound with each salt under similar circumstances; and in almost every case, particularly where the quantities operated on were minute, I obtained more satisfactory results with the dried ferrocyanide than with the cyanide.

The following were amongst my experiments:—I mixed the $\frac{1}{10}$th of a grain of sulphuret of arsenic with 3 grains of the dried ferrocyanide, and made a similar experiment, substituting the same quantity of cyanide; and on heating the mixtures in similar glass tubes, obtained almost identically fine and characteristic rings of metallic arsenic.

I then intimately mixed the same quantity of sulphuret of arsenic with 49·9 grains of very finely powdered glass, and taking 5 grains of this mixture, containing the $\frac{1}{100}$th part of a grain of the sulphuret, mixed it with 5 grains of the dried ferrocyanide, and made a comparative experiment with another 5 grains of the mixture, substituting the same quantity of cyanide; on heating both these mixtures in small reduction tubes, I got the characteristic metallic rings in both, but better defined in the case of the ferrocyanide.

I finally took 2·5 grains of the mixture of sulphuret and glass, containing about $\frac{1}{1000}$th parts of a grain of sulphuret of arsenic, and treated them in the same manner, using in one case 2·5 grains of ferrocyanide, and in the other 2·5 grains of cyanide, and obtained in each case a minute metallic ring, which, however, was much more distinct and satisfactory where the ferrocyanide had been used as the reducing agent.

The same comparative experiments were made with arsenious acid, when results similar to those in the case of the sulphuret of arsenic were obtained.

The ferrocyanide, therefore, is a most delicate reducing agent in the case of arsenical compounds, and where very minute quantities have to be detected, appears from my experiments to give more satisfactory results than the cyanide.

Whether the addition of dried carbonate of soda would improve the ferrocyanide for some cases of reduction, I am not at present able to say; but in one experiment which I made with the sulphuret of arsenic, I obtained as good results, using the ferrocyanide alone, as where it was mixed previously with its own weight of dried carbonate of soda. In many cases the ferrocyanide may be used as a reducing agent in a state of powder without separating its water of crystallization; but, in most cases, it will be rendered a far better reducing agent by being previously dried at 212° in a water-bath or oven; and in this dried condition it may be kept for any length of time in a good-stoppered or well-corked bottle.

Though as yet my experiments have been chiefly confined to the reduction of different compounds of arsenic and mercury, I entertain no doubt that the ferrocyanide of potassium will be found an equally effective reducing agent in the case of the combinations of other metals, and that it may with great advantage be substituted for the cyanide of potassium in many cases where the latter salt is used as a reducing agent.

XXXV. *Proceedings of Learned Societies.*

ROYAL SOCIETY.

[Continued from p. 153.]

April 26, 1860.—Sir Benjamin C. Brodie, Bart., Pres., in the Chair.

THE following communication was read:—

    "On the Effect of the Presence of Metals and Metalloids upon the Electric Conductivity of Pure Copper." By A. Matthiessen, Esq., and M. Holzmann, Esq.

After studying the effect of suboxide of copper, phosphorus, arsenic, sulphur, carbon, tin, zinc, iron, lead, silver, gold, &c., on the conducting power of pure copper, we have come to the conclusion *that there is no alloy of copper which conducts electricity better than the pure metal.*

May 3.—Sir Benjamin C. Brodie, Bart., President, in the Chair.

The following communications were read:—

"On the relations between the Elastic Force of Aqueous Vapour, at ordinary temperatures, and its Motive Force in producing Currents of Air in Vertical Tubes." By W. D. Chowne, M.D., F.R.C.P.

In 1853 the author of this communication made a considerable number of experiments which demonstrated that when a tube, open at both ends, was placed vertically in the undisturbed atmosphere of a closed room, there was an upward movement of the air within the tube of sufficient force to keep an anemometer of light weight in a state of constant revolution, though with a variable velocity. An abstract of the results of these experiments was printed in the Philosophical Magazine, vol. xi. p. 227.

In order to further investigate the immediate cause or nature of the force which set the machine in motion, the author instituted a series of fresh experiments.

These experiments were made in the room described in the former communication, guarded in the same manner against disturbing causes, and with such extra precautions as will be hereafter explained. The apparatus used was a tube 96 inches long and 6·75 inches uniform diameter, the material zinc. The upper extremity was open to its full extent; at the lower, the aperture was a lateral one only, into which a piece of zinc tube 3 inches in diameter, and bent once at right angles, was accurately fitted with the outer orifice upward. Within this orifice, which was about 5 inches above the level of the floor, an anemometer, described in the former paper, and weighing 7 grains, was placed in the horizontal position. About midway between the upper and the lower extremity of the tube, a very delicate differential thermometer was firmly and permanently fixed, with one bulb outside and the other inside, and the aperture through which the latter was inserted completely closed. The scale was on the stem of the outer bulb.

The results of a long series of observations were recorded. The state of the dry and the wet bulb of the hygrometer, as well as the indications of the differential thermometer, was noted, in connexion with the number of revolutions performed per minute by the anemometer. While the differential thermometer indicated the same relative differences between the heat of the atmosphere within and without the tube, the velocity of the revolutions was found to vary considerably. This variation was discovered to be chiefly, if not wholly, dependent on the *elasticity* of vapour, due to the hygrometrical state of the atmosphere, as estimated from the dry- and the wet-bulb thermometers, and calculated from the tables of Regnault.

240 observations were recorded and afterwards separated into groups, each group comprising those in which the differential thermometer gave the same indication.

If in either of these groups we separate into two classes the cases in which the elasticity was highest, from the cases in which it was. lowest, and multiply the mean of each with the corresponding mean of the number of the revolutions of the anemometer, their product is nearly a constant, thus showing that the velocity of ascent of the atmospheric vapour is inversely as its elasticity; and hence it follows that the velocity of the ascending current in the tube varies inversely as the density or elastic force of the vapour suspended in the atmosphere. This was rendered evident by the aid of Tables appended to the paper.

When the mean elastic force of vapour calculated from the dry and the wet bulbs is multiplied by the constant, 13·83, the result gives the whole amount of water in a vertical column of the atmosphere in inches; it follows therefore that when the difference of temperature between the external air and that in the tube, as shown by the differential thermometer, is constant, the velocity of the current in the tube varies inversely as the weight of the vapour suspended in the atmosphere.

In an Appendix the author describes some additional experiments, made with the view of ascertaining whether the readings of the differential thermometer were mainly due to actual changes of temperature within the tube, or to extraneous causes acting on the external bulb. He found that when the external bulb was covered with woollen cloth or protected by a zinc tube of about 4 inches diameter and 6 inches long, the temperature of the bulb was increased about 2° on the scale of the instrument, and that when they were removed the prior reading was restored, while the number of revolutions of the anemometer per minute was not appreciably affected by the change. This explains why the readings of the differential thermometer varied from 33°·0 to 33°·5 as described in the paper, without producing a corresponding change in the velocity of the anemometer.

For the purpose of obtaining a more correct estimate of the influence of a given increase of heat within the tube, the author introduced into the tube at its lowest extremity, a phial containing eight ounces of water at the temperature of 100° Fahr., corked so that no vapour could escape. The result showed that in thirteen observations a quantity of heat equal to an increase of one-tenth of a degree on the scale of the differential thermometer, was equivalent to a mean velocity of the anemometer of 3·6 revolutions per minute, the greatest number being 3·8, the least 3·3 per minute.

These observations render it still more evident, that if a higher temperature within the tube had been the main cause of the revolutions of the anemometer, the variations in their velocity would not have been in such exact relation to the elastic force of the atmospheric vapour, as has been shown to be the case. They also lead to the inference, that the apparent excess of heat within the tube alluded to by the author in his Paper read before the Society in 1855 did not really exist, and to the conclusion that, if such excess had been present, the anemometer would not have been brought to a state of rest by depriving the air of the room of a portion of the moisture ordinarily suspended in it.

"On the Relation between Boiling-point and Composition in Organic Compounds." By Hermann Kopp.

The author was the first to observe (in 1841) that, on comparing pairs of analogous organic compounds, the same difference in boiling-point corresponds frequently to the same difference in composition. This relation between boiling-point and composition, when first pointed out, was repeatedly denied, but is now generally admitted. The continued experiments of the author, as well as of numerous other inquirers, have since fixed many boiling-points which had hitherto remained undetermined, and corrected such as had been inaccurately observed. In the present paper the author has collected his experimental determinations, and has given a survey of all the facts satisfactorily established up to the present moment regarding the relations between boiling-point and composition.

The several propositions previously announced by the author were :—

1. An alcohol, $C_n H_{n+2} O_2$, differing in composition from ethylic alcohol ($C_4 H_6 O_2$, boiling at 78° C.) by $x C_2 H_2$, more or less, boils $x \times 19°$ higher or lower than ethylic alcohol.

2. The boiling-point of an acid, $C_n H_n O_4$, is 40° higher than that of the corresponding alcohol, $C_n H_{n+2} O_2$.

3. The boiling-point of a compound ether is 82° higher than the boiling-point of the isomeric acid, $C_n H_n O_4$.

These propositions supply the means of calculating the boiling-points of all alcohols, $C_n H_{n+2} O_2$; of all acids, $C_n H_n O_4$; of all compound ethers, $C_n H_n O_4$. The author contrasts the values thus calculated for these substances with the available results of direct observation. The Table embraces eight alcohols, $C_n H_{n+2} O_2$, nine acids, $C_n H_n O_4$, and twenty-three compound ethers, $C_n H_n O_4$; the calculated boiling-points agree, as a general rule, with those obtained by experiment, as well as two boiling-points of one and the same substance determined by different observers. We are thus justified in assuming that the calculated boiling-point of other alcohols, acids, and ethers belonging to this series will also be found to coincide with the results of observation.

The boiling-points of other monatomic alcohols, $C_n H_m O_2$, other monatomic acids, $C_n H_m O_4$, and other compound ethers, $C_n H_m O_4$, are closely allied with the series previously discussed. A substance containing $x C$ more or less than the analogous term of the previous class, in which the same number of oxygen and of hydrogen equivalents is present, boils $x \times 14°·5$ higher or lower ; or, what amounts to the same thing, a difference of $x H$ more or less of hydrogen lowers or raises the boiling-point by $x \times 5°$. Thus benzoic acid, $C_{14} H_6 O_4$, boils $8 \times 14°·5$ higher than propionic acid, $C_6 H_6 O_4$, or $8 \times 5°$ higher than œnanthylic acid, $C_{14} H_{14} O_4$; cinnamate of ethyl, $C_{22} H_{12} O_4$, boils $10 \times 14°·5$ higher than butyrate of ethyl, $C_{12} H_{12} O_4$, or $10 \times 5°$ higher than pelargonate of ethyl, $C_{22} H_{22} O_4$.

The author compares the boiling-points thus calculated for five alcohols, $C_n H_m O_4$; for six acids, $C_n H_m O_4$; and for sixteen compound ethers, $C_n H_m O_4$, with the results of observation. In almost all cases the concordance is sufficient.

The author demonstrates in the next place that in many series of compounds other than those hitherto considered, the elementary difference, $x C_2 H_2$, likewise involves a difference of $x \times 19°$ in the boiling-point. He further shows that on comparing the boiling-points of the corresponding terms in the several series of homologous substances hitherto considered, many other constant differences in boiling-point are found to correspond to certain differences in composition. Thus a monobasic acid is found to boil 44° higher than its ethyl compound, and 63° higher than its methyl compound ; and this constant relation holds good even for acids other than those previously examined, *e. g.* for the substitution-products of acetic acid. Also in substances which are not acids, the substitution of $C_4 H_5$ or $C_2 H_3$ for H, occasionally involves a depression of the boiling-points respectively of 44° and 63°; the relation, however, is by no means generally observed.

The author, in addition to the examples previously quoted, shows that compounds containing benzoyl ($C_{14} H_5 O_2$) and benzyl ($C_{14} H_7$) boil 78° ($= 4 \times 14^{0.5} + 4 \times 5°$) higher than the corresponding terms containing valeryl ($C_{10} H_9 O_2$) and amyl ($C_{10} H_{11}$), a relation, however, which is likewise not generally met with. He discusses, moreover, other coincidences and differences of boiling-points of compounds differing in a like manner in composition. Not in all homologous series does the elementary difference $x C_2 H_2$ involve a difference of $x \times 19°$ in boiling-point. The author shows that this difference is greater for the hydrocarbons, $C_n H_{n-6}$ and $C_n H_{n+2}$; for the acetones and aldehydes, $C_n H_n O_2$; for the so-called simple and mixed ethers, $C_n H_{n+2} O_2$; for the chlorides, bromides, and iodides of the alcohol radicals, $C_n H_{n+1}$, and for several other groups; that it is, on the contrary, smaller for the anhydrides of monobasic acids, $C_n H_{n-2} O_6$; for the ethers, $C_n H_{n-2} O_8$ (which may be formed either by the action of one molecule of a dibasic acid, $C_n H_{n-2} O_8$, upon two molecules of a monatomic alcohol, $C_n H_{n+2} O_2$, or by the action of two molecules of a monobasic acid, $C_n H_n O_4$, upon one molecule of a diatomic alcohol, $C_n H_{n+2} O_4$), and several other series.

The author thinks that the unequal differences in boiling-points corresponding in different homologous series to the elementary difference $x C_2 H_2$, are probably regulated by a more general law, which will be found when the boiling-points of many substances shall have been determined under pressures differing from those of the atmosphere.

"From the observations at present at our disposal it may be affirmed as a general rule, that in homologous compounds belonging to the same series, the differences in boiling-points are proportional to the differences in the formulæ. Exceptions obtain only in cases when terms of a particular group are rather difficult to prepare, or when the substances boil at a very high temperature, at which the observations now at our command are for the most part uncertain. Again, it may be affirmed that the difference in boiling-points, corresponding to the elementary difference $C_2 H_2$, is in a great many series $= 19°$; in some series greater, in some series less."

The author proceeds to discuss the boiling-points of isomeric compounds. He shows that in a great many cases isomeric compounds belonging to the same type, and exhibiting the same chemical character, boil at the same temperature, and that there is no reason why, for the class of bodies mentioned, this coincidence should not obtain generally. On the other hand, different boiling-points are observed in isomeric compounds possessing a different chemical character, although belonging to the same type (*e. g.* acids and compound ethers, $C_n H_n O_4$; alcohols and ethers, $C_n H_{n+2} O_2$), and in isomeric compounds belonging to different types (*e. g.* allylic alcohol and acetone).

The author shows that the determination of the boiling-point of a substance, together with an inquiry into the compounds serially allied with it by their boiling-points, constitutes a valuable means of fixing the character of the substance, the type to which it belongs, and the series of homologous bodies of which it is a term. He quotes as an illustration eugenic acid. The boiling-point of this acid, $C_{20} H_{12} O_4$, is 150°; and on comparing this boiling-point with the boiling-points of benzoic acid, $C_{14} H_6 O_4$ (boiling-point 253°), and of hydride of salicyl, $C_{14} H_6 O_4$ (boiling-point 196°), it is obvious that eugenic acid cannot be homologous to benzoic acid, whilst, on the other hand, it becomes extremely probable that it is homologous to hydride of salicyl, and consequently that it belongs rather to the aldehydes than to the acids proper.

The author, in conclusion, calls attention to the importance of considering the chemical character in comparing the boiling-points of the volatile organic bases, and shows the necessity of distinguishing between the primary, secondary, and tertiary monamines in order to exhibit constant differences of boiling-point for this class of substances. He discusses the boiling-points of the several bases, $C_n H_{n-5} N$ and $C_n H_{n+3} N$, and points out how in many cases the particular class to which a base belongs may be ascertained by the determination of the boiling-point.

The comprehensive recognition of definite relations between composition and boiling-point is for the present chiefly limited to organic compounds. But for the majority of these compounds, and indeed for the most important ones, this relation assumes the form of a simple law, which, more especially for the monatomic alcohols, $C_n H_m O_2$, for the monobasic acids, $C_n H_m O_4$, and for the compound ethers generated by the union of the two previous classes, is proved in the most general manner; so much so, indeed, that in many cases the determination of the boiling-point furnishes most material assistance in fixing the true position and character of a compound.

The author points out more especially that the simplest and most comprehensive relations have been recognized for those classes of organic compounds which have been longest known and most accurately investigated, and that even for those classes the generality and simplicity of the relation, on account of numerous boiling-points incorrectly observed at an earlier date, appeared in the commencement doubtful, and could be more fully acknowledged only after a considerable number of new determinations. Thus he considers himself justified in hoping that also in other classes of compounds, in which simple and comprehensive relations have not hitherto been traced, these relations will become perceptible as soon as the verification of the boiling-points of terms already known, and the examination of new terms, shall have laid a broader foundation for our conclusions.

May 10.—Sir Benjamin C. Brodie, Bart., President, in the Chair.

The Bakerian Lecture was delivered by Mr. Fairbairn, F.R.S. The Lecturer gave a condensed exposition of the experiments and results detailed in the following Paper. He also exhibited the apparatus employed, and explained the methods followed.

"Experimental Researches to determine the Density of Steam at all Temperatures, and to determine the Law of Expansion of Superheated Steam." By William Fairbairn, Esq., F.R.S., and Thomas Tate, Esq.

The object of these researches is to determine by direct experiment the law of the density and expansion of steam at all temperatures. Dumas determined the density of steam at 212° Fahr., but at this temperature only. Gay-Lussac and other physicists have deduced the density at other temperatures by a theoretical formula true for a perfect gas:

$$\frac{VP}{V_1 P_1} = \frac{459 + T}{459 + T_1}. \quad \cdot \quad \cdot \quad \cdot \quad \cdot \quad \cdot \quad \cdot \quad (1.)$$

On the expansion of superheated steam, the only experiments are those of Mr. Siemens, which give a rate of expansion extremely high, and physicists have in this case also generally assumed the rate of expansion of a perfect gas. Experimentalists have for some time questioned the truth of these gaseous formulæ in the case of condensable vapours, and have proposed new formulæ derived from the dynamic theory of heat; but up to the present time no *reliable direct experiments* have been made to determine either of the points at issue. The authors have sought to supply the want of data on these questions by researches on the density of steam upon a new and original method.

The general features of this method consist in vaporizing a known weight of water in a globe of about 70 cubic inches capacity, and devoid of air, and observing by means of a "*saturation gauge*" the exact temperature at which the whole of the water is converted into steam. The saturation gauge, in which the novelty of the experiment consists, is essentially a double mercury column balanced upon one side by the pressure of the steam produced from the weighed portion of water, and on the other by constantly saturated steam of the same temperature. Hence when heat is applied the mercury columns remain at the same level up to the point at which the weighed portion of water is wholly vaporized; from this point the columns indicate, by a difference of level, that the steam in the globe is superheating; for superheated steam increases in pressure at a far lower rate than saturated steam for equal increments of temperature. By continuing the process, and carefully measuring the difference of level of the columns, data are obtained for estimating the rate of expansion of superheated steam.

The apparatus for experiments at pressures of from 15 to 70 lbs. per square inch, consisted chiefly of a glass globe for the reception of the weighed portion of water, drawn out into a tube about 32 inches long. The globe was enclosed in a copper boiler, forming a steam-bath by which it could be uniformly heated. The copper steam-bath was prolonged downwards by a glass tube enclosing the globe stem. To heat this tube uniformly with the steam-bath, an outer oil-bath of blown glass was employed, heated like the copper bath by gas jets. The temperatures were observed by thermometers exposed naked in the steam, but corrected for pressure. The two mercury columns forming the saturation gauge were formed in the globe stem, and between this and the outer glass tube; so long as the steam in the glass globe continued in a state of saturation, the inner column in the globe stem remained stationary, at nearly the same level as that in the outer tube. But when, in raising the temperature, the whole of the water in the globe had been evaporated and the steam had become superheated, the pressure no longer balanced that in the outer steam-bath, and, in consequence, the column in the globe stem rose, and that in the outer tube fell, the difference of level forming a measure of the expansion of the steam. Observations of the levels of the columns were made by means of a cathetometer at different temperatures, up to 10° or 20° above the saturation point; and the maximum temperature of saturation was, for reasons developed by the experiments, deduced from a point at which the steam was decidedly superheated.

The results of the experiments, which in the paper are given in detail, show that the density of saturated steam at all temperatures, above as well as below 212°, is invariably greater than that derived from the gaseous laws.

The apparatus for the experiments at pressures below that of the atmosphere was considerably modified; and the condition of the steam was determined by comparing the column which it supported with that of a barometer. The results of these experiments, reduced in the same way, are extremely consistent.

As the authors propose to extend their experiments to steam of a very high pressure, and to institute a distinct series on the law of expansion of superheated steam, they have not at present given any elaborate generalizations of their results. The following formulæ, however, represent the relations of specific volume and pressure of saturated steam, as determined in their experiments, with much exactness.

Let V be the specific volume of saturated steam, at the pressure P, measured by a column of mercury in inches; then

$$V = 25\cdot62 + \frac{49513}{P + \cdot72}. \quad \ldots \ldots \quad (2.)$$

$$P = \frac{49513}{V - 25\cdot62} - 0\cdot72. \quad \ldots \ldots \quad (3.)$$

In regard to the rate of expansion of superheated steam, the experiments distinctly show that, for temperatures within about ten degrees of the saturation point, the rate of expansion greatly exceeds that of air, whereas at higher temperatures the rate of expansion approaches very near that of air. Thus in experiment 6, in which the maximum temperature of saturation is 174°·92, the coefficient of expansion between 174°·92 and 180° is $\frac{1}{150}$, or three times that of air; whereas between 180° and 200° the coefficient is very nearly the same as that of air (steam$=\frac{1}{437}$, air$=\frac{1}{439}$), and so on in other cases. The mean coefficient of expansion at zero of temperature from seven experiments below the pressure of the atmosphere, and calculated from a point several degrees above that of saturation, is $\frac{1}{135}$, whereas for air it is $\frac{1}{459}$. Hence it would appear that for some degrees above the saturation point the steam is not decidedly in an aëriform state, or, in other words, that it is watery, containing floating vesicles of unvaporized water.

Table of Results, showing the relation of density, pressure, and temperature of saturated steam.

Number of Exper.	Pressure		Max. temp. of saturation, Fahr.	Specific Volume.		Proportional error of formula (2).
	in lbs. per sq. in.	in inches of mercury.		From experiment.	By formula (2).	
1	2·6	5·35	136·77	8266	8183	$+\frac{1}{100}$
2	4·3	8·62	155·33	5326	5326	0
3	4·7	9·45	159·36	4914	4900	$-\frac{1}{340}$
4	6·2	12·47	170·92	3717	3766	$+\frac{1}{71}$
5	6·3	12·61	171·48	3710	3740	$+\frac{1}{123}$
6	6·8	13·62	174·92	3433	3478	$+\frac{1}{76}$
7	8·0	16·01	182·30	3046	2985	$-\frac{1}{50}$
8	9·1	18·36	188·30	2620	2620	0
9	11·3	22·88	198·78	2146	2124	$-\frac{1}{97}$
1'	26·5	53·61	242·90	941	937	$-\frac{1}{235}$
2'	27·4	55·52	244·82	906	906	0
3'	27·6	55·89	245·22	891	900	$+\frac{1}{100}$
4'	33·1	66·84	255·50	758	758	0
5'	37·8	76·20	263·14	648	669	$+\frac{1}{31}$
6'	40·3	81·53	267·21	634	628	$-\frac{1}{100}$
7'	41·7	84·20	269·20	604	608	$+\frac{1}{110}$
8'	45·7	92·23	274·76	583	562	$-\frac{1}{28}$
9'	49·4	99·60	279·42	514	519	$+\frac{1}{110}$
11'	51·7	104·54	282·58	496	496	0
12'	55·9	112·78	287·25	457	461	$+\frac{1}{114}$
13'	60·6	122·25	292·53	432	428	$-\frac{1}{108}$
14'	56·7	114·25	288·25	448	456	$+\frac{1}{56}$

Adopting the notation previously employed, and putting r for the rate or coefficient of expansion of an elastic fluid at t_1 temperature, we find

$$r = \frac{1}{\epsilon_1 + t_1} = \frac{\frac{V_2 p_2}{V_1 p_1} - 1}{t_2 - t_1}, \quad \ldots \ldots (4.)$$

where $\frac{1}{\epsilon_1} =$ the rate of expansion at zero of temperature. In the case of air $\epsilon_1 = 459$.

The following Table gives the value of the coefficient of expansion of superheated steam taken at different intervals of temperature from the maximum temperature of saturation.

Number of the Exper.	Max. temp. of saturation.	Temperatures between which the expansion is taken.		Coefficient of expansion of superheated steam.	Coefficient of expansion of air.
1	136·77	140	170	$\frac{1}{303}$	$\frac{1}{310}$
2	155·33	160	190	$\frac{1}{340}$	$\frac{1}{310}$
3	159·36 {	159·36	170·2	$\frac{1}{130}$	$\frac{1}{313}$
		170·2	209·9	$\frac{1}{314}$	$\frac{1}{300}$
5	171·48 {	171·48	180	$\frac{1}{300}$	$\frac{1}{310}$
		180	200	$\frac{1}{304}$	$\frac{1}{310}$
6	174·92 {	174·92	180	$\frac{1}{100}$	$\frac{1}{313}$
		180	200	$\frac{1}{307}$	$\frac{1}{310}$
7	182·30 {	182·3	186	$\frac{1}{310}$	$\frac{1}{317}$
		186	209·5	$\frac{1}{310}$	$\frac{1}{313}$
8	188·30	191	211	$\frac{1}{304}$	$\frac{1}{330}$
1′	242·9	243	249	$\frac{1}{317}$	$\frac{1}{303}$
4′	255·5 {	257	259	$\frac{1}{188}$	$\frac{1}{318}$
		257	264	$\frac{1}{300}$	$\frac{1}{313}$
6′	267·21 {	268	271	$\frac{1}{310}$	$\frac{1}{317}$
		271	279	$\frac{1}{310}$	$\frac{1}{310}$
7′	269·2 {	271	273	$\frac{1}{313}$	$\frac{1}{310}$
		273	279	$\frac{1}{317}$	$\frac{1}{310}$
9′	279·42 {	283	285	$\frac{1}{300}$	$\frac{1}{313}$
		285	289	$\frac{1}{303}$	$\frac{1}{313}$
13′	292·53 {	297	299	$\frac{1}{301}$	$\frac{1}{313}$
		299	302	$\frac{1}{317}$	$\frac{1}{313}$

Hence it appears, that as the steam becomes more and more superheated, the coefficient of expansion approaches that of a perfect gas. The authors hope that these experiments may be continued, and that the results obtained at greatly increased pressures will prove as important as those already arrived at.

GEOLOGICAL SOCIETY.

[Continued from p. 157.]

December 5, 1860.—L. Horner, Esq., President, in the Chair.

The following communication was read :—

" On the Structure of the North-west Highlands, and the Relations of the Gneiss, Red Sandstone, and Quartzite of Sutherland and Ross-shire." By Professor James Nicol, F.R.S.E., F.G.S.

The author first referred to his paper in the Quart. Journ. Geol. Soc., vol. xiii. pp. 17, &c., in which the order of the red sandstone on gneiss, and of quartzite and limestone on the sandstone, was established, and in which the relation of the eastern gneiss or mica-schist to the quartzite was stated to be somewhat obscure on account of the presence of intrusive rocks and other marks of disturbance. Having examined the country four times, with the view of settling some of the doubtful points in the sections, the author now offered the matured result of his observations. He agrees with Sir R. Murchison as far as the succession of the western gneiss, red sandstone, quartzites (quartzite and fucoid-bed), and limestone is concerned; but differs from him in maintaining that there is no upper series of quartzite and limestone, and that there is no evidence of an "upward conformable succession" from the quartzite and limestone into the eastern mica-slate or gneiss—the o-called "upper gneiss." The "upper quartzite" and "upper limestone" the author believes to be portions of the quartzite of the country, in some cases separated by anticlines and faults and cropping out in the higher ground, and in other instances inverted beds with the gneiss brought up by a contiguous fault and overhanging them. This latter condition of the strata, as well as other cases where the eastern gneiss is brought up against the quartzite series, have, according to the author, given rise to the supposed "upward conformable succession" above referred to. In some cases where "gneiss" is said to have been observed overlying the quartzite, Professor Nicol has determined that the overlying rock is granulite or other eruptive rock, not gneiss.

The sections described by the author in support of his views of the eastern gneiss not overlying the quartzite and limestone, but being the same as the gneiss of the west coast, and brought up by a powerful fault along a nearly north and south line passing from Whiten Head (Loch Erriboll) to Loch Carron and the Sound of Sleat, are chiefly those which had been brought forward as affording the proofs on which the opposite hypothesis is founded; and in all, the author finds irruptions of igneous rocks, and other indications of faults and disturbance, depriving them, in his opinion, of all weight as evidence of a regular order of "upward conformable succession."

Prof. Nicol further argues that the mode of the distribution of the rocks shows that there is through Sutherland and Ross-shire a real fault, and no overlap of eastern gneiss of more than a few feet or yards at most, and that the fact of different strata of the quartzite series being brought against the gneiss at different places supports this view, and points to a great denudation having taken place along the line of fault. Though the quartzite is here and there altered by the igneous rocks, yet it is truly a sedimentary rock, and so is the limestone; but the eastern gneiss or mica-schist is a crystalline rock throughout: this fact, according to the author, is inimical to the hypothesis of the eastern gneiss overlying the limestone and quartzite. It has been insisted upon, that the strike of the western gneiss is different from that of the east; but the author remarks that the strike is not persistent in either area, and that great movements subsequent to the deposition of the quartzite series have irregularly affected the whole region. With regard to mineralogical characters, Prof. Nicol insists that both the eastern and the western gneiss are essentially the same. Both are locally modified with granitic and hornblendic matter near igneous foci; but no proof of a difference of age in the two can be obtained therefrom. The alteration in bulk of the gneiss in the western area, by the intrusion of the vast quantities of granite now observable in it, may perhaps have caused the great amount of crumpling and faulting along the N. and S. line of fault, dividing the western from the eastern gneiss,—a fault comparable with and parallel to that running from the Moray Firth to the Linnhe Loch, and to the one passing along the south side of the Grampians.

December 19, 1860.—L. Horner, Esq., President, in the Chair.

The following communications were read :—

1. " On the Geological Structure of the South-west Highlands of Scotland." By T. F. Jamieson, Esq. Communicated by Sir R. I. Murchison, V.P.G.S.

In this paper the author attempts to throw light on the relations of those rocks which figure in geological maps as the mica-schist, clay-slate, the chlorite-slates, and the quartz-rock of the South-west-ern Highlands, which range N.E. through the middle of Scotland, forming an important feature in the geology of that country. An examination of these rocks, as displayed in Bute and Argyleshire, has led Mr. Jamieson to believe that, from the quartz-rock of Jura to the border of the Old Red Sandstone, there is a conformable series of strata, which, although closely linked together, may be classed into three distinct groups, namely, 1st, a set of lower grits (or quartz-rock), many thousand feet thick; 2ndly, a great mass of thin-bedded slates, 2000 feet or more thick; and 3rdly, a set of upper grits, with intercalated seams of slate of equal thickness. Beds of limestone occur here and there sparingly in all the three divisions; the thickest being deep down in the lower grits. All the limestones are thickest towards the west. The siliceous grits also appear to be freer from an admixture of green materials towards the west. All the members of the series (namely, the upper grits, slates, and lower grits) have a persistent S.W.–N.E. strike, sometimes in Bute approaching to due N. and S. They are conformable, and graduate one into another in such a way as to show that they belong to one continuous succession of deposits. The materials of which they have been formed seem to have been derived from very similar sources. The upper and the lower grits are very similar in composition, being made up of water-worn grains of quartz, many of which are of a peculiar semitransparent bluish tint.

The rocks of the district have been thrown into a great undula-tion, with an anticlinal axis extending from the north of Cantyre through Cowal by the head of Loch Ridun on to Loch Eck (and probably by the head of Loch Lomond on to the valley of the Tay, at Aberfeldy), and with a synclinal trough lying near the parallel of Loch Swen. The anticlinal fold is well seen in the hill called Ben-y-happel, near the Tighnabruich quay in the Kyles of Bute. Southward of this ridge, which is composed of the lower grits or quartzite, the thin-bedded greenish slates and the upper grits suc-ceed conformably; and the latter are separated by a trap-dyke from the Old Red Sandstone of Rothsay. This section the author de-scribed in detail; also the corresponding section to the north of the anticlinal axis, towards Loch Fyne, and along the west shore of Loch Fyne. The lower grits extend as far as Loch Gilp, and are then succeeded by the green slates and the upper grits, which falling in the synclinal trough are repeated through Knapdale towards Jura Sound, where the green slates again form the surface along the eastern coast of Jura, lying on the quartzite or grits of that island. Throughout the synclinal trough and the neighbouring district (that is, from

Loch Fyne to Jura Sound) the grits and slates are intimately mixed, with numerous intercalated beds of greenstone, some being of great thickness. Mr. Jamieson pointed out that this feature of the district has hitherto in great part been misunderstood, and that Macculloch was in error when he denominated these rocks "chlorite-schist."

The probable relationship of the rocks of the Islands of Shuna, Luing, and Scarba to those of Jura and Bute were then dwelt upon; the greenstones of Knapdale, &c., and their relation to the sedimentary rocks, were described in detail; and the limestones of the district briefly noticed. As no fossils have hitherto been found, palæontological evidence of the age of these rocks is wanting; but the author, regarding their general resemblance to the quartz-rocks, limestones, and mica-schists of Sutherlandshire, thinks them to be of the same date as those rocks of the North-west Highlands.

2. "On the position of the beds of the Old Red Sandstone in the Counties of Forfar and Kincardine, Scotland." By the Rev. Hugh Mitchell. Communicated by the Secretary.

In Forfar- and Kincardine-shire, south of the Grampians, the Old Red Sandstone is developed in the following series, with local modifications :—1st (at top), Conglomerate; 2nd, grey flagstone with intercalated sandstone (about 40 feet thick at Cauterland Den, 120 feet at Carmylie); 3rd, gritty ferruginous sandstone, with occasional thin layers of purplish flagstone. Of the last, 120 feet are seen at Cauterland Den; it occurs also at Ferry Den, &c. The flagstone of this third or lowest member of the group yields Ripple-marks, Rain-prints, Worm-markings, and Crustacean tracks (of several kinds, large and small). *Parka decipiens* has been found in the lowest grits; and *Cephalaspis* in the sandstone at Brechin, immediately under the grey flagstones.

In the second member, namely the grey flags, Fish-remains have of late been found more or less abundantly throughout the district, together with Crustacean fossils. *Cephalaspis Lyellii*, Ichthyodorulites, Acanthodian fishes, *Pterygotus*, *Eurypterus*, *Kampecaris Forfariensis*, *Stylonurus Powriensis*, *Parka decipiens*, and vegetable remains are the most characteristic fossils.

The author pointed out that some few genera of Fish and Crustaceans were present both in this zone and in the Upper Silurian formation, and that still fewer links existed to connect the fauna of the Forfarshire flags with the Old Red Sandstone north of the Grampians, with which it appears to have, in this respect, almost as little relation as with the Carboniferous system. With the Old Red of Herefordshire these flags appear to have some few fossils in common; but of about twenty *species* found in Forfarshire, only about four could be quoted from Herefordshire.

In conclusion, the author noticed the vast vertical development of the whole series, and its great geographical extent; and particularly dwelt upon the distinctness of the fauna of the flagstones of Forfarshire, as giving good grounds for the treatment of the Old Red Fauna as peculiar to a separate geological period, both as distinct

from the Silurian System, and in some degree as divisible into two
or more members of one group:—three members, if the upper or.
Holoptychius-beds of Moray, Perth, and Fife, the middle or Fish-
beds of Cromarty and Caithness, and the lowest or Forfarshire
beds be counted separately; but two, if we regard some of the Old
Red beds north of the Grampians as equivalent in time to those on
the south.

January 9, 1861.—L. Horner, Esq., President, in the Chair.

The following communications were read:—

1. " On the Distribution of the Corals in the Lias." By the Rev.
P. B. Brodie, M.A., F.G.S.

From observations made by himself and others, the author was
enabled to give the following notes. In the Upper Lias some Corals
of the genera *Thecocyathus* and *Trochocyathus* occur. The Middle
Lias of Northamptonshire and Somersetshire has yielded a few Corals.
The uppermost band of the Lower Lias, namely the zone with *Ammonites raricostatus* and *Hippopodium ponderosum*, contains an unde-
scribed Coral at Cheltenham and Honeybourn in Gloucestershire;
and a *Montlivaltia* in considerable abundance at Down Hatherly in
Gloucestershire, at Fenny Compton and Aston Magna in Worcester-
shire, and at Kilsby Tunnel in Northamptonshire. The middle members of the Lower Lias appear to be destitute of Corals. In the zone
with *Ammonites Bucklandi*, called also the Lima-beds, *Isastræa* occurs
in Warwickshire and Somersetshire. Dr. Wright states that *Isastræa
Murchisoni* occurs in the next lowest bed of the Lower Lias, namely
the White Lias with *Ammonites Planorbis*, at Street in Somerset;
and another Coral has been found in the same zone in Warwickshire. Lastly, in the " Guinea-bed " at Binton in Warwickshire
another Coral has been met with.

The *Montlivaltiæ* of the Hippopodium-bed and the *Isastræa* of the
Lima-beds appear to have grown over much larger areas in the
Liassic Sea than the other Corals here referred to.

2. " On the Sections of the Malvern and Ledbury Tunnels, on the
Worcester and Hereford Railway, and the intervening Line of Rail-
road." By the Rev. W. S. Symonds, A.M., F.G.S., and A. Lambert,
Esq.

In this paper the authors gave an account of the different strata
exposed by the cuttings of the Worcester and Hereford Railway
(illustrated by a carefully constructed section), including the different members of the New Red Sandstone (on the east of the Malvern Hills), the syenite and greenstone (forming the nucleus of the
Malverns), and the Upper Llandovery beds, the Woolhope shales,
the Woolhope limestone, Wenlock shales, Wenlock limestone, and
Lower Ludlow rock on the west side of the syenite, followed by some
beds of the Old Red series, violently faulted against the Ludlow rock
at the west end of the Malvern Tunnel. Then the open railway
passes over Upper Ludlow rocks and some lower beds of the Old Red
series, here and there covered by drift, until the Lower Ludlow rock
is again traversed at the east end of the Ledbury Tunnel, and is

shown ·to be much faulted and brought up against Upper Ludlow shales and Aymestry rocks. The Wenlock shales and the Wenlock limestone are then traversed; these are much faulted, the Lower Ludlow beds again coming in, followed by Aymestry rock, Upper Ludlow shales, Downton sandstone, and, at the east end of the tunnel, by red and mottled marls, grey shales and grits, purple shales and sandstones, with the Auchenaspis-beds, forming the passage-beds into the Old Red Sandstone, as described in a former paper (Quart. Journ. Geol. Soc. vol. xvi. p. 193).

In a note, Mr. J. W. Salter, F.G.S., described the great abundance of Upper Silurian fossils found in these cuttings, and now chiefly in the collection of Dr. Grindrod and other geologists at Malvern and the neighbourhood.

XXXVI. *Intelligence and Miscellaneous Articles.* ·

ON THE FIBROUS ARRANGEMENT OF IRON AND GLASS TUBES.

[Extract from a Letter from Dr. Debus to Professor Tyndall.]

Queenwood College, Feb. 17, 1861.

A FEW weeks ago Mr. E. brought me a piece of iron tube which had been exposed for several years to the action of moist air. Nearly the whole of the tube was converted into oxide of iron. This suggested certain thoughts, the results of which were new and interesting to me.

You know, when a piece of glass tube, sealed and filled with water, is heated, the tube is cracked, and cracked in a longitudinal direction. Why is this? Glass tubes are made by taking a piece of hollow glass in a viscous state and pulling it at both ends. Now, the particles of the glass do not adhere to each other on all sides with a force of the same strength, but in some directions this force is stronger, in others weaker. Under the strain produced by the pulling, they arrange themselves so as to offer to the pulling force the greatest resistance. Therefore the greatest cohesion of the particles is found, in the formed tube, parallel to the length of the tube, and the weakest cohesion in a direction perpendicular to this. This explains the cracking of the tube as mentioned above.

Mr. E. could not give me exact information as to how iron tubes are made, but he said they were passed through rollers. Now, if this is correct, the particles of iron ought to arrange themselves in a similar way to the particles in a glass tube. If such a tube oxidizes, the oxygen naturally would overcome the cohesion of the iron first in those places where this cohesion is weakest, that is, in lines parallel with the tube. Such is actually the case. The tube mentioned had deep furrows, so to say, hollowed out by the oxygen along its length, just in the same way as a glass tube would crack under pressure.

I need not mention to you other examples; but one case more, and then I have done. You gave some years ago an explanation of cleavage*. The paper wherein this explanation was given never came to my hands; and I do not remember that you explained the thing to me when we had personal intercourse. The principle alluded to appears to me to be the true cause of the phenomenon.

If a plastic mass is exposed to pressure, the particles turn until they are in such a position as to offer the greatest resistance to the pressure brought to bear upon them. But that direction wherein they offer the greatest resistance to pressure is also that where the cohesion is least. Consequently cleavage ought to take place in a direction perpendicular to the pressure exerted.

Am I right or wrong? Of course I could say more; but why should I carry water to the well?

H. Debus.

ON THE CALCIUM SPECTRUM.

To the Editors of the Philosophical Magazine and Journal.

Gentlemen,

In a communication published in the last Number of the Philosophical Magazine, we pointed out the appearance of a well-defined blue line in the spectrum produced by igniting the evaporated residue of a deep-well water. From the circumstance of this line, which is situated somewhat more towards the violet end of the spectrum than the line Sr δ, not being referred to in the paper of Profs. Kirchhoff and Bunsen, nor indicated in their beautiful representations of the spectra of the alkaligenous metals, we were induced to attribute its production to the probable existence of a previously unrecognized member of the calcium group of metals. Further experiments, however, have satisfied us that the blue line in question really belongs to the calcium spectrum. On finding this to be the case, we communicated with Professor Bunsen, who in return informed us that Professor Kirchhoff and himself had observed this line, but, not thinking it sufficiently bright to be suitable for the recognition of calcium, had not made reference to it in their paper. It can, however, be seen with a degree of brilliancy at least equal to that of many of the lines represented, when perfectly pure chloride of calcium is ignited in a somewhat darkened room.

We remain, Gentlemen,
Your obedient Servants,
F. W. and A. Dupré.

ON THE LUNAR TABLES AND THE INEQUALITIES OF LONG PERIOD DUE TO THE ACTION OF VENUS. BY M. DE PONTÉCOULANT.

In the Number of the *Comptes Rendus* of the labours of the Academy of the 12th of November last, M. Delaunay has inserted a memoir in which he gives an account of the researches in which he has been engaged, concerning the two lunar disturbances of long

* Fibrous iron was one of my illustrations.—J. T.

period depending on the action of the planet Venus, which Professor Hansen has proposed to introduce into the expressions for the moon's motions. According to M. Delaunay's calculations (the accuracy of which I have at present no intention of disputing) the value of the coefficient of the first of these disturbances—that, namely, whose period is about 273 years—is nearly the same as that attributed to it by the astronomer of Gotha; but the coefficient of the second, whose period is about 240 years (and which is the most important of the two, since its coefficient, calculated at first by Professor Hansen at 23″·2, is according to his account at least 21‴·47), ought to be considered, according to the researches of M. Delaunay, as altogether insensible, if not absolutely nothing.

This conclusion, which is moreover perfectly in accordance with the announcement of the illustrious geometrician Poisson, published more than twenty-seven years ago in his memoir of 1833, gives rise to several questions of extreme importance, not only, as it seems to me, with respect to the perfection of the lunar tables, but also on the subject of scientific priority, and even of national honour. In order that the members of the Academy may be able to judge of this for themselves, it will be sufficient to mention that the principal corrections which the astronomers of the Greenwich Observatory have thought it necessary to make in the precious tables of our fellow-countryman Damoiseau—tables which are so remarkable from the fact that they are the first that were constructed from theory alone, without recourse to observation,—and the preference which they have accorded to the new lunar tables of Professor Hansen, are principally founded on the existence (considered by them to be conclusively demonstrated) of the two inequalities arising from the action of Venus, which M. Delaunay has just calculated. To this it will be sufficient to add that it was on these grounds that the extraordinary prize of £1000 was allotted to the same Professor by the Lords of the Admiralty, at the suggestion of the learned director of the Greenwich Observatory, for the really marvellous addition, as Mr. Airy calls it, which he has made to the lunar theory. This assertion, if suffered to pass without refutation, might lead us to undervalue the labours of those astronomers, French and foreign, who have brought about the rapid progress of this difficult theory, and have advanced it to its present state of perfection.

I shall not now dwell further on these observations, which from their length would extend beyond the limits prescribed by the Academy to its own members, and still more to strangers whose claims it admits to the honour of an insertion in the *Comptes Rendus*; but I thought it advisable to lose no time in announcing that I am busily occupied in drawing up a memoir, in which all the observations called for by a question of gravity so great that the history of science has rarely furnished one similar to it will be detailed at length, so that it may not be supposed, either in France or abroad, that a memoir so important as that of M. Delaunay has escaped unnoticed or remained unanswered.—*Comptes Rendus*, December 1860.

THE

LONDON, EDINBURGH AND DUBLIN

PHILOSOPHICAL MAGAZINE

AND

JOURNAL OF SCIENCE.

[FOURTH SERIES.]

APRIL 1861.

XXXVII. *On a New Proposition in the Theory of Heat.* *By* Professor KIRCHHOFF*.

SOME few months ago I communicated to the Society certain observations, which appeared of interest because they give some information respecting the chemical composition of the solar atmosphere, and point the way to further knowledge on this subject. These observations led to the conclusion that a flame whose spectrum consists of bright lines is partially opake for rays of light of the colour of these lines, whilst it is perfectly transparent for all other light. In this statement we find the explanation of Fraunhofer's dark lines in the solar spectrum, and the justification of the conclusions regarding the composition of the sun's atmosphere; for we find that a substance which, when brought into a flame, produces bright lines coincident with the dark lines of the solar spectrum, must be present in the sun's atmosphere. The fact that a flame is partially opake solely for those rays which it emits, was, as I stated at the time, a matter of some surprise to me. Since that time I have arrived, by very simple theoretical considerations, at a proposition from which the above conclusion is immediately derived. As this proposition appears to me to be of considerable importance on other accounts, I beg to lay it before the Society. A hot body emits rays of heat. We feel this very perceptibly near a heated stove. The intensity of the rays of heat which a hot body emits, depends on the nature and on the temperature of the body, but is quite independent of the nature of the bodies on which the rays fall. We *feel* the rays of heat only in the case of very hot bodies; but they are

* Abstract of a Lecture delivered before the Natural History Society of Heidelberg. Communicated by Professor Roscoe.

emitted from a body, whatever be its temperature, although the amount diminishes with the temperature. In proportion as a body radiates, it loses heat, and its temperature must sink unless this loss is made up. A body surrounded by substances of the same temperature undergoes no change of temperature. In this case the loss of heat caused by its own radiation is exactly compensated by the rays which surrounding substances give out, a part of which the body absorbs. The quantity of heat which this body absorbs in a given time must be equal to that which in the same time it emits. This holds good whatever the nature of the body may be; the more rays a body emits, the more of the incident rays must it absorb. The intensity of the rays which a body emits has been called its *power of radiation or emission*; and the number denoting the fraction of the incident rays which is absorbed has been called the *power of absorption*. The larger the power of emission a body possesses, the larger must its power of absorption be.

A somewhat closer consideration shows that the relation between the powers of emission and absorption for *one* temperature is the same for all bodies. This conclusion has been verified in many special cases, both in the last ten years and in former times. The foregoing proposition requires, however, that all the rays of heat under consideration are of one and the same kind; so that these rays are not qualitatively so far different that one part of them are absorbed by the bodies more than another part; for, were this the case, we could not speak of the power of absorption of a body, simply because it would be different for different rays. Now we have long known that there really are different kinds of heating rays, and that in general they are unequally absorbed by bodies. There are both dark- and luminous-heating rays; the former are almost all absorbed by white bodies, whilst the latter rays are thus scarcely absorbed at all. Indeed the variety of the rays of heat is even greater than the variety of the coloured rays of light. The rays of heat, the dark as well as the luminous, are influenced in the same manner as the rays of light, by transmission, reflexion, refraction, double refraction, polarisation, interference, and diffraction. In the case of the luminous rays of heat, it is not possible to separate the light from the heat; when one is diminished in a given relation, the other is diminished in the same ratio. This has led to the conclusion that rays of light and heat are essentially of the same nature; that rays of light are simply a particular class of the heat-giving rays. The dark rays of heat are distinguished from the rays of light, just as the differently coloured rays are distinguished from each other, by their period of vibration, wave-length, and refractive index. They are not visible because the media of our eyes are not transparent to them.

A difference of quality is noticed amongst the rays of light, not only in respect to the colour, but also in respect to the state of polarisation. Hence not only have we to distinguish the heating rays according to the wave-lengths, but we have also to divide rays of one wave-length into those variously polarized. If we take into consideration these various kinds of rays of heat, the conclusions which we had drawn concerning the relation between the powers of absorption and emission cease to be binding.

Whether this relation is still found to exist when these variations are taken into consideration, is a question which has, as yet, not been decided either by theoretical considerations or by an appeal to experiment. I have succeeded in filling up this gap; and I have found that the proposition concerning the ratio between the power of emission and the power of absorption remains true, however different the rays which the bodies emit may be, as long as the notions of emissive and absorptive powers be confined to *one kind* of ray.

The proposition which I have discovered may be thus more precisely defined :—Let a body C be placed behind two screens S_1 and S_2, in which two small openings are made. Through these openings a pencil of rays proceeds from the body C. Of these rays we consider that portion which corresponds to a given wave-length λ, and we divide this into two polarised components, whose planes of polarization are two planes a and b at right angles to each other, passing through the axis of the pencil of rays. Let the intensity of the polarized component a be E (emissive power). Now suppose that a pencil of rays, having a wave-length $=\lambda$, and polarized in the plane a, falls through the openings 2 and 1 upon the body C. The fraction of this pencil which is absorbed by the body C is called A (absorptive power).

Then the relation $\dfrac{A}{E}$ is independent of the position, size, and nature of the body C, and is alone determined by the size of the openings 1 and 2, by the wave-length λ, and by the temperature. I will point out the way in which I have proved this proposition. I began by considering that bodies are conceivable which, although very thin, absorb all the rays which fall upon them, or which have the capacity of absorption $=1$. I call such bodies *perfectly black*, or simply *black*. I first investigated the radiation of such black bodies. Let C be a black body. The body C is supposed to be enclosed in a black envelope, of which the screens S_1 and S_2 are a part, and the two screens are supposed to be connected by a black surface surrounding all. Lastly, let the opening 2 be closed by a black surface, which I will call "surface 2." The whole system is to be considered to possess the same temperature, and to be protected against loss of heat from without by an absolutely non-conducting medium. Under these circumstances the temperature of the body C cannot alter; the sum of the intensities of the rays which it emits must therefore be equal to the sum of the intensities of the rays which it absorbs; and because it absorbs all those that fall upon it, the sum of the intensities of the rays it emits must be equal to the sum of the intensities of the rays which fall upon it. If, now, we suppose the following change: The "surface 2" is removed and replaced by a circular mirror which reflects all the rays falling upon it, and whose centre is in the middle of opening 1. The equilibrium of the heat must still be kept up; the sum of the rays which fall on the body C must still remain equal to the sum of the rays which it emits. But, as it emits just as much as before, the quantity of rays which the mirror reflects upon the body C must be equal to that which the surface 2 emitted. The mirror produces an image of opening 1, which is coincident with opening 1. For this reason, just those rays come back to the body C, after one reflexion from the mirror, as the body C would have emitted through the openings 1 and 2 if this last one had been open; and the intensity of these rays is equal to the intensity of the rays which the surface 2 sent back through the opening 1. This last intensity however, is, evidently independent of the nature of the body C; and hence it follows that the intensity of the pencil of rays which the body C radiates through the openings 1 and 2, is independent of the form, position, and constitution of the body C; supposing of course that this body is black, and that its temperature is a given one. According to this, however, the qualitative composition of the pencil of rays might become different if the body C were replaced by another black body of the same temperature. This is, however, not the case. If I call *e* the power of emission of this black body compared with a certain wave-length and a given plane of polarization—that, therefore, which I have called E under the supposition that C is a body of any kind—then *e* is independent of the nature of the body C, if it only be black. In order to render this evident, a further arrangement is necessary. Into the pencil of rays which passes from the opening 1 towards the surface 2, let us suppose a small plate placed, which is of so slight a thickness that it shows in the visible rays the colours of thin plates; let it be so placed that the pencil of rays is incident at the polarizing angle; let the material of the plate be so chosen that it neither absorbs nor emits a sensible amount of rays; let the envelope joining the screens S_1 and S_2 be so shaped that the image which the plate reflects of the surface 2 lies in the envelope. At the position, and of the size of this image, let an opening in the envelope be made; this I will call "opening 3." Let a screen be so placed that no straight line can be drawn from any point of opening 1 to any point of opening 3 without passing through the screen. Let the opening 3 be now closed with a black surface, which I will call "surface 3." The whole system is then supposed to possess the same temperature; there is therefore in this case equilibrium as regards the heat. This equilibrium is supported by rays which, proceeding from surface 3, suffer reflexion on the plate, pass through opening 1, and fall on the body C. These rays are polarized in the plane of incidence of the plate, and contain, according to the thickness of the plate, sometimes more of one, sometimes more of another kind of ray. Let the surface 3 be removed and replaced by a circular mirror whose centre is situated at the spot where the plate reflects an image of the centre of the opening 1; then the rays emitted by surface 3 will no longer fall on body C, but instead of them those reflected from the mirror will fall upon it, and the equilibrium of the temperature remains unchanged. If we reflect that it does not matter what thickness the plate possesses, or in what position we turn it round the axis of the pencil determined by passing through openings 1 and 2, we arrive, by means of similar considerations, at the conclusion that the power of emission of the black body C, considered with respect to a given wave-length and a given plane of polarization, is quite independent of the constitution of this body. A conclusion which naturally arises from this proposition is, that *all* rays which a black body emits are completely unpolarized.

If we imagine that in the foregoing arrangement the body C is not black, but of any other colour, the following equation is found by similar reasoning:—

$$\frac{E}{A} = e. \qquad \ldots \ldots \ldots \quad (1)$$

This equation indicates that the relation between emission and absorption remains constant for all bodies. The equation may obviously be written

$$E = Ae, \qquad \ldots \ldots \ldots \quad (2)$$

or

$$A = \frac{E}{e}. \qquad \ldots \ldots \ldots \quad (3)$$

I will now notice some remarkable conclusions derived from my proposition. If we heat any body, a platinum wire for example, gradually more and more, it first emits only dark rays; at the temperature at which it begins to glow, red rays begin to appear; at a certain higher temperature yellow rays are seen; then green rays, until at last it becomes white-hot, *i. e.* emits all the rays present in solar light. The power of emission (E) of the platinum wire is therefore equal to 0 for the red rays at all temperatures lower than that at which the wire begins to glow; for yellow rays it ceases to be equal to 0 at a rather higher temperature; for green at a still higher temperature, and so on. According to equation (1), the emission-power (e) of a completely black body must cease to be equal to 0 for red, yellow, green, &c. rays at the same temperatures at which the platinum wire began to emit red, yellow, green, &c. rays. Let us now consider the case of any other body which is gradually heated. According to equation (2), this body must begin to give off red, yellow, and green rays at the *same temperatures* as the platinum wire. All bodies must therefore begin to glow at the same temperature, or at the same temperature begin to give off red, and at the same temperature yellow rays, &c. This is the theoretical explanation of an experimental conclusion obtained by Draper thirteen years ago. The intensity of the rays of given colour which a body radiates at a given temperature may, however, be very different,—according to equation (2) it is proportional to the power of absorption (A). The more transparent a body is, the less luminous it appears. This is the reason why gases, in order to glow visibly, need a temperature so much higher than solid or liquid bodies.

A second deduction which I will mention brings me back to my special subject. The spectra of all opake glowing bodies are continuous; they contain neither bright nor dark lines. Hence we can conclude that the spectrum of a glowing *black* body (the term being used in the sense already defined) must also be a continuous one. The spectrum of an incandescent gas consists, at any rate most frequently, of a series of bright lines separated from each other by perfectly dark spaces. If the power of emission of such a gas be represented by E, the relation $\frac{E}{e}$ has an appreciable value for those rays which correspond to the bright lines of the spectrum of the gas, but it has an inappreciable value for all other rays. According to equation (3), however, this relation is equal to the absorptive power of the incandescent gas. Hence it follows that the spectrum of an incandescent gas *will be* the *converse* of this, as I express it, when it is placed before a source of light of sufficient intensity, which gives a continuous spectrum; i. e. the lines of the gas-spectrum, which before were bright, will be seen as dark lines on a bright ground. A remarkable deduction from my proposition which I will mention is, that, if the more remote source of light is an incandescent solid body, the temperature of this body must be higher than that of the incandescent gas in order that such a conversion of the spectrum may occur.

The sun consists of a luminous nucleus, which would by itself produce a continuous spectrum, and of an incandescent gaseous atmosphere, which by itself would produce a spectrum consisting of an immense number of bright lines characteristic of the numerous substances which it contains. The actual solar spectrum is the converse of this. Were it possible to observe the spectrum belonging to the solar atmosphere with all its attendant bright lines, no one would be surprised to hear that, from the existence of the characteristic bright lines of sodium, potassium, and iron in the solar spectrum, the presence of these bodies in the sun's atmosphere has been ascertained. According to the proposition which I have just laid down, there can, however, be just as little doubt concerning the truth of this assertion, as if we saw the real spectrum of the solar atmosphere.

I will, lastly, mention a phenomenon which, although apparently trivial, was of peculiar interest to me, because I foresaw it theoretically, and afterwards verified it by experiment. According to theory, a body which absorbs more rays polarized in *one* direction than in another, must also emit those rays in the same proportion. A plate of tourmaline cut parallel to the optical axis absorbs, at common temperatures, more of those rays falling perpendicularly, whose plane of polarization is parallel to the axis of the crystal, than of those whose plane is at right angles to the axis. At temperatures above a red heat, tourmaline also possesses this same property, although in a less marked degree. Hence the rays of light which the plate of tourmaline emits perpendicular to its surface must be partially polarized; and, moreover, they must be polarized in a plane perpendicular to the plane of polarization of the rays which have been transmitted by the tourmaline. This theoretical conclusion is borne out by experiment.

XXXVIII. *Remarks on* Ampère's *Experiment on the Repulsion of a Rectilinear Electrical Current on itself.* By Mr. JAMES CROLL, *Glasgow*[*].

IN reference to Dr. Forbes's "Notes on Ampère's Experiment on the Repulsion of a Rectilinear Electrical Current on itself," which appeared in the Philosophical Magazine for February, the following remarks may perhaps be acceptable.

I have long been under the impression that Ampère's experiment, although successful, does not prove the thing intended, namely, that the different parts of a rectilinear electrical current are mutually repulsive; for the motion of the wire is evidently

* Communicated by the Author.

due to the action of angular currents, and not to a repulsion existing between the current in the mercury and the current in the branch of the wire in the same straight line.

Let *a b c d* be the wire floating on the mercury, P the point where the current enters the mercury, and N the point where it leaves it, after passing through the wire in the direction indicated by the arrows. The common explanation is, that the movement of the wire is due to there being in each of the branches

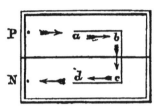

a b and *c d*, separately, a repulsion between the current which traverses them, and the current that is transmitted into the mercury before penetrating into the wire or after going out from it; and as the current of the mercury and that of the wire are only the prolongation of each other in a right line, this is considered sufficient proof that the one part of the rectilinear current repels the other part.

The following is, however, I think, the true explanation. The current P *a* in the mercury is at right angles to the current *b c* in the cross part of the wire. The former current is directed towards, and the latter current from the summit of the angle *a b c* formed by them. Now, according to Ampère's well-known law of angular currents, the two currents will repel each other. In this case the current *b c* being the moveable one, it will of course recede, maintaining a position parallel to itself. This cross current *b c* is also at right angles to the current N *d* on the other side of the glass partition; the former moving towards, and the latter from the summit of the angle *b c d* formed by them. These two currents for the same reason will repel each other. The combined influence of the currents P *a* and N *d* in the mercury will be to cause the cross section of the wire which is at *right angles* to them to recede, maintaining a position parallel to itself. It follows, therefore, according to the law of angular currents, that Ampère's experiment must be successful; but then its success does not prove that the parts of a rectilinear electrical current are repulsive of each other, but simply that a moveable current at right angles to a fixed one is repelled when the one current is directed towards, and the other from the summit of the angle formed by these two currents.

It would appear from Dr. Forbes's experiments, that the different sections of a rectilinear current are mutually attractive; not repulsive, as is generally supposed. To remove as much as possible all resistance to the motion of the wire, and also to simplify the conditions of the experiment, instead of floating the

moveable wire upon mercury, he placed it upon one of the arms of
a delicate torsion balance. The ends of the moveable wire were
made to bear slightly against the extremes of the two wires in
connexion with the poles of the pile. When the current was
established, the moveable wire placed upon the balance was
attracted by the wires proceeding from the pile, and not repelled
as in the case of Ampère's experiment; and the stronger the
current, and the more complete the contact of the ends of the
wires, the greater the attraction was found to be.

There is one objection which may be urged against the conclusiveness of the experiment. It is well known that the ends
of two wires connecting the poles of a voltaic pile before they
are brought into contact, are statically charged, the one with
positive, and the other with negative electricity, the intensity of
the charge depending upon the force of the pile. Now it is evident that these two wires being charged with different electricities, must attract each other. It is evident also that, however
close the two ends of the wires may be brought together, unless
they are in absolute contact in every part, a thing impossible
without soldering the ends together, the current will not pass
from the extremity of the one wire to that of the other through the
intermediate space, which is almost non-conducting, unless there
is an excess of positive electricity on the one wire and of negative
on the other; and the more so, considering the low tension of
voltaic electricity. This being the case, the ends of the conducting wire and those of the moveable branch will always be charged
with different quantities of electricity, and hence attraction will
be the consequence; yet one would suppose that, unless the
current is in reality self-attractive, the attraction arising from
this cause would not be able to overcome the repulsion which
must ensue from the action of angular currents just considered.

There is one objection to the common notion that the parts of
a rectilinear current are mutually repulsive, that I have never
seen adduced, which is as follows : It results from a law of Ohm,
which has been confirmed experimentally by Kohlrausch, that in
the conductor connecting the poles of a voltaic pile, while the
current is circulating, different sections of the conductor are differently charged. In any part of the conductor whatever, a
section taken towards the positive pole is always positive in relation to a section taken towards the negative pole; and *vice
versâ*, a section towards the negative pole is negative in relation
to any section taken towards the positive pole. It follows from
this that the different sections of the current must attract each
other.

The difference of the tension of any two sections depends upon
the resistance to conduction; that is to say, the force by which

any two sections of the current attract each other when the current is passing, is always as the amount of the resistance which opposes this attraction.   The attraction must always exceed the resistance, or else there can be no current.   What does all this mean if we do not admit that the sections of the current mutually attract each other ?

The same difference of electric state exists in the various sections of the pile itself; for we know that there is always an excess of positive electricity at the one pole, and of negative at the other, and these electricities must tend to unite through the pile itself.   But there is opposed to the attractive force of the electricities for uniting, not only the resistance in the pile itself, but also the electromotive force which decomposes the electricity. This electromotive force must always exceed the attraction of the electricities: this must be admitted; for unless the force which separates the electricities is greater than the attractive force which tends to unite them, there could be no decomposition of the electricity, and of course no current.

In the pile itself there are two forces—one tending to unite the various sections of the current, and the other tending to separate them; the latter force being the strongest, the sections of the current in the pile will mutually repel each other.   In the external conductor which unites the poles of the pile, the latter force has no existence; hence the various sections of the current in the conductor are mutually attractive.   By virtue of the repulsion in the pile and the attraction in the conductor, the electricity decomposes in the former and unites in the latter; and this constitutes what we call an electric current.

XXXIX. *On Theories of Magnetism and other Forces, in reply to Remarks by* Professor Maxwell. *By* J. CHALLIS, *F.R.S., F.R.A.S., Plumian Professor of Astronomy and Experimental Philosophy in the University of Cambridge* [*].

IN an article on "Molecular Vortices applied to Magnetic Phenomena," contained in the March Number of the Philosophical Magazine, Professor Maxwell has made, respecting certain points of the general physical theory which I have lately proposed, some remarks which call for a reply.   I refer chiefly to two paragraphs in p. 163, the first of which is as follows:—
"Currents, issuing from the north pole and entering the south pole of a magnet, or circulating round an electric current, have the advantage of representing correctly the geometrical arrangement of the lines of force, if we could account for the pheno-

* Communicated by the Author.

mena of attraction, or for the currents themselves, or explain
their continued existence." The generation of such currents I
have explained on hydrodynamical principles in my theories of
galvanism and magnetism, as also in that of electricity. They
are shown to be *secondary* currents, which are always produced
when a uniform primary current traverses a medium, in which
there is a gradation of density, such as that which must exist
from the top to the bottom of a heavy mass resting on a horizontal plane, in order that the force of gravity on the individual
particles may be counteracted. The *primary* currents are ascribed
exclusively to motions of the æther caused by the rotations of
the earth and of the other bodies of the solar system about their
axes. As this is a constant cause, the streams are constant.
The retention of an induced state of gradation of density from
end to end, is considered to be the distinctive property of a
magnetized bar. The observed attractions and repulsions are
satisfactorily accounted for by the variation of the fluid pressure
from point to point of space in the secondary currents considered
as instances of steady motion, such variation, together with the
dynamical effect of the currents, producing differences of pressure at different points of the surfaces of the atoms of which
the substances attracted or repelled are supposed to consist.
Thus the three explanations which Professor Maxwell considers
to be requisite respecting currents to which the phenomena of
galvanism and magnetism are attributed, are in fact given by
my general theory quite consistently with its original hypotheses.

The other paragraph commences thus :—" Undulations issuing
from a centre would, according to the calculations of Professor
Challis, produce an effect similar to attraction in the direction
of a centre." I consider that both central attraction and central
repulsion are accounted for by my calculations. Professor Maxwell then adduces the following objection :—" Admitting this to
be true, we know that two series of undulations traversing the
same space do not combine into one resultant as two attractions
do, but produce an effect depending on relations of *phase* as
well as intensity." This point I have considered in articles 2
and 5 of the Theory of Gravity contained in the Philosophical
Magazine for December 1859. There is no limitation as to
the function W, which expresses the velocity or condensation of
the ætherial waves, excepting that it must either be a single
circular function, or consist of the sum of several such functions.
Let it in general be represented by $\Sigma . m \sin (bt + c)$. Then, according to the theoretical calculation, the motion of translation
to an atom in the direction of the propagation of the
that is, *the repulsive action*, depends on the non-periodic
given part of the *square* of this function, the quantity which I have

called q being in this case insignificant. Now, whatever be the phases of the several component functions, the non-periodic part of the square of their sum is equal to the sum of the non-periodic parts of the squares of the separate functions. In the case of *attractive action*, according to the investigation given in the Mathematical Theory of Attractive Forces contained in the Philosophical Magazine for November 1859, the motion of translation depends on the value of q, and on the product of W and $\frac{d^2 W}{dt^2}$. But it is easily seen that if W have the form above assigned, the non-periodic part of the product for the sum of the terms is the sum of the non-periodic parts of the products for the component terms taken separately, independently of their phases. In short, as we are only concerned with *squares* of circular functions, the mutual interferences by difference of phase do not come under consideration. On this account the dynamical effects of two series of undulations from separate sources, take place independently of each other, and combine according to the laws of the composition of accelerative forces. To the additional objection, that, "if the series of undulations be allowed to proceed, they diverge from each other without any mutual action," I can make no reply, because I do not understand it. I can only conclude that it was written under some misapprehension.

Professor Maxwell goes on to assert that "the mathematical laws of attractions are not analogous in any respect to those of undulations." In making this assertion he must surely have overlooked the very remarkable analogies exhibited by the facts, that undulations, like *central* forces, diverge from a centre, and diminish in intensity according to some law of the distance. Each body in the universe on which a series of undulations is incident, becomes a centre of minor undulations, just as when waves on the surface of water encounter an insulated obstacle, the obstacle becomes a centre of subordinate waves. The continuous generation of subordinate undulations corresponds to the maintenance of the gravitating power of the body.

Perhaps, however, the assertion, although it is not limited, was intended to apply only to such attractions as are observed in a magnetic field. If so, it is in accordance with my general theory, which makes a distinction between attractions and repulsions by means of *undulations,* and attractions and repulsions due to *currents*. Electric, galvanic, and magnetic forces are of the latter kind. But, while it is admitted that the laws of these forces "have remarkable analogies with those of currents," I should not say that they are analogous "to the laws of the conduction of heat and electricity, and of elastic bodies," because, according to the views which I maintain, these are phenomena which ultimately may receive explanations by means of ætherial undulations and currents, and therefore ought not to be put in the same category.

As the article I have been referring to contains a theory of magnetic phenomena wholly different from that which I have advanced, it may be worth while to point out a difference in principle between the fundamental hypotheses of the two theories. Professor Maxwell assumes the existence of a magnetic medium, which is not fluid, but "mobile," and which acts along lines of magnetic force by *stress* combined with hydrostatic pressure; in other words, there is greater pressure in the equatorial, than in the axial, direction of the magnetic field. To account for this difference of pressure, it is assumed that "molecular vortices" circulate about axes parallel to the lines of magnetic force. Why they are called "molecular" is not expressly stated in this article; but it may be gathered from other of the author's writings, that he conceives the matter of the vortices to consist of molecules which by their motions may come into collision, the resulting dynamic effect depending on the number and frequency of such collisions. It is not my intention to criticise these hypotheses, which have been adduced merely for the sake of remarking that, as they are of a particular character, and have been framed apparently with special reference to observed laws of magnetic phenomena, the results of a mathematical investigation founded on them can hardly amount to more than an empirical expression of those laws. After all that can be done by this kind of research, an independent and *à priori* theory of the same kind as that which I have proposed, if not the very one, is still needed.

The hypotheses on which my investigations have been founded are these only. The physical forces are modes of action of the pressure of the æther, which is a continuous fluid medium, having the property of pressing in proportion to its density, and filling all space not occupied by the discrete atoms of sensible bodies, which atoms are inert, spherical, and of different, but constant, magnitudes. It may be remarked that these hypotheses include no ideas that are not intelligible by sensation and experience, and therefore conform to the rule of philosophy according to which our knowledge of natural operations must ultimately rest on such ideas. Also they are strictly related to antecedent and existing physical science. The prominent terms, *æther* and *atom*, which had their origin in ancient speculations, have obtained remarkable significance in modern science,—the first, by explanations of the phenomena of light, and the other, by aiding us to conceive of chemical analyses. The hypotheses under the above form are mainly due to Newton, who gave a definition of atoms, and suggested the dynamic action of the æther; but the state of mathematics in his day did not allow of investigating the consequences of the latter idea.   In attempting to do this in the present advanced state of mathematics, I have only added, for that purpose, to the Newtonian hypotheses, an exact definition of the æther, and the supposition that the atoms are spherical.   In the first instance I applied the hypotheses as a foundation for a theory of light, having long since seen that the theory which proposes to account for the phenomena of light by the oscillations of the discrete atoms of a medium having axes of elasticity, is contradicted by facts, and must therefore be abandoned.   This charge I have brought against it in an article on elliptically polarized light in the Philosophical Magazine for April 1859 (p. 288); and it has found no defender.   When this first application was made, I had no conception of any modes of applying the same hypotheses to explain the phenomena of gravity, electricity, galvanism, and magnetism.   If they are found to admit of applications so varied and extensive, the explanations are no longer *personal*, the hypotheses themselves explain, because they are true.   The only advantage I pretend to possess in these researches is, the discovery of the true principles of the application of partial differential equations to determine the motion and pressure of an elastic fluid.   But this kind of reasoning, though it is indispensable for the establishment of the truth of the general physical theory, may be tried on its own merits, quite independently of its application in that theory.   For this reason I expressed the intention of carefully revising the proofs of the propositions in hydrodynamics which have been already enunciated in this Journal, and the results of which have been applied in the physical theory.   But my occupations do not allow of entering on this task at present.

Cambridge Observatory,
    March 16, 1861,

XL. *On certain peculiar Forms of Capillary Action.*
By Thomas Tate, Esq.[*]

LIQUIDS rise in small tubes by what is called capillary action, that is, by the cohesion of the particles of the liquid for one another, as well as by their adhesion to the sides of the tube.   It has been ascertained that the height to which a liquid is raised by capillary action varies inversely as the diameter of the tube.   The chief object of this paper is to determine, by direct experiment, the law of capillary resistance exerted

* Communicated by the Author.

by a liquid filling small orifices or perforations made in rigid plates of different thicknesses.

Let A C represent a wide glass tube, closed at the top, and having a perforated plate E e, capable of being wetted, cemented to its lower extremity; E B a glass tube, about half an inch in diameter, cemented to this plate, open at its lower extremity, and communicating with the tube A C; e a small perforation made in the plate; D B a glass vessel containing water or any other liquid to be examined. The tube E B is graduated from the exterior surface of the plate into inches and decimal parts of an inch. The tubes being filled with water, and the extremity B inserted in the water contained in the vessel D B, it will be found that the orifice e may be raised for some inches above the level F D of the water in the vessel before the atmospheric air will enter the orifice. The height C D, at which the external

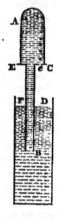

Fig. 1.

air enters the orifice, obviously gives us the measure of the capillary resistance of the liquid in the orifice. The following results of experiments show that, the temperature being constant, the height of the column C D, measuring the capillary resistance of the liquid, varies inversely as the diameter of the orifice.

The experiments recorded in the following Table of results, were made with the apparatus represented in fig. 1. The orifices were made in plates of gutta percha by means of fine steel wires, whose diameters had been previously determined. Slight oscillations and other extraneous causes having been found to affect the results of the experiments, each result here given is the mean of five experiments.

Table of results of experiments giving the columns of capillary resistance corresponding to different diameters of the orifices.

The liquid was water at the temperature of 56° F., and the thickness of the plate was ·05 of an inch.

Diameter of the orifice in parts of an inch, D.	Corresp. column C D of capillary resist. in inches, λ.	Value of λ by formula $\lambda = \frac{1}{22} \cdot \frac{1}{D}$
$\frac{1}{18}$	3·35	3·32
$\frac{1}{36}$	1·80	1·82
$\frac{1}{72}$	1·40	1·36
$\frac{1}{86}$	1·12	1·14

The near coincidence of the results of the second and third columns shows that, *other things being the same, the height of the column* C D, *measuring the capillary resistance, varies inversely. as the diameter of the orifice.*

The column of resistance, C D, is slightly affected by the thickness of the plate: thus with a half thickness of plate, and with an orifice of $\frac{1}{75}$th of an inch, the height of this column was found to be 3 inches nearly.

The number of perforations made in the plate does not affect the results.

An increase of temperature sensibly reduces the height of the column C D: thus at the temperature of 84°, with an orifice of $\frac{1}{75}$th of an inch, this column was found to be about $\frac{1}{10}$th of an inch less than it was at the temperature of 56°.

The column of capillary resistance, C D, for viscous liquids, such as diluted solutions of gum and sugar, was found to be considerably greater than the corresponding column for water.

Let the tube E B (fig. 2), closed at the top by the perforated plate E e, be depressed in the liquid, the orifice e being wet; it will be found that the liquid will not rise in the tube to the same level as the liquid in the vessel: the column of depression C D, in this case measuring the capillary resistance under the same

Fig. 2.   Fig. 3. 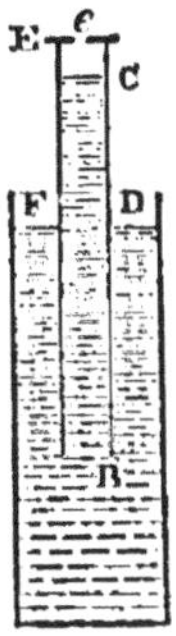   Fig. 4.

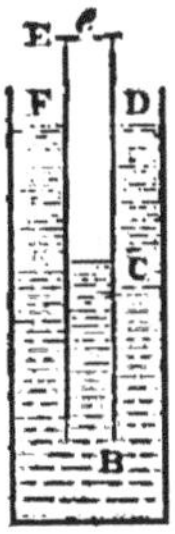

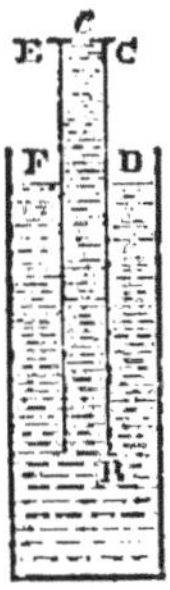

circumstances, is *nearly* equal to the elevation of the column C D of fig. 1. Here the resistance is simply due to the film of liquid filling the orifice e. It is scarcely necessary to state that, when the orifice e is dry, or nearly dry, the liquid will rise in the tube to the same level as the liquid in the vessel E B.

In like manner, if the tube, partially filled with liquid, be raised as represented in fig. 3, it will be found that the liquid will stand in the tube some distance higher than the level of the liquid in the vessel. The column of elevation, C D, in this case measures the capillary resistance to the pressure of the external air.

If the tube, completely filled with liquid, be raised as represented in fig. 4, it will be found that the liquid will stand in the tube some distance higher than the level of the liquid in the vessel. The column of elevation C D, measuring the capillary action, was found, under the same circumstances, to be nearly the same as in the preceding cases. Here the capillary action is restricted to the orifice of the thin plate, the diameter of the tube E B with which the experiment was performed being about one inch. This experiment strikingly shows that *the height of the column measuring capillary action for any given liquid at a constant temperature depends* (chiefly, if not entirely) *upon the diameter of that portion of the tube immediately in contact with the upper surface of the liquid.*

March 18, 1861.

XLI. *On a Theorem of* Abel's *relating to Equations of the Fifth Order. By* A. CAYLEY, *Esq.**

THE following is given (Abel, *Œuvres*, vol. xi. p. 253) as an extract of a letter to M. Crelle:—

"Si une équation du cinquième degré, dont les coefficients sont des *nombres rationnels*, est résoluble algébriquement, on peut donner aux racines la forme suivante,

$$x = c + A a^{\frac{1}{2}} a_1^{\frac{2}{3}} a_2^{\frac{2}{3}} a_3^{\frac{2}{3}} + A_1 a_1^{\frac{1}{2}} a_2^{\frac{2}{3}} a_3^{\frac{2}{3}} a^{\frac{2}{3}} + A_2 a_2^{\frac{1}{2}} a_3^{\frac{2}{3}} a^{\frac{2}{3}} a_1^{\frac{2}{3}} + A_3 a_3^{\frac{1}{2}} a^{\frac{2}{3}} a_1^{\frac{2}{3}} a_2^{\frac{2}{3}},$$

où

$$a = m + n\sqrt{1+e^2} + \sqrt{h(1+e^2 + \sqrt{1+e^2})},$$
$$a_1 = m - n\sqrt{1+e^2} + \sqrt{h(1+e^2 - \sqrt{1+e^2})},$$
$$a_2 = m + n\sqrt{1+e^2} - \sqrt{h(1+e^2 + \sqrt{1+e^2})},$$
$$a_3 = m - n\sqrt{1+e^2} - \sqrt{h(1+e^2 - \sqrt{1+e^2})},$$

$$A = K + K'a + K''a_2 + K'''aa_2, \quad A_1 = K + K'a_1 + K''a_3 + K'''a_1a_3,$$
$$A_2 = K + K'a_2 + K''a + K'''aa_2, \quad A_3 = K + K'a_3 + K''a_1 + K'''a_1a_3.$$

Les quantités c, h, e, m, n, K, K', K'', K''' sont des *nombres rationnels.*

"Mais de cette manière l'équation $x^5 + ax + b = 0$ n'est pas résoluble tant que a et b sont des quantités quelconques. J'ai trouvé de pareils théorèmes pour les équations du $7^{\text{ème}}$, $11^{\text{ème}}$, $13^{\text{ème}}$, &c. degré. Fribourg, le 14 Mars, 1826."

The theorem is referred to by M. Kronecker (Berl. *Monatsb.* June 20, 1853), but nowhere else that I am aware of.

It is to be noticed that in the expressions for a, a_1, a_2, a_3, the

* Communicated by the Author.

radicals are such that

$$\sqrt{1+e^2}\sqrt{h(1+e^2+\sqrt{1+e^2})}\sqrt{h(1+e^2-\sqrt{1+e^2})}=he(1+e^2),$$

a rational number.

The theorem is given as belonging to numerical equations; but considering it as belonging to literal equations, it will be convenient to change the notation; and in this point of view, and to avoid suffixes and accents, I write

$$x=\theta+A\alpha^{\frac{1}{2}}\beta^{\frac{3}{4}}\gamma^{\frac{1}{4}}\delta^{\frac{3}{4}}+B\beta^{\frac{1}{2}}\gamma^{\frac{3}{4}}\delta^{\frac{1}{4}}\alpha^{\frac{3}{4}}+C\gamma^{\frac{1}{2}}\delta^{\frac{3}{4}}\alpha^{\frac{1}{4}}\beta^{\frac{3}{4}}+D\delta^{\frac{1}{2}}\alpha^{\frac{3}{4}}\beta^{\frac{1}{4}}\gamma^{\frac{3}{4}},$$

where

$$\alpha=m+n\sqrt{\Theta}+\sqrt{p+q\sqrt{\Theta}},$$
$$\beta=m-n\sqrt{\Theta}+\sqrt{p-q\sqrt{\Theta}},$$
$$\gamma=m+n\sqrt{\Theta}-\sqrt{p+q\sqrt{\Theta}},$$
$$\delta=m-n\sqrt{\Theta}-\sqrt{p-q\sqrt{\Theta}};$$

the radicals being connected by

$$\sqrt{\Theta}\sqrt{p+q\sqrt{\Theta}}\sqrt{p-q\sqrt{\Theta}}=s,$$

and where

$$A=K+L\alpha+M\gamma+N\alpha\gamma,\quad B=K+L\beta+M\delta+N\beta\delta,$$
$$C=K+L\gamma+M\alpha+N\alpha\gamma,\quad D=K+L\delta+M\beta+N\beta\delta,$$

in which equations θ, m, n, p, q, Θ, s, K, L, M, N are rational functions of the elements of the given quintic equation.

The basis of the theorem is, that the expression for x has only the five values which it acquires by giving to the quintic radicals contained in it their five several values, and does not acquire any new value by substituting for the quadratic radicals their several values. For, this being so, x will be the root of a rational quintic; and conversely.

Now attending to the equation

$$\sqrt{\Theta}\sqrt{p+q\sqrt{\Theta}}\sqrt{p-q\sqrt{\Theta}}=s,$$

the different admissible values of the radicals are

$$\sqrt{\Theta}, \quad \sqrt{p+q\sqrt{\Theta}}, \quad \sqrt{p-q\sqrt{\Theta}},$$
$$-\sqrt{\Theta}, \quad \sqrt{p-q\sqrt{\Theta}}, \quad -\sqrt{p+q\sqrt{\Theta}},$$
$$\sqrt{\Theta}, \quad -\sqrt{p+q\sqrt{\Theta}}, \quad -\sqrt{p-q\sqrt{\Theta}},$$
$$-\sqrt{\Theta}, \quad -\sqrt{p-q\sqrt{\Theta}}, \quad \sqrt{p+q\sqrt{\Theta}},$$

corresponding to the systems

$$\alpha, \quad \beta, \quad \gamma, \quad \delta$$
$$\beta, \quad \gamma, \quad \delta, \quad \alpha$$
$$\gamma, \quad \delta, \quad \alpha, \quad \beta$$
$$\alpha, \quad \beta, \quad \gamma, \quad \delta$$

of the roots $\alpha, \beta, \gamma, \delta$; *i. e.* the effect of the alteration of the values of the quadratic radicals is merely to cyclically permute the roots $\alpha, \beta, \gamma, \delta$; and observing that any such cyclical permutation gives rise to a like cyclical permutation of A, B, C, D, the alteration of the quadratic radicals produces no alteration in the expression for x.

The quantities $\alpha, \beta, \gamma, \delta$ are the roots of a rational quartic. If, solving the quartic by Euler's method, we write

$$\alpha = m + \sqrt{F} + \sqrt{G} + \sqrt{H}, \qquad \sqrt{FGH} = \nu, \text{ a rational function,}$$
$$\beta = m - \sqrt{F} + \sqrt{G} - \sqrt{H},$$
$$\gamma = m + \sqrt{F} - \sqrt{G} - \sqrt{H},$$
$$\delta = m - \sqrt{F} - \sqrt{G} + \sqrt{H},$$

then the expressions for F, G, H in terms of the roots are

$$(\alpha + \gamma - \beta - \delta)^2, \quad (\alpha + \beta - \gamma - \delta)^2, \quad (\alpha + \delta - \beta - \gamma)^2,$$

which are the roots of a cubic equation

$$u^3 - \lambda u^2 + \mu u - \nu^2 = 0,$$

where λ, μ, ν are given rational functions of the coefficients of the quartic. We have

$$\sqrt{G} + \sqrt{H} = \sqrt{(\sqrt{G} + \sqrt{H})^2} = \sqrt{G + H + 2\sqrt{GH}}$$
$$= \sqrt{\lambda - F + \frac{2\nu}{F}\sqrt{F}}.$$

So that, taking $\Theta = F$, the last-mentioned expressions for $\alpha, \beta, \gamma, \delta$ will be of the assumed form

$$\alpha = m + \sqrt{\Theta} + \sqrt{p + q\sqrt{\Theta}}, \text{ \&c.}$$

The equation

$$\sqrt{\Theta}\sqrt{p + q\sqrt{\Theta}}\sqrt{p - q\sqrt{\Theta}} = s$$

thus becomes

$$\sqrt{F}\sqrt{(G - H)^2} = s, \text{ or } F(G - H)^2 = s^2;$$

that is,

$$-F^3 + F(F^2 + G^2 + H^2) - 2FGH = s^2;$$

S 2

or, what is the same thing, and putting Θ for F,

$$-\lambda\Theta^2 + (\lambda^2 - \mu)\Theta - 3\nu^2 = s^2.$$

Hence in order that the roots of the quartic may be of the assumed form,

$$\alpha = m + \sqrt{\Theta} + \sqrt{p + q\sqrt{\Theta}}, \text{ \&c.,}$$

where m, p, q, Θ are rational, and where also

$$\sqrt{\Theta}\sqrt{p + q\sqrt{\Theta}}\sqrt{p - q\sqrt{\Theta}} = s, \text{ a rational function,}$$

the necessary and sufficient conditions are that the quartic should be such that the reducing cubic

$$u^3 - \lambda u^2 + \mu u - \nu^2 = 0$$

(whose roots are $(\alpha + \beta - \gamma - \delta)^2, (\alpha + \gamma - \beta - \delta)^2, (\alpha + \delta - \beta - \gamma)^2$) may have *one rational root* Θ, and moreover that the function

$$-\lambda\Theta^2 + (\lambda^2 - \mu)\Theta - 3\nu^2$$

shall be the *square of a rational function s*. This being so, the roots of the quartic will be of the assumed form,

$$\alpha = m + \sqrt{\Theta} + \sqrt{p + q\sqrt{\Theta}}, \text{ \&c.}$$

And from what precedes, it is clear that any function of the roots of the quartic which remains unaltered by the cyclical substitution $\alpha\beta\gamma\delta$, or what is the same thing, any function of the form

$$\phi(\alpha, \beta, \gamma, \delta) + \phi(\beta, \gamma, \delta, \alpha) + \phi(\gamma, \delta, \alpha, \beta) + \phi(\delta, \alpha, \beta, \gamma)$$

will be a rational function of m, Θ, p, q, s, and consequently of the coefficients of the quartic. The above are the conditions in order that a quartic equation may be of the Abelian form.

It may be as well to remark that, assuming only the system of equations

$$\alpha = m + \sqrt{\Theta} + \sqrt{T},$$
$$\beta = m - \sqrt{\Theta} + \sqrt{T'},$$
$$\gamma = m + \sqrt{\Theta} - \sqrt{T},$$
$$\delta = m - \sqrt{\Theta} - \sqrt{T'},$$

then any rational function of $\alpha, \beta, \gamma, \delta$ which remains unaltered by the cyclical substitution $\alpha\beta\gamma\delta$ will be a rational function of $\Theta, T + T', TT', \sqrt{TT'}(T - T'), \sqrt{\Theta}(T - T'), \sqrt{\Theta}\sqrt{TT'}$. In fact, suppose such a function contains the term

$$(\sqrt{\Theta})^\alpha (\sqrt{T})^\beta (\sqrt{T'})^\gamma ;$$

then it will contain the four terms

$$(\quad \sqrt{\Theta})^{\alpha}(\quad \sqrt{T})^{\beta}(\quad \sqrt{T'})^{\gamma},$$
$$(-\sqrt{\Theta})^{\alpha}(\quad \sqrt{T})^{\beta}(-\sqrt{T'})^{\gamma},$$
$$(\quad \sqrt{\Theta})^{\alpha}(-\sqrt{T})^{\beta}(-\sqrt{T'})^{\gamma},$$
$$(-\sqrt{\Theta})^{\alpha}(-\sqrt{T})^{\beta}(\quad \sqrt{T'})^{\gamma},$$

which together are

$$\Theta^{\alpha}\{(1+(-)^{\beta+\gamma}1)(\sqrt{T})^{\beta}(\sqrt{T'})^{\gamma}+(-)^{\alpha}[(-)^{\beta}1+(-)^{\gamma}1](\sqrt{T})^{\gamma}(\sqrt{T}$$

an expression which vanishes unless $(-)^{\beta}$, $(-)^{\gamma}$ are both positive or both negative. The forms to be considered are therefore

$$(-)^{\alpha}, \quad (-)^{\beta}, \quad (-)^{\gamma}$$

$(-)^{\alpha}$	$(-)^{\beta}$	$(-)^{\gamma}$
+	+	+
−	+	+
+	−	−
−	−	−

The first form is

$$(\sqrt{\Theta})^{\alpha}\{(\sqrt{T})^{\beta}(\sqrt{T'})^{\gamma}+(\sqrt{T})^{\beta}(\sqrt{T'})^{\gamma},$$

which, α, β, γ being each of them even, is a rational function of Θ, $T+T'$, TT'.

The second form is

$$(\sqrt{\Theta})^{\alpha}\{(\sqrt{T})^{\beta}(\sqrt{T'})^{\gamma}-(\sqrt{T})^{\beta}(\sqrt{T'})^{\gamma}\},$$

which, α being odd and β and γ each of them even, is the product of such a function into $\sqrt{\Theta}(T-T')$.

The third form is

$$(\sqrt{\Theta})^{\alpha}\{(\sqrt{T})^{\beta}(\sqrt{T'})^{\gamma}-(\sqrt{T})^{\gamma}(\sqrt{T'})^{\beta}\},$$

which, α being even and β and γ each of them odd, is the product of such a function into $\sqrt{TT'}(T-T')$.

And the fourth form is

$$(\sqrt{\Theta})^{\alpha}\{(\sqrt{T})^{\beta}(\sqrt{T'})^{\gamma}+(\sqrt{T})^{\gamma}(\sqrt{T'})^{\beta}\},$$

which, α, β, γ being each of them odd, is the product of such a function into $\sqrt{\Theta}(T-T')$.

Hence if $T=p+q\sqrt{\Theta}$, $T'=p-q\sqrt{\Theta}$, and

$$\sqrt{\Theta}\sqrt{p+q\sqrt{\Theta}}\sqrt{p-q\sqrt{\Theta}}=s,$$

$$\Theta, \quad T+T'(=2p), \quad TT'(=p^2-q^2\Theta), \quad \sqrt{TT'}(T-T')\left(=\frac{2q}{s}\right),$$

$$\sqrt{\Theta}(T-T')(=2q\Theta), \text{ and } \sqrt{\Theta}\sqrt{TT'}(=s)$$

are respectively rational functions. This is the *à posteriori* verification, that with the system of equations

$$\alpha = m + \sqrt{\Theta} + \sqrt{p + q\sqrt{\Theta}}, \text{ &c., } \sqrt{\Theta}\sqrt{p + q\sqrt{\Theta}}\sqrt{p - q\sqrt{\Theta}} = s,$$

any function

$$\phi(\alpha, \beta, \gamma, \delta) + \phi(\beta, \gamma, \delta, \alpha) + \phi(\gamma, \delta, \alpha, \beta) + \phi(\delta, \alpha, \beta, \gamma)$$

is a rational function.

The coefficients of the quintic equation for x must of course be of the form just mentioned; that is, they must be functions of $\alpha, \beta, \gamma, \delta$, which remain unaltered by the cyclic substitution $\alpha\beta\gamma\delta$. To form the quintic equation, I write

$$\theta - x = a,$$

$$A\alpha^{\frac{1}{2}}\beta^{\frac{1}{2}}\gamma^{\frac{1}{2}}\delta^{\frac{1}{2}} = b, \quad D\delta^{\frac{1}{2}}\alpha^{\frac{1}{2}}\beta^{\frac{1}{2}}\gamma^{\frac{1}{2}} = c, \quad B\beta^{\frac{1}{2}}\gamma^{\frac{1}{2}}\delta^{\frac{1}{2}}\alpha^{\frac{1}{2}} = d, \quad C\gamma^{\frac{1}{2}}\delta^{\frac{1}{2}}\alpha^{\frac{1}{2}}\beta^{\frac{1}{2}} = e;$$

then we have

$$0 = a + b + c + d + e,$$

and the quintic equation is

$$f1 \, f\omega \, f\omega^2 \, f\omega^3 \, f\omega^4 = 0,$$

where ω is an imaginary fifth root of unity, and

$$f\omega = a + b\omega + c\omega^2 + d\omega^3 + e\omega^4.$$

We have

$$f\omega \, f\omega^4 = \Sigma a^2 + (\omega + \omega^4)\Sigma' ab + (\omega^2 + \omega^3)\Sigma' ac,$$

$$f\omega^2 f\omega^3 = \Sigma a^2 + (\omega^2 + \omega^3)\Sigma' ab + (\omega^2 + \omega^3)\Sigma' ac,$$

where Σ' is Mr. Harley's cyclical symbol, viz.

$$\Sigma' ab = ab + bc + cd + de + ea;$$

and so in other cases, the order of the cycle being always *abcde*. This gives

$$f\omega \, f\omega^2 f\omega^3 f\omega^4 = \Sigma a^4 + \Sigma a^2 b^2 - \Sigma a^3 b + 2\Sigma a^2 bc - \Sigma abcd$$
$$- 5\Sigma' a^2(be + cd);$$

and multiplying by $f1$, $= \Sigma a$, and equating to zero, the result is found to be

$$\Sigma a^5 - 5abcde - 5\Sigma' a^3(be + cd) + 5\Sigma' a(b^2 e^2 + c^2 d^2) = 0.$$

Or arranging in powers of a, this is

$$\left. \begin{array}{l} a^5 \\[4pt] + a^3 . \quad -5(be + cd) \\[4pt] + a^2 . \quad 5(bc^2 + ce^2 + ed^2 + db^2) \\[4pt] + a \cdot \left\{ \begin{array}{l} 5(b^3 c + c^3 e + e^3 d + d^3 b) \\ + 5(b^2 e^2 + c^2 d^2 - becd) \end{array} \right. \\[12pt] + \left\{ \begin{array}{l} b^5 + c^5 + e^5 + d^5 \\ -5(b^3 de + c^3 bd + e^3 cb + d^3 ec) \\ + 5 \, bd^2 e^2 + cb^2 d^2 + ec^2 b^2 + de^2 c^2 \end{array} \right. \end{array} \right\} = 0,$$

the several coefficients being, it will be observed, cyclical functions to the cycle b, c, e, d.

Putting for a its value $-(x-\theta)$, and for b, c, d, e their values, the quintic equation in x is

$$
\begin{aligned}
&(x-\theta)^5 \\
&x-\theta)^3. \quad -5(AC+BD)\alpha\beta\gamma\delta \\
&x-\theta)^2. \quad -5(A^2B\gamma\delta+B^2C\delta\alpha+C^2D\alpha\beta+D^2A\beta\gamma)\alpha\beta\gamma\delta \\
&(x-\theta). \left\{
\begin{array}{l}
-5(A^3D\beta\gamma^2\delta+B^3A\gamma\delta^2\alpha+C^3B\delta\alpha^2\beta+D^3C\alpha\beta^2\gamma)\alpha\beta\gamma\delta \\
+5(A^2C^2+B^2D^2-ABCD)\alpha^2\beta^2\gamma^2\delta^2
\end{array}\right. \\
&\left\{
\begin{array}{l}
(A^5\beta\gamma^3\delta^2+B^5\gamma\delta^3\alpha^2+C^5\delta\alpha^3\beta^2+D^5\alpha\beta^3\gamma^2)\alpha\beta\gamma\delta \\
-5(A^3BC\gamma\delta+B^3CD\delta\alpha+C^3DA\alpha\beta+D^3AB\beta\gamma)\alpha^2\beta^2\gamma^2\delta^2 \\
+5(AB^2C^2\alpha\delta+BC^2D^2\beta\alpha+CD^2A^2\gamma\beta+DE^2A^2\delta\alpha)\alpha^2\beta^2\gamma^2\delta^2
\end{array}\right.
\end{aligned}
$$

where, as before,

$$
\begin{aligned}
A &= K+L\alpha+M\gamma+N\alpha\gamma, \\
B &= K+L\beta+M\delta+N\beta\delta, \\
C &= K+L\gamma+M\alpha+N\gamma\alpha, \\
D &= K+L\delta+M\beta+N\delta\beta;
\end{aligned}
$$

and the coefficients of the quintic equation are, as they should be, cyclical functions to the cycle $\alpha\beta\gamma\delta$.

2 Stone Buildings, W.C.,
February 10, 1861.

XLII. *On the Stability of Satellites in small Orbits, and the Theory of Saturn's Rings.* By DANIEL VAUGHAN[*]

THE mysterious revolutions of planets and comets were not rendered intelligible to astronomers until mathematical investigations revealed the peculiar curves which moving bodies must describe when left to the exclusive control of solar gravity. The process of deductive inquiry, which proved so beneficial in this and other departments of celestial mechanics, may be successfully applied to another problem which the results of telescopic observation have forced on the attention of mathematicians. The physical constitution of Saturn's rings, the circumstances on which their stability depends, and the causes which prevent their conversion into satellites, have already been made the subject of many able essays; but, though regarding these productions as valuable contributions to science, I think it advisable to select a

[*] Communicated by the Author.

less difficult road to the solution of the curious problem, and to seek a clue to the stability of the annular appendage of Saturn by investigating the form which matter must necessarily assume in very great proximity to a central body.

In my communication published in the Philosophical Magazine for last December (1860), I treated on the equilibrium of satellites revolving extremely near to their primaries; and I endeavoured to give an estimate of the smallest orbits which they could describe in safety. In the cases I considered, the satellite was supposed to have its movements adjusted for keeping the same point of its surface always directed towards the primary, not merely because the hypothesis facilitated the investigation, but because observation lends it every support, and the principles of natural philosophy furnish most cogent reasons for its adoption. In describing a very small orbit without such an adjustment, a satellite must experience, not only excessive tides in its seas, but even incessant commotions in its solid matter; and the destruction of power by friction necessarily involves a continual change in the rotatory movement of the subordinate world, after a manner analogous to that which I described in a paper presented to the British Association for the Advancement of Science in 1857. This must have the ultimate effect of establishing a synchronism of the orbital and diurnal movements, together with a coincidence of the planes in which they are performed; so that the disturbing force may give the secondary planet a permanent elongation, without rendering it a prey to the effects of violent dynamic action. From late researches, however, I am convinced that a want of these peculiar conditions would not seriously affect the fate of a large satellite when brought into dangerous proximity to its primary; and would not change, to any great extent, the magnitude of the orbit in which its dismemberment must be inevitable.

A homogeneous fluid satellite, having its motions adapted for keeping one part of its surface in perpetual conjunction with the primary, must find repose in a form differing little from an ellipsoid. This proposition, which in my last article was assumed as true, may be proved by showing that the relation between the forces exerted on every part of the fluid mass is almost precisely such as is necessary for equilibrium when the figure is an ellipsoid, the dimensions being small compared with the diameter of its orbit. For this purpose put A, B, and C for the major, mean, and minor semiaxes of the ellipsoid, while P, Q, and R express the forces of attraction at their extremities in the absence of all disturbances. Now at any point in the surface, the coordinates of which referred to the centre are represented by a, b, and c, the components of the attraction in the direction of each axis will be expressed by

$$\frac{a\mathrm{P}}{\mathrm{A}}, \quad \frac{b\mathrm{Q}}{\mathrm{B}}, \quad \frac{c\mathrm{R}}{\mathrm{C}}*. \quad \cdots \quad \cdots \quad (1)$$

If N denote the centrifugal force at the extremities of the major axis, the intensity at the proposed point will be $\dfrac{\mathrm{N}\sqrt{a^2+b^2}}{\mathrm{A}}$, and the components in the direction of the three semiaxes will be

$$-\frac{a\mathrm{N}}{\mathrm{A}}, \quad -\frac{b\mathrm{N}}{\mathrm{B}}, \quad 0. \quad \cdots \quad \cdots \quad (2)$$

To find the components of the disturbing force of the primary when the major axis ranges with its centre, we may use methods analogous to those pursued in the lunar theory for estimating the amount of solar disturbance. Thus, putting x for the radius of the circular orbit which the satellite describes, and M for the attractive force of the primary at the distance x, the attraction which it exerts on the satellite at the point under consideration will be

$$\frac{\mathrm{M}x^2}{x^2-2ax+a^2+b^2+c^2}. \quad \cdots \quad \cdots \quad (3)$$

This is equivalent to two forces—one acting in the direction of the major axis and expressed by

$$\frac{\mathrm{M}x^3}{(x^2-2ax+a^2+b^2+c^2)^{\frac{3}{2}}}, \quad \cdots \quad \cdots \quad (4)$$

the other directed to the centre of the satellite and expressed by

$$\frac{\mathrm{M}x^2\sqrt{a^2+b^2+c^2}}{(x^2-2ax+a^2+b^2+c^2)^{\frac{3}{2}}}. \quad \cdots \quad \cdots \quad (5)$$

From the first, (4), arises a disturbance operating exclusively in the direction of the major axis, and represented by

$$\mathrm{M}-\frac{\mathrm{M}x^3}{(x^2-2ax+a^2+b^2+c^2)^{\frac{3}{2}}}, \text{ or } -3\frac{\mathrm{M}a}{x}, \quad \cdots \quad (6)$$

the squares and higher powers of $\dfrac{a}{x}, \dfrac{b}{x}$, and $\dfrac{c}{x}$ being rejected as too small to affect the result to any appreciable degree. Under the same conditions, the radial force (5) resolved with reference to the three axes gives the components

$$\frac{\mathrm{M}a}{x}, \quad \frac{\mathrm{M}b}{x}, \quad \frac{\mathrm{M}c}{x}. \quad \cdots \quad \cdots \quad (7)$$

Accordingly, if X represent the sum of the components acting in

* A demonstration of this theorem may be found in the article on "Attraction" in the eighth edition of the *Encyclopedia Britannica*.

the direction of A, Y the sum of those in the direction of B, and Z the sum of those in the direction of C, it appears from (1), (2), (6), and (7) that

$$X = \frac{aP}{A} - \frac{aN}{A} - \frac{2aM}{x}, \quad \ldots \ldots \quad (8)$$

$$Y = \frac{bQ}{B} - \frac{bN}{B} + \frac{bM}{x}, \quad \ldots \ldots \quad (9)$$

$$Z = \frac{cR}{C} + \frac{cM}{x}. \quad \ldots \ldots \quad (10)$$

Now it is well known that, to satisfy the conditions of equilibrium, or to make gravity perpendicular to the surface in all parts of the satellite, it is necessary that

$$X da + Y db + Z dc = 0. \quad \ldots \ldots \quad (11)$$

Substituting their values for X, Y, and Z, there results

$$\left(\frac{P-N}{A} - \frac{2M}{x} \right) a da + \left(\frac{Q-N}{B} + \frac{M}{x} \right) b db + \left(\frac{R}{C} + \frac{M}{x} \right) c dc = 0. \quad (12)$$

But the equation of an ellipsoid is $\frac{a^2}{A^2} + \frac{b^2}{B^2} + \frac{c^2}{C^2} = 1$, and its differential, multiplied by the constant quantity S, becomes

$$\frac{Sada}{A^2} + \frac{Sbdb}{B^2} + \frac{Scdc}{C^2} = 0. \quad \ldots \quad (13)$$

A comparison of the corresponding terms of equations (12) and (13) will enable us to fix the necessary relations of the constants for satisfying the former. It thus appears that

$$\frac{P-N}{A} - \frac{2M}{x} = \frac{S}{A^2}, \quad \ldots \ldots \quad (14)$$

$$\frac{Q-N}{B} + \frac{M}{x} = \frac{S}{B^2}, \quad \ldots \ldots \quad (15)$$

$$\frac{R}{C} + \frac{M}{x} = \frac{S}{C^2}. \quad \ldots \ldots \quad (16)$$

These relations being independent of the values of the coordinates *a*, *b*, and *c*, they will be the same for every part of the surface of the body; and it follows that an equilibrium established at any one locality must extend to every part of the entire mass. Accordingly the relative magnitudes which the axes A, B, and C must possess, to make gravity perpendicular to the surface at any intermediate point, must give gravity a like vertical direction in all places, and secure the same stability to every portion of the satellite which has an ellipsoidal form. This, however, would not appear to be rigorously correct if, in the expressions for the disturbances by the primary; the squares and higher powers of $\dfrac{a}{x}$, $\dfrac{b}{x}$, and $\dfrac{c}{x}$ were retained; and accordingly the very close approximation to a true ellipsoid can be exhibited only when the size of the satellite is very small compared with that of its orbit. If the disproportion between both were not very great, the form of the satellite would resemble that of an egg slightly flattened by lateral pressure; yet even in such extreme cases the hypothesis in regard to the ellipsoidal form can lead to no material error in estimating the intensity of gravity on its surface, and the dimensions of the smallest orbit in which its parts can be held together by their mutual attraction.

From equations (8) and (14), (9) and (15), and (10) and (16), the following are readily deduced :—

$$X = a\left(\frac{P-N}{A} - \frac{2M}{x}\right), \text{ or } = \frac{aS}{A^2}, \quad \cdots \quad (17)$$

$$Y = b\left(\frac{2-N}{B} + \frac{M}{x}\right), \text{ or } = \frac{bS}{B^2}, \quad \cdots \quad (18)$$

$$Z = c\left(\frac{R}{C} + \frac{M}{x}\right), \text{ or } = \frac{cS}{C^2}. \quad \cdots \quad (19)$$

But calling the force of gravity at the given locality F, it is evident that F is equal to $\sqrt{X^2 + Y^2 + Z^2}$. On substituting for X, Y, and Z their values given by the last equation, there results

$$F = \sqrt{\frac{a^2S^2}{A^4} + \frac{b^2S^2}{B^4} + \frac{c^2S^2}{C^4}}, \text{ or } = \frac{S}{C^2}\sqrt{\frac{C^4}{A^4}a^2 + \frac{C^4}{B^4}b^2 + c^2}. \quad (20)$$

The quantity under the radical in the last expression is the value of the normal of the ellipsoid; and hence the force of gravity everywhere on the surface is proportional to the length of the normal corresponding to the locality. At the extremity of each axis this gravitative power, like the normal, is inversely proportional to the lengths of the axes themselves—a result which might be more readily deduced from equations (17), (18), and (19). In the first, for instance, if the point be situated at the end of the major axis, a becomes equal to A and X, which then expresses that the entire gravity at the point is equal to $\dfrac{S}{A}$; while the two other equations, (18) and (19), treated in a similar manner, would give $\dfrac{S}{B}$ and $\dfrac{S}{C}$ for the values of the intensity of gravity at the terminations of the mean and minor axes.

The cases in which equilibrium is impossible will be indicated by the occurrence of imaginary radicals, when we determine the relation between the constant quantities in formulæ (14), (15), and (16); and as this may be found by simple equations for all except the semiaxes A, B, and C, it is to their values alone that we must look for imaginary expressions. The formulæ referred to give

$$A= \frac{x}{4M}(P-N)\pm \frac{1}{4M}\sqrt{x^2(P-N)^2-8SMx},$$

$$B=-\frac{x}{2M}(Q-N)\pm \frac{1}{2M}\sqrt{4MSx+x^2(2-N)^2}, \qquad . \ (21)$$

$$C=-\frac{Rx}{2M}\pm \frac{1}{2M}\sqrt{4MSx+R^2x^2}.$$

Now it is evident that none of the above radicals can become imaginary except the first; and the stability of the body ceases to be possible when $x^2(P-N)^2-8MSx$, in passing from a positive to a negative value, becomes equal to nothing. In this case

$$A= \frac{x}{4M}(P-N). \quad . \quad . \quad . \quad . \quad (22)$$

But by comparing the expressions given in my last article for centrifugal force and the disturbance of the primary at the extremity of the major axis of the satellite, it appears that the latter is double the former, or that N equals $\frac{MA}{x}$. We may deduce the same result by considering that the orbital velocity of the satellite's centre is equal to $\sqrt{Mx}$; and from this the rotatory velocity of the extremity of the greater axis is equal to $\frac{A}{x}\sqrt{Mx}$, or $A\sqrt{\frac{M}{x}}$. Calling this v,

$$N= \frac{v^2}{A}, \text{ or } N= \frac{MA}{x}. \quad . \quad . \quad . \quad . \quad (23)$$

This value being substituted for N in the last equation, gives

$$A= \frac{x}{4M}\left(P-\frac{MA}{x}\right); \text{ whence } P= \frac{5AM}{x}. \quad . \quad . \quad (24)$$

The diminished force of gravity which, at the extremity of A, is represented by $P-N-\frac{2MA}{x}$, thus becomes $\frac{2AM}{x}$; so that if the satellite were a homogeneous fluid, the stability must become impossible when more than three-fifths of the attraction along the major axis is neutralized by centrifugal force and the disturbing influence of the central sphere.

The cause of the unstable equilibrium in such cases will be rendered more intelligible by a further examination of equation (14), which, on multiplying its members by A, becomes

$$\cdots \quad \cdots \quad P - N - \frac{2MA}{x} = \frac{S}{A}. \quad \cdots \quad \cdots \quad (25)$$

The terms of the first member constitute the expression for the force of gravity at the extremity of A; and the impossible root merely shows that gravity at this point, after having lost over three-fifths of its intensity by the disturbances, cannot amount to $\frac{S}{A}$, and consequently can no longer maintain the inverse ratio to the length of the axis. This peculiar relation between the length of each axis and the gravity at its extremity has already been deduced from formula (20), and is indispensable to the equilibrium of similar columns of fluid extending from these points, either to the centre of the body, or through shells of matter equally dense, and bounded by the surfaces of concentric ellipsoids similar in position and dimensions. This leads to the conclusion maintained on different grounds in my last communication, in which I regarded the rupture of the satellite as inevitable, when an increase of elongation would fail to give any preponderance to the pressure along the greater axis, or when the ellipticity required to be increased to an infinite extent to counteract a very slight augmentation of the disturbing forces. My former estimates, indeed, do not agree very closely with the present investigation in determining the amount of disturbance necessary to bring stability to an end; but in these estimates the eccentricity of the elliptical section containing the mean and minor axes of the satellite was neglected; and from more exact calculations, which are not yet in a condition to be published, it appears that some reduction must be made in the value I first assigned to the smallest orbit in which a homogeneous satellite could be preserved.

To furnish another proof that the central and the superficial conditions of equilibrium necessarily lead to the same results in every respect, let us suppose a portion of the fluid to be enclosed in three tubes; two of which are connected at the centre and extend to the nearest and most distant part of the surface, while the third stretches along the surface to meet their extremities, while it coincides with the plane in which they are situated. Now the force of gravity being $\left(P - N - \frac{2MA}{x} \right)$ at the extremity of the major axis, it must be reduced to $\left(\frac{P-N}{A} - \frac{2M}{x} \right)_a$ along this line at a distance from the centre denoted by a, and the element of the pressure in the tube (taking the transverse section and density of the fluid as unity) will be

$$\left(\frac{P-N}{A}-\frac{2M}{x}\right)a\,da. \qquad (26)$$

The integral of this expression, taken within the limits of $a=0$ and $a=A$, gives for the central pressure of the fluid in the longer tube

$$\frac{A}{2}\left(P-N-\frac{MA}{x}\right). \qquad (27)$$

A similar process applied to the fluid in the tube coinciding with the minor axis, will give for the differential of pressure,

$$\left(\frac{R}{C}+\frac{M}{x}\right)c\,dc; \qquad (28)$$

and a similar integration will give for its pressure at the centre,

$$\frac{C}{2}\left(R+\frac{MC}{x}\right). \qquad (29)$$

For stability it is necessary that the contents of both tubes should press to the centre with the same amount of force, or that

$$\frac{A}{2}\left(P-N-\frac{MA}{x}\right)-\frac{C}{2}\left(R+\frac{MC}{x}\right)=0. \qquad (30)$$

Now from the peculiar position which the third tube is supposed to occupy on the surface, the general equation for the equilibrium of its contents will become $X\,da+Z\,dc=0$, or by substitution,

$$\left(\frac{P-N}{A}-\frac{2M}{x}\right)a\,da+\left(\frac{R}{C}+\frac{M}{x}\right)c\,dc=0. \qquad (31)$$

Integrating within the limits of $a=A$, $c=0$, and $a=0$, $c=C$, this becomes

$$\frac{A}{2}\left(P-N-\frac{2MA}{x}\right)-\frac{C}{2}\left(R+\frac{MC}{x}\right)=0. \qquad (32)$$

The identity of equations (30) and (32), and the relation between (26), (28), and (31), show that the equilibrium of the internal and external parts of the mass depend on precisely the same conditions, and that the fluid should rush to the most prominent parts of the satellite from the surface, as well as from its internal regions, whenever gravity along the major axis was diminished more than 60 per cent. by the disturbing forces. Brevity compels me to omit the more lengthy investigation which would be required to show that such consequences are not peculiar to special localities, but are the same on all parts of the surface of the body.

These results might lead us to infer that a satellite which had been introduced into the region of instability by the action of a resisting medium, must undergo a sudden and not a gradual dismemberment. Before embracing this opinion, however, a few modifying circumstances should be considered. The change in the figure of the body must increase the time of rotation, while the diminished size of the orbit calls for a shorter period of revolution; and the synchronism of the diurnal and progressive movements will be destroyed. But we may safely assert that the effects of the resisting medium in producing this change are exceedingly small compared with the influence of tidal action in keeping the same side of the satellite always turned to its primary, especially when the distance from the latter became very small. The result in such cases must be a little different for a solid satellite, which accommodates its form to the new conditions of equilibrium by a limited number of paroxysmal changes separated by intervals of many millions of years. On such occurrences, the reduction of the velocity of rotation, together with the tendency of the major axis to range with the primary*, would lead to a series of librations, which, in a dangerous proximity to the latter body, would tend more to promote than to prevent the final dismemberment.

It must be also recollected that our formulæ have been deduced on the supposition that all parts of the satellite are equally dense; and some modifications are therefore required in applying them to the cases likely to occur in the realms of Nature. If the density increased very rapidly from the surface to the centre, gravity might be entirely suppressed at the ends of the greater axis before it became incapable of maintaining the stability of the internal matter; and it would seem that in such a case the satellite might part successively with many layers of the fluid of which it is composed, before the increased disturbance called for a general disunion of the internal mass. If, however, the increasing density towards the centre merely results from the great pressure in these localities, the separation of matter from the surface must weaken the tie which holds the remainder of the satellite together; and the dismembering action, when once begun, will proceed without interruption until a dissolution of the entire mass is completed. But it is in cases where the satellite is solid that the mighty change in its condition assumes the most awful character, as the cohesion of its parts must prevent the gradual loss of matter from its surface, and keep the disturbing forces under restraint until they become capable of

---

* There is some inaccuracy in my last article in the incidental statement respecting the intensity of this directive force at different distances.

effecting a simultaneous dilapidation of the entire planetary structure.

I have regarded it as important to trace the precise manner in which these sublime catastrophes must take place; not so much on account of their connexion with the existence of planetary rings, as for the light which they throw on the nature of temporary stars. In an article published in the Supplemental Number of the Philosophical Magazine for December 1858, I maintained that these singular displays of stellar brilliancy were great meteoric displays in the atmospheres, or rather the dormant photospheres, of dark central bodies of space, as they were traversed by the wrecks of dilapidated worlds. The same theory has been set forth in my paper presented to the British Association in 1857; and I have endeavoured in other publications to support it with satisfactory proof. But the most conclusive evidence on which it depends, is to be derived from the instantaneous manner in which the attendant of a dark central body must undergo a total dismemberment, as it explains the sudden manner in which these celestial curiosities are ushered into existence with all the splendour of distant suns. Humboldt, in the third volume of his 'Cosmos,' calls special attention to the fact of the extreme brilliancy of the temporary stars in their incipient stages, regarding it as a remarkable peculiarity, and one well deserving of consideration.

Without adducing any further evidence on this subject, I shall now proceed to trace the condition which matter must assume in the region where such disturbing forces render it incapable of forming a single mass, held together by the power of gravity. On the dismemberment of a satellite on this dangerous ground, the resulting host of fragments would scatter into numberless orbits; and the wide range over which they must extend may be estimated from the greatest and least size of the elliptical paths which their velocities and positions should assign to them. For these, however, we can only give at present approximate values, taking no cognizance of the mutual disturbances of the fragmentary host. And in this case the matter from the most distant part of the satellite would describe an ellipse, the diameter of which is equal to

$$\frac{2x^3(x+A)}{2x^3-(x+A)^3}.$$

The fragments from the nearest point of the dismembering mass would describe an elliptical orbit the diameter of which is

$$\frac{2x^3(x-A)}{2x^3-(x-A)^3}.$$

But the size of the smallest orbits might fall considerably below this limit, in consequence of the rupture of many of the fragments at their least distances from the primary, either by the attraction of that body, or by the heat evolved when they are transformed into blazing meteoric masses.

The condition which matter must ultimately assume in the central zone, where it can no longer exist as one great satellite or in a limited number of smaller ones, must depend in some degree on the form of the primary planet. If this body be an oblate spheroid, considerably flattened by rapid rotation, as is the case with Saturn, the orbits of the several fragments must be subject to apsidal motion, to an extent depending on their transverse axes and excentricities. Accordingly those fragments describing the same track will be equally affected by it, and will form a line which remains unbroken during many revolutions. As one ring of fragments is thus made to roll within another, it is evident that both must ultimately become circular; and the fragmentary host will at length exhibit the nearest approximation to a state of repose, by moving in exact circles around the central planet.

There are even more cogent provisions for equalizing the distribution of the great ocean of disconnected matter over the vast zone in which it circulates. The attraction of the central body which led to the great dismemberment, must be adequate not only to forbid the reconstruction of a satellite, but even to prevent the parts of the mighty wreck from congregating to any point in an undue proportion. Whenever a preponderance of matter occurred at any locality, the impediments of friction would tend to equalize the angular velocity of the nearest and most distant fragments in the group; and the new relations between gravity and centrifugal force would immediately lead to their dispersion by the disturbing action of the primary. If the latter body were a very flattened spheroid, it would serve to confine the great annular ocean of fragments to the same plane, in opposition to small effects arising from the disturbances of distant spheres; and Laplace has shown that, supposing Saturn's ring to consist of numerous independent satellites, they will be prevented from departing from a common plane, in consequence of the action of his equatorial matter.

In addition to the foregoing agencies for securing the peculiar characters of the annular appendage, I must notice another which is inseparable from the movements of such collections of fluid or solid matter circulating in independent orbits. A vast amount of heat must be developed by their friction and their mutual collisions; while the calorific influence of such a mechanical action will be augmented by the slight excentricity impressed on their orbits by the disturbances of the external satellites. The increased temperature originating from this cause, must not only permit the existence of fluids in the extensive fields of floating matter, but also maintain an atmospheric covering of vapour, to give more continuity and symmetry to the annular appendage. If the physical characters of Saturn's rings be such as matter, not having an improbably great density, must necessarily assume in the region which they occupy, the independent movements of its parts may be regarded as a continual source of heat, which may perhaps in some degree mitigate the sway of intolerable cold in the frigid zone of the solar system.

Cincinnati, Feb. 19, 1861.

XLIII. *On the Principles of Energetics.*—Part I. *Ordinary Mechanics.* By J. S. STUART GLENNIE, *M.A., F.R.A.S.*[*]

1. IN the introductory paper, "On the Principles of the Science of Motion," I suggested that this name might be given to a new general science, ranking with the similar science of Growth, and not be used merely as a general name for the sciences of Ordinary[†] Mechanics (Stereatics and Hydratics) and Molecular Mechanics (Physics and Chemics[‡]). The science of Motion would, as a distinct science, or as a philosophy of the Mechanical Sciences, consider, first, the relations of motions, as motion, and without reference to their originating or determining causes or forces; and secondly, the conditions, and correlations of the conditions, of Pressure, bodily, or molecular, to which modern experiment and analysis give us the hope of being able to refer all the forces of motion. To the former section of General Mechanics I would give the name, coined or made current by Ampère, Kinematics[§]; for the latter section I would adopt the term Energetics[||], already introduced by Rankine with a similar meaning to that given by the above statement of its object.

2. It is proposed in the following papers partially to develope the conceptions of the introductory paper, by stating the fundamental principles of the proposed Science of Energetics, with their applications to the mechanical interpretation of phenomena. The justification and development of these principles is the work of the sciences of Mechanics, Physics, and Chemics respectively.

3. The science of Energetics may be defined as the theory of

* Communicated by the Author.
† Would there not be a more definite distinction between the two branches of Mechanics by means of the adjectives Corporal and Molecular?
‡ Phil. Mag. January 1861, p. 54.
§ *Essai sur la Philosophie des Sciences.*
|| "A science whose subjects are material bodies, and physical phenomena in general," Edinb. Phil. Journ. N. S. ii. 1855, p. 125.

Mechanical, as distinguished from Biological Forces. And without such a theory it is evident that no general and concurrent laws or relations can be established between phenomena of Motion, as distinguished from phenomena of Growth. Attractions (gravic, electric, and magnetic) and Waves (acoustic, optic, and thermotic) are the motions offered by Physics for explanation by mechanical forces, or conditions of pressure. The constitution and combination of bodies are the phenomena of which Chemics require a similar mechanical interpretation. These sciences may be distinguished from Mechanics, with its ordinary limitation of meaning, as forming together the science of Molecular Mechanics. But in Ordinary Mechanics also there are phenomena, the causal relations of which have been hitherto as little established as those of the phenomena of Attraction or of Affinity. Such unexplained mechanical phenomena are the uniform motion of the planets, and their velocities of rotation, as yet unconnected even by an empirical law.

The principles of Energetics more particularly belonging to Ordinary Mechanics will therefore be in this paper applied to the explanation of these mechanical facts.

4. (I.) A Force is the condition of difference between two pressures in relation to a third.

5. To establish the principles of Energetics applicable to the first part of Mechanics, it seems unnecessary to use the term pressure with other than its usual limited meaning as Statical pressure. Such a meaning would at least be wide enough for this first principle. For it might be otherwise expressed :—the general cause of the movement of a body is a difference between two (previously) equilibrating pressures upon it. But in order that it may be seen, at least generally, how I propose to bring the idea of Pressure into Physics and Chemics, and hence to make this principle the foundation of Molecular as well as of Ordinary Mechanics, it may be well to state at once that " under the term Pressure I shall include every kind of force which acts between elastic bodies, or the parts of an elastic body, as the cause (condition) or effect of a state of strain, whether that force is tensile, compressive, or distorting* ; " and that I consider elasticity to be " une des propriétés générales de la matière. Elle est en effet l'origine réelle, ou l'intermédiaire indispensable des phénomènes physiques les plus importans de l'univers. . . . . La gravitation et l'elasticité doivent être considérées comme les effets d'une même cause qui rend dépendantes ou solidaires toutes les parties matérielles de l'univers†."

* Rankine, Camb. and Dub. Math. Journ. 1851, vol. ii. p. 49.

† Lamé, *Théorie Mathématique de l'Elasticité des Corps Solides*, pp. 1 and 2.

6. The special application of this principle is less to phenomena than to physical hypotheses. For as Force is thus conceived, not as an absolute entity acting upon matter, but as a condition of the parts of matter itself, and as a condition determined by the relative masses and distances of these parts, any valid hypothesis of a force or of a motion to account for any set of phenomena is thus seen to imply an assertion as to relative masses and distances which can be more or less readily submitted to experiment or observation and analysis. And hypotheses of Forces which, like electric and magnetic fluids, or "uniform elastic ethers, the sole source of physical power*,", exist absolutely, and are not merely expressions of facts of mass- and distance-difference, are by this principle rejected as unscientific. "Lorsq'une branche de la Physique mathématique est ainsi parvenue à écarter tout principe douteux, toute hypothèse restrictive, elle entre réellement dans une phase nouvelle. Et cette phase paraît définitive, car la série historique, et en même temps rationnelle, des progrès accomplis, signale une tendance constante vers l'indépendance de toute loi préconçue†."

How the mutually pressing or repelling parts of matter are to be conceived in order that from facts of difference in relative masses and distances alone, the forces of Molecular may be referred to similar conditions with those of Ordinary Mechanics, has been in the introductory paper indicated, and will in the second part of this paper be more fully developed.

7. (II.) Motion, the effect of Force, whether mechanical, physical, or chemical, may be distinguished as beginning or continuous; and continuous motion as uniform or accelerated. The condition of the beginning of motion is a difference of pressure on the body that begins to move; the condition of a uniform continuous motion is a neutralization of the resisting pressure; the condition of an accelerated continuous motion is a uniform or varying resisting pressure.

8. This principle evidently embodies those of the Inertia of Matter, of the Composition of Motions, and of Accelerating Force.

The principle of Inertia is the fundamental scientific principle of Non-spontaneity, or the impossibility of a motion undetermined by a change in the previously existing relations of the body. The inertia of a body or molecule is simply the relation between its pressure and that of the bodies acting upon it. All the meaning of this principle is in the relativity of the conception it gives of the phenomena of matter; it appears, therefore, to betray

---

* Challis, "On a Theory of Magnetic Force," Phil. Mag. February 1861, p. 107.

† Lamé, *Théorie Analytique de la Chaleur, Discours préliminaire*, p. vi.

some obscurity of thought to speak of "intrinsic or absolute inertia."

The law of the Composition of Motions is but an extension of that of Inertia[*]. For the compounding of a motion is but the beginning of another motion; and the change in velocity and line of motion of the particle due to each force (difference of pressure) is the same as if the others did not act.

There seems to be a clearer conception afforded of uniform and accelerated motion by referring these phenomena, as by this principle, to their actual physical conditions.

9. The application of this principle leads to the following theorem suggestive of an explanation of the apparent effect and non-effect of the resisting medium on the comets and planets respectively.

According as the resultant of a resisting medium passes or not through the centre of gravity of a revolving body is it an accelerating force of revolution, or a partially neutralized accelerating force of rotation.

If the medium is uniform, or if—though it varies in density, according to some such law as that with so great probability assumed for the solar medium, viz. inversely as the square of the distance from the central body[†]—the face of the revolving body is so small that the resultant of the resisting pressures thereon passes infinitesimally near the centre of gravity of the whole body, it may be easily proved that such a resultant of resistance will act as an accelerating force, which, did the body move on a solid surface, would retard its revolution, but which, as it moves through a fluid medium, will, by the progressive decrease of its major axis and excentricity, cause its orbit to approach more and more to the circular form; and there will hence result, as in the case of Encke's comet, a secular inequality in the expression of the mean longitude, and consequently in the period.

But if, with the above law of decreasing density, the resultant of resistance to revolution falls at a finite distance below the centre of gravity of the body, it is clear that an unbalanced pressure thus applied will affect, not the revolution, but the rotation of the body; and that the tendency either to cause or accelerate rotation will be partly at least neutralized by the resistance of the medium to this new motion.

For let a be the direction of revolution, a' the resultant of the resistance thereto of a medium varying in density. It is evi-

---

[*] Price, 'Treatise on Infinitesimal Calculus,' vol. iii. p. 370.

[†] See Encke, *Ueber die Existenz eines widerstehenden Mittels im Weltraume*; and Pontécoulant, *Théorie Analytique du Système du Monde,* vol. iii. book 4, chap. 5.

dent that α' will act as au
accelerating force of rotation in the direction β;
and that this rotation will
be retarded, and α' partly
neutralized by the resistance in the direction β'.

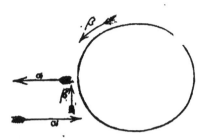

10. In the actual case
of the planets, their masses
and velocities of rotation
are such that the solar medium can be of course conceived, not
as causing, but only as tending slightly to accelerate their
rotations.   And a problem is by this theorem suggested of extreme interest, but also, in the present state of hydrodynamics,
of extreme difficulty, as to how far this accelerating force of rotation is neutralized.   The earth's rotation has been hitherto
considered and proved to be invariable, only in respect of the
action of the sun and moon; and it is to be remembered that
doubts have been thrown on its actual invariability even during
the short period of 4000 years; that one-tenth of a second in
10,000 years would be a large astronomical quantity; and that
their *actual* times are all that, at best, we know of the rotations
of the other planets.   I shall not at present offer any further
remarks on this problem, considered either as a purely hydrodynamical one, or with the data afforded by the planetary system,
except to note that nothing seems as yet to have been done towards
determining the relative effect of a resisting medium on (what
may at any moment be called) *the back* of a revolving and rotating body.   And it should seem that little further* can be done
towards the solution of this problem without experimental data on
this point especially.   The determination of the secular inequality,
the result of the variously directed and most improbably equilibrating forces of the medium, becomes still more complicated
when such a triple motion as that of a satellite is considered.

Such, then, is the theorem I would venture to offer as, if not
giving as yet the demonstrable explanation of the effect of the
resisting medium on the bodies of the solar system†, at least sug-
gesting new and very interesting experimental and analytical
problems in hydrodynamics.

11. (III.) The condition of Translation is a difference of
polar pressures on a point; the condition of Rotation is equal

* Stokes, ' On Fluid Friction.'
† I may refer to, though I cannot here discuss, the remarks on this subject of Sir John Herschel, ' Outlines of Astronomy,' 5th edit. p. 389 note;
and of Prof. Challis in his paper " On the Resistance of the Luminiferous
Medium," Phil. Mag. May 1859.

and opposite differences of polar pressures on two rigidly connected points; the condition of compound Translation and Rotation is unequal and opposite differences of polar pressures on two rigidly connected points; and the relation between the former and the latter depends on the distance between the centre of gravity of the body and the point of application of the resultant of such unequal opposite forces.

12. In explanation of this principle, it will be sufficient to remark that it is merely an expression of the ideas of a single force at the centre of gravity, a couple, and a single force not at the centre of gravity, in terms of the above general physical conception of a·Force.

13. The latter paragraph suggests the following general problem:—Given a force which, acting instantaneously at the centre of gravity of a body of a given mass in a vacuum, gives it a certain velocity: what are the different relations between the velocity of revolution and that of rotation when the same body is struck at certain different distances from the centre of gravity, and on any axis, by the same force?

The interest of this abstract problem is in the generality of its application to bodies which, while translated along *x*, rotate about *y* from a resultant of unequal pressures and resistances, having its point of application in or parallel to *z* (a wheel); and to bodies which, while translated along *x*, rotate about *z* from "a primitive impulse" applied at some point in *y* (a planet). For "le double mouvement de translation et de rotation des planètes, qui 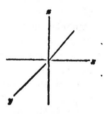 paraît au premier abord si compliqué, a pu résulter d'une seule impulsion primitive qui ne passait pas par leur centre de gravité*." From what previously existing conditions such a primitive impulse originated, whether from the rotation of a genetic ring, as in the Nebular Theory of Laplace, or otherwise, we have not here to inquire. But towards a mechanical explanation of the planetary elements, or a hydrodynamical theory of the formation of the solar system, the experimental† and analytical investigation of the above general problem seems to open the way.

14. In reference to the application of general or particular solutions of such a problem to the planets, it may be remarked that, as we are not of course here given the primitive impulse,

* Pontécoulant, *Théorie Analytique du Système du Monde*, vol. i. p. 144.
† Extend Plücker's experiments, for instance. See Taylor's 'Scientific Memoirs,' vol. iv. p. 16, and vol. v. pp. 584 and 621.

the first step towards a rational is the discovery of an empirical law of the rotations, in which such an element as the inclination of the rotation axis to the plane of revolution (easily calculable except for the two innermost and two outermost planets) would evidently be involved.

Such a law seems pointed to by the regularity of the decrease of the rotations, when the angular, instead of the linear velocities or times are considered. The respective angular velocities of rotation of the inner family are ·29811, ·26902, ·26181, and ·25879; and of Jupiter and Saturn, ·63313 and ·59907 respectively.

15. The attempts I have made to discover the law of the planetary rotations have had as yet no complete result[*]. But the following incidental observation with regard to the angular velocities of revolution and the distances may perhaps be worth noting towards such a theory of the formation of the system as above alluded to.

By Kepler's third law,

$$P = cD^{\frac{3}{2}}; \text{ whence } \frac{V}{D} \text{ or } \omega = c' \frac{1}{D^{\frac{3}{2}}}.$$

But under this law there might, in comparing successive velocities and distances, be found relations of inequality *ad infinitum*. The actual relations may, however, be thus expressed:—The angular velocities of revolution and the distances are in inverse geometrical progressions with inverse differences, except the innermost planet of each family.

To say that the distances are in geometrical progression, each nearer planet being half the distance of the next more remote[†], or that the angular velocities of revolution are in geometrical progression, each nearer planet revolving with twice the velocity of the next more remote, would be very far from accurate; but it seems interesting to observe, as by this law, that when the distance of a planet is *more* than twice that of the next inner, its angular velocity of rotation is *less* than half that of the next inner, and *vice versâ*. And that the only exceptions to this rule should be the innermost planet of each family, viz. Mercury and Jupiter, appears significant.

---

[*] The results of an approximative formula were given in a paper "On a general Law of Rotation applied to the Planets," read by me at the Oxford Meeting of the British Association, June 1860.

[†] See Humboldt's remarks on the Law of Bode, or rather of Titius. Cosmos, vol. iii. pp. 319, 320.

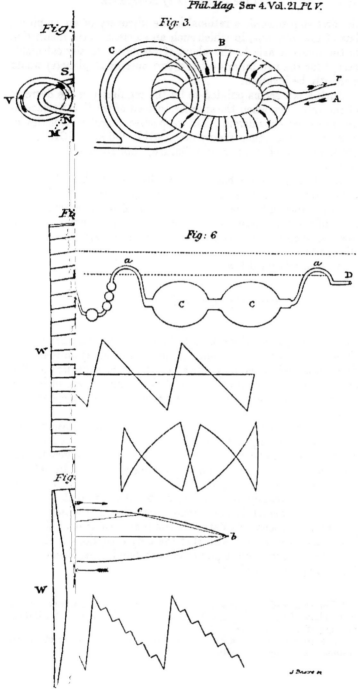

Fig.

Fig: 3.

Fig: 6

Fig.

J. Basire sc

Mercury and Venus.

$$D = 36,298051 = \tfrac{1}{2} \times 68,631843 + 1,982129,$$

$$\frac{V}{D} = 0.0029760 = 2 \times 0.0011651 + 0.0006468.$$

Venus and Earth.

$$D = 68,631843 = \tfrac{1}{2} \times 94,885000 + 21,189343,$$

$$\frac{V}{D} = 0.00116510 = 2 \times 0.00071676 - 0.00026842.$$

Earth and Mars.

$$D = 94,885000 = \tfrac{1}{2} \times 144,575333 + 22,587834,$$

$$\frac{V}{D} = 0.00071676 = 2 \times 0.00038108 - 0.00004540.$$

Jupiter and Saturn.

$$D = 493,654546 = \tfrac{1}{2} \times 905,087708 + 41,110692,$$

$$\frac{V}{D} = 0.000060411 = 2 \times 0.000024332 + 0.0000117470.$$

Saturn and Uranus.

$$D = 905,087708 = \tfrac{1}{2} \times 1820,020075 - 4,922329,$$

$$\frac{V}{D} = 0.000024332 = 2 \times 0.0000085313 + 0.0000072598.$$

Uranus and Neptune.

$$D = 1820,020075 = \tfrac{1}{2} \times 2849,991384 + 395,024383,$$

$$\frac{V}{D} = 0.0000085313 = 2 \times 0.0000043542 - 0.0000001871*.$$

6 Stone Buildings, Lincoln's Inn,
March 1861.

XLIV. *On Physical Lines of Force.* By J. C. MAXWELL, *Professor of Natural Philosophy in King's College, London†.*

[With a Plate.]

PART II.—*The Theory of Molecular Vortices applied to Electric Currents.*

WE have already shown that all the forces acting between magnets, substances capable of magnetic induction, and electric currents, may be mechanically accounted for on the sup-

---

* This fifteenth paragraph may be taken as an abstract of my paper "On the Revolutional Velocities and Distances of the Planets," read before the Royal Astronomical Society, Jan. 11, 1861.
† Communicated by the Author.

position that the surrounding medium is put into such a state that at every point the pressures are different in different directions, the direction of least pressure being that of the observed lines of force, and the difference of greatest and least pressures being proportional to the square of the intensity of the force at that point.

Such a state of stress, if assumed to exist in the medium, and to be arranged according to the known laws regulating lines of force, will act upon the magnets, currents, &c. in the field with precisely the same resultant forces as those calculated on the ordinary hypothesis of direct action at a distance. This is true independently of any particular theory as to the *cause* of this state of stress, or the mode in which it can be sustained in the medium. We have therefore a satisfactory answer to the question, " Is there any mechanical hypothesis as to the condition of the medium indicated by lines of force, by which the observed resultant forces may be accounted for ? " The answer is, the lines of force indicate the direction of *minimum pressure* at every point of the medium.

The second question must be, " What is the mechanical cause of this difference of pressure in different directions ? " We have supposed, in the first part of this paper, that this difference of pressures is caused by molecular vortices, having their axes parallel to the lines of force.

We also assumed, perfectly arbitrarily, that the direction of these vortices is such that, on looking along a line of force from south to north, we should see the vortices revolving in the direction of the hands of a watch.

We found that the velocity of the circumference of each vortex must be proportional to the intensity of the magnetic force, and that the density of the substance of the vortex must be proportional to the capacity of the medium for magnetic induction.

We have as yet given no answers to the questions, " How are these vortices set in rotation ? " and " Why are they arranged according to the known laws of lines of force about magnets and currents ? " These questions are certainly of a higher order of difficulty than either of the former ; and I wish to separate the suggestions I may offer by way of provisional answer to them, from the mechanical deductions which resolved the first question, and the hypothesis of vortices which gave a probable answer to the second.

We have, in fact, now come to inquire into the physical connexion of these vortices with electric currents, while we are still in doubt as to the nature of electricity, whether it is one substance, two substances, or not a substance at all, or in what way it differs from matter, and how it is connected with it.

We know that the lines of force are affected by electric currents, and we know the distribution of those lines about a current; so that from the force we can determine the amount of the current. Assuming that our explanation of the lines of force by molecular vortices is correct, why does a particular distribution of vortices indicate an electric current? A satisfactory answer to this question would lead us a long way towards that of a very important one, "What is an electric current?"

I have found great difficulty in conceiving of the existence of vortices in a medium, side by side, revolving in the same direction about parallel axes. The contiguous portions of consecutive vortices must be moving in opposite directions; and it is difficult to understand how the motion of one part of the medium can coexist with, and even produce, an opposite motion of a part in contact with it.

The only conception which has at all aided me in conceiving of this kind of motion is that of the vortices being separated by a layer of particles, revolving each on its own axis in the opposite direction to that of the vortices, so that the contiguous surfaces of the particles and of the vortices have the same motion.

In mechanism, when two wheels are intended to revolve in the same direction, a wheel is placed between them so as to be in gear with both, and this wheel is called an "idle wheel." The hypothesis about the vortices which I have to suggest is that a layer of particles, acting as idle wheels, is interposed between each vortex and the next, so that each vortex has a tendency to make the neighbouring vortices revolve in the same direction with itself.

In mechanism, the idle wheel is generally made to rotate about a *fixed* axle; but in epicyclic trains and other contrivances, as, for instance, in Siemens's governor for steam-engines*, we find idle wheels whose centres are capable of motion. In all these cases the motion of the centre is the half sum of the motions of the circumferences of the wheels between which it is placed. Let us examine the relations which must subsist between the motions of our vortices and those of the layer of particles interposed as idle wheels between them.

Prop. IV.—To determine the motion of a layer of particles separating two vortices.

Let the circumferential velocity of a vortex, multiplied by the three direction-cosines of its axis respectively, be a, β, γ, as in Prop. II. Let l, m, n be the direction-cosines of the normal to any part of the surface of this vortex, the outside of the surface being regarded positive. Then the components of the velocity of the particles of the vortex at this part of its surface will be

---

* See Goodeve's 'Elements of Mechanism,' p. 118.

$$n\beta - m\gamma \quad \text{parallel to } x,$$
$$l\gamma - n\alpha \quad \text{parallel to } y,$$
$$m\alpha - l\beta \quad \text{parallel to } z.$$

If this portion of the surface be in contact with another vortex whose velocities are α', β', γ', then a layer of very small particles placed between them will have a velocity which will be the mean of the superficial velocities of the vortices which they separate, so that if u is the velocity of the particles in the direction of x,

$$u = \tfrac{1}{2}m(\gamma' - \gamma) - \tfrac{1}{2}n(\beta' - \beta), \quad \dots \quad (27)$$

since the normal to the second vortex is in the opposite direction to that of the first.

Prop. V.—To determine the whole amount of particles transferred across unit of area in the direction of x in unit of time.

Let x_1, y_1, z_1 be the coordinates of the centre of the first vortex, x_2, y_2, z_2 those of the second, and so on. Let V_1, V_2, &c. be the volumes of the first, second, &c. vortices, and $\bar{V}$ the sum of their volumes. Let dS be an element of the surface separating the first and second vortices, and x, y, z its coordinates. Let ρ be the quantity of particles on every unit of surface. Then if p be the whole quantity of particles transferred across unit of area in unit of time in the direction of x, the whole momentum parallel to x of the particles within the space whose volume is $\bar{V}$ will be $\bar{V}p$, and we shall have

$$\bar{V}p = \Sigma u\rho dS, \quad \dots \quad (28)$$

the summation being extended to every surface separating any two vortices within the volume $\bar{V}$.

Let us consider the surface separating the first and second vortices. Let an element of this surface be dS, and let its direction-cosines be l_1, m_1, n_1 with respect to the first vortex, and l_2, m_2, n_2 with respect to the second; then we know that

$$l_1 + l_2 = 0, \quad m_1 + m_2 = 0, \quad n_1 + n_2 = 0. \quad \dots \quad (29)$$

The values of α, β, γ vary with the position of the centre of the vortex; so that we may write

$$\alpha_2 = \alpha_1 + \frac{d\alpha}{dx}(x_2 - x_1) + \frac{d\alpha}{dy}(y_2 - y_1) + \frac{d\alpha}{dz}(z_2 - z_1), \quad \dots \quad (30)$$

with similar equations for β and γ.

The value of u may be written:—

$$u = \frac{1}{2}\frac{d\gamma}{dx}\Big(m_1(x-x_1)+m_2(x-x_2)\Big)$$

$$+\frac{1}{2}\frac{d\gamma}{dy}\Big(m_1(y-y_1)+m_2(y-y_2)\Big)+\frac{1}{2}\frac{d\gamma}{dz}\Big(m_1(z-z_1)+m_2(z-z_2)\Big)$$

$$-\frac{1}{2}\frac{d\beta}{dx}\Big(n_1(x-x_1)+n_2(x-x_2)\Big)-\frac{1}{2}\frac{d\beta}{dy}\Big(n_1(y-y_1)+n_2(y-y_2)\Big)$$

$$-\frac{1}{2}\frac{d\beta}{dz}\Big(n_1(z-z_1)+n_1(z-z_2)\Big). \quad \ldots \ldots \ldots \quad (31)$$

In effecting the summation of $\Sigma u\rho dS$, we must remember that round any closed surface $\Sigma l dS$ and all similar terms vanish; also that terms of the form $\Sigma l y dS$, where l and y are measured in different directions, also vanish; but that terms of the form $\Sigma l x dS$, where l and x refer to the same axis of coordinates, do not vanish, but are equal to the volume enclosed by the surface. The result is

$$\nabla p = \frac{1}{2}\rho\left(\frac{d\gamma}{dy}-\frac{d\beta}{dz}\right)(V_1+V_2+\&c.); \quad \ldots \quad (32)$$

or dividing by $\overline{V}=V_1+V_2+\&c.$,

$$p=\frac{1}{2}\rho\left(\frac{d\gamma}{dy}-\frac{d\beta}{dz}\right). \quad \ldots \ldots \ldots \quad (33)$$

If we make

$$\rho = \frac{1}{2\pi}, \quad \ldots \ldots \ldots \ldots \ldots \quad (34)$$

then equation (33) will be identical with the first of equations (9), which give the relation between the quantity of an electric current and the intensity of the lines of force surrounding it.

It appears therefore that, according to our hypothesis, an electric current is represented by the transference of the moveable particles interposed between the neighbouring vortices. We may conceive that these particles are very small compared with the size of a vortex, and that the mass of all the particles together is inappreciable compared with that of the vortices, and that a great many vortices, with their surrounding particles, are contained in a single complete molecule of the medium. The particles must be conceived to roll without sliding between the vortices which they separate, and not to touch each other, so that, as long as they remain within the same complete molecule, there is no loss of energy by resistance. When, however, there is a general transference of particles in one direction, they must pass from one molecule to another, and in doing so, may ex-

perience resistance, so as to waste electrical energy and generate
heat.

Now let us suppose the vortices arranged in a medium in any
arbitrary manner. The quantities $\frac{d\gamma}{dy} - \frac{d\beta}{dz}$, &c. will then in
general have values, so that there will at first be electrical cur-
rents in the medium. These will be opposed by the electrical
resistance of the medium; so that, unless they are kept up by a
continuous supply of force, they will quickly disappear, and we
shall then have $\frac{d\gamma}{dy} - \frac{d\beta}{dz} = 0$, &c.; that is, $\alpha dx + \beta dy + \gamma dz$ will
be a complete differential (see equations (15) and (16)); so that
our hypothesis accounts for the distribution of the lines of force.

In Plate V. fig. 1, let the vertical circle E E represent an electric current flowing from copper C to zinc Z through the conductor E E', as shown by the arrows.

Let the horizontal circle M M' represent a line of magnetic
force embracing the electric circuit, the north and south directions being indicated by the lines S N and N S.

Let the vertical circles V and V' represent the molecular vortices of which the line of magnetic force is the axis. V revolves
as the hands of a watch, and V' the opposite way.

It will appear from this diagram, that if V and V' were contiguous vortices, particles placed between them would move downwards; and that if the particles were forced downwards by any
cause, they would make the vortices revolve as in the figure.
We have thus obtained a point of view from which we may regard
the relation of an electric current to its lines of force as analogous
to the relation of a toothed wheel or rack to wheels which it
drives.

In the first part of the paper we investigated the relations of
the statical forces of the system. We have now considered the
connexion of the motions of the parts considered as a system of
mechanism. It remains that we should investigate the dynamics
of the system, and determine the forces necessary to produce
given changes in the motions of the different parts.

Prop. VI.—To determine the actual energy of a portion of a
medium due to the motion of the vortices within it.

Let α, β, γ be the components of the circumferential velocity,
as in Prop. II., then the actual energy of the vortices in unit of
volume will be proportional to the density and to the square of
the velocity. As we do not know the distribution of density and
velocity in each vortex, we cannot determine the numerical value
of the energy directly; but since μ also bears a constant though
unknown ratio to the mean density, let us assume that the energy

in unit of volume is
$$E = C\mu(\alpha^2 + \beta^2 + \gamma^2),$$
where C is a constant to be determined.

Let us take the case in which
$$\alpha = \frac{d\phi}{dx}, \quad \beta = \frac{d\phi}{dy}, \quad \gamma = \frac{d\phi}{dz}. \quad \ldots \quad (35)$$

Let
$$\phi = \phi_1 + \phi_2, \quad \ldots \ldots \ldots \quad (36)$$
and let
$$\frac{\mu}{4\pi}\left(\frac{d^2\phi_1}{dx^2} + \frac{d^2\phi_1}{dy^2} + \frac{d^2\phi_2}{dz^2}\right) = m_1 \text{ and } \frac{\mu}{4\pi}\left(\frac{d^2\phi_2}{dx^2} + \frac{d^2\phi_2}{dy^2} + \frac{d^2\phi_2}{dz^2}\right) = m_2; \quad (37$$

then ϕ_1 is the potential at any point due to the magnetic system m_1, and ϕ_2 that due to the distribution of magnetism represented by m_2. The actual energy of all the vortices is
$$E = \Sigma C\mu(\alpha^2 + \beta^2 + \gamma^2)dV, \quad \ldots \quad (38)$$
the integration being performed over all space.

This may be shown by integration by parts (see Green's 'Essay on Electricity,' p. 10) to be equal to
$$E = -4\pi C\Sigma(\phi_1 m_1 + \phi_2 m_2 + \phi_1 m_2 + \phi_2 m_1)dV. \quad (39)$$
Or since it has been proved (Green's 'Essay,' p. 10) that
$$\Sigma\phi_1 m_2 dV = \Sigma\phi_2 m_1 dV,$$
$$E = -4\pi C(\phi_1 m_1 + \phi_2 m_2 + 2\phi_1 m_2)dV. \quad \ldots \quad (40)$$

Now let the magnetic system m_1 remain at rest, and let m_2 be moved parallel to itself in the direction of x through a space δx; then, since ϕ_1 depends on m_1 only, it will remain as before, so that $\phi_1 m_1$ will be constant; and since ϕ_2 depends on m_2 only, the distribution of ϕ_2 about m_2 will remain the same, so that $\phi_2 m_2$ will be the same as before the change. The only part of E that will be altered is that depending on $2\phi_1 m_2$, because ϕ_1 becomes $\phi_1 + \frac{d\phi_1}{dx}\delta x$ on account of the displacement. The variation of actual energy due to the displacement is therefore
$$\delta E = -4\pi C\Sigma\left(2\frac{d\phi_1}{dx}m_2\right)dV\,\delta x. \quad \ldots \quad (41)$$

But by equation (12), the work done by the mechanical forces on m_2 during the motion is
$$\delta W = \Sigma\left(\frac{d\phi_1}{dx}m_2 dV\right)\delta x; \quad \ldots \ldots \quad (42)$$

and since our hypothesis is a purely mechanical one, we must have by the conservation of force,

$$\delta E + \delta W = 0; \quad \dots \dots \dots (43)$$

that is, the loss of energy of the vortices must be made up by work done in moving magnets, so that

$$-4\pi C \Sigma \left(2\frac{d\phi_1}{dx} m_2 dV\right)\delta x + \Sigma \left(\frac{d\phi_1}{dx} m_2 dV\right)\delta x = 0,$$

or

$$C = \frac{1}{8\pi}; \quad \dots \dots \dots (44)$$

so that the energy of the vortices in unit of volume is

$$\frac{1}{8\pi}\mu(\alpha^2 + \beta^2 + \gamma^2); \quad \dots \dots (45)$$

and that of a vortex whose volume is V is

$$\frac{1}{8\pi}\mu(\alpha^2 + \beta^2 + \gamma^2)V. \quad \dots \dots (46)$$

In order to produce or destroy this energy, work must be expended on, or received from, the vortex, either by the tangential action of the layer of particles in contact with it, or by change of form in the vortex. We shall first investigate the tangential action between the vortices and the layer of particles in contact with them.

Prop. VII.—To find the energy spent upon a vortex in unit of time by the layer of particles which surrounds it.

Let P, Q, R be the forces acting on unity of the particles in the three coordinate directions, these quantities being functions of x, y, and z. Since each particle touches two vortices at the extremities of a diameter, the reaction of the particle on the vortices will be equally divided, and will be

$$-\frac{1}{2}P, \quad -\frac{1}{2}Q, \quad -\frac{1}{2}R$$

on each vortex for unity of the particles; but since the superficial density of the particles is $\frac{1}{2\pi}$ (see equation (34)), the forces on unit of surface of a vortex will be

$$-\frac{1}{4\pi}P, \quad -\frac{1}{4\pi}Q, \quad -\frac{1}{4\pi}R.$$

Now let dS be an element of the surface of a vortex. Let the direction-cosines of the normal be l, m, n. Let the coordinates of the element be x, y, z. Let the component velocities of the surface be u, v, w. Then the work expended on that element of surface will be

$$\frac{dE}{dt} = -\frac{1}{4\pi}(Pu + Qv + Rw)dS. \quad \ldots \quad (47)$$

Let us begin with the first term, $Pu\, dS$. P may be written

$$P_0 + \frac{dP}{dx}x + \frac{dP}{dy}y + \frac{dP}{dz}z, \quad \ldots \quad \ldots \quad (48)$$

and

$$u = n\beta - m\gamma.$$

Remembering that the surface of the vortex is a closed one, so that

$$\Sigma nx\, dS = \Sigma mx\, dS = \Sigma ny\, dS = \Sigma mz\, dS = 0,$$

and

$$\Sigma my\, dS = \Sigma nz\, dS = V,$$

we find

$$\Sigma Pu\, dS = \left(\frac{dP}{dz}\beta - \frac{dP}{dy}\gamma\right)V, \quad \ldots \quad \ldots \quad (49)$$

and the whole work done on the vortex in unit of time will be

$$\frac{dE}{dt} = -\frac{1}{4\pi}\Sigma(Pu + Qv + Rw)dS$$

$$= \frac{1}{4\pi}\left\{\alpha\left(\frac{dQ}{dz} - \frac{dR}{dy}\right) + \beta\left(\frac{dR}{dx} - \frac{dP}{dz}\right) + \gamma\left(\frac{dP}{dy} - \frac{dQ}{dx}\right)\right\}V. \ (50)$$

Prop. VIII.—To find the relations between the alterations of motion of the vortices, and the forces P, Q, R which they exert on the layer of particles between them.

Let V be the volume of a vortex, then by (46) its energy is

$$E = \frac{1}{8\pi}\mu(\alpha^2 + \beta^2 + \gamma^2)V, \quad \ldots \quad \ldots \quad (51)$$

and

$$\frac{dE}{dt} = \frac{1}{4\pi}\mu V\left(\alpha\frac{d\alpha}{dt} + \beta\frac{d\beta}{dt} + \gamma\frac{d\gamma}{dt}\right). \quad \ldots \quad (52)$$

Comparing this value with that given in equation (50), we find

$$\alpha\left(\frac{dQ}{dz} - \frac{dR}{dy} - \mu\frac{d\alpha}{dt}\right) + \beta\left(\frac{dR}{dx} - \frac{dP}{dz} - \mu\frac{d\beta}{dt}\right)$$

$$+ \gamma\left(\frac{dP}{dy} - \frac{dQ}{dx} - \mu\frac{d\gamma}{dt}\right) = 0. \quad \ldots \quad \ldots \quad (53)$$

This equation being true for all values of α, β, and γ, first let β and γ vanish, and divide by α. We find

$$\left.\begin{array}{l}\dfrac{dQ}{dz}-\dfrac{dR}{dy}=\mu\dfrac{d\alpha}{dt}.\\[2mm]\dfrac{dR}{dx}-\dfrac{dP}{dz}=\mu\dfrac{d\beta}{dt},\\[2mm]\dfrac{dP}{dy}-\dfrac{dQ}{dx}=\mu\dfrac{d\gamma}{dt}.\end{array}\right\}\quad\ldots\ldots\quad(54)$$

Similarly,

and

From these equations we may determine the relation between the alterations of motion $\dfrac{d\alpha}{dt}$, &c. and the forces exerted on the layers of particles between the vortices, or, in the language of our hypothesis, the relation between changes in the state of the magnetic field and the electromotive forces thereby brought into play.

In a memoir "On the Dynamical Theory of Diffraction" (Cambridge Philosophical Transactions, vol. ix. part 1, section 6), Professor Stokes has given a method by which we may solve equations (54), and find P, Q, and R in terms of the quantities on the right-hand of those equations. I have pointed out* the application of this method to questions in electricity and magnetism.

Let us then find three quantities F, G, H from the equations

$$\left.\begin{array}{l}\dfrac{dG}{dz}-\dfrac{dH}{dy}=\mu\alpha,\\[2mm]\dfrac{dH}{dx}-\dfrac{dF}{dz}=\mu\beta,\\[2mm]\dfrac{dF}{dy}-\dfrac{dG}{dx}=\mu\gamma,\end{array}\right\}\quad\ldots\ldots\quad(55)$$

with the conditions

$$\frac{1}{4\pi}\left(\frac{d}{dx}\mu\alpha+\frac{d}{dy}\mu\beta+\frac{d}{dz}\mu\gamma\right)=m=0,\quad(56)$$

and

$$\frac{dF}{dx}+\frac{dG}{dy}+\frac{dH}{dz}=0.\quad\ldots\ldots\quad(57)$$

Differentiating (55) with respect to t, and comparing with (54), we find

$$P=\frac{dF}{dt},\quad Q=\frac{dG}{dt},\quad R=\frac{dH}{dt}.\quad\ldots\quad(58)$$

* Cambridge Philosophical Transactions, vol. x. part 1. art. 3, "On Faraday's Lines of Force."

We have thus determined three quantities, F, G, H, from which we can find P, Q, and R by considering these latter quantities as the rates at which the former ones vary. In the paper already referred to, I have given reasons for considering the quantities F, G, H as the resolved parts of that which Faraday has conjectured to exist, and has called the *electrotonic state.* In that paper I have stated the mathematical relations between this electrotonic state and the lines of magnetic force as expressed in equations (55), and also between the electrotonic state and electromotive force as expressed in equations (58). We must now endeavour to interpret them from a mechanical point of view in connexion with our hypothesis.

We shall in the first place examine the process by which the lines of force are produced by an electric current.

Let A B, Pl. V. fig. 2, represent a current of electricity in the direction from A to B. Let the large spaces above and below A B represent the vortices, and let the small circles separating the vortices represent the layers of particles placed between them, which in our hypothesis represent electricity.

Now let an electric current from left to right commence in A B. The row of vortices $g\,h$ above A B will be set in motion in the opposite direction to that of a watch. (We shall call this direction $+$, and that of a watch $-$.) We shall suppose the row of vortices $k\,l$ still at rest, then the layer of particles between these rows will be acted on by the row $g\,h$ on their lower sides, and will be at rest above. If they are free to move, they will rotate in the negative direction, and will at the same time move from right to left, or in the opposite direction from the current, and so form an *induced* electric current.

If this current is checked by the electrical resistance of the medium, the rotating particles will act upon the row of vortices $k\,l$, and make them revolve in the positive direction till they arrive at such a velocity that the motion of the particles is reduced to that of rotation, and the induced current disappears. If, now, the primary current A B be stopped, the vortices in the row $g\,h$ will be checked, while those of the row $k\,l$ still continue in rapid motion. The momentum of the vortices beyond the layer of particles $p\,q$ will tend to move them from left to right, that is, in the direction of the primary current; but if this motion is resisted by the medium, the motion of the vortices beyond $p\,q$ will be gradually destroyed.

It appears therefore that the phenomena of induced currents are part of the process of communicating the rotatory velocity of the vortices from one part of the field to another.

[To be continued.]

U 2

XLV. *Chemical Notices from Foreign Journals.*
By E. ATKINSON, *Ph.D., F.C.S.*

[Continued from p. 126.]

BERNOULLI has published the result of an investigation of tungsten and some of its compounds*. With a view to a scientific investigation of the alloys of this metal, his first endeavour was to obtain the metal in a melted state; the result of his researches proves, however, that all previous statements as to the fusibility of pure tungsten are inaccurate. In his experiments pure tungstic acid was used; the experiments were made with the furnaces of the Royal Iron Foundry in Berlin, where he was able to command temperatures higher than any previously used in such experiments.

In one experiment a Hessian crucible was used lined with charcoal, in which there was a cavity to receive the tungstic acid, and over which there was a layer of charcoal powder. The crucible, provided with a cover, was kept at a white heat for nearly an hour. In this way a metallic mass was obtained free from carbon, but without any traces of fusion. In a subsequent experiment the Hessian crucible was completely fused, and accordingly they were replaced by the best American graphite crucibles. Even these did not resist the continuous heat of the furnace for $2\frac{1}{3}$ hours. A metallic mass was obtained, which was caked together and had some metallic lustre. This was heated again with charcoal in a crucible protected in the most complete manner; and the heat was greater than that ever observed in any puddling furnace, so much so that the slag from the coke dropped in a thin stream through the grates.

Notwithstanding this great heat the tungsten had not melted, although it had sintered to a tolerably compact mass. The metal thus obtained was heated for eighteen hours in a porcelain furnace without any change resulting.

Hence the author concludes that, with our present means, metallic tungsten is infusible.

Bernoulli also investigated the alloys of tungsten with metals, especially iron. Cast-iron turnings were intimately mixed with 1, 2, 3, 4, 5, 10, 15, and 20 per cent. of pure tungstic acid, in the idea that the carbon of the iron would reduce the acid to the state of metal. Some experiments were also made in which a larger per-centage of acid was taken; but in this case some powdered charcoal was added. The mixtures were heated in a graphite crucible to an intense white heat. With an addition of 10 per cent. of acid, the alloy had the properties of steel; it was very sonorous, had a clear grey colour, a pure fracture, and was

* Poggendorff's *Annalen*, December 1860.

malleable. The addition of 15 per cent. of acid yielded an alloy which might be considered as steel. It was very hard, but was not sufficiently malleable. With 20 per cent. the hardness was still greater, but the malleability much less.

The iron in these experiments was grey iron, and contained a considerable quantity of graphite, and it was found that with white iron a different result was obtained. The experiments were made in the same way as the previous ones; it was found that an alloy was only formed when charcoal dust was added; otherwise the tungstic acid sintered together, and very little tungsten combined with the iron. The alloy obtained with the addition of charcoal had none of the appearance of steel; it was white on fracture, had the structure of the iron used, and was imperfectly malleable.

With an addition of tungstic acid in the proportion of 75 per cent. no regulus was obtained. Analogous experiments were made with the minerals Wolfram and Scheelite, and similar results were obtained. The manganese present in Wolfram exerted a considerable influence on the result; and with Scheelite the lime combines with the silica to form a slag, so that the alloy is purer.

It follows from these experiments, that it is not the carbon present in the iron in a state of chemical combination which reduces the tungstic acid, but that which is mechanically intermingled. From white cast iron no carbon is withdrawn by the tungstic acid, and accordingly no steel is obtained if charcoal be not added.

The waste cast-iron turnings of the workshop may hence be used for preparing directly a cast steel, to which the tungsten imparts great hardness; or if the iron does not contain too much sulphur, phosphorus, or silica, a very useful rough cast steel may be obtained by fusing it directly with a quantity of powdered Wolfram proportionate to the per-centage of carbon which it contains.

The author determined the carbon in these alloys by three methods. In the first, a piece of the alloy was laid upon a fused cake of chloride of silver, and was left for several days, covered with distilled water. In this way the iron gradually dissolved; and when the decomposition was complete, the charcoal and tungsten were collected on an asbestos filter, dried and weighed, and then the carbon determined in the usual way by combustion with oxide of copper. In another case the alloy was decomposed by chloride of copper, and in a third case it was digested with iodine until the iron was dissolved. In these cases the carbon contained in the alloy amounted to about 1 per cent.

Experiments to alloy tungsten with other metals were also

made. With copper, reguli were obtained, which, however, were
not homogeneous; the individual particles could be distinctly
seen. In general it was also found that copper, lead, zinc,
antimony, bismuth, cobalt, and nickel only became alloyed with
tungsten when the reduction of the two metals took place simultaneously. The alloys are so infusible, that, with more than 10
per cent. of tungsten, no reguli are obtained, and at a higher
temperature the more volatile metals escape and metallic tungsten
is left behind. Iron differs in this respect from other metals.
It alloys in all proportions with tungsten; with above 80 per
cent., however, the alloys are infusible.

In order to prepare the tungstic acid used in these experiments, powdered Wolfram was fused with excess of carbonate of
soda in an iron crucible, the fused mass dissolved, and boiled to
reduce manganic acid, and filtered. The solution, which consisted of tungstate and carbonate of soda, was neutralized with
nitric acid, and the tungstate of soda crystallized out. This was
dissolved and treated with nitric acid, and the precipitate of hydrated tungstic acid was well washed. It was then dissolved in
ammonia, which left a residue of niobic and silicic acids; the
evaporated liquor deposited a fine crop of crystals of tungstate of
ammonia. This was well washed with water, and then repeatedly
treated with fresh quantities of nitric acid for some days to
remove nitrate of ammonia. The acid was ultimately washed
out and gently heated, by which it was obtained of a fine pure
sulphur-yellow colour.

The author found, in all these experiments, that it was not
possible to obtain pure yellow acid by directly heating the tungstate of ammonia, even when this was done under access of air.
It invariably became of a green colour. This has been observed
before, and has been differently interpreted, some ascribing it to
the formation of a suboxide, and some to an admixture of yellow
acid and the blue oxide $W^2 O^5$.

Bernoulli has found that it is a true compound. When the
yellow acid was heated to the highest temperature of a gas blowpipe, it gradually changed to a green colour, and from being
amorphous became crystalline. Similar results were obtained
by using the high temperature of a stoneware furnace, which
has an oxidizing flame. The tungstic acid was placed in suitably protected platinum crucibles, and heated for periods varying
from eighteen to seventy-two hours. A green crystalline mass
was obtained, while on the upper part of the crucible there were
smaller crystals of the same colour. When the caked mass was
divided and subjected to a further heat for eighteen hours, the
result was confirmed; part had sublimed in fine crystalline laminæ.

Bernoulli analysed these two modifications by reduction with hydrogen, and found that they had the same per-centage composition, WO^3. They must therefore be regarded as two isomeric modifications of the same acid, of which the *yellow* is formed both in the moist and in the dry way, but in the latter case only at a *low* temperature; while the green variety is only formed in the dry way and at a high temperature. The latter he proposes to call *pyrotungstic* acid, in antithesis to Scheibler's acid*, which is *metatungstic* acid. Including the ordinary acid, there are therefore three varieties.

Bernoulli has made a new determination of the equivalent of tungsten, both by oxidizing tungsten, and by reducing tungstic acid. He obtained results varying within very narrow limits, which lead to the number 93·4 as that of the equivalent. Dumas had obtained the number 92†.

Bernoulli finally discusses the formula of the natural tungstates, and attempts to show that the two modifications of tungstic acid also occur in nature. He adduces a great many analyses of Wolfram. Most of them contain a certain quantity of lime and magnesia, which he considers accidental constituents. But in most cases a niobic acid is present—in the tungsten from Zinnwald 1·1 per cent.; this is to be regarded as replacing part of the tungstic acid, with which it is therefore isomorphous.

The author finally describes the mode of analysing the tungsten minerals.

Engaged in investigating the methods of working up the platinum residues for the Russian Government, Deville and Debray had occasion to examine the different methods of preparing oxygen on a large scale. They find‡ that sulphate of zinc, which, as a waste product, is now so plentiful, furnishes an economical source of this gas. When calcined in an earthen vessel, it is converted into light white oxide, which, when the sulphate is pure, may be used for painting. The temperature required for its decomposition does not exceed that necessary for binoxide of manganese. The other products of the decomposition are sulphurous acid and oxygen, which may be separated by means of the solubility of the former in alkalies; or the following method may be used, which is employed by the authors for preparing oxygen by the decomposition of sulphuric acid.

This body at a red heat may be decomposed into sulphurous acid, water, and oxygen, by means of a very simple apparatus, consisting of a retort, of about 5 litres, filled with thin platinum foil, or, better, a serpentine tube filled with platinum sponge and heated to redness. A thin stream of sulphuric acid passes into

---

* Phil. Mag. vol. xx. p. 374.　　　† Ibid. vol. xvi. p. 211.
‡ *Comptes Rendus*, November 26, 1860.

this apparatus through an S tube; the products pass first through a cooler which condenses the water, and then through a washer of a special form.   In this way pure and inodorous gas is obtained, and a solution of sulphurous acid, which may be changed either into sulphite or hyposulphite of soda, or may be used in the sulphuric acid chambers.   The expense of oxygen prepared by this plan is very small; for the method consists essentially in abstracting oxygen from the atmosphere.   Even if the sulphurous acid were not utilized, sulphuric acid would still be the cheapest source of oxygen, cheaper even than binoxide of manganese.

M. Carré has applied[*] the great cold produced by the evaporation of condensed ammoniacal gas to the production of low degrees of temperature.

The apparatus he uses consists of two ordinary cylindrical metal receivers connected by a tube.   One of these is four times the size of the other, and is filled to three-quarters its capacity with the strongest solution of ammonia.   At the time of closing the vessel, care is taken to expel all air.   The largest vessel is placed over the fire, the smaller being immersed in cold water. The solution is heated to 130° or 140°, the temperature being indicated by a thermometer fitted in the larger vessel.   At this point nearly all the ammonia is expelled from the solution, and liquefies in the second retort.   When the separation is complete, the larger vessel is cooled down: the reabsorption of the liquefied gas commences immediately, and its volatilization produces a degree of cold which readily freezes the water surrounding it. The temperature sinks to −40°; and M. Balard, who tried the experiment at the Collége de France, was able to solidify mercury.

Besides this apparatus, M. Carré has devised another form of it, which is continuous in its action, but otherwise depends on the same principle.

M. Leroux, of the École Polytechnique, has made some determinations of the refractive indices of vapours at high temperatures, by means of an apparatus constructed for that purpose, on which M. Babinet[†] has reported to the Academy.

M. Dulong had found that the refractive index of oxygen was 1·000273; of hydrogen, 1·000138; of nitrogen, 1·000300; and of chlorine, 1·000772, that of air being 1·000294.   M. Leroux has found that the refractive indices of the vapours of the following substances, saturated at the ordinary pressure, are respectively—

[*] *Comptes Rendus*, December 24, 1860.
[†] Ibid. November 26, 1860.

Sulphur	1·001629
Phosphorus . . .	1·001364
Arsenic	1·001114
Mercury	1·000556

The apparatus by which these results were obtained consists of a very large furnace mounted on an axis, and provided at its lower part with a divided circle, by which it can be inclined at any angle. In the centre of the furnace there is a prism analogous to that of Borda, employed by Dulong. This prism is made of solid iron, and the rays enter and emerge through glass plates cemented by a method peculiar to M. Leroux. The exact measurement of the angle of the prism presents some ingenious points. As in Babinet's goniometer, the light is concentrated on the prism, while filled with the vapour, by a telescope, and the light on its emergence is received on another telescope provided at its focus with a micrometric wire. With the first telescope, its distance from the furnace, the size of the furnace, the distance of the second telescope and its focal distance, which is above 2 yards, the extent of this gigantic goniometer is about 23 feet. The means of verifying the results leave nothing to be desired.

Rose*, in a paper on the separation of tin from other metals, and on its quantitative estimation, describes the following method of effecting these objects.

The separation of tin from other metals is usually effected by oxidation with nitric acid, so as to convert the tin into stannic acid: in certain cases, however, this method gives inaccurate results, and can only be said to be quite successful in the case of the strongly basic oxides.

Another method consists in dissolving the binoxide of tin in hydrochloric acid, and precipitating by sulphuric acid. Both modifications of binoxide are precipitated in this manner, in the presence of a large excess of water. The precipitates require a long time in order to settle completely, more especially when there is a large quantity of free hydrochloric acid. The precipitate must be carefully washed free from hydrochloric acid otherwise when it is ignited some tin escapes in the form of chloride. The ignition is best effected with the addition of some carbonate of ammonia.

In the presence of certain substances, phosphoric acid for example, even this method gives inaccurate results. The presence of a large excess of hydrochloric acid does not hinder the precipitation of some phosphoric acid along with the binoxide of tin.

When tin is to be separated from other metals, it must be

* Poggendorff's *Annalen*, January 1861.

oxidized with nitric acid in the usual manner, and the residue digested with moderately strong hydrochloric acid. On the addition of a large quantity of water, all is dissolved, and the tin is then precipitated with sulphuric acid.

In the separation of copper and tin by the ordinary method, the binoxide always contains a trace of copper; but by the above process the separation of the two metals is complete, the binoxide is quite free from copper.

Tin and lead are best separated by fusing the alloy with sulphur and carbonate of potash. On treating the mass with water, the sulphide of tin dissolves. This is the best method of analysing the fusible alloy of tin, bismuth, and lead. Tin and silver are also best separated in this way.

In an acid solution, tin and bismuth are best separated by sulphide of ammonium.

The separation of iron from tin can only be completely effected by oxidizing the alloy with nitric acid, dissolving in hydrochloric acid, and saturating the hydrochloric solution with sulphuretted hydrogen.

Tin is best separated from titanium by means of sulphide of ammonium.

The separation of the oxides of tin from magnesia and the alkaline earths is best effected by igniting them with sal-ammoniac, by which the tin escapes as chloride. In general all the tin is expelled by one ignition; but two are always amply sufficient to remove the last traces of tin.

In the analysis of minerals, the separation and estimation of tin is best effected by converting it into the oxide. The volumetric analysis, however, is very convenient for the determination of the tin in tin-salts. The ordinary method is to convert the protochloride of tin in hydrochloric acid solution into the bichloride by means of oxidizing agents, such as permanganate or bichromate of potash. But protochloride of tin absorbs atmospheric oxygen, even during the determination, to such an extent as materially to vitiate the results of analyses.

Stromeyer[*] has observed that good results are obtained by the addition of sesquichloride of iron in excess to the freshly prepared solution of tin in hydrochloric acid. The reaction is as follows :—

$$\mathrm{Sn\,Cl + Fe^2\,Cl^3 = Sn\,Cl^2 + 2\,Fe\,Cl}.$$

The protochloride of iron formed is determined by permanganate; and as it is not nearly so susceptible to atmospheric oxygen as protochloride of tin, much more accurate results are obtained.

* Liebig's *Annalen*, February 1861.

. Tin may also be directly dissolved in sesquichloride of iron to which hydrochloric acid has been added,

$$Sn + 2 Fe^2 Cl^3 = Sn Cl^2 + 4 Fe Cl.$$

But this method is only available with tolerably pure tin; for many other metals reduce perchloride of iron, and consume solution of permanganate.

There are several compounds which have the formula $C^4 H^4 Cl^2$. One of these is obtained by the action of pentachloride of phosphorus on aldehyde, and another by the action of chlorine on chlorinated ethyle. Beilstein observed some time ago that these two bodies were identical, and he has recently proved[*] that the same is the case with two corresponding isomeric compounds of the benzoic acid series: the one, chlorobenzole, $C^{14} H^6 Cl^2$, is obtained by the action of pentachloride of phosphorus on oil of bitter almonds; and the other is the chlorinated chloride of benzyle, $C^{14} (H^6 Cl) Cl$.

The latter body is formed, as Cannizaro showed, by the action of chlorine on toluole; and Beilstein used this method of preparing it. He finds that it has all the physical properties of chlorobenzole. A careful comparison also of the chemical actions of the two substances, both by his own direct experiments and by those of other experimenters, leave no doubt as to the complete identity of the two substances.

Reboul has published[†] the results of a lengthened and important investigation on some derivatives of glycerine. The compounds which the author describes may be derived from an oxygenized body, *glycide*, $C^6 H^6 O^4$, which has not been isolated, and which might be considered as the anhydride of glycerine, to which it would bear the same relation as lactide does to lactic acid. It would play the part of a diatomic alcohol, and would yield a series of ethers, the general formation of which may be thus expressed:—

$$C^6 H^6 O^4 + A - H^2 O^2 = C^6 H^4 A O^2,$$

$$C^6 H^6 O^4 + A A' - 2 H^2 O^2 = C^6 H^2 A A',$$

A and A' representing the formulæ of monobasic acids.

The starting-point for his research is *hydrochloric glycide*, obtained by the action of potash on bihydrochloric glycerine, $C^6 H^6 Cl^2 O^2$, which simply removes hydrochloric acid,

$$C^6 H^6 Cl^2 O^2 - HCl = C^6 H^5 ClO^2.$$

Bihydrochloric glycerine itself is formed by saturating a mix-

* Liebig's *Annalen*, December 1860.
† *Annales de Chimie*, September 1860. *Répertoire de Chimie*, November 1860.

ture of glycerine and acetic acid with gaseous hydrochloric acid. The chief product of the action is bihydrochloric glycerine, which may be separated by fractional distillation; or, by heating the crude product with caustic potash, hydrochloric glycide is directly obtained. It is a colourless liquid, heavier than water, and smelling like chloroform. It boils at 118°—119°: it is metameric with Geuther's hydrochlorate of acroleine, with Riche's monochlorinated acetone, and with chloride of propionyle.

Monohydrochloric glycide combines directly with fuming hydrochloric acid to form *dihydrochloric glycide*, $C^6 H^4 Cl^2$: this body may also be formed by acting on hydrochloric glycide with pentachloride of phosphorus, an action which gives rise to the formation of trihydrochloric glycerine,

$$C^6 H^5 ClO^2 + PCl^5 = C^6 H^5 Cl^3 + PCl^3 O^2;$$
Hydrochloric   Trihydrochloric
glycide.   glycerine.

and this, decomposed by potash, loses hydrochloric acid, and the new body is formed,

$$C^6 H^5 Cl^3 - HCl = C^6 H^4 Cl^2.$$

This body is identical with what Berthelot has called epibromhydrine. It is metameric with bichlorinated propylene, and with Geuther's chloride of acroleine. Similar compounds containing bromine and iodine were also obtained.

By the action of ammonia on bihydrochloric glycide a base is formed, which the author considers identical with that obtained by the action of ammonia on terbromide of allyle.

Hydrochloric glycide unites directly with the oxyacids to form a glyceric ether, containing an equivalent of oxyacid and of hydracid. Thus,—

$$C^6 H^5 ClO^2 + C^4 H^4 O^4 = C^6 H^6 (C^4 H^3 O^2) ClO^4.$$
Hydrochloric   Acetic   Acetohydrochloric
glycide.   acid.   glycerine.

By the action of water, hydrochloric glycide fixes two equivalents, and monohydrochloric glycerine is formed:—

$$C^6 H^5 ClO^2 + H^2 O^2 = C^6 H^7 ClO^4,$$
Hydrochloric   Monohydrochloric
glycide.   glycerine.

By the action of alcohol the hydrochloric compounds of glycide are transformed into mixed glyceric ethers, containing an equivalent of acid and an equivalent of alcohol,

$$C^6 H^5 ClO^2 + C^4 H^6 O^2 = C^6 H^6 Cl (C^4 H^5) O^4.$$
Hydrochloric   Alcohol.   Hydrochloric
glycide.   ethylglycerine.

When these bodies are treated with potash, hydrochloric acid is removed, and the resultant compound, *ethylglycide*, is a glycide containing the alcohol radical in the place of an equivalent of hydrogen:—

$$C^6 H^6 (C^4 H^5) ClO^4 - HCl = C^6 H^5 (C^4 H^5) O^4.$$
Hydrochloric
ethylglycerine.   Ethylglycide.

Ethylglycide is a mobile liquid boiling at 128°. When treated with hydrochloric acid, it yields the compound $C^6 H^6 (C^4 H^5) ClO^4$. If the compound hydrochloric amylglycerine be treated with ethylate of soda, ethylamylglycerine is formed. Thus,—

$$C^6 H^6 (C^{10} H^{11}) ClO^4 + C^4 H^5 NaO^2 = C^6 H^6 (C^{10} H^{11})(C^4 H^5) O^6 + NaCl.$$
Hydrochloric   Ethylate   Ethylamylglycerine.
amylglycerine.   of soda.

The glyceric ethers containing two equivalents of the same acid are only a particular case of this reaction.

When hydrochloric glycide is acted upon by hydrosulphate of sulphide of potassium, a compound is obtained analogous to mercaptan,

$$C^6 H^5 ClO^2 + KS HS = KCl + C^5 H^6 S^2 O^2.$$
Hydrochloric   New body.
glycide.

A second mercaptan in this series is doubtless formed by the action of dihydrochloric glycide on hydrosulphate of sulphide of potassium, and which would have the formula $C^6 H^6 S^4$.

XLVI. *On the existence of a new Element, probably of the Sulphur Group.* By WILLIAM CROOKES, *F.C.S*[*].

IN the year 1850 Professor Hofmann placed at my disposal upwards of 10 lbs. of the seleniferous deposit from the sulphuric acid manufactory at Tilkerode in the Hartz Mountains, for the purpose of extracting from it the selenium, which was afterwards employed in an investigation upon the selenocyanides[†]. Some residues which were left in the purification of the crude selenium, and which from their reactions appeared to contain tellurium, were collected together and placed aside for examination at a more convenient opportunity. They remained unnoticed until the beginning of the present year, when, requiring some tellurium for experimental purposes, I attempted its extraction from these residues. Knowing that the spectra of the incandescent vapours of both selenium and tellurium were free from any

* Communicated by the Author.
† Chem. Soc. Quart. Journ. vol. iv. p. 12, and Gmelin's Handbook (Cavendish Soc. Translation), vol. viii. p. 122.

strongly-marked line which might lead to the identification of either of these elements, it was not until I had in vain tried numerous chemical methods for isolating the tellurium which I supposed to be present, that the method of spectrum-analysis was used. A portion of the residue, introduced into a blue gas-flame, gave abundant evidence of selenium; but as the alternate light and dark bands due to this element became fainter, and I was expecting the appearance of the somewhat similar, but closer, bands of tellurium, suddenly a *bright green line* flashed into view and as quickly disappeared. An isolated green line in this portion of the spectrum was new to me. I had become intimately acquainted with the appearances of most of the artificial spectra during many years' investigation, and had never before met with a similar line to this; and as from the chemical processes through which this residue had passed the elements which could possibly be present were limited to a few, it became of interest to discover which of them occasioned this green line.

After numerous experiments, I have been led to the conclusion that it is caused by the presence of a new element belonging to the sulphur group; but, unfortunately, the quantity of material upon which I have been able to experiment has been so small, that I hesitate to assert this very positively. I am, however, at work upon some of the seleniferous deposit itself, and hope shortly to be able to speak more confidently upon this point, as well as to give some account of its properties.

In the purest state that I have as yet succeeded in obtaining this substance, it communicates as definite a reaction to the flame as soda,—the smallest trace introduced into the burner of the spectrum apparatus giving rise to a brilliant green line, perfectly sharp and well-defined upon a black ground, and almost rivalling the Na line in brilliancy. It is not, however, very lasting:. owing to its volatility, which is almost as great as selenium, a portion introduced at once into a flame merely shows the line as a brilliant flash, remaining only a fraction of a second; but if it be introduced into the flame gradually, the line continues present for a much longer time.

The properties of the substance, both in solution and in the dry state, as nearly as I can make out from the small quantity at my disposal, are as follows :—

1. It is completely volatile below a red heat, both in the elementary state and in combination (except when united with a heavy fixed metal). 2. From its hydrochloric solution it is readily precipitated by metallic zinc in the form of a heavy black powder, insoluble in the acid liquid. 3. Ammonia added very gradually until in slight excess to its acid solution, gives no precipitate or coloration whatever, neither does the addition of car-

bonate or oxalate of ammonia to this alkaline solution. 4. Dry chlorine passed over it at a dull red heat unites with it, forming a readily volatile chloride soluble in water. 5. Sulphuretted hydrogen passed through its hydrochloric solution precipitates it incompletely, unless only a trace of free acid is present; but in an alkaline solution an immediate precipitation of a heavy black powder takes place. 6. Fused with carbonate of soda and nitre, it becomes soluble in water,—hydrochloric acid added in excess to this liquid producing a solution which answers to the above tests 2, 3, and 5.

An examination of these reactions shows that there are very few elements which could by the remotest possibility be mistaken for it.

The accompanying list includes every element, with the exception of the gases, bromine, iodine, and carbon. Opposite the name of each I have placed the number of the reaction which eliminates it from the list of possible substances, taking great care, in every case, to give the benefit of any doubt which might arise, on account of an imperfectly known or doubtful reaction, in favour of the opposite opinion to that which I desire to prove, and, in cases where several reactions would prove the same thing, only making use of the most trustworthy.

1, 5. Aluminium.	1. Iron.	Selenium.
Antimony.	1, 5. Lanthanium.	1, 5. Silicium.
Arsenic.	1. Lead.	1. Silver.
2, 3, 5. Barium.	2, 5. Lithium.	2, 5. Sodium.
2, 3, 5. Beryllium.	2, 5. Magnesium.	2, 3, 5. Strontium.
1. Bismuth.	1. Manganese.	5. Sulphur.
1, 2, 5. Boron.	3, 6. Mercury.	1. Tantalum.
6. Cadmium.	1. Molybdenum.	Tellurium.
2, 5. Cæsium.	1. Nickel.	1, 5. Terbium.
2, 3, 5. Calcium.	1. Niobium.	1, 5. Thorium.
1, 5. Cerium.	1. Norium.	1. Tin.
1. Chromium.	Osmium.	1. Titanium.
1. Cobalt.	1. Palladium.	1. Tungsten.
1. Copper.	5. Phosphorus.	1. Uranium.
1, 5. Didymium.	1. Platinum.	1. Vanadium.
1, 5. Erbium.	2, 5. Potassium.	1, 5. Yttrium.
1. Gold.	1. Rhodium.	2. Zinc.
1. Ilmenium.	1. Ruthenium.	1, 5. Zirconium.
1. Iridium.		

There are therefore left the following, amongst which, if already known, it must occur:—antimony, arsenic, osmium, selenium, and tellurium; and although, to my own mind, many of the reactions detailed above are sufficient proof that it cannot be one of the first three elements, yet I have thought it better to let them pass.

Each of the above five bodies, both in the elementary state and in combination, has been rigidly scrutinized in the spectrum

apparatus by myself and many friends. Not a trace of such a line
is shown by either of them in the green part of the spectrum,—
Antimony, arsenic, and osmium, in fact, giving continuous spectra,
in which every colour is visible. The remaining elements, selenium and tellurium, might almost be dismissed unchallenged,
inasmuch as I was first led to the examination by finding that it
was *not* either of these. Nevertheless I have, as stated at the
commencement of this paper, repeatedly examined their spectra,
and find no trace of such a line, the alternate light and dark
bands in the almost continuous spectra of selenium and tellurium
forming in fact so strong a contrast to the one single green ray
of the new substance, that the latter may readily be detected in
the presence of an enormous excess of either of the former.

In order to remove any remaining doubt which there might
be as to the green line being due to any of the elements mentioned in the above list, I have, moreover, specially examined the
spectra produced by each of these bodies in detail, either in their
elementary state, or in their most important compounds. Many
of them give rise to spectra of great and characteristic beauty,
but none give anything like the green line; nor, in fact, is there
any artificial spectrum, except that of sodium, which equals it in
simplicity.

There still may be urged the possibility of its being a compound
of two or more known elements, or an allotropic condition of one
of them; a moment's thought, however, will show that neither of
these hypotheses is tenable. They would in reality prove what
they are raised to oppose; for nothing less could follow than a
veritable transmutation of one body into another, and a consequent annihilation of all the groundwork upon which modern
science is based. If an element can be so changed as to have
totally different chemical reactions, and to have the spectrum of
its incandescent vapour (which is, *par excellence*, an elementary
property) altered to an appearance totally unlike that given by
its former self, it must have been changed into something which
it originally was not. This, in the present position of science, is
an absurdity.

The method of exhaustion which I have adopted to prove
the elementary character of the body which communicates this
green line to the spectrum of the blue gas-flame*, may seem
unnecessary as well as unchemical in the present state of the
science; I was obliged, however, to rely upon what I may call
circumstantial evidence of its not being a known element,
owing to the very small quantity of substance at my command

---

* I need scarcely add that the line is quite distinct from either of the
green or blue lines seen in a gas-flame which is undergoing complete combustion. It is moreover far more brilliant than these.

(I believe I overestimate the amount which I have as yet obtained, at two grains), which precluded me from trying many reactions. The method of spectrum-analysis adopted to prove the same fact, although perfectly conclusive to my own mind, might not have been so to others, unsupported by chemical evidence.

The following diagram will serve to show the position in the spectrum which the new green line occupies with respect to the two lithium and the sodium lines.

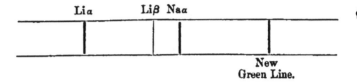

Li α   Li β Na α

New
Green Line.

For confirmatory experiments on many of the observations mentioned in this paper, I am indebted to my friend Mr. C. Greville Williams. The detailed examination of the various spectra are at present being jointly pursued by us, and will be published as soon as completed.

XLVII. *Proceedings of Learned Societies.*

GEOLOGICAL SOCIETY.

[Contined from p. 238.]

January 23, 1861.—L. Horner, Esq., President, in the Chair.

THE following communications were read:—

1. "On the Gravel and Boulders of the Punjâb." By J. D. Smithe, Esq., F.G.S.

In the Phimgota Valley (a continuation of the great Kangra or Palum Valley) the drift consists of sand and shingle with boulders of gneiss, schist, porphyry, and trap, from 6 inches to 5 feet in diameter. Some of the boulders, having a red vitreous glaze, occur in irregular beds. This moraine-like drift lies on the tertiary beds, which, here dipping gently towards the plains, gradually become vertical, and are succeeded by variegated compact sandstones, gradually inclining away from the plains; next come various slates, at a high angle; and gneissic rocks lie immediately over them.

2. "On *Pteraspis Dunensis* (*Archæoteuthis Dunensis*, Roemer)." By Prof. T. H. Huxley, F.R.S., Sec. G.S.

The fossil referred to in this communication is from Daun in the Eifel, and was described by Dr. Ferd. Roemer (in the 'Palæonto-

graphica,' vol. iv. p. 72, pl. 13) as belonging to the naked Cepha-
lopods, under the name of *Palæoteuthis Dunensis* (changed to *Archæoteuthis* in the 'Leth. Geogn.'); and in the Jahrb. 1858, p. 55, Dr.
F. Roemer described a second specimen from Wassennach on the
Laacher See. Prof. Huxley reproduced, with remarks, Dr. Roemer's
description of the specimens; and, after observing that Mr. S. P.
Woodward had already suggested (Manual of Mollusca, p. 417)
that Roemer's fossil was a fish, he stated his conviction that it was
really a *Pteraspis*, agreeing in all essential particulars with the
British *Pteraspides*, though possibly of a different species.

3. "On the ' Chalk-rock ' lying between the Lower and the Upper
Chalk in Wilts, Berks, Oxon, Bucks, and Herts." By W. Whitaker,
Esq., B.A., F.G.S.

The author has more particularly examined the band which he
terms "Chalk-rock" on the northern side of the western part of the
London Basin. Here it has its greatest thickness (12 feet) to the
west, gradually thinning eastward. It is a hard chalk, dividing into
blocks, by joints perpendicular to the bedding; and it contains hard
calcareo-phosphatic nodules. It contains no flints; and in the district referred to none occur below it; but there is often a band
of them resting on its upper surface. It seems to form an exact
boundary between the Upper and the Lower Chalk, being probably
the topmost bed of the latter. In this case it will often serve as an
index of the relative thickness of these divisions, or as a datum for
the measurement of the extent of denudation that the Upper Chalk
has suffered. North of Marlborough, where it is thick, the Chalk-
rock appears to have given rise to two escarpments (an upper and a
lower) to the western portion of the Chalk Range.

Fossils are usually rare in this bed; but Mr. J. Evans, F.G.S.,
collected several from it near Boxmoor; and amongst them the
genera *Belosepia* (hitherto known only as Tertiary), *Baculites, Nau-
tilus, Turrilites, Solarium, Inoceramus, Parasmilia,* and *Ventriculites*
are represented; and the following species have been identified—
Litorina monilifera and a new species, *Pleurotomaria* sp., *Myacites
Mandibula, Spondylus latus, Sp. spinosus, Rhynchonella Mantelliana,
Terebratula biplicata,* and *T. semiglobosa.*

February 6, 1861.—L. Horner, Esq., President, in the Chair.

The following communication was read :—

"On the Altered Rocks of the Western and Central Highlands."
By Sir R. I. Murchison, F.R.S., V.P.G.S., and A. Geikie, Esq.,
F.G.S.

In the introduction it was shown that the object of this paper was
to prove that the classification which had been previously established
by one of the authors in the county of Sutherland was applicable,
as he had inferred, to the whole of the Scottish Highlands. The
structure of the country from the borders of Sutherland down the
western part of Ross-shire was detailed, and illustrated by a large map

of Scotland coloured according to the new classification, and by numerous sections. Everywhere throughout this tract it could be proved that an older gneiss, which the authors called "Laurentian," was overlain unconformably by red Cambrian sandstones; these again unconformably by quartz-rocks, limestones, and a gneissose and schistose series of strata, as previously shown in the typical district of Assynt. From the base of these quartz-rocks a perfect conformable sequence was shown to exist upward into the gneissose rocks, which is not obliterated by granite or any similar rock.

The tract between the Atlantic and the Great Glen consists, according to the authors, of a series of convoluted folds of the upper gneissose rocks, until, along the line of the Great Glen, the underlying quartzose series is brought up on an anticlinal axis. A prolongation of this axis probably exists along part of the west coast of Islay and Jura, two islands which exhibit a grand development of the lower or quartzose portion of the altered Silurian rocks of the Highlands.

From the line of the Great Glen north-eastward to the Highland border, the country was explained as consisting of a great series of anticlinal and synclinal curves, whereby the same series of altered rocks which occurs on the north-west is repeated upon itself. One synclinal runs in a N.E. and S.W. direction across Loch Leven. The anticlinal of quartzose rocks that rises from under it to the S.E. spreads over the Breadalbane Forest to the Glen Lyon Mountains, where it sinks below the upper gneissose strata with their associated limestones. Ben Lawers occupies the synclinal formed by these upper strata; and the limestones and quartz-rock come up again in another anticlinal axis corresponding with the direction of Loch Tay. The continuity of these lines of axis was traced both to the N.E. and S.W.

It thus appeared that the crystalline rocks of the Highlands are capable of reduction to order; that the same curves and folds could be traced in them as in their less altered equivalents of the South of Scotland; and that in what had hitherto appeared as little else than a hopeless chaos, there yet reigned a regular and beautiful simplicity.

In conclusion, Sir Roderick Murchison vindicated the accuracy of his published sections in the N.W. of Sutherland, which had been approved after personal inspection by Professors Ramsay and Harkness; and he gave detailed reasons for disbelieving the accuracy of the sections recently put forth by Prof. Nicol, which were intended as corrections of his own. He concluded by affirming that, through the aid of Mr. Geikie, the proofs of the truthfulness of his own sections, showing a conformable ascending order from the quartz-rocks and limestones into crystalline and micaceous rocks, had now been extended over such large areas that there could no longer be any misgivings on the subject.

February 20, 1861.—L. Horner, Esq., President, in the Chair.

The following communications were read:—

1. "On the Coincidence between Stratification and Foliation in

the Crystalline Rocks of the Highlands." By Sir R. I. Murchison, V.P.G.S., and A. Geikie, Esq., F.G.S.

Allusion was, in the first place, made to the early opinions of Hutton and Macculloch, who regarded the gneissic and schistose rocks of the Highlands as stratified. Mr. Darwin's views of the nature of the "foliation" of gneiss and schist were then referred to ; and it was insisted that this condition was not to be found in the rocks of the Highlands,—the so-called "foliation" which the late Mr. D. Sharpe had described in 1846 as characterizing the crystalline rocks of that country being, according to the authors, really mineralized stratification. It was then pointed out that, as Prof. Sedgwick had previously insisted on the wide difference between "foliated " or " schistose" and "cleaved " or " slaty " rocks, and as Prof. Ramsay had in 1840 recognized interlaminated quartz as being parallel to stratification in the Isle of Arran, "foliation" should be regarded as coincident with stratification, and not with cleavage, in the Scottish Highlands.

After some observations on the occurrence of cleavage in slates at Dunkeld, Easdale, Ballahulish, and near the Spittal of Glenshee, the authors stated their belief that all the " foliation " of the crystalline rocks of the Highlands is nothing more than lamination due to the sedimentary origin of deposits, in which the sand, clay, lime, mica, &c. have subsequently been more or less altered, and that the "arches of foliation " described by Mr. D. Sharpe (Phil. Trans. 1852) correspond in a general way with the parallel anticlinal axes shown by the authors in a former paper to exist in the Highlands. They remarked that the synclinal troughs, however, are not expressed in Mr. Sharpe's figures, and that he has omitted the bands of limestone which they refer to as an important evidence of the stratification of the district. They also pointed to the acknowledged difficulty which the quartzites presented to Mr. Sharpe, but which readily fall into the system of undulated strata that they have described. One of the quartzites having yielded an Orthoceratite, and pebbles being present in one of the schists of Ben Lomond, these facts were adduced as further evidences of the real stratal condition of the schists and quartzites of the Highlands.

2. " On the Rocks of portions of the Highlands of Scotland South of the Caledonian Canal, and on their equivalents in the North of Ireland." By Professor R. Harkness, F.R.S., F.G.S.

The author, having had an opportunity of examining the geology of the North-west of Scotland in the year 1859, and more especially the arrangement of rocks described by Sir R. Murchison as "fundamental gneiss, Cambrian grits, lower quartz-rock, limestones, upper quartz-rock, and overlying gneissose flags," applied the results of his observations during last summer to portions of the Highlands lying south of the Caledonian Canal, and to the North of Ireland. Developed over a large portion of these districts are masses of gneissose rock, of varying mineral nature, and sometimes putting on the aspect of a simple flaggy rock. Where these gneissose masses come in contact with plutonic masses, they exhibit that highly crystalline aspect which induced Macculloch and others of the Scotch geologists to regard them as occupying an extremely low position among the sedimentary series, and to apply to them the Wernerian term "primitive." Many of Macculloch's descriptions, however, show that this assumed low position is not the true place of this gneiss among the sedimentary rocks which make up the Highlands of Scotland.

In a section from the southern flank of the Grampians to Loch Earn (and in other sections, from Loch Earn to Loch Tay, from Dunkeld to Blair Athol, in the Ben y Gloe Mountains, in Glen Shee, &c.), there is seen a sequence which indicates that this gneiss is the highest portion of the series of rocks, with underlying quartz-rock and limestone.

In the county of Donegal, Ireland, a like sequence is seen. A section from Inishowen Head to Malin Head, along the east side of Loch Foyle, presents us with gneissose rocks above limestone and quartz-rocks, exactly as in Scotland. In no portion of Scotland south of the Caledonian Canal, nor in the North of Ireland, did the author recognize any trace of the "fundamental gneiss."

March 6, 1861.—Leonard Horner, Esq., President, in the Chair.

The following communications were read :—

1. "On the Succession of Beds in the Hastings Sand in the Northern portion of the Wealden Area." By F. Drew, Esq., F.G.S., of the Geological Survey of Great Britain.

Having first referred to the division of the Wealden beds by former authors into the "Weald Clay," the "Hastings Sand," and the "Ashburnham Beds," and the subdivision of the "Hastings Sand" by Dr. Mantell into "Horsted Sands," "Tilgate Beds," and "Worth Sands," and having defined the district under notice as lying between and in the neighbourhood of the towns of Tenterden, Cranbrook, Tunbridge, Tunbridge Wells, East Grinstead, and Horsham, Mr. Drew proceeded to describe, first, the several beds in the meridian and vicinity of Tunbridge Wells. The Weald Clay is at least 600 feet thick in this district, and is underlain by sands and sandstones, termed by the author the "Tunbridge Wells Sand," on account of its being well exposed there. This subdivision is about 180 feet thick, and was described in detail,—an important feature being the "rock-sand," or massive sandstone forming the picturesque natural rocks of the neighbourhood. The shales and clays underlying these sands form the "Wadhurst Clay" of the author, and are at places 160 feet thick. This subdivision has yielded much ironstone in former times. It is underlain by other sand and sandstones, more than 250 feet thick, also yielding ironstone. These are termed "Ashdown Sand" by Mr. Drew on account of their forming the heights of Ashdown Forest.

Eastward of the meridian of Tunbridge Wells Mr. Drew has found the same sequence of beds, and he believes a similar succession to occur around Battle and Hastings. Westward of Tunbridge Wells as far as East Grinstead, the same beds occur, but beyond that the Weald Clay and Tunbridge Wells Sand alone are exposed; and the latter is here divided into upper and lower beds by shale and clay (termed "Grinstead Clay" by the author), which thicken westward to 50 feet and more. It is the "Lower Tunbridge Wells Sand" that forms natural rocks near Grinstead. Near Horsham the Weald Clay contains, at about 120 feet from its base, bands of stone known as the "Horsham Stone," used for roofing and paving.

The author then explained at large the grounds on which he proposed to replace Dr. Mantell's term "Horsted Sands" by "Upper Tunbridge Wells Sand," that of "Worth Sands" by "Lower Tunbridge Wells Sand,", and that of "Tilgate Beds" by "Wadhurst Clay," and his reason for proposing the name of "Ashdown" for the next lowest bed of the "Hastings Sand."

The paper concluded with a description of some of the chief lithological characters of the clays and sandstones of the Wealden area under notice.

2. "On the Permian Rocks of the South of Yorkshire; and on their Palæontological Relations." By J. W. Kirkby, Esq. Communicated by T. Davidson, Esq., F.G.S.

The author, after defining the area to be treated of, first noticed the results of the labours of former observers in this district; and then succinctly described the several strata, referring to Professor Sedgwick's Memoir on the Magnesian Limestone for descriptions of the physical geography and very much of the lithological characters of the country under notice. The strata treated of Mr. Kirkby recognizes (in descending order) as, 1. the Bunter Schiefer, about 50 feet thick; 2. the Brotherton Beds, 150 feet; 3. the small-grained Dolomite, 250 feet; 4. the Lower Limestone, 150 feet; 5. the Rothliegendes or Lower Red Sandstone, 100 feet. These were then compared and coordinated with the Permian strata of Durham, where the three limestone members are thus represented :—1. The Upper Limestone by the Yellow, Concretionary, and Crystaline Limestone (250 feet); 2. The Middle Limestone by the Shell- and Cellular Limestone (200 feet); and 3. The Lower Limestone by the Compact Limestone (200 feet) and the Marl-slate (10 feet),—the over- and under-lying sandstones being much alike as to thickness in the two areas.

After some remarks on the probable geographical conditions existing in the Permian epoch, the author proceeded to treat of the Permian fossils of South Yorkshire in detail. These belong to about thirty species, and are nearly all from the Lower Limestone,—three species only occurring in the Brotherton beds. With three exceptions they occur also in the several limestones of Durham; five of them are found in the lower part of the red marls of Lancashire; and six of them are found at Cultra and Tullyconnel in Ireland. The distribution of the species in the several beds at different locali ties having been fully treated of, the Permian fossils of South York-shire were compared, first, with those of Durham; next, with those of Lancashire; and thirdly, with those of Ireland. Remarks on the distribution of the Permian Fauna in time concluded the paper.

XLVIII. *Intelligence and Miscellaneous Articles.*

SOME RESULTS IN ELECTRO-MAGNETISM OBTAINED WITH THE BALANCE GALVANOMETER. BY GEORGE BLAIR, M.A.

SINCE bringing the balance galvanometer, along with some other apparatus, before the Society in the course of last session, the writer had made some experiments with this new galvanometer, which

Fig. 1.

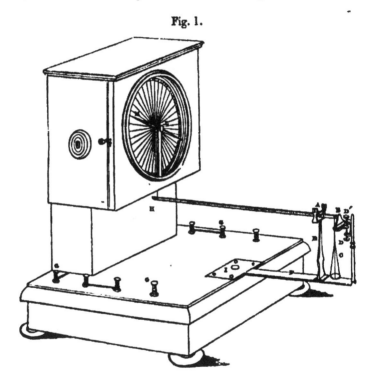

led to results that he did not anticipate, and which he considered to be of sufficient importance to justify him in presenting them to the Society. The object originally aimed at in its construction was to obtain an exact measure, by weight, of the actual amount of deflective force which the current exerts upon the magnetic needle. The instrument constructed for this purpose is represented in fig. 1. The coil consists of a total length of 1660 feet of No. 22 copper wire, weighing rather more than 6 lbs., and divided into four parts, the ends of which are brought out and connected with their respective terminals G G, so that they can be used separately or as one coil. The needle with which the first experiments were made consisted of a small rectangular steel bar, $1\frac{1}{2}$ inch in length, rather less than $\frac{1}{4}$th of an inch in breadth, and about half the thickness of a shilling. It weighed exactly 18 grains, and, when magnetized, its lifting power was 44 grains, or nearly $2\frac{1}{2}$ times its own weight. The index M, which is 9 inches in length, weighs only 2 grains. A small brass pulley, $\frac{1}{4}$th of an inch in diameter, is fixed upon the axis between the index and the supporting screw L. The balanced lever E H consists of a thin slip of hard spring-brass, placed edgewise for strength, and tapered, for lightness, towards the end of the long arm A H. The short arm A E is loaded to act as a counterpoise; and to this arm a scale-pan C is suspended, at a distance from the fulcrum equal to exactly $\frac{1}{10}$th of the length of the other arm. It carries also at its extremity a thin horizontal projection, which vibrates between two screw-points D, D', and by which, with the aid of the wooden foot-screws of the instrument, the lever can be always exactly levelled when balanced. The fulcrum B is supported on a stout brass bar F, which is firmly held in its place by means of the screw I, and can be removed at pleasure. When it is desired that the needle shall have liberty to move in both directions, the extremity H of the long arm of the lever is connected with the needle by a slender wire suspended from a very fine thread, fixed to the upper part of the pulley and carried down on both sides of it, as shown in fig. 2. The arm A H is divided into ten equal parts, each of which is subdivided into tenths; and, estimating the poles of the needle to be at a distance of about $\frac{1}{8}$th of its total length from the extremities, the diameter of the pulley is so adjusted that a weight of 100 grains, suspended at the distance of one of the large divisions from the fulcrum, acts with a force of 1 grain at the poles of the needle; suspended at division 2, it acts with a force of 2 grains; at 2·5, with a force of $2\frac{1}{2}$ grains, and so on.

Fig. 2.

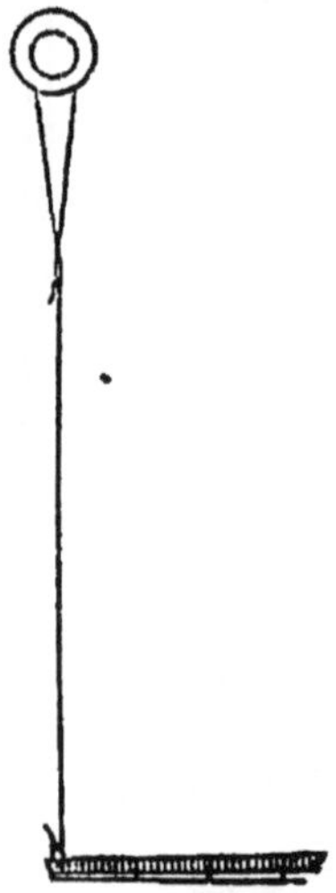

The following Table exhibits the results of the first series of experiments made with a small Grove's battery, the platinum plates of which expose only two inches of surface, and having the zinc plates immersed in a saturated solution of chloride of sodium. It is a striking characteristic of Grove's battery that it slightly increases in force after being some time in action; and it would have been preferable, therefore, to use a Daniell's, on account of its remarkable constancy; but the writer had not a sufficient number in series. The fourth column indicates the weight required to bring the index of the balance galvanometer back to zero; the fifth column expresses the same weight reduced to the force which it exerts at the poles of the needle, but increased in each case by half a grain, to compensate for the small preponderance given to the long arm of the lever in order to keep the needle vertical when not deflected by the current :—

Table I.

1.	2.	3.	4.	5.	6.
No. of pairs.	Angles on tangent galvanometer.	Angles on balance galvanometer, without weight.	Weights required to bring the needle to zero.	Force at poles of needle, in grains.	Ratio of tangents reduced.
1	4 30	66	100 grs. at 2·5	3·00	3·12
2	8 30	77 30	,, ,, 5·5	6·00	5·96
6	29 30	87	1000 grs. at 2·2	22·50	22·64
12	44 45	89	,, ,, 4·9	49·50	39·64

It will be seen that, up to six pairs, the numbers in the fifth column, expressing the force of the current in grains at the poles of the needle, vary very nearly in the same ratio as the tangents of the angles of deflection on the tangent galvanometer reduced to a comparable form in the sixth column. With twelve pairs, however, the weight required to balance the current is 49·5, or very nearly 50 grains; whereas, according to the tangent galvanometer, it should not have exceeded 40 grains. Reflecting on this anomaly, the writer could only arrive at the conclusion that the needle, surrounded by a very powerful current in such a large coil, ceased to act as a permanent magnet, and was temporarily charged with a higher magnetism induced by the current itself. Subsequent experiments completely confirmed this conclusion; and he was led to examine the subject more minutely by observing that the needle, after being several times subjected to the action of the current from twelve pairs, appeared to have lost its permanent magnetism; for he afterwards found repeatedly that when the index was brought back by successive increments of weight to 40°, the smallest possible additional weight sent it back to zero. Before taking out the needle to ascertain this, he submitted it a second time to the action of six and twelve pairs, with the following results :—

Number of pairs.	Current force by tang. galvan.	Weights supported.
6	22·64	7·5 grains.
12	39·64	24·5 ,,

whereas it will be seen, by referring to the preceding Table, that originally it supported 22·50 and 49·50 grains respectively. The needle was then taken out, and was found to have almost entirely lost its magnetism. It had originally lifted 44 grains; it was now only with the greatest precaution that it lifted 2 grains. But

$$7\cdot5 \quad : \quad 24\cdot5 \quad = 1 : 3\cdot27$$
$$\text{and } (22\cdot64)^2 : (39\cdot64)^2 = 1 : 3\cdot06.$$

The writer had therefore little doubt that, if not merely demagnetized, but formed of soft iron, the needle, when placed in a favourable position, would turn with a force proportional to the square of the current; whereas it plainly appears from Table I., that so long as its permanent magnetism is sufficient to resist the inducing action of the current, the needle is deflected with a force simply proportional to the current.

To determine this interesting question, two new needles were constructed, similar in shape and size to the former, but somewhat lighter, each weighing only 17·25 grains. The one was of steel, tempered to the hardness of glass; and was magnetized till it lifted with some difficulty 43 grains; the other was of soft hoop-iron, well annealed. With these needles the following results were obtained from experiments conducted very carefully, and using, for greater accuracy, a single-thread suspension :—

Table II.

	1.	2.	3.	4.	5.	6.
	No. of pairs.	Angles on tangent galvanometer.	Tangents to rad. 1.	Deflective force at poles of needle, in grains.	Ratio of tangents.	Ratio of squares of tangents.
With magnetized steel needle.	3	17 40	0·318	11·5	11·5	11·4
	6	31 20	0·609	22·0	22·0	42·1
	9	41 0	0·869	32·5	31·5	85·8
	12	47 0	1·072	43·5	38·8	130·6
With needle of soft iron	3	17 40	0·318	6·0	6·0	6·0
	6	31 0	0·601	20·5	11·4	21·5
	9	40 20	0·849	40·0	16·1	42·9
	12	47 0	1·072	57·0	20·4	68·4

A glance at the above Table will show that, with the permanently magnetized needle, the numbers in column 4, expressing the deflective force of the current in grains, are very nearly proportional to the numbers in column 5, expressing the *simple ratio of the tangents or quantities of current*; whereas with the soft iron needle they are nearly proportional to the *squares of the same quantities*, reduced to a comparable form in column 6. In both cases the only marked deviation coincides with the powerful current from twelve elements of the battery, in which case the steel needle, evidently acting under the superadded influence of induced temporary magnetism, is deflected with a force which exceeds the estimated amount by 4·7 grains, whereas the soft iron needle, under the same current force, falls short of the calculated amount by 11·4 grains. The last effect is probably attributable to the fact that, as the needle approaches saturation, the law of the squares gradually merges into the law of simple proportion. The writer regrets that he had not at command sufficient battery power to put this point to the test of decisive experiments, but hopes to do so shortly with a Daniell battery of 50 or 100 elements. In the meantime the results above given, having been arrived at with great care, and amply confirmed by experiments several times repeated, appear to establish very conclusively the following principles :—

1. A permanently magnetized steel needle, suspended in the middle of a galvanometer coil, is deflected with a force simply proportional to the quantity of current transmitted, so long as the current force which acts upon it is not sufficient to impart temporarily a higher magnetism than that which it permanently possesses. Beyond this point, the deflective force exerted on the magnetized steel needle increases in a somewhat higher ratio than the current, and therefore the accuracy of any form of galvanometer can be trusted only within certain limits of current force and of length and proximity of coil.

2. A pure soft iron needle, suspended at an angle of about 40° to the direction of the current (the angle varying according to the shape of the needle), is deflected with a force which, within certain limits of current power, is very exactly proportional to the squares of the quantities of current. Beyond these limits the deflective force exerted on the needle increases in some constantly diminishing ratio lower than that of the squares of the current.

3. The action of the current in deflecting a magnetic needle is precisely the same action, and follows the same law, as that which it exerts in magnetizing a bar of soft iron. The amount of magnetism actually imparted to a bar or needle of soft iron is directly proportional to the quantity of current; for the force with which a soft iron needle is deflected under different currents is not proportional to its temporary magnetism in each case, but to the product of its magnetism multiplied by the force of the current. By increasing the force of the current, two effects are produced; in the *first* place, the magnetism of the needle is increased in the same proportion; and *secondly*, the increased current acting upon this increased magnetism deflects the needle with a force proportional to the product of the two, or in other words, proportional to the square of the actual quantity of current.

It only remains to add the results of two series of experiments, showing the very striking difference between the deflective forces exerted upon the two needles at different angles of inclination. Table III. shows the increasing weights required to balance the needles at angles successively diminished by 10°; Table IV. exhibits the effect produced by successive additions of weights, equivalent to a force of ten grains acting at the poles of the needles. In both

cases the battery power employed was twelve small Grove's, but the
current declined, in the course of the experiments, from 47° to 45°
on the tangent galvanometer, which accounts for the fact that the
maximum weights supported are less than in the earlier experiments
recorded in Table II. In working out Table III., the weight employed (1000 grains) was simply advanced along the lever, and its
reduced amount at the poles of the needle noted when the index, in
gradually retreating, pointed to the successive angles specified. The
results in Table IV. were obtained by moving the weight from one
to another of the successive divisions, marked 1, 2, 3, &c. on the
lever; and the differences of the angles vary, as might be expected,
in nearly the reverse order of the differences of weights in Table
III. : —

Table III.

	Angles on balance galvanometer.	Weights reduced to force at needle, in grains.	Differences of successive weights.
With magnetized steel needle.	90	0·0	
	80	6·5	6·5
	70	15·5	9·0
	60	23·0	7·5
	50	30·0	7·0
	40	35·0	5·0
	30	39·4	4·4
	20	41·5	2·1
	10	42·0	0·5
With soft iron needle.	90	0·0	
	80	15·5	15·5
	70	29·5	14·0
	60	40·5	11·0
	50	50·0	9·5
	40	53·0	3·0

Table IV.

	Weights reduced to force at needle, in grains.	Angles on balance galvanometer.	Differences of successive angles.
With magnetized steel needle.	0·0	90 0	
	10·0	75 30	14 30
	20·0	63 0	12 30
	30·0	48 30	14 30
	40·0	26 0	22 30
	41·5		
With soft iron needle.	0·0	90 0	
	10·0	84 0	6 0
	20·0	76 30	7 30
	30·0	68 20	8 10
	40·0	59 10	8 10
	50·0	47 10	12 0
	52·0	0 0	

It will be observed from Table III. that a very small weight was sufficient to throw back the steel needle to 80°, and that, on the contrary, it is from 90° to 70° that the soft iron ·needle sustains itself with comparatively the greatest power, requiring very nearly double the weight which suffices for the steel needle to balance it at the latter angle. When reduced to 40°, however, the smallest possible additional weight throws back the iron needle to zero, and in every case it was necessary to move it aside with the finger to nearly that angle, before it would exhibit the slightest action under the influence of the current, or, in other words, any perceptible trace of *longitudinal* magnetization. In fact, being laterally magnetized when hanging in a vertical position, it necessarily offered a certain resistance to deflection.

On taking out the steel needle after these experiments, it was found to have retained its original magnetism unimpaired.

The tangent galvanometer by which the force of the current was

Fig. 3.

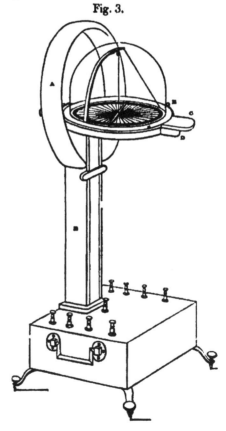

determined in the preceding experiments, and which the writer had also the honour of submitting to the Society last session, is represented in fig. 3. It is a very convenient modification of Gaugain's instrument, described in the *Annales de Chimie*, vol. xli. 1854. The circular frame A, containing a variety of coils of different lengths and sizes of covered wire, is 9·6 inches in external diameter; so that when the instrument is in use, the divided circle must be drawn out till its centre is 2·4 inches in front of the coil or coils through which the current is to be sent. To facilitate this operation, the horizontal bar D, upon which the disk slides, has the proper distances for each coil marked upon it, and these are successively exposed to view at the back of the instrument, in proportion as the disk is drawn smoothly forward by means of the handle C. The needle is only one inch in length, but carries parallel to itself a fine filament of glass for an index *; it is suspended by a silk fibre, and is raised so as to hang freely within its glass shade by turning the pin E. The ends of the coils are carried down through the hollow pillar B, and by connecting the electrodes of the battery with the proper terminals, the current can be sent through one or more of the coils. It can be sent through one convolution of No. 16, through 203 convolutions of No. 34, or any other of the intermediate lengths and sizes of wire; and in this way the resistance and the force exerted by the current upon the needle can be very exactly adapted to the character of the battery, or other rheomotor employed.—*From the Proceedings of the Glasgow Philosophical Society for* January 16, 1861.

ON THE PRESENCE OF ARSENIC AND ANTIMONY IN THE SOURCES AND BEDS OF STREAMS AND RIVERS. BY DUGALD CAMPBELL, ANALYTICAL CHEMIST TO THE HOSPITAL FOR CONSUMPTION, BROMPTON.

To the Editors of the Philosophical Magazine and Journal.

GENTLEMEN,

Since my communication upon the above subject, published in the Philosophical Magazine of October last, I have repeated my experiments upon several of the sands I then reported upon, and with the like results which I then gave. I have also made experiments upon other specimens since obtained, and in all I have hitherto examined I have found arsenic, and generally, if not always, accompanied with antimony. The process followed was the same as I formerly described, only I invariably used hydrochloric acid without the slightest trace of arsenic in it, as some doubts had been cast upon my former results, in a notice of my paper in the 'Chemical

---

* The thickness of the index is grossly exaggerated in the figure; it ought to be as fine as a hair, and short in one arm.

News' of the 20th of October last, because in my anxiety to admit of any one testing the accuracy of my results, I had described how the process might be conducted with what is generally sold as pure acid, but which, if properly tested, is rarely free from arsenic.

During these last experiments, it occurred to me to distil the sands with a second and a third dose of acid, and in most cases I have found the yield of arsenic and antimony to be much greater, say from two to five times, in the second distillate than in the first; and in some I have found the third distillate to give more than the first, but in others less.

These results induce me to say that, before a sand could be pronounced not to contain any arsenic or antimony, it should be distilled to dryness with at least three distinct doses of acid, each distillate being tested carefully in the manner described in my former communication.

I am, Gentlemen,
Your obedient Servant,
Dugald Campbell.

7 Quality Court, Chancery Lane,
    March 25th, 1861.

NOTE ON A MODIFICATION OF THE APPARATUS EMPLOYED FOR ONE OF AMPÈRE'S FUNDAMENTAL EXPERIMENTS IN ELECTRODYNAMICS. BY PROFESSOR TAIT.

My attention was recalled by Principal Forbes's note* (read to the Royal Society of Edinburgh on January 7), to his request that I should at leisure try to repeat Ampère's experiment for the mutual repulsion of two parts of the same straight conductor, by means of an apparatus which he had procured for the Natural Philosophy Collection in the University. Some days later I tried the experiment, but found that, on account of the narrowness of the troughs of mercury, it was impossible to prevent the capillary forces from driving the floating wire to the sides of the vessel. I therefore constructed an apparatus in which the troughs were two inches wide, the arms of the float being also at that distance apart. Making the experiment according to Ampère's method with this arrangement, I found one small Grove's cell sufficient to produce a steady motion of the float from the poles of the pile; in fact, the only difficulty in repeating the experiment lies in obtaining a perfectly clean mercurial surface.

Two objections have been raised against Ampère's interpretation of this experiment, one of which is intimately connected with the subject of Principal Forbes's note. This is, the difficulty of ascertaining exactly what takes place where a voltaic current passes from one conducting body to another of different material. It is known

* Phil. Mag. for February 1861.

that thermal and thermo-electric effects generally accompany such a passage. To get rid of this source of uncertainty, I have repeated Ampère's experiment in a form which excludes it entirely. In this form of the experiment the polar conductors and the float form one continuous metallic mass with the mercury in the troughs,—the float being formed of glass tube filled with mercury, with its extremities slightly curved downwards so as nearly to dip under the surface of the fluid, and the wires from the battery being plunged into the upturned outward extremities of two glass tubes, which are pushed through the ends of the troughs so as to project an inch or two inwards under the surface of the mercury. A little practice is requisite to success in filling the float and immersing it in the troughs without admitting a bubble of air. This float, being heavier than the ordinary copper wire, plunges deeper in the fluid, and encounters more resistance to its motion; but with two small Grove's cells only, Ampère's result was easily reproduced, even when the extremities of the float rested in contact with those of the polar tubes before the circuit was completed. It is obvious that here no thermo-electric effects can be produced in the mercury; and I have satisfied myself that the motion commences before the passage of the current can have sensibly heated the fluid in the tubes.

The other class of objections to Ampère's conclusion from this experiment, depending on the spreading of the current in the mercury of the troughs, is of course not met by this modification. I have made several experiments with a view to obviate this also; but my time has been so much occupied that I have not been able as yet to put them in a form suitable for communication to this Society.—*From the Proceedings of the Royal Society of Edinburgh*, vol. iv.

NOTE RESPECTING OZONE.

In the Philosophical Magazine, May 1860, page 403, is a short account of "the production of Ozone by means of a Platinum Wire made incandescent by an Electric Current," by M. Le Roux, which has just recalled to my memory the following fact.

I have frequently observed that a coil of platinum wire heated to *whiteness* in a strong jet of purified hydrogen, and then removed from the jet, imparted a feeble ozone-like odour to the ascending stream of hot air above the wire as long as the wire remained nearly white-hot, and ceased to impart this odour at a somewhat lower temperature.

G. GORE.

Birmingham.

THE

LONDON, EDINBURGH AND DUBLIN

PHILOSOPHICAL MAGAZINE

AND

JOURNAL OF SCIENCE.

[FOURTH SERIES.]

MAY 1861.

XLIX. *On the Determination of the Direction of the Vibrations of Polarized Light by means of Diffraction.* By L. Lorenz[*].

THE question whether the vibrations of polarized light are perpendicular to the plane of polarization, or in that plane, notwithstanding its great theoretical importance, is still undecided. On comparing the different arguments that may be advanced in favour of either hypothesis, only two will be found that have an essential bearing on the subject,—the experiments, namely, of Jamin on the reflexion of light by transparent media, and the polarization of light caused by diffraction.

The experiments of Jamin have hitherto been explained on the supposition that the vibrations of polarized light are perpendicular to the plane of polarization. This, however, is not decisive of the question, since it has hitherto been assumed that there is an instantaneous change of refractive power at the boundary of two transparent media; whereas I shall prove, in a subsequent essay[†] "On the Reflexion of Light," that Jamin's experiments can only be brought into complete accordance with Fresnel's formulæ for the reflexion and refraction of light, on the hypothesis that these formulæ hold good for infinitely small changes of the refractive index, that is to say, on the hypothesis that there is a gradual passage from one medium to the other.

It may be asked whether Fresnel's formulæ really hold good for an infinitely small change of refractive power, and whether these formulæ may be deduced on either hypothesis as to the direction of the vibrations of polarized light; two questions which I shall answer in a third essay.

* Translated by F. Guthrie, from Poggendorff's *Annalen*, vol. cxi. p. 315, 1860.
† See Poggendorff's *Annalen*, vol. cxi. p. 460: a translation of this paper will be given in a future Number of this Magazine.

The change of the plane of polarisation by diffraction conducts us by another road to the determination of the same question. Several years ago Mr. Stokes furnished a mathematical proof that the plane of polarization must be changed by diffraction. Doubts have, however, justly been entertained as to the accuracy of his conclusions, because he only succeeded in solving the problem of diffraction imperfectly; and I have therefore sought the complete solution of the problem by other methods, which I have found particularly applicable in the theory of elasticity.

When an undulation passes through an opening in a solid plane, waves proceed from the opening on both sides of the plane. The motion in the plane is not known, except insofar as it is determined by the fact that the sum of the components of the incident and reflected waves are equal to the components of those transmitted, and that the normal and tangential pressures on both sides of the plane of the opening are the same at every point. Let the components of the incident rays be denoted by u, v, and w; those of the transmitted rays by u_1, v_1, and w_1; of the reflected by u_2, v_2, and w_2; and let the plane of coordinates x, y, z coincide with the plane of the opening.

The first condition gives, for $x=0$,

$$u+u_2-u_1=0, \quad v+v_2-v_1=0, \quad w+w_2-w_1=0; \quad . \quad (1)$$

and by the help of these equations it may easily be deduced from the second condition for $x=0$, that

$$\frac{d(u+u_2-u_1)}{dx}=0, \quad \frac{d(v+v_2-v_1)}{dx}=0, \quad \frac{d(w+w_2-w_1)}{dx}=0. \quad (2)$$

If the incident waves are waves of light, then

$$\frac{du}{dx}+\frac{dv}{dy}+\frac{dw}{dz}=0;$$

and equations (1) and (2) may be satisfied by the supposition that

$$\left. \begin{array}{l} \dfrac{du_1}{dx}+\dfrac{dv_1}{dy}+\dfrac{dw_1}{dz}=0, \\[2mm] \dfrac{du_2}{dx}+\dfrac{dv_2}{dy}+\dfrac{dw_2}{dz}=0, \end{array} \right\} \quad \ldots \ldots \quad (3)$$

from which it is evident that no waves of condensation can be formed.

The law of the motion is expressed by the differential equation

$$\frac{d^2}{dx^2}+\frac{d^2}{dy^2}+\frac{d^2}{dz^2}=\frac{1}{w^2}\frac{d^2}{dt^2}, \quad \ldots \ldots \quad (4)$$

which must satisfy all the components, where w expresses the rate of propagation, and t the time. This equation will obviously be satisfied by the expression

$$\frac{\phi(wt-r)}{r},$$

where $r=\sqrt{x^2+(y-\beta)^2+(z-\gamma)^2}$, and therefore also by

$$\Phi=-\frac{1}{2\pi}\int d\beta \int d\gamma \frac{\phi(wt-r,\beta,\gamma)}{r};$$

which function Φ, when the limits of integration are determined by the boundaries of the opening, also possesses the property that its differential coefficient with respect to x becomes equal to $\phi(w, t, y, z)$ when x decreases to nothing, and the point yz is within the opening. If x increases to nothing, the value of the differential coefficient is $-\phi(wt, y, z)$; and if the point yz is without the opening, it becomes nothing when $x=0$; for by differentiating the integral with respect to x, $x=0$ enters as a factor, thus causing every element of the integral to disappear, except those in which $r=0$, that is to say, $y=\beta$, $z=\gamma$. Whence, if x is positive, and the point (yz) is within the limits of the integral,

$$\left[\frac{d\Phi}{dx}\right]^{x=0}=\left[\frac{1}{2\pi}\int d\beta \int d\gamma \frac{x}{r^3}\right]^{x=0}\phi(wt, y, z)=\phi(wt, y, z);$$

and if x is negative,

$$\left[\frac{d\Phi}{dx}\right]^{x=0}=-\phi(wt, y, z).$$

Introduce now other functions Ψ, X, Φ_1, Ψ_1, X_1, which are related to the respective functions ψ, χ, ϕ_1, ψ_1, χ_1 in the same way as Φ is to ϕ, and put

$$\left. \begin{aligned} u_1 &= \Phi + \frac{d\Phi_1}{dx} - \frac{d(F+F_1)}{dx}, \\ v_1 &= \Psi + \frac{d\Psi_1}{dx} - \frac{d(F+F_1)}{dy}, \\ w_1 &= X + \frac{dX_1}{dx} - \frac{d(F+F_1)}{dz}, \end{aligned} \right\} \quad \cdots \cdots \quad (5)$$

where the functions F, F_1 are so chosen that

$$\frac{du_1}{dx}+\frac{dv_1}{dy}+\frac{dw_1}{dz}$$

becomes equal to nothing, then

$$\Delta^2 F = \frac{d}{dx}\left[\Phi + \frac{d\Psi_1}{dy} + \frac{dX_1}{dz}\right], \left.\begin{array}{c}\\\\\end{array}\right\} \quad \ldots \quad (6)$$
$$\Delta^2 F_1 = \frac{d^2\Phi_1}{dx^2} + \frac{d\Psi}{dy} + \frac{dX}{dz}.$$

To the components u_2, v_2, w_2 of the reflected wave we give the same values as to u_1, v_1, w_1; only it must be observed that in the first x is always negative, in the latter positive.

Assuming now that

$$[F]^{x=0} = 0 \text{ and } \left[\frac{dF_1}{dx}\right]^{x=0} = 0, \quad \ldots \quad (7)$$

the truth of which will appear from what follows, we get by means of (1) for $x=0$,

$$\begin{array}{l}u + u_2 - u_1 = u - 2\phi_1(wt, y, z) = 0,\\v + v_2 - v_1 = v - 2\psi_1(wt, y, z) = 0,\\w + w_2 - w_1 = w - 2\chi_1(wt, y, z) = 0;\end{array}\left.\begin{array}{c}\\\\\\\end{array}\right\} \quad \ldots \quad (8)$$

and by means of (2) for $x=0$,

$$\begin{array}{l}\frac{d(u + u_2 - u_1)}{dx} = \frac{du}{dx} - 2\phi(wt, y, z) = 0,\\[2mm]\frac{d(v + v_2 - v_1)}{dx} = \frac{dv}{dx} - 2\psi(wt, y, z) = 0,\\[2mm]\frac{d(w + w_2 - w_1)}{dx} = \frac{dw}{dx} - 2\chi(wt, y, z) = 0.\end{array}\left.\begin{array}{c}\\\\\\\end{array}\right\} \quad \ldots \quad (9)$$

All the conditions are hereby fulfilled, and the functions ϕ, ϕ_1, ψ, &c. determined. The truth of equations (7) may now also be easily demonstrated.

The problem of diffraction is therefore completely solved by equations (5), (6), (8), and (9). If we now pass to the particular case in which the incident waves lie in a plane, the components are determined by the following equations,

$$u = \xi\mathfrak{b}, \quad v = \eta\mathfrak{b}, \quad w = \zeta\mathfrak{b},$$
where
$$\mathfrak{b} = \cos k(wt - ax - by - cz),$$
$$a\xi + b\eta + c\zeta = 0, \quad a^2 + b^2 + c^2 = 1.$$

Since, moreover, we only require to determine the motion at a point at a considerable distance behind, the opening in the screen must be very great; and putting ρ for the distance of the observed point from the origin of coordinates, we have

$$r = \sqrt{x^2 + (y-\beta)^2 + (z-\gamma)^2} = \rho - m\beta - n\gamma,$$

where

$$\rho = \sqrt{x^2 + y^2 + z^2}, \quad m = \frac{y}{\rho}, \quad n = \frac{z}{\rho};$$

and l, m, and n being the cosines of the angles made by the diffracted ray with the coordinate axes, $l^2 + m^2 + n^2 = 1$. Now from (9) we get

$$\phi(wt, y, z) = \tfrac{1}{2}ak\xi \sin k\,(wt - by - cz),$$

whence

$$\Phi = \tfrac{1}{2}ak\xi S,$$

where

$$S = -\frac{2}{2\pi\rho} \int d\beta \int d\gamma \sin k[wt - \rho + (m-b)\beta + (n-c)\gamma].$$

And the values of the functions that enter into (5) being found in a similar manner, we get

$$u_1 = \tfrac{1}{2}k(a+l)[\xi - l\,(l\xi + m\eta + n\zeta)]S,$$
$$v_1 = \tfrac{1}{2}k(a+l)[\eta - m(l\xi + m\eta + n\zeta)]S,$$
$$w_1 = \tfrac{1}{2}k(a+l)[\zeta - n(l\xi + m\eta + n\zeta)]S.$$

These expressions hold good also for the waves reflected from the opening, only that in this case l is negative. For a point in the direction opposite to that of the incident ray $a+l=0$, and therefore all the components of the motion are equal to nothing.

Mr. Stokes has arrived at the same result, although he did not regard the reflected wave, and has not completely solved the problem. If a plane be supposed to pass through the refracted and the incident ray, and if α denote the angle which the vibrations of the incident ray make with the normal to this plane, α_1 the angle made by the vibrations of the diffracted ray with the same normal, and β the angle of diffraction, then we easily find, as Stokes has done, that

$$\tan \alpha_1 = \cos \beta \tan \alpha,$$

which is independent of the form and position of the opening. The vibrations therefore become more nearly vertical after passing through a vertical slit or grating. Accordingly, therefore, as experiment shows that the plane of polarization is rendered more vertical or more horizontal by diffraction, so must the vibrations of polarized light be parallel or perpendicular to the plane of polarization. It must, however, be remembered that, mathematically speaking, the screen is supposed to be a plane which does not itself vibrate, and which reflects no light from its edges.

The experiments that have hitherto been tried leave this question still undecided; for while Stokes, from experiments made with glass gratings, found that the plane of polarization is rendered more horizontal, Holtzmann, from experiments made with a smoke-grating, came to the opposite conclusion. In order finally to settle this question, I have instituted a course of experiments with gratings of various kinds.

By means of a heliostat and collecting lens, I introduce a portion of the sun's rays into a chamber. At some distance from the focus of the first lens a smaller lens receives the rays, and transmits them, almost parallel, through a polarizing Nicol's prism, which is fastened in a tube to a vertical circular arc. An index with a vernier gives the angle which the plane of polarization of the transmitted light makes with the vertical line. At the distance of about 7 metres the light falls on a vertical grating, which is fastened to a small plate in the middle of a horizontal circular arc, which is provided with a moveable horizontal telescope. Before the object-glass is placed a doubly-refracting prism of rock-crystal, which divides the polarized pencil into two, polarized perpendicularly to each other. This prism can be turned about the axis of the telescope. In general, therefore, two horizontal bands of light of unequal brightness will be seen in the telescope; but by turning the Nicol's prism or the doubly-refracting crystal about its axis, the intensity of these two images can be rendered equal.

The experiments were generally conducted as follows:—The Nicol's prism was turned in such a direction that the plane of polarization made an angle of 45° with the vertical line, and the telescope was so placed that its vertical thread passed through the two illuminating points, while the horizontal thread lay between the two horizontal bands of diffracted light. When the telescope was turned through the angle β, both bands, owing to the position of the rock-crystal, were of equal intensity. The telescope was then put back into its original position, and the Nicol's prism was turned into such a position that one of the images entirely disappeared.

If the Nicol's prism required to be turned through the angle δ (or $\delta \pm 90°$ or $\delta + 180°$), the plane of polarization must have been turned through the same angle, provided that the light has not been elliptically polarized by diffraction; and if δ be positive, the plane of polarization has been rendered more horizontal.

Sometimes I first determined δ, and then the angle of diffraction β, for which the two bands are equally bright.

There is, however, a source of error in these experiments, of which I only became aware after some time. If, for example, the upper half of the grating produce a more brilliant diffracted image than the lower, then if the grating entirely cover the object-glass, as was always the case in the above experiments, the upper diffracted image will be too bright. And if this image be polarized horizontally, δ will be found too great; if vertically, too small. To render the experiment perfect, it is therefore necessary to turn the rock-crystal through an angle of 180°, and take the mean of the two values of δ so obtained. If this precaution be neglected, very considerable errors may be introduced, especially when smoke-gratings are employed, and I imagine it is this that has misled Holtzmann. He observed, in the case of a smoke-grating, that for a diffraction of 20° there was a very considerable difference in the brightness of the horizontal and vertically-polarized images. This is nearly always so with gratings of this description: the upper or the lower image appears the brightest, without reference to the position of the plane of polarization. With a perfectly accurate grating, M. Holtzmann would not have been able to distinguish the slight difference that really does exist.

My first experiments were made with a gold grating (1000 bars to the Paris inch). Light polarized at an angle of 45° with the vertical, when diffracted with this grating, gave two images, of which neither could be made entirely to disappear for any position of the rock-crystal; and this was still more evident when the grating was placed obliquely. The diffracted light must therefore either have been elliptically polarized, or have been partly converted into ordinary light. That the former was the case, I inferred from the fact that elliptically-polarized light could be converted by diffraction into circularly and plane-polarized. If, for example, I passed light polarized at the angle α through a Fresnel's parallelopiped, whose reflecting surfaces made the angle 45° with the vertical, the angle α could be so chosen that one image in the telescope could be made entirely to disappear, or, on the other hand, so that the two images, on turning the rock-crystal, always retained the same intensity.

By measurements made in this manner, I convinced myself that the phenomenon is essentially the same as that which accompanies the ordinary reflexion of light from polished metal surfaces, and that the effect of the diffracted light is imperceptible in comparison with that reflected from the edges.

I now provided myself with various smoke-gratings. Polished glass surfaces were smoked with burning camphor, and then treated with a few drops of oil of turpentine to fix the smoke to the glass. These were then divided by means of a machine into bars only an inch long (2, 5, 10, and 16 to the millim.).

With these gratings I no longer observed any elliptic polarization. I found no observable differences in the results for the different gratings, and content myself therefore with giving the mean results for them all.

When the grating was perpendicular to the incident ray, and on the side of the glass towards the telescope, as was the case in Holtzmann's experiments, I found the angle δ, through which the plane of polarization was turned, extraordinarily small, and therefore only determined it accurately for a single angle of diffraction (65°). The plane of polarization of the incident light was in all the following experiments inclined at an angle of 45° with the vertical.

The mean result for $\beta = 65°$ was

$$\delta = 1° \, 52'.$$

The plane of polarization therefore had become very slightly more horizontal. For greater values of β I found δ still smaller, which at first greatly perplexed me.

When the grating was turned round so as to be on the side towards the incident ray and perpendicular to it, the plane of polarization was turned through a greater angle in the same direction, and for $\beta = 65°$ I found

$$\delta = 12° \, 30'.$$

These results agree neither with Holtzmann's experiments, nor with the conclusions that seem to follow from theory. I think, however, they can be explained as follows.

When the light first passes through the glass and then through the grating, the circumstances are almost the same as when it encounters the grating in the substance of the glass, as may be concluded from the fact that there is no reflexion at the boundary between the smoke and the glass. The diffraction therefore takes place within the substance of the glass, and the diffracted light is *afterwards* refracted in passing out of the plate. If β_1 be the diffraction in the glass (β being the observed diffraction), and n the index of refraction of glass, then $\sin \beta = n \sin \beta_1$. In consequence of the refraction, the plane of polarization is again altered and becomes more *vertical*. Supposing now that the vibrations are perpendicular to the plane of polarization, the change of the plane of polarization δ_1, caused by the diffraction β_1, is determined by the equation

$$\tan (45° - \delta_1) = \cos \beta \, .$$

If therefore δ be the angular change of the plane of polarization after reflexion at the first surface of the glass, we have by Fresnel's formulæ,

$$\tan (45° - \delta) = \frac{\tan (45 - \delta_1)}{\cos (\beta - \beta_1)} = \frac{\cos \beta_1}{\cos (\beta - \beta_1)}.$$

The mean index of refraction n was determined by experiment as to the angle of polarization, and I found

$$\log n = 0 \cdot 18886.$$

For $\beta = 65°$ we should therefore have $\delta = 2° \ 11'$, which agrees pretty well with experiment, which gave $\delta = 1° \ 52'$. That δ should decrease as β increased, as appeared from experiment, follows also from this calculation.

If, on the contrary, the smoked side of the glass is turned towards the incident ray, the light is diffracted before it reaches the surface of the glass, and is afterwards twice refracted. We have therefore

$$\tan (45° - \delta) = \frac{\cos \beta}{\cos^2 (\beta - \beta_1)};$$

from which it follows that if $\beta = 65°$, $\delta = 16° \ 2' \ 30''$.

In this case experiment gave a decidedly less value for δ, which shows that the actual circumstances are only approximately those assumed, and that the diffraction of the light only takes place partially within the substance of the glass. This is still more evident when the grating is placed obliquely to the incident rays, so as to make equal angles with them and the axis of the telescope; since in that case for $\beta = 90°$ I found $\delta = 20°$ instead of 45°, which is given by calculation.

It is obvious, therefore, that the circumstances, though very complicated, are naturally accounted for in all essential particulars, on the supposition that the vibration of polarized light is perpendicular to the plane of polarization, whereas the other hypothesis is altogether irreconcileable with experiment.

In order to render these results less complicated and more susceptible of calculation, I contrived a different arrangement of the smoke-gratings. Canada balsam was melted over the surface of the glass, and a smooth glass plate was pressed down on it, which it was found could easily be done without injuring the grating. As Canada balsam has almost the same index of refraction as glass, all the circumstances could then easily be calculated. The grating was so placed that it made equal angles with the incident rays and the axis of the telescope. In this position of the apparatus the *vertically* polarized portion of the incident light was found to be weakened more than that polarized horizontally; and therefore the change of the plane of polarization was positive, *although* reflexion at the two glass surfaces tended to turn the plane of polarization in the opposite direction. As the mean of many experiments with several gratings (2, 5, 10 bars to the millim.), I found—

$\beta =$	40°	50°	60°	70°	80°	90°	100°
δ observed	2·24	3·0	4·54	6·36	7·42	9·6	12·3
Calculated	2·30	3·54	5·37	7·37	9·53	12·22	15·0

where δ is given by the equations

$$\tan(45°-\delta) = \frac{\cos\beta_1}{\cos^2\frac{1}{2}(\beta-\beta_1)}, \text{ or } \sin\tfrac{1}{2}\beta_1 = \sin\tfrac{1}{2}\beta,$$

$\frac{1}{2}\beta$ being the angle which the incident ray makes with the normal to the grating, $\frac{1}{2}\beta_1$ that made by the refracted ray with the same line, while β_1 is the diffraction within the substance of the glass itself.

It will be seen that experiment agrees very well with calculation, only that it gives results in all cases a little too small. Whatever, therefore, may be the cause of this difference, experiment most decidedly favours the hypothesis that the vibrations of polarized light are perpendicular to the plane of polarization, since in the opposite case δ would be negative and of much greater magnitude.

I next investigated the effect of diffraction by smoked metal gratings. When, however, these were of a perfect dull black, the diffraction images produced by them were far too feeble; they were therefore rendered smoother by passing a drop of oil of turpentine over them. Gratings of this description must, moreover, be rather fine and very accurately made. Some experiments made with a grating of 200 bars to the Paris inch (the thickness of the wire bars was $\frac{1}{800}$th of a millim.), the grating being equally inclined to the incident light and the axis of the telescope, gave approximately the following results:—

$\beta =$	25°	35°	40°	45°	50°	55°
δ =	10°	16°	20°	25°	30°	35°

The change of direction is positive, but much greater than it should be according to calculation. The polarization of the diffracted light was moreover slightly elliptical, from which it was evident that the reflexion from the metal surfaces of the wires was not entirely prevented by the smoke. On endeavouring to smoke the grating more perfectly, I partly destroyed its accuracy, and rendered it unfit for further experiments of this description. I have not succeeded in obtaining reliable results with other gratings of this description: there are peculiar difficulties in the way of checking the reflexion from the metal edges, and at the same time preserving the diffraction image sufficiently large and distinct.

The results obtained are, however, not unimportant, since the excessive values of δ can easily be accounted for on the ground of elliptic polarization.

If it be supposed, for example, that the difference of phase of the vertical and horizontal components is Δ, and that δ_1 is the change of direction when there is no elliptic polarization, an easy calculation gives

$$\tan 2\delta = \frac{\tan 2\delta_1}{\cos \Delta},$$

whence δ must always be greater than δ_1, their signs, however, being always the same.

As experiment gave δ positive, it confirms the result already obtained, *that the vibration of polarized light is perpendicular to the plane of polarization.*

Copenhagen, June 28, 1860.

L. *On certain Laws relating to the Boiling-points of different Liquids at the ordinary Pressure of the Atmosphere.* By THOMAS TATE, *Esq.*[*]

IT is well known that the boiling-point of water is raised by the addition of a soluble salt, or by the addition of a strong acid, and that this augmentation of the boiling-temperature depends upon the relative amount of salt or acid added, as the case may be; but, as far as I know, no general formulæ have hitherto been given to express the relation between the augmentation of boiling-temperature and the relative weight of the substance added to the water.

Different weights of anhydrous salt being dissolved in 100 parts of pure water, and the augmentation of boiling-temperature being observed, we obtain data for expressing the relation of the per-centage of the salt to the corresponding augmentation of the boiling-temperature of the solution. The salts which I have examined in this manner are as follows :—the chlorides of sodium, potassium, barium, calcium, and strontium; the nitrates of soda, potassa, lime, and ammonia; and the carbonates of soda and potassa. I have found for all these salts, that *the augmentation of boiling-temperature may be approximately expressed in a certain power of the per-centage of the salt dissolved*: thus, if k be put for the weight of dry salt in 100 parts of water, and T the corresponding temperature of ebullition above that of boiling water under the same atmospheric pressure, then

$$T = ak^x, \quad \ldots \ldots \ldots (1)$$

---

[*] Communicated by the Author.

where a is constant for each salt only, and the exponent α is constant for all the salts contained in certain special groups. The salts enumerated may be divided into four distinct groups, the salts in each possessing certain remarkable points of relationship with respect to their boiling-temperatures; viz., *the augmentations of boiling-temperature of the solutions in each group of salts have a constant ratio to one another for all equal weights of salt dissolved.* Thus if T and T′ be put for the augmentations of boiling-temperature corresponding to any equal portions of two salts dissolved, belonging to the same group, then

$$\frac{T}{T'} = \text{a constant quantity.}$$

The constant quantity here expressed is, in some cases, nearly equal to the inverse ratio of the combining equivalents of the bases of the salts.

Moreover, if the weights of two portions of one kind of salt, separately dissolved in 100 parts of water, be proportional to the two portions of another salt belonging to the same group and similarly dissolved, then the ratio of the augmented temperatures of ebullition of the former will be equal (approximately) to the ratio of the augmented temperatures of ebullition of the latter.

Thus if $\dfrac{k_1}{k_2} = \dfrac{k'_1}{k'_2}$, then $\dfrac{T_1}{T_2} = \dfrac{T'_1}{T'_2}$, and conversely;

where k_1 and k_2 are the respective weights of one kind of salt separately dissolved in 100 parts of pure water, T_1 and T_2 the respective augmented temperatures of ebullition; and k'_1, k'_2, T'_1, and T'_2 are the corresponding symbols for the other salt.

For the sake of conciseness of expression, I shall sometimes speak of this augmentation of temperature simply as the temperature of ebullition, which, in fact, it would be if the temperature of boiling water were taken as zero.

The first group of salts comprises the chlorides of sodium, potassium, and barium, together with carbonate of soda.

The second group comprises the chlorides of calcium and strontium, and probably other salts.

The third group comprises the nitrates of soda, potassa, and ammonia.

The fourth group comprises the carbonates of potassa, and nitrate of lime.

If $T = af(k)$ represent the relation between T and k of a salt in any group, a being constant, and $f(k)$ a known function of k, then $T = a'f(k)$ will be the formula of relation for any other salt in the same group. For let T and T′ be the temperatures of ebullition in each case respectively for equal values of k, then

$$\frac{T}{T'} = \frac{a}{a'} = \text{a constant for all values of T and T' corresponding to}$$

equal values of k.

The annexed diagram represents the apparatus with which these experiments were made. D B an oil-bath heated on a sand-bath; A B a wide tube of some length containing the solution of the salt; T a thermometer passing through a perforated cork C, and having its bulb immersed in the liquid to within about one-quarter of an inch of the bottom of the tube. A slit is made in the side of the cork to keep the vapour in the tube at the same pressure as the external air. Pieces of platinum-foil were put in the liquid to facilitate the discharge of vapour; and the oil in the bath was time after time agitated to keep all the parts of the liquid at a uniform temperature. The boiling-temperature of pure water corresponding to the atmospheric pressure was first determined; known weights of anhydrous salt, corresponding to 100 parts by weight of water, were time after time introduced into the tube, and the corresponding temperatures of ebullition were noted: the elevations of these temperatures above that of boiling water were entered in the following Tables of results. The process was continued, in some cases, until the solution of maximum salt was nearly attained. A correction for the observed temperatures was made on account of the column of mercury in the stem of the thermometer not in contact with the liquid.

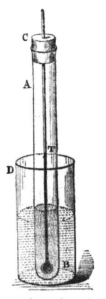

Augmentations of boiling-temperatures, in degrees Centigrade, of different solutions of the salts contained in Group 1.

Weight of salt in 100 parts of water, k.	Corresponding boiling-temperatures of the different solutions of salt above that of pure water.							
	Chloride of sodium, T.	Chloride of potassium, T'.	Chloride of barium, T''.	Carbonate of soda, T'''.	Value of T by formula $T = \frac{1}{12\cdot57}k^{\frac{4}{4}}$	Value of T' by formula $T' = \frac{2}{3}T$.	Value of T'' by formula $T'' = \frac{1}{2}T'$.	Value of T''' by formula $T''' = \frac{3}{7}T$.
0	0	0	0	0	0	0	0	0
8	1·1	·8	...	·5	1·07	·71	...	·47
16	2·5	1·86	·8	1·1	2·54	1·70	·93	1·07
24	4·2	2·93	1·4	1·8	4·22	2·80	1·46	1·80
32	6·0	4·10	2·0	2·5	6·05	4·00	2·05	2·57
40	8·0	5·34	2·6	3·3	8·00	5·33	2·67	3·14
48	...	6·60	3·3	4·5	......	......	3·30	
56	...	7·86	4·0	...	......	......	3·93	

Here the near coincidence of the results in the second and sixth columns shows that the experimental values of k and T may be approximately represented by the formula

$$T = \frac{1}{12\cdot 57} k^{\frac{4}{3}}. \quad \cdots \cdots \quad (2)$$

Again, the near coincidence of the results in the third and seventh, fourth and eighth, fifth and ninth columns shows that in this group of salts the temperatures of ebullition have a constant ratio to each other for all equal weights of salt.

The errors of these formulæ are probably within the limits of the errors of observation. Owing to the oscillations of the mercury column at the boiling-points of the liquids, the errors of the readings of the thermometer might amount to about one-fourth of a degree.

These experimental results for the most part agree with those given by Legrand (see Dr. Miller's 'Chemistry,' second edition, p. 247).

Augmentations of boiling-temperatures, in degrees Centigrade, of different solutions of the salts contained in Group 2.

If T and T′ be put for the augmentations of the boiling-temperatures of the solutions of the chlorides of calcium and strontium respectively for any equal portions of the salts dissolved, then it will be found that T′ = ·34T very nearly, as shown in the following examples:—

(1) For $k=16$, T$=2$; then by the formula, T′$=1\cdot 08$.
      By experiment, T′$=1$.
(2) For $k=64$, T$=16\cdot 8$; then by the formula, T′$=9\cdot 07$.
      By experiment, T′$=9$.
(3) For $k=96$, T$=27\cdot 2$; then by the formula, T′$=14\cdot 68$.
      By experiment, T′$=14\cdot 7$.
(4) For $k=112$, T$=32\cdot 3$; then by the formula, T′$=17\cdot 44$.
      By experiment, T′$=17\cdot 2$.

Augmentations of boiling-temperatures, in degrees Centigrade, of different solutions of the salts contained in Group 3.

If T, T′, T″ be put for the augmentations of the boiling-temperatures of the solutions of the nitrates of soda, potassa, and ammonia respectively, and k, $k′$, $k″$ the corresponding percentage of salts respectively dissolved, then the following formulæ will be found to represent the results of experiments very nearly:—

$$T = \frac{1}{8\cdot 1} k^{\cdot 945}; \quad \cdots \quad \cdots \quad (3)$$

and for all equal weights of salt,

$$T'' = \frac{6}{7}T, \quad \ldots \ldots \ldots \quad (4)$$

and

$$T' = \cdot 7\,T''. \quad \ldots \ldots \ldots \quad (5)$$

Applications of formula (3).

(1.) For $k = 24$; then by the formula, $T = 2\cdot57$.
By experiment, $T = 2\cdot6$.

(2.) For $k = 48$; then by the formula, $T = 4\cdot97$.
By experiment, $T = 5$.

(3.) For $k = 96$; then by the formula, $T = 9\cdot65$.
By experiment, $T = 9\cdot72$.

(4.) For $k = 120$; then by the formula, $T = 11\cdot93$.
By experiment, $T = 12$.

(5.) For $k = 168$; then by the formula, $T = 16\cdot47$.
By experiment, $T = 16\cdot3$.

(6.) For $k = 216$; then by the formula, $T = 20\cdot93$.
By experiment, $T = 20\cdot8$.

Applications of formula (4).

(1) For $T = 2\cdot6$, $T'' = 2\cdot23$. By experiment, $T'' = 2\cdot3$.
(2) For $T = 7\cdot45$, $T'' = 6\cdot38$. By experiment, $T'' = 6\cdot5$.
(3) For $T = 16\cdot3$, $T'' = 13\cdot97$. By experiment, $T'' = 13\cdot8$.
(4) For $T = 20\cdot3$, $T'' = 17\cdot4$. By experiment, $T'' = 17\cdot0$.

Applications of formula (5).

(1) For $T'' = 2\cdot3$, $T' = 1\cdot61$. By experiment, $T' = 1\cdot8$.
(2) For $T'' = 6\cdot5$, $T' = 4\cdot55$. By experiment, $T' = 4\cdot6$.
(3) For $T'' = 8\cdot5$, $T' = 5\cdot95$. By experiment, $T' = 5\cdot9$.
(4) For $T'' = 17$, $T' = 11\cdot9$. By experiment, $T' = 11\cdot3$.

Augmentations of boiling-temperatures, in degrees Centigrade, of different solutions of the salts in Group 4.

Weight of salt in 100 parts of water, k.	Corresponding boiling-temperatures of the different solutions of salt above that of pure water.			
	Nitrate of lime, T.	Carbonate of potassa, T'.	Value of T by formula $T = \frac{1}{33\cdot25}k^{1\cdot305}$.	Value of T by formula $T = \frac{7}{8}T'$.
0	0	0	0	0
16	1	1·33	1·12	1·16
32	2·7	3·12	2·76	2·73
48	4·7	5·30	4·70	4·63
64	7·0	7·8	6·97	6·82
80	9·2	10·5	9·15	9·18
96	11·63	13·5	11·62	11·81
112	14·2	17·0	14·20	14·87
176	25·0	29·5	25·61	25·81

Here the near coincidence of the results in the second and fourth columns shows that the relation of k and T may be very nearly represented by the formula :

$$T = \frac{1}{33 \cdot 25} k^{1 \cdot 305}. \quad . \quad . \quad . \quad . \quad . \quad (6)$$

Moreover, the near coincidence of the results in the second and fifth columns shows that $T = \frac{7}{8} T'$ very nearly, or that in this group of salts the temperatures of ebullition, T and T', have a constant ratio to each other for all equal weights of salt.

On a certain law connecting (approximately) *the boiling-temperatures of particular salts in the same group with the chemical equivalents of their bases, and in one instance with the equivalents of the entire salts.*

For the chlorides of sodium and barium we have found

$$\frac{T''}{T} = \frac{1}{3}, \text{ for all equal weights of the salts;}$$

but $$\frac{\text{Equiv. of sodium}}{\text{Equiv. of barium}} = \frac{23 \cdot 31}{68 \cdot 66} = \frac{1}{3} \text{ very nearly.}$$

Hence it follows that, *for all equal weights of salt, the boiling-temperatures,* T'' *and* T, *of these two chlorides are* (approximately) *in the inverse ratio of the equivalents of their bases.*

Again, for the chlorides of sodium and potassium we have

$$\frac{T}{T'} = 1 \cdot 5, \text{ for all equal weights of the salts;}$$

but $$\frac{\text{Equiv. of potassium}}{\text{Equiv. of sodium}} = \frac{39 \cdot 26}{23 \cdot 31} = 1 \cdot 68.$$

Hence it appears that the same law holds (approximately) true for these two salts.

In like manner, for the chlorides of calcium and strontium we have

$$\frac{T}{T'} = 1 \cdot 85, \text{ for all equal weights of the salts;}$$

but $$\frac{\text{Equiv. of strontium}}{\text{Equiv. of calcium}} = \frac{43 \cdot 75}{20} = 2 \cdot 18.$$

In this case the approximation is not so close.

For the nitrates of soda and potassa we have

$$\frac{T'}{T} = \cdot 60, \text{ for all equal weights of the salts;}$$

but $$\frac{\text{Equiv. of soda}}{\text{Equiv. of potassa}} = \frac{31 \cdot 31}{47 \cdot 26} = \cdot 66.$$

Hence it appears that the same law holds (approximately) true for these two salts.

For the nitrate of lime and the carbonate of potassa included in the fourth group, we have

$$\frac{T}{T'} = \frac{7}{8} = \cdot87, \text{ for all equal weights of the salts;}$$

but

$$\frac{\text{Equiv. carb. of potassa}}{\text{Equiv. nitrate of lime}} = \frac{69}{82} = \cdot84.$$

In this case, *for all equal weights of the salts, the boiling-temperatures,* T *and* T', *are* (approximately) *in the inverse ratio of the equivalents of the entire salts.*

How far these laws may be extended to other substances future researches will determine; at the same time it must be observed that it is quite consistent with analogy to suppose that the chemical composition of a substance should affect the boiling-temperature of its solution. Although the law here indicated is not strictly true, yet it is sufficiently exact to warrant further inquiry, and the cases to which it is found to apply are too numerous to be referred to accidental coincidence.

On the boiling-point of diluted sulphuric acid.

With the exception of the sixth, seventh, and ninth experiments, the following experimental results were given by Dalton.

The per-centages of concentrated acid in the liquids were calculated from the observed specific gravities of the liquids by means of Ure's Table, given at p. 801, fourth edition, of his work on the Arts and Manufactures.

Augmentations of the boiling-temperatures, in degrees Fahrenheit, of diluted sulphuric acid at mean atmospheric pressure, containing different proportions of concentrated acid in 100 parts, the specific gravity of the concentrated acid being 1·846.

Weight of concentrated acid in 100 parts of the liquid, k.	Corresponding excess of boiling temperature above 212°, T.	Value of k by formula $k = 14 \cdot 15\,T^{\frac{1}{4}}$.
100	363	100·90
98·21	333	98·08
93·66	289	93·55
90·53	261	90·43
86·82	223	85·81
76·88	150	75·18
48·00	40	48·39
41·00	28	42·96
34·00	16	35·66
0	0	0

· Here the near coincidence of the results in the first and third columns shows that the relation between k and T may be approximately expressed by the formula

$$k = 14.15\,\mathrm{T}^{\frac{1}{4}},\ \text{or}\ \mathrm{T} = \left(\frac{k}{14.15}\right)^{3}.\quad \cdots \quad (7)$$

H stings, April 1, 1861.

LI. *On Physical Lines of Force.* By J. C. MAXWELL, *Professor of Natural Philosophy in King's College, London.*

[With a Plate.]

PART II.—*The Theory of Molecular Vortices applied to Electric Currents.*

[Concluded from p. 291.]

AS an example of the action of the vortices in producing induced currents, let us take the following case:—Let B, Pl. V. fig. 3, be a circular ring, of uniform section, lapped uniformly with covered wire. It may be shown that if an electric current is passed through this wire, a magnet placed within the coil of wire will be strongly affected, but no magnetic effect will be produced on any external point. The effect will be that of a magnet bent round till its two poles are in contact.

If the coil is properly made, no effect on a magnet placed outside it can be discovered, whether the current is kept constant or made to vary in strength; but if a conducting wire C be made to *embrace* the ring any number of times, an electromotive force will act on that wire whenever the current in the coil is made to vary; and if the circuit be *closed*, there will be an actual current in the wire C.

This experiment shows that, in order to produce the electromotive force, it is not necessary that the conducting wire should be placed in a field of magnetic force, or that lines of magnetic force should pass through the substance of the wire or near it. All that is required is that lines of force should pass through the circuit of the conductor, and that these lines of force should vary in quantity during the experiment.

In this case the vortices, of which we suppose the lines of magnetic force to consist, are all within the hollow of the ring, and outside the ring all is at rest. If there is no conducting circuit embracing the ring, then, when the primary current is made or broken, there is no action outside the ring, except an instantaneous pressure between the particles and the vortices which they separate. If there is a continuous conducting circuit embracing the ring, then, when the primary current is made, there will be a current in the opposite direction through C; and when it is broken, there will be a current through C in the same direction as the primary current.

We may now perceive that induced currents are produced when the electricity yields to the electromotive force,—this force, however, still existing when the formation of a sensible current is prevented by the resistance of the circuit.

The electromotive force, of which the components are P, Q, R, arises from the action between the vortices and the interposed particles, when the velocity of rotation is altered in any part of the field. It corresponds to the pressure on the axle of a wheel in a machine when the velocity of the driving wheel is increased or diminished.

The electrotonic state, whose components are F, G, H, is what the electromotive force would be if the currents, &c. to which the lines of force are due, instead of arriving at their actual state by degrees, had started instantaneously from rest with their actual values. It corresponds to the *impulse* which would act on the axle of a wheel in a machine if the actual velocity were suddenly given to the driving wheel, the machine being previously at rest.

If the machine were suddenly stopped by stopping the driving wheel, each wheel would receive an impulse equal and opposite to that which it received when the machine was set in motion.

This impulse may be calculated for any part of a system of mechanism, and may be called the *reduced momentum* of the machine for that point. In the varied motion of the machine, the actual force on any part arising from the variation of motion may be found by differentiating the reduced momentum with respect to the time, just as we have found that the electromotive force may be deduced from the electrotonic state by the same process.

Having found the relation between the velocities of the vortices and the electromotive forces when the centres of the vortices are at rest, we must extend our theory to the case of a fluid medium containing vortices, and subject to all the varieties of fluid motion. If we fix our attention on any one elementary portion of a fluid, we shall find that it not only travels from one place to another, but also changes its form and position, so as to be elongated in certain directions and compressed in others, and at the same time (in the most general case) turned round by a displacement of rotation.

These changes of form and position produce changes in the velocity of the molecular vortices, which we must now examine.

The alteration of form and position may always be reduced to three simple extensions or compressions in the direction of three rectangular axes, together with three angular rotations about any set of three axes. We shall first consider the effect of three simple extensions or compressions.

Prop. IX.—To find the variations of α, β, γ in the parallelopiped x, y, z when x becomes $x+\delta x$; y, $y+\delta y$; and z, $z+\delta z$; the volume of the figure remaining the same.

By Prop. II. we find for the work done by the vortices against pressure,

$$\delta W = p_1 \delta(xyz) - \frac{\mu}{4\pi}(\alpha^2 yz\delta x + \beta^2 zx\delta y + \gamma^2 xy\delta z); \quad (59)$$

and by Prop. VI. we find for the variation of energy,

$$\delta E = \frac{\mu}{4\pi}(\alpha\delta\alpha + \beta\delta\beta + \gamma\delta\gamma)xyz. \quad \ldots \ldots (60)$$

The sum $\delta W + \delta E$ must be zero by the conservation of energy, and $\delta(xyz)=0$, since xyz is constant; so that

$$\alpha\left(\delta\alpha - \alpha\frac{\delta x}{x}\right) + \beta\left(\delta\beta - \beta\frac{\delta y}{y}\right) + \gamma\left(\delta\gamma - \gamma\frac{\delta z}{z}\right) = 0. \quad (61)$$

In order that this should be true independently of any relations between α, β, and γ, we must have

$$\delta\alpha = \alpha\frac{\delta x}{x}, \quad \delta\beta = \beta\frac{\delta y}{y}, \quad \delta\gamma = \gamma\frac{\delta z}{z}. \quad \ldots \ldots (62)$$

Prop. X.—To find the variations of α, β, γ due to a rotation θ_1 about the axis of x from y to z, a rotation θ_2 about the axis of y from z to x, and a rotation θ_3 about the axis of z from x to y.

The axis of β will move away from the axis of x by an angle θ_3; so that β resolved in the direction of x changes from 0 to $-\beta\theta_3$.

The axis of γ approaches that of x by an angle θ_2; so that the resolved part of γ in direction x changes from 0 to $\gamma\theta_2$.

The resolved part of α in the direction of x changes by a quantity depending on the second power of the rotations, which may be neglected. The variations of α, β, γ from this cause are therefore

$$\delta\alpha = \gamma\theta_2 - \beta\theta_3, \quad \delta\beta = \alpha\theta_3 - \gamma\theta_1, \quad \delta\gamma = \beta\theta_1 - \alpha\theta_2. \quad (63)$$

The most general expressions for the distortion of an element produced by the displacement of its different parts depend on the nine quantities

$$\frac{d}{dx}\delta x, \ \frac{d}{dy}\delta x, \ \frac{d}{dz}\delta x; \ \frac{d}{dx}\delta y, \ \frac{d}{dy}\delta y, \ \frac{d}{dz}\delta y; \ \frac{d}{dx}\delta z, \ \frac{d}{dy}\delta z, \ \frac{d}{dz}\delta z;$$

and these may always be expressed in terms of nine other quantities, namely, three simple extensions or compressions,

$$\frac{\delta x'}{x'}, \ \frac{\delta y'}{y'}, \ \frac{\delta z'}{z'}$$

along three axes properly chosen, x', y', z', the nine direction-cosines of these axes with their six connecting equations, which are equivalent to three independent quantities, and the three rotations θ_1, θ_2, θ_3 about the axes of x, y, z.

Let the direction-cosines of x' with respect to x, y, z be l_1, m_1, n_1, those of y', l_2, m_2, n_2, and those of z', l_3, m_3, n_3; then we find

$$\left. \begin{aligned} \frac{d}{dx}\delta x &= l_1^2\frac{\delta x'}{x'} + l_2^2\frac{\delta y'}{y'} + l_3^2\frac{\delta z'}{z'}, \\ \frac{d}{dy}\delta x &= l_1 m_1 \frac{\delta x'}{x'} + l_2 m_2 \frac{\delta y'}{y'} + l_3 m_3 \frac{\delta z'}{z'} - \theta_3, \\ \frac{d}{dz}\delta x &= l_1 n_1 \frac{\delta x'}{x'} + l_2 n_2 \frac{\delta y'}{y'} + l_3 n_3 \frac{\delta z'}{z'} + \theta_2 \end{aligned} \right\} \quad \cdots \quad (64)$$

with similar equations for quantities involving δy and δz.

Let α', β', γ' be the values of α, β, γ referred to the axes of x', y', z'; then

$$\left. \begin{aligned} \alpha' &= l_1\alpha + m_1\beta + n_1\gamma, \\ \beta' &= l_2\alpha + m_2\beta + n_2\gamma, \\ \gamma' &= l_3\alpha + m_3\beta + n_3\gamma. \end{aligned} \right\} \quad \cdots \cdots \quad (65)$$

We shall then have

$$\delta\alpha = l_1\delta\alpha' + l_2\delta\beta' + l_3\delta\gamma' + \gamma\theta_2 - \beta\theta_3, \quad \cdots \quad (66)$$

$$= l_1\alpha'\frac{\delta x'}{x'} + l_2\beta'\frac{\delta y'}{y'} + l_3\gamma'\frac{\delta z'}{z'} + \gamma\theta_2 - \beta\theta_3. \quad (67)$$

By substituting the values of α', β', γ', and comparing with equations (64), we find

$$\delta\alpha = \alpha\frac{d}{dx}\delta x + \beta\frac{d}{dy}\delta x + \gamma\frac{d}{dz}\delta x \quad \cdots \cdots \quad (68)$$

as the variation of α due to the change of form and position of the element. The variations of β and γ have similar expressions.

Prop. XI.—To find the electromotive forces in a moving body.

The variation of the velocity of the vortices in a moving element is due to two causes—the action of the electromotive forces, and the change of form and position of the element. The whole variation of α is therefore

$$\delta\alpha = \frac{1}{\mu}\left(\frac{dQ}{dx} - \frac{dR}{dy}\right)\delta t + \alpha\frac{d}{dx}\delta x + \beta\frac{d}{dy}\delta x + \gamma\frac{d}{dz}\delta x. \quad (69)$$

But since α is a function of x, y, z and t, the variation of α may be also written

$$\delta\alpha = \frac{d\alpha}{dx}\delta x + \frac{d\alpha}{dy}\delta y + \frac{d\alpha}{dz}\delta z + \frac{d\alpha}{dt}\delta t. \quad \cdots \cdots \quad (70)$$

Equating the two values of $\delta\alpha$ and dividing by δt, and remembering that in the motion of an incompressible medium

$$\frac{d}{dx}\frac{dx}{dt} + \frac{d}{dy}\frac{dy}{dt} + \frac{d}{dz}\frac{dz}{dt} = 0, \quad \ldots \ldots \quad (71)$$

and that in the absence of free magnetism

$$\frac{d\alpha}{dx} + \frac{d\beta}{dy} + \frac{d\gamma}{dz} = 0, \quad \ldots \ldots \ldots \quad (72)$$

we find

$$\frac{1}{\mu}\left(\frac{dQ}{dz} - \frac{dR}{dy}\right) + \gamma\frac{d}{dx}\frac{dx}{dt} - \alpha\frac{d}{dx}\frac{dz}{dt} - \alpha\frac{d}{dy}\frac{dy}{dt} + \beta\frac{d}{dy}\frac{dx}{dt}$$

$$+ \frac{dy}{dz}\frac{dx}{dt} - \frac{d\alpha}{dz}\frac{dz}{dt} - \frac{d\alpha}{dy}\frac{dy}{dt} + \frac{d\beta}{dy}\frac{dx}{dt} - \frac{d\alpha}{dt} = 0. \quad . \quad (73)$$

Putting

$$\alpha = \frac{1}{\mu}\left(\frac{dG}{dz} - \frac{dH}{dy}\right) . \quad \ldots \ldots \quad (74)$$

and

$$\frac{d\alpha}{dt} = \frac{1}{\mu}\left(\frac{d^2G}{dz\,dt} - \frac{d^2H}{dy\,dt}\right), \quad \ldots \ldots \quad (75)$$

where F, G, and H are the values of the electrotonic components for a fixed point of space, our equation becomes

$$\frac{d}{dz}\left(Q + \mu\gamma\frac{dx}{dt} - \mu\alpha\frac{dz}{dt} - \frac{dG}{dt}\right) - \frac{d}{dy}\left(R + \mu\alpha\frac{dy}{dt} - \mu\beta\frac{dx}{dt} - \frac{dH}{dt}\right) = 0. \quad (76)$$

The expressions for the variations of β and γ give us two other equations which may be written down from symmetry. The complete solution of the three equations is

$$P = \mu\gamma\frac{dy}{dt} - \mu\beta\frac{dz}{dt} + \frac{dF}{dt} - \frac{d\Psi}{dx},$$

$$Q = \mu\alpha\frac{dz}{dt} - \mu\gamma\frac{dx}{dt} + \frac{dG}{dt} - \frac{d\Psi}{dy}, \quad \left.\right\} \quad \ldots \quad (77)$$

$$R = \mu\beta\frac{dx}{dt} - \mu\alpha\frac{dy}{dt} + \frac{dH}{dt} - \frac{d\Psi}{dz}.$$

The first and second terms of each equation indicate the effect of the motion of any body in the magnetic field, the third term refers to changes in the electrotonic state produced by alterations of position or intensity of magnets or currents in the field, and Ψ is a function of x, y, z, and t, which is indeterminate as far as regards the solution of the original equations, but which may always be determined in any given case from the circumstances of the problem. The physical interpretation of Ψ is, that it is the *electric tension* at each point of space.

The physical meaning of the terms in the expression for the electromotive force depending on the motion of the body, may be made simpler by supposing the field of magnetic force uniformly magnetised with intensity α in the direction of the axis of x. Then if l, m, n be the direction-cosines of any portion of a linear conductor, and S its length, the electromotive force resolved in the direction of the conductor will be

$$e = S(Pl + Qm + Rn), \quad \ldots \ldots \quad (78)$$

or

$$e = S\mu\alpha\left(m\frac{dz}{dt} - n\frac{dy}{dt}\right), \quad \ldots \ldots \quad (79)$$

that is, the product of $\mu\alpha$, the quantity of magnetic induction over unit of area multiplied by $S\left(m\dfrac{dz}{dt} - n\dfrac{dy}{dt}\right)$, the area swept out by the conductor S in unit of time, resolved perpendicular to the direction of the magnetic force.

The electromotive force in any part of a conductor due to its motion is therefore measured by the *number* of lines of magnetic force which it crosses in unit of time; and the total electromotive force in a closed conductor is measured by the change of the number of lines of force which pass through it; and this is true whether the change be produced by the motion of the conductor or by any external cause.

In order to understand the mechanism by which the motion of a conductor across lines of magnetic force generates an electromotive force in that conductor, we must remember that in Prop. X. we have proved that the change of form of a portion of the medium containing vortices produces a change of the velocity of those vortices; and in particular that an extension of the medium in the direction of the axes of the vortices, combined with a contraction in all directions perpendicular to this, produces an increase of velocity of the vortices; while a shortening of the axis and bulging of the sides produces a diminution of the velocity of the vortices.

This change of the velocity of the vortices arises from the internal effects of change of form, and is independent of that produced by external electromotive forces. If, therefore, the change of velocity be prevented or checked, electromotive forces will arise, because each vortex will press on the surrounding particles in the direction in which it tends to alter its motion.

Let A, fig. 4, represent the section of a vertical wire moving in the direction of the arrow from west to east, across a system of lines of magnetic force running north and south. The curved lines in fig. 4 represent the lines of fluid motion about the wire, the wire being regarded as stationary, and the fluid as having a motion relative to it.   It is evident that, from this figure, we can trace the variations of form of an element of the fluid, as the form of the element depends, not on the absolute motion of the whole system, but on the relative motion of its parts.

In front of the wire, that is, on its east side, it will be seen that as the wire approaches each portion of the medium, that portion is more and more compressed in the direction from east to west, and extended in the direction from north to south; and since the axes of the vortices lie in the north and south direction, their velocity will continually tend to increase by Prop. X., unless prevented or checked by electromotive forces acting on the circumference of each vortex.

We shall consider an electromotive force as positive when the vortices tend to move the interjacent particles *upwards* perpendicularly to the plane of the paper.

The vortices appear to revolve as the hands of a watch when we look at them from south to north; so that each vortex moves upwards on its west side, and downwards on its east side.   In front of the wire, therefore, where each vortex is striving to increase its velocity, the electromotive force upwards must be greater on its west than on its east side.   There will therefore be a continual increase of upward electromotive force from the remote east, where it is zero, to the front of the moving wire, where the upward force will be strongest.

Behind the wire a different action takes place.   As the wire moves away from each successive portion of the medium, that portion is extended from east to west, and compressed from north to south, so as to tend to diminish the velocity of the vortices, and therefore to make the upward electromotive force greater on the east than on the west side of each vortex.   The upward electromotive force will therefore increase continually from the remote west, where it is zero, to the back of the moving wire, where it will be strongest.

It appears, therefore, that a vertical wire moving eastwards will experience an electromotive force tending to produce in it an upward current.   If there is no conducting circuit in connexion with the ends of the wire, no current will be formed, and the magnetic forces will not be altered; but if such a circuit exists, there will be a current, and the lines of magnetic force and the velocity of the vortices will be altered from their state previous to the motion of the wire.   The change in the lines of force is shown in fig. 5.   The vortices in front of the wire, instead of merely producing pressures, actually increase in velocity, while those behind have their velocity diminished, and those at the sides of the wire have the direction of their axes altered; so that the final effect is to produce a force acting on the wire as a resist-.

ance to its motion. We may now recapitulate the assumptions we have made, and the results we have obtained.

(1) Magneto-electric phenomena are due to the existence of matter under certain conditions of motion or of pressure in every part of the magnetic field, and not to direct action at a distance between the magnets or currents. The substance producing these effects may be a certain part of ordinary matter, or it may be an æther associated with matter. Its density is greatest in iron, and least in diamagnetic substances; but it must be in all cases, except that of iron, very rare, since no other substance has a large ratio of magnetic capacity to what we call a vacuum.

(2) The condition of any part of the field, through which lines of magnetic force pass, is one of unequal pressure in different directions, the direction of the lines of force being that of least pressure, so that the lines of force may be considered lines of tension.

(8) This inequality of pressure is produced by the existence in the medium of vortices or eddies, having their axes in the direction of the lines of force, and having their direction of rotation determined by that of the lines of force.

We have supposed that the direction was that of a watch to a spectator looking from south to north. We might with equal propriety have chosen the reverse direction, as far as known facts are concerned, by supposing resinous electricity instead of vitreous to be positive. The effect of these vortices depends on their density, and on their velocity at the circumference, and is independent of their diameter. The density must be proportional to the capacity of the substance for magnetic induction, that of the vortices in air being 1. The velocity must be very great, in order to produce so powerful effects in so rare a medium.

The size of the vortices is indeterminate, but is probably very small as compared with that of a complete molecule of ordinary matter*.

(4) The vortices are separated from each other by a single layer of round particles, so that a system of cells is formed, the partitions being these layers of particles, and the substance of each cell being capable of rotating as a vortex.

(5) The particles forming the layer are in *rolling contact* with both the vortices which they separate, but do not rub against each other. They are perfectly free to roll between the vortices

---

* The angular momentum of the system of vortices depends on their average diameter; so that if the diameter were sensible, we might expect that a magnet would behave as if it contained a revolving body within it, and that the existence of this rotation might be detected by experiments on the free rotation of a magnet. I have made experiments to investigate this question, but have not yet fully tried the apparatus.

and so to change their place, provided they keep within one *complete molecule* of the substance; but in passing from one molecule to another they experience resistance, and generate irregular motions, which constitute heat. These particles, in our theory, play the part of electricity. Their motion of translation constitutes an electric current, their rotation serves to transmit the motion of the vortices from one part of the field to another, and the tangential pressures thus called into play constitute electromotive force. The conception of a particle having its motion connected with that of a vortex by perfect rolling contact may appear somewhat awkward. I do not bring it forward as a mode of connexion existing in nature, or even as that which I would willingly assent to as an electrical hypothesis. It is, however, a mode of connexion which is mechanically conceivable, and easily investigated, and it serves to bring out the actual mechanical connexions between the known electro-magnetic phenomena; so that I venture to say that any one who understands the provisional and temporary character of this hypothesis, will find himself rather helped than hindered by it in his search after the true interpretation of the phenomena.

The action between the vortices and the layers of particles is in part tangential; so that if there were any slipping or differential motion between the parts in contact, there would be a loss of the energy belonging to the lines of force, and a gradual transformation of that energy into heat. Now we know that the lines of force about a magnet are maintained for an indefinite time without any expenditure of energy; so that we must conclude that wherever there is tangential action between different parts of the medium, there is no motion of slipping between those parts. We must therefore conceive that the vortices and particles roll together without slipping; and that the interior strata of each vortex receive their proper velocities from the exterior stratum without slipping, that is, the angular velocity must be the same throughout each vortex.

The only process in which electro-magnetic energy is lost and transformed into heat, is in the passage of electricity from one molecule to another. In all other cases the energy of the vortices can only be diminished when an equivalent quantity of mechanical work is done by magnetic action.

(6) The effect of an electric current upon the surrounding medium is to make the vortices in contact with the current revolve so that the parts next to the current move in the same direction as the current. The parts furthest from the current will move in the opposite direction; and if the medium is a conductor of electricity, so that the particles are free to move in any direction, the particles touching the outside of these vortices will

be moved in a direction contrary to that of the current, so that there will be an induced current in the opposite direction to the primary one.

If there were no resistance to the motion of the particles, the induced current would be equal and opposite to the primary one, and would continue as long as the primary current lasted, so that it would prevent all action of the primary current at a distance. If there is a resistance to the induced current, its particles act upon the vortices beyond them, and transmit the motion of rotation to them, till at last all the vortices in the medium are set in motion with such velocities of rotation that the particles between them have no motion except that of rotation, and do not produce currents.

In the transmission of the motion from one vortex to another, there arises a force between the particles and the vortices, by which the particles are pressed in one direction and the vortices in the opposite direction. We call the force acting on the particles the electromotive force. The reaction on the vortices is equal and opposite, so that the electromotive force cannot move any part of the medium as a whole, it can only produce currents. When the primary current is stopped, the electromotive forces all act in the opposite direction.

(7) When an electric current or a magnet is moved in presence of a conductor, the velocity of rotation of the vortices in any part of the field is altered by that motion. The force by which the proper amount of rotation is transmitted to each vortex, constitutes in this case also an electromotive force, and, if permitted, will produce currents.

(8) When a conductor is moved in a field of magnetic force, the vortices in it and in its neighbourhood are moved out of their places, and are changed in form. The force arising from these changes constitutes the electromotive force on a moving conductor, and is found by calculation to correspond with that determined by experiment.

We have now shown in what way electro-magnetic phenomena may be imitated by an imaginary system of molecular vortices. Those who have been already inclined to adopt an hypothesis of this kind, will find here the conditions which must be fulfilled in order to give it mathematical coherence, and a comparison, so far satisfactory, between its necessary results and known facts. Those who look in a different direction for the explanation of the facts, may be able to compare this theory with that of the existence of currents flowing freely through bodies, and with that which supposes electricity to act at a distance with a force depending on its velocity, and therefore not subject to the law of conservation of energy.

The facts of electro-magnetism are so complicated and various, that the explanation of any number of them by several different hypotheses must be interesting, not only to physicists, but to all who desire to understand how much evidence the explanation of phenomena lends to the credibility of a theory, or how far we ought to regard a coincidence in the mathematical expression of two sets of phenomena as an indication that these phenomena are of the same kind. . We know that partial coincidences of this kind have been discovered; and the fact that they are only partial is proved by the divergence of the laws of the two sets of phenomena in other respects. We may chance to find, in the higher parts of physics, instances of more complete coincidence, which may require much investigation to detect their ultimate divergence.

Note.—Since the first part of this paper was written, I have seen in Crelle's *Journal* for 1859, a paper by Prof. Helmholtz on Fluid Motion, in which he has pointed out that the lines of fluid motion are arranged according to the same laws as the lines of magnetic force, the path of an electric current corresponding to a line of axes of those particles of the fluid which are in a state of rotation.   This is an additional instance of a *physical analogy*, the investigation of which may illustrate both electro-magnetism and hydrodynamics.

LII. *Remarks on* Mr. Cayley's *Note.*   By G. B. Jerrard[*].

DESIGNATING by u, v two rational n-valued homogeneous functions of the roots of the equation

$$x^m + A_1 x^{m-1} + A_2 x^{m-2} + \ldots + A_m = 0,$$

we find by Lagrange's theory that

$$\left.\begin{aligned}
v &= \mu_{n-1} + \mu_{n-2} u + \mu_{n-3} u^2 + \ldots + \mu_0 u^{n-1} \\
u &= \nu_{n-1} + \nu_{n-2} v + \nu_{n-3} v^2 + \ldots + \nu_0 v^{n-1}
\end{aligned}\right\} ; \quad \ldots \quad (c)$$

in which $\mu_{n-1}, \mu_{n-2}, \ldots \mu_0, \nu_{n-1}, \nu_{n-2}, \ldots \nu_0$ are symmetrical functions of the roots of the original equation in x; and u, v depend separately on two equations of the nth degree

$$u^n + \alpha_1 u^{n-1} + \alpha_2 u^{n-2} + \ldots + \alpha_n = 0, \quad \ldots \quad (\text{U})$$

$$v^n + \beta_1 v^{n-1} + \beta_2 v^{n-2} + \ldots + \beta_n = 0, \quad \ldots \quad (\text{V})$$

$\alpha_1, \alpha_2, \ldots \alpha_n, \beta_1, \beta_2, \ldots \beta_n$ being, as well as $\mu_{n-1}, \ldots \nu_0$, symmetrical functions of the roots of the equation in x.

I ought to observe that any coefficient, μ_{n-r}, in the equation

---

[*] Communicated by the Author.

(e_1) may take the form

$$\frac{M_{n-s}}{D},$$

M_{n-s}, D being expressive of whole functions, and D, which remains constant while M_{n-s} successively becomes $M_{n-1}, M_{n-2}, .. M_0$, being such as not to vanish except when (U) has equal roots. We find in fact from the researches of Lagrange that

$$D = F(u_1)\, F(u_2) .. F(u_n),$$

where $F(u) = n u^{n-1} + (n-1)\alpha_1 u^{n-2} + (n-2)\alpha_2 u^{n-3} + .. + \alpha_{n-1};$ $u_1, u_2, .. u_n$ denoting the n roots of the equation (U).

Of the meaning of the analogous expression

$$\frac{N_{n-s}}{D'}$$

which obtains in (e_2) for ν_{n-s}, it is needless to speak. Indeed, having found one of the two equations (e), say (e_1), we may in general deduce the other, (e_2), from it by the method of the highest common divisor.

Let us now examine the following extract from Mr. Cayley's paper in the last Number (that for March) of the Philosophical Magazine.

"Writing," he says, "with Mr. Cockle and Mr. Harley,

$$\tau = x_a x_\beta + x_\beta x_\gamma + x_\gamma x_\delta + x_\delta x_\epsilon + x_\epsilon x_a,$$
$$\tau' = x_a x_\gamma + x_\gamma x_\epsilon + x_\epsilon x_\beta + x_\beta x_\delta + x_\delta x_a,$$

then $(\tau + \tau'$ is a symmetrical function of all the roots, and it must be excluded; but) $(\tau - \tau')^2$ or $\tau\tau'$ are each of them 6-valued functions of the form in question, and either of these functions is linearly connected with the Resolvent Product. In Lagrange's general theory of the solution of equations, if

$$f\iota = x_1 + \iota x_2 + \iota^2 x_3 + \iota^3 x_4 + \iota^4 x_5,$$

then the coefficients of the equation the roots whereof are $(f\iota)^5$, $(f\iota^2)^5$, $(f\iota^3)^5$, $(f\iota^4)^5$, and in particular the last coefficient $(f\iota f\iota^2 f\iota^3 f\iota^4)^5$, are determined by an equation of the sixth degree; and this last coefficient is a perfect fifth power, and its fifth root, or $f\iota f\iota^2 f\iota^3 f\iota^4$, is the function just referred to as the Resolvent Product.

"The conclusion from the foregoing remarks is, that *if the equation for* W *has the above property of the rational expressibility of its roots*, the equation of the sixth order resulting from Lagrange's general theory has the same property."

Here the question arises, Is it certain that $f\iota f\iota^2 f\iota^3 f\iota^4$ can, by means of Lagrange's theory, be expressed generally in rational terms of $(f_1 f_1^2 f_1^3 f_1^4)^5$?*

Denoting those functions by u, v respectively, we have in this case
$$v = u^{n-1},$$
$$n = 6, \quad m = 5.$$

Now on substituting u^{n-1} or u^5 for v in the equation (e_1), we see that (e_1) will merge into

$$\mu_5 + \mu_4 u + \mu_3 u^2 + \ldots + (\mu_0 - 1) u^5 = 0, \quad \ldots \quad (e'_1)$$

wherein, since (U) is in general irreducible†, we must have

$$\mu_5 = 0, \quad \mu_4 = 0, \ldots \mu_0 - 1 = 0.$$

Accordingly, on combining the equations (e_1), (U), that is to say, (e'_1), (U), we find

$$u = \frac{0}{0};$$

the equation (e'_1) being, as we see, illusory.

We are therefore not permitted to assume that the resolvent product can in general—that is, when (U) has no equal roots—be expressed rationally in terms of its fifth power.

Again, it is generally possible to establish a rational communication between that fifth power and the function W, as is evidenced in this latter case from the non-existence of any illusory equation corresponding to (e'_1).

We are thus furnished, as will be seen, with a new confirmation of the validity of my method of solving equations of the fifth degree.

April 1861.

[To be continued.]

LIII. *On the Principles of Energetics.*—Part II. *Molecular Mechanics.* By J. S. STUART GLENNIE, *M.A., F.R.A.S.*‡

SECTION I. *Physics.*

16. I PROCEED to consider the Principles of Energetics, or the science of Mechanical Forces, which seem to afford the bases of an explanation of physical motions. There is at present no attempt at a systematic elaboration of these principles, or mathematical application of them to the expression and expla-

---

* It will be understood that our present inquiry relates to the possibility of expressing either of the two quantities u, v as a rational function of the other and of the elements, $A_1, A_2, \ldots A_m$. Thus $R(v, \ldots)$ is supposed to mean the same thing as $R(v, A_1, A_2, \ldots A_m)$; R indicating a rational function.

† As defined by Abel.

‡ Communicated by the Author. In reference to the first part of this paper, note that the word "rotation" in the fourth line from the bottom of p. 280 of the last Number for "revolution."

nation of phenomena. Previous to such an attempt, it is thought advisable to enunciate these principles in their most general form, and give them merely experimental illustration.

The principles to be set forth in this paper will lead me to remark on the physical theories recently published in this Journal by Prof. Challis and Prof. Maxwell. It will be found that as my theory refers attractions to differential conditions of stress and strain, of pressure and tension, among elastic bodies, it agrees rather with the molecular theory of the latter, than with the hydrodynamical theory of the former; that the point of fundamental difference from both is in the conception offered of Matter; and that on this point my theory is a development of the views to which experiment has led Mr. Faraday.

17. (I.) Atoms are mutually determining centres of pressure.

18. If this idea of an atom, as a body of any size, acting and reacting on another similar body by the pressure of the continued, infinitesimal, but similar particles of which each centre is an aggregate, be clearly conceived, it may be expressed in many different forms. I have, for instance, in the introductory paper spoken of a body thus conceived as a Centre of Lines of Pressure, or an Elastic System with a centre of resistance. But here, more clearly to express the idea in contrast with the fundamental hypotheses of Prof. Challis, an atom may be defined as a centre of an *emanating* elastic æther, the pressure of which is directly as the mass of its centre, and the form of which depends on the relative pressures of surrounding atoms. Thus, if you will, matter may be said to be made up of particles in an elastic æther. But that æther is not a uniform circumambient fluid, but made up of the mutually determining æthers (if you wish to give the outer part of the atom a special name) emanating from the central particles. And these central particles are nothing but what (endeavouring to make my theory clear by expressing it in the language of the theories it opposes) I may call æthereal nuclei.

"Hence," according to the conception of Faraday, "matter will be continuous throughout, and in considering a mass of it we have not to suppose a distinction between its atoms and any intervening space . . . . The atoms may be conceived of as highly *elastic*, instead of being supposed excessively hard and unalterable in form . . . . . With regard also to the *shape* of the atoms . . . . . . That which is ordinarily referred to under the term *shape* would now be referred to the disposition and relative intensity of the forces *."

* Phil. Mag. 1844. vol. xxiv. p. 142; or Experimental Researches, vol. ii. p. 284. See also Phil. Mag. 1846, vol. xxviii. No. 188; or Experimental Researches, vol. iii. p. 447.

19. I venture to offer this conception of atoms, not as a mere hypothesis, but as a fundamental scientific principle. For there is this involved in it—that as a phenomenon is *scientifically* explained only when, and so far as, it is shown to be determined by other phenomena, the conception of Matter itself must be relative, and its parts be conceived as mutually determined.

Now Pressure is not only an ultimate idea, including all those qualities of Matter classed by the metaphysicians as the Secundo-primary, but is, unlike those, for instance, of Trinal extension, Ultimate incompressibility, Mobility, and Situation (the primary qualities), not an absolute, but a relative conception, and, as such, that on which alone can be founded a strictly scientific theory of material phenomena. For in the foundation of a theory based on the conception of the parts of matter as centres of pressure, there is nothing, properly speaking, hypothetical, as no absolute, intrinsic, or independent qualities of form, hardness, motion, &c. are postulated for atoms; and in their definition nothing more is done than an expression given to our ultimate and necessarily relative conception of matter.

In defining Atoms as Centres of Pressure, they are thus no less distinguished on the one hand from Centres of Force, than from the little hard bodies of the ordinary theories; for such Centres of Force are just as absolute and self-existent in the ordinary conception of them as are those little bodies. And in a scientific theory there can, except as temporary conveniences, be no absolute existences,—entities. Hence (Mechanical) Force, or the cause of motion, is conceived, not as an entity, but as a condition—the condition, namely, of a difference of Pressure *; and the figure, size, and hardness of all bodies are conceived as relative, dependent, and therefore changeable. There are thus no *absolutely* ultimate bodies.

20. But the full justification of advancing this conception of Atoms as a fundamental scientific principle, is found in the principles of the modern critical school of philosophy—in that especially of the relativity of knowledge. From such a point of view this principle cannot here be considered. I must limit myself, therefore, to a criticism of the opposed conception of atoms in a uniform æther, as developed by Prof. Challis, and to the attempt to show that, with the conception of atoms here offered, Prof. Maxwell's somewhat arbitrary hypothesis of vortices becomes unnecessary. For I agree with the former in thinking that, " after all that can be done by this kind of research, an independent and *à priori* theory . . . . . is still needed † ;" and I observe that

* See the first part of this paper, Phil. Mag. April, p. 275.
† "On Theories of Magnetism and other Forces, in reply to Remarks by Professor Maxwell," Phil. Mag. April, p. 253.

the object of the latter "is to clear the way for speculation*," rather than to advance a complete general theory.

In examining Professor Challis's "fundamental hypothetical facts," I hope to show that they are opposed (1) by Newton's Rules of Philosophizing; (2) by the principles of Metaphysics, as the modern Science of the Conditions of Thought; and (3) by the conceptions of matter, of the interaction of its different parts, and of motion and force, to which modern experimental researches have led.

21. "The fundamental hypothetical facts on which the [Prof. Challis's] theory rests are, that all substances consist of minute spherical atoms, of different, but constant, magnitudes, and of the same intrinsic inertia, and that the dynamical relations and movements of different substances are determined by the motions and pressures of a uniform elastic medium pervading all space not occupied by atoms, and varying in pressure in proportion to variations of its density†." Prof. Challis further says that he has "been guided by Newton's views on the ultimate properties of matter, especially as embodied in the *Regula Tertia Philosophandi* in the Third Book of the 'Principia';" and that he has merely "added to the Newtonian hypotheses two others, viz. that the ultimate atoms of bodies are spherical, and that they are acted upon by the pressure of a highly elastic medium‡." But on reference to the cited rule, no "Newtonian hypotheses" will be found, only a statement of the actual general qualities of matter. And, setting aside the "additional hypothesis" of sphericity, so far are the hypotheses of ultimate indivisible atoms, and these of an indeterminate number of different sizes, though of the same intrinsic inertia, Newtonian, that Newton says, "At si vel unico constaret experimento quod particula aliqua indivisa, frangendo corpus durum et solidum, divisionem pateretur; concluderemus vi bujus regulæ, quod non solum partes divisæ separabiles essent, sed etiam quod indivisæ in infinitum dividi possent." And Le Seur and Jacquier add in their note: "Hinc patet differentia Newtonianismi et hypotheseos Atomorum; Atomistæ necessario et metaphysicè atomos esse indivisibiles volunt, ut sint corporum unitates; Metaphysicam hanc quæstionem missam facit Newtonus . . . . omnem hac de re Theoriam Metaphysicam experimentis facile postponens." So that not only are Professor Challis's hypotheses as to "ultimate" bodies unwarranted by the rule he vouches, but he appears as of that very metaphysical school of Atomists, New-

---

* "The Theory of Molecular Vortices applied to Magnetic Phenomena," Phil. Mag. March, p. 162.
† Phil. Mag. Feb. 1861, p. 106.   ‡ Ibid. Dec. 1859, pp. 443 and 444.
Phil. Mag. S. 4. Vol. 21. No. 141. *May* 1861.   2 A

ton's opposition to which is implied in his *Regula Tertia Philo-sophandi.*

No less clear is it that the postulate of two different kinds of matter, one with the qualities of inertia and elasticity, the other without the second of these qualities, is opposed by the very terms, not only of the third rule, "Qualitates corporum quæ intendi et remitti nequeunt, quæque corporibus omnibus compe-tunt in quibus experimenta instituere licet, pro qualitatibus cor-porum universorum habendæ sunt," but by the terms of the first rule also, "causas rerum naturalium non plures admitti debere, quam quæ et *veræ sint* et earum phænomenis expli-candis sufficiant."

22. Consider, secondly, how such hypotheses are judged by the modern principles of Metaphysics. For it is evident that the theories of every science must ultimately be judged by the results of a science, τέχνη τεχνῶν καὶ ἐπιστήμη ἐπιστημῶν, which, defining the conditious of knowledge, gives canons for the criticism of hypotheses. As this is no place for a metaphy-sical discussion, let it suffice to say that the theory of the rela-tivity of cognition seems to justify the enouncement of this canon as a test of theories put forward as scientific. A scientific (phy-sical) theory is founded on postulates of Relations, not on postu-lates of absolutely existing Entities. According to this rule it is evident that if, for instance, a theory requires an atom of a cer-tain size or hardness, it can only be granted where it will stand as an expression of the relation between the forces distinguished at that point as internal and external; so if a certain elasticity, rotatory, or other motion of a body is required, the theory must take that elasticity or rotatory motion, not as an absolute pro-perty, but along with those relative conditions of other bodies which determine such elasticity or motion.

23. Without advancing any other defence, this canon may be justified by the consequence of its neglect. For a theory founded on postulates of absolute qualities—entities—must necessarily reason in a circle, accounting for phenomena by the same phe-nomena already assumed as ultimate.

Thus, though Professor Challis says that it would be contrary to principle "to ascribe to an atom the property of elasticity, because, from what we know of this property by experience, it is quantitative, and being most probably dependent on an *aggrega-tion* of atoms, may admit of explanation by a complete theory of molecular forces[*]," he has no hesitation in ascribing elasticity to the particles of the æther, which, if anything, are as much atoms of matter as the "hard" atoms. But further, as to hard-ness, is it not the case that, "from what we know of this pro-

[*] Phil. Mag. February 1860, p. 89.

perty by experience," it also "may admit of explanation by a complete theory of molecular forces?" Is it not therefore self-contradictory to attribute elasticity to one sort of matter, and justify its denial to another, on grounds which would equally apply to the quality by which this other sort of matter is distinguished from the former? And is not a theory fallacious which, if it attempts to explain relative elasticity or relative hardness, must do so by means of hypothetical and inconceivable, absolutely elastic, and absolutely hard entities?

24. But further, examining these fundamental facts by the results of the analysis of the qualities of matter, it will be seen that it is attempted to found a physical theory on the hypothesis of a physical matter acting on a mathematical matter. An elastic matter may be physically conceivable; but the interaction of such a matter and bodies without any physical quality, but mere abstractions of the metaphysical qualities deduceable from the respective conditions of occupying and being contained in space, cannot but be experimentally inconceivable.

25. Consider therefore, thirdly, the experimental conceptions to which these "hypothetical facts" are opposed; and (1) the conception of matter.

The conception of an absolute, or uniform, and universal elastic æther is opposed to the conception now formed of such similar entities as the old electric fluids, &c.; namely, that electricity is not an entity, but the expression of a certain physical relation between bodies, the electric state being kept up by, and entirely dependent upon, the bodies among which the electric body at any time is, or may be brought. Hence it should seem that if a theory requires an elastic æther, its elasticity must be conceived as relative or determined by the masses and distances of bodies, and hence evidently elasticity be conceived as "une des propriétés générales de la matière*"

And the notion of absolutely existing spherical atoms of different magnitudes not only begs as many separate creations of atoms as our fancy may suggest differences in their size, but is opposed to the conception of the transmutation of matter, generalized from the fact that we have in physics at least no creations, but perpetual changes dependent upon the ever-varying relations of bodies.

26. But (2) the idea of motions arising from the action of an elastic fluid on an inelastic absolutely hard and smooth body is opposed to all experimental conceptions of the interaction of the parts of matter. For not only do we seem to be led by experiment to a conception of the continuity of every part of matter by the cohesion of other bodies, so that it should seem to be impos-

* Lamé, quoted in Part I. of this paper, Phil. Mag. April, p. 275.

2 A 2

sible for a fluid to act on a solid except through a mediate or immediate cohesion, but we are led by the Mechanical Theory of Heat to conceive every impulse communicated to a body to be productive of internal as well as external motion. It is of course necessary to make abstraction, out of the infinite number of effects, of the particular effect we may desire to consider. But an hypothesis of infinitely hard atoms not merely requires, in the consideration of the motion of such an atom, abstraction to be made of the interior relative motions also consequent on that difference of pressures which causes its external relative motion, but explicitly denies any internal motion.

It may be here noted that the Mechanical Theory of Heat would· lead us to consider as "ultimate" no special class of bodies or molecules, except simply those, of the internal motions of which we do not in any particular theoretical, or cannot in an experimental, investigation take account. So any hardness may be called "infinite" if we do not consider the internal motions, or change of form, consequent on the application of a force which causes the translation of the body. But Professor Challis requires us to concede as physical facts what are properly but convenient mathematical abstractions.

27. Again (3), the conception of the origination of motion under such conditions as a uniform æther and discrete atoms therein, all of the same mass, is opposed to the experimental conception of motion as originating in difference in the mutual pressures of bodies. For these hypotheses give us the conditions of an eternal equilibrium. In the theory I propose, it is evident that anything short of an absolute equality in the masses and distances of the parts of matter implies infinite mutually determining motions.

And Professor Challis speaks of "the existence of the æther as the sole source of physical power*." But in a mechanical theory, as 1 have in the introductory, and in the first part of this, paper shown, nothing can be accurately spoken of as, of itself, "a source of power." "A source of power," a cause of motion, or a force, is simply the difference in relation to a third body of two resultant pressures upon it. And there can thus be no conceivable mechanical power in a fluid of which the elasticity is uniform, and on which the reaction of different solids within it should seem, by this theory, to be either nothing or the same.

28. Professor Challis further conceives the physical forces to be correlated as "modes of action of a single elastic medium." But I shall endeavour in the sequel to distinguish these correlations, and to show that they are either coexisting, mutually

* Phil. Mag. February 1861, p. 106; and December 1859, p. 444.

causative, or sequential *molecular* motions. For the true application of Hydrodynamics would seem to be rather to actual solids and fluids, than to such "hypothetical facts" as form the bases of Professor Challis's Theory of Physics. How much work remains to be done in that true direction is well known; and the greatness of the results in the knowledge that might thereby be given of the formation of the solar and other sidereal systems, makes every contribution to Hydrodynamics, whatever the immediate particular application of the theorems, of peculiar value.

29. If, therefore, the true application of Hydrodynamics has been mistaken, and if a Hydrodynamical Theory of Physics must be founded on entities, hypothetical solids and fluids, to which such objections as the foregoing can be urged, there remains for us only a Molecular Theory of Physics. It is because to such a Theory all the most important modern physical researches seem to point, that I have thought it necessary to examine at such length the "fundamental facts" of Professor Challis. For the great result of modern science may be said to be the relative conception it gives of every phenomenon, and hence the demand that the fundamental facts of any theory be conceived, not as absolute and independent existences, but as expressions of relations. Now a Molecular is distinguished from a Hydrodynamical Theory of Physics in this,—that while in the latter the states and motions of bodies are explained by the action on them of some hypothetical, uniform, absolute, and all-pervading entity, in the former theory, physical states and motions are referred to differential conditions of stress and strain among the actual constituent molecules of bodies. Hence the evident experimental advantage of a Molecular Theory is, that its hypotheses being as to relative conditions of Molecular pressure and tension, transmission of motion, &c., they are more or less capable of experimental proof or disproof; and such a theory will be at least prolific in the suggestion of experiment. But where one deals with the waves or currents of an absolute æther acting on absolute atoms, a plausible theory may indeed be made out, but it is because its conceptions are fundamentally opposed to, so that its minor hypotheses cannot be checked by, experiment.

30. It is as the new fundamental principle of a Molecular Theory of Physics that I venture to suggest the above conception of Atoms. It is because, however convinced of the soundness of this principle, I am very diffident of my own powers of applying it, that I have gone at such length into its illustration, and the criticism of the opposed conception, as developed by Professor Challis. For should the mechanical explanation which, by

means of this principle, I propose to give of physical and chemical phenomena be found liable to serious objection, I hope the above remarks will have made this conception of Atoms sufficiently clear to be applied with greater success by others.

6 Stone Buildings, Lincoln's Inn,
April 11, 1861.

[To be continued.]

LIV. *Chemical Notices from Foreign Journals.*
By E. ATKINSON, *Ph.D., F.C.S.*

[Continued from p. 301.]

SAWITSCH * found that, under certain circumstances, monobrominated ethylene, $C^2 H^3 Br$ (bromide of vinyle), parted with hydrobromic acid and became converted into acetylene, $C^2 H^2$, the gas discovered by Edmund Davy and investigated by Berthelot †. He was led to investigate this deportment more minutely, and having tried the action of monobrominated ethylene on amylate of sodium, has found in it a mode of preparing this gas.

About 45 grammes of brominated ethylene were heated in a closed vessel with amylate of sodium. An abundant precipitate of bromide of sodium was formed, and the mass became liquid from the regenerated amylic alcohol. The vessel was then carefully cooled down in a freezing mixture and opened, when about 4 litres of a gas escaped, which, agitated with an ammoniacal solution of chloride of copper, gave an abundant red precipitate. When this precipitate was treated with dilute hydrochloric acid, it gave off about a litre of a colourless gas with a peculiar odour, and which burned with a very fuliginous flame. The analysis of this gas, and its combination with copper, left no doubt that it was acetylene.

Its formation may be thus expressed :—

$$\left.\begin{array}{c} C^5 H^{11} \\ Na \end{array}\right\} \Theta + C^2 H^3 Br = C^2 H^2 + Na Br + C^5 H^{12} \Theta.$$

Amylate of sodium.   Monobrominated ethylene.   Acetylene.   Amylic alcohol.

This reaction is important, as it will probably lead to the formation of a new series of hydrocarbons of the general formula $C_n H_{2n-2}$, of which acetylene, $C^2 H^2$, is the first member, from the hydrocarbons of the general formula $C_n H_{2n}$. In fact Sawitsch has subsequently ‡ examined the action of monobromi-

* *Bulletin de la Société Chimique,* p. 7.
† Phil. Mag. vol. xx. p. 196.
‡ *Comptes Rendus,* March 4, 1861.

nated propylene on ethylate of sodium, and has obtained a second member of the series, *allylene*, $C^3 H^4$. The action is quite analogous to that in the former case : the gas is passed into an ammoniacal solution of copper, with which it forms a voluminous flocculent precipitate. This is decomposed, when heated, with the formation of a reddish flame ; with concentrated acids it disengages a gas even in the cold.

Allylene is best obtained from this precipitate by the action of dilute aqueous hydrochloric acid, from which, when heat is applied, there is given off a colourless gas of a strong and disagreeable odour, but less so than that of acetylene. It burns with a fuliginous flame, and precipitates silver and mercury salts, the former grey and the latter white. These compounds are analogous to the copper compound, and, like it, are very unstable. The property of combining with an ammoniacal oxide of copper appears to be characteristic of this group, and will probably lead to the discovery of the higher members.

The formation of allylene may be thus expressed :—

$$C^2 H^5 NaO + C^3 H^5 Br = Na Br + C^3 H^4 + C^2 H^6 O.$$
Ethylate of   Brominated   Allylene.   Alcohol.
sodium.   propylene.

Miasnikoff[*], in some experiments with monobrominated ethylene, has also noticed a mode of the formation of acetylene, which gives a simple and elegant method of preparing this gas.

When the vapours of crude monobrominated ethylene, prepared in the ordinary way, are passed into an ammoniacal solution of nitrate of silver, a precipitate forms, which is at first yellow, but which quickly becomes converted into grey ; at the same time an oil collects at the bottom of the vessel, which is brominated ethylene, and which can easily be separated by heating to 20°. When this bromide further acts upon a fresh portion of ammoniacal solution, it produces no change ; but after being passed through a boiling concentrated solution of potash, it again acquires the property of forming this grey pulverulent deposit. By arranging the apparatus so that the vapours of brominated ethylene pass more than once through solution of potash, a considerable quantity of the pulverulent deposit can be formed.

This powder detonates strongly on the application of heat, percussion, or friction, and also by the action of chlorine or of gaseous hydrochloric acid. Treated with dilute hydrochloric acid, it gives a gas which burns with a fuliginous flame, and which reproduces the pulverulent precipitate. The analysis and the properties of this gas show that it is acetylene ; and the analysis of the silver compound gives for it the formula

$$C^2 H^2 Ag^2.$$

[*] *Bulletin de la Société Chimique*, p. 12.

From the mode of its preparation, monobrominated ethylene, in passing through strong potash, is simply resolved into acetylene and hydrobromic acid,

$$C^2 H^3 Br - H Br = C^2 H^2.$$
Monobrominated   Acetylene.
ethylene.

By a similar series of actions, M. Morkownikoff has prepared what appears to be the gas *allylene,* described by Sawitsch.

According to Heinz[*], the best method of preparing glycolic acid is from monochloracetic acid, which, under the influence of alkalies, decomposes into an alkaline glycolate and into an alkaline chloride[†].

The hydrate of glycolic acid may readily be obtained by the following method :—To the solution of the mixture of glycolate of soda and chloride of sodium obtained in the above reaction, a sufficient quantity of solution of sulphate of copper is added. The glycolate of copper, $C^4 H^3 Cu O^6$, which forms, is a difficultly soluble salt; it precipitates as a crystalline mass, and is readily obtained pure by washing. This salt is then diffused in a large quantity of water, the mixture raised to boiling, and saturated with sulphuretted hydrogen. When all the copper is precipitated it is filtered; and as the filtrate is brownish, from the solution of a small quantity of sulphide of copper, it is evaporated to a small volume, while a slow stream of sulphuretted hydrogen is passed through, and when now filtered, is obtained quite colourless.

By a series of operations analogous to those by which an alcohol of the ethyle series may be transformed into the next higher acid, Cannizaro[‡] has obtained from anisic alcohol an acid homologous with anisic acid.

Anisic alcohol, $C^8 H^{10} O^2$[§], appears to contain the group $C^8 H^9 O$, which plays the part of a monoatomic radical. When the chloride of this group, $C^8 H^9 O$, Cl, was treated with cyanide of potassium, chloride of potassium was formed, and an oil which was the cyanide, $C^8 H^9 O C N$. This oil was obtained in the impure state, and was treated directly with potash at the temperature of ebullition, by which a large quantity of ammonia was disengaged; and on the subsequent addition of hydrochloric acid in excess, an oil deposited which solidified to a crystalline mass. This consisted of the new acid which Cannizaro names *homoanisic*

* Poggendorff's *Annalen*, January 1861.
† Phil. Mag. vol. xvi. p. 138.
‡ *Comptes Rendus*, vol. li. p. 606.
§ Phil. Mag. vol. xx. p. 294.

acid. It crystallizes in nacreous laminæ. Its formation may be thus expressed:—

$$C^9 H^9 \Theta C N + K H\Theta + H^2 \Theta = C^9 H^9 K\Theta^3 + NH^3.$$

Cyanide of   Homoanisate
anisyle.   of potash.

Rossi[*] has applied to cuminic alcohol the same series of transformations, and has obtained a new acid homologous with cuminic acid. He prepared the chloride of cumyle, $C^{10} H^{13} Cl$, by the action of hydrochloric acid on cuminic alcohol, and treated it with cyanide of potassium, by which means he obtained the corresponding cyanide. This crude cyanide of cumyle was boiled with strong caustic potash until its decomposition was complete; and to the mixture was then added hydrochloric acid in excess, by which the new acid was precipitated. On recrystallization it was obtained in small needles. Its formation is thus expressed:

$$C^{10} H^{13} C N + KH\Theta + H^2 \Theta = C^{11} H^{13} K\Theta^2 + NH^3.$$

Cyanide of   Homocuminate
cumyle.   of potash.

Homocuminic acid, $C^{11} H^{12} \Theta^2$, can be distilled without decomposition. It is difficultly soluble in cold water; but its solution reddens litmus, and decomposes the carbonates. Its salts are obtained by double decomposition: most of them crystallize well.

As the result of a lengthened investigation on filtration of the air in reference to fermentation, putrefaction, and crystallization, Schröder[†] is led to the following conclusions:—

1. A vegetable or animal can only be formed from living vegetable or animal organisms. *Omne vivum ex vivo.*

2. There is a series of phenomena of fermentation and putrefaction which arise solely from microscopic germs furnished by the atmosphere. These are more especially the formation of mould, of wine-yeast, of the lactic acid ferment, of the ferment which produces the decomposition of urine.

3. Vegetable or animal substances, boiled and closed while hot by means of cotton, remain in that condition quite protected against every kind of fermentation, putrefaction, or formation of mould, if all the germs in them capable of development are destroyed by boiling; for the germs which might reach them from the air are filtered out by the cotton.

4. The germs of most vegetable or animal substances are completely destroyed by simple boiling. A boiling for a short time at 100° C. is also sufficient to kill all germs furnished by the air.

[*] *Comptes Rendus,* March 4, 1861.
[†] *Liebig's Annalen,* March 1861.

5. But milk, the yellow of egg, and meat contain germs which are not completely destroyed by a short boiling at 100°. But boiling at a higher temperature under a pressure of two atmospheres in the digestor, or long-continued boiling at 100°, is sufficient to kill even these germs.

6. The germs of milk, yellow of egg, and of meat, even when they have been submitted to a boiling at 100°, not continued, however, too long, are capable of developing themselves as a specific putrefaction ferment, and not unfrequently, at least in the yelk of egg, in the form of long but inert fibrils.

7. This specific putrefying ferment is of animal nature. It developes and increases at the expense of all albuminous substances. It is, however, incapable of increase under conditions which are all that are necessary to vegetable formations.

8. The crystallisation of supersaturated solutions is commenced or induced by the action of the surface of solid bodies.

9. The induction necessary to set up the crystallization of the soluble hydrates from a supersaturated solution, is less than that necessary for the crystallization of the more difficultly soluble hydrates.

10. The surface of a crystal of the same nature exercises the strongest inducing action. Next to that comes the layer of air which forms on the surface of solid bodies. These coatings are destroyed by heating, continued wetting, or by cleaning, and are only formed slowly again in filtered air.

11. The crystallization of the more soluble hydrates from supersaturated solutions, which is set up even by a feeble induction, only experiences a feeble induction on the surface of the crystal of the same kind, and hence only progresses very slowly.

12. Supersaturated solutions closed with cotton keep for a long time unchanged, because the cotton filters all the solid particles from the air which gains access. Agitation has no action on the crystallization; it only induces it if supersaturated solutions are in contact with such places of the surface as are fitted to induce the crystallization.

By oxidizing cymole, $C^{10} H^{14}$, with dilute nitric acid, Noad found that it was converted into toluylic acid, $C^9 H^8 O^2$, and oxalic acid; from cumole, $C^9 H^{12}$, Abel similarly obtained benzoic acid, $C^7 H^6 O^2$. Hence it might have been expected that toluole, $C^7 H^8$, by analogous treatment would yield a new acid, $C^5 H^2 O^2$, homologous with these.

This is, however, not the case, as experiments by Fittig [*] have shown. When toluole is oxidized by means of nitric acid, the process is different. There is no formation either of oxalic or of carbonic acid. A colourless acid is formed, almost insoluble in

[*] Liebig's *Annalen*, February 1861.

cold water, and but slightly so in hot. It crystallizes from this solution in very small needles.

This body seems to have the composition of salicylic acid, $C^7 H^6 O^3$, without, however, being identical with it; for it does not give the well-known reaction with perchloride of iron, characteristic of salicylic acid. The baryta salt has the formula $C^7 H^5 Ba O^3$, and the silver salt $C^7 H^5 Ag O^3$.

As yet the glycols of the fatty acid series only are known. Wicke obtained a compound, $C^{11} H^{12} O^4$, which has the composition of an acetate of benzo-glycol, $\left. \begin{array}{c} C^7 H^{6\prime\prime} \\ (C^2 H^3 O)^2 \end{array} \right\} O^2$; but its properties differ from what might be expected of a glycol derivative, and it is rather analogous to the compounds which Geuther obtained by heating the aldehydes of the fatty acids with anhydrous acids. Beilstein and Seelheim[*] have made a series of experiments to prepare the glycol, $C^7 H^8 O^2$, of the aromatic series of acids, which stands to benzylic alcohol, $C^7 H^8 O$, and benzoic acid, $C^7 H^6 O^2$, in the same relation as ordinary glycol, $C^2 H^6 O^2$, does to alcohol, $C^2 H^6 O$, and acetic acid, $C^2 H^4 O^2$. By the action of sulphuric acid on benzylic alcohol, they hoped to obtain the bibasic radical $C^7 H^6$ in a manner analogous to that by which ethylene is obtained from ordinary alcohol. But the result of the above action was a resinous body, which, when treated with bromine in the expectation of obtaining $C^7 H^6 Br^2$, gave off hydrobromic acid and underwent a complete decomposition.

The composition of the desired body would be identical with that of saligenine, which in many points resembles a biatomic alcohol. An attempt was made to prepare from it the biatomic chloride, $C^7 H^6 Cl^2$, which did not give the desired result.

By the action of pentachloride of phosphorus, saligenine was resolved into saliretine, $C^7 H^6 O$, and water; at the same time a certain quantity of a chlorine compound was formed which seemed to contain the chloride $C^7 H^6 Cl^2$, but which could not be separated in the pure state.

Saligenine was heated with anhydrous acetic acid in the expectation of forming acetate of saligenine; but instead of this, saliretine and acetic acid were formed:—

$$C^7 H^8 O^2 + \left. \begin{array}{c} C^2 H^3 O \\ C^2 H^3 O \end{array} \right\} O = C^7 H^6 O + 2 C^2 H^4 O^2.$$

Saligenine. ·Acetic anhydride. Saliretine. Acetic acid.

When sodium was added to a solution of saligenine in pure ether, hydrogen was evolved, and a white pulverulent precipitate

[*] Liebig's *Annalen*, January 1861.

formed, which appeared to have the formula $C^{14} H^{13} Na \Theta^2$, Iodide of ethyle and chloride of acetyle act on this body; and form resinous bodies which could not be purified.

Saligenine dissolves in baryta water, forming a crystalline compound, $C^7 H^9 Ba \Theta^3 = C^7 H^7 Ba \Theta^2 + H^2 \Theta$.

By the action of pentasulphide of phosphorus, saligenine appears to become converted into an amorphous variety.

The law which regulates the contraction of solutions made it probable that the specific gravity of liquid ammonia must be less than that found by Faraday; and with a view of testing this point, Jolly* has made a redetermination of its specific gravity.

The liquid gas was prepared in the usual way, by heating ammoniacal chloride of silver in a bent closed tube, the empty end of which was much narrowed in one part and was provided with an arbitrary graduation. After the ammonia had been expelled from the ammoniacal chloride, and had been condensed in the cooled end of the tube, the part containing the liquefied gas was cooled down to a temperature of $-86°$ in a mixture of solid carbonic acid and ether, and was melted off, which at this temperature could be done without danger.

The tube was now immersed in pounded ice, and the height of the liquid read off. A subsequent weighing gave the weight of the tube, together with the liquid gas and the compressed gas above the liquid.

The tube was now cooled down to $-24°$ C., and the point softened, by which, as the tension of the gas at this temperature does not exceed two atmospheres, the gas escaped without danger or loss of glass. The opened tube was next transferred from the freezing mixture into pounded ice, upon which a violent ebullition commenced, after which, when terminated, and there was no more escape of NH^3, the tube was closed and weighed, by which the weight of the vapour at $0°$ was determined.

At this stage special experiments were made to ascertain if the ammonia was anhydrous and quite free from air, which was found to be the case. The tube was next weighed empty, and then calibrated.

From the data furnished by these various operations, Jolly found the specific gravity in three experiments to be

$$0·6239, \qquad 0·6261, \qquad 0·6193,$$

or in the mean 0·6231, which is one-sixth less than the number

* Liebig's *Annalen*, February 1861.

found by Faraday, 0·73, and agrees with a determination made by Andréeff[*], who obtained the number 0·6364.

Jolly also determined the coefficient of absorption of liquid ammonia, and found it to be for 1 degree 0·00155, which is about half as much as that of air.

LV. *On the Duration of the Spark which accompanies the discharge of an Electrical Conductor. By* P. L. Rijke, *Professor of Natural Philosophy at the University of Leyden*[†].

1. WHEN a Leyden jar is discharged in the ordinary way, the spark produced may be considered as instantaneous; its duration, at least, is so short that it has hitherto been found impossible to measure it even approximately. This, however, is no longer the case when the charge has to traverse a body which offers any considerable resistance, as, for instance, a copper wire half a mile long. In fact, Mr. Wheatstone found[‡] that the spark obtained through a copper wire of $\frac{1}{15}$th of an inch in diameter, and of the length above mentioned, lasted for about $\frac{1}{24.000}$th of a second.

2. This result, if Mr. Wheatstone had published it by itself, would probably have been explained simply on the ground that electricity required precisely this time, viz. $\frac{1}{24.000}$th of a second, to traverse the length of wire in question. This explanation would, however, have been, to say the least, incomplete, since in the same series of experiments Mr. Wheatstone proved that electricity really requires $\frac{1}{1.152.000}$th of a second to travel that distance; and in order to reconcile these results, he considered it necessary to have recourse to a new hypothesis, and to suppose "that the diameter of the wire was not sufficiently great to allow the charge to pass through it except in a *successive* manner."

3. In reflecting on this question, it appeared to me that the results obtained by the above-named illustrious physicist might be easily explained on known grounds; and that it was by no means necessary to have recourse to a hypothesis, in support of which it would be difficult to cite a single direct observation.

We shall see that it is, in fact, easy to prove *à priori* the following proposition :—

The time required by electricity to traverse a given conductor is much less than that required to discharge that conductor.

[*] Liebig's *Annalen*, vol. cx. p. 1.
[†] Communicated by the Author.
[‡] "On some Experiments to Measure the Velocity of Electricity and the duration of Electrical Light," Phil. Trans. 1834.

Let A B be an isolated conductor of such length that a charge of electricity requires a perceptible time, say t'', to traverse it.

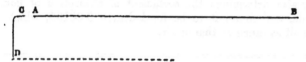

Let C D be another conductor much shorter than the first, and so placed that the extremities C and A of the two conductors are at the distance of several millims., the other extremity D of the second conductor being in communication with the ground.

.. Now suppose that at a given moment a certain quantity of electricity be communicated to the extremity B of the conductor A B. At the end of t'' this electricity will have spread over the whole surface of the conductor. At the same instant, supposing the tension sufficient, a luminous discharge will commence between A and C. We say will *commence*, admitting therefore that the discharge will occupy a certain time. In fact, in order that the conductor might discharge itself instantaneously, it would be necessary that all the electric fluid should be accumulated at A. This, however, is not the case; since at the moment when the discharge commences, the whole surface of the conductor is occupied, though unequally, with electricity. Now it is easy to see that the electricity, which at the moment we are considering is still at B, will only reach the other extremity t'' afterwards*; and even then it must be carefully noted that *all* the electricity at B will not have passed to A; only part will have done so, and the rest will have remained at B, of which a further portion will reach A $2t''$ after the discharge commences, and a third $3t''$ after the same epoch, and so on.

The above reasoning shows that, during the time occupied by the discharge, successive portions of electricity will arrive at A from B at equal intervals of time; and it is clear that the same will be the case with the electricity which at the beginning of the discharge was at any intermediate part of the conductor, only that the intervals of time will be shorter in proportion as the part considered is nearer A. It is therefore evident that there will be a *continuous* current of electricity towards this extremity, and that the discharge will therefore be equally *continuous*. The passage of the electricity from A to C will of course cease immediately the tension at A descends below a certain limit; but, on the other hand, it must not be forgotten

* If, as some physicists believe, the rapidity of the propagation of electricity diminishes with its density, then the electricity at B would require more than t'' to arrive at A.

that, in order to support the discharge already commenced, a much less considerable tension is required than is necessary to establish it in the first instance (Riess, *Die Lehre der Reibungs-elektricität*, vol. ii. p. 636). When electricity once commences to pass in the disruptive form from A to C, an expansion is produced in the intermediate layers of air, which amply suffices to explain the facility with which the same layers of air allow themselves to be traversed by electricity of much lower tension.

4. It is evident, however, that this expansion will depend on the quantity of electricity which in a given time passes from A to C. Consequently if the experiment be so conducted that, all other things being the same, a less quantity of the electric fluid passes from A to C, the expansion of the air will be less, and the discharge ought to cease sooner. But if the discharge ceases sooner, it will follow that the *residue* will be greater. Now it is easy to verify this theoretical conclusion.

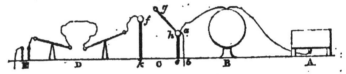

B is a metallic sphere of 0·31 metre diameter. This sphere is connected by means of metal wires, on one side with a sinus elec-trometer A (Pogg. *Ann.* vol. cvi. p. 438), and on the other with an apparatus C, called by Riess an "Entladungsapparat" (Riess, *Die Lehre der Reibungselektricität*, p. 365), and which consists of a moveable metal rod $g\,h$, turning at h on a horizontal axis provided with a ratchet and clapper, which can be moved by pulling the silk cord $a\,b$, so that the knob g can always be brought to a fixed distance from the metal ball f.

The ball f and the moveable rod $g\,h$ are both supported on glass pillars, the former being united by means of a metal wire with one branch of the universal excitator D, of which the other branch communicates by similar means with one of the knobs of the spark-micrometer E. The other knob of the spark-micro-meter is in connexion with the ground. Between the branches of the excitator D the substances are placed whose action on the residue we wish to determine.

I confined myself to the comparison of the actions of brass wire and a cord of hemp steeped in water. Both these conduc-tors were 0·3 metre long, the diameter of the brass wire being ·08 millim., that of the hempen cord 2 millims. In each experiment the sphere B received the same charge measured by the electro-meter. The clapper which supported the rod $g\,h$ was then pulled by means of the silk cord $a\,b$, the knob g descended

towards the ball *f*, establishing a metallic communication between
A B C D and E, and at the same instant a spark passed between
the two knobs of the spark-micrometer. The amount of residue
was then determined by the electrometer. The results I obtained are collected in the following Table, in which the amount
of residue is represented by R, that of the charge by L:—

Distance between the knobs of the micrometer.	Name of the substance placed between the branches of the excitator.	Deviation of the magnetic needle		Value of $\frac{R}{L}$
		Before the discharge.	After the discharge.	
0·88 millim.	Hempen cord.	5$\frac{3}{2}$ 20	5 18	0·117
„ „	Brass wire.	„	3 18	0·073
„ „	Hempen cord.	„	4 42	0·103
1 millim.	Hempen cord.	64 26	7 6	0·137
„ „	Brass wire.	„	3 56	0·076
„ „	Hempen cord.	„	6 56	0·134

From this Table it is obvious that the residue is more considerable when the body through which the discharge takes place
offers more resistance.

5. It is possible that the above theory may not at first meet
with universal assent. I trust, however, that those physicists
who hesitate to admit it will find their doubts dispelled on considering the following experiment:—A B C D is a hollow cylinder closed at the end C D and fitted at the other end A B with a
sucker-valve opening outwards. Near the end C D there is a
slide G H pierced with a large opening at I. The anterior por-

tion A B E F of the tube must be supposed exhausted of air,
while in the remaining part E F C D the air is in a state of great
compression. Now suppose the slide G H suddenly depressed
until the centre of the opening I coincides with the axis of the
cylinder. It is evident that the air enclosed in E F C D will immediately begin to spread throughout the anterior portion of the
tube until it reaches A B, where, if its tension be sufficiently
great, it will commence escaping from the tube. If the air has
taken *t''* to arrive at A B, it will of course *begin* to escape *t''* after
the slide has been depressed; but no one will imagine that the

escape of the air will *cease* in *t'*. In fact it would suffice to repeat, *mutatis mutandis*, the reasoning in (8) to see that the escape must necessarily last more than *t''*.

P.S. In my note on the Inductive Spark, which was inserted in the December Number of this Journal, the expression *point* of light should have been *streak* of light.

LVI. *Note on the Historical Origin of the unsymmetrical Six-valued Function of six Letters.* By J. J. SYLVESTER, *Professor of Mathematics at the Royal Military Academy, Woolwich* *.

THE discovery and first announcement of the existence of the celebrated function of six letters having six values, and not symmetrical in respect to all the letters, is usually assigned to my illustrious friend M. Hermite, to whom M. Cauchy expressly ascribes it in a memoir inserted in the *Comptes Rendus* of the Institut for December 8, 1845, p. 1247, and again, January 5, 1846, p. 30.

M. Cauchy adds that the conversation he held with M. Hermite on this subject excited in himself a lively desire to sound to its depths the question of permutations, and to develope the consequences to be deduced from the application of the principles relative thereto, which he had himself long previously laid down.

I was not at that date in the habit of consulting the *Comptes Rendus*, or I should at once have made the reclamation of priority which I now do, not from any unworthy motive of self-love in so small a matter, but out of regard to historic truth. It is a year or two since I first learnt that the origin of this function was usually referred to M. Cauchy or M. Hermite; but although aware that its existence was known to myself long previous to the dates quoted, I did not recollect that I had ever communicated it to the world through the medium of the press, and I therefore kept silence on the subject.

Turning over, a few days ago, for another purpose, the pages of a back volume of this Magazine, my eye chanced to alight on a footnote to a paper of my own inserted therein, under date of April 1844, "On the Principles of Combinatorial Aggregation," which I will take the liberty of quoting at length, as it proves incontestably the priority which I lay claim to.

"When the modulus is four, there is only one synthematic arrangement possible, and there is no indeterminateness of any kind; from this we can infer, *à priori*, the reducibility of a bi-

* Communicated by the Author.

Phil. Mag. S. 4. Vol. 21. No. 141. *May* 1861. 2 B

quadratic equation; for using $\phi, f,$ F to denote rational symmetrical forms of function, it follows that

$$F \begin{cases} f(\overline{\phi a, b}, \ \overline{\phi c, d}) \\ f(\overline{\phi a, c}, \ \overline{\phi b, d}) \\ f(\overline{\phi a, d}, \ \overline{\phi b, c}) \end{cases} \text{ is itself a rational symmetric function of } a, b, c, d.$$

Whence it follows that if a, b, c, d be the roots of a biquadratic equation, $f(\overline{\phi a, b}, \ \overline{\phi cd})$ can be found by the solution of a cubic: for instance, $(a+b) \times (c+d)$ can be thus determined, whence immediately the sum of any two of the roots comes out from a quadratic equation.

"To the modulus 6 there are fifteen different synthemes capable of being constructed. At first sight it might be supposed that these could be classed in natural families of three or of five each, on which supposition the equation of the sixth degree could be depressed; but on inquiry this hope will prove to be futile, not but what natural affinities do exist between the totals; but in order to separate them into families, each will have to be taken twice over; or in other words, the fifteen synthemes to modulus 6 being reduplicated, subdivide into six natural families of five each."

The six families above referred to (in which it is to be understood that $p.q$ and $q.p$ are identical in effect) are the following:—

a.b	c.d	e.f		a.c	d.e	f.b		a.d	e.f	b.c
a.c	b.e	d.f		a.d	c.f	e.b		a.e	d.b	f.c
a.d	b.f	c.e		a.e	c.b	d.f		a.f	d.c	e.b
a.e	b.d	c.f		a.f	c.e	d.b		a.b	d.f	e.c
a.f	b.c	d.e		a.b	c.d	e.f		a.c	d.e	f.b

a.e	f.b	c.d		a.f	b.c	d.e		a.b	c.d	e.f
a.f	e.c	b.d		a.b	f.d	c.e		a.c	b.e	d.f
a.b	e.d	f.c		a.c	f.e	b.d		a.d	b.f	c.e
a.c	e.b	f.d		a.d	f.c	b.e		a.e	b.d	c.f
a.d	e.f	b.c		a.e	f.b	c.d		a.f	b.c	d.e

And it will be observed that every two families have one, and only one, syntheme in common between them; and precisely in the same way as in the note above quoted, it is especially shown that the one single natural family

$$\begin{vmatrix} a.b & c.d \\ a.c & b.d \\ a.d & b.c \end{vmatrix}$$

gives rise to a function of four letters with only one value, so the six functions analogously formed with these six families obviously give rise to six functions, which change into one another when any interchange is effected between the letters which enter into them; so that any one of these is a function of six letters having only six values. I conceive that, after this reference, no writer on the subject wishing to specify the function in question would hesitate to call it after my name.

I may also take occasion to observe that, in connexion with my researches in combinatorial aggregation, long before the publication of my unfinished paper in the Magazine, I had fallen upon the question of forming a heptadic aggregate of triadic synthemes comprising all the duads to the base 15, which has since become so well known, and fluttered so many a gentle bosom, under the title of the fifteen school-girls' problem; and it is not improbable that the question, under its existing form, may have originated through channels which can no longer be traced in the oral communications made by myself to my fellow-undergraduates at the University of Cambridge long years before its first appearance, which I believe was in the 'Lady's Diary' for some year which my memory is unable to furnish.

In order to relieve this notice from the mere personal character which it may thus far appear to bear, I will state another question concerning the combinatorial aggregation of fifteen things which may serve as a pendant to the famous school-girl problem.

The number of triads to the base 15 is $\dfrac{15 \times 14 \times 13}{3 \cdot 2 \cdot 1} = 5 \times 91$.

Let it be required to arrange these into 91 synthemes, in other words, to set out the walks of 15 girls for 91 days (say a quarter of the year) in such a manner that the same three shall never *all* come together more than once in the quarter. Of the various ways in which it is probable this problem may be solved, the following deserves notice. Let 15 letters be arbitrarily divided into 5 sets, viz.

$$a_1\, b_1\, c_1;\quad a_2\, b_2\, c_2;\quad a_3\, b_3\, c_3;\quad a_4\, b_4\, c_4;\quad a_5\, b_5\, c_5.$$

The sets as they stand will represent one of the 91 arrangements sought for, which I call the basic syntheme. The remaining 90 may be obtained as follows in 10 batches of 9 each. Write down the 10 index distributions following:—

1 2 3; 4 5			1 4 5; 2 3
1 2 4; 3 5			2 3 4; 1 5
1 2 5; 3 4			2 3 5; 1 4
1 3 4; 2 5			2 4 5; 1 3
1 3 5; 2 4			3 4 5; 1 2

Take any one of these distributions, as for instance 2 3 5; 1 4, and proceed as follows:—In respect of 2, 3, 5, conjugate the three sets $\begin{array}{ccc} a_2 & b_2 & c_2 \\ a_3 & b_3 & c_3 \end{array}$; and in respect of 1, 4, conjugate the two

remaining sets $\begin{array}{ccc} a_5 & b_5 & c_5 \\ a_1 & b_1 & c_1 \\ a_4 & b_4 & c_4 \end{array}$.

From the ternary conjugation form the nine arrangements,

$a_2\ a_3\ a_5$	$b_2\ b_3\ b_5$	$c_2\ c_3\ c_5$
$a_2\ a_3\ b_5$	$b_2\ b_3\ c_5$	$c_2\ c_3\ a_5$
$a_2\ a_3\ c_5$	$b_2\ b_3\ a_5$	$c_2\ c_3\ b_5$
$a_2\ b_3\ a_5$	$b_2\ c_3\ b_5$	$c_2\ a_3\ c_5$
$a_2\ b_3\ b_5$	$b_2\ c_3\ c_5$	$c_2\ a_3\ a_5$
$a_2\ b_3\ c_5$	$b_2\ c_3\ a_5$	$c_2\ a_3\ b_5$
$a_2\ c_3\ a_5$	$b_2\ a_3\ b_5$	$c_2\ b_3\ c_5$
$a_2\ c_3\ b_5$	$b_2\ a_3\ c_5$	$c_2\ b_3\ a_5$
$a_2\ c_3\ c_5$	$b_2\ a_3\ a_5$	$c_2\ b_3\ b_5$

which call

$$\mathrm{L}_1\ \mathrm{L}_2\ \mathrm{L}_3 \qquad \mathrm{L}_4\ \mathrm{L}_5\ \mathrm{L}_6 \qquad \mathrm{L}_7\ \mathrm{L}_8\ \mathrm{L}_9.$$

Again, from the binary conjugation, form the nine arrangements,

$a_1\ b_1\ c_4$	$a_4\ b_4\ c_1$
$a_1\ b_1\ b_4$	$a_4\ c_1\ c_4$
$a_1\ b_1\ a_4$	$c_1\ b_4\ c_4$
$a_1\ c_1\ c_4$	$a_4\ b_4\ b_1$
$a_1\ c_1\ b_4$	$a_4\ b_1\ c_4$
$a_1\ c_1\ a_4$	$b_1\ b_4\ c_4$
$b_1\ c_1\ c_4$	$a_4\ b_4\ c_1$
$b_1\ c_1\ b_4$	$a_4\ a_1\ c_4$
$b_1\ c_1\ a_4$	$a_1\ b_4\ c_4$

which call

$$\mathrm{M}_1\ \mathrm{M}_2\ \mathrm{M}_3 \qquad \mathrm{M}_4\ \mathrm{M}_5\ \mathrm{M}_6 \qquad \mathrm{M}_7\ \mathrm{M}_8\ \mathrm{M}_9.$$

Now combine the L with the M system, each L with some M in any order whatever; the 9 combinations or appositions thus obtained will give a batch of 9 synthemes; and proceeding in like manner with each of the 10 distributions of the indices 1, 2, 3, 4, 5, we shall obtain 90 synthemes, which together with the basic syntheme complete the system required. The M system corresponding to any distribution of the indices is the system which contains the synthematic arrangement of the bipartite* triads which can be constituted out of six things, separated in two sets or parts, and is unique. The L system is *one* of those which represents the synthematic arrangement of the tripartite† triads of nine things separated into three sets or parts. I have set out above one in particular of these for the sake of greater clearness; but any other system having the same property will serve the same purpose, and a careful study will serve to show that the total number of L's corresponding to a given distribution of indices will be ( )†. Consequently the total number of LM's that we can form for a given distribution will be ( ) $\times 1.2.3.4.5.6.7.8.9$; and the number of *distinct* synthematic arrangements satisfying the given conditions corresponding to any assumed basic syntheme will be this number raised to the tenth power; and as this vastly exceeds the total number of permutations of fifteen things, we see, without even taking into consideration the diversity that may be produced by a change of the base, that this method must give rise to many distinct types of solution (arrangements being defined to belong to the same or different types, according as they admit or not of being deduced from each other by a permutation effected among their monadic elements). The common character of all these allotypical aggregations, and which serves to constitute them into a natural order or family, consists in their being derived from a base formed out of five sets, such that the monopartite triads corresponding to the base form one syntheme, and the other 90 synthemes each contain a conjugation of the tripartite triads belonging to three out of the five sets of the base with the bipartite triads belonging to the other two sets thereof. There is, moreover, no reason to suppose, or at all events no safe ground for affirming, that this family exhausts the whole possible number of types to which the arrangements satisfying the proposed condition admit of being reduced. A further question which I have somewhere raised, and which brings the two problems of the school-girls into *rapport*, is the following:—" To divide the system of 91 synthemes satisfying the conditions above stated into thirteen minor systems, each of which satisfies the conditions of the old problem, *i. e.* of containing all the duads that can be made out of the fifteen elements once and once only;" or to put the question in a more exact form, to exhibit thirteen systems,

* See note at end of paper.

† Some day or another a new combinatorial calculus must come into being to furnish general solutions to the infinite variety of questions of *multifariousness* to which the theory of combinatorial aggregation, *alias* compound permutations, gives rise.

each satisfying this last condition, which shall together include between them all the triads that can be made out of the fifteen elements.

The reader would have reason to be dissatisfied with the author's reticence, were he to leave altogether unmentioned the synthematic aggregation of the *binomial* triads appertaining to the same three triliteral sets or *nomes*; but space forbids my doing more at present than giving one of these aggregates, and indicating the number and mode of generation of all from this one. It will readily be seen that any such aggregate will be made up of two sub-aggregates, which I shall call A and B respectively; of which one bears the same relation to the disposition of the nomes in the order 123 456 789, as the other to their disposition in the order 123 789 456. Thus we may take for our A and B the following, which will each contain 9 synthemes, the total number of synthemes in the two together being 18* :—

(A.)			(B.)		
124	567	893	127	894	763
125	468	739	128	795	436
126	459	783	129	786	453
134	568	279	137	895	246
185	469	278	138	796	245
136	457	289	139	784	256
234	569	187	237	896	154
235	467	189	238	794	156
236	458	179	239	785	146

The system of triads contained in A may be arranged in twelve different aggregates similar to the one given, and the same will be true for the triads in the B; so that the total number of the combined systems will be 144. All the permutations which leave A or B (separately considered) unaltered will form a natural group,—the theory of groups in this, as in every other case, standing in the closest relation to the doctrine of combinatorial aggregation, or what for shortness may be termed syntax. I have elsewhere given the general name of Tactic to the third pure mathematical science, of which order is the proper sphere,

---

* Thus, since there is evidently one mononomial syntheme, the total number of synthemes of all three kinds will be $1+18+9=28=\frac{8\times7}{2}$, as it should be, the total number of triads being $\frac{9.8.7}{3.2}$, and $\frac{9}{3}$ of them going to a syntheme.

as is number and space of the other two. Syntax and Groups are each of them only special branches of Tactic. I shall on another occasion give reasons to show that the doctrine of groups may be treated as the arithmetic of ordinal numbers. With respect to the twelve varieties of the A or B aggregates, they may be obtained from the one given by combining the substitutions corresponding to the six permutations of the three constituents of one nome, as 7, 8, 9, with the permutation of any two constituents of another, as 5, 6. But I have said enough for my present purpose, which is to point out the boundless untrodden regions of thought in the sphere of order, and especially in the department of *syntax*, which remain to be expressed, mapped out, and brought under cultivation. The difficulty indeed is not to find material, of which there is a superabundance, but to discover the proper and principal centres of speculation that may serve to reduce the theory into a manageable compass.

I put on record (as a Christmas offering on the altar of science) for the benefit of those studying the theory of groups, or compound permutations (to which the prize shortly to be adjudicated by the Institute of France for the most important addition to the subject may tend to give a new impulse), and with an eye to the geometrical and algebraical verities with which, as a constant of reason, we may confidently anticipate it is pregnant, an exhaustive table of the monosynthematic aggregates of the trinomial triads that are contained in a system of three triliteral nomes. Let these latter be called respectively 1 2 3; 4 5 6; 7 8 9; then we have the annexed:—

Table of Synthemes of Trinomial Triads to Base 3.3.

(1.)	(2.)	(3.)	(4.)
147 258 369	147 258 369	147 258 369	147 258 369
148 259 367	148 259 367	148 259 367	148 259 367
149 257 368	149 257 368	149 257 368	149 267 358
157 268 349	157 268 349	157 269 348	157 268 349
158 269 347	158 269 347	158 267 349	158 269 347
159 267 348	159 267 348	159 268 347	159 247 368
167 248 359	167 249 358	167 248 359	167 248 359
168 249 357	168 247 359	168 249 357	168 249 357
169 247 358	169 248 357	169 247 358	169 257 348

(5.)	(6.)	(7.)	(8.)
147 258 369	147 258 369	147 258 369	147 258 369
148 267 359	148 267 359	148 269 357	148 269 357
149 268 357	149 257 368	149 257 368	149 267 358
157 249 368	157 268 349	157 268 349	157 268 349
158 269 347	158 269 347	158 249 367	158 249 367
159 248 367	159 248 367	159 267 348	159 247 368
167 259 348	167 259 348	167 248 359	167 248 359
168 257 349	168 249 357	168 259 347	168 259 347
169 247 358	169 247 358	169 247 358	169 257 348

The discussion of the properties of this Table, and the classification of the eight aggregates into natural families, must be reserved for a future occasion.

Note.—A triad is called tripartite if its three elements are culled out of three different parts or sets between which the total number of elements is supposed to be divided; bipartite if the elements are taken out of two distinct sets; unipartite if they all lie in the same set. The more ordinary method for the reduction of synthematic arrangements from a given base to a linear one which I employ, consists in the separate synthematization *inter se* of all the combinations of the *same* kind as regards the number of parts from which they are respectively drawn. Thus, *ex. gr.*, if the distribution of the $\frac{30 \times 29 \times 28}{6}$ triads to the base 30 into $\frac{29 \times 28}{2}$ synthemes be required, this may be effected by dividing the 30 elements in an arbitrary manner into 15 parts, each part containing 2 elements. These 15 parts being now themselves treated as elements, are first to be conjugated as in the old 15-school-girl problem, and each of these 7 conjugations can be made to furnish 6 synthemes containing exclusively bipartite triads. The same 15 parts are then to be conjugated as in the new school-girl problem, and the 91 conjugations thus obtained will each furnish 4 synthemes, containing exclusively the tripartite triads. These bipartite and tripartite synthemes will exhaust the entire number of triads of both kinds, and accordingly we shall find

$$7 \times 6 + 91 \times 4 = 406$$
$$= \frac{29 \times 28}{2}.$$

A syntheme, I need scarcely add, is an aggregate of combinations containing between them all the monadic elements of a given system, each appearing once only. In the more general theory of aggregation, such an aggregate would be distinguished by the name of a monosyntheme. A disyntheme would then signify an aggregate of combinations containing between them the duadic elements, each appearing once only, and so forth. Thus the old 15-school-girl question in my nomenclature would be enunciated under the form of a problem "to construct a triadic disyntheme, separable into monosynthemes to the base 15;" the new school question, as a problem "to divide the whole of the triads to base 15 into monosynthemes;" the question which connects the two, as a problem "to exhibit the whole of the triads to base 15 under the form of 13 disynthemes, each separated into 7 monosynthemes.

A question of a more general kind, and embracing this last, would be the problem of dividing the whole of the same system of triads into 13 disynthemes, without annexing the further condition of monosynthematic divisibility. So there is the simpler question of constructing a single disyntheme to the base 15 without any condition annexed as to its decomposability into 7 synthemes.

K, Woolwich Common,
  December 1860.

LVII. *On the Galvanic Polarization of buried Metal Plates. By* Dr. Ph. Carl[*].

LAST year, in consequence of the disturbances which were observed in the telegraphic wires during the appearance of the northern lights, Professor Lamont was induced to contrive an apparatus at the observatory of Munich in order to examine more closely into the occasional motion of the earth's electricity, and to determine its magnitude and direction. For this purpose large zinc plates were buried on the north, south, east, and west sides of the observatory garden; the north plate being connected with the south, and the east with the west, by means of copper wires, which were brought into the observatory and connected with galvanometers. As Professor Lamont, in testing this apparatus, remarked certain phenomena which he attributed to galvanic polarization, it appeared to me advisable to subject the matter to a more careful examination, and to obtain more accurate measurements.

Through the wire that connects two of the above-mentioned zinc plates, a current, which I shall call the terrestrial current, is perpetually circulating, the intensity of which is indicated by a fixed deviation of the galvanometer. If a galvanic element be inserted in these conducting wires and again removed, then, provided it has caused no modification in the conductor, the needle of the galvanometer will return to its former position. But if, on the other hand, a state of galvanic polarization has been produced in the zinc plates, then the deviation of the needle of the galvanometer, after the removal of the element, will be greater or less than that exhibited by it originally, accordingly as the direction of the galvanic current has been opposite to, or the same as that of the terrestrial current.

On trial, the latter result exhibited itself so unmistakeably that no further doubt could be entertained of the occurrence of galvanic polarization. In order to measure the magnitude of the effect produced, I made use of a weak Daniell's cell, which I inserted

[*] Translated by F. Guthrie from Poggendorff's *Annalen*, No. 10, 1860.

for five minutes in the wire conductor, and I thereby obtained the numbers exhibited in the following Table; in which G indicates the effect of the galvanic current, E that of the terrestrial current, and P that of the polarization expressed in divisions of the galvanometer. The effect of the polarization was observed 1 minute 30 seconds after the removal of the cell.

I. When the two currents passed in the same direction.

G+E.	P.	$\frac{P}{G+E}$.
129·8	3·4	0·026
129·3	3·4	0·026
123·0	2·0	0·021
108·2	2·4	0·022

II. When the current passed in opposite directions.

G−E.	P.	$\frac{P}{G-E}$.
124·5	5·4	0·043
122·7	5·1	0·041
107·4	3·8	0·035
109·2	3·5	0·031

From these experiments, it follows that the mean value of $\frac{P}{G+E}$ is 0·0237, that of $\frac{P}{G-E}$ 0·0375; so that when the galvanic and terrestrial currents pass in opposite directions, the polarization of the buried metal plates is greater than when they pass in the same direction.

Immediately on the removal of the cell, the effect of the galvanic polarization was greater by two or three divisions of the galvanometer scale than it was after the lapse of 1 minute 30 seconds; and after that period the effect still continued gradually to diminish. In order to exhibit the law of this diminution I subjoin the following Table:—

Time after the removal of the cell.	Deviation of the galvanometer needle.
1 minute.	23·0 divisions.
2 ,,	21·8 ,,
3 ,,	21·1 ,,
4 ,,	20·7 ,,
5 ,,	20·3 ,,
6 ,,	20·0 ,,
7 ,,	19·9 ,,
8 ,,	19·7 ,,
9 ,,	19·5 ,,
10 ,,	19·4 ,,

At the commencement of the experiment, the deviation of. the galvanometer caused by the terrestrial current was read off at 19·2 divisions.

The above experiments disclose nothing at variance with the known laws of galvanism; but it nevertheless appeared to me advisable to make them known, as they afford a simple explanation of certain phenomena which Professor Thomson has described*, and which he seems to attribute to entirely different causes.

LVIII. *On Transcendental and Algebraic Solution.* By JAMES
COCKLE, *M.A., F.R.A.S., F.C.P.S., Barrister-at-Law, of the
Middle Temple*†.

LET $fx=0$ be an algebraic equation of the nth degree, all the coefficients of which are functions of one parameter a. By differentiation we obtain a result of the form

$$\frac{dfx}{da}=\mathrm{F}x \cdot \frac{dx}{da}+fx=0.$$

But, since F and f are rational functions,

$$\frac{dx}{da}=-\frac{fx}{\mathrm{F}x}$$

$$=-\frac{fx \cdot \mathrm{F}x_2 \cdot \mathrm{F}x_3 \ldots \mathrm{F}x_n}{\mathrm{F}x \cdot \mathrm{F}x_2 \cdot \mathrm{F}x_3 \ldots \mathrm{F}x_n}=\mathrm{R}x,$$

where R is a rational and integral function of x. And

$$\mathrm{R}x=\mathrm{A}_1 x^{n-1}+\mathrm{A}_2 x^{n-2}+ \ldots +\mathrm{A}_n,$$

where A is a function of a. Moreover

$$\frac{d^2 fx}{da^2}=\frac{d\mathrm{R}x}{da}=\mathrm{F}_2 x \cdot \frac{dx}{da}+f_2 x$$

$$=\mathrm{F}_2 x \cdot \mathrm{R}x+f_2 x=\mathrm{R}_2 x.$$

Hence, repeating this process, we are conducted to the system

$$\frac{dx}{da}=\mathrm{R}x=\mathrm{A}_1 x^{n-1}+\mathrm{A}_2 x^{n-2}+ \ldots +\mathrm{A}_n,$$

$$\frac{d^2 x}{da^2}=\mathrm{R}_2 x=\mathrm{B}_1 x^{n-1}+\mathrm{B}_2 x^{n-2}+ \ldots +\mathrm{B}_n,$$

$$\vdots \qquad \vdots \qquad \qquad \vdots \qquad \quad \vdots$$

$$\frac{d^{n-1} x}{da^{n-1}}=\mathrm{R}_{n-1} x=\mathrm{G}_1 x^{n-1}+\mathrm{G}_2 x^{n-2}+ \ldots +\mathrm{G}_n;$$

and if we assign $n-2$ indeterminates $\lambda, \mu, \ldots \nu$ so as to satisfy

---

* Report of the Twenty-ninth Meeting of the British Association for the Advancement of Science (Transactions of the Sections), p. 26.
. † Communicated by the Author.

the $n-2$ conditions

$$\lambda A_1 + \mu B_1 + \ldots + \nu C_1 + G_1 = 0,$$
$$\lambda A_2 + \mu B_2 + \ldots + \nu C_2 + G_2 = 0,$$
$$\vdots \qquad \vdots$$
$$\lambda A_{n-2} + \mu B_{n-2} + \ldots + \nu C_{n-2} + G_{n-2} = 0,$$

we shall arrive at a linear differential equation,

$$\frac{d^{n-1}x}{da^{n-1}} + \nu \frac{d^{n-2}x}{da^{n-2}} + \ldots + \mu \cdot \frac{d^2 x}{da^2}$$
$$+ \lambda \cdot \frac{dx}{da} + L \cdot x = M.$$

For, when the above conditions are satisfied, $x^{n-1}, x^{n-2}, \ldots x^2$, all the powers of x in short, save x itself, disappear; and $\nu, \ldots \mu, \lambda, L, M$ are each of them known functions of a, inasmuch as $A, B, \ldots C, G$ are known functions of a.

Thus the roots of any equation whereof the coefficients are functions of only one parameter may be expressed in terms of algebraic, circular, or logarithmic functions, and of integrals of algebraic functions. These integrals depend upon the quantity M.

To a form involving only one parameter, Mr. Jerrard has shown that the general quintic may be reduced. Its resolvent sextic may also be reduced to the same form.

Mr. Jerrard's memorable discoveries also show that the general sextic may be regarded as involving two parameters only. The general sextic leads us to the consideration of the equation

$$\delta f x = F x \cdot \delta a + f x \cdot \delta b = 0,$$

where a and b are the independent parameters, of which the co-efficients may be considered as functions, and δ is the characteristic of the Calculus of Variations.

If, in art. 62 of my " Observations," &c. (vol. xviii. p. 342), we take the suffixes of θ to the modulus 6, the equations become

$$\theta_1 \theta_4 + \theta_2 \theta_0 + \theta_3 \theta_5 = \gamma_5 = r(x_5),$$
$$\theta_1 \theta_0 + \theta_2 \theta_5 + \theta_3 \theta_4 = \gamma_1 = r(x_1),$$
$$\theta_1 \theta_3 + \theta_2 \theta_4 + \theta_5 \theta_0 = \gamma_2 = r(x_2),$$
$$\theta_1 \theta_5 + \theta_2 \theta_3 + \theta_4 \theta_0 = \gamma_3 = r(x_3),$$
$$\theta_1 \theta_2 + \theta_3 \theta_0 + \theta_4 \theta_5 = \gamma_4 = r(x_4).$$

And this system has a certain relation to the formula

$$\theta_{a^2} \theta_{a^2+1} + \theta_{a^2+2} \theta_{a^2+3} + \theta_{a^2+3} \theta_{a^2+4},$$

which, taking the suffixes to the modulus 6, is, for all integral values of a, equal either to γ_1 or to γ_4. But all the values of γ are not thence evolved; and in order to obtain a convenient representation of the system, I avail myself of certain cyclical forms which may be given to it when one of the roots is supposed to become fixed.

The form here used will admit of an exceedingly simple representation if, throughout my "Observations," we replace θ_3 by θ_4, and *vice versâ*. This requires that in art. 48 (vol. xviii. p. 52) we write

$$\Theta_{f(14)}=\Theta'_3, \quad \Theta_{f(13)}=\Theta'_4,$$

a change of definition which I shall accordingly suppose to be made.

Further, I shall suppose that we replace θ_6 by θ_i, where i is an imaginary suffix defined by the congruence

$$i+a \equiv i \text{ (mod. 5)},$$

a being an integer. Or, if we agree to regard the infinite suffix ∞ as satisfying the congruence

$$\infty + a \equiv \infty \text{ (mod. 5)},$$

we may replace θ_6 by θ_∞. Lastly, I shall suppose the suffixes, after these changes, to be taken to the modulus 5.

The changes being made, it will be found that all the functions γ are deducible from the expression

$$\theta_i\theta_a+\theta_{a+1}\theta_{a+4}+\theta_{a+2}\theta_{a+3}$$

by writing, successively, 0, 1, 2, 3, 4 for a. In fact we have

$$\theta_i\theta_0+\theta_1\theta_4+\theta_2\theta_3=\gamma_2=r(x_2),$$
$$\theta_i\theta_1+\theta_2\theta_0+\theta_3\theta_4=\gamma_1=r(x_1),$$
$$\theta_i\theta_2+\theta_3\theta_1+\theta_4\theta_0=\gamma_0=r(x_0),$$
$$\theta_i\theta_3+\theta_4\theta_2+\theta_0\theta_1=\gamma_3=r(x_3),$$
$$\theta_i\theta_4+\theta_0\theta_3+\theta_1\theta_2=\gamma_4=r(x_4);$$

and if, in these equations, we change

$$\theta_i, \quad \theta_0, \quad \theta_3, \quad \theta_4, \quad \gamma_0, \quad x_0$$

into

$$\theta_6, \quad \theta_5, \quad \theta_4, \quad \theta_3, \quad \gamma_5, \quad x_5$$

respectively, we shall be reconducted to the system of art. 62 of my "Observations." More extensive changes in our fundamental formulæ and definitions would enable us to express the system with a greater concinnity between the suffixes of θ and those of x, and provided that, on the right of the last system, we interchange x_0 and x_2 the system may be deduced from the equation

$$\theta_i\theta_a+\theta_{a+1}\theta_{a+4}+\theta_{a+2}\theta_{a+3}=\gamma_a=r(x_a).$$

If, in the expression

$$\theta_i\theta_a+\theta_{a+1}\theta_{a+3}+\theta_{a+2}\theta_{a+4},$$

we make a equal to 0, 1, 2, 3, 4 successively, we find the following relations between it and the α and β functions of art. 70 of my "Observations" (vol. xviii. p. 508), viz. :—

$$\theta_i\theta_0 + \theta_1\theta_3 + \theta_2\theta_4 = \alpha_0,$$
$$\theta_i\theta_1 + \theta_2\theta_4 + \theta_3\theta_0 = \beta_1,$$
$$\theta_i\theta_2 + \theta_3\theta_0 + \theta_4\theta_1 = \beta_2,$$
$$\theta_i\theta_3 + \theta_4\theta_1 + \theta_0\theta_2 = \alpha_2,$$
$$\theta_i\theta_4 + \theta_0\theta_2 + \theta_1\theta_3 = \beta_0;$$

and in like manner from *

$$\theta_i\theta_s + \theta_{s+1}\theta_{s+2} + \theta_{s+3}\theta_{s+4}$$

we deduce

$$\theta_i\theta_0 + \theta_1\theta_2 + \theta_3\theta_4 = \alpha_4,$$
$$\theta_i\theta_1 + \theta_2\theta_3 + \theta_4\theta_0 = \alpha_1,$$
$$\theta_i\theta_2 + \theta_3\theta_4 + \theta_0\theta_1 = \beta_3,$$
$$\theta_i\theta_3 + \theta_4\theta_0 + \theta_1\theta_2 = \beta_4,$$
$$\theta_i\theta_4 + \theta_0\theta_1 + \theta_2\theta_3 = \alpha_3.$$

A contemplation of these systems of equations seems to lead to the conclusion that a certain equation of the fifteenth degree, which Mr. Jerrard supposes to be capable of decomposition into five cubics, is not irreducible, but composed of a quintic and a 10-ic factor.

I cannot think that Abel's conclusion is at all shaken by the researches of Mr. Jerrard. Of his cardinal proposition, the equation

$$\Xi_f - r\{P_{f(\beta_i)}\} = 0$$

of art. 106 of his 'Essay,' Mr. Jerrard gives no proof. Directions to compare (ab) and (ac) are insufficient instructions for attaining a result fraught with difficulties so serious. For reasons already assigned, I believe the proposition to be erroneous, and incapable of proof. And with it the whole argument of Mr. Jerrard falls.

I need not the warning from my own oversight to restrain me from dwelling unduly upon what I believe to be an error of Mr. Jerrard's. But, with every recognition of his great claims to

* Mr. Harley has completely determined these α and β functions; and I regret that he has not as yet published the whole of his investigations on quintics. The proposition of Mr. Jerrard, which Mr. Jerrard supposes that Mr. Harley has ignored, is, when considered in reference to processes which do not involve transformation to a soluble form, rather an axiom than a theorem. It may be stated thus : If Θ be equal to Θ', it is not in general equal to Θ''.

I would here add the expression of a hope that Mr. Harley, to whom I have communicated the above results on the theory of transcendental roots, may soon publish some developments of them to which he has been led.

the gratitude of mathematicians, it is scarcely possible to ignore the fact that Mr. Jerrard's hope (expressed at the conclusion of his paper of 1845), to discuss the resolution of the trinomial equation $x^5 + A_4 x + A_5 = 0$, has not been realized, and that little or no approach has yet been made towards its realization. Mr. Jerrard's subsequent researches on quintics seem to me, for reasons already adduced, to enhance rather than dispel any difficulties which arise upon the paper in question. It is perhaps to be desired that the mathematical world should be made acquainted with the whole of Mr. Jerrard's views on this important subject.

Does an absolutely impossible, or rootless, equation exist? MM. Terquem and Gilain have discussed this question in the *Nouvelles Annales de Mathématiques**, with reference to the equation

$$+ \sqrt{1+x} + \sqrt{1-x} = 1.$$

But this equation does not in reality raise the question under consideration. For (as I had occasion to write to Mr. Harley during last autumn) every one of the congeneric equations

$$\pm \sqrt{1+x} \pm \sqrt{1-x} = 1$$

is soluble. And some one of the four values of x given by

$$x = \pm (\pm 1)^2 \tfrac{1}{2} \sqrt{3}$$

will satisfy any one of the above four congeners. I shall therefore again (S. 3, vol, xxxvii. p. 281) have recourse, for illustration, to the equations

$$1 + \sqrt{x-4} + \sqrt{x-1} = 0,$$
$$1 - \sqrt{x-4} - \sqrt{x-1} = 0,$$

each of which must, I think, be deemed impossible or rootless.

The only gleam of a solution of the last is, so far as I can see, one which springs from the assumptions

$$\ldots x = 4(+1)^2 + (-1)^2,$$
$$4 = 4(+1)^2, \; 1 = (-1)^2,$$

while, for the first, we have the system

$$x = 4(-1)^2 + (+1)^2,$$
$$4 = 4(-1)^2, \; 1 = (+1)^2.$$

But how can that be called a solution which depends upon a modification of the constants of a problem (compare S. 4. vol. iii. p. 439)? The safer conclusion seems to be that the two equations are rootless.

Midland Circuit, at Lincoln,
    March 15, 1861.

* See Mr. Wilkinson's *Notæ Mathematicæ*, Mechanics' Magazine, vol. lxii. p. 582.

LIX. *Proceedings of Learned Societies.*

ROYAL SOCIETY.

[Continued from p. 233.]

May 24, 1860.—Sir Benjamin C. Brodie, Bart., Pres., in the Chair.

THE following communications were read :—
"On a new Method of Approximation applicable to Elliptic and Ultra-elliptic Functions." By C. W. Merrifield, Esq.

"On the Lunar-diurnal Variation of Magnetic Declination at the Magnetic Equator." By John Allan Broun, F.R.S., Director of the Trevandrum Observatory.

This variation, first obtained by M. Kreil, next by myself, and afterwards by General Sabine, presents several anomalies which require careful consideration, and especially a careful examination of the methods employed to obtain the results. The law obtained seems to vary from place to place even in the same hemisphere and in the same latitude, and this to such an extent, that, for example, when the moon is on the inferior meridian at Toronto it produces a minimum of westerly declination, while for the moon on the inferior meridian of Prague and Makerstoun in Scotland it produces a maximum of westerly declination. No two places have as yet given exactly the same result; though the result for each place has been confirmed by the discussion of different periods.

In order to obtain the lunar diurnal action, it has been usual to consider the magnetic declination at any time' as depending on the sun's and moon's hour-angles and on irregular causes. Thus, if at conjunction, H_0 be the variation due to the sun on the meridian, and h_0 be that due to the moon on the meridian, H_1 the variation for the sun at 1^h, h_1 for the moon on the meridian of 1^h, and so on; it is supposed that we may represent the variations for a series of days by the following expressions, where the nearest values of h to the whole hour-angles are given :—

1st day. $\quad H'_0 + h'_0 + x'_0, \quad H'_1 + h'_1 + x'_1 \ldots H'_{23} + h'_{23} + x'_{23}$

2nd day. $\quad H''_0 + h''_{23} + x''_0, \quad H''_1 + h''_0 + x''_1 \ldots H''_{23} + h''_{21} + x''_{23}$

$\vdots$

nth day. $\quad H^n_0 + h^n_1 + x^n_0, \quad H^n_1 + h^n_2 + x^n_1 \ldots H^n_{23} + h^n_{23} + x^n_{23},$

where x is due to irregular causes, and n is the number of days in a lunation nearly.

Summing these quantities we have approximately,

$$\Sigma H_0 + \Sigma^0_{23} h + \Sigma x_0, \quad \Sigma H_1 + \Sigma^0_{23} h + \Sigma x_1, \ldots \Sigma H_{23} + \Sigma^0_{23} h + \Sigma x_{23} \quad \text{(A)}$$

and the means are,

$$H_0 + \text{C} + \frac{\Sigma x_0}{n}, \quad H_1 + \text{C} + \frac{\Sigma x_1}{n}, \ldots H_{23} + \text{C} + \frac{\Sigma x_{23}}{n} \ldots \ldots \text{(B)}$$

Here the hourly means are affected by the constant due to the total action of the moon on all the meridians, and by variables depending on disturbing causes. If, on the other hand, we arrange the series as follows,

$$H'_0 + h'_0 + x'_0, \qquad H'_1 + h'_1 + x'_1, \ldots \quad H''_0 + h''_{23} + x''_0$$
$$H''_1 + h''_0 + x''_1, \qquad H''_2 + h'_1 + x'_2, \ldots \quad H'''_1 + h'''_{23} + x'''_1$$
$$\vdots \qquad\qquad \vdots \qquad\qquad \vdots$$
$$H^{n-1}_{23} + h^{n-1}_0 + x^{n-1}_{23}, \quad H^n_0 + h^n_1 + x^n_0, \ldots \quad H^n_{23} + h^n_{23} + x^n_{23}.$$

Summing these quantities we have,

$$\Sigma^0_{23} H + \Sigma h_0 + \Sigma^0_{23} x^{(1)}, \; \Sigma^0_{23} H + \Sigma h + \Sigma^0_{23} x^{(11)} \ldots \Sigma^0_{23} H + \Sigma h_{23} + \Sigma^0_{23} x^{(24)} \; \text{(C)}$$

and for the means,

$$\Theta + h_0 + \frac{\Sigma x^{(1)}}{n-1}, \; \Theta + h_1 + \frac{\Sigma x^{(11)}}{n-1} \ldots \Theta + h_{23} + \frac{\Sigma x^{(24)}}{n-1}. \quad \ldots \text{(D)}$$

In this case Θ is the mean of $n-1$ observations, of which 24 give the true means for the total solar influence, and the remaining $n-25$ being equally distributed through the hour-angles also give the mean approximately.

Instead, however, of combining the observations in this way, the following method has been preferred. Let, in the quantities (B),

$$(H_0) = H_0 + \mathfrak{C} + \frac{\Sigma x_0}{n}$$

$$(H_1) = H_1 + \mathfrak{C} + \frac{\Sigma x_1}{n}.$$

Then
$$H'_0 + h'_0 + x'_0 - (H_0) = h'_0 + (x'_0) = d'_0$$
$$H''_1 + h''_0 + x''_1 - (H_1) = h''_0 + (x''_1) = d''_0$$
$$\vdots \qquad \vdots \qquad \vdots$$
$$H^{n-1}_{23} + h^{n-1}_0 + x^{n-1}_{23} - (H_{23}) = h^{n-1}_0 + (x^{n-1}_{23}) = d^{n-1}_0.$$

Summing the last two columns, we have

$$\frac{\Sigma d_0}{n-1} = h_1 + \frac{\Sigma^0_{23}(x')}{n-1}.$$

Similarly we obtain

$$\frac{\Sigma d_1}{n-1} = h_1 + \frac{\Sigma^0_{23}(x'')}{n-1}, \text{ and so on.}$$

It will be observed that in these summations there are two assumptions; one, that the lunar diurnal law is constant throughout the lunation, or series of lunations, for which the means are obtained; or that the quantity $\mathfrak{C}$ in the expressions (B) is constant. If this be not exact, then the quantity $\frac{\Sigma x}{n}$ will contain the variation due to this cause, and depend in part on the lunar hour-angle; so that the mean (H) which is employed in taking the differences will eliminate part of the lunar action and partially distort the law. The other assumption is that the mean solar diurnal variation, represented by $(H_0),(H_1)$ ...., is nearly constant throughout the period; for, if not, the differences due to such changes might be sufficient to mask any lunar law, the latter having a small range compared with the former.

Also it should be remarked that the means h_0, h_1, &c. are combined with the irregular effect $\dfrac{\Sigma_{23}^0(x)}{n-1}$. This effect, as far as it is due to disturbance, we know obeys a solar diurnal law; and if independent of lunar action, a sufficiently large series of observations might suffice to eliminate it, as combining with and forming part of the regular solar diurnal variation. If, however, the series is not very large and the irregular disturbance considerable compared with the variation sought, it may be desirable to omit or modify the marked irregularities.

As regards the first assumption referred to above, the results obtained hitherto seem to show the error to be small, and the only way to determine its amount will be to consider it zero in the first instance, and thereafter a more accurate calculus may be employed. For the second assumption, it is certain that the solar diurnal law varies considerably in some cases within a lunation. At the magnetic equator, for example, the law of magnetic declination is inverted within a few weeks near the equinoxes. The attempt to correct the error due to considerable change in the solar diurnal variation by taking the means, as has been done, from shorter periods than a lunation, is liable to the serious objection that the resulting hourly means are affected unequally by the lunar action, so that the sums (A) take the form,

$$\Sigma H_0 + \Sigma_p^0 h + \Sigma x_0, \quad \Sigma H_1 + \Sigma_q^1 h + \Sigma x_1 \ldots \ldots \Sigma H_{23} + \Sigma^{22} h + \Sigma x_{23},$$

where the second term in each expression is a variable. In the discussion to which I am about to allude, the following plan has been followed. The hourly means for the following series of weeks were taken, namely—

> m_1 from 1st, 2nd, 3rd, and 4th weeks of the year.
> m_2 „ 2nd, 3rd, 4th, and 5th „ „
> m_3 „ 3rd, 4th, 5th, and 6th „ „.
> ⋮ ⋮

The means of m_1 and m_2 were then taken as normals for the 3rd or middle week, of m_2 and m_3 as normals for the 4th week, and so on: these means were then employed for the differences from the corresponding hourly observations of the weeks to which they belonged.

With reference to the irregular effect, it is evidently desirable that we should know in the first instance whether it may not be a function of the lunar, as well as of the solar, hour-angle; for this end it is essential in the first instance to obtain the result including all the supposed irregular actions, and afterwards to eliminate these in the best manner possible.

In the discussion of the Makerstoun Observations I had substituted for certain observations, which gave differences from the mean beyond a fixed limit, values derived by interpolation from pre-.

ceding and succeeding observations. General Sabine in his discussions
has rejected wholly the observations which exceeded the limit chosen
by him. The omission of observations accidentally or intentionally,
and the taking of means without any attempt to supply the omitted
observations by approximate values, require consideration.

Let m be the true hourly mean for an hour h, derived from the
complete series of n observations; let m' be the mean derived from
$n-1$ observations, one observation o being accidentally lost; then

$$m' = \frac{nm - o}{n-1},$$

$$m = m' - \frac{m' - o}{n} = m' - \frac{m - o}{n-1}.$$

If, however, we supply the omitted observation by an interpolation
between the preceding and succeeding observations, and if the interpolated value be $o + x$, we have

$$m'' = \frac{nm + x}{n},$$

$$m = m'' - \frac{x}{n}.$$

The comparative errors of m' and m'' are therefore

$$\frac{o - m}{n-1} \text{ and } \frac{x}{n}.$$

We may for any given class of observation determine the mean values
of these errors.

Example :—At Hobarton, in July 1846, the mean barometer
for 3^h (Hobarton mean time) was 29·848 in., and the mean difference of an observation at that hour from the mean for the hour was
0·403 in.; if an observation had been omitted with such a difference,
or for which $o - m = 0.403$ in., we should have an error in the resulting
mean of $\frac{0.403}{25} = 0.016$ in., and the error might have been twice as
great had the observation with the greatest difference been rejected.
If we now seek the error of m'', where the observation is interpolated,
we shall find for the same month that the mean value of $x = 0.005$ in.
nearly; whence the error $\frac{x}{n} = \frac{0.005}{26} = 0.0002$ in. only, and the error
would never exceed 0·001 in. A similar though less advantageous
result will be found in all classes of hourly observations.

In the case where observations are rejected which differ from the
mean for the corresponding hour more than a given quantity, let
us suppose, to simplify the question, that the sums of $n-1$ out of
n observations for each of two successive hours are each equal M,
and that the observations for the same hours of the nth day are
respectively $m' + l$ and $m' + l + x$, where $m' = \frac{M}{n-1}$, l is the limit
beyond which observations are rejected, and x is the excess of the
observation to be omitted. The means retaining all the observations

are,

$$m_1 = \frac{M + m' + l}{n} = m' + \frac{l}{n},$$

$$m_2 = \frac{M + m' + l + x}{n} = m' + \frac{l + x}{n};$$

but if we reject the observation $m' + l + x$, we have

$$m_2' = \frac{M}{n-1} = m'.$$

It is assumed that $m_1' - m_2' = 0$ (any other hypothesis of variation would give the same final result), and therefore the error of the change from the first hour to the second, when all the observations are retained, is $\frac{x}{n}$; but if the observation be rejected, the change is

$$m' + \frac{l}{n} - m' = \frac{l}{n}.$$

This error, therefore, will be greater than the other if $l > x$; so that the error in the resulting change from one hour to the next will be less by retaining an observation than by rejecting it, if the difference from the preceding observation be not greater than the difference from the hourly mean; that this will most frequently be the case will be obvious from the following fact:—At Makerstoun, in 1844, at 1 A.M. the number of observations which exceeded the monthly means by 3' and less than double that, or 6', was 99, while the whole number which exceeded by more than 6' was only 16.

It will be evident also that the difference l of an observation from the corresponding hourly mean may not be due to irregular causes, or to causes which affect the changes from one hour to the next in a perceptible manner, but to gradual and regular daily change. If we examine the *daily* means most free from irregular or intermittent disturbance, we shall find that they vary plus or minus of the monthly mean; if the difference amounts to l in any case, then the whole observations of the day may be rejected though they follow the normal law. By taking a proper value of l this case may not happen frequently, but cases like the following will. At Hobarton the daily means of magnetic declination differ in some months from the monthly means by 2'·0 nearly; as the limit chosen by General Sabine is 2'·4, any observation in such days differing by 0'·4 from the normal mean would be rejected. The 25th and 26th days of March 1844 had been chosen by me as days free from magnetic disturbance, and following the normal law at Makerstoun (Mak. Obs. 1844, p. 339), yet the means of horizontal force for these days differed 0·00064 and 0·00075 from the monthly means; had the former quantity been the limit, all the observations on these days might have been rejected.

Altogether it appears to me that the method of rejecting observations beyond certain limits should not be employed at all, or if employed, only when interpolated observations are substituted; and·

that this interpolation should constitute a second part of the discussion, the first including all the observations*.

These considerations may appear somewhat elementary, but it is essential that results which present such anomalies as the lunar diurnal variation of magnetic declination should be obtained in a manner the most free from objection, even though the objections should touch on quantities of a second order compared with those obtained.

The discussion of which I now proceed to note the results, includes all the hourly observations without exception, made in the Trevandrum Observatory (within a degree and a half of the magnetic equator) during the five years 1854 to 1858; the second part of the discussion, in which days of great magnetic irregularity have been wholly rejected, not being completed, I shall reserve the details for a more formal communication to the Royal Society. The results obtained are as follows :—

1st. At the magnetic equator the lunar diurnal law of magnetic declination varies with the moon's declination and with the sun's declination.

2nd. This variation is so considerable that the attempt to combine all the observations to form the mean law for the *year* gives results that are not true for any period. .Hence evidently the impossibility of relating the laws at different places. The so-called *mean* law for the year at Trevandrum obtained for the moon furthest north, on the equator going south, furthest south, and on the equator going north, consists of *three* maxima and *three* minima,—a result wholly false, excepting as an arithmetical operation due to combination of very different laws.

3rd. The lunar diurnal law varies chiefly with the position of the *sun*, the variation being comparatively small with the position of the *moon*.

4th. At the magnetic equator the range of the variations is *markedly* greatest in the months of January, February, November and December, or about perihelion.

The following results are derived after grouping the means for different positions of the moon in periods of six months, October to March, and April to September; they are therefore, for the reason given in the 2nd conclusion, not quite accurate; but the change of the law from month to month will be followed when the details are presented to the Society. The following will give a general idea of the changes :—

5th. *When the moon is furthest north.*

 a. About perihelion. The lunar diurnal law of magnetic declination consists of two *maxima*† when the moon is near the upper and lower meridians, the maximum for the latter being much the great-

---

* I should note here my belief that a peculiarity noticed by General Sabine in his discussions as requiring explanation, namely, that the excursions of the declination needle east and west in the lunar diurnal variation have very different magnitudes, is due to the rejection of observations, while the means by which the differences were obtained included the rejected quantities.

† The declination is easterly at Trevandrum, and the maxima indicate greater easterly declination.

est; of the two minima at intermediate epochs, that for the setting moon is the most marked.

b. About aphelion. The law consists of two nearly equal *minima* near the upper and lower transits: of the two intermediate maxima, that near the moonset is the most marked.

c. Thus the law about the winter solstice is inverted about the summer solstice, and the one law passes into the other at the epochs of the equinoxes, *exactly as for the solar diurnal variation.*

6th. *For the moon on the equator going south.*

a. About perihelion. The lunar diurnal law consists of two *nearly equal maxima* near the superior and inferior transits: of the two intermediate minima, the moonset minimum is by far the most marked.

b. About aphelion. The law consists of two *nearly equal minima* near the superior and inferior transits: of the two intermediate maxima, that near moonrise is by far the most marked.

c. In this case also the laws for the solstices are the opposite of each other, and the one law passes into the other near the epochs of the equinoxes.

7th. *For the moon furthest south.*

a. About perihelion. The lunar diurnal law consists of maxima near the upper and lower transits, that at the upper transit being by far the most marked: of the intermediate minima, that near moonset is the greater.

b. About aphelion. The law consists of two minima, the most marked at the inferior transit, the other about three hours before the superior transit; and of two equal maxima, one near moonrise, the other near the superior transit, but varying little till 3 hours before the inferior passage.

c. In this instance the inversion is not so complete as in the other cases; this, it is believed, will be found to be due to the fact that the change from one law to the other takes place after the vernal and before the autumnal equinox; so that in the means for six months, from which the above conclusions are drawn, the lunations following the law *a* are combined with those belonging to *b.*

8th. *The moon on the equator going north.*

a. About perihelion. The lunar diurnal law consists of two nearly equal maxima when the moon is near the superior and inferior meridians; of the two intermediate minima, that near moonrise is by far the most marked.

b. About aphelion. The law consists of two minima at the inferior and superior transits; and of two maxima, the greatest at moonset, the other between the meridians of 16^h and 21^h; between these points there is an inflexion constituting a slight minimum.

c. In this case also the opposition of the laws is sufficiently well marked; the only divergence from opposition being that due to the minor minimum about the meridian of 19^h, due, it is believed, as noted 7th *c*, to the partial combination of opposite laws in the aphelion half-year.

9th. It will be observed that the variations of the law with refer-

ence to the moon's declination *for any given period of the year*, consists chiefly in the difference of the relative values of the maxima and
minima, the differences of epochs being small.   Thus for perihelion,
the moon furthest north, the principal maximum occurs at the inferior passage ; the moon on the equator going south, the two maxima
are nearly equal ; the moon furthest south, the maximum at the
superior passage is by far the greatest : on the equator going north,
the two maxima are again nearly equal ; and so on for other epochs.

10th. The moon's action is chiefly, if not wholly, dependent on
the position of the sun, or (which is the same thing) on the position
of the earth relatively to the sun ; and the law of the lunar action at
the magnetic equator resembles in some points that for the solar
action at the same epochs. Thus about aphelion there is a *minimum*
of easterly (maximum of westerly) declination produced by the lunar
action, as well as by the solar action, for these two bodies near the
superior meridian ; whereas about perihelion both actions for the
sun and moon near the superior meridian produce *maxima* of easterly
declination. A like analogy holds for near the epochs of sunrise and moonrise.

June 14.—General Sabine, R.A., Treasurer and Vice-President, in
the Chair.

The following communication was read :—
"On the Nature of the Light emitted by heated Tourmaline." By
Balfour Stewart, Esq., M.A.

Some months ago I had the honour of submitting to the Royal
Society a paper on the light radiated by heated bodies, in which it
was endeavoured to explain the facts recorded by an extension of
the theory of exchanges.

Having mentioned the difficulty which I had in maintaining the
various transparent substances at a nearly steady red heat for a
sufficient length of time in experiments demanding a dark background, Professor Stokes suggested an apparatus by means of which
this difficulty might be overcome ; and it is owing to his kindness in
doing so that I have been enabled to lay these results before the
Society.

The apparatus consists of a thick, spherical, cast-iron bomb, about
5 inches in external and 3 inches in internal diameter—the thickness
of the shell being therefore 1 inch. It has a cover removeable at
pleasure.   There is a small stand in the inside, upon which the substance under examination is placed, and when so placed it is precisely at the centre of the bomb.   Two small round holes, opposite
to one another, viz. at the two extremities of a diameter, are bored in
the substance of the shell.   If, therefore, the substance placed upon
the stand be transparent, and have parallel surfaces, by placing these
surfaces so as to front the holes, we are enabled to see through the
substance, and consequently through the bomb,   Let the bomb with
the substance on the stand be heated to a good red heat, and then
withdrawn from the fire and allowed to cool.   It is evident that the
cooling of the substance on the stand will proceed very slowly, as it

is almost completely surrounded with a red-hot enclosure. It is also evident that, by placing the bomb in a dark room, we may view the transparent substance against a dark background. By this method of experimenting, therefore, the difficulty above alluded to is overcome.

Before describing the experiment performed on tourmaline, it may be well to state what result the theory of exchanges would lead us to expect when this mineral is heated, and we shall perceive at the same time the importance of the experiment with tourmaline as a test of the theory. When a suitable piece of tourmaline, with its faces cut parallel to the axis, is used to transmit ordinary light, the light which it transmits is nearly completely polarized, the plane of polarization depending on the position of the axis. The reason of this is, that if we resolve the incident light into two portions, one of which consists of light polarized in a plane perpendicular to the axis of the crystal, and the other of light polarized in a plane parallel to the same axis, nearly all the latter is absorbed, while a notable proportion of the former is allowed to pass.

Suppose now that such a piece of tourmaline is placed in a red-hot enclosure; the theory of exchanges, when fully carried out, demands that the light transmitted by the tourmaline, say in a direction perpendicular to its surface, *plus* the light radiated by the tourmaline in that direction, *plus* the small quantity of light reflected by the surface of the tourmaline in that direction, shall together equal in quantity and quality that which would have proceeded in the same direction from the wall of the enclosure alone, supposing the tourmaline to have been removed. Let us neglect the small quantity of light which is reflected from the surface of the tourmaline, and, standing in front of it, analyse with our polariscope the light which proceeds from it. This light consists of two portions, the transmitted and the radiated, both of which together ought to be equal in quality and intensity to that which would reach our polariscope from the enclosure alone were the tourmaline taken away. But the light which would fall on our polariscope from the enclosure alone would not be polarized; hence the whole body of light which falls upon it from the tourmaline, and which is similar in quality to the former, ought not to be polarized. Now part of this light, or that which is transmitted by the tourmaline, is polarized; hence it follows, in order that the whole be without polarization, that the light which is radiated should be partially polarized in a direction at right angles to that which is transmitted.

Another way of stating this conclusion is this. The light which the tourmaline radiates is equal to that which it absorbs, and this equality holds separately for light polarized in a plane parallel to the axis of the crystal, and for light polarized in a plane perpendicular to the same.

The experiment was made with a piece of brown tourmaline having a few opake streaks, procured from Mr. Barker of Lambeth. It was placed in a graphite frame between two circular holes made as above described in opposite sides of the bomb, the diameter of the holes being about $\frac{2}{10}$ths of an inch. On looking in at one of these holes you could thus see through the tourmaline and the opposite hole, or, in other words, see quite through the bomb. An arrangement was also made by which part of the tourmaline might be viewed with the graphite behind it.

The apparatus thus prepared was heated to a red or yellow heat in the fire, placed on a brick in a dark room, and the tourmaline viewed by a polariscope which Mr. Gassiot kindly lent me. The following was the appearance of the experiment :—

Without the polariscope the transparent parts of the tourmaline were slightly less radiant than the field around them. When the polariscope was used, the light from the transparent portions of the tourmaline was found to vary in intensity as the instrument was turned round. No change of intensity could be observed in the light radiated by the opake streaks of the tourmaline, or by the graphite.

The light from the transparent portions was therefore partially polarized. The polariscope was then brought to its darkest position, and a light from behind allowed to pass through the tourmaline. The light was distinctly visible in this position, but by turning round the polariscope about 90° it became eclipsed. The mean of four sets of experiments made the difference between the position of darkness for the two cases 88½°. It appears, therefore, that the light radiated by the tourmaline was partially polarized in a plane at right angles to that which was transmitted by it. It was also ascertained that the light from the tourmaline which had the graphite behind it gave no trace of polarization.

LX. *Intelligence and Miscellaneous Articles.*

ON THE MOTION OF THE STRINGS OF A VIOLIN.
BY PROFESSOR H. HELMHOLTZ.

I HAVE been studying for some time the causes of the different qualities of sound ; and as I found that those differences depended principally upon the number and intensity of the harmonic sounds accompanying the fundamental one, I was obliged to investigate the forms of elastic vibrations performed by different sounding bodies. Among such vibrations, the form of which is not yet exactly known, the vibrations of strings excited by the bow of a violin are peculiarly interesting. Th. Young describes them as very irregular; but I suppose that his assertion relates only to the motions which remain after the impulse of the bow has ceased. At least, I myself found the motion very regular as long as the bow is applied near one end of the string, in the regular way commonly followed by players of the violin. I used a method of observing very similar to that of Lissajoux. Already, without the assistance of any instruments, one can see easily that a string moved by the bow vibrates in one plane only—the same plane in which the string itself and the hairs of the bow are situated. This plane was horizontal in my experiments. The string was powdered with starch, and strongly illuminated. One of the little grains of starch, looking like a bright point, was observed by a vertical microscope, the object lens of which was fixed to one of the branches of a tuning-fork. The fork, making 120 vibrations in the second, was placed between the branches of a horseshoe electro-magnet, which was magnetized by an interrupted electric current, the number of interruptions being itself 120 in the second. In that way the fork was kept vibrating for as long a time as I desired. The lens of the microscope vibrated in a direction parallel to the string, and therefore perpendicular to its vibrations. The string I used was the second string of a violin, answering to the note A, tuned a little higher than common, to 480 vibrations, and therefore it performed four vibrations for every one of the tuning-fork. Looking through the microscope, I observed the grain of starch describing an illuminated curved line, the horizontal abscissæ of which corresponded to the deviations of the tuning-fork, and the vertical ordinates to the deviations of the string. I found it a matter of importance to use a violin of most perfect construction, and I was fortunate in getting a very fine instrument of Guadanini for these experiments. On the common instruments of inferior quality I could not keep the curve constante nough for numbering the little indentures which I shall describe afterwards, although the general character of the curve was the same on all the instruments I tried : the curve used to move by jerks along the line of abscissæ ; and every jerk was accompanied by a scratching noise of the bow. On the contrary, with the Italian instrument, and after some practice, I got a curve completely quiescent as long as the bow moved in one direction, the sound being very pure and free from scratching.

We may consider the motion of the string as being compounded of two different sets of vibrations, the first of which is the principal motion as to magnitude. Its period is equal to the period of the fundamental sound of the string, and it is independent of the situation of the point where the bow is applied. The second motion produces only very small indentures of the curve. Its period of vibration answers to one of the higher harmonics of the string. It is known that a string, when producing only one of its higher harmonics, is divided into several vibrating divisions of equal length, being separated by quiescent points, which are called nodes. In all the nodes of the second motion of the string in the compound result at present considered, the principal motion appears alone ; and also in the other points of the string the indentures corresponding to the second motion are easily obliterated, if the line of light is too broad.

The principal motion of the string is such that every point of it goes to one side with a constant velocity, and returns to the other side with another constant velocity.

Plate V. fig. 7 represents four such vibrations, corresponding to one vibration of the fork. The horizontal abscissæ are proportional to the time, the vertical ordinates to the deviation of the vibrating point. Every vibration is formed on the curve by two straight lines. The curve is not seen quite in the same way through the microscope, because there the horizontal abscissæ are not proportional to the time but to the sine of the time. It must be imagined that the curve (fig. 7) is wound up round a cylinder, so that the two ends of it meet together, and that the whole is seen in perspective from a great distance; thus it had the real appearance of the curve, as represented in two different positions in fig. 8. If the number of vibrations of the string is accurately equal to four times the number of the tuning-fork, the curve appears quietly keeping the same position. If there is, on the contrary, a little difference of tuning, it looks as if the cylinder rotated slowly about its axis, and by the motion of the curve the observer gets as lively an impression of a cylindrical surface, on which it seems to be drawn, as if looking at a stereoscopic picture. The same impression may be produced by combining, stereoscopically, the two diagrams of fig. 8.

We learn, therefore, by these experiments,—

1. That the strings of a violin, when struck by the bow, vibrate in one plane.

2. That every point of the string moves to and fro with two constant velocities.

These two data are sufficient for finding the complete equation of the motion of the whole string. It is the following;—

$$y = A\Sigma \left\{ \frac{1}{n^2}\sin\left(\frac{\pi n x}{l}\right)\sin\left(\frac{2\pi n t}{T}\right) \right\} \dots\dots\dots (1)$$

y is the deviation of the point whose distance from one end of the string is x; l, the length of the string; t, the time; T, the duration of one vibration; A, an arbitrary constant; and n, any whole number; and all values of the expression under the sign Σ, got in that way, are to be summed.

A comprehensive idea of the motion represented by this equation may be given in the following way:—Let $a\,b$, fig. 9, be the equilibrium position of the string. During the vibration its forms will be similar to $a\,c\,b$, compounded of two straight lines, $a\,c$ and $c\,b$, intersecting in the point c. Let this point of intersection move with a constant velocity along two flat circular arcs, lying symmetrically on the two sides of the string, and passing through its ends, as represented in fig. 9. A motion the same as the actual motion of the whole string is thus given.

As for the motion of every single point, it may be deduced from equation (1), that the two parts $a\,b$ and $b\,c$ (fig. 7) of the time of every vibration are proportional to the two parts of the string which are separated by the observed point. The two velocities of course are inversely proportional to the times $a\,b$ and $b\,c$. In that half of the string which is touched by the bow, the smaller velocity has the same direction as the bow; in the other half of the string it has the contrary direction. By comparing the velocity of the bow with the velocity of the point touched by it, I found that this point of the string adheres fast to the bow and partakes in its motion during the time $a\,b$, then is torn off and jumps back to its first position during the time $b\,c$, till the bow again gets hold of it.

With these principal vibrations smaller vibrations are compounded, the nature of which I can define accurately only in the case where the bow touches a point whose distance from the nearer end of the string is $\frac{1}{5}, \frac{1}{6}, \frac{1}{7}$, &c. of its whole length, or generally $\frac{1}{m}$, if m is a whole number. Because the point where the bow is applied is not moved by any vibration belonging to the mth, $2m$th, &c. harmonic, it is quite indifferent for the motion of that point, and for the impulses exerted by the bow upon the string, whether vibrations corresponding to the mth harmonic exist or not. Th. Young has proved that if we excite the vibrations of a string by bending it with the finger, as in the harp, or hit it with a single stroke, as in the piano, in the ensuing motion all those harmonics are wanting which have a node in the touched point. I therefore concluded also that the bow cannot excite those harmonics which have a node at the point where it is applied; and I found, indeed, that if this point is distant $\frac{1}{m}$ from the end, the ear does not hear the mth harmonic sound, although it distinguishes very well all the other harmonics. Therefore, in the equation (1), all those members of the sum will be wanting in which n is equal to m, or $2m$, or $3m$, &c. These members, taken together, constitute a vibration of the string with m vibrating divisions. Every such division performs the same form of vibration we have described as the principal vibration of the whole string. These small vibrations must be subtracted from the principal vibration of the whole string for getting its actual vibration. Curves constructed according to this theoretical view represent very well the really observed curves.

If $m = 6$ and the observed point is distant $\frac{1}{12}$ from the other end of the string, the motion is represented in fig. 10. Near the end of the string, where the bow is commonly applied by players, the nodes of different harmonics are very near to each other, so that the bow is nearly always at, or at least very near to, the place of a node. Striking in the middle between two nodes, I could not get a curve sufficiently constant for my observations. If I strike very near the end, the sound changes often between the fundamental and the second or third harmonic, which is indicated by gradual corresponding alterations of the curve.—*From the Proceedings of the Glasgow Philosophical Society* for Dec. 19, 1860.

ON CLAIRAUT'S THEOREM. BY PROF. HENNESSY, F.R.S.

Laplace has shown that this theorem follows, whatever may be the density of the interior parts of the earth, provided it consists of similar concentric strata, and that the form of the outer stratum is ellipsoidal. In the 'Philosophical Transactions' for 1826, Mr. Airy (the present

Astronomer Royal of England) has presented an equivalent result; more recently, Professor Stokes has shown that we can deduce the law of variation of terrestrial gravity without any hypothesis whatsoever as to the earth's interior structure. He assumes merely that its surface is spheroidal, and that the equation of fluid equilibrium holds good at that surface. In vol. vi. of the 'Cambridge Mathematical Journal,' Professor Haughton presented a demonstration, founded upon the same assumptions as those of Professor Stokes, and in which he uses certain propositions relative to attractions which had been enunciated by Gauss and Maccullagh. While studying the labours of those mathematicians, it appeared to me that the question could be entirely divested of the hydrostatical character, and that Clairaut's theorem may be directly deduced from the equations to the normal of any closed surface, without any considerations as to the physical condition of the matter forming that surface. Thus every surface concentric with the earth, and perpendicular to gravity, will possess the property of exhibiting this relation in the intensity of gravity at its various points.

Let X, Y, Z represent the components parallel to the rectangular axes of the forces by which a point is retained at rest on a given surface whose equation is L=0. Then from the equations of the normal we have

$$Y\frac{dL}{dx}-X\frac{dL}{dy}=0,\quad Z\frac{dL}{dx}-X\frac{dL}{dz}=0,$$

when the resultant of these forces is perpendicular to the given surface. If we represent by V the potential of the earth on the particle in question, by w the angular velocity of rotation, we have

$$X=\frac{dV}{dx}+w^2x,\quad Y=\frac{dV}{dy}+w^2y,\quad Z=\frac{dV}{dz},$$

and the above equations become

$$\frac{dV}{dy}\frac{dL}{dx}-\frac{dV}{dx}\frac{dL}{dy}=w^2\left(x\frac{dL}{dy}-y\frac{dL}{dx}\right),$$

$$\frac{dV}{dz}\frac{dL}{dx}-\frac{dV}{dx}\frac{dL}{dz}=w^2\,x\frac{dL}{dz}.$$

If, in conformity with General Schubert's[*] recent determinations, we assume the earth's surface to be that of an ellipsoid, with three unequal axes, we should substitute for L

$$\frac{x^2}{a^2}+\frac{y^2}{b^2}+\frac{z^2}{c^2}-1=0,$$

or

$$\frac{dL}{dx}=\frac{2x}{a^2},\quad \frac{dL}{dy}=\frac{2y}{b^2},\quad \frac{dL}{dz}=\frac{2z}{c^2};$$

[*] *Mémoires de l'Académie Impériale des Sciences de St. Pétersbourg*, sér. 7. tom. i.

whence we have

$$b^2 x \frac{dV}{dy} - a^2 y \frac{dV}{dx} = w^2 xy(a^2 - b^2), \quad c^2 x \frac{dV}{dz} - a^2 z \frac{dV}{dx} = w^2 a^2 xz.$$

Each of these partial differential equations can be easily integrated; and the value of V, finally obtained, is equivalent to the equation of fluid equilibrium, or

$$V + \frac{w^2}{2}(x^2 + y^2) = C.$$

Let θ represent the complement of the latitude, and ϕ the longitude, counted from the meridian of the greatest axis, then

$$x = r \cos \theta, \quad x = r \sin \theta \cos \phi, \quad y = r \sin \theta \sin \phi,$$

and

$$V + \frac{r^2 w^2}{2} \sin^2 \theta = C.$$

In the case of an ellipsoid having the ellipticity e, we have, neglecting small terms,

$$r = a (1 - e \cos^2 \theta).$$

From these equations, and from the properties of Laplace's functions into which V can be expanded, an expression can be obtained of the same kind as that deduced by Professor Stokes from his own and Gauss's theorems relative to attractions.—*Proceedings of the Royal Irish Academy*, Feb. 25, 1861.

ON A METHOD OF TAKING VAPOUR-DENSITIES AT LOW TEMPERATURES. BY DR. LYON PLAYFAIR, C.B., F.R.S., AND J. A. WANK-LYN, F.R.S.E.

The authors refer to Regnault's experiments, which have shown that aqueous vapour in the atmosphere has the same vapour-density at ordinary temperatures as aqueous vapour above 100° C.; and they bring forward fresh experiments upon alcohol and ether to show that when mixed with hydrogen these vapours preserve their normal density at 20° or 30° C. below the boiling-points of the liquids, and infer generally that vapours, when partially saturating a permanent gas, retain their normal densities at low temperatures.

From their researches the authors deduce the consequence—remarkable, but quite in harmony with theory—that permanent gases have the property of rendering vapour truly gaseous. Stated in more precise terms, the proposition maintained by the authors is, "The presence of a permanent gas affects a vapour, so that its expansion-coefficient at temperatures near its point of liquefaction tends to approximate to its expansion-coefficient at the highest temperatures."

The authors anticipate that admixture with a permanent gas may serve as a kind of reagent to distinguish between cases of unusually high expansion-coefficient in a vapour, and cases where chemical alteration takes place. It will also be possible, by the employment of a permanent gas, to obtain vapour-densities of compounds which will not bear boiling without undergoing decomposition.

In experimenting upon substances which may be heated above the boiling-point, the authors employ Gay-Lussac's process for taking the specific gravity of vapours. A slight modification, however, is necessary. Previous to the introduction of the bulb containing the weighed substance, dry hydrogen is introduced into the graduated tube and measured with all the precautions belonging to a gas analysis. It will be obvious that in the subsequent calculation the volume of hydrogen corrected at standard temperature and pressure must be subtracted from the volume of mixed gas and vapour, also corrected at standard temperature and pressure.

When the substance will not bear heating to its boiling-point, the authors employ a process resembling that of Dumas in principle, but differing very widely from it in detail. Dumas's flask with drawn-out neck is replaced by two bulbs, together of about 300 cub. cent. capacity, joined by a neck, and terminating on either side in a narrow tube. One of the narrow tubes has some very small dilatations blown upon it (*b*), the other is merely bent (D). (See Plate V. fig. 6.) The apparatus, whose weight should not exceed 70 grms., is weighed in dry air, then placed in a bath, being secured by a retort-holder grasping the neck joining the large bulbs C and C. The end A, projecting over the one side of the bath, is made to communicate with a hydrogen apparatus; the end D passes through a hole in the opposite side of the bath, which is plugged up water-tight by means of putty. Dry hydrogen is transmitted through the whole arrangement, and escapes at D through a long narrow tube joined to it by a caoutchouc connecter.

The bath is next filled with warm water until the bends *a* and *a* are covered. The connexion with the hydrogen apparatus is then for a moment interrupted, to allow of the introduction of a small quantity of the substance at A. The substance, which should not more than half-fill the small bulb *b*, is partially vaporized in the stream of hydrogen, and in that state passes into the part C C. All the while the temperature of the bath is kept uniform throughout by constant stirring, and made to rise very slowly. When within a few degrees of the temperature at which the determination is to be made, the current of hydrogen is almost stopped, so that the bulbs C and C may contain less vapour than will fully saturate the gas at the temperature of sealing. The water of the bath is then made to subside, by opening a large tap placed near the bottom. The bends *a* and *a* are thus exposed, the bulbs C C remaining covered. Immediately the current of hydrogen has been stopped, the flame is applied at *a a*, so as to seal the apparatus hermetically. The temperature of the bath, as well as the height of the barometer, must now be observed.

After being cleaned, the apparatus (which now consists of three portions, viz. the portion C C hermetically sealed and the two ends *b* and D) must be weighed.

The capacity of the apparatus is found by filling it completely with water and weighing; but previously to this operation the volume of hydrogen enclosed at the time of sealing must be found. On breaking one extremity under water, the water will rise in the bulbs, and, after a while, will have absorbed all the vapour, but will leave the hydrogen. The bulbs must then be lifted out of the water, without altering their temperature, and, with the water that has entered, weighed. The difference between the latter weighing and the weight of the bulbs quite full of water gives the weight in grammes, which expresses in cubic centimeters the volume of hydrogen enclosed; the pressure is the height of the barometer minus the column of water which had entered the bulbs; the temperature is that of the water.

An example of a determination of the vapour-density of alcohol at 30° C. below its boiling point is subjoined :—

Height of the barometer (at 0° C.)	763·09 millims.
Temperature of the balance case	7°·5 C.
Weight of apparatus in dry air..........	69·959 grms.
Temperature at time of sealing..........	48° C.
Weight of apparatus+hydrogen+vapour..	69·5275 grms.
Weight of apparatus+water (at 5°·2 C.)..	191·76 grms.
Weight of apparatus filled with water....	545·36 grms.
Height of water column	122 millims.

From which is deduced—

Volumes corrected at 0° C. and 760 millims. pressure, cubic centimeters.			Grm,
Hydrogen+vapour	406·43	weighing	0·1695
Hydrogen...............	341·27	„	0·0306
	65·16		0·1389

Therefore, 65·16 cub. cent. of alcohol-vapour weigh ·1389
but 65·16 cub. cent. of air weigh.......... ·0843

$$\text{Vapour-density of alcohol} = \frac{\cdot 1389}{\cdot 0843} = 1 \cdot 648.$$

The authors have extended their experiments to acetic acid and other substances. At low temperatures the vapour-density of acetic acid approximates to 4·00, no matter how much hydrogen be employed. At higher temperatures an approximation to 2·00 is obtained, but without heating so high as Cahours found necessary.

The authors are continuing these researches.—*From the Proceedings of the Royal Society of Edinburgh,* January 21, 1861.

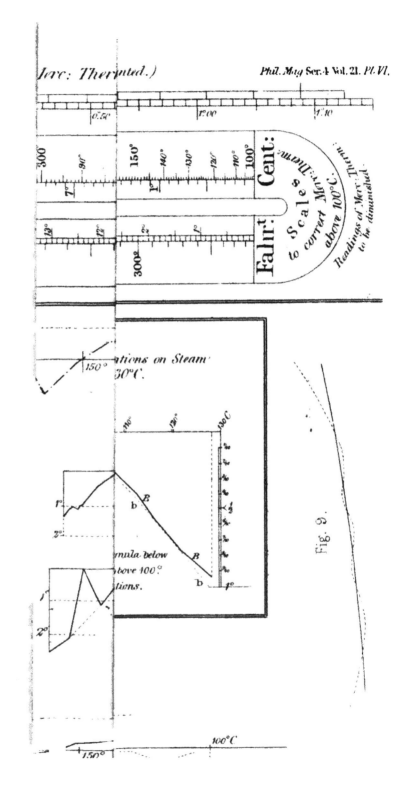

Fig. 9.

THE
LONDON, EDINBURGH AND DUBLIN
PHILOSOPHICAL MAGAZINE
AND
JOURNAL OF SCIENCE.

[FOURTH SERIES.]

JUNE 1861.

LXI. *On a Law of Liquid Expansion that connects the Volume of a Liquid with its Temperature and with the Density of its saturated Vapour.* By JOHN JAMES WATERSTON*, *Esq.*

[With a Plate.]

§ 1. IN the archives of the Royal Society for 1852, there is an account of observations on the density of liquids and their superjacent vapours at high temperatures, made in sealed graduated tubes filled with the same liquid in different proportions of their volumes. The general law of density in saturated vapours deduced from these observations, and from the various observations of vapour-tension already published by other physicists, is therein set forth, with the assistance of a chart (No. 2 chart) in which the observations are all projected, and lines drawn, that enables the eye to judge of the accordance between theory and observation. (See Note A.)

An account of this general law is also given in the Philosophical Magazine for March 1858, in a paper entitled "On the Evidence of a Graduated Difference between the Thermometers of Air and Mercury between 0° and 100° C."

By the same mode of observing in sealed graduated tubes, I afterwards extended the observations up to the transition-point of three of the liquids, viz. alcohol, ether, and sulphate of carbon, and found that the law of vapour-density was maintained in them without deviation to their extreme limiting temperatures. As these extensive series of observations supplied the curves of expansion of the three liquids, I have occasionally tried by means of them to discover a general law of liquid expansion. I submit the following account of the last attempt of this kind, as it appears to be successful.

* Communicated by the Author.

§ 2. The curves of expansion, drawn to a large and distinct scale, were examined by the following graphical process:—At four or five points nearly equidistant, tangents were drawn (carefully judging of the direction to be given to the straight edge by the sweep of the curve to the right and left of the point of contact). Thus were obtained several values of the quotient of the differential of volume by the volume or proportionate differentials for constant element of temperature $\left[\dfrac{dv}{dt \cdot v}\right]$. These were set off as ordinates to the temperatures, and the curve drawn through the points appeared to be the common equilateral hyperbola, having one asymptote coinciding with the axis of temperature and the other perpendicular to it, and intersecting it at a temperature $[\gamma]$ that evidently was above the transition-point. If this were the case, the product of the coordinates to each point of the curve, reckoning from the point γ as origin, ought to be constant $[=\rho]$; and accordingly it was found that when the inverse of the quotients were projected as ordinates to the temperatures, the points ranged in a straight line, which being produced, cut the axis in the point γ. The differential equation is thus

$$\frac{dv}{dt} \cdot \frac{\gamma - t}{v} = \rho,$$

the integration of which is

$$v = \left\{ \frac{k}{\gamma - t} \right\}^{\rho},$$

in which $k = (\gamma - t)$ when $v = 1$.   (See Note B.)

§ 3. There is a relation between this expression and that for saturated vapour-density which seems to prove that it is not empirical, but the true exponent of the physical condition of the molecules of a body in the liquid state. The following is a statement of it.

In the papers above referred to, there will be found an account of the law of saturated vapour-density, and the proofs on which it rests. It is expressed by the equation

$$\left\{ \frac{t - g}{h} \right\}^{6} = \mathrm{D}.$$

If we put $\Delta = \dfrac{1}{v}$, the law of liquid density is

$$\left\{ \frac{\gamma - t}{k} \right\}^{\rho} = \Delta.$$

On comparing the constants for different liquids, I find as a general rule that the quotient $\dfrac{h}{\rho}$ is a constant quantity $[\mathrm{E}]$.

Thus for Alcohol my observations give E=1717
„ Ether „ „ E=1789
„ Sulphate of carbon „ E=1725

M. Muncke's observations on sulphuric acid from 50° to 230° C. E=1600 to 1800. (See Note C.) MM. Dulong and Petit's observations on the expansion of mercury between 0° and 300° C. give $\rho = \dfrac{1}{1\cdot489}$. The line passing through M. Avogadro's observations on the vapour of mercury from 230° to 300° C. gives $h=1160$. These combined give E=1727. I have adopted 1717 as the nearest probable value at present, because the most labour was bestowed on the alcohol series of observations.

These values of E are derived from English measures of pressure and temperature, viz. inches of mercury and degrees of Fahrenheit scale. The value of this constant derived from French measures is F=504·44, which corresponds with E=1717. The ratio of reduction is $a = \dfrac{5}{9}\left\{ \dfrac{\frac{4}{5} \times 29\cdot922}{760} \right\}^{\frac{1}{4}}$, so that Ea=F,

$$\left[\begin{array}{l} \log F = 2\cdot70282 \\ \log \dfrac{1}{a} = 0\cdot53195 \end{array} \right]$$

§ 4. M. Regnault's observations on the tension of the vapour of mercury from low temperatures up to 200° C., respond to the same value of h as those at the higher temperatures by M. Avogadro, but with g augmented 12 degrees, showing a boiling-point 12 degrees higher. It is remarkable that M. Regnault's observations on the expansion of liquid mercury differs so far from those of MM. Dulong and Petit as to be represented with $\rho = \frac{4}{5}$ and E=870. At 300° C. this difference amounts to about the equivalent of $3\frac{1}{4}$° C. The acceleration of the rate of expansion is in M. Regnault's observations only one-half what is shown by those of MM. Dulong and Petit. (See Note D.) This is a remarkable discrepancy, both being so eminent in this class of observations.

§ 5. In all cases I have worked with temperatures reduced to the air-thermometer by scales of correction computed from the formula given in Appendix III. to the paper in the Philosophical Magazine for March 1858 above referred to.

The formula is founded on MM. Dulong and Petit's observations. I annex exact tracings of these scales (Plate VI.). In two cases (petroleum and sulphuric acid) the temperatures were taken uncorrected and compared with the results when corrected. In both, the differences between theory and observation were less when the temperatures were corrected.

It is not absolutely necessary to correct the temperatures in order to recognize these laws of density, and for practical purposes may not be required; but it seems best to accustom ourselves to do so, in order to be prepared for the recognition of any other relations of harmony that may exist in the thermo-molecular physics of different bodies.

§ 6. Water is, as might be expected, an exception to the law of liquid density, as it is to the law of capillarity and compressibility (see papers by M. Grassi and M. Simon in the *Annales de Chimie*). I have traced its curve of expansion by observations in sealed tubes up to 210° C. air-thermometer (see Note E), and projected the densities to the value of ρ required by its vapour-gradient; also those of M. Despretz from 0° to 100° C.; but they do not conform to the line required at any point of its range even at the highest temperature. These abnormal features in this first of liquids have had a prejudicial effect on the progress of science in this department. There is no other liquid as yet found with such point of maximum density that remains a liquid under its maximum; yet such a point seems invariably to be sought for. M. Muncke and M. Pierre have bestowed much unavailing labour on this question. (See Note F.)

§ 7. To determine the constants of these two equations for the density of the liquid and of its vapour, not more than four exact observations are strictly required; two of the vapour, and two of the liquid.

If the series of observations on the dilatation of a liquid extend over a considerable range of temperature, and have had their inequalities equalized by graphical processes equivalent to weighing by the method of least squares, the three constants of the equation may be directly determined.

Thus let t_0, t_1, t_2 be the three temperatures, and v_0, v_1, v_2 the corresponding volumes observed, to find ρ and k we have

$$\frac{t_1-t_0}{\left(\frac{1}{v_0}\right)^{\frac{1}{\rho}} - \left(\frac{1}{v_1}\right)^{\frac{1}{\rho}}} = \frac{t_2-t_0}{\left(\frac{1}{v_0}\right)^{\frac{1}{\rho}} - \left(\frac{1}{v_2}\right)^{\frac{1}{\rho}}} = k,$$

which may be solved by trial and error. But few observations as yet published will stand this test, the range of temperature being too small, and the irregularities proportionably too great.

§ 8. A simple and satisfactory way to test both of these laws of density by published observations, is to take two of the vapour-tensions not far from the boiling-point and compute the value of h. *Ex. gr.*, let e_0, e_1 be the two observations of the pressure of vapour in contact with its generating liquid; T_0, T_1 the corresponding temperatures by air-thermometer reckoned from the zero of gaseous tension [$-461°$ F. or $-273°·89$ C.]; then we have

$$h = \frac{T_1 - T_0}{\left(\dfrac{e_1}{T_1}\right)^t - \left(\dfrac{e_0}{T_0}\right)^t};$$

and since $\rho = \dfrac{h}{E}$, we obtain the index of the power of the density of the liquid, which being set off as ordinates to the temperature, ought to range in a straight line; and this line produced, cuts the axis of temperature at γ. If, when these points are connected by distinct lines, a general convexity in the range can be discovered, viewing it foreshortened with the eye close to the plane of projection, then we may infer that ρ requires to de diminished if the convexity is directed upwards from the axis of temperature, and *vice versâ*.

§ 9. Having found ρ and k and γ, the next step is to compute the values of t from the volumes by the equation, and tabulate the differences between the computed and observed temperatures. This will be found attended with but little additional labour. If we now project these differences as ordinates on an exaggerated scale to the temperatures, we obtain a distinct impression of how far theory and observation accord.

§ 10. Mercury and alcohol being the most important liquids for thermometric purposes, may serve as examples of the mode of computation.

I. *Mercury.*

M. Avogadro's observations on the tension of the vapour of mercury:—At $260°$ C. the observed tension was $133·62$ millims., at $290°$ it was $252·51$ millims. The correction to reduce the temperatures to the air-thermometer from the scale is $5°·60$ at $260°$, and $6°·91$ at $290°$. Hence—

$$260 - 5·60 + 273·89 = 528·29 = T_0, \quad 133·62 \text{ millims.} = e_0,$$
$$290 - 6·91 + 273·89 = 556·98 = T_1, \quad 252·51 \text{ millims.} = e_1;$$

and we arrive by computation at $\log h = 2·54796$ and $g = 247°·45$.

[I have computed the temperatures for M. Avogadro's other six observations. The computed, minus the observed, is, at

$$300 = +0·15$$
$$290 = 0$$
$$280 = +0·42$$
$$270 = -0·26$$
$$260 = 0$$
$$250 = -0·26$$
$$240 = -1·90$$
$$230 = -4·56$$

= difference in tension amounting to one-third inch mercury.]

This value of h being derived from French measures, is to be applied to F to find $\frac{1}{\rho}$, which thus comes out 1·4284.

The following are MM. Dulong and Petit's volumes of mercury computed with this index:—

Temp. Cent. Air.	Volumes.	Inverse of volumes raised to the power $\frac{1}{\rho}=1\cdot4284.$	First differences.	Second differences.
0	1·000000	1·000000		
100	1·0180180	·974815	·025185	·000028
200	1·0368664	·949602	·025213	·000023
300	1·0566037	·924366	·025236	

The second differences indicate a slight convexity in the line upwards. This shows that $\frac{1}{\rho}$ requires augmentation. The following is the result of computing the same observations with $\frac{1}{\rho} = 1\cdot489$:—

		First diff.	Second diff.
0	1·000000		
100	·973760	·026240	·000001
200	·947521	·026239	·000005
300	·921287	·026234	

To find γ, we have $1 - 0\cdot947521 : 200° :: 1 : 3811°\cdot05 = \gamma$, and $\left\{\dfrac{3811\cdot05 - t}{3811\cdot05}\right\}^{\frac{1}{1\cdot489}} = \Delta = \dfrac{1}{v}$, or $(3811\cdot05 - t)v^{1\cdot489} = 3811\cdot05$.

This expresses MM. Dulong and Petit's observations with a difference at 100° amounting to $\frac{1}{1000}$th of a degree, and at 300° the difference is $\frac{1}{30}$th of a degree.

§ 11. To bring the D of the vapour formula to the same standard as the Δ of the liquid formula, it is requisite to change the value of h in the one and k in the other, so that the weight in grains of a cubic inch of either may be indicated.

Let $\eta = \cdot0216216 =$ weight in grains of a cubic inch of hydrogen at the temperature 0° C. and pressure 760 millims.

$\delta =$ vapour-density of the body on the hydrogen scale.

$\tau =$ temperature (reckoned from the zero of gaseous tension) at which the pressure of the saturated vapour is 760 millims.

$\mu =$ required factor

$$\left\{\frac{\tau - g}{\mu h}\right\}^6 = \frac{273\cdot89}{\tau} \cdot \eta\delta = \beta^6.$$

By the formula,

$$\tau = 247°·45,$$
$$g = 611°·28,$$

also $\delta = 101$; hence $\mu\lambda = \dfrac{\tau - g}{A}$, and log $\mu\lambda = 2·56247$. Thus we obtain

$$\left\{ \frac{T - 247·45}{[2·56247]} \right\}^6 = \omega,$$

the general expression for the weight in grains of a cubic inch of the saturated vapour of mercury at T° (C. A. G.), temperature reckoned by Centigrade Air-thermometer from the sero of Gaseous tension.

§ 12. The weight of a cubic inch of mercury at 0° is 8754·4 grains, hence

$$\frac{3754·4}{v} = \left\{ \frac{3811·05 - (T - 278·89)}{M \times 3811·05} \right\}^{\frac{1}{1·409}};$$

and $M = \left(\dfrac{1}{3754·4} \right)^{1·409}$; hence

$$\left\{ \frac{3811·05 - (T - 273·89)}{[8·25857]} \right\}^{\frac{1}{1·409}} = W,$$

the weight of a cubic inch of mercury in grains at T° (C. A. G. temperature).

We may thus find the temperature at which $\omega = W$, or that at which liquid and vapour would be of equal density if the laws were maintained.

With alcohol, ether, and sulphate of carbon, transition occurs a few degrees below the theoretical temperature of equal density.

II. *Alcohol.*

§ 13. M. Regnault's observed tensions of saturated vapour :—
At 40° C. the tension is 134·1 millims.; at 70°, 589·2 millims.

$$40 + 0·48 + 273·89 = 314·37 \text{ C. A. G. } = T_0$$
$$70 + 0·41 + 273·89 = 344·30 \qquad\qquad = T_1$$

$$\log h = 2·15392$$
$$g = 190·70$$
$$\frac{1}{\rho} = \frac{F}{h} = 3·5392$$

This gives 78°·83 C. A. as boiling-point at pressure 760 millims. M. Pierre's observations on the expansion of alcohol were made on a specimen that boiled at 78°·63 C. A. under pressure 758 millims., and its specific gravity at 0° was 0·8151.

The following are his second series of observations computed with the above index, 3·5892 :—

	C.	C. A.	V.	$\left(\dfrac{1}{V}\right)^{3\cdot5392}$	Computed temp. minus observed temp.
(1)	$33\cdot46 + 0\cdot47 = 33\cdot93$		$1\cdot03714$	$\cdot87890$	$+0\cdot21$
(2)	$47\cdot52 + 0\cdot51 = 48\cdot03$		$1\cdot05356$	$\cdot83140$	0
(3)	$50\cdot33 + 0\cdot51 = 50\cdot84$		$1\cdot05676$	$\cdot82248$	$-0\cdot19$
(4)	$56\cdot26 + 0\cdot50 = 56\cdot76$		$1\cdot06416$	$\cdot80245$	$-0\cdot25$
(5)	$60\cdot41 + 0\cdot48 = 60\cdot89$		$1\cdot06989$	$\cdot78736$	$+0\cdot03$
(6)	$78\cdot70 + 0\cdot38 = 74\cdot08$		$1\cdot08780$	$\cdot74238$	0
(7)	$76\cdot73 + 0\cdot35 = 77\cdot08$		$1\cdot09168$	$\cdot78310$	$-0\cdot27$

The computed densities ($\cdot87890$, &c.) set off as ordinates to the temperatures, show a trend without any appearance of curvature. The straight line seems to pass exactly through the points of the second and fifth and sixth. Assuming it to pass through the second and sixth, we have

$$\cdot83140 - \cdot74238 = \cdot08902 : 74°\cdot08 - 48°\cdot03 = 26°\cdot05 : : \cdot74238 : 217\cdot24,$$

and $74\cdot08 + 217\cdot24 = 291°\cdot32 = \gamma$. This line gives unity volume at $-1°\cdot30$, hence $k = 292\cdot62 = \gamma - t$ when volume equal unity, and the equation is $(291°\cdot32 - t)v^{3\cdot539} = 292\cdot62$. The differences between the observed temperatures and those computed from the observed volumes by this equation are given in the last column.

§ 14. This equation answers well to the observations of M. Pierre above $10°$; also to those of M. Muncke (St. Petersburgh Memoirs) above the same temperature; but in both the trend of the points below this lies in a line inclined to that of the equation. The divergence is the greatest in M. Pierre's. In neither is it a general convexity, but distinctly the contour shows two lines diverging from about $15°$ to $20°$ C. This is most distinct in M. Muncke's observations. I have computed them by the above equation, and tabulated the differences between the observed and computed temperatures, which are set off in Pl. VI. fig. 3 as ordinates to the temperatures on ten times the natural scale. Above these, in fig. 2, M. Pierre's differences are set off to the same scale, and in fig. 4 the differences in my series of observations on alcohol, described in the paper above referred to as being in the archives of the Royal Society. This alcohol was not absolute; it had 19 per cent. water, and the index of its power derived from its line of vapour-density was $3\cdot60$; also $\gamma = 290\cdot89$, $k = 209\cdot81$, and its equation (the volume being reckoned as unity at the boiling point)

$$(290°\cdot89 - t)\ v^{3\cdot60} = 209°\cdot81\ \text{C. A. G.}$$

§ 15. The deflection in M. Pierre and M. Muncke's observations, it will be remarked, occurs in those below atmospheric tem-

peratures, where the reduction of temperature had to be artificially produced by mixtures of broken ice and muriate of lime, and
it represents the temperature of the mercury to be higher than
that of the alcohol. This is precisely what took place in some
observations I made on the contraction of ether about 20° below
the atmospheric temperature. A similar deflection, but in a
greater degree, appears in the ether observations of the same
authors. They are projected in figs. 5 and 6 to the same scale
as the others. (See Note G.)

The application of cold to maintain a constant temperature is
by no means under the same command as the application of
heat; and, besides, conductibility is very much reduced at low
temperatures. There is an evident dislocation, the law of continuity is broken, but it is at the part of the scale where the *mode*
of obervation underwent a change. I submit, therefore, that
the verdict should be against the observations at the lower temperatures, not against the law of expansion, which, if in fault,
would cause the trend of the points to have a general curvature
throughout the range.

In judging of the evidence afforded by these graphical projections, it should be kept in view that the vertical scale magnifies
the amount of the differences tenfold. The accordance of theory
with observation is in some cases remarkable. Thus, for 40° C.,.
Muncke's alcohol and Pierre's ether do not show a difference
greater than one-sixth of a degree. We have also to keep in
mind that the power $\frac{1}{\rho}$ that reduces the densities to a straight
trend, is not arbitrarily assumed to suit a particular series of observations, but that it is determined *à priori* from the vapour.
If we take any other value of $\frac{1}{\rho}$ but that which is thus determined, the graphical projection of the computed densities shows
a general bend. If $\frac{1}{\rho}$ is too great, the bow is turned downwards;
if too small, the bow is turned upwards. The string of the bow
only makes its appearance when the value of $\frac{1}{\rho}$ is that deduced
from the gradient of vapour-density above described.

§ 16. The time has not perhaps yet arrived for deducing these
laws of density from the dynamical theory of heat; but if we are
ever to arrive at a conception of the true ultimate nature of molecular force, it seems clear that the inductive path of least difficult
approach (if not the only one) is that which sets out from the study
of the gaseous state, and proceeds by way of that of the equili-
brated condition of saturated vapours in communication with their

generating liquids, to the molecular condition of the liquids, where the dynamic condition of the chemical element is constrained by the cohesive force; and the struggle in which this dynamic force is gradually subdued by the increase of temperature is, as now ascertained, represented by one quantitative relation throughout, that seems to indicate a certain simplicity in the ultimate recondite principle on which molecular force is based.

  · We know that the physics of gases conform to the physics of media that consist of perfectly free elastic projectiles*. Their free concourse and perfectly elastic recoil determines the resolution of their *vis viva* into the six rectangular directions of space; and it is this number that probably fixes the ratio of the proportionate increment of density in a saturated vapour to the corresponding proportionate increment of temperature reckoned from the fixed limit g. But the *absolute* increment of density corresponding to constant increment of temperature differs in different vapours, being ruled by a gradient, the sixth root of which has a constant ratio to the index of density of the generating liquid expressed as a function of the temperature.

  · The next step that seems within reach, if we had a few more observations to work from, is the discovery of the relation which no doubt exists between the increase of volume and decrease of latent heat or capillarity regarded as the integral of cohesion. The density and the capillarity both diminish as the temperature rises. (See Note H.) If there is a simple law of quantitative relation between them, its discovery would supply all that is now wanting to bring the dynamical theory of heat to bear upon the molecular physics of liquids.

Notes.

Note A. § 1.—The title of the paper is " On a General Law of Density in Saturated Vapours," illustrated by Chart No. 2. In the Philosophical Transactions for 1852 there is a paper with the same title, illustrated by a Chart No. 1. (This paper was originally sent to the British Association.) In Chart No. 1 the sixth root of density is laid off as ordinate to the square root of the temperature reckoned from the zero of gaseous tension. In Chart No. 2 the sixth root of density is laid off as ordinate to the temperatures simply. In Chart No. 1 the lines appear straight at the upper part of their course, but with an increasing flexure at the lower part of the range convex to axis of temperature. Also there is no relation of harmony apparent between them. In No. 2 Chart (of which a tracing is to be found in the archives of the Royal Society for 1852–53) the lines are straight

* See paper " On the Physics of Media that consist of perfectly elastic Molecules in a state of Motion," in the archives of the Royal Society, 1845–46.

throughout, and relations of paralellism appear; also several radiate from the same point in the axis of temperature, showing that, as a general law, *vapours in contact with their generating liquids have, at the same temperature, or at the same constant difference of temperature, densities that have a constant ratio.*

Note B. § 2.—This mode of graphical analysis seems a natural mode of operating when a law of nature has to be unmasked. If the proportionate differentials of volume had been laid off as ordinates, not to the temperature, but to the volume, the result would be the logarithmic curve, which might not be so easy to recognize with only a few points to lead from. We must be guided in the selection of the coordinate axis by the causal relation of dependent phenomena. Heat being, as it were, the instrument of action in molecular physics, claims the preference as a standard by which to measure the proportionate differentials, and to which other variables may be referred to as coordinate axis. As an example, the following is the analysis of the law of saturated vapours by this process.

The vapour-tensions being divided respectively by the corresponding temperatures reckoned from the zero of gaseous tension, the quotients represent densities of saturated vapour. Setting off these quotients as ordinates to the temperatures, we next draw the curve, and equalize the irregularities as far as possible; then take off the ordinates of the finished curve at equal intervals of temperature, say $10°$ or $5°$; next take the differences of adjacent ordinates and divide each by the intermediate ordinate. These quotients, $\frac{dD}{dtD}$, are to be laid off as ordinates to the temperatures. The points appear to range in a conic hyperbola, having the axis of temperature as an asymptote. This conjecture is to be tested by laying off the inverse of these quotients as ordinates to the temperatures. The conjecture is confirmed by the points ranging in a straight line which cuts the axis at a certain temperature g. Hence $(t-g)f = \frac{dtD}{dD}$ and $\frac{dD}{D} = \frac{dt}{f(t-g)}$. The integration of this gives $D = (t-g)^{\frac{1}{f}} \times \frac{1}{H}$, in which $H = (t-g)^{\frac{1}{f}}$ when $D =$ unity; or let $h^f = H$, then $D^f = \left(\frac{t-g}{h}\right)$. Comparing the value of f in different vapours, it is found to be constant for all and equal to $\frac{1}{4}$.

As another example, but unconnected with heat, we may inquire as to the possibility of ascertaining the law of gravitation from the changes in the moon's apparent size and motion, its actual distance and the earth's radius being supposed unknown, but assuming that the difficulty caused by an unknown parallax and augmentation of diameter might be evaded by taking lunar distances at equal altitudes on both sides of the meridian.

By observations on consecutive nights, while the diameter is increasing or diminishing at the maximum rate, we might obtain two angular velocities and two measurements of diameter; hence the proportionate differential of the diameter and the correction to be applied to the angular velocities to reduce them to the same radius. We might thus obtain the velocity, the increment of velocity, and increment of distance expressed in terms of r, the radial distance of the moon from the earth at the middle epoch. Now since $\frac{dr}{r^2} = 2vdv$, if we had these same quantities for different values of $\frac{1}{D}$ or r, and projected the different values of $\sqrt{\frac{1}{2vdv}}$ as ordinates to the corresponding values of r, the points would converge in a straight line to the zero of r; and if an approximate parallax was obtained, the point corresponding to the value of $2vdv$ at the earth's surface would fit in and confirm the propriety of the projection. If it is a question what should direct us to this particular projection, it might be answered the increment of square velocity is a square quantity, and the inverse form of function is applicable to a power depending on distance.

Note C. § 3.—The longest series of observations on the expansion of a liquid that I have met with is that of M. Muncke, on sulphuric acid from $-30°$ to $+230°$ C. I have been enabled to put them to the test by the following equation for the tension of its vapour, viz.
$\left(\frac{t-354\cdot7}{1288}\right)^6 t = p$ (English measures). In this the value of $g = 354\cdot7$ is assumed to be the same as that for steam, and for the vapours of several hydrates of sulphuric acid observed by M. Regnault, and referred to in § 1. of paper in the Philosophical Magazine for March 1858. The value of $h(=1288)$ is derived from the boiling-point. The value of $\frac{1}{\rho} = \frac{E}{h}$ is thus $1\cdot33$. The inverse volumes being computed to this power, and laid off as ordinates to the temperatures, were found to range well in a straight line above 30°. The line drawn through 45° and 220° is expressed by the equation

$$(1433°\cdot2 - t)v^{\frac{3}{4}} = 1436°\cdot1 \text{ C. A.}$$

The differences between the temperatures computed from the volumes by this equation and those observed are laid off in fig. 1 (Pl. VI.) as ordinates to the temperatures, the scale vertical to horizontal being 10 to 1.

The law of continuity is evidently broken at about 40°, the deflection being similar to the other cases referred to in §§ 14, 15, and probably due to the same cause.

Fig. 7 (Plate VI.) represents the differences of Muncke's observations on petroleum projected in the same way. The equation is

$$(489°\cdot5 - t)\, v^{2\cdot14} = 489°\cdot5 \text{ C.,}$$

in which $\frac{1}{\rho} = 2\cdot14$ has been deduced from Ure's observations on the vapour of petroleum. It is unlikely that the liquids were exactly the same. A slight convexity directed upwards is apparent in the trend of the points, which a small augmentation in $\frac{1}{\rho}$ would correct.

Note D. § 4.—M. Regnault stands so high as an authority, that an error in his observations that can be clearly demonstrated from internal evidence is of importance to science. Erroneous observations from eminent observers are serious obstacles to progress, as are unsound deductions from eminent men of science. They are weeds difficult to root up, and the attempt to do so is a task so ungracious and so irreverent as to incur every discouragement.

The projection of M. Regnault's observations on the tension of steam above and below 100° C. is given in fig. 8. The dotted line represents the empirical formula which had to be altered at 100°, the point at which the method employed in making the observations was changed. They are projected with temperatures uncorrected, in the manner described in § 1 of paper in the Philosophical Magazine for March 1858, and they are orthographically foreshortened as described in the latter part of § 4. See also § 6, and Appendix I. of the same paper.

The question to ask ourselves when looking at the figure 8 is, Do the points conform to the law of continuity? Is their trend not clearly broken at 100°? To put a series of observations to the test of this law can always be done, but it is attended with considerable labour, and seems to require a speciality different from that which characterizes the eminent observer and experimentalist.

Note E. § 6.—The following are those observations in series along with those of M. Despretz, both equalized graphically by elaborate processes, and the temperatures corrected and reduced to the air-thermometer. My observations from 100° up to 212° F. agree so well with M. Despretz's, that those at the higher temperatures will, I think, be found nearly correct, although there was some uncertainty in consequence of absorption by corrosion of glass.

C. A.	Volume.	C. A.	Volume.	C. A.	Volume.	C. A.	Volume.
-10	1·00184	55	1·01413	100	1·04333	155	1·09847
- 5	1·00068	60	1·01668	105	1·04743	160	1·10456
0	1·00013	65	1·01940	110	1·05172	165	1·11083
+ 5	1·00001	70	1·02230	115	1·05620	170	1·11731
10	1·00025	75	1·02538	120	1·06086	175	1·12400
15	1·00083	80	1·02863	125	1·06570	180	1·13093
20	1·00168	85	1·03205	130	1·07073	185	1·13811
25	1·00280	90	1·03563	135	1·07593	190	1·14553
30	1·00416	95	1·03939	140	1·08130	195	1·15323
35	1·00575	100	1·04333	145	1·08684	200	1·16124
40	1·00755			150	1·09257	205	1·16958
45	1·00955					210	1·17829
50	1·01175						

The following is an extract from note-book of the experiments :—

" The observations on pure distilled water could only be made up to 305° F., in consequence of the glass being corroded and becoming opake above that temperature. At the higher temperatures, five tubes were employed with water having $\frac{1}{23}$ of carbonate of soda in solution. Two only of these five were sufficiently transparent up to 413°. But on examining them next day, $\frac{1}{200}$th of the volume of liquid was absorbed. This allowed for.

" The expansion of the solution rather less than pure water. The corrosion of the glass began immediately above the surface of the liquid. The vapour was computed from formula, assuming the law of vapour-density maintained."

Note F. § 6.—M. Muncke and M. Pierre have employed the general formula $1 + \Delta_e = 1 + ax + bx^2 + cx^3$, &c. to represent their observations, and have computed the constants for each series. They have also sought, by means of the roots of this equation, to find points of maximum density of each liquid beyond the range of their observations. Thus M. Pierre, at p. 358, vol. xv. *Ann. de Chim.*, expresses himself as follows :—" .... puisque l'équation $\frac{d(1+\Delta_e)}{dx} = 0$, dont les racines *doivent* donner la température de ce maximum, a ses deux racines imaginaires."

I have traced graphically the curve of the equation and of the observations, and find that its course through them is similar to fig. 8, interlacing at the fixed points, and departing altogether from the line of observation beyond the extreme points to which it is bound down. The positive and negative differences at the loops sometimes amount to $\frac{1}{4}$ degree. A conic section may be drawn to represent almost perfectly a series of observations if the range is not great. The hyperbola answers well, and can be simply applied as the increasing rate of expansion adapts itself to the curve, referred to an asymptote parallel to the axis of temperature.

Note G. § 15.—The value $\frac{1}{\rho} = 3\cdot28$ is taken from Regnault's observations on the tension of its vapour at 0° and 20° C. The observations at 0° and 30° represent $\frac{1}{\rho} = 3\cdot25$. Dalton's observation on the vapour give it equal to $3\cdot2108$, which is probably the most correct, as h is thus represented to be the same for sulphuric ether and water, their lines of vapour-density being parallel.

Note H. § 16.—In a paper on Capillarity in the Philosophical Magazine for January 1858, the proofs are given in detail of a law that connects molecular volume with capillarity and latent heat. It is expressed by the equation $m = \frac{p}{L}$, in which m is the cube root of the molecular volume of a liquid, p the height of the same in a capillary tube of constant bore, and L the latent heat of the vapour of the same, all taken at the same temperature *t.* According to Wolf and Brunner (*Ann. de Chim.* vol. xlix.), $p=A-bt$ is the empirical equation for the capillary height in terms of the temperature.   According to the law of expansion, $m=\sqrt[3]{\left(\dfrac{k}{\gamma-t}\right)^{\rho}}$.   Hence

$$L \doteq \frac{p}{m} \doteq (A-bt)\sqrt[3]{\left(\frac{\gamma-t}{k}\right)^{\rho}}$$

is the equation which, at present partly empirical, it is so desirable to convert into one wholly expressive of a general natural law of quantitative relation between L and V.

Edinburgh, May 6, 1861.

LXII. *On a peculiar Acid (Dianic Acid) met with in the Group of Tantalum and Niobium compounds.* By Professor F. von Kobell*.

BEING engaged in preparing a new edition of my ‘Mineralogical Tables,’ I was anxious, among other matters, to arrive at as distinct chemical characters as possible for the tantalates and niobates with which we are familiar; and after various experiments, I came to the conviction that in several of these compounds there exists an acid different from the true tantalic acid which occurs in the tantalite from Kimito, and also from the niobic acid met with in the niobite from Bodenmais.

As from the previous labours of MM. Rose, Hermann, Wöhler, and others we are aware that in testing for these acids the one is very liable to be mistaken for the other, inasmuch as the tests give more or less various indications according to the manner of treating the substances and the quality of the reagents themselves, I have endeavoured in the first place to avert any possible error arising from those causes by conducting the whole of the assays in precisely the same manner, which I now proceed to describe in detail.

1·5 grm. of each assay was fused in a silver crucible with 12 grms. of hydrate of potash, and the mass, which melted quietly, was maintained for seven minutes longer in a state of fusion; hot water was then added till the fluid amounted to 20 cubic inches, and when cold it was filtered.   The filtrate was acidulated with hydrochloric acid, then neutralized with ammonia, the precipitate allowed to subside, and the liquor poured off; after this, the precipitate, which was frequently coloured by manganese, was shaken with caustic ammonia and filtered.   I had taken somewhat more ammonia than would have been required to remove from the precipitate an amount of 10 per cent. of tungstic acid.   By

* From the *Bulletin* of the Academy of Sciences of Munich, Meeting of March 10, 1860. Communicated by W. G. Lettsom, Esq.

this treatment any tungstic or molybdic acids which might have caused the reaction described below was got rid of.

In order to employ in my experiments as equal quantities as possible of the precipitates, which must be used when just thrown down, I made funnels of tinfoil, which I cut into the form of a filter one inch long in the side, and gave them the requisite shape in a small porcelain funnel. One of these funnels was filled with the freshly filtered pasty precipitate by means of a spatula, and then laid in a porcelain dish; the tinfoil having been opened out, one cubic inch of concentrated hydrochloric acid, of the specific gravity of 1·14, was added and heated to boiling, that temperature being maintained for three minutes, and the foil kept continually well stirred about in the fluid. Under this treatment the appearances observed were as follows.

1. The acid of the tantalite from Kimito, and of the niobite from Bodenmais, coloured the liquor bluish (smalt-blue); on adding half a cubic inch of water thereto, when poured into a glass the colour *disappeared rapidly*, the precipitate settled *without being dissolved*; on being filtered the liquor gave a *colourless* filtrate, and the precipitate, which at first was of a bluish tint, became speedily white on a further addition of water.

2. The acids of a so-called tantalite from Tammela, the powder of which was blackish grey, those of euxenite, æschynite, and samarskite, on being boiled with hydrochloric acid and tinfoil as above described, were dissolved to a dark blue cloudy fluid, which, when diluted with half a cubic inch of water or rather more, *became perfectly clear with a deep sapphire-blue colour*, and gave a transparent deep-blue filtrate. On being further diluted by the addition of a twofold or threefold quantity of water, the colour becomes indigo-blue and bluish green; and in open vessels, after some time, olive-green, maintaining that tint for several hours, but becoming paler. The fluid preserves its perfect transparency all the while, and in a closed vessel the colour remains unchanged for weeks.

Both with the assays under (1), and also with those under (2), I kept up the boiling for a longer time, indeed till the liquors were tolerably concentrated; I then added half the volume of water and poured the whole into a glass. The appearances observed were the same as before; the acids of (1) remained undissolved and gave a colourless filtrate, while those of (2) were dissolved and gave a transparent blue solution, the colour of the filtrate being also blue. When treating euxenite on one occasion, and concentrating the liquor by boiling, I obtained an olive-green fluid, which was, however, transparent; on the addition of concentrated hydrochloric acid, and boiling a second time with tinfoil, the blue colour was restored. If, on obtaining a green fluid of this nature, it is diluted with three times its volume of water and then slowly evaporated till it becomes turbid, on the addition of a suitable quantity of concentrated hydrochloric acid and boiling for a few minutes with tinfoil, the blue colour of the solution always appears on adding a little water.

It seems superfluous to say that a transparent blue fluid also gives a filtrate of the same colour; and yet the case occurs of such a fluid being coloured only from some substance being held suspended therein in a state of extreme subdivision, the filtrate being colourless. Such, for instance, is the behaviour of tungstic acid when it is precipitated from tungstate of potash with hydrochloric acid, and the precipitate boiled with concentrated hydrochloric acid and tinfoil. I obtained thus a dark-blue fluid, which, when considerably diluted, was quite transparent and of a bright sapphire-blue; but both the dark-blue and the light-blue diluted fluid gave a colourless filtrate; and when left to themselves, both these fluids also became colourless when the blue oxide of tungsten suspended therein, and which in that condition retains its blue colour, had settled to the bottom.

The tin contained in the blue solution of the acid in question is easily got rid of by a stream of sulphuretted hydrogen, and the acid is obtained again from the filtrate by precipitation with ammonia. The precipitate, on being boiled with hydrochloric acid and tinfoil, again produces the blue fluid. On evaporating slowly the liquor filtered from the sulphide of tin (which from its diluted state is colourless), it becomes turbid when considerably concentrated. On adding a little water the cloudiness disappears, and on the further addition of concentrated hydrochloric acid a white precipitate is produced. If the hydrochloric acid has been added in suitable quantity, and if the fluid is boiled with a slip of tinfoil placed in it, the appearance spoken of above is produced. The fluid becomes of a deep blue, and when poured into a glass appears turbid; but on the addition of half its vólume of water it becomes transparent, and presents itself in the glass like a clear sapphire. The original precipitate from the potash solution may be freed from any manganese it may contain by boiling it with a certain quantity of hydrochloric acid; this precipitate can further be boiled with tolerably strong sulphuric acid without being deprived of the property of being soluble in hydrochloric acid in the presence of tinfoil. The acid thus purified is white; on being heated, it assumes a very pale yellow colour, which it loses again on cooling, taking somewhat the appearance of porcelain.

Before the blowpipe it is dissolved in borax and salt of phosphorus to a colourless glass, both in the oxidating and the reducing flame. When the borax glass is saturated, it remains trans-parent on cooling after exposure to a good heat; on being warmed again it becomes cloudy, and assumes the appearance of enamel.

When the acid in question is boiled with zinc instead of tin, the blue solution is not obtained; the precipitate of the acid is blue, it is true, but the filtrate is colourless, and the acid loses its colour on the addition of water without being perceptibly dis-solved. It was only with a very large quantity of hydrochloric acid and zinc that I could obtain a dirty greenish solution, which, however, when diluted with half its amount of water, became speedily reduced in its colour, assuming a pale green hue with opalescence.

If equal quantities of the acid spoken of, of tantalic acid, and of hyponiobic acid, all three being measured in a platinum fun-nel, are boiled for three minutes with concentrated hydrochloric acid without tin in the manner above described, and are then poured out into a glass, they all three give yellow milky fluids. On the addition of a very moderate quantity of water the acid in question becomes perfectly transparent, whereas the tantalic acid, and also the hyponiobic acid, even on the addition of four or five times their volume of water, remain undissolved.

If the metallic acid in question, when freshly precipitated, is heated to boiling in diluted sulphuric acid (1 volume of concen-trated acid to 5 of water), it forms a cloudy fluid; and on this being poured into a glass with a few grains of distilled zinc, in the course of a few minutes the acid, which was previously white, becomes of a decided smalt-blue, even deep blue, and retains this colour for some time on the addition of water; the filtrate, however, is colourless. In this behaviour it resembles hyponiobic acid, whereas tantalic acid treated in the same manner is only coloured pale blue, which colour immediately disappears on the addition of water. The difference in the behaviour of tantalic and hyponiobic acids has been already mentioned by Heinrich Rose as character-istic; as I modified the experiment, by having recourse to a boil-ing temperature, the effect is not only produced more rapidly, but also in a more marked manner. I look on this reaction for distinguishing tantalic acid from other kindred acids as the most certain, that is to say, if one does not wish to investigate the be-haviour of the chlorides. For a qualitative testing of an acid of this class, the first step of the inquiry would be to precipitate it in the manner described from the solution of potash, and then to examine the solubility of the freshly obtained precipitate with hydrochloric acid and tinfoil, with due attention to the condi-tions above laid down. Should the acid not be dissolved to a blue fluid when, after three minutes' boiling, half a cubic inch or a cubic inch of water is added, it is tantalic or hyponiobic acid, and recourse must be had to an assay with sulphuric acid and zinc to determine between these two acids. I am at least inclined to think that our determinations, as far as correctness is concerned, will have as great a probability in their favour as with the other methods hitherto employed, which, as is shown by the constantly varying indications of the acids of euxenite, yttrotantalite, samarskite, &c., have not given any thoroughly reliable results.

With respect to the acid discovered by me, and which forms a blue solution with hydrochloric acid and tin with such remarkable facility, it is with certainty and ease distinguishable both from tantalic and from hyponiobic acid, and that evidently in a more marked manner than those acids mutually are; the method of its preparation, moreover, as above set forth, as well also as a comparison with kindred acids under closely similar circumstances, appear to me to exclude the idea of its being an allotropic state, or a not hitherto observed stage of oxidation of tantalum or niobium, and to claim for it an existence as a distinct acid. Hermann, as is well known, several years since assumed the occurrence in samarskite, formerly termed uranotantalite, of a peculiar acid which he termed ilmenic acid; he was, however, not enabled to characterize that acid with sufficient precision; and Heinrich Rose could at that time establish point by point for his own niobic acid, now termed hyponiobic, everything that Hermann sought to establish for ilmenic acid, so that at last Hermann ranged his acid under niobium, and has pronounced it to be a niobous-niobic combination*. That the acid discovered by me is an oxide of niobium is, as far as present experience goes, not to be assumed; for if it were a lower grade of oxidation than the hyponiobic acid we are acquainted with, it must, on being fused with potash in an open crucible, be converted to this hyponiobic acid, inasmuch as, according to Heinrich Rose, niobium itself is dissolved into hyponiobate of potash by boiling potash; and if it were a higher oxide than hyponiobic acid, it must, on being reduced with tin, be also converted into that acid, and consequently neither soluble in hydrochloric acid under the conditions spoken of, nor impart a blue colour to the solution, as is, however, the case.

The same argument holds good if it be regarded as an oxide of tantalum: under the treatment referred to, it must be converted into the tantalic acid with which we are familiar, and must agree with it in its reactions, which it does not. Heinrich Rose has duly established that the metallic acid of the tantalite of Bodenmais is distinct from that of certain Finland tantalites; and to mark the difference, he called the former

---

* According to Hermann, it colours salt of phosphorus dark brown before the blowpipe.

niobic acid, now termed hyponiobic acid, and has called the
mineral formerly designated as tantalite by the name of niobite.   According to my experiments, the same case occurs with
the acids of the tantalite from Kimito, and of that from Tammela
which I have examined.   I will therefore name the latter, in
which I first remarked the difference, after Diana, and term it
Dianic acid; the element I shall name *Dianium* (Di), and the
mineral from Tammela which contains this acid, *Dianite.*

.Besides occurring in the mineral here mentioned, this acid,
though in a less pure state, appears to occur in the Greenland
tantalite, in the pyrochlore from the Ilmen Mountains, and in
the brown Wöhlerite (I have not examined the yellow).   I could
only employ small quantities, however, of these minerals, and it
was not in my power to carry out the requisite investigations in
sufficient detail.   A small fragment of yttrotantalite, professedly
from Ytterby, gave the reaction of dianic acid; in another assay
of a specimen, the specific gravity of which I ascertained to be
5·5, from the collection of the late Duke of Leuchtenberg, the
acid proved to be tantalic acid.   The former assay refers therefore to a different species, the specific gravity of which I could
not determine.

When combinations of this nature contain at the same time
titanic acid, the latter is found in the residue of the potash-lye,
in which it can be easily detected even when this residue contains also a small portion of dianic acid.   The residue is boiled
with concentrated hydrochloric acid and filtered, the filtrate,
with a strip of tin laid in it, being then boiled longer.   If no
dianic acid, but tantalic acid is present, the.liquor on becoming
concentrated assumes a violet-blue colour, which on diluting
with water is changed very characteristically to pink.   The fluid
retains this latter colour for several days or longer.   When the
solution, in addition to titanic acid, contains a portion of dianic
acid as well, the blue colour of the latter predominates; on diluting it in an open glass, the pink colour due to titanic acid
makes its appearance in the course of a few hours, owing to the
colouring of the dianic acid disappearing gradually.   In this
way I recognized the presence of titanic acid (as it had been established previously by other methods) in æschynite, pyrochlore,
and euxenite.

I cannot of course say whether my dianic acid is contained in
all varieties, and from all the localities, of the above-named
species; with respect to the tantalites from Tammela, it is indeed
established that perhaps the majority of them contain tantalic
acid.   The specific gravity should probably be particularly attended to.   The mineral from Tammela examined by me (dianite) has a specific gravity of 5·5, while the tantalites from that

locality analysed by H. Rose, Weber, Jacobson, Brooke, Wornum, and Nordenskiöld, had a specific gravity of from 7·38 to 7·5 and more. The tantalite from Kimito, moreover, from which I procured the tantalic acid employed for my investigation, has a specific gravity of 7·06. The colour of the streak of dianite is, as before remarked, black-grey, while that of the Tammela tantalites analysed by Jacobson is stated to be dark brownish red, as is also the case with the Kimito tantalite.

In appearance dianite very closely resembles Finland tantalites. The assay analysed was taken from a large tabular broken crystal about 2 inches in size, on which, however, only two planes occur. Their angle of inclination, as measured by the hand-goniometer, amounts to about 151°; whether those planes are T and R of Naumann's tantalite, or T and G, or other ones, cannot of course be determined. Before the blowpipe, dianite affords no marked difference when compared with the Kimito tantalite.

The samarskite which I examined is from the Ilmen Mountains; I employed quite fresh, pure fragments, with a conchoidal fracture and strong, somewhat metallic, vitreous lustre. The euxenite is from Alva near Arendal (procured from Dr. Krantz); the æschynite, from the Ilmen Mountains, was from the Leuchtenberg collection.

·While preparing the above, I forwarded a portion of the dianic acid in question to Professor Heinrich Rose, and communicated to him the leading points of the paper, requesting his opinion on the matter. Professor Rose was so good as to prepare the chloride of this acid, and wrote to me that, in doing so, he had met with a trace of tungstic acid, adding that the reaction described by me might be brought about from that circumstance; and he advised me, as a first step, to purify the acid by the method suggested by him, namely, by fusing it with carbonate of soda and sulphur. The case might be similar to the one which had misled Hermann.

Now I had, it is true, established, by the very ready solubility of dianic acid in hydrochloric acid when compared with true tantalic and hyponiobic acids under similar conditions, a characteristic distinction for the first of these three acids; but it was none the less essential to prove that the property of becoming blue with hydrochloric acid and tin belonged to the acid in question, and is not attributable to tungstic acid. After the treatment with ammonia, to which reference has been made, but little tungstic acid could, it is true, contaminate the dianic acid; nevertheless the turning blue might be ascribed to that. A plan to clear this point up was soon formed. I first sought to impart to the non-colouring tantalic and hyponiobic acids the quality of becoming coloured by an addition of tungstic acid, and then endeavoured to ascertain how far an acid thus mixed was to be purified by ammonia, the process which I followed in my original experiments. I prepared tungstate of potash of a determined strength, and mixed it with a lye of tantalic acid in such proportions that 84 parts of tantalic acid were united to 16 parts of tungstic acid, corresponding, as it were, to a potash solution of a tantalite that consisted of tantalic and tungstic acids only. The mixture was divided into two portions (by means of a graduated glass), and precipitated with hydrochloric acid. The precipitate of one portion was decanted and filtered, and a tinfoil filter, one inch long in the side, being filled therewith, it was boiled for three minutes in 1 cubic inch of hydrochloric acid as described above; $1\frac{1}{2}$ cubic inch of water was added, and it was filtered. The filtrate was greenish yellow; on adding 1 cubic inch more water the fluid was yellowish, the precipitate was *not dissolved*, and after the lapse of twenty-four hours the fluid which was poured off deposited a dark blue precipitate. The same experiment, performed in the like manner with a similar quantity of the hyponiobic acid, gave an olive-green filtrate, which did not materially change in twenty-four hours. When boiled again it became of a blue colour, which was also the hue of the filtrate. The hyponiobic acid was a little dissolved in this experiment, as the tantalic acid was in the corresponding one. When, however, I agitated the precipitates of the mixed acids with ammonia (the approximate quantity required to dissolve the amount of tungstic acid contained therein having been ascertained by experiment), and then allowed them to subside, decanted and filtered them, the precipitates thus treated, on being boiled for three minutes, as above described, with hydrochloric acid and tinfoil, and diluted with half a cubic inch of water, behaved almost entirely like the acids prepared directly from the minerals themselves; the fluid of the tantalic acid passed through the filter colourless, that of the niobic acid had a slight greenish tint. These experiments prove that tantalic and hyponiobic acids, even when containing a great amount of tungstic acid, may be at least so far purified by ammonia as *not* to produce the deep blue colour given by dianic acid; and further, that the presence of tungstic acid in those acids, under the conditions spoken of, does *not* increase their solubility in hydrochloric acid. The latter circumstance, although to be anticipated, was of more importance to me than the absence of the blue colour; for similar experiments to these had already fully convinced me that the removal of tungstic acid by means of ammonia is not a perfect one. To remove, however, the last doubts as to the possible cooperation of the tungstic acid in the production of the blue colour alluded to, the acid of the dianite from Tammela was purified with the utmost care by the process suggested by H. Rose. The filtered acid was dried down till it was capable of being readily reduced to powder. I took 0·5 grm. of it, which was triturated with 1·5 grm of carbonate of potash and then with 0·5 grm. sulphur. I fused the mixture in a covered porcelain crucible over a gas-lamp, dissolved the mass in water, and after decantation transferred the acid remaining into a close glass vessel with sulphuretted hydrogen, which I agitated well, leaving it thus for twenty-four hours. The liquor was then decanted twice, and the residue boiled with diluted hydrochloric acid and well washed; and lastly, the metallic acid was attacked a second time with hydrate of potash in a silver crucible, precipitated with hydrochloric acid and filtered. A tin-foil funnel was filled, as above described, with the acid, and a cubic inch of concentrated hydrochloric acid having been poured into it, the fluid was maintained for three minutes at a boiling temperature. Having been poured into a glass, it proved to be deep blue and turbid; but on the addition of the necessary quantity of water, it afforded a splendid sapphire-blue solution, perfectly transparent, without a trace of undissolved precipitate. There is therefore no doubt, not only that by its very marked difference of solubility under similar conditions dianic acid is distinguishable from tantalic and hyponiobic acids, but also that the property of producing a blue colour, as above described, belongs to it essentially, a property which the other acids do not possess.

I purified in the same way the acid of euxenite and samarskite; and their behaviour was precisely the same as I observed it to be on my endeavouring to remove by agitation with ammonia any tungstic acid they might possibly contain. The blue solution of dianite and samarskite was of a peculiarly deep colour, almost black, so that twice or thrice its volume of water had to be added to it to recognize the blue colour distinctly, and to see that the solution was perfectly transparent. In a stoppered bottle the colour maintained itself quite unaltered for weeks.

The acid of æschynite I have not purified further than by agitation with ammonia; and as in two experiments, independent of the blue colour, it proves to be quite as thoroughly soluble as the acid of dianite, I have no doubt that it is dianic acid. The experiments here described have all been repeated several times, particularly those with true tantalic acid, with hyponiobic acid, and with the dianic acid of dianite.

The very peculiar behaviour of dianic acid above described with respect to zinc as compared with tin, and with hydrochloric acid alone, induced me to make some additional ex-

periments.   When the solution obtained by employing a large
quantity of hydrochloric acid and zinc is diluted with about
thrice its volume of water, it has a dirty yellowish colour, and
for the moment is tolerably transparent; in the course of a few
minutes it becomes turbid and loses its transparency, owing to
a finely-divided greenish-grey precipitate which forms after a
while; on the addition of more water it is deposited, so that the
liquor may be decanted.   An assay with hydrochloric acid and
tin shows this deposit to be hydrous dianic acid; for it is dissolved to the characteristic sapphire-blue fluid when treated in
the way so often referred to.   If this blue solution is boiled for
a few minutes with zinc, the dianic acid also is thrown down
with the tin; the precipitate is deposited with readiness in light
grey flakes on the zinc, which is coated with spongy tin,
and the supernatant fluid is transparent and colourless.   On
filtering the portion containing the flakes and boiling the
acid collected with hydrochloric acid and tinfoil, the blue solution is obtained again on the addition of a little water.   Thus
the behaviour of the acid of the Tammela dianite and that of
the acid of samarskite from the Ilmen Mountains is strictly identical.   The behaviour of zinc and tin therefore with reference
to the solutions of dianic acid in question, is, to an extent not
to be anticipated, entirely different, it may be said antagonistic.

Those who wish to repeat the investigations described would
do well to adhere to the quantities mentioned by me, or to employ them proportionally; for without this precaution it is possible
that the properties may not be as distinctly brought out as they
will be by adhering to them.

LXIII. *On the Partitions of a Close.*   By A. CAYLEY, *Esq.*[*]

IF F, S, E denote the number of faces, summits, and edges of
a polyhedron, then, by Euler's well-known theorem,

$$F + S = E + 2;$$

and if we imagine the polyhedron projected on the plane of any
one face in such manner that the projections of all the summits
not belonging to the face fall *within* the face, then we have a
partitioned polygon, in which, if P denote the number of component polygons, or, say, the number of parts, $F = P + 1$, or we
have

$$P + S = E + 1,$$

where S is the number of summits and E the number of edges
of the plane figure.   I retain for convenience the word *edge*, as
having a different initial letter from *summit*.

* Communicated by the Author.

The formula, however, excludes cases such as that of a polygon divided into two parts by means of an interior polygon wholly detached from it; and in order to extend it to such cases, the formula must be written under the form

$$P+S=E+1+B,$$

where B is the number of breaks of contour, as will be explained in the sequel.

The edges of a polygon are right lines: it might at first sight appear that the theory would not be materially altered by removing this restriction, and allowing the edges to be curved lines; but the fact is that we thus introduce closed figures bounded by two edges, or even by a single edge, or by what I term a mere contour; and we have a new theory, which I call that of the Partitions of a Close.

Several definitions and explanations are required. The words line and curve are used indifferently to denote any path which can be described *currente calamo* without lifting the pen from the paper. A closed curve, not cutting or meeting itself[*], is called a *contour*. An enclosed space, such that no part of it is shut out from any other part of it, or, what is the same thing, such that any part can be joined with any other part by a line not cutting the boundary, is termed a *close*. The boundary of a close may be considered as the limit of a single contour, or of two or more contours lying wholly within the close. The reason for speaking of a limit will appear by an example. Consider a circle, and within it, but wholly detached from it, a figure of eight; the space interior to the circle but exterior to the figure of eight is a close: its boundary may be considered as the limit of two contours,—the first of them interior to the close, and indefinitely near the circle (in this case we might say the circle itself); the second of them an hour-glass-shaped curve, interior to the close (that is, exterior to the figure of eight) and indefinitely near to the figure of eight. The figure of eight, as being a curve which cuts itself, is not a contour; and in the case in question we could not have said that the boundary of the close consisted of two contours. A similar instance is afforded by a circle having within it two circles exterior to each other, but connected by a line not cutting or meeting itself; or even two points, or, as they may be called, summits, connected by a line not cutting or meeting itself; or, again, a single summit: in each of these cases the boundary of the close may be considered as the limit of two contours. But this explanation once given, we may for shortness speak of the close as bounded by a

[*] It is hardly necessary to add, except in so far as any point whatever of the curve may be considered as a point where the curve meets itself.

single contour, or by two or more contours; and I shall throughout do so, instead of using the more precise expression of the
boundary being the limit of a contour, or of two or more contours.
The excess above unity of the number of the contours which
form the boundary of a close is the *break of contour* for such
close; in the case of a close bounded by a single contour, the
break of contour is zero.

Any point whatever on a curve may be considered as the point
of meeting of two curves, or, in the case of a closed curve, as the
point where the curve meets itself, but it is not of necessity so
considered. A point where a curve cuts or meets itself or any
other curve, is a *summit*; each point of termination of an un-
closed curve is also a *summit*; any isolated point may be taken
to be a *summit*. It follows that, in the case of a closed curve
not cutting or meeting itself (that is, a contour), any point
or points on the curve may be taken to be summits; but
the contour need not have upon it any summit: it is in this case
termed a *mere contour*. The curve which is the path from a
summit to itself, or to any other summit, is an *edge*: the former
case is that of a contour having upon it a single summit, the
latter that of an edge having, that is, terminated by, two summits, and no more. It is hardly necessary to remark that a
contour having upon it two or more summits consists of the
same number of edges, and, by what precedes, a contour having
upon it a single summit is an edge; but it is to be noted that a
contour without any summit upon it, or mere contour, is *not* an
edge. It may be added that an edge does not cut or meet itself
or any other edge except at the summit or summits of the edge
itself.

Consider now a close bounded by $\beta+1$ mere contours: if for
any partitioned close we have P the number of parts, S the
number of summits, E the number of edges, B the number
of breaks of contour; then, for the unpartitioned close, we have
$P=1$, $S=0$, $E=0$, $B=\beta$, and therefore

$$P+S+\beta=E+1+B;$$

and it is to be shown that this equation holds good in whatever
manner the close is partitioned. The partitionment is effected
by the addition, in any manner, of summits and mere contours,
and by drawing edges, any edge from a summit to itself or to
another summit. The effect of adding a summit is first to increase S by unity: if the summit added be on a contour, E will
be thereby increased by unity; for if the contour is a mere contour, it is not an edge, but becomes so by the addition of the
summit; if it is not a mere contour, but has upon it a summit
or summits, the addition of the summit will increase by unity

the number of edges of the contour. If, on the other hand, the summit added be an isolated one, then the addition of such summit causes a break of contour, or B is increased by unity. Hence the addition of a summit increases by unity S; and it also increases by unity E or else B, that is, it leaves the equation undisturbed. The effect of the addition of a mere contour is to increase P by unity, and also to increase B by unity: it is easy to see that this is the case, whether the new mere contour does or does not contain within it any contour or contours. Hence the addition of a mere contour leaves the equation undisturbed. The effect of drawing an edge is first to increase E by unity; if the edge is drawn from a summit to itself, or from a summit on a contour to another summit on the same contour, then the effect is also to increase P by unity; if, however, the edge is drawn from a summit on a contour to a summit on a different contour, then P remains unaltered, but B is diminished by unity. There are a few special cases, which, although apparently different, are really included in the two preceding ones: thus, if the edge be drawn to connect two isolated summits, these are in fact to be considered as summits belonging to two distinct contours, and the like when a summit on a contour is joined to an isolated summit. And so if there be two or more summits connected together in order, and a new edge is drawn connecting the first and last of them, this is the same as when the edge is drawn through two summits of the same contour. The effect of drawing a new edge is thus to increase E by unity, and also to increase P by unity, or else to diminish B by unity; that is, it leaves the equation undisturbed. Hence the equation $P + S + \beta = E + 1 + B$, which subsists for the unpartitioned close, continues to subsist in whatever manner the close is partitioned, or it is always true.

In particular, if $\beta = 0$, that is, if the original close be bounded by a mere contour, $P + S = E + 1 + B$; and if, besides, $B = 0$, then $P + S = E + 1$, which is the ordinary equation in the theory of the partitions of a polygon.

If we consider the surface of a plane as bounded by a mere contour at infinity, then for the infinite plane, $\beta = 0$, or we have $P + S = E + 1 + B$: in the case where the infinite plane is partitioned by a mere contour, $P = 2$, $S = 0$, $E = 0$, $B = 1$ (for the exterior part is bounded by the contour at infinity, and the partitioning contour, that is, for it, $B = 1$), and the equation is thus satisfied. And so for a contour having upon it n summits, $P = 2$, $S = n$, $E = n$, $B = 1$, and the equation is still satisfied: this is the case of the plane partitioned into two parts by means of a single polygon.

The case of a spherical surface is very interesting: the entire surface of the sphere must be considered as a close bounded by 0 contour, or we have have $\beta = -1$, and the equation thus becomes $P + S = E + 2 + B$. Thus, if the sphere be divided into two parts by a mere contour, $P = 2$, $S = 0$, $E = 0$, $B = 0$, and the equation is satisfied. And in general, when $B = 0$, then $P + S = E + 2$; or writing F for P, then $F + S = E + 2$, which is Euler's equation for a polyhedron.

2 Stone Buildings, W.C.,
March 8, 1861.

LXIV. *Some Notes on the Drift Deposits of Western Canada, and on the Ancient Extension of the Lake Area of that Region. By* E. J. CHAPMAN, *Professor of Mineralogy and Geology in University College, Toronto*.*

THE following notes and deductions are the result of a careful examination of the Drift deposits of Western Canada, undertaken during the last three or four summers in an unsuccessful search for marine post-tertiary fossils, such as occur so abundantly in many parts of Eastern Canada and throughout the New England States. The district more especially investigated extends from the Bay of Quinté westward to the mouth of the Saugeen on Lake Huron, and includes the line of country lying along, and immediately within, the outcrop of the Laurentian rocks north of that region. Detached observations have been made, moreover, at various points on the islands and north shore of Lake Huron; and also beyond the limits of the province, as in the district south of Lake Ontario, in Michigan, and along the southern shore of Lake Superior.

The notes recorded here are arranged under two sections, of which the first comprises a collection of data, and the second a corresponding series of deductions.

§ 1. *Data.*

1. The first point observable, with regard to our Drift deposits, is the very evident fact that the rock floor on which these accumulations are spread had been extensively denuded prior to their deposition upon it. They cover thus an undulating and more or less broken surface; and their thickness, consequently, apart from the denudation to which they have been themselves subjected, is exceedingly variable.

2. The lowest of these deposits appear to consist of dark-blue or greyish clays, with thin layers of yellowish or light-coloured clay

---

* Communicated by the Author.

in places. This deposit is often laminated horizontally, and is generally very calcareous. It appears also to be free from northern or large crystalline boulders. Pebbles of limestone and other fossiliferous rock, mixed with some small pebbles of water-worn gneiss, occur abundantly in it in many localities; but northern boulders, properly so called, are either absent or exceedingly rare. Amongst the localities in which these lower and boulder-free clay deposits are of marked occurrence, the district around Toronto and many parts of the valley of the Saugeen and western shores of Lake Huron may be especially mentioned; but wherever our Drift deposits are found to consist of clay and other materials, the clay-beds are almost invariably seen to occupy the lower place. At the same time, as described more fully in the sequel, beds of yellow and other-coloured clay, it should be observed, are occasionally found with northern boulders in a higher part of the series; but these are quite distinct from the lower clays now referred to. They are, moreover, of no great thickness, but alternate with, and are subordinate to, thick deposits of gravel and sand; whereas the lower clays attain in places to a thickness of over 100 feet, and present a general uniformity throughout. In these latter beds no traces of contemporaneous fossils have as yet been found.

3. It is generally assumed as an established fact, that the harder rocks beneath the Drift exhibit everywhere the marks of glacial action. Although we have numerous examples, throughout this section of the province, of polished and striated rock, I believe it to be still an open question as to whether the rocks which underlie these lower clays have been thus affected. I have not been able to discover any instances of it, nor can I find any recorded cases in our Geological Reports, or in other trustworthy sources. The question hitherto does not seem to have been mooted, the Drift accumulations generally being classed together by most observers under one common term. As the point is of much interest, however, it should be kept in view.

4. Above the lower clay deposits, or resting immediately (where these are absent) on the foundation rock of the country, we meet with a series of sands and gravels of evidently northern origin, containing boulders of gneissoid and other rock, and alternating occasionally with beds of clay, in which northern boulders are also frequently found. This clay, with scarcely an exception, is remarkably free from calcareous matter, the cause of which will be alluded to further on. In some places the clay and gravel are mixed up together, and present no signs of stratification; but more usually they are distinctly stratified, and the boulders are mostly accumulated towards the upper part of the series As a general rule, indeed, the boulders occur in by far the greatest

abundance, scattered, *per se*, over the surface of the gravels, or resting immediately on the underlying rocks where the clays and gravels are absent. This appears to have arisen in some cases from the subsequent removal or washing away of the looser materials in which the boulders were originally imbedded; but the greater number of these were evidently thrown down where they now lie, by melting or stranded icebergs, after the deposition of the other Drift materials. The boulders, whether of gneissoid or other fossiliferous rock, belong always to northern localities in relation to the spots on which they now occur. Here and there the infiltration of water containing bicarbonate of lime has cemented some of these upper Drift deposits into conglomerates of considerable solidity (Burlington Heights; vicinity of Niagara Falls; Georgetown, &c.).

5. Under the gravels and sands, or where the isolated boulders of this series are found, the rocks are always more or less marked by glacial action. The more common effects comprise a smoothed and polished surface, and a fine striation, the striæ running in long straight lines in a general N.E. and S.W. direction, although following to a certain extent, in hilly and broken districts, the natural windings of the rock-slopes on which they occur. These effects are seen in Western Canada, at various heights above the sea-level, up to an elevation of at least 1500 feet. They are well shown on the top of the Collingwood escarpment, at about 1000 feet above the level of Lake Huron; on the same line of escarpment near Niagara Falls; on many of the rock-exposures on the north shore of Lake Huron; and throughout the country at the junction of the Laurentian and Silurian formations, between the river Severn and the County of Frontenac; also in the vicinity of Belleville, Trenton*, &c.

The isolated boulders scattered over the country frequently exhibit in themselves a polished and striated surface; and the small boulders and pebbles imbedded in the gravel deposits often present the same effects (*e. g.* the pebbles found in the terraces north of Toronto; also those in Drift gravel in the environs of Belleville, Marmora, Guelph, Niagara Falls†, &c.).

6. The gravel and sand beds of this series occur in places in oblique stratification, or exhibit what is technically termed "false bedding." This occurs at or near the upper part of the series, and is evidently due to a rearrangement of the materials by the action of currents (*e. g.* Drift-bank seen in Great Western Rail-

---

* See a paper, by the writer, "On the Geology of Belleville and its Environs," in the 'Canadian Journal,' vol. v. (new series), pp. 41–48.

† The localities cited in this paper are those which have come more immediately under the author's observation. In most instances the lists given might be greatly added to.

way cutting at Toronto, and extending westward several miles; beds at Orillia, on Lake Couchiching; also near Collingwood, &c: A remarkable example, alluded to more fully in the second part of this paper, Deduction 3, occurs near the village of Lewiston, on the south shore of Lake Ontario). I think it will be rendered clear, by what follows, that the currents in question were not marine, but were produced in the lake waters when these stood at higher levels. In places, moreover, secondary ridges, or ancient spits, have been formed by the same action out of these drift materials (*e. g.* Ridge at Weston, near Toronto, described by Sandford Fleming, C.E., in the Canadian Journal, new series, vol. vi.; also the ridge at Craigleith, in Collingwood Township, mentioned by the writer in the same Journal, vol. v. p. 305). These secondary ridges, it should be observed, are altogether distinct from the terraces of the lake shores and intervening districts. A careful search would no doubt reveal their presence in very many localities.

7. We now come to a fact of great interest—the occurrence of shells of freshwater mollusca in the sands and gravels of these Drift deposits, at various levels above the present surface of our lakes. These shells belong to existing species, inhabitants of the surrounding waters. They must not be confounded with similar shells left in elevated spots by the drying up of streams and ponds, or by the cutting back and lowering of river-beds. As occurring in our modified Drift deposits, they are imbedded in sand or gravel containing northern pebbles and small boulders; and in situations, moreover, in which it is evident that no merely local causes could have been concerned in their deposition. The fragility of most freshwater shells necessarily operates against the preservation of these in the coarser sediments, and explains their absence, probably, as regards the upper Drift beds of many localities.

In some of these re-sorted beds the bones and teeth of both extinct and existing mammals are occasionally found. The extinct forms comprise a species of Mastodon (*M. Ohioticus?* see Canadian Journal, new series, vol. iii. p. 356), the *Elephas primigenius*, and apparently an extinct species of the Horse. The remains of existing species found in these deposits (always confining our remarks to Western Canada) include the Wapiti, the Moose, Beaver, Musk-rat, &c. These two classes of remains have been found together. In a railway-cutting through Burlington Heights near Hamilton, the tusk of a Mammoth (*Elephas primigenius*) and the horns of a Wapiti (*Elaphus Canadensis*) were met with at a depth of about forty feet below the present surface of the ground. I have also seen the lower jaw of a Beaver (*Castor fiber*) obtained from the same locality. The flint arrow-heads and other wrought implements of Amiens and Abbeville, occur apparently in deposits of the same kind and age.

I have discovered freshwater shells, under the conditions described above, in beds of stratified Drift consisting of coarse gravel filled with pebbles of gneiss and other northern rocks, on the Kingston road, about two miles east of Belleville, at an elevation, by rough measurement, of about 40 feet above the present level of Lake Ontario. These belong to *Planorbis trivolvis*, or to some closely related species. Other examples of the same shell were obtained from fine gravel in oblique stratification, near the village of Orillia, at a height of about 18 feet above the level of Lake Couchiching. This lake is about 120 feet higher than Lake Huron, and about 700 feet above the sea. Pieces of nacreous shell (belonging to a species of *Unio*?) were also found in gravel, in the vicinity of Barrie, at an estimated height of about 30 feet above Lake Simcoe. I have found lacustrine and terrestrial shells in many other places; but these I omit from mention, as the shells occurred on the sites of ancient swamps, in gullies, or in flat lands adjacent to running streams, or in other doubtful situations in which they may have been deposited by freshets and other agencies of comparatively recent date.

Mr. R. Bell, of the Geological Survey of Canada, has added greatly to our knowledge of the above localities, in a paper published in the 'Canadian Naturalist' for February of this year (1861). Amongst other spots in which he has discovered freshwater shells, the environs of Collingwood and Owen Sound may be cited. At the former, examples of *Planorbis trivolvis*, associated with several species of *Helix*, were found by him at an elevation of 78 feet above Lake Huron. Specimens of *Melania conica* have been obtained, according to Mr. Bell, from another spot in this locality. Dr. Benjamin Workman of Toronto, has also communicated the discovery of examples of a *Melania* and *Unio ellipsis*, on the high banks of the Don, about 30 feet above the lake. These may have been deposited by the river, however, when flowing at a higher level; but they were covered, according to Dr. Workman, by a considerable deposit of sand.

The upper deposits of the Drift period are separable with difficulty in many places from those of more recent age. As the one period merged gradually into the other, this must necessarily be the case. Among the more recent deposits of Western Canada, however, our river "flats" may be more especially cited, as those of the Grand River, filled with the remains of land mollusca. Also, the closely-similar deposits of the ancient bed of the Niagara, so high above the present level of that river; together with the shell-marls and calcareous tufas of our lakes and streams; and our deposits of bog-iron ore and iron ochres.

§ 2. *Deductions.*

The following deductions appear to flow naturally from the observations recorded above:—

1. A general depression of the land, at the commencement of the Drift period, must have taken place to such an extent as to admit of the deposition of the lower clays. These latter were evidently derived from the limestones and other Silurian and Devonian strata lying beneath and around them. Hence their generally calcareous nature. Their derivation from this source is proved, moreover, by the pebbles of Trenton limestone and other fossiliferous rocks which they frequently contain. Extensive denudation must thus have occurred both immediately prior to and during the deposition of these clays; but it may be questioned whether the bolder contours offered by the denuded rocks, such as the escarpment that sweeps from the Niagara river to Cabot's Head on Lake Huron, were not produced during the first uprise of the palæozoic strata from the earlier seas in which their materials were accumulated, ages before the period now under discussion. It appears, at least, to be a well-admitted point, that these rocks had been elevated into dry land before the deposition of the higher formations in the south and west.

2. After the deposition of the lower Drift clays, a sudden and abrupt change in the character of the sediments took place. A striking example of this may be seen in the natural sections about Hogg's Hollow, a few miles north of Toronto. The change in question must have been effected by a still further depression of the country, bringing the higher lands and gneissoid strata of the north within the influence of the waves, and yielding the sands, gravels, and boulders of the upper Drift accumulations.

This depression permitted an invasion and broad extension southwards of the ice-covered Arctic seas, the true cause, in all probability, of the cold of this epoch. The depression must have exceeded 1500 feet, since northern boulders are found at that height above the sea on the Collingwood escarpment. The gneissoid boulders there met with must at least have traversed the basin of Georgian Bay; but the glacial striæ which also occur there may have been produced by the action of ice originating at the spot itself: the three or four distinct sets of striæ observed at this locality, however, do not radiate from any fixed point, but run in the usual north and south direction, some being a little east and others a little west of north*.

3. At the close of this second series of phenomena, a gradual

---

* On a visit to this spot, since the publication of the "Note on the Geology of the Blue Mountain Escarpment" in the 'Canadian Journal,' vol. v. p. 304, some additional sets of striæ were observed.

uprise of the land appears to have taken place, and a vast area, extending over and around our present lake-basins, then became converted into a freshwater sea. This probably found its outlet to the ocean through what is now the broad valley of the Mississippi. Its waters stood at a great elevation above the waters of our present lakes, and were gradually lowered to these levels by physical changes in the surrounding country, and more especially by the depression of a higher region lying to the east. During this gradual fall and retrocession of the great lake-waters, the upper layers of the Drift were re-sorted, mixed with newer sediments, and thrown up here and there into secondary ridges; and the remarkable terraces which form so salient a feature in the general aspect of our lake shores and intervening districts, were then in chief part produced. The escarped faces of these Drift terraces, it should be observed, *always front the present lake-basins*, and thus look in some places towards the north, and in others towards the south, &c., according to the direction of the nearest shores. This would necessarily arise if they were produced, as here imagined, by a gradual lowering of the waters, with intervening periods of repose. The shells of freshwater mollusca buried in the modified Drift, at various levels above the existing lake-waters, and in localities so far apart—for these shells have been found throughout the region south of the lakes, in addition to the localities mentioned in this paper—prove incontestably the former expansion and union of our lakes, or, in other words, the presence in this part of Western America, of a widely-extended freshwater sea covering an enormous area. A curious circumstance, and one of great significance in its bearings on this question, is the fact that all the inclined layers of modified Drift (to the east, at least, of Lake Superior) appear to slope towards the west or south. A remarkable instance of this, hitherto, it is believed, unnoticed, may be seen near the mouth of the Niagara river, at Lewiston. At this spot, oblique layers of modified Drift, in beds made up of coarse gravel and pebbles, point nearly due south, and thus bear witness to the fact that the current which occasioned the inclined stratification must have set directly up the gorge, *or against the direction of the present stream.*

The assumption of an immense freshwater lake of this character gradually falling from a high level, necessarily involves the additional assumption of an eastern barrier, extending at one period between the lake-waters and the Atlantic. This view was maintained by some of the earlier investigators of our geology, and notably by Mr. Roy, in his well-known paper on the terraces of Lake Ontario, communicated to the Geological Society of London in 1837. The difficulty of finding a satisfactory location for a barrier of this kind led Sir Charles Lyell, however, to reject the idea of an original lake extension, and to refer the formation of our terraces entirely to the action of the sea during the slow uprise of the land at the commencement of the present epoch. In this he has been followed by all geologists who have subsequently examined these terraces. The difficulty may perhaps be surmounted by assuming the earlier and greater elevation of that portion of the country lying to the east of the gneissoid belt which connects our northern Laurentian district with the Adirondack Mountains of New York. The subsequent depression of this region would open an eastern outlet to the lake-waters, and gradually lower these to their present levels. But whatever the explanation, the undoubted fact remains, that, at the close of the Drift period, a vast freshwater sea extended over the greater portion of Western Canada, and at a level of at least 500 feet above the present surface of Lake Ontario.

Whilst the mollusca of this ancient lake were identical with existing species, its shores were peopled by the mastodon and mammoth, and probably by other extinct forms of life, together with various species that still survive. A great question remains to be solved. Our gravel beds may perhaps reply to this, and reveal to us that here, as in Europe, man and the departed mammoth once trod the earth together. Could this be established, the discovery would be fraught with even deeper interest than that which attaches itself to exhumed human relics of the ancient plains of Picardy and the gravel-beds of Suffolk. Our Indian arrow-heads are disentombed by hundreds: the connecting link of the extinct tooth or bone may not be long forthcoming[*].

University College, Toronto, Canada.
March 16, 1861.

LXV. *On the Neutralization of Colour in the mixtures of Solutions of certain Salts.* By FREDERICK FIELD, F.R.S.E.[†]

I BELIEVE Maumené first pointed out the fact that when nitrate of cobalt is added to nitrate of nickel in certain proportions, the green and pink colours of the solution entirely disappear, and the liquid becomes colourless, or assumes merely a pale neutral tint. Ever since the manufacture of the oxides of

---

[*] Since writing the above, Albert Koch's account of the discovery of the Missouri mastodon has come under the author's notice. In this account, published in 1841, it is stated that the mastodon bones were found in more or less immediate association with large arrow-heads. The same writer also attests to the discovery of wrought implements in connexion with Edentate remains in Gasconade county, Missouri.

[†] Communicated by the Author.

nickel and cobalt from the ordinary *speiss* by the wet method, carried on for many years in Birmingham, this phenomenon must have been observed by those employed in conducting the process; and to one ignorant of the fact, and who has devoted but little time to this particular ranch of chemical analysis, the solution of speiss must appear rather surprising, as, although very rich both in cobalt and nickel, in certain instances the solution appears almost destitute of colour, showing no trace either of the red of the cobalt or the green of the nickel. Professor Liebig (Liebig's *Annalen*, vol. xc. p. 112, and Chemical Gazette, vol. xii. p. 300), in remarking upon the decolorizing action of binoxide of manganese upon glass, does not impute the bleaching effect to the oxidation of the protoxide of iron by the reduction of the binoxide, as neither nitrate of potash nor other strongly oxidizing bodies effect the same change, but to the violet colour imparted to the glass by the manganese being complementary to the green produced by the iron, and hence the two affording a colourless mass. And that this is the case is evident, I imagine; for if borax be coloured by protoxide of iron, the resulting glass fused in a platinum crucible, and a little of the same salt (previously coloured by manganese) cautiously added, a point is arrived at where the mixture of the two has lost the individual tint of each and produced a nearly colourless glass. Liebig also mentions that a concentrated solution of sulphate of manganese having a slight rose colour, added to a solution of protosulphate of iron having a pale green tint, affords a perfectly colourless mixture.

A few experiments of my own have been extended to some other solutions and chemical compounds.

When nitrate of cobalt is gradually added to a cold solution of bicarbonate of soda, a beautiful amethyst-coloured liquid, at times almost approaching to violet, is produced. The colour has not the pure rose-red of the nitrate or sulphate of the metal, but has evidently a considerable portion of blue in its composition. If the fluid thus formed be divided into two equal parts, and to one of these a few drops of hypochlorite of soda be added, the liquid changes to an intense green colour, with no trace of blue, but a slightly yellow tinge, very much the colour of chloride of copper when dissolved in strong hydrochloric acid. If the violet and green liquids are united, the mixture becomes colourless, more strikingly so perhaps than by the union of the cobalt and nickel salts: the blue tint in the bicarbonate of cobalt forms green with the slight yellow shade of the peroxidized compound; and this, together with the remaining green of that liquid, is entirely neutralized by the pure rose colour.

Dilute sulphate of nickel (pale green) dissolves crystallized sulphate of manganese (pink), forming a colourless solution.

A solution of permanganate of potash evidently contains two colours, red and blue, forming by their union the magnificent violet tint so characteristic of that salt.   When a little chloride of sodium is added to a solution of sulphate of copper, chloride of copper is formed by double decomposition, and the liquid assumes a pure green colour.   Permanganate of potash carefully added to this compound changes it to a fine bright blue.   The red tint is neutralized by the green, and the visible blue remains. The experiment can also be performed by substituting the chloride of copper for the sulphate.   When the chloride, free from acid, is dissolved in water, the solution has a pale blue tint, and the addition of one drop of the permanganate causes a very dark blue shade.   If a little acid be introduced into the copper salt, and the permanganate added as before, a similar effect is produced, but in about half an hour the pure blue disappears and the solution becomes green.   The acid in this instance first changes the blue colour of the chloride into green, and subsequently decomposes the permanganate, and thus by destroying the red which was neutralized by the green, as well as the blue which remained intact, the original green tint becomes apparent.

When permanganate of potash is cautiously added to a solution of the bichromate of the same base, a bright-red liquid is produced.   The solutions, however, must be dilute, and carefully managed.   The yellow in the bichromate forms a green with the blue in the permanganate, which, neutralized by the red in both salts, would form a colourless solution, did not an excess of the latter tint prevail.

Most chemists must have observed that, in the estimation of iron by means of permanganate of potash, the last drop, which shows that the reaction is complete, imparts a rose-red to the liquid, differing somewhat from the bluish pink of the permanganate.   The pale yellow of the perchloride of iron has combined with the blue of the permanganate, and the resulting green, not sufficiently powerful to destroy the whole of the red, has left a portion of it visible.

M. Terreil estimates copper by the same reagent.   The cupreous salt is deoxidized by sulphite of ammonia, the sulphurous acid expelled by ebullition, and permanganate of potash added until the whole is converted into the protoxide.   The difference of tint which the last drop of permanganate imparts to this liquid and to that containing the iron salt, is very apparent.   In the case of the copper solution it is nearly blue; in the iron, pinkish red.   These facts are not without their significance in qualitative analysis.   Gibbs* informs us that the beautiful test for manganese, first proposed by Mr. Walter Crum, by the action of nitric

* Silliman's Journal, September 1852.

acid and peroxide of lead, fails to yield the characteristic tint if
nickel be present in large quantity, and the manganese only in
minute traces: the nickel salt destroys, or at all events modifies
the colour produced by the formation of permanganic acid.   If
cobalt is present, however, or a solution of a salt of that metal
is subsequently added, the colour of the nickel is nullified, and
that of the manganese becomes sensible.

When red fire, composed of nitrate of strontia, chlorate of
potash, &c., is mixed with green fire containing nitrate of baryta
and ignited, the red and green rays become invisible, and a white,
or, rather, a bluish-white flame is produced: the crimson of the
strontian flame has a dash of blue in its composition, and the
red being removed by the green, is clearly shown.   If a rose-red
fire be prepared by mixing thirty-four parts of carbonate of lime,
fifty-two of chlorate of potassa, and fourteen of sulphur, or still
better, perhaps, twenty-three of dry chloride of calcium, sixty-
one of chlorate of potassa, and sixteen of sulphur, and then
ignited with the ordinary green fire, pure white light is produced.

LXVI. *Researches on several Phenomena connected with the Polarization of Light.* By M. H. FIZEAU*.

IT has long been observed that when a ray of light is received
on a mirror whose surface, instead of being perfectly smooth,
is scratched with a fine point, it is no longer simply reflected.
Part indeed follows the ordinary course, but a considerable portion is dispersed in various directions, some of which deviate
greatly from that of the regularly reflected ray.   We may mention as an example of this species of phenomenon, the singular
reflexions produced by surfaces of metal, such as brass and steel,
&c., which have been artificially prepared by being rubbed continually in one direction with emery, the hard angular particles
of which produce fine parallel scratches on the surface of the
metal.   The amount of light dispersed by surfaces so prepared,
in directions other than that of the regularly reflected ray, is calculated to excite some surprise; and it is generally agreed to
attribute this effect partly to the variously inclined surfaces of
the minute furrows with which the mirror is covered, partly to
the diffraction caused by them in conjunction, acting as an irregular grating, and producing at the same time, and mingled
with each other, the phenomena observed by Fraunhofer with
simple gratings of various structure.

In studying the light dispersed in this manner with respect to

* Translated from the *Comptes Rendus* for February 18, 1861, by F.
Guthrie.

its polarization, I have observed several unexpected and, as I believe, hitherto unnoticed phenomena, which appear worthy of the attention of some of the learned members of the Academy, to whom I hasten to communicate them.

Many observers, among whom I may mention Fraunhofer, Sir D. Brewster, Foreign Associate of the Academy, and more recently Messrs. Stokes, Holtzmann, and Lorenz, have remarked certain phenomena of polarization in the light emitted by regular gratings. These phenomena, however, seem to be entirely distinct from those I am about to communicate.

On a plate of metal, say of silver, perfectly flat and polished, suppose a straight line to be drawn with a fine steel or diamond point, the surface of the metal being disturbed as little as possible, and the line so traced becoming finer and finer by degrees towards one end until it becomes imperceptible. If the plate on which the line has been so drawn be illuminated very obliquely, and in a direction perpendicular to the line, the latter will be visible in every position comprised in the common plane of incidence and reflexion, except its thin end, which, from its great tenuity, will be imperceptible. If the plate, illuminated in the same manner, be placed in the field of a microscope, a much larger portion of the thin end of the line will be visible, the amount depending on the power of the lenses employed.

Under these circumstances, the plate being viewed perpendicularly to its surface, if a doubly-refracting analysing prism be placed between the eye and the eyepiece of the microscope, the different intensity of the ordinary and extraordinary images at once indicates that the light emitted by the brilliant line thus drawn is more or less polarized,—the amount of polarization being greatest towards the fine end of the line, and the plane of polarization being parallel to the direction of the line. Lines produced in various ways were observed, and, provided they were sufficiently fine, always with the same result.

It was observed, moreover, that the plane of polarization, which was always parallel to the lines at their thinnest extremities, was perpendicular to them where they were a little broader, and became altogether imperceptible towards their thicker ends.

On varying the angle of incidence of the light on the plate and the angle at which the line was viewed, which could be done within certain limits, depending on the shape of the microscope, by inclining the plate and shifting the position of the source of light, certain changes were observed in these phenomena, of which the following are the principal.

The plate being still observed normally, if the line be illuminated less and less obliquely, the source of light being brought nearer to the eyepiece of the microscope, the amount of polarization rapidly decreases and soon becomes insensible.

. If the plate be illuminated perpendicularly, but observed very obliquely though perpendicularly to the line, the phenomena are the same as in the first case, the light reflected from the finest extremity of the line being still polarized in the direction of the line itself.

If the plate be kept in the same position as in the last case, but the source of light be brought nearer to the eyepiece, so that the plate is both observed and illuminated obliquely and nearly in the same direction, the polarization of the bright line is much increased, not only the fine end polarizing the light parallel to itself, but the same species of polarization being observed towards the thicker part of the line where the light was formerly polarized in the opposite direction, and even where the light was previously unaffected.

The distinct polarizing power of single lines traced on a silver surface having been thus demonstrated, it was easily foreseen that the same property would be possessed by the innumerable striæ produced on metallic surfaces by rubbing them with a body covered with some hard substance reduced to a fine powder; and it was anticipated that the large number of these lines would compensate for the want of brilliancy in each individually, so that the phenomena would be visible without the need of a microscope.

On a silver surface, a straight striated band about 2 centims. broad was accordingly traced by means of a ruler and a piece of cork charged with emery*. The striated band thus produced exhibits in the most striking manner, on account of the intensity of the light reflected, the phenomena of polarization before observed in the case of single lines.

If the striated band be placed under the microscope and illuminated obliquely, innumerable lines are observed of different degrees of brightness and of various colours, due no doubt to accidental phenomena of diffraction and interference : nearly all these lines polarize the light in their own direction, some of the thicker ones, however, produce the opposite effect.

If the striated band, instead of being straight, is in the form of an arc of a circle of about 50 centims. radius, and if the surface of the plate be horizontal, and the source of light be placed vertically above the centre of the band, at such a distance from the plate that the incident light may make an angle of from 60 to 80 degrees with the normal, then, the eye being placed behind the source of light a little on one side, and being

---

* Emery powder, No. 40 of the opticians, answers very well for these experiments, the mean diameter of its particles being about $\frac{1}{100}$ millim. Emery powder No. 20, ordinary Tripoli, or English red may also be employed.

shaded from the direct rays by means of a small screen, the striated band will appear illuminated throughout the entire arc, and the reflected light may be observed directly by means of an analyser, which causes it to appear alternately more or less bright.

The polarization thus produced does not seem to depend on the nature of the metal of which the plate is formed. Gold, platinum, copper, steel, brass, speculum metal, aluminium, tin, &c., have been substituted for silver; and all these metals, when suitably striated, have presented the same phenomena without sensible difference, except with regard to the colour of the light polarized by reflexion.

Metals which themselves possess least colour, give reflexions tinged with yellow, which, when most oblique, become bronze; while in the case of metals that have a marked colour of their own, the reflexions possess the same tinge, and sometimes, indeed, in a striking degree, as is the case with copper and gold.

I should mention here that the striated bands on silver and copper and some other metals are very brilliant when they have recently been made, but that they diminish in intensity on account of the action of the vapours accidentally present in the atmosphere. Gold and platinum are of course exempt from this inconvenience.

Non-metallic substances present similar phenomena, but their reflexions are so dull that the observation is often uncertain. Polarization parallel to the lines has nevertheless been clearly observed with a plate of specular iron and with one of obsidian, both of which I owe to the kindness of M. de Senarmont. With some precautions the phenomenon may even be observed with a plate of glass.

Finally, the groups of lines on silver and copper surfaces have been moulded with black wax, gum-lac, and even with galvanic copper; and the impressions so obtained presented to all appearance the same phenomena as the surface actually furrowed.

Among the various experiments that were tried, I may mention the case of a scratched metal surface which at first gave a reflexion distinctly polarized parallel to the lines, but which was afterwards covered with varnish. Under these circumstances the polarization became hardly sensible—a result which seems most naturally explicable on the ground of the change of the direction of the rays owing to the refraction of the varnish, which prevented the incident light from impinging at that angle which is required to produce the polarization in question. In fact, when glass plates differently cut were fastened to the striated plate by means of a varnish of turpentine or Canada balsam, the polarization was immediately reproduced in the refracting substance whenever the direction of the ray was such that the phenomenon would have been produced in the air. It is, however, well known that instruments constructed of brass, copper, and bronze, which are rendered brilliant by polishing, are generally covered with a coat of varnish for the purpose of preserving the brightness of the metal, and there is therefore no reason for surprise that the reflexions from these metals, however bright they may be, present no sensible polarization. When these surfaces have not been varnished, and when, moreover, they have been obtained by the action of polishing substances not too coarse, the phenomena are invariably observed.

No mention has yet been made of the light regularly reflected by a striated surface. In this direction the observation is not so easy, because the rays reflected by the furrows mingle with those reflected from the smooth surface of the mirror. Nevertheless, on observing with a microscope single lines suitably illuminated, a sensible polarization may be detected which varies little with the angle of reflexion; but what is strange is that the direction of this polarization is opposite to that which is produced in the former case, being *perpendicular* to the lines on the surface. The same fact may even be observed on regarding a striated surface directly by means of an analyser. In this case, to render the phenomenon sensible, it is sufficient to trace two bands of lines at right angles on a silver surface, and to observe the reflected rays almost normally, the source of light being a white surface equally illuminated. The direction of the lines being perpendicular in the two bands, and each band polarising the light perpendicularly to its direction, it results that the two bands give opposite polarizations, and that the phenomenon becomes more sensible by contrast.

The same effect may be observed, without much difference, whether the incident rays are normal or more or less oblique, the reflected rays invariably possessing a partial polarization sensibly perpendicular to the direction of the lines.

This phenomenon is rendered very obvious on striating a space of some centimetres diameter on a polished plate by means of a lathe; the portion covered with concentric lines, being regarded with an analyser, presents two dark tufts similar to the tufts discovered by M. Haidinger.

This phenomenon may be rendered still more visible by uniformly furrowing the entire surface of two mirrors, placing them parallel and opposite to each other, and causing light to be reflected repeatedly from one to the other: at each of these reflexions an additional portion of light is polarized; so that after several reflexions in a direction which may be as near the normal as possible, the polarized light greatly exceeds the non-polarized in quantity.

This polarization of the regularly reflected ray may be observed with different metals without sensible modification,—the metals which have the least effect in producing ordinary polarization, such as gold and silver, giving results sensibly the same as those produced by the metals that polarize the most, such as platinum and zinc. This result seems to preclude us from attributing these effects to partial polarization by the sides of the furrows, which is the first idea which naturally occurs to us. Such reflexions appear at least not to be the principal cause of the phenomenon.

In order to advance another step in the study of these singular phenomena, it was necessary to ascertain the exact dimensions of the small furrows which possess properties so singular. For this purpose recourse was had to exceedingly thin layers of silver deposited on glass from certain chemical solutions, and which may not only be used to replace the tin amalgam for silvering looking-glasses, but have even been employed with success by M. Foucault in the construction of a new species of telescope.

The first glass plate, A, was covered with a very thin, though still perfectly opake layer of silver, and a straight band of striæ was produced on the metallic surface by means of a piece of cork covered with very fine emery. This band presented the phenomena above described, viz. polarization of the dispersed light parallel to its own direction, and polarization of the reflected light perpendicular to the same. On examining, by means of the microscope, the state of the striated layer by transmitted light, it was easy to perceive that the furrows had not in general penetrated through the metallic layer, some deeper than the rest forming exceptions.

In order to ascertain the thickness of the metallic layer, a piece of iodine was placed on one point of the surface, and the coloured rings due to the film of iodide of silver produced by the evaporation of the iodine were suffered to spread from this point, which itself became completely transparent, owing to the change of the silver layer throughout its entire thickness into yellow iodide. From the part where the silver had not been in any degree affected by the emanations of iodine, to the transparent spot of iodide, there was a series of coloured rings commencing with white, and which, on being observed through a red glass, appeared to be nine in number. The series terminated in the middle of the ninth brilliant ring. The index of refraction of the iodide of silver being 2·246 (as deduced from the angle of polarization, which was 66°), the ninth bright ring gave for the iodide of silver a thickness of $\frac{1}{838}$ millim. From the thickness, composition, and known density of the iodide, the thickness of the layer of silver was calculated to be $\frac{1}{3500}$ millim.

A second glass plate, B, was covered with silver-leaf, opake like the former, except for rays of more than a certain degree of intensity; for on looking at the sun through this plate, the body of that luminary was visible, stripped of his rays, and of a rich blue colour. I am persuaded that this silver-leaf contained a small quantity of gold.

. The metallic surface so produced was found too unequal and too loosely adherent to the glass to be scratched with emery like the former. On being rubbed, however, with pure cotton, the polarization of the dispersed light was very distinct. Under the microscope it appeared to be torn in every direction, so that nothing could be inferred from this experiment; nevertheless, for the sake of comparison, the thickness of the metal was determined by the same means as before. In this case the rings were four in number, the layer becoming transparent in the middle of the fourth, whence it was concluded that the thickness of the metal was not more than $\frac{1}{90000}$ millim. This extreme thinness of beaten silver-leaf agrees with what we know with respect to the gold leaves used in gilding; these leaves, which, as is well known, are transparent and of a green tinge, have a medium thickness of about $\frac{1}{10.000}$ millim., as I ascertained by weighing measured quantities of them. Three different specimens gave the values 0·000108, 0·000095, and 0·000091 millim.

Returning then to the employment of glass plates covered with a layer of silver chemically deposited, one was obtained (which we will call C) not so thick as the former, as was proved by its greater transparency. The light transmitted by this plate was of a greyish blue, the reflected light of a yellowish blue, with this property, that towards the angle of maximum polarization it became of a pure blue for light polarized normally to the plane of reflexion. The rings of iodide formed on this surface terminated in the middle of the second bright line, beyond which the layer was completely transparent, from which it was concluded that the thickness of the silver layer was $\frac{1}{27,200}$ millim. This metallic film, when scratched with the same emery as the former, polarized the dispersed light very distinctly. On being examined microscopically, a very large number of the furrows seemed to have penetrated through the metal; but the majority did not, and consequently must have been less than $\frac{1}{27,200}$ millim. deep. It was ascertained also that the lines which, when illuminated obliquely, produced a distinct polarization in their own direction, were of the latter kind.

Lastly, a fourth layer of silver, D, was prepared thinner than all the others; it was more transparent than the last, and the light reflected by it possessed the same properties to a greater degree. Scratched like the former, the dispersed light was feebler, but still quite sensible and polarized as before. On being viewed with the microscope, it was found that the number of furrows that penetrated entirely through the metal was much greater than before; but that the polarization of the dispersed light was still 'due solely to those striæ which were merely superficial, and consequently were less than $\frac{1}{54.410}$ millim. deep,—such being the thickness of the silver, as estimated by the coloured rings of iodide, which terminated in the middle of the first dark line.

A general view of the results just described naturally leads one to anticipate that light may suffer changes of the same nature as that above described when it traverses extremely fine slits. The experiments, with a description of which I am about to close this memoir, prove that this is the case, and that under these circumstances also phenomena are produced closely related to the preceding.

It is well known that, to reproduce certain experiments of interference and refraction, small instruments are constructed which are to be found in all physical cabinets: these are slits with narrow walls and straight parallel edges, which can be approximated to each other from a distance of several millimetres to actual contact. If a pencil of light be suffered to pass through such a slit, the opening of which has been reduced to such a degree as to suffer but a trace of the light to pass, the emergent ray is invariably observed to be polarized in a direction perpendicular to the slit, the polarization being stronger in proportion as the slit is narrower.

If a very bright light be employed, and if the observation be conducted with a microscope, openings still narrower may be observed; and by slightly inclining one edge to the other, a slit may be obtained in the field of the microscope which gradually decreases until the edges actually meet. Now in this case the light which passes near the point of contact is almost entirely polarized in a direction perpendicular to the slit.

This phenomenon was at first attributed to the repeated reflexions of the light from one side of the slit to the other, reflexions which necessarily give rise to several phenomena of polarization; but we shall see directly that there are circumstances hard to reconcile with this explanation.

The first trials were made with a slit whose edges were of brass; steel, copper, and, lastly, silver were substituted for this metal: the phenomenon was but little modified, and that in a way which by no means agreed with the polarizing power of these different metals. Silver, for example, which by itself has but little polarizing power, polarizes almost completely the light which traverses a very narrow slit of which it forms the sides.

Moreover, in reducing the thickness of the sides so as to render them mere edges, the phenomenon is equally produced; and it then becomes difficult to imagine the existence of reflexions sufficiently numerous to cause the observed effect.

Bodies of the most dissimilar nature, so disposed as to present a narrow slit, give rise to the same phenomenon, provided the sides of the slit be well polished. Flint, glass, obsidian, ivory, fluor-spar, have in this respect afforded almost the same result.

Having observed that the edges of the slit require to be well polished, which agrees with the theory that the polarisation is due to repeated reflexions, the effect of removing this cause of polarization was tried, the edges of the slit being covered with lampblack. Under these circumstances the polarization was entirely destroyed, and the idea of repeated reflexions seemed thus to be confirmed, only, however, to be again overthrown; since on restoring the polish of one side only, the other being still covered with lampblack, the polarization reappeared distinctly, and in this case the existence of repeated reflexions seemed impossible. This experiment, taken in conjunction with the preceding, seems to show that there is some particular cause for the phenomenon.

In order to throw some light on this subject, the attempt was made to examine the phenomenon when the light undergoes total reflexion from the sides of the slit, such reflexion having, as is well known, no polarizing effect.

After several unsuccessful attempts, no better method was discovered than that of observing the edge of a soap-bubble formed in a straight tube. Under these circumstances the soap-bubble forms a very thin film of liquid at the centre, perpendicular to the axis of the tube, and terminated by two opposite concave menisci. Placing this under the microscope and illuminating it properly, it was easy to distinguish through the walls of the tube a bright line, formed by the light that had traversed the film, either directly, or having undergone none but total reflexions from its sides.

When the bubble was thick, the light thus transmitted exhibited no sensible polarization; but when the bubble was sufficiently thin to give rise to coloured rings by reflexion, partial polarization was invariably found present in the bright line, the plane of polarization being, as before, perpendicular to the direction of the slit. This observation was repeated in various ways, and always with the same result.

Finally, the slits in the thin layers of gold and silver mentioned above, were examined with regard to the above phenomena.

Leaves of beaten gold were at first examined with the microscope: they presented numerous rents and slits that offered no particular characteristics; but in some of the leaves, especially in their thicker parts, natural fissures were discovered extremely narrow, certainly less than $\frac{1}{5000}$ millim. across, which were clearly polarized, especially towards their narrowest ends, the plane of polarization being, as before, perpendicular to the direction of the slit. These polarized slits are rather rare in gold-leaf; and I never found them in the thinner and more transparent parts, but only in those whose thickness cannot be much less than $\frac{1}{5000}$ millim. This property seems to indicate that the phenomenon is only produced when the layers have a certain thickness, below which it is insensible.

The same conclusion was come to on observing the thin layers of silver already mentioned. In fact layers C and D, which were excessively thin and transparent, exhibited no sensible polarization even in their finest slits; while the somewhat thicker and more opake layer A, whose thickness amounted to $\frac{1}{5500}$ millim., exhibited the phenomenon in a remarkable manner, and with certain curious peculiarities. Some of the narrowest of the lines which penetrated through the entire thickness of the metal were polarized, some partially, others almost totally, the plane of polarization being still perpendicular to their length. On employing solar light, the phenomenon of coloured light appeared in conjunction with that of polarization; for on observing the polarized lines with a doubly-refracting prism, the two images appeared in certain cases of complementary colours.

I believe I have now communicated the principal facts which I have discovered relatively to the polarization,

1. Of light dispersed from furrowed metal surfaces,
2. Of the rays regularly reflected by such surfaces,
3. Of light which has traversed very narrow slits.

With the permission of the Academy I shall for the present content myself with the foregoing statement of facts, without entering on the premature and, as yet, uncertain question of the causes and connexion of the phenomena above detailed.

LXVII. *On the Cubical Compressibility of certain solid Homogeneous Bodies.* By M. G. WERTHEIM[*].

IN 1848 I published a memoir on the proportion between the elongation and transverse contraction of a homogeneous isotropic elastic bar when subjected to longitudinal traction. After having called attention to the fact that the value of this ratio as determined by Poisson's analysis, viz. $\frac{1}{4}$, had never been confirmed by any conclusive experiments, I shewed in the case of certain substances which I was enabled to submit to direct

[*] Translated from the *Comptes Rendus*, vol. li. p. 969.

experiment by means of a method susceptible of almost any degree of accuracy, that this ratio was really $\frac{1}{2}$. Subsequent experiments, less direct, but made on a greater variety of bodies, have confirmed this result.

These researches have given rise to numerous discussions. Several distinguished geometricians, without repeating my experiments, and without disputing their accuracy, have endeavoured to bring them into accordance with the ancient theory by various and, unfortunately also, very arbitrary hypotheses. I shall shortly mention and discuss these hypotheses before describing my new experiments on this subject.

In a memoir published shortly after mine, M. Clausius expressly acknowledged that the bodies on which I had experimented may be considered as truly homogeneous and isotropic; but he thinks that the secondary elasticity discovered by M. Weber in silk threads, and which I have observed in several organic bodies, may serve to explain this disagreement between experiment and the ancient theory. This secondary action being added to the true or primary elongation, will give rise to an excessive actual elongation, the numerator of the fraction being thus increased to such a degree that, though really equal to $\frac{1}{4}$ when only the primary elongation is considered, it in fact approximates to $\frac{1}{2}$.

Against this explanation it may be objected, in the first place, that it is founded on a fact absolutely hypothetical, no one having yet observed this secondary action either in metals or in glass, which are the only bodies I experimented on. Some experiments of M. Weber are, it is true, appealed to, according to which the transverse note produced by a metallic wire which has been suddenly stretched, becomes lower for several seconds. Seebeck has in fact endeavoured to explain this phenomenon on the ground of the secondary action above mentioned—contrary, however, to the very plausible opinion of M. Weber himself, who attributes it solely to the diminution of the temperature of the wire produced by its elongation, and its gradual return to the temperature of the surrounding atmosphere. But even admitting the hypothesis of M. Seebeck, this lowering of the note is at all events far too small for its corresponding secondary elongation to account for the numerical result of my experiments; and M. Clausius is therefore obliged to suppose that, in the case of metals, this species of elongation principally takes place during the first quarter of a second, and consequently produces no sensible effect on the note. But the primary action itself not being instantaneous, how is it possible to fix the limit of time beyond which the effect ought to be considered as secondary? It is thus that hypotheses accumulate.

A still graver objection to this hypothesis is that, contrary to all our theoretical notions and all the results of experience, we should be obliged to suppose that this secondary elongation produces no corresponding transverse contraction, since otherwise the ratio between the two observed quantities, namely the total longitudinal elongation and the total transverse contraction, would always remain that indicated by the ancient theory.

I have been obliged to enter into these details in consequence of the perseverance with which certain physicists have for twelve years opposed this theory to mine, and the manner in which they insist on treating as a demonstrated scientific truth that which M. Clausius himself only regarded as a hypothesis, and one indeed to which no great importance was to be attached.

MM. Lamé and Maxwell admit that the ratio above defined, or, what comes to the same thing, the ratio between the cubic and linear compressibilities, may vary in different substances. Experience alone can determine whether this is the case, as I have not failed to remark, both in my original memoir, and in several of those I have since published. M. Verdet is therefore wrong in asserting, as he does in the extract of a memoir which we shall have to discuss hereafter, and of which M. Kirchhoff is the author, that I have "endeavoured to show by numerous experiments that this ratio has in all bodies the same constant value $\frac{1}{5}$." On the contrary, while affirming and maintaining the exactitude of this value for those bodies which were the subject of my researches, I excluded those not yet submitted to experiment.

According to an interesting experiment made by M. Clapeyron on vulcanized caoutchouc, the fraction $\frac{\lambda}{\mu}$, instead of being equal to 1 according to the ancient theory, or 2 as my experiments require, attains in the case of this substance the enormous value of 2201; this fact, to which we shall return hereafter, seems to me to be explained by the results of the present memoir.

Contrary to the opinion of M. Clausius, M. de Saint-Venant attributes the disagreement between the results of my experiments and the ancient theory to a want of isotropism in the bodies on which I experimented: the author thinks "that there are as many species of mechanical homogeneousness as there are of possible curvilinear systems of coordinates, or of systems of conjugate orthogonal surfaces"—in fact, that we may imagine as many species as we please of non-isotropic homogeneousness; but what he has failed to show is that any such heterotropy really exists in the bodies I experimented on, and, which is absolutely incredible, that it exists to the same degree in them all.

But without going so far as this, and without comparing bodies chemically different, if we attribute to a body one of the species of homogeneousness imagined by M. de Saint-Venant, as for instance cylindrical or spherical homogeneousness, or any other, we ought at least to be able thus to explain the results of the various experiments to which these bodies may be submitted.

It would, for example, be easy to invent a molecular arrangement such that a cylindrical piezometer would possess a cubical compressibility conformable to that given by the ancient theory; but it would be necessary to show also that this cylinder, when subjected to longitudinal traction, would exhibit the elongation, and at the same time the transverse contraction, which is shown by experiment, that its resistance to torsion might be determined beforehand, &c.

As long as this demonstration has not even been attempted, all discussion on these hypotheses is necessarily futile.

Lastly, M. Kirchhoff has just published an important memoir on this point, which it will be necessary for me to examine with the degree of attention due to the name of the author and the interest of the subject. Instead of indulging in mere conjectures, M. Kirchhoff has devised the following experiment:—A weight applied to the end of a lever produces simultaneously the flexion and torsion of a homogeneous cylinder; these two displacements are measured exactly by means of an ingenious application of Gauss's method; and their ratio, which is independent of the modulus of elasticity and the radius of the cylinder, gives by known formulæ the required relation between the elongation and transverse contraction.

This method is liable to numerous objections. It would be difficult to imagine one more indirect and consequently more subject to error: the coefficient of the change of volume is determined by two distortions, which are themselves unaccompanied by any change of volume whatever; this at least is what is assumed in order to establish the formulæ, though it is not rigorously true; the experiment may be considered as the flexion of a cylinder which has been rendered non-homogeneous by torsion, or the torsion of a cylinder rendered heterogeneous by flexion; and the ordinary formulæ for torsion and flexion, which are inexact in themselves (as I think I have sufficiently shown in the case of the former, and as I shall hereafter endeavour to prove of the latter), become still less trustworthy in the present case.

M. Kirchhoff's apparatus is one of great delicacy, and does not seem as if it could possess sufficient stability for researches of this nature; the small dimensions of the cylinders subjected to experiment (less than 8 millims. in diameter, and only 145 millims. long), the sensible flexions produced originally by the mirrors and the levers supported by the cylinders, the necessity of soldering the latter in the middle, and lastly the complication of the calculations necessary to reduce the observations,´ are so many sources of inconvenience and error.

The following are, however, the results of these experiments: M. Kirchhoff found for yellow brass the value 0·387, for tempered steel 0·294: these numbers are, it will be seen, decidedly greater than $\frac{1}{4}$, while $\frac{1}{3}$ is very nearly equal to their mean value. M. Kirchhoff passes somewhat lightly over the former of these results, while he attaches great importance to the second: tempered steel appears to him to be a body eminently isotropic, while yellow brass is neither sufficiently homogeneous nor entirely free from the secondary effect before spoken of.

We have already done justice to the latter observation, which moreover applies as well to steel as it does to brass, since this secondary effect has in fact been observed in neither. As far as isotropism is concerned, it is very gratuitous to attribute that property peculiarly to a tempered body: the action that tempered glass exercises on polarized light abundantly proves this; and indeed if the least homogeneous among non-crystallized bodies were required, it would undoubtedly be on a tempered substance that the choice ought to fall.

· I am far, then, I repeat, from affirming that this ratio ought not to be somewhat less than $\frac{1}{3}$ in the case of homogeneous steel; but the present experiment does not seem to me to be conclusive on the subject.

The experiment, on the contrary, made on yellow brass is the first in which my results have been attempted to be verified on one of the substances I myself employed. The number 0·387 is, it is true, greater than $\frac{1}{4}$; but I shall prove, in a memoir on flexion I am about to produce, that the denominator of the fraction which represents in M. Kirchhoff's results the ratio of the torsion to the flexion, is too little, and that this fraction, properly corrected, approaches much more nearly to the value $\frac{1}{4}$.

In short, setting aside for the moment the experiment of M. Clapeyron, no fact has hitherto been advanced to show that the required relation is different in different bodies. The experiments that have been made, moreover, refer only to a small number of bodies; they have been made by means of methods all more or less indirect; and the cubical compressibility has never itself been the subject of any direct experiment; so that we know not whether the proportion supposed to exist between the pressures and the diminutions of volume really does exist for changes of pressure, however small. This research will be the subject of the memoir which I shall shortly have the honour of submitting to the Academy.

LXVIII. *On a New Electrometer (the Siphon Electrometer) for measuring the Electrical Charge of the Prime Conductor of a Machine; and on the Dispersion of different Liquids by Electrical Action.* By THOMAS TATE, *Esq.**

THE electrometer most commonly used by electricians, for ascertaining the intensity of the charge of the prime conductor of a machine, is Henley's. According to the experiments of Sir W. Snow Harris, it appears that the degrees of divergence of this instrument for high charges are nearly in proportion to the quantities of electricity generated; at the same time it must be observed that this result does not agree with what theoretical investigation would give. This instrument therefore would serve very well in all ordinary cases for indicating the power of any machine, provided that all the instruments used were constructed exactly alike; but this is practically impossible; and hence it follows that the degree of divergence of one instrument cannot be compared with that of another instrument, as regards its indication of electrical charge. The hydrostatic and balance electrometers are very complete instruments as regards scientific research; but it must be allowed that they are too delicate in their construction and mode of action to be used on ordinary occasions, when only an approximate value of an electric charge is required; moreover, the thermo-electrometer is only applicable to high charges of an electrical battery.

The siphon electrometer, represented in the annexed diagram, is sufficiently delicate and reliable in its indications, and admits of being constructed so that the results derived from one instrument may be fairly compared with those derived from another instrument. It depends on the principle, that different quantities of electricity discharge different quantities of liquid from a siphon-tube in which the liquid is suspended by capillary action.

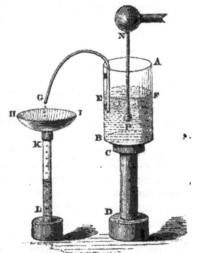

A B a glass jar, containing water, about 4 inches in diameter, placed upon the insulating stand C D of gutta percha; E G a small siphon about ·15 of an inch diameter, cemented to the side of the jar, as shown in the diagram; H I a funnel-shaped receiver

* Communicated by the Author.

about 8 inches diameter, connected with the ground by a damp cord, and placed directly below the orifice G of the tube, and connected with a glass tube, K L, divided into tenths and hundredths of a cubic inch; N P a conducting wire fixed to the prime conductor of the electrical machine and dipping into the liquid E F. The instrument is used in the following manner:—

A sufficient quantity of water is poured into the jar, so as to cause the siphon to act; the water then flows through the siphon until its pressure in the jar is balanced by the capillary action of the tube G E, when it will cease to flow; it will then be found that the level of the water in the jar stands somewhat above the orifice G of the siphon-tube: scarcely any amount of shaking or oscillation will now cause the water to flow from the orifice G. The graduated tube K L is then placed below the orifice G, the bottom of the funnel being from $2\frac{1}{4}$ to $2\frac{1}{2}$ inches from this orifice. The machine is then turned, and the electric action causes the water to flow in a continuous stream or jet from the orifice G, filling the tube K L; any proposed number of revolutions being given to the machine in a known time, the number of cubic inches of water discharged is taken as the measure of the efficiency of the machine.

It will be hereafter shown, the machine being in a fixed state of action, that in order to produce a given or constant discharge of liquid, the product of the number of revolutions of the machine by the time in which these revolutions are made must be a constant quantity. Thus, if n revolutions, performed in t seconds, produce a discharge of k cubic inches of water, and n_1 revolutions, performed in t_1 seconds, produce the same discharge of liquid, then

$$nt = n_1 t_1.$$

Hence it follows that (within certain limits) the relative efficiency of a machine (in different states of action, or of different machines) will be inversely as the product of the number of revolutions by the time requisite for producing a given or fixed amount of discharge. Thus, if a machine discharges one-half of a cubic inch of water in 20 revolutions per 60 seconds, and another machine discharges the same amount of water in 15 revolutions per 40 seconds, then the powers of the machines will be as

$$\frac{1}{20\times 60} \text{ to } \frac{1}{15\times 40}, \text{ or as 1 to 2.}$$

The following results of experiments show the uniformity of the action of the instrument.

24 revolutions of the machine, in 45 seconds, produced a discharge of ·61 of a cubic inch of water; and the experiment being repeated for two successive times, the discharges were found to be ·62 and ·61.

When the machine was in a higher state of action, 20 revolutions, in 60 seconds, produced a discharge of 1·06 cubic inch; and upon repeating the experiment the discharge was found to be 1·04 cubic inch.

Within certain limits, the quantity of liquid discharged is not sensibly affected by the diameter of the tube G E, or by the distance of the cup from the nozzle G : other things being the same, the diameter of the tube may vary from $\frac{3}{10}$ths to $\frac{4}{10}$ths of an inch; and the distance of the cup from the nozzle may vary from 2¼ to 2½ inches without sensibly affecting the amount of discharge. In like manner, slight variations in the diameter of the jar produced little or no sensible alteration in the amount of the liquid discharged.

At the commencement of the operation, the columns of liquid being in equilibrium, the electrical action has simply to overcome the cohesion of the particles of the liquid for one another; but as the process goes on, the water being more and more discharged, the equilibrium of the columns is destroyed, and the resistance to the discharge increases, so that, with the same electric force, the rate of discharge of the water becomes less and less; but when the section of the jar is ˙considerable, and the volume of the liquid displaced does not exceed a certain limit, the velocity with which the liquid is discharged is nearly uniform.

The following experimental results show that, for *equal quantities of water discharged* (*the machine being in a fixed state of action*), *the product of the number of revolutions by the corresponding time is* (*approximately*) *a constant quantity.*

Experiment I.

Number of revolutions of the machine, *n*.	Corresponding time in seconds, *t*.	Corresp. discharge in parts of a cubic inch, *k*.	Value of *n* × *t*.
15	70	·69	1050
25	42	·70	1050
18	60	·69	1080
12	90	·69	1080

The following experiment was made when the machine was in a different state of action.

Experiment II.

24 revolutions of the machine, performed in 80 seconds, produced a discharge of ·72 cubic inch of water; and 33 revolutions, in 60 seconds, produced the same amount of discharge.

In this case 24 × 80 = 1920, and 33 × 60 = 1980.

Experiment III.

10 revolutions, performed in 50 seconds, produced a discharge of ·41 ; and 20 revolutions, in 25 seconds, produced nearly the same discharge.

Here $10 \times 50 = 20 \times 25 = 500$.

Now, assuming that the quantity of electricity generated is in proportion to the number of revolutions of the machine, then it follows that a certain quantity of electricity, acting for 50 seconds, produces the same (or nearly the same) discharge as a double quantity of electricity acting for half the time; and so on, similarly to the results of the other experiments.

If ϵ be put for the quantity of electricity generated in t seconds, and ϵ_1 for the quantity generated in t_1 seconds, then for equal amounts of discharge we shall have

$$\epsilon \times t = \epsilon_1 \times t_1 ;$$

that is, *for equal amounts of discharge the quantities of electricity are in the inverse ratio of the times.*

In Experiment III. the angular velocities of the machine are as $1 : 4$, and therefore the intensities of the electricity generated would be in the same ratio, provided that no electricity had been carried off by the discharge of the liquid; but the deflections of Henley's electrometer indicated that the ratio of the intensities of the electricity in the two states of the conductor was only about 2 to 3.

Although equal volumes of water are discharged, we cannot infer that the dynamic effects are equal; for the liquids are respectively discharged with different velocities.

Let $k =$ the cubic inches of water discharged in each case, w being its weight in units of lbs.

$n =$ the corresponding number of revolutions in t seconds in the one case, ϵ being the amount of electricity generated, and v the velocity with which the liquid is discharged.

$n_1 =$ the corresponding number of revolutions in t_1 seconds in the other case, ϵ_1 being the quantity of electricity generated, and v_1 the velocity with which the liquid is discharged.

$u, u_1 =$ the accumulated work or dynamic effect in each case respectively.

Then $u = \dfrac{v^2 w}{2g}$, and $u_1 = \dfrac{v_1^2 w}{2g}$;

$$\therefore \frac{u}{u_1} = \left(\frac{v}{v_1}\right)^2 ;$$

but $\dfrac{v}{v_1} = \dfrac{t_1}{t}$, on the assumption that the velocities are uniform,

$$\therefore \frac{u}{u_1} = \left(\frac{t_1}{t}\right)^2.$$

Now by experiment we have

$$nt = n_1 t_1, \text{ and } \therefore \frac{n}{n_1} = \frac{t_1}{t};$$

$$\therefore \frac{u}{u_1} = \left(\frac{n}{n_1}\right)^2;$$

that is, *the discharge being constant, the dynamic effects are in the ratio of the squares of the number of revolutions of the machine.*

If $n = 20$, and $n_1 = 10$, as in Exp. III., then $\dfrac{u}{u_1} = 4$; that is, in this case the dynamic effects will be as 1 to 4, or a double number of revolutions produces a quadruple effect.

Let x be put for the number of revolutions of the machine (its state being constant) performed in 60 seconds, or 1 minute, to produce the same discharge as n revolutions in t seconds; then

$$x \times 60 = nt, \text{ or } x = \frac{nt}{60}.$$

Similarly, we have for the relation of equal discharge corresponding to any other state of the machine,

$$x_1 = \frac{n_1 t_1}{60}.$$

Now it may be presumed that the efficiency of the machine is inversely proportional to the number of revolutions per minute requisite to produce a given discharge; but we find

$$\frac{x_1}{x} = \frac{n_1 t_1}{nt};$$

that is, *the efficiency of a machine varies inversely as the product of the number of revolutions by the corresponding time requisite to produce a given discharge.*

If $t = t_1$, then

$$\frac{x_1}{x} = \frac{n_1}{n};$$

that is, in this case *the efficiency of a machine varies inversely as the number of revolutions* (the time being constant) *requisite to produce a given discharge.*

Thus, if a machine discharges half of a cubic inch of water in twenty revolutions in a certain time, and another machine discharges the same amount of water in ten revolutions in the same time, then the latter machine will have double the power of the former.

On the Dispersion of different Liquids by Electrical Action.

The siphon-electrometer enables us to determine the rate at which electrical charges will disperse different liquids. The liquid to be examined being placed in the jar A B, and the siphon being brought to act in the usual manner, the discharge produced by a given number of revolutions of the machine in a given time is determined; and having previously found the amount of pure water discharged by the same number of revolutions performed in the same time, we are enabled to estimate the dispersiveness of the particular liquid, as compared with that of pure water, under the same electrical action. In this manner, saturated solutions of chloride of sodium, carbonate of soda, and other salts were examined; and it was found that, *under the same electrical action, the volumes of the liquid discharged were in the inverse ratio of their specific gravities.* It will be observed that these liquids are all good conductors of electricity. But the case is very different with respect to liquids which are imperfect conductors of electricity, such as turpentine, fixed oils, and alcohol.

In twenty revolutions per minute the discharge of pure water was about three-fourths of a cubic inch; whereas with the same electrical action only about one-fourth of a cubic inch of turpentine was discharged, and not more than one-tenth of a cubic inch of fixed oil. Under the same electrical action, the volume of alcohol discharged did not exceed one-fifth of a cubic inch. Now although the specific gravities of these liquids are less than that of water, yet their dispersiveness, under the same electrical action, is considerably less than that of water.

Hastings, May 18.

LXIX. *Proceedings of Learned Societies.*

ROYAL SOCIETY.

[Continued from p. 393.]

June 14, 1860.—General Sabine, R.A., Treasurer and Vice-President, in the Chair.

THE following communications were read:—
"Notes of Researches on the Poly-Ammonias."—No. VIII. Action of Nitrous Acid upon Nitrophenylenediamine. By A. W. Hofmann, LL.D., F.R.S.

The experiments of Gottlieb have shown that dinitrophenylamine, when boiled with sulphide of ammonium, is converted into a remarkable base, crystallizing in crimson needles, generally known as nitrazophenylamine, and for which, in accordance with the views I entertain regarding its constitution, I now propose the name Nitrophenylenediamine. I owe to the kindness of Dr. Vincent Hall a considerable quantity of this substance, which is not quite easily procured.

I have made a few experiments with this compound in the hope of obtaining some insight into its molecular constitution. If, bearing in mind the numerous analogies between the radicals ethyle and phenyle, we assume that the latter, by the loss of hydrogen, may be converted into a diatomic molecule, phenylene $C_6 H_4$, corresponding to ethylene, the existence of a group of bases corresponding to the ethylene-bases cannot be doubted.

$$\text{Ethylamine} \quad \left. \begin{array}{c} C_2 H_6 \\ H \\ H \end{array} \right\} N^*. \qquad \text{Ethylenediamine} \quad \left. \begin{array}{c} (C_2 H_4)'' \\ H_2 \\ H_2 \end{array} \right\} N_2.$$

$$\text{Phenylamine} \quad \left. \begin{array}{c} C_6 H_6 \\ H \\ H \end{array} \right\} N. \qquad \text{Phenylenediamine} \quad \left. \begin{array}{c} (C_6 H_4)'' \\ H_2 \\ H_2 \end{array} \right\} N_2.$$

With the last-named body agrees in composition the compound known as semibenzidam, or azophenylamine, which Zinin obtained by exhausting the action of sulphide of ammonium on dinitrobenzole.

Those chemists, however, who have had an opportunity of becoming acquainted with the well-defined properties of ethylenediamine, will not be easily persuaded to consider the uncouth dinitrobenzol-product—sometimes appearing in brown flakes, sometimes as a yellow resin, rapidly turning green in contact with the air—as standing to smooth phenylamine in a relation similar to that which obtains between ethylenediamine and ethylamine; we much more readily admit a relation of this description between phenylamine and Gottlieb's crimson-coloured base, in which the clearly pronounced character of the former is still distinctly visible, although of necessity modified by the further substitution which has taken place in the radical.

$$\text{Phenylamine} \quad \left. \begin{array}{c} C_6 H_6 \\ H \\ H \end{array} \right\} N.$$

$$\text{Phenylenediamine} \quad \left. \begin{array}{c} (C_6 H_4)'' \\ H_2 \\ H_2 \end{array} \right\} N_2.$$

$$\text{Nitrophenylenediamine} \quad \left. \begin{array}{c} (C_6[H_2(NO_2)])'' \\ H_2 \\ H_2 \end{array} \right\} N_2.$$

Does the latter formula really represent the molecular constitution of the crimson needles? The degree of substitution of this body might have been determined by the frequently adopted process of ethylation. But even a simpler and a shorter method appeared to present itself in the beautiful mode of substituting nitrogen in the place of hydrogen, lately discovered by P. Griess. The red crystals

* $H = 1$; $O = 16$; $C = 12$, &c.

undergo, indeed, the transformation, which he has already proved for so many derivatives of ammonia, with the greatest facility.

On passing a current of nitrous acid into a moderately concentrated solution of the nitrate of the base, the liquid becomes slightly warm, and deposits on cooling a considerable quantity of brilliant white needles, the purification of which presents no difficulty: sparingly soluble in cold, readily soluble in boiling water, the new compound requires only to be once or twice recrystallized. Thus purified, this substance forms long prismatic crystals, frequently interlaced, white as long as they are in the solution, but assuming a slightly yellowish tint when dried, and especially when exposed to 100°: they are readily soluble both in alcohol and in ether. The new body exhibits a distinctly acid reaction; it dissolves on application of a gentle heat in potassa and in ammonia, without, however, neutralizing the alkaline character of these liquids; it also dissolves in the alkaline carbonates, but without expelling their carbonic acid. The new acid fuses at 211° C., and sublimes at a somewhat higher temperature, with partial decomposition. The sublimate consists of small prismatic crystals.

Analysis proves this substance to contain

$$C_6 H_4 N_4 O_2,$$

a formula which is confirmed by the analysis of a silver-compound,

$$C_6 (H_3 Ag) N_4 O_2,$$

and of a potassium-salt,

$$C_6 (H_3 K) N_4 O_2.$$

The analysis of the new compound shows that, under the influence of nitrous acid, nitrophenylenediamine exchanges three molecules of hydrogen for one molecule of nitrogen, three molecules of water being eliminated.

$$\underbrace{C_6 H_7 N_3 O_2}_{\text{Nitrophenylene-diamine.}} + H NO_2 = 2 H_2 O + \underbrace{C_6 H_4 N''' N_3 O_2}_{\text{New acid.}}.$$

I do not propose a name for the new compound, which can claim but a passing interest, as throwing, by its formation, some light on the constitution of nitrophenylenediamine.

The composition of the new acid, and of its salts, shows that in the crimson-red base four hydrogen molecules are still capable of replacement; in other words, that this body contains four extra-radical molecules of hydrogen. The result of these experiments appears to confirm the view which, in the commencement of this Note, I have taken of the constitution of this body; at all events, the mutual relation of the several bodies is satisfactorily illustrated by the formulæ—

$$\text{Nitrophenylenediamine} \quad \left. \begin{array}{c} (C_6[H_3(NO_2)])'' \\ H_2 \\ H_2 \end{array} \right\} N_2.$$

$$\text{New acid} \quad \left. \begin{array}{c} (C_6[H_3(NO_2)])'' \\ N''' \\ H \end{array} \right\} N_2.$$

$$\text{Silver-salt} \quad \left. \begin{array}{c} (C_6[H_3(NO_2)])'' \\ N''' \\ Ag \end{array} \right\} N_2.$$

If the admissibility of this interpretation be confirmed by further experiments, the reaction discovered by Griess furnishes a new and valuable method of recognizing the degree of substitution in the derivatives of ammonia.

The new acid differs in many respects from the substances produced from other nitrogenous compounds. As a class, these substances are remarkable for the facility with which they are changed under the influence of acids, and more especially of bases. The new acid exhibits remarkable stability; it may be boiled with either potassa or hydrochloric acid without undergoing the slightest change. Even a current of nitrous acid passed into the aqueous or alcoholic solution is without the least effect. The latter experiment appeared of some interest; for if the action of nitrous acid, in a second phase of the process, had assumed the form so frequently observed by Piria and others, it might have led to the formation of the diatomic nitrophenylene-alcohol, according to the equation

$$\left. \begin{array}{c} (C_6[H_3(NO_2)])'' \\ H_2 \\ H_2 \end{array} \right\} N_2 + 2H\,NO_2 = 2H_2O + N_4 + \left. \begin{array}{c} (C_6[H_3(NO_2)])'' \\ H_2 \end{array} \right\} O_2.$$

It deserves to be noticed that nitrophenylenediamine, although derived from two molecules of ammonia, is nevertheless a decidedly monacid base. Gottlieb's analyses of the chloride, nitrate, and sulphate left scarcely a doubt on this point. However, as some of the natural bases, quinine for instance, are capable of combining with either one or two molecules of acid, I thought it of sufficient interest to confirm Gottlieb's observations by some additional experiments. The crystals deposited on cooling from a solution of nitrophenylenediamine in concentrated hydrochloric acid, were washed with the same liquid and dried *in vacuo* over lime.

Analysis led to the formula

$$\left[H \left\{ \begin{array}{c} (C_6[H_3(NO_2)])'' \\ H_2 \\ H_2 \end{array} \right\} N_2 \right] Cl.$$

The dilute solution of this chloride is not precipitated by dichloride of platinum, nor can the double salt of the two chlorides be obtained by evaporating the mixed solutions, which, just as Gottlieb observed it, is readily decomposed with separation of metallic platinum. I had, however, no difficulty in preparing a platinum salt, crystallizing in splendid long brown-red prisms, by adding the dichloride of platinum to the *concentrated* solution of the hydrochlorate.

The platinum determination led to the formula

$$\left[H \left\{ \begin{array}{c} (C_6[H_2(NO_2)])'' \\ H_2 \\ H_2 \end{array} \right\} N_2 \right] Cl, PtCl_2.$$

These experiments prove that, even under the most favourable circumstances, nitrophenylenediamine combines only with 1 equiv. of acid, while the ethylene-derivatives are decidedly diacid. The diminution of saturating power in nitrophenylenediamine, at the first glance, seems somewhat anomalous, but the anomaly disappears if the constitution of the body be more accurately examined. It cannot be doubted that the diminution of the saturating power is due to the substitution which has taken place in the radical of the diamine. I have pointed out at an earlier period [*], that the basic character of phenylamine is considerably modified by successive changes introduced into the phenyl-radical by substitution. Chlorphenylamine, though less basic than the normal compound, still forms well-defined salts with the acids; the salts of dichlorphenylamine, on the other hand, are so feeble, that, under the influence of boiling water, they are split into their constituents; in trichlorphenylamine, lastly, all basic characters have entirely disappeared. Again, on examining the nitro-substitutes of phenylamine, we find that even nitrophenylamine is an exceedingly weak base, whilst dinitrophenylamine is perfectly indifferent. What wonder, then, that a molecular system, to which in the normal condition we attribute a diacid character, should, by the insertion of special radicals, be reduced to monoacidity? The normal phenylenediamine, which remains to be discovered, will doubtless be found to be diacid, like the diamines derived from ethylene. Even now the group of diacid diamines is represented in the naphtyl-series:

$$\text{Naphtylamine} \quad \left. \begin{array}{c} C_{10}H_7 \\ H \\ H \end{array} \right\} N \text{ monoacid.}$$

$$\text{Naphtylenediamine} \quad \left. \begin{array}{c} (C_{10}H_6)'' \\ H_2 \\ H_2 \end{array} \right\} N \text{ diacid.}$$

The body which I designate by the term Naphtylenediamine, is the base which Zinin obtained by the final action of sulphide of ammonium upon dinitronaphtaline. This substance, originally designated seminaphtalidam, and subsequently described as naphtalidine, combines, according to Zinin's experiments, with 2 equivalents of hydrochloric acid [†].

"On the Formula investigated by Dr. Brinkley for the general Term in the Development of Lagrange's Expression for the Summation of Series and for Successive Integration." By Sir J. F. W. Herschel, Bart., F.R.S. &c.

* Mem. of Chem. Soc. vol. ii. p. 298.
† Liebig's Annalen, vol. lxxxv. p. 328.

June 21.—Sir Benjamin C. Brodie, Bart., President, in the Chair.

The following communications were read:—

"Experimental Researches on various questions concerning Sensibility." By E. Brown-Séquard, M.D.

"On the Construction of a new Calorimeter for determining the Radiating Powers of Surfaces, and its application to the Surfaces of various Mineral Substances." By W. Hopkins, Esq., M.A., F.R.S.

When the author's Memoir on the Conductivity of various substances was presented to the Society, it was intimated to him on the part of the Council of the Society, that it might be advisable to determine absolute instead of relative conductivities, the latter only having been attempted in his previous experiments. It is partly in consequence of this intimation, and partly from the desire to make his former investigations more complete, that the author has given his attention to the construction of a calorimeter which might serve for this purpose. His present memoir contains a description of this instrument, with the results obtained from its application to the surfaces of various substances.

The apparatus used by Messrs. Dulong and Petit was more delicate and complete than the simpler instrument devised by the author of this paper, but it was calculated only to determine the radiating powers of substances of which the bulb of a thermometer could be constructed, or with which it could be delicately coated. The only substances to which, in fact, it was applied, were glass and silver, the radiation taking place, in the first case, from the naked bulb of the thermometer, and, in the second, from the same bulb coated with silver paper. In these cases, too, it was the whole heat radiating in a given time from the instrument, and not that which radiated from a given area, that was determined. For this latter purpose the apparatus was not well calculated, on account of the difficulty of obtaining with accuracy the area of the surface from which radiation took place. The instrument here described can be easily applied to any plane radiating surface, while the area of that surface can be easily determined to any required degree of accuracy. The quantity of heat radiating under given conditions, from a unit of surface in a unit of time, can thus be easily ascertained. The paper contains a detailed description of the instrument, and of the experiments made with it.

The following are experimental results thus obtained,—the unit of heat being that quantity of heat which would raise 1000 grs. of distilled water 1° Centigrade. The formula is that of Dulong and Petit, where

$\theta =$ temperature of the surrounding medium (the air in these experiments), expressed in Centigrade degrees;

$t =$ the excess of the temperature of the radiating surface above that of the surrounding medium, in Centigrade degrees;

$p =$ pressure of the surrounding medium (the atmosphere in these experiments), expressed by the height of the barometer in metres;

$a = 1.0077$, a numerical quantity which is always the same for all radiating surfaces and surrounding media.

Then if Q denote. the quantity of heat, expressed numerically, which radiates from a unit of surface (a square foot) in a unit of time (one minute), we have the following results for the substances specified :—

Glass.

$$Q= 9{\cdot}566\, a^\theta(a^t-1)+{\cdot}03720\left(\frac{p}{{\cdot}72}\right)^{{\cdot}45} t^{1{\cdot}233}$$

Dry Chalk.

$$Q= 8{\cdot}613\, a^\theta(a^t-1)+{\cdot}03720\left(\frac{p}{{\cdot}72}\right)^{{\cdot}45} t^{1{\cdot}233}$$

Dry New Red Sandstone.

$$Q= 8{\cdot}377\, a^\theta(a^t-1)+{\cdot}03720\left(\frac{p}{{\cdot}72}\right)^{{\cdot}45} t^{1{\cdot}233}$$

Sandstone (building stone).

$$Q= 8{\cdot}882\, a^\theta(a^t-1)+{\cdot}03720\left(\frac{p}{{\cdot}72}\right)^{{\cdot}45} t^{1{\cdot}233}$$

Polished Limestone.

$$Q= 9{\cdot}106\, a^\theta(a^t-1)+{\cdot}03720\left(\frac{p}{{\cdot}72}\right)^{{\cdot}45} t^{1{\cdot}233}$$

Unpolished Limestone (same block as the last).

$$Q=12{\cdot}808\, a^\theta(a^t-1)+{\cdot}03720\left(\frac{p}{{\cdot}72}\right)^{{\cdot}45} t^{1{\cdot}233}.$$

"On Isoprene and Caoutchine." By C. Greville Williams, Esq.

This paper contains the results of the investigation of the two principal hydrocarbons produced by destructive distillation of caoutchouc and gutta percha.

Isoprene.

This substance is an exceedingly volatile hydrocarbon, boiling between 37° and 38° C.; after repeated cohobations over sodium, it was distilled and analysed. The numbers obtained as the mean of five analyses were as follows :—

	Experiment.		Calculation.	
Carbon . . .	88·0	C^{10} 60	88·2
Hydrogen . .	12·1	H^8 8	11·8
		68	100·0

Three of the specimens were from caoutchouc and two from gutta percha. The vapour-density was found to be at 58° C. 2·40. Theory requires, for $C^{10} H^8 = 4$ volumes, 2·35. The density of the liquid was 0·6823 at 20° C.

Action of Atmospheric Oxygen upon Isoprene.

Isoprene, exposed to the air for some months, thickens and acquires powerful bleaching properties owing to the absorption of ozone. On distilling the ozonized liquid, a violent reaction takes place between the ozone and the hydrocarbon. All the unaltered hydrocarbon distils away, and the contents of the retort suddenly solidify to a pure, white, amorphous mass, yielding the annexed result on combustion :—

Experiment.		Calculation.		
Carbon	. . 78·8	C^{10}	60	78·95
Hydrogen	. . 10·7	H^8	8	10·52
Oxygen	. . 10·5	O	8	10·53
			76	100·00

This directly-formed oxide of a hydrocarbon is unique, as regards both its formula and mode of production.

Caoutchine.

Himly's analysis was correct. The mean results of three analyses are compared in the following Table with those of M. Himly :—

	Mean.	Himly.	Calculation.		
Carbon	. . 88·1	88·44	C^{20}	120	88·2
Hydrogen	. 11·9	11·56	H^{16}	16	11·8
				136	100·0

Two of the determinations, the results of which are incorporated in the above mean, were made on a substance from gutta percha. The vapour-density was :—

Experiment.	Himly.	Calculation = 4 vols.
4·65	4·46	4·6986

We now for the first time see the relation between the two hydro-carbons. It is the same as between amylene and paramylene. The author discusses the boiling-point of these bodies, and shows that they form most decided exceptions to Kopp's empirical law.

Action of Bromine on Caoutchine and its isomer Turpentine.

Caoutchine and turpentine act on bromine in precisely the same manner. One equivalent of the hydrocarbon decolorises four equivalents of bromine. To determine this point quantitatively, eight experiments were made, four with turpentine and four with caoutchine. The quantity of bromine-water employed was 20 cub. cents. =0·2527 gramme bromine.

Mean of four turpentine experiments.	Mean of four caoutchine experiments.
0·1074 grm.	0·1091 grm.

Conversion of Turpentine and Caoutchine into Cymole.

By the alternate action of bromine and sodium on caoutchine or turpentine, two equivalents of hydrogen are removed, the final result being cymole, having exactly the odour hitherto considered charac-teristic of the hydrocarbon obtained from oil of cumin, and quite distinct from that of camphogene. The liquid was identified by the annexed analyses. No. I. was from turpentine, II. and III. from caoutchine.

	Experiment.				Calculation.		
	I.	II.	III.	Mean.			
Carbon	89·2	89·5	89·5	89·4	C^{20}	120	89·6
Hydrogen	10·5	10·4	10·4	10·4	H^{14}	14	10·4
						134	100·0

Agreeing perfectly with the formula $C^{20} H^{14}$ *.

Paracymole.

At the same time that cymole is formed, there is a production of an oil having the same composition, but boiling about 300° C. The author has provisionally named it paracymole.

Action of Sulphuric Acid on Caoutchine.

Sulphuric acid acts on caoutchine, converting it almost entirely into a viscid fluid like hévéène, at the same time a very small quantity of a conjugate acid is formed, having the formula

$$C^{20} H^{18} S^2 O^4 ;$$

the composition was determined from that of the lime salt, which on ignition, &c., gave a quantity of sulphate of lime equal to 8·3 per cent. of calcium; $C^{20} H^{15} Ca S^2 O^4$ requires 8·5.

The author considers the action of heat on caoutchouc to be merely the disruption of a polymeric body into substances having a simple relation to the parent hydrocarbon. He deduces this view from the similarity in composition between pure caoutchouc, isoprene, and caoutchine.

The following Table contains the principal physical properties of isoprene and caoutchine:—

Table of the Physical Properties of Isoprene and Caoutchine.

Name.	Formula.	Boiling-point.	Specific gravity.	Vapour-density.	
				Expt.	Calculated.
Isoprene	$C^{10} H^8$	37°	0·6823 at 20°	2·44	2·349
Caoutchine	$C^{20} H^{16}$	171°	0·8420	4·65	4·699

In the calculations rendered necessary by the numerous vapour-density determinations contained in this paper, and more especially in those "On some of the products of the Destructive Distillation of Boghead Coal," the author has so repeatedly had to ascertain the value of the expression $\dfrac{1}{1+0\cdot00367\,T}$, that he was induced to calcu-

* (Note received July 27.) Both the cymole from turpentine and that from caoutchine were converted into insolinic acid by bichromate of potash and sulphuric acid. The quantitative determinations made on the silver salt of the acid were almost theoretically exact.

late it once for all for each degree of the Centigrade thermometer from 1° to 150°. As it is always easy so to manipulate as to prevent the value of T falling between the whole numbers, the Table proved a most valuable means of saving time; the author has therefore appended it to his paper in the hope of its proving equally useful to other working chemists.

"On the Thermal Effects of Fluids in Motion—Temperature of Bodies moving in Air." By J. P. Joule, LL.D., F.R.S., and Professor W. Thomson, LL.D., F.R.S.

An abstract of a great part of the present paper has appeared in the Phil. Mag. vol. xv. p. 477. To the experiments then adduced a large number have since been added, which have been made by whirling thermometers and thermo-electric junctions in the air. The result shows that at high velocities the thermal effect is proportional to the square of the velocity, the rise of temperature of the whirled body being evidently that due to the communication of the velocity to a constantly renewed film of air. With very small velocities of bodies of large surface, the thermal effect was very greatly increased by that kind of fluid friction the effect of which on the motion of pendulums has been investigated by Professor Stokes.

"On the Distribution of Nerves to the Elementary Fibres of Striped Muscle." By Lional S. Beale, M.B., F.R.S.

"On the Effects produced by Freezing on the Physiological Properties of Muscles." By Michael Foster, B.A., M.D. Lond.

"On the alleged Sugar-forming Function of the Liver." By Frederick W. Pavy, M.D.

"A new Ozone-box and Test-slips." By E. J. Lowe, Esq., F.R.A.S., F.L.S. &c.

The ordinary form of Ozone-box being very cumbersome, the present one has been contrived to supersede it*. The box is simple in construction, small in size, and cylindrical in form; the chamber in which the *test-slips* are hung is perfectly dark, and at the same time there is a constant current of air circulating through it, no matter from what quarter of the compass the wind is blowing. The air either passes in at the lower portion of the box and travels round a circular chamber twice; until it reaches the centre (where the test-slips are hung) and then out again at the upper portion of the box in the same circular manner, or in at the top and out again at the bottom of the box.

Fig. 1 represents a section of the upper portion of the box, showing the manner in which the air enters and moves along to the centre chamber (where the test-slip is hung at A), and figure 2 represents a section of the lower half of the box where the air circulates in the opposite direction, leaving the box on the side opposite to that on which it had entered.

* A specimen of the instrument was forwarded with the paper.

Fig. 1.   Fig. 2.

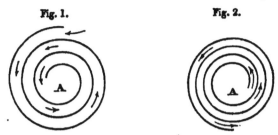

The box has been tested and found to work well.

On three different dates, when there was much ozone, test-slips were hung in one box, whilst others were hung in another which had the two entrances sealed up in order that no current of air should pass through; the result was satisfactory, viz. :—

Example.	New ozone-box.	New ozone-box sealed up.
1	10	0
2	9	0
3	9¼	0

Then again, in five examples of test-slips being exposed without any box, in comparison with those placed in this new box, the result was :—

Example.	In new ozone-box.	Exposed to north without a box.
1	10	9
2	9	9
3	7	7
4	10	5
5	2	0

The ozone-box is capable of being suspended at an elevation above the ground; and this appears to be a great advantage, because elevation seems necessary in order to get a proper current of air to pass across the test-slips; indeed as an instance it may be mentioned, that at an elevation of 20 feet there is almost always more indication of ozone than at 5 feet.

The plan adopted here is to suspend the box to a T support, it being drawn up to its proper place by means of a thin rope passing over a pulley; and there is less trouble in examining and changing the test-slips in this manner than there was in the old method.

The box, as described, is made by Messrs. Negretti and Zambra of Hatton Garden.

It has been urged that a box was scarcely necessary for ozone test-slips; but as the papers fade on exposure to light, it must be evident that in order to register the maximum amount of ozone a *dark* box is required.

Test-slips.—Paper-slips being so fragile, I have substituted others made of calico. The calico is to all intents and purposes chemically pure, containing only a few granules of starch, used in the first process of its manufacture, which it is very difficult to remove, being enveloped in the cotton fibre; it is, however, thought to be purer than the paper that is used for these test-slips, every precaution having been taken to make it so.

Results of observations.—The following Tables have been constructed from observations made between the 1st of May, 1859, and the 31st of March, 1860.

TABLE I.

Mean amount of Ozone observed from Test-slips hung for twelve hours, both at night and in the daytime, in comparison with others hung for twenty-four hours.

During the month of	Papers exposed for twelve hours.			Papers exposed for twenty-four hours.			Difference between twelve hours and twenty-four hours.	
	Day.	Night.	Differ-ence.	Day.	Night.	Differ-ence.		
1859. May	0·4	1·3	0·9	1·1	1·9	0·8	0·7	0·6
June	0·8	0·9	0·1	1·3	1·5	0·2	0·5	0·6
July	0·9	1·0	0·1	1·2	1·3	0·1	0·3	0·3
August......	0·7	1·4	0·7	1·2	1·8	0·6	0·5	0·4
September .	1·9	2·6	0·7	2·5	3·0	0·5	0·6	0·4
October ...	0·5	0·7	0·2	0·7	0·9	0·2	0·2	0·2
November ..	1·5	1·7	0·2	1·8	2·1	0·3	0·3	0·4
December...	1·7	2·0	0·3	2·1	2·5	0·4	0·4	0·5
1860. January ...	2·8	2·8	0·0	3·2	3·5	0·3	0·4	0·7
February ...	2·3	2·8	0·5	2·6	3·0	0·4	0·3	0·2
March	4·9	5·2	0·3	5·2	5·6	0·4	0·3	0·4
Mean	1·7	2·0	0·3	2·1	2·5	0·4	0·4	0·5

The ozone being always in excess in the night, and the tests exposed for twenty-four hours showing always an excess over those only exposed for twelve hours.

TABLE II.

Number of observations without any visible ozone.

Month.	During the night.		During the day.	
	Twelve hours' exposure.	Twenty-four hours' exposure.	Twelve hours' exposure.	Twenty-four hours' exposure.
1859. May	9	4	19	12
June.........	18	10	15	9
July	18	12	18	13
August......	10	4	15	9
September..	2	0	0	0
October ...	16	12	18	14
November..	10	7	10	10
December...	10	5	7	5
1860. January ...	8	6	7	5
February ...	12	6	9	9
March	0	0	0	0
Number of days...	113	66	118	86

Mean amount of ozone with the box suspended at the height of 25 feet.

1859. December 24 hours' exposure $=3.0$		48 hours' exposure $=5.0$
1860. January... 24 hours' exposure $=3.9$		48 hours' exposure $=4.5$
February 24 hours' exposure $=3.7$		48 hours' exposure $=5.4$
March ... 24 hours' exposure $=5.9$		48 hours' exposure $=6.4$

Mean amount of ozone with the box suspended at the height of 40 feet, March 1860, with twenty-four hours' exposure $=7.1$.

CAMBRIDGE PHILOSOPHICAL SOCIETY.

[Continued from vol. xviii. p. 316.]

October 31, 1859.—A communication was made by Mr. Hopkins "On the construction of a new Calorimeter for determining the Radiating Power of the Surfaces of Heated Bodies."

November 14.—A communication was made by the Master of Trinity College "On the Mathematical part of Plato's Meno."

November 28.—The Rev. Dr. Donaldson read a paper "On the Origin and proper value of the word 'Argument.'"

The author first investigated the etymology and meaning of the Latin verb *arguo*, and its participle *argutus*. He showed that *arguo* was a corruption of *argruo* = *ad gruo*; that *gruo* (in *argruo*, *ingruo*, *congruo*) ought to be compared with κρούω, which means "to dash one thing against another," especially for the purpose of making a shrill, ringing noise; that *arguo* means "to knock something for the purpose of making it ring, or testing its soundness," hence "to test, examine, and prove anything;" and that *argutus* signifies "made to ring," hence "making a distinct, shrill noise," or "tested and put to the proof." Accordingly *argumentum* means *id quod arguit*, "that which makes a substance ring, which sounds, examines, tests, and proves it."

It was then shown that these meanings were not only borne out by the classical usage of the word, but also by the technical application of "argument" as a logical term. For it is not equivalent to "argumentation," or the process of reasoning; it does not even denote a complete syllogism; though Dr. Whately and some other writers on logic have fallen into this vague use of the word, and though it was so understood in the disputations of the Cambridge schools. The proper use of the word "argument" in logic is to denote "the middle term," *i. e.* "the term used for proof." In a sense similar to this the word is employed by mathematicians; and there can be no doubt that the oldest and best logicians confine the word to this, which is still its most common signification.

. The author entered at some length into the Aristotelian definition of the *enthymeme*, which may be rendered approximately by the word "argument." He also explained how the words "topic" and "argument" came to denote the subject of a discourse or even of a picture. He showed, by a collection of examples from the best English poets, that the established meanings of the word "argument" are reducible to three: (1) a proof or means of proving; (2) a process of reasoning or controversy made up of such proofs; (3) the subject matter of any discourse, writing, or picture. And he maintained that the second of these meanings ought to be excluded from scientific language.

December 12.—The following paper from the Astronomer Royal was read, "Supplement to the proof of the Theorem that 'Every Algebraic Equation has a Root.'"

The author expressed his want of confidence in every result obtained by the use of imaginary symbols, and in this supplement demonstrated that the left-hand member of every algebraic equation of the form $\phi(x) = 0$ admitted of resolution, either into real linear factors, or into real quadratic factors.

Professor Miller also made a communication "On a new portable form of Heliotrope, and on the employment of Camera Lucida prisms and right-angled prisms in surveying."

February 13, 1860.—The Rev. H. A. J. Munro read a paper "On the Metre of an Inscription copied by Mr. Blakesley, and printed by him in his 'Four Months in Algeria,' p. 285."

February 27.—The Rev. Professor Sedgwick made the following communications :—

1. "An account of Mr. Barrett's progress in the Survey of Jamaica, with some remarks on the Distribution of Gold Veins."

2. "Some account of the Geological Discoveries in the Arctic Regions."

March 12.—The Rev. Professor Challis made a communication "On the Planet within the orbit of Mercury, discovered by M. Lescarbault."

By a recent comparison of the theory of Mercury's orbit with observation, M. Leverrier found that the calculated secular motion of the perihelion of that planet requires to be increased by 38″, and that this difference between observation and theory cannot be accounted for by the attractions of known bodies of the solar system. In a letter addressed to M. Faye, and published in the Paris Meteorological Bulletins of October 4, 5, and 6, 1859, he suggested that the difference might be due to the attraction of a group of small planets circulating between Mercury and the Sun. On December 22 of the same year, M. Lescarbault, a physician and amateur astronomer, residing at Orgères, about sixty miles south-west of Paris, announced in a letter to M. Leverrier that he had seen on March 26, 1859, a small round spot traversing the sun's disc, which he considered to be a planet inferior to Mercury. Naturally much interested by this information, M. Leverrier went to Orgères on December 31, and after closely interrogating M. Lescarbault respecting the particulars of the observation, and the instrumental means by which it was made, he returned with the conviction that the observation was trustworthy, and that a new planet had been discovered (*Comptes Rendus,* January 2, 1860, p. 40).

M. Lescarbault had long conceived the idea of detecting inferior planets by watching the sun's disk for transits, and in 1858 he put his project into execution. He was in possession of a good telescope of 3¼ inches aperture and 5 feet focal length, mounted with an altitude and azimuth movement, and provided with a finder magnifying 6 times. The power of the eyepiece employed in the observations of March 26 was 150. Not being furnished with a position-circle, he adopted the following means of obtaining angular measurements. The eyepiece of the telescope and the eyepiece of the finder each had at its focus two wires crossing at right angles, and the wires of the latter were so adjusted that a star seen at their intersection was seen at the same time at the intersection of the wires of the telescope. There were also in the eyepiece of the finder two wires parallel to, and on opposite sides of, each cross-wire, and distant by about 16′. A circular card about 6 inches in diameter, and graduated to half degrees, was placed concentric with the tube of the eyepiece of the finder, and apparently could be moved both about the tube and, with the tube, about the axis of the finder. A cross-wire of the telescope and a cross-wire of the finder were adjusted vertically by looking at a distant plumb-line, and the diameter of the card containing the zero of its graduation was placed vertically by means of a small plumb-line and eye-hole approximately arranged for that purpose. The mode of using this apparatus for angular measurements will be seen by the following account of the observations. The observer had also a small transit-instrument by which he obtained true time, using for timepiece his watch, which, as it only indicated minutes, required the supplement of a temporary seconds' pendulum.

In the account which M. Lescarbault gives of his observations, he says that it had been his practice to examine with the telescope the contour of the sun for a considerable interval on each day in which he had leisure, and that at length, on March 26, 1859, he saw a small round spot near the limb, which he immediately brought to the intersection of the wires of the telescope. Then, according to his statement, he quickly turned the graduated card till *two* of the wires of the finder were tangents to the sun's limbs, or equidistant from them. But it is evident that to effect an angular measurement in this way, *one* of the middle wires of the finder must have been placed tangentially to the sun's limb at the point of their intersection, to which point the spot had just been brought. Assuming that this operation was performed, the angular distance of the point from the vertical diameter of the sun might be read off, as the account states that it was, by applying the plumb-line apparatus to the graduated card. This method could only give a rough measure of the angular position of a point very near the sun's limb; and in fact M. Lescarbault does not appear to have attempted to determine the position of the spot during the interval between the beginning and the end of the transit. He states that the spot had entered a little way on the sun when he first saw it, and that the time and place of entrance were inferred by estimation.

The following are the immediate results of the observations:— The spot entered at $4^h 5^m 36^s$ mean time of Orgères at the angular distance of $57° 22'$ from the north point towards the west, and departed at $5^h 22^m 44^s$, at $85° 45'$ from the south point towards the west, occupying consequently in its transit $1^h 17^m 8^s$. The length of the chord it described was $9' 14''$, and its least distance from the sun's centre $15' 22''$. M. Lescarbault also states that he judged the apparent diameter of the spot to be at most one-fourth of that of Mercury, when seen by him with the same telescope and magnifying power during its transit across the sun on May 8, 1845. The latitude of Orgères is $48° 8' 55''$, and longitude west of Paris, $2^m 35^s$.

From these data M. Leverrier ascertained, by calculating on the hypothesis of a circular orbit, that the longitude of the ascending node is $12° 59'$, the inclination $12° 10'$, the mean distance $0·1427$, that of the earth being unity, and the periodic time $19·7$ days. Also he found that the greatest elongation of the body from the sun is $8°$, the inclination of its orbit to that of Mercury $7°$, the real ratio of its diameter to Mercury's 1 to $2·58$, and that its volume is one-seventeenth the volume of Mercury on the supposition of equal densities. This mass is much too small to account for the perturbation of Mercury's perihelion. According to these results, the periods at which transits may be expected are eight days before and after April 2 and October 5, the body being between the earth and sun near its descending node at the former period, and near its ascending node at the latter.

After the announcement of this singular discovery, it was found that other observations of a like kind had been previously made. Several instances are collected by Professor Wolf in the tenth number of his *Mittheilungen über die Sonnenflecken*, eight of which are quoted in vol. xx. (p. 100) of the Monthly Notices of the Royal Astronomical Society. Two of these, the observation of Stark on October 9, 1819, and that of Jenitsch on October 10, 1802, agree sufficiently well with the calculated position of the node of the object seen by Lescarbault. But the spot seen by Stark is stated to have been about the size of Mercury.

Capel Lofft saw at Ipswich, on January 6, 1818, at 11 A.M., a spot of a 'sub-elliptic form,' which advanced rapidly on the sun's disc, and was not visible in the evening of the same day (Monthly Magazine, 1818, part 1, p. 102).

Mr. Benjamin Scott, Chamberlain of London, saw about midsummer of 1847 a large and well-defined round spot, comparable in apparent size with Venus, which had departed at sunrise of the next day (Evening Mail, January 11, 1860).

Pastorff of Buchholz records that he saw on October 28 and November 1, 1836, and on February 17, 1837, *two* round black spots of unequal size, moving across the sun at the respective hourly rates of $14'$, $7''$, and $28'$. Also he announced, January 9, 1835, to the Editor of the *Astronomische Nachrichten*, that "six times in the previous year he had seen two new bodies pass before the sun in different directions and with different velocities. The larger was about 3" in diameter, and the smaller from 1" to 1"·25. Both appeared perfectly round. Sometimes the smaller preceded, and at other times the larger. The greatest observed interval between them was 1' 16": at times they were very near each other. Their passage occupied a few hours. Both appeared as black as Mercury on the sun, and had a sharp round form, which, however, especially in the smaller, was difficult to distinguish." Schumacher considered it his duty as editor to insert the communication, but evidently did not give credit to it (*Astron. Nachr.* No. 273).

In vol. ii. of the Correspondence between Olbers and Bessel, mention is made in p. 162 of an observation at Vienna by Steinhübel, of a dark and well-defined spot of circular form which passed over the sun's diameter in five hours. Olbers, from these data, estimates the distance from the sun to be 0·19, and the periodic time thirty days. It is remarkable that Stark saw about noon *of the same day* a singular and well-defined circular spot, which was not visible in the evening. This is one of the instances in vol. xx. of the Monthly Notices of the Astronomical Society.

These accounts appear to prove that transits of dark round objects across the sun are real phænomena; but it would perhaps be premature to conclude that they are planetary bodies. If the object observed by Lescarbault be a planet, it is certainly very surprising that it has not been often seen. Schwabe, after observations of the sun's face continued through thirty-three years, has recorded no instance of such a transit. It is probable that now attention has been especially drawn to the subject, future observations, accompanied by measures (of which Lescarbault's are the first instance), may throw light on the nature of these phænomena.

April 23.—Professor De Morgan read a paper " On the Syllogism, No. IV., and on the Logic of Relations."

In the third paper were presented the elements of a system in which only *onymatic* relations were considered; that is, relations which arise out of the mere notion of nomenclature—relations of name to name, as names. The present paper considers relation in general. It would hardly be possible to abstract the part of it which relates to relation itself, or to the author's controversy with the logicians, who declare all relations *material* except those which are onymatic, to which alone they give the name of *formal*. Mr. De Morgan denies that there is any purely formal proposition except " there is the probability *a* that X is in the relation L to Y;" and he maintains that the notion 'material' *non suscipit magis et minus*; so that the relating copula is as much materialized when for L we read *identical* as when for L we read *grandfather*.

Let X . . LY signify that X stands in the relation L to Y; and X . LY that it does not. Let LM signify the relation compounded of L and M, so that X . . LMY signifies that X is an L of an M of Y. In the doctrine of syllogism, it is necessary to take account of combinations involving a sign of *inherent quantity*, as follows:—

By X . . LM'Y is signified that X is an L of *every* M of Y.

By X . . L$_{\prime}$MY it is signified that X is an L of *none but* Ms of Y.

The *contrary* relation of L, not -L, is signified by *l*. Thus X . LY is identical with X . . *l*Y. The converse of L is signified by L^{-1}: thus X . . LY is identical with Y . . L^{-1}X. This is denominated the L-*verse* of X, and may be written LX by those who prefer to avoid the mathematical symbol.

The attachment of the sign of inherent quantity to the symbols of relation is the removal of a difficulty which, so long as it lasted, prevented any satisfactory treatment of the syllogism. There is nothing more in X . . LM'Y than in every M of Y is an L^{-1} of X, or MY))L^{-1}X, X and Y being individuals; and nothing more in X . . L$_{\prime}$MY than in L^{-1}X))MY, except only the attachment of the idea of quantity to the combination of the relation.

When X is related to Y and Y to Z, a relation of X to Z follows: and the relation of X to Z is compounded of the relations of X to Y and Y to Z. And this is syllogism. Accordingly every syllogism has its inference really formed in the first figure, with both premises affirmative. For example, Y . LX and Y . . MZ are premises stated in the third figure: they amount to X . . L^{-1}Y and Y . . MZ, giving X . . *l*$^{-1}$MZ for conclusion. This affirmative form of conclusion may be replaced by either of the negative forms X . L^{-1}M'Z or X . *l*$_{\prime}$$^{-1}$mZ.

The arrangement of all the forms of syllogism, the discussion of points connected with the forms of conclusion, the extension from individual terms in relation to quantified propositions, the treatment of the particular cases in which relations are convertible, or transitive, or both—form the bulk of the paper, so far as it is not controversially directed against those who contend for the confinement of the syllogism to what Mr. De Morgan calls the *onymatic* form.

An appendix follows the paper, on syllogism of transposed quantity, in which the number of instances included in one premise is equal to the whole number of existing instances of the concluding term in the other premise.

Mr. J. H. Röhrs also read a paper "On the Motion of Bows, and thin Elastic Rods."

May 7.—The Rev. Professor Sedgwick made a communication "On the Succession of Organic Forms during long geological periods; and on certain Theories which profess to account for the origin of new species."

May 21.—The Public Orator read a paper "On the Pronunciation of the Ancient and Modern Greek Languages."

He gave a rapid sketch of the "Reuchlin and Erasmus" controversy in the sixteenth century, especially the part taken in it at Cambridge by Cheke, Smith, Ascham, and Bishop Gardiner; and then proceeded to show how the proper sounds of the Greek letters may be determined from the following sources :—

1. Distinct statements of grammarians.
2. Incidental notices in other ancient authors.
3. Variations in writing of inscriptions and MSS.
4. Phonetic spelling of cries of animals.
5. Puns and riddles.
6. The value of the respective letters in other languages employing the same alphabet, especially Latin.
7. The way in which Latin proper names are spelt in Greek, and *vice versâ*.
8. The traditions of pronunciation preserved in modern Greek.

He concluded that, on the whole, the method of Erasmus approached more nearly to the ancient pronunciation than that of Reuchlin.

"But," he proceeded, "when we consider the untrustworthiness of each of these sources of evidence taken singly, and when moreover we find them often in conflict with one another, it cannot be expected that the result should be very certain or very satisfactory. There are also other considerations which enhance the difficulty of the inquiry. As there were very marked dialectic varieties in Greece, so there may have been local variations even in Attica itself.

"The pronunciation, too, changed from time to time. Plato gives us proof of this in the 'Cratylus.'"

After quoting several instances, and showing that great changes both in pronunciation and spelling had taken place in modern languages, French, Spanish, and English, "it would," he said, "be hopeless to attempt to determine the pronunciation of any language by a reference to its orthography at a time when both were perpetually changing. But in the history of every nation there arrives a time when the creative energy of its literature seems to have spent itself; when, instead of developing new forms, men begin to look back and not forward, to comment and to criticise. Then it is that a language begins to assume, even in minor and merely outward points, such as pronunciation and spelling, a fixity and rigidity which it retains with scarcely any change so long as the nation holds together. Such a period in Greek history was that which began with the grammarian sophists in the fifth century B.C., and culminated in Aristarchus and Aristophanes of Byzantium. In the spelling and pronunciation of Greek there was probably very little change from that time to the end of the third century A.D."

October 19.—Dr. Paget made a communication "On some Points in the Physiology of Laughter."

November 12.—The Public Orator read a paper (a sequel to that on May 21) "On the Accentuation of Ancient Greek."

The question of accents was not discussed in the Reuchlin and Erasmus dispute. At that time all pronounced according to the system of accents introduced by the Greeks of Constantinople, who first taught the ancient language to the Italians.

It was probably in Elizabeth's reign that we began to disuse the old pronunciation of vowels both in Greek and Latin; and concurrently with this change we, as well as the other nations of Europe, began to pronounce Greek, not with the modern Greek, but with the Latin accent. The reasons were:—

1. Teachers speaking the modern Greek were no longer required, so the tradition was not kept up.

2. It saved much trouble to pronounce both languages with the same accentuation.

3. The Greek accent perpetually clashes with quantity; the Latin much more rarely; never, indeed, in that syllable of which the quantity is most marked—the penultima.

Isaac Vossius (1650–60) advocated the disuse of accentual marks altogether, as the invention of a barbarous age to perpetuate a barbarous pronunciation.

After showing the meaning of the word 'accent' as applied to modern languages, and discussing the accentuation of the German, English, French, &c., he proceeded to say:

"There are three methods of emphasizing a syllable:—

1. By raising the note;
2. By prolonging the sound;
3. By increasing its volume.

"Scaliger, *De Causis Linguæ Latinæ*, lib. ii. cap. 52, recognizes this division when he says that a syllable may be considered of three dimensions in sound, having height, length, and breadth.

"Now in our own language, when we accent a syllable, which of these dimensions do we increase? Generally all three, but not necessarily; for when the prayers, for example, are intoned, *i. e.* read upon one note, the accent is marked by increasing the volume of sound (the third method), which involves also a longer time in utterance, *i. e.* a lengthening of quantity. In speaking, all three methods are employed, but one more prominently than the other, according to individual peculiarities of the speakers. What we blend, the Greeks kept distinct.

"We cannot understand the Greek system unless we bear this in mind. They never confounded accent with quantity. Ineradicable habit prevents us from reverting in practice to their method, just as they would have been unable to comprehend ours.

"It is clear from Dionysius, *De Comp. Verb.* lib. xi. cap. 75, that the dialogue in tragedy preserved the ordinary accentuation, which was disregarded only in choral passages set to music."

The practical conclusion was this: that while it would be desirable, if possible, to return to the Erasmian system of pronunciation, it would be extremely absurd to adopt the barbarous accentuation of modern Greek, which has quite lost the old essential distinction between accent and quantity. In this respect, as we cannot recover practically the ancient method, it is better to keep to our own system of the Latin accent, which does not confuse the learner's notion of quantity in verse as the modern Greek does.

An Athenian boy has the greatest difficulty in comprehending the rhythm of Homer or Sophocles. Hence it is not blind prejudice (as Professor Blackie asserts) which makes us keep to our old usage, but a well-grounded conviction that we should lose more by changing than we should gain.

· November 26.—Professor Challis made a communication "On the Solar Eclipse of July 18, 1860."

December 10.—Mr. Seeley read a "Notice of Opinions on the Red Limestone at Hunstanton."

Professor Miller also described "An Instrument for measuring the radii of arcs of Rainbows."

February 11, 1861.—Mr. H. D. Macleod read a paper "On the present State of the Science of Political Economy."

The writer took a general survey of the science as it at present exists, testing several generally received doctrines by the principles of inductive logic, and earnestly enforcing the necessity of a thorough reform of the whole science, which must be constructed on principles analogous to those of the other inductive sciences.

February 25.—Dr. Humphry made a communication "On the Growth of Bones."

March 11.—The Master of Trinity made a communication "On the Timæus of Plato."

LXX. *Intelligence and Miscellaneous Articles.*

ON THE OPTICAL PROPERTIES OF THE PICRATE OF MANGANESE.
BY M. CAREY LEA.

BREWSTER and Haidinger have described a remarkable property possessed by certain crystalline surfaces, of reflecting; besides the ray normally polarized in the plane of incidence and reflexion, another ray, polarized perpendicularly to that plane, and differing from the former in being coloured—a property rendered more conspicuous by the fact that the colour of the ray so polarized abnormally is either complementary to, or at least quite distinct from the colour of the crystal itself.

I find that this property is possessed to a remarkable degree by the picrate of manganese. This salt crystallizes in large and beautiful transparent right-rhombic prisms, sometimes amber-yellow, sometimes aurora-red, exhibiting generally the combination of principal prism, and macrodiagonal, brachydiagonal and principal end planes. In describing this substance in a paper on picric acid and the picrates[*], I mentioned that in a great number of specimens examined, no planes except those parallel with or perpendicular to the principal axis had been met with. Since then I have obtained in several crystallizations specimens exhibiting a brachydiagonal doma; but this appears to be rather unusual.

The optical properties of this salt are very interesting. It exhibits a beautiful dichroism. If the crystal be viewed by light transmitted in the direction of its principal axis, it appears of a pale straw-colour, in any other direction, rich aurora-red in some specimens, in others salmon-colour. A doubly refracting achromatized prism gives images of these two colours, unless the light be transmitted along the principal axis of the crystal of picrate, in which case both are pale straw-colour.

[*] Silliman's American Journal, Nov. 1858.

But it also possesses in a high degree the property of reflecting two oppositely polarized beams ; and the great size of the crystals in which it may readily be obtained renders it peculiarly fitted for optical examination. If one of these crystals be viewed by reflected light while it is held with its principal axis lying in the plane of incidence and reflexion, the reflected light is found to be not pure white, but to have a purple shade. Examined with a rhombohedron or an achromatized prism of Iceland spar, having its principal axis in the plane of incidence and reflexion, the ordinary image is white as usual, while the extraordinary is of a fine purple colour, the phenomenon having the greatest distinctness when the light is incident at the angle of maximum polarization.

The experiment may be varied and the purple light beautifully seen without the use of a doubly reflecting prism, by allowing only light polarized perpendicularly to the plane of incidence to fall on the crystal ; in this case the surface of the crystal appears rich deep purple, no white light reaching the eye.

This property is not possessed by all the planes of the crystal, but is limited to the principal prism and brachy- and macrodiagonal end planes, in other words, to the planes parallel with the principal axis of the crystal. The brachydiagonal doma and OP planes do not possess it. Nor is it exhibited by the first-mentioned planes when the crystal is turned with its prismatic axis at right angles to the plane of incidence.

All specimens of picrate of manganese do not possess this property to an equal extent: The crystals vary considerably in colour, and those which are full red exhibit it more strongly than the amber-coloured. Picric acid boiled with aqueous solution of cyanhydroferric acid and saturated with carbonate of manganese, gives crystals of a rich deep colour, which exhibit the purple polarized beam particularly well.

These properties are not possessed by the manganese salt alone, but also by the picrates of potash and ammonia (especially when crystallized by very slow spontaneous evaporation in prisms of sufficient size), and the picrates of cadmium and peroxide of iron—with this difference, however, that while the prismatic axis of the crystal in the case of the cadmium and manganese salts must be in the plane of incidence, in the alkaline salts it must be perpendicular to that plane. As they all crystallize in the right-rhombic system, it is probable that either the alkaline salts on the one hand, or the manganese and cadmium on the other, are prismatically elongated in the direction of a secondary axis.

It is convenient that distinct phenomena should have distinct names ; and none appears to have been assigned to this. Brewster speaks of it as a "property of light," and Haidinger uses the word "Schiller" for it. The terms dichroism, trichroism, and pleiochroism are limited to properties of transmitted light. I therefore suggest for the phenomenon here in question the name *catachroism*, using the preposition κατα in the same sense as in the word κατοπτρίζω, to reflect (as a polished surface), applying it to express the property of·

reflecting two beams—one normally polarized in the plane of incidence, and the other polarized in a plane perpendicular to it.

The chromatic properties exhibited by the picrates of ammonia and potash are very remarkable in their variety. Their crystals possess:—

1st. The well-known play of red and green light. If a little very dilute solution of pure picrate of potash be spontaneously evaporated in a hemispherical porcelain basin, so as to form a network of extremely slender needles, and these be viewed by gas-light, the play of colours is singularly brilliant.

2nd. Dichroism. When by spontaneous evaporation of large quantities of solution of potash, or, better, of ammonia salt, transparent prisms of $\frac{1}{15}$ to $\frac{1}{10}$ inch diameter are obtained; these, viewed with a doubly refracting prism by transmitted light, give two images—one pale straw-colour, and the other deep brownish red.

3rd. The above-described property of catachroism, or reflexion in the plane of incidence of oppositely polarized beams.—*Silliman's American Journal*, November 1860.

EXPERIMENTS ON THE POSSIBILITY OF A CAPILLARY INFILTRATION THROUGH POROUS SUBSTANCES, NOTWITHSTANDING A STRONG COUNTERPRESSURE OF VAPOUR. POSSIBLE APPLICATION TO GEOLOGICAL PHENOMENA. BY M. DAUBRÉE.

In the grand phenomena which are to us the principal manifestations of the activity of the interior of the globe, we see every day enormous quantities of water disengaged as steam from great depths. It may be asked if these incessant losses are not partially at least made up by a supply from this surface; and if so, in what way are these infiltrations effected?

It would be difficult to imagine that this supply was produced by a free circulation; for the way open for a descent would at the same time form an outlet also for the escape of vapour; and this objection would apply more especially to the volcanic regions, where the internal vapour has sufficient tension to send columns of lava with a density two or three times that of water, to great heights above the level of the sea. In trying to reconcile these apparent contradictions, I have been led to inquire if water could not reach the deep and heated reservoirs, which yield it in a variety of ways, not by means of extended *fissures*, as has hitherto been supposed, but also by the *porosity* and *capillarity* of rocks.

M. Jamin's ingenious experiments* have shown how considerable is the influence which capillarity exerts in changing the conditions of equilibrium, established through the intervention of a liquid column, between two opposite pressures.

But in previous experiments the temperature was the same in all parts of the capillary tube. It appeared important, more especially in reference to the geological problem which I have indicated, to see what would happen if the temperature was much higher at one part of

* Phil. Mag. vol. xix. p. 204.

the capillary passage, so as to convert the liquid into vapour, and thus change it into a state in which it would probably not be subject to the laws which at first had caused its infiltration.

I have constructed an apparatus, the principal object of which was to connect, by a porous plate of fine close-grained sandstone, on the one end a closed space in which the tension of vapour measured by a manometer was $1\frac{7}{8}$ atmosphere, and on the other a space in direct communication with the air, half-filled with water, which soon reached the boiling-point, but where the pressure could not exceed that of the atmosphere.

Although the thickness of the interposed plate was only 2 centimetres, the apparatus showed that the water is not driven back by the counterpressure of vapour; the difference of pressure on the two sides of the plate does not prevent the liquid from passing from the relatively cold region towards the relatively hot one, by a sort of capillary process, favoured by the rapid evaporation and drying of the latter.

The effects of this apparatus, which I cannot explain in detail, will manifestly be materially augmented by increasing the thickness of the porous plate, and working with vapour at a higher temperature.

But even these results prove that capillarity, acting in conjunction with gravity, can, in spite of very powerful internal counterpressures, force water from the superficial and cold regions of the globe to the deep and heated parts, where, in consequence of the temperature and pressure which it acquires, the vapour becomes susceptible of producing great mechanical and chemical effects*. Do not the preceding experiments thus touch the fundamental points of the mechanism of volcanos, and of the other phenomena generally attributed to the development of vapours in the interior of the globe, especially earthquakes, the formation of certain thermal springs, the filling metalliferous veins, as well as to various cases of the metamorphism of rocks? Without excluding the primitive water generally supposed to be incorporated in the internal melted masses, do not the same experiments show that infiltrations from the surface may also be operative, so that the deeper parts of the globe would be in a daily state of giving and taking, and that by a most simple process, although very different from the mechanism of the siphon and of ordinary springs? A slow, continuous, and regular phenomenon would thus become the cause of sudden and violent manifestations, like explosions and ruptures of equilibrium. — *Comptes Rendus*, January 28, 1861.

* It is known that water penetrates into the pores of most rocks, especially those belonging to the stratified formations, as is shown by the water which they generally contain in nature. Bischoff has long called the attention of geologists to this fact. Although the granite on which the sedimentary rocks rest is usually very impermeable, it has been traversed in many places by injections of eruptive rocks. Among the latter there are some, like the trachytes, so porous that they might well be particularly suspected of establishing a permanent capillary communication between the water of the surface and the heated masses which form the base of this kind of column.

THE

LONDON, EDINBURGH AND DUBLIN

PHILOSOPHICAL MAGAZINE

AND

JOURNAL OF SCIENCE.

SUPPLEMENT to VOL. XXI. FOURTH SERIES.

LXXI. *On the Reflexion of Light at the Boundary of two Isotropic Transparent Media. By* L. Lorenz[*].

JAMIN, as is well known, discovered that Fresnel's formulæ for the intensity of the rays reflected and refracted at the boundary of two isotropic transparent media do not perfectly agree with experiment, the difference being very considerable when the angle of incidence approaches to the angle of polarisation. Cauchy had already proved that, under these circumstances, waves must be produced with longitudinal vibrations; and having assumed that these waves were absorbed very rapidly (though not instantaneously, since in that case he would have returned to the formulæ of Fresnel), he now introduced a correction into the formulæ which caused them to agree with experiment.

All calculations, however, which have hitherto been made concerning the reflexion and refraction of light, have proceeded on the hypothesis of an instantaneous passage from one medium to the other, and a consequent instantaneous change of the index of refraction. Such a passage is, however, a mere metaphysical abstraction, which cannot possibly exist in nature; and the calculation would be more exact and more satisfactory if a gradual passage were admitted between the two media through a space which might afterwards be assumed to be as small as we please. It is, moreover, a fact that bodies are really surrounded by an atmosphere which must produce such a gradual change of refraction.

The object of this paper is to show that Jamin's experiments can only be reconciled with Fresnel's formulæ when the calculation is made on the above hypothesis.

In what follows, the case of total reflexion will not be considered.

If the incident light be polarized in the plane of incidence,

* Translated by F. Guthrie, from Poggendorff's *Annalen*, vol. cxi. p. 460.

Phil. Mag. S. 4. No. 143. *Suppl.* Vol. 21.   2 I

the angle of incidence being called x, and that of refraction x_1, the ratio of the amplitudes of the incident, refracted, and reflected light, according to Fresnel, is

$$1 : \frac{2 \cos x \sin x_1}{\sin (x+x_1)} : \frac{\sin (x-x_1)}{\sin (x+x_1)}. \quad \dots \quad (1)$$

For the light polarized perpendicularly to the plane of incidence, the ratio of the same three amplitudes is

$$1 : \frac{2 \cos x \sin x_1}{\sin (x+x_1) \cos (x-x_1)} : -\frac{\tan (x-x_1)}{\tan (x+x_1)}. \quad . \quad (2)$$

Assume now that these formulæ are correct when the difference between x and x_1 is infinitely small, so that $x_1 = x + dx$. Substituting this value of x_1, the above expressions become

$$1 : 1 + \frac{dx}{\sin 2x} : -\frac{dx}{\sin 2x}, \quad \dots \quad (3)$$

$$1 : 1 + \frac{dx}{\sin 2x} : \frac{dx}{\tan 2x}. \quad \dots \quad (4)$$

We suppose that the incident ray approaches the bounding surface of the media at an angle α, and that its direction is there gradually changed by having to traverse successive parallel refractive layers, until it emerges completely into the other medium at the constant angle β.

In order to simplify the calculation, we will in the first instance neglect the retardation of the ray.

Let A be the amplitude of the incident ray, and let this become χ and $\chi + d\chi$ for the refracted ray, when the angle of incidence, α, becomes x and $x + dx$. Then, whatever may be the polarization, we have, according to (3) and (4),

$$\frac{d\chi}{\chi} = \frac{dx}{\sin 2x};$$

from which, by integrating and determining the constants, we get

$$\chi = A \sqrt{\frac{\tan x}{\tan \alpha}}.$$

The ray reflected from this layer, if it be polarized in the plane of incidence, has, according to (3), the amplitude $-\chi \dfrac{dx}{\sin 2x}$; and if polarized perpendicularly, it has, according to (4), the amplitude $\chi \dfrac{dx}{\tan 2x}$. These two values we indicate by χdu, where, in the first case,

$$u = -\tfrac{1}{2} \log \tan x; \quad \dots \quad (5)$$

and in the second,

$$u = \tfrac{1}{2} \log \sin 2x. \quad . \quad . \quad . \quad . \quad . \quad . \quad (6)$$

The amplitude of the reflected ray is therefore

$$\chi du = A \sqrt{\frac{\tan x}{\tan \alpha}} du;$$

and when this ray encounters a layer whose angle of refraction is x_1, its amplitude becomes

$$\chi \sqrt{\frac{\tan x_1}{\tan x}} du = A \sqrt{\frac{\tan x_1}{\tan \alpha}} du.$$

At the boundary between this layer and the following, where the angle of refraction is $x_1 - dx_1$, a portion of the light is again reflected; and u_1 being the same function of x_1 that u is of x, the amplitude of the twice reflected ray is

$$-A \sqrt{\frac{\tan x_1}{\tan \alpha}} du \, du_1;$$

and when this ray has traversed all the layers until its angle of refraction has become constant and equal to β, its amplitude is

$$-A \sqrt{\frac{\tan \beta}{\tan \alpha}} du \, du_2.$$

The angle x_1 may now have all values between α and x, and x all values between α and β. The sum, therefore, of the amplitudes of all the twice reflected rays will be represented by the definite double integral

$$-A \sqrt{\frac{\tan \beta}{\tan \alpha}} \int_{u_\alpha}^{u_\beta} du \int_{u_\alpha}^{u} du_1;$$

where u_α and u_β indicate the values of u for x equal to α, and x equal to β.

In this manner the sum of the amplitudes of the rays reflected $4, 6, \ldots$ times can easily be calculated; and as the sum of the different rays that have been $0, 2, 4, 6 \ldots$ times reflected make up the whole of the refracted ray, the amplitude of the latter is

$$\frac{\tan \beta}{\tan \alpha} \left[1 - \int_{u_\alpha}^{u_\beta} du \int_{u_\alpha}^{u} du_1 + \int_{u_\alpha}^{u_\beta} du \int_{u_\alpha}^{u} du_1 \int_{u_1}^{u_\beta} du_2 \int_{u_\alpha}^{u_2} du_3 - \ldots \right], \; ($$

which we shall indicate by

$$A \sqrt{\frac{\tan \beta}{\tan \alpha}} f(u_\alpha),$$

where

$$f(u) = 1 - \int_u^{u_\beta} du \int_{u_\alpha}^u du_1 \, f(u_1).$$

From the last equation we get by differentiation

$$f'(u) = \int_{u_\alpha}^u du_1 \, f(u_1)$$

and

$$f''u = f(u),$$

which gives

$$f(u) = c e^u + c_1 e^{-u},$$

where the constants c and c_1 are to be determined by the equations

$$f(u_\beta) = 1, \text{ and } f'(u_\alpha) = 0.$$

Whence

$$f(u) = \frac{\epsilon^{u - u_\alpha} + \epsilon^{u_\alpha - u}}{\epsilon^{u_\beta - u_\alpha} + \epsilon^{u_\alpha - u_\beta}} ;$$

and the value of (7) or the amplitude of the refracted ray is

$$\frac{2 \mathrm{A} \sqrt{\dfrac{\tan \beta}{\tan \alpha}}}{\epsilon^{u_\beta - u_\alpha} + \epsilon^{u_\alpha - u_\beta}} \quad \cdot \quad \cdot \quad \cdot \quad \cdot \quad \cdot \quad \cdot \quad (8)$$

If now we wish to find the amplitude of the refracted ray polarized in the plane of incidence, which we will call B, we must in the above expression substitute

$$u = -\tfrac{1}{2} \log \tan x,$$

and we shall find

$$\mathrm{B} = 2\mathrm{A} \frac{\cos \alpha \sin \beta}{\sin (\alpha + \beta)}. \quad \cdot \quad \cdot \quad \cdot \quad \cdot \quad \cdot \quad (9)$$

If, on the other hand, in (8) we substitute

$$u = \tfrac{1}{2} \log \sin 2x,$$

we find for B′, the amplitude of the portion of the refracted ray polarized perpendicularly to the plane of incidence,

$$\mathrm{B'} = 2\mathrm{A} \frac{\cos \alpha \sin \beta}{\sin (\alpha + \beta) \cos (\alpha - \beta)}. \quad \cdot \quad \cdot \quad (10)$$

We return, therefore, exactly to Fresnel's formulæ, which is a remarkable property of those expressions. The calculation only assumes the relations indicated by (8) and (4); and these expressions might have been deduced from many other formulæ than Fresnel's.

The amplitude of the reflected ray, which is the sum of the

amplitudes of the rays reflected 1, 3, 5 ... times, may be similarly found, and may be expressed as follows :

$$-\int_{u_\alpha}^{u_\beta} du \int_{u_\alpha}^{u} du_1 \int_{u_1}^{u_\beta} du_2 + \int_{u_\alpha}^{'u_\beta} du \int_{u_\alpha}^{u} du_1 \int_{u_1}^{u_\beta} du_2 \int_{u_\alpha}^{u_2} du_3 \int_{u_3}^{u_\beta} du_4 - \ldots]$$

for which we will put

$$A \int_{u_\alpha}^{u_\beta} du (fu)$$

and

$$f(u) = 1 - \int_{u_\alpha}^{u} du_1 \int_{u_1}^{u_\beta} du_2 f(u_2)$$

From the last equation we get

$$f(u) = \frac{e^{u-u_\beta} + e^{u_\beta - u}}{e^{u_\alpha - u_\beta} + e^{u_\beta - u_\alpha}};$$

whence (11) or the amplitude of the reflected ray is

$$-A \frac{e^{u_\alpha - u_\beta} - e^{u_\beta - u_\alpha}}{e^{u_\alpha - u_\beta} + e^{u_\beta - u_\alpha}}. \quad \ldots \ldots \quad (12)$$

And substituting $u = -\frac{1}{2} \log \tan x$, we get for the amplitude of the portion polarised in the plane of incidence, which we will call R,

$$R = A \frac{\sin (\alpha - \beta)}{\sin (\alpha + \beta)}. \quad \ldots \ldots \quad (13)$$

If R′ is the amplitude of the ray polarized perpendicularly to the plane of incidence, and if in (12) we substitute $u = \frac{1}{2} \log \sin 2x$, we get

$$R' = -A \frac{\tan (\alpha - \beta)}{\tan (\alpha + \beta)}. \quad \ldots \ldots \quad (14)$$

In this case also, therefore, we return to Fresnel's formulæ.

The result is, that even if there be a gradual change of the index of refraction between the two media, and consequently an infinite number of reflexions at the boundary, Fresnel's formulæ nevertheless remain true so long as the thickness of the intermediate layers is infinitely small as compared with the length of a wave. If this be not the case, then the retardations of the different rays must be taken into consideration.

For the refracted light this correction is very small, and could hardly be confirmed by experiment. We shall therefore proceed to calculate it in the case of the reflected light.

. A wave which is reflected by the layer whose angle of refraction is x, or $x_1, x_2, \ldots$, and afterwards interferes with the wave reflected by the first layer, will be retarded relatively to the latter; and we may indicate the successive retardations of phase by the letters $\delta, \delta_1, \delta_2, \ldots$. These quantities are functions of $x, x_1, x_2 \ldots$; but may also be regarded as functions of $u, u_1, u_2, \ldots$

We may therefore represent the amplitude of the ray reflected once at the layer whose angle of refraction is x, by

$$A \cos (kt - \delta)\, du,$$

where t is the time, and k a constant.

For reasons analogous to those stated above, it is easy to see that the amplitude of all the reflected rays may be expressed by

$$\int_{u_\alpha} du \cos (kt - \delta) - \int_{u_1}^{u_\beta} du \int_{u_\alpha}^{u} du_1 \int_{u_1}^{u_\beta} du_2 \cos (kt - \delta + \delta_1 - \delta_2) + \ldots \Big] . \ ($$

This series is the real part of

$$A \int_{u_\alpha}^{u} du\, \epsilon^{(kt - \delta)\sqrt{-1}} f(u),$$

where

$$f(u) = 1 - \int_{u_\alpha}^{u} du_1 \int_{u_1}^{u_\beta} du_2\, \epsilon^{(\delta_1 - \delta_2)\sqrt{-1}} f(u_2).$$

From this last equation we may deduce the differential equation

$$\frac{d}{du}\Big[\epsilon^{-\delta\sqrt{-1}} f'(u) \Big] = \epsilon^{-\delta\sqrt{-1}} f(u);$$

from which we obtain another expression for series (15), since it is the real part of

$$A \Big[\epsilon^{(kt - \delta)\sqrt{-1}} f'(u) \Big]_{u_\alpha}^{u_\beta};$$

or since for $u = u_\alpha$, we have $f'(u_\beta) = 0$ and $\delta = 0$, the real part of

$$-A \epsilon^{kt\sqrt{-1}} f'(u_\alpha). \quad \cdots \cdots \cdots \quad (15')$$

If in the above differential equation we substitute

$$f(u) = \epsilon^{\lambda\sqrt{-1}} \frac{\epsilon^{u - u_\beta} + \epsilon^{u_\beta - u}}{\epsilon^{u_\alpha - u_\beta} + \epsilon^{u_\beta - u_\alpha}},$$

λ will be determined as a function of u by the following equation:

$$\frac{d^2\lambda}{du^2} + \frac{\epsilon^{u - u_\beta} - \epsilon^{u_\beta - u}}{\epsilon^{u - u_\beta} + \epsilon^{u_\beta - u}} \cdot \frac{d(2\lambda - \delta)}{du} + \frac{d\lambda}{du} \cdot \frac{d(\lambda - \delta)}{du} \sqrt{-1} = 0,$$

$\lambda = 0$ for $u = u_\alpha$, and $\frac{d\lambda}{du} = 0$ for $u = u_\beta$.

We confine ourselves now to the case in which the values of $\frac{d\lambda}{du}$ and $\frac{d\delta}{du}$ are very small. The differential equation for λ then gives by integration, since the last member vanishes,

$$\frac{d\lambda}{du} = -\frac{1}{(e^{u-u_\beta}+e^{u_\beta-u})^2}\int_u^{u_\beta}\left(e^{2(u-u_\beta)}-e^{2(u_\beta-u)}\right)\frac{d\delta}{du}\,du.$$

If in this expression we substitute the value of u, it is obvious that for all angles of incidence $\frac{d\lambda}{du}$ is small so long as $\frac{d\delta}{du}$ is so, which is the only hypothesis.

If now we substitute for $f(u)$ its values as found in (15'), then series (15) becomes the real part of

$$-Ae^{kt\sqrt{-1}}\left[\frac{e^{u_\alpha-u_\beta}-e^{u_\beta-u_\alpha}}{e^{u_\alpha-u_\beta}+e^{u_\beta-u_\alpha}}+\left[\frac{d\lambda}{du}\right]^{u=u_\alpha}\sqrt{-1}\,\right],$$

and its sum is therefore

$$-A\frac{e^{u_\beta-u_\beta}-e^{u_\beta-u_\alpha}}{e^{u_\alpha-u_\beta}+e^{u_\beta-u_\alpha}}\left[\cos kt+\sin kt\int_{u_\alpha}^{u_\beta}\frac{e^{2(u-u_\beta)}-e^{2(u_\beta-u)}}{e^{2(u_\alpha-u_\beta)}-e^{2(u_\beta-u_\alpha)}}\frac{d\delta}{du}\,du\right]$$

And substituting again in this expression for u its value $-\frac{1}{2}\log\tan x$, we get for R the amplitude of the reflected ray polarized in the plane of incidence,

$$\left.\begin{aligned}&\mathrm{R}=A\frac{\sin(\alpha-\beta)}{\sin(\alpha+\beta)}\left[\cos kt+\sin kt\tan\Delta\right]\\[2mm]&\tan\Delta=\frac{\sin\alpha\cos\alpha}{\sin^2\alpha-\sin^2\beta}\int_\alpha^\beta\left[\cos^2\beta\tan x-\sin^2\beta\cot x\right]\frac{d\delta}{dx}\,dx.\end{aligned}\right\}(16)$$

If, on the other hand, for u we substitute $\frac{1}{2}\log\sin 2x$, we get for the amplitude of the reflected ray polarized-perpendicularly to the plane of incidence,

$$\left.\begin{aligned}&\mathrm{R}'=-A\frac{\tan(\alpha-\beta)}{\tan(\alpha+\beta)}\left[\cos kt+\sin kt\tan\Delta'\right]\\[2mm]&\tan\Delta'=\frac{\sin 2\alpha\sin 2\beta}{\sin^2 2\alpha-\sin^2 2\beta}\int_\alpha^\beta\left[\frac{\sin 2x}{\sin 2\beta}-\frac{\sin 2\beta}{\sin 2x}\right]\frac{d\delta}{dx}\,dx.\end{aligned}\right\}\cdot(17)$$

From these equations it may be seen that $\tan\Delta$, for all angles of incidence, is small provided $\frac{d\delta}{du}$ is so also; while, on the contrary, $\tan\Delta'$ may be infinite, as when $\sin 2\alpha=\sin 2\beta$; that is, when the angle of incidence is equal to the angle of polarization, in which case α and β are complementary.

If Δ', for a given angle of incidence, is a small positive quantity, it gradually approaches $\frac{\pi}{2}$ as the angle of incidence approximates to the angle of polarization, and afterwards approaches π. If, on the contrary, Δ' is a small negative quantity, on changing the angle of incidence Δ' approaches $-\frac{\pi}{2}$ and $-\pi$.

The retardation of the phase of the reflected ray R', compared with the other polarized in the plane of incidence R, may be expressed by $\Delta'-\Delta$ if the coefficients of $\cos kt$ have the same sign for both rays, that is to say, when the angle of incidence is *greater* than the angle of polarization. If, however, Δ' and Δ be always taken in the first positive or negative quadrant, we can introduce any multiple we please of 2π, and therefore express the retardation by $\Delta'-\Delta+2p\pi$, where p is a whole number.

If now Δ is *positive* for this angle of incidence, it will increase as the angle of incidence diminishes; and when the angle of incidence becomes less than the angle of polarization, and Δ' is taken in the first quadrant, the retardation of phase will be expressed by

$$\Delta'-\Delta+(2p+1)\pi.$$

If, on the other hand, Δ' is *negative* for an angle of incidence greater than the angle of polarization, the retardation of phase will become $\Delta'-\Delta+(2p-1)\pi$, if the angle of incidence is made less than that of polarization. These results agree with those of Jamin. In the first case Jamin puts $p=-1$, whereby the retardation of phase becomes

$$\Delta'-\Delta-2\pi \text{ for } \alpha+\beta>\frac{\pi}{2},$$

$$\Delta'-\Delta-\pi \text{ for } \alpha+\beta<\frac{\pi}{2};$$

and bodies in which this is the case ho calls "bodies of *negative* reflexion."

In the second case (Δ' negative for $\alpha+\beta<\frac{\pi}{2}$) he puts $p=1$; and the retardation of phase of these bodies, which he calls "bodies of *positive* reflexion," then becomes

$$\Delta'-\Delta+2\pi \text{ for } \alpha+\beta>\frac{\pi}{2},$$

$$\Delta'-\Delta+\pi \text{ for } \alpha+\beta<\frac{\pi}{2}.$$

Jamin found, moreover, that most bodies whose index of refraction is less than a given amount (about 1·46) give negative reflexions, while those whose index of refraction is greater than that amount give positive reflexions. Between these are bodies which at the angle of polarization produce a sudden change of phase from 0 to π. These remarkable relations between the difference of phase and index of refraction could not have been anticipated from Cauchy's theory, while on the other hand they can be immediately deduced from the theory above enunciated.

Let ρ be the distance of the first intermediate layer from that in which the angle of refraction is x, and $d\rho$ the distance of the latter from the next whose angle of refraction is $x + dx$. It will easily be seen that if a wave be reflected at these two consecutive layers, the difference of path of the two reflected waves will be equal to $2d\rho \cos x$.

If then l be the wave-length in the first medium, and therefore $l \frac{\sin x}{\cos x}$ the wave-length in the layer under consideration, we have for the difference of phase corresponding to this difference of path,

$$\frac{2\pi \sin \alpha}{l \sin x} . 2d\rho \cos x = \frac{d\delta}{dx} dx.$$

Instead of x we might introduce, as a new variable, the square of the index of refraction. If this variable be called v, we have

$$v = \frac{\sin^2 \alpha}{\sin^2 x}.$$

The limits of the variables ρ and v are, for $x = \alpha$, 0 and 1; and for $x = \beta$ they may be denoted by ρ_1 and v_1, where ρ_1 is the thickness of all the intermediate layers, and v_1 the square of the observed index of refraction.

If now the new values of $\frac{d\delta}{dx}$ and x be substituted in (16) and (17), and these expressions be integrated by parts, we have

$$\tan \Delta = \frac{4\pi}{l} \cdot \frac{\cos \alpha}{v_1 - 1} \int_1^{v_1} \rho \, dv, \quad \ldots \ldots \ldots \quad (18)$$

$$\tan \Delta' = \frac{4\pi}{l} \cdot \frac{v_1^2}{v_1 \cos^2 \alpha - \cos^2 \beta} \int_1^{v_1} \rho \left[\frac{\cos^2 \beta}{v_1^2} - \frac{\sin^2 \beta}{v^2} \right] dv. \quad (19)$$

From which it is obvious that $\tan \Delta$ is always positive, whereas $\tan \Delta'$ may either be positive or negative.

In one particular case Δ' passes suddenly from 0 to $\pm \pi$ at the angle of polarization; this is when for this angle

$$\int_1^{v_1} \rho \left[\frac{\cos^2 \beta}{v_1^2} - \frac{\sin^2 \beta}{v^2} \right] dv = 0;$$

or, as $\dfrac{\cos^2 \beta}{\sin^2 \beta} = v_1$,

$$\int_1^{v_1} \rho \left[1 - \frac{v_1}{v^2} \right] dv = 0.$$

As this integral includes positive as well as negative elements, it is evident that this equation is possible for some values of v_1. And if it be now supposed that for different bodies ρ is generally only approximately the same function of v, when v_1 increases by the positive increment dv_1, this integral increases by

$$\left[\rho_1 \left(1 - \frac{1}{v_1} \right) - \int_1^{v_1} \rho \frac{dv}{v^2} \right] dv_1,$$

which is less than

$$\left[\rho_1 \left(1 - \frac{1}{v_1} \right) - \rho_1 \int_1^{v_1} \frac{dv}{v^2} \right] dv_1 = 0.$$

This increase is therefore *positive*; and it follows that the definite integral in (19), when v_1 exceeds a certain amount, is positive, and consequently Δ' *negative* for $v_1 \cos^2 \alpha - \cos^2 \beta < 0$ or for $\alpha + \beta > \dfrac{\pi}{2}$, if the light, as in Jamin's experiments, passes from a less refracting to a more refracting body $(\alpha > \beta)$. But we have, however, seen above that this case answers to that of bodies with a positive reflexion.

Calculation therefore, like experiment, proves that positive reflexion occurs in the case of bodies with a greater index of refraction, negative reflexion in the case of bodies with a smaller index, while the difference of phase in the case of bodies which lie between these passes suddenly from 0 to $\pm \pi$.

The intensity of the reflected light polarized perpendicularly to the plane of incidence is, according to (17),

$$A^2 \frac{\tan^2 (\alpha - \beta)}{\tan^2 (\alpha + \beta)} (1 + \tan^2 \Delta') ;$$

the intensity of the reflected light polarized in that plane is, according to (16),

$$A^2 \frac{\sin^2 (\alpha - \beta)}{\sin^2 (\alpha + \beta)} (1 + \tan^2 \Delta).$$

If the ratio of these two intensities be expressed by k^2, we get

$$k = \frac{\cos (\alpha + \beta)}{\cos (\alpha - \beta)} \frac{\cos \Delta}{\cos \Delta'}. \quad \cdots \quad (20)$$

These results agree with those of Cauchy.

As Jamin has in his experiments determined these magnitudes directly, it is possible from his experiments, that is, from the principal angle of incidence, and the index of refraction, to deduce the value of Δ, and consequently by (18) the value of $\int_{1}^{v_1} \rho \, dv$. If now it be supposed that ρ can be approximately determined by $\rho_1 \frac{v-1}{v_1-1}$, the thickness of the intermediate layers can be deduced from the experiments.

In this way we have a new means of testing the theory, since this requires that the thickness in question should be small and positive. The calculation from Jamin's experiments shows that this is actually the case, though, as might be expected, no great degree of exactness can be thus attained in the determination of these quantities. I have found that the thickness of the layers in the case of the bodies experimented on lies between $\frac{1}{10}$ and $\frac{1}{100}$ of the length of a wave.

It appears, therefore, that the result of Jamin's experiments can be completely explained on the simple supposition of an exceedingly thin stratum of intermediate layers in which there is a gradual change of refractive power, a supposition which we have obviously more right to make than to omit.

Copenhagen, June 28, 1860.

LXXII. *On a Surface of the Fourth Order.*
By A. CAYLEY, *Esq.*[*]

LET A, B, C be fixed points; it is required to investigate the nature of the surface, the locus of a point P such that

$$\lambda AP + \mu BP + \nu CP = 0,$$

where λ, μ, ν are given coefficients; the equation depends, it is clear, on the ratios only of these quantities.

The surface is easily seen to be of the fourth order; it is obviously symmetrical in regard to the plane ABC; and the section by this plane, or say the principal section, is a curve of the fourth order, the locus of a point M such that

$$\lambda AM + \mu BM + \nu CM = 0.$$

The curve is considered incidentally by Mr. Salmon, p. 125 of his 'Higher Plane Curves;' and he has remarked that the two circular points at infinity are double points on the curve, which is therefore of the eighth class. Moreover, that there are two double foci, since at each of these circular points there are two tangents, each tangent of the one pair intersecting a tangent of the other pair in a double focus; hence, further, that there are

* Communicated by the Author.

four other foci, the points A, B, C, and a fourth point D lying in a circle with A, B, C, and which are such that, selecting any three at pleasure of the points A, B, C, D, the equation of the curve is in respect to such three points of the same form as it is in regard to the points A, B, C.,

Consider a given point M, on the principal section, then the equations

$$\frac{BP}{BM} = \frac{CP}{CM}, \quad \frac{CP}{CM} = \frac{AP}{AM}, \quad \frac{AP}{AM} = \frac{BP}{BM}$$

belong respectively to three spheres: each of the spheres passes through the point M. The first of the spheres is such that, with respect to it, B and C are the images each of the other; that is, the centre of the sphere lies on the line BC, and the product of its distances from B and C is equal to the square of the radius; in like manner the second sphere is such that, with regard to it, C and A are the images each of the other; and the third sphere is such that, with regard to it, A and B are the images each of the other. The three spheres intersect in a circle through M at right angles to the principal plane (that is, the three spheres have a common circular section), and the equations of this circle may be taken to be

$$\frac{AP}{AM} = \frac{BP}{BM} = \frac{CP}{CM}.$$

It is clear that the circle of intersection lies wholly on the surface.

The spheres meet the principal plane in three circles, which are the diametral circles of the spheres; these circles are related to each other and to the points A, B, C, in like manner as the spheres are to each other and to the same points. The circles have thus a common chord; that is, they meet in the point M and in another point M′. And MM′ is the diameter of the circle, the intersection of the three spheres.

It may be shown that M, M′ are the images each of the other in respect to the circle through A, B, C. In fact, consider in the first place the two points A, B, and a circle such that, with respect to it, A, B are the images each of the other; take M a point on this circle, and let O be any point on the line at right angles to AB through its middle point, and join OM cutting the circle in M′; then it is easy to see that M, M′ are the images each of the other, in regard to the circle, centre O and radius OA (=OB). Hence starting with the points A, B, C and the point M, let O be the centre of the circle through A, B, C, and take M′ the image of M in respect to this circle; then considering the circle which passes through M, and in respect to which B, C are images each of the other, this circle passes through M′; and so the circle through M, in respect to which C, A are images each of the other, and the circle through M, in respect to which A, B are images each of the other, pass each of them through M'; that is, the three circles intersect in M'.

It is to be noticed that M', being on the surface, must be on the principal section; that is, the principal section is such that, taking upon it any point M, and taking M' the image of M in regard to the circle through A, B, C, then M' is also on the principal section. It is very easily shown that the curve of the fourth order possesses this property; for M, M' being images each of the other in respect to the circle through A, B, C, then A, B, C are points of this circle, or we have

$$\frac{MA}{M'A} = \frac{MB}{M'B} = \frac{MC}{M'C};$$

that is, the equation

$$\lambda AM + \mu BM + \nu CM = 0$$

being satisfied, the equation

$$\lambda AM' + \mu BM' + \nu CM' = 0$$

is also satisfied.

The points M, M' of the curve, which are images each of the other in respect to the circle through A, B, C, may be called conjugate points of the curve. The above-mentioned circle, the intersection of the three spheres, is the circle having MM' for its diameter; hence the required surface is the locus of a circle at right angles to the principal plane, and having for its diameter MM', where M and M' are conjugate points of the curve.

In the particular case where the equation of the surface is

$$BC \cdot AP + CA \cdot BP + AB \cdot CP = 0,$$

the principal section is the circle through A, B, C, twice repeated. Any point on the circle is its own conjugate, and the radius of the generating circle of the surface is zero; that is, the surface is the annulus, the envelope of a sphere radius 0, having its centre on the circle through A, B, C. Or attending to real points only, the surface reduces itself to the circle through A, B, C. But this last statement of the solution is an incomplete one. The equation of an annulus, the envelope of a sphere radius c, having its centre on a circle radius unity, is

$$\sqrt{x^2 + y^2} = 1 \pm \sqrt{c^2 - z^2};$$

and hence putting $c = 0$, the equation of the surface is,

$$\sqrt{x^2 + y^2} = 1 \pm zi$$

(if, as usual, $i = \sqrt{-1}$), or, what is the same thing, it is

$$x^2 + y^2 + (z \pm i)^2 = 0 ;$$

that is, the surface is made up of the two spheres, passing through the points A, B, C, and having each of them the radius zero; or say the two *cone-spheres* through the points A, B, C. In other words, the equation

$$BC . AP + CA . BP + AB . CP = 0$$

is the condition in order that the four points A, B, C, P may lie on a sphere radius zero, or cone-sphere. Using 1, 2, 3, 4 in the place of A, B, C, P to denote the four points, the last-mentioned equation becomes

$$12 . 34 + 13 . 42 + 14 . 23 = 0 ;$$

and considering $\overline{12}$, &c. as quadratic radicals, the rational form of this equation is

$$\square = \begin{vmatrix} 0 , & \overline{12}^2, & \overline{13}^2, & \overline{14}^2 \\ \overline{21}^2, & 0 , & \overline{23}^2, & \overline{24}^2 \\ \overline{31}^2, & \overline{32}^2, & 0 , & \overline{34}^2 \\ \overline{41}^2, & \overline{42}^2, & \overline{43}^2, & 0 \end{vmatrix} = 0.$$

In my paper "On a Theorem in the Geometry of Position," Camb. Math. Journ. vol. ii. pp. 267–271 (1841), I obtained this equation, the four points being there considered as lying in a plane, as the relation between the distances of four points in a circle, in addition to the relation

$$\begin{vmatrix} & 1 , & 1 , & 1 , & 1 \\ 1, & 0 , & \overline{12}^2, & \overline{13}^2, & \overline{14}^2 \\ 1, & \overline{21}^2, & 0 , & \overline{23}^2, & \overline{24}^2 \\ 1, & \overline{31}^2, & \overline{32}^2, & 0 , & \overline{34}^2 \\ 1, & \overline{41}^2, & \overline{42}^2, & \overline{43}^2, & 0 \end{vmatrix} = 0,$$

which exists between the distances of any four points in a plane. The present investigation shows the signification of the equation $\square = 0$ between the distances of four points in space; viz. it expresses that the four points lie in a sphere radius zero, or cone-sphere. But the formula in question is in reality included in that given in the paper for the distances of five points in space. For calling the points 0, 1, 2, 3, 4, the relation between the distances of these five points is

$$
\begin{vmatrix}
0, & 1, & 1, & 1, & 1, & 1 \\
1, & 0, & \overline{01}^2, & \overline{02}^2, & \overline{03}^2, & \overline{04}^2 \\
1, & \overline{10}^2, & 0, & \overline{12}^2, & \overline{13}^2, & \overline{14}^2 \\
1, & \overline{20}^2, & \overline{21}^2, & 0, & \overline{23}^2, & \overline{24}^2 \\
1, & \overline{30}^2, & \overline{31}^2, & \overline{32}^2, & 0, & \overline{34}^2 \\
1, & \overline{40}^2, & \overline{41}^2, & \overline{42}^2, & \overline{43}^2, & 0
\end{vmatrix} = 0.
$$

Hence if 1, 2, 3, 4 are the centres of spheres radii α, β, γ, δ, and if 0 is the centre of a tangent sphere radius r, we have

$$\overline{01} = r \pm \alpha, \quad \overline{02} = r \pm \beta, \quad \overline{03} = r \pm \gamma, \quad \overline{04} = r \pm \delta;$$

so that, for any given combination of signs, it would at first sight appear that r is determined by a quartic equation; but by means of a simple transformation (indicated to me by Prof. Sylvester) it may be shown that the equation for r is really a quadratic one; moreover, the equation remains unaltered if the signs of α, β, γ, δ and of r, are all reversed; and r^2 has thus in the whole sixteen values. In particular, if α, β, γ, δ are each equal 0, then r^2 is determined by a simple equation (r the radius of the sphere through the four points); and if, moreover, $r = 0$, then we have for the relation between the distances of the four points, the foregoing equation $\square = 0$.

2 Stone Buildings, W.C.,
   March 25, 1861.

LXXIII. *Chemical Notices from Foreign Journals.*
By E. ATKINSON, *Ph.D., F.C.S.*

[Continued from p. 365.]

MM. LOIR and Drion* describe the following method of obtaining solid carbonic acid, merely requiring for its preparation apparatus within the ordinary reach of the laboratory. It depends on the great cold produced by the evaporation of liquid sulphurous acid. Liquid ammonia is placed in a glass vessel, and connected with the receiver of an air-pump by a vessel containing pumice impregnated with sulphuric acid. On exhausting, the temperature of the liquid ammonia rapidly sinks, and it commences to solidify at $-81°$ C.; when the pressure is reduced to 1 millim., the temperature of the liquid ammonia is $-89°\cdot5$. This is sufficient for the liquefaction of carbonic acid under the ordinary atmospheric pressure; for when a current of dry carbonic

* *Comptes Rendus,* April 15, 1861.

acid is passed through a U-tube dipping in the ammonia, a small portion of it liquefies.

By increasing the pressure to some extent, considerable quantities of carbonic acid may be readily solidified. About 150 cubic centims. of liquid ammonia are introduced into an inverted bell-jar provided with a collar, on which a plate perforated with two apertures is hermetically fitted. In the central aperture there is a tube closed at one end and reaching to the bottom of the jar; the other aperture serves to connect the apparatus with the air-pump. The carbonic acid is produced by heating dried bicarbonate of soda to reduess in a copper flask. This flask is connected with the tube dipping in the liquid ammonia, and also with a small air-manometer. All the air having been expelled from the apparatus, and the temperature of the liquid ammonia reduced to near solidification, the flask is heated until the manometer indicates a pressure of 3 to 4 atmospheres. Crystals of carbonic acid soon begin to form on the inside of the tube, and in half an hour about 25 grammes of solid carbonic acid are obtained, forming a thick layer on the inside of the tube which dips in the liquid ammonia.

This solid carbonic acid is a colourless mass, as transparent as glass; it may be detached from the tube by touching it with a glass rod, and is seen to consist of small cubical crystals. Exposed to the air, these crystals slowly evaporate without leaving any residue; they may be placed on the hand without producing any sensation either of heat or of cold: they can be scarcely seized between the fingers. Mixed with ether and exposed to the air, they form a freezing mixture, the temperature of which is −81° C.

The temperatures were observed by MM. Loir and Drion, by means of an alcohol thermometer on which two fixed points had been marked; that is, 0° the temperature of melting ice, and −40° the temperature of melting mercury. The liquid ammonia was prepared by Bussy's method[*], of passing gaseous ammonia into a flask surrounded by liquid sulphurous acid, the evaporation of which was promoted by the air-pump. In this way 6 to 7 fluid ounces may be obtained without difficulty in the course of two hours.

The following experiments by Deville[†] throw considerable light on the formation of some native minerals.

When fluoride of silicon was passed over calcined alumina heated to whiteness in a porcelain tube, fluoride of aluminium was disengaged, and staurotide formed analogous in all its pro-

---

[*] Phil. Mag. vol. xx. p. 202.
[†] *Comptes Rendus,* April 22, 1861.

perties to the natural mineral. This experiment was repeated in a modified manner.

In a porcelain tube placed vertically, a series of alternate layers of alumina and quartz were arranged, the alumina being at the bottom, and the quartz at the top; fluoride of silicon was then passed through the tube at a white heat. In this way the fluoride of silicon meeting alumina was decomposed, and staurotide formed; but the fluoride of aluminium which was formed at the same time was decomposed on coming in contact with the layer of quartz, with the formation also of staurotide and regeneration of fluoride of silicon. The same process followed with all the successive layers; so that the quartz and alumina were both converted into staurotide, and, as the last layer was quartz, as much fluoride of silicon left the apparatus as entered it. None of the fluorine was fixed, and it served no other purpose than to cause the combination of two of the most stable bodies in nature.

From the formula of topaz, which is a silicate of alumina and fluoride of silicon, it was probable that it might be formed in a similar way. But direct experiments showed that this is not the case; and Deville is inclined to think that it is formed in the moist way.

In the expectation of obtaining phenakite, Deville heated glucina in fluoride of silicon. He obtained a mineral which crystallizes well, and consists of silica and glucina, but could not be identified with any known mineral species.

When fluoride of silicon was passed over zirconia, beautiful octahedral crystals were obtained which had all the characters of the native zircon. An experiment of Deville's seems to show that a very small quantity of fluorine can produce an indefinite quantity of this mineral.

Alternate layers of zirconia and quartz were placed in a porcelain tube, commencing with the former and ending with the latter, and a current of fluoride of silicon was passed through the tube at a white heat. The zirconia in contact with fluoride of silicon was changed into zircon and volatile fluoride of zircon; the latter meeting quartz, gave zircon also and fluoride of silicon; and so on with the whole of the layers. The contents of the tube were entirely mineralized, and the quantity of fluoride of silicon which left the tube was equal to that which entered it. No fluorine had been fixed.

In a subsequent communication Deville will describe a method for obtaining metallic sulphurets by the dry way.

Schützenberger[*] has described a new class of salts, in which the electro-negative elements chlorine, bromine, iodine, &c.

---

[*] *Comptes Rendus*, January 23, 1861.

are substituted' for the basic hydrogen, the metals, &c. He
has effected this by acting on salts with such compounds as
chloride of iodine and iodide of cyanogen. The formation of
acetate of iodine will illustrate this class of actions, which is susceptible of great extension.

$$C^4 H^3 NaO^4 + I Cl = Na Cl + C^4 H^3 I O^4.$$

Acetate of	Chloride	Acetate of
soda.	of iodine.	iodine.

These bodies, as may be expected, are endowed with special
properties, and especially are very unstable.

Anhydrous hypochlorous and acetic acids mixed in equivalents
form a red mixture, which soon becomes decolorized. A slight
excess of hypochlorous acid imparts to it a red tint, which is
removed by heating the mixture to a temperature not exceeding
30°. This body is the acetate of chlorine, $C^4 H^3 ClO^4$; its composition is that of chloracetic acid, but it differs greatly in properties. It dissolves immediately in water, producing hypochlorous and acetic acids, and explodes at 100°, with formation
of chlorine, oxygen, and anhydrous acetic acid. Singularly
enough it is attacked by mercury even in the cold, with liberation of chlorine, and formation of acetate of mercury and a little
calomel—

$$C^4 H^3 ClO^4 + Hg = Cl + C^4 H^3 Hg O^4,—$$

Acetate of	Acetate of
chlorine.	mercury.

a curious instance of the replacement of chlorine by a metal.

It dissolves iodine instantaneously without becoming coloured,
and disengages chlorine; acetate of iodine is formed, a white
crystalline solid isomeric with iodacetic acid.

$$C^4 H^3 ClO^4 + I = C^4 H^3 I O^4 + Cl.$$

Acetate of	Acetate of
chlorine.	iodine.

Another mode of forming this body has been given above. It
is decomposed at 100° into iodine, oxygen, and acetate of
methyle.

$$2(C^4 H^3 IO^4) = I^2 + C^2 O^4 + C^4 H^3 (C^2 H^3) O^4.$$

Acetate of		Acetate of
iodine.		methyle.

It is decomposed by water into iodic acid, iodine, and acetic acid.

Butyrate of iodine is formed by the action of chloride of iodine
on butyrate of soda. Acetate of bromine is obtained by the
action of bromine on acetate of chlorine.

Sulphur dissolves in acetate of chlorine with disengagement
of chlorine; but the acetate of sulphur which forms is very unstable, for it soon decomposes into anhydrous acetic acid, sul-

phurous acid, sulphur, and chlorine. The action of iodide of cyanogen on acetate of silver forms iodide of silver, and apparently iodide of cyanogen.

These interesting facts are capable, as Wurtz suggests[*], of another interpretation, by assuming that the bodies are mixed anhydrous acids. Thus the acetate of chlorine is hypochlor. acetic anhydride, and Wurtz expresses its formation in the following manner :—

$$\left.\begin{matrix} Cl \\ Cl \end{matrix}\right\}\Theta + \left.\begin{matrix} C^2H^3\Theta \\ C^2H^3\Theta \end{matrix}\right\}\Theta = 2\left[\left.\begin{matrix} C^2H^3\Theta \\ Cl \end{matrix}\right\}\cdot\Theta\right].$$

Anhydrous   Anhydrous   Anhydrous hypo-
hypochlorous acetic acid.   chloracetic acid.
acid.

Lourenço has described[†] a series of new compounds, the polyglyceric alcohols. They bear the same relation to glycerine that the polyethylenic alcohols[‡] do to glycol. These latter bodies were obtained by the action of hydrochloric glycol on glycol in excess: the formation of the new bodies is quite analogous; it ensues when hydrochlorate of glycerine acts on glycerine in excess. Lourenço saturated a portion of glycerine with hydrochloric acid gas, and having added to it an equal quantity of glycerine, he heated the whole to 180° C. for several hours in a flask connected with a condenser so that the distillate fell again into the flask. The result of this action was a very thick brown liquid, which was distilled under a pressure of 10 millims. A body was obtained boiling at 220°—230° under this pressure. It had the composition $C^6H^{14}\Theta^5$, and its formation may be thus expressed :—

$$\left.\begin{matrix} C^3H^5 \\ H^3 \end{matrix}\right\}\Theta^3 + \left.\begin{matrix} C^3H^5 \\ H^2 \end{matrix}\right\}\Theta^2 = 2\left.\begin{matrix} (C^3H^5) \\ H^4 \end{matrix}\right\}\Theta^5 + HCl.$$

Glycerine.   Cl   New body.

Hydrochloric
glycerine.

This body, which Lourenço names pyroglycerine, or diglyceric alcohol, is formed by the condensation of two molecules of glycerine and elimination of one molecule of water. It is analogous in its chemical composition to Graham's pyrophosphoric acid.

$$\left.\begin{matrix} C^6H^{5'''} \\ C^3H^{5'''} \\ H^4 \end{matrix}\right\}\Theta^5 \qquad \left.\begin{matrix} P\Theta^{'''} \\ P\Theta_4^{'''} \\ H^4 \end{matrix}\right\}\Theta^5.$$

Pyroglycerine.   Pyrophosphoric acid.

* *Répertoire de Chimie*, April 1861.
† *Comptes Rendus*, February 25, 1861.
‡ Phil. Mag. vol. xix. p. 124.

Besides this there was formed at the same time another body of analogous properties, but possessing a greater viscosity. It boils at 275° to 285° under the pressure of 10 millims.; it has the composition $C^9 H^{20} O^7$, and is derived from three molecules of glycerine with elimination of two molecules of water. It is analogous to triethylenic alcohol in the series of condensed glycols.

In the crude product from which these bodies were obtained, there were several chlorine compounds which distilled at the ordinary atmospheric pressure, and were separated by fractional distillation. A portion of this, boiling between 230° and 270°, which chiefly consisted of hydrochlorate and dibydrochlorate of pyroglycerine, was treated with potash, by which chloride of potassium was formed, and a body obtained which, on purification and analysis, was found to have the composition

$$C^6 H^{12} O^4 = \left. \begin{array}{c} C^3 H^5 \\ C^3 H^5 \\ H^2 \end{array} \right\} O^4.$$

It is metameric with glycide, the existence of which has been placed out of doubt by Reboul's researches[*]. Lourenço names it *pyroglycide*; it stands in the same relation to pyroglycerine that glycide does to glycerine, being formed from it by the elimination of water.

$$\left. \begin{array}{c} C^3 H^5 \\ H^3 \end{array} \right\} O^3 - H^2 O = \left. \begin{array}{c} C^3 H^5 \\ H \end{array} \right\} O^2.$$
Glycide.

$$\left. \begin{array}{c} 2(C^3 H^5) \\ H^4 \end{array} \right\} O^5 - H^2 O = \left. \begin{array}{c} 2(C^3 H^5) \\ H^2 \end{array} \right\} O^4.$$
Pyroglycide.

There is another way of obtaining these polyglyceric alcohols, which throws some light on their formation. When glycerine was heated, and the part collected which distilled between 130° and 266°, and this portion treated with ether, an insoluble residue was left. This body gave distillates up to 300°, under a pressure of 10 millims., consisting of pyroglyceric alcohols. It is highly probable that in this decomposition, glycerine losing one molecule of water forms glycide, and this combining with one, two, or three equivalents of glycerine, forms polyglyceric compounds; just as oxide of ethylene, in acting upon one, two, or three equivalents of glycol, forms polyethylenic alcohols.

Lourenço points out that the formation of these polyethylenic alcohols suggests a plausible explanation of the formation of the different modifications of metaphosphoric acid, which are ob-

[*] Phil. Mag. April 1861.

tained by heating microcosmic salt, or acid phosphate of soda. Graham's metaphosphate of soda acts like glycide or oxide of ethylene, and by successive condensations gives rise to the different modifications of the acid.

$$\left.\begin{array}{l} P\Theta \\ Na \\ H^2 \end{array}\right\}\Theta^3 - H^2\Theta = \left.\begin{array}{l} P\Theta \\ Na \end{array}\right\}\Theta^2, \text{corresponding to } \left.\begin{array}{l} \Theta^3 H^5 \\ H \end{array}\right\}\Theta^2.$$

Acid phosphate   Metaphosphate   Glycide.
of soda.   of soda.

$$\left.\begin{array}{l} \\ \\ \end{array}\right\}\Theta^3 + \left.\begin{array}{l} P\Theta \\ Na \end{array}\right\}\Theta^2 - H^2\Theta = \left.\begin{array}{l} 2P\Theta \\ Na^2 \end{array}\right\}\Theta^4, \text{ corresponding to } \left.\begin{array}{l} 2(\Theta^3 H^5) \\ H^2 \end{array}\right\}$$

                                     Maddrell's meta-   Pyroglycide.
                                  phosphate of soda.

Freund* has made a series of experiments on the preparation of the oxygen radicals of the formic acid series by the action of metals on the chloride of these acids, a method analogous to that of the preparation of the ether radicals.

When chloride of acetyle was treated with sodium-amalgam no action ensued in the cold, and the action set up at a higher temperature produced a complete decomposition with formation of empyreumatic substances. The action of chloride of butyryle, $\Theta^4 H^7 \Theta Cl$, on sodium-amalgam gave better results. When slightly warmed together, an action was induced which disengaged heat sufficient to continue it; the flask was connected with a Liebig's condenser, so that the distillate flowed back. After a short time the mixture was distilled; the unaltered chloride passed off; the residue, consisting of mercury, chloride of sodium, and butyryle, was digested with water, and the butyryle which rose to the surface removed. The chloride of butyryle which had distilled off was treated again with sodium-amalgam, and the process repeated until a large quantity of butyryle had been accumulated. This was digested with carbonate of potash, washed, dried over chloride of calcium, and rectified. No product of constant boiling-point was obtained; but a portion distilling between 260° and 280° gave on analysis numbers agreeing with the formula of butyryle, or rather dibutyryle, $\Theta^8 H^{14}\Theta^2 = \left.\begin{array}{l} \Theta^4 H^7\Theta \\ \Theta^4 H^7\Theta \end{array}\right\}$. Its formation may be thus expressed:—

$$2\left.\begin{array}{l} \Theta^4 H^7\Theta \\ Cl \end{array}\right\} + 2Na = \left.\begin{array}{l} \Theta^4 H^7\Theta \\ \Theta^4 H^7\Theta \end{array}\right\} + 2NaCl.$$

Chloride of   Sodium. Butyryle.
butyryle.

* Liebig's *Annalen,* April 1861.

.. The action of strong potash on butyryle is very energetic; butyrate of potash is formed as well as a substance of a pleasant odour, which has the composition of the ketone of butyric acid, but in properties appears to be quite different.

Martius[*] has published an investigation on the cyanides of the metals associated with platinum. In their preparation he used the residues obtained from the manufacture of Russian platinum. The method of separating the metals which he adopted is a combination of several methods, and presents some interesting points.

The residues were finely powdered, and the larger grains of osmium-iridium separated by decantation. The residue having been dried and heated, was fused with a mixture of lead and oxide of lead, by which all the silicates and other similar impurities passed into the slag, and a lead regulus was obtained containing all the platinum metals. When this was treated with diluted nitric acid, a residue was left consisting principally of iridium and osmium-iridium. The latter was separated by decantation. To bring it into a state of fine powder, which could not be effected in the ordinary way on account of its hardness, it was melted with zinc in a carbon crucible, by which it was dissolved; when this mass was afterwards heated in a wind furnace, the zinc was expelled and the mineral left in a state of fine powder.

. The osmium-iridium was then heated in a current of oxygen; some osmic acid was formed, which volatilized, and was collected in a well-cooled receiver. The residue and the iridium were then mixed with an equal weight of common salt, and heated in a current of chlorine; the mass was dissolved in water, and the solution which contained the double chlorides was boiled with aqua regia, by which osmium was removed as osmic acid, and was received in a solution of ammonia. The residual solution was then mixed with sal-ammoniac, which precipitated everything, excepting a little rhodium, as ammonium double salt. The precipitate consisting principally of iridium, but containing also some platinum and ruthenium, was fused with cyanide of potassium to convert it into cyanides; this was boiled with hydrochloric acid to decompose excess of cyanide of potassium, and then sulphate of copper added, which gave a red precipitate of the copper salt.

By digesting this precipitate with baryta water, oxide of copper was formed and the barium double cyanides. They were easily separated by crystallization, the platinocyanide of barium being more insoluble than the iridiocyanide of barium. The small quantity of ruthenium was contained in the mother-liquor of this latter salt.

[*] Liebig's *Annalen*, March 1861.

them. He has not even alluded to my mathematics. I say, therefore, that there is no argument which I have to meet.

In prosecuting physical inquiry, it appears to be necessary to proceed by way of hypotheses. But hypotheses of themselves teach nothing: we *learn* by mathematics, as the very name implies, because by mathematics the truth of a hypothesis may be tested or established. The existence of gravity as a *force*, and the law of gravity, are truths which could not be ascertained by observation alone; but being taken to be true hypothetically, they are proved to be actually true, by the aid of mathematics.

Hence hypotheses respecting the physical forces are deserving of consideration only so far as they afford a basis for mathematical reasoning. In fact this quality of a hypothesis is a criterion of its truth, because all quantitative laws are deducible mathematically from *true* hypotheses. In selecting hypotheses for the foundation of a general theory of the physical forces, I had regard, in the first place, to their conformity with the antecedents of physical science, and then to the possibility of arguing from them mathematically. I have not met with any which in the latter respect are preferable to those I have selected, which, consequently, I have good reason to adhere to.

To assume that an atom is of constant form and magnitude, is, I admit, virtually to call it an indivisible particle, and not, as Newton does, an undivided particle, "particula indivisa." But as the hypothesis is expressed in perfectly intelligible terms, it is open to no objection, provided we admit with Newton, that if *by a single experiment* it can be shown that the supposed indivisible particle is divided when a solid mass is broken, the theory of atoms is untenable. When a physical hypothesis satisfies the condition of being expressed in terms which common experience renders intelligible, special observation and experiment, or comparisons of its mathematical consequences with facts, alone determine whether or not it be true. I do not admit that any *metaphysical* argument can be adduced either in support of, or against, a physical hypothesis. *Meta*-physics come after physics. If a general physical theory should be established on verified hypotheses, we should have a secure basis for metaphysical reasoning; and possibly it might then appear that some of the speculative metaphysics which have prevailed during the last century are without foundation.

By the same mode of reasoning, the hypothesis of a universal fluid æther, the pressure of which varies proportionally to its density, is unobjectionable *as a hypothesis*, simply because it is expressed in terms which experience has made intelligible. Whether it be a true hypothesis, that is, whether such an æther be a *reality,* is another question. It does not admit of *à priori* proof or disproof, but may be disproved by a single contradictory fact, or may receive accumulative evidence by the agreement of its mathematical results with many facts. Now I venture to assert respecting this particular hypothesis, that the mathematical evidence of its truth and of the reality of a fluid æther is so varied and comprehensive, that it may be pronounced to be all but conclusive. My reasons for this assertion are the following:—When a mathematical inquiry is made into the laws of the motion and pressure of a fluid constituted as above supposed, certain results are obtained by the formation and solution of partial differential equations which correspond to various phenomena of light. The difference of the intensities of different rays, the variation of intensity with the distance from a centre, and the law of the variation, the coexistence at the same instant of different portions of light in the same portions of space, the interference and non-interference of different rays, the composite character of light, its colour, results of compounding colours, and lastly the polarization of light, are all phenomena which have their exact analogues in the motions, as mathematically deduced, of a fluid medium whose pressure varies as its density. When the number, variety, and speciality of these analogies are considered, it seems difficult to resist the conclusion that properties of the fluid æther *explain* phenomena of light, and that the phenomena reciprocally give evidence of the reality of the æther. Some of the properties—for instance, that of transverse vibration, which accounts for polarization—have been deduced by mathematical reasoning for which I am responsible. I have, however, given to mathematicians the fullest opportunity of discussing these parts of the general argument; and when, as I hope to be able to do, I go through a revision of the propositions, further opportunity will be given. The proof of the reality of the ætherial medium, drawn from the explanations which the hypothesis of such a medium gives of phenomena of light, is an essential preliminary of my general theory of physical force, and I am well aware that on this ground the truth of the theory must be contested. If this point be carried, the rest, I think, must follow.

There is already evidence from *experiment* that the action of physical force may be explained hydrodynamically. In the Philosophical Magazine for May (p. 348), Professor Maxwell has referred to a paper by M. Helmholtz on Fluid Motion, in which the author points out that lines of fluid motion are arranged according to the same laws as the lines of magnetic force. This, which Prof. Maxwell chooses to call a "physical analogy," I of course take to be confirmatory of the hydrodynamical theory of magnetic force. In the same light I regard the experiments of Professor Wiedemann, mentioned in vol. xi. No. 42 of the ' Proceedings of the Royal Society,' the results of which point to the same conclusion.

Cambridge Observatory,
   May 22, 1861.

LXXIV. *On Phenomena which may be traced to the Presence of a Medium pervading all Space.* By DANIEL VAUGHAN.

IF the permanent change which seems to have been detected in the revolution of Encke's comet be not sufficient to establish the doctrine of a space-pervading æther, it may afford reasonable motives for examining other indications of the impediments of such a fluid to celestial motion. The direct information which can be obtained on this subject is at present very limited and uncertain. The approximate investigations hitherto given by mathematicians of the cause of the perturbation of the planets, necessarily overlook many slight effects of their mutual attraction ; and we are thus prevented from discovering the unperiodical changes which a small resistance to their movements might occasion. In addition to this, we are incommoded by the want of observations made during very long periods of time ; for these are as necessary in tracing the course of remote physical events, as an extensive base-line is in determining the distances of the fixed stars. But by investigating the necessary consequence of a resisting medium, and testing the result by a comparison with observed facts, we may be enabled to base our conclusions respecting this important question on evidence no less satisfactory than that which has already served to establish many of the received doctrines of physical science.

As there has prevailed among some astronomers an impression, not unwholly unfounded, in regard to a modification which the sun's attractive power is supposed to experience from the emission of his light, it seems advisable to give special attention to cases in which the central body is not luminous ; and certain phenomena, observed in the secondary systems and in the dark systems of space, afford evidence not vitiated by any effects which light might be expected to produce. In my communication in the Philosophical Magazine for last April, I showed that a satellite impeded by the resistance of a medium would, by an imperceptibly slow diminution of its orbit, be finally introduced into the region of instability, where its dismemberment must be inevitable, and where it must be transformed into a ring, similar

* Communicated by the Author.

in all respects to those of Saturn. · But it might be premature to suppose that the annular appendage of Saturn has originated in this manner, or that it is to be regarded as an index of mutability in the heavens, if the conclusion were not supported by investigations of a different character. Were the rings two integral solid masses, the inner one, even with the most favourable velocity of rotation, would require to be composed of materials having over two hundred times the tenacity of wrought iron to escape being ruptured, in consequence of the enormous strain arising from a preponderance of centrifugal force on one part, and of gravity on the other. Even if this danger were removed, solid rings could not be prevented from striking the planet, unless each were loaded with some inequality; and, according to the investigations of Professor Maxwell, the load must contain about four and a half times as much matter as the remainder of the ring. A slight excess or deficiency in the amount of this load would be fatal to stability; and the tendency of any fluid or loose solid matter to the locality where it occurs must add much to the serious perils and the infirmities of the annular structure.

Regarding the hypothesis of two solid rings as untenable, Professor Maxwell considers the case of their fluidity, and he arrives at the conclusion that the fluid composing them would break up into satellites, unless its density were less than $\frac{1}{17}$ of that of the primary. But, from the result deduced in my articles in the Philosophical Magazine for December 1860 and April 1861, it is evident that, in so great a proximity to the central body, any liquid matter would require a far greater density to exist in the form of independent satellites. In investigating the case of a ring of numerous solid satellites, or fragments, he finds a combination of very extraordinary conditions necessary to prevent the derangements and permanent changes which collisions and friction are expected to occasion. The bodies are to be all equal in mass, and placed in regular array around Saturn; but the intervals between them must be very great compared with the linear dimensions; and the ratio between the planet and the ring must, according to his formulæ, be greater than ·4352 multiplied by the square of the number of satellites composing the latter. When we consider the vast number of such bodies required to maintain the continuity of the ring, and the great improbability that all the immense group should have the peculiar conditions for preventing one from striking another, we may regard the essay of the eminent mathematician as a proof that the disconnected matter composing the annular appendage, whether it be fluid or solid, cannot be maintained in its present condition without the occurrence of friction and collisions between its parts.

Besides the valid objections which Professor Maxwell urges against the common idea which regards the rings as two flat solids, others of a somewhat different character have been suggested by Mr. Bond, who has embraced the opinion that the ring is fluid. But whatever be its composition, or whatever proportions of fluid and solid matter it may consist of, all its parts must have independent movements around Saturn, and velocities depending on their distance from his centre. The attraction of the planet will be an insurmountable obstacle to their conversion into satellites, and will even prevent them from concentrating in excessive numbers in any locality; but their incessant action must be attended with a constant development of heat and a gradual destruction of motion. In consequence of the necessary alteration in the orbit of its parts from this cause, the dimensions of the ring cannot always remain the same; and though it is not likely that the nearest edge is approaching the planet so rapidly as the researches of Struve and Hansen would indicate, yet, as some change of this nature is unavoidable, we cannot resist the conclusion that the rings have been introduced into the zone which they now occupy, from one in which their matter could only exist in the form of two satellites. Accordingly there appears to be no ground for any other inference than that I have adopted, in regard to the imperceptible diminution of the orbits of secondary planets by the action of a resisting medium.

In tracing the ultimate effects of a similar impediment to motion in the dark systems of remote space, we deduce so satisfactory an explanation of the temporary stars, that we may regard these celestial apparitions as indicating the existence of the same æthereal fluid, and manifesting the great revolutions to which it leads in the condition of the heavenly bodies. In my last article, I have shown that the instantaneous manner in which a secondary or a primary planet must undergo a total dismemberment on coming into fatal proximity with the central sphere harmonizes in a very decided manner with the astonishing rapidity with which temporary stars attain their greatest brilliancy. This peculiarity, taken in connexion with the comparatively slow and gradual decline, is sufficient to set aside the theory which ascribes such ephemeral exhibitions of light to the rotation of great orbs, self-luminous on one side and dark on the other. But this theory, though adopted by Arago and other eminent astronomers, is liable to a more fatal objection. This will be apparent when we investigate the circumstances necessary to make a partially luminous sphere or spheroid display its brilliancy to the inhabitants of the earth for only seventeen months, while its period of rotation has been estimated at

309 or 318 years.   Under the most favourable circumstances for manifesting such an extraordinary inequality between its periods of light and darkness, the surface of the supposed distant sphere must be nearly 200,000,000 times as great as the part of it sending light to our planet during the period of maximum brightness.   The light, moreover, must have proceeded from the verge of the invisible disk; and this circumstance, taken in connexion with the surprising brilliance of the star of 1572, together with the invariability of its position, will compel us to ascribe to the spectral orb in question a diameter far exceeding that of Neptune's orbit.   We must also regard these vast bodies as solid; for, if composed of liquid or gaseous matter, they could not have the luminosity confined to particular localities.   Even if stellar movements could permit us to suppose the existence of such stupendous spheres, the explanation would be applicable to one or two cases only; and we must therefore reject a hypothesis whose claims rests solely on the greater imperfections of others proposed to account for the same phenomena.

But investigations respecting the necessary course of physical events in the dark systems afford still more important evidence in regard to the æthereal contents of space.   Were the central body composed of solid matter, or surrounded with an atmosphere of oxygen, nitrogen, or carbonic acid, a development of heat and light might be expected to attend the dilapidation of one of the satellites, or the ultimate incorporation of its matter with the great orb; but the appearance would not correspond to that exhibited by the temporary stars.   Admitting that a solid globe, almost as large as the sun, may be rendered so highly incandescent as to shine like the star of 1572 in its greatest brilliancy, it would be impossible for it to cool so rapidly as to become invisible in the course of seventeen months. Besides this, it may be easily shown that, if our earth had a diameter of 80,000 miles, with its present density and superficial temperature, our atmosphere would have its density reduced a millionfold with an elevation of six or seven miles.   Thus, the greater mass we assign to the central body, the more narrow must we regard the atmospheric region where light can be developed by aërial compression; and the less display of lustre could we expect from this cause when a satellite fell from its stage of planetary existence.   But this difficulty will disappear when we suppose that the æther of space forms for the several great celestial bodies extensive atmospheres, which are rendered luminous by adequate compression, or rather by the chemical action it induces—a theory which becomes necessary to account for the luminosity of meteors and the perpetual brilliancy of suns.

The theory which ascribes the sun's light to the incessant fall of meteors to his surface, and which I have controverted in my article in the Philosophical Magazine for December 1858, appears to have been suggested by the recently discovered relation between heat and mechanical energy. From this it may be estimated that a pound of solid matter, falling to the sun from a distance of 35,000,000 miles, is capable of generating at his surface an amount of heat about 4000 times as great as could be developed by the combustion of a pound of coal in oxygen gas. But the large amount of heat arising from the combustion of hydrogen, and other facts and principles connected with thermal agencies, give support to the opinion that the development of heat must be proportional to the intensity of the chemical forces by which it is produced; and these must be commensurate with the degree of elasticity between the elements concerned in the calorific or the illuminating action. We have therefore no grounds for supposing that the accession of temperature imparted to the solar orb by the fall of a body, even from an infinite distance, is necessarily greater than that originating from the chemical action of an equal amount of matter, the elements of which were so elastic as to diffuse themselves into space, in opposition to the attractive power of suns and planets. If the undulatory theory of light be admitted, the medium which conveys it with nearly a million times the rapidity of sound, must have a modulus of elasticity almost 1,000,000,000,000 times as great as that of common air. But though not regarding the æther which gives birth to solar light as so inconceivably elastic, we may safely presume that no matter is better adapted for sustaining the great fountain of brilliancy by energetic chemical action, than that whose particles are associated with forces sufficiently powerful to cause its diffusion through universal space.

That the fall of meteors is far more frequent and more conspicuous on the sun than on the earth cannot be questioned. If these small bodies are to be regarded as independent occupants of space, two large spheres, moving with the same velocity through the region in which they are located, would each be likely to receive a number of them proportional to its mass multiplied by its diameter. The circumstances in which the earth and sun are placed will change, to some extent, their relative capabilities of receiving these foreign bodies; but the facts which Mr. Carrington's observations have made known in regard to meteoric phenomena on the solar disk are not inconsistent with what might be reasonably expected, and do not indicate any special provision for feeding our central luminary with regular supplies of meteorites.

The definite information which Arago was enabled to furnish respecting the sun by means of his polariscope, has recently received an important accession from the labours of Bunsen and Kirchhoff. A comparison of the spectrum of the sun with that of various metallic vapours in a state of incandescence, enabled these chemists to show that potassium, sodium, calcium, and other elements widely diffused on our globe, enter into the composition of the solar atmosphere. Their observations proved, however, that these substances, while abundant in the sun's envelope, instead of being concerned in producing his light, only exerted a negative influence—absorbing certain rays, and causing dark lines to replace the bright ones peculiar to their luminous vapours. It is therefore evident that the light from the vapours of the elements alluded to must be overpowered by the rays emanating from some other source. Professor Kirchhoff has embraced the opinion that the solid globe of the sun must have a far higher temperature and a greater illuminating power than his atmosphere; but the observations with Arago's polariscope have afforded positive evidence that solar light does not emanate from an incandescent solid or liquid body. It appears more philosophical to conclude that the light of our great luminary originates, not from the vapours discoverable in its atmosphere, but from a more subtle æthereal medium combined with them, and possessed of far greater illuminating power.

The periodical changes recently discovered in the sun's spots seem to furnish a fatal objection to the idea that the self-luminous condition results from the high temperature of his solid nucleus, or from the heat developed by its compression. It can scarcely be doubted that the periodicity of the spots is dependent on the movements of the planets; and the position of Jupiter seem to exert the greatest influence on their occurrence; as recent observations show, the period of his revolution agrees very closely with the interval between the times at which the spots are most numerous. Although it may be premature to express a decided opinion on so obscure a subject, there seem to be legitimate motives to justify an examination of the more obvious ways in which it would be possible for a planet to affect the luminous condition of the solar disk. If, as Helmholtz contends, the ocean of heat and light be maintained by the compression of the sun, the planets can only exert their influence on his spots by diminishing the weight and pressure of his materials, in the same manner in which the moon acts to raise tides on our oceans. But the alteration in the weight of terrestrial matter from lunar attraction, though extremely small, is about 80,000 times as great as that which the component parts of the sun experience from the attractive force of Jupiter. This planet holds the highest place in its capability of affecting the pressure of the solar matter: it is almost equalled by Venus, but it is far superior to the other members of our system. If, however, the planets moved in circles, the peculiar action alluded to could only increase the tendency of the spots to appear on certain sides of the sun, without materially increasing the numbers visible during the year. When we take into consideration the changes occasioned by the eccentricity of their orbits, the greatest effect must be ascribed to Mercury; Jupiter holds the next place; after which we must rank Saturn and the earth. But even admitting the compressibility of the sun's materials, his mean diameter could not be altered more than the $\frac{1}{10}$th of an inch by the attraction of any of his planetary attendants; and the variation of temperature from this peculiar action cannot exceed $\frac{1}{50,000}$th part of a degree (F.). That so small a variation could be manifested in the appearance of the sun's disk, seems wholly improbable, especially if we adopt the estimate of Mr. Waterston, which assigns to the great orb a mean temperature of one thousand million degrees.

Any effect which the planets may be supposed to occasion by their electric or magnetic forces must be also rendered extremely feeble in consequence of their great distance. If the great ocean of solar light is sensitive to the electricity or magnetism of Jupiter, the nearest satellite of this planet must feel the power of his mysterious influence to an extent several million times as great; yet no indications of such a fact have been observed. But supposing the sun's motion through space to be concerned in maintaining his effulgence, the planets would derive a far more considerable influence from the general movement around the centre of gravity of our system. The position of Jupiter would change the progressive motion of the great luminary about twenty-four miles an hour; and the other planets will be attended with results proportional to their masses multiplied by the square roots of their distances. Now the amount of æther which the sun collects from space, and the density it attains on his surface, will depend on the rapidity of his translatory motion; but I have shown in the Philosophical Magazine for May 1858 another way in which the position of the planets would increase or diminish the supply of æthereal fuel which sustains the great solar conflagration.

The idea that the space-pervading medium is condensed by the attraction of the celestial orbs, is not to be considered a new hypothesis, but rather a necessary inference from that of Professor Encke. Were the density of the subtle fluid uniform, small and large planets would be so unequally affected by its resistance, that their orbits could not retain the relation necessary for their stability, and they must be destroyed by collisions long

before the natural term of their existence. This difficulty will,
however, disappear when the effects of planetary attraction in
condensing the medium are taken into consideration. How far
observation of primary or secondary worlds give evidence of unperiodical changes in our system has not been yet determined
with positive certainty. The constant acceleration of the moon's
orbital velocity, during the past 2000 years, has been traced by
Laplace to a periodical change in the eccentricity of the earth's
orbits. But an error in his investigations being lately pointed
out by Mr. Adams, there appears to be some definite ground for
regarding the lunar orbit as subject to a very slow permanent
diminution, which, after some allowance for the effects of tidal
action, we may consider as depending, to some extent, on the
resistance of a medium. The great oblateness which Arago and
Sir William Herschel assign to Mars would indicate that the
time of its rotation has been considerably lengthened, since the
remote period at which it was moulded into its present form ;
and this may be looked upon as circumstantial evidence of the
effects of an æthereal resistance in changing the diurnal motion
of planets. It is only to smaller worlds that we could look for
such results, for in larger orbs the strength of their solid matter
can have little influence in preventing alterations of form to correspond with the relations of gravity and centrifugal force.

As doubts are entertained by some eminent astronomers as to
the sensible ellipticity of Mars, it may be well to refer to certain
appearances which show a slight deviation, at least, in his form
from a figure of equilibrium. The marked indications of atmospheric phenomena around his poles, while they are either wholly
absent or only faintly exhibited in the vicinity of his equator, is so
much opposed to everything we might expect from the condition
of our own globe and the belted appearance of Jupiter, that we
cannot avoid concluding that the aërial ocean of Mars is much
deeper in his polar than in his equatorial regions. Perhaps this
may account for the very discordant results of observations in
determining the extent of the atmosphere of this planet by oc-
cultations of the fixed stars.

It is to the revolutions of comets that astronomical curiosity
has chiefly turned for evidence of the contents of interplanetary
space ; but the advantages of low density in these bodies have
been counterbalanced by the great elongation of their orbits,
which exposes them to very great disturbances from the planets.
During the past eighteen centuries Halley's comet has occupied
in its revolution a period varying from 74·88 to 79·34 years, according to the Table in Mr. Hind's work on Comets (page 57).
Of the twenty-four consecutive revolutions here recorded, the
first eight average 77·59 years, the next eight 76·84, and the last

76·54 years. This would seem to favour the idea of a permanent diminution of the orbit; and I understand that De Vico regards the movements of his own comet as indicative of a similar result from the widely diffused æther. But the information on this subject hitherto deemed worthy of the most confidence, has been derived from the successive returns of Encke's comet. The advantages which this body affords for such inquiries depend chiefly on the moderate eccentricity of its orbit and the position of the transverse axis, which is nearly perpendicular to the line of the sun's progressive motion. This arrangement must give a more decided preponderance to the perihelion resistance, which has the greatest influence in diminishing the size of the orbit.

As shooting stars are now regarded as small bodies describing very elongated ellipses around the sun, they seem calculated to furnish perhaps the most satisfactory means of testing the perfection of the celestial vacuum. Supposing these bodies to be more sensitive to the resistance of the medium than to planetary disturbances, the transverse axes of their orbits will have a tendency to assume a uniform direction in consequence of the sun's progressive motion. From the same cause the planes in which they move will have their intersections confined for the most part to a very limited range, and will also exhibit, though in a less degree, a tendency to coincidence. This peculiar arrangement of their orbits must cause vast swarms of these minute cosmical bodies to congregate from the most distant parts of the solar domain to a comparatively narrow region at their perihelion passage. For the appearance of the zodiacal light and the periodical fall of meteors, I have endeavoured to account in this manner in a paper sent to the meeting of the British Association and published in the Sections (1854, p. 26). My late researches on the subject exhibit a closer accordance with observed facts than I could then obtain, and they give much support to the ideas very prevalent in scientific circles, with regard to the agency of meteors in reflecting the zodiacal light.

Cincinnati, May 11th, 1861.

LXXVI. *On a Problem in Tactic which serves to disclose the existence of a Four-valued Function of three sets of three letters each. By* J. J. SYLVESTER, *M.A., F.R.S.; Professor of Mathematics at the Royal Military Academy, Woolwich*[*].

AT page 375 of the May Number of the Magazine (in that paragraph commencing at the middle of the page) I gave a Table of Synthemes, correct as far as it went, but left in a very

* Communicated by the Author.

imperfect state. It was intended to be supplemented with a material addition which escaped my recollection when, after a long delay, the proofs of the paper passed through my hands. The question to which this Table refers is the following :—

Three *nomes*, each containing three elements, are given; the number of *trinomial* triads (i. e. ternary combinations, composed by taking one element out of each nome) will be 27, and these 27 may be *grouped* together into 9 synthemes (each syntheme consisting of 3 of the triads in question, which together include between them all the 9 elements). It is desirable to know :—
1st. How many distinct *groupings* of this kind can be formed.
2nd. Whether there is more than one, and, if so, how many distinct types of groupings. The criterion of one grouping being cotypal or allotypal to another is its capability or incapability of being transformed into that other by means of an interchange of elements. Be it once for all stated that the question in hand is throughout one of combinations, and not of permutations; the order of the elements in a triad, of a triad in a syntheme, of a syntheme in a grouping is treated as immaterial. As we are only concerned with the elements as distributed into *nomes*, the number of interchanges of elements with which we are concerned is 6×6^3 or 1296; the factor 6^3 arises from the permutability of the elements of each nome *inter se*, the remaining factor 6 from the permutability of any nome with any other. I find, by a method which carries its own demonstration with it on its face, that the number of distinct *groupings* is 40, of which 4 belong to one *type* or *family*, and 36 to a second type or family.

Let the nomes be $1.2.3, 4.5.6, 7.8.9$, and let

c_1 denote	1.4, 2.5, 3.6		$\dot{c}_1$ denote	1.4, 2.6, 3.5
c_2 „	1.5, 2.6, 3.4		$\dot{c}_2$ „	1.5, 2.4, 3.6
c_3 „	1.6, 2.4, 3.5		$\dot{c}_3$ „	1.6, 2.5, 3.4

$$\gamma \text{ denote } \begin{matrix} 7, 8, 9 \\ 8, 9, 7 \\ 9, 7, 8 \end{matrix} \qquad \gamma' \text{ denote } \begin{matrix} 7, 9, 8 \\ 9, 8, 7 \\ 8, 7, 9 \end{matrix}$$

b_1 denote	1.7, 2.8, 3.9		$\dot{b}_1$ denote	1.7, 2.9, 3.8
b_2 „	1.8, 2.9, 3.7		$\dot{b}_2$ „	1.8, 2.7, 3.9
b_3 „	1.9, 2.7, 3.8		$\dot{b}_3$ „	1.9, 2.8, 3.7

$$\beta \text{ denote } \begin{matrix} 4, 5, 6 \\ 5, 6, 4 \\ 6, 4, 5 \end{matrix} \qquad \beta' \text{ denote } \begin{matrix} 4, 6, 5 \\ 6, 5, 4 \\ 5, 4, 6 \end{matrix}$$

a_1 denote   4.7, 5.8, 6.9 $\dot{a}_1$ denote   4.7, 5.9, 6.8

a_2 „   4.8, 5.9, 6.7 $\dot{a}_2$ „   4.8, 5.7, 6.9

a_3 „   4.9, 5.7, 6.8 $\dot{a}_3$ „   4.9, 5.8, 6.7

$$
\alpha \text{ denote }
\begin{matrix}
1, & 2, & 3\\
2, & 3, & 1\\
3, & 1, & 2
\end{matrix}
\qquad
\alpha' \text{ denote }
\begin{matrix}
1, & 3, & 2.\\
2, & 3, & 1\\
3, & 1, & 2
\end{matrix}
$$

I take first the larger family of 36 groupings; these may be represented as follows:

$$
\begin{array}{c|c|c|c|c|c|c|c|c|c|c|c}
a_1\alpha & a_1\alpha & a_1\alpha' & a_1\alpha & a_1\alpha' & a_1\alpha' & \dot{a}_1\alpha & \dot{a}_1\alpha & \dot{a}_1\alpha' & \dot{a}_1\alpha & \dot{a}_1\alpha' & \dot{a}_1\alpha' \\
a_2\alpha & a_2\alpha' & a_2\alpha & a_2\alpha' & a_2\alpha & a_2\alpha' & \dot{a}_2\alpha & \dot{a}_2\alpha' & \dot{a}_2\alpha & \dot{a}_2\alpha' & \dot{a}_2\alpha & \dot{a}_2\alpha' \\
a_3\alpha' & a_3\alpha & a_3\alpha & a_3\alpha' & a_3\alpha' & a_3\alpha & \dot{a}_3\alpha' & \dot{a}_3\alpha & \dot{a}_3\alpha & \dot{a}_3\alpha' & \dot{a}_3\alpha' & \dot{a}_3\alpha
\end{array}
$$

$$
\begin{array}{c|c|c|c|c|c|c|c|c|c|c|c}
b_1\beta & b_1\beta & b_1\beta' & b_1\beta & b_1\beta' & b_1\beta' & \dot{b}_1\beta & \dot{b}_1\beta & \dot{b}_1\beta' & \dot{b}_1\beta & \dot{b}_1\beta' & \dot{b}_1\beta' \\
b_2\beta & b_2\beta' & b_2\beta & b_2\beta' & b_2\beta & b_2\beta' & \dot{b}_2\beta & \dot{b}_2\beta' & \dot{b}_2\beta & \dot{b}_2\beta' & \dot{b}_2\beta & \dot{b}_2\beta' \\
b_3\beta' & b_3\beta & b_3\beta & b_3\beta' & b_3\beta' & b_3\beta & \dot{b}_3\beta' & \dot{b}_3\beta & \dot{b}_3\beta & \dot{b}_3\beta' & \dot{b}_3\beta' & \dot{b}_3\beta
\end{array}
$$

$$
\begin{array}{c|c|c|c|c|c|c|c|c|c|c|c}
c_1\gamma & c_1\gamma & c_1\gamma' & c_1\gamma & c_1\gamma' & c_1\gamma' & \dot{c}_1\gamma & \dot{c}_1\gamma & \dot{c}_1\gamma' & \dot{c}_1\gamma & \dot{c}_1\gamma' & \dot{c}_1\gamma' \\
c_2\gamma & c_2\gamma' & c_2\gamma & c_2\gamma' & c_2\gamma & c_2\gamma' & \dot{c}_2\gamma & \dot{c}_2\gamma' & \dot{c}_2\gamma & \dot{c}_2\gamma' & \dot{c}_2\gamma & \dot{c}_2\gamma' \\
c_3\gamma' & c_3\gamma & c_3\gamma & c_3\gamma' & c_3\gamma' & c_3\gamma & \dot{c}_3\gamma' & \dot{c}_3\gamma & \dot{c}_3\gamma & \dot{c}_3\gamma' & \dot{c}_3\gamma' & \dot{c}_3\gamma
\end{array}
$$

An example of the development of any one of the above symbolisms into its correspondent grouping will serve to render perfectly intelligible the whole Table.

Let it be required to develope

$$
\begin{aligned}
\dot{b}_1\,\beta\\
b_2\,\beta'\\
b_3\,\beta'.
\end{aligned}
$$

Since

$$
\begin{array}{lllll}
\dot{b}_1 = 1.7 & 2.9 & 3.8 & 4, 5, 6 & 4, 6, 5\\
\dot{b}_2 = 1.8 & 2.7 & 3.9 & \beta = 5, 6, 4 & \beta' = 6, 5, 4\\
\dot{b}_3 = 1.9 & 2.8 & 3.7 & 6, 4, 5 & 5, 4, 6,
\end{array}
$$

the development required is the following:—

$$
\begin{array}{ccc}
1.7.4 & 2.9.5 & 3.8.6 \\
1.7.5 & 2.9.6 & 3.8.4 \\
1.7.6 & 2.9.4 & 3.8.5 \\
1.8.4 & 2.7.6 & 3.9.5 \\
1.8.6 & 2.7.5 & 3.9.4 \\
1.8.5 & 2.7.4 & 3.9.6 \\
1.9.4 & 2.8.6 & 3.7.5 \\
1.9.6 & 2.8.5 & 3.7.4 \\
1.9.5 & 2.8.4 & 3.7.6 \\
\end{array}
$$

The whole of this family of 36 may be represented under the following condensed form, according to the notation usual in the theory of substitutions.

$$
\left(\begin{vmatrix} a_1\alpha \\ a_2\alpha \\ a_3\alpha \end{vmatrix}\right) \times \begin{pmatrix} 123 & 123 & 123 \\ 123 & 231 & 312 \end{pmatrix} \times \begin{pmatrix} a & a \\ a & a \end{pmatrix} \times \begin{pmatrix} a\alpha' & a\alpha' \\ a\alpha' & a'\alpha \end{pmatrix} \times \begin{pmatrix} a\alpha & b\beta & c\gamma \\ a\alpha & a\alpha & a\alpha \end{pmatrix}
$$

It remains to describe the principal and most symmetrical family. This contains only 4 groupings, and may be represented indifferently under any of the three following forms :

$$
\begin{array}{lcl}
a_1\alpha\ \ a_1\alpha'\ \ \dot a_1\alpha\ \ \dot a_1\alpha' & b_1\beta\ \ b_1\beta'\ \ \dot b_1\beta\ \ \dot b_1\beta' & c_1\gamma\ \ c_1\gamma'\ \ \dot c_1\gamma\ \ \dot c_1\gamma' \\[4pt]
a_2\alpha\ \ a_2\alpha'\ \ \dot a_2\alpha\ \ \dot a_2\alpha' \ \text{or}\ & b_2\beta\ \ b_2\beta'\ \ \dot b_2\beta\ \ \dot b_2\beta' \ \text{or}\ & c_2\gamma\ \ c_2\gamma'\ \ \dot c_2\gamma\ \ \dot c_2\gamma' \\[4pt]
a_3\alpha\ \ a_3\alpha'\ \ \dot a_3\alpha\ \ \dot a_3\alpha & b_3\beta\ \ b_3\beta'\ \ \dot b_3\beta\ \ \dot b_3\beta' & c_3\gamma\ \ c_3\gamma'\ \ \dot c_3\gamma\ \ \dot c_3\gamma'
\end{array}
$$

In developing, it will be found that each of these three representations gives rise to the *same* family of groupings, which from its importance it is proper to set out in full as follows :—

```
4.7 2.5.8 3.6.9 | 1.4.7 2.5.9 3.6.8 | 1.4.7 2.6.8 3.5.9 | 1.4.7 2.6.9 3.5
4.8 2.5.9 3.6.7 | 1.4.9 2.5.8 3.6.7 | 1.4.8 2.6.9 3.5.7 | 1.4.9 2.6.8 3.5
4.9 2.5.7 3.6.8 | 1.4.8 2.5.7 3.6.9 | 1.4.9 2.6.7 3.5.8 | 1.4.8 2.6.7 3.5
5.7 2.6.8 3.4.9 | 1.5.7 2.6.9 3.4.8 | 1.5.7 2.4.8 3.6.9 | 1.5.7 2.4.9 3.6
5.8 2.6.9 3.4.7 | 1.5.9 2.6.8 3.4.7 | 1.5.8 2.4.9 3.6.7 | 1.5.9 2.4.8 3.6
5.9 2.6.7 3.4.8 | 1.5.8 2.6.7 3.4.9 | 1.5.9 2.4.7 3.6.8 | 1.5.8 2.4.7 3.6
6.7 2.4.8 3.5.9 | 1.6.7 2.4.9 3.5.8 | 1.6.7 2.5.8 3.4.9 | 1.6.7 2.5.9 3.4
6.8 2.4.9 3.5.7 | 1.6.9 2.4.8 3.5.7 | 1.6.8 2.5.9 3.4.7 | 1.6.9 2.5.8 3.4
6.9 2.4.7 3.5.8 | 1.6.8 2.4.7 3.5.9 | 1.6.9 2.5.7 3.4.8 | 1.6.8 2.5.7 3.4
```

It follows at once from the above Table, that if 3 cubic equations be given, we may form a function of the 9 roots, which, when any of the roots of any of the equations are interchanged *inter se*, or all the roots of one with all those of any other, will receive only *four distinct values.*

It also follows that we may form with 9 letters an intransitive group (of Cauchy) containing $\dfrac{216}{4}$, *i. e.* 54, or a transitive group containing $\dfrac{1296}{4}$, or 324 substitutions. So the family of 36 groupings lead to the formation of an intransitive substitution group of $\dfrac{216}{12}$, i. e. 18, and of a transitive group of $\dfrac{1296}{36}$, or 16 substitutions.

Since 9 letters may be thrown, in $\dfrac{8.7}{2} \times \dfrac{5.4}{2}$, i. e. 280 different ways, into nomes of 3 letters each, it further follows that by repeating each of the above two families 280 times we shall obtain new families remaining unaltered by any substitution of any of the nine elements *inter se*, and consequently indicating the existence of substitution-groups containing

$$\frac{1.2.3.4.5.6.7.8.9}{280 \times 36} \text{ and } \frac{1.2.3.4.5.6.7.8.9}{280 \times 4},$$

i. e. 36 and 324 substitutions respectively.

In the above solution a little consideration will show that the method is essentially based on the solution of a *previous* question, viz. of grouping together the synthemes of *binomial duads* of *two* nomes of three letters each, which can be done in two distinct modes, which (if, *ex. gr.*, we take $1.2.3, 4.5.6$ as the two nomes in question) are represented in the notation used above by $\begin{smallmatrix} c_1 \\ c_2 \end{smallmatrix}$ and $\begin{smallmatrix} \dot{c}_1 \\ \dot{c}_2 \end{smallmatrix}$ respectively. So, more generally, the groupings of the q-nomial q-ads of r nomes of s elements may be made to depend on the groupings of the $(q-1)$-nomial $(q-1)$-ads of $(r-1)$ nomes of s elements each. The more general question is to discover the groupings and their families of the synthemes composed of p-nomial q-ads of r nomes of s elements, of which the simplest example next that which has been considered and solved is to discover the groupings of the synthemes composed of 54 *binomial triads* of 8 nomes of 3 elements each [*].

The chief difficulty of calculating *à priori* the number of such groupings is of a similar nature to that which lies at the bottom of the ordinary theory of the partition of numbers, namely, the liability of the same groupings to make their appearance under distinct symbolical representations. Of this we have seen an example in the threefold representation of the principal family of 4 groupings just treated of. But for the existence of this

---

[*] I have ascertained, by a direct analytical method, since the above was written, that the number of different groupings of the synthemes composed of these binomial triads is 144. The number of distinct types or families is *three*, one containing 12, another 24, and the third 108 groupings.

multiform representation of the same grouping we could have affirmed *à priori* the number of groupings to be $2 \times 8 \times 2^3$ or 48, whereas the true number is only 40. I believe that the above is the first instance of the doctrine of types making its appearance explicitly, and illustrated by example in the theory of tactic. It were much to be desired that some one would endeavour to collect and collate the various solutions that have been given of the noted 15-school-girl problem by Messrs. Kirkman (in the Ladies' Diary), Moses Ansted (in the Cambridge and Dublin Mathematical Journal), by Messrs. Cayley and Spottiswoode (in the Philosophical Magazine and elsewhere), and Professor Pierce, the latest and probably the best (in the American Astronomical Journal), besides various others originating and still floating about in the fashionable world (one, if not two, of which I remember having been communicated to me many years ago by Mr. Archibald Smith, F.R.S.), with a view to ascertain whether they belong to the same or to distinct types of aggregation.

LXXVII. *Notices respecting New Books.*

A History of the Progress of the Calculus of Variations during the Nineteenth Century. By I. TODHUNTER, *M.A., Fellow and Principal Mathematical Lecturer of St. John's College, Cambridge.* Cambridge: Macmillan and Co. 1861.

MR. TODHUNTER, whose name is already so familiar to the mathematical student, has at length produced a work of much greater originality and research than any of his former and more elementary treatises.

The "Calculus of Variations," one of the most difficult branches of pure mathematics, has been the subject of the labours of several eminent mathematicians, Euler, Lagrange, Gauss, Poisson, &c., whose successive researches and improvements form an exceedingly interesting department of scientific history, which, however, has hitherto been specially treated by only one writer in our own language, viz. Woodhouse, whose 'Treatise on Isoperimetrical Problems and the Calculus of Variations' was published in 1810, and is now an extremely scarce book.

Woodhouse's work has always received very high praise by such competent judges as Messrs. Peacock, Herschel, and Babbage, in their 'Examples;' Professor De Morgan, in his 'Differential and Integral Calculus;' and Professor Jellett, in his 'Calculus of Variations.' But since its publication the calculus has been greatly advanced and improved; and it is to record this progress that Mr. Todhunter has written the volume before us, which commences where Woodhouse left off. It is evidently the work of one who thoroughly understands the science itself, and who has most conscientiously and laboriously consulted and *studied* all the available materials and sources of information. He unites the qualifications of a sound mathematician and a good linguist—a rare combination. The Memoirs and Treatises in the German and Italian languages, as well as those in the French and Latin, have been completely mastered and analysed: and some account is given even of a dissertation in the Russian language. Of those works which are difficult of access to the English student, a more copious account is given; and throughout the whole history, "numerous remarks, criticisms, and corrections are suggested relative to the various treatises and memoirs which are analysed. The writer trusts that it will not be supposed that he undervalues the labours of the eminent mathematicians in whose works he ventures occasionally to indicate inaccuracies or imperfections, but that his aim has been to remove difficulties which might perplex a student" (Preface). We would specially point out the last chapter in the book (pp. 505–530) as deserving attention in this respect.

Fully agreeing with Mr. Todhunter as to the "value of a history of any department of science, when that history is presented with accuracy and completeness," we congratulate him on having produced a History which so well merits this character of "accuracy and completeness;" and we sincerely hope that the success of his present contribution to scientific history may induce him to carry out the intention expressed in the conclusion of his Preface, viz. "to undertake a similar survey of some other department of science."

LXXVIII. *Proceedings of Learned Societies.*

ROYAL SOCIETY.

[Continued from p. 469.]

June 21, 1860.—Sir Benjamin C. Brodie, Bart., Pres., in the Chair.

THE following communications were read :—

"On the Sources of the Nitrogen of Vegetation ; with special reference to the Question whether Plants assimilate free or uncombined Nitrogen." By J. B. Lawes, Esq., F.R.S.; J. H. Gilbert, Ph.D., F.R.S.; and Evan Pugh, Ph.D., F.C.S.

After referring to the earlier history of the subject, and especially to the conclusion of De Saussure, that plants derive their nitrogen from the nitrogenous compounds of the soil and the small amount of ammonia which he found to exist in the atmosphere, the Authors preface the discussion of their own experiments on the sources of the nitrogen of plants, by a consideration of the most prominent facts established by their own investigations concerning the amount of nitrogen yielded by different crops over a given area of land, and of the relation of these to certain measured, or known sources of it.

On growing the same crop year after year on the same land, without any supply of nitrogen by manure, it was found that wheat, over a period of 14 years, had given rather more than 30 lbs.—barley, over a period of 6 years, somewhat less—meadow-hay, over a period of 3 years, nearly 40 lbs.—and beans, over 11 years, rather more than 50 lbs. of nitrogen, per acre, per annum. Clover, another Leguminous crop, grown in 3 out of 4 consecutive years, had given an average of 120 lbs. Turnips, over 8 consecutive years, had yielded about 45 lbs.

The Graminaceous crops had not, during the periods referred to, shown signs of diminution of produce. The yield of the Leguminous crops had fallen considerably. Turnips, again, appeared greatly to have exhausted the immediately available nitrogen in the soil. The amount of nitrogen harvested in the Leguminous and Root-crops was considerably increased by the use of "mineral manures," whilst that in the Graminaceous crops was so in a very limited degree.

Direct experiments further showed that pretty nearly the same amount of nitrogen was taken from a given area of land in *wheat* in 8 years, whether 8 crops were grown consecutively, 4 in alternation with fallow, or 4 in alternation with beans.

Taking the results of 6 separate courses of rotation, Boussingault obtained an average of between one-third and one-half more nitrogen in the produce than had been supplied in manure. His largest yields of nitrogen were in the Leguminous crops; and the cereal crops were larger when they next succeeded the removal of the highly nitrogenous Leguminous crops. In their own experiments upon an actual course of rotation, without manure, the Authors had obtained, over 8 years, an average annual yield of 57·7 lbs. of nitrogen per acre; about twice as much as was obtained in either wheat or barley, when these crops were, respectively, grown year after year on the same land. The greatest yield of nitrogen had been in a clover crop, grown once during the 8 years; and the wheat crops grown after this clover in the first course of 4 years, and after beans in the second course, were about double those obtained when wheat succeeded wheat.

Thus, Cereal crops, grown year after year on the same land, had given an average of about 30 lbs. of nitrogen, per acre, per annum; and Leguminous crops much more. Nevertheless the Cereal crop was nearly doubled when preceded by a Leguminous one. It was also about doubled when preceded by fallow. Lastly, an entirely unmanured rotation had yielded nearly twice as much nitrogen as the continuously grown Cereals.

Leguminous crops were, however, little benefited, indeed frequently injured, by the use of the ordinary direct nitrogenous manures. Cereal crops, on the other hand, though their yield of nitrogen was comparatively small, were very much increased by direct nitrogenous manures, as well as when they succeeded a highly nitrogenous Leguminous crop, or fallow. But when nitrogenous manures had been employed for the increased growth of the Cereals, the nitrogen in the immediate increase of produce had amounted to little more than 40 per cent. of that supplied, and that in the increase of the second year after the application, to little more than one-tenth of the remainder. Estimated in the same way, there had been in the case of the meadow grasses scarcely any larger proportion of the supplied nitrogen recovered. In the Leguminous crops the proportion so recovered appeared to be even less; whilst in the root-crops it was probably somewhat greater. Several possible explanations of this real or apparent loss of the nitrogen supplied by manure are enumerated.

The question arises—what are the sources of all the nitrogen of our crops beyond that which is directly supplied to the soil by artificial means? The following actual or possible sources may be enumerated:—the nitrogen in certain constituent minerals of the soil; the combined nitrogen annually coming down in the direct aqueous depositions from the atmosphere; the accumulation of combined nitrogen from the atmosphere by the soil in other ways; the formation of ammonia in the soil from free nitrogen and nascent hydrogen; the formation of nitric acid from free nitrogen; the direct absorption of combined nitrogen from the atmosphere by plants themselves; the assimilation of free nitrogen by plants.

A consideration of these several sources of the nitrogen of the vegetation which covers the earth's surface showed that those of them which have as yet been quantitatively estimated are inadequate to account for the amount of nitrogen obtained in the annual produce of a given area of land beyond that which may be attributed to supplies by previous manuring. Those, on the other hand, which have not yet been even approximately estimated as to quantity —if indeed fully established qualitatively—offer many practical difficulties in the way of such an investigation as would afford results applicable in any such estimates as are here supposed. It appeared important, therefore, to endeavour to settle the question whether or not that vast storehouse of nitrogen, the atmosphere, affords to growing plants any measurable amount of its *free* nitrogen. Moreover, this question had of late years been submitted to very extended and laborious experimental researches by M. Boussingault, and M. Ville, and also to more limited investigation by MM. Mène, Roy, Cloez, De Luca, Harting, Petzholdt and others, from the results of which diametrically opposite conclusions had been arrived at. Before entering on the discussion of their own experimental evidence, the Authors give a review of these results and inferences; more especially those of M. Boussingault who questions, and those of M. Georges Ville who affirms the assimilation of *free* nitrogen in the process of vegetation.

The general method of experiment instituted by Boussingault, which has been followed, with more or less modification, in most subsequent researches, and by the Authors in the present inquiry, was—to set seeds or young plants, the amount of nitrogen in which was estimated by the analysis of carefully chosen similar specimens; to employ soils and water containing either no combined nitrogen, or only known quantities of it; to allow the access either of free air (the plants being protected from rain and dust)— of a current of air freed by washing from all *combined* nitrogen—or of a limited quantity of air, too small to be of any avail so far as any compounds of nitrogen contained in it were concerned; and finally, to determine the amount of combined nitrogen in the plants produced and in the soil, pot, &c., and so to provide the means of estimating the gain or loss of nitrogen during the course of the experiment.

The plan adopted by the Authors in discussing their own experimental results, was—

To consider the conditions to be fulfilled in order to effect the solution of the main question, and to endeavour to eliminate all sources of error in the investigation.

To examine a number of collateral questions bearing upon the points at issue, and to endeavour so far to solve them, as to reduce the general solution to that of a single question to be answered by the results of a final set of experiments.

To give the results of the final experiments, and to discuss their bearings upon the question which it is proposed to solve by them.

Accordingly, the following points are considered :—

1. The preparation of the soil, or matrix, for the reception of the plants and of the nutriment to be supplied to them.

2. The preparation of the nutriment, embracing that of mineral constituents, of certain solutions, and of water.

3. The conditions of atmosphere to be supplied to the plants, and the means of securing them ; the apparatus to be employed, &c.

4. The changes undergone by nitrogenous organic matter during decomposition, affecting the quantity of combined nitrogen present, in circumstances more or less analogous to those in which the experimental plants are grown.

5. The action of agents, as ozone ; and the influence of other circumstances which may affect the quantity of combined nitrogen present in connexion with the plants, independently of the direct action of the growing process.

In most of the experiments a rather clayey soil, ignited with free access of air, well-washed with distilled water, and re-ignited, was used as the matrix or soil. In a few cases washed and ignited pumice-stone was used.

The mineral constituents were supplied in the form of the ash of plants, of the description to be grown if practicable, and if not, of some closely allied kind.

The distilled water used for the final rinsing of all the important parts of the apparatus, and for the supply of water to the plants, was prepared by boiling off one-third from ordinary water, collecting the second third as distillate, and redistilling this, previously acidulated with phosphoric acid.

Most of the pots used were specially made, of porous ware, with a great many holes at the bottom and round the sides near to the bottom. These were placed in glazed stone-ware pans with inward-turned rims to lessen evaporation.

Before use, the red-hot matrix and the freshly ignited ash were mixed in the red-hot pot, and the whole allowed to cool over sulphuric acid. The soil was then moistened with distilled water, and after the lapse of a day or so the seeds or plants were put in.

Very carefully picked bulks of seed were chosen ; specimens of the average weight were taken for the experiment, and in similar specimens the nitrogen was determined.

The atmosphere supplied to the plants was washed free from

ammonia by passing through sulphuric acid, and then over pumice-stone saturated with sulphuric acid. It then passed through a solution of carbonate of soda before entering the apparatus enclosing the plant, and it passed out again through sulphuric acid.

Carbonic acid, evolved from marble by measured quantities of hydrochloric acid, was passed daily into the apparatus, after passing, with the air, through the sulphuric acid and the carbonate of soda solution.

The enclosing apparatus consisted of a large glass shade, resting in a groove filled with mercury, in a slate or glazed earthenware stand, upon which the pan, with the pot of soil, &c., was placed. Tubes passed under the shade, for the ingress and the egress of air, for the supply of water to the plants, and, in some cases, for the withdrawal of the water which condensed within the shade. In other cases, the condensed water was removed by means of a special arrangement.

One advantage of the apparatus adopted was, that the washed air was forced, instead of being aspirated, through the enclosing vessel. The pressure upon it was thus not only very small, and the danger from breakage, therefore, also small, but it was exerted upon the inside instead of the outside of the shade; hence, any leakage would be from the inside outwards, so that there was no danger of unwashed air gaining access to the plants.

The conditions of atmosphere were proved to be adapted for healthy growth, by growing plants under exactly the same circumstances, but in a garden soil. The conditions of the artificial soil were shown to be suitable for the purpose, by the fact that plants grown in such soil, and in the artificial conditions of atmosphere, developed luxuriantly, if only manured with substances supplying combined nitrogen.

Passing to the subjects of collateral inquiry, the first question considered was, whether plants growing under the conditions stated would be likely to acquire nitrogen from the air through the medium of ozone, either within or around the plant, or in the soil; that body oxidating free nitrogen, and thus rendering it assimilable by the plants.

Several series of experiments were made upon the gases contained in plants or evolved from them, under different circumstances of light, shade, supply of carbonic acid, &c. When sought for, ozone was in no case detected. The results of the inquiry in other respects, bearing upon the points at issue, may be briefly summed up as follows:—

1. Carbonic acid within growing vegetable cells and intercellular passages suffers decomposition very rapidly on the penetration of the sun's rays, oxygen being evolved.

2. Living vegetable cells, in the dark, or not penetrated by the direct rays of the sun, consume oxygen very rapidly, carbonic acid being formed.

3. Hence, the proportion of oxygen must vary greatly according to the position of the cell, and to the external conditions of light, and it will oscillate under the influence of the reducing force of carbon-matter (forming carbonic acid) on the one hand, and of that of the

sun's rays (liberating oxygen) on the other. Both actions may go on simultaneously according to the depth of the cell; and the once outer cells may gradually pass from the state in which the sunlight is the greater reducing agent to that in which the carbon-matter becomes the greater.

4. The great reducing power operating in those parts of the plant where ozone is most likely, if at all, to be evolved, seems unfavourable to the oxidation of nitrogen; that is under circumstances in which carbon-matter is not oxidized, but on the contrary, carbonic acid reduced. And where beyond the influence of the direct rays of the sun, the cells seem to supply an abundance of more easily oxidised carbon-matter, available for oxidation should free oxygen or ozone be present. On the assumption that nitrates are available as a direct source of nitrogen to plants, if it were admitted that nitrogen is oxidated within the plant, it must be supposed (as in the case of carbon) that there are conditions under which the oxygen compound of nitrogen may be reduced within the organism, and that there are others in which the reverse action, namely, the oxidation of nitrogen, can take place.

5. So great is the reducing power of certain carbon-compounds of vegetable matter, that when the growing process has ceased, and all the free oxygen in the cells has been consumed, water is for a time decomposed, carbonic acid formed, and hydrogen evolved.

The suggestion arises, whether ozone may not be formed under the influence of the powerful reducing action of the carbon-compounds of the cell on the oxygen eliminated from carbonic acid by sunlight, rather than under the direct action of the sunlight itself —in a manner analogous to that in which it is ordinarily obtained under the influence of the active reducing agency of phosphorus? But, even if it were so, it may be questioned whether the ozone would not be at once destroyed when in contact with the carbon-compounds present. It is more probable, however, that the ozone said to be observed in the vicinity of vegetation, is due to the action of the oxygen of the air upon minute quantities of volatile carbo-hydrogens emitted by plants.

Supposing ozone to be present, it might, however, be supposed to act in a more indirect manner as a source of combined and assimilable nitrogen in the Authors' experiments, namely,—by oxidating the nitrogen dissolved in the condensed water of the apparatus—by forming nitrates in contact with the moist, porous, and alkaline soil—or by oxidating the free nitrogen in the cells of the older roots, or that evolved in their decomposition.

Experiments were accordingly made to ascertain the influence of ozone upon organic matter, and on certain porous and alkaline bodies, under various circumstances. A current of ozonous air was passed over the substances for some time daily, for several months, including the whole of the warm weather of the summer; but in only one case out of eleven was any trace of nitric acid detected, namely, that of garden soil; and this was proved to contain nitrates before being submitted to the action of ozone.

It is not, indeed, hence inferred that nitric acid could under no circumstances be formed through the influence of ozone on certain nitrogenous compounds, on nascent nitrogen, on gaseous nitrogen in contact with porous and alkaline substances, or even in the atmosphere. But, considering the negative result with large quantities of ozonous air, acting upon organic matter, soil, &c., in a wide range of circumstances, and for so long a period, it is believed that no error will be introduced into the main investigation by the cause referred to.

Numerous experiments were made to determine whether free nitrogen was evolved during the decomposition of nitrogenous organic compounds.

In the first series of 6 experiments, wheat, barley, and bean-meal were respectively mixed with ignited pumice, and ignited soil, and submitted for some months to decomposition in a current of air, in such manner that any ammonia evolved could be collected and estimated. The result was, that, in 5 out of the 6 cases, there was a greater or less evolution of free nitrogen—amounting, in two of the cases, to more than 12 per cent. of the original nitrogen of the substance.

The second series consisted of 9 experiments; wheat, barley, and beans being again employed, and, as before, either ignited soil or pumice used as the matrix. In some cases the seeds were submitted to experiment whole, and allowed to grow, and the vegetable matter produced permitted to die down and decompose. In other cases, the ground seeds, or "meals," were employed. The conditions of moisture were also varied. The experiments were continued through several months, when from 60 to 70 per cent. of the carbon had disappeared.

In 8 out of the 9 experiments, a loss of nitrogen, evolved in the free state, was indicated. In most cases, the loss amounted to about one-seventh or one-eighth, but in one instance to 40 per cent. of the original nitrogen. In all these experiments the decomposition of the organic substance was very complete, and the amount of carbon lost was comparatively uniform.

It thus appeared that, under rare circumstances, there might be no loss of nitrogen in the decomposition of nitrogenous organic matter; but that, under a wide range of circumstances, the loss was very considerable—a point, it may be observed, of practical importance in the management of the manures of the farm and the stable.

Numerous direct experiments showed, that when nitrogenous organic matter was submitted to decomposition in water, over mercury, in the absence of free oxygen, there was no free nitrogen evolved. In fact, the evolution in question appeared to be the result of an oxidating process.

Direct experiments also showed, that seeds may be submitted to germination and growth, and that nearly the whole of the nitrogen may be found in the vegetable matter produced.

It is observed that, in the cases referred to in which so large an evolution of free nitrogen took place, the organic substances were submitted to decomposition for several months, during which time they lost two-thirds of their carbon. In the experiments on the question of assimilation, however, but a very small proportion of the total organic matter is submitted to decomposing actions apart from those associated with growth, and this for a comparatively short period of time, at the termination of which the organic form is retained, and therefore but very little carbon is lost. It would appear, then, that in experiments on assimilation no fear need be entertained of any serious error arising from the evolution of free nitrogen in the decomposition of the nitrogenous organic matter necessarily involved, so long as it is subjected to the ordinary process of germination, and exhaustion to supply materials for growth. On the other hand, the facts adduced afford a probable explanation of any small loss of nitrogen which may occur when seeds have not grown, or when leaves, or other dead matters, have suffered partial decomposition. They also point out an objection to the application of nitrogenous organic manure in such experiments.

. Although there can be no doubt of the evolution of hydrogen during the decomposition of organic matter under certain conditions, and although it has long been admitted that nascent hydrogen may, under certain circumstances, combine with gaseous nitrogen and form ammonia—nevertheless, from considerations stated at length in the paper, the Authors infer that there need be little apprehension of error in the results of their experiments, arising from an unaccounted supply of ammonia, formed under the influence of nascent hydrogen given off in the decomposition of the organic matter involved.

Turning to their direct experiments on the question of the assimilation of free nitrogen, the Authors first consider whether such assimilation would be most likely to take place when the plant had no other supply of combined nitrogen than that contained in the seed sown, or when supplied with a limited amount of combined nitrogen, or with an excess of combined nitrogen? And again— whether at an early stage of growth, at the most active stage, or when the plant was approaching maturity? Combinations of these several circumstances might give a number of special conditions, in perhaps only one of which assimilation of free nitrogen might take place, in case it could in any.

It is hardly to be supposed that free nitrogen would be assimilated if an excess of combined nitrogen were at the disposal of the plant. It is obvious, however, that a wide range of conditions would be experimentally provided, if in some instances plants were supplied with no more combined nitrogen than that contained in the seed, in others brought to a given stage of growth by means of limited extraneous supplies of combined nitrogen, and in others supplied with combined nitrogen in a more liberal measure. It has been sought to provide these conditions in the experiments under consideration.

In the selection of plants, it was sought to take such as would be adapted to the artificial conditions of temperature, moisture, &c. involved in the experiment, and also such as were of importance in an agricultural point of view—to have representatives, moreover, of the two great Natural Families, the Graminaceæ and the Leguminosæ, which seem to differ so widely in their relations to the combined nitrogen supplied within the soil—and finally, to have some of the same descriptions as those experimented upon by M. Boussingault, and M. G. Ville, with such discordant results.

Thirteen experiments were made (4 in 1857 and 9 in 1858), in which the plants were supplied with no other combined nitrogen than that contained in the original seed. In 12 of the cases prepared soil was the matrix, and in the remaining one prepared pumice.

Of 9 experiments with Graminaceous plants, 1 with wheat and 2 with barley were made in 1857. In one of the experiments with barley there was a gain of 0·0016, and in the other of 0·0026 gramme of nitrogen. In only two cases of the experiments with cereals in 1858, was there any gain of nitrogen indicated; and in both it amounted to only a small fraction of a milligramme. Indeed, in no one of the cases, in either 1857 or 1858, was there more nitrogen in the *plants themselves*, than in the seed sown. A gain was indicated only when the nitrogen in the soil and pot—which together weighed about 1500 grammes—was brought into the calculation. Moreover, the gain only exceeded 1 milligramme in the case of the experiments of 1857, when slate, instead of glazed earthenware stands were used as the lute vessels; and there was some reason to believe that the gain indicated was due to this circumstance. In none of the other cases was the gain more than would be expected from error in analysis.

The result was then, that in no one case of these experiments was there any such gain of nitrogen as could lead to the supposition that *free* nitrogen had been assimilated. The plants had, however, vegetated for several months, had in most cases more than trebled the carbon of the seed, and had obviously been limited in their growth for want of a supply of available nitrogen in some form. During this long period they were surrounded by an atmosphere containing free nitrogen; and their cells were penetrated by fluid saturated with that element. It may be further mentioned, that many of the plants formed glumes and paleæ for seed.

It is to be observed that the results of these experiments with cereals go to confirm those of M. Boussingault.

The Leguminous plants experimented upon did not grow so healthily under the artificial conditions as did the cereals. Still, in all three of the cases of these plants in which no combined nitrogen was provided beyond that contained in the original seed, the carbon in the vegetable matter produced was much greater than that in the seed—in one instance more than 3 times greater. In no case, however, was there any indication of assimilation of free nitrogen, any more than there had been by the Graminaceous plants grown under similar circumstances.

One experiment was made with buckwheat, supplied with no other combined nitrogen than that contained in the seed. The result gave no indication of assimilation of free nitrogen.

In regard to the whole of the experiments in which the plants were supplied with no combined nitrogen beyond that contained in the seed, it may be observed that, from the constancy of the amount of combined nitrogen present in relation to that supplied, throughout the experiments, it may be inferred, as well that there was no evolution of free nitrogen by the growing plant, as that there was no assimilation of it; but it cannot hence be concluded that there would be no such evolution if an excess of combined nitrogen were supplied.

The results of a number of experiments, in which the plants were supplied with more or less of combined nitrogen, in the form of ammonia-salts, or of nitrates, are recorded. Ten were with Cereals; 4 in 1857, and 6 in 1858. Three were with Leguminous plants; and there were also some with plants of other descriptions—all in 1858.

In the case of the cereals more particularly, the growth was very greatly increased by the extraneous supply of combined nitrogen; in fact, the amount of vegetable matter produced was 8, 12, and even 30 times greater than in parallel cases without such supply. The amount of nitrogen appropriated was also, in all cases many times greater, and in one case more than 30 times as great, when a supply of combined nitrogen was provided. The evidence is therefore sufficiently clear that all the conditions provided, apart from those which depended upon a supply of combined nitrogen, were adapted for vigorous growth; and that the limitation of growth where no combined nitrogen was supplied was due to the want of such supply.

In 2 out of the 4 experiments with cereals in 1857, there was a slight gain of nitrogen beyond that which should occur from error in analysis; but in no one of the 6 in 1858, when glazed earthenware instead of slate stands were used, was there any such gain. It is concluded, therefore, that there was no assimilation of free nitrogen. In some cases the supply of combined nitrogen was not given until the plants showed signs of decline; when, on each addition, increased vigour was rapidly manifested. In others the supply was given earlier and was more liberal.

As in the case of the Leguminous plants grown without extraneous supply of combined nitrogen, those grown with it progressed much less healthily than the Graminaceous plants. But the results under these conditions, so far as they go, did not indicate any assimilation of free nitrogen.

The results of experiments with plants of other descriptions, in which an extraneous supply of combined nitrogen was provided, also failed to show an assimilation of free nitrogen.

Thus, 19 experiments with Graminaceous plants, 9 without and 10 with an extraneous supply of combined nitrogen—6 with Leguminous plants, 3 without and 3 with an extraneous supply of combined nitrogen, and also some with other plants, have been made. In none of the experiments, with plants so widely different as the Graminaceous and the Leguminous, and with a wide range of conditions of growth, was there evidence of an assimilation of free nitrogen.

The conclusions from the whole inquiry may be briefly summed up as follows:—

The yield of nitrogen in the vegetation over a given area, within a given time, especially in the case of Leguminous crops, is not satisfactorily explained by reference to the hitherto quantitatively determined supplies of *combined* nitrogen.

The results and conclusions hitherto recorded by different experimenters on the question whether plants assimilate *free* or *uncombined* nitrogen, are very conflicting.

The conditions provided in the experiments of the Authors on this question were found to be quite consistent with the healthy development of various Graminaceous Plants, but not so much so for that of the Leguminous Plants experimented upon.

It is not probable that, under the circumstances of the experiments on assimilation, there would be any supply to the plants of an unaccounted quantity of combined nitrogen, due to the influence either of ozone, or of nascent hydrogen.

It is not probable that there would be a loss of any of the combined nitrogen involved in an experiment on assimilation, due to the evolution of free nitrogen in the decomposition of organic matter, excepting in certain cases when it might be presupposed.

It is not probable that there would be any loss due to the evolution of free nitrogen from the nitrogenous constituents of the plants during growth.

In numerous experiments with Graminaceous plants, under a wide range of conditions of growth, in no case was there any evidence of an assimilation of free nitrogen.

In experiments with Leguminous plants the growth was less satisfactory, and the range of conditions was, therefore, more limited. But the results with these plants, so far as they go, do not indicate any assimilation of free nitrogen. It is desirable that the evidence of further experiments with such plants, under conditions of more healthy growth, should be obtained.

Results obtained with some other plants, are in the same sense as those with Graminaceous and Leguminous ones, in regard to the question of the assimilation of free nitrogen.

In view of the evidence afforded of the non-assimilation of *free* nitrogen by plants, it is very desirable that the several actual or possible sources whence they may derive *combined* nitrogen should be more fully investigated, both qualitatively and quantitatively.

If it be established that plants do not assimilate free or uncombined nitrogen, the source of the large amount of combined nitrogen known to exist on the surface of the globe, and in the atmosphere, still awaits a satisfactory explanation.

" Reduction and Discussion of the Deviations of the Compass observed on board of all the Iron-built Ships and a selection of the Wood-built Steam-ships in Her Majesty's Navy, and the Iron Steamship ' Great Eastern'." By Frederick J. Evans, Esq.

The analysis of the deviations of the compass in this paper comprises the observations made in forty-two iron ships, varying in size from 3400 to 165 tons, a selection of wood-built screw and paddle-wheel steam-vessels, as also the steam-ship 'Great Eastern' at various times prior to her departure from England.

The observations made in the iron-built ships extend over periods varying between thirteen and five years; and having been made with the same description of compass—the Admiralty standard—and under similar conditions of arrangement and situation, in accordance with the system carried out in Her Majesty's Navy, details of which are given, the general results are strictly comparable.

In the analysis of the Tables, amounting to nearly 250 in number, of deviations observed in various parts of both hemispheres, the formula deduced from Poisson's General Equations by Mr. Archibald Smith, given in the Philosophical Transactions for 1846, p. 348, has been employed.

In this formula, the deviation of the compass on board ship, reckoned positive when the north point of the needle deviates to the east, is given by the following expression:—

$$\text{Deviation } (\delta) = A + B \sin \zeta' + C \cos \zeta' + D \sin 2\zeta' + E \cos 2\zeta',$$

ζ' being the azimuth (by compass) of the ship's head, reckoned from the magnetic north towards the east;

A, D, E being constant coefficients depending only on the amount, quality, and arrangement or position of the iron in the ship: B and C, coefficients depending on these, and also on the magnetic dip and horizontal intensity, are each consisting of two parts; one caused by the permanent magnetism of the hard iron, the deviation produced by which varies inversely as the horizontal force at the place; and the other, caused by the vertical part of the earth's force inducing the soft iron in the ship, the deviation produced by which varies as the tangent of the dip: B representing that part of the combined attraction acting in a fore-and-aft direction, C that acting in a transverse, or athwart-ship direction

From the equation $\tan^{-1}\dfrac{C}{B}$, the direction of the ship's force, and $\sqrt{B^2 + C^2}$, the total magnetic force of the ship in proportion to the horizontal force at the place of observation is obtained: for convenience, 1000 has been adopted to represent the value of the earth's horizontal force at the English ports of observation, in order, by an easy comparison, to note the changes on foreign stations.

By comparison of the coefficients of the several descriptions of ships, it is observed that in wood-built steam-vessels, the coefficients B and C vary nearly as the tangent of the dip; from whence it may be inferred, as a general rule, that in steam machinery permanent magnetism bears but a small proportion to induced; but in iron-built ships, B and C generally vary more nearly as the inverse horizontal force, showing that they depend more on the permanent magnetism of the iron of the ship, and thus confirming the view of the Astronomer Royal, given in his earliest deductions (Phil. Trans. 1839), that the effect of transient induced magnetism is in these ships small comparatively. Numerous examples are given in detail of this permanency of magnetism, as also of the gradual diminution of the ship's force resulting from time.

An investigation of the coefficient D, which is caused entirely by the horizontal induction of the soft iron in the ship, and which is known as the "quadrantal" deviation, shows, that while in wood-built steam-ships it seldom exceeds 1° or 1½°, it rises in iron-built ships from 1½° to 6° and 7°; the Liverpool Compass-Committee recording even a point of the compass.

The chief characteristics of the quadrantal deviation, as developed in this investigation, are—

1. That it has invariably a positive sign, causing an easterly deviation in the N.E. and S.W. quadrants; and a westerly deviation in the S.E. and N.W. quadrants.

2. Its amount does not appear to depend on the size, or mass of the vessel, or direction when building; or on the existence of iron beams.

3. That a gradual decrease in amount has occurred, after the lapse of a number of years, in nearly every vessel that has been observed.

4. That the value remains unchanged in sign and amount, on changes of geographic position.

5. That a value not exceeding 4°, and ranging between that amount and 2°, may be assumed to represent the average or normal amount in vessels of all sizes.

Numerous examples are given in support of these propositions, as also of the uniformity of the amount of quadrantal deviation when determined in various parts of the ship; and, assuming the normal amount in iron steam-ships as from 2° to 4°, an analysis is given by which it is seen that 75 per cent. of the iron ships of the Royal Navy are included in this condition.

Two questions of importance here arise; are the results of this analysis conclusive, and if so, under what conditions do large quadrantal deviations occur? Reverting to the Astronomer Royal's early experiments in 1838–39, in the iron ships 'Rainbow' and 'Ironsides,' whose values were very small, and presuming that those vessels were built of good material—from their then experimental character—as also that similar conditions of material of good quality exist in the iron ships of the Royal Navy, it is assumed that the value (2° to 4°) represents the average condition of a ship built of the best or superior iron.

On the other hand, can the inference be drawn that large quadrantal deviation in an iron ship implies that inferior material has been used in her construction? Attention is here directed to the ships 'Birkenhead' and 'Royal Charter,' which from their well-known magnetic coefficients may be regarded as the types respectively of "hard" and "soft" iron constructed vessels, and from their consideration, as also from a review of the general results, these conclusions are derived:—

1. That in an iron ship of ordinary dimensions, a standard compass can be placed, the deviations of which will but little exceed those obtaining in wood-built steam-ships; and further, that on changes of geographic position, however distant, these deviations will be within smaller limits, and can be approximately predicted.

2. A divergence from these conditions will arise when the inductive

magnetism of the hull or machinery predominates; and it is inferred, especially from the example of the 'Royal Charter,' that large qua-drantal deviation and fluctuating sub-permanent magnetism (due to hull alone) are co-existent, and give rise to conditions of compass disturbance which are beyond prediction, and which have hitherto baffled inquiry and given a complexion to theoretical deductions varying as regarded from different points of view.

In order to examine the change which the original magnetism of an iron ship undergoes after launching, a series of compass observa-tions were made in the steam-ship 'Great Eastern' prior to her quit-ting the River Thames in 1859, and subsequently at Portland, Holy-head, and Southampton*—at the three first-named places within short periods of time of each other.

The results, from an Admiralty Standard Compass placed in a position the least subject to influence from local masses of iron, were as follows:—In the first five days, from Deptford to Portland, the ship's force had diminished from 0·585 to 0·480 [the earth's force $=1\cdot000$], or nearly one-fifth; representing a decrease in the "semi-circular" deviation from $35° 50'$ to $28° 45'$; the direction of the force, or neutral points, approaching the fore-and-aft line by $10°$, or changing from $47°$ on the starboard bow to $37°$.

At the expiration of the next six weeks, the ship in the interim having made the passage to Holyhead, the ship's force diminished from 0·480 to 0·390, or about one-sixth, corresponding to a decrease of "semicircular" deviation from $28° 45'$ to $23° 0'$, the direction of the force changing from $37°$ to $32°$.

At Southampton, in June 1860, or nearly eight months after the experiments made at Holyhead, the force had further diminished from 0·390 to 0·235, or by one-half, corresponding to a decrease in the "semicircular" deviation from $23° 0'$ to $13° 30'$; whilst the direction of the force approached the fore-and-aft line $25°$, or from $32°$ to $7°$; the quadrantal deviation remaining nearly constant $[+4\frac{1}{4}°]$ the whole time included in the various observations.

The unvarying tendency of the direction of the ship's force in the 'Great Eastern' to assume a fore-and-aft line, supports the view that time, with the vibrations and concussions due to sea service, leads to a distribution of the magnetic lines, of the nature of a stable equili-brium depending on the average of the inducing forces to which the ship is exposed; the respective sections of the hull having north and south polarity, being separated by lines approximating more nearly a horizontal plane and vertical axis through the body of the ship; instead of the inclined axis and equatorial plane of separation due to the magnetic dip of the locality, and divergence from the magnetic meridian, of the hull while building.

The practical information resulting from the example of the 'Great Eastern' is, that prior to a newly built iron ship being sent to sea, her head during equipment should be secured in an opposite direction to that in which she was built; and that the magnetic lines should

---

* The observations at Southampton were made after the paper was communi-cated to the Royal Society, and are introduced by way of supplement.

be assisted to be "shaken down" by the vibrations of the machinery in a short preparatory trip prior to the determination of her compass errors, or their compensation; but especially that in the early voyages vigilant supervision should be exercised in the determination of the compass disturbances.

Another important point, generally neglected when compasses are adjusted by the aid of magnets in a newly built iron ship, is rendered manifest by the results of this investigation; namely, the necessity of the errors of the compass being determined and placed on record prior to the adjustment. Without the knowledge to be derived from these observations of the magnetic force of the ship, all future changes of magnetism and consequent errors of the compass are mere guesswork both to those who adjust, and those in charge of the navigation of the ship.

It is recommended that, in any future legislation for the security of the navigation of our mercantile marine with reference to iron-built ships, the determination and record of these preliminary observations should be secured.

The paper concludes by directing attention to the general principles of practical import which result from the investigation, viz. as to the best direction with reference to the magnetic meridian for the keel and head of an iron ship to be placed in building, to ensure the least compass disturbance; the best position and arrangement for a compass to ensure small deviations, and permanency on changes of geographic position; and the changes to which the compass is liable from various causes on the foregoing conditions being fulfilled.

For the best direction in building, it is shown that, from the nature of the polarity of the hull, and especially of the top sides in the after section of the ship and adjoining the compass, where usually placed, the latter is least affected in those vessels built in the line of the magnetic meridian.

For iron steam-vessels engaged in the home or foreign trades in the northern hemisphere, it is recommended, from the then antagonistic magnetic influence of the hull and machinery, to build them head to the north: for iron sailing vessels, from the top sides, in the usual position of the compass, being magnetically weak if built head to the south, the latter direction is to be preferred.

The selection for the position of the compass depends on the direction of the ship during building; in those built head to north, it must be removed as far from the stern as convenience will permit; in those built head to south, as near to the stern as convenient, but avoiding especially, in all cases, proximity to vertical masses of iron. In ships built head east or west, there is little choice of position: in those built on the intercardinal points, a position approximating to the stern when the action from the top-sides—to be determined experimentally—is at a minimum, is to be preferred.

Ample elevation above the deck and exact position in the middle line of the ship, are primary conditions to be observed; and no compass should be nearer iron deck beams than 4 feet. As every piece of iron not forming a part of, or hammered in the fabrication of the hull, such as the rudder, funnel, fastenings of deck houses, &c., is of a magnetic character differing from the hull of the ship, proximity to any such should be avoided, and, as far as possible, the compass should be so placed that they may act as correctors of the general magnetism of the hull.

As most compasses are affected by the magnetism of the ship to an amount depending on their elevation, and the direction of the ship in building, the disturbances will be large comparatively, except in those vessels built head east or west.

A series of Tables is appended, wherein the magnetic coefficients and ship's force and direction of the various classes of vessels are given, the ships being classed according to the nature of their material and machinery.

GEOLOGICAL SOCIETY.

[Continued from p. 311.]

March 20, 1861.—L. Horner, Esq., President, in the Chair.

The following communications were read :—

1. "On a Collection of Fossil Plants from the Nagpur Territory, Central India." By Sir C. Bunbury, Bart., F.R.S., F.G.S. &c.

The specimens examined by the author were collected by the Rev. Messrs. S. Hislop and R. Hunter, and presented to the Geological Society in 1854 and since. The vegetable remains described in this paper are :—1. *Glossopteris Browniana*, var. *Australasica*, Ad. Brongn. *G. Browniana*, var. *Indica*, Ad. Brongn. By much the most abundant plant in the collection. 2. *G. musæfolia*, sp. nov. 3. *G. leptoneura*. 4. *G. stricta*, sp. n. 5. *Pecopteris*, sp. 6. *Cladophlebis* (?). 7. *Tæniopteris danæoides*, M'Celland (?). 8 and 9. *Filicites* : possibly *Glossopteris*. 10. *Nœggerathia* (?). 11. *Phyllotheca Indica*, sp. n. 12. *Vertebraria* (?). Different from the true *Vertebraria*, and probably *roots*. 13. *Knorria* (?). 14. *Stigmaria* (?). 15. Part of a stem, somewhat Sigillarian in appearance. 16. Yuccites (?). The fruits and seeds are reserved for further examination. On a general survey of all these plant-remains, the author for the present considers the *facies* of the fossil flora under notice to be Mesozoic rather than Palæozoic, but he regards the question as an open one, and requiring much further light for its perfect elucidation.

2. "On the Age of the Fossiliferous thin-bedded Sandstones and Coal-beds of the Province of Nagpur, Central India." By the Rev. Stephen Hislop. Communicated by the President:

The author first pointed out the places near the city of Nagpur where the *plant-bearing sandstone* has been best studied. He next noticed the carbonaceous shales underlying thick sandstones, at the foot of the Mahádewa Hills and the coal-seams of Barkoi, near Umret, 80 miles and more N.W. of Nagpur ; and pointed out their relationship to the plant-bearing sandstone near Nagpur, as shown by the *Glossopteris* and other fossils found in each locality.

· At Mángali, between 50 and 60 miles S. of Nagpur, dark red sandstones are found, rich in *Estheria*, and containing remains of Plants, Ganoid Fishes, and Reptiles (*Brachyops laticeps*, Owen). These beds Mr. Hislop thinks to be of the same age as those of Nagpur and Chanda. Still further S. (170 miles from Nagpur), at Kotá, there are (under thick sandstones) limestones and shales, containing fishes of the genera *Æchmodus* and *Lepidotus*, Teleosaurian remains, Coprolites, fossil Insects, *Cypridæ*, and *Estheriæ*, with obscure plant-remains. These beds are also regarded by the author as equivalent in age to the plant-bearing sandstones of Nagpur; whilst the sandstone above them may be equal to the sandstone of the Mahádewas; and the red clay beneath them may be the same as that of Maledi 30 miles off (to the N.E.), where *Ceratodus* teeth and Coprolites have been found in abundance.

Mr. Hislop then compared in detail, 1. the fossil flora of the coal-fields of New South Wales with that of Central India; 2. the fossil plants of Western Bengal with those of Central India; and 3. the fossil fauna of these two regions; and came to the conclusion that, on the whole, they probably represent the Jurassic (or possibly the Triassic) period,—at all events some portion of the Lower Mesozoic epoch.

3. "On the Geological Age of the Coal-bearing Rocks of New South Wales." By the Rev. W. B. Clarke, F.G.S.

The author first referred to his report, in 1847, of the occurrence of *Lepidodendron, Sigillaria*, and *Stigmaria* in the coal-fields of Australia; and advanced proofs of the occurrence of *Lepidodendron (Pachyphlæus* (?), Gœppert) over a region extending from 23° to 37° S. lat., and at least 1000 miles long. After some observations on the association of Carboniferous and Devonian fossils with the coal-beds of Australia and Tasmania, Mr. Clarke stated that in 1859, at Stony Creek, near Maitland, Mr. B. Russell, having sunk two pits in search of coal, found four or five coal-seams lying between beds containing *Pachydomi, Spiriferi, Orthoceratites, Conulariæ*, &c.; and beneath them a shale containing *Nœggerathia, Glossopteris, Cyclopteris*, &c. From this and other evidence the author is induced to believe that the beds are of palæozoic age, in spite of the "Jurassic" appearance of the plant-remains.

4. "On some Reptilian Remains from North-western Bengal." By Prof. T. H. Huxley, F.R.S., Sec.G.S.

Some bones, found by Mr. Blanford in the uppermost portion of the "Lower Damūda" group of strata in the Ranigunj coal-field, and forwarded to the author by Professor Oldham, have proved to belong to Labyrinthodont Amphibia and Dicynodont Reptiles, hereby affording new and interesting links with the fossil fauna of the Karoo-beds of South Africa, and largely increasing the probability that the rocks in which they were found are of Triassic, or perhaps Permian, age.

April 10.—Sir R. I. Murchison, V.P.G.S., in the Chair.

The following communications were read :—

1. "On the Geology of the Country between Lake Superior and

the Pacific Ocean (between 48° and 55° parallels of latitude), explored by the Government Exploring Expedition under the command of Captain J. Palliser (1857–60)." By James Hector, M.D. Communicated by Sir R. I. Murchison, V.P.G.S.

This paper gave the geological results of three years' exploration of the British Territories in North America along the frontier-line of the United States, and westward from Lake Superior to the Pacific Ocean.

· It began by showing that the central portion of North America is a great triangular plateau, bounded by the Rocky Mountains, Alleghanies, and Laurentian axis, stretching from Canada to the Arctic Ocean, and divided into two slopes by a watershed that nearly follows the political boundary-line, and throws the drainage to the Gulf of Mexico and the Arctic Ocean. The northern part of this plateau has a slope, from the Rocky Mountains to the eastern or Laurentian axis, of six feet in the mile, but is broken by steppes, which exhibit lines of ancient denudation at three different levels : the lowest is of freshwater origin; the next belongs to the Drift-deposits ; and the highest is the great Prairie-level of undenuded Cretaceous strata. This plateau has once been complete to the eastern axis, but is now incomplete along its eastern edge, the soft strata having been removed in the region of Lake Winipeg.

The eastern axis sends off a spur that encircles the west shore of Lake Superior, and is composed of metamorphic rocks and granite of the Laurentian Series. To the west of this follows a belt where the floor of the plateau is exposed, consisting of Lower Silurian and Devonian rocks. On these rest Cretaceous strata, which prevail all the way to the Rocky Mountains, overlain here and there by detached tertiary basins.

The Rocky Mountains are composed of Carboniferous and Devonian limestones, with massive quartzites and conglomerates, followed to the west by a granitic tract which occupies the bottom of the great valley between the Rocky and the Cascade Mountains. The Cascade chain is volcanic, but the volcanos are now inactive; to the west of it, along the Pacific coast, Cretaceous and Tertiary strata prevail. The description of these rocks was given with considerable detail, on account of their containing a lignite which for the first time has been determined to be of Cretaceous age. This lignite, which is of very superior quality, has been worked for some years past by the Hudson Bay Company, and is in great demand for the steam-navy of the Pacific station, and for the manufacture of gas. Extensive lignite-deposits in the Prairie were also alluded to; and, like those above mentioned, were considered to be of Cretaceous age ; but, besides these, there are also lignites of the Tertiary period.

The general conclusion was that the existence of a supply of fuel in the Islands of Formosa and Japan, in Vancouver's Island, in the Cretaceous strata of the western shores of the Pacific, but principally within the British territory, and in the plains along the Saskatchewan, will exercise a most important influence in considering the practicability of a route to our Eastern possessions through the Canadas, the Prairies, and British Columbia.

2. "On Elevations and Depressions of the Earth in North America." By Dr. A. Gesner, F.G.S.

After some observations on the differences between volcanic uplifts of the land and the slow upward and downward shiftings produced by changes in the position of great parallel areas during long periods of time, the author proceeds to enumerate evidences of local elevation and subsidence that he has observed along the coast from the northern part of Labrador to New Jersey.

In the south-eastern part of New Jersey, at Nantucket, Martha's Vineyard, and Portland, submergence of the land is proceeding, locally at the rate of probably four feet in sixty years. In New Brunswick, at St. John's the land has been elevated; at the Great Manan Island and the Great Tantaman Marsh there has been subsidence. At Bathurst and on the opposite coast of Lower Canada the land seems to be rising. In Nova Scotia, near the Bay of Fundy and Mines Basin there is subsidence; on the southern side, however, there are signs of elevation. The sea rapidly encroaches upon Louisberg in Cape Breton; and in Prince Edward's Island, also, at Cascumpec, submergence of the land is taking place.

LXXIX. *Intelligence and Miscellaneous Articles.*

ON THE THEORY OF CYLINDRICAL CONDENSERS.

BY M. J. M. GAUGAIN.

I HAVE in a former note * called attention to the fact that it is very difficult to analyse the phenomena of condensation produced in submarine telegraphic cables, the gutta percha which envelopes them being only imperfectly non-conducting, so that there is at one and the same time propagation by conductibility and condensation. To study the latter phenomenon by itself, I substituted for gutta percha dielectrics, which isolate much more perfectly; I employed for this purpose gum-lac and air; with gum-lac the absorption is very small, with air it is nothing, or altogether imperceptible.

The laws which I have succeeded in establishing are very simple, and may be of some practical utility, since they afford a solution of the different questions which relate to electric condensation in submerged cables; it is, however, from a philosophical point of view that they seem to be of the greatest interest, since they confirm in a remarkable manner the views of Faraday. This illustrious physicist, in a memoir published in 1837 (Experimental Researches, Series XI. No. 1320), expressed himself nearly as follows:—"The power of isolating and that of conducting are only two extreme degrees of the same property, and ought to be considered as being of the same nature in any satisfactory mathematical theory." Now it will be seen that, in the case at least of cylindrical condensers, the laws which regulate the propagation of electricity by *excitation* do not differ from those which Ohm has established for propagation by conductibility. The general results of my researches may be stated shortly as follows:—

1. When the internal cylinder is the collector, that is to say, when it communicates with the source, and the external cylinder commu-

* *Comptes Rendus*, 28th October, 1860.

nicates with the ground, the *excited* charge of the external cylinder is equal to the *exciting* charge of the internal cylinder.

2. When the external cylinder is the collector, the *excited* charge of the internal cylinder is precisely the same as it would have been if it had been put in direct communication with the same source.

3. When the external cylinder is the collector, its charge may be considered as consisting of two parts, one of which is equal to the *excited* charge of the internal cylinder, the other representing the quantity of electricity which the external cylinder would take up by itself under the influence only of the medium in which it is placed.

The latter law enables us to foresee what would happen in the case of a condenser formed of three concentric cylinders. The charge which the middle cylinder would take up when put in communication with the source, the other two being connected with the ground, must be equal to the charge which would be excited in it by the two other cylinders. I have found by experiment that this is really the case.

It follows that condensers arranged in a spiral form, may serve to collect in a small volume a large quantity of electricity. -

4. If we agree to call by the name of *resistance to excitation* a quantity inversely proportional to the charge received by either armature, when the tension of the internal cylinder is maintained at unity, and that of the external cylinder at zero, then this resistance, which I shall call ρ, is expressed by the formula

$$\rho = k \log \frac{R}{r},$$

R and r representing the respective radii of the external and internal cylinders, and k being a constant which depends on the inductive capacity of the dielectric, and on the length of the cylinder employed.

This formula is remarkable, since it might have been deduced à *priori* from the ordinary theory of propagation by conductibility. Suppose, in fact, that the substance which separates the two cylindrical armatures of the condensers possesses a certain conductibility, and let us call by the name of *resistance to conductibility*, a quantity inversely proportional to the amount of electricity which, in a unit of time, traverses the annular space between the two cylinders, the tension of the internal cylinder being maintained at unity, that of the external cylinder at nothing. This *resistance to conductibility* may be calculated according to the principles established by Ohm, and it will be found that it is expressed by the same formula as the *resistance to excitation.* ·To pass from one formula to the other, it is only necessary to change the meaning of k. We may say therefore that the same theory, that namely of Ohm, regulates propagation by excitation and propagation by conductibility, at least when we confine ourselves to the consideration of spaces bounded by concentric cylinders. I propose to verify this principle under other circumstances, and especially in the case of spherical condensers.—*Comptes Rendus,* Feb. 18th, 1861.

ON THE PRODUCTION OF GRAPHITE BY THE DECOMPOSITION OF CYANOGEN COMPOUNDS.   BY DR. P. PAULI.

The mother-liquors obtained from the evaporation of a solution of the so-called black ash are now commercially worked for caustic alkali.   These liquors contain the following compounds :—

1. Chiefly hydrate of soda.
2. Some quantity of carbonate of soda.
3. Several sulphur compounds of sodium : viz. sulphide of sodium, hyposulphite of soda, sulphite of soda, and sulphate of soda.
4. Sulphide of iron held in solution by the sulphide of sodium.
5. Chloride of sodium.
6. Several cyanogen compounds of sodium, and especially ferro-cyanide of sodium.

These liquors are evaporated in large cast-iron pots, and in order to destroy or oxidize the sulphides of sodium and iron, as also the cyanogen compounds, an equivalent quantity of soda-saltpetre is added.   All the oxidable sulphur compounds, together with the small quantity of sulphide of iron, are changed to sulphate of soda and peroxide of iron by the nitrate of soda in the boiling liquor, at a temperature not below 260° to 270° F.   The cyanogen compounds, on the other hand, are not decomposed by the nitre until the liquor begins to pass from the watery into the dry fusion, and the uncombined water of the hydrate of soda has been driven off.   When the whole mass of alkali (generally about four tons) reaches a low red heat, a regular evolution of gas is observed : this is evidently owing to the oxygen produced by the decomposition of the nitrate, and to the nitrogen from the decomposition of the cyanides; at the same time a plentiful liberation of graphite is observed, covering the whole surface of the liquor with a bright layer of graphite.   This liberation of graphite is still more plainly seen if no nitre be added to the liquor at first, or only so much as is sufficient to oxidize the sulphur compounds : but if a few pounds of nitrate of soda be added when the water has been driven off, and the mass is allowed to become red-hot, a violent reaction takes place, and a large quantity of graphite is set free.   This sudden liberation of graphite proves that this substance cannot be derived from the cast iron of the pot in which the fusion is made.   So violent is the evolution of gas, that a complete cloud of fine particles of caustic soda is carried up into the air, rendering it almost impossible to remain in the neighbourhood of the operation. In this way all the cyanogen compounds are completely decomposed, the iron in the ferrocyanide of sodium becomes peroxide, and in a few hours falls to the bottom of the pot.   If the right quantity of saltpetre has been added, a colourless mass of fused caustic soda remains; but if too large an amount of nitre has been added, the liquor becomes coloured deep green, owing to the formation of manganate of soda.   It is remarkable that, in the absence of nitrate of soda, the cyanogen compounds act reducing upon the sulphide of sodium; this is seen from the fact that a portion of the soda-lye, which gives no sulphide reaction with a lead salt, produces a blackening after the caustic alkali has been heated to redness.

The graphite may be skimmed off the surface of the fused alkali;

and when washed with water and hydrochloric acid, it appears in the form of an extremely fine bright powder. If allowed to swim on the top of the almost red-hot fused soda, the graphite is oxidized gradually, and after the lapse of about three or four hours it altogether disappears. Heated in a platinum crucible by itself it is incombustible; but it generally contains small particles of charcoal mixed with it, and these undergo oxidation.

The temperature at which this evolution of graphite takes place is a very low one, compared with that at which graphite is liberated from cast iron; for a thin iron wire can scarcely be brought to a visible red heat by dipping it into the fused alkali.

From this peculiar decomposition it would appear that we have good reason to assume that the carbon contained in cyanogen is present in the graphite modification; for if this be not the case, how is it that the easily combustible charcoal can withstand the oxidizing action of the saltpetre, whilst none of the iron of the ferrocyanide of sodium is reduced to the metallic state.

It has, besides, been lately shown by M. Caron, that the formation of steel, *i. e.* the combination of iron with carbon in the graphite modification, can only take place in presence of cyanogen compounds, and that no carbon whatever is taken up by the iron when this metal is heated with other carboniferous gases. The mode of production of graphite noticed in this communication appears to be an intermediate reaction between that from the carbide of iron and from the nitride of carbon.

As in the process of cementation it is seen that the carbon of the cyanogen is taken up by the iron without being set free, so this reaction proves that cyanogen can be split up into its constituent parts without either of them combining with a third body.

Despretz asserts that the carbonization of iron is always preceded by a combination of this metal with nitrogen, a process which makes it porous and more fit for uniting with carbon. The correctness of this supposition has, however, become rather doubtful, by Caron's recently published experiments (*Comptes Rendus*, Nos. 15 and 24, 1860).

To conclude, I beg to say some words about the formation of native graphite; I do not think that this body has been formed from coal or diamond, but I rather believe it has been separated out of carbon compounds as graphite, by processes perhaps analogous to those above described.—*Proc. Manchester Phil. Soc.*, April 16, 1861.

ON ELECTRICAL PARTIAL DISCHARGES. BY P. RIESS.
To the Editors of the Philosophical Magazine and Journal.
GENTLEMEN,

Professor Rijke has met with a difficulty in the explanation of Wheatstone's experiments on the discharge of a Leyden jar, and has himself instituted experiments*, from which it seems that more electricity remains on a conductor when it is discharged through a wet string than when it is discharged through a metallic wire.

I do not think that this difficulty exists, nor that these experiments have any claim to novelty. Twenty years ago I showed that when

* Phil. Mag. for May, p. 365.

a battery is discharged through a column of water, three-eighths of the charge remain on the battery, but that two-thirteenths remain when the discharge is effected through a metallic wire; and from this I concluded that the electricity of the battery is discharged in a successive manner (Pogg. *Ann.* vol. liii. p. 14. " Lehre von der Rei-bungselectricität," § 634).

Since then I have always assumed that the spark accompanying the discharge consists of several sparks; and the discharge of an electrified body, of several individual discharges, which I have called partial discharges. By this assumption Wheatstone's and many other electrical experiments have become capable of explanation.

I am, Gentlemen,
Yours truly,

Berlin, May 10, 1861.   P. RIESS.

ON THE FREEZING OF WATER AND THE FORMATION OF HAIL.
BY M. L. DUFOUR.

When water is preserved from contact with solid bodies by placing it in a mixture which has the same density, and which does not form aqueous mixtures, its congelation may be materially retarded. Water placed in a mixture of chloroform and oil (the best is oil of sweet almonds) takes the form of perfect globules, and remains at rest in the interior of the mixture. If this mixture be cooled, the water in this condition scarcely ever freezes at $0°$ C.; its temperature sinks to $-6°$, $-10°$ before this change takes place. Globules have in this way been even reduced to $-20°$ while still liquid.

The globules either change into globules of ice, or they simply freeze on the surface, according to their dimensions and the diminution of temperature. They persist in the liquid state with remarkable stability. In this mixture of chloroform and oil they may be shaken, and foreign bodies introduced, without solidifying; but solidification immediately ensues when they are touched with a piece of ice. The discharge of a Leyden jar or a galvanic current may traverse these globules without their solidifying; but the powerful discharge of a Ruhmkorff's coil causes their immediate solidification.

When an ice-sphericle formed in the mixture of chloroform and oil is surrounded by other spheres which still remain liquid, the congelation of the latter may be effected by bringing them in contact with the first. Different effects are obtained according to the temperature and dimensions of the globules. Sometimes (with small globules and low temperatures) the spheres touched solidify suddenly, and remain separate; sometimes (with larger globules and somewhat higher temperatures) they coalesce more or less completely; they stretch out on each other at the moment of solidification. In this way pieces of ice of the most varied shapes may be obtained—irregular spheres formed of concentric layers (each layer consisting of a globule which enveloped the nucleus at the moment of its formation), spheres with protuberances, &c. These varied forms would have but a subordinate interest, if they did not recall the concentric zones and the irregular shapes observed in hailstones. This resemblance is evident in these experiments; and the question naturally arises whether hailstones are not formed under similar circumstances. In a memoir which I shall publish in the *Bibliothèque Universelle*, I examine this analogy more closely, and endeavour to show that it is not superficial, but extends into numerous details. I attempt to show that this particular case of freezing gives a suitable explanation of the general phenomena, as well as of the accidental peculiarities of hailstones: I attempt to show that these aqueous globules may also be cooled below 0° in the atmosphere; that they can then freeze and unite just as in the mixture of chloroform and oil, and that the grains of ice thus formed, increased by the condensation of atmospheric vapour on their surface, may be hailstones.—*Comptes Rendus*, April 15, 1861.

ON AN APPARATUS FOR EXPERIMENTS ON RESPIRATION AND PERSPIRATION IN THE PHYSIOLOGICAL INSTITUTE AT MUNICH. BY PROFESSOR PETTENKOFER.

In order to determine the quantities of carbonic acid and of water which are eliminated through the skin and lungs, numerous methods have been proposed, and the processes and results of Scharling. Vierordt, Valentin and Brunner, Regnault and Reiset, Smith, and others, are well known to every physiologist and chemist. The objections to all previous methods are of two kinds: first, that the accuracy of the method cannot be determined by control experiments with known quantities of carbonic acid; and second, that the men and animals were compelled to breathe, in these experiments, under more or less unusual and burdensome, and therefore unnatural conditions. The experiments of Bischoff and Voit on the food of carnivora have further shown that the carbonic acid eliminated by the lungs and skin, cannot be obtained by taking the difference between that administered as food, and that secreted in the urine and fæces, allowing for the difference in weight of the body; because two unknown quantities of carbonic acid and water are eliminated from the lungs simultaneously, and in varying quantities. As it was necessary to determine directly at least one of these magnitudes, the author endeavoured to construct an apparatus by which a constant current of air could be passed over a man, and the increase of carbonic acid and water of this air determined.

Pettenkofer's apparatus consists of a small sheet-iron chamber (which will be called the *saloon*) 8 Bavarian feet in every dimension, with an iron door, a light at the top, and windows at the sides. The windows were cemented, and the sides and cover riveted as air-tight as possible. The door had moveable openings in order to ensure access of air at other points besides the joints of the door. On the side opposite to the door there are two apertures, one below and the other above, which, by means of two tubings, are connected with a single wide tube outside, in which the air flows towards that part of the apparatus which serves as an aspirator. This piece of the apparatus, which is placed in a different part of the house, consists of two suction cylinders with valves, which can be uniformly worked to the same height of stroke by means of a powerful clockwork motion. The weight of the clockwork is continually raised in proportion as it sinks, and in this way is obtained a continuous current of air through the doors of the iron room towards the suction cylinders. But the air cannot reach the suction cylinder without first passing through a continuously acting measuring apparatus. For this purpose the author selected a large gas-meter, of such dimensions that 3000 English cubic feet could be measured with it in an hour.

In order to investigate a portion of the air entering by the apertures in the door and any accidental leakages in the apparatus, as well as of the air flowing towards the gas-meter, and to calculate from the observed difference in the proportion of water and carbonic acid the quantities which had entered from the apparatus, there are two aspirators, each of which simultaneously withdraws an equal quantity of air. The water of the air is absorbed by sulphuric acid and weighed, and the carbonic acid is determined by allowing the air to pass in fine bubbles through a determinate quantity of lime-water of known strength, and the strength of the lime-water finally determined by means of dilute oxalic acid.

In order to take a specimen of the air remaining in the saloon, a forcing and suction pump is connected with the outlet pipe, by which flasks holding 6 or 8 litres may be filled with air, and the quantity of carbonic acid determined by means of lime-water. The same pump serves to determine at any time the variations in the carbonic acid during the progress of an experiment. There is an arrangement by which test-quantities of any amount may be taken out without causing any loss in measuring the whole current. For this purpose a flask is connected air-tight with the pump, and by continuous pumping its air is completely replaced by air from the outlet pipe. The air pressed out of the flask is not allowed to escape, but is passed by means of a caoutchouc tube into the current which goes to the gas-meter; of course, in a place where it cannot affect the determinations of carbonic acid.

In order that the air-current may take no water from the large gas-meter by evaporation, the air before entering the gas-meter first passes through an upright cylinder filled with pieces of pumice kept moist.

Where the air issues from this apparatus, there is in the tube a psychrometer, in order to measure the temperature and moisture of the air which passes into the gas-meter. In the tube which leads to the moistening apparatus there is a psychrometer, and several tubes for taking out specimens of air, &c.

The apparatus has been examined since May last in every particular, and the author recommends the methods of investigation as in every way convenient. It was above all important to prove that the carbonic acid disengaged in the saloon could actually be found again and determined—a control which has been omitted in all previous experiments on respiration. After the author, by numerous experiments, had investigated all the influences of the apparatus and of the methods on the accuracy of the results, he took a stearine candle and determined its carbon by elementary analysis. When the suction cylinders of the apparatus, and simultaneously the apparatus for analysis of the air, were at work, a weighed candle was lighted in the saloon from without, and, before the experiment terminated, again extinguished from without and weighed.

The carbonic acid formed by combustion of the candle must be partially contained in the air which had passed through the large gas-meter, and partly in that contained in the saloon. The carbonic acid in the air which had passed through the gas-meter was determined by allowing a continually equal proportion (about 100 cubic centims. in a minute) of the current from the saloon to the gas-meter to bubble through lime-water. The carbonic acid of the air remaining in the saloon was determined, after mixing the different layers of air well together, by filling two or three vessels of 6 to 8 litres capacity, determining by lime-water, and calculating this upon the known capacity of the saloon. It is only after these flasks have been filled that the saloon may be entered to take out the candle and weigh it.

Since the air which passed into the gas-meter, and that which remains in the saloon, contained not only the carbonic acid formed in the experiment, but that already contained in the air as it passed into the saloon, the quantity of the carbonic acid of the air entering must be subtracted. This is obtained from the experiment where the air which enters is withdrawn and investigated in just the same manner and the same quantity as the emergent air. In this way the difference in the quantity of carbonic acid inside and outside is determined; and this ensures the exactitude of the determinations, because all constant errors of the method are thereby eliminated. Of course all the measurements are made with allowances for the tension of aqueous vapour, of temperature, and pressure of the atmosphere.

The author adduces some experiments which he made with stearine candles, and states his reasons for believing that, during an experiment in which more than four-fifths of the disengaged carbonic acid pass into the current between the saloon and the gas-meter, no greater errors than at most 1 or 2 per cent. are to be feared. Inasmuch as the duration of experiments with men and animals can be extended to twelve or twenty-four hours, it is to be hoped that even greater accuracy may be attained.

This is the first apparatus of its kind in which living is possible under normal conditions. Men can live in it just as well as in any well-ventilated room, and can move about, eat, and drink in the ordinary manner. By a moveable window in the door of the saloon, food and other articles can be supplied or taken out—just as, in a small room, provided the draught in the chimney is in order, a stove-door can be opened without admitting smoke. The observer outside the saloon who has charge of the experiment does not by his respiration affect the result in the least; for the carbonic acid in the air passing into the saloon can be continually checked by one of the control apparatus, and can be allowed for.—*Journal für Prakt. Chemie*, vol. lxxxii. p. 40.

INDEX to VOL. XXI.

END OF THE TWENTY-FIRST VOLUME.

PRINTED BY TAYLOR AND FRANCIS,
RED LION COURT, FLEET STREET.

ALERE FLAMMAM.

THE

LONDON, EDINBURGH, AND DUBLIN

PHILOSOPHICAL MAGAZINE

AND

JOURNAL OF SCIENCE.

CONDUCTED BY

SIR DAVID BREWSTER, K.H. LL.D. F.R.S.L. & E. &c.

SIR ROBERT KANE, M.D., F.R.S., M.R.I.A.

WILLIAM FRANCIS, Ph.D. F.L.S. F.R.A.S. F.C.S.

JOHN TYNDALL, F.R.S. &c.

"Nec aranearum sane textus ideo melior quia ex se fila gignunt, nec noster vilior quia ex alienis libamus ut apes." Just. Lips. *Polit.* lib. i. cap. 1. Not.

VOL. XXII.—FOURTH SERIES.

JULY—DECEMBER, 1861.

LONDON.

TAYLOR AND FRANCIS, RED LION COURT, FLEET STREET,
Printers and Publishers to the University of London;
SOLD BY LONGMAN, GREEN, LONGMANS, AND ROBERTS; SIMPKIN, MARSHALL
AND CO.; WHITTAKER AND CO.; AND PIPER AND CO., LONDON:—
BY ADAM AND CHARLES BLACK, AND THOMAS CLARK,
EDINBURGH; SMITH AND SON, GLASGOW; HODGES
AND SMITH, DUBLIN; AND PUTNAM,
NEW YORK.

"Meditationis est · perscrutari occulta; contemplationis est admirari perspicua . . . . . Admiratio generat quæstionem, quæstio investigationem, investigatio inventionem."—*Hugo de S. Victore.*

—"Cur spirent venti, cur terra dehiscat,
Cur mare turgescat, pelago cur tantus amaror,
Cur caput obscura Phœbus ferrugine condat,
Quid toties diros cogat flagrare cometas;
Quid pariat nubes, veniant cur fulmina cœlo,
Quo micet igne Iris, superos quis conciat orbes
Tam vario motu."

J. B. Pinelli ad Mazonium.

CONTENTS OF VOL. XXII.

(FOURTH SERIES.)

NUMBER CXLVI.—SEPTEMBER.

NUMBER CXLVII.—OCTOBER.

ERRATA IN VOL. XXI.

Page 407, line 2, *for* $\tau=247^{\circ}\cdot45$ *read* $g=247^{\circ}\cdot45$.
— line 3, *for* $g=611^{\circ}\cdot28$ *read* $\tau=611^{\circ}\cdot28$.

— line 4, *for* hence $\mu h = \dfrac{\tau - g}{A}$ *read* hence $\mu h = \dfrac{\tau - g}{\beta}$.

415, last line of Note H, *for* between L and V *read* between L and m^{3}.

PLATES.

I. Illustrative of Prof. Magnus's Paper on the Propagation of Heat in Gases.

II. Illustrative of Mr. F. Galton's Paper on Meteorological Charts.

III. Illustrative of Prof. Tyndall's Paper on the Absorption and Radiation of Heat by Gases and Vapours.

IV. Illustrative of Mr. C. Tomlinson's Paper on the Cohesion-Figures of Liquids.

V. Illustrative of MM. Kirchhoff and Bunsen's Paper on Chemical Analysis by Spectrum-observations.

VI. Illustrative of Mr. W. S. Jevons's Paper on the Deficiency of Rain in an elevated Rain-gauge; of Dr. Lamont's Paper on the Form of Magnets; and MM. Kirchhoff and Bunsen's Paper on Chemical Analysis by Spectrum-observations.

Fig. 2

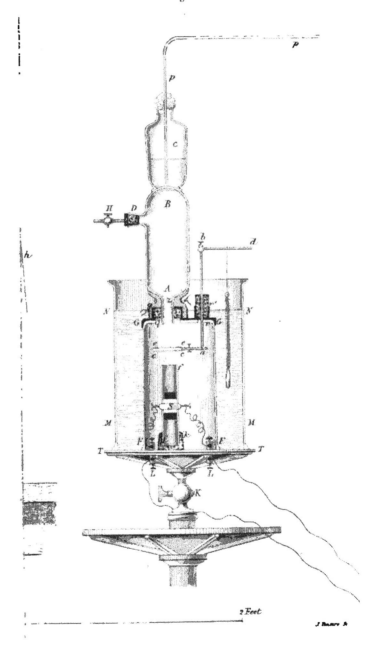

2 Feet

J. Basire Sc

THE

LONDON, EDINBURGH AND DUBLIN

PHILOSOPHICAL MAGAZINE

AND

JOURNAL OF SCIENCE.

[FOURTH SERIES.]

JULY 1861.

I. *On the Propagation of Heat in Gases.* By G. MAGNUS[*].

[With a Plate.]

Conduction of Heat by Gases.

THE cooling of a body, when it takes place *in vacuo*, simply depends on the exchange of heat by radiation between the body and the surrounding envelope. If, however, the space in which the cooling takes place is filled with a gas, an ascending current is formed which accelerates the process. The cooling is likewise promoted by the capacity of the gas to transmit heat (or its diathermancy), as well as by its conductibility, assuming that gases can conduct heat. Dulong and Petit, in enunciating their laws of the loss of heat in their comprehensive memoir *Sur la Mesure des Températures et sur les Lois de la Communication de la Chaleur*, have disregarded the latter actions, manifestly because they could neglect them as being infinitely small in comparison with the influence of the ascending current. Accordingly, since the appearance of their memoir, it has been universally assumed that the differences of cooling in various gases depend on the different mobility of their particles. This was the more justifiable, because almost simultaneously with Dulong and Petit's investigation Sir H. Davy's celebrated memoir[†] on Flame appeared, in which he says, "It appears that the property of elastic fluids to withdraw caloric from the surface of solid bodies increases as their density decreases, and that there is something in the con-

[*] Read before the Academy of Sciences of Berlin, July 30, 1860, and February 7, 1861. Translated by Dr. E. Atkinson. [A short abstract of a portion of the results has already appeared in this Journal, vol. xx. p. 510.]

[†] Phil. Trans. 1817, part. I, p. 61.

stitution of the light gases which makes them capable of withdrawing caloric from these surfaces in a manner different to what
other gases would do, which doubtless depends on the mobility
of their particles." Such a mobility is met with in other cases.
I have shown* that hydrogen passes through fine clefts and
apertures more easily than atmospheric air; and in fact the
different degree of diffusion in gases mainly depends on their
greater or less readiness to penetrate into capillary apertures.
Along with this difference in the mobility of the particles, which,
as follows from experiments on the transpiration of gases through
narrow apertures, depends on the specific gravity of the gas,
there may also be a different degree of friction within the gas
itself. It is, however, difficult to assume that this friction alone
occasions the great differences observed on cooling in different
gases. I hope to show in the following pages that the conductibility of heat by gases exercises an essential influence on the
internal cooling.

In 1792 Count Rumford published a memoir on heat†, in which
he contended that the propagation of heat in gases and vapours
is only produced by the motion of the gaseous particles, and
that a communication of heat from particle to particle—a conduction—does not take place in gases. Subsequently, in the
seventh of his 'Essays‡,' he has extended his researches to
liquids, and has denied that they possess the property of conducting heat.

This assertion was soon opposed by John Dalton§, while
Murray‖ and Dr. Thomas Thomson¶ afterwards endeavoured
to refute it by comprehensive investigations. Biot also**, in
reporting on the memoir of the latter, observed that Rumford's
experiments only justified the conclusion that liquids conduct
heat to a very small extent, and not that the property is entirely
absent.

It is only necessary to dip the hand in mercury to be convinced that this liquid is a good conductor; for the insupportable
cold which the hand experiences, and which is also perceptible
when it is laid on the surface, in which case certainly no currents
are formed, can only depend on conduction. But if this fluid
can conduct heat, may not others do the same, even though in
a smaller degree?

* Poggendorff's *Annalen*, vol. x. p. 153.
† Phil. Trans. for 1792, part 1. p. 48.
‡ Experimental Essays, vol. iii. p. 1.
§ Memoirs of the Literary and Philosophical Society of Manchester,
vol. v. part. 2. p. 372.
‖ Nicholson's Journal, vol. i. pp. 165 and 241. Gilbert's *Annalen*,
vol. xiv. p. 158.
¶ Nicholson's Journal, vol. iv. p. 529. Gilbert's *Annalen*, vol. xiv. p. 129.
** *Bulletin des Sciences par la Société Philomatique de Paris*, vol. iii. p. 36.

Despretz has since shown[*] that the conduction of heat in water follows the laws given by Fourier for its conduction in metals. A conductibility of heat in gases has never been imagined to exist. Although it is in any case very small, it appeared interesting to investigate what influence it might exert, and whether differences existed in the conductibility of different gases; for the deportment of gases is of especial importance, not only for the laws propounded by Dulong and Petit, but also for any theory of the nature of heat.

The more immediate inducement to this investigation was a repetition of Grove's[†] interesting experiment, that a platinum wire is less strongly heated by a galvanic current when surrounded by hydrogen than when it is in atmospheric air or any other gas.

On the first publication of these experiments, M. Poggendorff [‡] expressed the opinion that they depended on the laws which Dulong and Petit had established for the cooling of a body heated in the ordinary manner. Clausius § has since shown the concordance between Grove's results and the numbers obtained by Dulong and Petit.

In repeating Grove's experiments, I found that hydrogen exerts its preventive action even when only a very thin layer surrounds the platinum wire. Two very thin, equally long platinum wires were enclosed in tubes of 1 millim. diameter, one of the tubes being filled with atmospheric air and the other with hydrogen. On passing the current through both wires in succession, the one surrounded by atmospheric air became strongly incandescent, while the other did not even reach a red heat. It is scarcely necessary to say that the result was always the same, whichever of the two tubes was filled with hydrogen. Even when the tube filled with hydrogen was quite horizontal, 'the wire did not become incandescent. As the existence of currents in such a narrow horizontal tube can scarcely be assumed, it appeared improbable that the mobility of the particles of hydrogen was the cause of the strong cooling of the wire.

It is also impossible to conceive why currents produced by differences of temperature should be stronger in hydrogen than in other gases. This gas, it is well known, instead of being more, is even somewhat less expansible than atmospheric air.' Hence the same differences in temperature produce in hydrogen less change in the specific gravity than in atmospheric air. But it is by these changes alone that currents are produced in gases. Even if the friction of the particles exercises any influence, and

---

[*] *Ann. de Chim.* S. 2. vol. lxxi. p. 206.

[†] Phil. Mag. vol. xxvii. p. 445; vol. xxxv. p. 114. Pogg. *Ann.* vol. lxxviii. p. 366.

[‡] Pogg. *Ann.* vol. lxxi. p. 197. § Ibid. vol. lxxxvii. p. 501.

offers a greater hindrance to motion in other gases than in hydrogen, still this hindrance is in any case so small as not to cancel the influences of the greater expansibility of other gases, such as carbonic acid or sulphurous acid. .

But if the expansion in hydrogen can produce no stronger currents than in other gases, there remains no other assumption to explain the more rapid cooling in it than that this gas can conduct heat—that is, can give it from particle to particle, as is the case with metals—and that it possesses this property in a higher degree than other gases. The small density of hydrogen militates against this assumption; and it appeared necessary to decide by a few experiments how far it was correct. Accordingly in a glass tube 2 centims. broad and 10 centims. long, and closed at one end, a thermometer was fitted air-tight, so that the bulb was in the middle of the tube, while the graduation was above the cork. In order to fill the tube with different gases, there were two narrow glass tubes fitted into holes near the thermometer. Outside the tube they were bent at right angles, and could be closed by stopcocks. After water had been boiled in a capacious flask until all air had been expelled, the tube which had been previously filled with gas, was introduced into the flask, during the ebullition, in such a manner that it was entirely surrounded by vapour. The time was then measured which elapsed before the thermometer rose from 20° to 80° C. or 90° C.

The following results were obtained for the time necessary to heat the thermometer in the different gases:—

In	20° to 80°.	20° to 90°.
Atmospheric air .	3·5 minutes.	5·25 minutes.
	3·5 ,,	5·25 ,,
	3·5	5·2
Hydrogen . . .	1·0 ,,	1·5
	1·0	1·25
	1·1 ,,	1·4
	1·0 ,,	1·5
Carbonic acid . .	4·25 ,, ·	6·5
	4·25 ,,	6·25
Ammonia . . .	3·5 ,,	5·5
	3·5	5·5
	3·5	5·25 ,,
	3·5 ,,	5·25 ,,

Analogous experiments have been made by Leslie[*], Dalton[†],

---

[*] Inquiry into the Nature of Heat, p. 483.
[†] Memoirs of the Manchester Literary and Philosophical Society, vol. v. part 2. p. 379.

Davy* and others; but they, like Dulong and Petit, introduced the heated thermometer into a space which was successively filled with different gases, and observed the times which were necessary for the same cooling in these gases. Currents were thereby produced within the gas, which in my experiments were very small if not entirely absent, for the tube containing the gas was heated almost equally on all sides, above as well as below. But as the times which the thermometer required to become heated varied considerably, it appeared probable that the heating in the gases was not produced by currents alone, but that there was also a propagation of heat from particle to particle—in other words, a conduction. Accordingly I next made a series of experiments in which the gases were only heated from above, and observed the temperatures which a thermometer placed in them ultimately assumed. As in this case also the temperature was always higher in hydrogen than in other gases, and was also very different in them, I was confirmed in the conclusion that gases can conduct heat. It might still be objected that in the heating from above currents were formed, which caused the differences in temperature. There was a ready means of testing this objection. For if gases actually can conduct heat, the temperature which a thermometer assumes in a space heated from above, must be lower when the conducting substance is absent, that is, if the space is exhausted. In order to investigate whether this is the case I made use of the following apparatus.

Experiments on Conduction.

On a very thin glass vessel AB, fig. 1, Plate I., 56 millims. wide and 160 millims. in height, a second vessel C was fixed by fusion, of the same diameter, but only 100 millims. in height. A B is provided with a lateral tubulure D, in which a thermometer fg is hermetically fitted in such a manner that its bulb is in the axis of A B, and 35 millims. under the bottom of l, while the horizontal scale is outside A B. The lower end of A B is closed by means of a cork, in which are two narrow glass tubes provided with stopcocks, which serve to fill A B with different gases. Boiling water was poured into C, and then, from a flask at some distance in which water boiled, steam was passed into this water by the glass tube pp, so as to keep it in a state of ebullition. A plug of cotton wool prevented spirting.

In order to compare the thermometer-indications, obtained in using different gases, it was necessary to ensure that the space surrounding the vessel A B was always at the same temperature. For this purpose the vessel A B, with its thermometer,

* Philosophical Transactions for 1817, part 1. p. 60.   Schweigger's Journal, xx. p. 154.

was placed in a glass cylinder P Q, 235 millims. wide and
400 millims. in height. This stood in a second similar cylinder
X Y, so that there was a space of 30 millims. on every side,
which was filled with water. In order that the internal cylinder
P Q might not be raised by this water, flat leaden weights were
placed on its base, which loaded it so that it rested on the cork
supports U U. This cylinder was closed at the top by a hollow
metallic cylindrical cover, E E, 75 millims. deep, in which water
was poured. In this cover there was a cylindrical aperture *d d' s s'*,
55 millims. in diameter. The vessel C which received the boiling
water was fitted in this aperture by means of a slit cork; it was
prevented from falling by a couple of metal slides, *s s'*, fitted on
the under surface of the cover. This arrangement served at the
same time to fasten the whole apparatus A B C.

The heat which the vessel radiated laterally, heated the water
in the cover E E. In order to keep it at an invariable temperature, cold water continually flowed from a high reservoir
through the tube *r r*, while the heated water escaped by a siphon
h h. To observe the temperature on the inner cylinder there
were several thermometers, one of which, *k k*, was placed in a
horizontal position right under the cover E E; a second, *m*, was
suspended in the middle of the space, and a third, *l*, near the
bottom. During the experiment all these thermometers were
kept at the same temperature, 15° C. For this it was necessary
to have the room in which the experiments were performed at
about this temperature.

The bulb of the inner thermometer *f g* was protected by a
screen *o o* from direct radiation from above. At first I used a
cork screen, but afterwards one of silvered copper foil. According to the kind of gas contained in A B, the thermometer reached
its highest point and kept it unchanged, in from twenty to forty
minutes from the time at which hot water was poured into the
vessel C and steam passed into it. In the same kind of gas,
the density being constant, the maximum was always reached
in the same time; and provided that the thermometer *f g*, with
its screen, always remained in a fixed position and distance from
the vessel C, the temperature did not vary more than 0°·1 to 0°·2 C.
Under these conditions also the gradual increase of the temperature, up to the maximum, took place in such a manner that
after the same time the thermometer always indicated the same
temperature. This concordance furnished a proof of the accuracy
of the method.

Before turning to a few of the conclusions from the numbers
obtained, it will be convenient to enter upon the circumstances
which influence the maxima of temperature.

The heat proceeding from the lower surface of the vessel C is

propagated either by radiation alone, or by radiation and conduction. The thermometer is indeed protected from direct radiation by the cork screen; but this screen itself becomes heated by
a long-continued action of the rays, and then gives part of its
heat to the thermometer. I confess I at first believed that
the heat transferred in this manner to the thermometer would
be scarcely perceptible with a screen of 2 millims. thickness, and
would in any case be less than with a metal screen. Hence the
greater part of the experiments were made with a cork screen.
Afterwards, however, I found that a metal screen, although six
times as thin as a cork screen, is a better protection against
radiation. This doubtless depends upon the fact that a metal
screen absorbs fewer of the rays, and also radiates worse than
the cork screen; for when the silvered copper foil was blackened
on both sides by a tallow candle, the thermometer was heated
more than by the cork screen. Hence the metal screen was
never used blackened. But whatever the nature of the screen,
even when it consisted of two metal plates with an interposed
layer of air, the thermometer after the lapse of a sufficient time
always attained an invariable temperature, just as it did when
without a screen. Other circumstances being the same, this was
highest when the thermometer was without a screen. In an
apparatus similar to that represented in A B C, Pl. I. fig. 1,
but in which the thermometer was somewhat more distant from
the vessel of boiling water, the temperatures which it indicated
in atmospheric air under a pressure of 1 atmosphere were as
follows :—

Cork screen 2 millims. thick.	Two copper foils 1 millim. distant.	No screen.
23° C.	21·5	25·5.

It might be thought that the temperatures obtained in different
gases, with the use of different screens, would be proportional to
one another, since the different screens would absorb proportional quantities of the heat incident upon them, and would again
part with proportional quantities. But the result has shown
that, although these temperatures do follow the same series (that
is, if in one gas the temperature with the use of one screen is
higher than in another, it is also higher when another screen
is used), yet that there is no proportion between the two cases.
This arises from the fact that, besides the screen, the side of the
vessel A B which becomes heated during the experiment, also
acts on the thermometer. Although the vessel is surrounded on
the outside with air at 15°, it continually receives heat on the
inside, partly from the air in contact, partly by radiation from
the thermometer *fg*, and partly from the heated base of the
vessel C. In consequence of this, the side, although of very thin

glass in the neighbourhood of C, assumes temperatures which are higher than 15° C. Although the thermometer is protected against rays which proceed from the side above, it is directly exposed to the rays which come from those parts of the side which are lower than the screen. · And as these parts of the side receive more rays when the vessel A B is filled with a gas which readily transmits heat (that is, with a better-conducting gas) than if it contains a gas which possesses these properties in a lower degree, the influence of the side must change with the kind of gas, and cannot therefore be proportional to the indications of the thermometer.

In the following Tables the temperatures are given which have been obtained by the use of two different screens in different gases, at different densities. In obtaining these results, I have had the advantage of the careful help of M. Rüdorff, who for some time has been engaged in my laboratory. The temperatures are counted from that of the surrounding medium, that is, from 15° C.

Gas.	Pressure.	Thermometer with	
		Cork screen.	Metal screen.
	mm		
Atmospheric air . . .	759·4	9·6 C.	
	753·2		7·0 C.
	741·5	9·5	
	738·0	9·5	
	553·3	9·6	
	373·0	10·0	
	356·0	10·1	
	194·7	11·0	
	15·3	11·5	
	11·6	11·7	
	11·6		7·8
Oxygen	771·2	9·6	
	10·0	11·6	
Hydrogen	760·0	13·0	
	763·5		12·0
	517·7	12·5	
	195·4	12·1	
	11·7	11·8	
	9·6	11·6	
	13·8		8·6
Carbonic acid . . .	750·4	8·2	
	765·3	8·2	
	309·1	9·3	
	16·4	11·3	

Gas	Pressure.	Thermometer with	
		Cork screen.	Metal screen
	mm		
Carbonic oxide . . .	760·0	9·5 C.	
	758·9		8·8 C.
	14·4		7·8
	11·0	11·6	
Protoxide of nitrogen .	760·0	8·8	
	752·5		6·3
	289·0		6·5
	17·7		7·5
	12·0	11·5	
Marsh-gas	771·3	9·4	
	764·2		7·0
	306·8		7·3
	13·3		7·8
	12·0	11·6	
Olefiant gas	749·1	9·0	
	319·2	9·9	
	268·8	10·0	
	19·8	11·7	
Ammonia	770·3	8·1	
	746·5	8·3	
	267·7	9·4	
	63·3	10·8	
	18·7	10·9	
	15·4	11·0	
Cyanogen	760·0	8·8	
	14·0	11·4	
Sulphurous acid . .	757·3	7·8	
	763·3	8·0	
	301·1	9·1	
	11·4	11·0	

The hydrogen used in the experiments was prepared from zinc and sulphuric acid; it was dried by chloride of calcium, but not further purified.

The oxygen was prepared from chlorate of potass and binoxide of manganese.

The carbonic acid was liberated from marble by dilute hydrochloric acid, and then passed through a tube containing bicarbonate of soda.

The carbonic oxide was obtained by heating formiate of soda with sulphuric acid, and

The protoxide of nitrogen was obtained from nitrate of ammonia.

The marsh-gas was obtained from acetate of soda with lime and caustic soda.

The olefiant gas was obtained partly by Wöhler's method, from a mixture of alcohol, sulphuric acid, and sand; and partly by Mitscherlich's method, of passing alcohol vapour with sulphuric acid at 165° C. All these gases were dried by chloride of calcium.

The ammonia was prepared from sal-ammoniac and burnt marble, and dried by passing through a tube of caustic soda.

The sulphurous acid was generated from sulphuric acid and mercury, and dried by chloride of calcium and sulphuric acid.

The cyanogen was prepared from dry cyanide of mercury.

The great concordance between the heating of the thermometer in oxygen and in atmospheric air, shows that this would be the case in nitrogen, and makes a determination in this gas unnecessary.

If the temperatures obtained in the different gases with the use of a screen and under the pressure of an atmosphere are compared, we obtain for

Atmospheric air . .	9·6	or	100·0
Oxygen	9·6	„	100·0
Hydrogen . • . .	13·0	„	135·4
Carbonic acid . .	8·2	„	85·4
Carbonic oxide . .	9·5	„	98·9
Protoxide of nitrogen	8·8	„	91·6
Marsh-gas . . .	9·4	„	97·9
Olefiant gas . . .	9·0	„	93·7
Ammonia. . . .	8·1	„	84·3
Cyanogen. . . .	8·8	„	91·6
Sulphurous acid. .	7·8	„	81·2

The temperatures obtained in these gases when greatly rarefied, are not very concordant, because the small quantity of gas still present doubtless exercises an influence; but if a temperature of 26°·7 − 15° = 11°·7 C. be assumed as the most likely for vacuum, and if this be put = 100, the proportion between the temperatures obtained in the other gases under the pressure of an atmosphere, are as—

	Temperature.
Vacuum	100
Atmospheric air . . .	82·0
Oxygen	82·0
Hydrogen	111·1
Carbonic acid	70·0
Carbonic oxide	81·2
Protoxide of nitrogen . .	75·2
Marsh-gas	80·3
Olefiant gas	76·9
Ammonia	69·2
Cyanogen	75·2
Sulphurous acid . . .	66·6

It follows from these numbers that hydrogen really conducts heat in a manner similar to the metals; for the temperature which a thermometer placed in it ultimately assumes is higher as the gas is denser.

Of all gases this is only the case with hydrogen, with all the others the temperature is higher when they are more rarefied. It follows therefore that these gases oppose a hindrance to the transmission of radiant heat, and that they are athermanous to such an extent that their athermancy exercises a greater resistance than their capacity to conduct heat. This property is, however, not entirely absent; for, apart from other reasons already adduced by Dalton and Biot, which speak for a conductibility of heat by gases, it would be contrary to all other known laws if we assumed that the capacity of conducting heat was confined to hydrogen. But it is certainly very remarkable that this gas, the lightest of all, possesses the greatest conductibility.

This surprising result has led me to undertake a few experiments with a view of removing, as far as possible, any doubt as to conductibility. For as the upper part of the side of the vessel A B in the neighbourhood of C also became gradually heated, it might be supposed that although the temperature of this side decreases from above downwards, yet that currents are produced in the gases contained in A B, and that the differences of temperature observed only arise from these currents. This assumption is indeed refuted by the fact that, owing to currents, the temperatures in an exhausted space cannot be higher than in one filled with air. In order to remove every objection, I repeated the experiments just mentioned in such a manner that the apparatus A B, or an entirely-similar one, was filled with a light substance, with feathers or eider-down, or with cotton wool. It then appeared that the denser the light substance, the higher was the temperature which the thermometer assumed. This higher temperature was therefore certainly not produced by a motion of the air. When the air among the cotton was removed as completely as possible, the thermometer did not attain the same temperature as before, when the interstices were filled with air. The small difference might be caused by an alteration in the density which the cotton had experienced on exhausting the air in the neighbourhood of the thermometer; for the various observations made with the same quantity of cotton gave similar deviations. But when hydrogen was introduced among the cotton, the thermometer always rose higher than when the space was filled with atmospheric air. Hydrogen produced the same effect, whether eider-down or cotton wool was used. The following are a few of the numbers :—

The apparatus contained.	Atmospheric air	Hydrogen	Vacuum.
	under a pressure of 1 atmosphere.		
Loose cotton	7·2° C.*	11·0° C.	7·0° C.
,, ,,	7·7	11·0	
,,	7·2		
,,	7·5	11·0	7·0
,, ,,	7·5		7·1
,, ,,	6·0		
Eider-down	6·0		

After these results it cannot be doubted that hydrogen conducts heat, and that in a higher degree than all other gases. This is the more unexpected, since although the conductibility is not directly dependent on the density of bodies (for example, platinum conducts worse than copper or silver), yet the metals, the densest of all bodies, are the best conductors, and in general the looser and less dense substances conduct worse than the denser ones. If hydrogen exhibits in this respect a deviation, a new proof is afforded of that similarity to the metals, so often maintained from its chemical relations.

[To be continued.]

II. *On a supposed Failure of the Calculus of Variations.*
  By G. B. AIRY, *Esq., Astronomer Royal*†.

PROFESSOR JELLETT, in his comprehensive Treatise on the Calculus of Variations, has alluded twice (pages 161 and 365) to the problem "To construct upon a given base A B a curve such that the superficial area of the surface generated by its revolution round A B may be given, and that its solid content may be a maximum." The curve found by Professor Jellett's treatment is a semicircle, and the solid therefore is a sphere. On this he remarks, page 365, "The solution is not given by the sphere, inasmuch as its superficial area is a determinate function of A B [that is, supposing the sphere of the solution to have its diameter equal to and coinciding with A B], and cannot therefore be made equal to any other given quantity." And in page 366 he concludes, "The method therefore fails altogether."

Mr. Todhunter has cited this solution and remark of Professor Jellett, in his invaluable 'History of the Calculus of Variations,' page 410. Mr. Todhunter points out the form of the solution when the solid required has circular ends, but does not allude further to the case considered by Professor Jellett. And thus the matter is left, as an apparent failure of the Calculus.

* The temperatures are counted from 15° C. upwards.
† Communicated by the Author.

I submit the following solution, as what I believe to be the real interpretation of the formulæ given by the Calculus. It is founded upon these three principles :—

(1) There is nothing to prevent us from accepting as solution of the problem a discontinuous curve, provided the different parts meet in a way which is suitable to the conditions of the problem. Mr..Todhunter in several places has alluded to such disconti-·nuity (see pages 19 and 174).

(2) If, in the solution given immediately by the Calculus of Variations, we are certain that no accidental or adventitious factor has been introduced; and if we find that the solution, expressed under the form $\phi\left(x,\ y,\ \dfrac{dy}{dx},\ \&c.\right)=0$, is the product of two factors, then we are bound to consider each of the curves represented by the two factors as a good and sufficient solution of the problem.

(3) And, to exhibit the solution in its utmost generality, we must use both the solutions given by these curves, in such combination as the circumstances of the problem indicate to be proper.

I now proceed to apply these principles to the problem before us.

Since $\pi \int dx \,.\, y^2$ is to be maximum, while $2\pi \int dx \,.\, y\sqrt{(1+p^2)}$ is given, then if a be a constant to be determined hereafter, the value of V will be

$$y^2 + 2ay\sqrt{(1+p^2)}\,;$$

and treating this in the usual way, we find

$$\frac{2ay}{\sqrt{(1+p^2)}} = b - y^2,$$

where b is another constant produced by integration. It is certain here that no factor has been introduced.

Since the curve is to meet the axis, $y=0$ at certain points, and $\sqrt{(1+p^2)}$ is never $=0$. Hence b must $=0$; and our equation becomes

$$\frac{2ay}{\sqrt{(1+p^2)}} = -y^2,$$

or

$$y\left\{\frac{2a}{\sqrt{(1+p^2)}} + y\right\} = 0\,;$$

which is satisfied by either of the following,

$$\frac{2a}{\sqrt{(1+p^2)}} + y = 0,$$

$$y \qquad\quad = 0.$$

The first of these denotes a sphere of radius $-2a$, the first or last limit upon the axis of x being arbitrary. The second denotes a cylinder whose radius is indefinitely small. And the union of the two, which gives the complete solution of the problem, is a sphere of such a radius that its surface has the prescribed value, connected by indefinitely small cylinders or pipes with the points adopted as the limits of x, that is, with A and B.

The following diagrams may be conceived to represent the various forms which the solution takes. The first is peculiar to the case when the diameter of the sphere which has the given superficies is less than A B; the second is peculiar to the case when the diameter is greater than A B; the third and fourth both apply to both cases. The diameter of the small pipe is made finite, to be visible to the eye.

Form 1.   Form 4.

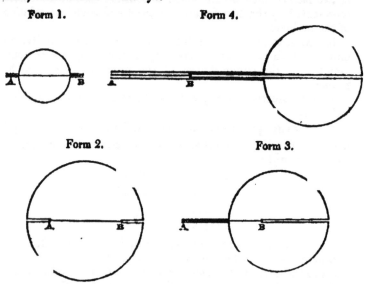

Form 2.   Form 3.

The practical solution evidently is, that a sphere is to be constructed, at any part of the axis of x, whose diameter is such that its surface will be equal to the surface prescribed in the data of the problem. But the metaphysical solution, containing the idea of the tubular connexions with A and B, enables us also to satisfy the condition of terminating the integrations at any points that we may select, not necessarily defined by the position of the sphere.

Royal Observatory, Greenwich,
        June 5, 1861.

III. *On the Theory of Types in Chemistry.*
By T. Sterry Hunt, *M.A. F.R.S.**

IN the *Annalen der Chemie und Pharmacie* for March 1860 (vol. cxiii. p. 293), M. Kolbe has given a paper on the natural relations between mineral and organic compounds, considered as a scientific basis for a new classification of the latter. He objects to the four types admitted by Gerhardt, namely, hydrogen, hydrochloric acid, water, and ammonia, that they sustain to organic compounds only artificial and external relations, while he conceives that between these and certain other bodies there are natural relations having reference to the origin of the organic species. Starting from the fact that all the bodies of the carbon series found in the vegetable kingdom are derived from carbonic acid with the concurrence of water, he proceeds to show how all the compounds of carbon, hydrogen, and oxygen may be derived from the type of an oxide of carbon, which is either $C^2 O^4$, $C^2 O^2$, or the hypothetical $C^2 O$.

When in the former we replace one atom of oxygen by one of hydrogen, we have $C^2 O^3 H$, or anhydrous formic acid; the replacement of a second equivalent would yield $C^2 O^2 H^2$, or the unknown formic aldehyde; a third, $C^2 O H^3$, the oxide of methyle; and a fourth, $C^2 H^4$, or formene. By substituting methyle for one or more atoms of hydrogen in the previous formula, we obtain those of the corresponding bodies of the vinic series; and it will be readily seen that by introducing the higher alcoholic radicals, we may derive from $C^2 O^4$ the formulas of all the alcoholic series. A grave objection to this view is, however, found in the fact that, while this compound may be made the type of the aldehydes, acetones, and hydrocarbons, it becomes necessary to assume the hypothetical $C^2 O^2$, $H O$ as the type of the acids and alcohols. Oxide of carbon, $C^2 O^2$, is, according to Kolbe, to be received as the type of hydrocarbons, like olefiant gas ($C^2 H Me$), while $C^2 O$, in which ethyle replaces oxygen, is $C^6 H^5$, or lipyle, the supposed triatomic base of glycerine.

The monobasic organic acids are thus derived from one atom of $C^2 O^4$, while the bibasic acids, like the succinic, are by Kolbe deduced from a double molecule, $C^4 O^8$, and tribasic acids, like the citric, from a triple molecule, $C^6 O^{12}$. He moreover compares sulphuric acid to carbonic acid, and derives from it by substitution the various sulphuric organic compounds. Ammonia, arseniuretted and phosphuretted hydrogen, are regarded as so many types; and by an extension of his view of the replacement of oxygen by electro-positive groups, the ethylides $ZnEt$, $PbEt^2$,

* Communicated by the Author.

and BiEt^3, are by Kolbe assimilated to the oxides of ZnO, PbO^2, and BiO^3.

Ad. Wurtz, in the *Répertoire de Chimie Pure* for October 1860, has given an analysis of Kolbe's memoir (to which, not having the original before me, I am indebted for the preceding sketch), and follows it by a judicious criticism. While Kolbe adopts as types a number of mineral species, including the oxides of carbon, of sulphur and the metals, Wurtz would maintain but three, hydrogen (H^2), water (H^{2}O^2), and ammonia (N H^3); and these three types, as he endeavoured to show in 1855, represent different degrees of condensation of matter. The molecule of hydrogen, H^2 (M^2), corresponding to four volumes, combines with two volumes of oxygen (O^2) to form four volumes of water, and may thus be regarded as condensed to one-half in its union with oxygen, and derived from a double molecule, M^{2}M^2. In like manner four volumes of ammonia contain two volumes of nitrogen and six of hydrogen, which, being reduced to one-third, correspond to a triple molecule, M^{3}M^3, so that these three types and their multiples are reducible to that of hydrogen more or less condensed *.

As regards the rejection of water as a type of organic compounds, and the substitution of carbonic acid, founded upon the consideration that these in nature are derived from C^{2}O^4, Wurtz has well remarked that water, as the source of hydrogen, is equally essential to their formation, and indeed that the carbonic anhydride, C^{2}O^4, like all other anhydrous acids, may be regarded as a simple derivative of the water type. Having then adopted the notion of referring a great variety of bodies to a mineral species of simple constitution, water is to be preferred to carbonic anhydride,—first, because we can compare with it many mineral compounds which can with difficulty be compared with carbonic acid; and secondly, because, the two atoms of water being replaceable singly, the mode of derivation of a great number of compounds (acids, alcohols, ethers, &c.) is much more simple and natural than from carbonic acid. As Wurtz happily remarks, Kolbe has so fully adopted the theory of types, that he wishes to multiply them, and even admits condensed types, which are, however, molecules of carbonic acid, and not of water; "he combats the types of Gerhardt, and at the same time counterfeits them."

Thus far we are in accordance with M. Wurtz, who has shown himself one of the ablest and most intelligent expounders of this doctrine of molecular types, as above defined, now almost universally adopted by chemists. He writes, "To my mind this idea of referring to water, taken as a type, a very great

* Wurtz, *Ann. de Chim. et de Phys.* [3.] vol. xliv. p. 304.

number of compounds, is one of the most beautiful conceptions of modern chemistry[*];" and again, he declares the idea of regarding both water and ammonia as representatives of the hydrogen type, more or less condensed, to be so simple and so general in its application, that it is worthy "to form the basis of a system of chemistry[†]."

We have in this theory two important conceptions: the first is that of hydrogen and water regarded as types to which both mineral and organic compounds may be referred; and the second is the notion of condensed and derived types, according to which we not only assume two or three molecules of hydrogen or water as typical forms, but even look on water as the derivative of hydrogen, which is itself the primal type.

As to the history of these ideas, Wurtz remarks that the proposition enunciated by Kolbe, that all organic bodies are derived by substitution from mineral compounds, is not new, but has been known in the science for about ten years. "Williamson was the first who said that alcohol, ether, and acetic acid were comparable to water—organic waters. Hofmann and myself had already compared the compound ammonias to ammonia itself. * * * * * To Gerhardt belongs the merit of generalizing these ideas, of developing them, and supporting them with his beautiful discovery of anhydrous monobasic acids. Although he did not introduce into the science the idea of types, which belongs to M. Dumas, he gave it a new form, which is expressed and essentially reproduced by the proposition of Kolbe. Gerhardt reduced all organic bodies to four types—hydrogen, hydrochloric acid, water, and ammonia[‡]."

The historical inaccuracies of the above quotation are the more surprising, since in March 1854 I published in the American Journal of Science (vol. xvii. p. 194) a concise account of the progress of these views. This paper was republished in the 'Chemical Gazette' (1854, p. 181), and copies of it were by myself placed in the hands of most of the distinguished chemists of England, France, and Germany. In this paper I have shown that the germ of the idea of mineral types is to be found in an essay of Auguste Laurent[§], where he showed that alcohol may be looked upon as water (H^2O^2) in which ethyle replaces one atom of hydrogen, and hydric ether as the result of a complete substitution of the hydrogen by a second atom of ethyle. Hence he observed that while ether is neutral, alcohol is monobasic and the type of the monobasic vinic acids, as water is the type of

---

[*] *Répertoire de Chimie Pure*, 1860, p. 359.   [†] Ibid. p. 356.
[‡] Ibid. p. 355.
[§] "Sur les Combinaisons Azotées," *Ann. de Chim. et de Phys.* November 1846.

bibasic acids. In extending and developing this idea of Laurent's, I insisted in March 1848, and again in January 1850, upon the relation between the alcohols and water as one of homology, water being the first term in the series, and H^2 being in like manner the homologue of acetene and formene, while the bases of Wurtz were said to "sustain to their corresponding alcohols the same relation that ammonia does to water *."

In a notice of his essay, published in September 1848[†], I endeavoured to show that Laurent's view might be further extended, so as to include in the type of water "*all those saline combinations (acids) which contain oxygen;*" and in a paper read before the American Association for the Advancement of Science at Philadelphia, in September 1848, I further suggested that as many neutral oxygenized compounds which do not possess a saline character are derivatives of acids which are referable to the type $H^2 O^2$, "*we may regard all oxygenized bodies as belonging to this type,*" which I further showed in the same essay is but a derivative of the primal type H^2, to which I referred all hydrocarbons and their chlorinized derivatives, as also the volatile alkaloids, which were regarded "as amidized species" of the hydrocarbons, in which the residue amidogen, $N H^2$, replaced an atom of H or Cl, or what is equivalent, the residue N H was substituted for O^2 in the corresponding alcohols[‡].]

In the paper published in September 1848, I showed that while water is bibasic, the acids which, like hypochlorous and nitric acids, were derived from it by a simple substitution of Cl and NO^4 for H, were necessarily monobasic; and I then pointed out the possible existence of the nitric anhydride $(NO^4)^2 O^2$, which was soon after discovered by Deville. Gerhardt at this time denied the existence of anhydrides of the monobasic acids, while he regarded anhydrides as characteristic of polybasic acids, and indeed was only led to adopt my views by the discovery of the very anhydrides whose formation I had foreseen[§].

In explaining the origin of bibasic acids, I described them as produced by the replacement, in a second equivalent of water, of an atom of hydrogen by a monobasic saline group; thus sulphuric acid would be $(S^2 HO^6 H) O^2$. Tribasic acids, in like manner, are to be regarded as derived from a third equivalent of

---

* American Journal of Science [2], vol. v. p. 265; vol. ix. p. 65; vol. xiii. p. 206.

 † Ibid. vol. vi. p. 173.   ‡ Ibid. vol. viii. p. 92.

 § The anhydrides of the monabasic acids correspond to two equivalents of the acid, minus one of water, as $2(C^4 H^1 O^4) - H^2 O^2 = C^8 H^4 O^6$, while one equivalent of a bibasic acid (itself derived from $2(H^2 O^2)$) loses one of water, and becomes an anhydride, as $C^2 H^2 O^6 - H^2 O^2 = C^2 O^4$. So that both classes of anhydrides are to be referred to the type of one molecule of water, $H^2 O^2$.

water in which a bibasic residue replaces an atom of hydrogen. The idea of polymeric types was further illustrated in the same paper, where three hydrogen types were proposed, (HH), (H^2H^2), and (H^3H^3), corresponding to the chlorides MCl, MCl³, and MCl⁵. It was also illustrated by sulphur in its ordinary state, which I showed is to be regarded as a triple molecule S³ (or S⁶ = 4 volumes), and referred sulphurous acid SO² to this type, to which also probably belongs selenic oxide. (At the same time I suggested that the odorant form of oxygen or ozone was possibly O³.) Wurtz, in his memoir published in 1855, adopts my view, and makes sulphur vapour at 400° C. the type of the triple molecule. I further suggested* that gaseous nitrogen is NN, an anhydride amide or nitryle, corresponding to nitrite of ammonia, (NO^3, NH^4O) — H^4O^4 = NN. This view a late writer attributes to Gerhardt, who adopted it from me†. May not nitrogen gas, as I have elsewere suggested, regenerate under certain conditions ammonia and a nitrite, and thus explain not only the frequent formation of ammonia in presence of air and reducing agents, but certain cases of nitrification‡?

I endeavoured still further to show that hydrogen is to be looked upon as the fundamental type, from which the water type is derived by the replacement of an atom of H by the residue HO^2§. In the same way I regarded ammonia as water in which the residue NH replaced O².

I have always protested against the view which regards the so-called rational formulæ as expressing in any way the real structure of the bodies which are thus represented. These formulæ are invented to explain a certain class of reactions, and we may construct, from other points of view, other rational formulæ which are equally admissible. As I have elsewhere said, "the various hypotheses of copulates and radicals are based upon the notion of dualism, which has no other foundation than the observed order of generation, and can have no place in a theory of

---

* American Journal of Science, vol. v. p. 408; vol. vi. p. 172.

† *Ann. de Chim. et de Phys.* vol. lx. p. 381.

‡ The formation of a nitrite in the experiments of Cloez appears to be independent of the presence of ammonia, and to require only the elements of air and water (*Comptes Rendus*, vol. lxi. p. 135). Some experiments now in progress lead me to conclude that the appearance of a nitrite in the various processes for ozone is due to the power of nascent oxygen to destroy by oxidation the ammonia generated by the action of water on nitrogen, the nitrous nitryle; so that the odour and many of the reactions assigned to ozone or nascent oxygen are really due to the nitrous acid which is set free when the former encounters nitrogen and moisture. On the other hand, nascent hydrogen, which readily reduces nitrates and nitrites to ammonia, by destroying the regenerated nitrite of the nitryle, produces ammonia in many cases from atmospheric nitrogen.

§ American Journal, vol. viii. p. 93.

science." All chemical changes are reducible to union (identification), and division (differentiation). When in these changes only one species is concerned, we designate the process as metamorphosis, which is either by condensation or by expansion (homogeneous differentiation). In metagenesis, on the contrary, unlike species may unite, and by a subsequent heterogeneous differentiation give rise to new species, constituting what is called double decomposition, the results of which, differently interpreted, have given origin to the hypothesis of radicals and the notion of substitution by residues, to express the relations between the parent bodies and their progeny. The chemical history of bodies is then a record of their changes; it is, in fact, their genealogy; and in making use of typical formulæ to indicate the derivation of chemical species, we should endeavour to show the ordinary modes of their generation*.

Keeping this principle in mind, let us now examine the theory of the formation of acids. As we have just seen, I taught in 1848 that the monobasic, bibasic, and tribasic acids are derived respectively from one, two, and three molecules of water, $H^2 O^2$. M. Wurtz, seven years later (in 1855), put forth a similar view. He supposes a monatomic radical $PO^{4'}$, a diatomic radical $PO^{3''}$, and a triatomic radical $PO^{2'''}$, replacing respectively one, two, and three atoms of hydrogen in $H^2 O^2$, $H^4 O^4$, and $H^6 O^6$, thus $(PO^{4'} H) O^2$, $(PO^{3''} H^2) O^4$, and $(PO^{2'''} H^3) O^6$. These radicals evidently correspond to PO^5 which has lost one, two, and three atoms of oxygen in reacting upon the hydrogen of the water type; and these acids may be accordingly represented as formed by the substitution of the residue $PO^5 - O$ for H, &c.

To this manner of representing the generation of polybasic acids we object that it encumbers the science with numerous hypothetical radicals, and that it moreover fails to show the actual successive generation of the series of acids in question.

When phosphoric anhydride, $P^2 O^{10} = (PO^4)^2 O^2$, is placed in contact with water, it combines with one equivalent, $H^2 O^2$. The union is followed by homogeneous differentiation, and two equivalents of metaphosphoric acid result;

$$(PO^4)^2 O^2 + H^2 O^2 = 2(PO^4 H) O^2.$$

Two equivalents of this acid with one of water at ordinary temperatures are slowly transformed into two of pyrophosphoric acid by a reaction precisely similar to the last,

$$2(PHO^6) [= (PHO^5)^2 O^2] + H^2 O^2 = 2(PHO^5 H) O^2;$$

* See "On the Theory of Chemical Changes," Amer. Journ. of Science, vol. xv. p. 226; Lond. Edinb. and Dub. Phil. Mag. [4] vol.v. p. 526; and *Chem. Centralblatt*, 1853, p. 849. Also, "Thoughts on Solution," Amer. Journ. of Science, vol. xix. p. 100; and 'Chemical Gazette,' 1855, p. 92;

and two equivalents of pyrophosphoric acid, when heated with a third equivalent of water, yield in like manner two of tribasic phosphoric acid,

$$2(PH^2O^7) = [(PH^2O^6)^2O^2] + H^2O^2 = 2(PH^2O^6H)O^2 = 2PH^3O^8.$$

Gerhardt long since maintained that we cannot distinguish between polybasic salts and what are called subsalts; which are as truly neutral salts of a particular type. Thus the bibasic and tribasic phosphates are to be looked upon as subsalts which sustain the same relation to the monobasic phosphates that the basic nitrates bear to the neutral nitrates. He succeeded in preparing two crystalline subnitrates of lead and copper, having the formulæ NO^5, M^2O^2, HQ (tribasic), and NO^5, M^4O^4, H^3O^3 (quadri or heptabasic), both of which retain their water of composition at 392° F. The compounds of sulphuric acid are,—1st. the true monobasic sulphate, S^2O^6MO, corresponding to the Nordhausen acid and the anhydrous bisulphates; 2nd. the ordinary neutral sulphates, S^2O^6, M^2O^2; 3rd. the so-called disulphates, S^2O^6, M^4O^4, corresponding to the glacial acid density 1·780; 4th. the type S^2O^6, M^6O^6, represented by turpeth mineral; and 5th. the so-called quadribasic sulphates, $S^2O^6M^8O^8$. The copper salt of this type, according to Gerhardt, retains, moreover, $6HO$ at 392° F.*

Without counting the still more basic sulphates of zinc and copper, described by Kane and Schindler, we have the following salts, which, in accordance with Wurtz's notation, correspond to the annexed radicals:—

1. Unibasic	. . . .	$S^2HO^7 = S^2O^5$	monatomic.
2. Bibasic	. . . .	$S^2H^2O^8 = S^2O^4$	diatomic.
3. Quadribasic	. . .	$S^2H^4O^{10} = S^2O^2$	tetratomic.
4. Sexbasic	. . .	$S^2H^6O^{12} = S^2$	hexatomic.
5. Octobasic	. . . .	$S^2H^8O^{14} = S^2 - O^2$	octatomic.

It is easy to apply a similar *reductio ad absurdum* to the radical theory in the case of the oxychlorides and other basic salts, and to show that the radicals of the dualists are often merely algebraic expressions. (See further my remarks in the American Journal of Science, vol. vii. pp. 402–404†.)

The above, which we conceive to be a simple statement of the

* Gerhardt "On Salts," *Journ. de Pharm.* 1848, vol. xii. American Journal of Science, vol. vi. p. 337.

† Those who are familiar with chemical literature will remember an amusing *jeu d'esprit* of Laurent's, in which he invited the attention of the advocates of the radical theory to a newly invented electro-negative radical, *Eurhizene* (*Comptes Rendus des Travaux de Chimie* for 1850, pp. 251 and 376). We observe a late writer in the 'Chemical News' (vol. i. p. 326) proposing, as a new electro-negative radical, under the name of hydrine, the peroxide of hydrogen, HO, the eurhizene of Laurent!

process as it takes place in nature, dispenses alike with hypothetical radicals and residues, both of which are, however, convenient for the purposes of notation. In the selection of a typical form, to which a great number of species may be referred, hydrogen or water merits the preference from its simplicity, and from the important part which it plays in the generation of species. Water and carbonic anhydride are both so directly concerned in the generation of the bodies in the carbon series, that either may be assumed as the type; but we prefer to regard $C^2 O^4$, like the other anhydrides, as only a derivative of the type of water, and eventually of the hydrogen type.

These views were first put forward by myself in 1848, when I expressed the opinion that they were destined to form "the basis of a true natural system of chemical classification;" and it was only after having opposed them for four years to those of Gerhardt, that this chemist, in June 1852, renounced his views, and without any acknowledgment adopted my own*. Already in 1851, Williamson, in a paper read before the British Association, had developed the ideas on the water type to which Wurtz refers above; and to him the English editor of Gmelin's 'Handbook' ascribes the theory. The notion of condensed types, and of H^2 as the primal type, was not, so far as I am aware, brought forward by either of these, and remained unnoticed until resuscitated by Wurtz in 1855, seven years after I had first announced it, and one year after my reclamation, published in the American Journal of Science, in March 1854.

My claims have not, however, been overlooked by Dr Wolcott Gibbs. In an essay on the polyacid bases, he remarks that in a previous paper he had attributed the theory of water types to Gerhardt and Williamson, and adds, "In this I find I have not done justice to Mr. T. Sterry Hunt, to whom is exclusively due the credit of having first applied the theory to the so-called oxygen acids and to the anhydrides, and in whose earlier papers may be found the germs of most of the ideas on classification usually attributed to Gerhardt and his disciples†." It will be seen, from what precedes, that I not only applied the theory, as Dr. Gibbs remarks, but, except so far as Laurent's suggestion goes, invented it and published it in all its details some years before it was accepted by a single chemist.

In conclusion, I have only to ask that future historians will do justice to the memory of Auguste Laurent, and will ascribe to whom it is due the credit of having given to the science a theory which has exercised such an important influence on modern che-

* *Ann. de Chim. et de Phys.* [3] vol. xxxvii. p. 285.
† Proceedings of the American Association, Baltimore, May 1858, p. 197.

mical speculation and research, remembering that my own publications on the subject, which cover the whole ground, were some years earlier than those of Williamson, Gerhardt, Wurtz, or Kolbe.

Montreal, January 1861.

IV. *On the Reduction of Observations of Underground Temperature; with Application to* Professor Forbes's *Edinburgh Observations, and the continued Calton Hill Series.* By Professor WILLIAM THOMSON, F.R.S.*

I. *Analysis of Periodic Variations.*

1. EVERY purely periodical function is, as is well known, expressible by means of a series of constant coefficients multiplying sines and cosines of the independent variable with a constant factor and its multiples. This important truth was arrived at by an admirable piece of mathematical analysis, called for by Daniel Bernoulli, partially given by La Grange, and perfected by Fourier.

2. To simplify my references to the mathematical propositions of this theory, I shall commence by laying down the following definitions :—

Def. 1. A simple harmonic function is a function which varies as the sine or cosine of the independent variable, or of an angle varying in simple proportion with the independent variable. The harmonic curve is the well-known name applied to the graphic representation, on the ordinary Cartesian system, of what I am now defining as a simple harmonic function. It is the form of a string vibrating in such a manner as to give the simplest and smoothest possible character of sound; and, in this case, the displacement of each particle of the string is a harmonic function of the time, besides being a harmonic function of the distance of its position of equilibrium from either end of the string. The sound in this case may be called a perfect unison.

Def. 2. The argument of a simple harmonic function is the angle to the sine or cosine of which it is proportional.

Cor. The argument of a harmonic function is equal to the independent variable multiplied by a constant factor, with a constant added; that is to say, it may be any linear function of the independent variable.

Def. 3. When time is the independent variable, the epoch is

---

* From the Transactions of the Royal Society of Edinburgh, vol. xxii. part 2. Communicated by the Author.

the interval which elapses from the era of reckoning till the function first acquires a maximum value. The augmentation of argument corresponding to that interval will be called "the epoch in angular measure," or simply "the epoch" when no ambiguity can exist as to what is meant.

Def. 4. The period of a simple harmonic function is the augmentation which the independent variable must receive to increase the argument by a circumference.

Cor. If c denote the coefficient of the independent variable in the argument, the period is equal to $\dfrac{2\pi}{c}$. Thus if T denote the period, ϵ the epoch in angular measure, and t the independent variable, the argument proper for a cosine is

$$\frac{2\pi t}{T} - \epsilon;$$

and the argument for a sine,

$$\frac{2\pi t}{T} - \epsilon + \frac{\pi}{2}.$$

3. *Composition and Resolution of Simple Harmonic Functions of one Period.*

Prop. The sum of any two simple harmonic functions of one period is equal to one simple harmonic function whose amplitude is the diagonal of a parallelogram described upon lines drawn from one point to lengths equal to the amplitudes of the given functions, at angles measured from a fixed line of reference equal to their epochs, and whose epoch is the inclination of the same diagonal to the same line of reference.

Cor. 1. If A, A' be the amplitudes of two simple harmonic functions of equal period, and ϵ, ϵ' their epochs, that is to say, if $A \cos (mt - \epsilon)$, $A' \cos (mt - \epsilon')$ be two simple harmonic functions, the one simple harmonic function equal to their sum has for its amplitude and its epoch the following values respectively:—

(amplitude) $\{(A \cos \epsilon + A' \cos \epsilon')^2 + (A \sin \epsilon + A' \sin \epsilon')^2\}^{\frac{1}{2}},$

or $\{A^2 + 2AA' \cos (\epsilon' - \epsilon) + A'^2\}^{\frac{1}{2}};$

(epoch) $\tan^{-1} \dfrac{A \sin \epsilon + A' \sin \epsilon'}{A \cos \epsilon + A' \cos \epsilon'}.$

Cor. 2. Any number of simple harmonic functions, of equal period, added together, are equivalent to a single harmonic function of which the amplitude and epoch are derived from the amplitude and epochs of the given functions, in the same manner as the magnitude and inclination to a fixed line of reference, of the resultant of any number of forces in one plane, are derived from the magnitudes and the inclinations to the same line of reference of the given forces.

Cor. 3. The physical principle of the superposition of sounds being admitted, any number of simple unisons of one period co-existing, produce one simple unison of the same period, of which the intensity (measured by the square of the amplitude) and the . epoch are determined in the manner just specified.

Cor. 4. The sum of any number of simple harmonic functions of one period vanishes for every argument, if it vanishes for any two arguments not differing by a semicircumference, or by some multiple of a semicircumference.

Cor. 5. The co-existence of perfect unisons may constitute perfect silence.

Cor. 6. A simple harmonic function of any epoch may be resolved into the sum of two whose epochs are respectively zero and a quarter period, and whose amplitudes are respectively equal to the value of the given function for the arguments zero and a quarter period respectively.

4. *Complex Harmonic Functions.*—Harmonic functions of different periods added can never produce a simple harmonic function. If their periods are commensurable, their sum may be called a complex harmonic function.

Cor. A complex harmonic function is the proper expression for a perfect harmony in music.

5. *Expressibility of Arbitrary Functions by Trigonometrical series.*

Prop. A complex harmonic function, with a constant term added, is the proper expression, in mathematical language, for any arbitrary periodic function.

6. *Investigation of the Trigonometrical Series expressing an Arbitrary Function.*—Any arbitrary periodic function whatever being given, the amplitudes and epochs of the terms of a complex harmonic function, which shall be equal to it for every value of the independent variable, may be investigated by the "method of indeterminate coefficients," applied to determine an infinite number of coefficients from an infinite number of equations of condition, by the assistance of the integral calculus as follows:—

Let $F(t)$ denote the function, and T its period. We must suppose the value of $F(t)$ known for every value of t, from $t=o$ to $t=T$. Let M_0 denote the constant term, and let $M_1, M_2, M_3,$ &c. denote the amplitudes, and $\epsilon_1, \epsilon_2, \epsilon_3,$ &c. the epochs of the successive terms of the complex harmonic functions by which it is to be expressed; that is to say, let these constants be such that

$$(Ft) = M_0 + M_1 \cos\left(\frac{2\pi t}{T} - \epsilon_1\right) + M_2 \cos\left(\frac{4\pi t}{T} - \epsilon_2\right)$$

$$+ M_3 \cos\left(\frac{6\pi t}{T} - \epsilon_3\right) + \&c.$$

Then, expanding each cosine by the ordinary formula, and assuming

$$M_1 \cos \epsilon_1 = A_1, \quad M_2 \cos \epsilon_2 = A_2, \quad \&c.,$$

$$M_1 \sin \epsilon_1 = B_1, \quad M_2 \sin \epsilon_2 = B_2, \quad \&c.,$$

we have

$$F(t) = A_0 + A_1 \cos\frac{2\pi t}{T} + A_2 \cos\frac{4\pi t}{T} + A_3 \cos\frac{6\pi t}{T} + \&c.,$$

$$+ B_1 \sin\frac{2\pi t}{T} + B_2 \sin\frac{4\pi t}{T} + B_3 \sin\frac{6\pi t}{T} + \&c.$$

Multiplying each member by $\cos \dfrac{2i\pi t}{T} dt$, where i denotes o or any integer, and integrating from $t = o$ to $t = T$, we have

$$\int_0^T F(t) \cos\frac{2i\pi t}{T} dt = A_i \int_0^T \left(\cos\frac{2i\pi t}{T}\right)^2 dt,$$

$$= A_i \times \tfrac{1}{2}T, \quad \text{when } i \text{ is any integer;}$$

or

$$= A_0 \times T, \quad \text{when } i = 0.$$

Hence

$$A_0 = \frac{1}{T}\int_0^T F(t)dt,$$

$$A_i = \frac{2}{T}\int_0^T F(t) \cos\frac{2i\pi t}{T} dt;$$

and similarly we find

$$B = \frac{2}{T}\int_0^T F(t) \sin\frac{2i\pi t}{T} dt :$$

equations by which the coefficients in the double series of sines and cosines are expressed in terms of the values of the function supposed known from $t = o$ to $t = T$. The amplitudes and epochs of the single harmonic terms of the chief period and its submultiples are calculated from them, according to the following formula :—

$$\tan \epsilon_i = \frac{B}{A_i}; \quad M_i = (A_i^2 + B_i^2)^{\frac{1}{2}}$$

(or for logarithmic calculation,

$$M_i = A_i \sec \epsilon_i).$$

The preceding investigation is sufficient as a solution of the problem, to find a complex harmonic function expressing a given arbitrary periodic function, when once we are assured that the problem is possible; and when we have this assurance, it proves that the resolution is *determinate*, that is to say, that no other complex harmonic function than the one we have found can satisfy the conditions. For a thorough and most interesting analysis of the subject, supplying all that is wanting to complete the investigation, and giving admirable views of the problem from all sides, the reader is referred to Fourier's delightful treatise. A concise and perfect synthetical investigation of the harmonic expression of an arbitrary periodic function is to be found in Poisson's *Théorie Mathématique de la Chaleur*, chap. vii.

II. *Periodic Variations of Terrestrial Temperature.*

7. If the whole surface of the earth were at each instant of uniform temperature, and if this temperature were made to vary as a perfectly periodic function of the time, the temperature at any internal point must ultimately come to vary also as a periodic function of the time, with the same period, whatever may have been the initial distribution of temperature throughout the whole. Fourier's principles show how the periodic variation of internal temperature is to be conceived as following, with diminished amplitude and retarded phase, from the varying temperature at the surface supposed given : and by his formulæ the precise law according to which the amplitude would diminish and the phase would be retarded, for points more and more remote from the surface, if the figure were truly spherical and the substance homogeneous, is determined.

8. The largest application of this theory to the earth as a whole is to the analysis of imaginable secular changes of temperature, with at least thousands of millions of years for a period. In such an application, it would be necessary to take into account the spherical figure of the earth as a whole. Periodic variations at the surface with any period less than a million* of years will,

* A periodic variation of external temperature of one million years' period would give variations of temperature within the earth sensible to one thousand times greater depths than a similar variation of one year's period. Now the ordinary annual variation is reduced to $\frac{1}{10}$th of its superficial

at points below the surface, give rise to variations of temperature
not appreciably influenced by the general curvature, and sensibly
agreeing with what would be produced if the surface were an
infinite plane, except insofar as they are modified by superficial
irregularities.   Hence Fourier's formulæ for an infinite solid,
bounded on one side by an infinite plane, of which the temperature is made to vary arbitrarily, contain the proper analysis for
diurnal or annual variations of terrestrial temperature, unless a
theory of the effect of inequalities of surface (upon which no investigator has yet ventured) is aimed at.

9. The effect of diurnal variations of temperature becomes
insensible at so small a distance below the surface, that in most
localities irregularities of soil and drainage must prevent any very
satisfactory theoretical treatment of their inward progression and
extinction from being carried out.   At depths exceeding three
feet below the surface, all periodic effects of daily variations of
temperature become insensible in most soils, and the observable
changes are those due to a daily average, varying from day to
day.   If now the annual variation of temperature were truly
periodic, a complex harmonic function could be determined to
represent for all time the temperature at three feet or any greater
depth.   But in reality the annual variation is very far from
recurring in a perfectly periodic manner, since there are both
great differences in the annual average temperatures, and never-
ceasing irregularities in the progress of the variation within each
year.   A full theory of the consequent variations of temperature
propagated downwards, must include the consideration of
non-periodic changes; but the most convenient first step is
that which I propose to take in the present communication, in
which the average annual variations for groups of years will be
discussed according to the laws to which periodic variations are
subject.

10. The method which Fourier has given for treating this and
other similar problems is founded on the principle of the independent superposition of thermal conductions.   This principle
holds rigorously in nature, except insofar as the conductivity or

---

amount at a depth of 25 French feet, and is scarcely sensible at a depth of
50 French feet (being there reduced, in such rock as that of Calton Hill,
to $\frac{1}{100}$).   Hence, at a depth of 50,000 French feet, or about ten English
miles, a variation having one million years for its period would be reduced
to $\frac{1}{100}$.   If the period were ten thousand million years, the variation would
similarly be reduced to $\frac{1}{100}$ at 1000 miles' depth, and would be to some
appreciable extent affected by the spherical figure of the whole earth,
although to only a very small extent, since there would be comparatively but
very little change of temperature (less than $\frac{1}{10}$ of the superficial amount)
beyond the first layer of 500 miles' thickness.

the specific heat of the conducting substance may vary with the changes of temperature to which it is subjected; and it may be accepted with very great confidence in the case with which we are now concerned, as it is not at all probable that either the conductivity or the specific heat of the rock or soil can vary at all sensibly under the influence of the greatest changes of temperature experienced in their natural circumstances; and, indeed, the only cause we can conceive as giving rise to sensible change in these physical qualities is the unequal percolation of water, which we may safely assume to be confined in ordinary localities to depths of less than three feet below the surface.   The particular mode of treatment which I propose to apply to the present subject consists in expressing the temperature at any depth as a complex harmonic function of the time, and considering each term of this function separately, according to Fourier's formulæ for the case of a simple harmonic variation of temperature, propagated inwards from the surface.   The laws expressed by these formulæ may be stated in general terms as follows.

11. *Fourier's Solution stated*.*—If the temperature at any point of an infinite plane, in a solid extending infinitely in all directions, be subjected to a simple harmonic variation, the temperature throughout the solid on each side of this plane will follow everywhere according to the simple harmonic law, with epochs retarded equally, and with amplitudes diminished in a constant proportion for equal augmentations of distance.   The retardation of epoch expressed in circular measure (arc divided by radius) is equal to the diminution of the Napierian logarithm of the amplitude; and the amount of each per unit of distance

is equal to $\sqrt{\dfrac{\pi c}{Tk}}$, if c denote the capacity for heat of a unit bulk

of the substance, and k its conductivity †.

12. Hence, if the complex harmonic functions expressing the varying temperature at two different depths be determined, and each term of the first be compared with the corresponding term

of the second, the value of $\sqrt{\dfrac{\pi c}{Tk}}$ may be determined either by

dividing the difference of the Napierian logarithms of the amplitudes, or the difference of the epochs by the distance between the points.   The comparison of each term in the one series with the

---

* For the mathematical demonstration of this solution, see Note appended to Professor Everett's paper, which follows the present article in the Transactions.

† That is to say, the quantity of heat conducted per unit of time across a unit area of a plate of unit thickness, with its two surfaces permanently maintained at temperatures differing by unity.

corresponding term in the other series gives us, therefore, two determinations of the value of $\sqrt{\dfrac{\pi c}{k}}$, which should agree perfectly, if (1) the data were perfectly accurate, if (2) the isothermal surfaces throughout were parallel planes, and if (3) the specific heat and conductivity of the soil were everywhere and always constant.

As these conditions are not strictly fulfilled in any natural application, the first thing to be done in working out the theory is to test how far the different determinations agree, and to judge accordingly of the applicability of the theory in the circumstances. If the test thus afforded prove satisfactory, the value of the conductivity in absolute measure may be deduced from the result with the aid of a separate experimental determination of the specific heat.

18. The method thus described differs from that followed by Professor Forbes, in substituting the separate consideration of separate terms of the complex harmonic function for the examination of the whole variation unanalysed, which he conducted according to the plan laid down by Poisson.

This plan consists in using the formulæ for a simple harmonic variation, as approximately applicable to the actual variation. At great depths the amplitudes of the second and higher terms of the complex harmonic function become so much reduced as not sensibly to influence the variation, which is consequently there expressed with sufficient accuracy by a single harmonic term of yearly period; but at even the greatest depths for which continuous observations have actually been made, the second (or semi-annual) term has a very sensible influence, and the third and fourth terms are by no means without effect on the variations at three feet and six feet from the surface. A close agreement with theory is therefore not to be expected, until the method of analysis which I now propose is applied. It may be added that in the theoretical reductions hitherto made, either by Professor Forbes or others, the amplitudes of the variations for the different depths have alone been compared, and the very interesting conclusion of theory, as to the relation between the absolute amount of retardation of phase and the diminution of amplitude for any increase of depth, has remained untested.

14. In Professor Forbes's paper [*], the very difficult operations which he had performed for effecting the construction and the sinking of the thermometers, and the determination of the cor-

* "Account of some Experiments on the Temperature of the Earth at different Depths and in different Soils near Edinburgh," Transactions of the Royal Society of Edinburgh, vol. xvi. part 2. Edinburgh, 1846.

rections to be applied to obtain the true temperatures of the earth at the different depths from the readings of the scales graduated on their stems protruding above the surface, are fully described. The results of five years' observations—1837 to 1842—are given, along with most interesting graphical representations and illustrations. A process of graphic interpolation, for estimating the temperatures at times intermediate between those of the observations, is applied for the purpose of obtaining data from which the complex harmonic functions expressing the temperatures actually observed for the different depths are determined. I am thus indebted to Professor Forbes for the mode of procedure (described below) which I have myself followed in expressing the variations of temperature during the succeeding thirteen years for the Calton Hill station (where alone the observations were continued). The only variation from his process which I have made is, that, instead of taking twelve points of division for the yearly period, I have taken thirty-two, with a view to obtaining a more perfect representation of all the features of the observed variations, and a more exact average for the principal terms, especially the annual and the semi-annual terms of the complex harmonic function expressing them.

15. *Application of the General Theory to Five Years' Observations—1837 to 1842—at* Professor Forbes's *three Thermometric Stations.*—The first application which I made of the analytical theory explained above, was to the harmonic terms which Professor Forbes had found for expressing the average annual progressions of temperature during the five years' term of observations at the three stations. These terms (which I have recalculated to get their values true to a greater number of significant figures), with alterations of notation which I have found convenient for the analytical expressions, are as follows :—

Three Feet below Surface.

Observatory . . . $45 \cdot 49 + 7 \cdot 39 \cos 2\pi(t - \cdot 63\ ) + 0 \cdot 362 \cos 2\pi(2t - \cdot 669)$

Experimental Gardens. $46 \cdot 13 + 9 \cdot 00 \cos 2\pi(t - \cdot 616) + 0 \cdot 737 \cos 2\pi(2t - \cdot 183)$

Craigleith . . . . $45 \cdot 88 + 8 \cdot 16 \cos 2\pi(t - \cdot 617) + 0 \cdot 284 \cos 2\pi(2t - \cdot 154)$

Six Feet below Surface.

Observatory . . . $45 \cdot 86 + 5 \cdot 06 \cos 2\pi(t - \cdot 686) + 0 \cdot 433 \cos 2\pi(3t - \cdot 731)$

Experimental Gardens. $46 \cdot 42 + 6 \cdot 66 \cos 2\pi(t - \cdot 665) + 0 \cdot 501 \cos 2\pi(2t - \cdot 182)$

Craigleith . . . . $45 \cdot 92 + 6 \cdot 16 \cos 2\pi(t - \cdot 649) + 0 \cdot 368 \cos 2\pi(2t - \cdot 305)$

Twelve Feet below Surface.

Observatory . . . $46 \cdot 36 + 2 \cdot 44 \cos 2\pi(t - \cdot 799) + 0 \cdot 075 \cos 2\pi(2t - \cdot 833)$

Experimental Gardens $46 \cdot 76 + 3 \cdot 38 \cos 2\pi(t - \cdot 782) + 0 \cdot 230 \cos 2\pi(2t - \cdot 390)$

Craigleith . . . . $45 \cdot 92 + \cdot 4 \cdot 22 \cos 2\pi(t - \cdot 713) + 0 \cdot 067 \cos 2\pi(2t - \cdot 819)$

Twenty-four Feet below Surface.

Observatory . . . $46\cdot87+0\cdot655\cos2\pi(t-1\cdot013)$
ExperimentalGardens $47\cdot09+0\cdot920\cos2\pi(t- \cdot986)$
Craigleith . . . . $46\cdot07+1\cdot940\cos2\pi(t- \cdot849)$

The semi-annual terms in these equations present so great
irregularities (those for the Calton Hill station, for instance,
showing a greater amplitude at 6 feet depth than at 3 feet), that
no satisfactory result can be obtained by including them in the
theoretical discussion on which we are now about to enter. We
shall see later, however, that when an average for the whole period
of eighteen years for the Calton Hill station is taken, the semiannual terms are, for the 3 feet and 6 feet depths, in fair agreement with theory; and for the two greater depths are as small as
is necessary for the verification of the theory, and so small as not
to be much influenced by errors of observation and of reduction,
or of "corrections" for temperature of the thermometer tubes.
For the present, we attend exclusively to the annual terms. The
amplitudes and epochs of these terms, extracted from the preceding equations, are shown in the following Table :—

TABLE I. Annual Harmonic Variations of Temperature.

Depths below surface in French feet.	Calton Hill.			Experimental Garden.			Craigleith Quarry.		
	Amplitudes in degrees Fahr.	Epochs of maximum.		Amplitudes in degrees Fahr.	Epochs of maximum.		Amplitudes in degrees Fahr.	Epochs of maximum.	
		In degrees and minutes.	In months and days.		In degrees and minutes.	In months and days.		In degrees and minutes.	In months and days.
Feet. 3	7·386	226 52	Aug. 19	9·063	221 40	Aug. 13	8·069	220 0	Aug. 14
6	5·063	247 5	Sept. 8	6·661	239 20	31	6·148	233 43	26
12	2·455	287 30	Oct. 19	3·408	281 27	Oct. 13	4·216	256 42	Sept. 17
24	0·655	365 6	Jan. 6	0·920	355 0	Dec. 27	1·836	305 46	Nov. 7

By taking the differences of the Napierian logarithms of the
amplitudes, and the differences of epochs reduced to circular
measure (arc divided by radius), thus shown for the different
depths, and dividing each by the corresponding difference of
depths, we find the following numbers :—

TABLE II.—Rates of Logarithmic Diminution in Amplitude, and of Retardation in Epoch, of Annual Harmonic Variations Downwards.

Depths below surface in French feet.	Calton Hill.		Experimental Garden.		Craigleith Quarry.	
feet.	Rate of Diminution of Amplitude.	Rate of Retardation of Epoch.	Rate of Diminution of Amplitude.	Rate of Retardation of Epoch.	Rate of Diminution of Amplitude.	Rate of Retardation of Epoch.
3 to 6	·1250	·1176	·1004	·1163	·09372	·06599
6 to 12	·1206	·1176	·1130	·1193	·06304	·06690
12 to 24	·1101	·1129	·1064	·1062	·06476	·06690
3 to 24	·1154	·1149	·1082	·1114	·06841	·06648

16. All the numbers here shown for each station would be equal, if the conditions of uniformity supposed in the theoretical solution were fulfilled. The discrepancies are, with the exception of one of the numbers for Craigleith Quarry, on the whole small; smaller, indeed, than might be expected when the very notable deviations of the true circumstances from the theoretical conditions are considered. The mean results over the 21 feet, shown in the last line, present very remarkable agreements,—the numbers derived from amplitudes being identical with that derived from epochs for the Calton Hill station, while the differences between the corresponding numbers for the two other stations are in each case only about three per cent. Taking that one number for the first station, and the mean of the slightly differing numbers derived from amplitudes and from epochs respectively for the second and third, we have undoubtedly very accurate determinations of the value of $\sqrt{\frac{\pi c}{k}}$ for the three stations, which are as follows:—

Calton Hill trap rock.	Experimental Garden sand.	Craigleith Quarry sandstone.
$\sqrt{\frac{\pi c}{k}} = \cdot 1154$	$\sqrt{\frac{\pi c}{k}} = \cdot 1098$	$\sqrt{\frac{\pi c}{k}} = \cdot 06744$

A continuation of the observations at Calton Hill not only leads, as we shall see, to almost identical results, both by diminution of amplitude and by retardation, on the whole 21 feet, but also reproduces some of the features of discrepance presented by the progress of the variation through the intermediate depths, and therefore confirms the general accuracy of the preceding results, for all the stations, so far as it might be questioned because of only five years' observations having been available. Further consideration of these results, and deduction of the conductivities of the different portions of the earth's crust involved, are deferred until after we have taken into account the further data for Calton Hill, to the reduction of which we now proceed.

[To be continued.]

V. *Meteorological Charts.*　By Francis Galton, *Esq.**

[With a Plate.]

WHEN contemporary meteorological reports from numerous stations are printed one after another in a column (such as we may see in newspapers and certain foreign publications), they present no picture to the reader's mind. Lists of this description are therefore insufficient to do more than supply data which meteorological students must protract as they best can, upon a map, in some notation intelligible to themselves, at a considerable expense of labour and artistic skill.

It is needless to enlarge upon the serious obstacle which the necessity of doing this opposes to the pursuit of meteorology. It has sufficed to convert what might be a very popular science into a laborious and difficult study. We require means of printing, not lists of dry figures, but actual charts which should record meteorological observations pictorially and geographically, without sacrificing detail. It is then in the belief that an attempt I have just made to supply this desideratum might interest some of your readers, and perhaps lead to useful suggestions, that I forward the accompanying chart. (Plate II). It has been printed with moveable types, which I designed and caused to be cast; and I am much indebted to Mr. W. Spottiswoode, who printed it, for his aid in carrying out my ideas. The map simply incorporates the newspaper data of the day to which it refers, and was printed, not with any scientific object, but solely for the purpose of experiment.

Explanation of the Symbols.

The shade signifies cloud, of an amount proportional to its depth. The types with lines round them, , stand for rain.

Cloud types have been interpolated where observations were

* Communicated by the Author.

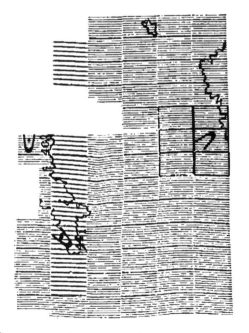

29·6.
rise.

29·5.
rise.

29·4.
rise.

Digitized

wanting. The horseshoes show the direction of the wind current: thus, $\supset$ means wind *from* the west. An included spot $\ni$, or line $\ni$, or cross $\ni$, respectively signify that the wind is gentle, moderate, or strong; where neither dot, line, nor cross are inserted, the force of the wind is unknown. Thermometrical data are expressed by figures, printed below the wind symbols. The first two figures of each set stand for the height of the ordinary thermometer, and the last figure (in a different type) for the difference between this and the thermometer with a wetted bulb. To save confusion of figures, barometer heights are not inserted on the face of the present map; but lines of equal barometric pressure have been deduced from the existing observations, and the places where lines corresponding to each integral one-tenth of an inch cut the marginal columns, have been marked. Thus a straight line joining the pair of figures, 29·7, is approximately the line of that pressure.

I do not consider the types here employed as forming a complete series. An additional shade for cloud is especially wanted.

It will be observed that no space would be lost by this mode of representation, supposing we possessed observations corresponding to every type space of the map.

42 Rutland Gate, S.W.

VI. *On the Curves situate on a Surface of the Second Order.*
By A. CAYLEY, *Esq.*[*]

A SURFACE of the second order has on it a double system of generating lines, real or imaginary; and any two generating lines of the first kind form with any two generating lines of the second kind a skew quadrangle. If the equations of the planes containing respectively the first and second, second and third, third and fourth, fourth and first sides of the quadrangle are $x=0$, $y=0$, $z=0$, $w=0$, and if the constant multipliers which are implicitly contained in x, y, z, w respectively are suitably determined, then the equation of the surface of the second order (or say for shortness the quadric surface) is $xw - yz = 0$.

Assume $\frac{y}{x} = \frac{\mu}{\lambda}$, $\frac{z}{x} = \frac{\nu}{\rho}$, then $\frac{\mu}{\lambda}$, $\frac{\nu}{\rho}$, or say $(\lambda, \mu, \nu, \rho)$, may be regarded as the coordinates of a point on the quadric surface; we in fact have $x : y : z : w = 1 : \frac{\mu}{\lambda} : \frac{\nu}{\rho} : \frac{\mu\nu}{\lambda\rho}$, or what is the same

* Communicated by the Author.
D 2

thing, $=\lambda\rho:\mu\rho:\nu\lambda:\mu\nu$. The four quantities (λ,μ,ν,ρ) are for symmetry of notation used as coordinates; but it is to be throughout borne in mind that the absolute magnitudes of λ and μ, and of ν and ρ are essentially indeterminate; it is only the ratios $\lambda:\mu$ and $\nu:\rho$ that we are concerned with.

An equation of the form

$$(*\!\!\!\;)\!(\lambda,\mu)^p(\nu,\rho)^q=0,$$

that is, an equation homogeneous of the degree p as regards (λ,μ), and homogeneous of the degree q as regards (ν,ρ), represents a curve on the quadric surface; and this curve is of the order $p+q$. In fact, combining with the equation of the curve the equation of an arbitrary plane

$$Ax+By+Cz+Dw=0,$$

this equation, expressed in terms of the coordinates (λ,μ,ν,ρ), is

$$A\lambda\rho+B\mu\rho+C\nu\lambda+D\mu\nu=0;$$

or, as it is more conveniently written,

$$\begin{vmatrix}C, & D\\A, & B\end{vmatrix}\!(\lambda,\mu)(\nu,\rho)=0;$$

and if from this and the equation of the curve we eliminate $\lambda:\mu$ or $\nu:\rho$, say the second of these quantities, we obtain

$$(*\!\!\!\;)\!(\lambda,\mu)^p(-A\lambda-B\mu,C\lambda+D\mu)^q=0,$$

which is of the order $p+q$ in (λ,μ); and $\lambda:\mu$ being known, $\nu:\rho$ is linearly determined. There are thus $p+q$ systems of values of the coordinates, or the plane meets the curve in $p+q$ points; that is, the curve is of the order $p+q$.

A linear equation $A\lambda+B\mu=0$ gives a generating line, say of the first kind, of the quadric surface, and a linear equation $C\nu+D\rho=0$ gives a generating line of the second kind. And by combining the one or the other of these equations with the equation of the curve, it is at once seen that the curve meets each generating line of the first kind in q points, and each generating line of the second kind in p points.

Consider the curves of the order n: the different solutions of the equation $p+q=n$ give different species of curves. But the solution $(n, 0)$ gives only a system of n generating lines of the first kind, and the solution $(0, n)$ gives only a system of generating lines of the second kind. And in general the solutions (p, q) and (q, p) give species of curves which are related, the one of them to the generating lines of the first and second kinds, in the same way as the other of them to the generating lines of the second and first kinds; and they may be considered as correlative members of the same species. The number of distinct species is thus $\dfrac{n-1}{2}$ or $\dfrac{n}{2}$, according as n is odd or even; for $n=3$ we have the single species (2, 1) or (1, 2); for $n=4$, the two species (1, 3) or (3, 1), and (2, 2); for $n=5$, the two species (4, 1) or (1, 4), and (3, 2) or (2, 3); and so on. Thus for $n=3$, the species (2, 1) is represented by an equation of the form

$$(a,\ b,\ c\,\rangle\!\langle\lambda, \mu)^2\nu + (a',\ b',\ c'\,\rangle\!\langle\lambda,\ \mu)^2\rho = 0,$$

which belongs to a cubic curve in space. To show *à posteriori* that this is so, I observe that the equation expressed in terms of the original coordinates $(x,\ y,\ z,\ w)$ is

$$x(a,\ b,\ c\,\rangle\!\langle x,\ y)^2 + z(a',\ b',\ c'\,\rangle\!\langle x,\ y)^2 = 0,$$

which by means of the equation $xw - yz = 0$ of the quadric surface is reduced to

$$(a,\ b,\ c\,\rangle\!\langle x,y)^2 + a'xz + 2b'yz + c'yw = 0;.$$

and this is the equation of a quadric surface intersecting the quadric surface $xw - yz = 0$ in the line $x=0$, $y=0$; and therefore also intersecting it in a cubic curve.

For $n=4$, I take first the species (2, 2) which is represented by an equation of the form

$$(a,\ b,\ c\,\rangle\!\langle\lambda,\ \mu)^2\nu^2 + 2(a',\ b'\ c'\,\rangle\!\langle\lambda,\ \mu)^2\nu\rho + (a'',\ b'',\ c''\,\rangle\!\langle\lambda,\ \mu)^2\rho^2 = 0,$$

which in fact belongs to a quartic curve, the intersection of two quadric surfaces. For, reverting to the original coordinates, the equation becomes

$$(a,\ b,\ c\,\rangle\!\langle x,y)^2x^2 + 2(a',\ b',\ c'\,\rangle\!\langle x,y)^2xz + (a'',\ b'',\ c''\,\rangle\!\langle x,y)^2z^2 = 0,$$

which by means of the equation $xw - yz = 0$ of the quadric surface is at once reduced to

$$(a,\ b,\ c\,\rangle\!\langle x,\ y)^2 + 2a'xz + 4b'yz + 2c'yw + a''z^2 + 2b''zw + c''w^2 = 0,$$

which is the equation of a quadric surface intersecting the given quadric surface $xw - yz = 0$ in the curve in question.

Consider next the species (3, 1) represented by an equation of the form

$$(a,\ b,\ c,\ d\,\rangle\!\langle\lambda,\ \mu)^3\nu + (a',\ b',\ c',\ d'\,\rangle\!\langle\lambda,\ \mu)^3\rho = 0,$$

which is the other species of quartic curve situate on only a single quadric surface. Reverting to the original coordinates, the equation becomes

$$(a, b, c, d\chi x, y)^3 x + (a', b', c', d'\chi x, y)^3 z = 0.$$

And by means of the equation $xw - yz = 0$ of the quadric surface this is reduced to

$$(a, b, c, d\chi x, y)^3 + d'x^2 z + 3b'xyz + 3c'y^2 z + d'y^2 w = 0,$$

which is the equation of a cubic surface containing the line ($x=0$, $y=0$) twice, and therefore along this line touching the quadric surface $xw - yz = 0$; and consequently intersecting it besides in a quartic curve. And in like manner for the curves of the fifth and higher orders which lie upon a quadric surface.

The combination of the equations

$$(*\chi\lambda, \mu)^p(\nu, \rho)^q = 0,$$

$$(*'\chi\lambda, \mu)^{p'}(\nu, \rho)^{q'} = 0,$$

shows at once that two curves on the same quadric surface of the species (p, q) and (p', q') respectively intersect in a number $(pq' + p'q)$ of points. Thus if the curves are $(1, 0)$ and $(1, 0)$, or $(0, 1)$ and $(0, 1)$, *i. e.* generating lines of the same kind, the number of intersections is $1.0 + 0.1 = 0$; but if the curves are $(1, 0)$ and $(0, 1)$, *i. e.* generating lines of different kinds, the number of intersections is $1.1 + 0.0 = 1$.

The notion of the employment of hyperboloidal coordinates presented itself several years ago to Prof. Plücker (see his paper "Die analytische Geometrie der Curven auf den Flächen zweiter Ordnung und Classe," Crelle, vol. xxxiv. pp. 341–359 [1847]); but the systems made use of, *e. g.* $\xi = -\dfrac{d}{\mu}\dfrac{z}{y}$, $\eta = -\dfrac{d}{\mu}\dfrac{z}{x}$, with $z(z+d) + \mu xy = 0$ for the equation of the surface of the second order, is less simple; and the question of the classification of the curves on the surface is not entered on.

2 Stone Buildings, W.C.,
    May 24, 1861.

VII. *On some Experiments connected with* Dove's *Theory of Lustre. By* Prof. O. N. Rood, *of Troy*.*

IN the *Farbenlehre*, p. 177, Prof. Dove writes, "In every case where a surface appears lustrous, there is always a transparent or translucent reflecting stratum of minor intensity, through which we see another body. It is therefore externally

* From Silliman's American Journal for May 1861.

reflected light in combination with internally reflected or dispersed light, whose combined action produces the idea of lustre."

Thus by combining in the stereoscope two projections of a pyramid, one drawn in black lines on a white ground, the other in white lines on black ground, Dove found that the pyramid appeared lustrous as though made of graphite. [To me it, recalls rather the idea of highly polished glass.] He found also that a yellow and blue surface, when combined in the stereoscope and viewed through a plate of violet glass, produced, in the act of combination, the idea of a polished metal.

Similar to Dove's theory of lustre is that of Prof. Reute*.

This view of the nature of lustre opens to us the possibility of reproducing by the stereoscopic combination of suitably coloured surfaces, the individual lustre and appearance of gold, copper, brass, &c.; it also affords us a means of examining separately the components which *may produce* the appearances peculiar to each.

1. I combined in the stereoscope on white or on black grounds, a piece of tinfoil one inch square with a piece of yellow paper of the same size. The value of the tint on the chromatic circles of Chevreul was, 1st circle, orange-yellow, No. 4. When the field containing the tinfoil was somewhat shaded by the hand or otherwise, the surface seen in the stereoscope could not be distinguished from gold-leaf. The union of the images took place as readily and the illusion was as strong with persons unaccustomed to the use of the instrument.

2. By combining in the same way tinfoil with orange-tinted paper (1st circle, orange), the lustre and appearance of copper is imitated.

3. Tinfoil in the act of combination with Nos. 14 and 15 of the red and black scale imitate bismuth.

4. Tinfoil or silver-foil in the act of combination with ultramarine paper appears scarcely blue, rather black like foliated graphite.

5. Gold-leaf in combination with paper of a tint nearly that of the green of the 1st circle imitated murexide.

6. Gold-leaf in combination with ultramarine paper resembled a surface of graphite.

Upon substituting dark grey paper for the tinfoil the same effects in degree were not produced, owing, as it seemed to me, to the fact that the well-known texture and appearance of the paper forcing itself on the attention, precluded the idea of anything metallic. To remove this difficulty I employed two means:—

* *Das Stereoscop;* C. G. Th. Reute, Leipzig, 1860.

1. A crumpled sheet of tinfoil was photographed, and from the negative, prints were taken by the "ammonia-nitrate process," which were toned to the so-called black of the photographers. This furnished dark paper upon whose surface was an accurate drawing of the irregularities characteristic of metallic foil; the surface of the paper was of course wholly without lustre.

(*a*) Upon combining, in black or white fields, a square inch of one of those photographs with the above-mentioned yellow paper, and shading the photograph a little, a representation of gold was obtained but little inferior to that given by the use of the real tinfoil.

(*b*) This photographic paper in combination with orange paper (1st circle, orange) made an imitation of metallic copper.

(*c*) The ultramarine paper in combination with the photograph of tinfoil gave a striking imitation of foliated graphite. The blue colour is perceived much less than would be expected.

2. The surface of a plate of brass 1 inch square was polished, and then rather heavily scratched by a coarse file. Into the scratches a small amount of yellow or white oil paint was rubbed, and upon this prepared surface dark grey or black paper was laid, and the whole submitted to the action of a press as in copper-plate printing. By this means a drawing of a scratched metallic surface was transferred to paper. These markings serve also to enable the observer much more easily to direct his attention simultaneously to the two impressions presented.

(*a*) Upon combining dark grey paper (black and white scale, Nos. 18, 19, 20) prepared in this way with the above-mentioned yellow paper, the appearance of a polished, scratched plate of gold was obtained.

(*b*) When these dark prepared papers were combined with yellow paper coloured by gamboge (yellow and black scale, No. 9), the appearance and lustre of brass were obtained.

According to Dove's theory the darker surface in the stereoscope represents the dispersed light, the brighter, that regularly reflected. As the polish of a metallic surface is proportional to the smallness in amount of the light it disperses, we should be led to expect that by varying the shade of the black paper, we should be able to alter the apparent degree of polish of these imitated metallic surfaces.

This is the case: yellow paper (1st circle, orange-yellow No. 4), in combination with black (No. 21), gives the idea of a very highly polished golden surface; as we descend in the scale, the lustre and resemblance to polished metal regularly diminishes till at grey, No. 8, almost no effect like gold is to be perceived.

On the other hand, by diminishing the brightness of the yellow paper, the black tint remaining constant, the idea of a polished golden plate in the shade, or so placed as to reflect the image of some dark object, is produced. Thus we may descend through the circles of Chevreul to the 7th, when by combining the orange-yellow of that circle with No. 21 of the grey scale, the idea of a golden plate much shaded is produced. I constructed tables expressing the effects produced by varying the intensity of the two components; but it is not worth while to introduce them here.

As we are accustomed to see gold tinted variously from nearly a yellow as in gold-leaf, to almost a copper hue as in some specimens of our American coin, so the tint of the paper placed in the stereoscope may be varied within certain limits, without greatly affecting the results.

Prof. Helmholtz, in his admirable work on Physiological Optics*, mentions that by a peculiar arrangement he was able to cause the homogeneous golden-yellow light of the spectrum to appear brown, proving thus that the tint brown is only weak yellow light. These stereoscopic experiments give us, on the other hand, the means of apparently converting brown into a metallic golden yellow; for many specimens of even brown wrapping paper, when combined in the stereoscope with very black prepared paper, acquire the lustre and appearance of yellow plates in the shade, and reflecting images of dark objects.

In the same manner, and corresponding to the investigations of Helmholtz, I found that the stereoscopic union of black glazed paper with red (No. 14, red and black scale), imitated with surprising perfection the appearance of a glazed plate of chocolate.

The chromatic scales of Chevreul furnish us with a ready means of combining in rapid succession in the stereoscope a great number of definite tints; thus by cutting in a card-board two parallel apertures $\frac{3}{10}$ inch broad and 1 inch long, their distance apart being 2·6 inches, and pasting under one of them black prepared paper, the other can be brought over any desired tint and the effect noted.

1. In this way I found that a pretty good representation of the appearance of slightly tarnished lead was produced by the stereoscopic union of grey No. 18 and No. 4 on the blue-violet and black scale.

2. A somewhat inferior imitation of antimony was given by No. 1 blue and black scale, with grey Nos. 18 to 20, or by using No. 17 blue and black scale with white.

* P. 281. *Physiologische Optik* (*Encyklopädie der Physik.* Leipzig, 1860).

8. Tarnished zinc surfaces may be imitated by the use of grey No. 5 with No. 18 blue and black scale.

4. Ultramarine paper, with some of the lighter violet-blues, gave an imitation of blue glass. The idea of blue polished glass was also obtained by using in combination with the ultramarine paper No. 1 of the yellow and black scale.

I will mention here that the stereoscopic union of this blue with yellow paper, never induced in my mind the idea of green.

I made some experiments to ascertain how far the *stereoscopic* mixture of two masses of different coloured light corresponded to their true mixture by the method of rapid rotation, use being made of the imitations above described. It is however so diffi-- cult to compare a varying with a fixed tint that I will not record. the results obtained; in many cases a certain moderate amount of agreement in the resultant tints was observed. Brücke found, when a deeply-coloured yellow glass was held before one eye and a blue cobalt glass before the other, that a landscape viewed through this combination was simply darkened in appearance. I repeated this experiment with similar lasses, and obtained a like result; objects appeared darkened but in their natural colours, though sometimes the blue or yellow tint predominated a little. But when I presented to a *single* eye these two masses of light, a very different result was obtained; the plates of glass were attached to a blackened disk opposite suitable perforations, and it was set in rapid rotation; a landscape viewed through it appeared deep purple, though not a trace of this colour was to be perceived in the binocular use of these glasses.

When these two glasses were held before the same eye, a landscape viewed through them was very much darkened but scarcely coloured.

Sir David Brewster's Theory of Lustre.

Sir David Brewster opposes Dove's theory of lustre, as he has found that when black and white surfaces without drawings are combined in the stereoscope, no lustre is produced. The lustre, then, according to this philosopher, is due not to one mass of light passing through another, but to the effort of the eyes to combine the stereoscopic pictures.

Admitting the correctness of Sir David's experiment, Dove has shown that the objection founded on it is without weight (p. 3, Optical Studies).

In repeating Brewster's experiment I always obtain the *opposite result*; in combining uniform black and white surfaces, without drawings, I always obtain a distinct impression of lustre, like that of the blackened mirror of a polariscope, and in strict accordance with Dove's theory; when the black field is so darkened that *no light* is sent from it to the eye, this lustre vanishes, and the white paper alone is perceived. This disagreement is not a cause of astonishment when we reflect that De Haldat's original experiment waited nearly half a century for confirmation.

To Brewster's own theory, the simple objection, which has already been made by others, that we daily perceive lustre plainly with one eye, would seem sufficient.

Production of Lustre in Monocular Vision.

I proceed now to describe some experiments where by the action upon a single eye of two masses of light of unequal intensity, the idea of lustre is produced.

1. If a disk of coloured card-board, out of which a number of sectors has been removed, be made to rotate rapidly, and an object be viewed through it by a single eye, two masses of light will reach the eye, which apparently proceed from the object; one is reflected from the surface of the disk, the other emanates from the object behind the disk, and passes through the first mass of light. Dark objects viewed in this way assume to me, to a small extent, an appearance like that of blackened glass. The effect is not at all striking, and would be overlooked by many persons; I therefore prepared paper in a peculiar way, so as to imitate distantly the appearance of foliated graphite or crumpled mica.

White smooth drawing-paper was rubbed over irregularly with a brush *slightly* moistened with a weak wash of India ink or lampblack; when dry, another wash of a deeper hue applied as before, care being taken to leave many small spots untouched. The final wash was laid on with pure black. If the brush be kept nearly dry and passed only lightly over the paper, it is easy to obtain a surface bearing some very distant resemblance to the minerals above mentioned; it is of course without lustre. Similar papers were prepared with red and blue water colours.

When these papers were held behind disks of ultramarine or orange-tinted paper, from which equal alternate sectors had been removed, and which were revolving at such rates that their surfaces seemed uniform, or at lower rates, they often appeared, to a single eye, highly lustrous. This was true of the prepared paper in a *state of rest;* when moved slightly by the hand it glittered strongly. Dark photographs of tinfoil held behind a revolving disk of ultramarine paper and viewed by a single eye, assume often to a striking degree the lustre and appearance of foliated graphite.

2. If a piece of this peculiarly blackened paper $\frac{1}{4}$ an inch square be placed in a blue field (rather light ultramarine paper)

and he steadily regarded for some minutes by one eye, it assumes a red-orange hue, and appears suspended over the blue paper and nearer to the eye than the latter; at the same instant it appears lustrous like crumpled mica. The illusion with me often lasts half a minute in great perfection; this is particularly the case when the eye is not quite accurately focused on the paper.

3. If a sheet of this prepared paper be brightly illuminated by light from a window, and be held so near one eye as to produce indistinct vision, it often apparently becomes highly lustrous. In this case enlarged images of the white and grey points are formed on the retina, which overlap, so that again we have two masses of light, one passing through the other.

4. If a *roll* of black paper like the above, but coarser in its markings, be brightly illuminated on one side and viewed through deeply coloured plates of glass (red, green, blue), in a few seconds it appears lustrous, resembling a roll of polished zinc which has been irregularly and deeply corroded by an acid. Upon removing the glass, the surface of the paper appears lustrous for an instant.

5. A sheet of the finer variety of this prepared paper viewed through a large rhomb of calc-spar, gives often in spots the appearance of lustre, particularly when the head of the observer, or the rhomb, is slightly moved. Some persons compared this to the appearance of water.

It would seem probable that in all cases where two masses of light reach a single eye, one passing through the other, particularly when there is any *perception of their individuality*, that the appearance of more or less lustre is produced, though from habit we often overlook it. Thus Helmholtz remarks * (upon the combination of two coloured surfaces in monocular vision by means of a simple instrument he figures), " It is particularly favourable when the drawings, or spots on the two surfaces, are made to shift their position. Then we often believe that we see both colours simultaneously in the same place, the one through the other. We have an impression in such cases of seeing objects through a coloured veil or reflected from a coloured surface."

I found, in fact, that by placing stereographs consisting of coloured paper for one eye and a photographic drawing of tinfoil for the other in this instrument, that lustre could be perceived, particularly with the imitations of copper.

The diagram represents the instrument referred to; it consists of a plate of glass, P, with parallel sides, which is properly supported over a blackened board B. Differently coloured papers are placed at K and Y; one is seen through the plate, and the

* *Physiologische Optik*, p. 273.

other by reflexion from it. The images are
made to overlap, and their intensity is regulated by altering their distance from F.

Analogous to this is the observation of
Brewster*. Speaking of uniting similar
pictures (patterns on hanging-paper) in bin-
ocular vision, he remarks, "The surface of it (the wall) seems
slightly curved. It has a *silvery transparent aspect.*" Here the
images (though of the same intensity, &c.) moving with each
slight movement of the head induces in the mind the idea of one
object seen through another.

In closing, I will remark that while many of the experiments
above mentioned are easily repeated, others require considerable
practice in this kind of observation.

VIII. *Concluding Paper on Tactic.* By J. J. Sᴙʟᴠᴇsᴛᴇʀ, *M.A.,*
F.R.S., Professor of Mathematics at the Royal Military Aca-
demy, Woolwich†.

IN my tactical paper in the May Number of the Magazine, I
considered the number of groupings and of types of groupings of *synthemes* formed out of triads of three *nomes* of three
elements each. The first example of considering the *ensemble* of
the groupings of a defined species of synthemes (each of such
groupings being subjected to satisfy a certain exhaustive condition) was, as already stated, furnished by me in this Magazine,
April 1844. In that case the synthemes consisted of duads
belonging to a single nome of 6 elements, and the total number
of the groupings was observed to be 6, all contained in one type
or family. The total number of synthemes in that instance being
15, and there being 6 groupings of 5 synthemes each, it followed that in the whole family every syntheme is met with twice
over; once in one grouping, and once in another. In the case
treated of in my last communication to the Magazine, the total
number of the synthemes of the kind under consideration is 36
(for it may easily be shown that the number of synthemes of
n-nomial n-ads of n nomes of q elements each is $(1.2.3\ldots q)^{n-1}$);
and as each grouping contains 9 synthemes, these 36 are distributed *without repetition* between the 4 groupings of the smaller
of the two natural species,—a phenomenon of a kind here met
with for the first time in the study of *syntax.* If now we go on
(as a natural and irrepressible curiosity urges) to ascertain the
groupings of the synthemes of *binomial* triads of the same 3

---

* The Stereoscope, p. 91. London, 1856.
† Communicated by the Author.

nomes of three elements each, we advance just one step further in the direction of type-complexity; that is to say, we meet with the existence of 3, and not more than 3, types or species in which all such groupings are comprised. The investigation by which this is made out appears to me well worthy to be given to the world, as affording an example of a new and beautiful kind of analysis proper to the study of *tactic*, and thus lighting the way to the further opening up of this fundamental doctrine of mathematic, the science of necessary relations, of which, combined with logic (if indeed the two be not identical), tactic appears to me to constitute the main stem from which all others, including even arithmetic itself, are derived and secondary branches. The key to success in dealing with the problems of this incipient science (as I suppose of most others) must be sought for in the construction of an apt and expressive notation, and in the discovery of language by force of which the mind may be enabled to lay hold of complex ,operations and mould them into simple and easily transmissible .forms of thought. I must then entreat the indulgence of the reader if, in this early grappling with the difficulties of a new .language and a new notation, I may occasionally appear wanting in absolute clearness and fullness of expression.

Let us, as before, represent the nine elements by the numbers from 1 to 9, and suppose the *nomes* to be $1, 2, 3 : 4, 5, 6 : 7, 8, 9$.

If we take any syntheme formed out of the *binomial* triads belonging to the above nomes, and if out of such syntheme we omit the elements 1, 2, 3 (belonging to the 1st nome) wherever they occur, the slightest consideration will serve to show that the synthemes thus denuded will assume the *form $l . m . r, p . q, n$*, where l, m, r may be regarded as belonging to one of the remaining nomes, and p, q, n to the other. The total number of synthemes in a grouping which contains all the binomial triads is 18, because the total number of these triads is 54; and consequently it will be seen that every grouping will in fact consist of the same *framework*, so to say, of combinations of elements belonging to the second and third nomes variously compounded . with the elements of the first nome.

This framework may be advantageously divided into two parts, 'each containing nine terms, and which I shall call respectively U and U̇. Thus by U I shall understand the nine arrangements following:—

$$4.5.7, 8.9, 6; \quad 4.5.8, 7.9, 6; \quad 4.5.9, 7.8, 6$$
$$5.6.7, 8.9, 4; \quad 5.6.8, 7.9, 4; \quad 5.6.9, 7.8, 4$$
$$6.4.7, 8.9, 5; \quad 6.4.8, 7.9, 5; \quad 6.4.9, 7.8, 5$$

each *imperfect* or defective syntheme being separated from the .next by a semicolon, or else by a change of line. So by U̇

I shall understand the *complementary* part of the framework, viz. :—

$$8.9.6, 4.5, 7; \quad 7.9.6, 4.5, 8; \quad 7.8.6, 4.5, 9$$
$$8.9.4, 5.6, 7; \quad 7.9.4, 5.6, 8; \quad 7.8.4, 5.6, 9$$
$$8.9.5, 6.4, 7; \quad 7.9.5, 6.4, 8; \quad 7.8.5, 6.4, 9$$

It is of cardinal importance to notice that the order in which the *imperfect synthemes* are arranged in U and U̇ is one of absolute reciprocity. It is in this reciprocity, and in the fact of U or U̇ being each in *strict regimen* (so to say) with the other, that the cause of the success of the method about to be applied essentially resides.

The slightest reflection will serve to show that every *complete* syntheme of the kind required will be of the form

$$\left| \begin{array}{c} U \times P \\ \dot{U} \times \dot{P} \end{array} \right|$$

where the symbolical multipliers P and Ṗ are each of them some one of the forms (by no means necessarily the *same*) represented generally by the framework of defective synthemes hereunder written (defective in the sense that all the elements of the second and third nomes are supposed to be omitted),

$$, a, bc; \quad , b, ca; \quad , c, ab$$
$$, b, ca; \quad , c, ab; \quad , a, bc$$
$$, c, ab; \quad , a, bc; \quad , b, ca$$

or else by the cognate framework

$$, a, bc; \quad , c, ab; \quad , b, ca$$
$$, b, ca; \quad , a, bc; \quad c, ab$$
$$, c, ab; \quad , b, ca; \quad a, bc$$

where a, b, c are identical in some order or another with the elements of the first nome, viz. 1, 2, 3; so that there are six different systems of a, b, c in each of these two frameworks.

No other combination of the elements in U or U̇ (all of which belong to the second and third nomes) with the elements in the first nome is possible; for any such combination would involve the fact of a *repetition* of the same *triad* or triads in the same grouping, contrary to the nature of a grouping. Hence, then, the number of forms of P and of Ṗ being twice six, or 12, we at once perceive that the total number of groupings is 12×12, or 144.

But now comes the more difficult question of ascertaining between how many distinct species or types these groupings are distributed. If we study the form of P or Ṗ, it is obvious that

it will be completely and distinctively *denoted in brief* by the
twelve forms arising from the development of

$$
\begin{array}{ccc}
a & b & c \\
b & c & a \\
c & a & b
\end{array}
\quad \text{and} \quad
\begin{array}{ccc}
a & c & b \\
b & a & c \\
c & b & a
\end{array}
\quad ; \text{videlicet}
$$

(1)	(2)	(3)	(4)	(5)	(6)
1 2 3	2 3 1	3 1 2	2 1 3	1 3 2	3 2 1
2 3 1	3 1 2	1 2 3	1 3 2	3 2 1	2 1 3
3 1 2	1 2 3	2 3 1	3 2 1	2 1 3	1 3 2

(7)	(8)	(9)	(10)	(11)	(12)
1 3 2	2 1 3	3 2 1	2 3 1	1 2 3	3 1 2
2 1 3	3 2 1	1 3 2	1 2 3	3 1 2	2 3 1
3 2 1	1 3 2	2 1 3	3 1 2	2 3 1	1 2 3

which we may for facility of future reference denote by

$$ \pi_1\, \pi_2\, \pi_3\, \pi_4\, \pi_5 \ldots \pi_{12}. $$

Now as regards the types :· since the order of the elements in
one nome is entirely independent of the order of the elements in
any other, it is obvious that it is not the particular form of P or
of $\dot{\mathrm{P}}$ which can have any influence on the form of the type, but
solely the relation of P and $\dot{\mathrm{P}}$ to one another. In order then to
fix the ideas, I shall for the moment consider P equal to

$$
\begin{array}{ccc}
1 & 2 & 3 \\
2 & 3 & 1 \\
3 & 1 & 2
\end{array}
$$

This at once enables us to fix *a limit* to the number of distinct
types. In the first place, the essentially distinct FORMS of the
first column in $\dot{\mathrm{P}}$, with respect to that of P, may be sufficiently
represented by taking the two columns identical, or differing by
a single interchange, or else having no two elements in the same
place. Hence $\dot{\mathrm{P}}$, so far as the ascertainment of types is concerned, may be limited to the six forms following :—

(α)	(γ)	(ε)
1 2 3	2 1 3	2 3 1
2 3 1	1 3 2	3 1 2
3 1 2	3 2 1	1 2 3

(β)	(δ)	(η)
1 3 2	2 3 1	2 1 3
2 1 3	1 2 3	3 2 1
3 2 1	3 1 2	1 3 2

But again, since (β) and (η) are each derivable from (α) (the
assumed form of P) by an interchange of two columns *inter se,*

it is clear that, as regards distinction of type, $\eta = \beta$, and conse-quently there are only *at utmost* five types remaining, which may be respectively described by the symbols

$$\left| \begin{array}{c} U\alpha \\ \dot{U}\alpha \end{array} \right| \quad \left| \begin{array}{c} U\alpha \\ \dot{U}\beta \end{array} \right| \quad \left| \begin{array}{c} U\alpha \\ \dot{U}\gamma \end{array} \right| \quad \left| \begin{array}{c} U\alpha \\ \dot{U}\delta \end{array} \right| \quad \left| \begin{array}{c} U\alpha \\ \dot{U}\epsilon \end{array} \right|$$

It must be noticed that α comprehends or typifies the squares numbered 1; β those numbered 7, 8, 9; γ those numbered 4, 5, 6; δ those numbered 10, 11, 12; ϵ those numbered 2, 3.

I say designedly that the number of types is *at utmost* limited to these five. But it by no means follows that the number is so great as five; for it will not fail to be borne in mind that these differences have reference to the peculiar mode in which we have chosen to decompose in idea each syntheme, by viewing it as a symbolical product of an arrangement containing only the ele-ments of the second and third nomes by an arrangement con-taining only those of the first nomes. But the nomes are inter-changeable, and therefore it may very well be the case that two types which appear to be distinct are in reality identical, their elements in the groupings appertaining to such types having absolutely analogous relations to different orderings of the nomes, so that the groupings will be convertible into each other by per-mutations among the given elements. We must therefore ascer-tain how the above types, or any specific forms of them, come to. be represented when we interchange the first nome with either of the other two, or, to fix the ideas, let us say with the second.

To effect this, let $U\alpha$, $\dot{U}\alpha$, $\dot{U}\beta$, $\dot{U}\gamma$, $\dot{U}\delta$, $\dot{U}\epsilon$ be actually ex-panded; by the performance of the symbolical multiplications we obtain—

$$U_a = \left| \begin{array}{llll} 4.5.7 & 8.9.1 & 6.2.3; & 4.5.8 \ 7.9.2 \ 6.1.3; & 4.5.9 \ 7.8.3 \ 6.2.1 \\ 5.6.7 & 8.9.2 & 4.1.3; & 5.6.8 \ 7.9.3 \ 4.1.2; & 5.6.9 \ 7.8.1 \ 4.2.3 \\ 6.4.7 & 8.9.3 & 5.1.2; & 6.4.8 \ 7.9.1 \ 5.2.3; & 6.4.9 \ 7.8.2 \ 5.1.3 \end{array} \right|$$

$$\dot{U}_a = \left| \begin{array}{llll} 8.9.6 & 4.5.1 & 7.2.3 & 7.9.6 \ 4.5.2 \ 8.1.3 & 7.8.6 \ 4.5.3 \ 9.2.1 \\ 8.9.4 & 5.6.2 & 7.1.3 & 7.9.4 \ 5.6.3 \ 8.1.2 & 7.8.4 \ 5.6.1 \ 9.2.3 \\ 8.9.5 & 6.4.3 & 7.2.1 & 7.9.5 \ 6.4.1 \ 8.2.3 & 7.8.5 \ 6.4.2 \ 9.1.3 \end{array} \right|$$

$$\dot{U}\beta = \left| \begin{array}{llll} 8.9.6 & 4.5.1 & 7.2.3 & 7.9.6 \ 4.5.3 \ 8.1.2 & 7.8.6 \ 4.5.2 \ 9.1.3 \\ 8.9.4 & 5.6.2 & 7.1.3 & 7.9.4 \ 5.6.1 \ 8.2.3 & 7.8.4 \ 5.6.3 \ 9.2.1 \\ 8.9.5 & 6.4.3 & 7.2.1 & 7.9.5 \ 6.4.2 \ 8.1.3 & 7.8.5 \ 6.4.1 \ 9.2.3 \end{array} \right|$$

$$\dot{U}\gamma = \left| \begin{array}{llll} 8.9.6 & 4.5.2 & 7.1.3 & 7.9.6 \ 4.5.1 \ 8.2.3 & 7.8.6 \ 4.5.3 \ 9.1.2 \\ 8.9.4 & 5.6.1 & 7.2.3 & 7.9.4 \ 5.6.3 \ 8.1.2 & 7.8.4 \ 5.6.2 \ 9.1.3 \\ 8.9.5 & 6.4.3 & 7.1.2 & 7.9.5 \ 6.4.2 \ 8.1.3 & 7.8.5 \ 6.4.1 \ 9.2.3 \end{array} \right|$$

$$\dot{U}\delta = \left| \begin{array}{llll} 8.9.6 & 4.5.2 & 7.1.3 & 7.9.6 \ 4.5.3 \ 8.1.2 & 7.8.6 \ 4.5.1 \ 9.2.3 \\ 8.9.4 & 5.6.1 & 7.2.3 & 7.9.4 \ 5.6.2 \ 8.1.3 & 7.8.4 \ 5.6.3 \ 9.1.2 \\ 8.9.5 & 6.4.3 & 7.1.2 & 7.9.5 \ 6.4.1 \ 8.2.3 & 7.8.5 \ 6.4.2 \ 9.1.3 \end{array} \right|$$

$$\dot{U}\epsilon = \left| \begin{array}{llll} 8.9.6 & 4.5.2 & 7.1.3 & 7.9.6 \ 4.5.3 \ 8.1.2 & 7.8.6 \ 4.5.1 \ 9.2.3 \\ 8.9.4 & 5.6.3 & 7.1.2 & 7.9.4 \ 5.6.1 \ 8.2.3 & 7.8.4 \ 5.6.2 \ 9.1.3 \\ 8.9.5 & 6.4.1 & 7.2.3 & 7.9.5 \ 6.4.2 \ 8.1.3 & 7.8.5 \ 6.4.3 \ 9.1.2 \end{array} \right|$$

. Let us form a *framework* with the nomes $1.2.3, 7.8.9$ exactly similar to that which we formed before with $4.5.6$, $7.8.9$, and let V, $\dot{V}$ be its two parts respectively analogous to U, $\dot{U}$, we thus obtain for $\dot{V}$,

$$
\begin{array}{lll}
1.2.7, 8.9, 3; & 7.2.8, 7.9, 3; & 1.2.9, 7.8, 3 \\
2.3.7, 8.9, 1; & 2.8.8, 7.9, 1; & 2.3.9, 7.8, 1 \\
3.1.7, 8.9, 2; & 3.1.8, 7.9, 2; & 3.1.9, 7.8, 2
\end{array}
$$

and for V,

$$
\begin{array}{lll}
8.9.3, 1.2, 7; & 7.9.8, 1.2, 8; & 7.8.3, 1.2, 9 \\
8.9.1, 2.3, 7; & 7.9.1, 2.3, 8; & 7.8.1, 2.3, 9 \\
8.9.2, 3.1, 7; & 7.9.2, 3.1, 8; & 7.8.2, 3.1, 9
\end{array}
$$

We must now perform the unwonted process of symbolical division, and obtain the quotients of $U\alpha$ by V, and of $\dot{U}\alpha$, $\dot{U}\beta$, $\dot{U}\gamma$, $\dot{U}\delta$, $\dot{U}\epsilon$ by $\dot{V}$ (it will of course be perceived that it is known *à priori* that the dividend forms of arrangement are actual multipliers of the divisors V and $\dot{V}$). In writing down the results of these divisions, which will consist exclusively of elements belonging to the nome $4.5.6$, and of which each term will be of the form $d, e.f$, we may, analogously to what we have done before for greater brevity, write down only the single element (d), and omit the residue (ef), which is determined when (d) is determined. We shall thus obtain the quotients following :—

$$
\frac{U\alpha}{V} = \begin{array}{ccc} 5 & 4 & 6 \\ 6 & 5 & 4 \\ 4 & 6 & 5 \end{array}
$$

$$
\frac{\dot{U}\alpha}{\dot{V}} = \begin{array}{ccc} 5 & 4 & 6 \\ 6 & 5 & 4 \\ 4 & 6 & 5 \end{array} \qquad
\frac{\dot{U}\beta}{\dot{V}} = \begin{array}{ccc} 5 & 6 & 4 \\ 6 & 4 & 5 \\ 4 & 5 & 6 \end{array} \qquad
\frac{\dot{U}\gamma}{\dot{V}} = \begin{array}{ccc} 5 & 4 & 6 \\ 4 & 6 & 5 \\ 6 & 5 & 4 \end{array}
$$

$$
\frac{\dot{U}\delta}{\dot{V}} = \begin{array}{ccc} 5 & 6 & 4 \\ 4 & 5 & 6 \\ 6 & 4 & 5 \end{array} \qquad
\frac{\dot{U}\epsilon}{\dot{V}} = \begin{array}{ccc} 4 & 6 & 5 \\ 5 & 4 & 6 \\ 6 & 5 & 4 \end{array}
$$

It may be observed that these divisions may be effected with great rapidity; because when three out of the nine figures (in any quotient) not in the same line or column are known, all the rest are known. Thus, for example, to find $\dfrac{\dot{U}\epsilon}{\dot{V}}$ it is only necessary to seek in $\dot{U}\epsilon$ the syntheme which contains $1.2.7$, and then to take out the figure in that syntheme associated with 8.9 in that line, viz. 4; then again to seek the syntheme which contains $1.2.8$, and to take out the figure in that syntheme associated with 7.9, which is 6; and finally to seek the syntheme which contains $2.3.7$, and then to take out the figure associated with 8.9, viz. 5; we thus obtain the three corner figures of the square which represents $\dfrac{\dot{U}\epsilon}{V}$ as thus:

$$
\begin{array}{ccc}
4 & 6 & . \\
5 & . & . \\
. & . & .
\end{array}
$$

of which the six remaining figures are given by the condition that in no line and in no column must the same two figures be found. In order to compare these quotients, or rather the relations of the first of them to the remaining five with those of α to $\alpha, \beta, \gamma, \delta, \epsilon$, it will be convenient to subtract the constant number 3 from each figure, and to transpose the first and second columns; we thus obtain

$$
\frac{U\alpha}{V} \equiv \begin{vmatrix} 1 & 2 & 3 \\ 2 & 3 & 1 \\ 3 & 1 & 2 \end{vmatrix} \equiv \pi_1 \equiv \alpha,
$$

$$
\frac{\dot{U}\alpha}{V} \equiv \begin{vmatrix} 1 & 2 & 3 \\ 2 & 3 & 1 \\ 3 & 1 & 2 \end{vmatrix} \equiv \pi_1 \equiv \alpha, \qquad
\frac{\dot{U}\beta}{V} \equiv \begin{vmatrix} 3 & 2 & 1 \\ 1 & 3 & 2 \\ 2 & 1 & 3 \end{vmatrix} \equiv \pi_9 \equiv \beta,
$$

$$
\frac{\dot{U}\gamma}{V} \equiv \begin{vmatrix} 1 & 2 & 3 \\ 3 & 1 & 2 \\ 2 & 3 & 1 \end{vmatrix} \equiv \pi_{11} \equiv \delta, \qquad
\frac{\dot{U}\delta}{V} \equiv \begin{vmatrix} 3 & 2 & 1 \\ 2 & 1 & 3 \\ 1 & 3 & 2 \end{vmatrix} \equiv \pi_6 \equiv \gamma,
$$

$$
\frac{\dot{U}\epsilon}{V} \equiv \begin{vmatrix} 3 & 1 & 2 \\ 1 & 2 & 3 \\ 2 & 3 & 1 \end{vmatrix} \equiv \pi_3 \equiv \epsilon.
$$

Thus, for greater brevity, considering the five types to be represented by

$$
\begin{array}{ccccc}
\alpha & \alpha & \alpha & \alpha & \alpha \\
\alpha & \beta & \gamma & \delta & \epsilon,
\end{array}
$$

or still more briefly by

$$
\alpha \quad \beta \quad \gamma \quad \delta \quad \epsilon;
$$

and calling the nomes N_1, N_2, N_3, we find that the effect of interchanging N_1 and N_2 with each other is to change

$$
\alpha \quad \beta \quad \gamma \quad \delta \quad \epsilon
$$

into

$$
\alpha \quad \beta \quad \delta \quad \gamma \quad \epsilon.
$$

In like manner it may be ascertained (and the student is ad-

vised to satisfy himself by actual trial of the fact) that the effect
of interchanging N_1 and N_3 with each other is to convert

$$\alpha \quad \beta \quad \gamma \quad \delta \quad \epsilon$$

into

$$\alpha \quad \delta \quad \gamma \quad \beta \quad \epsilon.$$

From these two calculations it follows that the effect of any
permutation between N_1, N_2, N_3 is to produce a permutation in
β, γ, δ *inter se*, but will leave α and ϵ unaltered*. Hence then
we have arrived at the goal of our inquiry, having demonstrated
that

$$\left| \begin{matrix} V\alpha \\ \dot{V}\alpha \end{matrix} \right|$$

indicates one type,

$$\left| \begin{matrix} V\alpha \\ \dot{V}\beta \end{matrix} \right|, \quad \left| \begin{matrix} V\alpha \\ \dot{V}\gamma \end{matrix} \right|, \quad \left| \begin{matrix} V\alpha \\ \dot{V}\delta \end{matrix} \right|$$

each of them another *the same* type, and

$$\left| \begin{matrix} V\alpha \\ \dot{V}\epsilon \end{matrix} \right|$$

a third type,—and bearing in mind that

     (α) belongs to π_1 exclusively,

     (ϵ)   ,, π_2, π_3 ,,

     (β)   ,, π_7, π_8, π_9 ,,

     (γ).   ,, π_4, π_5, π_6 ,,

     (δ)   ,, π_{10}, π_{11}, π_{12} ,,

and that each form of π comprehends 12 groupings due to the
12 forms of $V\alpha$, we are enabled to affirm that the total number
of groupings of the binomial triads of 3 nomes of 3 elements
each is 144, and that the number of types or species between
which these 144 are distributed is 3, comprising 12, 24, and
108 respectively,—a conclusion which it would almost have
exceeded the practical limits of human labour and perspi-
cuity to have established by the direct comparison of the 144

---

* This result, by the aid of a fine observation, may be more rapidly
established *uno ictu* (I mean by *one* calculation instead of two) as follows
Let $N_1 N_2 N_3$ be made to undergo a *cyclical* interchange, then it will be
found that β, γ, δ also undergo a cyclical interchange, whilst α and ϵ
remain unchanged. This proves that β, γ, δ are only different phases of
the same type, which is *sufficient*; for as regards α and ϵ, the fact of the
number of individuals which they represent being unequal *inter se*, and
also unequal to the number contained in β, γ, δ, renders it *à priori* impossible to allow that they can either pass into each other or into the forms
β, γ, δ, by virtue of any interchange among the elements.

groupings of 18 synthemes each with each other, with a view to ascertain which admit of being permutable into each other, and which not.

The largest species of 108 groupings, it may be observed, is subdivisible into 3 *varieties*, not really allotypical, of 36 each,—the characteristic of those groupings which belong to the same variety being that they permute *exclusively* into each other when the permutations of the elements are confined to perturbations of the order of the elements in the same nome or nomes, and the different nomes are subject to no interchange of elements between themselves.

Just so the species of 36 groupings of trinomial triads, treated of in my preceding paper, subdivides into 3 varieties or subfamilies characterized by a similar property.

The total number of modes of subdivision of 9 elements between 3 nomes being 280, it follows, from considerations of the same kind as stated in the May Number of the Magazine, that there exist transitive substitution-groups belonging to 9 elements of

$$\frac{\pi(9)}{280 \times 12}, \quad \frac{\pi(9)}{280 \times 24}, \quad \frac{\pi(9)}{280 \times 108},$$

that is, 108, 24 and 12 substitutions respectively.

Again, let us consider the question of forming the synthemes of the triads of a *single nome* of 9 elements into groupings where *every* triad shall be found without repetition. We may obtain such groupings by choosing arbitrarily any one of the 280 sets of 3 nomes into which the 9 elements may be segregated*, and then forming one syntheme with the three monomial triads (corresponding to such set so chosen), 18 synthemes (in any one of the 144 possible ways) of exclusively binomial triads, and 9 synthemes (in any one of the 40 possible ways) of exclusively trinomial triads; we shall thus obtain in all $280 \times 144 \times 40$, or 1,612,800 solutions of the question proposed; I mean 1,612,800 groupings, all satisfying the imposed condition, and reducible to 6 *genera*†, comprising respectively

$$4 \times 12 \times 280 \quad 4 \times 24 \times 280 \quad 4 \times 108 \times 280 \quad 36 \times 12 \times 280$$
$$36 \times 24 \times 280 \quad 36 \times 108 \times 280,$$

---

* 280 is also evidently the number of synthemes of triads belonging to one nome of 9 elements. In general the number of r-ads belonging to one nome of mn elements is

$$\frac{\pi(mn-1)\pi\big((m-1)n-1\big)\pi\big((m-2)n-1\big)\ldots\pi(n-1)}{\big(\pi(n-1)\big)^m \pi\big((m-1)n\big)\pi\big((m-2)n\big) \ldots \pi(n)}.$$

† The above *genera* must not be confounded with types or species. (In my preceding communications I may inadvertently have used the word

i. e. 18,440, 26,880, 120,960, 120,960, 241,920, 1,088,640 individual groupings. I conclude with putting a grand question, more easy to propose than to answer, viz. are these one million six hundred thousand (and upwards) groupings (classifiable under six distinct genera) all the possible modes and types of grouping which will satisfy the conditions of the question? and if not, what other mode or type of grouping can be found? Were I compelled to give an answer to this question, I would say that the balance of my mind leans to the opinion that the six types in question are the sole possible types of solution; but I do not pretend to rest this judgment upon any solid grounds of demonstration, nor to entertain it with any strong degree of assurance. It is a question which the effort to resolve cannot but react powerfully on our knowledge of the principles of tactic in general, and of the theory of substitution-groups in particular; and as such I submit it to the consideration of the rising chivalry of analysis, seeking myself meanwhile fresh fields and pastures new of meditation.

K, Woolwich Common,
    June 6, 1861.

---

family as coincident with type: *species* is the proper term.) The type of a total grouping in the problem referred to in the text will depend not only on the particular combination of the types of the binomial and trinomial partial groupings which give rise to these 6(=2×3) genera, but also on the relative *phases* of the types so combined. The number of groupings in one type or species is always a submultiple of the number of permutations of the elements; whereas it will be seen that the number of groupings in one of the above genera greatly exceeds that number, which in the present case is only

$$1.2.3.4.5.6.7.8.9, \text{ or } 362,880.$$

Whatever may be the case in natural history, the nature of a type or species, as distinguished from a genus, family, or any other higher kind of aggregation of individuals, in *pure syntax* is perfectly clear and unambiguous; those groupings form a species which are commutable into one another by an interchange of elements: thus the different *phases* of the same type or species are in analogy with the different values of the same function arising out of a change in a constant parameter. If it should turn out that the above sixteen hundred thousand and odd groupings are not the sole solutions of which the question admits, then it will follow that even in this early instance we shall have an example not only of species and genera, but of distinct families of genera, for it is certain that the above six genera constitute within themselves a complete natural family. It will form an interesting subject of inquiry to ascertain how many types are included within each of the six genera belonging to this family; and be it never forgotten that to each species corresponds, and from it is, so to say, capable of being extracted or sublimated, a Cauchian substitution-group.

IX. *Chemical Notices from Foreign Journals.* By E. ATKINSON,
Ph.D., F.C.S., *Teacher of Physical Science in Cheltenham
College.*

[Continued from vol. xxi. p. 504.]

IN the investigation of the new metal cæsium (Cs), which stands
nearest potassium, Bunsen[*] has found that, besides cæsium,
there exist another metal previously unknown, and which seems
to resemble potassium as closely as does cæsium.

The platinum salt of cæsium is more difficultly soluble in water
than that of potassium. On trying to separate the latter from
the former by repeated boilings with water, in proportion as the
quantity of potassium decreases, the continuous potassium spectrum between Kα and Kβ becomes fainter, and new lines appear,
more especially two very intense ones in the violet between Srδ
and Kβ. A limit is soon reached at which the quantity of potassium cannot be further diminished. This is the case when the
sum of the atomic weights of the metals, combined with platinum
and chlorine, has reached 109 (H=1). If from the platinum
compound thus obtained the mixture of the hydrated oxides of
potassium and cæsium is prepared, and if about the fifth part of
this is converted into carbonate, absolute alcohol will extract
from the dried mixture of the salts principally the hydrated
oxide of cæsium. If this operation be repeated, a limit is ultimately attained at which the part dissolved in alcohol has a
constant composition. This limit is reached when the atomic
weight has risen from 109 to 123·4. The substance which has
this enormous weight (next to gold and iodine the highest
known) forms a deliquescent hydrate, as caustic as hydrate of
potass. It also forms a deliquescent, strongly alkaline carbonate, of which about 10 parts are soluble in 100 parts of absolute alcohol at the ordinary temperature, and an anhydrous
nitrate, which, unlike nitre, is not rhombic but is hexagonal,
and by a hemihedral form is isomorphous with nitrate of soda,
&c. The spectrum of the substances purified up to the atomic
weight 123·4, shows the blue cæsium lines in the most brilliant
lustre, but the violet lines of the unpurified mixture (of the
atomic weight 109) in so feeble a degree, that a small addition
of chloride of potassium, which is almost without perceptible
action on the lines Csα, causes them entirely to disappear in
consequence of the brightness of the ground produced by potassium. The material for this investigation, only amounting to a
few grammes, was prepared from 44,000 kilogrammes of Dürckheim mineral water. On repeating the preparation from

[*] *Bericht der Akad. der Wissenschaften zu Berlin,* 1861.

about 150 kilogrammes of Saxon lepidolite, a product was obtained on the first treatment with bichloride of platinum which showed the violet lines between Srδ and Kβ in the most intense manner, but not a trace of the lines Csα. If this platinum double salt obtained from lepidolite had been a mixture, the blue line Csα must have been visible, together with the violet ones; for with the product obtained from the Dürckheim mineral water, on increasing the quantity of chloride of potassium the violet lines always disappear first, but the cæsium lines much later, and indeed only with a great excess of the potassium salt. Hence, besides potassium, sodium, lithium, and cæsium there must be a fifth alkali metal, which is present in small quantities in Dürckheim, Kreuznach and other similar mineral springs, but in lepidolite in larger quantity.

M. Ste.-Claire Deville and Troost, in continuation of their investigation on the reproduction of the natural minerals, have described the preparation of some sulphurets*.

Sulphuret of zinc is very easily prepared by melting together equal parts of sulphate of zinc, of fluoride of calcium, and of sulphide of barium. A fusible gangue of sulphate of baryta and of fluoride of calcium is obtained, in which are found beautiful crystals of sulphuret of zinc, either imbedded or arranged in geodes. Analysis proved them to be identical in composition with the natural blende; but they have an entirely different form. Instead of belonging to the monometric system, they crystallize in a regular double hexagonal prism, which is precisely the form of the crystals of sulphuret of cadmium. This singular observation supplies a link in the analogies of sulphur and cadmium, and establishes the dimorphism of sulphuret of zinc. It has received a timely confirmation in a discovery which Friedel has made† of the existence of a natural sulphide of zinc which crystallizes in the hexagonal system. On examining an argentiferous antimonio-sulphide of lead, he found imbedded certain crystals which had all the chemical reactions and composition of ordinary zinc blende, but was entirely different in crystalline form. The crystals consisted of a double hexagonal pyramid, with occasionally the faces of the hexagonal prism. These faces are strongly striated parallel to the base, and the angle between the adjacent faces of the pyramid was found to be about 129°, which is very near that of one of the pyramids of Greenockite. It has four easy cleavages parallel to the base and to the faces of the hexagonal prism. Its action on polarized light further establishes the crystalline form of the

* *Comptes Rendus,* May 6, 1861.   † Ibid. May 13.

substance. Friedel proposes to give the name *Wurtzite* to this dimorphous variety of zinc blende.

Deville and Troost have also obtained this hexagonal blende by a kind of sublimation. Some sulphuret of zinc placed in trays in a porcelain tube, was heated to bright redness in a current of hydrogen. No hydrogen was absorbed, and no trace of sulphuretted hydrogen produced. Notwithstanding this the sulphuret of zinc appeared as if volatilized, and was removed to the cooler parts of the tube in the form of transparent crystals of the greatest regularity. Hexagonal blende had been formed, as was seen by a powerful action on polarized light. The reaction had doubtless taken place in the following manner. The sulphuret of zinc at a red heat had been reduced by hydrogen, forming a mixture of zinc vapour and sulphuretted hydrogen. Arrived slowly in the cooler parts of the tube, an inverse reaction occurred; the zinc again took up sulphur to form hexagonal blende, and hydrogen became free. It served as mineralizing agent; and the native sulphide may have been formed in the same manner. That the volatilization of zinc was only apparent, was proved by heating sulphide of zinc to a very high temperature in a current of sulphuretted hydrogen. No trace of sublimation was obtained in the porcelain tube.

A number of crystallized oxides may be obtained by heating in a platinum crucible a mixture of the sulphates of these oxides and of alkaline sulphates. The oxide thus liberated at a very high temperature in melted sulphate of potass or soda crystallizes. Debray, who had previously obtained glucina in this way, has also succeeded in preparing magnesia (periclase) and oxide of nickel [*]. With sulphate of manganese pretty large crystals of red oxide of manganese, $Mn^3 O^4$, were obtained, but they were so interlaced that it was impossible to measure their angles with sufficient accuracy to be enabled to identify them with Hausmannite. They have the same composition and hardness; the colour of their powder is the same, but the artificial crystals are transparent.

Alumina, magnetic oxide of iron, and green oxide of uranium may be obtained by an analogous method, based on the decomposition of certain phosphates by alkaline sulphates at a high temperature.

Gorup-Besanez [†] recommends the use of ozone for cleaning and restoring the colour of old spotted and soiled books and prints.

[*] *Comptes Rendus,* May 13, 1861.
[†] *Liebig's Annalen,* May 1861.

Ozone completely removes writing ink; but printing ink is not attacked by it, at any rate to no perceptible extent: grease spots and mineral colours also remain unchanged, but vegetable colours are completely removed. The method used is as follows:—The air in a sulphuric-acid carboy is ozonized by Schönbein's method, which consists in placing in it a piece of phosphorus 8 inches long and $\frac{1}{2}$ an inch thick, and pouring into the carboy as much water at $30°$ C. as will half cover the phosphorus; the carboy is loosely corked and allowed to stand in a moderately warm place until the air is charged with ozone, which generally requires from twelve to eighteen hours. Without removing the phosphorus and water, the article to be bleached is uniformly moistened with distilled water, and after being rolled up is suspended by a platinum wire in about the centre of the carboy. The roll of paper is soon seen to be continually surrounded by the column of vapour rising from the surface of the phosphorus. The time required for the bleaching depends on the nature of the substance, but never requires more than three days; paper brown with age and coloured with coffee spots, in two days was quite white and clean. If the paper were now dried, it would not only be very brittle, but would also rapidly become brown; hence the acid must be completely removed. The paper is immersed in water, which is frequently renewed, until it only gives a very feeble acid reaction with litmus. It is next placed in water to which a few drops of solution of soda have been added, and then, being spread on a piece of glass and placed in an inclined position, is exposed to a thin stream of water for twenty-four hours. After being allowed to stand till nearly dry, it is carefully removed and dried between blotting-paper.

Gorup-Besanez found that ozone was not well adapted for cleaning oil colours.

Pohl has communicated* a research on the white gunpowder invented by Augendre, which consists of prussiate of potash, white sugar, and chlorate of potash. Pohl finds that the following mixture gave a very good burning powder:—

> Prussiate of potash 28 parts.
> Sugar . . . . 23 „
> Chlorate of potash 49 „
> ———
> 100

which is very nearly in the relation

$$K^2 Cfy, 3HO + C^{12} H^{11} O^{11} + 3KO ClO^5.$$

Of the products formed by the combustion of this powder, it

---

* *Sitzungsberichte der Wiener Akademie,* vol. xli.

would be difficult to state any thing with accuracy without very numerous analyses, and they would differ according as the combustion was free or in a closed space, and whether it was slow or rapid. Assuming that the possible products of decomposition of the ferrocyanide are nitrogen, cyanide of potassium, and a carbide of iron of the formula FeC^2, the decomposition might be represented in the following manner:—

$$K^2 Fe Cy^3 + C^{12} H^{11} O^{11} + 3(KO ClO^5) = 2K Cy + 3K Cl + Fe C^2.$$
$$+ N + 6CO + 6CO^2 + 14HO;$$

according to which 100 parts of the powder would yield 52·56 parts of non-volatile, and 47·44 of gaseous substances.

The decomposition may take place in conformity with other reactions, but from a preliminary investigation this appears the most probable.

In accordance with this, 100 parts of the powder would yield—

Nitrogen	1·86
Carbonic oxide	11·19
Carbonic acid	17·59
Water	16·79
	47·43

and

Cyanide of potassium	17·88
Chloride of potassium	29·84
Carbide of iron	5·33
	52·55

hence, reduced to volumes at 0° and 760 millims., 100 parts would yield—

Nitrogen	1927	cub. centims.
Carbonic oxide	8943	,,
Carbonic acid	8943	,,
Aqueous vapour	20867	,,
	40680	,,

Pohl calculates from this that the quantity of heat furnished by the combustion of this substance would be equal to 506·3 thermal units. The temperature of the combustion, is obtained by dividing the number of thermal units by the sum of the specific heats of the products of combustion, which amounts to 0·2636; and this gives 1920° C. as the temperature. From Bunsen and Schischkoff's research[*], it appears that the heating effect of ordinary gunpowder is 619·5 thermal units, and that the temperature of its free combustion is 2993° C. It furnishes in

[*] Phil. Mag. vol. xv. p. 489.

100 parts 68·06 of solid residue, and 31·38 of gaseous products, corresponding to 19310 cubic centims. From these data the relation between the two substances is—

		Black powder.		White powder.
The quantity of gas	. . .	1	:	2·107
The temperature of flame	.	1	:	0·641
The residues	. . . . .	1	:	0·77

But for the respective temperatures of combustion the reduced volumes of gas would be for black powder 231411 cubic cen_tims., and for white powder 300798 cubic centims., and hence the quantities of gas would be as 1 : 1·13.

In the combustion in a confined space the temperature of the combustion would be altered, for there would be a great difference in the specific heat of the products of combustion. Hence the volume of gas, when reduced to the normal temperature and pressure, would vary. For white gunpowder, Pohl calculates the temperature of combustion in a closed space at 2604° C., and the volume of gas furnished by 100 parts at 431162 cubic centims. Under similar circumstances Bunsen and Schischkoff found that the temperature was 3340° C., and the volume of gas 258420 cubic centims.

Hence the relation between the products of combustion in confined space would be—

		Black powder.		White powder.
The temperature of the flames	. .	1	:	0·779
The volumes of gas	. . . . .	1	:	1·669

As the action of an explosive powder principally depends on the volume of the gases formed, for equal weights the new white powder would produce 1·67 times the action of the other. But for equal volumes of the powder the ratio would be different. Pohl found that a vessel which held 102·542 grms. of white powder, held 132·355 grms. of ordinary black powder. Hence the density of the new powder in reference to the other would be as 0·774 : 1, and the work performed by equal volumes would be as 1·292 : 1.

In order to produce the same effect on projectiles, in firing mines, &c., 60 parts by weight of the new, would be required for 100 parts of the old. The weights of the residues in the two cases are respectively 31·53 and 68 parts. Another advantage of the white powder is, that the temperature of the flame is much lower; a greater number of shots could be fired without heating the projectile too much. The new powder is more energetic in its action than the old, and in this respect stands nearest gun-cotton. It has the advantage over this substance of being cheaper and easier to prepare, and it can be kept for a long time without undergoing any change.

The new powder contains chlorate of potash; and this, in all substitutes for gunpowder of which it is a constituent, forms products of combustion at a high temperature, which attack the firearms.   If the decomposition of the white powder takes place in accordance with the equation already given, it is not easy to see why this evil is to be feared.   It could be most simply decided by firing off a certain number of shots with a given weapon. Another advantage of the new powder is its difficult explosibility by pressure and percussion.   Explosion is only produced by the heaviest blow of iron upon iron; it is not produced by the friction of wood upon metal, or between stones, &c.   The new powder is also far easier of preparation than the old; and if the raw materials are at hand, a large quantity of it may be prepared in a few hours with no other apparatus than a stamping-mill and mixing tub.

The following observations have been made by Ste.-Claire Deville* on the influence exerted by the sides of certain vessels on the motion and composition of gases traversing them.

In laboratories, earthen and stoneware vessels are often used for reaction with gases at a high temperature.   They are suitable for most gases; but they are permeable to hydrogen, and they absorb water.

1. A rapid current from hydrogen is passed through one of these tubes.   The tube is closed by two corks, in which are fitted glass tubes; one of these tubes admits hydrogen; the other, which dips in water, serves for the escape of the gas.   On closing the stopcock by which hydrogen enters, not only does the gas cease to be liberated, but the water rises to a height of 60 to 70 centimetres above its level, as if the hydrogen had been drawn into the interior of the apparatus.   With coal-gas the aspiration is less, and appears to depend on the density of the gas; and there is none at all in the case of carbonic acid.

2. If the air be passed more slowly into the interior of the tube, but still more rapidly than in the majority of operations, the gas collected in the trough is no longer hydrogen, but pure air.

3. If an earthen tube be made red-hot in a furnace, and a current of hydrogen be passed through it, a mixture of carbonic acid and nitrogen (and also sulphurous acid if the combustible contains pyrites) is obtained—that is, the gases of combustion by which the tube is surrounded.   The experiment succeeds

* *Comptes Rendus*, March 1861.

even when the gases in the interior are under a pressure of 7 to 8 centimetres of mercury.

4. This experiment may be made more striking by the following method:—the earthen tube is enclosed in a larger glass tube, and carbonic acid is passed into the annular space between the two tubes, while hydrogen traverses the earthen tube. The two gases emerge by two distinct delivery tubes. One of the two currents of gas is inflammable, and it is precisely that which proceeds from the end of the apparatus communicating with the source of carbonic acid. The gases change their places during this short and rapid passage.

X. *On the Principles of Energetics.* Part II. *Molecular Mechanics.* By J. S. STUART GLENNIE, *M.A., F.R.A.S.*

[Concluded from vol, xxi. p. 358.]

31. THE misconception of my theory of material forces displayed in the remarks of Professor Challis *, seems to require a brief and, I trust, clearer restatement of the proposed first principle of Molecular Mechanics, along with its second and third principles, before proceeding to their application and development.

32. (I.) Matter is conceived as made up, not of an elastic æther and inelastic atoms, but of elastic molecules of different orders as to size and density.

If a rough physical conception of these molecules be required, they may be conceived as æthereal nuclei, the æther of the nuclei of a lower being made up of nuclei of a higher order, and so on *ad infinitum.*

I shall, perhaps, best defend this principle by restating the *experimental* objections to that for which it is substituted:—(1) We are led by experiment to conclude that *all* matter is elastic, and hence we are not justified in assuming two kinds of matter, an elastic and inelastic. (2) Not to speak of the inconsistency of denying to the atoms the elasticity which is attributed to the æther, which must be made up of atoms, it is impossible to conceive that, from any arrangement of inelastic atoms, the elasticity of the bodies they constitute should arise; though it is at once admissible that *degrees* of elasticity may depend on the arrangement of elastic atoms. (3) The action of an elastic æther, or anything else, on an "absolutely hard, ultimate atom" is experimentally inconceivable; for all known action of one body on another implies motion of the particles of that other body. If a body is struck, it is heated, and moves; if it moves little, it is

* Phil, Mag. p. 504, June 1861,

much heated. (4) The analogy "to the production of secondary waves, when a small obstacle is encountered by primary waves, propagated on the surface of water," is fallacious. For all such obstacles are elastic, and it seems a prodigious assumption to imagine that unknown inelastic, would have the same effect as known elastic bodies. (5) Not only the generation of secondary waves, but the origination of primary waves is experimentally inconceivable. For the hypothesis is—inelastic atoms of the same inertia in a *uniform* æther; and the experimental condition of motion is—*difference* of pressure.

33. (II.) Physical phenomena are to be explained from the conception of motions of different orders of molecules.

This conception has been forced on me by the impossibility of reconciling the notion of Electricity, Light, Heat, Actinism, &c. as states of molecular strain or motion, with the experimental facts of their coexistence, and mutual modification.

34. (III.) Chemical phenomena are to be explained from the conception of systems of different orders of molecules in dynamical equilibrium.

Bodies are thus conceived as systems of molecular motion; their sensible differences as dependent on differences of the orders, and motions, of their constituent molecules; their permanence as dependent on the continuance of dynamic equilibrium, that is, of the same state of motion at every point; and dissolution, combination, or the formation of new bodies as the result of difference between two or more systems of molecular motion, in mediate or immediate contact. As an illustration of this and the preceding principle, take the explanation afforded of Dr. Tyndall's discovery of the greater absorption and radiation of compound gases*. A compound gas will, in this theory, be conceived as a system, the moving molecules of which are of a relatively low order. Now it is clear that, suppose, for instance, the molecules, whose motions determine the chemical character of the gas, are of order (1), and that the molecules, whose vibrations give the sensation of heat, are of order (6), there are the motions of five, instead of, as in a simple gas, a smaller number of orders of molecules to be affected. Hence, degree of absorption of motion is seen to depend on the number of motions to be affected.

But I must reserve a fuller explanation of this and connected phenomena to its proper place in the development of the theory; for my object in these papers has merely been to give a general introductory statement and explanation of the Principles of Energetics, forming the basis of Ordinary Mecha-

---

* Bakerian Lecture, 1860,

nics (Stereatics and Hydratics) and Molecular Mechanics (Physics and Chemics).

.35. As to Analytical Investigations on the basis of these principles, there does not seem to be any peculiar difficulty raised when any separate order of molecules is considered. But that there will be found great analytical difficulty in passing from a higher to a lower order of molecules, and in expressing the correlations of their motions, is not to be concealed. And that little help is to be found in existing analytical investigations, except, perhaps, those relative to the conduction of heat in crystalline media, is not to be wondered at, seeing how recent is the conception of the Correlation of Forces.

6 Stone Buildings, Lincoln's Inn,
　25th June, 1861.

XI. *Proceedings of Learned Societies.*

ROYAL SOCIETY.

[Continued from vol. xxi. p. 536.]

November 22, 1860.—Major-General Sabine, R.A., Treasurer and Vice-President, in the Chair.

THE following communications were read :—
　"On Boric Ethide." By Edward Frankland, Ph.D., F.R.S., and B. Duppa, Esq. Received July 7, 1860.

When zincethyle in excess is brought into contact with tribasic boracic ether, $\left(B \begin{cases} C_4 H_5 O_2 \\ C_4 H_5 O_2 \\ C_4 H_5 O_2 \end{cases} \right)$, the temperature of the mixture gradually rises for about half an hour. If it be now submitted to distillation, it begins to boil at 94° C., and between this temperature and 140° a considerable quantity of a colourless liquid distils over. The distillation then suddenly stops, the thermometer rises rapidly, and, to avoid secondary products of decomposition, the operation should now be interrupted. The materials remaining in the retort solidify, on cooling, into a mass of large crystals, which are a compound of ethylate of zinc with zincethyle. On rectification, the distillate began to boil at 70°, but the thermometer rapidly rose to 95°, at which temperature the last two-thirds of the liquid passed over and were received apart. The product thus collected exhibited a constant boiling-point on redistillation. Submitted to analysis, it yielded results agreeing with the formula

$$B \begin{cases} C_4 H_5 \\ C_4 H_5 \\ C_4 H_5 \end{cases}$$

This body, for which we propose the name *boric ethide*, is produced by the following reaction :—

$$2B\begin{cases}C_4H_5O_2\\C_4H_5O_2\\C_4H_5O_2\end{cases}+3Zn_2\begin{cases}C_4H_5\\C_4H_5\end{cases}=2B\begin{cases}C_4H_5\\C_4H_5\\C_4H_5\end{cases}+6\,{}^{C_4H_5}_{Zn}\Big\}O_2$$

<u>Boracic ether.</u>   <u>Zincethyle.</u>   <u>Boric ethide.</u>   <u>Ethylate of zinc.</u>

The ethylate of zinc thus formed combines with zincethyle to form the crystalline compound above alluded to.

Boric ethide possesses the following properties :—It is a colourless mobile liquid of a pungent odour; its vapour is very irritating to the mucous membrane, and provokes a copious flow of tears. The specific gravity of boric ethide at 23° C. is ·6961; it boils at 95° C., and the results of several determinations of its vapour-density give the number 3·4006. The calculated vapour-density of boric ethide, volumetrically composed like terchloride of boron, is 3·3824.

Boric ethide is insoluble in water, and is very slowly decomposed by prolonged contact with it. Iodine has scarcely any action upon it, even at 100° C. It floats upon concentrated nitric acid for several minutes without change; but suddenly a violent oxidation takes place, and crystals of boracic acid separate. When boric ethide vapour comes in contact with the air it produces slight bluish-white fumes, which have a high temperature. The liquid is spontaneously inflammable in air, burning with a beautiful green and somewhat fuliginous flame. In contact with pure oxygen it explodes. Placed in a flask and allowed to oxidize gradually, first in dry air and finally in dry oxygen, it forms a colourless liquid, which boils at a higher temperature than boric ethide, but cannot be distilled under atmospheric pressure without partial decomposition. In a stream of dry carbonic acid this product of oxidation evaporates without residue. By distillation *in vacuo* it is obtained pure, and it then exhibits a composition expressed by the formula

$$B\begin{cases}C_4H_5\\C_4H_5O_2\\C_4H_5O_2\end{cases}$$

The product of the oxidation of boric ethide is therefore the *diethylate* of a body which may be conveniently named *boric dioxyethide,*

$\left(B\begin{cases}C_4H_5\\O\\O\end{cases}\right)$. The formation of diethylate of boric dioxyethide from boric ethide may be thus represented :

$$B\begin{cases}C_4H_5\\C_4H_5\\C_4H_5\end{cases}+O_4=B\begin{cases}C_4H_5\\C_4H_5O_2\\C_4H_5O_2\end{cases}$$

<u>Boric ethide.</u>   <u>Diethylate of boric dioxyethide.</u>

This compound dissolves instantly in water, and is resolved into alcohol and a volatile white crystalline body, which may be sublimed without change, at a gentle heat, in a stream of carbonic acid, and then condenses in magnificent crystalline plates like naphthaline.

The analytical results yielded by this body agree closely with the formula

$$B \left\{ \begin{matrix} C_4 H_5 \\ H O_2 \\ H O_3 \end{matrix} \right.$$

It is therefore obviously produced by the substitution of two atoms of hydrogen for two of ethyle in diethylate of boric dioxyethide:

$$B \left\{ \begin{matrix} C_4 H_5 \\ C_4 H_5 O_2 \\ C_4 H_5 O_3 \end{matrix} \right. + 2 \left. \begin{matrix} H \\ H \end{matrix} \right\} O_2 = B \left\{ \begin{matrix} C_4 H_5 \\ HO_2 \\ HO_3 \end{matrix} \right. + 2 \left. \begin{matrix} C_4 H_5 \\ H \end{matrix} \right\} O_2.$$

Diethylate of boric   Dihydrate of boric   Alcohol.
dioxyethide.   dioxyethide.

Dihydrate of boric dioxyethide possesses an agreeable etherial odour and a most intensely sweet taste. Exposed to the air it evaporates slowly at ordinary temperatures, undergoing at the same time partial decomposition, and invariably leaving a slight residue of boracic acid. Its vapour tastes intensely sweet. It reddens litmus paper, although in other respects its acid qualities are very obscure. It is very soluble in water, alcohol, and ether. Exposed to a gentle heat it fuses, and at a higher temperature boils with partial decomposition.

We are at present engaged with the further study of these bodies, and with the corresponding reactions of zincethyle upon silicic, carbonic, and oxalic ethers.

"On Fermat's Theorems of the Polygonal Numbers." First Communication. By the Right Hon. Sir Frederick Pollock, F.R.S., Lord Chief Baron. Received July 11, 1860.

"On Cyanide of Ethylene and Succinic Acid." Preliminary Notice. By Maxwell Simpson, Ph.D. Received August 1, 1860.

Succinic acid bears the same relation to the diatomic alcohol (glycol) that propionic acid bears to ordinary alcohol. Propionic acid can be obtained by treating the cyanide of the alcohol radical with potash. Can succinic acid be obtained by treating the cyanide of the glycol radical with the same reagent, or is it an isomeric acid that is formed under these circumstances?

$$C_4 H_5, Cy + O_2 \left\{ \begin{matrix} K \\ H \end{matrix} \right. + 2HO = O_2 \left\{ \begin{matrix} C_6 H_5 O_3 \\ K \end{matrix} \right. + NH_3.$$

Cyanide of ethyle.   Propionate of potash.

$$C_4 H_4 2Cy + 2 \left(O_2 \left\{ \begin{matrix} K \\ H \end{matrix} \right. \right) + 4HO = O_4 \left\{ \begin{matrix} C_6 H_4 O_4'' \\ K_2 \end{matrix} \right. + 2NH_3.$$

Cyanide of ethylene.   Succinate of potash?

The following experiments were performed with the view of determining this point:—

Preparation of Cyanide of Ethylene.—As a preliminary step to the formation of succinic acid in this way, it became of course necessary to prepare the cyanide of ethylene. This body I obtained by submitting bromide of ethylene to the action of cyanide of potassium.

The process was thus conducted:—A mixture of two equivalents of the cyanide and one of the bromide was introduced into a large balloon, together with a considerable quantity of alcohol, sp. gr. ·840, and exposed to the temperature of a water-bath, a Liebig's condenser having been previously attached to the balloon in such a manner as to prevent the alcohol from distilling off the reacting ingredients. As soon as all the cyanide of potassium had been converted into bromide, the alcohol was separated and distilled. A semifluid residue was thus obtained, which was filtered at the temperature of 100° Cent. On treating the filtrate with a saturated solution of chloride of calcium, a reddish oil rose to the surface, which was well washed with ether, and exposed for some time to the temperature of 140°, in order to remove any bromide of ethylene that might have escaped the solvent action of the ether. This body proved, on analysis, to be cyanide of ethylene. It was not, however, quite pure. There are difficulties attending its complete purification which I have not yet overcome.

At the temperature of the air, cyanide of ethylene is a semisolid crystalline mass of a brownish colour. It melts under 50° Cent. It is very soluble in water and alcohol, and sparingly soluble in ether. It cannot be distilled. Nevertheless it bears a tolerably high temperature without suffering much decomposition. Heated with an alcoholic solution of potash, it gives off ammonia. Treated with nitric acid, it forms a body which crystallizes from alcohol in long needles. This and some other reactions I am at present engaged in studying.

Preparation of Succinic Acid.—Bromide of ethylene and cyanide of potassium were made to react upon each other in the same manner as in the preparation of the cyanide of ethylene. As soon as the reaction was complete, the alcohol was separated from the bromide of potassium, some sticks of caustic potash were added to it, and the whole heated for several days by means of a water-bath. Torrents of ammonia were given off on applying the heat. As soon as the evolution of this gas had ceased, the alcohol was distilled off and the residue treated with a considerable excess of hydrochloric acid. This was then heated gently as long as acid vapours continued to be evolved, digested with absolute alcohol, and filtered, and then the filtrate was evaporated to dryness. The dry mass thus obtained was treated several times with alcohol in a similar manner. The result of these repeated digestions was then dissolved in water, and a few drops of a solution of nitrate of silver were added to it, which occasioned a slight precipitate of chloride of silver. This was separated by filtration, and the filtrate was exactly neutralized with ammonia. On adding excess of nitrate of silver to this, an abundant white precipitate was obtained, very soluble in nitric acid and ammonia. This gave, on analysis, numbers agreeing very well with the composition of succinate of silver. The acid itself possessed also all the properties of succinic acid. It sublimed on the application of heat, was soluble in water, alcohol, and ether, and gave, when neutralized, a reddish-brown precipitate with per-chloride of iron. Moreover, on digesting this precipitate with ammonia, an acid could be detected in the filtered liquor, which gave white precipitates with nitrate of silver, and with a mixture of chloride of barium and alcohol.

Succinic acid *can* then be obtained from glycol in the same manner as propionic acid from ordinary alcohol; the bromide of ethylene, the point from which I started, being capable of derivation from the diatomic alcohol.

I propose extending this investigation to some other hydrocarbons of the series $C_n H_n$, with the view of ascertaining whether or not the homologues of succinic acid can be obtained from these bodies by a similar process.

"Results of Researches on the Electric Function of the Torpedo." By Professor Carlo Matteucci of Pisa. In a Letter to Dr. Sharpey, Sec. R.S. Received August 3, 1860.

"It has hitherto been believed that the action of the electric organ of the Torpedo was momentary only;—that it becomes charged, under the influence of nervous action and discharged immediately that action ceases, somewhat like soft iron under the influence of an electric current. Such, however, is not the real state of the case. The electric organ is always charged. It may be conclusively shown by experiment that the action of that organ never ceases, and that round the body of a Torpedo, and probably of every other electric fish, there is a continual circulation of electricity in the liquid medium in which the animal is immersed. In fact, when the electric organ, or even a fragment of it, is removed from the living fish and placed between the ends of a galvanometer, the needle remains deflected at a constant angle for twenty or thirty hours, or even longer.

"I must here explain that in electro-physiological experiments it is highly advantageous to employ, as extremities of the galvanometer, plates of amalgamated zinc immersed in a neutral saturated solution of sulphate of zinc. This arrangement, which can be worked with the greatest facility, gives a perfectly homogeneous circuit, leaving the needle at zero in an instrument of 24,000 coils; the liquid in contact with the animal part experimented on has the greatest possible conductibility while it does not act chemically on the tissue, and the apparatus is entirely free from secondary polarity.

"To return to the Torpedo. The electric organ, or a portion of it, detached from the fish and kept at the temperature of freezing, preserves its electromotive properties for four, six, or even eight days; and an organ which has been kept for twenty-four hours in a vessel surrounded with a frigorific mixture of ice and salt, is found to possess an electromotive power as great as that of the organ recently detached from the living fish. Thus the electric organ retains its functional activity long after both muscular and nervous excitability have been extinguished.

"What then is the action of the nerves on this apparatus? Here again experiment affords a very distinct and conclusive answer. Detach the organ of a live torpedo and cut it into two equal portions, in such a way as to leave each half in connexion with one of the large nervous trunks; place the two halves on a plate of gutta percha, with electric couples opposed; that is, with the similar surfaces (say the dorsal) in contact; and connect the two free (ventral) surfaces with the extremities of the galvanometer. There will usually be no deflection of the needle, or, at most, a very slight effect which will soon disappear. Now, after having opened the circuit of the galvanometer, irritate the nerve of one of the segments, by pinching, by the interrupted electric current, or in any other way; or prick the piece itself with a needle. The portion of organ thus stimulated will give several discharges in succession, and a rheoscopic frog's limb with its nerve applied to the part will each time be thrown into violent convulsions. If, after this, the galvanometer be applied as before, there will be a very strong deflection in a direction answering to the segment stimulated. This deviation endures for a short time, but gradually becomes less, so that in a few minutes the effect of the two segments is equal. Stimulation now of the other segment will in like manner render its electricity predominant. These alternations may be repeated several times, but naturally the effect becomes less and less marked.

"Thus the electromotive apparatus becomes charged and acts independently of the influence of the nerves, but that influence renews and renders persistent the activity of the apparatus. We know, moreover, that the discharge, which is only a state of temporary increased activity of the organ, is brought on by an act of the will in the live animal, or by the excitation of the nerves of the organ.

"I shall not enter now into further details respecting my recent experiments on the Torpedo, but I venture to think that we have really made a step towards clearing up the theory of the animal electromotive apparatus. The organ of the Torpedo does not, under the influence of the nerves, act as an induction apparatus; the operation seems more analogous to that of a 'secondary pile,' created, through the influence of the nerves, in each constituent cell of the organ.

"The case is very different in muscular action, the changes occurring in which are better understood now that we know the phenomena of muscular respiration. I do not here refer to the variation of the muscular current which takes place at the moment of contraction. In that case it would appear from experiment, as I lately showed, that there are indications of a current in an opposite direction; but the conditions of the animal structure in action are so complex that no inference can be drawn as to the intimate nature of the phenomenon. It is otherwise, however, in comparing muscles which have been left at rest with muscles which have been fatigued by repeated contraction. Being still engaged in the investigation of this matter, I shall content myself now with mentioning one result of my inquiry, which I consider as well established; the result, in fact, of performing on muscles the same kind of experiment as the one above described on the organ of the Torpedo. The experiment is as follows:—Having selected a series of muscles, entire or divided, which have been proved (by my method of opposed muscular piles) to be equal in electromotive power ; subject a certain number of them to repeated stimulation, and then, by means of the method of opposed couples, compare the muscles which have been exercised with those which have been left at rest, and it will be found that the latter will manifest a much greater degree of electromotive power than the former. The nervous excitation, which causes muscular contraction, developes heat, generates mechanical force. and consumes chemical affinity ; and as the electromotive apparatus of muscle operates through means of that affinity, it must get weakened, like a pile in which the acid has become weaker. In the Torpedo, on the other hand, there is neither heat nor mechanical force produced, and the electromotive apparatus is set up again, as it were, through the influence of the nerves, after the manner of a secondary pile."

"Natural History of the Purple of the Ancients." By M. Lucase Duthiers, Professor of Zoology in the Faculty of Sciences of Lille. Received March 22, 1860.

"Contributions towards the History of Azobenzol and Benzidine." By P. W. Hofmann, Ph.D.   Received July 24, 1860.

Among the numerous compounds into which benzol, when submitted to reagents, is converted, *azobenzol* and its derivatives have as yet received but limited attention. Although more than twenty-five years have elapsed since this interesting body was discovered by Mitscherlich, both its formation and its constitution remain still doubtful.

Mitscherlich [*], who discovered azobenzol in 1834, when submitting nitrobenzol to the action of an alcoholic solution of potassa, represented this compound by the formula

$$C_6 H_5 N †,$$

but left the reaction which gives rise to the formation of azobenzol unexplained. In 1845 this body was reprepared by Hofmann and Muspratt[‡], who observed among the collateral products of the reaction *aniline* and *oxalic acid*. They represent the formation of azobenzol by the equation

$$2C_6 H_5 NO_2 + C_2 H_6 O = C_6 H_5 N + C_6 H_7 N + C_2 H_2 O_4 + H_2 O,$$

Nitrobenzol.   Alcohol.   Azobenzol. Phenylamine. Oxalic acid.

adding at the same time that they are far from considering this equation as more than the representation of *one* phase of the transformation of nitrobenzol, since several other rather indefinite compounds or products are formed simultaneously.

At about the same period Zinin made the interesting observation that azobenzol is capable of fixing hydrogen and of being thereby

[*] Pogg. Ann. xxxii. p. 224.
[†] H = 1, O = 16, C = 12, &c.
[‡] Mem. of the Chem. Soc. vol. iii. p. 113.

converted into a well-defined base, benzidine, which he represented by the formula

$$C_6 H_4 N.$$

Considering the physical characters both of azobenzol and of benzidine, especially the high boiling-points of these substances, and the ratio of hydrogen and nitrogen in the latter compound, the sum of the number of equivalents of these two elements not being divisible by 2, many chemists were inclined to double the formulæ of both bodies, and to represent them by the following expressions :—

$$\text{Azobenzol} \ldots \ldots \ C_{12} H_{10} N_2$$
$$\text{Benzidine} \ldots \ldots \ C_{12} H_{12} N_2$$

This view received the first experimental confirmation in the formation of the nitro-derivatives of azobenzol, which were examined in 1849 by Gerhardt and Laurent. The formation of

$$\text{Nitrazobenzol} \ldots \ C_{12} H_9 N_2 O_2 = C_{12} (H_9 \ NO_2) \ N_2, \ \text{of}$$
$$\text{Dinitrazobenzol} \ldots \ C_{12} H_8 N_4 O_4 = C_{12} \big[H_8 (NO_2)_2 \big] N_2,$$

and of several derivatives of these bodies, having established the C_{12}-formula of azobenzol, but little doubt could be entertained regarding the formula of benzidine, which is as readily obtained from azobenzol by reducing agents, as it may be reconverted into azobenzol by nitric acid[*].

The molecular value of benzidine being thus almost exclusively fixed by the determination of the formula of the compound from which it originates, it was of some interest to obtain additional experimental evidence for the molecular weight of azobenzol.

With this view I have determined the vapour-density of azobenzol. This body boiling at a rather high temperature, I have availed myself of the method of displacement lately proposed by Professor Hofmann. Experiment proved the density of the azobenzol-vapour to be 94 referred to hydrogen as unity, or 6·50 referred to air. The theoretical vapour-density of azobenzol, assuming that one molecule of this compound furnishes, like the rest of well-examined substances, 2 vols. of vapour[†], is $\dfrac{182}{2} = 91$ referred to hydrogen, and 6·32 referred to air.

The determination of the vapour-density, then, plainly confirms the higher molecular weights proposed for azobenzol and for benzidine.

When determining the vapour-density of azobenzol, I had occasion to observe that, probably in consequence of a typographical error, the boiling-point of this compound is misstated in all the manuals which I could consult, and even in the original memoirs of Mitscherlich himself. The boiling-point is stated to be 193° C., whilst it is in reality 293° C.

Benzidine, when expressed by the formula

$$C_{12} H_{12} N_2,$$

---

[*] Noble, Journal of the Chem. Soc. vol. viii. p. 293.　　[†] $H_2 O = 2$ vols.

presents itself as a well-defined diacid diamine. The molecular construction of the diatomic base remained to be decided.

I have endeavoured to solve this problem by the process of ethylation, as yet the simplest and the best guide in determining questions of this kind. Benzidine in the presence of alcohol is rapidly attacked by iodide of ethyle. After two hours' digestion at 100° C. in sealed tubes, the reaction is complete. The solution on evaporation yields a crystalline iodide,

$$C_{16} H_{22} N_2 I_2 = C_{12} H_{12} (C_2 H_5)_2 N_2 I_2,$$

from which ammonia separates a solid crystalline base very similar to benzidine. This compound, which fuses at 65° C., and resolidifies at 60° C., is *diethylbenzidine* :

$$C_{16} H_{20} N_2 = C_{12} H_{10} (C_2 H_5)_2 N_2,$$

which forms well-crystallizable salts with the acids, and yields with dichloride of platinum a difficultly soluble crystalline platinum-salt containing

$$C_{16} H_{22} N_2 Cl_2 2PtCl_2.$$

When diethylbenzidine is treated again with iodide of ethyle, the phenomena previously observed repeat themselves. The iodide

$$C_{20} H_{30} N_2 I_2 = C_{12} H_{10} (C_2 H_5)_4 N_2 I_2$$

is formed, which when decomposed by ammonia yields *tetrethylbenzidine*

$$C_{20} H_{28} N_2 = C_{12} H_8 (C_2 H_5)_4 N_2.$$

Tetrethylbenzidine resembles the diethylated and the non-ethylated base. It fuses at 85° C., resolidifying at 80° C., produces with the acids crystalline compounds, and furnishes with dichloride of platinum a platinum-salt of the formula

$$C_{20} H_{30} N_2 Cl_2, 2PtCl_2.$$

The further action of iodide of ethyle upon tetrethylbenzidine is extremely slow. After 12 hours' digestion at 100° C. only a very minute quantity of the base had been transformed into an iodide. Iodide of methyle, on the other hand, acts with greate nergy. An hour's digestion is sufficient to produce the final diammonium-compound.

The iodide

$$C_{22} H_{34} N_2 I_2 = C_{12} H_8 (C_2 H_5)_4 (CH_3)_2 N_2 I_2$$

is very difficultly soluble in absolute alcohol, but dissolves with facility in boiling water, from which it is deposited on cooling, in long beautiful needles. The solution of this iodide is no longer precipitated by ammonia, but yields with oxide of silver a powerfully alkaline solution, exhibiting all the characters of the completely substituted ammonium- and diammonium-bases discovered by Professor Hofmann. The solution of this dimethyl-tetrethylated base, which contains

$$C_{22} H_{34} N_2 O_2 = \left. \begin{matrix} C_{12} H_8 (C_2 H_5)_4 (CH_3)_2 N_2 \\ H_2 \end{matrix} \right\} O_2,$$

is not further acted upon by either iodide of ethyle or methyle. With acids it forms a series of salts which are remarkable for the beauty with which they crystallize. The platinum-salt is almost insoluble in water, but soluble with difficulty in concentrated boiling hydrochloric acid, crystallizing from this solution on cooling in beautiful needles. This salt contains

$$C_{22} H_{34} N_2 Cl_2, 2 PtCl_2.$$

The above experiments appear to establish the molecular construction of benzidine in a satisfactory manner. This base is obviously a primary diamine, in which the molecular group $C_{12} H_8$, whatever its nature may be, functions as a diatomic radical. A glance at the subjoined Table exhibits the construction of benzidine and of the several compounds which I have described.

Diamines.

$$\text{Benzidine} \ldots \ldots \left. \begin{array}{l} (C_{12} H_8)'' \\ H_2 \\ H_2 \end{array} \right\} N_2,$$

$$\text{Diethylated benzidine} \ldots \ldots \left. \begin{array}{l} (C_{12} H_8)'' \\ (C_2 H_5)_2 \\ (H_2)_2 \end{array} \right\} N_2,$$

$$\text{Tetrethylated benzidine} \ldots \ldots \left. \begin{array}{l} (C_{12} H_8)'' \\ (C_2 H_5)_2 \\ (C_2 H_5)_2 \end{array} \right\} N_2.$$

Iodides of Diammoniums.

Primary $\ldots \ldots$	$[(C_{12} H_8)''$	H_4		$N_2]'' I_2,$
Secondary $\ldots \ldots$	$[(C_{12} H_8)''$	H_4	$(C_2 H_5)_2 N_2]'' I_2,$	
Tertiary $\ldots \ldots$	$[(C_{12} H_8)''$	H_2	$(C_2 H_5)_4 N_2]'' I_2,$	
Quartary $\ldots \ldots$	$[(C_{12} H_8)'' (CH_3)_2 (C_2 H_5)_4 N_2]'' I_2.$			

"On Bromphenylamine and Chlorphenylamine." By E. T. Mills. Received July 24, 1860.

Nitrophenylamine, when prepared from dinitrobenzol (*i. e.* by the indirect method), differs in so many respects from the isomeric base which is obtained from phenyle-compounds (*i. e.* by the direct method), that chemists have distinguished these two bodies as alpha- and beta-nitrophenylamine [*]—Bromphenylamine and chlorphenylamine

---

[*] The alpha-nitrophenylamine (nitraniline) was formed about sixteen years ago by Dr. Muspratt and myself (Chem. Soc. Mem. vol. iii. p. 112), by the action of reducing agents on dinitrobenzol. The beta-nitraniline was discovered by Arppe (Chem. Soc. vol. viii. p. 175), who obtained this compound when distilling pyrotartronitrophenylamide with potash. The two bases resemble each other in a remarkable manner; but there are differences in their physical and chemical characters which leave no doubt as to the fact of their having different constitutions. I may here remark that I have repeated Arppe's experiments, the results of which I can confirm in every particular. Since the phenyle-compound from which Arppe obtained his substance is accessible only with difficulty, I have endeavoured to nitronate a more easily procurable phenyle-compound. Acetyl-

have hitherto been produced only by the action of potash upon bromisatine and chorisatine, the indirect method, by which they were originally obtained by Dr. Hofmann; it appeared therefore of some interest to ascertain whether the bodies generated directly from compounds of phenylamine would exhibit differences in their properties similar to those which distinguish alpha- and beta-nitrophenylamine.

With the view of deciding this question experimentally, I have submitted acetylphenylamide to the action of bromine and chlorine, in the hope of thus forming directly from phenylamine the brominated and chlorinated compounds in question.

Action of Bromine on Acetylphenylamide.

A cold aqueous solution of acetylphenylamide, when agitated with bromine gradually added in small quantities until the yellow colour imparted to the liquid no longer disappears, furnishes a crystalline compound difficultly soluble in cold, but easily recrystallizable from boiling water. The substance consists chiefly of monobrominated acetylphenylamide

$$(C_6 H_6 Br N O = \left. \begin{array}{c} (C_6 H_4 Br) \\ (C_2 H_3 O) \\ H \end{array} \right\} N,$$

which is however invariably mixed with small quantities of dibrominated acetylphenylamide

$$(C_6 H_7 Br_2 N O = \left. \begin{array}{c} (C_6 H_3 Br_2) \\ (C_2 H_3 O) \\ H \end{array} \right\} N.$$

I have not been able to find a method of separating these two bodies perfectly.

The brominated compound is readily attacked by potash. On distilling the mixture, the vapour of water carries over a volatile

---

phenylamide may be used for this purpose with considerable advantage. A solution of the compound in cold fuming nitric acid yields, on the addition of water, a crystalline difficultly soluble precipitate, which is easily obtained pure by recrystallization. This substance contains

$$C_3 H_8 N_2 O_4 = \left. \begin{array}{c} [C_6 (H_4 NO_2)] \\ (C_2 H_3 O_2) \\ H \end{array} \right\} N,$$

and yields, when heated with potassa, the beta-nitrophenylamine of Arppe with all its properties. I may here recall a former observation, which has now become perfectly intelligible. When studying the action of nitric acid upon melaniline, I found (Chem. Soc. Mem. t. i. 305) that the dinitromelaniline, which is thus formed, essentially differs from the dinitromelaniline obtained by submitting nitrophenylamine (alpha-) to the action of chloride of cyanogen. The two nitrobases, which are both expressed by the formula

$$C_{13} H_{11} N_5 O_4 = C_{13} [H_{11} (NO_2)_2] N_5,$$

stand to each other in the same relation which obtains between alpha-nitrophenylamine and beta-nitrophenylamine. In fact, I have since found that the distillation of the nitro-base, obtained by treating alpha-nitrophenylamine with chloride of cyanogen, furnishes alpha-nitrophenylamine; whilst beta-nitrophenylamine may be detected amongst the products of the distillation of the dinitromelaniline which is formed directly from melaniline by means of nitric acid.—A. W. H.

body which solidifies in the condenser into beautiful acicular crystals, acetate of potassium remaining in the retort.

The solidified distillate was purified by recrystallization from boiling water, and submitted to analysis. Both the combustion of the base itself and the platinum-determination of the beautiful golden-yellow platinum-salt proved this body to be bromphenylamine

$$C_6 H_6 BrN = \left.\begin{array}{c} (C_6 H_4 Br) \\ H \\ H \end{array}\right\} N.$$

In its appearance, odour, and taste, as likewise in its deportment with acids and with solvents generally, the brominated base obtained from acetylbromophenylamide resembles perfectly the bromphenylamine produced from bromisatine, a specimen of which I obtained from Dr. Hofmann's collection. There is only one point in which a slight difference was observed. Both compounds are capable of crystallizing either in needles or in well-defined octahedra, the former being generally obtained from water, and the latter from alcohol. The bromphenylamine, obtained from the acetyle-compound, appears to be more inclined to crystallise in needles than in octahedra. Circumstances have prevented me from entering into an examination of the products of decomposition of the two bromphenylamines; and the question whether these two bodies are really identical, or similarly related as the two nitro-compounds, must be decided by further experiments[*].

Action of Chlorine on Acetylphenylamide.

The phenomena observed in the action of chlorine on a cold saturated solution of the phenyle-compound are perfectly similar to those presented in the corresponding reaction with bromine. A crystalline compound immediately separates from the solution; as soon as the crystals cease to augment, the current of chlorine is interrupted. Washed with cold, and once recrystallized from boiling water, the chlorinated body is found to be nearly perfectly pure monochlorinated acetylphenylamide

$$C_8 H_8 Cl N O = \left.\begin{array}{c} C_6 H_4 Cl \\ C_2 H_3 O \\ H \end{array}\right\} N,$$

which, when distilled with potash, furnishes abundance of chlorphenylamine, resembling in a marked manner the chlorphenylamine obtained by the action of potash upon chlorisatine.

"New Compounds produced by the substitution of Nitrogen for Hydrogen." By P. Griess, Esq. Received July 24, 1860.

In several previous notes I have called attention to a peculiar double acid which is formed by the action of nitrous acid upon amidobenzoic acid,

$$C_{14} H_{14} N_2 O_4 + H NO_2 = C_{14} H_{11} N_3 O_4 + 2 H_2 O \dagger,$$

[*] These experiments have since been made by Mr. P. Griess, whose results are given in the next abstract.—A. W. H.

$\dagger$ H = 1; O = 16; C = 12, &c.

the constitution of which, as far as my experiments go, may be represented by the formula

$$[C_7 (H_3 N_2') O][C_7 (H_4 H_2 N) O] \atop H_2 \Big\} O_2.$$

There are not less than three other compounds known which empirically may be represented by the same formula as amidobenzoic acid, viz. nitrotoluol, salicylamide, and anthranilic acid. The two former substances differ from amidobenzoic acid both physically and chemically in a marked manner; anthranilic acid, on the other hand, is so closely allied to the benzoic derivative, that special experiments were required to distinguish these two bodies. Gerland, when he submitted the two acids to Piria's well-known reaction, observed that both are converted by nitrous acid into non-nitrogenated acids, which, although still isomeric, essentially differ in their properties; amidobenzoic acid being transformed into a new acid,—oxybenzoic acid, whilst anthranilic acid yields salicylic acid.

It appeared of some interest to try whether the substitution of nitrogen for hydrogen in anthranilic acid would furnish a compound isomeric with the double acid obtained from amidobenzoic acid. A current of nitrous acid, when passed into a cold alcoholic solution of anthranilic acid, rapidly transforms this substance into a compound crystallizing in white prisms, which is easily obtained by allowing the alcohol to evaporate at the common temperature. The new body is extremely soluble in water, insoluble in ether. By analysis it was proved to contain

$$C_{14} H_9 N_3 O_7.$$

The new compound is thus seen to be far from isomeric with the derivative of amidobenzoic acid produced under similar circumstances, with which, in fact, it shows no analogy whatever. I have not yet arrived at a definite view regarding the molecular construction of this body; nevertheless its deportment with water shows even now that the nitrogen in it exists in two different forms. Gently heated with water, the new compound disengages torrents of nitrogen; on cooling, the liquid solidifies into a crystalline mass of salicylic acid, free nitric acid remaining in solution. This metamorphosis is represented by the equation

$$\underbrace{C_{14} H_9 N_3 O_7}_{\text{New body.}} + 2 H_2 O = N_4 + HNO_3 + \underbrace{2 C_7 H_6 O_3,}_{\text{Salicylic acid.}}$$

which has been controlled by quantitative experiments. The idea suggests itself to assume one-fifth of the nitrogen in the form of nitric acid, when the new body might be viewed as a salt-like compound of the formula

$$C_7 H_4 N_2' O_2 \atop C_7 H_4 N_2' O_2 \Big\} HNO_3;$$

the action of the water consisting simply in the replacement of the monatomic nitrogen by the elements of water, which would produce salicylic acid, nitric acid being liberated.

I avail myself of this opportunity of mentioning the deportment of several other isomeric bodies under the influence of nitrous acid. There are two basic compounds,

$$C_6 (H_4 NO_2) N,$$

known; the one is the alphaphenylamine of Hofmann and Muspratt, the other the betaphenylamine observed by Arppe. When submitted to the action of nitrous acid, these two isomeric bodies yield two perfectly different nitrogen-substituted derivatives. The substance obtained from alphaphenylamine (the base formed by the reduction of dinitrobenzol) has been already mentioned in one of my previous notes, the body derived from betaphenylamine is still under examination.

The action of nitrous acid proves that there are also two bromphenylamines similar to the two nitrophenylamines. The original bromphenylamine discovered by Hofmann, and which is formed by the distillation of bromisatine with hydrated potash, yields with nitrous acid a compound,

$$C_{12} H_8 Br_2 N_3 = \left. \begin{matrix} (C_6 H_4 Br)_2 \\ N''' \\ H \end{matrix} \right\} N_2,$$

crystallizing in beautiful golden-yellow needles, insoluble in water, and difficultly soluble in alcohol and ether. The bromphenylamine, on the other hand, which was lately prepared by Mills * from acetylbromphenylamide, exhibits with nitrous acid a perfectly different deportment, being transformed into a yellow scarcely crystalline compound, easily soluble in alcohol and ether, but insoluble in water. I have not as yet analysed this compound; its formation, however, and its properties render it probable that it will be found to be isomeric with the product of decomposition previously mentioned. I am engaged in a more minute examination of this compound, which I hope may assist in explaining the cause of the still enigmatical isomerism exhibited by the derivatives of phenylamine.

I have already repeatedly called attention to the different atomicity exhibited by nitrogen under different conditions. In the derivatives of amidobenzoic and of anthranilic acids, it can be proved that 1 equiv. of nitrogen replaces 1 equiv. of hydrogen; while in the derivatives of phenylamine, the nitrogen is present with the value of three molecules of hydrogen.

GEOLOGICAL SOCIETY.

[Continued from vol. xxi. p. 539.]

April 24, 1861.—Leonard Horner, Esq., President, in the Chair.

The following communications were read:—

1. "On the 'Symon Fault' in the Coalbrook Dale Coal-field." By Marcus W. T. Scott, Esq., F.G.S.

This communication was based on observations made during many years on a section through a part of the Shropshire Coal-field in

-----

* See the previous abstract.

nearly a straight line from north to south, commencing at the Greyhound Pit, near Oakengates Tunnel of the Shrewsbury and Birmingham Railway, and terminating at John Anstice and Co.'s Halesfield Pits near Madely. Particular reference was made to the explanation of the nature of the Great East or Symon Fault. The author commenced making his observations on the Malinslee and Stirchlee Royalties in 1843; and in 1845 he came to the conclusion that what the miners termed in this locality the "Symon Fault," that is the successive dying out of certain coal-seams, ironstones, &c. at various depths underground, was due to an old denudation which had produced an inclined surface at the expense of some of the beds before the upper measures were deposited. Having obtained, in course of time, correct sections of several pits situated in the N.–S. line above mentioned, the author, taking the "Little Flint" (the lowest workable coal) as a base-line, plotted the several shifted segments of the coal-field in a vertical plan, and thus restored the original outline of the denuded area (one side of a valley) as seen in a transverse section. Six sinkings in the N.–S. line having indicated the successive disappearance of five workable coal-beds in a distance of 2484 yards, a seventh pit, 2000 yards further south, was found to yield all the coals again ; and the author thinks that between the 6th (the Grange) and the 7th (Halesfield) pit the coals re-occur successively on the opposite side of the old valley of denudation, and that they may here be sought for and worked advantageously. The line of the old valley of denudation apparently strikes the Great East fault, as laid down on the Geological Survey Map, at a considerable angle.

2. "On the Occurrence of *Cyrena fluminalis* associated with Marine Shells in Sand and Gravel above the Boulder-clay at Kelsey Hill near Hull." By Joseph Prestwich, Esq., F.R.S., Treas.G.S. &c.

The author's observations tended to show that the *Cyrena fluminalis*, instead of being limited, in its occurrence, to beds beneath the Boulder-clay (under which circumstance it is found in Norfolk), occurs in deposits of newer date, and that the argument, that the well-known beds at Grays, in Essex, are older than the Boulder-clay, depending much on the presence of this shell, would lose much of its force if this *Cyrena* were proved to belong also to the newer geological horizon. The question is now the more important, as this shell has been found by Mr. Prestwich in the beds that contain flint implements at Abbeville.

The author proceeded to show that some gravels and sands near Hull in Yorkshire, formerly described by Professor Phillips, contain abundance of the *Cyrena fluminalis*, associated with twenty-two species of marine shells, two of which have Arctic characters, the others being common littoral forms. These gravels and sands were proved, by well-sections and other exposures, especially by borings and trenches made by the author and Mr. T. J. Smith, of Hull, to overlie the Boulder-clay.

XII. *Intelligence and Miscellaneous Articles.*

ON THE SOLIDIFICATION OF CERTAIN SUBSTANCES.

BY M. L. DUFOUR.

IN a preceding communication it has been shown that water, kept in suspension in a liquid of its own density, could be cooled much below 0° without solidifying. It was probable that other bodies placed in similar conditions would experience a similar retardation of solidification. The following are three examples:—

Sulphur.—The persistence of this body in the fluid state below 115° has been already noted (M. Person and Prof. Faraday), but it is an exception not frequently mentioned.

It is easy to prepare a solution of chloride of zinc which has the same or a little higher density than that of liquid sulphur. This solution can be heated to 115° without boiling; sulphur may be melted in it, and then floats in spheres. In order to keep the spheres surrounded by liquid, a layer of oil may be poured on the solution. On cooling, the solidification scarcely ever takes place at the melting-point. The liquid globules usually sink to 70°, 50°, &c. before solidifying. The solidification is spontaneous, or it may be provoked by the contact of a solid body, especially of a fragment of sulphur; but in the special conditions of these experiments the liquid state presents a remarkable stability. At 60°, salts, metallic wires, &c. may occasionally be introduced into globules 6 millims. in diameter without inducing an immediate solidification. Globules of $\frac{1}{2}$ a millim. in diameter frequently remain fluid at 5°, and persist in that state for several days.

When the spheres of sulphur remain liquid at 50° or 60° below the ordinary temperature of solidification, it is truly interesting to see their change of state. The fluid mass, which is transparent and of a deep red, suddenly changes into a hard, opake yellow fragment. This experiment, which is very pretty and easily performed, is exceedingly well adapted to exhibit the curious phenomenon of superfusion.

Phosphorus.—M. Desains has already noted the conservation of the liquid state by this body below 44°. The method which serves for sulphur is perfectly applicable to phosphorus. The solution of chloride of zinc of a suitable density is covered with a layer of oil in order to avoid contact of the air. The liquid transparent globules of phosphorus are easily seen, and their solidification does not take place till far below 44°. Globules, $\frac{1}{2}$ to 2 millims. in diameter, frequently sink to 5° or even to 0°. The liquid state is also remarkably stable, and the change of condition gives occasion for observations analogous to those relative to sulphur.

Naphthaline.—The fusion and solidification of this body usually take place at 79°. It has almost exactly the same density as water, but is somewhat less dense in the liquid state. With suitable precautions the phenomenon of superfusion may be easily produced. It is merely requisite to melt the body in a flask filled to the neck with boiled water, and then to incline the flask so that the liquid napthaline lodges in the upper part of the flask, pressed, but feebly so, against the side of the glass. In virtue of the slight difference in density, this liquid assumes the spheroidal form, and does not adhere to the glass. I have seen globules 8 millims. in diameter retain the liquid state to 55°.

It is probable that other bodies, if placed in suitable conditions, would also present the phenomenon which the preceding bodies manifest in such a pronounced degree. Unfortunately it is difficult with a large number to realize the essential condition, which is to pass the ordinary temperature of change of state while the body floats in equilibrium in a liquid of the same density. The liquid selected must, in point of fact, fulfil the four following conditions: it must have the same density as the body under experiment, retain the fluid state above and below its melting-point, and not exert any chemical action. Spite of these requirements, I do not doubt that chemistry will furnish the means of successfully applying to other substances the method by which the retardation of the solidification of water, sulphur, and phosphorus is so easily and certainly effected.— *Comptes Rendus*, April 29, 1861.

ON THE CHANGES PRODUCED IN THE POSITION OF THE FIXED
LINES IN THE SPECTRUM OF HYPONITRIC ACID BY CHANGES
IN DENSITY. BY M. WEISS.

Weiss has found by actual measurement that the distance between the dark lines in the spectrum of hyponitric acid diminishes as the density of the gas increases. The measurements were made with an Oertling's circle reading directly to two seconds of arc, and, by a filar micrometer in the ocular, to a single second. The same phenomenon occurs with the spectrum of chlorophyll. The stronger the extract in ether, the less is the distance of the absorption-bands. Thus the absorption-band in the red, in the case of a strong extract, corresponds quite well with Fraunhofer's line C; in the case of a weak extract it stands at some distance from it. The other absorption-bands in this spectrum undergo similar dislocations.

These changes in the distances of the dark lines are very sensible, even in the spectrum of hyponitric acid, when the changes in the density of the gas are considerable; they are not, however, equal for all the dark lines.

The cause of these dislocations is to be sought, according to Weiss, in a one-sided absorption which each line undergoes toward the violet end of the spectrum when the density of the body is increased. This is shown by direct observations and comparisons with the solar spectrum as well as by numerous measurements. There is no specific absorption upon both sides of each line, but only an absorption upon the side of the line which lies toward the violet end of the spectrum. In this manner the bands become broader, and the distance between them less. The author has observed similar changes in the breadth of Fraunhofer's lines at sunset. In this case also the absorption was only upon one side. From this it appears that the lines of hyponitric acid cannot be used as standards in determinations of indices, &c.—Poggendorff's *Annalen*, vol. cxii. p. 153, Jan. 1861; and Silliman's *Journal* for May.

THE

LONDON, EDINBURGH AND DUBLIN

PHILOSOPHICAL MAGAZINE

AND

JOURNAL OF SCIENCE.

[FOURTH SERIES.]

AUGUST 1861.

XIII. *On the Klaprothine or Lazulite of North Carolina.* By E.
J. CHAPMAN, *Professor of Mineralogy and Geology in University College, Toronto*[*].

THE Klaprothine or lazulite is comparatively a rare mineral.
It appears to have been first recognized by Widenmann
in 1791, in the valley of the Muhr, near Krieglach in Upper Styria.
By Werner it was mistaken for felspar; and, although examined
by Klaproth, its true nature was not detected until the analysis
by Fuchs of specimens afterwards discovered near Werzen in
Salzburg. Brandes then examined the Krieglach specimens,
and showed their identity in composition with the examples
analysed by Fuchs[†]. The other known localities of this mineral
comprise Vorau near Gratz in Styria (examples from which spot
have been analysed by Rammelsberg); the foot of the Wechsels
near Therenberg in Lower Austria; Minas Geraes in Brazil; and
Sinclair County in North Carolina. Specimens from this latter
locality have been very carefully analysed by Professor J. Lawrence
Smith and George J. Brush (now Professor of Metallurgy in Yale
College); but I have failed to discover in any publication a cry-
stallographic or mineralogical description of this North American
lazulite. A specimen, however, consisting of numerous small
crystals imbedded in fine granular quartz or sandstone, having
been kindly presented to me within the few last months, by

---

[*] Communicated by the Author.
[†] Brandes appears, however, to have missed the water present in this
substance, unless there be a typographical error in his recorded numbers.
If we transpose these numbers, as regards the silica (an impurity) and the
half per cent. of water said to have been obtained, his analysis will agree
closely with those of other chemists.

Prof. T. Sterry Hunt, of the Geological Survey of Canada, I propose in the present place to offer a brief notice of its leading mineralogical characters.

All the earlier determinations of lazulite crystals referred the mineral to the trimetric or rhombic system. Prüfer of Vienna was the first to maintain its monoclinic character; and the angles given in the more recent works on mineralogy are adopted from his measurements. The European crystals present in general a somewhat complicated aspect, although certain combinations closely resemble those of the trimetric system. Two "augite pairs" are always present. These, according to Prüfer, measure respectively over a front edge 100° 20′ and 99° 40′, the difference being but little more than half a degree. According to the same observer, moreover, the inclination of the base on the prism-plane (0 P : ∞ P, in the notation of Naumann) only differs from a right angle by 23′. Were these values consequently all that we had to depend upon, it would be manifestly unsafe to rely upon them as proofs of the monoclinic crystallization of lazulite. But in some combinations the forms below the middle zone of the crystal are less numerous than those above this zone, or otherwise differ from the latter in their measurements. Nevertheless in certain trimetric minerals, and notably in datolite and Wolfram, we have the same peculiarity, and we might therefore look upon these lazulite crystals as trimetric combinations hemihedrally modified. From my examination of the North Carolina specimens, I cannot but think that this view will in the end prevail. It is supported by the fact that in many combinations the upper and lower forms do actually correspond in number and character, and that practised crystallographers like Phillips and Lévy, skilled in the use of the goniometer, were unable to detect in their measurements the differences announced by Prüfer*.

The North Carolina crystals (presuming those in my possession to represent the generality of crystals obtained at this locality), although usually distorted, are of extreme simplicity, contrasting remarkably in this respect with the majority of European examples. At first sight they resemble a monoclinic prism terminated by a single "augite pair" or hemi-pyramid; but they really consist (if monoclinic) of two hemi-pyramids, the four planes of one of which are greatly elongated; or if trimetric (as I conceive them to be), they form a rhombic octahedron in which four planes, in opposite sets of two, are thus lengthened beyond

---

* These observers appear to be the only crystallographers who have practically examined crystals of lazulite. Thus the measurements of Phillips are followed by Hausmann, Breithaupt, and others; those of Lévy, by Dufrénoy; and those of Prüfer, by Naumann, Dana, Quenstedt, and Miller.

the others. Fig. 1 represents this distorted aspect; fig. 2 the same form (or combination, if monoclinic) in symmetrical proportions. These symmetrical crystals are of smaller size and less numerous than the distorted forms.

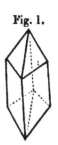

Fig. 1.

Fig. 2.

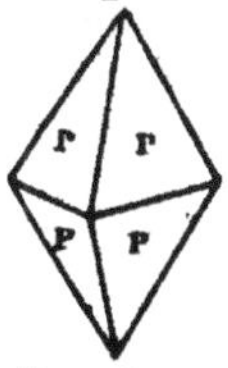

Although the edges of these crystals are sharply defined, the planes are unfortunately without lustre. The most careful measurements of five crystals, by means of a fixed or Adelmann's goniometer, gave me the same angles for both the upper and lower faces. The difference found by Prüfer is too slight, however, to be satisfactorily detected by any kind of application goniometer. I attached, therefore, thin films of mica as carefully as possible to the planes of one of the crystals, and measured the angles by reflected light with a Wollaston goniometer of the best construction. The following Table (sustaining the apparently trimetric character of these crystals) shows the measurements thus obtained :—

	Upper planes over front edge.	Lower planes over front edge.
1st measurement . . .	100 · 4	100 0
2nd „ . . .	99 99	99 99
3rd „ . . .	99 99	· 100 2

	Upper planes over side edge.	Lower planes over side edge.
1st measurement . . .	97 27	97 28
2nd „ . . .	97 30	97 26
3rd „ . . .	97 26	97 27

	Front planes over middle edge.	Back planes over middle edge.	
1st measurement	134 10	134 7	Whether monoclinic or trimetric, these measurements should of course correspond.
2nd „	134 10	134 12	The two sets were taken, however, for greater satisfaction.
3rd „	134 8	134 10	

Adelmann's goniometer gave me 100°—100° 30′ over a front edge, 97°—97° 30′ over a side edge, and 134°—134° 30′ over a middle edge. If we look upon the mineral as trimetric, and adopt the angle of 100° as the mean inclination over a front edge, with 91° 30′ for the value of the prism-angle (according to general adoption), the following angles and axial relations are obtained by calculation :—

$$P : P \text{ (over a front edge)} = 100 \; \overset{\circ}{} \; 0$$
$$P : P \text{ (over a side edge)} = 97 \; 24\tfrac{1}{2}$$
$$P : P \text{ (over a middle edge)} = 134 \; 12$$

$$x \text{ (vertical axis)} = 1\cdot652$$
$$\bar{x} \text{ (macrodiagonal)} = 1$$
$$\breve{x} \text{ (brachydiagonal)} = 0\cdot9741$$

The measurements of Phillips give for the octahedral angles, as deduced by Hausmann, 99° 16′ (over front edge), 96° 39′ (over side edge), and 136° 20′ (over middle edge). The position of the crystals, as adopted by Phillips, is here changed, however, his middle edge being made a front polar edge, and the reverse.

Many of these North Carolina crystals appear to possess another form in addition to those enumerated above. This is the front polar or macrodome, occurring generally on two opposite edges only, and thus presenting a monoclinic character, but lying sometimes on only one edge, and being consequently (if the mineral be trimetric) a tetartohedral modification. It is a mere line, dull like the other planes, and too narrow to admit of satisfactory measurement. The crystals are sometimes implanted in one another; but I have not detected any definite twin-combinations. The crystals extracted from my specimen, together with those exposed on the surface of this, do not amount, however, to more than ten or twelve in number. The hardness of these crystals is equal to 5·75, or very nearly to 6·0. The specific gravity (one determination only) I found to equal 3·108, a value corresponding sufficiently with that obtained by Smith and Brush (3·122). The cleavage I have not been able to determine in a satisfactory manner. The blowpipe reactions are as follows:—

In the closed tube the assay gives off water and loses its colour, becoming yellowish or greyish white.

Per se, it exfoliates and expands greatly in bulk, changes colour, tinges the flame green, and crumbles away without fusing.

In borax it dissolves very easily, imparting to the glass a pale ferruginous tinge.

In salt of phosphorus it dissolves also very readily, and with slight effervescence.

In carbonate of soda it dissolves partially, but the dissolved portion is in great part precipitated as the glass cools, forming a white enamel. If the bead be dissolved in a little boiling water, a drop of nitric acid added to decompose the excess of carbonate of soda, and the clear supernatant liquid be then poured upon a small crystal of nitrate of silver, a yellow precipitate of phosphate is at once obtained. In employing this test for phosphates, the beginner should be cautioned, however, that silicates (if decomposable by carbonate of soda) will produce the same reaction, but the silica may be eliminated by adding several drops of acid and evaporating to dryness. By treatment with salt of phosphorus, moreover, silicates are at once recognized. If the solution of our mineral, as obtained above, be treated with acetate of lead, the precipitate presents the well-known blowpipe reaction of phosphate of lead, *i. e.* the formation of a faceted globule without reduction.

Two analyses of the North Carolina lazulite are given by Professors Smith and Brush in the 'American Journal of Science and Arts' for September 1853. These exhibit the following results :—

	I.	II.
Phosphoric acid	43·38	44·15
Alumina	31·22	32·17
Protoxide of iron	8·29	8·05
Magnesia	10·06	10·02
Water	5·68	5·50
Silica (an impurity)	1·07	1·07
	99·70	100·96

From the above values, Messrs. Smith and Brush have deduced the annexed formula :—

$$2[3(\text{MgO, FeO}), \text{PO}^5] + 5\text{Al}^2\text{O}^3, 3\text{PO}^5 + 5\text{HO}.$$

The true position of lazulite, in a natural classification, appears to be amongst a group of phosphates containing both anhydrous and hydrous species (the distinction between these being entirely artificial), and in some of which fluorine is also present. In this group I would place the following minerals :—Childrenite, Wavellite, Fischerite, turquoise, lazulite, Wagnerite, Herderite, amblygonite, monazite, xenotime, and cryptolite.

Toronto, Canada,
June 20, 1861.

XIV. *On the Propagation of Heat in Gases.* By G. Magnus.
[Concluded from p. 12.]

Passage of the Rays of Heat through Gases*.

A N objection might still be raised against any conductibility in gases. It might be maintained that the stronger heating of the thermometer in hydrogen depended on the fact that it permits the passage of heat-rays more easily than all other gases.

* Read before the Academy of Berlin, February 7, 1861.

The above experiments with cotton and eider-down speak against this; for it can scarcely be assumed that the heating takes place through radiation. Moreover, the experiments which Dr. Franz has published * on the radiation through hydrogen, show that more heat-rays do not pass through hydrogen than through at-. mospheric air. It appeared, however, necessary, before main-. taining that gases can conduct heat, to determine by new experiments how far that doubt was founded. Hence the deter-minations on the passage of heat-rays through various gases were concluded when I communicated to the Academy, in July of last year (1860), the investigation on the conduction of heat.

As far as I know, Dr. Franz's are the only experiments which have hitherto been published on the diathermancy of the gases. These, which moreover only refer to atmospheric air, hydrogen, and carbonic acid, could not be sufficient for the present pur-pose, because an argand lamp was used as a source of heat. But it was not merely possible, but even probable that the transmission of thermal rays would differ with the source whence they came. If therefore the experiments were to be conclusive, the transmission must be investigated for rays proceeding from the same source of heat, that of boiling water.

Boiling water as a source of Heat.

Dr. Franz in his experiments enclosed the gases in tubes closed at both ends by pieces of plate glass. Now from Mel-loni's experiments†, rays from so low a source of heat penetrate plate glass in scarcely perceptible quantities. Even when I used plates of rock-salt to close a tube a metre in length, the action which the rays of boiling water produced upon the thermo-pile were so small, that a comparison of the different gases furnished no satisfactory results. It further appeared desirable to avoid any kind of plates, even of rock-salt; for although, from Melloni's experiments, the rays which have passed through this substance comport themselves exactly like those which proceed directly from the source of heat and have only passed through air, yet the rock-salt might possibly alter the rays, and exert an influence on their subsequent passage through different gases. I have accordingly undertaken new experiments on the diathermancy of gases for obscure heat, in which it was my object to allow the rays of heat to pass through the gases without necessitating their passage through any plate.

When these experiments were finished, I saw from the 'Pro-

---

* Poggendorff's *Annalen*, vol. xciv. p. 337.
† Ibid. vol. xxxv. p. 393.

ceedings of the Royal Society * ' that Dr. Tyndall in London was engaged with an investigation on the transmission of heat through gases. As Dr. Tyndall, whose research is only just announced, has experimented with tubes which were closed by plates of rock-salt, I considered that the following investigation was independent of that of Dr. Tyndall.

The apparatus which I used was constructed as follows. Upon the plate of an air-pump T T, fig. 2. Plate I., which could be placed apart from the pump on a separate foot, a thermo-electric pile was firmly fixed by means of a cork ring cemented on the plate. The mounting of this pile was of brass, and had an internal diameter of 24 millims. and a length of 118 millims. The pile itself was only 30 millims. long. It contained 56 pairs of antimony and bismuth, which together formed a section of 30 millims. square. The wires from this pile to the galvanometer passed through the plate insulated at L L. Over the pile was a glass vessel, F G, with a broad ground edge air-tight upon the plate of the pump. This was 175 millims. high and 100 millims. in diameter.

At the upper part there were two apertures, q and r, to which corresponded the two tubes $q q_1$ and $r r_1$ of the brass cover G G, which was fastened on the top of this vessel. These tubes were 30 millims. in height. In the tube $q q_1$ right over the thermo-pile S, the glass vessel A B, upon which the vessel C was fused, was firmly fitted by means of a cork, and made air-tight by means of caoutchouc. In the tubulure D of this vessel there was a cork, through which a glass tube passed, which could be closed with a stopcock H. The tube $r r_1$ contained a stuffing-box, through which the round brass rod $a b$ passed. Inside the vessel F G, this rod was provided with a horizontal arm $a c$, at whose end c, two circular pieces of tinplate 34 millims. in diameter were fitted parallel above one another at a distance of 3 millims. They served as a screen, and when the thermo-pile was to be exposed, could be easily moved aside. This could be effected by a horizontal arm $b d$, fitted at $b b$ on the brass rod $a b$ outside the vessel F G. In order to protect the pile as far as possible from the influence of external sources of heat, the vessel F G was surrounded by a wide glass cylinder N M, which with its broad edge was pressed on the plate. The space between both vessels was filled up to N M with water, which was kept at a temperature of 15° C.

The vessel C contained boiling water, which by passing steam into it was kept at a temperature of 100°; and this formed the source of heat. Its action upon the thermo-pile was indicated

* Proceedings of the Royal Society, vol. x. p. 37. Phil. Mag. vol. xix. p. 60.

by means of a very delicate multiplier with a double needle, which was connected by copper wires with the binding screws L L. . The wire of the multiplier consisted of copper which had been galvanoplastically deposited, and was free from iron. It is the multiplier which I used in my investigation on thermo-electric currents. I tried to use in its stead a multiplier with a steel mirror, which was read by means of a telescope and scale; but spite of this mode of observation I found it less delicate, manifestly because the steel mirror was not astatic. Whether a multiplier with mirror and astatic needle would not be better for these observations I have not tried; but I doubt it; for the reading off by means of a mirror is only suited for small differences of angles, while in using an astatic needle greater deviations are observed. The multiplier used was placed upon a firm stand separate from the rest of the apparatus.

In investigating the diathermancy of a gas, water at 100° C. was poured into the vessel C, and kept at that temperature by means of steam passed into it from a flask in the neighbourhood. The moment the screen *c c e e* was displaced, the needle began to move slowly, and after it had reached its greatest deflection, it assumed a fixed position in the course of about two minutes, after a few very small oscillations. This was read off partly directly, and partly by means of a telescope; in the latter case a rectangular prism was placed directly over that part of the divided circle which was to be observed, so that the position of the needle could be seen by reflexion. When this was effected, the screen was replaced over the pile, upon which the needle reverted to its original position. It was never, however, exactly over the null-point of the scale, either because the torsion of the wire had changed a little, or because there had been a slight difference of temperature in the pile. As in the multiplier used the coils of wire were quite free from magnetism, the replacement at 0° could easily have been effected by turning the divided circle independently of the magnetic needle. But this might easily have produced fluctuations of the entire apparatus, and it therefore appeared better to take the mean of the positions of equilibrium before and after the deflection, and to subtract this from the observed deflection. The observations thus obtained agreed very well with each other when several were made successively. It was only after the experiments had been continued some time that the numbers somewhat disagreed, because the pile became a little warmer at one end. Four to six of such observations were always made in succession, and the mean of these taken.

The values corresponding to the deflections of the galvanometer were determined by the method given by Melloni in his

'Thermochrose,' page 59. As far as 14°·5 the strengths of the current were proportional to the deviations.

In order to fill the apparatus with any given gas, it was first exhausted, the gas admitted by the stopcock H, then again exhausted and filled a second time with the gas, and so on for four times, upon which the atmospheric air could be considered to be completely removed. With gases like cyanogen and ammonia which attack the pump, the filling was performed by displacement, the gas being admitted by the stopcock C, and escaping by the stopcock K, under the plate TT. For this purpose the whole apparatus, with the plate TT, was removed from the air-pump and placed on a tripod. On the lower part of the stopcock K a glass tube was fitted, through which the gas passed into an absorbent liquid. If the gas was to be used in a rarefied state, the rarefaction was effected, after the filling was complete, by means of the air-pump. The rarefaction thus produced was either directly observed by the barometer, or a manometer was introduced, which was read off by means of a cathetometer. Thus the most different gases could be examined as to their capacity of transmitting heat, with the exception of those which attacked the metal of the pile. This excluded, to my great regret, all coloured gases.

The gases were prepared exactly as in the experiments on conduction.

As the intensity of the galvanometer might have changed in the course of time, in almost every case before the apparatus was filled with a new gas the radiation through atmospheric air was determined. In this way the relation of the radiation in the particular gas to that of atmospheric air was obtained. This method of comparison I have always retained, for it ensures great certainty. In experiments with boiling water this comparison was ultimately found superfluous, for the galvanometer remained so unchanged that the values obtained at different times for atmospheric air agreed very closely. Nevertheless in the following Table the observed deflections are so arranged that the control experiments with atmospheric air are found in one column, and the gases examined either directly before or after are placed opposite, being separated from the rest by a horizontal line.

For the gases which exhibit the greatest deviation, the radiation has been determined at different times. Since the numbers obtained agreed as closely as could be expected with such experiments and with such angles, I have only adduced one series.

Transmission of Heat by a Glass surface at 100° C.

Atmospheric air under 1 atm. pressure.

Position of equilibrium of the needle.	Mean.	Observed deflection.	Difference.
−1	−0.5	14.0	14.5
0	+0.2	14.5	14.3
+0.5	+0.5	15.0	14.5
0.5	+0.5	15.0	14.5
0.5			
Mean 14.5			

Atmos. air under 22.5 millims. pressure.

Position of equilibrium of the needle.	Mean.	Observed deflection.	Difference.
+0.5	+0.5	18.0	15.5
0.5	0.6	16.0	15.4
0.7	0.7	16.2	15.5
0.7	0.7	16.2	15.5
0.7			
Mean 15.5			

Atmospheric air under 1 atm. pressure

Position of equilibrium of the needle.	Mean.	Observed deflection.	Difference.
+3.5	+2.5	17.0	14.5
1.5	1.7	16.0	14.3
2.0	2.0	16.2	14.2
2.0	2.0	16.5	14.5
2.0	2.0	16.5	14.5
Mean 14.4			

Atmospheric air under 18.0 millims. pres.

Position of equilibrium of the needle.	Mean.	Observed deflection.	Difference.
−0.5	−0.15	14.7	14.85
+0.2	+0.7	15.7	15.0
+1.2	1.1	16.2	15.1
+1.0	1.0	16.2	15.2
+1.0	1.0	16.2	15.2
+1.0			
Mean 15.1			

Atmospheric air under 1 atm. pressure.

Position of equilibrium of the needle.	Mean.	Observed deflection.	Difference.
−0.0	+0.25	14.7	14.5
+0.5	0.5	15.0	14.5
+0.5	0.5	15.0	14.5
+0.5	0.5	15.0	14.5
+0.5			
Mean 14.5			

Atmospheric air under 8 millims. pres.

Position of equilibrium of the needle.	Mean.	Observed deflection.	Difference.
0.0	+0.1	16.0	15.9
+0.2	0.35	16.0	15.65
+0.5	0.6	16.5	15.9
+0.7	0.7	16.5	15.8
+0.7			
Mean 15.8			

Atmospheric air under 9 millims. pres.

Position of equilibrium of the needle.	Mean.	Observed deflection.	Difference.
0.0	0.0	15.7	15.7
0.0	−0.1	15.7	15.8
−0.2	−0.2	15.7	15.9
−0.2	−0.2	15.5	15.7
−0.2			
Mean 15.8			

Atmospheric air under 1 atm. pressure.

Position of equilibrium of the needle.	Mean.	Observed deflection.	Difference.
+0.5	+0.7	15.2	14.5
+1.0	+1.1	15.5	14.4
1.2	1.2	15.7	14.5
1.2	1.5	16.0	14.5
1.7	1.5	16.0	14.5
1.2			
Mean 14.5			

Hydrogen under 1 atm. pressure.

Position of equilibrium of the needle.	Mean.	Observed deflection.	Difference.
+1.5	+1.2	15.0	13.8
+1.0	1.1	15.0	13.9
1.2	1.2	15.0	13.8
1.2	1.0	15.0	14.0
0.7	0.6	14.7	14.1
0.5			
Mean 13.9			

Transmission of Heat by a Glass surface (*continued*).

Position of equilibrium of the needle.	Mean.	Observed deflection.	Difference.

Hydrogen under 18 millims. pressure.

Position of equilibrium of the needle.	Mean.	Observed deflection.	Difference.
−0·5	−0·4	15·0	15·4
−0·3		15·0	15·2
−0·1	−0·2	15·2	15·2
+0·2	+0·3	15·5	15·2
+0·5		15·7	15·2
+0·5	+0·5		

Mean........ 15·2

Atmospheric air under 1 atm. pressure

Position	Mean.	Observed.	Difference.
−1·0	−0·6	13·5	14·1
−0·2	0·0	14·2	14·2
+0·1	+0·3	14·7	14·4
+0·5	+0·3	14·7	14·4
+0·2	+0·2	14·7	14·5
+0·2			

Carbonic acid under 1 atm. pressure.

Position	Mean.	Observed.	Difference.
0	+0·2	13·2	13·0
+0·5	0·7	13·7	13·0
+1·0	1·0	14·0	13·0
1·0	1·0	14·0	13·0
1·0	1·0	14·0	13·0
1·0			

Atmospheric air under 1 atm. pressure.

Position	Mean.	Observed.	Difference.
−0·2	+0·4	15·0	14·6
+1·0	+1·2	15·7	14·5
+1·5	1·7	16·2	14·5
+2·0	2·4	17·0	14·6
+2·7	2·5	17·0	14·5
+2·2			

Mean........ 14·5

Mean.	Observed.
−2·8	3·5
−2·2	4·0
−1·6	4·7
−1·0	4·0
−0·3	6·0

Atmospheric air under 1 atm. pressure.

Position	Mean.	Observed.	Difference.
+0·5	+0·5	15·0	14·5
0·5	+0·6	15·2	14·6
0·7	1·1	15·7	14·6
1·5	1·5	15·7	14·2
1·5	1·6	16·0	14·4
1·7			

Mean........ 14·4

Marsh-gas under 1 atm. pressure.

Position	Mean.	Observed.	Difference.
+2·2	+2·3	14·0	11·7
2·5	2·7	14·5	
	3·1	15·0	
	3·2	15·0	
	3·1	14·7	
	2·8	14·0	
	2·6	14·5	

Mean........ 11·7

Position	Mean.	Observed.	Difference.
0·8	+0·9	12·2	11·3
	0·9	12·8	11·9
	1·0	12·5	11·5
	1·0	12·6	11·6
	1·0	12·6	11·6

Mean........ 11·6

Position of equilibrium of the needle.	Mean.	Observed deflection.	Difference.

Atmospheric air under 1 atm. pressure.

+2·0	+2·1	16·5	14·4
+2·2	2·3	17·0	14·7
2·7	2·6	17·0	14·4
2·5	2·3	16·7	14·4
2·2	2·2	16·7	14·4
2·2			14·5

Mean .... 14·4

Atmospheric air under 1 atm. pressure.

0·0	+0·4	14·7	14·3
+0·7	0·6	15·0	14·4
+0·5	0·6	15·0	14·4
+0·7	0·6	15·0	14·4
+0·5			

Mean .... 14·4

Atmospheric air under 1 atm. pressure.

0·0	+0·1	14·5	14·4
0·2	0·2	14·7	14·5
0·2	0·3	14·7	14·4
0·5	0·5	15·0	14·5
0·5			

Mean .... 14·45

Atmospheric air under 1 atm. pressure.

0·0	+0·1	14·5	14·4
+0·2	0·1	14·5	14·4
0·0	0·0	14·4	14·4
0·0	0·0	14·4	14·4
0·0			

14·4

Protoxide of nitrogen under 1 atm. pres.

+3·0	+3·2	15·0	11·8
3·5	3·6	15·5	11·9
3·7	3·7	15·7	12·0
3·7	3·6	15·7	12·1
3·5	3·5	15·5	12·0
3·5			

Mean .... 12·0

Olefiant gas under 1 atm. pressure.

+1·2	+1·2	8·5	7·3
1·2	1·0	8·5	7·5
0·7	0·7	8·2	7·5
0·7	0·7	8·2	7·5
0·7	0·7	8·5	7·8
0·7			

0·0	+0·1	12·7	12·6
+0·2	0·1	13·0	12·9
0·2	0·0	13·0	13·0
0·0	−0·1	12·7	12·8
−0·2			

Carbonic oxide ui

+0·5	+0·25	15·5	15·25
0·0	−0·25	15·2	15·45
−0·5	−0·5	15·2	15·70
−0·5	−0·5	15·0	15·50
−0·5			

Oxygen under 1 atm. pressure.

0·0	+0·1	14·5	14·4
+0·2	0·2	14·5	14·3
0·2	0·35	14·7	14·35
0·5	0·5	15·0	14·5
0·5			

0·0	+0·25	12·0	11·75
+0·5	0·35	12·0	11·65
+0·2	0·35	12·0	11·65
0·5	0·50	12·2	11·70
0·5			

Mean .... 11·7

If the deflections obtained with atmospheric air under a pressure of 1 atmosphere are collated, we obtain

$$14°\cdot5,\ 14°\cdot4,\ 14°\cdot5,\ 14°\cdot5,\ 14°\cdot3,\ 14°\cdot5,\ 14°\cdot4,\ 14°\cdot4,\ 14°\cdot4,$$
$$14\cdot45,\ 14°\cdot4,$$

the mean being $14°\cdot4$.

For all other gases the deflections are less. But as the deflections are proportional to the intensities of the current up to $14°\cdot4$, and as these are proportional to the increase in temperature of the pile, the quantities of heat which pass through different gases under the same pressure are as follows :—

Atmospheric air . .	14·4 or	100
Oxygen	14·4 „	100
Hydrogen ·. . . .	13·9 „	96·5
Carbonic acid . . .	13·0 „	90·3
Carbonic oxide. . .	12·8 „	88·8
Protoxide of nitrogen.	12·0 „	83·3
Marsh-gas	11·7 „	81·2
Cyanogen	11·7 „	81·2
Olefiant gas . . .	7·5 „	52·1
Ammonia	6·3 „	43·7

As oxygen gave exactly the same value as atmospheric air, it was unnecessary to examine nitrogen.

For atmospheric air the deflections were for a pressure of—

$$8\ \text{millims.} = 15°\cdot8$$
$$9\ \text{millims.} = 15\ \cdot8$$

This deflection is no longer proportional to the intensity of the current, but corresponds to a value of $16\cdot2$, the value for $1°$ between $0°$ and $14°$ being placed equal to 1. If it be assumed that the radiation through vacuum would produce the same deflection, the heat which passes through vacuum would be to that which passes through atmospheric air under a pressure of 1 atmosphere, as

$$16\cdot2 : 14\cdot4 = 100 : 88\cdot88.$$

In order to obtain greater certainty for this proportion, I determined the radiation through rarefied air by interposing in the conduction a wire which offered considerable resistance, in order that the deflections might be smaller, and proportional to the intensities of the current. In the three following determinations the air was under a pressure of 4 millims. Directly before and after each of them the radiation through atmospheric air under a pressure of one atmosphere was determined. The corresponding determinations are indicated by the same number.

No.	Position of equilibrium of the needle.	Mean.	Observed deflection.	Difference.	No.	Position of equilibrium of the needle.	Mean.	Observed deflection.	Difference.
Atmospheric air under 1 atm. pressure.					Atmospheric air under 4 millims. pres.				

Atmospheric air under 1 atm. pressure. | Atmospheric air under 4 millims. pres.

No.	Position	Mean.	Obs. defl.	Diff.	No.	Position	Mean.	Obs. defl.	Diff.
I	+0·2 +0·5 +0·5 0·0 0·0 0·0	+0·35 +0·5 +0·25 0·0 0·0	11·2 11·7 11·2 11·0 11·0 11·0	10·85 11·2 10·95 11·0 11·0	Ia	0·0 −0·25 0·0 0·0	−0·1 −0·1 0·0	12·0 12·0 12·0	12·1 12·1 12·1
		Mean .... 11·0					Mean .... 12·1		

Ia : I = 12·1 : 11·0 = 100 : 90·9.

| II | 0·0
+0·25
0·0
0·0
+0·25 | +0·1
+0·1
0·0
+0·1 | 11·0
11·0
11·0
11·0 | 10·9
10·9
11·0
10·9 | IIa | 0·0
−0·25
−0·25
0·0 | −0·1
−0·25
−0·1
0·0 | 12·0
12·25
12·5
12·5 | 12·1
12·5
12·6
12·5 |
| | | Mean .... 10·9 | | | | | Mean......... 12·4 | | |

IIa : II = 12·4 : 10·9 = 100 : 87·9.

| III | 0·0
0·0
0·0
0·0
0·0 | 0·0
0·0
0·0
0·0 | 11·0
11·0
11·0
12·0 | 11·0
11·0
11·0
12·0 | IIIa | −1·75
−1·5
−1·5
−2·0
−1·5
−1·5 | −1·6
−1·5
−1·75
−1·75
−1·5 | 10·5
11·25
11·0
11·0
11·0 | 12·1
12·75
12·75
12·75
12·5 |
| | | Mean......... 11·2 | | | | | Mean .... 12·5 | | |

IIIa : III = 12·5 : 11·2 = 100 : 89·5.

After the observations I a and II a were complete, the air was still under a pressure of 10 millims.; after III a it was under 7 millims.

The relation thus ascertained between radiation through rarefied air to that through atmospheric air under a pressure of 1 atmosphere, agrees so far with that previously given, that I have taken the former as a basis for calculating the relation of the radiation through other gases to that through vacuum. Hence of 100 rays which pass through vacuum, the following quantities pass through the different gases, all under the pressure of one atmosphere :—

	Deflection.	Rays.
Vacuum	15·8°	
corresponding to . .	16·2 $=$	100
Atmospheric air . . .	14·4	88·88
Oxygen	14·4	88·88
Hydrogen	13·9	85·79
Carbonic acid . . .	13·0	80·23
Carbonic oxide . . .	12·8	79·01
Protoxide of nitrogen .	12·0 .	74·06
Marsh-gas	11·7	72·21
Cyanogen	11·7	72·21
Olefiant gas	7·5	46·29
Ammonia	6·8	38·88

Although these values cannot be looked upon as quite reliable, inasmuch as variations may occur from imperfect purity of the gas, or from other almost unavoidable impurities, they yet show how considerable are the differences which perfectly transparent gases exhibit in reference to the property of transmitting heat. This surprising deportment, which I had already established before I laid the first part of this treatise "On the Conduction of Heat" before the Academy, led me to make a separate investigation of the radiation through gases, and first of all to ascertain whether similar differences prevailed when another source of heat was used.

A Gas-flame as source of Heat.

I desired first of all to use a source of heat at a higher temperature, for which purpose the apparatus depicted in fig. 2, Plate I. was unfitted. I was accordingly compelled to use the gases in a tube closed at both ends by plates. In testing this method, I had occasion to make some observations which have probably also been made by others, but which I have nowhere found mentioned.

Influence of the side of the Tube.

If the rays from any source of heat be allowed to act upon a thermo-pile without having passed through any tube, a smaller deflection is obtained than when the rays from the same source of heat placed at the same distance from the thermo-pile are allowed to pass through a tube open at both ends, that is, not closed by any kind of plate. This increased action is obviously caused by the rays reflected from the inner side of the tube, so that the thermo-pile is acted on, not only by the rays which come directly from the source of heat, but also by those which fall obliquely into the tube and are again reflected. Even if the tube were blackened on the inside, or if, as was usually the case in the following.

experiments, the inside was lined with a black, rough, non-lustrous paper, the action was likewise stronger than without a tube, although the increase was not so considerable as if the glass was without this coating.

The influence of the tube can, it is true, be diminished by introducing diaphragms, which hinder the irradiation of the inside; but I have not succeeded in entirely obviating it, for the edges of the diaphragm likewise reflect heat. But the significance of the action of the sides of the tube in investigating the diathermancy of gases is best seen from the following experiments.

In these experiments a strong gas-flame with a double draught, surrounded by a glass cylinder, was used as a source of heat. It was provided with a small parabolic metallic mirror, which reflected the rays of the lamp in such a manner that they passed into a tube 1 metre long and 35 millims. in internal diameter, at the other end of which was the above-named thermo-electric pile. Between the tube and the lamp, and somewhat nearer the latter, there was a screen consisting of two metal plates at a distance of 12 millims. from each other. This could be removed when the rays were to fall on the pile, and replaced as soon as this was finished. Between this screen and the tube was a second similar screen, which had an aperture 30 millims. square, the centre of which was in a line with the axis of the tube. This screen, which was always in a fixed position, protected the rays of the lamp from the outside of the tube when the other was removed. There was another screen with a similar aperture close to the thermo-pile and between it and the tube, the object of which was to protect the pile from all external rays.

When using this arrangement, the rays were allowed to pass through a tube open at both ends, and lined with rough black paper; the deflection of the needle amounted to 24°·7, corresponding to 32·2 units. On removing the tube the deflection was only 10°, corresponding to 10. If, after removing the tube, the rays were allowed to pass through two glass plates 4 millims. thick, placed at the same distance at which they would be if they closed the tube, the deflection of the needle would only be 1° to 2°. If, on the contrary, the blackened tube, as the tube lined with black paper will for the future be called, was between the glass plates, the deflection increased to 12°·6, corresponding to 12·6. If the tube closed with glass plates was not blackened on the inside, the deflection increased to 64°, corresponding to 320.

From this it will be seen how greatly investigations on the passage of heat-rays will be affected by the nature of the tubes in which the gases are experimented on.

Transmission of the Heat of a Gas-flame (*continued*).

	Blackened tube.					Unblackened tube.			
No.	Position of equilibrium of the needle.	Mean.	Observed deflection.	Difference.	No.	Position of equilibrium of the needle.	Mean.	Observed deflection.	Difference.
			Oxygen under 1 atm. pressure.						
IIIa	+0·5	+0·3	11·5	11·2	IIIa	+1·	+1·1	59·0	57·9
	+0·2	+0·1	11·0	10·9		1·25	1·0	59·25	58·25
	0·0	0·0	11·0	11·0		0·75	0·8	58·75	57·9
	0·0	+0·1	11·0	10·9		1·0	0·5	58·5	58
	+0·2					0·0			
	Mean .... 11·0					Mean .... 58·0			
	Corresponding to ... 11·0					Corresponding to ... 193·0			
	III : IIIa = 11·0 : 11·0 = 100 : 100					III : IIIa = 193 : 193 = 100 : 100			
			Atmospheric air under 1 atm. pressure.						
IV	+0·5	+0·5	16·7	16·2	IV	+0·5	+0·75	61·0	60·25
	+0·5	0·6	16·7	16·1		+1·0	1·0	61·2	60·2
	0·7	0·7	16·7	16·0		1·0	1·0	61·2	60·2
	0·7	0·7	17·0	16·3		1·0	1·0	61·2	60·2
	0·7					1·0			
	Mean .... 16·15					Mean .... 60·2			
	Corresponding to ... 17·40					Corresponding to ... 230·0			
			Hydrogen under 1 atm. pressure.						
IVa	+0·7	+0·6	16·7	16·1	IVa	+1·0	+0·85	61·0	60·15
	0·5	0·5	16·5	16·0		0·7	+0·85	61·0	60·15
	0·5	0·5	16·7	16·2		1·0	+0·85	60·7	59·85
	0·5	0·5	16·5	16·0		0·7	0·70	60·5	59·8
	0·5					0·7			
	Mean .... 16·07					Mean .... 60·0			
	Corresponding to ... 17·2					Corresponding to ... 226·0			
	IV : IVa = 17·4 : 17·2 = 100 : 98·85					IV : IVa = 230 : 226 = 100 : 98·26			
			Atmospheric air under 1 atm. pressure.						
V	−1·2	−1·1	17·7	18·8	V	+0·5	+0·6	61·0	60·4
	−1·0	−0·8	18·0	18·8		+0·7	+0·6	61·0	60·4
	0·7	−0·6	18·2	18·8		0·5	0·5	61·2	60·7
	0·5	−0·5	18·5	19·0		0·5	0·5	61·2	60·7
	0·5					0·5			
	Mean .... 18·8					Mean .... 60·55			
	Corresponding to ... 22·1					Corresponding to ... 233·0			

H 2

Transmission of the Heat of a Gas-flame (*continued*).

	Blackened tube.					Unblackened tube.			
No.	Position of equilibrium of the needle.	Mean.	Observed deflection.	Difference.	No.	Position of equilibrium of the needle.	Mean.	Observed deflection.	Difference.

Carbonic acid under 1 atm. pressure.

No.	Position	Mean.	Observed	Difference.	No.	Position	Mean.	Observed	Difference.
Va	−0.5 −0.2 −0.5 0.0 0.0	−0.35 −0.35 −0.25 0.0	17.7 17.7 17.7 18.0	18.0 18.0 17.95 18.0	Va	+0.2 0.5 0.5 0.7 0.7	+0.35 0.5 0.6 0.7	60.0 60.2 80.2 60.2	59.65 59.7 59.6 59.5

Mean 18.0
Corresponding to ... 20.8
V:Va=22.1:20.8=100:94.11

Mean 59.6
Corresponding to ... 218.0
V:Va=238:218=100:91.59

Atmospheric air under 1 atm. pressure.

No.	Position	Mean.	Observed	Difference.	No.	Position	Mean.	Observed	Difference.
VI	0 0 +0.2 0 0	0.0 +0.1 0.1 0.0	23.75 24.0 24.2 24.0	23.75 23.9 24.1 24.0	VI	+0.7 +1.2 1.2 1.2 1.2	+1.0 1.2 1.2 1.2	63.0 63.0 63.0 63.0	62.0 61.8 61.8 61.8

Mean..................... 23.9
Corresponding to...... 30.8

Mean 61.8
Corresponding to ... 265.0

Carbonic oxide under 1 atm. pressure.

No.	Position	Mean.	Observed	Difference.	No.	Position	Mean.	Observed	Difference.
VIa	+0.2 0.2 0.2 0.0 0.0	+0.2 0.2 0.1 0.0	23.0 23.0 23.0 22.5	22.8 22.8 22.9 22.5	VIa	0.0 +0.5 0.5 1.0 0.7	+0.25 +0.50 0.75 0.85	60.25 60.3 60.75 60.75	60.0 59.8 60.0 59.9

Mean 22.75
Corresponding to ... 29.00
VI:VIa=30.8:29.0=100:94.15

Mean 59.9
Corresponding to ... 224.0
VI:VIa=265:224=100:84.52

Atmospheric air under 1 atm. pressure.

No.	Position	Mean.	Observed	Difference.	No.	Position	Mean.	Observed	Difference.
VII	+0.7 0.7 0.5 0.5 0.5	+0.7 0.6 0.5 0.5	16.7 16.7 16.5 16.5	16.0 16.1 16.0 16.0	VII	−0.5 −0.2 −0.2 −0.2 −0.2	−0.35 −0.2 −0.2 −0.2	60.5 60.7 60.7 60.7	60.85 60.9 60.9 60.9

Mean 16.0
Corresponding to ... 17.1

Mean 60.9
Corresponding to ... 245.0

Transmission of the Heat of a Gas-flame (*continued*).

	Blackened tube.					Unblackened tube.			
No.	Position of equilibrium of the needle.	Mean.	Observed deflection.	Difference.	No.	Position of equilibrium of the needle.	Mean.	Observed deflection.	Difference.
Protoxide of nitrogen under 1 atm. pressure.									
VIIa	+0.5	+0.35	15.2	14.85	VIIa	0.0	+0.25	60.0	59.75
	0.2	0.1	15.2	15.1		+0.5	+0.6	60.2	59.6
	0.0	0.0	15.2	15.2		+0.7	+0.7	60.5	58.8
	0.0	0.0	15.2	15.2		0.7	+0.7	60.7	60.0
	0.0	0.0	15.2	15.2		0.7			
	Mean 15.1					Mean 59.5			
	Corresponding to ... 15.4					Corresponding to ... 217.0			
	VII : VIIa = 17.1 : 15.4 = 100 : 90.05					VII : VIIa = 245 : 217 = 100 : 88.57			
Atmospheric air under 1 atm. pressure.									
VIII	0	0	22.5	22.5	VIII	+0.75	0.85	65.2	64.35
	0	0	22.7	22.7		+1.0	1.0	65.2	64.2
	0	+0.1	23.0	22.8		1.0	1.1	65.2	64.1
	+0.2	+0.35	23.0	22.65		1.2	1.2	65.5	64.3
	+0.5					1.2			
	Mean 22.7					Mean 64.2			
	Corresponding to ... 29.0					Corresponding to ... 226.0			
Marsh-gas under 1 atm. pressure.									
VIIIa	0	0	22.5	22.5	VIIIa	+1.0	+1.25	64.0	62.75
	0	+0.1	22.5	22.4		1.5	1.5	64.5	63.0
	+0.2	0.35	22.7	22.35		1.5	1.6	64.5	62.9
	0.5	0.5	22.7	22.2		1.7	1.6	64.7	63.1
	0.5					1.5			
	Mean 22.35					Mean 62.95			
	Corresponding to ... 28.5					Corresponding to ... 293.0			
	VIII : VIIIa = 29.0 : 28.5 = 100 : 98.27					VIII : VIIIa = 326 : 293 = 100 : 89.87			
Atmospheric air under 1 atm. pressure.									
IX	0	0	18.0	18.0	IX	+0.5	+0.6	62.5	61.9
	0	0	18.2	18.2		0.7	0.7	63.0	62.3
	0	0	18.2	18.2		0.7	0.7	62.7	62.0
	0	0	18.2	18.2		0.5	0.6	63.0	62.4
	0					0.7			
	Mean 18.15					Mean 62.15			
	Corresponding to ... 21.0					Corresponding to ... 273.0			

Transmission of the Heat of a Gas-flame (*continued*).

	Blackened tube.					Unblackened tube.			
No.	Position of equilibrium of the needle.	Mean.	Observed deflection.	Difference.	No.	Position of equilibrium of the needle.	Mean.	Observed deflection.	Difference.
			Olefiant gas under 1 atm. pressure.						
IX a	0	−0·1	14·0	14·1	IX d	+0·5	+0·5	58·5	58·0
	−0·2	−0·2	13·7	13·9		0·5	0·5	58·5	58·0
	−0·2	−0·1	13·5	13·6		0·5	0·5	58·2	57·7
	0	0	13·7	13·7		0·5	0·5	58·2	57·7
	0					0·5			
	Mean .... 13·8					Mean .... 57·85			
	Corresponding to ... 13·8					Corresponding to ... 192·0			
	IX : IX a = 21·0 : 13·8 = 100 : 65·71					IX : IX a = 273 : 192 = 100 : 70·33			
			Atmospheric air under 1 atm. pressure.						
X	+1·0	+1·1	27·5	26·4	X	+0·75	+0·6	65·0	64·4
	+1·2	+0·6	27·0	26·4		0·5	0·25	65·0	64·75
	0	0	26·2	26·2		0·0	0·25	64·7	64·45
	0	0	26·5	26·5		0·5	0·5	64·7	64·2
	0					0·5			
	Mean .... 26·4					Mean .... 64·45			
	Corresponding to ... 35·8					Corresponding to ... 334·0			
			Olefiant gas under 1 atm. pressure.						
X a	+0·25	+0·12	20·0	19·9	X a	−1·0	−1·1	60·0	61·1
	0	0·25	20·5	20·25		−1·2	−0·6	60·2	60·8
	+0·5	0·5	20·5	20·0		0	0	60·75	60·75
	0·5	0·5	20·25	19·75		0	0	60·75	60·75
	0·5					0			
	Mean .... 20·0					Mean .... 60·6			
	Corresponding to ... 24·0					Corresponding to ... 239·0			
	X : X a = 35·8 : 24·0 = 100 : 67·03					X : X a = 334 : 239 = 100 : 71·55			
			Atmospheric air under 1 atm. pressure.						
XI	+0·7	+0·85	18·0	17·15	XI.	0	+0·25	62·0	61·75
	+1·0	0·75	17·7	17·0		+0·5	0·5	62·5	62·0
	+0·5	0·50	17·7	17·2		0·5	0·5	62·2	61·7
	0·5	0·5	17·5	17·0		0·5	0·5	62·2	61·7
	0·5					0·5			
	Mean .... 17·1					Mean .... 61·8			
	Corresponding to ... 19·3					Corresponding to ... 265·0			

Transmission of the Heat of a Gas-flame (*continued*).

	Blackened tube.					Unblackened tube.			
No.	Position of equilibrium of the needle.	Mean.	Observed deflection.	Difference.	No.	Position of equilibrium of the needle.	Mean.	Observed deflection.	Difference.
	Ammonia under 1 atm. pressure.								
XIa	0· +0·5 0·7 0·7 0·7	+0·25 0·6 0·7 0·7	11·7 12·0 12·2 12·2	11·45 11·4 11·5 11·5	XIa	+0·5 0·7 0·7 0·7 0·7	+0·6 0·7 0·7 0·7	56·5 56·7 56·5 56·7	55·9 56·0 55·8 56·0
	Mean .... 11·5 Corresponding to ... 11·5 XI : XIa = 19·3 : 11·5 = 100 : 59·58					Mean .... 55·9 Corresponding to ... 224·0 XI : XIa = 265 : 224 = 100 : 84·52			

The relation of the radiation through the various gases is therefore the following :—

	Blackened tube.	Unblackened tube.
Atmospheric air under 1 atm. . . .	100	100
,, ,, 4 millims. . .	102·6	117·21
,, ,, 6 millims. . .	102·3	
,, ,, 8 millims. . .		117·37
Oxygen under 1 atm. . . .	100	100
Hydrogen ,, ,,	98·85	98·26
Carbonic acid ,,	94·11	91·59
Carbonic oxide ,,	94·15	84·52
Protoxide of nitrogen ,,	90·05	88·57
Marsh-gas ,,	98·27	89·87
Olefiant gas ,,	{ 65·71 67·03	70·83 71·55
Ammonia* ,, ,,	59·58	84·52

Since oxygen gave the same values as atmospheric air, it appeared superfluous to investigate nitrogen.

The great difference between the radiation in rarefied space, according as it is investigated in the blackened or in the unblackened tube, led me to determine it once more. This was done with the unblackened tube in such a manner that the deflections were smaller than before, which was effected by remo-

* As the tube had to be filled with ammonia without using the air-pump, it is not improbable that small quantities of atmospheric air may have been left; for in passing the gas through a tube 35 millims. wide, it is very difficult to expel the air completely. For this reason I have not examined cyanogen.

ving the mirror with which the lamp was provided, and placing the lamp itself at greater distances. For it appeared not improbable that the values of the galvanometer corresponding to the deflections, which for greater deflections are not capable of such exact determination as for smaller ones, might have caused the great difference in the radiation through rarefied air, and through air at the normal pressure. With the blackened tube it was not possible to dispense with the mirror; and hence, in order to produce a smaller deflection, the resistance of the wire connected with the galvanometer was increased. In this way the following values were obtained :—

		Blackened tube.					Unblackened tube.		
No.	Position of equilibrium of the needle.	Mean.	Observed deflection.	Difference.	No.	Position of equilibrium of the needle.	Mean.	Observed deflection.	Difference.
			Atmospheric air under 1 atm. pressure.						
I	0·0 0·0 +0·5 +0·5 +0·5	0·0 +0·25 +0·5 +0·5	11·0 11·25 11·5 11·5	11·0 11·0 11·0 11·0	I	+0·75 +2·0 +2·0 +1·75 +2·0	+1·75 +2·0 +1·75 +1·75	28·25 28·75 28·50 28·50	26·50 26·75 26·75 26·75
	Mean .... 11·0 Corresponding to ... 11·0					Mean .... 26·70 Corresponding to ... 36·3			
		Under 4 millims. pressure.					Under 9 millims. pressure.		
Ia	+0·5 +0·5 +0·5 +0·5 +0·5	+0·5 +0·5 +0·5 +0·5	12·0 11·75 12·0 12·0	11·5 11·25 11·5 11·5	Ia	0·0 0·0 −0·2 −0·2 −0·2	0·0 −0·1 −0·2 −0·2	29·0 29·0 29·0 29·0	29·0 29·1 29·2 29·2
	Mean .... 11·5 Corresponding to ... 11·5 I : Ia = 11·0 : 11·5 = 100 : 103·8					Mean .... 29·1 Corresponding to ... 41·2 I : Ia = 36·3 : 41·2 = 100 : 113·5			
			Unblackened tube.						
		Atmospheric air under 1 atm. pres.				Atmospheric air under 6 millims. pres.			
II	−1·0 −1·0 −1·0 −1·0 −1·0	−1·0 −1·0 −1·0 −1·0	13·0 13·0 13·0 13·0	14·0 14·0 14·0 14·0	IIa	−1·5 −1·5 −1·5 −1·5 −1·5	−1·5 −1·5 −1·5 −1·5	14·7 14·7 14·5 15·0	16·2 16·2 16·0 16·5
	Mean .... 14·0 Corresponding to ... 14·0					Mean .... 16·2 Corresponding to ... 16·4 II : IIa = 14 : 16·4 = 100 : 117·1			

Table (*continued*).

	Unblackened tube.								
No.	Position of equilibrium of the needle.	Mean.	Observed deflection.	Difference.	No.	Position of equilibrium of the needle.	Mean.	Observed deflection.	Difference.
	Atmospheric air under 1 atm. pressure.				Under 6 millims. pressure.				
III	0·0	+0·1	11·7	11·6	IIIa	0·0	0·0	12·7	12·7
	+0·2	+0·2	11·5	11·3		0·0	0·0	13·0	13·0
	+0·2	+0·1	11·5	11·4		0·0	0·0	13·0	13·0
	+0·0	+0·1	11·5	11·4		0·0	0·0	13·2	13·2
	+0·2					0·0			
	Mean .... 11·4				Mean .... 13·2				
	Corresponding to ... 11·4				Corresponding to ... 13·2				
					III : IIIa = 11·4 : 13·2 = 100 : 113·2				

As in these determinations differences were obtained similar to those already mentioned, in calculating the relation between the radiation through vacuum and through different gases, I have taken as a basis the values previously found and detailed in page 103. In so doing I have assumed that, if 100 rays pass through atmospheric air under 1 atm. pressure, by using the blackened tube 102·5, and the unblackened tube 117·3, would pass through vacuum. Hence of 100 rays from a gas-flame which pass through vacuum, the following quantities pass through the various gases under the pressure of one atmosphere :—

	Blackened tube.	Unblackened tube.
Vacuum	100	100
Atmospheric air . .	97·56	85·25
Oxygen	97·56	85·25
Hydrogen	96·43	83·77
Carbonic acid . . .	91·81	78·08
Carbonic oxide . . .	91·85	72·05
Protoxide of nitrogen .	87·85	75·50
Marsh-gas	95·87	76·61
Olefiant gas . .	64·10 / 65·39	59·96 / 60·99
Ammonia	58·12	55·00

Influence of Aqueous Vapour on Radiation.

Although it might with certainty be predicted that the small quantity of aqueous vapour which air can take up at the ordinary temperature (not 2 per cent. of its volume at 16° C.) could exercise no influence on the radiation, it appeared desirable to determine experimentally that this supposition was correct. With this view I made comparative determinations of the radiation through perfectly dry air and through air entirely saturated with moisture. The air was passed through several chloride of calcium tubes, and afterwards, by means of a respirator, was drawn through the unblackened tube in such quantity that all the air previously in the tube might be considered as displaced. After the radiation was determined, the air was exhausted by means of the air-pump, and fresh air admitted, which before entering had passed through water slowly and in small bubbles. This air was then exhausted, and a fresh quantity admitted under the same circumstances. After moist air had been thrice successively admitted, it could be assumed that the whole of the air contained in the tube at the temperature 16° C., and pressure 764·6 millims., was saturated with aqueous vapour.

The capacity of dry air and moist air to transmit heat-rays of 100° C. was investigated in exactly the same way by means of the apparatus described in page 87. In this way the following results were obtained:—

Dry air.				Air saturated with aqueous vapour.			
Position of equilibrium of the needle.	Mean.	Observed deflection.	Difference.	Position of equilibrium of the needle.	Mean.	Observed deflection.	Difference.
With the Gas-lamp.							
0·0	−0·25	11·5	11·75	−1·5	−1·35	10·75	12·1
−0·5	−0·25	11·5	11·75	−1·25	−1·25	10·5	11·75
0·0	0·0	12·0	12·0	−1·25	−1·25	10·75	12·0
0·0	0·0	12·0	12·0	−1·25	−1·25	10·5	11·75
0·0				−1·25			
Mean.......... 12·0				Mean .... 12·0			
With the source of heat at 100° C.							
0·0	......	12·5	......	+0·5	+0·5	13·25	12·75
0·0	......	12·75	......	+0·5	+0·5	13·25	12·75
0·0	......	12·5	......	+0·5	+0·5	13·0	12·5
0·0	......	12·5	......	+0·5	+0·75	13·5	12·75
				+1·0			
Mean ...... 12·5				Mean ...... 12·6			

These experiments show that the water present in the atmosphere at 16° C. exercises no perceptible influence on the radiation. That such an influence should be felt as soon as part of the vapour separates as fog, appears very probable.

XV. On New Falls of Meteoric Stones.

To the Editors of the Philosophical Magazine and Journal.

Gentlemen,

SOME weeks ago I received a letter from Professor Joaquin Balcells of Barcelona, stating that he had heard of a large fall of meteorites, accompanied by tremendous detonations, said to have taken place at Cañellas, near Villanova in Catalonia, at some distance from Barcelona, on the 14th of May this year. I have just received from him another letter dated the 27th of June, enclosing an account of his expedition to Cañellas for the purpose of procuring additional information, and also, if possible, some specimens. I give the following translation of this part of his letter:—

"There is no doubt that stones really fell on May 14 at about 1 P.M.; but the greater number are lost, from having fallen with such violence upon the arable land that they could not be found. Two or three fell, however, upon rocks, which they penetrated and cut up to a depth of 5 inches (*pouces*) in a direction towards N.E. at an angle of 45°. They broke into pieces with a tremendous noise and great light. The largest specimen only weighed 18 ounces, and is already destined for the Natural History Museum at Madrid. The second specimen which I saw, was destined for the Professor of Physics, Señor Arbá of Barcelona. I likewise saw other specimens of from 5 to 9 grammes in weight, which were in the hands of the peasants, who would not part with them at any price, because they fancied that these stones, coming from heaven, would bring them good luck. From this cause I was only able to procure for myself one small fragment of 5 grammes weight."

An aërolitic fall is mentioned in 'Cosmos' for April 26, 1861, as having taken place at Tocane St. Apre in Dordogne, France; an aërolite fell on the 14th of February, 1861, with a streak of fire (without noise apparently), in the market place of that town; it weighed only 7 grammes, and is now deposited in the museum of the department at Dordogne.

Another meteoric stone in all probability fell last year on the 8th or 9th of June, about two miles from Raphoe in Co. Donegal, Ireland, on the farm of Dr. M'Clintock of Raphoe, about 2 P.M. It was about the size of a hen's egg, and fell during a storm of thunder, lightning, and hail. It resembled a friable sandstone; but it does not appear there was either any black crust to it, or that there was any fire-ball seen at the time. This fall is mentioned in the 'Londonderry Sentinel' of June 15, 1860. It appears that one portion of this stone has been lost or mislaid, and the remainder had crumbled into sand and has been thrown away. When it fell it broke into three pieces, and was cold and saturated with wet; it was seen to fall by a ploughman of Dr. M'Clintock's, and immediately afterwards picked up by him.

R. P. GREG.

Manchester, July 12, 1861.

XVI. *Solution of a Problem in the Calculus of Variations.* By Professor CHALLIS[*].

IN the July Number of this Magazine the Astronomer Royal has called attention to a problem in the calculus of variations, the solution of which presents some difficulty. The method of solution I am about to propose, which appears to meet the difficulty, is, I believe, new.

The following is the enunciation of the problem as given in Mr. Todhunter's 'History of the Calculus of Variations':—To construct upon a given base AB a curve, such that the superficial area of the surface generated by its revolution round AB may be given, and that its solid content may be a maximum. By the rules of the calculus of variations, the ordinary notation being adopted, the solution of the problem is given by the equation

$$\delta . \int (y^2 + 2ay \sqrt{1+p^2}) dx = 0.$$

Integrating from $x = 0$ to $x = x_1$, and equating separately to zero the parts outside and those under the sign of integration, we have

$$\frac{ap_1 y_1 \delta y_1}{\sqrt{1+p_1^2}} - \frac{ap_0 y_0 \delta y_0}{\sqrt{1+p_0^2}} = 0,$$

$$y + a \sqrt{1+p^2} - \frac{d}{dx} \frac{ayp}{\sqrt{1+p^2}} = 0.$$

The first equation is evidently satisfied by the hypothesis that $y_0 \delta y_0 = 0$ and $y_1 \delta y_1 = 0$. The integration of the other gives

$$\frac{2ay}{\sqrt{1+p^2}} = b - y^2, \quad \cdots \cdots \quad (1)$$

b being the arbitrary constant introduced by the integration.

The next step usually taken in solving this problem is to put $b = 0$, because at the fixed points A and B y vanishes. This appears to have been done previous to the second integration solely because the equation (1) is not integrable unless $b = 0$. I

---

[*] Communicated by the Author.

shall now show that the effect of this step is to restrict the generality of the solution.

Although the above differential equation is not generally integrable, we can obtain from it an exact expression for the length s of the curve. For we have

$$\frac{ds}{dx} = p\frac{ds}{dy} = \frac{2ay}{b-y^2},$$

and

$$p^2 = \left(\frac{2ay}{b-y^2}\right)^2 - 1;$$

$$\therefore ds = \frac{2ay\,dy}{\sqrt{4a^2y^2 - (b-y^2)^2}}.$$

The integration of this equation gives

$$\cos\left(\frac{s}{a}+k\right) = \frac{2a^2+b-y^2}{2a\sqrt{a^2+b}}, \quad \ldots \quad (2)$$

k being a new arbitrary constant. Now at the point A, $y=0$ and $s=0$. Hence

$$\cos k = \frac{2a^2+b}{2a\sqrt{a^2+b}} = \frac{2a^2+b}{\sqrt{(2a^2+b)^2-b^2}}. \quad \ldots \quad (3)$$

Thus the denominator of this value of $\cos k$ is less than the numerator; which is impossible. If s_1 be the length of the arc, it would similarly be found, for the point B, that $\cos\left(\frac{s_1}{a}+k\right)$ is greater than unity. The inference to be drawn from these results is, that when $s=0$ and $s=s_1$, y cannot be equal to zero, but must have some other values, which it is required to find before the solution of the question can proceed. This may be done as follows.

Since the coordinates of the extremities of the arc must have certain values y_0 and y_1 different from zero, it is *necessary* that the circular areas generated by the revolution of these coordinates should be taken into account in the expression for the total surface of the solid; that is, the reasoning must be conducted in the manner proposed by Mr. Todhunter (p. 410). Hence, if r and r' be distances of points of the circular areas from A and B respectively, there will be the two additional terms,

$$\delta\int ar\,dr + \delta\int ar'\,dr',$$

the integrations being taken from $r=0$ to $r=y_0$, and from $r'=0$ to $r'=y_1$. Consequently the total quantity outside the sign of integration will be

$$\left(\frac{p_1}{\sqrt{1+p_1^2}}+1\right)ay_1\,\delta y_1 - \left(\frac{ap_0}{\sqrt{1+p_0^2}}-1\right)ay_0\,\delta y_0.$$

Now, since y_0 and y_1 are quantities to be determined, neither δy_0 nor δy_1 is equal to zero.   Hence

$$\frac{p_1}{\sqrt{1+p_1{}^2}}+1=0 \quad \text{and} \quad \frac{p_0}{\sqrt{1+p_0{}^2}}-1=0;$$

that is, $p_0=+\infty$ and $p_1=-\infty$.   Hence if C and D be the extremities of the ordinates, the curve is at these points continuous with the ordinates; also the equation (1) shows, by putting $p=\infty$, that for each of the points $y^2=b$.

The equation (2) between y and s being put under the form

$$y^2=2a^2+b-2a\,\sqrt{a^2+b}\,\cos\left(\frac{s}{a}+k\right), \quad . \quad . \quad . \quad (4)$$

it will be seen that in the case in which $b=0$, and consequently by (3) $\cos k=1$ and $k=0$, we shall have

$$y^2=2a^2\left(1-\cos\frac{s}{a}\right),$$

$$\text{or } y=2a\sin\frac{s}{2a}.$$

The curve is therefore a semicircle, the radius of which is $2a$. Thus it appears that this is a particular instance contained in the general solution.

The equation (4) shows that for any other value of b the curve differs from a semicircle.   Its length from C to D is readily found.   For since $y_0=y_1$, it follows from (3) that

$$\cos k=\cos\left(\frac{s_1}{a}+k\right).$$

Hence $\frac{s_1}{a}=2\pi$, or $s_1=2\pi a=$ half the circumference of a circle

whose radius is $2a$.

Also the area enclosed by the curve, the ordinates AC and BD, and the base AB, can be exactly determined.   For from the equation

$$p^2=\left(\frac{2ay}{b-y^2}\right)^2-1$$

we obtain

$$ydx=\frac{(b-y^2)y\,dy}{\sqrt{4a^2y^2-(b-y^2)^2}},$$

which being integrated from A to B, gives for the above-mentioned area

$$2a\,\sqrt{b}+2\pi a^2.$$

If the equation (1) were integrable, the values of the three arbitrary constants might be found from the given value of the enclosing surface, and the given condition that the surface passes through A and B.

The solid resulting from the foregoing investigation possesses the characteristic of a maximum, and is the only solution which the problem admits of. It is antecedently evident that the conditions of the question must admit of being satisfied by a *surface* of some kind passing through the given points, and that consequently the calculus of variations could not fail to give such a solution.

Cambridge Observatory,
    July 19, 1861.

XVII. *On the Action of certain Vapours on Films; on the Motions of Creosote on the surface of Water, and other phenomena. In a Letter addressed to* W. A. MILLER, *Esq., M.D., F.R.S. &c., Professor of Chemistry, King's College, London. By* CHARLES TOMLINSON, *Lecturer on Science, King's College School*[*].

MY DEAR MILLER,

A FEW days ago, after a lecture at College on *Cohesion and Adhesion,* one of my pupils asked me, "What is the cause of the remarkable agitation that takes place when sulphuric ether is dropped on the surface of water?" I put that same question to myself more than five and twenty years ago while studying chemistry, and made a large number of experiments on the subject, some of which I have lately had the pleasure of showing you. As you were kind enough to express great interest in them, and a desire that I would complete the inquiry by pushing it to a definite conclusion, I have endeavoured to do so, and will with your permission submit the whole inquiry to you from my own point of view.

But in order to do this I must go back to the years 1837–38, when I obtained a large number of results, and embodied them in three Articles which are now before me in MS. I did not publish them, because the conclusions were not quite satisfactory to my own mind. But being engaged about that time in seeing my 'Students' Manual of Natural Philosophy' through the press, I included the principal experimental results in that work, where you will find them at pages 545–549, and again at pages 553–555. The popular nature of this work doubtless caused these experiments to remain unknown to scientific men; and I venture to think that they will even now strike many with an air of novelty. This is my excuse for a short summary of old results by way of introduction to new ones, or, at least, to such as have not been published.

* Communicated by Dr. Miller.

One of my first experiments, made with the view to ascertain what takes place when ether is brought into contact with water, was the following:—A goblet being quite filled with water was placed in a good light, and the finger well wetted with ether was brought down very near the surface. On looking along this surface in the direction of the light, a cup-shaped depression was evident. I then dusted the surface of the water with a light powder, such as lycopodium, and on presenting the finger wet with ether there was a strong repulsive action; the powder was forcibly driven aside, and the surface of the water was laid bare, evidently in a state of agitation under the influence of the vapour of ether. I tried many sorts of powders with a similar result, but none answered better than lycopodium.

My next proceeding was to try to represent the action of the ether by means of films of oily compounds formed by the spreading of an oily drop on the surface of water. A large soda-water glass was first employed, but a common white dinner-plate showed the effects best. Oil of turpentine, many of the turpentine varnishes, such as gold-size, black Japan, carriage, copal, &c., make admirable films. Some of the fixed and essential oils also answer very well. In experiments of this kind, a single drop of the oily substance must be gently placed on, or rather *delivered* to the water without any fall or disturbance; otherwise the varnish, &c. may sink below the surface in the form of a perfectly spherical bead, and so remain as a good example of cohesion. The best method of obtaining a film is to dip a glass rod into one of the oily liquids, and allow it to drain so that it may deliver only a single drop to the surface of the water. The plate filled with water should be placed before a good light, when a drop of the oily substance, being gently placed on the centre, usually spreads out with a beautiful exhibition of colour, or the film may be quite colourless. Take the latter case. The finger, or, what is better, a flat piece of sponge tied with thread over the rounded end of a glass stirring-rod, wetted with ether and held over the film, produces a cup-shaped cavity, within which a beautiful set of Newton's rings may be seen so long as the sponge is wet with ether. In this case the vapour of ether attenuates the film; the point immediately below the sponge is the point of greatest action, and here the black of the centre of the first series of rings is seen: the action diminishes from this depressed point, where the film is thinnest, and it gradually increases in thickness until it unites with the rest of the film, where colour ceases to be displayed. The film is in fact under tension so long as the ether vapour is acting upon it; and the tension is greatest in the direction of a vertical ray from the sponge to the water, and gradually diminishes as the rays increase in length from the sponge to the

water. Now as these thicknesses vary, for water, from about the
0·38 to the 57·75 millionths of an inch, the film gives all or some
of the series of colours known as Newton's rings. In the second
place, supposing the drop of oil, &c. to form a coloured film (and
most of the turpentine varnishes do so to perfection), the ether-
sponge still developes a system of rings, not always beginning with
those of the first series, but exhibiting the colours of the second,
third, fourth, &c. The diameter of the coloured rings on the oil-
film may vary from ⅓ of an inch to 2 or 3 inches, and in general
they close up and disappear as soon as the ether-sponge is removed or gets dry.

But not only was ether found to produce these effects, but also
liquor ammoniæ, wood-spirit, alcohol, and napbtha, and, as I
afterwards found, chloroform, benzole, bisulphide of carbon—in
fact any substance that throws off vapour with facility—when a
sponge wetted with one of these fluids was held over the film.
The effects were not always so good as with ether, but each substance had peculiar features of its own, and no two films of different oils, &c. exhibited the same phenomena; indeed the films of
the same substance would vary from day to day with varying temperatures of the air and other causes. It may be remarked that
a sponge wetted with ammonia and held over the film often produced so violent an action as to break it up and scatter it about.
It also forms with it a soapy compound which dissolves in the
water. Ether-vapour may also produce as violent an action as
ammonia. For example, a drop of oil of cinnamon produces on
water a mottled film, reminding one of marbled paper. A sponge
dipped in chloroform and held over the film, spreads it with a
development of colour and incipient rings. The ether-sponge is
then powerfully repulsive, spreading, breaking up and scattering
the numerous discs into which a single drop of the oil forms.
But the ammonia-sponge occasions a remarkable spreading, rapid
motion, producing first coloured rings, and then a granular soapy
structure, after which all further action ceases, from the film combining with the water.

In this way I accumulated a large number of results, which
did not appear to throw much light upon the question as to what
takes place between ether and water. There seemed to be a
repulsive action of some kind, and I tried to measure it by means
of a delicately hung torsion-balance of straw, making the straw
carry a piece of filtering-paper which was saturated with water,
while another piece of paper saturated with ether was brought
up near to it; but I obtained no results in this way. I therefore tried the action of ether on a thin sheet of water just sufficient to cover the surface of clean mercury in a wine glass, or
spread over a glass disc 5 or 6 inches in diameter with a ledge of

bees'-wax run round it. On presenting the ether-sponge to the
centre of this sheet. of water, the vapour drove away the water
and left a circular dry space in the centre of the mercury or of
the disc. This is a striking experiment, especially on the surface of the mercury, which shows the effect very well, and allows
the thickness of the sheet of water to be somewhat greater than
on the glass. The cohesion of the water is also well shown by
its not closing up again when the ether-vapour is removed; but
it forms a beautiful circular pupil with a convex surface towards
the mercury. If, instead of the sponge, a dropping-tube containing ether be brought down to the surface of the water on the
glass disc, the water will open as before; and on letting the ether
fall the water will be still further repelled, so as to form a more
convex ring round the liquid ether than it did around its vapour.
Other volatile liquids produce different effects on this sheet of
water. A single drop of creosote placed on the centre disperses
the water, and leaves a long irregular portion of the glass dry.
Several hundred drops of naphtha form a lenticular disc on
the water without displacing it. A single drop of ether brought
down upon it disperses both naphtha and water, and finds its
way to the glass, leaving a convex-bounded ring which slowly
closes in upon the dry space of glass. But the most remarkable
result is with benzole: a disc of this being formed on the sheet
of water, and the ether-sponge held over it, hollows it out
into a thick ring and holds it in that state for some time. In
fact there is a thick convex ring of benzole on water, the force
which holds it open being the vapour of ether. Chloroform
dropped on the sheet of water displaces it, and forms as it were
a cavity, which it occupies by itself as in a pit of solid matter.
The ammonia-sponge, when presented to the chloroform, drives
small globules of the latter out of the cavity, and forms with the
remainder a soapy looking compound which permanently excludes
the water.

Results of this kind, however curious, only served to convince me that it is far more easy to multiply phenomena than
to discover laws. Being strongly impressed with the idea of
repulsion which these results seemed to favour, I tried the effects
of heat; and instead of obtaining a clue to the explanation I was
in search of, I extended the phenomena which we are now accustomed to call the "spheroidal condition" of matter. Boutigny's
striking experiments had not then been contrived; and my first
acquaintance with that class of phenomena was derived from
Dumas's *Chimie Appliquée*, vol. i. p. 31, where is described the
experiment of dropping water into a red-hot platinum crucible.
I varied the experiment by dropping ether into it, and found it
possible to accumulate a considerable quantity. I did the same

with spirits of wine, saline solutions, and even mercury. I changed the nature of the hot surface, and found that ether would roll about on hot mercury, hot oil, and hot water. I also found it perfectly easy to place a drop of water on rape or olive-oil heated to about 400° or 500°. All that was necessary was to deliver the water gently from a dropping-tube to the oil without any fall or splashing; it would then roll about for a considerable time. If ether were also placed on the hot oil, it would unite with the water and form a shell about it.

When M. Boutigny showed his remarkable experiments in your laboratory in 1845, you informed him of my results, and he admitted that they were quite new to him.

On trying some of the fixed oils on the surface of hot water and mercury, turpentine on hot sulphuric acid, &c., the single drop used for each experiment either became spheroidal, or flattened into a disc, the latter rotating on a vertical axis. Experiments of this kind were connected in my mind with the motions of camphor on the surface of water, as well as the agitation of ether and other liquids. I tried the effect of various vapours on camphor while rotating on water; and the results first suggested to me what I think is the key, if not the master-key, to these experiments; for, as I shall hereafter endeavour to show, electricity has some share in these results. A pellet of sponge saturated with benzole held over a rotating piece of camphor, had the effect of increasing the rotations of the smaller fragments to such an extent that the form of the camphor often became quite indistinct, and appeared as a mere cloudy haze. After an experiment of this kind, the morsel of camphor displays one or two brilliant points where the structure is altered and the light abundantly reflected. These points are the effect of solution. The benzole vapour seizes the camphor and begins to dissolve it; and during this action there is a contest between the cohesion of the camphor and the film formed by the condensed vapour of benzole, and the diffusive tendency of the water: there is a contest, in fact, between cohesion and adhesion. The formation of this film about the camphor may be plainly seen by holding another sponge dipped in chloroform instead of benzole over camphor: it first produces a rapid spinning, the effect of solution; but nearly as fast as the solution is formed the camphor is displaced by the water, and a solid opake crust of camphor is formed. Bisulphide of carbon held over the spinning camphor drives it about; and when a drop of that substance is placed on the water, it does not arrest the motions of the camphor, but follows it about. Persian naphtha causes the camphor to spin more rapidly; and a drop of that substance placed on the water will pursue the camphor with great swiftness, combine with it, and form a film which sometimes displays colour.

Taking *solution* as one of the keys, if not the master-key, to these experiments, and defining it as you do in your 'Chemical Physics' as a case of *adhesion* of a liquid to a solid (often overcoming cohesion), or of a liquid to a liquid, and moreover defining *saturation* as an equilibrium between the forces of adhesion and cohesion, I began to see more clearly the *rationale* of my numerous experiments. In order to test the adhesion view of the case, I looked about me for some fluid of nearly the same density as water, with but slight adhesion to it (that is, very sparingly soluble in it), but one which would readily saturate a moderate quantity of water (that is, a liquid whose cohesion would soon balance the adhesion of the water), so that whatever visible action might take place between the two would admit of being renewed from time to time by increasing the quantity of the water. After many trials I found exactly what was required in creosote. Although this substance is slightly heavier than water (spec. grav. 1·059), yet by carefully delivering a drop to the surface of water from the end of a glass rod it will not sink; the under surface of the drop will, however, present a convex bulge below the general surface of the water.

I wonder whether it ever occurred to a chemist to place a drop of creosote on the surface of water. It presents a most singular appearance: it flattens out into a disc with a silvery reflexion of the light, and sails about on the water with some speed, while it is all the time rapidly agitated with a motion that gives it the appearance of a living creature. Its edge vibrates with rapid crispations; it darts out small globules, which immediately begin a series of motions of rotation and translation on their own account. In the mean time a silvery film of creosote spreads over the surface of the water: the parent globule and the smaller globules become less energetic; they perform a number of motions among themselves, moving about in circular or curved paths, carefully avoiding each other, and reminding one of the water insects which may be seen sporting on the surface of a pond in summer. Sometimes the larger globules will remain still, and the smaller ones will rotate in little lakes, which they seem to clear out for themselves in the film to disport in. After some time they all come to rest; but again begin to move for a time, once more to come to rest, and, it may be, again to rotate.

Now there is evidently a struggle going on between the cohesion of the drop and the adhesion of the water. These two forces are so nicely balanced that it seems doubtful for a time which will prevail. The water tends to adhere to and diffuse the creosote; the cohesion of the creosote tends to prevent this action, and the struggle between the two is manifested by a series of vibrations which take place at the edge of the disc; the creosote tends to spread, its cohesive force struggles to prevent the spreading. Small globules, however, are constantly being torn away from the parent drop, and as these spin round and disappear, they leave a film which gradually covers more or less of the surface of the water. The motions of the parent disc and of the globules cease; but as the film becomes dissolved by the water, the motions (now very slow ones) set in again with the formation of another film, which in its turn is dissolved. But if the quantity of water be small, the globules soon cease to be disturbed, since the water has become saturated, or the adhesion of the water balances the cohesion of the globule, and hence the quiescence.

When I showed you this experiment, it naturally struck you as a case of *solution,* and you inquired whether the motions of the disc would take place in a saturated solution. I had already ascertained that if, when one drop, or rather disc of creosote, is in rapid agitation, and also moving about on the water, a second drop be placed by its side, it stops the motions of the first disc, and is itself soon brought to rest. In other words, the water is sooner saturated. I further ascertained that if, when the discs of creosote are at rest in a small quantity of water—a wine-glass full, for example—the contents of such wine-glass were transferred to a half-pint goblet nearly full of water (the transfer being gently made, so as to keep the creosote on the surface), the motions of the discs will begin again with as much energy as before. When this half-pint of water is saturated and all motion has ceased, the motions will be renewed if the half-pint of water be poured into a large soda-water glass, two-thirds filled with water.

Thus, by increasing the quantity of water, we remove it further from the point of saturation, and favour the gyrations and other motions of the creosote. The same effect may be produced if (the quantity of water being fixed) we increase its solvent power. For example, a disc of creosote is very lively for some minutes on the surface of water; but as the latter becomes saturated the motions decline, and then cease. If we now touch the water with a glass rod wet with acetic acid, a new solvent power is conferred on the water, and the motions of the creosote set in again. So also these and other motions may be produced if we hold over the quiescent globule the vapours of substances in which creosote is soluble. The ether-sponge will cause the disc to display its crispations, and to dart out numerous globules. The ammonia sponge restores motion to

the globule while the latter is under the immediate influence of
the gas. . The benzole sponge acts with remarkable energy,
causing the creosote to spread to the utmost verge of its cohesion, and then to split with a jerking kind of motion. Bisulphide of carbon has also a powerful action. The motions are
also exceedingly curious when a drop of one of these substances is
placed on the surface of water with the creosote, and about half
an inch away from it. There is an interchange of action between them, an apparent repulsion, but in fact a contest between the solvent powers of the water and of the naphtha, &c.
for the creosote. If a drop of bisulphide of carbon be placed
near the creosote, the former remains lenticular, and does not
evaporate so quickly as when placed alone on water; the creosote bombards it with a number of small globules, and is active
only on the side nearest to it. If a drop of bisulphide be
placed on either side of the creosote, the latter will carry on the
bombardment from two sides. A drop of benzole is, in certain
states of saturation of the water, so exceedingly active that it
pursues the creosote, and attacks it with life-like motions.
The latter darts about as if seeking to escape from it, and re-
minds one of an aquatic beetle pursuing its prey on the surface
of a pond. These globules of creosote, benzole, &c., have thus
a decided action on each other, but the lenticular discs which
they form on water do not coalesce; they often attract each
other with increasing velocity up to a certain point, and then
repel each other, sailing slowly away until the force of attraction
brings them near together again.

The phenomena may be further complicated by the action of
vapours on the two dissimilar lenses. Thus creosote in the
presence of a naphtha globule may be very lively, and the
ether-sponge held over the creosote may break it up with a
very decided action, and yet have little or no action on the
naphtha.

I should detain you too long were I to describe the varied
phenomena of this kind. They have an especial interest to me,
because they led me to explain some of the other results partly
by the same laws of solution. Thus one of my earliest experiments—the repulsion of lycopodium dust on the surface of
water by ether—was not altogether a case of repulsion, but also
of attraction—the attraction, in fact, of ether for water. But
first, as to repulsion. That many of these phenomena display
repulsion cannot be denied by any one who has witnessed them.
The very circumstance of ether assuming so readily the vaporous
state implies a strong degree of repulsion. As the ether leaves
the saturated sponge, its comparatively feeble cohesion becomes
changed into repulsion, *i. e.* the liquid becomes vapour, which

vapour being very heavy, falls down upon the excessively attenuated film, whose thickness must be measured by millionths of an inch. This heavy repulsive vapour then sweeps aside the film in a regular manner, producing in some cases a large central opening, and then a thinning off of the film, sufficient to allow of the interference of the light required for the phenomena of coloured rings. This descending vapour, moreover, meets with an ascending vapour from the film, and the contact of the two produces further complications in the way of chemical and electrical effects which I will not ask you to consider at present. I will only remark, *first,* that the electrical condition of these vapours is very decided, and must be considered before the explanation of these phenomena is at all complete; and *secondly,* that during solution, as of a salt in water, current electricity is (as I have reason to believe) largely concerned in the action and in the motions of the solid in the solvent. But my immediate point is with the mechanical action of this repulsive vapour from the wet sponge upon the film, the light powders, and the mobile discs of creosote, &c. There is a mechanical action about this vapour which goes some way to explain the production of the rings. A stream of gas from a gas-bottle produces them, and, what is equally remarkable, the vapour of oil of turpentine will repel a turpentine film. A drop of turpentine on water forms a good film, often at first without colour, but as it evaporates it displays colour. As soon, however, as the film is formed, if a sponge dipped into the same bottle from which the drop was taken be held over the film, it will spread out into very beautiful rings.

But now let us consider the influence of attraction. Admitting for the moment that water becomes saturated with one-eighth of its bulk of ether, it will be found that there is a very strong attraction between ether and its vapour for water up to this point. The water quickly becomes saturated, but the combination up to saturation is very energetic, especially at first. The great density of ether-vapour also assists the attractive force of the water; it falls down, the water seizes it, and disperses any dust or powder that may be in its way. If a film of oil or varnish be interposed, it begins to dissolve that, and thins it out to the gradually decreasing thicknesses necessary to the display of Newton's rings, often making a complete perforation, half an inch in diameter, in the film to get to the water, and showing the rings of colour at the inner edge of this perforation.

That this explanation is likely to be true the following experiment will prove:—Seven parts of water and one part of ether were shaken up together and poured out into a small porcelain dish; the surface was dusted with lycopodium, and the ether-sponge presented to it: there was no action; the powder was not displaced or disturbed. The solution of ether was then boiled and filtered, and, when cold, the surface was again dusted with the powder. The ether-sponge now produced a repulsion of the powder, not so decidedly as with plain water, but still a good repulsion.

The solution of ether was also made to carry an oil-film. A drop of varnish formed an exquisite series of coloured rings, and the ether-sponge also displayed some very beautiful rings; but after a minute or two, when the adhesion between the solution and the film was complete, the ether-sponge was powerless.

It may also be mentioned that a vapour acts differently on the film according as it has a greater attraction for the water or for the film. If it has a strong attraction for the water, it will thin out and disperse the film. If its attraction is strong for the film, it will gather it up, thicken it, and deprive it of colour. Thus with a film of oil of lavender the ether scatters and disperses, while the benzole sponge thickens and attracts; in fact the benzole vapour condenses into little discs, which unite with the film. So also if a drop of oil of peppermint be placed on water, it spreads out into a honeycombed film displaying colour. If the ether-sponge be presented, the vapour pours down in a cataract and powerfully displaces the film (a very common effect of ether-vapour on films of the essential oils); whereas, if the turpentine sponge be held over it, the scattered parts of the film sail up to it, gather themselves together, and form a number of thickening lenticules.

I do not like to intrude further on your patience at present. Should this letter not disappoint the interest you have kindly expressed in this inquiry, I will trouble you with a second, and in the mean time subscribe myself,

Your attached friend,

CHARLES TOMLINSON.

King's College, London,
June 22, 1861.

XVIII. *On the Reduction of Observations of Underground Temperature; with Application to* Professor Forbes's *Edinburgh Observations, and the continued Calton Hill Series.* By Professor WILLIAM THOMSON, *F.R.S.*

[Concluded from p. 34.]

17. *APPLICATION to Thirteen Years' Observations* (1842-54) *at the Thermometric Station, Calton Hill.*—The observations on thermometers fixed by Professor Forbes at the different depths in the rock of Calton Hill, have been regularly continued weekly till the present time by the staff of the Royal Edinburgh Observatory, and regularly corrected to reduce to true temperatures of the bulbs, on the same system as before. Tables of these corrected observations, for the twelve years 1842 to 1854 inclusive, having been supplied to me through the kindness of Professor Piazzi Smyth, I have had the first five terms of the harmonic expression for each year determined in the following manner[*]:— In the first place, the observations were laid down graphically, and an interpolating curve drawn through the points, according to the method of Professor Forbes. The four curves thus obtained represent the history of the varying temperature, at the four different depths respectively, as completely and accurately as it can be inferred from the weekly observations. The space corresponding to each year was then divided into thirty-two equal parts (the first point of division being taken at the begining of the year), and the corresponding temperatures were taken from the curve. The coefficients of the double harmonic series (cosines and sines) for each year were calculated from these data, with the aid of the forms given by Mr. Archibald Smith, and published by the Board of Admiralty, for deducing the harmonic expression of the error of a ship's compass from observations on the thirty-two points. The general form of the harmonic expression being written thus—

$$V = A_0 + A_1 \cos 2\pi t + B_1 \sin 2\pi t + A_2 \cos 4\pi t + B_2 \sin 4\pi t + \&c.,$$

where V denotes the varying temperature to be expressed, and t the time, in terms of a year as unit. The following Table shows the results which were obtained, with the exception of the values of A_0:—

---

[*] The operations here described, involving, as may be conceived, no small amount of labour, were performed by Mr. D. M'Farlane, my laboratory assistant, and Mr. J. D. Everett, now Professor of Mathematics and Natural Philosophy in King's College, Windsor, N.S.

A₁.	B₁.	A₂.	B₂.	A₃.	B₃.	A₄.	
	−5·00	+·01	+·25	+·60	+·06	+·23	−·71
	−4·80	−·15	+·03	+·10	+·10	+·12	−·26
+·34	−2·73	−·12	−·13	−·08	−·04	+·01	−·04
+·68	−·14	·00	−·07	−·02	−·04	−·01	−·02
−4·75	−5·11	+·17	+·91	+1·23-	+·30	+·79	−·17
−1·63	−4·38	−·20	+·61	+·45	+·42	+·32	+·30
	−2·04	−·18	−·08	−·05	+·17	−·03	+·10
+·62	+·12	·00	−·02	−·01	−·01	·00	·00
−5·29	−4·53	−·05	+·70	+·74	+·71	+·08	+·49
−2·11	−4·09	+·22	+·50	+·20	+·50	−·06	+·20
	−2·15	+·18	+·05	+·11	+·13	−·05	−·01
	−·02	−·03	−·02	·00	−·03	−·01	−·02
5·17	−5·01	−·17	+·56	+·67	+·29	−·28	+·02
	−4·38	+·07	+·30	·00	+·18	−·04	−·08
	−2·15	+·12	+·06	−·01	−·03	·00	+·02
	+·13	+·04	·00	+·01	+·02	+·01	+·02
	−5·17	+·03	+1·05	−·86	+·64	+·00	−·49
	−4·64	−·38	+·44	−·63	−·39	−·11	−·22
	−2·70	−·30	−·17	−·14	−·45	·00	−·07
	−·22	−·02	−·17	+·03	−·11	−·03	−·06
5·36	−5·31	+·69	+·24	−·18	−·81	−·02	−·14
	−4·58	+·18	+·32	+·11	−·39 ·	−·05	−·04
	−2·37	−·03	+·17	+·12	+·14	+·03	+·02
	+·16	−·01	+·04	+·01	+·03	+·01	+·03
	−4·46	+·33	+·27	+·29	+·35	+·45	−·30
−2·32	−4·16	+·13	+·27	+·02	+·23	+·28	+·09
	−2·15	+·04	+·16	−·01	+·09	+·04	+·11
	+·10	−·01	+·03	·00	+·02	−·01	+·01
	−4·44	+·05	+1·14	−·66	−·10	−·48	−·69
−1·85	−3·97	−·20	+·45	−·28	−·15	+·01	−·25
	−2·06	−·23	+·04	+·04	−·06	+·09	−·05
+·57	+·03	·00	−·02	+·01	+·02	·00	+·01
−5·40	−4·50	−·12	+·70	−·54	−·82	−·15	−·42
	−4·15	−·22	+·31	+·03	−·47	+·11	−·17
	−2·27	−·15	−·04	−·10	−·05	+·04	+·01
	−·04	+·01	−·03	+·01	·00	−·01	−·01
4·18	−4·53	+·12	+·96	−·09	+·31	+·22	+·18
	−3·92	−·19	+·53	−·18	+·07	−·03	+·14
	−1·99	−·22	+·01	−·04	−·06	−·05	−·02
	+·02	+·01	−·05	·00	−·01	−·14	−·01
−4·92	−4·80	+·20	+1·32	+·64	−·24	−·46	+·31
−1·87	−4·25	−·23	+·71	+·15	+·10	−·31	−·02
+·54	−2·24	−·26	+·05	+·01	+·09	−·01	−·07
+·61	−·03	−·12	−·07	+·01	−·04	·00	−·02
	−5·43	+·83	+·30	+·11	+·27	+·18	+·19
−1·92	−4·57	+·38	+·41	−·05	+·17	+·06	+·13
+·76	−3·15	−·01	+·21	−·01	·00	−·01	+·03
+·62	+·18	−·39	+·03	·00	+·10	+·01	+·03
−5·69	−4·56	−·61	+·53	·00	−·15	+·15	−·20
−2·48	−4·27	−·50	−·01	·00	−·13	+·08	−·03
	−2·31	−·12	+·21	+·02	−·03	+·02	+·01
	−·03	+·02	−·02	·00	−·01	−·01	−·01
	−4·835	+·114	+·687	+·150	+·0778	+·05462	−·149
−2·122	−4·320	−·0838	+·375	−·00615	+·0185	+·02923	−·016
+·5415	−2·332	−·0985	+·00923	−·01846	−·00778	+·006154	+·003
+·6231	−·0200	−·0385	−·0285	−·00231	−·00462	−·01462	−·003

The values which were found for A_0 should represent the annual mean temperatures. They differ slightly from the annual means shown in the Royal Observatory Report, which, derived as they are from a direct summation of all the weekly observations, must be more accurate. The variations, and the final average values of these annual means, present topics for investigation of the highest interest and importance, as I have remarked elsewhere (see British Association's Report, Section A, Glasgow, 1855); but as they do not belong to the special subject of the present paper, their consideration must be deferred to a future occasion.

18. *Theoretical Discussion.*—The mean value of the coefficients in the last line of the Table being obtained from so considerable a number of years, can be but very little influenced by irregularities from year to year, and must therefore correspond to harmonic functions for the different depths, which would express truly periodic variations of internal temperature consequent upon a continued periodical variation of temperature at the surface.

19. According to the principle of the superposition of thermal conductions, the difference between this continuous harmonic function of five terms for any one of the depths, and the actual temperature there at the corresponding time of each year, would be the real temperature consequent upon a certain real variation of superficial temperature: Hence the coefficients shown in the preceding Table afford the data, first by their mean values, to test the theory explained above for simple harmonic variations, and to estimate the conductivity of the soil or rock, as I propose now to do; and secondly, as I may attempt on a future occasion, to express analytically the residual variations which depend on the inequalities of climate from year to year, and to apply the mathematical theory of conduction to the non-periodic variations of internal temperature so expressed.

20. Let us accordingly now consider the complex harmonic functions corresponding to the mean coefficients of the preceding Table; and in the first place, let us reduce the double harmonic series in each case to series in each of which a single term represents the resultant simple harmonic variation of the period to which it corresponds, in the manner shown by the proposition and formulæ of § 3 above.

21. On looking to the annual and semiannual terms of the series so found, we see that their amplitudes diminish, and their epochs of maximum augment, with considerable regularity from the less to the greater depths. The following Table shows, for the annual terms, the logarithmic rate of diminution of the amplitudes, and the rate of retardation of the epoch between the points of observation in order of depth:—

TABLE IV.—Average of Thirteen Years, 1842 to 1854; Trap
Rock of Calton Hill.

Depths below surface, in French feet.	Diminution of Napierian logarithm of amplitude, per French foot of descent.	Retardation of epoch in circular measure, per French foot of descent.
3 to 6 feet	·1310	·1233
6 to 12 ,,	·1163	·1140
12 to 24 ,,	·1121	·1145
3 to 24 ,,	·1160	·1156

22. The numbers here shown would all be the same if the
conditions of uniformity supposed in the theoretical solution
were fulfilled. Although, as in the previous comparisons, the
agreement is on the whole better than might have been expected,
there are certainly greater differences than can be attributed to
errors of observation. Thus the means of the numbers in the two
columns are for the three different intervals of depth in order as
follows :—

Mean deductions from
amplitude and epoch.

   3 to 6 feet . . . . . ·127
   6 to 12 ,, ... ·115
   12 to 24 ,, ... ·113

numbers which seem to indicate an essential tendency to diminish at the greater depths. This tendency is shown very decidedly in each column separately; and it is also shown in each of
the corresponding columns, in tables given above, of results derived from Professor Forbes's own series of a period of five years.

23. There can be no doubt that this discrepance is not attributable to errors of observation, and it must therefore be owing
to deviation in the natural circumstances from those assumed for
the foundation of the mathematical formula. In reality, none
of the conditions assumed in Fourier's solution is rigorously fulfilled in the natural problem; and it becomes a most interesting
subject for investigation to discover to what particular violation
or violations of these conditions the remarkable and systematic
difference discovered between the deductions from the formula
and the results of observation is due. In the first place, the
formula is strictly applicable only to periodic variations, and the
natural variations of temperature are very far from being precisely periodic; but if we take the average annual variation
through a sufficiently great number of years, it may be fairly
presumed that irregularities from year to year will be eliminated :
and that the discrepance we have now to explain does not de-

pend on residual inequalities of this kind seems certain, from the fact that it exists in the average of Professor Forbes's first five years' series no less decidedly than in that of the period of thirteen years following.

24. For the true explanation we must therefore look either to inequalities (formal or physical) in the surface at the locality, or to inequalities of physical character of the rock below. It may be remarked, in the first place, that if the rates of diminution of logarithmic amplitude and of retardation of epoch, while less, as they both are, at the greater depths, remained exactly equal to one another, the conductivity must obviously be greater, and the specific heat less in the same proportion inversely, at the greater depths. For in that case, all that would be necessary to reconcile the results of observation with Fourier's formula, would be to alter the scale of measurement of depths so as to give a nominally constant rate of diminution of the logarithmic amplitude and of the retardation of epoch; and the physical explanation would be, that thicker strata at the greater depths, and thinner strata at the less depths (all of equal horizontal area), have all equal conducting powers and equal thermal capacities *.

25. Now in reality, a portion, but only a portion, of the discrepance may be done away with in this manner; for while the logarithmic amplitudes and the epochs each experience a somewhat diminished rate of variation per French foot of descent at the greater depths, this diminution is much greater for the former than for the latter; so that, although the mean rates per foot on the whole 21 feet are as nearly as possible equal for the two (being ·1160 for the logarithmic amplitudes, and ·1156 for the epoch), the rate of variation of the logarithmic amplitude exceeds that of the epoch by about 6 per cent. on the average of the stratum of 3 to 6 feet; and falls short of it by somewhat more than 2 per cent. in the lower stratum, 12 to 24 feet. To find how much of the discrepance is to be explained by the variation of conductivity and specific heat in inverse proportion to one another at the different depths, we may take the mean of the

---

* The "conducting power" of a solid plate is an expression of great convenience, which I define as the quantity of heat which it conducts per unit of time when its two surfaces are permanently maintained at temperatures differing by unity. In terms of this definition, the specific conductivity of a substance may be defined as the conducting power per unit area of a plate of unit thickness. The conducting power of a plate is calculated by multiplying the number which measures the specific conductivity of its substance by its area, and dividing by its thickness.

The *thermal capacity of a body* may be defined as the quantity of heat required to raise its mass a unit (or one degree) of temperature. The specific heat of a substance is the thermal capacity of a unit quantity of it, which may be either a unit of weight or a unit of bulk.

rates of variation of logarithmic amplitude and of epoch at each depth, and alter the scale of longitudinal reckoning downwards, so as to reduce the numerical measures of these rates to equality. This, however, we shall not do in either the five years' or the thirteen years' term, which we have hitherto considered separately, but for a harmonic annual variation representing the average of the whole eighteen years 1837 to 1854.

26. By taking for each depth the coefficients A_1, B_1 (not explicitly shown above), derived from the first five years' average, and multiplying by 5; taking similarly the coefficients A_1, B_1 for the succeeding thirteen years' average, and multiplying by 13; adding each of the former products to the corresponding one of the latter, and dividing by 18; we obtain, as the proper average for the whole eighteen years, the values shown in the following Table, in the columns headed A_1, B_1. The amplitudes and epochs shown in the next columns are deduced from these by the formulæ $\sqrt{(A_1{}^2 + B_1{}^2)}$ and $\tan^{-1}\dfrac{B_1}{A_1}$ respectively:—

TABLE V.—Annual Harmonic Variation of Temperature in Calton Hill, from 1837 to 1844 inclusive.

Depths.	A_1 in degrees Fahr.	B_1 in degrees Fahr.	Amplitudes in degrees Fahr.	Epochs in degrees and minutes.
3 feet	−5·184	−4·989	7·1949	223 54
6 ,,	−2·080	−4·416	4·8812	244 47
12 ,,	+ ·5961	−2·3345	2·4094	284 19
24 ,,	+ ·6311	+ ·0306	·6319	362 47

From these, as before, for ten terms of five years and of thirteen years separately, we deduce the following:—

TABLE VI.—Average of Eighteen Years, 1837 to 1844; Trap Rock of Calton Hill.

Depths below surface in French feet.	Diminution of logarithmic amplitude, per French foot of descent.	Retardation of epoch in circular measure, per French foot of descent.
3 to 6 feet	·1286	·1215
6 to 12 ,,	· ·1177	·1150
12 to 24 ,,	·1115	·1141
3 to 24 ,,	·1157	·1154

27. Hence we have as final means, of effects on logarithmic amplitudes and on epochs, for the average annual variation on the whole period of eighteen years,—

1. From depth 3 feet to 6 feet . . . ·1250
2.   „   6 „   12 „ . . . ·1163
3.   „   12 „   24 „ . . . ·1128

If now, in accordance with the proposed plan, we measure depths, not in constant units of length, but in terms of thicknesses corresponding to equal conducting powers and thermal capacities, and if we continue to designate the thickness of the first stratum by its number 3 of French feet, our reckoning for the positions of the different thermometers will stand as follows :—

TABLE VII.

Thermometers numbered downwards.	Depths in true French feet below No. 1.	Depths in terms of conductive equivalents.
I.	0	0
II.	3	3
III.	9	$3 + \dfrac{\cdot1163}{\cdot1250} \times 6 = 8\cdot58$
IV.	21	$8\cdot58 + \dfrac{\cdot1128}{\cdot1250} \times 12 = 19\cdot41$

According to this way of reckoning depths, we have the following rates of variation of the logarithmic amplitudes, and of the epochs separately, reduced from the previously stated means for the whole period of eighteen years :—

TABLE VIII.

Portions of rock.	Rates of diminution of logarithmic amplitude per French foot, and conductive equivalents.	Rates of retardation of epoch per French foot, and conductive equivalents.
Between thermometers Nos. I. and II.	·1286	·1215
„ „ II. and III.	·1265	·1236
„ „ III. and IV.	·1236	·1264
Between thermometers Nos. I. and IV.	·1252	·1248

28. Comparing this Table with the preceding Table VI., we see that the discrepancies are very much diminished; and we cannot doubt that the conductive power of the rock is less in the lower parts of the rock, and that the amount of the variation is approximately represented by Table VII. We have, however, in Table VIII. still too great discrepancies to allow us to consider variation in the value of kc as the only appreciable deviation from Fourier's conditions of uniformity.

29. In endeavouring to find whether these residual discre-

pancies are owing to variations of k and c not in inverse proportion one to the other, I have taken Fourier's equation

$$c\frac{dv}{dt} = k\frac{d^2v}{dx^2} + \frac{dk}{dx}\frac{dv}{dx},$$

where v denotes the temperature at time t, and at a distance x from an isothermal plane of reference (a horizontal plane through thermometer No.' 1., for instance); k the conductivity, varying with x; and c the capacity for heat of a unit of volume, which may also vary with x. In this equation I have taken

$$v = a\epsilon^{-P}\cos\left(\frac{2\pi t}{T} - Q\right),$$

where P and Q are functions of x, assumed so as to express, as nearly as may be, the logarithmic amplitudes, and the epochs, deduced from observation.   I have thus obtained two equations of condition, from which I have determined k and c, as functions of x. . The problem of finding what must be the conductivity and the specific heat at different depths below the surface, in order that, with all the other conditions of uniformity perfectly fulfilled, the annual harmonic variation may be exactly that which we have found on the average of the eighteen years' term at Calton Hill, is thus solved. The result is, however, far from satisfactory. The small variations in the values of P and Q which we have found in the representation of the observed temperatures require very large and seemingly unnatural variations in the values of k and c.

30. I can only infer that the residual discrepancies from Fourier's formula shown in Table VIII. are not with any probability attributable to variations of conductivity and specific heat in the rock, and conclude that they are to be explained by irregularities, physical and formal, in the surface. It is possible, indeed, that thermometric errors may have considerable influence, since there is necessarily some uncertainty in the corrections estimated for the temperatures of the different portions of the columns of liquid above the bulbs; and before putting much confidence in the discrepancies we have found as true expressions of the deviations in the natural circumstances from Fourier's conditions, a careful estimate of the probable or possible amount of error in the observed temperatures should be made. That even with perfect *data* of observation as great discrepancies should still be found in final reductions such as we have made, need not be unexpected when we consider the nature of the locality, which is described by Professor Forbes in the following terms :—

The position chosen for placing the thermometer was below the surface " in the Observatory enclosure on the Calton Hill, at a height of 350 feet above the sea. The rock is a porphyritic trap, with a somewhat earthy basis, dull and tough fracture. *The exact position is a few yards east of the little transit house.* There are *also other buildings in the neighbourhood.* The ground rises slightly to the east, and *falls abruptly to the west at a distance of fifteen yards.* The immediate surface is flat, *partly covered with grass, partly with gravel*."*

I have marked by italics those passages which describe circumstances such as it appears to me might account for the discrepancies in question.

31. *Application to Semiannual Harmonic Terms.*—The harmonic expressions given above (§ 15) for the average periodic variations for the three stations of Professor Forbes's original series of five years' observations, contain semiannual terms which are obviously not in accordance with theory. The retardations of epochs and the diminutions of amplitudes are, on the whole, too irregular to be reconcileable by any supposition as to the conductivities and specific heat of the soils and rocks involved, or as to the possible effects of irregularity of surface; and in two of the three stations the amplitude of the semiannual term is actually greater as found for the six-feet deep than for the three-feet deep thermometer, which is clearly an impossible result. The careful manner in which the observations have been made and corrected seems to preclude the supposition that these discrepancies, especially for the three-feet and six-feet thermometers, for which the amplitudes of the semiannual terms are from ·28° to ·74° (corresponding to variations of double those amounts, or from ·56° to 1°·48), can be attributed to errors in the *data.* It must be concluded, therefore, that the semiannual terms of those expressions do not represent any truly periodic elements of variation, and that they rather depend on irregularities of temperature in the individual years of the term of observation. Hence, until methods for investigating the conduction inwards of non-periodic variations of temperature are applied, we cannot consider that the special features of the progress of temperature during the five years' period at the three stations, from which our apparent semiannual terms have been derived, have been theoretically analysed. But, as we have seen, every irregularity depending on individual years is perfectly eliminated when the average annual variation over a sufficiently great number of years is taken. Hence it becomes interesting to examine particularly the semiannual terms for the eighteen years' average of the Calton Hill thermometers, which we now proceed to do.

* Professor Forbes "On the Temperature of the Earth," Trans. Roy. Soc. Edinb. 1846, p. 194.

32. Calculating as above (§ 26), for the coefficients A_1, B_1, the average values of A_2 and B_2, from Professor Forbes's results for his first five years' term, and from the averages for the next thirteen years shown in Table III. above, we find the values of A_2 and B_2 shown in the following Table. The amplitudes and epochs are deduced as usual by the formulæ $\sqrt{(A_2^2 + B_2^2)}$ and $\tan^{-1} \dfrac{B_2}{A_2}$. These reductions I only make for the three-feet deep and the six-feet deep thermometers, since, for the two others, as may be ju.ged by looking at the thirteen years' average shown in the former Table, the amounts of the semiannual variation do not exceed the probable errors in the data of observation sufficiently to allow us to draw any reliable conclusions from their apparent values.

TABLE IX.—Average Semiannual Harmonic Term, from Eighteen Years' Observations at Calton Hill.

Depths below surface, in French feet.	A_2 in degrees Fahr.	B_2 in degrees Fahr.	Amplitudes in degrees Fahr.	Epochs in degrees and minutes.
3 feet.	°·1518	°·5842	°·604	75° 26′
6 feet.	·0461	·3911	·394	96 43

The ratio of diminution of the amplitude here is $\dfrac{·604}{·394}$, or 1·53, of which the Napierian logarithm is ·426. Dividing this by 3, we find

$$·142$$

as the rate of diminution of the logarithmic amplitude per French foot of descent.

The retardation of epoch shown is $21° 17'$; and therefore the retardation per French foot of descent is $7° 6'$, or, in circular measure,

$$·1239.$$

If the data were perfect for a periodical variation, and the conditions of uniformity supposed in Fourier's solution were fulfilled, these two numbers would agree, and each would be equal to $\sqrt{\dfrac{2\pi k}{c}}$. Hence, dividing them each by $\sqrt{2}$, we find

Apparent values of $\sqrt{\dfrac{\pi c}{k}}$.

·100   (by amplitudes).
·0877   (by epochs).

The true value of $\sqrt{\frac{\pi c}{k}}$ must, as we have seen, be ·116, to a very close degree of approximation.

33. When we consider the character of the reduction we have made, and remember that the data were such as to give no semblance of a theoretical agreement when the first five years' term of observations was taken separately, we may be well satisfied with the approach to agreement presented by these results, depending as they do on only eighteen years in all, and we may expect that, when the average is of a still larger term of observation, the discrepancies will be much diminished. In the mean time we may regard the semiannual term we have found for the three-feet deep thermometer as representing a true feature of the yearly vicissitude; and it will surely be interesting to find whether it is a constant feature for the locality of Edinburgh, to be reproduced on averages of subsequent terms of observation.

34. It may be remarked that the nearer to the equator is the locality, the greater relatively will be the semiannual term; that within the tropics the semiannual term may predominate, except at great depths; and that at the equator the tendency is for the annual term to disappear altogether, and to leave a semiannual term as the first in a harmonic expression of the yearly vicissitude of temperature. The facilities which underground observation affords for the analysis of periodic variations of temperature when the method of reduction which I have adopted is followed, will, it is to be hoped, induce those who have made similar observations in other localities to apply the same kind of analysis to their results; and it is much to be desired that the system of observing temperatures at two, if not more depths below the surface may be generally adopted at all meteorological stations, as it will be a most valuable means for investigating the harmonic composition of the annual vicissitudes.

III. *Deduction of Conductivities.*

35. Notwithstanding the difficulty we have seen must attend any attempt to investigate all the circumstances which must be understood in order to reconcile perfectly the observed results with theory, the general agreement which we have found is quite sufficient to allow us to form a very close estimate of the ratio of the conductivity of the rock to its specific heat per unit of bulk. Thus, according to the means deduced from the whole period of eighteen years' observation, the average rate of variation of the logarithmic amplitude of the annual term through the whole space of twenty-one feet is ·1157, and of the epoch of the same term, ·1154. The mean of these, or ·1156, can differ but very little

from the true average value of $\sqrt{\dfrac{\pi c}{k}}$ for the portion of rock between the extreme thermometers.

36. Dividing π by the square of the reciprocal of this number, we find 235·1 as the value of $\dfrac{k}{c}$, or, as we may call it, the conductivity of the rock in terms of the thermal capacity of a cubic foot of its own substance. In other words, we infer that all the heat conducted in a year (the unit of time) across each square foot of a plate one French foot thick, with its two sides maintained constantly at temperatures differing by 1°, would, if applied to raise the temperature of portions of the rock itself, produce a rise of 1° in 235 cubic feet. As it is difficult (although by no means impossible) to imagine circumstances in which the heat, regularly conducted through a stratum maintained, with its two sides, at perfectly constant temperatures, could be applied to *raise* the temperatures of other portions of the same substance, we may vary the statement of the preceding result, and obtain the following completely realizable illustration.

37. Let a large plate of the rock, everywhere one French foot thick, have every part of one of its sides (which, to avoid circumlocution, we shall call its lower side) maintained at one constant temperature, and let portions of homogeneous substance, at a temperature 1° lower, be continually placed in contact with the upper surface, and removed to be replaced by other homogeneous portions at the same lower temperature, as soon as the temperature of the matter actually thus applied rises in temperature by $\frac{1}{1000}$ of a degree. If this process is continued for a year, the whole quantity of the refrigerating matter thus used to carry away the heat conducted through the stratum must amount to 235,000 cubic feet for each square foot of area, which will be at the rate of ·00745 of a cubic foot per second. We may therefore imagine the process as effected by applying an extra stratum ·00745 of a foot thick every second of time. This extra stratum, after lying in contact for one second, will have risen in temperature by $\frac{1}{1000}$ of a degree. By means of the information contained in this apparently unpractical statement, many interesting problems may be practically solved, as I hope to show in a subsequent communication.

38. The value of $\sqrt{\dfrac{\pi c}{k}}$, derived from the whole eighteen years' period of observation (·1156), differs so little from that (·1154) found previously (§ 16) from Professor Forbes's observations and reductions of the first five of the years, that we may feel much confidence in the accuracy of the values ·1098 and

·06744, which, from his five years' data alone, we found (§ 16) for the corresponding constant with reference to the sand at the Experimental Garden and the sandstone of Craigleith Quarry. From them, calculating as above (§ 36), we find 260·5 and 690·7 as the values of $\frac{k}{c}$ for the terrestrial substances of these localities respectively,—results of which the meaning is illustrated by the statements of §§ 36 and 37.

39. To deduce the conductivities of the strata in terms of uniform thermal units, Professor Forbes had the "specific heats" of the substances determined experimentally by M. Regnault. The results, multiplied by the specific gravities, gave for the thermal capacities of portions of the three substances, in terms of that of an equal bulk of water, the values ·5283, ·3006, and ·4623 respectively. Now these must be the values of c if the thermal unit in which k is measured is the thermal capacity of a French cubic foot of water. Multiplying the values of $\frac{k}{c}$ found above by these values of c, we find for k the following values:—

Trap-rock of Calton Hill.	Sand of Experimental Gardens.	Sandstone of Craigleith.
124·2	78·31	319·3

The values found by Professor Forbes were—

111·2	82·6	298·3

Although many comparisons have been made between the conducting powers of different substances, scarcely any data as to thermal conductivity in absolute measure have been hitherto published, except these of Professor Forbes, and probably none approaching to their accuracy. The slightly different numbers to which we have been led by the preceding investigation are no doubt still more accurate.

40. To reduce these results to any other scale of linear measurement, we must clearly alter them in the inverse ratio of the square of the absolute lengths chosen for the units*. The

---

* Because the absolute amount of heat flowing through the plate across equal areas will be inversely as the thickness of the plate, and the effect of equal quantities of heat in raising t e temperature of equal areas of the water will be inversely as the depth of the water. The same thing may be perhaps more easily seen by referring to the elementary definition of thermal conductivity (footnote to § 11 above). The absolute quantity of heat conducted across unit area of a plate of unit thickness, with its two sides maintained at temperatures differing by always the same amount, will be directly as the areas, and inversely as the thickness, and therefore simply as the absolute length chosen for unity. But the thermal unit in which these quantities are measured, being the capacity of a unit bulk of water, is

length of a French foot being 1·06575 of the British standard foot, we must therefore multiply the preceding numbers by 1·18581 to reduce them to convenient terms.

41. We may, lastly, express them in terms of the most common unit, which is the quantity of heat required to raise the temperature of a grain of water by 1°; and to do this we have only to multiply each of them by 7000 × 62·447, being the weight of a cubic foot in grains.

42. The following Table contains a summary of our results as to conductivity expressed in several different ways, one or other of which will generally be found convenient :—

TABLE X.—Thermal Conductivities of Edinburgh Strata, in British Absolute Units [Unit of Length, the English Foot].

Description of terrestrial substance.	Conductivities in terms of thermal capacity of unit bulk of substance $\left(\dfrac{k}{c}\right)$.			Conductivities in terms of thermal capacity of unit bulk of water (k).			Conductivities in terms of thermal capacity of one grain of water.
	Per ann.	Per 24 h.	Per second.	Per ann.	Per 24 h.	Per second.	Per second.
Trap-rock of Calton Hill.	267·0	·7310	·000008461	141·1	·3863	·000004471	1·9544
Sand of Experimental Gardens...	295·9	·8100	·000009375	88·9	·2435	·000002818	1·2319
Sandstone of Craigleith Quarry ...	784·5	2·1478	·00002486	362·7	·9929	·00001149	5·0225

43. The statements (§§ 36 and 37) by which the signification of $\dfrac{k}{c}$ has been defined and illustrated, require only to have *cubic feet of water* substituted for *cubic feet of rock*, in their calorimetric specifications, to be applicable similarly to define and illustrate the meaning of the conductivity denoted by k. The fluidity of the water allows a modified and somewhat simpler explanation, equivalent to that of § 36, to be now given as follows :—

44. If a long rectangular plate of rock one foot thick, in a position slightly inclined to the horizontal, have water one foot deep flowing over it in a direction parallel to its length, and if the lower surface of the plate be everywhere kept 1° higher in temperature than the upper, the water must flow at the rate of k times the length of the plate per unit of time in order that the heat conducted through the plate may raise it just 1° in tempe-

directly as the cube of the unit length, and therefore the numbers expressing the quantities of heat compared will be inversely as the cubes of the lengths chosen for unity, and directly as these simple lengths; that is to say, finally, they will be inversely as the squares of these lengths.

rature in its flow over the whole length. [It must be understood here that the plate becomes warmer, on the whole, under the lower parts of the stream of water, its upper surface being everywhere at the same temperature as the water in contact with it, while its lower surface is, by hypothesis, at a temperature 1° higher.] If, for instance, the plate be of Calton Hill trap-rock, the water must, according to the result we have found, flow at the rate of 141·1 times its length in a year, or of ·3863 of its length in twenty-four hours, to be raised just 1° in temperature in flowing over it. Thus, water one French foot deep, flowing over a plane bed of such rock at the rate of ·3863 of a mile in twenty-four hours, will in flowing one mile have its temperature raised 1° by heat conducted through the plate. The rates required to fulfil similar conditions for the sand of the Experimental Gardens and the sandstone of Craigleith Quarry are similarly found to be ·2435 of the length and ·9929 of the length in twenty-four hours.

XIX. *Chemical Notices from Foreign Journals.* By E. Atkinson, *Ph.D., F.C.S.*

[Continued from p. 62.]

LOURENCO* has succeeded in converting glycerine into propylic glycol, and glycol into ordinary alcohol. The formula of monohydrochloric glycerine only differs from that of propylic glycol by containing chlorine in the place of an atom of hydrogen. This relation, as well as that between monohydrochloric glycol and the corresponding monoatomic alcohol, is indicated in the following formulæ:—

$$C^3 H^7 Cl \Theta^2$$
Monohydrochloric glycerine.

$$C^3 H^7 Cl \Theta^4$$
Monohydrochloric glycol.

$$C^2 H^5 Cl \Theta^4$$
Hydrochloric glycol.

$$C^3 H^8 \Theta^3$$
Propylic glycol.

$$C^3 H^8 \Theta^4$$
Propylic alcohol.

$$C^2 H^6 \Theta^4$$
Alcohol.

By treating these hydrochloric ethers with nascent hydrogen, this chlorine is removed and replaced by hydrogen.

When monohydrochloric glycerine, diluted with its volume of water, was placed in contact with excess of sodium-amalgam, and the mixture left at the ordinary temperature, the amalgam was slowly decomposed with a slight disengagement of hydrogen, and formation of an abundant deposit of chloride of sodium.

* *Comptes Rendus,* May 20, 1861.

The reaction was terminated in two or three days; the contents were digested with strong alcohol, filtered, and the alkaline liquor neutralized with acetic acid and distilled. When the water and alcohol had passed over, the thermometer rapidly rose, and between 180° and 190° a colourless oily liquid distilled over, which was found to have all the properties, physical and chemical, of Wurtz's propylglycol.

Ordinary glycol was converted into alcohol by an analogous process. Hydrochloric glycol, diluted with half its volume of water, was mixed with excess of sodium-amalgam. When left at the ordinary temperature, it became converted into ordinary alcohol; at a higher temperature the character of the reaction was different, some oxide of ethylene being formed. The product of the reaction was distilled off in the water-bath, and the distillate, dried by carbonate of potash and by caustic baryta, was found to have the composition and properties of ordinary alcohol.

It is exceedingly probable that the transformation of propylic glycol into propylic alcohol would take place just in the same way.

Strecker has published* a very interesting investigation on the relations between guanine, xanthine, caffeine, theobromine, and creatinine.

He describes a modification of the method of preparing guanine, and also describes some compounds which it forms with nitrate of silver and with baryta, and which are analogous to the compounds with sarcine and xanthine.

Unger found, by oxidizing guanine with chlorate of potash, that an acid was formed which had the formula $C^{10} H^5 N^4 O^9$, and which he named *peruric* acid. These experiments have been repeated by Strecker, who has found that the acid in question is parabanic acid, $C^6 H^2 N^2 O^6$. The mother-liquor contains, in addition to this, the hydrochlorate of a new organic base, which he calls *guanidine*. It is a strongly alkaline body, which forms neutral crystalline salts with most acids. The free base is a crystalline mass with a caustic taste, which rapidly attracts water and carbonic acid from the air, and thus is unfitted for direct analysis. Its formula, $C^2 H^5 N^3$, was determined from the analysis of its platinum-salt, $C^2 H^5 N^3 HCl, PtCl^2$; its carbonate, $C^2 H^5 N^3 HO, CO^2$; and its oxalate, $C^2 H^5 N^3 2 HO, C^4 H^2 O^6$.

The decomposition of guanine may be thus expressed:—

$$C^{10} H^5 N^5 O^2 + 2 HO + 6 O = C^6 H^2 N^2 O^6 + C^2 H^5 N^3 + 2 CO^2.$$

Guanine.   Parabanic acid.   Guanidine.

* Liebig's *Annalen*, May 1861.

When treated with nitric acid, guanidine is oxidized into urea:

$$C^2 H^5 N^3 + 2 HO = C^2 H^4 N^2 O^2 + N H^3.$$
$$\text{Guanidine.} \qquad\qquad \text{Urea.}$$

When heated, guanidine yields products analogous to those from mellone; it stands in close relation to cyanamide, and may be considered as *cyanodiamine*, $C^2 H^5 N^3 = \left. \begin{array}{l} C^2 N \\ H^2 \\ H^3 \end{array} \right\} N$ $\}N.$

In its decompositions guanine is closely allied to creatine; for Dessaignes found that this body, when oxidized, is resolved into oxalic acid and methyluramine, $C^4 H^7 N^3$, which is nothing more than methylguanidine, $C^2 H^3 \left. \begin{array}{l} C^2 N \\ H^4 \end{array} \right\} N^2$. And under other circum-

stances Dessaignes found that creatine yields an acid, $C^8 H^4 N^2 O^6$, which appears to be a methylparabanic acid, judging from a comparison of the properties and of the formulæ of the two acids,

$$\left. \begin{array}{l} C^4 O^4 \\ C^2 O^2 \\ H^2 \end{array} \right\} N^2 \qquad\qquad \left. \begin{array}{l} C^4 O^4 \\ C^2 O^2 \\ C^2 H^3.H \end{array} \right\} N^2$$
$$\text{Parabanic acid.} \qquad\qquad \text{Methylparabanic acid.}$$

From its decompositions, creatine may be regarded as composed of cyanamide and of methylglycocol (sarcosine); and its relations to creatinine and guanine are evident from a comparison of the formulæ

$$\left. \begin{array}{l} C^2 N \\ H^2 \\ C^4 H^2 O^2 \\ C^2 H^3 \\ H^2 \end{array} \right\} N \, \left. \begin{array}{l} \\ N \\ O^2 \end{array} \right. = \left. \begin{array}{l} C^2 N \\ C^4 H^2 O^2 \\ C^2 H^3 \\ H^4 \end{array} \right\} N^2 \, \left. \begin{array}{l} \\ O^2 \end{array} \right. \quad \left. \begin{array}{l} C^2 N \\ C^4 H^2 O^2 \\ C^2 H^3 \\ H^2 \end{array} \right\} N^2 \quad \left. \begin{array}{l} C^4 H^2 \\ C^2 N \\ C^4 H^2 O^2 \\ H^3 \end{array} \right\} N^2;$$
$$\text{Creatine.} \qquad\qquad \text{Creatinine.} \qquad \text{Guanine.}$$

that is, guanine may be regarded as creatinine containing an atom of hydrogen in the place of methyle, and containing besides 2 equivs. of cyanogen, in which it is analogous to a number of organic bases, such as cyaniline, cyanocodeine, &c.

Strecker has also investigated several points in reference to the artificial xanthine which he prepared[*], which prove that it is identical with the natural product, and not merely isomeric, as had been suggested. He finds that xanthine is soluble in 570 parts of boiling, and in 2120 parts of cold water.

From its formula, xanthine may be considered as belonging to an homologous series along with theobromine and caffeine, as

---

[*] Phil. Mag. vol. xviii. p. 135.

their formulæ only differ by $nC^2 H^2$,

$C^{10} H^4 N^4 O^4$	$C^{14} H^8 N^4 O^4$	$C^{16} H^{10} N^4 O^4$
Xanthine.	Theobromine.	Caffeine.

Strecker has indeed shown that theobromine may be converted into caffeine, to which it bears the same relation as aniline to methylaniline. When theobromine is dissolved in ammonia, and nitrate of silver added, a precipitate forms which readily dissolves in warm ammonia. On boiling this, a granular crystalline precipitate is deposited, which is theobromine-silver. When this is treated with iodide of methyle, iodide of silver is formed, and a crystalline body, which is caffeine:

$$C^{14} H^7 Ag N^4 O^4 + C^2 H^3 I = C^{16} H^{10} N^4 O^4 + Ag I.$$

Theobromine-silver.   Iodide of   Caffeine.
                        methyle.

Strecker attempted a similar transformation of xanthine into theobromine by treating the silver compound of xanthine, which contains 2 equivs. of silver, $C^{10} H^2 Ag^2 N^2 O^4$, with 2 equivs. of iodide of methyle. He obtained a body of the same composition as theobromine, but which is simply isomeric, and not identical with it.

Gerhardt had already pointed out that *cholestrophane*, a product of the decomposition of caffeine, might be regarded as a bimethylated parabanic acid; and from this point of view Hlasiwetz endeavoured, but without success, to convert parabanic acid into this body. Strecker, by treating parabanate of silver with iodide of methyle, has succeeded in effecting the change; that is, by replacing 2 equivs. of hydrogen in parabanic acid by 2 equivs. of methyle, cholestrophane is obtained, as is indicated in the formula

$\left.\begin{array}{c} C^4 O^4 \\ C^2 O^2 \\ H^2 \end{array}\right\} N^2$	$\left.\begin{array}{c} C^4 O^4 \\ C^2 O^2 \\ 2C^2 H^3 \end{array}\right\} N^2$
Parabanic acid.	Cholestrophane.

Hence, while the oxidation of uric acid, of guanine, and of xanthine gives parabanic acid, from caffeine, dimethylparabanic acid is obtained; and from creatine, methylparabanic acid.

Stas* has published the results of an investigation of the atomic weights of the elements which has occupied him during several years. His object was to subject Prout's hypothesis, that all the atomic weights of the elements are multiples by whole numbers of that of hydrogen as unity, to a more rigorous scrutiny, and to ascertain whether there was in fact a common divi-

* *Bulletin de l'Académie Royale de Belgique*, sér. 2. vol. x. No. 8. Liebig's *Annalen*, Supplement, May 1861.

sor for the atomic weights. The original memoir must be consulted for an account of the great labour and pains taken to
ensure the purity of the substances used, accuracy in the weigh-
ings, and the exclusion of all sources of error from the apparatus
employed. In these respects the memoir is probably one of the
most important which has ever appeared. In one point his determinations differ from those of preceding chemists—that is, in
the quantities taken, which are very considerably larger than
those usually employed; the balances used were on a corresponding scale, and of the most perfect construction. It may be
mentioned that he found any kind of glass attacked when heated
directly by flame; but that, when protected by a coating of charcoal or of magnesia, it could be heated to softening without undergoing any alteration in weight. In all cases, where practicable, Stas used vessels of platinum.

The present communication refers to the atomic weights of
nitrogen, chlorine, sulphur, potassium, sodium, and silver, and
their relation to the atomic weight of hydrogen. The author
made for this purpose the synthesis of the following substances—
chloride of silver, sulphide of silver, nitrate of silver, nitrate of
lead, sulphate of lead; the analysis of chlorate of potash, and
sulphate of silver; and he determined the relations between the
atomic weights of the following substances:—silver and chloride
of potassium, silver and chloride of ammonium, silver and chloride of sodium, nitrate of silver and chloride of potassium, nitrate
of silver and chloride of ammonium.

A most essential point was the preparation of pure metallic
silver. Several different methods were used, one of which consisted in digesting a dilute solution of silver with finely divided
phosphorus. This action is slow, but the metal, after having
been digested with ammonia, is quite pure.

The synthesis of chloride of silver was effected by heating
pure silver in a current of chlorine, and expelling the excess of
chlorine by dry air. In three experiments there were obtained
for 100 parts of silver, 132·841, 132·843, and 132·843 respectively of chloride of silver. Chloride of silver was also prepared
by passing hydrochloric acid gas over the surface of a solution of
silver in acid. The precipitate was dried in the same vessel, and
fused in an atmosphere of hydrochloric acid. Two experiments
of this kind, one of which was made with 400 grammes of silver,
gave respectively 132·849 and 132·846 of chloride of silver from
100 parts of silver. In another series the silver was dissolved
in nitric acid, and precipitated by a feeble excess of hydrochloric
acid, washed, and fused in a current of the gas. This gave for
100 of silver 132·846 of chloride. In the last series, solution
of silver was precipitated by chloride of ammonium, the precipi-

tate washed, dried, and fused in hydrochloric acid. From 100 of silver there were obtained 132·848 and 132·8417 of chloride of silver.

The synthesis of nitrate of silver was effected by dissolving pure silver in nitric acid, evaporating to dryness in a Bohemian-glass vessel, heating in a current of dry air, and fusion till constant. The mean of eight experiments gave for the relation between silver and nitrate of silver

$$100 : 157\text{·}492.$$

Sulphuret of silver was prepared by heating a known weight of silver, either in the vapour of pure sulphur or in sulphuretted hydrogen gas. In the mean of five experiments, 100 parts of silver gave 114·8522 of sulphuret.

The method for estimating the relation between the equivalents of silver and of the chlorides of sodium, ammonium, and potassium, was that of Gay-Lussac. It consisted in dissolving a known weight of silver in nitric acid, and adding an equivalent of the chloride in question; the equivalents being calculated according to Prout's law. The excess of silver after precipitation, estimated by standard solutions, gave the required relations.

The chloride of potassium was prepared either from carefully purified chlorate of potash, from platinochloride of potassium, from nitrate of potash, or from tartrate of potash. The mean of 19 experiments with quantities of pure silver varying from 3 to 32 grms., gave for the relation,

$$Ag : K\,Cl = 100 : 69\text{·}103.$$

Fourteen similar experiments with chloride of sodium led to the relation

$$Ag : Na\,Cl = 100 : 54\text{·}2078.$$

The relation between the equivalent of silver and that of chloride of ammonium, as obtained from ten experiments, was

$$Ag : NH^4\,Cl = 100 : 49\text{·}5944.$$

For the relation between nitrate of silver and chloride of potassium the numbers were

$$AgO\,NO^5 : K\,Cl = 100 : 43\text{·}8758;$$

and for the relation between nitrate of silver and chloride of ammonium,

$$AgO\,NO^5 : NH^4\,Cl = 100 : 31\text{·}488.$$

The preparation of pure lead was attended with greater difficulties even than that of pure silver. It was effected by digesting solution of acetate of lead in a leaden vessel with thin lead-foil until all copper and silver were precipitated. Sulphuric acid was then added so as to form sulphate of lead, which was converted into carbonate by digestion with ammonia and carbonate of ammonia. This carbonate of lead was dried and reduced by cyanide of potassium. Pure lead was also obtained by the reduction of the chloride, either by fusion with cyanide of potassium, or with a mixture of cyanide and black flux.

The lead was converted into nitrate by heating it with strong nitric acid until it was completely converted into nitrate, and then evaporating to dryness in the same vessel in a current of dry air at 140°.

100 parts of lead gave 159·974 of nitrate of lead.

The synthesis of sulphate of lead gave the relation

$$Pb : PbO\ SO^3 = 100 : 146\cdot4275.$$

The analysis of chlorate of potash, that is, the determination of the quantity of oxygen, was effected either by heating chlorate of potash alone, or by heating it with hydrochloric acid. The first of these methods gave the relation

$$KO\ ClO^5 : KCl = 100 : 60\cdot8428;$$

and the second

$$KO\ ClO^5 : KCl = 100 : 60\cdot849.$$

The analysis of sulphate of silver was effected by reduction with hydrogen.

100 pure sulphate gave 69·208 silver.

The relations of the atomic weights investigated are seen in the following Table, in which the numbers obtained are compared with those which would be required on the hypothesis of Prout:—

				Prout.	Stas.
Ag	:	Cl . . .	= 100 :	32·87	= 100 : 32·8445
Ag	:	S . . .	= 100 :	14·814	= 100 : 14·852
Ag	:	AgO NO⁵.	= 100 :	157·404	= 100 : 157·473
Ag	:	K Cl . .	= 100 :	68·981	= 100 : 69·103
Ag	:	Na Cl . .	= 100 :	54·166	= 100 : 54·2078
Ag	:	NH⁴ Cl .	= 100 :	49·537	= 100 : 49·5944
AgO NO⁵	:	K Cl . .	= 100 :	43·823	= 100 : 43·878
AgO NO⁵	:	NH⁴ Cl	= 100 :	31·470	= 100 : 31·488
Pb	:	PbO NO⁵.	= 100 :	159·903	= 100 : 159·969
Pb	:	PbO SO³ .	= 100 :	146·376	= 100 : 146·427
KO ClO⁵	:	KCl . .	= 100 :	60·816	= 100 : 60·846
AgO SO³	:	Ag . . .	= 100 :	69·23	= 100 : 69·203

The following are the atomic weights deduced from Stas's experiments as compared with those usually admitted, and taking oxygen at 8 :—

	Stas.	
Chloride of potassium . .	74·59	74·5
Silver	107·943	108
Chlorine	35·46	35·5
Potassium	39·18	89
Sodium	23·05	23
Ammonium	18·06	18
Nitrogen	14·041	14
Sulphur	16·037	16
Lead (synthesis of PbO SO³)	103·458	103·5
Lead (synthesis of PbO NO⁵)	103·460	

It will thus be seen that the atomic weights of nitrogen and ammonium differ by 4·02 instead of by 4. Hence if the synthesis of nitrate of silver is correct, the atomic weight of hydrogen cannot be exactly one-eighth that of water. The author proposes to return to the synthesis of water, for his researches lead him to believe that the error will be found there rather than in the synthesis of nitrate of silver.

Stas, who at the commencement of his researches had confidence in the hypothesis of Prout, has been led to the conclusion that it is untenable, as well as the modification introduced by Dumas[*]. He says, in conclusion, " So long as, in establishing the laws of matter, we are to adhere to experiment, we must consider Prout's law as an entire delusion, and must regard the undecomposable bodies of our globe as distinct beings having no relation to each other. The undoubted analogy of properties observed in certain elements must be sought for in other causes than those derived from the ratio of the weight of their acting masses."

In reference to Stas's investigation, Marignac[†], in giving an abstract of it, objects to the conclusion as too absolute. He says " he can only form a clear idea of the degree of confidence which the determination of an atomic weight deserves when this weight has been obtained by several methods absolutely independent of each other, based on the analysis of several distinct compounds."

When M. Stas, as a control of the synthesis of nitrate of silver, refers to the experiments by which he determined the proportion between this nitrate and the chloride of potassium, which latter is directly connected with silver, M. Marignac only sees a confirmation of the exactitude of the experiments themselves, but by no means a control of the method.

---

[*] Phil. Mag. vol. xvi. p. 209.

[†] *Bibliothèque Universelle,* vol. ix. p. 202. *Répertoire de Chimie,* May 1861.

If, for instance, we suppose that nitrate of silver does not contain its elements exactly in the proportion of their atomic weights, even the best methods for its analysis or synthesis will give with the same inaccuracy the relation of this weight.

In developing his ideas on this point, M. Marignac refers to his own experiments on monohydrated sulphuric acid. He has shown that this compound, which was always considered very stable, is really very unstable; it is only when it contains a slight excess of water (1 per cent.) that it is quite stable, otherwise the least increase of temperature causes it to give off vapours of anhydrous sulphuric acid.

Who could say *à priori* that the sulphuret and nitrate of silver are not capable of retaining at high temperatures a trace of sulphur or nitric acid, seeing that sulphuric acid can retain a slight excess of water far above 100°?

Causes of error of this kind, to which others might be added, lead M. Marignac to doubt whether the differences between experiment and Prout's law do not arise from the imperfection of experimental methods.

M. Marignac has another objection against Stas's conclusion. "If the numbers of M. Stas do not absolutely coincide with those of Prout, they approximate to such an extent that they in fact cannot be considered accidental. What has been said of Mariotte's and Gay-Lussac's laws may be applied to Prout's laws. These laws, long considered absolute, were found to be inexact when the experiments were made with the accuracy attained by M. Regnault and by M. Magnus. Nevertheless they will always be considered as expressing natural laws, whether from a practical point of view (for by their means the changes of volume in gases may be calculated with sufficient accuracy) or from the point of view of theory; for they probably express the normal law of these changes of volume abstracted from some disturbing influences, the effects of which may some day be calculated. The same may be believed of Prout's law."

M. Marignac terminates his remarks by an observation due to Dumas :—The fundamental principle which led Prout to propound his law, that is to say, the idea of the unity of matter and all the conceptions which have been based on this principle, is quite independent of the magnitude of the unit which might serve as common divisor of the atomic weights. Whether this weight be that of an atom of hydrogen, of half or a quarter of an atom, or whether it be any infinitely small fraction, all these considerations would nevertheless retain the same degree of probability. The relations between the constitution merely become somewhat less simple between the different elements.

XX. *Remark on the Tactic of* 9 *Elements. By* J. J. SYLVESTER, *M.A., F.R.S., Professor of Mathematics at the Royal Military Academy, Woolwich**.

AT the end of my preceding paper in this Magazine for July, I hazarded an opinion that any grouping of 28 synthemes comprising the 84 triads belonging to a system of 9 elements, might be regarded as made up of 1 syntheme of monomial triads, 18 synthemes of binomial triads, and 9 of trinomial triads, the denominations (monomial, binomial, and trinomial) having reference to a duly chosen distribution of the 9 elements into 3 nomes of 3 elements each. This conjecture is capable of being brought to a very significant, although not decisive test, by examining a peculiar and important distribution of the 28 synthemes into 7 sets of 4 synthemes each, the property of each set being that its 12 triads contain amongst them all the 36 *duads* appertaining to the 9 elements. I discovered this mode of distribution very many years ago; but it was first published independently by a mathematician whose name I forget, either in the Philosophical Magazine or in the Cambridge and Dublin Mathematical Journal, I think at some time between the years 1847–53. A similar mode of distribution exists for any system of elements of which the number is a power of 3. Without pausing to give the law of formation, I shall simply observe that for 9 elements we may take as a basic arrangement the square

$$
\begin{array}{ccc}
1 & 2 & 3 \\
4 & 5 & 6 \\
7 & 8 & 9
\end{array}
$$

and form from this, by a symmetrical method, the annexed six derived arrangements :—

$$
\begin{array}{ccc}
\begin{array}{ccc} 7 & 1 & 2 \\ 3 & 5 & 6 \\ 4 & 8 & 9 \end{array} &
\begin{array}{ccc} 7 & 2 & 3 \\ 1 & 4 & 5 \\ 6 & 8 & 9 \end{array} &
\begin{array}{ccc} 9 & 2 & 3 \\ 1 & 5 & 6 \\ 4 & 7 & 8 \end{array} \\
\\
\begin{array}{ccc} 4 & 3 & 1 \\ 7 & 5 & 6 \\ 2 & 8 & 9 \end{array} &
\begin{array}{ccc} 5 & 2 & 3 \\ 7 & 6 & 4 \\ 1 & 8 & 9 \end{array} &
\begin{array}{ccc} 4 & 2 & 3 \\ 8 & 5 & 6 \\ 1 & 9 & 7 \end{array}
\end{array}
$$

and reading off each of these squares in lines, in columns, and in right and left diagonal fashion, we obtain the 7 sets of 4 synthemes each referred to, viz.

$$
\begin{array}{ccc}
123 & 456 & 789 \\
\hline
147 & 258 & 369 \\
\hline
159 & 267 & 348 \\
\hline
168 & 249 & 357
\end{array}
$$

* Communicated by the Author.

712 356 489	·723 145 689	923 156 478
734 158 269	716 248 359	914 257 368
759 164 238	749 256 318	988 264 317
768 139 254	758 219 346	967 218 354
431 756 289	523 764 189	423 859 197
472 358 169	571 268 349	481 259 367
459 362 178	569 241 378	457 261 889
468 879 152	548 279 361	469 287 351

If, now, we take any distribution of the 9 elements into nomes other than 123, 456, 789, we shall find that some of the synthemes will contain trinomials, some binomials only, but others (in number either 9 or 18, according to the distribution chosen) will contain binomials and trinomials mixed; but if we adopt 123, 456, 789 as the nomes, then it will be found that the remaining 27 synthemes (after excluding the monomial syntheme 123, 456, 789) will consist of 18 purely binomial triads, and 9 purely trinomial triads. The former will consist of the first, second, and fourth synthemes of the 6 derived groups; the latter of the second, third, and fourth of the basic group, and of the second synthemes of each of the 6 derived groups.

It may be remembered that there are two types or species of groupings of trinomial triad synthemes appertaining to 3 nomes of 3 elements; one of these species contains 4, the other 36 individual groupings. It may easily be ascertained that the grouping above indicated belongs to the first (the less numerous) of these species. Again, there are 3 types or species of groupings of binomial triad synthemes appertaining to the same system of nomes; one containing 12, one 24, and the third 108 groupings. The grouping with which we are here concerned will be found to belong to the *second* of these species,—that denoted by the symbols $\begin{vmatrix} a \\ e \end{vmatrix}$ in my paper of last month. Hence, then, we derive a very considerable presumption in favour of the opinion which I advanced at the close of my preceding paper on Tactic, and derived, too, from a case apparently unfavourable to the verisimilitude of the conjecture; for a natural subdivision of 28 things into 7 sets of 4 each seems at first sight hardly compatible with another natural division into 3 sets of 1, 18, and 9 respectively. Notwithstanding this seeming incompatibility, we have found that the two methods of decomposition do coexist, owing essentially to the fact that the 7 sets (of 4 synthemes each) stand not in a relation of indifference set to set, but are to be considered as composed of a base and 6 derivatives indifferently related to the base and to each other. The theory of these 7 sets is extremely curious, and well worthy of being fully investigated by the student of tactic, but cannot be gone into within the limits suitable to the pages of · a philosophical miscellany.

Before taking final leave of the subject (at all events for the present, and in the pages of this Magazine), as I have been questioned as to the meaning of the important word "syntheme," derived from συν θημα, I repeat that a "syntheme" is the general name for any consociation of the single or combined elements of a given system of elements in which each element is once and once only contained. A *nome*, from νεμω (*to divide*), means a consociation of a certain number out of a given system of elements; and a binomial, trinomial, or r-nomial combination of any specified sort, means a combination whose elements are dispersed between 2, 3, or r of the nomes between which the entire system of elements is supposed to have been divided.

K, Woolwich Common,
    July 14, 1861.

P.S. I have found the date and place of the resolution into 7 sets referred to in the text; it is given in a paper by Mr. Kirkman, vol. v. p. 261 of the Cambridge and Dublin Mathematical Journal for 1850. His 7 squares, whose horizontal, vertical, and two diagonal readings (like mine) constitute the 7 sets in question, are substantially as follows:—

$$\begin{array}{ccc} 1 & 2 & 3 \\ 4 & 5 & 6 \\ 7 & 8 & 9 \end{array}$$

$$\begin{array}{ccc} 1\ 2\ 4 & 4\ 5\ 7 & 7\ 8\ 1 \\ 5\ 6\ 7 & 8\ 9\ 1 & 2\ 3\ 4 \\ 8\ 9\ 3 & 2\ 3\ 6 & 5\ 6\ 9 \end{array}$$

$$\begin{array}{ccc} 7\ 1\ 2 & 1\ 4\ 5 & 4\ 7\ 8 \\ 3\ 4\ 5 & 6\ 7\ 8 & 9\ 1\ 2 \\ 6\ 8\ 9 & 9\ 2\ 3 & 3\ 5\ 6* \end{array}$$

On assuming 123, 456, 789 as the three nomes, the 28 syn-

* By changing the positions of the lines and columns of the six derivative squares, which may be done without affecting the value of their *readings*, they may be represented under the form following, which will be seen to render much clearer their relation to the primitive square:—

412	623	423	‖	239	129	127
756	745	956	‖	451	563	453
389	189	178	‖	786	784	896

themes contained in the sets will be found to consist of purely monomial, binomial, and trinomial synthemes.

Thus there would be an additional presumption in favour of the supposed law of *homonomial resolubility*, provided that Mr. Kirkman's solution were essentially distinct in type from my own; his binomial and trinomial systems, taken separately, coincide in type with those afforded by my solution, notwithstanding which it would not be lawful to assume (indeed I had at first some reasons for doubting) the identity of type of the total groupings of which these systems form part; all we could have positively inferred from that fact would have been, that these two groupings both belong to the same class or genus containing 26,880 individuals, the second of the six referred to at the close of my last paper; a comparison of the two solutions has, however, satisfied me that they are absolutely identical in form.

XXI. *Proceedings of Learned Societies.*

ROYAL INSTITUTION OF GREAT BRITAIN.

June 7, 1861.

" ON the Physical Basis of Solar Chemistry." By John Tyndall, Esq., F.R.S., Professor of Natural Philosophy, Royal Institution.

Omitting all preface, the speaker drew attention to an experimental arrangement intended to prove that gaseous bodies radiate heat in different degrees. Behind a double screen of polished tin was placed an ordinary ring gas-burner; on this was placed a hot copper ball, from which a column of heated air ascended: behind the screen, but so placed that no ray from the ball could reach the instrument, was an excellent thermo-electric pile, connected by wires with a very delicate galvanometer. The thermo-electric pile was known to be an instrument whereby heat was applied to the generation of electric currents; the strength of the current being an accurate measure of the quantity of the heat. As long as both faces of the pile were at the same temperature no current was produced; but the slightest difference in the temperature of the two faces at once declared itself by the production of a current, which, when carried through the galvanometer, indicated by the deflection of the needle both its strength and its direction.

The two faces of the pile were in the first instance brought to the same temperature, the equilibrium being shown by the needle of the galvanometer standing at zero. The rays emitted by the current of hot air already referred to were permitted to fall upon one of the faces of the pile; and an extremely slight movement of the needle showed that the radiation from the hot air, though sensible, was extremely feeble. Connected with the ring-burner was a holder containing oxygen gas; and by turning a cock, a stream of this gas was

permitted to issue from the burner, strike the copper ball, and ascend
in a heated column in front of the pile.   The result was that oxygen
showed itself, as a radiator of heat, to be quite as feeble as atmospheric air.

A second holder containing olefiant gas was also connected by
its own system of tubes with the ring-burner.   Oxygen had already
flowed over the ball and cooled it in some degree.   Hence, as a
radiator in comparison with oxygen, the olefiant gas laboured under a
disadvantage.   It was purposely arranged that this should be the
case; so that if, notwithstanding its being less hot, the olefiant gas
showed itself a better radiator, its claim to superiority in this respect
would be decisively proved.   On permitting the gas to issue upwards,
it cast an amount of heat against the adjacent face of the pile sufficient
to impel the needle of the galvanometer almost to its stops at 90°.
This experiment proved the vast difference between two equally
transparent gases with regard to their power of emitting radiant
heat.

The converse experiment was now performed.   The thermo-electric
pile was removed and placed between two cubes filled with water kept
in a state of constant ebullition; and it was so arranged that the quantities of heat falling from the cubes on the opposite faces of the pile
were exactly equal, thus neutralizing each other.   The needle of the
galvanometer being at zero, a sheet of oxygen gas was caused to issue
from a slit between one of the cubes and the adjacent face of the pile.
If this sheet of gas possessed any sensible power of intercepting the
thermal rays from the cube, one face of the pile being deprived of the
heat thus intercepted, a difference of temperature between its two
faces would instantly set in, and the result would be declared by the
galvanometer.   The quantity absorbed by the oxygen under those
circumstances was too feeble to affect the galvanometer; the gas, in
fact, proved sensibly transparent to the rays of heat.   It had but a
feeble power of radiation; it had an equally feeble power of absorption.

The pile remaining in its position, a sheet of olefiant gas was
caused to issue from the same slit as that through which the oxygen
had passed.   No one present could see the gas; it was quite invisible,
the light went through it as freely as through oxygen or air, but its
effect upon the thermal rays emanating from the cube was what
might be expected from a sheet of metal.   A quantity so large was
cut off that the needle of the galvanometer, promptly quitting the zero
line, moved with energy to its stops.   Thus the olefiant gas, so light
and clear and pervious to luminous rays, was a most potent destroyer
of the rays emanating from an obscure source.   The reciprocity of
action established in the case of oxygen comes out here; the good
radiator is found by this experiment to be the good absorber.

This result, which was exhibited before a public audience this
evening for the first time, was typical of what had been obtained with
gases generally.   Going through the entire list of gases and vapours
in this way, we should find radiation and absorption to be as rigidly
associated as positive and negative in electricity, or as north and south

polarity in magnetism. The gas which, when heated, is most competent to generate a calorific ray, is precisely that which is most competent to stop such a ray. If the radiation be high, the absorption is high; if the radiation be moderate, the absorption is moderate; if the radiation be low, the absorption is low; so that if we make the number which expresses the absorptive power the numerator of a fraction, and that which expresses its radiative power the denominator, the result would be that, on account of the numerator and denominator varying in the same proportion, the value of that fraction would always remain the same, whatever might be the gas or vapour experimented with.

But why should this reciprocity exist? What is the meaning of absorption? what is the meaning of radiation? When you cast a stone into still water, rings of waves surround the place where it falls; motion is radiated on all sides from the centre of disturbance. When the hammer strikes a bell, the latter vibrates; and sound, which is nothing more than an undulatory motion of the air, is radiated in all directions. Modern philosophy reduces light and heat to the same mechanical category. A luminous body is one with its particles in a state of vibration; a hot body is one with its particles also vibrating, but at a rate which is incompetent to excite the sense of vision; and as a sounding body has the air around it, through which it propagates its vibrations, so also the luminous or heated body has a medium called æther, which accepts its motions and carries them forward with inconceivable velocity. Radiation, then, as regards both light and heat, is *the transference of motion from the vibrating body to the æther in which it swings*; and, as in the case of sound, the motion imparted to the air is soon transferred to the surrounding objects, against which the aërial undulations strike, the sound being, in technical language, *absorbed*, so also with regard to light and heat, absorption consists in *the transference of motion from the agitated æther to the particles of the absorbing body.*

The simple atoms are found to be bad radiators; the compound atoms good ones: and the higher the degree of complexity in the atomic grouping, the more potent, as a general rule, is the radiation and absorption. Let us get definite ideas here, however gross, and purify them afterwards by the process of abstraction. Imagine our simple atoms swinging like single spheres in the æther; they cannot create the swell which a group of them united to form a system can produce. An oar runs freely edgeways through the water and imparts far less of its motion to the water than when its broad flat side is brought to bear upon it. In our present language the oar, broad side vertical, is a good radiator; broad side horizontal, it is a bad radiator. Conversely, the waves of water, impinging upon the flat face of the oar-blade, will impart a greater amount of motion to it than when impinging upon the edge. In the position in which the oar radiates well it also absorbs well. Simple atoms glide through the æther without much resistance; compound ones encounter this, and yield up more speedily their motion to the æther. Mix oxygen and nitrogen mechanically, they absorb and radiate a certain amount.

Cause these gases to *combine* chemically and form nitrous oxide, both the absorption and radiation are thereby augmented 250 times!

In this way we look with the telescope of the intellect into atomic systems, and obtain a conception of processes which the eye of sense can never reach. But gases and vapours possess a power of choice as to the rays which they absorb. They single out certain groups of rays for destruction, and allow other groups to pass unharmed. This is best illustrated by a famous experiment of Sir David Brewster's, modified to suit the requirements of the present discourse. Into a glass cylinder, with its ends stopped by discs of plate glass, a small quantity of nitrous acid gas was introduced, the presence of the gas being indicated by its rich brown colour. The beam from an electric lamp being sent through two prisms of bisulphide of carbon, a spectrum 7 feet long and 18 inches wide was cast upon a screen. Introducing the cylinder containing the nitrous acid into the path of the beam as it issued from the lamp, the splendid and continuous spectrum became instantly furrowed by numerous dark bands, the rays answering to which were struck down by the nitric gas, while it permitted the light which fell upon the intervening spaces to pass with comparative impunity.

Here also the principle of reciprocity, as regards radiation and absorption, holds good; and could we, without otherwise altering its physical character, render that nitrous gas luminous, we should find that the very rays which it absorbs are precisely those which it would emit. When atmospheric air and other gases are brought to a state of intense incandescence by the passage of an electric spark, the spectra which we obtain from them consist of a series of bright bands. But such spectra are produced with the greatest brilliancy when, instead of ordinary gases, we make use of metals heated so highly as to volatilize them. This is easily done by the voltaic current. A capsule of carbon was filled with mercury, which formed the positive electrode of the electric lamp; a carbon-point was brought down upon this; and on separating one from the other, a brilliant arc containing the mercury in a volatilized condition passed between them. The spectrum of this arc was not continuous like that from the solid carbon points, but consisted of a series of vivid bands, each corresponding in colour to that particular portion of the spectrum to which its rays belonged. Copper gave its system of bands; zinc gave its system; and brass, which is an alloy of copper and zinc, gave a splendid spectrum made up of the bands belonging to both metals.

Not only, however, when metals are united like zinc and copper to form an alloy is it possible to obtain the bands which belonged to them. No matter how we may disguise the metal—allowing it to unite with oxygen to form an oxide, and this again with an acid to form a salt; if the heat applied be sufficiently intense, the bands belonging to the metal reveal themselves with perfect definition. Holes were drilled in a cylinder of retort carbon, and, these being filled with pure culinary salt, the carbon was made the positive electrode of the lamp; the resultant spectrum showed the brilliant yellow lines of the metal sodium. Similar experiments were made with the chlorides of strontium, calcium, lithium*, and other metals; each salt gave the bands due to the metal. Different salts were then mixed together and rammed into the holes in the carbon; a spectrum was obtained which contained the bands of them all.

The position of these bright bands never varies; and each metal has its own system. Hence the competent observer can infer from the bands of the spectrum the metals which produce it. It is a language addressed to the eye instead of the ear; and the certainty would not be augmented if each metal possessed the power of audibly calling out, "I am here!" Nor is this language affected by distance. If we find that the sun or the stars give us the bands of our terrestrial metals, it is a declaration on the part of these orbs that such metals enter into their composition. Does the sun give us any such intimation? Does the solar spectrum exhibit bright lines which we might compare with those produced by our terrestrial metals, and prove either their identity or difference? No. The solar spectrum, when closely examined, gives us a multitude of fine dark lines instead of bright ones. They were first noticed by Dr. Wollaston, were investigated with profound skill by Fraunhofer, and named from him Fraunhofer's lines. They have been long a standing puzzle to philosophers. The bright lines which the metals give us have been also known to us for years; but the connexion between both classes of phenomena was wholly unknown, until Kirchhoff, with admirable acuteness, revealed the secret, and placed it at the same time in our power to chemically analyse the sun.

We have now some hard work before us; hitherto we have been delighted by objects which addressed themselves rather to our æsthetic taste than to our scientific faculty. We have ridden pleasantly to the base of the final cone of Etna, and must now dismount and march wearily through ashes and lava, if we would enjoy the prospect from the summit. Our problem is to connect the dark lines of Fraunhofer with the bright ones of the metals. The white beam of the lamp is refracted in passing through our two prisms, but its different components are refracted in different degrees, and thus its colours are drawn apart. Now the colour depends solely upon the rate of oscillation of the particles of the luminous body,—red light being produced by one rate, blue light by a much quicker rate, and the colours between red and blue by the intermediate rates. The solid incandescent coal-points give us a continuous spectrum; or, in other words, they emit rays of all possible periods between the two extremes of the spectrum. They have particles oscillating so as to produce red; others to produce orange; others to produce yellow, green, blue, indigo, and violet respectively. Colour, as many of you know, is to light what *pitch* is to sound. When a violin-player

---

* The vividness of the colours of the lithium spectrum is extraordinary: it contained a blue band of indescribable splendour. It was thought by many, during the discourse, that I had mistaken strontium for lithium, as this blue band had never before been seen. I have obtained it many times since; and my friend Dr. Miller, having kindly analysed the substance made use of, pronounces it chloride of lithium.—J. T.

presses his finger on a string he makes it shorter and tighter, and, thus, causing it to vibrate more speedily, augments the pitch.   Imagine such a player to move his finger slowly along the string, shortening it gradually as he draws his bow, the note would rise in pitch by a regular gradation; there would be no gap intervening between note and note.   Here we have the analogue to the continuous spectrum, whose colours insensibly blend together without gap or interruption, from the red of the lowest pitch to the violet of the highest.   But suppose the player, instead of gradually shortening his string, to press his finger on a certain point, and to sound the corresponding note; then to pass on to another point more or less distant, and sound its note; then to another, and so on, thus sounding particular notes separated from each other by gaps which correspond to the intervals of the string passed over; we should then have the exact analogue of a spectrum composed of separate bright bands with intervals of darkness between them.   But this, though a perfectly true and intelligible analogy, is not sufficient for our purpose; we must look with the mind's eye at the very oscillating atoms of the volatilized metal.   Figure these atoms connected by springs of a certain tension, and which, if the atoms are squeezed together, push them asunder, or, if the atoms are drawn apart, pull them together, causing them, before coming to rest, to quiver at a certain definite rate determined by the strength of the spring.   Now the volatilized metal which gives us one bright band is to be figured as having its atoms united by springs all of the same tension, its vibrations are all of one kind.   The metal which gives us two bands may be figured as having some of its atoms united by springs of one tension, and others by a second series of springs of a different tension.   Its vibrations are of two distinct kinds; so also when we have three or more bands, we are to figure as many distinct sets of springs, each set capable of vibrating in its own particular time and at a different rate from the other.   If we seize this idea definitely, we shall have no difficulty in dropping the metaphor of springs, and substituting for it mentally the forces by which the atoms act upon each other.   Having thus far cleared our way, let us make another effort to advance.

Here is a pendulum,—a heavy ivory ball suspended from a string. I blow against this ball; a single puff of my breath moves it a little way from its position of rest; it swings back towards me, and when it reaches the limit of its swing I puff again.   It now swings further; and thus by timing my puffs I can so accumulate their action as to produce oscillations of large amplitude.   The ivory ball here has absorbed the motions which my breath communicated to the air.   I now bring the ball to rest.   Suppose, instead of my breath, a wave of air to strike against it, and that this wave is followed by a series of others which succeed each other exactly in the same intervals as my puffs; it is perfectly manifest that these waves would communicate their motion to the ball and cause it to swing as the puffs did. And it is equally manifest that this would not be the case if the impulses of the waves were not properly timed; for then the motion imparted to the pendulum by one wave would be neutralized by another, and there could not be that accumulation of effect which we have when the periods of the waves correspond with the periods of the pendulum. So much for the kind of impulses absorbed by the pendulum. But such a pendulum set oscillating in air produces waves in the air; and we see that the waves which it produces must be of the same period as those whose motions it would take up or absorb most copiously if they struck against it. Just in passing I may remark that, if the periods of the waves be double, treble, quadruple, &c. the periods of the pendulum, the shocks imparted to the latter would also be so timed as to produce an accumulation of motion.

Perhaps the most curious effect of these timed impulses ever described was that observed by a watchmaker named Ellicott, in the year 1741. He set two clocks leaning against the same rail; one of them, which we may call A, was set going; the other, B, not. Some time afterwards he found, to his surprise, that B was ticking also. The pendulums being of the same length, the shocks imparted by the ticking of A to the rail against which both clocks rested were propagated to B, and were so timed as to set B going. Other curious effects were at the same time observed. When the pendulums differed from each other a certain amount, A set B going, but the reaction of B stopped A. Then B set A going, and the reaction of A stopped B. If the periods of oscillation were close to each other, but still not quite alike, the clocks mutually controlled each other, and by a kind of mutual compromise they ticked in perfect unison.

But what has all this to do with our present subject? They are mechanically identical. The varied actions of the universe are all modes of motion; and the vibration of a ray claims strict brotherhood with the vibrations of our pendulum. Suppose æthereal waves striking upon atoms which oscillate in the same periods as the waves succeed each other, the motion of the waves will be absorbed by the atoms; suppose we send our beam of white light through a sodium flame, the particles of that flame will be chiefly affected by those undulations which are synchronous with their own periods of vibration. There will be on the part of those particular rays a transference of motion from the agitated æther to the atoms of the volatilized sodium, which, as already defined, is absorption. We use glass screens to defend us from the heat of our fires; how do they act? Thus:—The heat emanating from the fire is for the most part due to radiations which are incompetent to excite the sense of vision; we call these rays obscure. Glass, though pervious to the luminous rays, is opake in a high degree to those obscure rays, and cuts them off, while the cheerful light of the fire is allowed to pass. Now mark me clearly. The heat cut off from your person is to be found in the glass, the latter becomes heated and radiates towards your person; what, then, is the use of the glass if it merely thus acts as a temporary halting-place for the rays, and sends them on afterwards? It does this:—it not only sends the heat it receives towards you, but

scatters it also in all other directions round the room. Thus the
rays which, were the glass not interposed, would be shot directly
against your person, are for the most part diverted from their original
direction, and you are preserved from their impact.

· Now for our experiment. I pass the beam from the electric lamp
through the two prisms, and the spectrum spreads its colours upon
the screen. Between the lamp and the prism I interpose this snap-
dragon light. Alcohol and water are here mixed up with a quantity
of common salt, and the metal dish that contains them is heated by
a spirit-lamp. The vapour from the mixture ignites, and we have
this monochromatic flame. Through this flame the beam from the
lamp is now passing; and observe the result upon the spectrum.
You see a dark band cut out of the yellow,—not very dark, but sufficiently so to be seen by everybody present. Observe how the
band quivers and varies in shade as the amount of yellow light cut
off by the unsteady flame varies in amount. The flame of this monochromatic lamp is at the present moment casting its proper yellow light
upon that shaded line; and more than this, it casts in part the light
which it absorbs from the electric lamp upon it; but it scatters the
greater portion of this light in other directions, and thus withdraws
it from its place upon the screen; as the glass, in the case above supposed, diverted the heat of the fire from your person. Hence the
band appears dark; not absolutely, but dark in comparison with the
adjacent brilliant portions of the spectrum.

But let me exalt this effect. I place in front of the electric lamp
the intense flame of a large Bunsen's burner. I have here a platinum
capsule into which I put a bit of sodium less than a pea in magnitude. The sodium placed in the flame soon volatilizes and burns
with brilliant incandescence. Observe the spectrum. The yellow
band is clearly and sharply cut out, and a band of intense obscurity
occupies its place. I withdraw the sodium, the brilliant yellow of
the spectrum takes its proper place : I reintroduce the sodium, and
the black band appears.

· Let me be more precise :—The yellow colour of the spectrum extends over a sensible space, blending on one side into orange and on
the other into green. The term "yellow band" is therefore somewhat indefinite. I want to show you that it is the precise yellow
band emitted by the volatilized sodium which the same substance
absorbs. By dipping the coal-point used for the positive electrode
into a solution of common salt, and replacing it in the lamp, I obtain
that bright yellow band which you now see drawn across the spectrum. Observe the fate of that band when I interpose my sodium
light. It is first obliterated, and instantly that black streak occupies
its place. See how it alternately flashes and vanishes as I withdraw
and introduce the sodium flame!

And supposing that instead of the flame of sodium alone I introduce into the path of the beam a flame in which lithium, strontium,
magnesium, calcium, &c. are in a state of volatilization, each metallic
vapour would cut out its own system of bands, each corresponding
exactly in position with the bright band which that metal itself would

cast upon the screen. The light of our electric lamp then shining through such a composite flame would give us a spectrum cut up by dark lines, exactly as the solar spectrum is cut up by the lines of Fraunhofer.

And hence we infer the constitution of the great centre of our system. The sun consists of a nucleus which is surrounded by a flaming atmosphere. The light of the nucleus would give us a continuous spectrum, as our common coal-points did; but having to pass through the photosphere, as our beam through the flame, those rays of the nucleus which the photosphere can itself emit are absorbed, and shaded spaces, corresponding to the particular rays absorbed, occur in the spectrum. Abolish the solar nucleus, and we should have a spectrum showing a bright band in the place of every dark line of Fraunhofer. These lines are therefore not absolutely dark, but dark by an amount corresponding to the difference between the light of the nucleus intercepted by the photosphere, and the light which issues from the latter.

The man to whom we owe this beautiful generalization is Kirchhoff, Professor of Natural Philosophy in the University of Heidelberg; but, like every other great discovery, it is compounded of various elements. Mr. Talbot observed the bright lines in the spectra of coloured flames. Sixteen years ago Dr. Miller gave drawings and descriptions of the spectra of various coloured flames. Wheatstone, with his accustomed ingenuity, analysed the light of the electric spark, and showed that the metals between which the spark passed determined the bright bands in the spectrum of the spark. Masson published a prize essay on these bands; Van der Willigen, and more recently Plücker, have given us beautiful drawings of the spectra obtained from the discharge of Ruhmkorff's coil. But none of these distinguished men betrayed the least knowledge of the connexion between the bright bands of the metals and the dark lines of the solar spectrum. The man who came nearest to the philosophy of the subject was Angström. In a paper translated from Poggendorff's *Annalen* by myself, and published in the Philosophical Magazine for 1855, he indicates that the rays which a body absorbs are precisely those which it can emit when rendered luminous. In another place he speaks of one of his spectra giving the general impression of *reversal* of the solar spectrum. Foucault, Stokes, and Thomson have all been very close to the discovery; and, for my own part, the examination of the radiation and absorption of heat by gases and vapours, some of the results of which I placed before you at the commencement of this discourse, would have led me in 1859 to the law on which all Kirchhoff's speculations are founded, had not an accident withdrawn me from the investigation. But Kirchhoff's claims are unaffected by these circumstances. True, much that I have referred to formed the necessary basis of his discovery; so did the laws of Kepler furnish to Newton the basis of the theory of gravitation. But what Kirchhoff has done carries us far beyond all that had before been accomplished. He has introduced the order of law amid a vast assemblage of empirical observations, and has ennobled our previous knowledge by showing its relationship to some of the most sublime of natural phenomena.

Postscript, July 24.—As far back as the year 1822 Sir John Herschel described the spectra of various coloured flames in the Transactions of the Royal Society of Edinburgh. In his "Treatise on Light" in the *Encycl. Metropol.*, published in 1827, he describes the spectra derived from the introduction of various salts into flames, and finds exactly as Bunsen and Kirchhoff have recently found, the muriates most suitable for such experiments, on account of their volatility. He also adds the distinct statement, that "the colours thus communicated by different bases to flame afford in many cases a ready and neat way *of detecting extremely minute quantities of them.*"

ROYAL SOCIETY.

[Continued from p. 77.]

November 22, 1860.—Major-General Sabine, R.A., Treasurer and Vice-President, in the Chair.

The following communications were read :—

"Contributions towards the History of the Monamines."—No. III. Compound Ammonias by Inverse Substitution. By A. W. Hofmann, LL.D., F.R.S. &c. Received July 24, 1860.

Many years ago I showed that the bromide or iodide of a quartary ammonium splits under the influence of heat into the bromide or iodide of an alcohol-radical on the one hand, and a tertiary monamine on the other.

Having lately returned to the study of this class of substances, I was led to examine the deportment, under the influence of heat, of the tertiary, secondary, and, lastly, of the primary monammonium-salts.

Experiment has shown that these substances undergo an analogous decomposition. The chloride of a tertiary monammonium when submitted to distillation yields, together with the chloride of an alcohol-radical, a secondary monamine; the chloride of a secondary monammonium, together with an alcohol-chloride, a primary monamine; lastly, the chloride of a primary monammonium, the chloride of an alcohol-radical and ammonia.

Exactly, then, as my former experiments show that we may rise in the scale by replacing the four equivalents of hydrogen in ammonium one by one by radicals, so it is obvious from these new experiments that we may also step by step descend, by substituting again hydrogen for the radicals in succession.

To take as an illustration the monammonium-salts of the ethyle-series which as yet I have chiefly examined :

Ascent.

$$H_3 N + (C_2 H_5) Br = [(C_2 H_5) H_3 N] Br$$

Ammonia.   Bromide of ethyle.   Bromide of Ethylammonium.

$$(C_2 H_5) H_2 N + (C_2 H_5) Br = \cdot [(C_2 H_5)_2 H_2 N] Br$$

Ethylamine.   Bromide of Diethylammonium.

$$(C_2 H_5)_2 H N + (C_2 H_5) Br = [(C_2 H_5)_3 H N] Br$$

Diethylamine.   Bromide of Triethylammonium.

$$(C_2 H_5)_3 N + (C_2 H_5) Br = [(C_2 H_5)_4 N] Br$$

Triethylamine.   Bromide of Tetrethylammonium.

Note.—H = 1 ; C = 12.

Descent.

$$[(C_2 H_5)_4 N] Cl = (C_2 H_5) Cl + (C_2 H_5)_3 N$$

Chloride of Tetrethylammonium.   Chloride of Ethyle.   Triethylamine.

$$[(C_2 H_5)_3 H N] Cl = (C_2 H_5) Cl + (C_2 H_5)_2 H N$$

Chloride of Triethylammonium.   Diethylamine.

$$[(C_2 H_5)_2 H_2 N] Cl = (C_2 H_5) Cl + (C_2 H_5) H_2 N$$

Chloride of Diethylammonium.   Ethylamine.

$$[(C_2 H_5) H_3 N] Cl = (C_2 H_5) Cl + H_3 N$$

Chloride of Ethylammonium.   Ammonia.

The above reactions, interesting when regarded from a scientific point of view, admit of but limited application in practice. The purity of the result is disturbed by several circumstances, which it is difficult to exclude. Unless the temperature be sufficiently high, a small portion of the ammonium-salt submitted to distillation sublimes without change; again, a portion of the same salt is reproduced in the neck of the retort and in the receiver[*], from the very constituents into which it splits; lastly, if the temperature be too high, the chloride of ethyle is apt to be decomposed into ethylene and hydrochloric acid, the latter producing, with the monamine liberated in the reaction, a salt which in its turn is likewise decomposed.

Thus the chloride of diethylammonium, for instance, together with chloride of ethyle and ethylamine, yields ethylene and chloride of ethylammonium which splits into chloride of ethyle and ammonia.

The idea naturally suggested itself, to attempt, by means of this

[*] This inconvenience may be partly obviated by distilling into an acid.

reaction, the formation of the primary and secondary monophosphines, which are at present unknown. Experiments made with the view of transforming triethylphosphine into diethylphosphine have as yet remained unsuccessful, the chloride of triethylphosphonium distilling without alteration.

"Notes of Researches on the Poly-Ammonias."—No. IX. Remarks on *anomalous* Vapour-densities. By A. W. Hofmann, LL.D., F.R.S. Received July 24, 1860.

In a note addressed to the Royal Society* at the commencement of this year, I have shown that the molecules of the diamines, like those of all other well-examined compounds, correspond to two volumes of vapour†, and I have endeavoured to explain the apparent anomalous vapour-densities of the hydrated diamines by assuming that the vapour-volume experimentally obtained was a mixture of the vapour of the anhydrous base and of the vapour of water. Thus, hydrated ethylene-diamine was assumed to split under the influence of heat into anhydrous ethylene-diamine (2 vols. of vapour) and water (2 vols. of vapour).

$$C_2H_{10}N_2O = \left. \begin{matrix} (C_2H_4)'' \\ H_2 \\ H_2 \end{matrix} \right\} N_2 + \left. \begin{matrix} H \\ H \end{matrix} \right\} O\ddagger.$$

The vapour-density of ethylene-diamine referred to hydrogen being 30, and that of water-vapour 9, the vapour-density of a mixture of equal volumes of ethylene-diamine and water-vapour $= \dfrac{30+9}{2} = 19 \cdot 5$, which closely agrees with the result of experiment.

In continuing the study of the diamines, I have expanded these experiments. Without going into the detail of the inquiry, I beg leave to record an observation which appears to furnish an experimental solution to the question.

Ethylene-diamine, when submitted to the action of iodide of ethyl, yields a series of ethylated derivatives, amongst which the diethylated compound has claimed my particular attention. This body in the anhydrous state is an oily liquid containing

$$C_6H_{16}N_2 = \left. \begin{matrix} (C_2H_4)'' \\ (C_2H_5)_2 \\ H_2 \end{matrix} \right\} N_2.$$

With water it forms a beautiful crystalline very stable hydrate §, of the composition

$$C_{6}H_{14}N_2O = \left. \begin{matrix} (C_2H_4)'' \\ (C_2H_5)_2 \\ H_2 \end{matrix} \right\} N_2 + \left. \begin{matrix} H \\ H \end{matrix} \right\} O.$$

---

* Phil. Mag. vol. xx. p. 66.
† $H_2O = 2$ vols.   ‡ $H = 1$; $O = 16$; $C = 12$, &c.
§ Phil. Mag. vol. xix. p. 232.

The vapour-density of the anhydrous base was found by experiment to be 57·61, showing that the molecule of diethyl-ethylene-diamine corresponds to 2 vols. of vapour, the theoretical density being $\frac{114}{2} = 57$.

On submitting the crystalline hydrate to experiment, I arrived at the vapour-density 33·2. This number is in perfect accordance with the result obtained in the case of ethylene-diamine. The legitimate interpretation of this number is that here again the hydrated base splits into the anhydrous diamine and water, and that the density observed is that of a mixture of equal volumes of diamine-vapour and of water-vapour, the theoretical density of which is $\frac{57+9}{2} = 33$.

The correctness of this interpretation admits of an elegant experimental demonstration.

Having observed that the hydrate loses its water when repeatedly distilled with a large excess of anhydrous baryta, the idea suggested itself, to attempt the decomposition of the hydrate in the state of *vapour.* If the vapour obtained by heating this hydrate to a temperature 15° or 20° higher than its boiling-point actually consisted of a mixture of equal volumes of its two proximate constituents in a state of *dissociation* (to use a happy term proposed by Deville), it appeared very probable that the volume would be halved by the introduction of anhydrous baryta. Experiment has verified this anticipation.

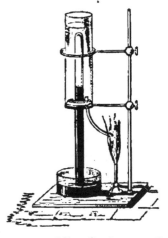

The upper half of a glass tube filled with, and inverted over, mercury, was surrounded by a second glass tube open at both ends and of a diameter about treble that of the former, the annular space between the two being closed at the bottom of the outer tube by a well-fitting cork. The vessel thus formed round the upper part of the inner tube was moreover provided with a small bent copper tube open at the top and closed at the bottom, which was likewise fixed in the cork. The vessel being filled with paraffin and a lamp being applied to the copper tube, the upper part of the mercury-tube could be conveniently kept at a high and constant temperature, whilst the lower end, immersed in the mercury-trough, remained accessible. A glance at the figure explains the disposition of the apparatus. A small quantity of the hydrated base was then allowed to rise on the top of the mercury in the tube; and the paraffin bath having been heated to 170°, the volume of the vapour was observed. Several pellets of anhydrous baryta were then allowed to ascend into the vapour-volume, while the temperature was maintained constant. The mercury began immediately to rise, becoming stationary again, when a fraction of the vapour had disappeared, which amounted, the necessary corrections being made, to half the original volume.

"Notes of Researches on the Poly-Ammonias."—No. X. On Sulphamidobenzamine, a new base ; and some Remarks upon Ureas and so-called Ureas. By A. W. Hofmann, LL.D., F.R.S. Received July 24, 1860.

Among the numerous compounds capable of the metamorphosis involved in Zinin's beautiful reaction, the nitriles have hitherto escaped the attention of chemists. This is the more remarkable, since some of these bodies are easily converted into crystalline nitro-compounds.

When examining several of the diamines which I have lately submitted to the Royal Society*, I was induced to study the transformation which benzonitrile undergoes under the successive influence of nitric acid and reducing agents.

Benzonitrile, when treated with a mixture of sulphuric and fuming nitric acid, furnishes, as is well known, a solid nitro-substitute which crystallizes from alcohol in beautiful white needles, containing

$$C_7 H_4 N_2 O_2 = C_7 (H_4, NO_2) N†.$$

In order to obtain this body, it is desirable to perform the operation with small quantities, and to cool the liquid carefully, otherwise the formation of appreciable proportions of nitrobenzoic acid can scarcely be avoided.

The nitro-compound is readily attacked by an aqueous solution of sulphide of ammonium ; sulphur is abundantly precipitated, and on evaporating the liquid, a yellowish red oil is separated, which gradually and imperfectly solidifies. This substance possesses the characters of a weak base, dissolving with facility in acids, and being again precipitated by the addition of ammonia and the alkalies. The preparation in the state of purity, both of the base itself and of its compounds, presents some difficulty. This circumstance has prevented me from analysing the base. I have, however, examined one of its products of decomposition, which leaves no doubt that nitrobenzonitrile, under the influence of reducing agents, undergoes the well-known transformation of nitro-compounds, and that the composition of the oily base is represented by the formula

$$C_7 H_6 N_2 = C_7 H_4 (NH_2) N.$$

The oily base, when left in contact with sulphide of ammonium, is gradually changed, a crystalline compound being formed, which is easily soluble in alcohol and in ether, but difficultly soluble in water, and which may be purified by several crystallizations from boiling

---

* Phil. Mag. vol. xix. p. 232.   † H=1; O=16; C=12, &c.

water, being deposited on cooling in white brilliant needles. This compound is a well-defined organic base; it dissolves with facility in acids, and is precipitated from these solutions by the addition of potassa or of ammonia. With hydrochloric acid it forms a crystallizable salt, which yields, with dichloride of platinum, an orange-yellow crystalline precipitate.

On analysis, the new base was found to have the composition

$$C_7 H_8 N_2 S,$$

explaining its formation, in which evidently two phases have to be distinguished:

(1) $C_7 (H_4 NO_2) N + 3H_2 S = 2H_2 O + 3S + C_7 H_6 N_2,$

(2) $C_7 H_6 N_2 + H_2 S = C_7 H_8 N_2 S.$

The new sulphuretted base has the same composition as sulpho-carbonyl-phenyldiamide, a feebly basic compound which I obtained some time ago by the action of ammonia on sulphocyanide of phenyle *.

$$C_7 H_6 NS + H_2 N = C_7 H_8 N_2 S.$$

A superficial comparison of the properties of the two bodies shows, however, that they are only isomeric, the constitution of the latter compound being represented by the expression

$$\left.\begin{array}{l}(C S)'' \\ (C_6 H_5)' \\ H_2\end{array}\right\} N_2,$$

whilst the constitution of the former may be expressed by the formula

$$\left.\begin{array}{l}C_7 (H_4 H_2 N) S \\ H \\ H\end{array}\right\} N = \left.\begin{array}{l}(C_7 H_4 S)'' \\ H_2 \\ H_2\end{array}\right\} N_2.$$

The new sulphuretted base is closely connected with an interesting compound which Chancel obtained some years ago, when he submitted nitrobenzamide to the action of reducing agents. The crystalline base produced in this reaction contains

$$C_7 H_8 N_2 O,$$

and differs from the body which forms the subject of this note only by having oxygen in the place of sulphur.

The formation of this oxygenated compound has given rise to some misconceptions, which I take this opportunity to elucidate. A short time before the discovery of the body in question, I had obtained a compound of exactly the same composition by the action of the vapour of cyanic acid upon aniline,

$$C_6 H_7 N + C H N O = C_7 H_8 N_2 O.$$

The mode of producing this substance pointed it out as an ana-

* Phil. Mag. vol. xvii. p. 65.

logue of urea, and hence the designation *aniline-urea,* under which I described the new body as the first of the group of compound ureas, which has since been so remarkably enriched by Wurtz and several other chemists.

The aniline-urea, or phenyl-urea as it is more appropriately called, differs from ordinary urea in its deportment with acids, being, in fact, no longer capable of producing saline compounds. The absence of basic properties in the new phenyl-compound was sufficient to throw some doubt upon its ureic character, and this doubt appeared to receive additional support by Chancel's subsequent discovery of a compound possessing not only the composition of phenyl-urea, but forming likewise well-defined saline combinations. This compound is, however, the amide of amidobenzoic acid, its constitution being interpreted by Chancel, in accordance with its formation:

$$\text{Benzamide} \qquad \left.\begin{array}{c} C_7\,H_5\,O \\ H \\ H \end{array}\right\} N.$$

$$\text{Nitrobenzamide} \qquad \left.\begin{array}{c} C_7\,(H_4\,NO_2)\,O \\ H \\ H \end{array}\right\} N.$$

$$\text{Amidobenzamide} \qquad \left.\begin{array}{c} C_7\,(H_4\,NH_2)\,O \\ H \\ H \end{array}\right\} N.$$

Nevertheless chemists, by silent but general consent, began to look upon this compound as the *true* phenyl-urea; and in most manuals, even Gerhardt's 'Traité de Chimie' not excepted, it figures under this appellation.

Let us see how far this view is supported by the deportment of this substance. Compound ureas, as I conceive the character of this class, must imitate the deportment of urea *par excellence,* both in their mode of formation and their products of decomposition. Urea is formed whenever cyanic acid or cyanates come in contact with ammonia or ammoniacal salts. These are precisely the conditions under which the substance which I have described as phenyl-urea is generated. This compound is obtained by the union of cyanic acid with phenylamine, or of ammonia with cyanate of phenyle.

$$\left.\begin{array}{c} C_6\,H_5 \\ H \\ H \end{array}\right\} N + \left.\begin{array}{c} (CO)'' \\ H \end{array}\right\} N = \left.\begin{array}{c} (CO)'' \\ (C_6\,H_5) \\ H_3 \end{array}\right\} N_2.$$

$$\left.\begin{array}{c} H \\ H \\ H \end{array}\right\} N + \left.\begin{array}{c} (CO)'' \\ (C_6\,H_5) \end{array}\right\} N = \left.\begin{array}{c} (CO)'' \\ (C_6\,H_5) \\ H_3 \end{array}\right\} N_2.$$

On the other hand, no cyanogen-compound is involved in the formation of amidobenzamide, or amidobenzamine, as it might be more appropriately called, on account of its basic properties.

Not less decisive is the evidence furnished by the products of decomposition of the two bodies. The most characteristic transformation of urea is its decomposition into ammonia and carbonic acid when it is submitted to the action of the alkalies. A compound urea thus treated should yield, together with carbonic acid and ammonia, the monamine from which it has arisen. Phenyl-urea should furnish carbonic acid, ammonia, and phenylamine: these are precisely the products observed in the decomposition of the compound which is formed by the action of cyanic acid on phenylamine.

$$\left.\begin{array}{c}(CO)''\\ C_6H_5\\ H_2\end{array}\right\}N_2+\left.\begin{array}{c}H\\ H\end{array}\right\}O=(CO)''O+\left.\begin{array}{c}H\\ H\\ H\end{array}\right\}N+\left.\begin{array}{c}C_6H_5\\ H\\ H\end{array}\right\}N.$$

Amidobenzamine, on the other hand, exhibits with potassa the deportment of an amidated amide. The reaction presents two distinct phases, ammonia and amidobenzoic acid being formed in the first phase, and ammonia and benzoic acid in the second:

$$\left.\begin{array}{c}C_7(H_4H_2N)O\\ H\\ H\end{array}\right\}N+\left.\begin{array}{c}H\\ H\end{array}\right\}O=\left.\begin{array}{c}H\\ H\\ H\end{array}\right\}N+\left.\begin{array}{c}C_7(H_4H_2N)O\\ H\end{array}\right\}O$$

$$\left.\begin{array}{c}C_7(H_4H_2N)O\\ H\end{array}\right\}O+\left.\begin{array}{c}H\\ H\end{array}\right\}O=\left.\begin{array}{c}H\\ H\\ H\end{array}\right\}N+\left.\begin{array}{c}C_7H_5O\\ H\end{array}\right\}O.$$

No trace of carbonic acid and no trace of phenylamine are eliminated by potassa. It is only by fusing with soda-lime that a perfect destruction of the compound ensues, when, as Chancel has distinctly observed, in the first place ammonia, and ultimately carbonic acid and phenylamine are evolved.

What I have said respecting phenyl-urea applies with equal force to diphenyl-urea. Gerhardt describes as diphenyl-urea the compound obtained by Laurent and Chancel when they examined the action of reducing agents upon nitrobenzophenone, and which, on account of its yellow colour, was originally described as *flavine*. This body contains

$$C_{13}H_{12}N_2O,$$

which is certainly the formula of diphenyl-urea. But here again chemists have been misled by the basic properties of the substance.

It is not my object at present to dwell on the constitution of flavine, which I intend to examine in a subsequent note; suffice it to say that this substance is not diphenyl-urea.

The true diphenyl-urea is the substance commonly called carbanilide, or carbophenylamide.

$$C_{13}H_{12}N_2O=\left.\begin{array}{c}(CO)''\\ (C_6H_5)_2\\ H_2\end{array}\right\}N_2.$$

Both the conditions under which this body forms, and the products into which it is decomposed, leave no doubt regarding its position in the system.

This compound is formed by the action of cyanate of phenyle upon either water or phenylamine.

$$2\left[\begin{matrix}(CO)'' \\ (C_6H_5)\end{matrix}\right\}N\right] + \begin{matrix}H \\ H\end{matrix}\right\}O = (CO)''O + \begin{matrix}(CO)'' \\ (C_6H_5)_2 \\ H_2\end{matrix}\right\}N_2,$$

$$\begin{matrix}(CO)'' \\ (C_6H_5)\end{matrix}\right\}N + \begin{matrix}C_6H_6 \\ H \\ H\end{matrix}\right\}N = \begin{matrix}(CO)'' \\ (C_6H_5)_2 \\ H_2\end{matrix}\right\}N_2.$$

When boiled with potassa, it splits into carbonic acid and phenyl-amine.

$$\begin{matrix}(CO)'' \\ (C_6H_5)_2 \\ H_2\end{matrix}\right\}N_2 + \begin{matrix}H \\ H\end{matrix}\right\}O = (CO)''O + 2\left[\begin{matrix}C_6H_5 \\ H \\ H\end{matrix}\right\}N\right].$$

These are the characters of *true* diphenyl-urea.

GEOLOGICAL SOCIETY.

[Continued from p. 78.]

May 8, 1861.—Leonard Horner, Esq., President, in the Chair.

The following communications were read :—

1. "Description of two Bone-caves in the Mountain of Ker, at Massat, in the Department of the Arriège." By M. Alfred Fontan. Communicated by M. E. Lartet, For. Mem. G.S.

The valley of Massat, on the northern side of the Pyrenees, is of a triangular shape, its northern angle being narrowed by the projecting limestone mountain of Ker. Among the fissures and grottos that traverse this mountain in every direction are two caves in particular : one is situated near the top, at about 100 metres above the valley; the other is near the base, at about 20 metres above the river. They both open towards the north. In the upper cave M. Fontan found a sandy loam with pebbles (the pebbles being of rocks different from that of the mountain), extending inwards for 100 metres, and containing a large quantity of bones of *Carnivora, Ruminantia,* and *Rodentia,*—those of the great Cave-bear, a large *Hyæna,* and a large *Felis* being the most numerous. On the surface some fragments of pottery, an iron poignard, and two Roman coins were found, with a quantity of cinders and charcoal; and at a depth of more than 3 feet in the ossiferous loam another bed of cinders and charcoal was met with, and in this M. Fontan found a bone arrow-head and two human teeth; the latter were at a distance of 5 or 6 metres one from the other.

In the lower cavern a blackish earth, with large granitic and other pebbles, was found to contain bones of the Red Deer, Antelope.

Aurochs, and Lynx; also worked flints and numerous utensils of bone (of deer chiefly), such as bodkins and arrows; the latter have grooves on their barbs, probably for poison. Some of the bones bear marks made of incisions by sharp instruments in flaying or cutting up the carcases. In each cavern a chasm crosses the gallery and terminates the deposits—in the upper cave at 100 metres, in the lower one at about 7 metres from the entrance.

The author argues that, from the facts which he has noticed, these caverns must have been subjected simultaneously to the effects of a great transient diluvial cataclysm coming from the N.N.W. or West, in the opposite direction to the present course of the waters of that region; that man and all the other animals the remains of which are buried in these caves existed in the valley before this inundation; and that the greater part of the animals inhabited the caves, but that man was not contemporary with all of them.

2. "Notes on some further Discoveries of Flint Implements in the Drift; with a few suggestions for search elsewhere." By J. Prestwich, Esq., F.R.S., Treas. G.S.

Since the author's communication to the Royal Society last year on the discovery of Flint Implements in Pleistocene beds at Abbeville, Amiens, and Hoxne, similar implements have been found in some new localities in this country.

In Suffolk, between Icklingham and Mildenhall, Mr. Warren has met with some specimens in the gravel of Rampart Hill in the valley of the Lark. This gravel is of later date than the Boulder-clay of the neighbourhood. In Kent, Mr. Leech, Mr. Evans, and the author found some specimens at the foot of the cliffs between Herne Bay and the Reculvers. The author believes them to have been derived from a freshwater deposit that caps the cliff, and which has been found by Mr. Evans and himself to yield similar specimens at Swale Cliff near Whitstable. In Bedfordshire, Mr. J. Wyatt, F.G.S., has found some specimens in the gravel at Biddenham, near Bedford; this gravel also is of freshwater origin, and is younger than the Boulder-clay. In Surrey, a specimen found in the gravel of Peasemarsh twenty-five years ago has been brought forward by its discoverer, Mr. Whitburn of Guildford. In Herts, Mr. Evans has found a specimen in the surface-drift on the Chalk Hills near Abbots Langley. Lastly, the author recommended that diligent search be made in the gravel and brick-earth at Copford and Lexden near Colchester, at Grays and Ilford in Essex, at Erith, Brentford, Taplow, Hurley, Reading, Oxford, Cambridge, Chippenham, Bath, Blandford, Salisbury, Chichester, Selsea, Peasemarsh, Godalming, Croydon, Hertford, Stamford, Orton near Peterborough, &c.

3. "On the *Corbicula* (or *Cyrena fluminalis*) geologically considered." By J. Gwyn Jeffreys, F.R.S., F.G.S.

Mr. Jeffreys has identified the species of *Corbicula*, found by Mr. Prestwich in a raised sea-beach at Kelsey Hill in Yorkshire, with that of the Grays deposit, as well as with the recent species from the

Euphrates and the Nile, He mentioned the great tendency to variation in freshwater shells, and the distribution of the same species
throughout different and widely separated parts of the world; and
he therefore considered that there was no difficulty in supposing that
the *Corbicula* was contemporaneous in this country with Arctic shells
found with it at Kelsey Hill. According to Mr. Jeffreys, specimens
of Testacea from the north are larger than those of the same species
from southern localities.

XXII. *Intelligence and Miscellaneous Articles.*

PHOTOGRAPHIC MICROMETER.

To the Editors of the Philosophical Magazine and Journal.

GENTLEMEN,   Parsonstown, July 1861.

I SAW last week in some newspaper a notice that in America mi-
crometers have lately been made by means of photography. I do
not know from what publication this notice has been taken; but as I
have been for the last month or so engaged occasionally in making
experiments with the same intention as an original idea, I send
you a short account of what I have done, for publication if you think
it sufficiently interesting.

My endeavour was to get a glass slide for a microscope marked so
as to measure very minute objects; and as the micrometer I have
(measuring $\frac{1}{100}$th of an inch) was useless for the purpose I had in
view, it occurred to me that by the diminishing power of the camera
I might succeed in obtaining smaller divisions. I tried first for pictures of dark lines, $\frac{1}{8}$th of an inch in breadth, on a white ground,
reduced to a small compass, but I did not succeed even with a very
small aperture to the lens. I then substituted lines $\frac{1}{4}$th of an inch in
breadth removed to a greater distance, and I got a pretty sharp picture; but I found that the sharpest and best-marked picture of distant
lines I obtained was given by opake bars, placed so that the light
from a clear sky came to the camera between them.

By nailing rods of blackened wood, $\frac{1}{4}$th of an inch broad and $\frac{1}{4}$th of
an inch asunder, across a frame, and placing this at a suitable distance
with a clear light behind, and using an aperture of about $\frac{1}{4}$th of an
inch in diameter, I easily obtained well-marked and sharp lines the
$\frac{1}{1000}$th of an inch apart and the $\frac{1}{1000}$th of an inch in breadth, sufficiently accurate for all the purposes of a micrometer. The picture
of the lines requires to be covered with transparent varnish to prevent rubbing. I have taken the picture on very thin talc, and cemented it to glass with the collodion between the plates; and for
object-glasses of small power I have found it answer; but the thickness of the talc is too much for the higher powers, as the object
viewed and the lines do not sufficiently agree in focus.

I suppose the reason why lines with spaces between them give a
better picture than black lines ruled on a white ground is because

there is no irradiation of light from behind, at least not nearly so much from the spaces as from the white ground. At all events, whatever the cause may be, I have found the lines with the spaces give a much better and sharper impression.

The picture of the lines should be a positive one, and very clear. I found the collodion prepared with the iodide of iron, according to the formula given in this Magazine, July 1854, to act admirably. It must be very sensitive, on account of the smallness of the aperture necessary for the required sharpness.

I have no doubt that much finer lines than these I have got might be obtained by the same process.

Your obedient Servant,

THOMAS WOODS, M.D.

ON THE BOILING OF LIQUIDS. BY M. L. DUFOUR.

The ebullition of liquids, instead of taking place under normal circumstances of temperature and pressure, varies, as is well known, with the vessel in which the liquid is contained. In an earthen vessel, for instance, the ebullition is at a higher point than in one of metal; and Marcet has shown that the treatment the glass has experienced, washing with sulphuric acid, &c., often modifies the boiling-point to the extent of several degrees. Water deprived of air and placed in the conditions of a *water-hammer*, may be heated several degrees above 100° C. without passing into the gaseous state, but it then boils violently. Donny has shown that water free from air and carefully heated, may be raised to 135° without assuming the gaseous state. This retardation of ebullition is further found in other liquids; and the violent production of vapour is a frequent indication of it in glass vessels.

The ebullition not being produced except at a temperature higher than that at which the elastic force of the vapour is equal to the external pressure, is due to two causes,—first, the adhesion of the liquid to the substance of the vessel; and secondly, the absence of air in solution.

There are nevertheless some curious cases in which the retardation of boiling cannot be explained by the adhesion to a solid, and the absence of air, yet where the contact of a solid produces a sudden formation of vapour. If linseed oil be heated in a dish to 105° or 110° and a few drops of water be allowed to fall, they will sink to the bottom of the vessel. The moment they touch there is a sudden formation of vapour; the globule of water, slightly diminished, is repelled a few millimetres from the bottom; it again sinks, giving rise to a fresh disengagement of vapour, which raises it again, and so on. There is no perceptible evaporation from the globules of water so long as they float on the oil, and are not in contact with the side of the vessel; and it is only on the sudden contact of the solid that a bubble of vapour is suddenly produced. It is natural to inquire what would take place if the water during its being heated was kept from the side of the vessel, and floated in a medium of the same density as its own. The medium to be employed ought to exceed 100° without boiling, have the same density as water, and not form aqueous mixtures. Oils are unsuitable, but certain essences realize these conditions.

Essence of cloves, to which a small quantity of oil has been added, forms a liquid in which water remains in equilibrium in round spheres, and perfectly moveable in the interior. If heated carefully, a temperature far above 100° may frequently be attained before the ebullition of the water ensues. A temperature of 120° and 130° is frequently reached; and I have often had these aqueous spheres 10 millims. in diameter at 140° to 150°. Smaller spheres, 1 to 2 millims. in diameter, have often been raised to 170°, and even 175°; that is, to temperatures at which the tension of aqueous vapour is 8 atmospheres. The water had undergone no preparation; it was neither distilled nor free from air. At these high temperatures there is not, as might be thought, a slow and continuous ebullition. The spheres are as limpid and calm at 150° as at 10°.

Ebullition ensues when the globules come in contact with a solid. If, carried by the currents which are produced during the heating, they strike against the side of the vessel or the bulb of the thermometer, there is a sudden production of vapour. The globule, which has become somewhat smaller, is driven to some distance from the point at which the explosion is produced, but it continues to float. If, when the temperature exceeds 115° to 120°, the aqueous globule is touched with a glass or metal rod, a similar effect is produced; an explosion takes place at the point of contact, a bubble of vapour is disengaged which traverses the essence, and the globule is driven away as if the point had exerted on it a sudden repulsion. The solids best fitted for producing this effect are a pointed piece of wood or of charcoal. Glass or metal rods occasionally fail; the contact of saline crystals is generally successful.

The preceding phenomena may also be produced with other liquids when heated under suitable conditions. Chloroform, for example, heated in a solution of chloride of zinc, may be raised to 90° or 100°.

It is natural to connect these phenomena with those in which the contact of a solid induces the crystallization of supersaturated saline solutions, as well as with the sudden solidification of water, sulphur, &c., reduced below the ordinary temperature of solidification. They are also intimately connected with the phenomenon of liquids resisting solidification when they are immersed in a fluid medium. It appears as if the contact of solids were a determining cause for the change of condition in liquids; and it may be that the limits of temperature which we have assigned to the different conditions of bodies are less absolute than they appear.—*Comptes Rendus,* May 13, 1861.

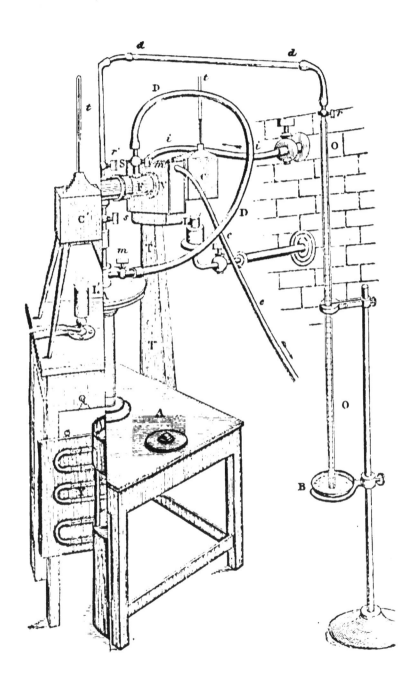

THE

LONDON, EDINBURGH AND DUBLIN

PHILOSOPHICAL MAGAZINE

AND

JOURNAL OF SCIENCE.

[FOURTH SERIES.]

SEPTEMBER 1861.

XXIII. *On the Absorption and Radiation of Heat by Gases and Vapours, and on the Physical Connexion of Radiation, Absorption, and Conduction.—The Bakerian Lecture.* By JOHN TYNDALL *Esq., F.R.S. &c.**

[With a Plate.]

§ 1. THE researches on glaciers which I have had the honour of submitting from time to time to the notice of the Royal Society, directed my attention in a special manner to the observations and speculations of De Saussure, Fourier, M. Pouillet, and Mr. Hopkins, on the transmission of solar and terrestrial heat through the earth's atmosphere. This gave practical effect to a desire which I had previously entertained to make the mutual action of radiant heat and gases of all kinds the subject of an experimental inquiry.

Our acquaintance with this department of Physics is exceedingly limited. So far as my knowledge extends, the literature of the subject may be stated in a few words.

From experiments with his admirable thermo-electric apparatus, Melloni inferred that for a distance of 18 or 20 feet the absorption of radiant heat by atmospheric air is perfectly insensible†.

With a delicate apparatus of the same kind, Dr. Franz of Berlin found that the air contained in a tube 3 feet long absorbed 3·54 per cent. of the heat sent through it from an Argand lamp; that is to say, calling the number of rays which passed through the exhausted tube 100, the number which passed when the tube was filled with air was only 96·46‡.

---

* From the Philosophical Transactions, Part I. for 1861, having been read at the Royal Society February 7, 1861.

† *La Thermochrose*, p. 136.   ‡ Pogg. *Ann.* vol. xciv. p. 342.

In the sequel I shall refer to circumstances which induce me to conclude that the result obtained by Dr. Franz is due to an inadvertence in his mode of observation. These are the only experiments of this nature with which I am acquainted, and they leave the field of inquiry now before us perfectly unbroken ground.

§ 2. At an early stage of the investigation, I experienced the need of a first-class galvanometer. My instrument was constructed by that excellent workman, Sauerwald of Berlin. The needles are suspended independently of the shade; the latter is constructed so as to enclose the smallest possible amount of air, the disturbance of aërial currents being thereby practically avoided. The plane glass plate, which forms the cover of the instrument, is close to the needle; so that the position of the latter can be read off with ease and accuracy either by the naked eye or by a magnifying lens.

The wire of the coil belonging to this instrument was drawn from copper obtained from a galvano-plastic manufactory in the Prussian Capital; but it was not free from the magnetic metals.

In consequence of its impurity in this respect, when the needles were perfectly astatic they deviated as much as 30° right and left of the neutral line. To neutralize this, a "compensator" was made use of, by which the needle was gently drawn to zero in opposition to the magnetism of the coil.

But the instrument suffered much in point of delicacy from this arrangement, and accurate quantitative determinations with it were unattainable. I therefore sought to replace the Berlin coil by a less magnetic one. Mr. Becker first supplied me with a coil which reduced the lateral deflection from 30° to 3°.

But even this small residue was a source of great annoyance to me; and for a time I almost despaired of obtaining pure copper wire. I knew that Professor Magnus had succeeded in obtaining it for his galvanometer, but the labour of doing so was immense[*]. Previous to undertaking a similar task, the thought occurred to me, that for my purpose a magnet furnished an immediate and perfect test as to the quality of the wire. Pure copper is *diamagnetic*; hence its repulsion or attraction by the magnet would at once declare its fitness or unfitness for the purpose which I had in view.

Fragments of the wire first furnished to me by M. Sauerwald were strongly attracted by the magnet. The wire furnished by Mr. Becker, when covered with its green silk, was also attracted, though in a much feebler degree.

I then removed the green silk covering from the latter and tested the naked wire. *It was repelled.* The whole annoyance

* Pogg. *Ann.* vol. lxxxiii. p. 489; and Phil. Mag. 1852, vol. iii. p. 82.

was thus fastened on the green silk; some iron compound had been used in the dyeing of it; and to this the deviation of my needle from zero was manifestly due.

I had the green coating removed and the wire overspun with white silk, clean hands being used in the process. A perfect galvanometer is the result. The needle, when released from the action of a current, returns accurately to zero, and is perfectly free from all magnetic action on the part of the coil. In fact while we have been devising agate plates and other elaborate methods to get rid of the great nuisance of a magnetic coil*, the means of doing so are at hand. Nothing is more easy to be found than diamagnetic copper wire. Out of eleven specimens, four of which were furnished by Mr. Becker, and seven taken at random from our laboratory, nine were found diamagnetic and only two magnetic.

Perhaps the only defect of those fine instruments with which Du Bois Raymond conducts his admirable researches in animal electricity is that above alluded to. The needle never comes to zero, but is drawn to it by a minute magnet. This defect may be completely removed. By the substitution of clean white silk for green, however large the coil may be, the compensator may be dispensed with, and a great augmentation of delicacy secured. The instrument will be rendered suitable for quantitative measurements; effects which are now beyond the reach of experiment will be rendered manifest; while the important results hitherto established will be obtained with a fraction of the length of wire now in use†.

§ 3. Our present knowledge of the deportment of liquids and solids, would lead to the inference that, if gases and vapours exercised any appreciable absorptive power on radiant heat, the absorption would make itself most manifest on heat emanating from an obscure source. But an experimental difficulty occurs at the outset in dealing with such heat. How must we close the receiver containing the gases through which the calorific rays are to be sent? Melloni found that a glass plate one-tenth of an inch in thickness intercepted all the rays emanating from a source of the temperature of boiling water, and fully 94 per cent. of the rays from a source of 400° Centigrade. Hence a tube closed with glass plates would be scarcely more suitable for the purpose now under consideration, than if its ends were stopped by plates of metal.

* See Melloni upon this subject, *Thermochrose*, pp. 31–33.

† Mr. Becker, to whose skill and intelligence I have been greatly indebted, furnished me with several specimens of wire of the same fineness as that used by Du Bois Raymond, some covered with green silk and others with white. The former were invariably attracted, the latter invariably repelled. In all cases the *naked* wire was repelled.

Rock-salt immediately suggests itself as the proper substance; but to obtain plates of suitable size and transparency was exceedingly difficult. Indeed, had I been less efficiently seconded, the obstacles thus arising might have been insuperable. To the Trustees of the British Museum I am indebted for the material of one good plate of salt; to Mr. Harlin for another; while Mr. Lettsom, at the instance of Mr. Darker*, brought me a piece of salt from Germany from which two fair plates were taken. To Lady Murchison, Sir Emerson Tennant, Sir Philip Egerton, and Mr. Pattison my best thanks are also due for their friendly assistance.

The first experiments were made with a tube of tin polished inside, 4 feet long and 2·4 inches in diameter, the ends of which were furnished with brass appendages to receive the plates of rock-salt. Each plate was pressed firmly against a flange by means of a bayonet joint, being separated from the flange by a suitable washer. Various descriptions of leather washers were tried for this purpose and rejected. The substance finally chosen was vulcanized india-rubber very lightly smeared with a mixture of bees-wax and spermaceti. A T-piece was attached to the tube, communicating on one side with a good air-pump, and on the other with the external air, or with a vessel containing the proper gas.

The tube being mounted horizontally, a Leslie's cube containing hot water was placed close to one of its ends, while an excellent thermo-electric pile, connected with its galvanometer, was presented to the other. The tube being exhausted, the calorific rays sent through it fell upon the pile, a permanent deflection of 30° being the consequence. The temperature of the water was in the first instance purposely so arranged as to produce this deflection.

Dry air was now admitted into the tube, while the needle of the galvanometer was observed with all possible care. Even by the aid of a magnifying lens I could not detect the slightest change of position. Oxygen, hydrogen, and nitrogen, subjected to the same test, gave the same negative result. The temperature of the water was subsequently lowered so as to produce a deflection of 20° and 10° in succession, and then heightened till the deflection amounted to 40°, 50°, 60° and 70°; but in no case did the admission of air, or any of the above gases into the exhausted tube, produce any sensible change in the position of the needle.

It is a well-known peculiarity of the galvanometer, that its higher and lower degrees represent different amounts of calorific

---

* During the course of the inquiry, I have often had occasion to avail myself of the assistance of this excellent mechanician.

action. In my instrument, for example, the quantity of heat
necessary to move the needle from 60° to 61° is about twenty
times that required to move it from 11° to 12°. Now in the
case of the small deflections above referred to, the needle was, it
is true, in a sensitive position; but then the total amount of
heat passing through the tube was so inconsiderable that a small
per-centage of it, even if absorbed, might well escape detection.
In the case of the large deflections, on the other hand, though
the total amount of heat was large, and though the quantity
absorbed might be proportionate, the needle was in such a position as to require a very considerable abstraction of heat to produce any sensible change in its position. Hence arose the
thought of operating, if possible, with large quantities of heat,
while the needle intended to reveal its absorption should continue to occupy its position of maximum delicacy.

The first attempt at solving this problem was as follows :—My
galvanometer is a differential one—the coil being composed of
two wires wound side by side, so that a current could be sent
through either of them independent of the other. The thermo-
electric pile was placed at one end of the tin tube, and the ends
of one of the galvanometer wires connected with it. A copper
ball heated to low redness being placed at the other end of the
tube, the needle of the galvanometer was propelled to its stops
near 90°. The ends of the second wire were now so attached
to a second pile that when the latter was caused to approach the
copper ball, the current thus excited passed through the coil in
a direction opposed to the first one. Gradually, as the second
pile was brought nearer to the source of heat, the needle descended from the stops, and when the two currents were nearly
equal the position of the needle was close to zero.

Here then we had a powerful flux of heat through the tube;
and if a column of gas four feet long exercised any sensible
absorption, the needle was in the position best calculated to
reveal it. In the first experiment made in this way, the neutralization of one current by the other occurred when the tube was
filled with air; and after the exhaustion of the tube had commenced, the needle started suddenly off in a direction which
indicated that a *less* amount of heat passed through the partially
exhausted tube, than through the tube filled with air. The
needle, however, soon stopped, turned, descended quickly to zero,
and passed on to the other side, where its deflection became permanent. The air made use of in this experiment came direct
from the laboratory, and the first impulsion of the needle was
probably due to the aqueous vapour precipitated as a cloud
by the sudden exhaustion of the tube. When, previous to its
admission, the air was passed over chloride of calcium, or

pumice-stone moistened with sulphuric acid, no such effect was observed. The needle moved steadily in one direction until its maximum deflection was attained, and this deflection showed that in all cases radiant heat was absorbed by the air within the tube.

These experiments were commenced in the spring of 1859, and continued without intermission for seven weeks. The course of the inquiry during this whole period was an incessant struggle with experimental difficulties. Approximate results were easily obtainable; but I aimed at exact measurements, which could not be made with a varying source of heat like the copper ball. I resorted to copper cubes containing fusible metal, or oil, raised to a high temperature; but was not satisfied with their action. I finally had a lamp constructed which poured a sheet of gas-flame along a plate of copper; and to keep the flame constant, a gas regulator specially constructed for me by Mr. Hulet was made use of. It was also arranged that the radiating plate should form one of the walls of a chamber which could be connected with the air-pump and exhausted, so that the heat emitted by the copper plate might cross a vacuum before entering the experimental tube. With this apparatus I determined approximately the absorption of nine gases and twenty vapours during the summer of 1859. The results would furnish materials for a long memoir; but increased experience and improved methods have enabled me to substitute for them others of greater value; I shall therefore pass over the work of these seven weeks without further allusion to it.

On the 9th of September of the present year (1860) I resumed the inquiry. For three weeks I worked with the plate of copper as my source of heat, but finally rejected it on the score of insufficient constancy. I again resorted to the cube of hot oil, and continued to work with it up to Monday the 29th of October. During the seven weeks just referred to, I experimented from eight to ten hours daily; but these experiments, though more accurate, must unhappily share the fate of the former ones. In fact the period was one of discipline—a continued struggle against the difficulties of the subject and the defects of the locality in which the inquiry was conducted.

My reason for making use of the high sources of heat above referred to was, that the absorptive power of some of the gases which I had examined was so small that, to make it clearly evident, a high temperature was essential. For other gases, and for *all* the vapours that had come under my notice, a source of lower temperature would have been not only sufficient, but far preferable. I was finally induced to resort to boiling water, which, though it gave greatly diminished effects, was capable of being preserved at so constant a temperature that deflections which, with the other sources, would be masked by the errors of observation, became with it true quantitative measures of absorption.

§ 4. The entire apparatus made use of in the experiments on absorption is figured on Plate III. S S′ is the *experimental tube*, composed of brass, polished within, and connected, as shown in the figure, with the air-pump, A A. At S and S′ are the plates of rock-salt which close the tube air-tight. The length from S to S′ is 4 feet. C is a cube containing boiling water, in which is immersed the thermometer *t*. The cube is of cast copper, and on one of its faces a projecting ring was cast to which a brass tube of the same diameter as S S′, and capable of being connected air-tight with the latter, was carefully soldered. The face of the cube within the ring is the radiating plate, which is coated with lampblack. Thus between the cube C and the first plate of rock-salt there is *a front chamber* F, connected with the air-pump by the flexible tube D D, and capable of being exhausted independently of S S′. To prevent the heat of conduction from reaching the plate of rock-salt S, the tube F is caused to pass through a vessel V, being soldered to the latter where it enters it and issues from it. This vessel is supplied with a continuous flow of cold water through the influx tube *i i*, which dips to the bottom of the vessel; the water escapes through the efflux tube *e e*, and the continued circulation of the cold liquid completely intercepts the heat that would otherwise reach the plate S.

The cube C is heated by the gas-lamp L. P is the thermo-electric pile placed on its stand at the end of the experimental tube, and furnished with two conical reflectors, as shown in the figure. C′ is the *compensating cube*, used to neutralize by its radiation* the effect of the rays passing through S S′. The regulation of this neutralization was an operation of some delicacy; to effect it the double screen H was connected with a winch and screw arrangement, by which it could be advanced or withdrawn through extremely minute spaces. For this most useful adjunct I am indebted to the kindness of my friend Mr. Gassiot. N N is the galvanometer, with perfectly astatic needles and perfectly non-magnetic coil; it is connected with the pile P by the wires *w w*; Y Y is a system of six chloride-of-calcium tubes, each 32 inches long; R is a U-tube containing fragments of pumice-stone, moistened with strong caustic potash; and Z is a second similar tube, containing fragments of pumice-stone wetted with strong sulphuric acid. When *drying* only was aimed at, the potash tube was suppressed. When, on

---

* It will be seen that in this arrangement I have abandoned the use of the differential galvanometer, and made the thermo-electric pile the differential instrument:

the contrary, as in the case of atmospheric air, both moisture and carbonic acid were to be removed, the potash tube was included. G G is a holder from which the gas to be experimented with was sent through the drying-tubes, and thence through the pipe *p p* into the experimental tube S S'. The appendage at M and the arrangement at O O may for the present be disregarded; I shall refer to them particularly by and by.

The mode of proceeding was as follows:—The tube S S' and the chamber F being exhausted as perfectly as possible, the connexion between them was intercepted by shutting off the cocks *m, m'*. The rays from the interior blackened surface of the cube C passed first across the vacuum F, then through the plate of rock-salt S, traversed the experimental tube, crossed the second plate S', and being concentrated by the anterior conical reflector, impinged upon the adjacent face of the pile P. Meanwhile the rays from the hot cube C' fell upon the opposite face of the pile, and the position of the galvanometer needle declared at once which source was predominant. A movement of the screen H back or forward with the hand sufficed to establish an approximate equality; but to make the radiations perfectly equal, and thus bring the needle exactly to 0°, the fine motion of the screw above referred to was necessary. The needle being at 0°, the gas to be examined was admitted into the tube; passing, in the first place, through the drying apparatus. Any required quantity of the gas may be admitted; and here experiments on gases and vapours enjoy an advantage over those with liquids and solids, namely, the capability of changing the density at pleasure. When the required quantity of gas had been admitted, the galvanometer was observed, and from the deflection of its needle the absorption was accurately determined.

Up to about its 36th degree, the degrees of my galvanometer are all equal in value; that is to say, it requires the same amount of heat to move the needle from 1° to 2° as to move it from 35° to 36°. Beyond this limit the degrees are equivalent to larger amounts of heat. The instrument was accurately calibrated by the method recommended by Melloni (*Thermochrose*, p. 59); so that the precise value of its larger deflections are at once obtained by reference to a table. Up to the 36th degree, therefore, the simple deflections may be regarded as the expression of the absorption; but beyond this the absorption equivalent to any deflection is obtained from the table of calibration.

§ 5. The air of the laboratory, freed from its moisture and carbonic acid, and permitted to enter until the tube was filled, produced a deflection of about . . . . 1°.

Oxygen obtained from chlorate of potash and peroxide of manganese produced a deflection of about . . . . . . 1°.

One specimen of nitrogen, obtained from the decomposition of nitrate of potash, produced a deflection of about . . .. 1°.

Hydrogen from zinc and sulphuric acid produced a deflection of about . . . . 1°.

Hydrogen obtained from the electrolysis of water produced a deflection of about . . . . 1°.

Oxygen obtained from the electrolysis of water, and sent through a series of eight bulbs containing a strong solution of iodide of potassium, produced a deflection of about . . 1°.

In the last experiment the electrolytic oxygen was freed from its ozone. The iodide of potassium was afterwards suppressed, and the oxygen, plus its ozone, admitted into the tube; the deflection produced was . . . . 4°.

Hence the small quantity of ozone which accompanied the oxygen in this case trebled the absorption of the oxygen itself*.

I have repeated this experiment many times, employing different sources of heat. With sources of high temperature the difference between the ozone and the ordinary oxygen comes out very strikingly. By careful decomposition a much larger amount of ozone might be obtained, and a corresponding large effect on radiant heat produced.

In obtaining the electrolytic oxygen, I made use of two different vessels. To diminish the resistance of the acidulated water to the passage of the current, I placed in one vessel a pair of very large platinum plates, between which the current from a battery of ten of Grove's cells was transmitted. The oxygen bubbles liberated on so large a surface were extremely minute, and the gas thus generated, on being sent through iodide of potassium, scarcely coloured the liquid; the characteristic odour of ozone was also almost entirely absent. In the second vessel smaller plates were used. The bubbles of oxygen were much larger, and did not come into such intimate contact with either the platinum or the water. The oxygen thus obtained showed the characteristic reactions of ozone; and with it the above result was obtained.

The total amount of heat transmitted through the tube in these experiments produced a deflection of . . . . 71°·5.

Taking as unit of heat the quantity necessary to cause the needle to move from 0° to 1°, the number of units expressed by the above deflection is . . . . 808.

Hence the absorption by the above gases amounted to about 0·33 per cent.

I am unable at the present moment to range with certainty oxygen, hydrogen, nitrogen, and atmospheric air in the order of

* It will be seen further on that this result is in harmony with the supposition that ozone, obtained in the manner described, is a *compound* body.

their absorptive powers, though I have made several hundred experiments with the view of doing so.   Their proper action is so small that the slightest foreign impurity gives one a predominance over the other.· In preparing the gases, I have resorted to the methods which I found recommended in chemical treatises, but as yet only to discover ·the defects incidental to these methods. Augmented experience and the assistance of my friends will, I trust, enable me to solve this point by and by. An examination of the whole of the experiments induces me to regard hydrogen as the gas which exercises the lowest absorptive power.

We have here the cases of minimum gaseous absorption. It will be interesting to place in juxtaposition with the above results some of those obtained with olefiant gas—the most highly absorbent permanent gas that I have hitherto examined.   I select for this purpose an experiment made on the 21st of November.

The needle being steady at zero in consequence of the equality of the actions on the opposite faces of the pile, the admission of olefiant gas gave a permanent deflection of . . 70°·3.

The gas being completely removed, and the equilibrium re-established, a plate of polished metal was interposed between one of the faces of the pile and the source of heat adjacent. The total amount of heat passing through the exhausted tube was thus found to produce a deflection of . . . . . 75°.

Now a deflection of 70°·3 is equivalent to 290 units, and a deflection of 75° is equivalent to 360 units; hence more than seven-ninths of the total heat was cut off by the olefiant gas, or about 81 per cent.

The extraordinary energy with which the needle was deflected when the olefiant gas was admitted into the tube, was such as might occur had the plates of rock-salt become suddenly covered with an opake layer.   To test whether any such action occurred, I polished a plate carefully, and projected against it for a considerable time a stream of the gas; there was no dimness produced.   The plates of rock-salt, moreover, which were removed daily from the tube, usually appeared as bright when taken out as when they were put in.

The gas in these experiments issued from its holder, and had there been in contact with cold water.   To test whether it had chilled the plates of rock-salt, and thus produced the effect, I filled a similar holder with atmospheric air and allowed it to attain the temperature of the water; but its action was not thereby sensibly augmented.

In order to subject the gas to ocular examination, I had a glass tube constructed and connected with the air-pump.   On permitting olefiant gas to enter it, not the slightest dimness or opacity was observed. To remove the last trace of doubt as to the possible action of the gas on the plates of rock-salt, the tin tube referred to at the commencement was perforated · at its centre and a cock inserted into it; the source of heat was at one end of the tube, and the thermo-electric pile at some distance from the other. The plates of salt were entirely abandoned, the tube being open at its ends and consequently full of air. On allowing the olefiant gas to stream for a second or two into the tube through the central cock, the needle flew off and struck against its stops. It was held steadily for a considerable time between 80° and 90°.

A slow current of air sent through the tube gradually removed the gas, and the needle returned accurately to zero.

The gas within the holder being under a pressure of about 12 inches of water, the cock attached to the cube was turned quickly on and off; the quantity of gas which entered the tube in this brief interval was sufficient to cause the needle to be driven to the stops, and steadily held between 60° and 70°.

The gas being again removed, the cock was turned once half round as quickly as possible. The needle was driven in the first instance through an arc of 60°, and was held permanently at 50°.

The quantity of gas which produced this last effect, on being admitted into a graduated tube, was found not to exceed one-sixth of a cubic inch in volume.

The tube was now taken away, and both sources of heat allowed to act from some distance on the thermo-electric pile. When the needle was at zero, olefiant gas was allowed to issue from a common argand burner into the air between one of the sources of heat and the pile. The gas was invisible, nothing was seen in the air, but the needle immediately declared its presence, being driven through an arc of 41°. In the four experiments last described, the source of heat was a cube of oil heated to 250° Centigrade, the compensation cube being filled with boiling water*.

Those who like myself have been taught to regard transparent gases as almost perfectly diathermanous, will probably share the astonishment with which I witnessed the foregoing effects. I was indeed slow to believe it possible that a body so constituted, and so transparent to light as olefiant gas, could be so densely opake to any kind of calorific rays; and to secure myself against error, I made several hundred experiments with this single substance. By citing them at greater length, however, I do not think I

---

* With a cube containing boiling water I have since made this experiment visible to a large audience.

could add to the conclusiveness of the proofs just furnished, that the case is one of true calorific absorption*.

§ 6. Having thus established in a general way the absorptive power of olefiant gas, the question arises, "What is the relation which subsists between the density of the gas and the quantity of heat extinguished?"

I sought at first to answer this question in the following way:— An ordinary mercurial gauge was attached to the air-pump; the experimental tube being exhausted, and the needle of the galvanometer at zero, olefiant gas was admitted until it depressed the mercurial column 1 inch, the consequent deflection being noted; the gas was then admitted until a depression of 2 inches was observed, and thus the absorption effected by gas of 1, 2, 3, and more inches tension was determined. In the following Table the first column contains the tensions in inches, the second the deflections, and the third the absorption equivalent to each deflection.

TABLE I.—Olefiant Gas.

Tensions in inches.	Deflections.	Absorption.
1	56	90
2	58·2	123
3	59·3	142
4	60·0	157
5	60·5	168
6	61·0	177
7	61·4	182
8	61·7	186
9	62·0	190
10	62·2	192
20	66·0	227

No definite relation between the density of the gas and its absorption is here exhibited. We see that an augmentation of the density *seven times* about *doubles* the amount of the absorption; while gas of 20 inches tension effects only 2¼ times the absorption of gas possessing 1 inch of tension.

But here the following reflections suggest themselves:—It is evident that olefiant gas of 1 inch tension, producing so large a deflection as 56°, must extinguish a large proportion of the rays which are capable of being absorbed by the gas, and hence the succeeding measures having a less and less amount of heat to act upon must produce a continually smaller effect. But sup-

---

* It is evident that the old mode of experiment might be applied to this gas. Indeed, several of the solids examined by Melloni are inferior to it in absorptive power. Had time permitted, I should have checked my results by experiments made in the usual way; this I intend to do on a future occasion.

posing the quantity of gas first introduced to be so inconsiderable that the number of rays extinguished by it is a vanishing quantity compared with the total number capable of absorption, we might reasonably expect that in this case a double quantity of gas would produce a double effect, a treble quantity a treble effect, or in general terms, that the absorption would, for a time, be proportional to the density

To test this idea, a portion of the apparatus, which was purposely omitted in the description already given, was made use of. O O, Plate III., is a graduated glass tube, the end of which dips into the basin of water B. The tube can be stopped above by means of the stopcock r; dd is a tube containing fragments of chloride of calcium. The tube O O being first filled with water to the cock r, had this water displaced by olefiant gas; and afterwards the tube S S', and the entire space between the cock r and the experimental tube, was exhausted. The cock n being now closed and r' left open, the cock r at the top of the tube O O was carefully turned on and the gas permitted to enter the tube S S' with extreme slowness. The water rose in O O, each of whose smallest divisions represents a volume of $\frac{1}{50}$th of a cubic inch. Successive measures of this capacity were admitted into the tube and the absorption in each case determined.

In the following Table the first column contains the quantity of gas admitted into the tube; the second contains the corresponding deflection, which, within the limits of the Table, expresses the absorption; the third column contains the absorption, calculated on the supposition that it is proportional to the density.

TABLE II.—Olefiant Gas.

Unit-measure $\frac{1}{50}$th of a cubic inch.

Measures of gas.	Absorption.	
	Observed.	Calculated.
1	2·2	2·2
2	4·5	4·4
3	6·6	6·6
4	8·8	8·8
5	11·0	11·0
6	12·0	13·2
7	14·8	15·4
8	16·8	17·6
9	19·8	19·8
10	22·0	22·0
11	24·0	24·2
12	25·4	26·4
13	29·0	28·6
14	30·2	29·8
15	33·5	33·0

This Table shows the correctness of the foregoing surmise, and proves that for small quantities of gas the absorption is exactly proportional to the density.

Let us now estimate the tensions of the quantities of gas with which we have here operated. The length of the experimental tube is 48 inches, and its diameter 2·4 inches; its volume is therefore 218 cubic inches. Adding to this the contents of the cocks and other conduits which led to the tube, we may assume that each fiftieth of a cubic inch of the gas had to diffuse itself through a space of 220 cubic inches. The tension, therefore, of a single measure of the gas thus diffused would be $\frac{1}{11,000}$th of an atmosphere,—a tension capable of depressing the mercurial column connected with the pump $\frac{1}{367}$th of an inch, or about $\frac{1}{15}$th of a millimetre!

But the absorptive energy of olefiant gas, extraordinary as it is shown to be by the above experiments, is far exceeded by that of some of the vapours of volatile liquids. A glass flask was provided with a brass cap furnished with an interior thread, by means of which a stopcock could be screwed air-tight on to the flask. Sulphuric ether being placed in the latter, the space above the liquid was completely freed of air by means of a second air-pump. The flask, with its closed stopcock, was now attached to the experimental tube; the latter was exhausted and the needle brought to zero. The cock was then turned on so that the ether-vapour slowly entered the experimental tube. An assistant observed the gauge of the air-pump, and when it had sunk an inch, the stopcock was promptly closed. The galvanometric deflection consequent on the partial cutting off of the calorific rays was then noted; a second quantity of the vapour, sufficient to depress the gauge another inch, was then admitted, and in this way the absorptions of five successive measures, each possessing within the tube 1 inch of tension, were determined.

In the following Table the first column contains the tensions in inches, the second the deflection due to each, and the third the amount of heat absorbed, expressed in the units already referred to. For the purpose of comparison I have placed the corresponding absorption of olefiant gas in the fourth column.

TABLE III.—Sulphuric Ether.

Tensions in inches.	Deflections.	Absorption.	Corresponding absorption by olefiant gas.
1	64·8	214	90
2	70·0	282	123
3	72·0	315	142
4	73·0	330	154
5	73·0	330	163

For these tensions the absorption of radiant heat by the vapour of sulphuric ether is more than twice the absorption of olefiant gas. We also observe that in the case of the former the successive absorptions approximate more quickly to a ratio of equality. In fact the absorption produced by 4 inches of the vapour was sensibly the same as that produced by 5.

But reflections similar to those which we have already applied to olefiant gas are also applicable to ether. Supposing we make our unit-measure small enough, the number of rays first destroyed will vanish in comparison with the total number, and for a time the fact will probably manifest itself that the absorption is directly proportional to the density. To examine whether this is the case, the other portion of the apparatus, omitted in the general description, was made use of. K is a small flask with a brass cap, which is closely screwed on to the stopcock c'. Between the cocks c' and c, which latter is connected with the experimental tube, is the chamber M, the capacity of which was accurately determined. The flask k was partially filled with ether, and the air above the liquid removed. The stopcock c' being shut off and c turned on, the tube S S' and the chamber M are exhausted. The cock c is now shut off, and c' being turned on, the chamber M becomes filled with pure ether vapour. By turning c' off and c on, this quantity of vapour is allowed to diffuse itself through the experimental tube, and its absorption determined; successive measures are thus sent into the tube, and the effect produced by each is noted. Measures of various capacities were made use of, according to the requirements of the vapours examined.

In the first series of experiments made with this apparatus, I omitted to remove the air from the space above the liquid; each measure therefore sent in to the tube was a mixture of vapour and air. This diminished the effect of the former; but the proportionality, for small quantities, of density to absorption exhibits itself so decidedly as to induce me to give the observations. The first column, as usual, contains the measures of vapour, the second the observed absorption, and the third the calculated absorption. The galvanometric deflections are omitted, their equivalents being contained in the second column. In fact as far as the eighth observation, the absorptions are merely the record of the deflections.

TABLE IV.—Mixture of Ether Vapour and Air.

Unit-measure $\frac{1}{50}$th of a cubic inch.

| Measures. | Absorption. | |
	Observed.	Calculated.
1	4·5	4·5
2	9·2	9·0
3	13·5	13·5
4	18·0	18·0
5	22·8	23·5
6	27·0	27·0
7	31·8	31·5
8	36·0	36·0
9	39·7	40·0
10	45·0	45·0
20	81·0	90·0
21	82·8	95·0
22	84·0	99·0
23	87·0	104·0
24	88·0	108·0
25	90·0	113·0
26	93·0	117·0
27	94·0	122·0
28	95·0	126·0
29	98·0	131·0
30	100·0	135·0

Up to the 10th measure we find that density and absorption augment in precisely the same ratio. .While the former varies from 1 to 10, the latter varies from 4·5 to 45. At the 20th measure, however, a deviation from proportionality is apparent, and the divergence gradually augments from 20 to 30. In fact 20 measures tell upon the rays capable of being absorbed,—the quantity destroyed becoming so considerable, that every additional measure encounters a smaller number of such rays, and hence produces a diminished effect.

With ether vapour alone, the results recorded in the following Table were obtained. Wishing to determine the absorption exercised by vapour of very low tension, the capacity of the unit-measure was reduced to $\frac{1}{100}$th of a cubic inch.

TABLE V.—Sulphuric Ether.

Unit-measure $\frac{1}{100}$th of a cubic inch.

Measures.	Absorption.		Measures.	Absorption.	
	Observed.	Calculated.		Observed.	Calculated.
1	5·0	4·6	17	65·5	77·2
2	10·3	9·2	18	68·0	83·0
4	19·2	18·4	19	70·0	87·4
5	24·5	23·0	20	72·0	92·0
6	29·5	27·0	21	73·0	96·7
7	34·5	32·2	22	73·0	101·2
8	38·0	36·8	23	73·0	105·8
9	44·0	41·4	24	77·0	110·4
10	46·2	46·2	25	78·0	115·0
11	50·0	50·6	26	78·0	119·6
12	52·8	55·2	27	80·0	124·2
13	55·0	59·8	28	80·5	128·8
14	57·2	64·4	29	81·0	133·4
15	59·4	69·0	30	81·0	138·0
16	62·5	73·6			

We here find that the proportion between density and absorption holds sensibly good for the first eleven measures, after which the deviation gradually augments.

I have examined some specimens of ether which acted still more energetically on the thermal rays than those above recorded. No doubt for smaller measures than $\frac{1}{100}$th of a cubic inch the above law holds still more rigidly true; and in a suitable locality it would be easy to determine with perfect accuracy $\frac{1}{10}$th of the absorption produced by the first measure; this would correspond to $\frac{1}{1000}$th of a cubic inch of vapour. But on entering the tube the vapour had only the tension due to the temperature of the laboratory, namely 12 inches. This would require to be multiplied by 2·5 to bring it up to that of the atmosphere. Hence the $\frac{1}{1000}$th of a cubic inch, the absorption of which I have affirmed to be capable of measurement, would, on being diffused through a tube possessing a capacity of 220 cubic inches, have a tension of $\frac{1}{220} \times \frac{1}{2·5} \times \frac{1}{1000} = \frac{1}{500,000}$th part of an atmosphere!

I have now to record the results obtained with thirteen other vapours. The method of experiment was in all cases the same as that just employed in the case of ether, the only variable element being the size of the unit-measure; for with many substances no sensible effect could be obtained with a unit volume so small as that used in the experiments last recorded. With bisulphide of carbon, for example, it was necessary to augment the unit-measure 50 times to render the measurements satisfactory.

TABLE VI.—Bisulphide of Carbon.'

Unit-measure ½ a cubic inch.

| Measures. | Absorption. | |
	Observed.	Calculated.
1	2·2	2·2
2	4·9	4·4
3	6·5	6·6
4	8·8	8·8
5	10·7	11·0
6	12·5	13·0
7	13·8	15·4
8	14·5	17·6
9	15·0	19·0
10	15·6	22·0
11	16·2	24·2
12	16·8	26·4
13	17·5	28·6
14	18·2	30·8
15	19·0	33·0
16	20·0	35·2
17	20·0	37·4
18	20·2	39·6
19	21·0	41·8
20	21·0	44·0

As far as the sixth measure the absorption is proportional to the density; after which the effect of each successive measure diminishes. Comparing the absorption effected by a quantity of vapour which depressed the mercury column half an inch, with that effected by vapour possessing one inch of tension, the same deviation from proportionality is observed.

By mercurial gauge.

Tension.	Absorption.
½ inch	14·8
1 inch	18·8

These numbers simply express the galvanometric deflections, which, as already stated, are strictly proportional to the absorption as far as 36° or 37°. Did the law of proportion hold good, the absorption due to 1 inch of tension ought of course to be 29·6 instead of 18·8.

Whether for equal volumes of the vapours at their maximum density, or for equal tensions as measured by the depression of the mercurial column, bisulphide of carbon exercises the lowest absorptive power of all the vapours which I have hitherto examined. For very small quantities, a volume of sulphuric ether vapour, at its maximum density in the measure, and expanded thence into the tube, absorbs 100 times the quantity of radiant heat intercepted by an equal volume of bisulphide of carbon vapour at its maximum density. These are the extreme limits of the scale, as far as my inquiries have hitherto proceeded. The action of every other vapour is less than that of sulphuric ether, and greater than that of bisulphide of carbon.

A very singular phenomenon was repeatedly observed during the experiments with bisulphide of carbon. After determining the absorption of the vapour, the tube was exhausted as perfectly as possible, the trace of vapour left behind being exceedingly minute. Dry air was then admitted to cleanse the tube. On again exhausting, after the first few strokes of the pump a jar was felt and a kind of explosion heard, while dense volumes of blue smoke immediately issued from the cylinders. The action was confined to the latter, and never propagated backwards into the experimental tube.

It is only with bisulphide of carbon that this effect has been observed. It may, I think, be explained in the following manner:—To open the valve of the piston, the gas beneath it must have a certain tension, and the compression necessary to produce this appears sufficient to cause the combination of the constituents of the bisulphide of carbon with the oxygen of the air. Such a combination certainly takes place, for the odour of sulphurous acid is unmistakeable amid the fumes.

To test this idea I tried the effect of compression in the air-syringe. A bit of tow or cotton wool moistened with bisulphide of carbon, and placed in the syringe, emitted a bright flash when the air was compressed. By blowing out the fumes with a glass tube, this experiment may be repeated twenty times with the same bit of cotton.

It is not necessary even to let the moistened cotton remain in the syringe. If the bit of tow or cotton be thrown into it, and out again as quickly as it can be ejected, on compressing the air the luminous flash is seen. Pure oxygen produces a brighter flash than atmospheric air. These facts are in harmony with the above explanation.

TABLE VII.—Amylene.

Unit-measure $\frac{1}{10}$th of a cubic inch.

Measures.	Absorption.	
	Observed.	Calculated.
1	3·4	4·3
2	8·4	8·6
3	12·0	12·9
4	16·5	17·2
5	21·6	21·5
6	26·5	25·8
7	30·6	30·1
8	35·3	34·4
9	39·0	38·7
10	44·0	43·0

For these quantities the absorption is proportional to the density, but for large quantities the usual deviation is observed, as shown by the following observations:—

By mercurial gauge.

Tension.	Deflection.	Absorption.
¼ inch	60°	157
1 inch	65	216

Did the proportion hold good, the absorption for an inch of tension ought of course to be 314 instead of 216.

TABLE VIII.—Iodide of Ethyle.

Unit-measure $\frac{1}{10}$th of a cubic inch.

	Absorption.	
Measures.	Observed.	Calculated.
1	5·4	5·1
2	10·3	10·2
3	16·8	15·3
4	22·2	20·4
5	26·6	25·5
6	31·8	30·6
7	35·6	35·9
8	40·0	40·8
9	44·0	45·9
10	47·5	51·0

By mercurial gauge.

Tension.	Deflection.	Absorption.
¼ inch	56°·3	94
1 inch	58·2	120

TABLE IX.—Iodide of Methyle.

Unit-measure $\frac{1}{10}$th of a cubic inch.

	Absorption.	
Measures.	Observed.	Calculated.
1	3·5	3·4
2	7·0	6·8
3	10·3	10·2
4	15·0	13·6
5	17·5	17·0
6	20·5	20·4
7	24·0	23·8
8	26·3	27·2
9	30·0	30·6
10	32·3	34·0

By mercurial gauge.

Tension.	Deflection.	Absorption.
¼ inch	48·5°	60
1 inch	56·5	96

TABLE X.—Iodide of Amyle.

Unit-measure $\frac{1}{10}$th of a cubic inch.

Measures.	Absorption.	
	Observed.	Calculated.
1	0·6	0·57
2	1·0	1·1
3	1·4	1·7
4	2·0	2·3
5	3·0	2·9
6	3·8	3·4
7	4·5	4·0
8	5·0	4·6
9	5·0	5·1
10	5·8	5·7

The deflections here are very small; the substance, however, possesses so feeble a volatility, that the tension of a measure of its vapour, when diffused through the experimental tube, must be infinitesimal. With the specimen which I examined, it was not practicable to obtain a tension sufficient to depress the mercury gauge ½ an inch; hence no observations of this kind are recorded.

TABLE XI.—Chloride of Amyle.

Unit-measure $\frac{1}{10}$th of a cubic inch.

Measures.	Absorption.	
	Observed.	Calculated.
1	1·3	1·3
2	3·0	2·6
3	3·8	3·9
4	5·1	5·2
5	6·8	6·5
6	8·5	7·8
7	9·0	9·1
8	10·9	10·4
9	11·3	11·7
10	12·3	13·0

By mercurial gauge.

Tension.	Deflection.	Absorption.
¼ inch	59°	137
1 inch	not practicable.	

TABLE XII.—Benzole.

Unit-measure $\frac{1}{10}$th of a cubic inch.

Measures.	Absorption.	
	Observed.	Calculated.
1	4·5	4·5
2	9·5	9·0
3	14·0	13·5
4	18·5	18·0
5	22·5	22·5
6	27·5	27·0
7	31·6	31·5
8	35·5	36·0
9	39·0	40·0
10	44·0	45·0
11	47·0	49·0
12	49·0	54·0
13	51·0	58·5
14	54·0	63·0
15	56·0	67·5
16	59·0	72·0
17	63·0	76·5
18	67·0	81·0
19	69·0	85·5
20	72·0	90·0

Up to the 10th measure, or thereabouts, the proportion between density and absorption holds good, from which onwards the deviation from the law gradually augments.

By mercurial gauge.

Tension.	Deflection.	Absorption.
¼ inch	54	78
1 inch	57	103

TABLE XIII.—Methylic Alcohol.

Unit-measure $\frac{1}{10}$th of a cubic inch.

Measures.	Absorption.	
	Observed.	Calculated.
1	10·0	10·0
2	20·0	20·0
3	30·0	30·0
4	40·5	40·0
5	49·0	50·0
6	53·5	60·0
7	59·2	70·0
8	71·5	80·0
9	78·0	90·0
10	84·0	100·0

By mercurial gauge.

Tension.	Deflection.	Absorption.
½ inch	58·8	133
1 inch	60·5	168

TABLE XIV.—Formic Ether.

Unit-measure $\frac{1}{10}$th of a cubic inch.

	Absorption.	
Measures.	Observed.	Calculated.
1	8·0	7·5
2	16·0	15·0
3	22·5	22·5
4	30·0	30·0
5	35·2	37·5
6	39·5	45·0
7	45·0	52·5
8	48·0	60·0
9	50·2	67·5
10	53·5	75·0

By mercurial gauge.

Tension.	Deflection.	Absorption.
½ inch	58·8	133
1 inch	62·5	193

TABLE XV.—Propionate of Ethyle.

Unit-measure $\frac{1}{10}$th of a cubic inch.

	Absorption.	
Measures.	Observed.	Calculated.
1	7·0	7·0
2	14·0	14·0
3	21·8	21·0
4	28·8	28·0
5	34·4	35·0
6	38·8	42·0
7	41·0	49·0
8	42·5	56·0
9	44·8	63·0
10	46·5	70·0

By mercurial gauge.

Tension.	Deflection.	Absorption.
½ inch	60·5	168
1 inch	not practicable.	

Table XVI.—Chloroform.

Unit-measure $\frac{1}{10}$th of a cubic inch.

Measures.	Absorption.	
	Observed.	Calculated.
1	4·5	4·5
2	9·0	9·0
3	13·8	13·5
4	18·2	18·0
5	22·3	22·5
6	27·0	27·0
7	31·2	31·5
8	35·0	36·0
9	39·0	40·5
10	40·0	45·0

Subsequent observations lead me to believe that the absorption by chloroform is a little higher than that given in the above Table.

Table XVII.—Alcohol.

Unit-measure $\frac{1}{2}$ a cubic inch.

Measures.	Absorption.	
	Observed.	Calculated.
1	4·0	4·0
2	7·2	8·0
3	10·5	12·0
4	14·0	16·0
5	19·0	20·0
6	23·0	24·0
7	28·5	28·0
8	32·0	32·0
9	37·5	36·0
10	41·5	40·0
11	45·8	44·0
12	48·0	48·0
13	50·4	52·0
14	53·5	56·0
15	55·8	60·0

By mercurial gauge.

Tension.	Deflection.	Absorption.
$\frac{1}{2}$ inch	60°	157
1 inch	not practicable.	

The difference between the measurements when equal *tensions* and when equal *volumes* at the maximum density are made use of is here strikingly exhibited.

In the case of alcohol I was obliged to resort to a unit-measure of $\frac{1}{2}$ a cubic inch to obtain an effect about equal to that produced by benzole with a measure possessing only $\frac{1}{10}$th of a cubic inch in capacity; and yet for equal tensions of 0·5 of an inch, alcohol cuts off precisely twice as much heat as benzole. There is also an enormous difference between alcohol and sulphuric ether when equal measures at the maximum density are compared; but to bring the alcohol and ether vapours up to a common tension, the density of the former must be many times augmented. Hence it follows that when *equal tensions* of these two substances are compared, the difference between them diminishes considerably. Similar observations apply to many of the substances whose deportment is recorded in the foregoing Tables; to the iodide and chloride of amyle, for example, and to the propionate of ethyle. Indeed it is not unlikely that with equal tensions the vapour of a perfectly pure specimen of the substance last mentioned would be found to possess a higher absorptive power than that of ether itself.

It has been already stated that the tube made use of in these experiments was of brass ‚polished within, for the purpose of bringing into clearer light the action of the feebler gases and vapours. Once, however, I wished to try the effect of chlorine, and with this view admitted a quantity of the gas into the experimental tube. The needle was deflected with prompt energy; but on pumping out *, it refused to return to zero. To cleanse the tube, dry air was introduced into it ten times in succession; but the needle pointed persistently to the 40th degree from zero. The cause of this was easily surmised: the chlorine had attacked the metal and partially destroyed its reflecting power; thus the absorption by the sides of the tube itself cut off an amount of heat competent to produce the deflection mentioned above. For subsequent experiments the interior of the tube had to be repolished.

Though no other vapour with which I had experimented produced a permanent effect of this kind, it was necessary to be perfectly satisfied that this source of error had not vitiated the experiments. To check the results, therefore, I had a length of 2 feet of similar brass tube coated carefully on the inside with lampblack, and determined by means of it the absorptions of all the vapours which I had previously examined, at a common tension of 0·3 of an inch. A general corroboration was all I sought, and I am satisfied that the few discrepancies which the measurements exhibit would disappear, or be accounted for, in a more careful examination.

In the following Table the results obtained with the blackened and with the bright tubes are placed side by side, the tension

---

* Dense dark fumes rose from the cylinders on this occasion; a similar effect was produced by sulphuretted hydrogen.

in the former being three-tenths, and in the latter five-tenths of
an inch.

TABLE XVIII.

Absorption.

Vapour.	Bright tube, 0·5 tension.	Blackened tube, 0·3 tension.	Absorption with bright tube proportional to
Bisulphide of Carbon . .	5·0	21	23
Iodide of Methyle . . .	15·8	60	71
Benzole	17·5	78	79
Chloroform	17·5	89	79
Iodide of Ethyle . . .	21·5	94	97
Wood-spirit	26·5	123	120
Methylic Alcohol . . .	29·0	133	131
Chloride of Amyle . . .	30·0	137	135
Amylene	31·8	157	143

The order of absorption is here shown to be the same in both
tubes, and the quantity absorbed in the bright tube is, in
general, about 4½ times that absorbed in the black one. In the
third column, indeed, I have placed the products of the numbers
contained in the first column by 4·5. These results completely
dissipate the suspicion that the effects observed with the bright
tube could be due to a change of the reflecting power of its
inner surface by the contact of the vapours.

With the blackened tube the order of absorption of the following substances, commencing with the lowest, stood thus:—

> Alcohol,
> Sulphuric ether,
> Formic ether,
> Propionate of ethyle;

whereas with the bright tube they stood thus:—

> Formic ether,
> Alcohol,
> Propionate of ethyle,
> Sulphuric ether.

As already stated, these differences would in all probability
disappear, or be accounted for on re-examination. Indeed very
slight differences in the purity of the specimens used would be
more than sufficient to produce the observed differences of absorption*.   [To be continued.]

* In illustration of this I may state, that of two specimens of methylic
alcohol with which I was furnished by two of my chemical friends, one gave
an absorption of 84 and the other of 203. The former specimen had been
purified with great care, but the latter was not pure. Both specimens,
however, went under the common name of methylic alcohol. I have had
a special apparatus constructed with a view to examine the influence of
ozone on the interior of the experimental tube.

XXIV. *Some Remarks on* Dr. Siemens's *Paper " On Standards of Electrical Resistance, and on the Influence of Temperature on the Resistance of Metals."* By A. MATTHIESSEN, F.R.S.*

1. IN the above paper †, page 92 (2nd paper), M. Siemens states, " *It may be asserted without all doubt, that the most experienced and skilful physicists, even with the best instruments and most appropriate localities, are not able to determine resistances in absolute measure which do not vary several per cent. A standard of so little accuracy would not even answer the requirements of technical purposes.*" M. Siemens, however, does not give the grounds on which he bases the above assertion.

Prof. W. Thomson, in a paper published in the 'Proceedings of the Royal Society' (vol. viii. p. 555), says, " *It is impossible to over-estimate the great practical value of this system of absolute measurement carried out by Weber into every department of electrical science.*" I have always understood that the determinations of resistances in absolute measure by Weber's methods were most accurate; and in order to be able to answer this point more definitely, I wrote to Prof. W. Thomson and asked him to give me his opinion on the subject, knowing that the opinion of such a distinguished physicist would have great weight—in fact, would settle the question.

Prof. Thomson's answer was the following:—

" *There can scarcely be a doubt but that Weber's original determination of resistance in absolute measure* (Pogg. Ann. vol. lxxxii. p. 33) *was considerably within one half per cent. of the truth. He used two remarkably different methods, and obtained by means of them* 190·8 *and* 189·8 *respectively for the absolute measure of the resistance of one of his conductors. The details of the application of each of the two methods separately present so much consistence, that the possibility of so great an error as one half per cent. could not be admitted in the mean result of either considered alone, unless through some error in the corrections directly applied to it. Any such doubt seems perfectly removed by the close agreement between the two results derived from the two different methods, with different instruments, very dissimilar experimental operations, and perfectly distinct reductions and corrections to reduce to absolute measure. The mean of the two numbers quoted above, being* 190·05, *differs by less than* 0·14 *per cent. from each of them. It is not improbable that this mean may be within* 0·1 *per cent. of the truth: it is*

---

* Communicated by the Author.

† Pogg. *Ann.* vol. cxiii. p. 91. In order to prevent mistakes I will call this the 2nd paper; the 1st paper being the original one, where M. Siemens proposes mercury as a standard (Phil. Mag. Jan. 1861).

improbable that it differs by 0·2 *per cent. from the truth, and it is scarcely possible that it is wrong by* 0·5 *per cent.**"

2. M. Siemens states, page 93 (2nd paper), "Because the differences found in the conducting powers of the gold-silver alloys I had made in different places amount to 1·5 per cent., the alloy is useless for the purpose proposed by me (Phil. Mag. Feb. 1861), namely, the reproduction of a resistance by means of which the observations of different experimenters may be compared with each other, or the reproduction of a resistance in absolute measure.   For if two alloys are made and their resistances determined, we should certainly come within one half per cent. of the true value, six out of the eight alloys tested agreeing within that limit."   Let us now for a moment see what M. Siemens says of his proposed mercury standard ; and on referring to his first paper we find a Table, where he gives the resistances of six tubes filled with mercury.   The values found by him for $\frac{w}{w_1}$, where w is the calculated and w_1 the observed resistances, are given in the following Table, together with those found by myself for the conducting power of the gold-silver alloy.

TABLE I.

No. of tube.	Values found for $\frac{w}{w_1}$.	Conducting power of hard-drawn alloys.
1	1·008	1 = 1·003
2	1·000	2 = 1·002
3	1·0008	3 = 0·988
4	0·992	5 = 1·004
5	0·994	6 = 0·997
6	1·005	7 = 1·003
		8 = 1 001

M. Siemens, when speaking of the differences he found, says *they are not greater than were to be expected*; and further on, he continues, *the temperatures of the* étalon (*copper*) *and the mercury varied* 2—3° C. *during the experiments* ; but does not state which of the determinations were made at the higher or lower temperature, so that the differences he finds may be greater or smaller, as the case may be.   Now on comparing the above values it will be seen that the maximum differences are in each case the same. If, therefore, in the opinion of M. Siemens, the gold-silver alloy is useless as a standard, how much more must his mercury standard be so, when, according to his own determinations with

* Prof. Weber, in a letter written a short time since to Prof. Thomson, states, when speaking of some new determinations of resistance in absolute measure he is about to undertake, that by some improvements in the method and apparatus he hopes to arrive *at a still greater accuracy* than that which he formerly obtained.

the same mercury in tubes carefully picked from a large quantity, he does not arrive at a greater accuracy than I did with alloys made in *different* places, by *different* persons, of *different* gold and silver, and drawn by *different* wire-drawers. If, on the contrary, I had made and drawn the eight alloys myself of the same gold and silver, I should undoubtedly have obtained results not varying 0·1 per cent. If now different experimenters determine the conducting power of mercury, is it not probable that much greater differences would be found between their results than those obtained by M. Siemens *himself*? Now it so happens that different observers have already determined the conducting power of mercury. Let us compare their results; and we will first compare the conducting powers of the metals, taking silver = 100, and afterwards taking mercury = 100. Now I maintain that if the values obtained for one and the same metal by these different observers agree better when compared with silver than with mercury as unit, then M. Siemens's proposed standard must be useless as such.

TABLE II.—Conducting Power of Metals.   Silver = 100.

	Siemens.	Lenz.	Becquerel.	Matthiessen.
Silver*	100	100	100	100
Copper*......	96·9†	73·4	95·3	99·5
Gold*	......	58·5	68·9	78
Cadmium ...	......	......	26·3	23·8
Zinc	......	......	25·7	29·2
Tin	......	22·6	15·0	12·3
Iron	......	13·0	13·1	14·4 at 20·4
Lead	......	10·7	8·8	8·3
Platinum* ...	14·2	10·4	8·6	10·5 at 20·7
Mercury......	1·72	3·42 at 18·9	1·86	1·65

TABLE III.—Conducting Power of Metals.   Mercury = 100.

	Siemens.	Lenz.	Becquerel.	Matthiessen.
Silver*	5820	2924	5376	6060
Copper*......	5640	2146	5123	6030
Gold*	......	1710	3704	4727
Cadmium ...	......	......	1414	1442
Zinc	......	......	1382	1770
Tin	......	659	810	745
Iron	......	380	704	872 at 20·4
Lead	......	312	473	503
Platinum* ...	825	304	462	636 at 20·7
Mercury......	100	100 at 18·7	100	100

* Hard-drawn. All temperatures 0° C., except when the contrary is stated.

† Value given in 2nd paper 100.

One glance at the foregoing Table will suffice to show how very badly Lenz's series agrees with the rest when mercury is taken as unit; and comparing Becquerel's and my own, our values differ for—

TABLE IV.

	When silver = 100.	When mercury = 100.
Copper	4·3 per cent.	15 per cent.
Gold	11·6 ,,	21·6 ,,
Cadmium . . .	9·0 ,,	2·0
Zinc	11·9 ,,	22·1
Tin	18·0 ,,	8·0
Iron	9·0 ,,	19·2
Lead	5·7 ,,	5·9
Platinum . . .	18·1 ,,	27·3 ,,

These results prove that the mercury standard proposed by M. Siemens cannot be a useful and good one; for, in fact, we obtain more concordant results if we take in the above series any other metal as unit. The mercury employed by three of the observers was stated by them to have been pure.

3. Page 93 (2nd paper), M. Siemens states, *"German-silver wire is much better for resistance coils than the gold-silver alloy, on account of its high conducting power and expense."* I quite agree with him. I only proposed the gold-silver alloy to be used for the same purpose as he does mercury.

4. Page 93 (2nd paper), M. Siemens states, *" Even if the con- ducting power of the gold-silver alloys were the same, yet small resistances cannot be accurately compared with them, as there would always be a slight difference in the resistance at points where the alloy is connected with the connectors of the apparatus."* I may, however, mention that I always solder the ends of the normal wire to two thick copper wires (of 2–3 millims. diameter and about 35 millims. long), the free ends of which are carefully amalgamated by dipping them into a solution of nitrate of mer- cury in dilute nitric acid; and the connexions are made by means of mercury cups, the bottoms of which are amalgamated copper plates. These can be removed, and are of course from time to time reamalgamated. The free copper ends of the normal wire are reamalgamated every time before use. This arrangement gives most satisfactory results; not the slightest change in the resist- ance is observed when the normal wire is taken out of the mer- cury cups and put in again. If, however, a wire of the gold- silver alloy has once been made and arranged for use, when wanted it is only necessary to reamalgamate the ends, and it may then be used without further loss of time. On the contrary, for M. Siemens's proposed unit there must be a great deal of time

spent in cleaning the tube (in which operation the tube is liable to be broken) and in purifying the mercury.

5. Page 95 (2nd paper), M. Siemens gives a table, by which he wishes to prove that he is able to reproduce resistances of exactly the same values. He, however, only proves that he is able to fill the same tubes with different mercury, and that their resistances only vary 0·05 per cent.; for he compared three unknown resistances with two equal ones (when reduced to equal lengths and diameters), and obtained very nearly the same values. Now if, instead of taking normal tubes, called 3 and 7, he had taken those called No. 1 and 4 (1st paper), would his results have been the same? No; they would have varied 1·5 per cent. (See his results given in Table I.)

6. Page 96 (2nd paper), M. Siemens says, the statement I made *that the traces of foreign metals cause a decrement in the conducting power of mercury, and not, as stated by* Siemens, *an increment, is incorrect.* In this M. Siemens is perfectly right. I was misled by the fact that when mercury is alloyed with several per cent. of foreign metal, a smaller conducting power is observed than the mean of the conducting powers of the relative volumes of the metals employed; and as in no case I had found an increment in the conducting power of a metal when alloyed with a trace of another, I concluded that traces (0·1 or 0·2 per cent.) of foreign metals would also cause a decrement in the conducting power of mercury.

As mercury behaves in this respect differently from the other metals, instead of assuming, as I did in my paper on the conducting power of alloys*, that the metals may be classed under two heads, viz.,—

I. Those metals which, when alloyed with each other, conduct electricity in the ratio of their relative volumes;

II. Those metals which, when alloyed with one of the metals belonging to the first class or with one another, *do not* conduct electricity in the ratio of their relative volumes, but *always* in a lower degree than the mean of their volumes,—

we must now have three classes of metals, the third probably being—

Those metals which, when alloyed with very small per-centages of another, have a *greater* conducting power, but when alloyed with larger per-centages, have a *lower* conducting power than the mean of their volumes. I am at present investigating how far this may be true; and it will be very interesting to see whether pure metals, such as bismuth, tin, &c., in a liquid state behave like mercury; that is to say, if, when melted, traces of other metals be added, an increment in the conductor will be observed. I also

* Phil. Trans. 1860, p. 161.

intend trying whether the conducting power of mercury when solid is increased or decreased by the addition of traces of other metals.

To prove the assumption I have made as to the behaviour of the third class of metals is probably correct, I have given in Table V. some experiments.

TABLE V.

Taking the conducting power of the hard-drawn gold-silver alloys at $0° = 100$,—

			Calculated conducting power.
Pure mercury conducts	24·47 at 18° C.	
„ alloyed with 0·1 per cent. pure } bismuth .... }	24·58 at 18·6	24·46
„ 0·01 per cent. pure tin	24·51 at 18·4	24·50
„ 0·02 „	24·54 at 18·0	24·52
„ 0·05 „	24·63 at 18 2	24·61
„. 0·1 „	24·76 at 18·8	24·75
„ 0·2 „	25·04 at 19·0	25·02
„ 0·5 „	25·86 at 18·4	25·83
„ 1·0 „	26·62 at 18·6	27·19
„ 2·0 „	27·66 at 18·8	29·19
„ „ 4·0 „	29·69 at 19·0	35·09

For the calculations, the conducting power of tin was taken at 172·09, that of bismuth 17·88; the specific gravity of mercury 13·573, that of bismuth 9·823, and that of tin 7·294.

The resistances of the amalgams were determined in the same tube as the mercury, so that any error in the measurement of the length or diameter will not have any influence on the relative values obtained.

From the above Table we see that even bismuth, a worse conductor than mercury, increases the conducting power of mercury, as would be expected from the above assumption. The experiments with the amalgams show how important it would be, if mercury were to be taken as unit for determinations of resistances, that it should be *absolutely chemically pure.* We cannot be surprised to find discrepancies in the values obtained for mercury by different observers, when such small traces of impurity so materially affect its conducting power.

7. Page 103 (2nd paper), M. Siemens gives a Table, from which he deduces that the increase in the resistance of mercury between 0° and 100° C. is in direct ratio with the increase of temperature. In other words, M. Siemens assumes that the formula

$$w = 1 + at$$

expresses the resistance of mercury at any temperature between 0° and 100°. Let us now calculate from his results the values of " *a,*" the temperatures at which the resistances have been observed. These will be found by using the formula $a = \dfrac{w-1}{t}$.

In Table VI. I have given M. Siemens's Table of the resistances of mercury for different temperatures, together with the value of the coefficient " *a* " for each of the observations. The resistance of mercury at 0° C. is taken = 1.

TABLE VI.

T.	Resistance.	Coefficient " *a.*"
0	1·00	
18·51	1·0166	0·000897
28·19	1·0263	0·000933
41·29	1·0391	0·000947
57·34	1·0548	0·000956
97·29	1·0959	0·000986

If the formula $w = 1 + at$ were correct, the values found for " *a* " ought all to be equal; but as there is a gradual increment in the values there can be no doubt that a formula with two terms, as $w = 1 + at + bt^2$, will express the resistances for different temperatures much better.

The increase of resistance of mercury between 0° and 100° is, ·according to

Becquerel.	Siemens.	Matthiessen and von Bose.	Schröder van der Kolk[*].
10·3 per cent.	9·85 per cent.	9·0 per cent.	8·6 per cent.

M. Siemens's value is deduced from 12 observations; Schröder van der Kolk's from 29; von Bose's and my own from 96. Again, M. Siemens deduces from 14 observations that the resistance of copper between 0° and 100° increases in direct ratio with the increase of temperature; whereas von Bose and myself deduce from 332 observations that the formula for the resistance of copper must be $w = 1 + at + bt^2$. ·Our experiments are almost finished; we hope that they will be published before the end of the year.

8. Page 105 (2nd paper), M. Siemens states, " *What induced Mr. Matthiessen to make at the end of his paper the following asser- tion I am not able to judge, as he does not give the grounds on which he bases it :—' it has been generally assumed that the conducting power of all copper wire, whether pure or commercial, varies with an increase of temperature to the same degree, which, however, is far from the truth.'* " Two reasons for my having made the above

---

[*] Poggendorff's *Annalen*, vol. cx. p. 452.

assertion were, (1) M. Siemens himself assumes in his first paper that the conducting power of his copper (étalon) varies 0·1 per cent. with each degree Centigrade; and (2) M. C. W. Siemens*, in describing his resistance thermometer, assumes also the same; in fact he bases his calculation on Arndtsen's formula without stating the sort of copper he uses.

That my statement regarding the difference of the coefficients of the increase of resistance for different temperatures of coppers is correct, may be deduced from the following data:—M. Siemens finds (2nd paper) the resistances of a commercial copper he tested to vary between 0° and 100° C. 32·9 per cent.; Arndtsen finds copper containing traces of iron to vary 36 per cent.; von Bose and myself have found pure copper to vary 42 per cent.; and lastly, one commercial copper I have tested varies only about 8 per cent.

XXV. *On the True and False Discharge of a Coiled Electric Cable.* By Professor W. Thomson, *LL.D., F.R.S.,* and Mr. Fleeming Jenkin, *C.E.*†

IN an article in the last May Number of this Magazine, "On the Galvanic Polarization of buried Metal Plates," translated from Poggendorff's *Annalen*, No. 10, 1860, Dr. Carl describes certain interesting experiments on the electro-polarization produced between two large zinc plates buried in the garden of the Observatory of Munich, by opposing and by augmenting the natural earth-current between them by the application of a single element of Daniell's; and concludes with the following remark:—

"The above experiments disclose nothing at variance with the known laws of galvanism; but it nevertheless appeared to me advisable to make them known, as they afford a simple explanation of certain phenomena which Professor Thomson has described (Report of the Twenty-ninth Meeting of the British Association, Aberdeen, 1859, Trans. of Sections, p. 26), and which he seems to attribute to entirely different causes."

In the report of Prof. Thomson's communication to the British Association here referred to, it is stated that (after mentioning certain experiments by Mr. F. Jenkin on submarine cables coiled in the manufactory of Messrs. Newall and Co., Birkenhead, in which one end of the battery used, and one end of the cable experimented on, in each case was kept in connexion with the earth while the other end of the cable, after having been for a time in

* Phil. Mag. January 1861.
† Communicated by the Authors.

connexion with the insulated pole of the battery, was suddenly
removed from the battery and put in connexion with the earth
through the coil of a galvanometer) Prof. Thomson and Mr.
Fleeming Jenkin remarked "that the deflections recorded in these

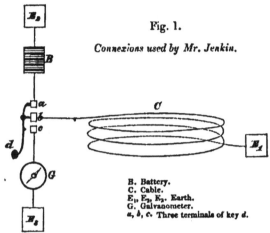

Fig. 1.

Connexions used by Mr. Jenkin.

B. Battery.
C. Cable.
E₁, E₂, E₃. Earth.
G. Galvanometer.
a, b, c. Three terminals of key d.

experiments were in the contrary direction to that which the true
discharge of the cable would give;" and at Prof. Thomson's
request "Mr. Jenkin repeated the experiments, watching carefully for indications of reverse currents to those previously noted.
It was thus found that the first effect of pressing down the key
[to throw the cable from battery to earth through galvano-
meter] was to give the galvanometer a deflection in the direction
corresponding to the true discharge current, and that this was
quickly followed by a reverse current generally greater in degree,
which gave a deflection corresponding to a current in the same
direction as that of the original flow through the cable.

"Professor Thomson explained this second current, or false
discharge, as it has since been sometimes called, by attributing
it to mutual electro-magnetic induction between different parts of
the coil, and anticipated that no such reversal could ever be
found in a submerged cable. The effect of this induction is to
produce in those parts of the coil first influenced by the motion
of the key, a tendency for the electricity to flow in the same
direction as that of the decreasing current flowing through the
remoter parts of the coil. Thus, after the first violence of the
back flow through the key and galvanometer, the remote parts
of the cable begin, by their electro-magnetic induction on the
near parts, to draw electricity back from the earth through the
galvanometer into the cable again, and the current is once more
in one and the same direction throughout the cable."

P 2

The phenomena thus described and explained are entirely different from any that could result from the galvanic polarization supposed by Dr. Carl to account for them*. It is true that the discharging earth-plate might become polarized by the discharge in certain cases sufficiently to cause a slight reversal in the current through the galvanometer coil, after the subsidence of the violent discharge current through it. But in no case could the whole quantity of electricity flowing in this supposed polarization current be more than a very small fraction of the quantity which previously flowed in the true discharge current, of which it is a feeble electro-chemical reflexion. Its effect on the galvanometer needle must in every case be as nothing in comparison to the great impulsive deflection produced by the true discharge current ; and there is no combination of circumstances, as to size of the earth-plates, amount of the battery power, and rapidity or sensibility of the galvanometer needle, in which the cause supposed by Dr. Carl could possibly be adequate to explain the phenomena described in Prof. Thomson's communication.

In point of fact, all effects of polarization of the earth-plates were extremely small in comparison with the main currents observed, which in the experiments on cables with one end kept to earth, consisted of (1) the constant *through-current*, produced by a battery of 72 elements Daniell's in series ; (2) the true discharge through the galvanometer to be observed instantly after breaking the battery connexion of the end of the cable to which the battery was applied, and making instead a connexion, through the galvanometer coil, between the same end of the cable and the earth ; and (3) the " false discharge, " so called because it must have been often mistaken for the true discharge, which almost necessarily escapes notice altogether when short lengths of coiled cable are tested with slow galvanometer nee-

* They are also different from any effects which could result from polarization of the plate connecting the far end of the cable with earth—a cause suggested by Prof. Wheatstone in a report published by the Committee appointed by the Board of Trade to inquire into the Construction of Submarine Cables. In support of his opinion, Prof. Wheatstone quotes some experiments in which he could observe only the well-known effects due to polarization, which on the short pieces of wire at his command quite overpowered both the true and false discharge. The current from the polarized end of a cable is always in the direction of the true discharge when the battery has been long enough applied : it is observed on both straight and coiled cables, and is capriciously variable. The details given in the present paper show that the currents due to electro-magnetic induction, called false discharge currents, are on the contrary always in the opposite direction to that of the true discharge, that they can only be observed on coiled cables, and that they are in each case sensibly constant. The galvanometer used by Mr. Jenkin would not have been deflected half a degree by the current from a polarized earth-plate at the end of cables from 300 to 500 knots in length.

dles. The *through-current* (1) was measured at the beginning of the discharge experiments by introducing the galvanometer into the circuit of cable and battery. Neither the whole amount of the true and false discharges, nor the rapidly varying strength of the current from instant to instant, could be distinctly observed, because the period of vibration of the galvanometer needle, being about $4\frac{1}{2}$ seconds each way, was neither incomparably greater nor incomparably smaller than the duration of the current in either direction. Thus the back-flow, or true discharge, which was of comparatively short duration, first gave the needle an impulse to the left (let us suppose); but before its natural swing, from even an instantaneous impulse, could have allowed it to begin to return, it was caught by the reverse current of false discharge and turned and thrown to the other side of zero through an angle to the right, which, except in the cases of the longest lengths of cable experimented on, was much greater than the angle of the first deflection to the left. It is obvious from what has been stated, that the durations of these deflections of the needle on the two sides do. not even approximately coincide with the times during which the current flowed in the directions of the true and false discharges respectively, but that they depend in a complicated manner on the inertia of the needle and the varying forces to which it is subjected. The general character of the phenomena will be made sufficiently clear by the following examples, which are quoted from letters of Mr. Jenkin's to Prof. Thomson, of dates April 9 and April 22, 1859.

TABLE I.

Lengths of cable in nautical miles*,—the first being for the Dardanelles, and the other three, of a different gauge, for the Alexandria and Candia telegraph.	Remote end of cable kept insulated.	Remote end of cable kept to earth.
	First throw of needle.	First observed throw of needle.
123	$1\frac{3}{4}$ left	$3\frac{3}{4}$ right
$137\frac{3}{4}$	$15\frac{1}{2}$ „	37 „
$261\frac{1}{2}$	$28\frac{1}{2}$ „	31 „
$399\frac{1}{4}$	$41\frac{1}{2}$ „	21 „

To explain the cause of the deflections to the right recorded in the last column of this Table, the following observations were made, with care that the first motion of the needle in either direction, however slight or rapid, should not escape notice.

* A nautical or geographical mile, or a knot as it is generally called in nautical language, is taken as 6087 feet.

TABLE II.—455 nautical miles of Alexandria and Candia Cable.

Remote end of cable kept	First throw of needle.	Recoil or second throw.	Excess of recoil above first throw.
1. To earth direct	$2\frac{1}{2}^\circ$ right	$24\frac{1}{2}^\circ$ left	22°
2. To earth through 10 German miles, resistance units*	5 „	22 „	17
3. To earth through 50 „	$11\frac{1}{2}$ „	$18\frac{1}{2}$ „	7
4. To earth through 90 „	$16\frac{1}{2}$ „	21 „	$4\frac{1}{2}$
5. Insulated	$44\frac{1}{2}$ „	not observed	
6. To earth direct, and key "pressed very sharp home"	$3\frac{1}{4}$ „	24 „	$20\frac{3}{4}$

If the whole duration of current, with or without reversal, through the galvanometer coil had been infinitely small in comparison with the natural time of oscillation of the needle (which, reckoned in one direction, was about $4\frac{1}{2}$ seconds), the recoils would have been sensibly equal to the first throws in the contrary direction, being only less by the effect of resistance of the air, &c. to the motion of the needle. Hence the numbers in the last column of the preceding Table prove that at some interval of time, not incomparably less than $4\frac{1}{2}$ seconds, after the first motion of the needle, there was a current through the galvanometer coil opposite in direction to that which produced the first or *right* deflection, in each case except No. 5, or that in which the remote end of the cable was insulated. It may be safely assumed that the conductors used in cases 2, 3, and 4 to give the stated resistances between the remote end of the cable and the earth, exercised no sensible electro-magnetic influence, and held no sensible charge, in the actual circumstances; and it is interesting to see how the greater the resistance thus introduced, that is to say the more nearly the remote end is insulated, the greater is the first throw (due, as explained above, to true discharge), and the less is the excess of the recoil above it.

This excess, shown in the last column of the Table, exhibits the effect of the electro-magnetic induction from coil to coil which stops short the true discharge, and produces after it a reverse current constituting the "false discharge." The following experiments, performed by Mr. Jenkin on the 19th of April, 1859, on different lengths of the Red Sea cable, illustrate the relations between true and false discharge.

* The resistance of this unit was found by experiment to be equal to about 190×10^6 British absolute units of feet per second, or to $6\frac{1}{4}$ nautical miles of the Alexandria and Candia cable, or to 4·39 of the Dardanelles, or to 7·44 of the Red Sea.

TABLE III.

Lengths of Red Sea cable.	Remote end of length used kept insulated. Discharge from electrification of 36 cells.			Remote end of length used kept to earth. True and false discharge from electrification and current of 72 cells.		
	First throw.	Recoil.	Excess of first throw above recoil.	First throw.	Recoil.	Excess of recoil above first throw.
312 nautical miles	} 20° left	19° right	1	1¼ left	18° right	16¾
546 ,,	29½ ,,	27 ,,	2½	5¼ ,,	15 ,,	9¼
858 ,,	35 ,,	14 ,,	21	17 ,,	22 ,,	5
Col. 1.	Col. 2. True discharge.	Col. 3. Inertia of needle.	Col. 4. Effect of duration of discharge.	Col. 5. True discharge.	Col. 6. "False discharge" and inertia.	Col. 7. "False discharge," or effect of electro-magnetic induction.

The great increase of the numbers in column 4, for the longer portions of cable, illustrates the fact first demonstrated by Prof. Thomson in 1854[*], that, when undisturbed by electro-magnetic induction, the discharge of a cable takes place at a rate inversely proportional to the square of the length. The duration of the discharge, which, when the remote end is kept insulated, is probably much increased by electro-magnetic induction, must be very considerable in the case of the 858 miles length, to produce so great a diminution as 21° in the recoil, from a throw of 35°, on a needle whose period of vibration was 4½ seconds. The diminution of 1° from the throw of 20°, as observed in the case of the 312 miles length, may be to some considerable proportion of its amount due to resistance of the air, although, as this is probably scarcely sensible on a single swing of the needle, it may be supposed that it is chiefly the effect of the duration of the discharge current. From column 7 it is clear that nearly all trace of the electro-magnetic influence would be lost sight of in comparison with the greater effect of true discharge, in the method of experimenting that was followed, if applied to lengths exceeding 1000 knots, in a coil or coils of similar dimensions to those actually used; while for the 546 knots, and shorter lengths, the effect of electro-magnetic induction is greater than that of the true discharge. It is remarkable that the effect of electro-magnetic induction is absolutely greatest for the shortest of the three lengths. These relations between the different lengths must of course, according to the explanation we have given, depend on the plan of coiling, whether in one coil or in several coils, and on the dimensions of the coil or coils, as

[*] Proceedings of the Royal Society, 1855; and Phil. Mag. vol. xi. p. 146.

well as on the dimensions of the conductor, the gutta percha, and the outer iron sheath of the cable. The magnetic properties of the iron sheath must greatly influence the false discharge; and it would be interesting to compare the discharge from a plain gutta-percha-covered wire coiled under water with that from an iron-sheathed cable.

The following set of experiments, the last which we at present adduce, illustrate the influence of less or greater intervals of time during which the near end of the cable remains insulated, after removal from the battery but before application to earth through the galvanometer coil.

TABLE IV.—455 nautical miles of Alexandria and Candia Cable, remote end kept to earth.   Battery of 72 cells Daniell's.

Experiment.		Throw of needle by true discharge.	Recoil, if any, and throw by false discharge.
No. 1	Key struck down	3 left	27 right
2	Key pressed down as usual .	2¾ ,,	26 ,,
3	Key pressed very gently ...	2¼ ,,	20½ ,,
4	Key held 5 seconds half-way.	0 ,,	14 ,,
5	,, 10 ,, ,,	0 ,,	17 ,,
6	,, 15 . ,, ,,	0 ,,	4 ,,

In order to detect whether there might not have been "a slight hesitation in these three last instances, a much more delicate instrument was taken, but no such hesitation could be detected." These results are very remarkable, especially as regards the duration of the electro-magnetic influence. If the conductor of the cable were circumstanced like that of a common electro-magnet, and had no sensible electrostatic capacity, the "mechanical value* of the current in it" at the instant of the connexion between its near end and the battery being broken, would be spent in a spark, or electric arc of sensible duration between the separated metal surfaces. But in the cable, the electrostatic capa-

* See a paper "On Transient Electric Currents," by Prof. W. Thomson, Phil. Mag. June 1853, where it is shown that, like the mechanical value of the motion of a moving body, which is equal to half the square of its velocity, multiplied by its mass, the mechanical value of a current at any instant, in a coiled conductor, depending on electro-magnetic induction, is equal to half the square of the strength of the current through it, multiplied by a constant which the author defined as the "electrodynamic capacity of the conductor," and which he showed how to calculate according to the form and dimensions of the coil. Additional explanations and illustrations will be found in Nichol's 'Cyclopædia of Physical Science,' second edition, 1859, under the heads "Magnetism—Dynamical Relations of," and "Electricity—Velocity of."

city of the near portions of the conductor has an effect analogous to that of Fizeau's condenser in the Ruhmkorff coils; and there was little or no spark (none was observed, although it was looked for, in the key) on breaking the battery circuit, and consequently, as nearly as may be, the whole mechanical value of the current left by the battery must have been expended in the development of heat in the conductor itself, and by induced currents in the iron of the sheath; and therefore we need not wonder at the great length of time during which electric motion remains in the cable.

The first column of results for experiments Nos. 1, 2, and 3, and the two columns for Nos. 4, 5, and 6, show that the continued flow of the main current through the cable, after the near end is removed from the battery and kept insulated, is to reduce its potential gradually from that of the battery (which for the moment we may call positive), through zero, to negative, in some time less than five seconds, and to keep it negative ever after, if it is kept insulated, as long as any trace of electro-dynamic action remains*. It is probable that, at the same time, there may be oscillations of current backwards and forwards again†, and of potential to negative, and positive again, in some parts, especially towards the middle, of the cable. The mathematical theory of the whole action is very easily reduced to equations; but anything like a complete practical analysis of these equations presents what may be safely called insuperable difficulties, because of the mutual electro-magnetic influence of the different parts of the cable with differently varying current through them. These peculiar difficulties do not, theoretically viewed, present any specially interesting features; and the problem is of little practical importance when once practical electricians are warned to avoid being misled by electro-magnetic induction, in testing by discharge during either the manufacture, the submergence, or lifting of a cable, and not to *under-estimate* the rate of signalling through a long submarine cable to be attained when it is laid, from trials through the same cable in coils, when electro-magnetic induction must embarrass the signalling more or less according to the dimensions and disposition of the coils, and probably does so in some cases to such an ex-

---

* After what has been said in the text above, it is scarcely necessary to point out that this effect is both opposed to, and much greater than, anything producible by polarization of the earth-plates.

† As in the oscillatory discharge of a Leyden phial, investigated mathematically by Prof. Thomson ("Transient Electric Currents," Phil. Mag. June 1853), and actually observed by Feddersen, in his beautiful photographic investigation of the electric spark (Poggendorff's *Ann.* vol. cviii. p. 497, probably year 1860; also second paper, year 1861).

tent as to necessitate a considerably slower rate of working than will be found practicable after the cable is laid.

The theoretical conclusion that the "false discharge" would not be observed in submerged cables, has been recently verified by Mr. Jenkin on various lengths of Bona cable up to 100 miles, which he was engaged in recovering, and which, under careful tests, never gave the slightest indication of "false discharge," although, even when the remote end had completely lost insulation, they gave not only polarization effects *, but also, in the same direction as these, but distinguishable from them, indications of true discharge.   But, in fact, a fortnight before the theoretical conclusion was published by Prof. Thomson at the Aberdeen meeting, a most remarkable and decisive experimental demonstration of it was published by Mr. Webb, Engineer to the Electric and International Telegraph Company, who had independently discovered the phenomena which form the subject of this paper, and given substantially the same explanation as that which we now maintain.   If there could be a doubt as to the electro-magnetic theory, the following extract from a letter of Mr. Webb's, published in 'The Engineer' of August 26, 1859, is decisive:—

"It is, however, on making contact at F with earth [that is to

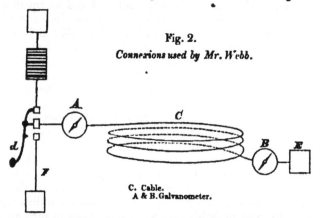

Fig. 2.

Connexions used by Mr. Webb.

C. Cable.
A & B. Galvanometer.

say, putting what we have called the near end of the cable to earth] that the greatest and most singular difference occurs [between straight and coiled cables].   It will then be seen that the needle at A [that is to say, the needle of a galvanometer in circuit between key and cable instead of between key and earth, as in our experiments], instead of being reversed will continue

* Of the same nature as those observed by Prof. Wheatstone on his short cables.

deflected in the original direction, and both needles will very gradually resume the perpendicular."

"There is a most marked difference between the effect produced between a coiled and a straight cable. The return current appears obliterated, or rather it is overpowered by the effects of the inductive action which takes place from coil to coil. The deflection thus produced is much greater than that produced by the return current. I have had perhaps peculiar facilities for observing this striking phenomenon. Whilst picking up a cable at sea, I frequently test the length I am operating on for return current ; *and as the cable becomes coiled into the ship the deflection of the needle, when testing for return current, becomes reversed.*

"It is also my practice to cut the cable at certain distances as it is picked up, and then test such sections separately. On these occasions, sections which, when one end is insulated, will give a charge and discharge of 5°, will when that end is to earth, give a current at the battery end, after contact, of 90°, but in the reverse direction to that in which the discharge or return current would be if the cable were laid out straight."

XXVI. *On the Movements of Gases.* By J. A. WANKLYN, *Demonstrator of Chemistry in the University of Edinburgh*[*].

WHEN a gas heavier than air is placed in a cylindrical vessel closed at the top and open at the bottom, it does not descend rapidly. In like manner, a gas lighter than air contained in a cylinder closed below, but freely communicating with the atmosphere above, does not move upwards with rapidity.

By simply placing a gas in a vertical cylinder shut at one end, the ordinary course of gravitation is disturbed—to how great an extent few people would anticipate.

The following experiments show how remarkably the fall of gases is retarded by such an arrangement.

A tube[†] filled with carbonic acid was allowed to remain with its mouth open and directed downwards for the space of five seconds. After the lapse of that time, the gaseous contents of the tube were analysed in order to ascertain how much carbonic acid had made its escape.

The composition of the gas was—

Air	26·3
Carbonic acid	73·7
	100·0

---

* Communicated by the Author.
† Dimensions of the tube :—Diameter, 14·5 millims.; length, 232 millims.; capacity, 37 cubic centims.

A second experiment with the same tube, and the same length of exposure, gave—

Air . . . . . . .20·3
Carbonic acid . . . 79·7
                    ———
                    100·0

A third experiment, also with the same tube and same exposure, gave—

Air . . . . . . 26·6
Carbonic acid . . . 73·4
                    ———
                    100·0

Three more experiments, where the time of exposure was twenty seconds, other conditions remaining unaltered, gave—

	I.	II.	III.
Air	49·67	60·68	51·34
Carbonic acid	50·33	39·32	48·66
	100·00	100·00	100·00

The differences observed between the quantities of gas which escaped in the same times in the different experiments are no doubt caused by the action of currents of air which are produced by the act of unclosing the tube, which currents are necessarily variable in extent and direction. Though it would be idle to draw any inference as to the precise numerical relation subsisting between times of exposure and quantities of gas fallen, yet these oscillations in our different experiments do not at all affect the certainty of the general result.

In five seconds about one quarter of the carbonic acid escaped; in twenty seconds about one half.

Let us translate this into retardation of the fall of the gas.

Our tube was 232 millimetres long. Therefore, in five seconds three-fourths of the gas cannot have fallen more than 60 millimetres in vertical distance. By comparing this with the distance through which carbonic acid, contained in a balloon, would fall in five seconds, we arrive at an estimate of the retardation which we seek to measure.

At first sight, the cause of this retardation would appear to be friction between the carbonic acid and air which must enter to supply the place of the carbonic acid. But if we carefully consider the conditions under which we find the gases in these experiments, we shall see that another explanation is possible.

For, notwithstanding the absence of cohesion in gases, the carbonic acid in our experiment seems to be very much in the condition of water in the two arms of an equal-armed siphon; the essential difference between the two cases being that, whereas the cohesion between the particles of water hinders all movement whatever, the want of cohesion in the gas permits movement, which accordingly takes place, but with exceeding slowness, and indeed (if certain theoretical conditions could be realized) with infinite slowness.

If we could place our gases in contact without occasioning any current by the act of making them communicate, and besides could realize—

(1) The inferior surface of the gas a mathematically horizontal superficies.

(2) A vessel mathematically cylindrical and vertical;

(3) Molecules of the gas infinitely little and absolutely non-adherent either to one another or to the glass;

then, these conditions being granted, an infinitely prolonged time would be required for any finite fraction of the gas to fall.

At the beginning of the experiment it would be the lowest stratum of molecules alone whose gravitation would tend to cause motion. All the molecules, situated above the *plane* of *contact* between our gas and the air, would be in equilibrium, as the descent of one of them would involve the ascent of another.

During the first instant the lowest plane of carbonic acid would change place with the uppermost plane of air. Thus a plane of air-molecules would be interposed between the mass of carbonic acid above and a plane of carbonic acid-molecules below.

During instant the second, the isolated plane of carbonic acid molecules would change places with the adjacent air immediately below it, while simultaneously the lowermost stratum in the mass of carbonic acid would change with the isolated stratum of air.

We should thus have an isolated stratum of carbonic acid-molecules of infinitely small thickness travelling downwards through the air; and if it could be shown that this isolated lowest stratum would require eternity to traverse a finite vertical distance, it will follow, *à fortiori*, that a finite fraction of the carbonic acid would require eternity to fall a finite distance.

That a body of infinitely small vertical diameter requires an infinite time to fall through a finite portion of a medium may be thus proved.

Assign any finite time, *e. g.* a second. In a second a body falling *in vacuo* acquires a velocity of 32 feet per second. Let our body be conceived to enter a medium being charged with a velocity of 32 feet per second (which is consequently the product of a greater force than the gravitation during a second). In moving through any finite portion of the medium, the body would encounter an infinite number of times its weight of the medium. It would therefore have to communicate its motion to its weight multiplied by infinity.

Hence in traversing a finite space its velocity would become 32 feet divided by infinity. It would therefore require an infinite time to traverse a finite space. *A fortiori*, in a second it could not traverse a finite space.

In like manner, any finite time being assigned, it can be shown that in that time no finite space can be traversed.

We are thus led to expect that carbonic acid should not escape from a tube more rapidly when its mouth is turned downwards than when it is turned upwards. Nor does the fact, that in the actual experiment the escape was more rapid in the former position than in the latter, disprove the proposition; for in the experiment there is a very great imperfection : viz., it is impossible to open a tube without creating a current. When a current is set up, the gas moves *en masse*, and then it is quite conceivable how gravitation can increase the movement; so that the descent of a quarter of the gas in our experiments is no proof that if the tube could be opened without disturbance there would be a higher rate of egress than there is when simple diffusion acts.

In order to show experimentally that it is the upward current of air which produces the retardation, the following experiment was devised and executed.

A tube 9 to 10 millimetres in diameter and 242 millimetres long was provided with ground-glass plates, closing both top and bottom. The top was fixed on with tallow, and the tube used as though it consisted of a single piece. It was filled with mercury and inverted in the mercurial trough, when it proved to be tight. Carbonic acid was then introduced in the usual way. The tube charged with that gas was then closed with the other ground-glass plate and removed from the trough. The top was taken off. The bottom was then removed and replaced in five seconds. The top was then put on, and the tube taken back to the trough, and its gas passed into a graduated tube and examined. It consisted of almost pure air: 91·111 vols. left 90·665 vols. not capable of absorption by potash.

This shows clearly that, however we may account for the mode of action of the upward current of air, it is the upward current which produces the remarkable retardation forming the subject of this paper.

Here it may be well to mention that a trial was made to ascertain the extent of movement produced by simple diffusion. The tube employed in the six first experiments was filled with carbonic acid and exposed, mouth opening upwards, for five seconds.

Only about 3 per cent. o   ·nic acid had left the tube, and probably even that small amount was chiefly due to the disturbance on opening the tube.

The following experiment was made with hydrogen (not dried). The same tube as was used in the six first experiments was filled with that gas and exposed for five seconds, with the open end upwards.   The residual gas after the experiment contained—

$$
\begin{array}{lr}
\text{Air} \cdot \cdot \cdot \cdot \cdot \cdot \cdot \cdot & 38\cdot8 \\
\text{Hydrogen} \cdot \cdot \cdot \cdot \cdot \cdot & 61\cdot2 \\
\hline
& 100\cdot0
\end{array}
$$

It will be obvious, on a little consideration, that the same causes are in operation in this instance as in the former instances where carbonic acid refused to descend.

To show that the same phenomena occur in very wide tubes, the following experiments may be cited.

A tube, 38 millimetres in diameter, 256·5 cubic centimetres in capacity, but of the same length (232 millim.) as the former tube, was filled with carbonic acid.

The residues in different experiments consisted of—

	Exposure of 5 seconds.		Exposure of 10·5 seconds.
	I.	II.	
Air	43·4	52·5	76·1
Carbonic acid	56·6	47·5	23·9
	100·0	100·0	100·0

The loss of carbonic acid is therefore greater than when the narrow tube is employed.   We may explain this by the greater extent of current, which is of necessity produced on unclosing a wider tube.

Lastly, an experiment may be brought forward in which a mixture of gases was used, and in which the point aimed at was, whether or not relative change takes place in the composition of a mixed gas.

The mixed gas was prepared by heating oxalic acid with sulphuric acid*.   An analysis previously to the experiment gave—

$$
\begin{array}{lr}
\text{Carbonic acid} \cdot \cdot \cdot \cdot & 50\cdot62 \\
\text{Carbonic oxide} \cdot \cdot \cdot \cdot & 49\cdot38 \\
\hline
& 100\cdot00
\end{array}
$$

After exposure in the narrow tube opening downwards for 60 seconds, the product contained—

* No doubt a trace of SO_2 was present, but this it was not deemed necessary to remove.

$$\begin{array}{llll}\text{Carbonic acid} & . & . & . & . & 17\cdot91 \\ \text{Carbonic oxide} & . & . & . & . & 18\cdot94 \\ \text{Air} & . & . & . & . & . & . & 63\cdot15 \\ \hline & & & & & 100\cdot00 \end{array}$$

From which we see that little or no change in the relative proportions of GO^2 and GO had taken place,—a circumstance tending to prove that the exit of gas, *en masse*, is due to currents and not molecular.

XXVII. *On some Cerium Compounds.*
By M. Holzmann, *Ph.D.*[*]

ON continuing my former researches on the cerium compounds[†], I found a new class of double nitrates of cerium which do not contain the cerium in the state of protosesquioxide, but simply as protoxide. I prepared the cerium double salts of ammonium, potassium, strontium, magnesium, zinc, manganese, nickel, cobalt, and uranium, partly by dissolving the metal in a solution of nitrate of protosesquioxide of cerium containing a considerable quantity of free acid, partly by mixing the solutions of the two nitrates. The deoxidation of the protosesquioxide of cerium was effected in the first case by the hydrogen generated in dissolving the metal; in the second by boiling the nitrate of protosesquioxide of cerium with alcohol. If the solution of the cerium-salt contains an excess of nitric acid, the alcohol must be added in small quantities, as the disengagement of gas causes a violent ebullition. The analyses of the ammonium, magnesium, zinc, manganese, nickel, and cobalt salts were already finished, when a paper was published by L. Th. Lange[‡], in which the same salts are described,—in consequence of which I have discontinued my research, and will now only state those of my observations which do not agree with Lange's.

The double salt of *nitrate of cerium* and *nitrate of magnesium*, prepared by myself, is not of a pale pink colour, but perfectly colourless, and only contains *six* atoms of water of crystallization. As this composition differs from that of the other salts belonging to the same group, I analysed the products of several preparations, but always obtained the same results. The salt was obtained by mixing equal parts of concentrated solutions of nitrate of magnesium and nitrate of protoxide of cerium, and leaving the mixture to crystallize over caustic lime and chloride of calcium. The crystals, representing perfectly developed hex-

---

* Communicated by the Author.
† *Journal für Praktische Chemie,* vol. lxxv. p. 321.
‡ Ibid. vol. lxxxii. p. 129.

agonal plates, frequently of one or two centimetres in diameter, were recrystallized three or four times, and the formation of large crystals prevented by stirring. For analysis, the salt was dried over caustic lime and chloride of calcium.

(1) 0·6942 grm., treated with recently precipitated oxide of silver, gave, after precipitating the filtered liquid with hydrochlo‑ ric acid, 0·4156 Ag Cl. The liquid filtered off from the chloride of silver gave with ammonia and phosphate of sodium, 0·1690 $(MgO)^2 PO^5$. After treating the oxide of silver on the filter with hydrochloric acid, the liquid gave, on precipitation with oxalate of ammonium and ignition of the oxalate of cerium, 0·1647 $Ce^3 O^4$ *.

(2) 0·52 grm., dissolved in water and precipitated by oxalate of ammonium, gave 0·1235 $Ce^3 O^4$: the filtrate gave with am‑ monia and phosphate of sodium 0·1237 $(MgO)^2 PO^5$.

(3) 0·4527 grm. gave, after Dumas's method (the substance in a platinum tray), 43·09 cubic centims. nitrogen of 0° C., and 760 millims. pressure. The residue, treated with concentrated nitric acid and precipitated by oxalate of ammonium, gave 0·1078 $Ce^3 O^4$; and the filtrate with ammonia and phosphate of sodium 0·1093 $(MgO)^2 PO^5$.

(4) 0·9543 grm., dissolved in water and precipitated with oxalate of ammonium, gave 0·2307 $Ce^3 O^4$, and the filtrate on evaporation and ignition 0·0778 MgO.

(5) 0·5343 grm., treated in the same manner, gave 0·1285 $Ce^3 O^4$ and 0·045 MgO.

These numbers lead to the formula

$$CeO, NO^5 + Mg O, NO^5 + 6 HO.$$

	Theory.		(1)	(2)	(3)	(4)	(5)
					Experiment.		
CeO	54	22·88	22·61	22 63	22·69	23·04	22·92
MgO	20	8·48	8 77	8·58	8·70	8·15	8·42
NO⁴	54	22·88	22·53	......	} 46·13		
NO⁵	54	22·88	......	......			
6 HO	54	22·88					
	236	100·00					

The *double nitrate of cerium and ammonium* has not been prepared by Lange. It is obtained by mixing equal parts of rather con‑ centrated solutions of the two salts, concentrating the liquid on the water‑bath, and allowing it to cool over chloride of calcium and caustic lime. If the liquid cools gradually, it soli‑ difies to a radiated crystalline mass; but if cooled quickly by

---

* It appears from this that only the nitrate of cerium gives up its acid to the oxide of silver; the quantity of chloride of silver found corresponds therefore only to half the quantity of nitric acid contained in the salt.

stirring, the crystalline powder may easily be separated from the mother-liquor. · The salt is perfectly colourless, very soluble in water and alcohol, and exceedingly deliquescent in moist air.

(1) 0·5585 grm., three times recrystallized from water and dried over caustic lime and chloride of calcium, gave, after evaporation with hydrochloric · acid and bichloride of platinum and ignition of the ammonio-chloride of platinum, 0·1485 Pt. After treating the filtrate with sulphuretted hydrogen and precipitating it with oxalate of ammonium, 0·1702 $Ce^3 O^4$ were left on the ignition of the oxalate of cerium.

(2) 0·4325 grm., treated in the same way, gave 0·112 Pt.

(3) 0·3138 grm., precipitated with oxalate of ammonium, gave 0·0977 $Ce^3 O^4$.

(4) 0·5905 grm. gave, after Dumas's method, 74·44 cubic centims. nitrogen of 0° C., and 760 millims. pressure.

These numbers are represented by the formula

$$2(CeO, NO^5) + NH^4 O, NO^5 + 8HO.$$

	Theory.			Experiment.			
				(1)	(2)	(3)	(4)
2CeO . .	108	29·35		29·04	......	29·67	
NH⁴O . .	26	7·06	} N² 15·22	6·98	6·80	....	} 15·84
3NO⁵ . .	162	44·02		....	....	....	
8HO . .	72	19·57					
	368	100·00					

In addition to the double salts of the *nitrate of proto-sesquioxide of cerium,* formerly described, I have prepared the *ammonium-salt,* which corresponds in composition and properties to the potassium-salt.

A mixture of the solutions of the two salts crystallizes, when left over caustic lime and chloride of calcium, in orange-red crystals, which have the appearance, under the microscope, of hexagonal prisms; this double salt is exceedingly deliquescent. The salt, recrystallized several times from water and dried over lime and chloride of calcium, gave on analysis the following results:—

(1) 0·4285 grm., treated as the ammonium double salt of the protoxide of cerium, gave 0·1477 Pt, and 0·1265 $Ce^3 O^4$.

(2) 0·5442 grm., treated in the same manner, gave 0·1832 Pt, and 0·16 $Ce^3 O^4$.

(3) 0·4645 grm., ignited alone, gave 0·1385 $Ce^3 O^4$.

(4) 0·5684 grm. gave, after Dumas's method, 91·52 cubic centims. N of 0° C., and 760 millims. pressure.

(5) 0·5633 grm., dissolved in water and precipitated with oxalate of ammonium, gave 0·1658 $Ce^3 O^4$.

I thought it superfluous to determine the degree of oxidation of the cerium, as the values obtained agree exactly with the formula $\left. \begin{array}{c} Ce\,O \\ 2\,NH^4\,O \end{array} \right\} 3\,NO^5 + Ce^2\,O^3,\ 3\,NO^5 + 3\,HO.$

Theory.			Experiment.				
			(1)	(2)	(3)	(4)	(5)
Ce³O⁴	170	29·67	29·52	29·40	29·32	......	29·43
2NH⁴O	52	9·08 ⎫	9·05	8·91	......	⎫ 20·23	
6NO⁵	324	56·54 ⎬ N³ 19·55	......	......	......	⎭	
3HO	27	4·71 ⎭					
	573	100·00					

On trying to prepare a double salt of the nitrate of the proto-sesquioxide of cerium with nitrate of aniline, the latter was instantaneously oxidized, and at the same time a dirty green precipitate was formed. In this way even a very small quantity of the proto-sesquioxide may be detected; for the liquid, when dilute, directly assumes a red colour.

The nitrate of the protoxide of cerium seems to form double salts with the nitrates of some of the organic bases, with the investigation of which I am now engaged.

A mixture of the solutions of *protochloride of cerium* and *bichloride of platinum*, when highly concentrated, deposits on cooling orange-coloured crystals, easily soluble in water and alcohol, but insoluble in ether. They fuse in the water-bath, and are deliquescent in moist air. An alcoholic solution, when slowly evaporated over chloride of calcium, often furnishes perfectly developed rectangular prisms. For analysis, the salt was twice recrystallized from water and dried over caustic lime and chloride of calcium.

(1) 0·9615 grm., treated with sulphuretted hydrogen, gave, after filtering and igniting the sulphide of platinum, 0·2375 Pt. The filtrate, boiled and precipitated with oxalate of ammonium, gave 0·2671 Ce³O⁴. The remaining liquid, when treated with nitrate of silver, gave 1·3485 Ag Cl.

(2) 0·782 grm., dissolved in alcohol and mixed with chloride of ammonium, gave after ignition of the ammonio-chloride of platinum 0·1904 Pt. The filtrate, precipitated by oxalate of ammonium, gave 0·22 Ce³O⁴.

(3) 0·6215 grm., treated in the same manner, gave 0·151 Pt, and 0·1724 Ce³O⁴.

These numbers lead to the formula

$$(Ce\,Cl)^2\,Pt\,Cl^2 + 8\,HO.$$

Theory.			Experiment.		
			(1)	(2)	(3)
2Ce	92	22·72	22·55	22·84	22·53
Pt	99	24·44	24·70	24·35	24·30
4Cl	142	35·06	34·69		
8HO	72	17·78			
	405	100·00			

The double chlorides of manganese or magnesium with platinum differ from this composition; for their formula, according to Bonsdorff*, is $M\,Cl,\,Pt\,Cl^2+6HO$.

When mixed solutions of *protochloride of cerium* and *iodide of zinc* are left for some time over chloride of calcium, a syrupy mass is generally obtained; very rarely a crystalline double salt is deposited from the solution. I have not succeeded in purifying this compound, as it attracts water with great avidity, and can hardly be recrystallized in consequence of its extreme solubility in water and alcohol. On concentrating a solution of the salt in the water-bath, iodine is liberated.

In conclusion I may mention that *oxalate of cerium, lanthanium,* or *didymium* may be obtained in perfectly developed rhombohedrons, attaining often a diameter of 2 or 3 millims., when dissolved in moderately concentrated nitric acid and allowed to evaporate slowly over caustic lime. An acid salt, however, is not obtained in this way, even when free oxalic acid is dissolved together with the oxalate: for 0·7968 grm. of oxalate of cerium, dried over caustic lime and chloride of calcium, left, on ignition, $0·3605\ Ce^3\,O^4$, corresponding to 36·70 per cent. of cerium; and 1·3862 grm., burnt with oxide of copper, gave $0·4784\ CO^2$ and $0·4175\ HO$, corresponding to 9·41 per cent. of carbon, and 3·35 per cent. of hydrogen. The formula $C^4\,O^8\,Ce^2+8HO$ requires 36·51 of cerium, 9·52 of carbon, and 3·18 per cent. of hydrogen. When the nitric acid is employed in a too concentrated state, and when the solution is heated to ebullition, a partial decomposition takes place, and a mixture of crystals of the oxalate and of free oxalic acid is obtained.

New Lodge, August 1, 1861.

XXVIII. *Some Observations on the Sensibility of the Eye to Colour.* By JOHN Z. LAURENCE, *F.R.C.S., M.B. Lond., Surgeon to the South London Ophthalmic Hospital†.*

IF, closing one eye—say the right—any highly luminous white ground, such as some portions of the sky on a sunny day, is viewed with the left through a dark tube so as to exclude all extraneous light, after a little the eye will begin to feel fatigued, and a vibrating circular smoky spectrum will be perceived at the end of the tube. When the tube is laid aside and

* Gmelin's *Handbuch*, vol. iii. p. 765 and 767.

† From the Glasgow Medical Journal, July 1, 1861. Communicated by the Author.

[Since writing this paper, my attention has been directed to a series of elaborate disquisitions by Brücke and Fechner in Poggendorff's *Annalen der Physik und Chemie*, vols. xliv., l., and lxxxiv., to which I beg to refer my readers.—J. Z. L.]

both eyes are directed to the sky, a similar spectrum will be observed, projected, as it were, on the surface of the heavens, but much darker. But if after a time each eye is alternately opened and closed, a rose-coloured spectrum is seen with the left eye, a pale green one with the right. These appearances are seen still better if, instead of the sky, a white screen is used as the plane of projection in the second part of the experiment. At first an almost black circular disc is seen; this becomes lighter and lighter, till it is finally succeeded in the left eye by a bright rose-colour disc, surrounded by a violet border; in the right eye by an equally bright green with a rose border. These spectra sometimes appear as if upon the surface of the screen, sometimes, on the contrary, as if originating within the eyeball itself, and indeed may be even seen with both eyes closed. To see the above phenomena in all their intensity, a slightly different plan must be adopted. As the field of projection, a sheet of dead black paper in a dark room is to be used; the spectra then seen with either eye are the same, and their colours most splendid, both as regards brightness and tint. At first an emerald-green disc appears, surrounded by a narrow carmine, or perhaps, more accurately, magenta border; the magenta tint is then seen to encroach more and more upon the green, till the whole disc is of the former colour, surrounded by a bluish-violet border; this last, in its turn, invades the magenta, till the final spectrum is of one uniform indigo-violet colour.

The above is the general sequence of colours which I, and other persons whom I have asked to perform the experiment, have observed; but these are liable to exceptions. Occasionally, the librating spectrum observed at the end of the tube in the first part of the experiment, acquires a faint rose, green, or violet tint. Sometimes I have seen the spectra of the right and left eyes, in the second part of the experiment, reversed as regards colour.

These facts appear to prove the following propositions :—

1. That colour sensations may be excited in the retina, or brain, altogether independently of any external colour-stimulus.

2. That as an *optical* analysis of white light may be effected by a prism, so with the eye we possess the power of effecting, what may be called, its *physiological* analysis.

3. The last proposition tends to the conclusion that white light consists of three fundamental colours—magenta, emerald-green, and indigo-violet—corroborating in a remarkable manner the opinions of Professor Maxwell and Dr. Young on the same subject.

4. That a colour sensation excited in one eye is generally felt in the other, although this latter has not been exposed to the

influence of light in any part of the experiment; that, in a word, a very close sympathy exists in the two retinæ, of which the consentaneous action of the two irides is probably but a reflex nervous consequence.

I may here allude to a distinction in ocular spectra which has, I believe, not been taken much account of by observers of these phenomena. Some spectra seem as if projected on the plane to which we direct the eye, and in that case appear, as I have found from numerous measurements, linearly magnified in proportion to the distance of the eye from the plane of projection. Other spectra, on the contrary, are perceived, so to say, in the eyeball itself, and are of a subjective nature. Independent of the differences of their apparent seats, the two classes of spectra present certain other well-defined distinctions. Projected spectra are only perceived with the eyes open, and are generally but faint in colour; while subjective ones may be seen with the eyes shut, and are always intense in colour. At the same time I am disposed to ascribe the differences of colour, in a certain degree, to the diluting influence of extraneous light; for projected spectra are always seen more vivid in a dark room than in daylight.

The green spectrum observed on a sheet of white paper, after prolonged contemplation of a red wafer, has been commonly explained thus :—"When the eye has been for some time fixed on the *red* wafer, the part of the retina occupied by the red image is deadened by its continued action, and insensible to the red rays which form part of the white light from the paper; consequently will see the paper of that colour which arises from all the rays in the white light of the paper, but the red; that is, of a *bluish-green* colour, which is therefore the true complementary colour of the *red* wafer*."

That this explanation is not correct seems to me to be proved by the following experiment :—

I, at night, made a room (which is provided with thick American-leather blinds for ophthalmoscopic purposes), to all appearance, absolutely dark, then viewed with the left eye a small aperture in a dark box covered with a piece of emerald-green glass, behind which was the nearly white flame of a lamp. The right eye was kept closed, and covered with a thick handkerchief. After a time I blew out the light in the box, and looked at a screen covered with a sheet of dead black paper. With the left eye a large carmine-coloured projected spectrum of the flame could be seen; with the right eye I generally perceived no spectrum at all, or if any, but of a very faint tint. But if the latter eye was exposed to a white light during the first part of the ex-

* Brewster's 'Optics,' 1831, p. 305.

periment, I invariably perceived the same spectrum with this eye as I did with the left one.

This experiment shows that the presence of white light is not necessary for the perception of complementary ocular spectra, and further, would appear to indicate that for a sympathetic spectrum to be excited in the eye which has not been exposed to the colour-stimulus, the excitation of some light is necessary.

M. Plateau painted one half of a piece of paper red, the other green; and after alternately directing the eyes to each half, covered them with a handkerchief, and observed a black image, having on each side a complementary-coloured image*. He hence inferred that "the combination of accidental colours produces black." Sir D. Brewster very properly objects to this conclusion, "because the eye has been in succession rendered insensible to the two colours which compose white light itself†." Elsewhere the same author says, "If we take the two complementary colours, namely, the *red* and the *green* tints forming the *ordinary* and *extraordinary* pencils in the polarized ring, which, by overlapping, form *white light*, then it is manifest that the accidental colour of the overlapping part is *black*, and hence the sum of the action of the *red* and *green* acting separately must also be black‡."

Notwithstanding, however, the authority of Sir D. Brewster, the following experiment which I have performed appears to me rather to corroborate Plateau's view. If the two halves of a card painted red and green respectively be illuminated by a green or red light, they appear black. In the same way, but depending on a different cause, the two halves of the card, if viewed through green or red glass, appear black.

Another set of observations, connected in a degree with the preceding, may be here noticed. Chevreul§ distinguishes two chief species of contrast of colours, *simultaneous* and *successive* contrasts. But an examination of these distinctions shows them in my judgment to be more apparent than real, and but the expression of one fundamental fact, viz. that the eye on perceiving any one colour acquires a tendency to see its complementary. Thus, to take an example of Chevreul's simultaneous contrast:—If a slip of red and one of yellow paper be viewed side by side, near the line of contact the red paper inclines to violet, the yellow to green. The *rationale* of this is at once obvious: the red mingling with the complementary of yellow,

* *Annales de Chimie* for 1833.
† Lond. and Edinb. Phil. Mag. for May 1839, p. 335.
‡ *Op. cit.* for December 1839, p. 437.
§ The Principles of Harmony and Contrasts of Colour, by M. E. Chevreul.

i. e. blue, produces the violet tint; whilst the yellow mingling with the complementary of red, *i. e.* green, produces a light green; and this same law holds good in the juxtaposition of any two colours whatever. By the term successive contrast Chevreul designates the familiar phenomena of complementary ocular spectra, of which a most comprehensive history has been given by Darwin in the Philosophical Transactions, vol. lxxvi. p. 33 *et seq.* Du Tour* thought that the two eyes cannot perceive each a separate colour at once. He says that if, *e. g.*, a blue disc be presented to one eye and a yellow one to the other, the result is that the mind perceives alternately the one or the other colour, but not the two at once. But I would submit that these two statements do not include the whole facts of the case. I took two tubes, each 10½ inches long, and applying the end of one to each eye, viewed the sky through them. I found that when the contiguous edges of the tubes at their further ends were some inches apart, two distinct white circles of sky were seen; these circles touched when the edges of the tubes were from $2\frac{1}{4}$ to $2\frac{1}{8}$ inches apart, and, when closer, the two circles appeared as one. If now the further end of one tube was covered with a piece of green glass, the end of the other with a piece of red, as long as the ends of the tubes were kept not closer than $2\frac{1}{4}$ to $2\frac{1}{8}$ inches asunder the two coloured discs were perceived perfectly distinct from one another; no alternation of either colour to the exclusion of the other, as in Du Tour's experiment, ensued, so long as the tubes were inclined to each other at this or any greater degree of divergence.

Another very interesting series of phenomena depending on the intrinsic sensibility of the eye to the impressions of colours, are those of coloured shadows. The first exact observations on these were made by Count Rumford†. He observed that the two shadows of an object placed in front of a white ground, from a white and a coloured light, were of the two colours complementary to the latter. I have investigated this fact a little more closely. The method adopted has been to throw a white and a coloured (red) circle of light from two magic lanterns on a white screen, before which a slender wooden rod was placed. It is easy to satisfy ourselves that the red shadow is produced by the (otherwise colourless) shadow cast from the interception of the white light being simply illuminated by the other red light. The green shadow is the shadow produced by the interception of the red light, illuminated by the white light. These coloured sha-

* *Mémoires de Mathématique et de Physique présentés à l'Académie Royale des Sciences,* vol. iii. p. 514; iv. p. 499. Paris, 1760–63.

† Philosophical Papers by Benjamin, Count of Rumford. London, 1802, vol. i. p. 333.

dows have, by Rumford and many subsequent observers, been ascribed to the effect of contrast. But this appears an inadequate explanation; for if, with one magic lantern, a half-white and a half-red circle of light be thrown on a screen, a shadow thrown across the two fields is simply dark, without any colour at all. If, again, a red and a white disc of light be thrown from two magic lanterns respectively on a screen, so as partially to overlap, where the overlapping takes place two complementary shadows of any object are seen, but in the other two parts of the field only one colourless dark shadow is seen.

The following facts seem to form the basis of the explanation of coloured shadows:—First, the experiment of Rumford[*],— that a piece of grey paper placed next to a piece of coloured paper, both on a black ground with the exclusion of extraneous light, appears tinged with the complementary colour. Secondly. I found by my own experiments that if, in a dark room, the screen is illuminated with a red circle of light from a magic lantern, the greenness of the shadow and the redness of the ground on which it appears are inversely proportional to one another. By approximating the red light to the screen this becomes redder, whilst the shadow of the rod placed before it becomes less green and darker, till it becomes an ordinary black shadow; that, on the other hand, removing the red light till it leaves the white screen but faintly tinged with red, brings out the green shadow very prominently, and on admission of light into the room, a second faint red shadow comes out.

Meusnier observed "that when the sun shone through a hole a quarter of an inch in diameter on a red curtain, the image of the luminous spot was green." Another observer, Mr. Smith of Fochabers[†], states, "If we hold a narrow strip of white paper vertically, about a foot from the eye, and fix both eyes upon an object at some distance beyond it, so as to see it double, then if we allow the light of the sun, or a light from a candle, to act strongly upon the right eye without affecting the left, which may be easily protected from its influence, the left-hand strip of paper will be seen of a bright *green* colour, and the right-hand strip of a *red* colour."

From all these facts, I think the conclusion arrived at by Sir David Brewster appears highly probable, that "as in acoustics, where every fundamental sound is actually accompanied with its harmonic sound, so in the impressions of light, the sensation of one colour is accompanied by a weaker sensation of its accidental

* *Op. cit.* p. 336.
† Brewster's 'Optics,' p. 405. Lond. and Edinb. Phil. Mag. for October 1832, vol. i. p. 249.

or harmonic colour*." To this might perhaps be added, that there is a tendency in the eye to, as it were, decompose white light into two complementary colours; and further, that the predominant decomposition is into red and green.

Applying this theory to the phenomena of coloured (*e. g.* red and green) shadows, the red shadow has already been shown to be simply due to the illumination of a colourless shadow by a red light; whilst on the whole of the rest of the field of the white screen, the red tint cast from the magic lantern is sufficiently powerful to overcome the green tint which the eye would otherwise perceive, excepting at one spot, namely, that which does not receive any red light on account of the interposition of the opake rod. Here the green (harmonic) colour, having no antagonistic red to overcome it, is rendered sensible to the eye.

XXIX. *On the Measurement of Electric Resistance according to an absolute Standard.* By WILHELM WEBER†.

§ 1. *Explanation of the absolute unit of measure for Electric Resistances.*

IF there are measures for time and space, a special *fundamental measure* for *velocity* is not necessary; and in like manner no special *fundamental measure for electric resistance* is needed if there are measures for electromotive force and for intensity of the current; for then *that resistance can be taken as unit of measure*, which a closed *conductor possesses in which the unit of measure of electromotive force produces the unit of measure of intensity.* Upon this depends the reduction of the measurements of electric resistance to an absolute standard.

It might be thought that this reduction would be more simply effected by reverting to the special dimensions, length and section, and adhering to that metal (*copper*) which is best fitted and is most frequently used for such conductors. In that case the absolute unit of measure of resistance would be that resistance which a copper conductor possesses whose length is equal to the measure of length, and whose section is equal to the measure of surface, in which, therefore, besides measure of length and surface, *the specific resistance of copper must be given as unit* for the specific resistance of conducting substances. Thus a special

* Brewster's 'Optics,' p. 309.
† Translated from Poggendorff's *Annalen*, vol. lxxxii. p. 337, by Dr. E. Atkinson. [From the great scientific and practical importance which the determination of electric resistances has of late acquired, it has been thought advisable to give a translation of Weber's original paper published in 1851, containing the method of referring these resistances to an absolute standard.—EDS.]

fundamental measure for specific resistances would be necessary, the introduction of which would be open to question. *First,* because there would be no saving in the number of the fundamental measures if, in order to do without a fundamental measure for the absolute resistance, another fundamental measure must be introduced which is otherwise superfluous. And *secondly,* neither copper nor any other metal is fitted for use in establishing a fundamental measure for resistances. Jacobi says that there are differences in the resistances of even the chemically purest metals, which cannot be explained by a difference in the dimensions; and that, accordingly, if one physicist referred his rheostat and multiplicator to copper wire a metre in length and 1 millimetre thick, other physicists could not be sure that his copper wire and theirs had the same *coefficient of resistance*, that is, whether the *specific* resistance of all these wires was the same. The reduction of measurements of galvanic resistances to an absolute measure can therefore only have an essential importance, and find a practical application, if it takes place in the first mentioned way, in which no other measures are presupposed than those for *electromotive force* and for *intensity.*

The question then arises, as to what are the measurements of *electromotive forces* and *intensities* ? In measuring these magnitudes, no specific *fundamental measures* are requisite, but they can be referred to *absolute measure* if the magnetic measures for *bar magnetism* and *terrestrial magnetism*, as well as *measure of space and time*, are given.

As an absolute unit of measure of electromotive force, may be understood *that electromotive force which the unit of measure of the earth's magnetism exerts upon a closed conductor, if the latter is so turned that the area of its projection on a plane normal to the direction of the earth's magnetism increases or decreases during the unit of time by the unit of surface.* As an absolute unit of intensity, can be understood *the intensity of that current which, when it circulates through a plane of the magnitude of the unit of measure, exercises, according to electro-magnetic laws, the same action at a distance as a bar-magnet which contains the unit of measure of bar magnetism.* The absolute measures of *bar magnetism* and of *terrestrial magnetism* are known from the treatise of Gauss, "Intensitas Vis Magneticæ Terrestris ad mensuram absolutam revocata," Göttingæ, 1833 (Poggendorff's *Annalen*, vol. xxviii. pp. 241 and 591).

From this statement it is clear that the measures of electric resistances can be referred to an absolute standard, provided measures of *space, time,* and *mass* are given as fundamental measures; for the absolute measures of *bar magnetism* and of *terrestrial magnetism* depend simply on these three fundamental measures. A closer consideration shows that even of these three fundamental measures, the *measure of mass* does not come into consideration, as follows from the following summary of the simple relations which are established by the determination of the absolute measures of these various kinds of magnitude.

As fundamental measures, there are to be considered the *measure of length* R, and the *measure of time* S; as *absolute measures*, the superficial measure F, and the units of measure of *bar magnetism* M, of *terrestrial magnetism* T, of *electromotive force* E, of *intensity* I, and of *resistance* W.

Hence, *first*, if wW is the resistance of any closed circuit, eE the electromotive force acting upon this conductor, and iI the intensity of the current produced by this electromotive force, we have the relation between the three numbers

$$w = \frac{e}{i};$$

from which it is clear that if the numbers e and i are determined, the number w is also indirectly obtained without needing a special determination.

Secondly, let eE stand for the electromotive force which acts upon any closed (plane) conductor, fF the area of the plane enclosed by this conductor, tT the earth's magnetism on which the electromotive force depends; and let sS express the space of time in which the plane of that conductor is moved by rotation from a position parallel to the direction of the earth's magnetism to a position at right angles to it, in such a manner that the limited surface produced by its projection on a plane at right angles to this direction of the earth's magnetism increases by the unit of measure during the unit of time proportional to the time. We shall then have between these four numbers e, f, t, s, the following relations,

$$e = \frac{ft}{s};$$

and hence it is clear that if the three numbers f, t, s are determined, the number e is also thereby directly given without necessitating a special measurement.

If, *thirdly*, iI is the intensity of the current in any closed conductor, fF the area of the plane enclosed by this conductor, and mM the magnetism of a bar which, when substituted for that conductor (its magnetic axis at right angles to the plane of the conductor), exercises the same actions at a distance, according to electro-magnetic laws, as that conductor, the following relation obtains between the three numbers i, f, and m,

$$i = \frac{m}{f};$$

from which it follows that if the numbers f and m are determined by measurement, i can be directly obtained without a special measurement.

From these three relations we get, finally,

$$w = \frac{e}{i} = \frac{fft}{sm}:$$

hence if the four numbers f, s, m, t are determined, the number w is also directly obtained. The number f is obtained by measuring the area of the plane embraced by the conductor; s is found by measuring the time; and there only remain the numbers m and t, which are obtained by measuring the bar magnetism by the method described by Gauss in the above paper. The unchangeability of the unit of measure for electric resistance can accordingly be guaranteed so long as the four given measures (space, time, and the units of measure for the earth's magnetism and for bar magnetism) are obtained unchanged. But it by no means follows that the maintenance of these four given measures is a necessary condition for the unchangeability of the unit of measure of electric resistances; the simple maintenance of that unit of measure for *velocities* is sufficient for the purpose.

For if tT is the earth's magnetism, on which the electromotive force depends, which acts upon the closed conductor whose resistance has been measured; if, further, m'M is the magnetism of a bar (whose magnetic axis is parallel to the direction of the earth's magnetism, while the straight line drawn from its centre to the centre of the plane enclosed by the conductor is normal thereto) which, according to magnetic laws, would, from a great distance, exert the same action as tT the earth's magnetism; and, finally, if Rr is the length of the straight line drawn from the middle of this bar to the middle of the plane enclosed by the conductor, we have, according to the *Intensitas,* the simple relation

$$t = \frac{m'}{r^3}.$$

Substituting this value of t in the equation for w, we have

$$w = \frac{ft}{r^3} \cdot \frac{m'}{m} \cdot \frac{1}{s}.$$

If, finally, r'R is the side of a square whose area is equal to the area of the plane enclosed by the conductor, from which is obtained the relation

$$f = r'r',$$

and substituting this value of f in the above equation, we have

$$w = \frac{r'^3}{r^3} \cdot \frac{m'}{m} \cdot \frac{r'}{s}.$$

It is self-evident that a change of the given measures has no influence on the value of the factor $\left(\dfrac{r'^3}{r^3}\cdot\dfrac{m'}{m}\right)$; but a change of the given measures of time and space does influence the value of the factor $\dfrac{f'}{s}$, and accordingly the value of the number w, if both measures are not simultaneously increased or diminished in proportion. The value of the number w is hence quite independent of all alterations of the given measures, so long as there is no change in the *measure of velocity*. But if, by an alteration of the given measures, the standard of velocity is increased or diminished n times, an n times larger or smaller value is obtained for the factor $\dfrac{r'}{s}$, and therefore also for the number w, which is as much as to say that the resistance in this case is expressed according to an n times smaller or larger standard. The unchangeability of the unit of measure for resistance merely depends therefore on the unchangeability of the given measure of velocity. But if the measure of velocity is taken n times larger or smaller, the unit of measure for resistance becomes simultaneously n times larger or smaller.

§ 2. *Method of measuring Electric Resistance according to an absolute standard.*

The measurements of length and of time, which, according to the preceding paragraph, are adequate for the determination of electric resistance, presuppose circumstances on the convenient arrangement of which the practical execution and accuracy of such a determination depend. The following arrangement may serve as a simple summary of the essential circumstances.

Out of the galvanic conductor whose resistance is to be determined, two circular rings, A and B, are formed, which are connected in the manner represented in the figure. The whole conductor, consisting of the two circles A, B, and the junctions form a continuous line, of which it may be assumed, for the sake of simplicity, that it is situate in one plane, and that the straight line connecting the centres of both circles coincides with the direction of the earth's magnetism. Let T be the force of the earth's magnetism as determined according to an absolute standard by magnetometric measurements; let r be the diameter of the circles, which, for simplicity sake, are assumed to be equal. If now the circle A is projected in the direction of the earth's magnetism AB on a plane normal to AB, the area of the projected plane is 0. From the flexibility of the wires connecting the two circles, let it be supposed that the circle A is so twisted as to be at right angles to AB, in which case the area of the plane of the projection is πrr. Let this rotation take place in a short time s, in such a manner that the area of the plane of the projection of the circle increases uniformly in this time from 0 to πrr. From the *magneto-electrical laws*, an *electromotive force* results which the terrestrial magnetism T exerts upon the rotated circular conductor A during the time s, and which, according to the unit of measure explained in the preceding paragraph, is expressed by Ee, in which the number e is determined by the equation

$$e = \frac{\pi rr}{s} . \text{T.}$$

By this electromotive force a current is produced in the time s passing through the whole closed conductor, whose *intensity*, according to the unit explained in the preceding paragraph, is expressed by iI. This current passes also through the circle B, and acts from here on a distant magnetic needle in C, whose axis of rotation lies in the plane of the circle at right angles to the direction of the earth's magnetism. Let C lie in the produced AB (that is, the line joining the centres of the circles A and B). It follows now from *electro-magnetic* laws, that the momentum of rotation exerted on the needle at C by a current passing through the circle B, is equal to the rotation exerted by a bar-magnet placed in the centre of the circle in such a manner that its magnetic axis is at right angles to the plane of the circle, if its magnetism M, expressed according to absolute measure, is

$$\text{M} = \pi rri.$$

If, further, the magnetism of the needle in C expressed in the same measure $= m$, and Bc = R, and ϕ the angle which the magnetic axis of the needle in C makes with the direction of the earth's magnetism AB, the momentum of rotation exerted by the bar magnetism M on the bar magnetism m is expressed, according to known magnetic laws, by

$$\frac{\text{M}m}{\text{R}^3} . \cos \phi = \frac{\pi rr}{\text{R}^3} . im \cos \phi.$$

From which it follows that if K is the inertia of the needle, the *acceleration* of the rotation is

$$\frac{dd\phi}{ds^2} = \frac{\pi rr}{\text{R}^3} . \frac{im}{\text{K}} . \cos \phi ;$$

and therefore that if the needle were previously at rest, and $\phi = 0$, the velocity of rotation at the end of the short time s is

$$\frac{d\phi}{ds} = \frac{\pi rr}{\text{R}^3} . \frac{im}{\text{K}} . s.$$

The greatest deflection α of the needle set in oscillation is known by *direct observation*; and the following expression is obtained for it from the above velocity, from known laws of oscillation, by multiplying by the length of oscillation t and dividing by the number π:

$$\alpha = \frac{rr}{R^3} \cdot \frac{im}{K} \cdot st.$$

For the length of oscillation we have the known equation

$$mT = \frac{\pi\pi K}{tt};$$

from which

$$\frac{mt}{K} = \frac{\pi\pi}{tT};$$

and thus

$$\alpha = \frac{\pi\pi rr}{R^3} \cdot \frac{is}{tT}.$$

Now α is obtained by direct observation; and hence for determining i we have

$$i = \frac{R^3}{\pi\pi rr} \cdot \frac{t}{s} \cdot T\alpha.$$

Remembering that the current passing through the circle B also traverses the circle A, we might also calculate the action of the circular current A upon the needle in C; but, for the sake of simplicity, it may be assumed that the distance AC is so great that this action vanishes in comparison with the action of the circular current B; in that case the *actually observed* deflection of the needle in C gives directly the value of α.

Consequently, by the *electromotive force* eE, expressed in an absolute measure, for which has been found the expression

$$e = \frac{\pi rr}{s} \cdot T,$$

a current is produced, in the whole closed conductor whose space is to be measured, the *intensity* of which is expressed in an absolute measure by iI, in which

$$i = \frac{R^3}{\pi\pi rr} \cdot \frac{t}{s} \cdot T\alpha$$

has been found. But, according to the unit explained in the preceding paragraph, the *desired resistance* of the whole closed conductor is expressed by wW, in which w is determined by the relation of the numbers e and i; for

$$w = \frac{e}{i} = \frac{\pi^3 r^4}{R^3 t\alpha}.$$

Hence the execution of the measurement of an electric resistance depends on the measurement of the magnitudes

$$r, \text{R}, t, \alpha;$$

in other words, the resistance of the whole closed conductor can be expressed in an absolute measure, if by observations, *first,* the number α has been found which the deflection of the needle gives in parts of the diameter; *secondly,* the number $\frac{r}{\text{R}}$, which gives the diameter of both circles in parts of the distance BC; *thirdly,* the velocity $\frac{r}{t}$, with which the diameter of those circles is traversed during one rotation of the needle. Hence it appears that the *measure of velocity* is the only measure which must be given if the resistance of a conductor is to be determined according to an absolute standard.

§ 3. *Observations.*

Of the four magnitudes which, according to the preceding paragraph, are to be found by *observation* for the purpose of determining electric resistances according to an absolute standard, three can readily be measured, namely, the diameter r of the two circles, the distance $BC = R$ of the circle B from the needle at C, and the time of oscillation of the needle t. There only remains the fourth magnitude, that is the deflection of the needle α expressed in parts of the diameter, and this is usually so small that it cannot be observed. This is the reason why, in actually making the observations, a slight deviation must be made from the arrangement described in the previous paragraph. For in order to obtain a value of α large enough for accurate observation, it is *first* necessary that the magnetic needle, upon which the circular current B is to act, instead of being at a great distance $BC = R$, be suspended in the centre of the circular current itself, in which case the action is the greater the smaller is the diameter r in comparison with R. Care must also be taken that the length of the needle is much smaller than the diameter of the circle, in order that the peculiar distribution of the magnetism in the needle need not be taken into account, because the investigation of this distribution is attended with difficulties. It is further necessary that both circles, instead of one, shall consist of several windings of the conductor, by which they become changed into rings of large diameter. In that case, however, the influence of all the windings must be individually taken into account, because they have different diameters, and are not all on the same plane as the needle.

For the conductor whose resistance was to be measured, a very long thick copper wire was chosen which weighed 169 kilogrammes. Of this 16 kilogrammes were used for the ring A, which consisted of 145 windings; enclosing altogether a surface of nearly 105 square metres. This ring was placed vertically, and by means of a winch could be rapidly rotated in a semi-circle, so that the perpendicular upon the plane of the ring at the commencement and at the end of the rotation coincided with the magnetic meridian. The other 153 kilogrammes were used for the ring B, which consisted of 1854 windings, giving together a section 202 millims.in breadth, and 70·9 millims. in height: the internal diameter of this ring was 303·51, and the external 374·41 millims. This second ring was firmly fixed, and its plane coincided with that of the magnetic meridian. In the centre of this second ring B, a small magnetic needle 60 millims. long, provided with a mirror, was suspended by a filament of silk, as in a small magnetometer; and the oscillations and deflections of the needle were observed with a telescope, directed to the mirror, on a scale about 4 metres from the mirror.

The observations were made in the following manner. The ring A was first so placed that its plane coincided with the magnetic meridian, and the needle in the middle of the ring was thereby brought to rest; thereupon the ring A was suddenly turned 90°. By this means the needle in the middle of the ring was set in rotation, and by means of the telescope the position of the needle was observed on the scale at its greatest (positive) deflection after half an oscillation. After a complete oscillation, and therefore an oscillation and a half after the beginning,·the needle attained its greatest deflection on the opposite side, which was also observed on the scale. In the moment at which the needle passed its original position of rest, and therefore two oscillations after the beginning of the experiments, the ring A was rotated 180°. The oscillating needle was thereby arrested in the middle of its motion, and thrown backwards, upon which its greatest negative and greatest positive deflections were observed on the scale. After the expiration of four oscillations from the commencement, that is, at the moment at which the needle returning from its last deflection passed its original position of rest, the ring was again turned forwards by 180°, and then the same oscillation observed as in the first case, and in this manner the experiments were continued until a sufficient series of observations was obtained. For each series, in the *first* column of the following Table are given the deflections observed on the scale and arranged in order under one another; in the *second* column the mean between two successive positive or negative deflections are added. In the *third* column are the differences of the means referring to positive and negative deflection, that is, the magnitude of the whole arc.

First Series			Second Series			Third Series			Fourth Series		
467·1			467·1			463·0			462		
540·7	543·70	80·10	540·5	543·65	79·65	536·7	539·65		534·7	538·20	80·00
516·7			546·8			542·6			541·7		
461·4	463·6	79·75	461·3	464·00	79·55	456·6	459·25		455·3	458·20	79·75
465·8			466·7			401·9			401·1		
540·6	543·85	79·25	540·8	543·55	79·80	537·6	539·60		535·1	537·95	79·50
546·1			546·3			541·6			540·8		
462·3	464·10	79·45	461·8	463·65	80·00	458·3	460·05		456·0	458·45	79·50
465·9			465·5			461·8			400·9		
541·4	543·55	79·75	542·1	543·65	79·70	537·7	539·75		535·3	537·95	80·05
545·7			545·2			541·8			540·6		
462·3	463·80	79·70	462·8	463·95	79·85	457·9	459·80		456·0	457·90	79·85
465·3			465·1			461·7			459·8		
542·0	543·50	79·45	542·3	543·80	80·10	537·6	539·65		536·1	537·75	79·55
545·0			545·3			541·7			539·4		
462·4	464·05	79·45	462·7	463·70	79·80	458·2	459·95		456·8	458·20	79·65
465·3			464·7			461·7			459·6		
542·0	543·50	79·65	542·3	543·5	79·75	537·6	540·05		536·0	537·85	79·70
543·0			544·7			542·5			539·7		
462·9	463·85	79·85	462·8	463·75	79·60	457·3	460·00		456·5	452·15	79·60
464·8			464·7			462·7			459·8		
543·7	543·70	79·45	541·9	543·35	79·75	536·6	539·5		535·8	537·75	79·55
544·8			544·8			542·4			539·7		
463·4	464·25	79·70	462·3	463·6	79·85	457·2	459·75		456·4	458·20	79·55
465·1			464·9			462·3			460·0		
542·6	543·95	79·75	541·3	543·45		462·3			535·7	537·75	
545·3			545·6						539·8		
462·8	464·20										
465·6											
Mean 79·64			**Mean 79·79**			**Mean 79·90**			**Mean 79·69**		

The mean value of these four series is 79·755 parts of the scale =79·4 millims., which must be increased by $\frac{1}{2}$ a millim. if we are to take into account the influence of the fact that the rotation of the ring A cannot be effected in a time so small that it can be neglected in comparison with the time of oscillation of the needle. From this we obtain for α the value

$$\alpha = \frac{79\cdot9}{8175},$$

inasmuch as double the horizontal distance of the mirror from the scale is exactly 8175 millims.

The time of oscillation of the needle was found from 300 oscillations to be

$$t = 10''\cdot2818,$$

in which the part of the directive force arising from the elasticity of the thread was the 1770th part of the magnetic directive force, and hence

$$\frac{1}{1+\theta} = \frac{1770}{1771}.$$

Finally, on account of the great distance of the two rings in a room not free from iron, the time of oscillation of the same needle was compared for the position of both rings, and their ratio found to be as 2·9126 : 2·9095 ; from which it follows that if T' is the terrestrial magnetism for A, T'' for B, we have

$$T' : T'' = 470 : 471.$$

These observations are sufficient for determining the resistance of the whole closed conductor; and by accurate calculation we get the value

$$w = 2166\cdot10.$$

§ 4. *Application of the principle of Deadening.*

Instead of using *terrestrial magnetism* to obtain an electromotive force which can be referred to an absolute measure, *bar magnetism* may be employed; in that case it is obvious that the most convenient position for the bar-magnet whose magnetism is to be used, will be in the centre of the ring formed by the closed conductor. The magnet may then either be fixed, and the ring moveable about its diameter at right angles to the magnetic axis of the bar; or inversely, the ring may be fixed and the magnet moved backwards and forwards about that diameter. In the latter case a strong *oscillating magnetic needle* may be used, suspended in the centre of the ring.

The current produced in the closed conductor by the electromotive force arising from the bar magnetism of a magnetic needle oscillating in the centre of the ring, itself reacts according to the principle of *deadening* on the oscillating needle, and produces a diminution in the amplitude of its oscillations which can be observed with great accuracy; and the *intensity* of this current may also, from these observations, be determined according to an absolute standard with great accuracy. It is then evident that the current does not need to be passed through a second ring serving as galvanometer, in order to measure the intensity of the current. Hence the whole conductor, whose resistance is to be measured, can be used to form a single ring which serves at once for indicator and multiplicator.

According to this simplification, the *observation of the arcs of oscillation of a magnetic needle oscillating in the centre of the ring* is sufficient: by their *magnitude* the strength of the electromotive force, and by their *decrease* the intensity of the current produced in the closed conductor by that electromotive force, can be determined.

In executing the observations according to this principle of *deadening*, it is of prime importance that the magnetism of the needle oscillating in the centre of the ring be very powerful; and also that the length of the needle be very small as compared with the diameter of the ring, in order that, in calculating the resistance, there shall be no necessity for an accurate knowledge of the distribution of the magnetism in the needle, the determination of which would be difficult. In the ring now solely used, which is that previously called B, and which has 303·51 millims. internal, and 374·41 millims. external diameter, and is 202 millims. in height, a magnetic needle 90 millims. long, and as strong as possible, was suspended. The experiment was commenced by *detaching* from each other the ends of the wire forming the ring. The needle was then·set in oscillation, and its time of oscillation and the decrease of its amplitude, or the logarithmic decrement of this decrease, was determined according to the method given by Gauss in the 'Results of the Observations of the Magnetic Verein in the year 1837*.' Thereupon the annular conductor was *closed*, and the same observations repeated. The results of these observations are given in the following Table, in which the logarithmic decrement of the diminution of the arc of oscillation with a *closed* conductor, stands in the first column under A, the same with an open conductor stands under B, while in the third column under *t* is given the observed time of oscillation. The mean values are indicated underneath :—

* See Taylor's Scientific Memoirs, Part VI. Vol. II.

A.	B.	t.
0·028645	0·000460	9·1128
0·027955	0·000360	9·1148
0·028565	0·000380	9·1107
0·028388	0·000400	9·1128

From this we obtain, according to Brigg's system, for that part of the logarithmic decrement arising from the deadening,

$$0·028388 - 0·000400 = 0·027988,$$

or according to the natural system,

$$\lambda = 0·064445.$$

The bar magnetism of the oscillating needle M, determined from magnetometric measurements, was found, according to absolute standard as compared with the horizontal part of the earth's magnetism T,

$$\frac{M}{T} = 20733000.$$

That part of the directive force of the needle arising from the elasticity of the thread was found to be 68 times less than that arising from the magnetism, or

$$\frac{1}{1+\theta.} = \frac{68}{69}.$$

For the calculation of the resistance from these observations, executed on the principle of deadening, we have the following rules.

According to the law of magnetic induction, the *electromotive force* of a small magnet oscillating in the centre of a circular conductor, whose magnetic axis makes the angle ϕ with the plane of the circle, is directly proportional to its magnetism M, to the cosine of the angle ϕ, and to the velocity of rotation $\frac{d\phi}{dt}$, and inversely proportional to the diameter of the circle r; and if M is expressed according to an absolute measure, is determined by

$$e = \frac{2\pi M}{r} \cdot \cos\phi \frac{d\phi}{dt}.$$

On the contrary, according to electro-magnetic laws the *momentum of rotation* which the induced current in the circular conductor exerts upon the small magnet oscillating in the centre is directly proportional to the magnetism M, to the cosine of the angle ϕ, and to the intensity, and is inversely proportional to the diameter r; and if i is expressed in absolute measure, is determined by

$$D\frac{d\phi}{dt} = \frac{2\pi M}{r} \cdot i \cos\phi.$$

For small oscillations in which ϕ differs little from O, we have

$$e = \frac{2\pi M}{r} \cdot \frac{d\phi}{dt},$$

$$D\frac{d\phi}{dt} = \frac{2\pi M}{r} \cdot i.$$

If K is the inertia of the oscillating magnet, upon which the directive force MT, arising from the horizontal part of the terrestrial magnetism, acts, the equation of its motion becomes

$$0 = \frac{dd\phi}{di^2} + \frac{MT}{k}\phi + \frac{D}{k}\frac{d\phi}{dt},$$

and hence by integration,

$$\phi = p + Ae^{-\frac{Dt}{2t}}\sin(t-B)\sqrt{\left(\frac{MT}{K} - \frac{1}{4}\frac{DD}{KK}\right)}.$$

$\frac{D}{2K}$ is the logarithmic decrement on the natural system of the diminution of the amplitude of oscillation reduced to the unit of time: hence if τ is the time of oscillation under the influence of deadening,

$$\lambda = \frac{D\tau}{2K} = \frac{\pi M}{rK} \cdot \frac{dt}{d\phi} \cdot \tau i;$$

and the *intensity of the current* is

$$i = \frac{rK\lambda}{\pi M\tau} \cdot \frac{d\phi}{dt}.$$

From this we obtain for calculating the resistance,

$$w = \frac{e}{i} = \frac{2\pi\pi MM}{rrK\lambda} \cdot \tau.$$

From the above equation for ϕ we get for the determination of the time of oscillation under the influence of the deadening,

$$\tau\sqrt{\left(\frac{MT}{K} - \frac{1}{4}\frac{DD}{KK}\right)} = \pi = \tau\sqrt{\left(\frac{MT}{K} - \frac{\lambda\lambda}{\tau\tau}\right)},$$

from which

$$MT = \frac{\pi\pi + \lambda\lambda}{\tau T};$$

hence

$$w' = \frac{2\pi\pi}{rr} \cdot \frac{\pi\pi + \lambda\lambda}{\lambda\tau} \cdot \frac{M}{T}.$$

From this, taking into account the correction arising from the deadener as being made up of several windings, and the correction for the elasticity of the thread, we find from the above observations

$$w' = 1898 \cdot 10^8.$$

[To be continued.]

XXX. *Notices respecting New Books.*

An Elementary Treatise on Trilinear Coordinates, the Method of Reciprocal Polars, and the Theory of Projections. By the Rev. N. M. Ferrers, *M.A., Fellow and Mathematical Lecturer of Gonville and Caius College, Cambridge.* Cambridge: Macmillan and Co., 1861.

IN the researches of the ancient geometers a problem presented itself to them in an almost tangible shape; the eye was a most important auxiliary to the brain; and, without questioning the truth of the old French definition, "La géométrie est une science par laquelle on raisonne droit sur des figures faites de travers," there is no doubt that a well-drawn figure would often suggest a property or method of investigation which might otherwise have escaped; and at any rate the ancients never contemplated reasoning on symbols which bore no resemblance whatever to the figure. The moderns, however, without any loss of distinctness of conception, have, by the introduction of symbols, gained important advantages. Among others, they have freed themselves from the necessity of verifying their results in every variation of case arising from a mere change of position in the data of a problem, and they have acquired an almost unlimited power of generalization.

The coordinate geometry was one of the first grand steps in this direction; but many important additions have been made since Descartes; and of late years new methods of investigation have been pursued which bid fair to carry science onwards with a speed and safety hitherto undreamt of.

The book before us makes known in a simple and intelligent manner the characteristic features of these new methods; it seems especially prepared with reference to the wants of students in the University of Cambridge, and will prove a valuable complement to the works now in use there as text-books. In his preface, the author says that his object in writing on the subject of trilinear coordinates has mainly been to place it on a basis altogether independent of the Cartesian system; but as several results of that system are assumed, as, for instance, in the definition of a conic, p. 33, and in the means of determining the centre of a conic, p. 35, it is obviously not intended to be throughout a perfectly independent work which may be studied without any previous knowledge of any other;

for a student who should attempt this would find his progress stayed at the beginning of the second chapter. This, however, does in no way detract from 'the merit of the book, which, we repeat, must be considered as a complement and a valuable step in advance.

The terms Trilinear coordinates, Anharmonic ratio, Involution, Reciprocal Polars, &c., have been for some years familiar terms in the studies of the University of Cambridge; and those who have read Salmon's 'Conic Sections' or Todhunter's 'Coordinate Geometry,' know the immense power they confer as a means of investigation; but we meet here with a term which to many students will be a new one, although the subject owes its existence and vitality chiefly to the labours of our own countrymen, Sylvester, Salmon, Boole, Spottiswoode, Cayley, and others. What is a determinant? Answer: Write down n rows of symbols with n symbols in each row, and enclose the whole between two vertical lines. That is a determinant. It is a conventional form of expressing in a concise manner a complicated function of these n symbols; and these same functions are so frequently recurring, not only in investigations concerning curves, but in almost every branch of mathematical inquiry, that an abridged notation for them was absolutely needed. A determinant of 5 rows, and therefore containing 25 symbols, would, if written at full length, contain 120 terms with 5 symbols in each term, *i. e.* 600 symbols instead of 25. Chapter III. of the book is devoted to a clear exposition of the simplest laws of combination of these functions, and will serve as a most useful introduction to the study of many modern scientific memoirs. We are only sorry that Mr. Ferrers does not dwell at greater length on them, and give us exact proofs of some of the remarkable results to be found in Spottiswoode, Salmon, Brioschi, Crelle's Journal, &c.

We cordially recommend this little work to those of our readers who have mastered the ordinary coordinate geometry.

XXXI. *Proceedings of Learned Societies.*

ROYAL SOCIETY.

[Continued from p. 164.]

November 22, 1860.—Major-General Sabine, R.A., Treasurer and Vice-President, in the Chair.

THE following communications were read:—

"Researches on the Phosphorus-Bases."—No. VIII. Oxide of Triethylphosphine. By A. W. Hofmann, LL.D. Received July 24, 1860.

In our former experiments *, Cahours and myself had often observed this substance, but we did not succeed in obtaining it in a state of purity fit for analysis. Nevertheless, founding our conclusion on the composition of the corresponding sulphur-compound, and having regard to the analogies presented by the corresponding terms of the arsenic- and antimony-series, we designated this body as

* Phil. Trans. 1857, p. 575.

the oxide of the phosphorus-base

$$C_6 H_{15} PO = (C_2 H_5)_3 PO *.$$

I have since confirmed this formula by actual analysis.

The difficulties which in our former experiments opposed the preparation of this compound in the pure state, arose entirely from the comparatively small quantity of material with which we had to work. Nothing is easier than to obtain the oxide in a state of purity, provided the available quantity of material is sufficient for distillation. In the course of a number of preparations of triethylphosphine for new experiments, a considerable quantity of the oxide had accumulated in the residues left after distilling the zinc-chloride-compound with potash. On subjecting these residues to distillation in a copper retort, a considerable quantity of the oxide passed over with the aqueous vapours, and a further quantity was obtained, as a tolerably anhydrous but strongly coloured liquid, by dry distillation of the solid cake of salts which remained after all the water had passed over. The watery distillate was evaporated on the water-bath as far as practicable, with or without addition of hydrochloric acid; and the concentrated solution was mixed with solid hydrate of potassium, which immediately separated the oxide in the form of an oily layer floating on the surface of the potash. The united products were then left in contact with solid potash for twenty-four hours and again distilled. The first portion of the distillate still contained traces of water and a thin superficial layer of triethylphosphine. As soon as the distillate solidified, the receiver was charged, and the remaining portion—about nine-tenths —collected separately as the pure product. To prevent absorption of water, the quantity required for analysis was taken during the distillation.

With reference to the properties of oxide of triethylphosphine, I may add the following statements to the description formerly given†. This substance crystallizes in beautiful needles, which, if an appreciable quantity of the fused compound be allowed to cool slowly, frequently acquire the length of several inches. I have been unable to obtain well-formed crystals; as yet I have not found a solvent from which this substance could be crystallized. It is soluble in all proportions, both in water and alcohol, and separates from these solvents on evaporation in the liquid condition, solidifying only after every trace of water or alcohol is expelled. Addition of ether to the alcoholic solution precipitates this body likewise as a liquid. The fusing-point of oxide of triethylphosphine is 44°; the point of solidification at the same temperature. It boils at 240° (corr.).

As no determination of the vapour-density of any member of the group of compounds to which oxide of triethylphosphine belongs has yet been made, it appeared to me of some interest to perform this experiment with the oxide in question. As the quantity of material at my disposal was scarcely sufficient for the determination by Dumas's method, and Gay-Lussac's was inapplicable on account of the high boiling-point of the compound, I adopted a modifica-

* H=1; O=16; C=12, &c.   † Phil. Trans. 1857, p. 575.

tion of the latter, consisting essentially in generating the vapour in the closed arm of a U-shaped tube immersed in a copper vessel containing heated paraffin, and calculating its volume from the weight of the mercury driven out of the other arm. Since I intend to publish a full description of this method, which promises to be very useful in certain cases, I shall here content myself with stating the results obtained in one of the experiments.

Substance .... 0·150 grm.
Volume of vapour .... 49·1 cub. cent.
Temperature (corrected) .... 266·6
Barometer at 0° .... 0·7670 metre.
Additional mercury column at 0° 0·1056 ,,

These numbers prove the vapour-density of oxide of triethylphosphine to be 66·30, referred to hydrogen as unity, or 4·60 referred to atmospheric air. Assuming that the molecule of oxide of triethylphosphine corresponds to 2 volumes of vapour[*], the spec. grav. of its vapour $= \frac{134}{2} = 67$, when referred to hydrogen, and 4·63 when referred to air. Hence we may conclude that in oxide of triethylphosphine the elements are condensed in the same manner as in the majority of thoroughly investigated organic compounds.

From the facility with which triethylphosphine is converted into the oxide by exposure to the air, even at ordinary temperatures, and the very high boiling-point of the resulting compound, in consequence of which the vapour of the latter can exert but a very slight tension at the common temperature, I am induced to think that the phosphorus-base may be used in many cases for the volumetric estimation of oxygen. When a paper ball soaked in triethylphosphine is passed up in a portion of air confined over mercury, the mercury immediately begins to rise, and continues to do so for about two hours, after which the volume becomes constant, the diminution corresponding very nearly to the proportion of oxygen in the air. To obtain very exact results, however, it would be probably necessary in every case to remove the residual vapour of triethylphosphine by means of a ball saturated with sulphuric acid.

Oxide of triethylphosphine exhibits in general but a small tendency to unite with other bodies. Nevertheless it forms crystalline compounds with iodide and bromide of zinc. I have examined more particularly the iodine-compound.

Oxide of Triethylphosphine and Iodide of Zinc.—On mixing the solutions of the two bodies, the compound separates, either as a crystalline precipitate or in oily drops which soon solidify with crystalline structure. It is easily purified by recrystallization from alcohol, when it is deposited in often well-formed monoclinic crystals containing

$$C_6 H_{15} PO, ZnI = (C_2 H_5)_3 PO, ZnI.$$

It is remarkable that this compound formed in presence of a large excess of hydriodic and even of hydrochloric acid.

Oxide of Triethylphosphine and Dichloride of Platinum.—No precipitate is formed on mixing the aqueous solutions of the two com-

[*] $H_2O = 2$ vols. vapour.

pounds, however concentrated. But on adding the anhydrous oxide to a concentrated solution of dichloride of platinum in absolute alcohol, a crystalline platinum-compound is deposited after a few moments. This compound is exceedingly soluble in water, easily soluble in alcohol, insoluble in ether. On adding ether to the alcoholic solution, the salt is precipitated, although with difficulty, in the crystalline state. The alcoholic solution, when evaporating spontaneously, yields beautiful hexagonal plates of the monoclinic system, frequently of very considerable dimensions. The crystals have the rather complex formula

$$C_{24} H_{80} P_4 O_3 Pt_2 Cl_6 = 3[(C_2 H_5)_3 PO] + (C_2 H_5)_3 PCl_2, 2Pt Cl_2.$$

On mixing the concentrated solution of the oxide with trichloride of gold, a deep yellow oil is separated, which crystallizes with difficulty after considerable standing. This compound is exceedingly soluble in water and in alcohol. When the aqueous solution is heated, the gold is reduced ; the transformation which the oxide of triethylphosphine undergoes in this reaction is not examined.

Chloride of tin forms likewise an oily compound with the oxide : I have not succeeded in crystallizing this compound.

Chloride of mercury is without any action on oxide of triethyl-phosphine.

Oxychloride of Triethylphosphine.—On passing a current of dry hydrochloric acid through a layer of oxide of triethylphosphine which is fused in a U-shaped tube surrounded by boiling water, brilliant crystals are soon deposited. These crystals disappear, how-ever, rapidly, the compound formed in the commencement of the reaction uniting with an excess of hydrochloric acid. The viscous liquid which ultimately remains behind, when heated loses the excess of hydrochloric acid, leaving an exceedingly deliquescent crystalline mass, very soluble in alcohol, insoluble in ether.

For analysis, the new compound was washed with absolute ether and dried over sulphuric acid *in vacuo*, either at the common tem-perature or at 40°. Three chlorine-determinations in specimens of different preparations, which, owing to the extraordinary avidity of this compound for moisture, exhibit greater discrepancies than are generally observed in experiments of this description, lead to the formula $\quad C_{12} H_{30} P_2 O Cl_2 = (C_2 H_5)_3 PO, (C_2 H_5)_3 PCl_2.$

The dichloride of triethylphosphine cannot be formed by the action of hydrochloric acid upon the oxide.

The oxychloride exhibits with other compounds the deportment of the oxide. It furnishes with dichloride of platinum the same platinum-salt which is obtained with the oxide. In a similar man-ner it gives with iodide of zinc the iodide of zinc-compound of the oxide previously described. Only once—under conditions not sharply enough observed at the time, and which I was afterwards unable to reproduce in repeated experiments—a compound of the oxy-chloride with iodide of zinc was formed. This substance, readily soluble in water and alcohol, crystallized from the latter solvent in beautiful colourless, transparent octahedra of the composition

$$C_{12} H_{30} P_2 O Cl_2 Zn_2 I_2 = (C_2 H_5)_3 PO, (C_2 H_5)_3 PCl_2, 2ZnI.$$

"Researches on the Phosphorus-Bases."—No. IX. Phosphar-sonium Compounds.   By A. W. Hofmann, LL.D.   Received July 24, 1860.

The facility with which the bromide of bromethyl-triethylphosphonium furnishes, when submitted to the action of ammonia and monamines, the extensive and well-defined group of phosphammonium-compounds, induced me to try whether similar diatomic bases containing phosphorus and arsenic might be formed by the mutual reaction between the bromethylated bromide and *monarsines.* There was no necessity for entering into a detailed examination of this class of compounds. I have, in fact, been satisfied to establish by a few characteristic numbers the existence of the phospharsonium-group.

Action of Triethylarsine on Bromide of Bromethyl-triethyl-phosphonium.

On digesting the two·substances in sealed tubes at $100°$, the usual phenomena are observed; the reaction being complete after the lapse of twenty-four hours. The saline mass which is formed yields with oxide of silver in the *cold*, a powerfully alkaline solution, containing the hydrated oxide of ethylene-hexethylphospharsonium,

$$C_{10} H_{36} P As O_2 = \frac{[(C_2 H_4)''(C_2 H_5)_6 P As]''}{H_2} O_2.$$

It is thus obvious that the arsenic-base imitates triethylphosphine in its deportment with the brominated bromide. The two substances simply combine to form the dibromide of the phospharsonium,

$$[(C_2 H_4 Br)(C_2 H_5)_3 P] Br + (C_2 H_5)_3 As = \left[(C_2 H_4)'' \frac{(C_2 H_5)_3 P}{(C_2 H_5)_3 As} \right]'' Br_2.$$

The alkaline solution of the oxide of the phospharsonium exhibits the leading characters of this class of bases; I may therefore refer to the account which I have given of the oxide of diphosphonium. The saline compounds likewise resemble those of the diphosphonium. The dichloride and the di-iodide were obtained in beautiful crystalline needles, exhibiting a marked tendency to form splendidly.crystallized double-compounds. I have prepared the compounds of the dichloride with chloride of tin, bromide of zinc, trichloride of gold, and lastly with dichloride of platinum. The latter compound was analysed in order to fix the composition of the series.

Platinum-salt.—The product of the reaction of triethylarsine upon the bromethylated bromide was treated with oxide of silver in the *cold*, and the alkaline solution thus obtained, saturated with hydrochloric acid and precipitated with dichloride of platinum. An exceedingly pale-yellow, apparently amorphous precipitate of diphosphonic appearance was thrown down, almost insoluble in water, but dissolving in boiling concentrated hydrochloric acid. The hydrochloric solution deposited, on cooling, beautiful orange-red crystals, resembling those of the diphosphonium-platinum-salt. The crystals, according to the measurement of Quintino Sella, belong to the trimetric system. The analysis of the platinum-salt led to the formula

$$C_{14} H_{34} P As Pt_2 Cl_6 = \left[(C_2 H_4)'' {}^{(C_2 H_5)_3 P}_{(C_2 H_5)_3 As} \right]'' Cl_2, 2 Pt Cl_2.$$

The phospharsonium-compounds, and more especially the hydrated oxide of the series, are far less stable than the corresponding terms of the diphosphonium- and even of the phosphammonium-series. If the product of the action of triethylarsine upon the brominated bromide be *boiled* with oxide of silver instead of being treated in the cold, not a trace of the phospharsonium-compound is obtained. The caustic solution which is formed, when saturated with hydrochloric acid and precipitated with dichloride of platinum, furnishes only the rather soluble octahedral crystals of the oxethylated triethylphosphonium-platinum-salt. The nature of this transformation- is clearly exhibited when a solution of the dioxide of phospharsonium is submitted to ebullition. Immediately the clear solution is rendered turbid from separated triethylarsine, which becomes perceptible, moreover, by its powerful odour, the liquid then containing the oxide of the oxethylated triethylphosphonium,

$$\left. {}^{[(C_2 H_4)''_2 (C_2 H_5)_6 PAs]''}_{H_2} \right\} O_2 = (C_2 H_5)_3 As + \left. {}^{[(C_2 H_5 O)(C_2 H_5)_3 P]}_{H} \right\} O.$$

GEOLOGICAL SOCIETY.
[Continued from p. 166.]
May 22, 1861.—Leonard Horner, Esq., President, in the Chair.
The following communications were read :—

1. "On the Geology of a part of Western Australia." By F. T. Gregory, Esq.

The author first described the granitic and gneissose tract of the elevated table-land ranging northwards from Cape Entrecasteaux and comprising the Darling Downs. The igneous rocks and quartz-dykes were next referred to, and also the clays, sandstones, and conglomerates capping the table-land. Carboniferous, cretaceous, and pleistocene rocks were also alluded to; and some evidences of the recent elevation of the coast were brought forward. Besides specimens of rocks and minerals, the following fossils from Western Australia were exhibited: Carboniferous fossils and cannel-coal from the Irvin River; Fossils of secondary age (*Trigoniæ, Ammonites,* and fossil wood) from the Moresby Range; fossil wood from the Stirling Range and from the Upper Murchison River; Ventriculites in flint from Gingin, and Brown-coal from the Fitzgerald River. The author's views of the geology of the district were shown by an original map and accompanying sections.

2. "On the Zones of the Lower Lias and the *Avicula contorta* Zone." By Charles Moore, Esq., F.G.S.

Referring to a paper on this subject, by Dr. Wright, which appeared in the sixteenth volume of the Society's Journal, the author stated that details of the section at Beer-Crowcombe (near Ilminster) in Somersetshire are now more fully known than they were when the Rev. P. B. Brodie, after having been taken to see that section by the author, communicated to Dr. Wright the notes on it that are published in the paper above referred to. In the first place, Mr. C. Moore described the characters of the Liassic beds at Ilminster, and their relations to the *Avicula contorta* beds and the Keuper as seen in passing from Ilminster through Beer-Crowcombe to Curry-Rival and North Curry,—a distance of ten miles. He then treated of the subdivisions of the Lower Lias and the true position of the "White Lias;" and stated that, although Dr. Wright had proposed the folꞁ lowing classification—5. *Ammonites Bucklandi* zone; 6. *A. Planorbis* zone (including the White Lias and the *Ostrea* beds); and 7. *Avicula contorta* zone, yet he preferred to group them thus—5. *A. Buck-landi* zone; 6. *A. Planorbis* zone; 7. Enaliosaurian zone; 8. White Lias; 9. *Avicula contorta* zone: 8 and 9 being equivalent to the "Kössener Schichten" or "Rhætic beds" of Gümbel and other Continental geologists.

The arguments in favour of his views the author based chiefly on observations made at Beer-Crowcombe, Stoke St. Mary, Pibsbury, Long Sutton, and other places in Somersetshire; and on a critical examination of the sections at Street, Saltford, &c. as given by Dr. Wright.

The communication concluded with descriptions of upwards of sixty species of fossils belonging to the Rhætic beds of England (including their thin representatives discovered by the author in the Vallis near Frome). Twenty-eight of these species are new.

XXXII. *Intelligence and Miscellaneous Articles.*

ADDITIONAL NOTE ON THE CRYSTALS OF LAZULITE DESCRIBED IN THE AUGUST NUMBER OF THIS JOURNAL.

To Dr. William Francis.

Dear Sir,

SINCE the transmission of my paper on the American crystals of Klaprothine or Lazulite, I have received a communication from Professor George J. Brush of Yale College, New Haven, informing me that the crystals in question do not come from North Carolina, but from Georgia. They occur at Graves' Mountain in Lincoln County of that State. The North Carolina examples, analysed by Smith and Brush, do not appear to have been met with in crystals.

Prof. Brush also informs me that these Georgian crystals have been described and figured in a paper by Prof. Shepherd, in the American Journal of Science and Arts, vol. xxvii. (2nd series). This paper had quite escaped my notice, and I have at present no means of referring to it. I hasten, however, in apologizing for past negligence, to point out the fact of its publication. As regards the assumed Trimetric character of these crystals, my views, I may venture to observe, remain unchanged.

Trusting that you will allow this explanation an early place in the pages of the Philosophical Magazine,

I am, dear Sir,

Sault de Ste. Marie, Lake Huron, .   Yours very truly,
July 29, 1861.   E. J. Chapman.

ON OZONE, NITROUS ACID, AND NITROGEN.

BY T. STERRY HUNT, F.R.S.

The formation of a nitrite when moist air is ozonized by means of the electric spark (the old experiment of Cavendish) or by phosphorus, was shown by Rivier and de Fellenberg, who concluded that the reactions ascribed by Schönbein to ozone were due to traces of nitrous acid. The subsequent experiments of Marignac and Andrews have, however, established that ozone is really a modification of oxygen, which Houzeau has shown to be identical with the so-called nascent oxygen, which is evolved, together with ordinary oxygen, when peroxide of barium is decomposed by sulphuric acid at ordinary temperatures. The spontaneous decomposition of a solution of permanganic acid also evolves a similar product having the characters of ozone.

Believing that the nitrous acid in the above experiments is not an accidental product of electric or catalytic action, but dependent upon the formation of active or nascent oxygen, I caused a current of air to pass through a solution of permanganate of potash mixed with sulphuric acid. The air, which had thus acquired the odour and other reactions of ozone, was then passed through a solution of potash; by which process it lost its peculiar properties, while the potash solution was found to contain a salt having the reactions of a nitrite.

As I suggested in this Journal in 1848, I conceive gaseous nitrogen to be the anhydride amide or nitryle of nitrous acid, which in contact with water might under certain circumstances generate nitrous acid and ammonia. From the instability of the compound of these two bodies, however, it becomes necessary to decompose one at the instant of its formation in order to isolate the other. Certain reducing agents which convert nitrous acid into ammonia may thus transform nitrogen (NN) into $2NH^3$. In this way I explain the action of nascent hydrogen in forming ammonia with atmospheric nitrogen in presence of oxidizing metals and alkalies. (Zinc in presence of a heated solution of potash readily reduces nitrates and nitrites with the evolution of ammonia.)

Now an agent which, instead of attacking the nitrous acid would destroy the newly formed ammonia, would permit us to isolate the nitrous acid. Houzeau has shown that nascent oxygen is such an agent, at once oxidizing ammonia with formation of nitrate (nitrite?) of ammonia; and thus when ozone (nascent oxygen) is brought in contact with moist air, both of the atoms of nitrogen in the nitryle (NN) appear in the oxidized state.

From this view it follows that the odour and most of the reactions ascribed to ozone are due to nitrous acid which is liberated by the decomposition of atmospheric nitrogen in presence of water and nascent oxygen. We have thus a key to a new theory of nitrification, and an explanation of the experiments of Cloez on the slow formation of nitrite by the action of air exempt from ammonia upon porous bodies moistened with alkaline solutions.—Silliman's *American Journal* for July 1861.

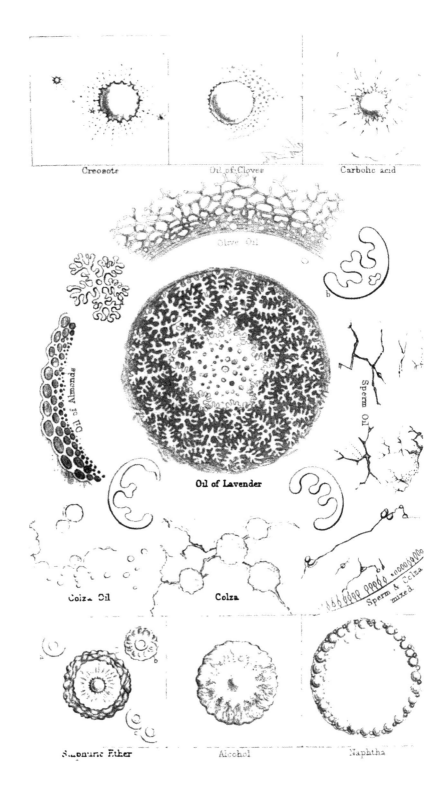

Creosote

Oil of Cloves

Carbolic acid

Olive Oil

Oil of Almonds

Oil of Lavender

Sperm Oil

Coiza Oil

Colza

Sperm & Colza mixed

Sulphuric Ether

Alcohol

Naphtha

THE

LONDON, EDINBURGH AND DUBLIN

PHILOSOPHICAL MAGAZINE

AND

JOURNAL OF SCIENCE.

[FOURTH SERIES.]

OCTOBER 1861.

XXXIII. *On the Cohesion-Figures of Liquids.* By CHARLES TOM-
LINSON, *Lecturer on Science, King's College School, London**.

[With a Plate.]

WE are accustomed to consider a solution as an example
of adhesion, as when water adheres to and dissolves
a salt, or mercury a metal. In such cases the adhesion is sufficiently powerful to overcome the cohesion of the solid. This
process continues until the adhesion of the liquid and the cohesion of the solid counterbalance each other, and we then get
what is called *saturation*. The solution of one liquid in another
is also a case of adhesion overcoming cohesion. The solution of
a gas or of a vapour in a liquid may also be regarded as a case of
adhesion; but often accompanied by this additional phenomenon,
that the particles of the gas or vapour reassume the cohesive
states of their liquids. For example, if we hold a pellet of sponge
saturated with sulphuric ether about half an inch over the surface of water, a portion of the vapour of the ether will be condensed upon the surface in the form of a film with a sharp,
well-defined edge; and this will continue so long as the sponge
is wet, but diminishing in size as the ether evaporates. So
powerful is the adhesion between the water and the ether, that,
if the surface of the former be dusted with lycopodium or tripoli,
or any loose dry powder, the ether vapour will sweep it aside,
and it will be seen, in a state of agitation, outside the edge of the
ether film.

When one liquid is added to another, and solution takes place
between them, there is always a breaking up of the cohesion of

* Communicated by the Author, having been read at the British Association at Manchester, September 1861.

one or other liquid: where there is no solution, there may be simply adhesion. In both cases, whether there be solution or not, one of the liquids displays the phenomena of cohesion in a characteristic manner. For example, the essential oils are but slightly soluble in water. If we place a drop of oil of lavender on the surface of water, the adhesion of the water will cause it to spread out into a film; but the cohesion of the oil immediately begins to reassert itself; the film opens in a number of places, forming long irregular arms or processes resembling the pattern assumed by wood when it has been much worm-eaten. These processes tend to gather up into separate discs or lenticules; the adhesion of the water spreads them out, the cohesion of the oil struggles to prevent this, and soon prevails; the almost immediate issue being the formation of the original drop into a number of discs with sharp, well-defined outlines and convex surfaces. The action is often so rapid, and the pattern so complicated, that it requires repeated observation to become master of all the phe-nomena. (See Plate IV. principal figure, and the subsidiary figures *a*, *b*, *c*, *d*.)

Now this struggle on the part of the oil of lavender to pre-serve its cohesion gives rise to a figure which is characteristic of the substance, and which I propose to name its *cohesion-figure*. It may be regarded as the resultant of the cohesive force of the substance, its density, and the adhesion of the surface on which it is placed. I believe that every independent liquid has its own cohesion-figure. By an independent liquid, I mean not a solu-tion; for in the solutions of solids and liquids cohesion has been already overcome.

The cohesion-figures of liquids can be conveniently studied by gently placing on the surface of water, of mercury, &c. a drop of the substance in question, which we will suppose exerts no chemical action on the receiving surface. Now the cohesion-figures of liquids will be more or less permanent in the inverse ratio of the solubility of the substance. A drop of one of the fixed oils placed on the surface of water will spread out into a film, which is characteristic of the substance, and may last some minutes or even hours, according to the degree of force with which cohesion reasserts itself. A drop of one of the essential oils will also give a characteristic film or cohesion-figure which may change every moment from evaporation and display some beautiful effects of colour; but all these phenomena will be cha-racteristic of the substance in question, and will enable it to be recognized. A drop of a substance like creosote, which is slightly soluble in water, may continue five minutes; a drop of ether or of alcohol may last only a fraction of a second; but whether the time be long or short, these figures are typical of the substances that produce them; and so sensitive are they to any variations in the conditions under which they are produced, that a slight alteration in one of those conditions leads to a marked change in the cohesion-figure. Thus the cohesion-figure of wood-spirit on water is very different from what it is on mercury, since the surface attraction or adhesion of mercury is very different from that of water.

Now let us examine the cohesion-figure of a liquid that is sparingly soluble in water, such as creosote. If we deliver a drop of this substance from the end of a glass rod to the surface of one ounce of water, we may witness a struggle between cohesion and adhesion that will last about five minutes. The creosote sails about on the surface of the water in a state of considerable agitation, discharging a number of small globules on all sides, which, in their turn, are greatly agitated; they rotate and disappear, leaving behind them a thin silvery film. Meanwhile the parent globule diminishes in size, but preserves all the characteristics of its cohesion-figure, until at length it disappears in the form of a film. If a second drop of creosote be now placed on the water, its behaviour will resemble that of the first in a mitigated form; it will be much less energetic, and will last a much longer time before it disappears in the form of a film. A third drop will remain on the surface in the form of a double convex lens sharply defined, showing that the cohesion of the creosote exactly balances the adhesion of the water; or, in other words, that saturation has been attained. If we now add more water, or increase its solvent power by the addition of a drop of acetic acid, the action will set in again, and the lens will change into the cohesion-figure. (See Plate.)

The following comparative experiment was made with fresh colourless creosote (Morson's) in two exactly similar shallow glasses, one containing one ounce, and the other two ounces of New River water. In such water adhesion is diminished by its mineral contents. In distilled water the phenomena are the same, but the time is diminished, a drop of creosote disappearing in five minutes instead of seven :—

Glass No. 1.—1 oz. *water.*		*Glass No.* 2.—2 ozs. *water.*	
	min.		min.
First drop of creosote disappeared in......	7	First drop of creosote disappeared in......	7
Second drop disappeared in..	20	Second drop disappeared in..	12¼
But two or three minute specks of creosote remained, moving in circular orbits in the film.		The disc was flatter and more vigorous than in No. 1; towards the end the disc broke up into separate portions, which rotated with immense rapidity and disappeared.	

S 2

Glass No. 1.—1 oz. water.

Third drop had not disap-} min. 135
peared after...........}

When this drop was first placed on the water, it repelled the film, and the crispations set in although sluggishly. No vollies of small globules. The edge like that of window glass. After 1 minute it became still. After 2 minutes lenticular, slowly sailing about, with occasional jerking of the edge. After 7 minutes, slowly revolving on vertical axis. After 13 minutes, slowly sailing about. After 25 minutes, a very convex lenticule. After 60 minutes, at rest and slowly evaporating.

Glass No. 2.—2 ozs. water.

Third drop disappeared in} min. 25
about ....}

The disc active and vigorous, and of good figure, with vollies of minute globules. After 20 minutes, broke up into three portions, two of which were active; then one split into three or four, which were scattered to a distance; then all still: crispations slowly resumed, and after 25 minutes only a few globules, scarcely visible, remained.

Fourth drop had not disap-} min. 110
peared after ....}

The drop was active for a few minutes, then subsided into a well-shaped lenticule, which slowly disappeared by evaporation.

In connexion with creosote, the cohesion-figure of carbolic acid is interesting. (See Plate.) It is an exaggerated form of the figure of creosote; the water seems to tear it to pieces; the crispations are amazingly active, and the disc quickly breaks up and disappears. Indeed, while a drop of creosote will endure five minutes in an ounce of distilled water, a drop of carbolic acid will last only a few seconds in the same quantity of water. The cohesion-figure is, however, quite characteristic of the substance, and cannot be for a moment mistaken for any other substance that I have examined.

. In cases of this kind, where the conditions are different, we get different cohesion-figures. It has already been stated that, by changing the receiving surface, as by substituting mercury for water, we get a new figure from the same liquid. So, also, if we change the character of the liquid, we vary the figure. The figure given by the unwashed sulphuric ether of the shops is very different from that afforded by rectified ether. Let us take up a quantity of the former in a dropping tube, and gently deliver it, drop by drop, to the surface of about 2 ozs. of water in a clean foot-glass. The very act of gently placing a drop of ether on water leads to the formation of a disc of condensed vapour, just as in pouring ether from a bottle we must first pour a quantity of vapour. As the drop of ether is hanging over the water, it forms a well-defined circular disc or film of condensed ether-vapour on the surface of the water immediately below the drop of ether. But as soon as the drop is delivered to the water, it combines with this disc, and spreads into another disc to the utmost limit of its cohesion: it forms, in fact, a circular or centrifugal wave of such extent that there is not matter enough to prevent the centre from opening and following the general impulse outwards. We thus get a perforated disc: the disc itself is in rapid motion and agitation, but the water seen through the central perforation is tranquil. Both the outer and the inner edges of this disc are perfectly sharp and well defined. The cohesive force of the ether prevents it from breaking up, and even produces a rebound: the disc closes in upon itself, becomes smaller and smaller, still preserving its central perforation and well-defined form, until at length it vanishes under the influence of evaporation, adhesion of the water, and probably electrical action; or, as is generally the case, the attraction of the sides of the vessel causes the mobile body to disappear by dashing up against the glass. If two or three drops be allowed to fall in quick succession, the perforated discs become partially superposed, but still preserve their distinctive features. (See figs. *a*, *b*, *c* in the sulphuric-ether figure.)

On placing a drop of rectified ether on the surface of water, it is evident that it has a much stronger cohesion than when adulterated with alcohol, or rather the water has a less adhesion for it. The drop of ether becomes lenticular, and in doing so discharges from all sides a portion of its substance, which assumes the form of a tolerably smooth flat ring: this in its turn discharges a portion of ether into the water, which, in seizing it, produces a troubled motion. Hence we have the true ether figure, consisting of a central lenticule, surrounded by a nearly smooth flat ring with radiated markings, and this by an agitated ring with curved markings, as if minute globules of the liquid were diffusing. I have attempted, with the assistance of an artist, to represent this in the Plate. It is tolerably accurate; but I need hardly suggest that a chemist's eye retains such figures better than an artist's; for to the one they are expressions of natural truths—additional exponents, in fact, of those endless properties which he endeavours to frame into laws; while to the artist these things are mere forms—if beautiful, so much the better—but still only forms, containing no latent truths.

I may also remark that, in order to get the typical form of each substance, certain precautions are necessary. The water may be distilled; but this is not absolutely necessary, provided it be chemically clean. It should be in sufficient quantity to prevent its becoming quickly saturated; for as the water approaches saturation, the figure becomes slightly modified, although it presents the no small advantage of greater persistence. The ether figure will remain for about a second in a nearly saturated solution. But what is in many cases absolutely necessary to success is, that the glass containing the water be quite clean; it should be purified from the organic film which covers most matter exposed to the air, by washing it in strong sulphuric acid, and then in a solution of caustic potash.   If, after this, the water completely wet the glass, it may be rinsed and used.   It had better not be wiped.   Having been once washed with sulphuric acid, it will serve for a great number of experiments, provided it be washed in a weak solution of caustic potash after each experiment, and rinsed with clean water.   The substance of which the figure is to be determined must, as already remarked, be pure : the figure given by pure washed ether becomes changed into the perforated discs by the addition of a few drops of absolute alcohol to a small quantity of the washed ether; and the unwashed ether, if exposed in an open vessel for a few minutes, will throw off its ethereal portion, and the cohesion-figure will quickly pass into the alcohol figure.   It is quite remarkable how rapidly this change takes place.   I had poured some unwashed ether into a test-glass, from which I fed the dropping-tube, and in about ten minutes the ether figure was completely superseded by the alcohol one.   I do not pretend to say that all the substances made use of in this inquiry are pure.   I have taken pains to procure them from the best sources, such as the manufacturers themselves; but some of the essential oils, for example, are prepared on the Continent, and may possibly not be quite pure.

There is not so marked a difference between the cohesion-figures produced by spirits of wine and absolute alcohol as between unwashed and washed ether : nor is this to be wondered at, seeing that spirits of wine only differ from absolute alcohol in having already received a portion of the water which the absolute alcohol takes up in forming its peculiar figure on water. The figure of spirits of wine consists of a central disc with a foliated outline surrounded by a tolerably smooth disc.   The figure of absolute alcohol (Pl. IV.) has the central disc more minutely foliated than in the former case, and it has a greater tendency to a stellar arrangement.

Without further multiplying these examples, I may once more recur to the law on which they seem to rest—viz. that each figure is the resultant of the cohesive force of the liquid, its density, and the adhesion of the receiving surface.   If this be true, it follows that two liquids although of very different chemical character, yet being of the same density, similarly cohesive, whether viscid or fluid, and the adhesion of the receiving surface being the same (*i. e.* having the same degree of solubility), we get precisely the same cohesion-figures for both liquids.   Now creosote and oil of cloves are chemically two very dissimilar liquids.   It is true that both are hydro-carbons, and that each consists of two distinct bodies ; but their points of difference are numerous and important.   Nevertheless their physical resem-·

blances are striking : they are of about the same density (*i. e.* a little heavier than water) ; they are about equal in liquidity or cohesion, and they are both sparingly soluble in water. Now the cohesion-figures of these two substances are so much alike that a casual observer would declare them to be identical. (See Plate.) There are the same crispations in the oil of cloves as in the drop of creosote; the same flattened, indented, waving, agitated outline; the same sailing about ; a similar film, and the same repulsion of the film ; the same whirling off of small globules, and the rapid rotation and disappearance of those small bodies. Like the creosote, too, the second drop of oil of cloves reproduces the phenomena of the first in a mitigated form, and is much longer in disappearing. But now for the differences. The film formed by oil of cloves is more like smoke, more dense, persistent, and plicated than that of creosote; and being constantly driven about by the parent disc, it becomes powdery, like fine flour, on the surface of the water. In the midst of this film, the parent globule will sometimes remain for several minutes, keeping a clear space of considerable extent all around it, pulsating in a regular manner, and flashing out lines which are visible only by the motions of the water. But the most considerable difference between oil of cloves and creosote is in their respective duration. We have seen that 2 ozs. of New River water will dissolve three drops of creosote ; the same quan ity of river water will not dissolve so much as two drops of oil of cloves. After the first drop has disappeared, a second will be lively at intervals during nearly an hour, but after twenty-four hours some small lenticules of the oil will remain on the water. Hence oil of cloves has only about half the solubility of creosote in water—only half the adhesion, in fact; but being denser than water, it tends to sink, and thus appears to be more adhesive than it really is. An interesting result may also be obtained with ol. pimentæ, which is a little heavier than water (1·021 to 1·044). It is much more sluggish than creosote and oil of cloves, but exhibits similar phenomena on a small scale. If, however, the water be heated to about 110° F., we get a large crispating figure of great beauty. There are special characters about it which I do not stop to describe.

Should any one wish to repeat these observations on oil of cloves, he may have some difficulty in doing so on account of the difficulty of obtaining the pure oil. The oil of the shops is commonly adulterated with the cheaper oils, such as oil of olives or of almonds, or of turpentine ; and I have ascertained that a single drop of oil of olives to twenty drops of oil of cloves is sufficient to prevent the formation of the cohesion-figure, and the display of those curious and interesting motions of the pure oil. Even in cases where the oil is not adulterated, the fixation of oxygen, simple or ozonized, may prevent the display of these characteristic phenomena; but at the same time it will introduce other results which are equally characteristic.

Oil of turpentine is also used as an adulterant of the essential oils. Its presence can be detected by the brilliant iridescent colours which it imparts to films that are otherwise colourless. It also makes many oils more limpid, and thus renders them more active in the display of their peculiar phenomena. Oil of turpentine alone gives a delicate film with iridescent rings and an outer border of minute globules, with bosses within the edge: these flatten into discs surrounded by small dots. Iridescent colours now set in and cover the film. Suddenly the whole film opens into holes, which, in the midst of the colours, have a beautiful effect. The film slowly disappears, leaving an outline lace-pattern which lasts for hours.

Now when oil of lavender is adulterated with turpentine in the proportion of 5 to 1, the film spreads with a brilliant display of colour, which is characteristic of the turpentine, the lavender being colourless; at the same time the peculiar worm-eaten pattern of lavender is more minute, and its action much more rapid than in the case of the pure oil. By increasing the proportion of the turpentine, the characteristics of the latter film override those of the former.

Now this brings me to speak of the use to which these cohesion-figures may be applied in detecting adulteration. It is perfectly easy to distinguish unwashed ether from rectified ether, alcohol from spirits of wine, &c., by their respective cohesion-figures. It is also equally easy to name a varnish, a fixed, or an essential oil, from the characters of the film which a drop of each substance forms on water. Having become acquainted with the characters of each film, it is not difficult to detect the films formed by mixtures, and even to name the component parts of a mixture. For example, oil of cinnamon is now worth about 5s per oz., so that there is an inducement to adulterate it. The readiest means of adulteration is with oil of olives or oil of sweet almonds. To be able to detect the adulteration, we must become acquainted with the characters of the films of all three oils. Now, to begin with oil of cinnamon:—As soon as a drop of this substance is delivered to the surface of the water, it spreads out into a film, but the more fluid portion of the oil (the elæopten) precedes the film in radial lines of minute globules, and these form an outer boundary line of detached spots to the film. The film itself even on a large surface of water is not more than about an inch in diameter; it is of a beautiful delicate structure and silvery reflection; its edge is well defined, and it has small bosses just within it. Almost immediately after its formation, holes open near the edge, starting into existence and altering rapidly, and the film separates into a kind of network and two or three well-defined flat discs.

When a drop of oil of olives is placed on water, the first thing that strikes the eye is a beautiful widening rainbow, which seems to deposit the film and then disappear. The film itself is colourless, and it has an indented edge displaying a very light and elegant kind of lace-pattern, similar to what, I believe, is called *guipure*, in which a raised thread traces the outline. (Plate IV.) After a few minutes the pattern vanishes, and the oil collects into an irregular trail with ragged edges, surrounded by numerous small globules.

A drop of oil of almonds on water spreads into a large film with a beautiful lace-like edge, which soon disappears by the holes opening into each other. (See Plate.) The edge separates from the parent film, and forms small lenticules outside it. The edge of the film appears a little raised; the holes in it continue to open and widen, and the detached pieces shrink up into lenticules; and in a few minutes the parent film has diminished to the size of a shilling, surrounded by a number of lenticules of various degrees of smallness.

Now when olive oil is used to adulterate oil of cinnamon, its presence may be detected by some of the characters which the oil-of-olives film exhibits alone, and especially by its iridescence. I added one drop of oil of olives to ten drops of oil of cinnamon; and the film formed by one drop of the mixture on water exhibited the following characters:—1st. A display of beautiful iridescent rings, which shrank into angular masses and so disappeared. 2nd. A considerable portion of the film gathered itself up into a central disc about the size of a wafer. 3rd. This disc was surrounded by a delicate perforated silvery film which quickly evaporated, leaving some minute lenticules which became fringed with a kind of frill. And then, 5th. These lenticules exhibited minute systems of iridescent rings. These iridescent effects at the commencement and close of the observation do not belong to oil of cinnamon, but are characteristic of oil of olives. These and the other phenomena are sufficient to detect the presence of a small portion of oil of olives in oil of cinnamon. By increasing the proportion of the adulterating oil, the properties of the oil of olives are displayed more strikingly.

When one drop of oil of almonds is added to ten drops of oil of cinnamon, and one drop of the adulterated oil is placed on water, there is a shooting out of minute globules in radial lines, which is characteristic of oil of cinnamon; a delicate film is formed with holes in the edge which close and open again, and in a few minutes cohesion gathers up the film in the form of long, ragged, irregular, oily-looking smears—an effect which is characteristic of oil of almonds.

Alcohol is also used to adulterate the essential oils: its presence can be detected by phenomena which vary with the proportions used. There are also special phenomena with each oil, that would take a long time to describe.

The films formed by such common oils as sperm and colza are also characteristic. Sperm oil forms a smooth large film, which occupies the whole surface and is accompanied by iridescent rings, which disappear when the film is formed. Minute and nicely perforated holes open in the film, and after a short time long, thin, narrow cracks open in it, darting out from the holes and often connecting them together like beads on a thread. These cracks are characteristic of sperm oil. (See Plate.)

Colza oil forms a large smooth film, accompanied by iridescent rings, which immediately disappear. Minute holes open at the edge at intervals, three or four together, sharp and clean as if punched. Similar distinct perforations are also formed in other parts of the film; and these widen and thicken at the edges until the surface is covered with a kind of honeycomb-pattern, the holes pressing together in twos and threes. The characteristic feature of the colza film is to be found in these large holes with thickened edges grouped together and opening into each other. After about an hour the film becomes whitish and greasy-looking, and the holes are surrounded by dark rings. It may be remarked, however, that an increase of temperature quickens and exalts the phenomena of this and other films. Thus the effects are more numerous and more quickly brought about on a fine warm day than on a dull and cloudy one. The films should also be formed on a given fixed area of water, or the film of the same liquid may vary in thickness at different times and thus disturb the phenomena. I have found a conical foot-glass nearly 4 inches in diameter at the mouth, nearly filled with water, answer well.

A mixture of sperm and colza in various proportions forms a good film, in which may be recognized the cracks of the sperm and the peculiar holes of the colza. I think it would be easy for any one to detect the mixture of these two oils by the character of the film. (See Plate.)

I have thus briefly indicated the mode of obtaining these cohesion-figures, and their value in determining the nature and purity of various liquids in common use and which are liable to adulteration, such as sulphuric ether, the essential and fixed oils, &c. By simply noticing the cohesion-figure of sulphuric ether for example, we can decide whether it contains alcohol or not. I believe it would be easy in many cases to decide by this mode whether liquids professedly delivered according to sample had been tampered with between the delivery of the sample and the goods.

A considerable amount of labour requires to be expended on these cohesion-figures before the subject can be said to be ripe for extended practical application; and this labour, if health and opportunity be granted me, I intend to bestow on this most interesting subject. It will be necessary to procure many specimens from different markets of the pure liquid, to take the cohesion-figure of each sample many times under varying temperatures and hygrometric conditions of the air, to make drawings of these figures, and to decide after repeated trials what is the characteristic feature of each liquid as constantly exhibited in its cohesion-figure. The next step will be to ascertain the characteristics of mixtures of certain liquids made with a view to adulteration, and by means of cohesion-figures to enable the observer to state not only whether a costly liquid be adulterated, but what it is adulterated with, and the relative proportions of the adulterated mixture. I have done all this for a moderate number of liquids, but very much remains to be done before the method can be considered as complete. There are also certain scientific questions to be answered, such as the relations between the figures of isomeric liquids.

I am aware that some of the phenomena of films on the surface of water have, on a few occasions, attracted the attention of philosophers; but no one, so far as I am informed, has considered films to differ from each other, or to be characteristic of the substances which produce them. The behaviour of films has generally had reference, in the labours of others, to some point irrespective of the films themselves—as in the repulsion-theory of B. Prevost[*] and the epipolic theory of Dutrochet[*].

It was many years ago observed by Ermann[†] that a drop of sulphuric acid deposited on the clean surface of pure mercury spreads out into a film. M. Dutrochet finds that many liquids also form films on the surfaces of glass, metal, &c., provided they be chemically clean; and he attributes all these phenomena "au développement subit et toujours de courte durée de la force épipolique centrifuge." It seems to me (with great submission to so distinguished an observer as the discoverer of endosmose and exosmose) that many of the phenomena described as the effects of the epipolic force are simple results of cohesion and adhesion. M. Dutrochet repeatedly states that his phenomena cannot be produced unless the surfaces be absolutely clean. That, I am quite sure, is a correct observation; but the impurities, according

---

[*] Quoted in M. Dutrochet's work entitled *Recherches physiques sur la Force Epipolique,* Paris, 1842, and 2nd part, 1843.

[†] *Annales de Physique de Gilbert,* vol. xxxii.

to my view, do not act in preventing the exhibition of a new force, but simply by preventing adhesion. Many of the phenomena of cohesion-figures I have been unable to produce away from home in vessels which have been cleaned and wiped in the usual manner; but they have succeeded perfectly when the glasses were washed in a weak solution of caustic potash and rinsed in clean water.

. I have taken advantage of Ermann's observation to obtain on mercury the cohesion-figures of sulphuric acid and other substances which act chemically when brought into contact with water. In the case of sulphuric acid, the drop spreads instantly and covers the surface of the mercury; but cohesion immediately begins to reassert its claims, and forms the acid near the edge into large flat bosses, each of which becomes a centre of action; minute globules pass in and out of it; similar small globules also move to and fro over the rounded edge of the mercury. After a few minutes all action ceases: the film contracts with a smooth surface and a well-defined edge. A drop of alcohol, or of ether held over the sulphuric acid film when at its widest, gathers it up in an instant into a small disc. The reason for this is, that there is a much stronger adhesion between the acid and the vapour of ether or alcohol, than between the acid and the mercury.

It was observed by Dutrochet that a drop of water placed on mercury remains globular, an effect which he explains according to his epipolic theory. I explain it by the cohesion of the water being stronger than the adhesion of the mercury; and I imagine that the absorption of water by the sulphuric acid film lessens the adhesion of the mercury, and enables cohesion to reassert its claims with more effect. If the vessel be covered up, the diffused sulphuric acid film is much more persistent. The superior cohesive force of water is also shown by placing a drop of it on the sulphuric acid film. It does not spread, but remains in a very convex lenticular state, at the same time repelling, apparently, the sulphuric acid all around it, so that the lenticule of water remains on a dry disc of mercury. These effects I attribute to the stronger adhesion between the acid and the mercury than between water and mercury. The water does not repel the acid, as Dutrochet supposed, but simply absorbs a portion around it, sufficient to allow it to rest on the mercury, and to prevent all further action of the acid.

The liquids whose cohesion-figures on water have been determined, present, of course, different figures on mercury, because one of the conditions in the production of these figures (viz. the adhesion) is no longer the same. Thus wood spirit, which on water forms a figure something like that of alcohol, produces on mercury a lens which flattens with a well-defined edge; then a rapid motion sets in from the edge and spreads all over the surface; bosses form and disappear; the film becomes divided into two or three parts by lines, but without separating; the agitation ceases; the film spreads more and more; but at a certain point cohesion begins to reassert itself, and the film gradually contracts and at length becomes a perfect circular disc.

King's College, London,
26th July, 1861.

XXXIV. *On the Measurement of Electric Resistance according to an absolute Standard.* By WILHELM WEBER.

[Concluded from p. 240.]

§ 5. *Comparison of the Resistance determined according to absolute measure with Jacobi's Standard of Resistance.*

TO compare the resistance of two conductors, there are different methods which need no explanation. The resistances considered in the preceding paragraphs have been compared according to the method examined in this memoir, and it has been found that

$$w : w' = 1138 : 1000.$$

If the first resistance be reduced to the second according to this proportion, we obtain

$$w' = \frac{1000}{1138} w = 1903 \cdot 10^8,$$

while the direct determination in the preceding paragraph gave

$$w' = 1898 \cdot 10^8.$$

From both these closely agreeing values, determined according to entirely different methods, the number $19 \cdot 10^8$ will in future be assumed as the mean value of this resistance.

Jacobi has dwelt on the importance of introducing a definite measure for resistance to be accepted by all physicists, especially at the present time, when so many voltaic investigations are being made with the most varied instruments, the comparison of which is often of great importance. For this purpose he has proposed as a *standard measure* a copper wire, which he has sent to several physicists who are engaged with voltaic measurements, and has requested them to compare this standard with theirs, and for the future to give their measurements in this measure.

This standard is a copper wire $7169\frac{3}{4}$ millims. in length, and $\frac{3}{4}$ millim. in thickness, which weighs $22449\frac{5}{10}$ milligrammes.

The standard introduced by Jacobi, which, it is to be hoped, will find general acceptance, is by no means supplanted by the

absolute measure here discussed; for it is not possible to compare
every resistance directly according to this measure, while every
resistance can be directly compared with Jacobi's standard. But
considering the importance which absolute determinations of
measure have in many investigations, it is desirable to be able
to reduce all the values, made according to Jacobi's standard, to
an absolute measure, which can be easily effected by comparing
the resistance determined as above according to an absolute measure with the resistance of Jacobi's standard.

Such a comparison has been made; and it has been found that
the two resistances are nearly as 32 : 10, or, more accurately, as
19000 : 5980. But as the first resistance has been found in absolute measure to represent 19000 million units, Jacobi's standard represents 5980 million units; or the resistance determined
according to Jacobi's measure can be reduced to absolute measure by multiplication by 6 milliards. By this determination
it would be possible to reproduce approximately Jacobi's standard, even if it were lost.

§ 6. *On the value of the constants found by Kirchhoff, on which
the intensity of induced electric currents depends.*

The *induction-constant* which Neumann, in his development
of the mathematical laws of induced electric currents, calls ϵ,
has the following meaning. If W be the absolute unit of measure proposed as above for electric resistances, and W' that measure of resistance which is actually used; if, further, C be the
measure of velocity which forms the basis in establishing the
above absolute measure (1 millimetre in a second); if, on the
contrary, C' be the measure of velocity actually used in measuring the induced motions and actions of the induced currents
(1 Prussian inch = 26·154 millims. in a second, according to
Kirchhoff), we have

$$\epsilon = 2\,\frac{C'W}{CW'}$$

It follows from this, that if the value of this induction-constant
is once determined, any resistance given according to the measure chosen can be referred to an absolute measure.

In the determination of the induction-constant ϵ given by
Kirchhoff in the seventy-sixth volume of Poggendorff's *Annalen,*
the resistance of a *copper wire* has been chosen as a standard, the
length of which was 1 Prussian inch = 26·154 millims., and the
section 1 Prussian square inch = 684 square millims. Here
unfortunately there is no determinate measure of resistance; for
different pieces of copper of the same dimensions have different
resistance; and it follows, therefore, that the value of the induc-

tion-constant ϵ is left undetermined within the limits of that variability of the resistance of copper. Kirchhoff himself says, "Since the conductibility of copper varies within certain limits, in giving the value of ϵ, only a limited accuracy is of interest." Kirchhoff wished to give only an approximate value of ϵ, which would be sufficient for his purpose; and he was the more content therewith because the methods and instruments which he used would scarcely have permitted a better determination of ϵ if he proposed a perfectly definite measure of resistance.

The interest which an accurate determination of the value ϵ has, is lost in consequence of that uncertainty in the choice of the measure of resistance; and it is important to restore it by the removal of that uncertainty. This may be accomplished by keeping, not to *copper in general*, but to the piece of copper actually used by Kirchhoff in his investigations, and by choosing the resistance of a wire of this copper 26·154 millims. in length, and with a section of 684 square millims. as a measure of resistance. It is thus only necessary to reduce the result found by Kirchhoff, as well as the measures made therewith or referred thereto, to the measure thus accurately determined in this manner. Kirchhoff took one Prussian inch in a second as a measure of velocity, and found in this way

$$\epsilon = \frac{1}{192};$$

from which it follows (since $C' = 26·154$ C) that that resistance which amounts to 52·808 units of the above absolute measure is the $\frac{1}{192}$ of the resistance of a wire of Kirchhoff's copper the length of which is 26·154 millims. and the section 584 square. millims.; in other words, that the measure of resistance chosen by Kirchhoff is 10043 times that of the above absolute measure.

Although this value of ϵ can only be considered as approximative, it is interesting to compare it with other values which have been found by entirely different methods and with different instruments, because an examination of the various natural laws brought thereby into operation is obtained. Kirchhoff's measurements refer to currents produced by *voltaic induction*, and hence in his case it is the laws of voltaic induction which have been used in determining the value of ϵ. My measurements, on the contrary, refer to currents produced by *magnetic induction*, and hence in this case it is the laws of magnetic induction which lead to the value of ϵ.

First of all, the value of ϵ shall be given which is obtained from my measurements. It is clear that the value of ϵ can be determined from these measurements, if only the resistance of

Kirchhoff's copper wire is compared with the resistance of Jacobi's standard. I have made that comparison by means of the wire which Kirchhoff kindly sent to me, and can here give the result of the comparison: it is as follows.

A piece of Kirchhoff's wire which was 13·573 Prussian inches in length and 0·4061 square line in section, had a resistance which was to the resistance of Jacobi's standard as

$$1 : 106.$$

From this we get the relation of the resistance of the measure chosen by Kirchhoff to that of Jacobi's standard as

$$1 : 106 \times 13 \cdot 573 \times \frac{144}{0 \cdot 4061}.$$

If J be the resistance of Jacobi's standard, and W' that of Kirchhoff's, we have

$$\frac{J}{W'} = 510180.$$

Now the resistance of Jacobi's standard is equal to 5980 million units of the absolute measure found above; hence, if W be the absolute resistance,

$$\frac{J}{W} = 5980000000 ;$$

hence

$$\frac{W'}{W} = 11720.$$

But now

$$\frac{C'}{C} = 26 \cdot 154 ;$$

hence

$$\epsilon = 2 \frac{C'W}{CW'} = \frac{J}{224} ;$$

that is, one-seventh less than Kirchhoff had found. A closer agreement was not to be expected, inasmuch as only an approximate value was claimed for Kirchhoff's statement.

I may give here a determination *of the specific resistance of the different kinds of copper* which have been used for Jacobi's standard, for Kirchhoff's wire, and for the damper which I used.

The *specific resistance* of a body is usually given according to an absolute unit by taking for this unit the specific resistance of a body whose absolute resistance with a length $= 1$ and a section $= 1$ is equal to the fixed measure of resistance. But the determination of specific resistance according to this unit meets with a practical difficulty in the accurate measurement of the

section, especially in fine wires, and hence, to obviate this difficulty, Kirchhoff has indirectly ascertained the section of the wire
by determining its absolute and specific gravity.

Now the determination of specific resistance according to this
unit, presupposes that the resistance of a wire whose length
remains unchanged, but the thickness of which is increased or
diminished, varies inversely as the section. This has not, however, been proved, and, with the small alterations of section
which are produced by pressure, can scarcely be proved. There
is just as much reason for assuming that, if the mass and the
length of the wire remain unchanged, the resistance does not
alter even with a changing section. On this assumption the
absolute unit would have to be fixed in another way than as
being the specific resistance of a body whose absolute resistance
for the length $=1$ and for a mass $=1$ is equal to the fixed measure of resistance. According to this, the specific resistance of a
body would be determined by multiplying the resistance of a wire
formed of that substance expressed according to the fixed measure
of resistance by its mass, and dividing by the square of its length.

The specific resistances of the wires used by Jacobi, Kirchhoff,
and myself will be determined according to the unit thus fixed;
for apart from the above considerations, this determination is in
any case the most applicable and capable of execution.

The following Table exhibits the results of these determinations :—

Quality of copper in	Length in millime-tres.	Mass in milli-grammes.	Resistance in abso-lute measure.	Specific resistance.	ϵ.
Jacobi's wire ...	7620	22435	5980000000	2310000	$\frac{1}{810}$
Kirchhoff's wire	355	4278	58500000	1916000	$\frac{1}{1143}$
Weber's wire ...	3946000	152890000	190000000000	1865600	$\frac{1}{710}$

It will be seen that there is only a small difference between
my copper and Kirchhoff's; while the difference in the case of
Jacobi's is far more considerable, as the latter possesses a far
smaller conductibility. In the supposition that Jacobi may have
used galvanoplastic copper for his standard, I examined a wire
of that material which I procured through the kindness of Professor Schellbach in Berlin, and found the following result,
which proves, contrary to the above supposition, that galvanoplastic copper is a somewhat better conductor.

Wire of galvano-plastic copper.	Length in millime-tres.	Mass in milli-grammes.	Resistance in abso-lute measure.	Specific resistance.	ϵ.
	12780	221295	1243000000	.1684000	$\frac{1}{166}$

In the last column here and in the upper Table are given the different values of ϵ which were obtained for the Neumann's induction-constant by adhering to the measure chosen by Kirchhoff, but using the different kinds of copper which have been mentioned. Adhering, however, to the absolute measure fixed as above, $C'=C$, $W'=W$, and ϵ has always the value 2.

§ 7. *On the constants of the electric laws which depend on the choice of measures.*

The law of induced currents propounded by Neumann represents the intensity of these currents as dependent on a constant the value of which must be determined from the measures according to which the magnitudes taken into consideration are to be determined. This constant Neumann has called the *induction-constant*. Such a constant occurs in the general expression of any natural law which states how one magnitude is determined by another. I may here give a summary of these constants for all the fundamental laws which refer to *electromotive force, intensity,* and *electric resistance.* Each of these laws represents the desired magnitude as an expression of other measurable magnitudes, which has a constant as a factor the value of which is to be determined from the measures chosen.

1. The fundamental law of the voltaic circuit represents the intensity of the current i as an expression of the electromotive force e, and of the resistance w; for, if the constant whose value is to be determined is called α,

$$i = \alpha \cdot \frac{e}{w}.$$

This constant α has the following meaning. If J, E, W are the absolute measures fixed as above for intensities, electromotive forces, and resistance; and if J', E', W' are the measures actually used, we have

$$\alpha = \frac{JEW}{J'E'W'}.$$

Hence using the absolute measure itself, $\alpha = 1$.

2. The fundamental law of electro-magnetism represents the electromotive force F as an expression of the quantity of magnetic fluid μ, of the length ds, and of the intensity i of the element, of their distance from one another r, and of a number which is given by the angle ϕ which r makes with ds; that is, if the constant whose value is to be determined from the measures chosen is β,

$$F = \beta \cdot \frac{\mu i ds}{rr} \sin \phi.$$

The constant β has the following signification:—If P is the absolute unit of measure of the momentum of rotation (the product of a millimetre into that force which in one second imparts to the mass a milligramme, the absolute unit of measure of velocity), if M is the absolute unit of measure of the magnetic fluid, and J is the absolute measure for intensities; if, further, P', M', and J' are the measures actually used,

$$\beta = \frac{PM'J'}{P'MJ};$$

consequently, using the absolute measure, $\beta = 1$.

3. Ampère's fundamental law of electrodynamics represents the electrodynamic force of attraction F as an expression of the intensities of two elements i and i', and of a number which is fixed by the relations of the lengths of the two elements to their distance $\frac{ds}{r}$, $\frac{ds'}{r}$; and by the three angles ϵ, θ, θ', which ds and ds' form with one another and with r; that is, if the constant whose value is to be determined from the given measures is designated by γ,

$$F = \gamma \cdot ii' \cdot \frac{ds\, ds'}{rr} \left(\cos \epsilon - \frac{3}{2} \cos \theta \cos \theta' \right).$$

The constant γ has the following signification:—If F is the absolute measure of force (that force which in a second imparts to the mass of a milligramme a velocity of a millimetre in a second), if J is the absolute measures for intensities, and F', J' the measures actually used, we get

$$\gamma = \frac{FJ'J'}{F'JJ};$$

hence using the absolute measure, $\gamma = r$.

4. The fundamental law of magneto-induction represents the electromotive force e as an expression of the mass of magnetic fluid μ, of the velocity of induced motion c, of the length of the induced element ds, and of its distance r from μ, and of a number given by the two angles ϕ, ψ which ds makes with r and c with the normal to the plane rds; that is, if the constant whose value is to be determined from the measures chosen is called δ, 1

$$e = \delta \frac{\mu c ds}{rr} \sin \phi \cos \psi.$$

The constant δ has the following signification:—If E is the absolute unit of measure of electromotive force, M the absolute unit of measure of magnetic fluid, S the seconds of time, and E', M', S' the measures actually used, we get

$$\delta = \frac{\text{EM}'\text{S}}{\text{E}'\text{MS}'} \; ;$$

hence using the absolute measure, $\delta = 1$.

5. The fundamental law of voltaic induction represents the electromotive force e as an expression of the intensity i and of its change $\frac{di}{dt}$, of the velocity of the inducing motion c, and of the distance r of the induced from the inducing element, and of several numbers which are given by the relations of the lengths of the two elements to their distance $\frac{ds}{r}, \frac{ds'}{r}$, and by the four angles e, θ, θ', ϕ which ds and c form with each other and with r, and which ds' forms with r; that is to say, if the constant whose value is to be determined from the measures chosen is called ζ,

$$e = \zeta \left[ci \cdot \frac{ds\, ds'}{rr} \left(\cos e - \frac{3}{2} \cos \theta \cos \theta' \right) \cos \phi + \frac{1}{2} \frac{di}{dt} \cdot \frac{ds\, ds'}{r} \cos \theta \cos \phi \right].$$

The constant ζ has the following significance:—If E and I are the absolute units for electromotive forces and for intensities, and C the absolute measure of velocity (a millimetre in a second), and E', I', C' the measures actually used, we have

$$\zeta = 2 \frac{\text{EI}'\text{C}'}{\text{E}'\text{IC}} ;$$

hence using the absolute measure itself,

$$\zeta = 2.$$

6. The general fundamental law of electric action represents the electric force F as an expression of the electric masses v, v', of their distance r, their relative velocity $\frac{ds}{dt}$, and their change $\frac{ddr}{dt^2}$; that is, if the constant whose value is to be determined from the given measures is called η, we have

$$\text{F} = \eta \cdot \frac{vv'}{rr} \left[1 - \frac{1}{aa} \left(\frac{dr^2}{dt^2} - 2r \frac{ddr}{dt^2} \right) \right] :$$

a stands for the number indicating the relation of that velocity with which two electric masses must be moved against each other in order that they exert no force on each other, to the velocity of a millimetre in a second.

The constant η has the following signification:—If F is the absolute measure of force, N the absolute unit of electric fluid (that mass of electric fluid which at a distance of a millimetre exerts upon a similar mass the absolute unit of force), if R is a millimetre, and F', N', R' the measures actually used, we have

$$\eta = \frac{FN'N'RR}{F'NNR'R'};$$

hence using the absolute measure itself, $\eta = 1$

Every electric force can act, however, as electromotive force; and this latter e is represented, according to the general fundamental law of electric action, as an expression of the electric mass v, of the length of the element in which is contained the quantity of electricity acted upon; further, of the distance r of both from each other, of their relative velocity $\frac{dr}{dt}$, and their change $\frac{d\,dr}{dt^2}$, and of the angle ϕ which ds forms with r; that is, if the constant whose value is to be determined from the measures chosen is called k, we have

$$e = k \cdot \frac{vds}{rr}\left[a - 1\frac{1}{a}\left(\frac{dr^2}{dt^2} - 2r\frac{ddr}{dt}\right)\right]\cos\phi.$$

The constant k has the following meaning:—If E is the absolute unit of measure of electromotive forces, N the absolute unit of measure of the electric fluid, C the absolute unit of velocity (a millimetre in a second), R a millimetre, and E', N', C', R' the measures actually used, we have

$$k = \frac{1}{2\sqrt{2}}\frac{EN'C'R}{E'NCR'};$$

hence using the absolute measure,

$$k = \frac{1}{2\sqrt{2}}.$$

XXXV. *On the Action of Uncrystallized Films upon Common and Polarized Light.* By Sir DAVID BREWSTER, *K.H., F.R.S.*[*]

IN a paper "On the Polarization of Light by Refraction," published in the Philosophical Transactions for 1814, I have shown that when a pencil of light is incident on a number of uncrystallized plates, inclined at the same or different angles to the incident ray, all their surfaces being perpendicular to the plane of the first incidence, the transmitted pencil will be wholly polarized when the sum of the tangents of the angle of incidence upon each plate is equal to a constant quantity depending upon

* Communicated by the Author. This paper was read at the meeting of the British Association held at Aberdeen in Sept. 1859.

the refractive power of the plates and the intensity of the incident pencil.

This law, though admitted by M. Arago in his article on Polarization in the *Encyclopædia Britannica**, was called in question by Dr. Young†, on the ground that no finite number of plates could polarize the whole transmitted beam, as a small portion of light must always remain unpolarized, or in the state of natural light. This is doubtless true; but, as Sir John Herschel has shown, it does not affect the truth of the law, which involves the intensity of the incident pencil. According to the law of geometrical progression, indeed, a small portion of unpolarized light exists mathematically in the transmitted beam; but a beam of light may be said to be completely polarized when the unpolarized portion is invisible, vanishing entirely in certain positions of the analysing prism.

Neither M. Arago nor Dr. Young has made the slightest reference to that portion of the refracted light which is reflected at the surfaces of each plate and returned into the transmitted beam. Sir John Herschel, however, has distinctly referred to it, and remarks that "it mixes with the transmitted beam, and, being in an opposite plane, destroys a part of its polarization ‡."

Although the law of the tangents which I have mentioned refers only to the transmitted pencil, yet, in the paper which contains it, I have shown that the light reflected back into that pencil is distinctly visible, not as ordinary light, as Sir John Herschel maintains, but as light polarized in an opposite plane to the refracted pencil.

When the angle of incidence is considerable, this oppositely polarized light appears as a nebulous mass, like the nebulous image in the agate; and after examining it, I found it to have the same relation to the refracted pencil "*as the nebulous image has to the bright image of the agate, or as the first has to the second pencil of doubly refracting crystals §.*"

In making the experiment with a small bright image of a candle, and using plates of parallel glass, I found that the reflected images *a, a, a, a* were distinctly separated from the bright or refracted image A, and were all polarized by reflexion in a plane opposite to that of A ‖.

Although these two facts, which have much theoretical importance, were not only minutely described, but represented in diagrams, in my paper of 1814, yet they escaped the notice of the

---

* *Encyc. Brit.* vol. xviii. part 1. sect. v.
† Ibid., in the passage within brackets.
‡ Treatise on Light, art. 868.
§ Phil. Trans. 1814, p. 226, and plate 8. figs. 2, 3.
‖ Ibid.

three distinguished philosophers I have named, and of all subsequent writers; and the consequence of this has been that the
true action of a pile of plates or films has never been the subject of research during the last forty-six years, though such
piles have been used in some of the most delicate and important
researches in physical optics.

The difficulty of procuring transparent plates with parallel
surfaces, and of sufficient thinness, would have prevented the
most skilful observer from making any progress in the inquiry;
and had I not been fortunate enough to obtain, from the museum
of the Marquis Campana in Rome, a large quantity of glass in
different stages of decomposition, I could hardly have done more
than confirm the result which I obtained in 1813, that the light
transmitted by a pile of transparent plates consists of two portions of light polarized in opposite planes.

In submitting the films of decomposed glass to the polarizing
microscope, I observed a number of polygonal portions, approaching more or less to circles, but often perfectly circular, and exhibiting the black cross with coloured sectors and rings analogous to those produced by uniaxal crystals. This observation,
which was made with decomposed glass given me forty years
ago by the late Marquis of Northampton, was communicated to
the British Association at Glasgow in 1855; but at that time I
regarded the black cross and its accompanying tints, as shown
in the drawings on the table, as produced by the refraction and
polarization in different azimuths of the light transmitted
through the spherical shells, like a group of watch-glasses, of
which the circular portions were composed. The light surrounding the black cross was so highly coloured with the colours of
the thin plates which composed the film of glass, that I failed in
every attempt to analyse it. After examining, however, many
hundreds of these films from the new specimens which I have
mentioned, I succeeded in finding a few in which there were no
such colours, and which
enabled me to arrive at
results that could not
have been obtained from
the finest and the thin-
nest plates of glass artificially produced.

These results will be
understood from the an-
nexed diagram, in which
M N is a thin plate and
A B a ray of common
light incident perpen-

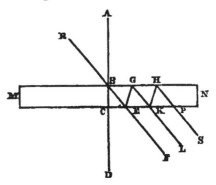

dicularly at B, and emerging at C in the direction C D. As a portion of the ray B C is reflected at C, and again reflected at B and transmitted at C, the pencil C D will consist of two distinct portions, one of which has been twice refracted, and the other and much feebler portion twice reflected. As neither of these portions are polarized, no physical change is produced by their combination, unless when the plate M N is so extremely thin as to produce the colours of thin plates by the interference of the reflected with the refracted portions.

When a ray R B is incident obliquely at B, it suffers refraction at B and E, and the emergent pencil E F contains a portion of light polarized by refraction. This ray, in passing through other plates or films parallel to M N, is at last completely polarized in one plane, having grown feebler in intensity by the abstraction of the light reflected at the two surfaces of each plate.

The portion of the refracted pencil B E which is reflected at E and G, and a portion of it polarized, emerges at K as a pencil K L, partly polarized by reflexion. A portion of G K is again reflected at K and H, and emerges at P as a pencil P S, more polarized by reflexion than K L. Hence the principal or refracted pencil E F is combined with the pencils K L, P S (and others by reflexions at P, &c.), polarized in an opposite plane, so that with a certain number of plates, varying with the angle of incidence, the emergent pencil E F, K L, and P S consists of two oppositely-polarized portions of light approximately equal.

When polarized light is incident upon a pile of these thin and colourless films, and subsequently analysed, it exhibits all the properties of a plate cut perpendicularly to the axis of a uniaxal crystal. The line A D corresponds with the axis of the crystal; and the different azimuths in which the polarized ray may be inclined to this axis correspond with the principal sections of a uniaxal crystal. The polarized tints have the same value in every azimuth at the same angle of incidence, and therefore form rings which, when crossed with plates of sulphate of lime, descend in Newton's Scale like the tints of negative uniaxal crystals.

Out of hundreds of specimens now on the table, I have found a few so colourless and so perfect as to produce, at different incidences, all the polarized tints or rings up to the *blue* of the second order of Newton's Scale. These colours are so pure, and so regularly developed by the inclination of the plate, that the most skilful observer could not fail to pronounce it to be a portion of a doubly refracting crystal.

The production of the leading phenomena of doubly refracting crystals, namely, two oppositely polarized pencils, and the system of coloured rings by the interference of these pencils, is certainly one of the most remarkable facts in physical optics; and, in a theoretical point of view, no less remarkable is the fact that one of the interfering portions is a fasciculus of pencils returned into the refracted beam by different routes, and having different origins.

Owing to the extreme thinness of the combined films, we cannot, as with thick plates of uniaxal crystals, see at once the black cross and its attendant rings; but in numerous specimens of decomposed glass to which I have already referred, the films are spherical shells of different diameters and thicknesses, and exhibit the black cross with the greatest sharpness and beauty. In many specimens these circular combinations are perfectly colourless, and the colours of the four luminous sectors which embrace the black cross rise only to the *white* of the first order.

When the films are so thin as to give the colours of thin plates, the colour of the luminous sectors is generally the same as that of the film in which the circular portions occur, and the rings or bands which surround them have a very peculiar character, owing to the manner in which the spherical shells are joined to the films which compose the plate.

How far these results may lead to new views of the structure which produces double refraction, it would be unprofitable to inquire in the present state of our knowledge of the atomical constitution of transparent bodies.

XXXVI. *On the Absorption and Radiation of Heat by Gases and Vapours, and on the Physical Connexion of Radiation, Absorption, and Conduction.—The Bakerian Lecture.* By JOHN TYNDALL, *Esq., F.R.S. &c.*

[Concluded from p. 194.]

§ 7. *ACTION of permanent Gases on Radiant Heat.*—The deportment of oxygen, nitrogen, hydrogen, atmospheric air, and olefiant gas has been already recorded. Besides these I have examined carbonic oxide, carbonic acid, sulphuretted hydrogen, and nitrous oxide. The action of these gases is so much feebler than that of any of the vapours referred to in the last section, that, in examining the relationship between absorption and density, the measures used with the vapours were abandoned, and the quantities of gas admitted were measured by the depression of the mercurial gauge.

TABLE XIX.—Carbonic Oxide.

Tension in inches.	Absorption.	
	Observed.	Calculated.
0·5	2·5	2·5
1·0	5·6	5·0
1·5	8·0	7·5
2·0	10·0	10·0
2·5	12·0	12·5
3·0	15·0	15·0
3·5	17·5	17·5

Up to a tension of 3½ inches the absorption by carbonic oxide is proportional to the density of the gas. But this proportion does not obtain with large quantities of the gas, as shown by the following Table :—

Tension in inches.	Deflection.	Absorption.
5	18·0°	18
10	32·5	32·5
15	41·0	45

TABLE XX.—Carbonic Acid.

Tension in inches.	Absorption.	
	Observed.	Calculated.
0·5	5·0	3·5
1·0	7·5	7·0
1·5	10·5	10·5
2·0	14·0	14·0
2·5	17·8	17·5
3·0	21·8	21·0
3·5	24·5	24·5

Here we have the proportion exhibited, but not so with larger quantities.

Tension in inches.	Deflection.	Absorption.
5	25·0°	25
10	36·0	36
15	42·5	48

TABLE XXI.—Sulphuretted Hydrogen.

Tension in inches.	Absorption.	
	Observed.	Calculated.
0·5	7·8	6
1·0	12·5	12
1·5	18·0	18
2·0	24·0	24
2·5	30·0	30
3·0	34·5	36
3·5	36·0	42
4·0	36·5	48
4·5	38·0	54
5·0	40·0	60

The proportion here holds good up to a tension of 2·5 inches, when the deviation from it commences and gradually augments.

Though these measurements were made with all possible care, I should like to repeat them. Dense fumes issued from the cylinders of the air-pump on exhausting the tube of this gas, and I am not at present able to state with confidence that a trace of such in a very diffuse form within the tube did not interfere with the purity of the results.

TABLE XXII.—Nitrous Oxide.

| Tension in inches. | Absorption. | |
	Observed.	Calculated.
0·5	14·5	14·5
1·0	23·5	29·0
1·5	30·0	43·5
2·0	35·5	58·0
2·5	41·0	71·5
3·0	45·0	87·0
3·5	47·7	101·5
4·0	49·0	116·0
4·5	51·5	130·5
5·0	54·0	145·0

Here the divergence from proportionality makes itself manifest from the commencement.

I promised at the first page of this memoir to allude to the results of Dr. Franz, and I will now do so. With a tube 3 feet long and blackened within, an absorption of 3·54 per cent. by atmospheric air was observed in his experiments. In my experiments, however, with a tube 4 feet long and polished within, which makes the distance traversed by the reflected rays more than 4 feet, the absorption is only one-tenth of the above amount. In the experiments of Dr. Franz, carbonic acid appears as a feebler absorber than oxygen. According to my experiments, for small quantities the absorptive power of the former is about 150 times that of the latter; and for atmospheric tensions, carbonic acid probably absorbs nearly 100 times as much as oxygen.

The differences between Dr. Franz and myself admit, perhaps, of the following explanation. His source of heat was an argand lamp, and the ends of his experimental tube were stopped with plates of glass. Now Melloni has shown that fully 61 per cent. of the heat-rays emanating from a Locatelli lamp are absorbed by a plate of glass one-tenth of an inch in thickness. Hence in all probability the greater portion of the rays issuing from the lamp of Dr. Franz was expended in heating the two glass ends of his experimental tube. These ends thus became secondary sources of heat which radiated against his pile. On admitting air into the tube, the partial withdrawal by conduction and convection of the heat of the glass plates would produce an effect exactly the same as that of true absorption. By allowing the air in my tube to come into contact with the radiating plate, I have often obtained a deflection of twenty or thirty degrees,—the effect being due to the cooling of the plate, and not to absorption. It is also certain that had I used heat from a luminous source, I should have found the absorption of 0·33 per cent. considerably diminished.

§ 8. I have now to refer briefly to a point of considerable interest as regards the effect of our atmosphere on solar and terrestrial heat. In examining the separate effects of the air, carbonic acid, and aqueous vapour of the atmosphere, on the 20th of last November, the following results were obtained:—

Air sent through the system of drying-tubes and through the caustic-potash tube produced an absorption of about   .   .   1.

Air direct from the laboratory, containing therefore its carbonic acid* and aqueous vapour, produced an absorption of . 15.

Deducting the effect of the gaseous acids, it was found that the quantity of aqueous vapour diffused through the atmosphere on the day in question, produced an absorption at least equal to thirteen times that of the atmosphere itself.

It is my intention to repeat and extend these experiments on a future occasion†; but even at present conclusions of great importance may be drawn from them. It is exceedingly probable that the absorption of the solar rays by the atmosphere, as established by M. Pouillet, is mainly due to the watery vapour contained in the air. The vast difference between the temperature of the sun at midday and in the evening, is also probably due in the main to that comparatively shallow stratum of aqueous vapour which lies close to the earth. At noon the depth of it pierced by the sunbeams is very small; in the evening very great in comparison.

The intense heat of the sun's direct rays on high mountains is not, I believe, due to his beams having to penetrate only a small depth of air, but to the comparative absence of aqueous vapour at those great elevations.

But this aqueous vapour, which exercises such a destructive action on the obscure rays, is comparatively transparent to the rays of light. Hence the differential action, as regards the heat coming from the sun to the earth and that radiated from the earth into space, is vastly augmented by the aqueous vapour of the atmosphere.

* And a portion of sulphurous acid produced by the two gas-lamps used to heat the cubes.

† The peculiarities of the locality in which this experiment was made render its repetition under other circumstances necessary.

De Saussure, Fourier, M. Pouillet, and Mr. Hopkins regard this interception of the terrestrial rays as exercising the most important influence on climate. Now if, as the above experiments indicate, the chief influence be exercised by the aqueous vapour, every variation of this constituent must produce a change of climate. Similar remarks would apply to the carbonic acid diffused through the air, while an almost inappreciable admixture of any of the hydrocarbon vapours would produce great effects on the terrestrial rays and produce corresponding changes of climate. It is not, therefore, necessary to assume alterations in the density and height of the atmosphere to account for different amounts of heat being preserved to the earth at different times; a slight change in its variable constituents would suffice for this. Such changes in fact may have produced all the mutations of climate which the researches of geologists reveal. However this may be, the facts above cited remain; they constitute true causes, the *extent* alone of the operation remaining doubtful.

The measurements recorded in the foregoing pages constitute only a fraction of those actually made; but they fulfil the object of the present portion of the inquiry. They establish the existence of enormous differences among colourless gases and vapours as to their action upon radiant heat; and they also show that, when the quantities are sufficiently small, the absorption in the case of each particular vapour is exactly proportional to the density.

These experiments furnish us with purer cases of molecular action than have been hitherto attained in experiments of this nature. In both solids and liquids the cohesion of the particles is implicated; they mutually control and limit each other. A certain action, over and above that which belongs to them separately, comes into play and embarrasses our conceptions. But in the cases above recorded the molecules are perfectly free, and we fix upon them individually the effects which the experiments exhibit; thus the mind's eye is directed more firmly than ever on those distinctive physical qualities whereby a ray of heat is stopped by one molecule and unimpeded by another.

§ 9. *Radiation of Heat by Gases.*—It is known that the quantity of light emitted by a flame depends chiefly on the incandescence of solid matter,—the brightness of an ignited jet of ordinary gas, for example, being chiefly due to the solid particles of carbon liberated in the flame.

Melloni drew a parallel between this action and that of radiant heat. He found the radiation from his alcohol lamp greatly augmented by lunging a spiral of platinum wire into the flame. He also found that a bundle of wire placed in the current of hot air ascending from an argand chimney gave a copious radiation, while when the wire was withdrawn no trace of radiant heat could be detected by his apparatus. He concluded from this experiment that air possesses the power of radiation in so feeble a degree, that our best thermoscopic instruments fail to detect this power*.

These are the only experiments hitherto published upon this subject; and I have now to record those which have been made in connexion with the present inquiry. The pile furnished with its conical reflector was placed upon a stand, with a screen of polished tin in front of it. An alcohol lamp was placed behind the screen so that its flame was entirely hidden by the latter; on rising above the screen, the gaseous column radiated its heat against the pile and produced a considerable deflection. The same effect was produced when a candle or an ordinary jet of gas was substituted for the alcohol lamp.

The heated products of combustion acted on the pile in the above experiments, but the radiation from pure air was easily demonstrated by placing a heated iron spatula or metal sphere behind the screen. A deflection was thus obtained which, when the spatula was raised to a red heat, amounted to more than sixty degrees. This action was due solely to the radiation of the air; no radiation from the spatula to the pile was possible, and no portion of the heated air itself approached the pile so as to communicate its warmth by contact to the latter. These effects are so easily produced that I am at a loss to account for the inability of so excellent an experimenter as Melloni to obtain them.

My next care was to examine whether different gases possessed different powers of radiation; and for this purpose the following arrangement was devised. P (fig. 1) represents the thermo-electric pile with its two conical reflectors; S is a double screen of polished tin; A is an argand burner consisting of two concentric rings perforated with orifices for the escape of the gas; C is a heated copper ball; the tube $t\,t$ leads to a gas-holder containing the gas to be examined. When the ball C is placed on the argand burner, it of course heats the air in contact with it; an ascending current is established, which acts on the pile as in the experiments last described. It was found necessary to neutralize this radiation from the heated air, and for this purpose a large Leslie's cube L, filled with water a few degrees above the temperature of the air, was allowed to act on the opposite face of the pile.

When the needle was thus brought to zero, the cock of the gas-holder was turned on; the gas passed through the burner, came into contact with the ball, and ascended afterwards in a heated column in front of the pile. The galvanometer was now

* La Thermochrose, p. 94.

Fig. 1.

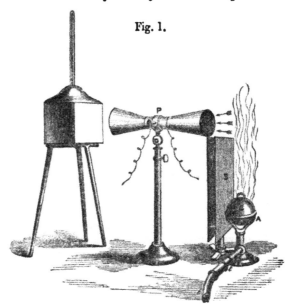

observed, and the limit of the arc through which its needle was urged was noted. It is needless to remark that the ball was entirely hidden by the screen from the thermo-electric pile, and that, even were this not the case, the mode of neutralization adopted would still give us the pure action of the gas.

The results of the experiments are given in the following Table, the figure appended to the name of each gas marking the number of degrees through which the radiation from the latter urged the needle of the galvanometer*:—

Air	$\overset{\circ}{0}$
Oxygen	0
Nitrogen	0
Hydrogen	0
Carbonic oxide . .	12
Carbonic acid . . .	18
Nitrous oxide . . .	29
Olefiant gas . . .	53

The radiation from air, it will be remembered, was neutralized by the   merely denotes that the propulsion of air from the gas-holder through the argand burner did not augment the effect. Oxygen, hydro-

* I have also rendered these experiments on radiation visible to a large audience. They may be readily introduced in lectures on radiant heat.

gen, and nitrogen, sent in a similar manner over the ball, were equally ineffective. The other gases, however, not only exhibit a marked action, but also marked differences of action. Their radiative powers follow precisely the same order as their powers of absorption. In fact, the deflections actually produced by their respective absorptions at 5 inches tension are as follow :—

Air . . . . . A fraction of a degree	
Oxygen . . . „ „	
Nitrogen . . „ „	
Hydrogen . . „ „	
Carbonic oxide . . . 18°	
Carbonic acid . . . . 25°	
Nitrous oxide . . . . 44°	
Olefiant gas . . . . 61°	

It would be easy to give these experiments a more elegant form, and to arrive at greater accuracy, which I intend to do on a future occasion; but my object now is simply to establish the general order of their radiative powers. An interesting way of exhibiting both radiation and absorption is as follows :—When the polished face of a Leslie's cube is turned towards a thermo-electric pile the effect produced is inconsiderable, but it is greatly augmented when a coat of varnish is laid upon the polished surface. Instead of the coat of varnish, a film of gas may be made use of. Such a cube, containing boiling water, had its polished face turned towards the pile, and its effect on the galvanometer neutralized in the usual manner. The needle being at 0°, a film of olefiant gas, issuing from a narrow slit, was passed over the metal. The increase of radiation produced a deflection of 45°. When the gas was cut off, the needle returned accurately to 0°.

The absorption by a film may be shown by filling the cube with cold water, but not so cold as to produce the precipitation of the aqueous vapour of the atmosphere. A gilt copper ball, cooled in a freezing mixture, was placed in front of the pile, and its effect was neutralized by presenting a beaker containing a little iced water to the opposite face of the pile. A film of olefiant gas was sent over the ball, but the consequent deflection proved that the absorption, instead of being greater, was less than before. The ball, in fact, had been coated by a crust of ice, which is one of the best absorbers of radiant heat. The olefiant gas, being warmer than the ice, partially neutralized its absorption. When, however, the temperature of the ball was only a few degrees lower than that of the atmosphere, and its surface quite dry, the film of gas was found to act as a film of varnish; it augmented the absorption.

A remarkable effect, which contributed at first to the complexity of the experiments, can now be explained. Conceive the experimental tube exhausted and the needle at zero; conceive a small quantity of alcohol or ether vapour admitted; it cuts off a portion of the heat from one source, and the opposite source triumphs. Let the consequent deflection be 45°. If dry air be now admitted till the tube is filled, its effect of course will be slightly to augment the absorption and make the above deflection greater. But the following action is really observed :—when the air first enters, the needle, instead of ascending, descends; it falls to 26°, as if a portion of the heat originally cut off had been restored. At 26°, however, the needle stops, turns, moves quickly upwards, and takes up a permanent position a little higher than 45°. Let the tube now be exhausted, the withdrawal of the mixed air and vapour ought of course to restore the equilibrium with which we started; but the following effects are observed :—When the exhaustion commences, the needle moves upwards from 45° to 54°; it then halts, turns, and descends speedily to 0°, where it permanently remains.

After many trials to account for the anomaly, I proceeded thus:—A thermo-electric couple was soldered to the external surface of the experimental tube, and its ends connected with a galvanometer. When air was admitted, a deflection was produced, which showed that the air, on entering the vacuum, was heated. On exhausting, the needle was also deflected, showing that the interior of the tube was chilled. These are indeed known effects; but I was desirous to make myself perfectly sure of them. I subsequently had the tube perforated and thermometers screwed into it air-tight. On filling the tube the thermometric columns rose, on exhausting it they sank, the range between the maximum and minimum amounting in the case of air to 5° Fahr.

Hence the following explanation of the above singular effects. The absorptive power of the vapour referred to is very great, and its radiative power is equally so. The heat generated by the air on its entrance is communicated to the vapour, which thus becomes a temporary source of radiant heat, and diminishes the deflection produced in the first instance by its presence. The reverse occurs when the tube is exhausted; the vapour is chilled, its great absorptive action on the heat radiated from the adjacent face of the pile comes more into play, and the original effect is augmented. In both cases, however, the action is transient; the vapour soon loses the heat communicated to it, and soon gains the heat which it has lost, and matters then take their normal course.

§ 10. *On the Physical Connexion of Radiation, Absorption, and*

Conduction.—Notwithstanding the great accessions of late years to our knowledge of the nature of heat, we are as yet, I believe, quite ignorant of the atomic conditions on which radiation, absorption, and conduction depend. What are the specific qualities which cause one body to radiate copiously and another feebly? Why, on theoretic grounds, must the equivalence of radiation and absorption exist? Why should a highly diathermanous body, as shown by Mr. Balfour Stewart, be a bad radiator, and an adiathermanous body a good radiator? How is heat conducted? and what is the strict physical meaning of good conduction and bad conduction? Why should good conductors be, in general, bad radiators, and bad conductors good radiators? These, and other questions, referring to facts more or less established, have still to receive their complete answers. It is less with a hope of furnishing such than of shadowing forth the possibility of uniting these various effects by a common bond, that I submit the following reflections to the notice of the Royal Society.

In the experiments recorded in the foregoing pages, we have dealt with *free* atoms, both simple and compound, and it has been found that in all cases absorption takes place. The meaning of this, according to the dynamical theory of heat, is that no atom is capable of existing in vibrating ether without accepting a portion of its motion. We may, if we wish, imagine a certain roughness of the surface of the atoms which enables the ether to *bite* them and carry the atom along with it. But no matter what the quality may be which enables any atom to accept motion from the agitated ether, the same quality must enable it to impart motion to still ether when it is plunged in the latter and agitated. It is only necessary to imagine the case of a body immersed in water to see that this must be the case. There is a polarity here as rigid as that of magnetism. From the existence of absorption, we may on theoretic grounds infallibly infer a capacity for radiation; from the existence of radiation, we may with equal certainty infer a capacity for absorption; and each of them must be regarded as the measure of the other*.

This reasoning, founded simply on the mechanical relations of the ether and the atoms immersed in it, is completely verified by experiment. Great differences have been shown to exist among gases as to their powers of absorption, and precisely similar differences as regards their powers of radiation. But what specific property is it which makes one free molecule a strong absorber, while another offers scarcely any impediment to the passage of radiant heat? I think the experiments throw

* This was written long before Kirchhoff's admirable papers on the relation of emission to absorption were known to me.

some light upon this question. If we inspect the results above recorded, we shall find that the *elementary* gases hydrogen, oxygen, nitrogen, and the *mixture* atmospheric air, possess absorptive and radiative powers beyond comparison less than those of the *compound* gases. Uniting the atomic theory with the conception of an ether, this result appears to be exactly what ought to be expected. Taking Dalton's idea of an elementary body as a single sphere, and supposing such a sphere to be set in motion in still ether, or placed without motion in moving ether, the communication of motion by the atom in the first instance, and the acceptance of it in the second, must be less than when a number of such atoms are grouped together and move as a system. Thus we see that hydrogen and nitrogen, which, when *mixed* together, produce a small effect, when *chemically united* to form ammonia, produce an enormous effect. Thus oxygen and hydrogen, which, when mixed in their electrolytic proportions, show a scarcely sensible action, when chemically combined to form aqueous vapour exert a powerful action. So also with oxygen and nitrogen, which, when mixed, as in our atmosphere, both absorb and radiate feebly, when united to form oscillating systems, as in nitrous oxide, have their powers vastly augmented. Pure atmospheric air, of 5 inches tension, does not effect an absorption equivalent to more than the one-fifth of a degree, while nitrous oxide of the same tension effects an absorption equivalent to fifty-one such degrees. Hence the absorption by nitrous oxide at this tension is about 250 times that of air. No fact in chemistry carries the same conviction to my mind, that air is a *mixture* and not a *compound*, as that just cited. In like manner, the absorption by carbonic oxide of this tension is nearly 100 times that of oxygen alone; the absorption by carbonic acid is about 150 times that of oxygen; while the absorption by olefiant gas of this tension is 1000 times that of its constituent hydrogen. Even the enormous action last mentioned is surpassed by the vapours of many of the volatile liquids, in which the atomic groups are known to attain their highest degree of complexity.

I have hitherto limited myself to the consideration, that the compound molecules present broad sides to the ether, while the simple atoms with which we have operated do not,—that in consequence of these differences the ether must swell into billows when the former are moved, while it merely trembles into ripples when the latter are agitated,—that, in the interception of motion also, the former, other things being equal, must be far more influential than the latter; but another important consideration remains. All the gases and vapours whose deportment we have examined are transparent to light; that is to say,

284 Prof. Tyndall *on the Absorption and*

the waves of the visible spectrum pass among them without sensible absorption. Hence it is plain that their absorptive power depends on the periodicity of the undulations which strike them. At this point the present inquiry connects itself with the experiments of Nièpce, the observation of Foucault, the surmises of Ångstrom, Stokes, and Thomson, and those splendid researches of Kirchhoff and Bunsen which so immeasurably extend our experimental range. By Kirchhoff it has been conclusively shown that every atom absorbs in a special degree those waves which are synchronous with its own periods of vibration. Now, besides presenting broader sides to the ether, the association of simple atoms to form groups must, as a general rule, render their motions through the ether more sluggish, and tend to bring the periods of oscillation into isochronism with the slow undulations of obscure heat, thus enabling the molecules to absorb more effectually such rays as have been made use of in our experiments.

Let me here state briefly the grounds which induce me to conclude that an agreement in period alone is not sufficient to cause powerful absorption and radiation—that in addition to this the molecules must be so constituted as to furnish *points d'appui* to the ether. The heat of contact is accepted with extreme freedom by rock-salt, but a plate of the substance once heated requires a great length of time to cool. This surprised me when I first noticed it. But the effect is explained by the experiments of Mr. Balfour Stewart, by which it is proved that the radiative power of heated rock-salt is extremely feeble. Periodicity can have no influence here, for the ether is capable of accepting and transmitting impulses of all periods; and the fact that rock-salt requires more time to cool than alum, simply proves that the molecules of the former glide through the ether with comparatively small resistance, and thus continue moving for a longer time; while those of the latter presenting broad sides to the ether, speedily communicate to it the motion which we call heat. This power of gliding through still ether possessed by the rock-salt molecules, must of course enable the moving ether to glide round them, and no coincidence of period could, I think, make such a body a powerful absorber.

Many chemists, I believe, are disposed to reject the idea of an atom, and to adhere to that of equivalent proportions merely. They figure the act of combination as a kind of interpenetration of one substance by another. But this is a mere masking of the fundamental phenomenon. The value of the atomic theory consists in its furnishing the physical explanation of the law of equivalents: assuming the one, the other follows; and assuming the act of chemical union as Dalton figured it, we see that it blends harmoniously with the perfectly independent conception of an ether, and enables us to reduce the phenomena of radiation and absorption to the simplest mechanical principles.

Considerations similar to the above may, I think, be applied to the phenomena of *conduction.* In the Philosophical Magazine for August 1853, I have described an instrument used in examining the transmission of heat through cubes of wood and other substances. When engaged with this instrument, I had also cubes of various crystals prepared, and determined with it their powers of conduction. With one exception, I found that the conductivity augmented with the diathermancy. The exception was furnished by a cube of very perfect rock-crystal, which conducted slightly better than my cube of rock-salt. The latter, however, had a very high conductive power; in fact rock-salt, calcareous spar, glass, selenite, and alum stood in my experiments, as regards conductivity, exactly in their order of diathermancy in the experiments of Melloni. I have already adduced considerations which show that the molecules of rock-salt glide with facility through the ether; but the ease of motion which these molecules enjoy must facilitate their mutual collision. Their motion, instead of being expended on the ether which exists between them, and communicated by it to the external ether, is in great part transferred directly from particle to particle, or in other words, is freely conducted. When a molecule of alum, on the contrary, approaches a neighbour molecule, it produces a swell in the intervening ether, which swell is in part transmitted, not to the molecules, but to the general ether of space, and thus lost as regards conduction. This lateral waste prevents the motion from penetrating the alum to any great extent, and the substance is what we call a bad conductor*.

Such considerations as these could hardly occur without carrying the mind to the kindred question of electric conduction; but the speculations have been pursued sufficiently far for the present, and must now abide the judgment of those competent to decide whether they are the mere emanations of fancy, or a fair application of principles which are acknowledged to be secure.

The present paper, I may remark, embraces only the first section of these researches.

* In the above considerations regarding conduction, I have limited myself to the illustration furnished by two compound bodies; but the elementary atoms also differ among themselves as regards their powers of accepting motion from the ether and of communicating motion to it. I should infer, for example, that the atoms of platinum encounter more resistance in moving through the ether than the atoms of silver. It is needless to say that the physical texture of a substance also has a great influence.

XXXVII. *Experimental and Theoretical Researches on the Figures of Equilibrium of a Liquid Mass devoid of Weight.*—Fifth Series *. *By* M. J. PLATEAU †.

New process for the production of figures in a state of equilibrium.—Pressure exerted by a liquid spherical film on the air which it contains.—Investigation of the very small limit within which, in a particular liquid, the value of radius of appreciable molecular attraction varies.

IN the Second and Fourth Series of this investigation I have applied my process of the immersion of a mass of oil in a mixture of water and alcohol to the production of some of the figures in a state of equilibrium which pertain to a liquid mass, supposed to be devoid of gravity and in a state of repose. This process, so simple in principle, presents in practice certain difficulties, and it required a certain cleverness to arrive at perfectly regular results. In the present series, I shall point out a process wholly different, far more simple and more convenient, and entirely exempt from the inconveniences of the previously described plan; I shall demonstrate afterwards some of the numerous results which the employment of the new method has furnished me, and the theoretical principles on which it rests.

I may remark in the first place, that oil immersed in the alcoholic mixture is easily converted into thin films; I shall show, for example, that, with a number of precautions which I describe, one can obtain, in the mixture in question, a hollow bubble of oil more than 12 centimetres in diameter, by inflating it with the same alcoholic mixture, just as one obtains in air a soap-bubble filled with air itself.

It must be remembered, with regard to these films of oil, that in the experiment in my First Series where a ring of oil is formed, this ring remains at first united to the central apparatus by a thin film; and starting with that fact, I shall show once more the incorrectness of every deduction, derived from this experiment, in favour of a cosmogonic hypothesis.

After having thus established the facilities for the production of liquid films removed from the action of gravity, I shall demonstrate that the figures in a state of equilibrium which appertain to the liquid films devoid of weight, are identically the same as those of full liquid masses, likewise deprived of weight.

---

* For the preceding Series see Taylor's Scientific Memoirs, Parts XIII. and XXI; and Phil. Mag. (S. 4), vol. xiv. p. 1, and vol. xvi. p. 23.

† The original memoir will be found in the thirty-third volume of the *Mémoires de l'Académie de Bruxelles.* The abstract, of which a translation is here given, appeared in the *Annales de Chimie et de Physique* for June 1861.

Besides, it is possible, without having recourse to mathematical analysis, sufficiently to account for this identity. Let me repeat, for this purpose, a principle on which I have many times dwelt in the preceding series. When a surface fulfils the general condition of equilibrium, it is indifferent whether the liquid be on one side or on the other of this surface; in other words, to each figure in a state of equilibrium, which is in relief, corresponds a figure in a state of equilibrium, identical with the same, only in depression. Now, the two faces of a liquid film, on account of the thinness of the latter, being capable of being considered as though they were two identical surfaces, the one in relief and the other depressed with regard to the liquid which forms the film, it follows, from the principle in question, that if one of these two faces constitutes a surface of equilibrium, it is the same with the other face, and that thus equilibrium exists throughout the entire film.

Suppose, however, it was possible to form in air liquid films devoid of weight; these films would necessarily take the same form as the films of oil formed in the alcoholic mixture. Now liquid films formed in air (films of soapy water, for example) are so thin that the action of gravity upon them can generally be regarded as inappreciable in comparison with that of molecular forces; we should therefore obtain in air, with films of soapy water or of an analogous liquid, the same figures in a state of equilibrium as with films of oil in the alcoholic mixture, and consequently, after what I have said above, figures which would belong to a full liquid mass devoid of weight. Therein consists the process which I have mentioned.

Thus we arrive at the curious result, that, with a liquid acted upon by gravity and in a state of repose, one can produce on a large scale all forms of equilibrium which belong to a liquid mass without weight and likewise at rest.

Soap-bubbles offer the first example for the employment of the process under consideration; floating in air, they are spherical, just as a full liquid mass would be if devoid of weight and freed from all adhesion.

The films, however, which are obtained from common solution of soap have but a very short existence, unless they be in a close vessel; a soap-bubble of one decimetre diameter, formed in the open air of a room, rarely lasts two minutes; it was therefore essential to find out some better liquid; and I have been happy enough to discover one which furnishes in the open air, whilst preserving its liquid nature, films of great durability. This liquid is formed by mixing, in proper proportions, glycerine, water, and soap. A glycerine which seems very pure and very concentrated can be easily procured at no great cost in London, at Mr. Bolton's, 146 Holborn Bars, for instance. I shall point out in a note at the end of the memoir, the proper way to obtain sufficiently good results with the ordinary glycerine of commerce.

. The mixture must be prepared in summer, and when the temperature out-of-doors is at least 19° Cent. Dissolve at a gentle heat one part by weight of Marseilles soap, previously cut into thin shavings, in 40 parts of distilled water; and when the solution is cold, filter it. That done, carefully mix in a flask by violent and continual agitation, 2 vols. of glycerine with 3 vols. of the above-mentioned solution, and then allow it to stand. The mixture, limpid at the time of its formation, begins after some hours to grow turbid; a slight white precipitate is produced, which rises with great slowness, and after some days forms a distinct layer at the top of the liquid; the limpid portion is then collected by means of a siphon, which draws off by a lateral tube, and the preparation is at an end.

The liquid thus obtained, and which I name *glyceric liquid*, gives films of great durability; for instance, if with this liquid, by means of a common clay pipe, a bubble 1 decimetre in diameter be inflated, and then placed in the open air of a room upon a ring of iron wire 4 centimetres in diameter and previously moistened with the same liquid, this bubble, provided it is perfectly at rest, will remain entire for three hours.

The glyceric liquid can be preserved about a year, after which time it rapidly decomposes. I have not observed any disengagement of gas; however, as the liquid is of an organic nature, it would not be unlikely that such might be produced sometimes; and it would be prudent, to prevent a possible explosion, to only close the flask with a cork which does not fit very firmly.

Just as the films of soapy water last very much longer in a closed vessel than in the open air, the endurance of the films of glyceric liquid, already so great in the open air, becomes still much more considerable when these films are enclosed in a vessel, especially if certain precautions are taken. I shall quote an example of this further on.

Having thus obtained a liquid easily furnishing films both large and very durable, I employ it in order to produce by means of them all the figures, in a state of equilibrium, of revolution. In order not to give this analysis too great length, I shall limit myself here to succinctly describing the formation of the cylinder.

For this purpose, use is made of an apparatus of two rings of iron wire 7 centimetres in diameter, similar to those mentioned by me in the preceding series, that is to say, the lower ring on a tripod, and the upper ring, supported by a fork, fixed into the two extremities of its diameter; the end of this fork is attached to a support, fixed in such a manner that the ring can be raised or lowered by a gentle movement. Place the first ring upon its stand on the table, sustain the second at a convenient height over it, and well moisten both with the glyceric liquid; then inflate a bubble of about 10 centimetres diameter, place it upon the lower ring and withdraw the pipe; now lower the upper ring until it comes in contact with the bubble, which immediately attaches itself to it; at last gradually raise this ring, and the bubble, which thus drawn out loses more and more its spherical curvature, is converted, by a certain separation of the rings, into a perfectly regular cylinder, having convex bases like the full cylinders of oil.

A rather larger diameter can be given to the bubble; but when it is too large, the cylindrical form is no longer obtained, either because the cylinder which it is desired to obtain exceeds its limit of stability*, or because, if it be still within this limit, it begins to approach it; in this last case, in fact, the figure-producing forces becoming very little intense, the small weight of the film exerts an appreciable influence, and the figure appears more or less swollen at the lower half, and compressed at the upper half. The tallest regular cylinder which can be formed with the rings, before pointed out, has a height of about 17 centimetres. Let us state in this place, that, for the complete success of experiments of this kind, the rings should have undergone a little preparation: when they leave the hands of the workman they should be slightly oxidized on their surface by dipping them for two minutes into nitric acid diluted with four times its volume of water; afterwards wash them in pure water.

In the memoir will be found the way to produce, in the laminated condition as well, the other forms of equilibrium of revolution, namely, those to which I have given† the names of *catenoids, onduloids,* and *nodoids.*

These experiments are very curious; there is a peculiar charm in the contemplation of these figures, so slender, almost reduced to mathematical surfaces, which make their appearance tinted with the most brilliant colours, and which, in spite of their extreme frailness, endure for such a time. These same experiments can be readily performed, and in the most convenient manner.

I now pass on to another application of my new process. Procure a collection of frames of iron wire, each one of which exhibits all the edges of a polyhedron—for example, of a cube, a regular octahedron, of prisms with triangular, pentagonal, and other bases. Each of these frames is to be fixed like the upper ring in the before-mentioned experiment, by a fork attached to two of its edges; they ought also to be oxidized by nitric acid.

* See Second Series in Scientific Memoirs, Part XXI.
† See the abstract of the Fourth Series in Phil. Mag. vol. xvi. p. 23.

In order to give an idea of the most convenient dimensions for this apparatus, I will just say that the edges of my cúbical frame are 7 centimetres in length, and that the iron wire, of which it is formed, is a little less than 1 millimetre in thickness. I have already employed similar frames in the experiments (mentioned in my second memoir) for the formation of liquid polyhedra.

If one of these frames were completely dipped (with the exception of the upper part of the fork) into the glyceric liquid and then withdrawn, it would be expected that the adhesion of this liquid to the solid frame would cause the formation of a set of films, occupying the interior of the frame; and this *does* take place; but a most remarkable thing it is, that the arrangement of these films is not a matter of chance; it is, on the contrary, perfectly regular and perfectly constant for each frame. In the cubical frame, for instance, is invariably obtained a collection of twelve films, starting respectively from the twelve wires, and all converging on a much smaller thirteenth one of quadrangular form and occupying the centre of the apparatus.

These systems of films, thus prepared in these polyhedral frames, have excited the admiration of all to whom I have shown them; they have a perfect regularity; the liquid edges that join among them the films of which they are composed are extremely fine, and the films themselves after some time exhibit the richest colours; again, the arrangement of these same films is regulated by simple and uniform laws, which I shall examine from a theoretical point of view in the next series, and of which here are the two principles:—

I. At one and the same liquid edge never more than three films can meet, and these same are inclined to each other at equal angles.

II. When several liquid edges meet at one and the same point in the interior of a system, these edges are always four in number, and are inclined to each other, at the point in question, at equal angles.

I had already obtained, by totally different means, these systems of films with oil immersed in the alcoholic liquid, as will be seen in my second series; but they are far less perfect and far less easily produced than by my present process.

· We now pass on to another subject. It is well known that a soap-bubble exerts a pressure on the air which it contains. Mr. Henry, in an oral communication made in 1844 to the American Society, has described experiments by means of which he measured this pressure by the height of the column of water with which it is in equilibrium; but I believe that his numbers have not been published. I have looked at the question in a general way from a theoretical point of view, and have arrived at the following result:—Let ρ stand for the density of

the liquid of which the film is formed, h the height to which
the same liquid rises in a capillary tube of 1 millimetre internal
diameter, d the diameter of the bubble, and lastly, let p denote
the pressure which this bubble exerts, or, more precisely speaking, the height of the column with which it would be in equilibrium ; then this pressure is expressed by the formula

$$p = \frac{2h\rho}{d}.$$

The product $h\rho$ is, as can easily be shown, proportional to the
cohesion of the liquid; the pressure exerted by a bubble upon
its enclosed air is consequently in direct ratio to the cohesion of
the liquid, and in inverse ratio to the diameter of the bubble.

I verify my formula by the experiment with the glyceric liquid.
By means of my apparatus, which is merely Mr. Henry's slightly
modified, a bubble is inflated at the orifice of a small inverted
funnel which communicates with a water-manometer. The
difference of the level in the two branches of this instrument is
measured by means of a cathetometer ; and the latter is likewise
employed to measure the diameter of the bubble, for which purpose it is placed in a horizontal position on suitable supports.

The formula gives $hd = 2h\rho,$

which shows that the product of the pressure by the diameter
must be constant for the same liquid and at the same temperature, since under these conditions h and ρ do not vary : it is this
constant which I first of all sought to verify. These measurements have been made by means of ten bubbles, of which the
smallest had a diameter of 7·55 millimetres, and the longest a
diameter of 48·1 millimetres, and consequently within limits
which were to one another nearly as 1 to 6 ; the temperature
ranged between 18°·5 and 20°.

The mean of the ten values obtained for the product pd is
22·75. Except in the case of the two largest diameters, there
was very little difference from the general mean ; and if the
results are arranged in order, with the diameters increasing,
it will be perceived that these small discrepancies are irregularly
distributed. The two values which form exceptions are 20·57
and 26·45, and it is seen that the first is under the mean, whilst
the second is over it. As the other eight values presented a
remarkable agreement, I have deemed it allowable to reject, as
spoiled by errors of accident, the two that I have mentioned,
and I thought I might take, in order to estimate the product
pd as regards the glyceric liquid, the mean of the eight agreeing
determinations, which mean is 22·56.

It remains to compare the value of the product pd, thus deduced by experiment, with that which our formula gives ; and

for this purpose it was necessary to determine, at the temperature of the preceding experiment, the density ρ and the height h
in regard to the glyceric liquid. This I have done, employing
every known precau ion, and I have found $\rho = 1\cdot 1065$ and
$h = 10\cdot 018$ millimetrès. One has consequently $2\, h\rho = 22\cdot 17$, a
number that differs but little from $22\cdot 56$, which experiment has
furnished me; and the agreement appears still more satisfactory
when it is remembered that these two numbers are respectively
deduced from elements totally different. The formula

$$p = \frac{2\,h\rho}{d}$$

may therefore be regarded as clearly verified by experiment.

. The accuracy of this formula requires, however, that the film
which constitutes the bubble should not have in any of its points
a thickness less than twice the radius of appreciable molecular
attraction. In fact, the pressure exerted upon the enclosed air
is the sum of the two actions due to the curvatures of the faces
of the film; and, on the other hand, it is known that in the case
of a full liquid mass the capillary pressure of the liquid upon
itself emanates from all the points of a superficial stratum having
for thickness the radius of activity in question. If, then, in all its
points the film has a thickness less than twice this same radius,
the superficial layers of its two faces have no longer their complete thickness, and, the number of molecules contained in one
of these layers being thus diminished, these same layers must
necessarily exert a weaker action; hence the sum of the latter,
that is to say the pressures on the enclosed air, must be less than
is indicated by the formula.

I shall thence deduce a convenient method which furnishes an
approximate value for the radius of activity now under consideration, or at least within a limit extremely little below which this
radius is found. If, having inflated a small bubble in the orifice
of the funnel of my apparatus, it is enclosed in a small glass
globe, it exhibits a remarkable phenomenon; for, after some time,
by placing the eye on a level with its centre, one sees a large
space, perceptibly circular, coloured with a uniform tint, and surrounded by narrow concentric rings of other colours. One would
infer from this that the point has been reached at which the film
has thickness appreciably uniform throughout the whole extent
of the bubble, except of course the lowest part, where there is
always a small accumulation of liquid: the colours of the rings
which surround the central part evidently arise from the oblique-
ness of the rays from them to the eye. This fact respecting
thickness has already been noticed by Newton, but only as occurring by chance, in the hemispherical bubbles of soapy water.
From the moment the bubble assumes this appearance it main-

tains it till it bursts; the respective tints of the central space and of the rings, however, vary progressively, changing in the order of the colours of Newton's rings, whence it follows that the film becomes thinner and thinner but equally all over, always excepting the very lowest portion of the bubble.

Now, after the film has acquired a uniform degree of thinness, if the pressure exerted on the enclosed air experienced a diminution, it would be rendered apparent by the manometer, and it would be seen to proceed in a regular manner and in proportion to the further weakening of the film. In this case the thickness of the film, when the diminution of the pressure commenced, could be determined by means of the colour which the central space at that moment presented, and half of this thickness would be the value of the radius of appreciable molecular attraction. If, on the contrary, the pressure continued constant until the bursting of the bubble, one would infer, from the colour of the central space, the final thickness of the film, and the half of this thickness would at least constitute the limit but a very little below that in which is found the radius in question.

I have tried the application of this method. By means of a number of precautions, which I have pointed out in the memoir, a bubble, 2 centimetres in diameter, inflated in the orifice of a small funnel and enclosed in a glass globe, existed for nearly three days, and at the time it burst it had reached the state of transition from yellow to white of the first order. The levels of water in the manometer had made little oscillations during this period, sometimes in one direction, sometimes in another; still the last was indicative of an increase of pressure. For reasons mentioned in the memoir, these oscillations could not be attributed—at least entirely—to variations of temperature, and I have thought it admissible that the continual diminution of the thickness of the film had not brought about any decrease of pressure; consequently the final thickness was most likely more than twice the radius of molecular attraction.

Calculating the final thickness of the film by means of Newton's numbers and the index of refraction of glyceric liquid, an index whose value, previously determined, was 1·377, I have found the thickness in question to be $\frac{1}{8811}$ of a millimetre. Half of this quantity, or $\frac{1}{17622}$ of a millimetre consequently constitutes the limit furnished by my experiment; but, to be on the safe side, I prefer $\frac{1}{17000}$.

I have thus arrived at a very probable conclusion, that in the glyceric liquid the radius of appreciable molecular attraction is less than $\frac{1}{17000}$th of a millimetre.

I propose to continue this research in order to investigate the black colour, and to throw light on the question of the variations of the manometer.

XXXVIII. *On the Amount of the direct Magnetic Effect of the Sun or Moon on Instruments at the Earth's Surface. By* G. JOHNSTONE STONEY, *M.A., F.R.S., Secretary to the Queen's University in Ireland**.

IN the Philosophical Magazine for March 1858, Dr. Lloyd showed that the observed disturbances of the magnetic needle, depending on the hours of solar or lunar time, follow laws inconsistent with their being due to the *direct* magnetic attraction of the sun or moon. Hence it might be too hastily concluded, from the absence of observed effects following the proper laws, that these luminaries are not magnetic. An inquiry into the amount of this influence, however, shows that, though the sun or moon were as highly magnetized as the earth, their direct effects would be so small as to be masked by the more powerful unknown perturbating causes which the observations prove to be at work.

In fact let O and O′ be the centres of a distant magnet and of a needle acted on. Let x, y, z be the coordinates of dm, a molecule of the distant magnet referred to O as origin, and rectangular coordinates so taken that the axis of x may pass through O′. Let also x', y', z' be the coordinates of dm', a molecule of the needle acted upon, referred to parallel coordinates passing through O′. Then using ρ for the distance betwen dm and dm', and D for the distance between the centres of the magnets, the components of the action of dm on dm' will be

$$dX = \frac{dm\,dm'}{\rho^2} \cdot \frac{D + x' - x}{\rho},$$

$$dY = \frac{dm\,dm'}{\rho^2} \cdot \frac{y' - y}{\rho},$$

$$dZ = \frac{dm\,dm'}{\rho^2} \cdot \frac{z' - z}{\rho}.$$

Therefore the elementary moments turning dm' round O′ will be

$$dP = \frac{dm\,dm'}{\rho^3} \cdot (yz' - zy'),$$

$$dQ = \frac{dm\,dm'}{\rho^3} \cdot (zx' - xz' + z'D),$$

$$dR = \frac{dm\,dm'}{\rho^3} \cdot (xy' - yx' - y'D).$$

But $\rho^2 = (D + x' - x)^2 + (y' - y)^2 + (z' - z)^2$. Therefore, expanding

---

* Communicated by the Author. An abstract of this paper was read at the recent Manchester Meeting of the British Association.

in inverse powers of D,

$$\frac{1}{\rho^3} = \frac{1}{D^3} \cdot \left(1 - 3\frac{x'-x}{D} + \therefore .\right).$$

Hence, expanding, rejecting terms in which D^{-4} occurs, and those into which coordinates of both dm and dm' do not enter (since they would disappear in integrating, from the fundamental property of magnetism that $\int dm = 0$),

$$d\mathrm{P} = + \frac{dm\,dm'}{D^3} \cdot (yz' - zy') + \text{small terms},$$

$$d\mathrm{Q} = + \frac{dm\,dm'}{D^3} \cdot (zx' + 2\,xz') + \text{rejected terms},$$

$$d\mathrm{R} = - \frac{dm\,dm'}{D^3} \cdot (2xy' + yx') + \text{rejected terms}.$$

Let M and M' be the magnetic moments, and $\alpha\beta\gamma$, $\alpha'\beta'\gamma'$, the directions of the magnetic axes, so that

$$M = (\textstyle\int x\,dm)^2 + (\int y\,dm)^2 + (\int z\,dm)^2,$$

$$M'^2 = (\textstyle\int x'\,dm')^2 + (\int y'\,dm')^2 + (\int z'\,dm')^2;$$

$$\cos\alpha = \frac{\int x\,dm}{M}, \quad \cos\beta = \frac{\int y\,dm}{M}, \quad \cos\gamma = \frac{\int z\,dm}{M},$$

$$\cos\alpha' = \frac{\int x'\,dm'}{M'}, \quad \cos\beta' = \frac{\int y'\,dm'}{M'}, \quad \cos\gamma' = \frac{\int z'\,dm'}{M'}.$$

Then integrating, the components of the moment turning the needle round O' will be

$$\mathrm{P} = + \frac{M\,M'}{D^3} (\cos\beta \cos\gamma' - \cos\gamma \cos\beta') + \text{small terms},$$

$$\mathrm{Q} = + \frac{M\,M'}{D^3} (\cos\gamma \cos\alpha' + 2\cos\alpha \cos\gamma') + \text{small terms},$$

$$\mathrm{R} = - \frac{M\,M'}{D^3} (2\cos\alpha \cos\beta' + \cos\beta \cos\alpha') + \text{small terms}.$$

Hence the resultant moment tending to turn the needle round O'

$$= \frac{MM'}{D^3} \sqrt{(\cos\beta\cos\gamma' - \cos\gamma\cos\beta')^2 + (\cos\gamma\cos\alpha' + 2\cos\alpha\cos\gamma')^2}$$
$$\overline{+ (2\cos\alpha\cos\beta' + \cos\beta\cos\alpha')^2} + \text{small terms}.$$

The maximum value of this (neglecting the small terms) is

$2\dfrac{MM'}{D^3}$, and arises when $\cos\alpha=1$ and $\cos\alpha'=0$*; that is, when the magnetic axis of the distant magnet is pointed towards the needle, and at the same time the needle stands in a perpendicular direction.

Now if H signify the horizontal, and T the total intensity of the earth's magnetism at any station expressed in Gauss's absolute units, $M'H\sin h$ and $M'T\sin t$ will be the moments by which the earth tends to restore the declination and dipping needles respectively, when displaced from their positions of rest through the angles h and t. If we suppose then that the moon is brought successively into the positions in which it will most deviate the two needles, we find that

$M'H\sin h=2\dfrac{MM'}{D^3}$ is the condition of rest for the declination needle, and

$M'T\sin t=\dfrac{2MM'}{D^3}$ is the condition of rest for the dipping needle.

Hence the greatest deviations which the moon can produce on the declination and dipping needles respectively will be

$$h=\frac{2M}{D^3H}, \qquad t=\frac{2M}{D^3T},$$

writing the small angles h and t instead of their sines.

In order to arrive at numerical values, it will be necessary to remember that the magnetic moment M, or

$$\sqrt{\{(\textstyle\int xdm)^2+(\int ydm)^2+(\int zdm)^2\}},$$

is independent of the position of the origin of coordinates. In fact the moment referring to a new origin abc is

$$\sqrt{\{(\textstyle\int(x-a)dm)^2+(\int(y-b)dm)^2+(\int(z-c)dm)^2\}},$$

which $=M$, since, from the fundamental property of magnetism, $\int dm=0$. It is obvious that it is also independent of the direction of the coordinate axes. From this we conclude that in magnetic bodies, since they consist of parts throughout each of

---

* This may be easily seen by conceiving the force of which the components are $-2\cos\mu$, $\cos\beta$, and $\cos\gamma$, applied to the point of which the coordinates are $\cos\alpha'$, $\cos\beta'$, $\cos\gamma'$. The radical in the text will then represent the moment of this force round the origin: and bearing in mind that $\cos\alpha'$, $\cos\beta'$, $\cos\gamma'$ are coordinates of a point at a unit distance from the origin, and that $-2\cos\alpha$, $2\cos\beta$, $2\cos\gamma$ would be components of a force equal to 2, it is obvious that the maximum moment of the given force will amount to 2, and will arise when $\cos\alpha=1$ $\cos\alpha'=0$.

which $\int dm = 0$*, the magnetic moment of the whole is the sum
of the magnetic moments of its parts; from which it follows that
the magnetic moments of similar bodies, if equally magnetized in
corresponding parts, *are proportional to their volumes.* Therefore, as the action of the moon on our instruments varies as
$\dfrac{M}{D^3}$, we may substitute for the moon a hypothetical globe sub-
tending at our instruments the same angle as the moon, and
equally magnetized bulk for bulk. If, then, the moon be as
magnetic as the earth, its maximum effect will equal that of a
globe one metre in diameter, of materials as magnetic as the
earth, and placed at such a distance from the instrument as to
subtend an angle of $2043''$, which is the greatest apparent
diameter of the moon as seen from the *surface* of the earth.

Now Gauss found the magnetic moment of a steel magnet
bar one pound in weight, referred to his absolute unit, to be
100,877,000, and he has shown† that the moment of the earth's
magnetism is equal to what would be produced by 7·831 such
bars placed parallel to one another in each cubic metre of its
volume. Hence the magnetic moment of a cubic metre magnetized in proportion to its bulk as much as the earth is
$7·831 \times 100,877,000$; and multiplying this by ·5236, the ratio
of the contents of a sphere to the cube of its diameter, we find
for the moment of the globe a metre in diameter, expressed in
Gauss's absolute units,

$$M = 0·5236 \times 7·831 \times 100,877,000.$$

Again, 206264·8 being the number of seconds in radius unity,
the distance of the globe, in order to subtend the same angle as
the moon when nearest, will be $\dfrac{206264·8}{2043}$ metres, or (to express
it in Gauss's unit of length) $\dfrac{206264800}{2043}$ millimetres.

Also, expressed in the same units, Gauss found the horizontal
intensity at Göttingen on the 19th July, 1834, $H = 1·7748$, and
the total intensity $T = 4·7414$. Therefore, finally, the maximum
deviations, expressed in seconds, which the moon, if as magnetic
bulk for bulk as the earth, could produce at Göttingen were

---

* This is equivalent to requiring that the *parts* spoken of in the text be
formed by divisions so disposed as not to split any magnetic molecule in
such a way as would place the north magnetism it contains in one part and
the south magnetism in another. All fractures which can in practice be
effected, fulfil this condition.

† See Gauss's Memoir "On the General Theory of Terrestrial Magnetism," translated in 'Taylor's Scientific Memoirs.'

$$h = \frac{2 \times 0{\cdot}5236 \times 7{\cdot}831 \times 100877000 \times 206264{\cdot}8}{\left(\dfrac{206264800}{2043}\right)^3 \times 1{\cdot}7748},$$

which is less than $0''{\cdot}094$, and

$$t = \frac{2 \times 0{\cdot}5236 \times 7{\cdot}831 \times 100877000 \times 206264{\cdot}8}{\left(\dfrac{206264800}{2043}\right)^3 \times 4{\cdot}7414},$$

which is less than $0''{\cdot}036$. *Hence at Göttingen the direct dis-*
turbance of the instrument of declination does not amount, at its
maximum, to one-tenth of a second of arc, and that of the dip
circle does not reach even one twenty-seventh of a second.

Now the observations with which these should be compared
have been made at several stations. The principal part of the
observed lunar diurnal variation consists of a term depending
on twice the lunar hour-angle, but there is also a small term
containing the simple hour-angle. This latter is the one which,
as Dr. Lloyd has shown, the direct action of the moon would
affect, and General Sabine* has determined the following values
for its coefficient, in calculating the formulæ which would best
represent the observations at the several stations:—

$-\overset{_{\prime\prime}}{1}{\cdot}05$ at Toronto,
$+0{\cdot}88$ at St. Helena,
$+1{\cdot}21$ at the Cape,
$+0{\cdot}97$ at Hobarton,
$-0{\cdot}81$ at Pekin, and
$-2{\cdot}04$ at Kew

for the declination; and—

$-\overset{_{\prime\prime}}{1}{\cdot}14$ at Toronto,
$+1{\cdot}32$ at St. Helena,
$-0{\cdot}94$ at the Cape, and
$-0{\cdot}48$ at Hobarton

for the inclination.

There is then no ground for presuming, from the minuteness of
the coefficient, that the moon is not of as magnetic or even much
more magnetic materials than the earth. On the contrary, the
actual magnitudes of the coefficients are *too large* to be with
probability attributed solely to the direct effect of the moon, even
if it were not evident from other considerations, that some cause
acting by different laws has contributed the greater part to them.

If the comparison with the earth be made mass for mass
instead of bulk for bulk, the above disturbances must be reduced

* See p. cxlvi of the Introduction to the 2nd volume of the St. Helena
Observations.

in the ratio of the moon's density to that of the earth, that is, to about $\frac{2}{3}$rds of the values already given.

The same method of course applies equally to the sun; and whether his magnetism be regarded as exceeding that of the earth in proportion to his mass or to his bulk, his maximum influence will be even less than that of the moon; for he never attains an apparent size as great as the maximum of the moon, and his density is only about half that of the moon.

XXXIX. *Chemical Notices from Foreign Journals.* By E. ATKINSON, *Ph.D., F.C.S.*

[Continued from p. 143.]

FROM the readiness with which, in the vegetable kingdom, the oxygen in carbonic acid is replaced by hydrogen, it was highly probable that carbonic acid could similarly be reduced artificially, and that the first product of substitution (formic acid) could be prepared from carbonic acid. Led by these considerations, Kolbe and Schmitt* undertook an investigation on the direct conversion of carbonic acid into formic acid, and their first experiments have been successful. The change succeeds so easily and in such a simple manner as to make it surprising that it has not been previously observed. When potassium was spread out in a thin layer on a flat dish, and this was placed under a bell-jar standing over milk-warm water, and kept continually filled with carbonic acid, the potassium was found in twenty-four hours to be converted into a mixture of bicarbonate and of formiate of potash. The reaction may be thus written:—

$$2\,K + 2C^2O^4 + 2HO = KO, C^2HO^3 + \left.\begin{matrix} KO \\ HO \end{matrix}\right\} C^2O^4.$$

The above mixture was supersaturated in the cold with sulphuric acid, the acid liquor poured off from the bisulphate of potash distilled, and the distillate neutralized with carbonate of lead. On evaporating the hot filtered solution, chemically pure formiate of lead was obtained.

Sodium exposed for twenty-four hours to the action of carbonic acid and aqueous vapour, also gives rise to the formation of formic acid, but in smaller quantity than potassium.

Schischkoff, in continuing his researches on nitroform†, has obtained results of which he communicates a preliminary notice‡.

---

* Liebig's *Annalen*, August 1861.
† Phil. Mag. vol. xv. p. 302.
‡ Liebig's *Annalen*, August 1861.

He finds that nitroform, $G(NO^2)^3 H$, is a strong acid, and readily exchanges its atom of hydrogen for metals, forming true salts. This hydrogen can also be replaced by bromine and by hyponitrous acid. When nitroform, mixed with bromine, was exposed to the sunlight, hydrobromic acid was formed, and the mixture became decolorized. The resultant product was washed with water, in which it is somewhat soluble; it is liquid at temperatures above $+12°$ C., but below that point solidifies to a crystalline mass. It has the formula $G(NO^2)^3 Br$.

In order to replace the hydrogen in nitroform by hyponitrous acid, a current of air was passed through a mixture of nitroform with sulphuric and nitric acids heated to 100°. A liquid distilled over, from which, on the addition of water, an insoluble oily liquid was precipitated. This substance boils at 126° C. without any decomposition; it is colourless, mobile, and fluid at ordinary temperatures, but solidifies at $+13°$ C. to a white crystalline mass. It has the composition $G(NO^2)^4$, and singularly enough, although it contains an atom more hyponitrous acid, it is more stable than nitroform; it does not explode when rapidly heated, but decomposes, giving off nitrous vapours.

Schischkoff had found that trinitroacetonitrile, $G^2(NO^2)^3 N$, was decomposed by sulphuretted hydrogen, yielding a body $G^2(NO^2)^2(NH^4) N$, which he called binitroammonyle. This he has since found[*] to be the ammonium-salt of the body *binitroacetonitrile*, $G^2(NO^2)^2 H N$, which has strongly acid properties. Binitroacetonitrile is obtained by treating an aqueous solution of binitroammonyle with sulphuric acid and agitating the mixture with ether. On evaporating the etherial solution, the acid crystallizes in large colourless plates. The silver and potassium salts of this body were prepared. The silver-salt has the formula $G^2(NO^2)^2 Ag N$; when treated with bromine in presence of water, bromide of silver is formed, and an oily product, which is probably bromobinitroacetonitrile, $G^2(NO^2)^2 Br N$.

A series of experiments by Wurtz and Friedel on lactic acid[†] confirm the conclusion[‡] that lactic acid contains a diatomic radical, and that its two equivalents of replaceable hydrogen are not of identical value.

There are two ethers of lactic acid which are isomeric, but completely different in properties. One of them, ethylactic acid, is obtained by treating dilactate of ethyle with caustic potash; the other is neutral, and was first obtained by Strecker in distilling lactate of lime with sulphovinate of potash. Wurtz

* Liebig's *Annalen*, August 1861.
† *Comptes Rendus*, May 27, 1861.
‡ Phil. Mag. vol. xviii. p. 287.

and Friedel have found that it is also obtained by heating lactic acid with alcohol in closed vessels to a temperature of 170°. The former of these compounds is a true acid, and readily forms salts. When monoethylic lactate, as the latter compound is called, is treated with potassium, hydrogen is disengaged, and a compound is obtained isomeric with ethylactate of potash, and which, when treated with iodide of ethyle, forms dilactic ether.

These two ethers present a most curious example of isomerism. They are formed by the same acid, both contain the same group ethyle, and yet one of them is an energetic acid, while the other is perfectly neutral. This is accounted for by the different parts that the two atoms play in lactic acid. One of them is strongly basic, and can be replaced by a metal or by an organic group, such as ethyle; in both cases a neutral compound is obtained. The other atom can be easily replaced by oxygen groups, such as the radicals of monobasic acids. If replaced by an indifferent group, such as ethyle, it is still acid, because the atom of basic hydrogen has not been touched.

The authors have noted similar isomeric relations between lactamethane, $C^5 H^{11} NO^2$, and a new amide produced by the action of ethylamine on lactide, $C^3 H^4 O^2$. By potash they undergo a different decomposition—the latter into lactic acid and ethylamine, and the former into ammonia and ethylactic acid.

Lactyle, the radical of lactic acid, has the property of multiplying itself in one and the same body so as to form compounds which may be referred to condensed types analogous to the polyethylenic compounds.

Dilactic Ether.—When chlorolactic ether* acts upon lactate of potash, chloride of potassium is formed, and a dilactic ether, according to the equation

$$C^5 H^9 O^2 Cl + C^3 H^5 K O^3 = K Cl + C^8 H^{14} O^5.$$

Chlorolactic ether.	Lactate of potassium.		New body.

The formula of the new body, *monoethylic dilactate,* may be written thus, $\left. \begin{array}{c} (C^3 H^4 O'')^2 \\ (C^2 H^5) H \end{array} \right\} O^3$. Besides this there is another lactic ether, *diethylic dilactate,* $\left. \begin{array}{c} (C^3 H^4 O')^2 \\ (C^2 H^5)^2 \end{array} \right\} O^3$, obtained by the action of chlorolactic ether on ethylolactate of potassium. These compounds are the ethers of the anhydrous lactic acid of Pelouze, and can be regarded as containing two equivalents of the radical lactyle according to the formula $\left. \begin{array}{c} C^3 H^4 O'' \\ H^2 \end{array} \right\} O^3$. There is a lac-

* Phil. Mag. vol. xviii. p. 287.

tosuccinic ether, $C^9 H^{18} O^6$, a mixed ether obtained by treating chlorolactic ether with ethylosuccinate of potash. It has the

formula $\left.\begin{array}{c} C^3 H^4 O^{\prime\prime} \\ C^4 H^4 O^{2\prime\prime} \\ (C^2 H^5)^2 \end{array}\right\} O^2.$ There is, further, a trilactic ether, formed by the direct union of lactide with lactic ether.

By the action of sodium-alcohol on iodoform, Boutlerow obtained, among other products, an acid which he believed was valerolactic acid. He has since found* that this acid is ethylactic acid; for when treated with hydriodic acid it is decomposed into lactic acid and iodide of ethyle,

$$C^5 H^{10} O^3 + H I = C^9 H^6 O^3 + C^2 H^5 I.$$
New acid.   Lactic acid.   Iodide of
                                            ethyle.

He has also proved by direct experiments that this acid is identical in properties with Wurtz's ethylactic acid, obtained by the decomposition of lactic ether by potash.

Vogt has given a fuller account† of the preparation and properties of the new benzylic mercaptan which he discovered, and of which a preliminary notice has already appeared‡; he has also described a series of its compounds. The sodium-benzylic mercaptan, $C^{12} H^5 Na S^2$, is obtained as a white saline mass by the addition of sodium to benzyle-mercaptan and subsequent evaporation to dryness. The lead compound, $C^{12} H^5 Pb S^2$, a yellow crystallized body, is obtained by adding an alcoholic solution of acetate of lead to an alcoholic solution of the mercaptan. The mercury compound, $C^{12} H^5 Hg S^2$, crystallizes in very fine white needles, and is obtained by the action of oxide of mercury on the mercaptan.

Nitric acid acts with considerable energy on benzyle-mercaptan. The result of this action is a body crystallizing in white lustrous needles, which has the formula $C^{12} H^5 S^2$, and is accordingly *bisulphide of benzyle*; its formation may be thus expressed:—

$$C^{12} H^6 S^2 + NO^5 HO = C^{12} H^5 S^2 + NO^4 + 2 HO.$$
Benzylic   Bisulphide
mercaptan.   of benzyle.

It has a faint but not unpleasant smell, and melts at 60° to a yellowish oil, which can be distilled at a high temperature without decomposition.

* Liebig's *Annalen*, June 1861.
† Ibid. August 1861.
‡ Phil. Mag. vol. xx. p. 522.

Bisulphide of benzyle can readily be reduced to benzyle-mercaptan by nascent hydrogen.

Another remarkable and hitherto unexplained mode of forming bisulphide of benzyle, by which it may be obtained in large transparent crystals, consists in dissolving the mercaptan in alcoholic ammonia and exposing the solution to spontaneous evaporation.

By the further oxidation of benzylic mercaptan, benzyle-sulphuric acid is formed, $HO \, C^{12} \, H^5 \, S^2 \, O^5$.

Mosling has investigated* the action of hydrochloric acid and of sulphuretted hydrogen on benzoic anhydride. The action of the former substance is simply in accordance with the equation

$$\left.\begin{array}{c} C^7 H^5 \Theta \\ C^7 H^5 \Theta \end{array}\right\} \Theta + \left.\begin{array}{c} H \\ Cl \end{array}\right\} = \left.\begin{array}{c} C^7 H^5 \Theta \\ H \end{array}\right\} \Theta + \left.\begin{array}{c} C^7 H^5 \Theta \\ Cl \end{array}\right\}$$

Benzoic anhydride.   Benzoic acid.   Chloride of benzoyle.

Benzoic anhydride was heated with sulphuretted hydrogen to a temperature of 130° for twenty hours. Some benzoic acid sublimed, and the residue in the retort, when crystallized from alcohol and bisulphide of carbon, was found to consist of a new body, which had the formula

$$C^7 H^5 \Theta S \text{ or } \left.\begin{array}{c} C^7 H^5 \Theta \\ C^7 H^5 \Theta \end{array}\right\} S^2,$$

and is therefore the *persulphide of benzoyle*. It is not soluble in water, and difficultly so in alcohol.

It readily dissolves in ether, and especially in bisulphide of carbon, from which it crystallizes in colourless plates which appear to be rhombic columns. It melts at 123°, and decomposes at a somewhat higher temperature. It is the first member of a new series of sulphur-compounds, and corresponds to Brodie's peroxide of benzoyle and acetyle.

It is probably formed in accordance with the following reaction:—

$$3 \left.\begin{array}{c} C^7 H^5 \Theta \\ C^7 H^5 \Theta \end{array}\right\} \Theta + 2 \left.\begin{array}{c} H \\ H \end{array}\right\} S = 3 \left.\begin{array}{c} C^7 H^5 \Theta \\ H \end{array}\right\} \Theta + \left.\begin{array}{c} C^7 H^5 \Theta \\ C^7 H^5 \Theta \end{array}\right\} S^2 + \left.\begin{array}{c} C^7 H^5 \Theta \\ H \end{array}\right.$$

Benzoic anhydride.   Benzoic acid.   Persulphide of benzoyle.   Hydride of benzoyle.

In a preliminary notice, Kalle† announced that, by the action of zinc-ethyle on chloride of sulphon-benzyle, he had obtained a new body which was a mixed acetone belonging to the benzyle

---

* Liebig's *Annalen*, June 1861.
† Phil. Mag. vol. xx. p. 522.

series, but containing sulphur in the place of some of the carbon. A subsequent examination* of the reaction has shown that chloride of ethyle is formed at the same time, along with the zinc-salt of a new acid. The reaction may be thus expressed:—

$$(C^{12} H^5) [S^2 O^4] Cl + Zn C^4 H^5 = ZnO C^{12} H^5 S^2 O^3 + C^4 H^5 Cl.$$
$$\text{Chloride of} \qquad \text{Zinc-ethyle.} \qquad \text{Benzyle-sulphite} \qquad \text{Chloride of}$$
$$\text{sulphon-benzyle.} \qquad\qquad\qquad \text{of zinc.} \qquad\qquad \text{ethyle.}$$

Benzyle-sulphurous acid, the product of this reaction, stands to sulphurous acid in the same relation as benzyle-sulphuric acid to sulphuric acid; it is sulphurous acid $(S^2 O^2)O^2$ in which an atom of oxygen is replaced by benzyle. It crystallizes in large prisms, often an inch long, and mostly occurring in stellate groups.

The author describes several of the salts of the new acid, and also a series of experiments made with the view of finding a more productive method of its preparation.

Wurtz has continued† his researches on the oxyethylenic bases formed by the action of oxide of ethylene on ammonia. The product of this action, when treated by hydrochloric acid, consists mainly of the hydrochlorates of trioxethylenamine, $(C^2 H^4 \Theta)^3 NH^3$, and dioxethylenamine, $(C^2 H^4 \Theta)^2 NH^3$. The former is insoluble; and from the alcoholic mother-liquor, the platinum-salt of the second, $C^4 H^{11} N\Theta^2$, HCl PtCl^2, is precipitated on the addition of bichloride of platinum. On the addition of ether to the mother-liquor, the platinum-salt of a third base, monoxethylenamine, is precipitated. It has the formula $(C^2 H^4 \Theta)$ HCl, PtCl^2, and crystallizes in golden-yellow nacreous laminæ.

The hydrochlorates of monoxethylenamine and of dioxethylenamine are formed by the action of ammonia on hydrochloric glycol when these substances, enclosed in strong vessels, are heated in the water bath.

$$C^2 H^5 ClO + NH^3 = (C^2 H^4 \Theta) NH^3, HCl.$$
$$\text{Hydrochlorate of}$$
$$\text{monoxethylenamine.}$$

$$2 (C^2 H^5 Cl\Theta) + 2 NH^3 = 2 (C^2 H^4 \Theta^2) NH^3, HCl + NH^4 Cl$$
$$\text{Hydrochlorate of}$$
$$\text{dioxethylenamine.}$$

The base trioxethylenamine may be isolated and obtained as a thick syrup by the action of oxide of silver on its hydrochlorate.

---

* Liebig's *Annalen*, August 1861.
† *Comptes Rendus*, August 19, 1861, Phil. Mag. vol. xix. p. 125.

When this base is heated with hydrochloric glycol, the hydrochlorate of the base tetroxethylenamine is formed.

$$(\mathrm{C^2\,H^4\,\Theta})^3\,\mathrm{N\,H^3} + \mathrm{C^2\,H^5\,Cl\Theta} = (\mathrm{C^2\,H^4\,\Theta})^4\,\mathrm{N\,H^3,\,HCl.}$$

Trioxethylenamine.   Hydrochloric   Hydrochlorate of
                        glycol.   tetroxethylenamine.

With reference to the constitution of these bases, if it be assumed that the diatomic oxide of ethylene, by fixing an atom of hydrogen in ammonia, may become monatomic,

$$\mathrm{C^2\,H^4\,\Theta'' + H = C^2\,H^5\,\Theta',}$$

they may be referred to the type ammonia, and the formulæ of their hydrochlorates become,—

$$\left.\begin{array}{c}\mathrm{C^2\,H^5\,\Theta'}\\ \mathrm{H^3}\end{array}\right\}\mathrm{NCl}$$ Hydrochlorate of monoxethylenamine.

$$\left.\begin{array}{c}(\mathrm{C^2\,H^5\,\Theta})^2\\ \mathrm{H^2}\end{array}\right\}\mathrm{NCl}$$ Hydrochlorate of dioxethylenamine.

$$\left.\begin{array}{c}(\mathrm{C^2\,H^5\,\Theta})^3\\ \mathrm{H}\end{array}\right\}\mathrm{NCl}$$ Hydrochlorate of trioxethylenamine.

$$(\mathrm{C^2\,H^5\,\Theta})^4\,\}\,\mathrm{NCl}$$ Hydrochlorate of tetroxethylenamine.

But they may also be referred to the mixed type, water and ammonia, $\left.\begin{array}{c}n\mathrm{H^2}\\ \mathrm{H^3}\end{array}\right\}\begin{array}{c}\Theta^n\\ \mathrm{N}\end{array}$; and the author prefers this view.

The union of anhydrous trioxethylenamine with ammonia can take place in several proportions. One, two, three, or four molecules of oxide of ethylene can unite with one molecule of the anhydrous base, forming oxygenated bases more and more complex, but in which the basic power is also feebler. They nevertheless have an alkaline reaction, combine with hydrochloric acid, and form double salts with bichloride of platinum. These latter do not crystallize, and are very difficult to purify and separate. The analysis of some of these bases gave results agreeing with the formulæ

$$(\mathrm{C^2\,H^4\,\Theta})^5\,\mathrm{N H^3\,.\,HCl,\,PtCl^2,}$$
$$(\mathrm{C^2\,H^\bullet\,\Theta})^7\,\mathrm{N H^3\,.\,HCl,\,PtCl^2.}$$

These bodies, although containing nitrogen, and being distinctly alkaline, are not compound ammonias, and cannot be referred to that type. It is accordingly probable that among natural oxygen bases there are some which are not compound ammonias, that is, cannot be regarded as derived from ammonia by substitution.

Wurtz[*] tried the action of aldehyde on glycol, expecting to

[*] *Comptes Rendus,* August 26, 1861.

obtain a series of bodies isomeric with the polyethylenic alcohols; the reaction, however, is quite different: the aldehyde dehydrates glycol, and unites with the oxide of ethylene thus formed :—

$$C^2 H^6 \Theta^2 + C^2 H^4 \Theta = C^4 H^8 \Theta^2 + H^2 O.$$
Glycol.   Aldehyde.   New body.

It is a colourless limpid liquid, with an agreeable penetrating odour, resembling that of aldehyde. It boils at $82^\circ\cdot5$.

If aldehyde is the oxide of ethylidene, the compound is a mixed oxide of ethylene-ethylidene. The body slowly reduces alcoholic solution of nitrate of silver. Heated with acetic acid, it regenerates diacetate of glycol.

Kekulé has published* an interesting communication on fumaric and some allied organic acids. When malic acid, $C^4 H^6 \Theta^5$, is heated, it loses water, and gives two isomeric bodies, fumaric and maleic acids, $C^4 H^4 \Theta^4$.

When fumaric acid is treated with bromine in the presence of water, no action takes place in the cold, but at the temperature of the water-bath the bromine rapidly disappears, and a quantity of perfectly white crystals are obtained, which are dibromosuccinic acid, $C^4 H^4 Br^2 \Theta^4$. The formation of this body is interesting, inasmuch as it takes place by a simple *addition* of the elements, and not, as is usually the case in the action of bromine on organic substances, by *substitution*: thus

$$C^4 H^4 \Theta^4 + Br^2 = C^4 H^4 Br^2 O^4$$
Fumaric acid.   Dibromosuccinic acid.

Hydrobromic acid also, when heated with fumaric acid, yields some monobromosuccinic acid, but the action is very slow.

$$C^4 H^4 \Theta^4 + HBr = C^4 H^5 Br \Theta^4$$
Fumaric acid.   Monobromosuccinic acid.

Fumaric acid can also be converted into succinic acid by the action of hydrogen. The experiment succeeds by means of hydriodic acid, but is most easily effected by means of nascent hydrogen. It is simply necessary to add sodium-amalgam to a solution of fumaric acid in water to convert it entirely into succinic acid.

$$C^4 H^4 \Theta^4 + H^2 = C^4 H^6 \Theta^4$$
Fumaric acid.   Succinic acid.

This action of nascent hydrogen is as unusual as that of bromine. Hydrogen in the nascent state can reduce organic substances by taking away oxygen; but there are few cases in which an organic substance unites directly with hydrogen.

* Liebig's *Annalen*, Supplement, July 1861.

Kekulé has found that maleic acid, when acted upon by the same reagents, yields the same bodies.

In conclusion, he developes his views as to the relations between fumaric acid and its allied substances. He establishes a close and interesting analogy between fumaric acid and ethylene. Fumaric acid stands in the same relation to malic acid as ethylene does to alcohol; it stands to dibromosuccinic acid as ethylene does to bromide of ethylene; and to monobromosuccinic acid as ethylene to bromide of ethyle, and so on. Tartaric acid is to fumaric acid what glycol is to ethylene. In fact tartaric acid is obtained when the bromide of fumaric acid, that is, dibromosuccinic acid, is heated with oxide of silver, just as an ether of glycol is obtained when a silver salt acts upon bromide of ethylene.

In Poggendorff's *Annalen* for 1855, Kessler described a volumetric method of estimating arsenic and antimony, by which he made a determination of their atomic weights. The method consisted in oxidizing these substances, which were employed in the form of arsenious and antimonious acids, to arsenic and antimonic acids by means of a standard solution of bichromate of potash; the excess of bichromate of potash was determined by means of a standard solution of protochloride of iron. The applicability of this method depends upon the fact that protochloride of iron reduces bichromate of potash, but does not affect arsenic or antimonic acid. Kessler has since then made [*] some additional experiments, partly confirming and partly rectifying previous results.

From these experiments he concludes that the atomic weight of arsenic is 75·15.

For the atomic weight of antimony, Kessler's previous experiments led to the number 123·78. In his recent experiments, in which some sources of error, to which his previous methods were liable, have been avoided, he has obtained different results.

Pure oxide of antimony was prepared, and was further purified by sublimation in a porcelain tube in a current of carbonic acid; a given weight of this was partially oxidized by a given weight of pure chlorate of potash, and the oxidation completed by means of a standard solution of bichromate of potash; the excess of the latter was estimated by a standard solution of protochloride of iron. In this way six experiments gave numbers for the equivalent of antimony, varying between 121·67 and 122·58, the mean being 122·16.

In another case, in which pure metallic antimony was oxi-

[*] Poggendorff's *Annalen,* vol. cxiii. p. 134.

dized to antimonic acid by bichromate of potash, the number found was 122·34.

In a third case, a double determination of terchloride of antimony was made, by oxidation to pentachloride, and by directly determining the quantity of chlorine in the ordinary way. This gave the number 122·37 for the equivalent of antimony.

The mean of these results obtained by different methods is 122·29; they furnish a remarkable confirmation of the excellent determinations of Dexter. The method employed by this chemist was that originally used by Berzelius, and consisted in the direct oxidation of pure antimony to antimoniate of oxide of antimony, SbO^4. By numerous very careful experiments he obtained the mean number 122·33.

In the above series of experiments Kessler obtained the number 26·1 for the atomic weight of chromium.

De Luca describes[*] the following method of preparing oxygen which he has used for some time; it only differs in manipulatory details from that of Deville and Debray[†]. A tubulated retort is filled three-quarters full with pumice and concentrated sulphuric acid, and luted on to a porcelain tube by means of a mixture of asbestos and clay. The tube also contains pumice; it is heated to redness, and the vapour of sulphuric acid passed over it. The oxygen is disengaged with regularity, and is easily purified; in one operation 2 ounces of acid furnished about a gallon and a quarter of gas. The process is analogous to that in which hydrogen is prepared by decomposing water by iron; and it is not more difficult.

Lapschin and Tichanowitsch[‡] have made a series of experiments on the electrolysis of organic and other substances, in which they had at their disposal a battery of 1000 elements.

Salicine is decomposed by the battery; the first stage appeared to be its decomposition into grape-sugar and saligenine. On the zinc pole gases were disengaged which were not collected; the next stage appeared to be that the saligenine was oxidized successively to hydride of salicyle and to salicylic acid.

The action of a battery of 900 elements produced in *crystallized acetic acid* a rapid disengagement of gas at the carbon pole, consisting of carbonic acid and carbonic oxide. A very slight quantity of gas was disengaged at the zinc pole, which, however, was lost; at the same time an amorphous mass of carbon was deposited.

[*] *Comptes Rendus,* July 22.
[†] Phil. Mag. vol. xxi. p. 295.
[‡] *Bulletin de l'Académie de St. Petersbourg,* vol. iv. p. 80.

Absolute alcohol was almost unaffected by a battery of 900 elements; the quantity of gas collected after seven hours was so small that it could not be analysed. At first there was no action, and a slight action was only set up after some time, when the alcohol had attracted moisture by standing. It may be assumed, therefore, that absolute alcohol offers a complete resistance to the current.

Ether was unattacked by the action of a battery of 900 elements, even when the poles were only a millimetre apart. There was no disengagement of gas; and the boiling-point remained the same as before. When the electrodes were 20 millims. asunder, there was an undulatory motion of the liquid from the carbon to the zinc pole.

With *amylic alcohol* 900 elements were used, the electrodes being 1 millim. apart. The multiplier stood at 20°. There was an undulatory motion from the carbon to the zinc pole, and after some time a yellowish deposit was formed on the zinc pole, which, under the microscope, was seen to consist of a pulverulent mass of yellow colour. When, subsequently, the electrodes were brought nearer, they melted together, the liquid became heated, and a black carbonaceous mass was deposited.

Valerianic acid, turpentine, and *anhydrous boracic acid* were unacted upon by a battery of 900 elements.

950 elements produced no action on *bisulphide of carbon*; the multiplier stood at 0°. A previous experiment in 1858 with 800 elements gave an equally negative result.

Silicic acid in the pulverulent form was placed in a clay crucible and exposed to the action of the current. At first there was no action, but afterwards the whole mass became ignited; the side of the crucible nearest the zinc pole was perforated, and a platinum globule melted through, which was found to contain silicon.

800 elements produced no action on dry powdered *oxide of antimony*, nor did 370 on *oxychloride of antimony*.

With dried powdered *oxide of zinc* 370 elements produced an energetic action, and the reduced zinc became ignited. The decomposition also ensued with 60 elements; even with 20 there was a slight action.

40 elements acted strongly on *sulphuret of antimony*; sulphur was liberated at the charcoal pole and became ignited.

Realgar required 260 elements for its decomposition; the products of the action, sulphur and arsenic, immediately took fire, and were converted into sulphurous and arsenious acids.

XL. *Proceedings of Learned Societies.*

ROYAL SOCIETY.

· [Continued from p. 246.]

November 15, 1860.—Major-General Sabine, R.A., Treasurer and
Vice-President, in the Chair.

THE following communication was read :—
"On the Laws of the Phenomena of the larger Disturbances
of the Magnetic Declination in the Kew Observatory : with notices
of the progress of our knowledge regarding the Magnetic Storms."
By Major-General Edward Sabine, R.A., Treas. and V.P.

The laws manifested by the mean effects of the larger magnetic
disturbances (regarded commonly as effects of magnetic storms)
have been investigated at several stations on the globe, being chiefly
those of the British Colonial Observatories ; but hitherto there has
been no similar examination of the phenomena in the British Islands
themselves. The object of the present paper is to supply this deficiency, as far as one element, namely the declination, is concerned,
by a first approximation derived from the photographs in the years
1858 and 1859, of the self-recording declinometer of the observatory
of the British Association at Kew ; leaving it to the photographs
of subsequent years to confirm, rectify, or render more precise the
results now obtained by a first approximation. The method of investigation is simple, and may be briefly described as follows :—
The photographs furnish a continuous record of the variations
which take place in the direction of the declination-magnet, and admit of exact measurement in the two relations of time, and of the
amount of departure from a zero line. From this automatic record,
the direction of the magnet is measured at twenty-four equal intervals of time in every solar day, which thus become the equivalents
of the "hourly observations" of the magnetometers in use at the
Colonial Observatories. These measures, or hourly directions of the
magnet, are entered in monthly tables, having the days of the month
in successive horizontal lines, and the hours of the day in vertical
columns. The "means" of the entries in each vertical column indicate the mean direction of the magnet at the different hours of the
month to which the table belongs, and have received the name of
"First Normals." On inspecting any such monthly table, it is at
once seen that a considerable portion of the entries in the several
columns differ considerably from their respective means or first normals, and must be regarded as "disturbed observations." The laws
of their relative frequency, and amount of disturbance, in different
years, months and hours, are then sought out, by separating for that
purpose a sufficient body of the most disturbed observations, computing the amount of departure in each case from the normal of the
same month and hour, and arranging the amounts in annual, monthly,
and hourly tables. In making these computations, the first normals
require to be themselves corrected, by the omission in each vertical
column of the entries noted as disturbed, and by taking fresh means,
representing the normals of each month and hour after this omission,
and therefore uninfluenced by the larger disturbances. These new

means have received the name of " Final Normals," and may be defined as being the mean directions of the magnet in every month and every hour, after the omission from the record of every entry which differed from the mean a certain amount either in excess or in defect.

In this process there is nothing indefinite; and nothing arbitrary save the assignment of the particular amount of difference from the normal which shall be held to constitute the measure of a large disturbance, and which, for distinction sake, we may call "*the separating value.*" It must be an amount which will separate a sufficient body of disturbed observations to permit their laws to be satisfactorily ascertained; but in other respects its precise value is of minor significancy; and the limits within which a selection may be made, without materially affecting the results, are usually by no means narrow; for it has been found experimentally on several occasions, that the *Ratios* by which the periodical variations of disturbance in different years, months and hours are characterized and expressed, do not undergo any material change by even considerable differences in the amount of the separating value. The separating value must necessarily be larger at some stations than at others, because the absolute magnitude of the disturbance-variation itself is very different in different parts of the globe, as well as its comparative magnitude in relation to the more regular solar-diurnal variation; but it must be a *constant* quantity throughout at one and the same station, or it will not truly show the relative proportion of disturbance in different years and different months.

The strength of the Kew establishment being insufficient for the complete work of a magnetic observatory, the tabulation of the hourly directions from the photographic records has been performed by the non-commissioned officers of the Royal Artillery, employed under my direction at Woolwich, where this work has been superintended by Mr. John Magrath, the principal clerk, as have been also the several reductions and calculations, which have been made on the same plan as those of the Colonial Observatories.

In the scale on which the changes of direction of the declination-magnet are recorded in the Kew photographs, one inch of space is equivalent to 22'·04 of arc. On a general view and consideration of the photographs during 1858 and 1859, 0·15 inch, or 3'·31 of arc appeared to be a suitable amount for the separating value to be adopted at that station; consequently every tabulated value which differed 3'·31 or more, either in excess or defect from the final normal of the same month and hour, has been regarded as one of the larger disturbances, and separated accordingly. The number of disturbed observations in the two years was 2424 (viz. 1211 in 1858, and 1213 in 1859), being between one-seventh and one-eighth of the whole body of hourly directions tabulated from the photographs, of which the number was 17,319. The aggregate value of disturbance in the 2424 observations, was 14,901 minutes of arc; of which 7207 minutes were deflections of the north end of the magnet to the west, and 7694 to the east; the easterly deflections thus having a slight preponderance. The number of the disturbed observations, as well as their aggregate values, approximated very closely in each of the two years, 1859 being very slightly in excess. The *decennial period* of the magnetic storms, indicated by the observations at the British Colonial Observatories between 1840 and 1850, had led to the anticipation that the next epoch of maximum of the cycle might take place in the years 1858–1859. The nearly equal proportions in which the numbers and aggregate values of the larger disturbances took place in 1858 and 1859 are so far in accordance with this view. Should the records of the succeeding years at Kew, made with the same instruments, and examined by the same method, show decreasing disturbance in 1860 and 1861, the precise epoch of the maximum indicated by the records of the Kew declinometer will be " the end of 1858 or commencement of 1859."

In Table I. are shown the aggregate values of disturbance in the two years, arranged under the several hours of solar time in which they occurred. They are also divided into the two categories of westerly and easterly deflections, since the experience gained at other stations has now fully established that the westerly and easterly disturbance-deflections are characterized in all parts of the globe by distinct and dissimilar laws. The Ratios are also shown which the aggregate values at the different hours, both of the westerly and the easterly deflections, bear to their respective mean values,—or, in other words, to the sums respectively of the westerly and easterly deflections at all the hours, divided by 24, and taken as the respective units.

TABLE I.—Showing the aggregate values of the larger disturbances of the Declination at the different hours of solar time in 1858 and 1859, derived from the Kew Photographs ; with the Ratios of disturbance at the several hours to the mean hourly value taken as the Unit.

Mean astronomical hours.	Westerly deflections.		Easterly deflections.		Mean civil hours.
	Aggregate values. (Minutes of arc.)	Ratios.	Aggregate values. (Minutes of arc.)	Ratios.	
18	553·9	1·85	118·9	0·37	6 A.M.
19	549·3	1·83	120·9	0·38	7 A.M.
20	442·9	1·48	115·2	0·36	8 A.M.
21	370·1	1·23	121·2	0·38	9 A.M.
22	376·9	1·26	104·6	0·33	10 A.M.
23	361·8	1·21	125·8	0·39	11 A.M.
0	413·7	1·38	173·0	0·54	Noon.
1	431·1	1·44	153·3	0·48	1 P.M.
2	459·8	1·53	173·0	0·54	2 P.M.
3	513·0	1·71	108·4	0·34	3 P.M.
4	403·9	1·35	141·0	0·44	4 P.M.
5	343·8	1·15	164·8	0·51	5 P.M.
6	282·5	0·94	291·1	0·91	6 P.M.
7	110·7	0·37	381·8	1·19	7 P.M.
8	65·6	0·22	499·0	1·56	8 P.M.
9	88·2	0·29	572·9	1·79	9 P.M.
10	59·0	0·20	724·3	2·25	10 P.M.
11	35·7	0·12	767·8	2·38	11 P.M.
12	146·7	0·49	709·5	2·21	Midnight.
13	141·8	0·47	634·8	1·98	1 A.M.
14	146·7	0·49	577·2	1·80	2 A.M.
15	151·5	0·51	464·8	1·45	3 A.M.
16	289·5	0·97	305·8	0·95	4 A.M.
17	458·9	1·53	144·9	0·45	5 A.M.
Mean hourly value 299·9 = 1·00			Mean hourly value 320·6 = 1·00		

The westerly and easterly deflections in the British Islands, as represented by the automatic records at Kew, are obviously governed, as in all other parts of the globe where the phenomena have been analysed, by distinct laws. The westerly deflections have their chief prevalence from 5 A.M. to 5 P.M., or during the hours of the *day*; the easterly deflections, on the other hand, prevail chiefly during the hours of the *night*, the ratios being above unity from 7 P.M. to 3 A.M., and below unity at all other hours. The easterly have one decided maximum, viz. at 11 P.M., towards which they steadily and continuously progress from 5 P.M., and from which they as steadily and continuously recede until 5 A.M. the following morning. The westerly deflections appear to have two epochs of maximum, one from 6 to 7 A.M., the other about 3 P.M., progressing regularly towards the first named from 3 A.M., and receding from it to 9 A.M.; at 9, 10, and 11 A.M. the ratios remain almost sensibly the same, but towards noon they begin to increase afresh, and continue to do so progressively to the second maximum at 3 P.M., from which hour they progressively decrease to 7 P.M. Those ratios which are less than unity, viz. those of the westerly deflections from 6 P.M. to 4 A.M., and of the easterly from 4 A.M. to 6 P.M., do not in either case exhibit the same decided tendency to one or two well-marked minima, as the ratios which are above unity do in both cases towards their maxima. It is possible, however, that this may in some degree be explained by the following consideration :—

The aggregate values of the disturbances prevailing at the different hours, as stated in the Table, are those which have prevailed, not only over the forces which would retain the magnet in its *mean* position, but also over any disturbing influences in an opposite direction, which may be conceived to have existed contemporaneously; and we cannot but suppose that as both westerly and easterly disturbances do record themselves as prevailing at the same hours on different days, that these opposite influences may sometimes *coexist*, neutralizing each other and not appearing in the record. We may reasonably suppose that the degree in which the aggregate values in the Table, both westerly and easterly, may be diminished thereby at the different hours, may be in some measure indicated by the disparity, or the reverse, in the amount of the aggregate values of disturbance in the opposite directions at those hours. Thus we may suppose that at a particular hour, 11 P.M. for example, when the amount of westerly deflections is very small, and of easterly very great, the diminution of the aggregate values of either by mutual counterbalance may be extremely small, while of equal absolute amount in both. Now a very small amount deducted from the large aggregate easterly value will scarcely have any effect whatsoever on the ratio at that hour to its unit or mean hourly value; whereas the same small amount deducted from the far less aggregate westerly value at the same hour would have a far more sensible effect upon its ratio. Assuming, therefore, the probability that westerly and easterly disturbing influences do sometimes coexist and neutralize each other in the record, and that we may in some degree judge of the respective amounts of the conflicting influences at the several hours by the means above stated, we should be prepared to expect that the ratios which are below unity do not represent the actual variations of the disturbing influences at those hours quite so purely as do the ratios which are above unity ; and that they are liable to be affected, though in a very subordinate degree, by the abstraction of the neutralized portion, when the aggregate values which they represent are very small.

Without, however, resting undue weight upon this suggestion, we may safely say that the hours, when the ratios are below unity, are hours of comparative tranquillity, and that their variations from hour to hour are of a far less marked character than during the hours when the ratios exceed unity. Thus viewed, the character of the disturbance-diurnal variations may be conceived to have some analogy with that of the phenomena of the regular solar-diurnal variation. We may imagine the disturbance-variation (either the westerly or the easterly, it is indifferent which is taken),—divided as it is into two portions, by the ratios being in the one case above, and in the other below unity,— to correspond in one of its divisions to the hours when the sun is above the horizon, in the part of the hemisphere where the disturbance may be imagined to originate, whilst the other division, or that in which the ratios are below unity, and manifest hours of comparative tranquillity, may be viewed as the hours of night at the same locality. The solar hours at a station of observation which are characterized, by disturbance ratios above unity, will in such case correspond in absolute time with the hours of the *day* at the supposed originating locality, modified (it may be) by a more or less rapid transmission of the disturbance. It will be understood, that in this hypothetical suggestion, the purpose in view is to aid the imagination, if it may be so, in apprehending the *ensemble* of the phenomena as far as they are yet known to us, rather than to advance a theoretical explanation, when we have not yet sufficient facts before us by which it may be judged ; it may be remarked, however, that the conception of a double locality of origination of the disturbances (easterly and westerly) in the one hemisphere will present no especial difficulty to those who are conversant with the general facts of terrestrial magnetism.

If our attention be limited to the consideration of the facts observed at a single station, unaccompanied by a view of corresponding phenomena elsewhere, we might be in danger of regarding some of the features, particularly perhaps those which are not the most prominent, as having an accidental rather than a systematic origin ; and we might thus lose a portion of the instruction which they may otherwise convey. On this account it has appeared desirable to exhibit the phenomena as observed at a second station, in comparison with those at Kew ; and I have selected for this purpose the results of a similar investigation to the present at Hobarton in Tasmania ; not only because the facts have been remarkably well determined there, but also because, though it is a very distant station, differing widely in geographical latitude and longitude, and situated indeed in a different hemisphere, there is a striking resemblance in the laws of the magnetic storms experienced at both. This resemblance, which is not only general, but extends to very minute particulars, is such that it seems impossible to resist the impression that the accordance cannot be accidental ; and that the methods of observation and of analysis which have been pursued, have proved themselves well adapted to open to us the knowledge of the existence of systematic laws, pervading and regulating the action of the forces which are in daily operation around us, and are *at least* co-extensive with the limits of our globe ; and thus to lead us ultimately to the correct theory of these forces. I have placed therefore beside each other in the next Table the Ratios of Disturbance at the different hours of local solar time at each of the two stations, separating them as before, into westerly and easterly deflections, and placing the westerly deflections at Kew in immediate juxtaposition with the easterly at Hobarton, and *vice versâ*, as that obviously constitutes the just comparison. The Hobarton Ratios exhibit the relative prevalence of disturbance at the several hours, derived from hourly observations continued for seven years and nine months, viz. from January 1, 1841, to September 30, 1848 ; a series unparalleled in duration at any other of the Colonial Observatories, and which has borne admirably, as I shall hope to have a future opportunity of explaining to the Society, an unquestionable test of its substantial accuracy and fidelity. The number of recorded hourly observations was 56,202, of which 7638 differed from their respective normals of the same month and hour by an amount equalling or exceeding 2'·13 of arc, and constituted the body of separated observations from which the aggregate values of disturbance at the different hours and their ratios have been obtained. The proportion of disturbed observations thus separated, to the whole body of observations, is about 1 in 7·35 ; differing very little from the proportion already noticed as obtained at Kew by a separating value of 5'·3. The disturbing effects due to magnetic storms are therefore somewhat greater at Kew than at Hobarton, though some portion of the difference may be ascribed to the circumstance, that the terrestrial horizontal force, antagonistic to the disturbing forces and tending to retain the magnet in its mean position, is less at Kew than at Hobarton, in the proportion, approximately, of 3·7 to 4·5.

TABLE II.—Showing the comparison of the Ratios of the larger Disturbances of the Declination at the different hours of local solar time at Kew and Hobarton.

Local astronomical hours.	KEW. Westerly deflection.	HOBARTON. Easterly deflection.	KEW. Easterly deflection.	HOBARTON. Westerly deflection.	Local civil hours.
18	1·85	1·18	0·37	0·42	6 A.M.
19	1·83	1·75	0 38	0·44	7 A.M.
20	1·48	1·76	0·36	0·62	8 A.M.
21	1·23	1·47	0·38	0·60	9 A.M.
22	1·26	1·38	0·33	0·54	10 A.M.
· 23	1·21	1·31	0·39	0·53	11 A.M.
0	1·38	1·17	0·54	0·67	Noon.
1	1·44	1·44	0·48	0·56	1 P.M.
2	1·53	1·31	0·54	0·68	2 P.M.
3	1·71	1·56	0·34	0·60	3 P.M.
4	1·35	1·58	0·44	0·50	4 P.M.
5	1·15	1·41	0·51	0·42	5 P.M.
6	0·94	1·10	0·91	0·68	6 P.M.
7	0·37	0·62	1·19	0·90	7 P.M.
8	0·22	0·37	1·56	1·50	8 P.M.
9	0·29	0·22	1·79	1·87	9 P.M.
10	0·20	0·17	2·25	2·20	10 P.M.
11	0·12	0·22	2·38	2·43	11 P.M.
12	0·49	0·38	2·21	2·15	Mid.
13	0·47	0·41	1·98	1·74	1 A.M.
14	0·49	0·53	1·80	1·35	2 A.M.
15	0·51	0·71	1·45	1·25	3 A.M.
16	0·97	1·01	0·95	0·85	4 A.M.
17	1·53	0·96	0·45	0·48	5 A.M.

For the convenience of those who prefer graphical illustration, I have represented on an accompanying woodcut the results to which I have referred. The curves drawn in unbroken black lines, in figures 1 and 2, show the phenomena at Kew; those in dotted lines in the same figures, the phenomena at Hobarton. Fig. 1 presents westerly disturbances at Kew, and easterly at Hobarton in comparison with each other; they are obviously allied phenomena. Fig. 2 presents easterly disturbances at Kew and westerly at Hobarton; these are also, obviously, allied phenomena, but are as obviously governed by distinct laws from those in fig. 1.

Had the phenomena at Kew and Hobarton been the only ones known to us, we might have inferred that we had obtained the characteristic forms of the diurnal variations due to the action of two distinct and independent forces; and we might have expected with some degree of confidence to have found curves of corresponding form by a similar analysis elsewhere;—and so far experience has been in accord with expectation. But, as the *forms* of these two pair of curves are not only respectively similar, but as they also correspond in the *hours* at which their chief characteristic features occur, we might also have formed an inference which would have proved erroneous, viz. that the hours as well as the forms would be the same at other stations. Now this is so far from being in accordance with the facts which we already possess, that whilst the

Mean Diurnal Disturbance Variation of the Magnetic Declination.
Figs. 1 and 2, Kew and Hobarton. Fig. 3, St. Helena.

Local Astronomical Hours.

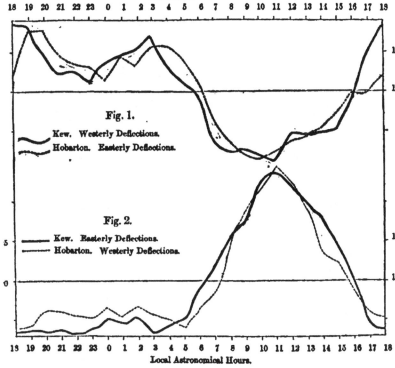

Fig. 1.

Kew. Westerly Deflections.
Hobarton. Easterly Deflections.

Fig. 2.

Kew. Easterly Deflections.
Hobarton. Westerly Deflections.

Local Astronomical Hours.

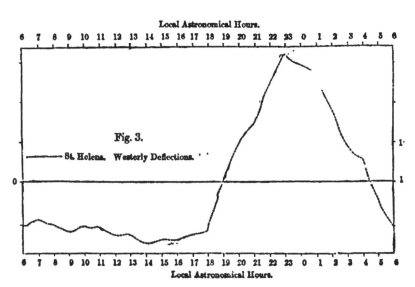

Local Astronomical Hours.

Fig. 3.

St. Helena. Westerly Deflections.

Local Astronomical Hours.

forms present generally a marked resemblance, the hours at different stations exhibit every variety.   To exemplify this I have given in a third figure the curve of the westerly disturbance-diurnal variation at St. Helena, of which the form is manifestly the same as that of the two curves in fig. 2, whilst the hours of its most marked features exhibit a difference of nearly 12 hours of local time from those in fig. 2.

It may not be unsuitable on the present occasion to take a brief retrospective view of the progress of our knowledge respecting these remarkable phenomena, videlicet, the *casual magnetic disturbances*, or *magnetic storms*.   Antecedently to the formation of the German Magnetic Association and the publication of its first Annual Report in 1837, our information concerning them went no further than that there occurred at times, apparently not of regular recurrence, extraordinary agitations or perturbations of the magnetic needle, which had been noticed in several instances to have taken place *contemporaneously* in parts of the European continent distant from each other; and to have been accompanied by remarkable displays of Aurora, seen either at the locality itself where the needle was disturbed, or observed contemporaneously elsewhere.   The opinion which appears to have generally prevailed at this time, was that the Aurora and the magnetic disturbances were kindred phenomena, originating probably in atmospherical derangements, or connected at least in some way with disturbances of the atmospherical equilibrium.   They were classed accordingly as " Meteorological Phenomena," and were supposed to have a local, though it might be in some instances a wide, extension and prevalence.

The special purpose of the German Magnetic Association was to subject the *"irregular magnetic disturbances"* (as they were then called in contradistinction to the regular periodical and secular variations) to a more close examination, by means of systematized observations made simultaneously in many parts of Germany.   With this view, six concerted days in each year were set apart in which the direction of the declination-magnet should be observed with great accuracy, by methods then for the first time introduced, at successive intervals of five minutes for twenty-four consecutive hours ; the meteorological instruments being observed at the same time.   The clocks at all the stations were set to Göttingen mean time (Göttingen being the birth-place of the Association), and the observations were thus rendered strictly simultaneous throughout.   The high respect entertained for the eminent persons with whom the scheme of the Association originated, obtained for it a very extensive cooperation, not limited to Germany alone, but extending over a great part of the European Continent.   The observations of the "Term-days," as they were called, were maintained until 1841, and were all transmitted to Göttingen for coordination and comparison.

The principal results of this great and admirably conducted cooperative undertaking were published in works well known to magneticians.   They may be summed up as follows :—The phenomena which were the subjects of investigation were shown to be of casual and not regular occurrence; to prevail contemporaneously everywhere within the limits comprehended by the observations; and to exhibit a correspondence surprisingly great, not only in the larger, but even in almost all the smaller oscillations; so that, in the words of the Reporters, MM. Gauss and Weber, "nothing in fact remained which could justly be ascribed to *local* causes."

Equally decided were the conclusions drawn against the previously imagined connexion between the magnetic disturbances and derangements of the atmosphere, or particular states of the weather. No perceptible influence whatsoever on the needle appeared to be produced either by wind-storms or by thunder-storms, even when close at hand.

The correspondence in the simultaneous movements of the declination-magnet, so strikingly manifested over an area of such wide extent, was however more remarkable in respect to the *direction* of a perturbation than to its *amount*. The disturbances at different stations, and even, as was expressly stated, at *all* the stations, coincided, even in the smaller instances, in time and in direction, but *with dissimilar proportions of magnitude*. Thus it was found generally that by far the greater number of the anomalous indications were smaller at the southern stations and larger at the northern; the difference being greater than would be due to the difference in the antagonistic retaining force (*i.e.* the horizontal force of the earth's magnetism, which is greater at the southern than at the northern stations). The generality of this occurrence led to the unavoidable inference, that, in Europe, the energy of the disturbing force must be regarded weaker as we follow its action towards the south.

A close and minute comparison of the simultaneous movements at stations in near proximity to each other led to the further conclusion, also stated to be unavoidable, that "various forces must be admitted to be contemporaneously in action, being probably quite independent of each other, and having very different sources; the effects of these various forces being intermixed in very dissimilar proportions at various places of observation according to the directions and distances of these from the sources whence the perturbations proceed." (Resultate aus den Beob. des Mag. Vereins, 1836. pp. 99, 100.) The difficulty of disentangling the complications which thus occur at every individual station was fully foreseen and recognized; and the Report, which bears the initial of M. Gauss, concludes with the remark that "it will be a triumph of science, if at some future time we should succeed in reducing into order the manifold intricacies of the combinations, in separating from each other the several forces of which they are the compound results, and in assigning the source and measure of each."

Such was the state of the inquiry when it was entered upon by the Royal Society. The Report of the Committee of Physics drawn up (*inter alia*) for the guidance of the Magnetic Observatories established by H.M. Government for a limited period in four of the British Colonies, bears date in 1840. The objects proposed by this Report were a very considerable enlargement upon those of the German Association, as well as an extension of the research to more distant parts of the globe. The German observations had been

limited for the most part to one only of the three elements required
in a complete investigation. When the German Association commenced its operations, the Declination was the sole element for which
an apparatus had been devised capable of recording its variations
with the necessary precision. To meet the deficiency in respect to
the horizontal component of the magnetic force, M. Gauss constructed
in 1837 his bifilar magnetometer, which was employed at Göttingen
and at some few of the German stations, concurrently with the
Declinometer, in the term observations of the concluding years of
the Association. But an apparatus for the corresponding observation of the vertical portion of the Force was as yet wholly wanting ;
without such an apparatus as a companion to the bifilar, no determination could be made of the perturbations or momentary changes
of the magnetic Dip and Force : and without a knowledge of these no
satisfactory conclusion in regard to the real nature, amount and
direction of the perturbing forces could be expected. The ingenuity
of Dr. Lloyd supplied the desideratum by devising the vertical force
magnetometer, which, with adequate care, has been found scarcely, if
at all, inferior to the bifilar in the performance of its work. The
scheme of the British Observatories was thus enabled to comprehend
all the data required for the investigation of the casual disturbances,
whether that investigation was to be pursued as before by concerted
simultaneous observations at different stations, or, as suggested in the
Report, *by the determination of the laws, relations and dependencies
of the disturbances at individual stations obtained independently and
without concert with other observers or other stations.* Thus, in
reference to these particular phenomena, the British system was both
an enlargement and an extension of the objects of the German Association ; but it also embraced within its scope the determinations
with a precision, not previously attempted, of the *absolute values* of
the three elements, and of *the periodical and progressive changes* to
which they are subject ; premising however, and insisting with a
sagacity which has been fully justified by subsequent experience, on
the necessity of eliminating in the first instance the effects of the
casual and transitory variations, as an indispensable preliminary to a
correct knowledge and analysis of the progressive and periodical
changes. A further prominency was given to investigations into the
particular class of phenomena which form the subject of this paper,
by the declaration that "the theory of the transitory changes is in
itself one of the most interesting and important points to which the
attention of magnetic inquirers can be turned, as they are no doubt
intimately connected with the general causes of terrestrial magnetism,
and will probably lead us to a much more perfect knowledge of these
causes than we now possess."

The instructions contained in the Royal Society's Report for the
adjustments and manipulation of the several instruments provided for
these purposes were clear, simple and precise. In looking back upon
them after the completion of the services for which they were
designed, it is impossible to speak of the instructions otherwise than
with unqualified praise. But the guidance afforded by the instructions terminated with the completion of the observations. To have

attempted to prescribe the methods by which conclusions, the nature of which could not be anticipated, should be sought out from observations not yet made, would have been obviously premature. Yet without some discussion of the results, the mere publication of unreduced observations is comparatively valueless. It has been well remarked by an eminent authority, whose opinions expressed in the Royal Society's Report have been frequently referred to in the course of this paper, that " a*man may as well keep a register of his dreams, as of the weather, or *any other set of daily phenomena,* if the spirit of grouping, combining, and eliciting results be absent." It was indispensable that the attempt should be made to gather in at least the first fruits of an undertaking on which a considerable amount of public money and of individual labour had been expended; and the duty of making the attempt might naturally be considered to rest on the person who had been entrusted with the superintendence of the Government Observatories. The methods and processes adopted for reducing, combining, eliminating, and otherwise eliciting results were necessarily of a novel description; they were in fact an endeavour to find a way by untrodden paths to simple and general phenomenal laws where no definite knowledge of the origin or mode of causation of the phenomena previously existed. Happily it is not necessary to trespass on the time or attention of the Society by a description of the methods and processes which have been employed to elucidate some of the leading features of the magnetic storms, as these are fully described in the discussions prefixed to the ten large volumes in which the observations at the Colonial Observatories have been printed. It will be only necessary to advert, and that very briefly, to some of the principal conclusions which may be supposed to throw most light on the theory of these phenomena.

The results of the extension of the term-day comparisons to the American Continent, and to the Southern Hemisphere and the Tropics, may first be disposed of in a very few words. The contemporaneous character of the disturbances, which had been shown by the German term-observations to extend over the larger portion of the European Continent, manifested itself also in the comparisons of the term-days in 1840, 1841, and 1842 at Prague and Breslau in Europe, and Toronto and Philadelphia in America, published in 1845; and the same conclusion was obtained by comparing with each other the term-days at the Colonial Observatories, situated in parts of the globe most distant from one another. The days of disturbance still appeared to be of casual occurrence, but were now recognized as affections common to the *whole globe,* showing themselves simultaneously at stations most widely removed from each other. When distant stations were compared, as for example stations in Europe with those in America, and either or both with Tasmania, discrepancies in the amount of particular perturbations, similar to those which had been found in comparing the European stations with each other, presented themselves, but larger and more frequent, and extending occasionally even to the *reversal* of the *direction* of the simultaneous disturbance. Instances were not unfrequent of the same element, or of different elements, being disturbed at the same observation-instant in Europe and

America; and on the other hand, there were perturbations, sometimes of considerable magnitude, on the one continent, of which no trace was visible on the other. Hence it was concluded, with the increased confidence due to this additional and more extensive experience, that various forces proceeding from different sources were contemporaneously in action; and it was further inferred that the most suitable and promising mode of pursuing the investigation was by an endeavour to analyse the effects produced at individual stations, and to resolve them if possible into their respective constituents.

The hourly observations which had been commenced at the Colonial stations in 1841 and 1842, and continued through several subsequent years, furnished suitable materials for this investigation, the first fruits of which were the discovery, that the disturbances, though casual in the times of their occurrence, and most irregular when individual perturbations only were regarded, were, in their *mean* effects, *strictly periodical phenomena*; conforming in each element, and at each station, on a mean of many days, to a law dependent on the solar hour; thus constituting a systematic mean diurnal variation distinct from the regular daily solar-diurnal variation, and admitting of being separated from it by proper processes of reduction. This conformity of the disturbances to a law depending on the solar hours was the first known circumstance which pointed to the sun as their primary cause, whilst at the same time a difference in the *mode* of causation of the regular- and of the disturbance-diurnal variations seemed to be indicated by the fact, that in the disturbance-variation the local hours of maximum and minimum were found to vary (apparently without limit) in different meridians, in contrast to the general uniformity of those hours in the previously and more generally recognized regular solar-diurnal variation.

This first reference of the magnetic storms to the sun as their primary cause, was soon followed by a far more striking presumptive evidence of the same, by a further discovery of the existence of a periodical variation in the frequency of occurrence, and amount of aggregate effects, of the magnetic storms, corresponding in period, and coincident in epochs of maximum and minimum, with the decennial variation in the frequency and amount of the spots on the sun's disk, derived by Schwabe from his own systematic observations commenced in 1826 and continued thenceforward. The decennial variation of the magnetic storms is based on the observations of the four widely distributed Colonial Observatories, and is concurred in by all. This remarkable correspondence between the magnetic storms and physical changes in the sun's photosphere, of such enormous magnitude as to be visible from the earth even by the unassisted eye, must be held to terminate altogether any hypothesis which would assign to the cause of the magnetic disturbances a local origin on the surface or in the atmosphere of our globe, or even in the terrestrial magnetism itself, and to refer them, as cosmical phenomena, to direct solar influence; leaving for future solution the question of the *mode* in which that influence produces the effects which we believe we have thus traced to their source in the central body of our system *.

* The existence of a decennial period of the magnetic storms was not, as some

We may regard as a step towards this solution the separation of the disturbances of the declination into two distinct forces acting in different directions and proceeding apparently from different foci; the phenomena of distinct (though in so many respects closely allied) variations exhibit the same peculiar features at all the stations to which the analysis has hitherto extended, and have been exemplified by the observations at Kew, as shown in the early part of this paper. A similar separation into two independent affections, each having its own distinct phenomenal laws, has followed from an analysis of the same description applied to the disturbances of the magnetic dip and force at the Colonial stations; thus placing in evidence, and tracing the approximate laws of the effects of six distinct forces (two in each element) contemporaneously in action in all parts of the globe, and pointing in no doubtful manner to the existence of two terrestrial foci or sources in each hemisphere from which the action of the forces emanating from the sun and communicated to the earth may be conceived to proceed. Such an ascription naturally suggests to those conversant with the facts of terrestrial magnetism the possibility that Halley's two terrestrial magnetic foci in each hemisphere may be either themselves the localities in question, or may be in some way intimately connected with them. The important observations which we owe to the zeal and devotion of Captain Maguire, R.N. and the Officers of H.M.S. 'Plover,' have made us acquainted with Point Barrow as a locality where the magnetic disturbances prevail with an energy far beyond ordinary experience, indicating the proximity of that station to the source or sources from which the action of the forces may proceed. Now Point Barrow is situated in a nearly intermediate position between what we believe to be the present localities of Halley's northern foci, and at no great distance from either : in such a situation the exposure to disturbing influences proceeding from both might well be supposed to be very great. The displays of Aurora at Point Barrow exceed also in numerical frequency any record received from any other part of the globe.

The further prosecution of this investigation appears to stand in need of some more systematic proceeding than would be supplied by the uncombined efforts of individual zeal. Observations similar to those of the Kew Observatory, made at a few stations in the middle latitudes of the hemisphere, distributed with some approach to symmetry in their longitudinal distances apart, would probably fur-

have supposed, a fortuitous discovery; but a consequence of a process of exami-nation early adopted and expressly devised, by the employment of a *constant* sepa-rating value, to make known any period of longer or shorter duration which might fall within the limits comprised by the observations. The period being decennial, and the epoch of minimum occurring at the end of 1843 or beginning of 1844, the epoch of maximum was necessarily waited for in order to ascertain the precise duration of the cycle. The maximum took place in 1848–1849, the observations in 1850 and 1851 showing that the aggregate value of the annual disturbances was again diminishing as it had been in 1842 and 1843. The process of deter-mining the proportion of disturbance in different years is a somewhat laborious one, and requires time; but in March 1852, I was able to announce to the Royal Society the existence of a decennial variation, based on the concurrent testimony of the observations at Toronto and Hobarton; deeming it proper that so remark-able a fact should not be publicly stated until it had been thoroughly assured by independent observations at two very distant parts of the globe.

nish data, which by their combination might serve to assign the localities from whence the disturbances are propagated—contribute still further to disentangle the complications of the forces which produce them,—and thus hasten the attainment of that "triumph of science" foreseen and foreshadowed by the great geometrician of the last age. Of such a nature was the scheme contemplated by the Joint Committee of the Royal Society and British Association, and submitted to H.M. Government in the hope of obtaining their aid in the execution of such part of it as fell within British dominion; and of thus "maintaining and perpetuating our national claim to the furtherance and perfecting of this magnificent department of physical inquiry." (Herschel in 'Quarterly Review' September 1840, p. 277.) The scheme was no unreasonable one: probably eight or nine stations in the contour of the hemisphere might suffice; and of these we already possess the observations at Toronto; those at Kew are in progress; and self-recording instruments, similar to those at Kew, are now under verification at Kew preparatory to being employed on the Western or Pacific side of the United States Territory, at a point not far from the previously desired Station of Vancouver Island, for which a substitute is thus provided. This Observatory, as well as one at Key West on the southern coast of the United States, in which self-recording instruments are already at work, will be maintained under the authority and at the expense of the American Government, and both have been placed under the superintendence of the able and indefatigable director of the "Coast Survey," Dr. Alexander Dallas Bache. The Russian Observatory at Pekin, the trustworthy observations of which are already known to the Society, is understood to have recommenced its hourly observations, and stands only in need of an apparatus for the vertical force (which might be readily supplied from this country), to contribute its full complement to the required data. More than half the stations may therefore be regarded as already provided for, and there are other Russian observatories in the desired latitudes and longitudes which might be completed with instruments for a full participation.

It would be wrong to conclude these imperfect notices without recognizing how greatly the researches have been aided in their progress by the united and unfailing countenance and support of the Royal Society and of the British Association. The Kew Observatory owes its existence and maintenance to funds most liberally supplied from year to year by the British Association; and the cost of the self-recording magnetic instruments, of which the first instalment of the results has formed the early part of this paper, was supplied from funds at the disposal of the Council of the Royal Society.

GEOLOGICAL SOCIETY.

[Continued from p. 247.]

June 5, 1861.—Leonard Horner, Esq., President, in the Chair.

The following communications were read:—

1. "On the Occurrence of some large Granite Boulders, at a

great depth, in West Rosewarne Mine, Gwinear, Cornwall." By H. C. Salmon, Esq., F.G.S.

The boulders of granite referred to were found in the 50-fathom level below the adit, the adit being 24 fathoms from the surface. One of the boulders was 4 feet 2 inches, and another 3 feet 10 inches in diameter; there were five other smaller boulders or pebbles also met with in the level. The boulders are in the killas close to the lode, and both the lode and the "country" near the lode are made up of brecciated killas. After quoting the details of somewhat similar phenomena formerly observed at Relistian and Herland Mines, the author treated of the probable origin of the boulders in question; and although lodes are regarded by some as having been formed from below upwards, yet in this case the author thinks that the boulders must have had a common origin with the lode, and have been introduced by a fissure from the surface.

2. " On an erect Sigillaria from the South Joggins, Nova Scotia." By Dr. J. W. Dawson, F.G.S.

This specimen, presenting the external markings of leaf-scars and ribs with more than usual clearness and with some instructive peculiarities, has afforded to the author the type of a new species, *Sigillaria Brownii*. Observations on the probable mode of growth, on the structure, and on the classification of *Sigillariæ*, were also given in this paper, together with a *résumé* of the observations previously published regarding *Sigillaria* by Brongniart, Corda, and others.

3. "On a Carpolite from the Coal-formation of Cape Breton." By Dr. J. W. Dawson, F.G.S.

Numerous *Trigonocarpa* belonging to a new species (*Trigonocarpum Hookeri*) occur in a thin calcareous layer in the coal-measures near Port Hood, Cape Breton. The author thinks it highly probable that though some *Trigonocarpa* may have belonged to Conifers, yet in this case they were the seeds of *Sigillaria*.

4. "On a Reconstructed Bed on the top of the Chalk." By W. Whitaker, Esq., B.A., F.G.S.

At some places near Reading (Maidenhatch Farm, about six miles to the W.; and Tilehurst, two miles to the S.W.), and also near Maidenhead, from 18 to 20 feet of broken chalk overlies the true chalk; and in places is overlain by the bottom-bed of the Reading Beds, and therefore must have been reconstructed before the deposition of the Tertiary strata. For the most part, however, in Berkshire the Woolwich and Reading Beds rest on an undisturbed surface of the Chalk. In Wiltshire also the author has observed similarly reconstructed chalk, probably there also underlying Tertiary beds; and he suggests that possibly the local reconstruction of the Chalk may have been contemporaneous with the formation of the Thanet Sands further to the east.

5. "On some of the Higher Crustacea from the British Coal-measures." By J. W. Salter, Esq., F.G.S.

In this paper were described, (1) a new Macrurous Crustacean, under the name of *Anthrapalæmon Grossarti*, from the slaty band of the black-band ironstone of the coal-measures, Goodhock Hill,

Shotts, Lanarkshire. (2) The Macrurous Crustacean of which an imperfect specimen was figured in Mr. Prestwich's memoir on the Coalbrook Dale Coal-field (plate 41, fig. 9, *Apus dubius*): this is referred to a subgenus (*Palæocarabus*) of the genus *Anthrapalæmon*; and another specimen from Ridgeacre Colliery was referred to. (3) A specimen from the Carboniferous Limestone of Derbyshire. (4) A small Crustacean, from the Mountain-limestone of Fifeshire, figured and described by the author in the 'Transactions of the Royal Society of Edinburgh,' vol. xxii. p. 394, as *Uronectes socialis*, but now regarded by him as belonging to the Macrura.

XLI. *Intelligence and Miscellaneous Articles.*

ANALYSIS OF GYROLITE. BY HENRY HOW, PROFESSOR OF CHE-
MISTRY, KING'S COLLEGE, WINDSOR, NOVA SCOTIA.

THE mineral gyrolite was first described by Professor Anderson of Glasgow[*] as a new species from the Isle of Skye; it is stated by Greg and Lettsom[†] to occur without doubt at two localities in Greenland, and, according to Heddle, at Faröe. The only other notice of it that I am acquainted with is by L. Sæmann[‡], who mentions that he examined a specimen, no locality being given, mixed or interlaminated with pectolite, and suggests that this mineral losing its alkali becomes gyrolite, and losing its lime becomes Okenite. No other analysis than the original one of Professor Anderson has, I believe, been published; the following account of its occurrence among the minerals of Nova Scotia shows it in such association as affords a mode of explaining its origin by change in apophyllite:—I met with it in Anapolis County, N. S., some twenty-five miles south-west of Cape Blomidon, between Margaretville and Port George, on the surface of fractured crystalline apophyllite; and on further breaking the mass a good many spherical concretions of pearly lustrous plates were observed in the interior, of sizes varying from that of a pin's head to nearly half an inch in diameter: their outline was well defined, and the external characters as given by Anderson were recognized on examination; it afforded the following results on analysis:—The mineral was ignited for water, and the residue treated with HCl; the resulting dried silica was weighed, and then fused with carbonated alkali; and the weight of the small quantities of alumina, &c. so separated was deducted from that of the first silica. I place my numbers by the side of those of Anderson, and give the calculated per-centages for his formula:—

	How.	Anderson.	Calculation.		
Potassa	1·60	....			
Magnesia	0·08	0·18			
Alumina	1·27	1·48			
Lime	29·95	33·24	32·26	2CaO	= 56
Silica	51·90	50·70	52·18	2SiO'	= 90·6
Water	15·05	14·18	15·55	3HO	= 27·0
	99·85	99·78	99·99		173·6

* Trans. Roy. Soc. Edinb., and Phil. Mag. Feb. 1851.
† Manual of Mineralogy, p. 217.
‡ First Supp. to Dana's Mineralogy, p. 9.

A general accordance is observed sufficient to show the identity of chemical composition in the minerals examined; the small quantity of potassa present in my specimen probably modified the blowpipe character a little, as I found it not to exfoliate completely, and it fused without any difficulty, and even with some ebullition.

Some of the numerous cavities in the apophyllite were empty, some entirely filled with gyrolite, and in others separate plates of this mineral were standing edgewise, leaving vacant spaces, while upon and by the side of the plates were in some cases rhombohedral crystals which proved to consist of calcite, and were sometimes present alone in the cavities, which varied from being quite shallow to half an inch in depth. It is mentioned by Anderson that gyrolite occurs associated with stilbite, Laumonite, and other zeolites, and is sometimes found coating crystals of apophyllite.

The difference in chemical composition between apophyllite and gyrolite is very well seen on comparing the respective theoretical per-centages of their constituents, thus :—

	SiO³.	CaO.	KO.	HO.
Apophyllite	= 52·70	26·00	4·40	16·70+HF variable.
Gyrolite	= 52·18	32·26		15·50

and the existence of the calcite in the cavities seems clearly to show that the gyrolite is formed from the apophyllite by the waters which deposited the carbonate of lime reacting on the silicate of potash, and dissolving out at the same time the fluorine as fluoride of calcium* : trial was made for fluorine on two fragments of the gyrolite, but no evidence of its existence obtained.—Silliman's *American Journal*, July 1861.

PRODUCTION OF THE GREEN MATTER OF LEAVES UNDER THE
INFLUENCE OF THE ELECTRIC LIGHT. BY M. HERVÉ MANGON.

It appeared interesting to ascertain whether the green matter developed so readily in young leaves exposed to the sun, was also produced under the influence of the bright light of the electric lamp. This experiment has been tried by the kind aid of M. Allard, chief engineer of lighthouses, who has allowed me to use for several days the powerful apparatus under his control.

The electricity was produced by a powerful electro-magnetic machine driven by a steam-engine. The light was that of a charcoal lamp.

The lamp was lit for eleven hours on the 30th of July, twelve hours on July 31, Aug. 1, and Aug. 2, and eleven hours and a half on Aug. 3. The temperature of the air varied from 22° to 25° C., and that of the earth from 19° to 21° C.

On the 30th of July, at 8 in the morning, small flower-pots, each containing four grains of rye, sown respectively on the 24th, 26th, 27th, and 28th of July, were placed in a perfectly dark room, about a yard from the lamp, and about 2 feet below the luminous focus, and without the interposition of any glass.

The grains sown on the 24th and 26th had sprouted; the stalks

* See Dana's Mineralogy, vol. i. pp. 232, 233.

were 0·005 metre to 0·012 metre in length. There was a slight green tint on the top of one of these plants; the other was quite white. The grains sown on the 27th and 28th of July had not sprouted on the 31st of July at 2 o'clock; the plants sown on the 24th and 26th of July were 0·010 metre to 0·060 metre in length; they were all *very green*, and strongly turned towards the light. The grain sown on the 27th of July had sprouted; the plants were 0·020 metre to 0·030 metre high, and there was a little green on the top of one of them.

At 1 o'clock on the 1st of August the plants continued to grow just as in the light. The rye sown on the 28th of July had sprouted, but showed no green.

On the 2nd of August, at 2 o'clock, all the plants continued to grow; the rye which had sprouted on the night before was decidedly green.

The seeds kept in the dark for the sake of comparison, gave plants which were completely yellow.—*Comptes Rendus*, Aug. 5, 1861.

NATURE OF THE DEPOSIT WHICH FORMS UPON THE COPPER EMPLOYED IN REINSCH'S TEST FOR ARSENIC.

Lippert has made a careful examination of the crust which forms upon bright metallic copper when this is placed in a solution of arsenic acidified with chlorhydric acid. This coating had been pretty generally mistaken for metallic arsenic until Fresenius (in his *Anleitung zur qualitativen Analyse*, 10te Aufl., Braunschweig, 1860, p. 141) called attention to the fact that it contained a large quantity of copper. From the experiments of Lippert, it now appears that the crust in question contains only 32 per cent. of arsenic, 68 per cent. of its weight being copper. This composition having been nearly constant in several specimens which he analysed, Lippert maintains that the compound is a definite alloy, $AsCu^5$. When ignited, at the temperature of a combustion furnace, in a current of hydrogen, the compound lost only 7 per cent. of its weight, an alloy of the composition $AsCu^6$ (same as that of the mineral Domeykite of F. Field) being formed.

The delicacy of Reinsch's test is evidently directly referable to the large amount of copper which the characteristic coating contains; for a proportionally small quantity of arsenic is thus obtained in an enlarged and, as it were, more tangible form. But, on the other hand, it is not easy to prove in a simple manner the presence of arsenic in this crust; for only a small portion of the arsenic can be volatilized in a current of hydrogen, and even if the alloy be first oxidized in a current of air and then reduced in a current of hydrogen, the per-centage of arsenic only falls from 32 to 20. By far the largest portion of the arsenic is therefore kept out of sight.

For the details of this interesting research, and the author's discussion of the proposition of Reinsch and v. Kobell to estimate arsenic quantitatively by determining the amount of copper which dissolves while the arsenic is being precipitated, we must refer to the original article.—*Journ. für Prakt. Chemie*, vol. lxxxi. p. 168.

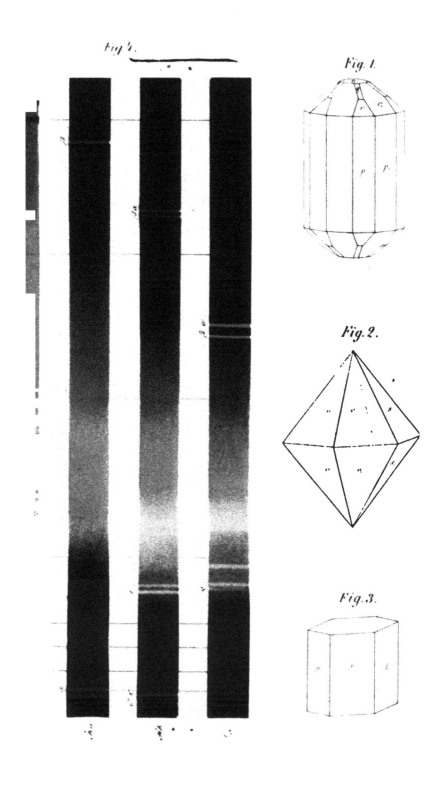

Fig. 4.

Fig. 1.

Fig. 2.

Fig. 3.

THE

LONDON, EDINBURGH and DUBLIN

PHILOSOPHICAL MAGAZINE

AND

JOURNAL OF SCIENCE.

[FOURTH SERIES.]

NOVEMBER 1861.

XLII. *Chemical Analysis by Spectrum-observations.—Second Memoir.* By G. Kirchhoff *and* R. Bunsen*.

[With a Plate.]

IN our first memoir on this subject, a translation of which appeared at page 89 of the 20th volume of this Magazine, we showed that the bright lines observed in the spectra of the incandescent vapour of certain metallic compounds may be employed as the most sure and delicate tests for ascertaining the presence of these metals. The analytical method founded upon such observations is of special importance when applied to the examination of groups of substances either occurring in very small quantities or possessing nearly identical chemical characters, because in these special cases this mode of examination introduces a whole series of most delicate distinctive reactions, hitherto wholly unknown. Considering how much more delicate the spectrum reactions are than the ordinary chemical tests, it appeared to us that this method would serve especially well for discovering the presence of substances which might have been overlooked by the rough methods previously employed, either because the bodies occurred in very small quantities, or because they were not distinguishable by the ordinary tests from other well-known bodies. This assumption was verified by the first appeal to experiment; for we have succeeded in discovering the presence of two new alkaline metals in addition to potassium, sodium, and lithium, notwithstanding that they both give all the characteristic precipitates of the potash salts, and occur in such minute quantities that, in order to obtain a few grammes of these substances, or a sufficient amount for investigation, we had to operate upon 44,000 kilogrammes (about 40 tons) of the mineral water of Dürkheim, and upon 180 kilogs. of Lepidolite.

* From Poggendorff's *Annalen*, No. 7, 1861. Communicated by Professor Roscoe.

Phil. Mag. S. 4. Vol. 22. No. 148. *Nov.* 1861.   Z

If a drop of the mother-liquor of the Dürkheim water be brought into the flame of the spectrum-apparatus, the characteristic lines of sodium, potassium, lithium, calcium, and strontium are at once seen. If the lime, strontia, and magnesia be separated according to well-known processes, and if the residual alkaline bases in the form of nitrates be washed out with alcohol and the lithium removed as completely as possible by precipitation with carbonate of ammonium, a mother-liquor is obtained which in the spectrum-apparatus shows the lines of sodium, potassium, and lithium, but besides these, two splendid blue lines situated close together, and almost coinciding with the blue strontium line Sr δ.

As no known elementary body produces two blue lines in this portion of the spectrum, we may consider the existence of this hitherto unknown alkaline element as thus placed beyond a doubt.

The facility with which a few thousandths of a milligramme of this body may be recognized by the bright blue light of its incandescent vapour, even when mixed with large quantities of the more common alkalies, has induced us to propose for it the name *Cæsium* (and the symbol Cs), derived from the Latin "cæsius," used to designate the blue of the clear sky*.

If Saxony lepidolite be treated by any of the known plans for separating the alkalies from the other constituents, and if the solution of the alkalies thus obtained be precipitated with bichloride of platinum, an abundant precipitate is formed, which, when examined in the spectrum-apparatus, shows only the bright potassium lines. If this precipitate be repeatedly washed with boiling water, and the residual salt occasionally examined in the apparatus, two splendid violet lines, lying between the strontium line Sr δ and the blue potassium line Kβ, will be noticed on the gradually fading continuous background of the potassium spectrum. These new lines increase in brilliancy as the washing is continued, and a number more appear in the red, yellow, and green portions of the spectrum.

None of these lines belong to any previously known body. Amongst them are two which are especially remarkable, as lying beyond Fraunhofer's line A and the potassium line Ka coincident with it, and therefore situated in the outermost portion of the red solar rays. Hence we propose for this new metal the name *Rubidium* (and the symbol Rb), from the Latin "rubidus," which was used to express the darkest red colour†.

Before describing the special spectra of cæsium and rubidium,

---

* Aulus Gellius, in the *Noctes Atticæ*, ii. 26, quotes Nigidius Figulus as follows :—Nostris autem veteribus cæsia dicta est, quæ a Græcis γλαυκῶπις, ut Nigidius ait, de colore cœli quasi cœlia.

† Aulus Gellius, *Noctes Atticæ*, ii. 26. Rubidus autem est rufus atrior et nigrore multo inustus.

we proceed to recount the experiments which one of us has conducted for the purpose of establishing the properties of the two new elements, and their more important compounds.

I. *Of the Preparation, Atomic Weight, and occurrence of the Rubidium Compounds.*

The pure chloride of rubidium was procured from the saline residue obtained by fusing a mass of about 150 kilogrammes of Saxony lepidolite, from which the alkaline earths and lithium salts had been removed. The separation of the new element, and the preliminary determination of its atomic weight, were effected as follows :—

The saline residue was dissolved in water, and treated with about 100 grms. of bichloride of platinum, a quantity, however, quite insufficient to precipitate all the potassium ; the double platinum salt was then boiled out twenty times with a small volume of water, and the boilings added to the original solution of the saline residue, whereby a precipitate again occurred, which was treated exactly as the former. In the course of the process of continued boiling with small quantities of water, the solution, which originally was of a dark yellowish-brown colour, becomes gradually lighter, so that it is easy to see, by the light colour of the precipitate remaining unchanged, the point at which the boiling-out has been continued long enough. The extraction is carried on until the whole of the precipitate formed by the saline residue dissolves on repeated boiling with small quantities of water. The several platinum precipitates, after having been again purified by treating them altogether with boiling water, are dried and reduced in a current of hydrogen gas, by which means a mixture of metallic platinum and impure chloride of rubidium is obtained, the latter being extracted by water. This aqueous solution is diluted, and, whilst boiling, again precipitated by chloride of platinum, and the insoluble double salt reduced, as before, in a current of hydrogen.

Of the chloride of rubidium thus prepared, which we will designate as portion A, 2·2496 grammes gave on precipitation 2·7688 grms. chloride of silver. A portion of this same preparation A was dissolved in about thirty times its weight of water, and precipitated whilst hot with a solution of chloride of platinum so diluted that the precipitate appeared only after the lapse of a few minutes. As the liquid cooled, the precipitate became more dense ; and when the temperature had sunk to about 40° C., it was filtered off, dried, and reduced in hydrogen as described. The chloride thus prepared we will call portion B ; of this preparation 0·9022 grm. gave 1·0712 grm. chloride of silver. A similar mode of separation was adopted in the case of the salt B, and another salt, which we call portion C, obtained ; 1·3540 grm. of this portion yielded 1·6076 grm. chloride of silver. By a repetition of this process on the salt C, a product D was prepared, of which 1·9486 grm. gave 2·3091 grms. chloride of silver. The quantities of chloride of silver obtained from one part by weight of chloride of rubidium after each of these purifications, are therefore,—

A . . . . . . 1·2308
B . . . . . . 1·1873
C . . . . . . 1·1873
D . . . . . . 1·1850

These numbers prove that the products of the three last preparations possess a constant composition. The bright spectrum-lines of cæsium and lithium were almost invisible in the last of these preparations; and the line Kα of potassium could not be seen at all in our spectrum-apparatus; so that we may fairly conclude that the product of the last preparation is pure chloride of rubidium.

In order to obtain a still further proof of the purity of the chloride thus prepared, a method was employed of which we shall again have to speak when discussing the mode of separation of cæsium from potassium and rubidium. This consists in treating the three caustic alkalies with carbonic acid until about one-fifth by weight of the whole mass is converted into carbonates, and then extracting the anhydrous salt with alcohol. If an alkali possessing greater or less basic properties than rubidium, and having a different atomic weight, were present together with this metal, the alcoholic solution must possess a composition differing from that of the residue; the portion of caustic alkali which dissolved in alcohol yielded, however, a chloride possessing a composition identical with that obtained from the portion of alkali undissolved by the alcohol, 0·5116 grm. of the former yielding 0·6078 grm. chloride of silver, or 1 part of chloride giving 1·1830 of chloride of silver, closely corresponding with the previous results. If we only consider the precipitation of the products possessing constant composition, and if we take, according to Stas, the atomic weight of silver to be 107·94, and that of chlorine 35·46, we obtain the following numbers for the atomic weight of rubidium on the hydrogen scale :—

B . . . . . . 85·31
C . . . , . . 85·32
D . . . . . . 85·55
E . . . . . . 85·24

or a mean of Rb = 85·36. The weight of the atom of the new metal is therefore more than twice as great as that of potassium. Although the numbers thus obtained do not coincide with the degree of accuracy which may be desirable in determinations of atomic weights, we believe that the mean experimental number does not differ from the true combining proportion more than is the case with a large proportion of the atomic weights at present considered as correct, and received without question.

It is almost needless to add that the residues obtained by the treatment above described, when operated upon again, yielded a considerable uantity of chloride of rubidium.

Although impossible to determine with exactitude the quantity of rubidium contained in lepidolite, it appeared of interest to ascertain this as accurately as possible. For this purpose a specimen of lepidolite was employed, found at Rozena near Hradisko, which was seen by means of spectrum-analysis to contain traces of cæsium as well as rubidium. The solution obtained from 13·509 grms. of this lepidolite fused with lime, was precipitated in the usual manner with chloride of platinum, and the precipitated double chlorides of potassium, rubidium and platinum reduced with hydrogen, yielding 2·0963 grms. of the chlorides of potassium and rubidium. These salts were again precipitated by chloride of platinum, and the precipitate thus obtained, boiled out with *small* quantities of water until the solution appeared of a light yellow colour. The united wash waters, on evaporation and cooling, deposited a second crop of crystals, which were treated in a similar way to the first precipitate. The platinum double salt which separated out a third time, was likewise submitted to the same treatment, and the operation repeated until the precipitate formed, on concentrating the wash-waters, easily dissolved on boiling in a small quantity of water without leaving any residue. The whole of the insoluble platinum double salt yielded, after reduction in a current of hydrogen, 0·0421 grm. of chloride of rubidium, corresponding to 0·24 part of oxide of rubidium in 100 parts of the lepidolite in question. This determination, together with an analysis of the other constituents, made by Mr. Cooper in my laboratory, gives the following composition for the lepidolite from Rozena:—

Silicic acid	50·32
Alumina	28·54
Peroxide of iron	0·73
Lime	1·01
Magnesia	0·51
Oxide of rubidium	0·24
Oxide of cæsium	trace.
Lithia	0·70
Fluoride of lithium	0·99
Fluoride of sodium	1·77
Fluoride of calcium	12·06
Water	3·12
	99·99

We have assured ourselves in a series of spectrum-analytical investigations, which we here omit, as we shall return to the subject when discussing the properties of the cæsium compounds, that in almost all mineral springs containing chloride of sodium, traces of rubidium compounds accompany the salts of potassium and sodium; so that, although rubidium is found in but small quantities, it is by no means a body of rare occurrence.

II. *Of Metallic Rubidium and some of its compounds.*

a. *The Metal.*

As the total quantity of pure material which we possessed scarcely exceeded one ounce, it would have been unwise to waste the whole of this in one experiment upon the reduction of the metal from the carbonate, and we therefore for the present confined ourselves to separating the metal by means of electrolysis. If a current be passed through fused chloride of rubidium, the positive pole consisting of a rod of graphite, and the negative pole being formed of an iron wire, the metallic rubidium, is seen to rise to the surface of the liquid from the latter, and burns with reddish colour on coming in contact with the air. When the iron wire is surrounded by a small glass bell through which a current of pure dry hydrogen is led, the metal does not burn, but it does not collect in the hollow bell, as it disappears as soon as it is liberated, uniting with the chloride to form a subchloride, which dissolves in the fused mass. This subchloride imparts a deep blue colour to the salt in the neighbourhood of the iron pole; and although this blue mass is perfectly transparent and does not exhibit, either when examined by the naked eye or with the microscope, any trace of a metallic substance, it decomposes water with evolution of hydrogen and with formation of a colourless solution having a strong alkaline reaction; chloride of potassium also forms, under similar circumstances, a blue subchloride. If the reduction be repeated with a mixture containing an equal number of atoms of chlorides of calcium and rubidium at the temperature, almost below redness, at which this mixture fuses, a mass is obtained which evolves large quantities of hydrogen when thrown into water, and from which small grains of metal are thrown out, which ascending take fire spontaneously on coming into contact with the air. The metal cannot, however, be obtained in this way in sufficient quantity to be properly investigated. The amalgam of rubidium can, on the contrary, be very easily prepared from a concentrated solution of chloride of rubidium when metallic mercury is used as the negative pole, and a platinum wire is employed as the positive pole. The mercury is thus quickly changed into an amalgam of rubidium, which on cooling appears as a solid crystalline brittle mass of a silver-white colour. This amalgam decomposes water at the ordinary temperature, absorbs oxygen from the air, becoming hot, and being covered with a white coating of caustic hydrated oxide of rubidium. Rubidium amalgam is strongly electro-positive in respect to potassium-amalgam when a circuit is completed with both by means of the chlorides of rubidium and potassium.

Potassium, therefore, can no longer be considered as the most electro-positive element, for in the foregoing experiment it has been shown to be more electro-negative than rubidium.

b. *Hydrated Oxide of Rubidium.*

This substance is best prepared from the sulphate; the latter salt is dissolved in 100 parts of water, and the solution boiled for some time to free it from air; to the boiling solution hydrate of baryta is cautiously added; the sulphate separates quickly out, so that the point of complete precipitation is easily and accurately reached. If the liquid be now quickly evaporated in a silver basin, the hydrated oxide is obtained as a white, or greyish-white, porous mass, which melts and fuses quickly almost below a red heat. It does not lose its water of hydration at a red heat; on cooling, it solidifies to a brittle though not easily breakable mass, which does not exhibit any crystalline structure. This substance is completely and quickly volatilized when placed in a flame; and placed in contact with water, it dissolves with evolution of great heat. Placed on the skin it acts as a powerful caustic, resembling the hydrate of potash. Exposed to the air it rapidly deliquesces, forming a syrupy liquid which possesses the peculiar oily feeling, when placed on the finger, characteristic of the common alkalies; and it gradually absorbs carbonic acid, at first becoming carbonate, and at last forming bicarbonate of rubidium. Alcohol dissolves this substance as easily as it does caustic potash, and a thick oily liquid is produced. As regards alkaline reaction and alkaline taste, it is not surpassed by potash. The alkali cannot be evaporated in platinum vessels, as it attacks this metal as strongly as caustic potash.

0·7200 grm. of this hydrated oxide of rubidium yielded 0·9286 grm. of sulphate. Hence it consists of—

			Calculated.	Found.
RbO .	. .	93·36	91·21	90·29
HO .	. .	9·00	8·79	9·71
		102·36	100·00	100·00

The somewhat large excess of water here found is explained by the difficulty of obtaining the salt perfectly free from carbonic acid. We have, as yet, not made experiments to determine whether rubidium possesses any higher or lower oxides.

c. *Monocarbonate of Rubidium.*

This salt is best prepared from the sulphate of rubidium by precipitating with baryta water, and evaporating the solution of the caustic alkali to dryness with carbonate of ammonium. The excess of baryta added remains behind on treating the mass with water. The solution yields on concentration indistinct crystals and crystalline crusts of hydrated carbonate of rubidium, which, on heating strongly, melt in their water of crystallization, and leave at last a porous mass, melting at a red heat, and solidifying on cooling to an opake white crystalline salt. The anhydrous salt is strongly hygroscopic, and dissolves in water with evolution of heat. It has a caustic and corrosive action upon the skin. The alkaline reaction of the salt is so powerful that boiled water, to which only $\frac{2}{10,000}$ths of the salt has been added, imparts a distinct alkaline reaction to red litmus paper. The salt is almost insoluble in boiling absolute alcohol, 100 parts of alcohol only dissolving 0·74 of the salt. When fused in a platinum crucible, it does not lose its carbonic acid, even at very high temperatures. 1·4632 grm. of the salt which had been fused for some time, lost 0·2748 grm. of carbonic acid upon treatment with sulphuric acid. Hence the composition of the salt is as follows :—

		Calculated.	Found.
RbO . . .	93·36	80·93	81·22
CO^2 . . .	22·00	19·07	18·78
	115·36	100·00	100·00

d. *Bicarbonate of Rubidium.*

The aqueous solution of the monocarbonate is easily converted into the acid salt when placed in contact with an atmosphere of carbonic acid. If the solution be allowed to evaporate at the ordinary atmospheric temperature over sulphuric acid, the salt forms shining crystals, permanent in the air, possessing a prismatic form, but of which no sample sufficiently well crystallized for exact measurement could be obtained. The crystals give a very slightly alkaline reaction, and they possess a cooling, non-caustic, agreeable taste, similar to that of saltpetre. On heating, they easily lose the second atom of carbonic acid. They are very soluble in water; and the aqueous solution gives off carbonic acid on boiling, probably owing to the formation of a sesquicarbonate?

0·5416 grm. of monocarbonate of rubidium was dissolved in water in a weighed platinum crucible, and left for fourteen days in an atmosphere of carbonic acid, which was slowly from time to time renewed. After the solution had been evaporated at the ordinary temperature over sulphuric acid, the mass was again moistened with a solution of carbonic acid in water, and again dried in the same way until no further loss of weight occurred. The salt then weighed 0·6878 grm. Hence it is seen that the bicarbonate of rubidium has a composition analogous to bicarbonate of potassium, or is represented by the formula—

		Calculated.	Found.
RbO . . .	93·36	63·79	63·72
$2CO^2$ . . .	44·00	30·06	
HO . . .	9·00	6·15	
	146·36	100·00	

e. *Nitrate of Rubidium.*

This salt crystallizes from aqueous solution, when quickly cooled, in long indistinct crystals. When the crystallization is conducted more slowly, double hexagonal prisms terminated with less distinct double hexagonal pyramids are obtained in a state fit for measurement. The crystals invariably incline to a prismatic form, and belong to the hexagonal system, corresponding to a ratio of the axes of

$$1 : a \text{ as } 1 : 0.7097.$$

This ratio belongs to an obtuse hexagonal dodecahedron, having polar angles of 78° 40', and basal angles of 143° 0'. The pyramidal faces were very imperfectly formed, so that the measurement of the angles could not be very exactly made. The faces P . ∞ P . P2 . ∞ P2, Plate V. fig. 1, were the only ones observed.

	Found.	Calculated.
$p-p_1$ . . .	149 49	150
p_1-p . . .	149 53	150
$r-p^*$ . . .	129 20	

The nitrate of potassium crystallizes, as is well known, in the rhombic system, but according to Frankenheim it occurs sometimes as a secondary hexagonal form whose hemihedral form corresponds to a hexagonal dodecahedron having polar angles of 106° 40'. This form corresponds to a hexagonal dodecahedron of another order than that in which nitrate of rubidium crystallizes; to this, however, we will recur in speaking of the nitrate of cæsium.

The nitrate of rubidium is anhydrous, but, like saltpetre, it contains water enclosed in the pores of the crystals, which therefore decrepitate on heating. Near red heat it fuses without decomposition to a clear liquid, and on cooling solidifies to a striated crystalline mass. When heated to a higher point it loses oxygen, and forms nitrite together with caustic oxide of

---

* This angle served for the calculation of the ratio of the axes.

rubidium, which acts rapidly upon the platinum vessels. Brought into the colourless gas flame on a platinum wire, the salt is completely volatilized.   It is much more soluble in water than saltpetre; 100 parts of water at 0° C. dissolve 20·1, and at 10° C. 43·5 parts of the salt.   Under the same circumstances water only dissolves 13·8 and 20·4 parts of saltpetre.

2·3543 grms. of the salt, when decomposed by sulphuric acid, yielded 2·1306 grms. of sulphate of rubidium.   Hence the salt consists of—

		Calculated.	Found.
RbO . . .	93·36	63·35	63·36
NO⁵ . . .	54·00	36·65	36·64
	147·36	100·00	100·00

f. *Sulphate of Rubidium.*

The acid salt having the formula $RbO, 2SO^3$ fuses like the corresponding potassium salt near redness, and when more strongly heated froths considerably, losing half its sulphuric acid, and leaving a solid residue, fusible only at a white heat.   If the aqueous solution of this neutral salt be slowly evaporated, fine large hard brilliant crystals are obtained, which belong to the rhombic system, and possess a ratio of the axes of $a : b : c$ as 0·5723 : 1 : 0·7522, corresponding to a rhombic octahedron whose basal angles are 113° 6′, and whose polar angles are 131° 6′ and 87° 8′.   The crystal represented in fig. 2 gave. the following surfaces:—

	$P.\infty.\bar{P}2.$	
	Found.	Calculated.
$o - o$ . . .	131 6	
$o - o_1$ . . .	113 6	
$s - o$ . . .	130 36	130 42′
$s - s_1$ . . .	112 43	112 46

This salt is therefore isomorphous with sulphate of potassium, which, according to Mitscherlich, possesses the following ratio of the axes:—$a : b : c$ as 0·5727 : 1 : 0·7464.   The sulphate of rubidium is anhydrous, perfectly unalterable in the air, and it possesses a peculiar taste, resembling that of sulphate of potassium.   On heating, it decrepitates, and loses its tranparency. Placed on a platinum wire in the flame, it is completely volatilized. 100 parts of water at +70° C. dissolve 42·4 parts of this salt; under the same conditions only 9·58 parts of sulphate of potassium are dissolved.

1·0098 grm. of this salt yielded 0·8872 grm. of sulphate of barium.   Hence it consists of—

		Calculated.	Found.
RbO . . .	93·36	70·01	69·86
SO³ . . .	40·00	29·99	30·14
	133·36	100·00	100·00

With sulphate of alumina this salt forms rubidium alum, $RbO\,SO^3 + Al^2O^3\,3SO^3 + 24HO$, which can be obtained very easily in large, bright, transparent crystals belonging to the regular system. Besides the prominent faces 0, the following, both ∞0 and ∞0∞, are seen to occur. The crystals are unalterable in the air, and in other respects closely resemble those of potash alum.

Sulphate of rubidium also forms with the sulphates of the magnesian class of bases a series of double salts corresponding to the formula $KO\,SO^3 + NiO\,SO^3 + 6HO$, and isomorphous with the respective potassium salts. These double rubidium salts are more difficultly soluble than the sulphate of rubidium itself, and can be easily obtained in large well-developed crystals. They generally exhibit the following faces :—

$$\infty P.0P. + P.P \,\infty. + 2P \,\infty.$$

h. *Chloride of Rubidium.*

This compound crystallizes indistinctly from aqueous solution upon quick evaporation or cooling; but on allowing the solution to evaporate slowly, cubic crystals are obtained. No other combination besides the cubic faces ∞0∞ were noticed. The crystals are unalterable in the air; they decrepitate on warming, and they fuse when heated to a temperature just below a red heat. Brought into the flame on a platinum wire, the salt volatilizes quickly and completely. 100 parts of water at +1°C. dissolve 76·38 parts, and at +7°C. 82·89 parts of this salt. Under similar circumstances 29·47 and 31·12 parts of chloride of potassium are dissolved.

0·9740 grm. of this chloride of rubidium yielded 1·1541 grm. of chloride of silver, hence the salt consists of—

		Calculated.	Found.
Rb . . .	85·36	70·65	70·30
Cl . . .	35·46	29·35	29·70
	120·82	100·00	100·00

i. *Double Chloride of Platinum and Rubidium.*

To obtain this compound a solution of a salt of rubidium is precipitated with bichloride of platinum.

The precipitate is of a light yellow colour, immediately depo-

On treatment with strong alcohol, this mass yielded a residue weighing 6·5 kilogrammes, which was tolerably rich in cæsium salts, and was subjected to a series of processes which will be described under the head of "Residue No. 1."

To the alcoholic extract from the original mass a concentrated aqueous solution of carbonate of ammonium was added in order to precipitate the greater portion of the lithium-salts. The solution was then evaporated to dryness in an iron vessel and heated to expel all ammoniacal salts, the brown mass (containing much oxide of iron) dissolved in water, and the aqueous solution evaporated to dryness. On extraction with alcohol, this salt yielded a residue which we will call "Residue No. 2," to the treatment of which we shall recur.

This second alcoholic extract gave with bichloride of platinum a yellow precipitate, which weighed 8·5134 grms. after washing with water. The precipitate did not undergo any change of composition on boiling with water; and when it was placed in the spectrum-apparatus, showed the cæsium and rubidium lines with great intensity. It therefore consisted almost entirely of a mixture of the double chlorides of cæsium and rubidium, and platinum.

These 8·5134 grms. $=$ A lost by reduction in a current of hydrogen 1·8719 grm. $=$ B. The residue contained metallic platinum, and the neutral chlorides of rubidium and cæsium. If we call x the quantity of the double chloride of rubidium and platinum, and y the quantity of the chloride of cæsium* and platinum, we have then

$$x + y = A,$$
$$\frac{2Cl}{Pt + Rb + 3Cl} x + \frac{2Cl}{Pt + Cs + 3Cl} y = B;$$

hence we find

$$x = 35 \cdot 4975 \, B - 7 \cdot 65588 \, A,$$
$$y = 8 \cdot 6559 \, A - 35 \cdot 4975 \, B.$$

By substitution of the values of A and B, we have the following as the composition of the precipitate :—

Double chloride of platinum and cæsium . . 1·2701
Double chloride of platinum and rubidium . . 7·2433
                                                ————
                                                8·5134

Hence 100 parts of the alkaline chlorides, combined with the chloride of platinum, are composed of—

Chloride of cæsium . . . 16·93
Chloride of rubidium . . 83·07
                          ———
                          100·00

* The atomic weight of the cæsium is taken to be 123·35, according to determinations which will be found in the sequel.

Residue No. 2 from the second extraction with alcohol, when dissolved in water and treated with bichloride of platinum, gave a yellow precipitate, which, after being boiled out from ten to twelve times, weighed 23 grms. 13·83 grms. = A of this precipitate lost, on reduction in hydrogen, 3·182 grms. =B. Hence the whole 23 grms. consisted of,—

Double chloride of platinum and cæsium . . 11·76 grms.
Double chloride of platinum and rubidium . . 11·24 „
                                                            ———
                                                            23·00 „

100 grms. of the alkaline chlorides contained in the precipitate were, therefore, made up of—

Chloride of cæsium . . . 54·89 grms.
Chloride of rubidium . . . 45·11 „
                                           ———
                                           100·00 „

The residue No. 1 weighed 6·5 kilogrammes, and for the most part consisted of the chlorides of potassium and sodium. In order to obtain the cæsium-salts still contained in it, the mass was dissolved in water, and the boiling solution precipitated with a quantity of bichloride of platinum amounting to only from 8 to 10 thousandths of the weight of the total mass. By boiling out the platinum precipitate fifteen to twenty times with water, and by adding the boilings to the original solution until they become slightly yellow-coloured, a second platinum precipitate is obtained, which must be treated in the same way as the first. The platinum precipitates are thus boiled out until no further deposit of a light-yellow insoluble crystalline mass is observed, and then all the precipitates, thus purified by washing, are reduced in a current of hydrogen, and the soluble salts extracted with water. The aqueous solution contains a mixture of chloride of cæsium and chloride of rubidium.

A kilogramme of the residue thus treated yielded 1·0348 grm. of such a mixture of the chlorides of rubidium and cæsium, from which nitrate of silver precipitated 1·1404 grm. of chloride of silver. Let A_1 signify the mixture of x_1 parts of chloride of rubidium and y_1 parts of chloride of cæsium; and let B_1 signify the weight of chloride of silver obtained from the salt A_1; we then find the values of x_1 and y_1 from the following equations:—

$$x_1 = 3.50963\,B_1 - 3.16906\,A_1,$$
$$y_1 = 4.16906\,A_1 - 3.50963\,B_1.$$

By help of these equations, and of the values of A_1 and B_1, it is easy to see that the residue No. 1, weighing 6·5 kilogrammes, contains—

Chloride of cæsium . . 2·0267 grms.
Chloride of rubidium . . 4·6995 „
                                           ———
                                           6·7262 „

Or 100 parts of the mixed chlorides contain—

Chloride of cæsium . . . 30·13
Chloride of rubidium . . 69·87
        100·00

Taking a mean of all these experiments, we find that the mother-liquor from 44,200 kilogrammes of the Dürkheim water yielded altogether—

9·237 grms. of chloride of rubidium,
7·272 grms. of chloride of cæsium.

These determinations do not, of course, profess to be very accurate. The numbers thus obtained are, however, correct enough for us to be able to give an approximate value for the quantity of the rubidium and cæsium compounds contained in the Dürkheim mineral water. The following numbers express the composition of 1000 parts of this remarkable mineral water, according to analyses made in the Heidelberg laboratory:—

Mineral Water of Dürkheim.

Bicarbonate of calcium	0·28350
Bicarbonate of magnesium	0·01460
Ferrous-bicarbonate	0·00840
Manganous-bicarbonate	trace
Chloride of calcium	3·03100
Chloride of magnesium	0·39870
Chloride of strontium	0·00810
Sulphate of strontium	0·01950
Chloride of sodium	12·71000
Chloride of potassium	0·09660
Bromide of potassium	0·02220
Chloride of lithium	0·03910
Chloride of rubidium	0·00021
Chloride of cæsium	0·00017
Alumina	0·00020
Silica	0·00040
Free carbonic acid	1·64300
Nitrogen	0·00460
Traces of sulphuretted hydrogen	0·00000
Traces of phosphates	0·00000
Traces of ammoniacal salts	0·00000
Traces of indeterminate organic bodies.	0·00000
	18·28028

The mother-liquors obtained from the Dürkheim salt-works, and sold for the purpose of manufacturing brine-baths, were found to contain the new alkaline chlorides in a more concentrated form, as is seen from the following analysis:—

Mother-liquor from Dürkheim Waters, composition in 1000 *parts.*

Chloride of calcium	296·90
Chloride of magnesium	41·34
Chloride of strontium	8·00
Sulphate of strontium	0·20
Chloride of sodium	20·98
Chloride of potassium	16·13
Bromide of potassium	2·17
Chloride of lithium	11·09
Chloride of cæsium	0·03
Chloride of rubidium	0·04
	396·88

The mother-liquors of the brine-springs of Kreuznach, Kissingen, and Nauheim were likewise found to contain evident traces of rubidium and cæsium compounds, as is seen by the following analyses made in the Heidelberg laboratory:—

Mother-liquor of Brine from Kissingen, composition in 1000 *parts.*

Chloride of magnesium . .	189·59
Sulphate of magnesium . .	36·01
Chloride of sodium . . .	41·37
Chloride of potassium . . .	18·72
Bromide of potassium . . .	10·62
Chloride of lithium . . .	12·85
Chloride of cæsium . . .	trace
Chloride of rubidium . . .	trace
	309·16

Mother-liquor of Brine from Theodorshall near Kreuznach, composition in 1000 *parts.*

Chloride of calcium	332·89
Chloride of magnesium . . .	32·45
Chloride of strontium . . .	2·86
Chloride of sodium	3·44
Chloride of potassium . . .	17·12
Bromide of potassium . . .	6·89
Iodide of potassium	0·08
Chloride of lithium	14·53
Chloride of cæsium . . .	considerable trace
Chloride of rubidium . . .	trace
	409·76

The salt which crystallizes from this mother-liquor appears to be free from cæsium and rubidium; it is, however, remarkable for the large quantity of chloride of strontium which it contains. According to an analysis made in the Heidelberg laboratory by

M. Sieber, the salt possesses the following per-centage composition :—

Chloride of calcium	54·28
Chloride of magnesium . . .	2·76
Chloride of strontium . . .	11·19
Chloride of sodium	2·01
Chloride of potassium . . .	7·98
Water	21·78
	100·00

From the foregoing analyses, it would appear that cæsium and rubidium occur pretty generally in the water of brine-springs. These metals, however, are likewise found to be present in the waters of the non-alkaline springs containing but small quantities of soluble salts. Thus, we have proved the presence of the salts of these two new alkalies in two of the hot springs at Baden-Baden, namely in the Ungemach and Höllenquelle.

The former of these springs was found to possess the following composition in 1000 parts :—

Water of the Ungemach hot spring at Baden-Baden.

Bicarbonate of calcium . .	1·475
Bicarbonate of magnesium .	0·712
Ferrous bicarbonate . . .	0·010
Manganous bicarbonate . .	trace
Sulphate of calcium . . .	2·202
Sulphate of strontium . . .	0·023
Sulphate of barium . . .	slight trace
Chloride of calcium . . .	0·463
Chloride of magnesium . .	0·126
Chloride of sodium . . .	20·834
Chloride of potassium . . .	1·518
Bromide of potassium . . .	trace
Chloride of lithium . . .	0·451
Chloride of rubidium . . .	0·0013
Chloride of cæsium . . .	trace
Silica	1·230
Alumina	0·001
Combined nitric acid . . .	0·030
Combined ammonia . . .	0·008
Combined arsenic acid . .	trace
Combined phosphoric acid .	trace
Combined oxide of copper .	trace
Indeterminate organic bodies	trace
Free carbonic acid	0·456
	29·552

The salts of cæsium and rubidium, as well as those of lithium and strontium, have been likewise found in the water of the Wiesbaden springs, as also in that of the newly-bored artesian well at Soden near Frankfurt. In order to obtain evidence of the presence of the new alkalies in this water, it is only necessary to boil out the platinum precipitate obtained from the mother-liquor of 6 to 8 litres of the water; the cæsium and rubidium lines are then easily recognised in the spectrum-apparatus.

We have examined small quantities of the ashes of land-and sea-plants, as well as Chili saltpetre and other alkaline salts occurring in commerce, for the compounds of the new alkaline metals, but we have not succeeded in detecting, in these substances, the presence of the salts of either metal.

Having thus considered the occurrence and diffusion of cæsium, we pass on to the consideration of the methods of separation, by means of which the compounds of this metal can be obtained in a state of purity. If, as is almost always the case, potassium, rubidium, and cæsium occur together with sodium and lithium, the first three metals can be separated from the two latter by means of bichloride of platinum. The double chloride of platinum and potassium can be separated from the platinum compound of the two new alkaline chlorides, as has been described, by boiling the double salts out about twenty times with small quantities of water; and thus the more soluble potassium-salt may be almost entirely removed. The double platinum compounds, which now contain but traces of potassium-salts, are next heated to redness in a current of hydrogen, at which temperature the chlorides of cæsium and rubidium do not fuse. The mass is then treated with about seventy times its weight of water, and the alkaline chlorides thus dissolved. The residual platinum is again converted into chloride, which is diluted with water to the same bulk as the solution of the alkaline chlorides; both solutions are then heated to boiling and mixed together. As soon as the precipitate which forms on mixing the solutions has collected in sufficient quantity by the cooling of the liquid, it is thrown on a filter, dried, and again subjected to the same treatment of reduction in hydrogen, &c., until a small portion brought into the spectrum-apparatus shows at most a faint trace of the potassium line $K\alpha$. The precipitate then contains solely the chlorides of cæsium and rubidium. For the purpose of separating these two bodies, the solubility of the carbonate of cæsium in absolute alcohol, and the insolubility of carbonate of rubidium in the same liquid, is made use of.

The separation of the carbonate of cæsium by repeated extractions with alcohol is, however, a difficult operation, as a double

discharges from heavy ordnance. The ground shook under the observer's feet, as was corroborated by three other gentlemen in company with Dr. S., who likewise heard the four detonations.

I have received some other observations concerning the same phenomenon; but I consider Dr. Scheffczik's observation most valuable, from the excellent account it gives of the first approach of the "star-like" luminous body, and of its subsequent progression.

Scarcely any fall of aërolites has ever been so exactly and fully observed as that which fell at New Concord, Muskingum Co., Ohio, on May 1st, 1860*. Professor Evans, of Marietta College, Ohio, calculated several elements of the orbit. The meteor, first seen as a fiery globe at a horizontal distance of 20 to 30 (English) miles, appeared like the full moon. An altitude of 40 miles, derived from other observations, would give to it a real diameter of ⅗ths of a mile. It moved from S.E. to N.W. The final velocity was about 4 miles a second. Nearly thirty stones, of about 700 lbs. total weight, were found to have fallen; the largest of them, weighing 103 lbs., is now in the Museum of Marietta College. All these stones taken together would fall far short of the apparent size of the meteor, as is the case with many other observations of a similar nature, especially with that of Agram in Croatia, May 26th, 1751, where two masses of native iron, the one of 71 lbs., the other of 16 lbs., were the only material residuum of a meteor whose apparent diameter was scarcely under 3000 feet†! At first the New Concord stones were warm, so that particles of the moist ground on which one of them had fallen, soon dried up, at least in the case of one weighing 71 lbs.

Its greatest heat was not more than that which the stone would have had if exposed for some time to the natural heat of the sun's rays. The largest of the stones (103 lbs.) was found about three weeks after the fall, beneath the root of an oak tree. It had gone through another root in an oblique direction, and had penetrated to the depth of nearly 8 feet into a hard argillaceous ground: no mention is made of its probable temperature at the time of falling. Those who witnessed the fall, only perceived that the stones were "black," they did not mention the appearance of any fireball ‡. At the moment of the fall they

* See Silliman's American Journal, vol. xxx. for July 1860.

† See "Der Meteorstein-Fall v. Hrashina bei Agram, 26 Mai, 1751," by W. Haidinger, Proceedings of the Vienna Imperial Academy, Class of Mathematics and Natural Sciences, 1859, vol. xxxv. p. 361 (283).

‡ "No one of the many persons who saw the stones fall, and who were in the immediate vicinity at the time, noticed anything of the luminous appearance described by those who saw it from a distance."—*Silliman's Journal.*

heard hissing sounds, and that before the chief detonation had attracted their attention. All the stones were covered with a black crust bearing evidence of fusion, and presented angular and fragmentary shapes; their interior resembled grey solid rocks[*]. The American naturalists inferred from the collated accounts about the igneous globe, the acoustic phenomena, and the fall itself, that the first and chief detonation took place at an altitude of about 40 miles (English) above the southern portion of Noble County, at a distance of about 30 miles from New Concord; and that the fall of the stones themselves commenced about one mile S.E. of that place, extending over an area of 10 miles in length by 2 to 3 in breadth, the largest ones falling last. The sound perceived was supposed to have been explosive in its nature; and the meteor, after having ceased to be visible, must have continued its course towards the Northwest. These are some of the most important facts relating to the phenomenon. Desirable as it is to pursue induction step by step, it is impossible to give a clear exposition without sketching previously the succession in time of each event as they are observed and perceived by our senses. Nobody who has ever examined meteorites with more than superficial attention, can have doubted that their interior and their exterior present two different periods of formation. The general form of meteorites is that of *fragments*, the constitution of their external crust is the consequence of *superficial fusion*. They are fragments coming from remote cosmical regions, which having entered the earth's atmosphere, are first perceived by us as stars, increasing in size as they come nearer to us. Great care should be taken to observe and note the moment or time with as much exactitude as possible, as, combined with the time of the year and the hour of the day, it gives us the direction of the meteor. The direction and the velocity of our globe in its circum-solar orbit (19 English miles per second, while a point on the equator by diurnal rotation moves 1464·7 Vienna feet in a second, or 900 nautical miles per hour), are well known.

Many observations have proved meteorites to travel 16 to 40 English miles in a second. Humboldt, in his 'Cosmos,' has even, from the observations of J. Schmidt, Heis, and Houzeau, calculated a velocity of 95 miles a second.

These orbits cross and oppose each other in every conceivable direction. Important consequences may be deduced from these enormous velocities, as compared with what takes place on the surface of our globe. A devastating hurricane takes place in

---

[*] " Viewed from most positions, this stone (that of 103 lbs. at Marietta College) is angular, and appears to have been quite recently broken from a larger mass."

our atmosphere whenever an air-current is progressing at the rate of 92 miles (English) per hour. A point on the equator, by diurnal rotation progresses at the rate of 1464·7 Vienna feet per second without disturbance of the atmospheric equilibrium, on account of the general atmospheric pressure being nearly equal in places lying very near each other. According to Sir John Herschel, the movement of a "devastating hurricane" is equivalent to a pressure of 37·9 lbs. Vienna (32·81 lbs. English) weight, on one square foot Vienna measure. The atmospheric pressure (=32 feet of water) on one Vienna square foot is 1804·8 lbs., or, compared with that of the most powerful hurricane, as 55 to 1.

I am glad to see these details, as I give them from sources most within reach, confirmed by Prof. E. E. Schmidt's, of the University of Jena, in his copious and excellent 'Manual of Meteorology' (vol. xxi. of the *Allgemeine Encyklopädie der Physik*), edited by Dr. G. Kersten and other eminent physicists). The following synoptic Table, calculated by Mr. Rouse (Report of the Tenth Meeting of the British Association, held at Southampton in Sept. 1846, p. 344) is found on page 483 of Prof. Schmidt's Manual :—

Velocity of wind.		Pressure per square foot in lbs. avoirdupois.	Character of wind.
English miles per hour.	English feet per second.		
60	88·02	17·715	Great storm.
80	177·36	31·490	Hurricane.
100	146·70	49·200	Destructive hurricane.
913–916	1340·0	1 atmosphere.	

The wind-scale of the Smithsonian Institute (published by the Smithsonian Institute, Nov. 1853, Washington, p. 173) offers analogous results.

It is the best proof in favour of the use of such extensive and elaborate synoptic works as Prof. Schmidt's Manual, that they gave me complete confidence in the *data* I had so laboriously compiled from other sources, and this, thanks to the author's kind attention to me, just at the moment when I felt most in want of such.

What is the state of the single particles of air composing our atmosphere in the elevated regions, where meteorites, first entering it, are capable of producing luminous phenomena as intense as observed at New Concord, even at the enormous altitude of 40 English miles? In these elevated regions the temperature may probably not exceed that of the interplanetary space, *i. e.* 100° Reaumur. Movements of particles may be supposed indeed to take place in the higher regions of the atmosphere, as on these depend the changes of atmospheric pressure nearer to the earth's surface, the causes of the winds, &c. Whenever solid bodies move through them, so abnormal an event goes on with such enormous rapidity, that these particles, quite isolated from each other, must be positively pushed aside. In the van of the progressing meteorite a stratum of atmospheric particles is formed, having no time to escape before the progressing body, but by streaming back alongside of it. The velocity of a meteorite, supposed on an average to be seven German miles (24,000 feet) per second, is to that of a hurricane of 134·72 feet per second as 124·4 to 1. Suppose the pressure to increase in the same proportion, it would be per square foot, for the hurricane =32·8 lbs., and for the meteorite 4080·32 lbs., or more than 22 atmospheres.

It may be supposed that such a sudden compression (action and reaction continually remaining equal) must have the same effect as the compression of air in the old tinder-boxes alluded to by Prof. Benzenberg. It might not here be out of place to quote *in extenso* a passage from a book published fifty years ago[*], expounding views still far from being cleared up:—

"The incandescence perceived around fireballs in a state of ignition may be t e result either of *combustion*, although with difficulty admissible in air so very rarefied, or of *friction*, as generally believed. I think it results still more from the *compression* of air, as in our newly invented tinder-boxes air produces fire by mere compression. Could not *electricity* become free in the same way? Suppose a cubic mile of air to be suddenly compressed to a volume of one cubic foot, would not then the electricity originally contained in it be set at liberty? The circumstances attending the explosion of igneous globes seem to be in accordance with this supposition. These globes, when first seen, do not appear larger than bright stars; as they approach the terrestrial surface (generally in an oblique direction) they increase to the size of the full moon, and at last, when at a few miles distant, explode with a violent detonation. The cause of this explosion is probably an excessive accumulation of electric matter, streaming from compressed air into the igneous globe of about 3000 feet in size of (?) metallic substance. The distance being still too considerable to admit of a discharge to the earth, this takes place in the open air, or within a cloud.

"Probably the place of the discharge depends less on the proximity of the terrestrial surface than on the density of air regulating the *maximum* of compression and accumulation of electricity. Subsequently to the explosion, the single fragments

[*] *Briefe geschrieben auf einer Reise durch die Schweiz im Jahre* 1810, von J. F. Benzenberg. 1 vol. Düsseldorf, 1811.

fall to the surface of the earth, with a velocity probably inferior
to that of a bullet shot from a gun, the air increasing in density
as it becomes nearer the surface of the globe, resistance increasing proportionally, so that there may be but a slight difference in the final velocity, whether the body fall from a height of
one or of five miles."

I have here quoted views somewhat opposed to those which I
myself intend to propose (as those relating to the explanation of
the explosion of fireballs); yet some of the above-quoted assertions
may perhaps be worth further consideration.

The following exposition of the way in which this may occur
may not be altogether devoid of probability. Compression, first
of all, developes heat and light. Immediately in front of the
meteorite is formed a centre of expansion, from which the compressed air tends to expand in every direction. Whatever lies
in the direction of the orbit, is left in the rear of the progressing
meteorite; whatever lies opposite to it, contributes to the fusion
of the superficial crust, or by its resistance either retards its
progress, or gives rise to a rotatory movement around an axis
coinciding with the meteorite's orbit, even if it should have undergone such a motion only on entering the terrestrial orbit. A
part of the air made luminous by compression, is forced out as
at C, in every direction perpendicularly to the orbit AB (fig. 1).
Resistance continues against this luminous disc, forces it backwards, overcomes it
gradually at some
distance from the
centre towards EE',
and rounds it off
behind the meteor-
ite in the shape of
an igneous globe,
either round, or as
frequently happens
oviform; occasion-
ally extended so far
back as to form
even an actual tail.

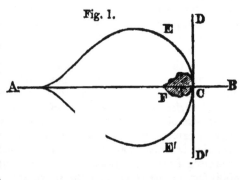

Fig. 1.

Instances of two or more luminous bodies behind each other
have been observed, as those seen at Elmira, Long Island, United
States, July 20, 1860; at Littau in Moravia, end of August 1848
or 1849; at Collioure in France, February 21st, 1846. In
these cases we may suppose that the single fireball first seen
contained already a certain number of fragments, acted upon
differently by the resistance of the air, according to their differences in size, shape, and perhaps specific gravity, so that the

heavier among them found less obstruction in pursuing their way than the lighter ones.

M. Julius Schmidt observed at Athens, July 27th, 1859, a magnificent green meteor, moving slowly in twelve seconds through an arc of 28°, commencing with a faint light, and ending as faintly, while about the middle of its course it expanded into a ball of 8–10 minutes in diameter, casting an intense light over the whole town and neighbouring hills.

An orbit having its convexity turned towards the earth's surface, as that of the meteor of 20th July, 1860, seen in the United States, may be indicative of a degree of specific gravity inferior to that generally the case in meteorites. In this case the motion of meteorites may become slower and slower, and at last be completely stopped; while there is little chance of their again returning into the cosmical space, from whence they entered our atmosphere.

The meteorite in question evidently entered the more rarefied strata of the atmosphere, and, perhaps influenced by the short duration of the igneous globe surrounding it, continued on its course into space. Its speed, though somewhat diminished, was certainly not annihilated.

Hitherto we have left out of consideration the altitude of the atmospheric strata in which a meteor is supposed to move; nor is this omission objectionable: suppose the meteorite moved along close to the surface of the earth under the pressure of a whole atmosphere, answering to a column of mercury 30 inches in height, and at the rate of seven miles per second, it would act on every square foot of resisting air with a pressure of 22 atmospheres[*]; this pressure would only amount to 11 atmospheres at a height of between 18,000 and 19,000 feet, where the barometer would indicate an atmospheric pressure of only 15 inches. It must, however, not be lost sight of, that under such circumstances the resistance of the surrounding air is also notably diminished, and that consequently the atmospheric particles forcibly driven out before the centre of elasticity will find the same facility for streaming along, or flowing in the directions C D, C D′ (fig. 1).

If electric tension in the extremely rarefied strata of our atmosphere is really as energetic as it is generally admitted to be, we are entitled to suppose a high development of electrical light. The expressions used by Benzenberg in the above-quoted passage, suggest no idea adequate to our present mode of viewing this subject. The view recently enounced in a totally different direction by one of the first of living physicists, Professor Plücker, seems to be in exact accordance with the subject considered here.

[*] See E. E. Schmidt's *Lehrbuch der Meteorologie*, page 913.

In his paper on the constitution of the electric spectra of certain
gases and vapours (see Poggendorff's *Annalen*, 1859, vol. cvii.
p. 505), the illustrious Professor says, " What is the thing
emitting light when an electrical discharge takes place through
the narrow passage of a Geisslerian tube, as much exhausted of
air as possible, and including gas or vapour? There is no light
unless some ponderable substratum emits it; *there is consequently no electrical light in the abstract sense of the word*. All
my observations have confirmed me in this persuasion. But
how is electricity acting here on the gaseous particles? In my
opinion only as an exciter of heat. *The gaseous particles become
incandescent.* The thick glass in the narrower portion of the
Geisslerian tube is very notably heated when the discharge from
Ruhmkorff's apparatus passes through the gas contained in it.
If, then, the heat transmitted to the glass from the dispersed
gaseous particles, whose tension is often measurable by fractions
of millimetres, increases to a notable degree, to what a degree
of intensity must these particles be heated !"

The cosmical orbit of the meteorite M entering the terrestrial
atmosphere   A
(fig. 2) termi-
nates at C; from
this moment the
meteorite   be-
longs to our earth,
and falls straight
down   from   C
(where, after ex-
ploding, its light

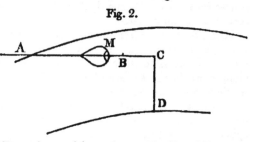

Fig. 2.

is extinguished) to D on the earth's surface. The line C D represents its terrestrial orbit.

Terrestrial attraction is quite an insignificant element compared to the planetary or cosmical impulsion peculiar to any
meteorite; and but for the resistance offered by the atmosphere,
few or no meteorites would reach the surface of our earth, except
those whose orbits were directly aimed towards it.

Meteoric stones after falling appear black; and their ena-
melled crust proves them to have undergone superficial fusion
from exposure to high temperature. Their interior is frequently
not more heated than would permit of their being held in the
hand without inconvenience.

This is no matter for surprise, as the stone having quite recently come from the cold regions of interplanetary space, a
compensation may be supposed to take place between the outside and the inside.

Fragments of the Dhurmsala meteorite (Punjab, 14th July,

1861), probably from the extreme cold of the interior, showed, at the moment of their fall, a temperature notably below congelation*. Meteoric *iron*, however, being a good conductor of heat, comes down far more heated, and even in a state of intense incandescence, as was the case with the iron of Corrientes in Caritas Paso, near the river Mocorita, in January 1844, mentioned by Mr. Greg†, which fell at 2 A.M. in the shape of a lengthened globe, a fiery streak marking its passage through the air. The mass fell down at a distance of about 1200 feet from Mr. Symonds, who indeed first made known this fall. Later in the morning it could not be approached nearer than ten or twelve yards, on account of the heat emanating from the mass, which projected several feet out of the ground. During the fall the atmosphere was evidently in a state of motion, as if repelled by the falling body, producing a whirlwind of short duration.

This description is quite consistent with the facts to be derived from the preceding considerations. In this case the meteoric iron-mass struck the earth nearly point blank, falling under an acute angle of 60°.

Another very characteristic phenomenon connected with the vanishing of meteoric light, is the accompaniment of intense explosive *sound*, resembling the ignition of gunpowder fired from guns or mines. Generally one detonation is strikingly loud, frequently followed by others of a "rattling" character. The meteor "explodes," as it is commonly called, and lets fall from it one or more stones, disproportionally small in quantity as compared with the probable size of the fireball itself.

What could then have become of a body so luminous as that of a large meteor, which, according to Prof. Laurence Smith's experiments, might indeed appear far larger than the solid matter contained in it could justify one in supposing possible?

According to Prof. Smith's experiments—made, 1st, with the electric light between carbon points; 2nd, with the oxy-hydrogen light falling on lime; and 3rd, with the light from steel burning in oxygen,—the irradiation of a luminous point gives the following numbers for the apparent size at four different distances:—

Distance ...	10 inches.	600 feet.		1320 feet or ¼ mile.		2640 feet or ½ mile.	
1. Carbon	0·3 line	½	Diameter of the moon.	3	Diameter of the moon.	3½	Diameter of the moon.
2. Lime ...	0·4 ,,	½		2		2	
3. Steel....	0·2 ,,	¼		1		1	

Though persons struck by any uncommon sight are generally

* Proceedings of the Imperial Academy of Vienna, sitting of the 29th Nov. 1860.

† Philosophical Magazine for July 1855.

inclined to overrate the size of an object, the reports or accounts of fireballs showing a half or the whole of the full-moon's diameter, when seen from a distance of 20, 40, 60, or 100 miles, cannot, however, entirely rest on self-delusion*. A large space may be occupied by the igneous globe, surrounding a far smaller nucleus, consisting of one or more fragments.

On coming with enormous velocity from planetary space into our atmosphere, the acoustic phenomenon may be accounted for by supposing that the fireball includes, as we have attempted to explain, a real vacuum maintained by the *resistance* of the *atmosphere* against it. The original velocity having at length been sufficiently retarded by the air, the meteor becomes almost stationary; at this moment the vacuum suddenly collapses in the already rather dense air, and detonation ensues from repercussion of the air filling up the vacuum. The intensity of the sound ceases to be a matter of wonder when we consider the explosions caused by setting fire to bubbles filled with oxy-hydrogen gas suspended in the air. The so-called "consecutive explosions," or series of smaller detonations, may depend on the more or less gradual diminution of the cosmical velocity†.

Hitherto only *one* solid body has come into question. When, however, meteorites arrive in flocks or groups, as when 3000 stones (the largest 17 lbs.) fell from a detonating meteor at L'Aigle in France, on the 26th of April, 1803, nearly 200 at Stannern in Moravia, 22nd May, 1808, and some 30 or 40 near New Concord, Ohio, on May 1st, 1860, it may be supposed that even if one principal explosion "had commenced the action," subsequent detonations of the several isolated portions could likewise have taken place. I do not, however, believe that in the above-

---

* The meteor of Feb. 11, 1850, seen in England at a distance subsequently calculated at 50 miles, appeared, as at Hartwell, as large as the full moon; at places 100 miles distant from its vertical passage, as large as Venus. That of July 20, 1860, seen in the United States, had a decided apparent diameter nearly equal to that of the full moon, when at a height of 41 miles. That of October 18, 1783, at a height of 60 miles, over Lincolnshire, presented a similar appearance.—R. P. G.

† It may be here fair to mention that Mr. Benj. W. Marsh, of the United States, considers (in his Report in the Journal of the Franklin Institute, on the daylight meteor of Nov. 15, 1859, seen in New Jersey) that the aërolitic *detonation* arises from a series of decrepitations caused by the sudden expansion of the surface of the stony fragment, the whole time of flight not being sufficient to penetrate the mass. At the forward end these explosions would take place under great pressure, which might account for the loudness of the sound. The force of these explosions, directed backwards, would likewise tend to check the forward velocity of the mass. He also considers that the audible explosion, often lasting several minutes, is the result of the actual bursting of the meteor; for though the explosion might only occupy in reality half a second of time, yet in that interval the noise might be distributed over a distance of twenty or thirty miles.—R. P. G.

cited instances the stones were formed by the bursting or explosion of one large stone, but that they actually entered the atmosphere as a group or swarm of separate individuals, surrounded, as I have ventured to suggest, by what appears to us the luminous fireball*. I must here shortly allude to some peculiarities common both to *stone* and iron meteorites. One is the "pitted" or indented appearance usually presented on their surfaces. This "pitted" surface is particularly evident on the meteoric stone of Gross-Divina, which fell July 24th, 1837, in Hungary; and in the meteoric iron of Nebraska (Transactions of the Acad. of Sciences of St. Louis, vol. i. no. 4, plate 21). They are best developed on the side supposed to have lain backwards (see F in fig. 1). The side turned towards C is constantly more uneven and rough, as though it had pressed against a homogeneous mass of air, while air-currents may, like pointed flames, turn alternately towards the plane F. Marginal seams, as on the stones of Stannern, owing to the fusibility of the crust, give place to similar conjectures. As for the general form, the centre of gravity must have been in the forepart or front, as long as the meteorite was moving through space. When rotation round an axis had once commenced, and become accelerated in consequence of the propulsory movement diminishing, the point next in gravity must have taken its place in the plane of rotation, so that an iron mass of a flat form, as that of Agram is, could be propelled lying on its flat side. This iron is indeed of very different aspect on each of its broader planes; the rougher of them was certainly directed forward, as long as propulsion continued, the smoother surface remained turned backward, and not acted upon by external agents. The flat shape of the whole characterizes the Agram iron as having originally filled up a vein-like narrow cavity.

A disruptive explosion is only indubitable where, as in the stone-fall of Pegu (December 27, 1857), two fragments of the same stone, fitting each other exactly, have been found at a certain distance (in the case in question, 10 English miles!)†; such a disruption may cause a sound, as would a millstone under analogous circumstances, but certainly of less intensity

---

* See Haidinger on "eine Leitform [typical form] der Meteoriten," Vienna Acad. Proceedings, vol. xl. 1860, page 525, note. It yet by no means seems proved that meteorites do enter our atmosphere in groups, and that then an explosion again scatters them as they fall to the earth; it seems more probable, and certainly as possible, on the other hand, that one large friable mass, constituting probably the nucleus of the *single* fireball, as that of L'Aigle, bursts into many pieces, sometimes, no doubt, into hundreds of small fragments, as well as occasionally into the finest dust.—R. P. G.

† Haidinger, vol. xlii. p. 301 of the Proceedings of the Imperial Academy of Vienna, "Die Meteoritenfälle von Quenggouk bei Bassein in Pegu."

than that caused by the sudden collapse of the vacuum within a large fireball.

I have left unnoticed many other particulars concerning this class of phenomena, as well as attempts at explanations, and the views of others respecting them; and I even abstain from mentioning their connexion with M. Coulvier-Gravier's long-continued and accurate investigations. Meantime I have received through the editor's particular kindness, a copy of Dr. Laurence Smith's paper on the late fall of stones at New Concord, before referred to in this paper, and published in Prof. Silliman's American Journal (Jan. 1861, vol. xxxi. p. 87). In a letter addressed to me, Dr. Smith, for a long time a most careful investigator of meteorites, writes as follows :—" The method hitherto used in studying meteorites is still very deficient. To obtain tolerably accurate notions concerning their nature and origin, it would be necessary to submit to stricter criticism than is generally done the phenomena attending their fall, together with their physical properties, mineralogical as well as chemical. We have no right to speak of the explosion of large bodies within our atmosphere, while the so-called fragments of them show no marks of any explosion; nor should we speak of superficial heating to fusion in our atmosphere, while masses of 50 lbs. weight were found, ten minutes after their fall, not warmer than any stone exposed to the sun's rays, while others fell on dry leaves without leaving on them any traces of combustion or heating. So I could point out several other erroneous views relative to the fall of meteorites, and fully refuted by the chemical and physical facts proved by the stones themselves, and about which my account of the Ohio fall in Silliman's Journal is to give some hints."

I have overcome, I believe, this difficulty by placing in the first period, viz. that of cosmical motion within the atmosphere, the formation of the crust by superficial fusion, and in the second period (that of telluric motion, or simple falling to the earth) the compensation between the internal and external temperatures. At all events, I may feel satisfied to see my own views to some extent corroborated by the independent assent of such a distinguished and competent observer as Prof. Laurence Smith.

l'articles separated from the surface of meteorites, appearing perhaps to observers in the shape of sparks, may again be covered with a thinner crust, and belong to a later but still cosmical portion of the orbit, as B C (fig. 2).

It would be desirable to ascertain in new cases, and as far as possible in those of older date, what is the direction of the line C D with respect to the diurnal movement (west to east) of the

terrestrial surface, as the supposition of a tangential force adequate to the elevation required for experiments on free fall close
to the earth's surface would prove inadmissible in the present
case. At all events, observations on such fugitive phenomena
require an uncommon amount of manifold circumspection.

Professor Laurence Smith concludes his above highly important memoir with the following propositions :—

1st. " The *luminous* phenomena attending the appearance of
meteorites are not caused by *incandescence*, but rather by electricity, or some other agent.

2nd. " The sound comes not from the explosion of any solid
body, but rather from *concussion* caused by its rapid movement
through the atmosphere, partly also from electric discharge.

3rd. " Meteoric *showers* owe not their existence to fragments
caused by the rupture of a single solid body, but to the division
of *smaller* aërolites entering the atmosphere in groups.

4th. " The *black crust* is not of atmospheric origin, but is
already formed in cosmical space, before the meteorites enter our
atmosphere."

I think (says M. Haidinger) I have now given some explanation applicable to each of these four propositions; some in
the same sense (2nd and 3rd), the others (1st and 4th) in a
somewhat different sense, without actually excluding mutual
compromises. At all events, I would recommend the utmost
accuracy in the observations of future meteoric falls, as well as
in all investigations concerning those already known.

[To be continued.]

XLIV. *On the Silicates of Copper from Chile.* By FREDERICK FIELD, *F.R.S.E., M.R.I.A.*[*]

ALTHOUGH I believe that crystallized silicate of copper has
not yet been found in Chile, several varieties of this mineral
exist in very large quantities, generally in amorphous masses,
with various shades of colour, some of which are of considerable
beauty. These silicates, owing to the great difference in their
composition and to the entire absence of crystallization, have
not excited the same amount of interest which has been attached
to other species,—mineralogists supposing that they are not true
minerals, but simply consist of oxide of copper in combination
with silica in greater or less proportion, the varieties containing
the most oxide being comparatively soft and friable, and the
poorer kinds, having but little metal, being exceedingly hard and
brittle, resembling in many respects masses of partially fused

[*] Communicated by the Author.

Phil. Mag. S. 4. Vol. 22. No. 148. *Nov.* 1861. 2 B

translucent glass. Some short account of those minerals which are found most frequently may perhaps not be entirely devoid of interest.

Green and Blue Silicates.—The Chilian miners frequently meet with veins of hard blue or green mineral, which they term *Llanca*, consisting, with the exception of small quantities of lime, alumina, and oxide of iron, entirely of oxide of copper, silicic acid, and water. The following is the composition of one of these llancas, which was found coating thin veins of suboxide of copper and the native metal, in the mines of Andacollo, Chile; the analysis was made by M. Domeyko:—

Oxide of copper . . .	29·50
Silica	52·20
Water	16·70
Alumina	1·20
	99·60

Another specimen of a pure green colour, analysed by the same chemist, yielded—

Oxide of copper . . .	12·00
Silica	75·90
Water	10·10
Alumina	2·00
	100·00

showing a very great difference in composition.

I obtained from a mine in the neighbourhood of Tambillos near Coquimbo, a considerable quantity of very fine silicate of copper, having a pure turquoise-blue colour, with little or no shade of green, perfectly amorphous and opake, and which appears, as the analysis will show, to have a far more definite composition than either of the samples quoted above:—

Oxide of copper . . .	39·50
Silica	28·21
Water	24·52
Oxide of iron	2·80
Alumina	4·97
	100·00

Regarding the alumina and oxide of iron as foreign to the mineral, we have in every 100 parts,—

Oxide of copper . . .	42·83
Silica	30·59
Water	26·58
	100·00

Silicate of copper, consisting of one equivalent of silicic acid, one of water, and one of oxide of copper, would require the following numbers * :—

Oxide of copper . . .	41·23
Silica	30·94
Water	27·83
	100·00

The mineral may thus be regarded as nearly approaching in composition to CuO SiO², 3HO, and we have—

Dioptase	CuO SiO², HO.
Chrysocolla . . .	CuO SiO², 2HO.
Blue silicate . . .	CuO SiO², 3HO.

Black Silicate of Copper.—This mineral is of a dense black colour, of compact structure, conchoidal fracture with a glassy lustre, very much resembling obsidian. Before the blowpipe it does not change colour, and only fuses round the edges with difficulty. It gives off water when heated, and is easily attacked by hydrochloric acid. This silicate is found in some few mines in Chile, particularly those of the Higuera in the province of Coquimbo, and always in very narrow veins, which are generally found associated with the red oxide of copper, and the blue and green silicates. An analysis by Domeyko gave the following numbers :—

Oxide of copper . . .	50·10
Silica	28·20
Water	19·10
Oxide of iron	2·60
	100·00

In the year 1858 I published a short account of a double silicate of copper and manganese in the Chemical Gazette. The mineral, which in an impure state is found in considerable abundance, has a deep black colour, vitreous lustre, and is immediately decomposed by hydrochloric acid in the cold. A pure specimen yielded on analysis,—

Oxide of copper	24·71
Silica	18·90
Water	15·52
Oxide of iron	·23
Peroxide of manganese . . .	40·28
	99·64[†]

This mineral, as can be imagined, would prove highly valu-

---

* Silica is taken here at 30·20, Si = 14·2, 2O = 16.

[†] Chemical Gazette, vol. xvi. p. 105.

able could it be obtained of sufficient richness in copper to warrant its exportation to Europe, as, by the action of hydrochloric acid, large quantities of chlorine are evolved, which might be made available for the production of chloride of lime, and the residual solution of chloride of copper would yield the metal in a very pure state upon the introduction of metallic iron. Unfortunately, although the ore is plentiful, but small quantities are obtained having the composition given above, the average yield of copper scarcely exceeding 5 per cent.

The following Table may serve to illustrate the composition of some of the principal silicates of copper :—

Composition.	1.	2.	3.	4.	5.	6.	7.
Oxide of copper	50·00	44·94	29·50	12·00	50·10	39·50	24·71
Silica ....:	38·70	34·83	52·20	75·90	28·20	28·21	18·90
Water................	11·30	20·23	16·70	10·10	19·10	24·52	15·52
Oxide of iron	....	....	1·20	......	} 1·46	2·80	·23
Alumina	......	......	......	2·00		4·97	
Lime and magnesia ..							
Oxide of manganese.	......	....	....	....	....	......	40·28
	100·00	100·00	99·60	100·00	98·86	100·00	99·64

(1) Dioptase.
(2) Chrysocolla
(3) Bluish-green amorphous silicate. Chile (Domeyko).
(4) Green silicate (Domeyko).
(5) Black silicate (Domeyko).
(6) Blue silicate. Chile (Field).
(7) Double silicate of copper and manganese. Chile (Field).

It may be mentioned that these silicates are, under certain circumstances, very advantageous to the smelter, especially when he has to operate upon highly ferruginous ores, by combining with the oxide of iron, and thus saving the sides of the furnace, which would otherwise be much injured. None of them contain either antimony or arsenic; and the copper therefore, after the necessary fusions, is exceedingly pure.

St. Mary's Hospital Medical School,
    London, October 11, 1861.

XLV. *Note on the Readings of the Graduated Arc in Spectrum-Analysis, and Distortion of the Spectrum. By* J. M. WILSON, *B.A., Fellow of St. John's College, Cambridge, and Natural Philosophy Master in Rugby School**.

IN the ordinary apparatus for spectrum-analysis, the rays passing along the axis of a fixed telescope and incident on

* Communicated by the Author.

the prism are there refracted, and a portion of them pass down another telescope, and are brought to the vertical wire of its eye-piece. By moving either the telescope by which the rays are viewed, or the position of the prism, all the different lines of the spectrum can be seen in succession. In the first case the changes in the refrangibility of the rays and the requisite angular motion of the telescope are nearly proportional; in the second case, in which the prism is moved, it will be found that the angular changes in the position of the prism are by no means proportional to the changes in the indices of refraction of the rays corresponding to those positions.

The following investigation arose from a suggestion I made in the summer to Mr. Becker, that the readings on the graduated arc should be either the refractive indices of the lines corresponding, or should give the principal lines of the solar spectrum. Mr. Becker then requested me to examine how the scale might be so graduated.

1. Let ABC be a prism whose angle is α; DE a ray from the slit of the first telescope; ϕ, ϕ', ψ', ψ the angles of incidence and refraction at the first and second surfaces of the prism; μ_E the index of refraction for the line E; D the deviation for the line whose index is μ. Then

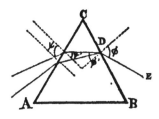

$$\sin \phi = \mu \sin \phi' \quad \cdots \quad (1)$$

$$\sin \psi = \mu \sin \psi' \quad \cdots \quad (2)$$

$$\phi' + \psi' = \alpha \quad \cdots \quad (3)$$

$$D = \phi + \psi - \alpha \quad \cdots \quad (4)$$

Let the angle between the telescopes, or the deviation, be fixed by the condition that the line X shall have a minimum deviation. Then

$$\phi = \psi, \ \phi' = \psi' = \frac{\alpha}{2},$$

and

$$D = 2 \sin^{-1}\left(\mu_X \sin \frac{\alpha}{2}\right) - \alpha.$$

The direction of DE being fixed, an angular change in the position of the prism is equivalent to a change in ϕ. It is required therefore to investigate a relation between μ and ϕ when the total deviation is fixed.

2. Differentiating the four equations given,

$$\cos\phi\,\frac{d\phi}{d\mu}=\mu\cos\phi'\cdot\frac{d\phi'}{d\mu}+\sin\phi',$$

$$\cos\psi\,\frac{d\psi}{d\mu}=\mu\cos\psi'\,\frac{d\psi'}{d\mu}+\sin\psi',$$

$$\frac{d\phi'}{d\mu}+\frac{d\psi'}{d\mu}=0=\frac{d\phi}{d\mu}+\frac{d\psi}{d\mu}.$$

Eliminating $\dfrac{d\phi'}{d\mu}$, $\dfrac{d\psi}{d\mu}$, $\dfrac{d\psi'}{d\mu}$,

$$\frac{d\phi}{d\mu}=\frac{\sin\alpha}{\cos\phi\cos\psi'-\cos\psi\cos\phi'}.$$

This result indicates that when $\phi=\psi$ and $\psi'=\phi'$ nearly, the change in ϕ is very large for a small change in μ.

3. Using special values of μ and α, let the angle of the prism be 60°, and let $\mu_x=1\cdot6801330$. (The reason of selecting this value for μ_x will appear presently; and it cannot differ by any appreciable error from the value given by Müller for μ_R.)   Then

$$D=2\sin^{-1}.\,(\cdot8400665)-60°$$
$$=54°\,17'\,42'',$$

and we have therefore the equations

$$\sin\phi=\mu\sin\phi', \quad\text{. . . . . (1)}$$
$$\sin(114°\,17'\,42''-\phi)=\mu\sin(60°-\phi'). \quad\text{. . . (2)}$$

Eliminating ϕ', we obtain

$$\frac{\sqrt{3}}{2}\,\sqrt{\mu^2-\sin^2\phi}=\sin(114°\,17'\,42''-\phi)+\frac{1}{2}\sin\phi.$$

Now the sine and cosine of the $\angle\ 114°\,17'\,42''$ differ by $\cdot5$, and therefore the left-hand side of the equation becomes

$$\cdot9114369\,(\sin\phi+\cos\phi),$$

and therefore using logarithms,

$$\log(\mu^2-\sin^2\phi)=\cdot8454309+2\log\sin(45°+\phi). \quad\text{. . (A)}$$

4. It will now be easy to calculate the value of μ for any values of ϕ.

When the deviation is a minimum,

$$\phi=\frac{D}{2}+30°,\text{ and }\phi=57°\,7'\,51''.$$

I shall proceed to calculate the value of μ for all values of ϕ from 57° at intervals of half a degree, as far as is required by the limits of the spectrum, by means of the formula (A). The subjoined Table gives the results:—

φ.	μ.	Diff. of μ.
57 0′ ″	1·680148	
57 7 51	1·680133	
57 30	1·680125	
58	1·679993	132
58 30	1·679750	243
59	1·679398	352
59 30	1·678934	464
60	1·678361	573
60 30	1·677678	683
61	1·676884	794
61 30	1·675982	902
62	1·674969	1013
62 30	1·673847	1122
63	1·672623	1224
63 30	1·671259	1374
64	1·669824	1435
64 30	1·668277	1547
65	1·666596	1681
65 30	1·664796	1800
66	1·662935	1861
66 30	1·660941	
67	1·658840	
67 30		
68	1·654313	
68 30		
69	1·649359	
69 30		
70	1·643999	

From this Table it is easy to construct a curve which shall represent to the eye the relations of μ and·φ; and either from the curve, or directly from the Table, to show to what extent the spectrum is distorted. The scale given on the following page is intended to illustrate this. The left column of the scale represents an arc graduated to half degrees, along which moves the index attached to the handle by which the prism is turned round. The degrees marked on the scale indicate the angle of incidence of the light on the first face of the prism. It might be convenient, however, to graduate this arc from the central line of minimum deviation (very near the line E) as zero. The right column of the scale gives the corresponding values of the refractive indices. It would be convenient to mark on one of the two scales the positions of the principal lines in the spectrum. This might be done either by experiment, or from the value of the refractive indices into bisulphide of carbon of the fixed lines, if they have been determined.

This distortion of the spectrum, or apparent exaggeration of its central portions, may be made evident to the eye by turning the prism slowly round with a uniform motion by moving the

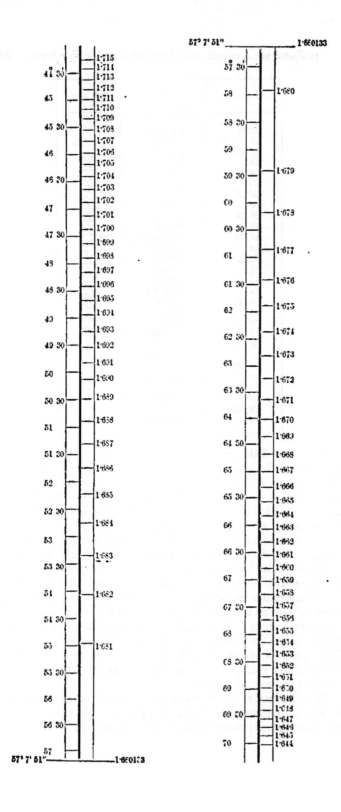

index from one end of the scale to the other, while the eye is applied to the telescope. The spectrum will then be seen to move with varying velocity. It will seem at first to move fast, and then gradually to diminish in velocity till the middle of the scale is reached, when it will be for the moment stationary, and then begin to move slowly in the same direction as before and with ever-increasing velocity till the end of the scale is reached. This also accounts for a fact which every one who has used the instrument must have observed, viz. the great preponderance of green in the spectrum thrown by the sun or common gaslight. It is obvious that the knowledge of these results gives a method of magnifying any portion of the spectrum, and of determining its limits with greater accuracy.

XLVI. *On the most advantageous Form of Magnets.*
By Dr. Lamont*.

[With a Plate.]

THE further the science of magnetism advances in its development, the more important becomes the decision of the question *what form should be given to the magnets in order to obtain the most advantageous effect.* If we at present confine ourselves to those magnets which have been employed in the investigation of the magnetism of the earth, we find that by some observers very acute sharp-pointed needles, by others flat prismatic needles, have been pronounced to be the best adapted to the purpose; solid or hollow cylinders also have been recommended. I am not aware, however, that experiments of a discriminating character have as yet been carried out; nor, so far as I know, have even the principles been established according to which the preference of one form over the others is to be determined. Nevertheless, as regards the latter point, a nearer consideration will show that scarcely an uncertainty or difference of opinion can exist, as there are in general only *three* subjects of observation which come into consideration in magnets, and it cannot be doubtful in what relations to the result these determinations stand.

The three determinations here referred to are—the *magnetic moment*, the *weight* or the *mass*, and the *moment of inertia*; and that form is to be recognized as the most advantageous, in which are united the greatest possible magnetic moment, with the smallest possible mass, and the smallest possible moment of inertia.

The direct way to decide upon the most advantageous form of

* Translated from Poggendorff's *Annalen*, vol. cxiii. pp. 239-249. Communicated by the Astronomer Royal.

the magnets, would consist in procuring hardened steel bars of different forms, magnetizing them to saturation, and investigating by measurement for every form the above-mentioned determinations. By this method I have instituted different experiments, but have given to them no great extension, because I have found.another way which attains the desired end more simply and more surely.

A magnet is composed of *magnetic molecules.* If the molecules were separated, it would appear that every molecule forms a small magnet with a determinate quantity of positive and negative magnetism; and this is what I denote by *independent magnetism.* As soon as the molecules are put together, each induces in the rest new magnetism, and to the independent magnetism of each molecule there is added a more or less considerable quantity of *induced magnetism,* according to the situation which the molecule occupies in the magnet.

The whole effect of a magnet is regulated by the *independent* and *induced* magnetism of the molecules.

A magnet is *then* magnetized to saturation when every molecule possesses the *greatest possible independent* magnetism; from which of course it follows that in a bar magnetized to saturation all the molecules have *equal* independent magnetism.

Now if we introduce a soft iron core of tolerable length into a very long spiral, through which a galvanic current passes, it is known that the same magnetizing force is exerted upon every molecule of the iron core; that is to say, equal independent magnetism is imparted to each molecule, and the mutual induction of the molecules comes then into operation as in the molecules of a magnet. From this it results that the distribution of the magnetism in an iron core placed within a long spiral, and that in a magnet which is magnetized to saturation, will be *the same;* and the laws under the limitations above-mentioned can be determined quite as well by iron cores as by magnets. But by substitution of iron cores, the great advantage is obtained, that the investigation is not only more easily executed, but also the disturbing influences which arise from the unequal or disproportionate hardness of different bars, and the consequent uncertainty whether, in the magnetizing, the point of saturation* is really reached, totally disappear.

---

* If a needle is rubbed with a pair of magnetic bars which are somewhat larger than the needle itself, and this rubbing is continued till the needle no longer receives additional magnetism, it is said to be " magnetized to saturation," although it is not proved whether a far greater magnetism might not be imparted by more powerful instrumental means. On the magnitude of the means which is required to communicate the maximum of force, no satisfactory investigations have hitherto been instituted; and

The principles above stated are here only mentioned *en passant*, as they have been already on an earlier occasion [*] stated, at least in outline, and will be hereafter more fully explained in a more detailed memoir. It has been already stated above, that the more or less advantageous form of a magnet is to be decided by the proportion of the magnetic moment to the mass, and to the moment of inertia: now, as to the last, it only comes into consideration in the oscillations, and it is of more trifling significance, on which account we will first investigate the proportion of the magnetic moment to the mass.

1st Series of Experiments.—In order to ascertain the dependence of the magnetism upon the diameter, I caused to be prepared four pieces of iron (Plate VI. fig. 4) of equal length $= 48'''\cdot2$ (Paris measure), and equal weight, but different transverse sections; the sections were,—

Of A, an equilateral triangle; length of one side $= 7'''\cdot5$.

Of B, a circle; diameter $= 5'''\cdot7$.

Of C, a square; length of one side $= 5'''\cdot3$.

Of D, a parallelogram; sides $= 6'''\cdot0$ and $4'''\cdot1$.

Of E, a parallelogram; sides $= 12'''\cdot4$ and $2'''\cdot1$.

In a long spiral of 212 turns, these pieces of iron, inserted as cores, gave the following magnetic moments (reduced to equal strength of current):—

	Magnetic moment.	Mass.	Proportion.
A	7·255	1·00	7·255
B	6·806	0·99	6·875
C	7·300	1·14	6·404
D	6·952	1·05	6·621
E	8·248	1·13	7·299

how little foundation there is for the ordinary opinions may be seen from the following statement.

In the mechanical workshop of the observatory of this city are two magnetizing apparatus, of which the one consists of two 25-pound bars, and the other is an electro-magnetic apparatus of great energy. Amongst the experiments which were carried out to prove the relation of the two apparatus appears the following case.

Two prismatic magnets, length $56'''\cdot0$ and $56'''\cdot6$, breadth $6'''\cdot8$ and $4'''\cdot9$, thickness $1'''\cdot5$ and $1'''\cdot0$, perfectly hard, were magnetized with the 25-pound bars, and the magnetic moment was determined by means of the deviation, whereby I obtained—

Greater Magnet, deviation ......116·3 scale-divisions.

Smaller Magnet, deviation ...... 81·7 scale-divisions.

Afterwards, when the two needles had been magnetized by means of the electro-magnetic apparatus, there resulted—

Greater Magnet, deviation ......177·8 scale-divisions.

Smaller Magnet, deviation ......112·4 scale-divisions.

From this we see that the magnetizing by the 25-pound bars, in regard to the degree of saturation, was deficient, in the greater magnet by somewhat more, and in the smaller by somewhat less, than a third part.

[*] *Jahresbericht der Münchener Sternwarte für* 1854, p. 27.

The mass is here, as well as in the following series of experiments, determined by means of the balance, not deduced from the dimensions above given, which are only approximate.

The most disadvantageous forms are the prism with square section and the cylinder, in which the mass is collected as much as possible about the axis of the figure, while, on the other hand, the greater widening of the mass in the other forms apppears to possess considerable advantage.

2nd Series of Experiments.—Twelve equal laminæ of iron plate (fig. 5), length $43'''\cdot2$, breadth $5'''\cdot3$, thickness $0'''\cdot4$, were so managed that first a single one, then two, three, &c. laid together, or rather joined together, were brought into the above-mentioned spiral.   When the twelve laminæ were laid together, they formed a prism of equal magnitude with C in the first series of experiments, and had a weight of 94·8 grms. The results were,—

	Magnetic moment.	Proportion to the mass.
1 lamina . . .	3·53	3·53
2 laminæ . . .	4·11	2·05
3 „ . . .	4·36	1·45
4 „ . . .	4·65	1·16
5 „ . . .	4·94	0·99
6 „ . . .	5·15	0·86
7 „ . . .	5·39	0·77
8 „ . . .	5·61	0·70
9 „ . . .	5·83	0·65
10 „ . . .	6·05	0·60
11 „ . . .	6·27	0·57
12 „ . . .	6·44	0·54

Here is shown in a striking manner how disadvantageous it is to increase the thickness.

As a deduction from the results given above, 14·4 parallelograms would, according to the weight, be equal to the prism C (Series of Experiments 1), and the whole magnetism of the same would have amounted to 6·874; but a double comparison gave 7·194—without doubt a consequence of this, that the parallelograms had been covered in the heating with tinder [Zunder].

3rd Series of Experiments.—Six parallelograms (fig. 6) of $45'''\cdot6$ length, $0'''\cdot3$ thickness, and breadths $2'''\cdot3$, $4'''\cdot6$, $6'''\cdot8$, $9'''\cdot1$, $11'''\cdot4$, $13'''\cdot7$, were cut out of an iron plate, and after they had been carefully heated were brought into the above-mentioned spiral; the result was as follows:—

	Magnetic moment.	Mass.	Proportion to the mass.
A . . .	2·69	2·8	0·961
B . . .	4·05	5·8	0·699
C . . .	5·04	9·0	0·560
D . . .	5·77	11·7	0·498
E . . .	6·52	14·3	0·454
F . . .	7·12	16·7	0·425

From this it is deducible that the augmentation of the breadth is also to be considered as disadvantageous, but in a more trifling proportion than we have found that of the thickness in the second series of experiments.

4th Series of Experiments.—Four needles (fig. 7) contracting from the middle to sharp points at the two ends (rhomboids) were cut out of an iron plate. They had all the same length $=39^{m}{\cdot}6$; the breadths in the middle were very nearly in the proportion of 1, 2, 3, 4, and amounted in the broadest needle to $19^{m}{\cdot}5$. The observation gave the following numbers :—

	Magnetic moment.	Mass.	Proportion to the mass.
A . . .	4·304	4·95	0·870
B . . .	5·313	9·84	0·539
C . . .	5·944	14·45	0·412
D . . .	6·595	19·45	0·339

It appears hereby that the proportion of the magnetism to the weight is the more advantageous the more sharply the needles are pointed, that is, the smaller the breadth is in the middle.

5th Series of Experiments.—Three equal needles (fig. 8) were made in form similar to those of the fourth series; length $46^{m}{\cdot}0$, breadth in the middle $13^{m}\,3$; from two of them a part was taken out of the middle, so that they had the appearance of perforated rhomboids, and the part cut out was similar to the whole figure. The magnitude of the part cut out amounted in B to one-third, and in C to two-thirds of the whole figure. The observation gave—

	Magnetic moment.	Mass.	Proportion to the mass.
A . . .	3·46	1·02	3·39
B . . .	3·47	0·85	4·08
C . . .	3·17	0·52	6·04

It is therefore very advantageous to take out a part of the mass in the middle.

6th Series of Experiments.—In the fourth and fifth series of experiments the needle contracted to a point from the middle towards the two ends; in the present series of experiments it is to be ascertained what difference depends on the circumstance whether the breadth begins to diminish directly from the middle or nearer to the ends. For this purpose flat pieces of steel were employed of $43^{m}{\cdot}1$ length, $1^{m}{\cdot}0$ thickness and $10^{m}{\cdot}0$ breadth (in the middle), whose figure is represented in fig. 9; the part $a\,b$ amounted in B to a sixth, in C to a third, and in D to a half of the length. The results were—

	Magnetic moment.	Mass.	Proportion to the mass.
A	44·6	37·2	1·20
B	34·3	28·8	1·19
C	27·7	23·6	1·17
D	23·6	18·0	1·32

This series of observations is not very decisive; nevertheless it shows distinctly that the pointing of the ends of the magnets is not advantageous, except when the diminution of the breadth begins from the middle. A flat needle contracting its breadth from the middle to a point is, by the above measures, more advantageous by one-tenth than one of the form of a parallelogram; from other far more decisive series of experiments I have found a somewhat greater proportion, *i. e.* one-eighth.

7th Series of Experiments.—It is known that magnetism shows its strength in corners and points, and it appeared proper to investigate what result would be obtained if a magnet had several points. With this view, three parallelograms of $47'''·0$ length, $9'''·0$ breadth, $0'''·4$ thickness, were cut out of a plate of iron, and triangular notches were cut out of the ends, so that one piece received two, the other three points at each end, whilst in the third piece no cut was made. The form of the pieces is seen in fig. 10; the depth of the cuts amounted to a fourth of the length. The observation gave—

	Magnetic moment.	Mass.	Proportion to the mass.
A	5·075	1·00	4·659
B	4·908	1·10	4·462
C	6·005	1·41	4·259

According to this it is advantageous to cut notches in the ends of flat magnets, and the proportion increases with the number of notches.

As a consequence of the determination given in the sixth series of experiments, the proportion-number would be 4·79 for a needle contracting to a point from the middle; it is not improbable that by increase of the number of notches this proportion could be exceeded, nevertheless the figure recommended itself, as to what is here in question, so little in other respects, that it will scarcely find practical application. ,

From the preceding determinations it results—

(1) That *narrower* magnets are more advantageous than *broader*.

(2) That *thinner* magnets are more advantageous than *thicker*.

(8) That consequently the most advantageous form is that in which breadth and thickness disappear, and the magnet is

transformed into a mathematical line, *i. e.* into a so-called *linear magnet.*

The most advantageous form of a magnet, so far as the proportion of the magnetism to the weight is considered, is therefore an *imaginary* one; practically, however, there are two forms which appear advantageous, namely the *flat, contracting to a point from the middle,* and the *flat prismatic* : and indeed in the former form the proportion of the magnetism to the weight is more advantageous by one-eighth part than in the latter; so that it must always hold as a rule that the thickness and breadth must be as far diminished as the other necessary conditions permit.

We should still have to investigate in what proportion in the above-mentioned forms the magnetism stands to the moment of inertia; but I consider it superfluous to annex here the tabular exhibitions relative to this, since without such it is easy to see that the form which we have pronounced as disadvantageous in reference to the weight, must also be disadvantageous as regards the moment of inertia. But as respects the flat form contracting to a point from the middle, and the flat prismatic form, which have been noted above as the only appropriate forms, the weights are, with equal length, and equal breadth in the middle, as 1 to 2, and the moments of inertia as 1 to 3·75, so that the form contracting to a point must be recognized as by far the best.

In regard to the preceding investigation, it ought yet to be mentioned that it must prove always too much dependent on circumstantial details, and too little satisfactory, as long as we are not in a position to lay down the laws of the distribution of the magnetism and of the dependence of the magnetic moment upon the dimensions. In this latter point of view the labours hitherto employed have had only very trifling success. From numerous observations which I have made with the prismatic bars, it results that with equal thickness the magnetic moments are in the proportion of the square roots of the thickness; nevertheless this law only obtains for greater transverse sections, and is perfectly unavailable for smaller dimensions. I have now made substitutions in the formula

$$\sqrt{\frac{ax+b}{x+c}}\,x,$$

where x is the variable dimension, and a, b, c constants; and I find that it very accurately corresponds with observation in small as in great dimensions. Even when laminæ are laid together, this formula represents very well the result, as will be proved by the following Table, in which the second series of experiments is calculated by the formula

$$\sqrt{\frac{12 \cdot 89 + 2 \cdot 46\, n}{n + 0 \cdot 218}}\; n.$$

Number of laminæ $= n$.	Magnetic moment. Observed.	Calculated.	Difference.
1 . . .	3·53	3·54	+0·01
2 . . .	4·11	4·00	−0·11
3 . . .	4·36	4·34	−0·02
4 . . .	4·65	4·63	−0·02
5 . . .	4·94	4·90	−0·04
6 . . .	5·15	5·16	+0·01
7 . . .	5·39	5·40	+0·01
8 . . .	5·61	5·62	+0·01
9 . . .	5·83	5·84	+0·01
10 . . .	6·05	6·05	0·00
11 . . .	6·27	6·25	−0·02
12 . . .	6·44	6·45	+0·01

A practical inference results from the preceding investigation, which I believe deserves to be carefully·considered on the part of those who concern themselves with the manufacture of magnetic instruments. A freely moveable magnet is to be employed with advantage only so far as the magnetic moment is as large as possible in proportion to the weight. But the more the transverse size is augmented, the greater is the departure from the fulfilment of this condition, and consequently the use of massive magnetic bars must be pronounced inadmissible. There is only one means of obtaining great magnetic strength with trifling weight; namely, by firmly connecting several thin and flat magnets near or upon one another in one system without their touching each other. Many years ago I began in magnetic variation-instruments, later also in magnetic theodolites, to unite several magnets; and at present I use universally systems of three laminæ, which are laid upon each other and held separated in the middle by small pieces of brass of about the thickness of three-quarters of a line. Also in ships' compasses several needles near each other are at present continually used with the best result. Hollow cylindrical magnets, to which some artists have given a great preference in regard to strength and lightness, remain, as can be proved even from theoretical considerations, very far behind in comparison with a single flat needle; and with this agree also the experiments which I have made.

XLVII. *Remarks on Radiation and Absorption.*

To Sir John F. W. Herschel, Bart. &c. &c. &c.

DEAR SIR JOHN,

I AM anxious to address this note to you upon a subject which you have in great part made your own, because I fear that neither in my book upon the Alps, nor in my recently published papers, have I made due reference to your estimable researches on Solar Radiation. I have been for some time experimenting on the permeability of our atmosphere to radiant heat, and have arrived at the conclusion that true air, that is to say, the mixture of oxygen and nitrogen which forms the body of our atmosphere, is, as regards the transmission of radiant heat, a practical vacuum. The results from which the opacity of air has been inferred are all to be ascribed to the extraneous matters diffused in the atmosphere, and mainly to the aqueous vapour. The negative results recently obtained by that eminent experimenter, Professor Magnus of Berlin, have induced me to reinvestigate this point; and the experiments which I have made not only establish the action of aqueous vapour, but prove this action to be comparatively enormous. Here is a typical case:—
On the 10th of this month I found the absorptive action of the common air of our laboratory to be made up of three components, the first of which, due to the pure air, was represented in magnitude by the number 1; the second, due to the transparent aqueous vapour, was represented by the number 40; while the third, due to the effluvia of the locality and the carbonic acid of the air, was represented by the number 27. The total action of its foreign constituents on the day in question was certainly sixty-seven times that of the atmosphere itself; while the aqueous vapour alone exerted an action at least forty times that of the air.

I have also to communicate to you some results of lunar radiation which connect themselves with your speculations. On Friday the 18th of this month, I made a series of observations on the moon from the roof of the Royal Institution. From six concurrent experiments, I was compelled to infer that my thermoelectric pile lost more heat when presented to the moon than when turned to any other portion of the heavens of the same altitude. The effect was equivalent to a radiation of *cold* from our satellite. I was quite unprepared for this result, which, however, you will at once perceive, may be an immediate consequence of the moon's *heat*. On the evening in question a faint halo which surrounded the moon, and which was only visible when sought for, showed that a small quantity of precipitated vapour was afloat in the atmosphere. Such precipitated particles, in virtue of their multitudinous reflexions, constitute a powerful screen to intercept the terrestrial rays, and any agency that removes them and establishes the optical continuity of the atmosphere must assist the transmission of terrestrial heat*. I think it may be affirmed that no sensible quantity of the obscure heat of the moon, which, when she is full, probably constitutes a large proportion of the total heat emitted in the direction of the earth, reaches us. This heat is entirely absorbed in our atmosphere; and on the evening in question it was in part applied to evaporate the precipitated particles, hence to augment the transparency of the air round the moon, and thus to open a door in that direction for the escape of heat from the face of my pile. The instrument, I may remark, was furnished with a conical reflector, the angular area of which was very many times that of the moon itself.

I remain,

Yours very faithfully,

October 21, 1861.   JOHN TYNDALL.

XLVIII. On a Generalization of a Theorem of Cauchy on Arrangements. By J. J. SYLVESTER, M.A., F.R.S., Professor of Mathematics at the Royal Military Academy, Woolwich†.

IN a paper "On the Theory of Determinants" in the Philosophical Magazine for March in this year, Mr. Cayley has referred and added to a theorem of Cauchy deduced from the latter's method of *arrangements*, viz. that if we resolve an integer n in every possible way into parts, to wit α parts of a, β parts of $b, \ldots$ of l, $(a, b, c \ldots l$ being all distinct integers), then

$$\Sigma \frac{1}{\Pi a a^\alpha \, \Pi \beta b^\beta \ldots \Pi \lambda l^\lambda} = 1.$$

Now both Cauchy's theorem and Mr. Cayley's addition to it (which essentially consists in the observation, that if before the numerator 1 in the above quantity under the sign of summation we write $(-)^{\alpha + \beta + \cdots + \lambda}$, the sum becomes zero) are no more than particular cases of the following theorem: viz. that if

---

* I was going to add "into space;" but the expression might lead to misapprehension. My experiments indicate that the absorption of water is a *molecular* phenomenon. If we suppose the aqueous vapour of the atmosphere to be condensed to a liquid shell enveloping the earth, the experiments of Melloni would lead us to conclude that such a shell would completely intercept the obscure terrestrial rays. And if the vapour be equally energetic, our atmosphere would prevent the *direct* transmission of the obscure heat of the earth into space. On this point, however, I wish to make some further observations.

† Communicated by the Author.

instead of 1 we write $\rho^{\alpha+\beta+\cdots+\lambda}$ in the numerator of the quantity under the sign of summation (ρ being any quantity whatever), the sum becomes expressible as a known function of ρ. Nothing can be easier than the proof.

Let the α, β, γ, ... λ in the preceding statement be supposed subject to the further condition that their sum is r; then for any assigned value of r (a positive integer) it is easy to see that the sum of the terms within the sign of summation in Cauchy's theorem is

$$\mathrm{S}\left(\frac{1}{x_1\,x_2\ldots x_r}\cdot\frac{\rho^r}{\Pi(r)}\right),$$

where x_1, x_2, ... x_r mean *every* system of values of x_1, x_2, ... x_r (permutations *admitted*) which satisfy the equation

$$x_1+x_2+\ldots+x_r=n.$$

(It should here be observed that α, β, γ, ... λ; a, b, c, ... l are the systems which satisfy $\alpha a+\beta b+\gamma c+\ldots+\lambda l=n$, permutations being *excluded*; that is to say, if, for example, α, β, γ should happen to be equal for any partition of n, the values α, a; α, b; α, c would figure only *once*, and not *six* times, among the systems included under the sign of Σ.) Hence then we see that

$$\Sigma\frac{\rho^{\alpha+\beta+\gamma+\ldots+\lambda}}{\Pi\alpha.a^\alpha\ \Pi\beta.b^\beta\ldots\Pi\lambda.l^\lambda}=\Sigma_{r=\alpha}^{r=1}\mathrm{S}_r\frac{\rho^r}{\Pi(r)}\,*.$$

where S_r is the coefficient of t^n in $\left(\dfrac{t}{1}+\dfrac{t^2}{2}+\dfrac{t^3}{3}+\&\mathrm{c.}\ ad\ infin.\right)^r$,

i. e. in $\left(\log\left(\dfrac{1}{1-t}\right)\right)^r$; and the total sum designated by Σ will be consequently the coefficient of t^n in

$$\log\left(\frac{1}{1-t}\right)\rho+\left(\log\frac{1}{1-t}\right)^2\frac{\rho^2}{1.2}+\&\mathrm{c.},$$

i. e. in $e^{\rho\log\left(\frac{1}{1-t}\right)}$, *i. e.* in $\left(\dfrac{1}{1-t}\right)^\rho$.

---

* For if we take a system of values satisfying the above equation, consisting of α equal values a, β equal values b ... λ equal values l, such a system will give rise in $\Sigma\dfrac{1}{x_1\,x_2\ldots x_r}$ to $\dfrac{\pi(r)}{(\pi\alpha)\,(\pi\beta)\ldots(\pi\lambda)}$ repetitions of the term $\dfrac{1}{a^\alpha.b^\beta\ldots l^\lambda}$, and consequently in $\Sigma\dfrac{1}{x_1\,x_2\ldots x_r}\cdot\dfrac{1}{\pi r}$ to a total value $\dfrac{1}{(\pi\alpha)a^\alpha\,(\pi\beta)b^\beta\ldots(\pi\lambda)l^\lambda}$, condensed into a single term in Cauchy's theorem.

Thus if $\rho=1$, we have Cauchy's theorem, viz. $\Sigma=1$;

„ $\rho=-1$,   „   Cayley's theorem, viz. $\Sigma=0*$;

and in general for any value of ρ,

$$\Sigma=\frac{\rho(\rho+1)\ldots(\rho+n-1)}{1.2\ldots n}\dagger.$$

In this theorem is in fact included another, viz. that if

$$\alpha a+\beta b+\ldots+\lambda l=n \text{ and } \alpha+\beta+\ldots+\lambda=r$$

(permutations *not* admissible), then

$$\Sigma\frac{\Pi n}{\Pi\alpha.a^{\alpha}.\Pi\beta.b^{\beta}\ldots\Pi\lambda.c^{\lambda}}$$

is equal to the coefficient of ρ^{r-1} in

$$(\rho+1)(\rho+2)\ldots(\rho+n-1).$$

This coefficient is accordingly (to return to Cauchy's theory of arrangements) the number of substitutions of n elements capable of being expressed by the product of r cyclical substitutions. As, for instance, the number of substitutions of four elements a,b,c,d capable of expression by the product of two cyclical substitutions

---

* Provided, however, that n exceeds 1, a limitation accidentally omitted in Mr. Cayley's paper; and so in general

$$\Sigma\frac{(-\rho)^{\alpha+\beta+\ldots+\lambda}}{\Pi\alpha.a^{\alpha}\ldots\Pi\lambda.l^{\lambda}}=0,$$

ρ being any positive integer provided n is greater than ρ.

$\dagger$ If $\rho=\frac{1}{2}$, we obtain

$$\Sigma=\frac{1.3.5\ldots(2n-1)}{2.4.6\ldots2n};$$

from which it is easy to infer that the number of substitutions of $2n$ things representable by the product of cyclical substitutions, all of an even order, is $(1.3.5\ldots(2n-1))^2$. If $\rho=-\frac{1}{2}$, we obtain

$$\Sigma=\frac{1.1.3\ldots(2n-1)}{2.4.6\ldots(2n)},$$

combining which with the preceding result, it is easy to infer that the number of substitutions of $2n$ things representable by the product of an odd number of cyclical substitutions, all of an even order, is to the number of such representable by the product of an even number of cyclical substitutions, all of an even order, in the ratio of n to $(n-1)$. The former of these two theorems is intimately allied with Mr. Cayley's celebrated theorem on "skew," or what, for good reasons hereafter to be alleged, I should prefer to call *polar* determinants, viz. that every such of the $2n$th order is the square of a Pfaffian. A Pfaffian is in fact a sum of quantities typifiable completely, both as to sign and magnitude, by a duadic *syntheme* of $2n$ elements, the number of which is readily seen to be $1.3.5\ldots(2n-1)$. I believe I shall soon be in a condition to announce a remarkable extension of this theory to embrace the case of Polar *Commutants* and *Hyperpfaffians.*

ought to be the coefficient of λ in $(\lambda+1)(\lambda+2)(\lambda+3)$, *i. e.*
11, which is right; for the number of substitutions of the
form (a, b) (c, d) will be 3, and of the form (a, b, c) (d) 8. In
conclusion, I may notice that by an obvious deduction from
this last theorem, we are led to the well-known one in the theory
of numbers, that every coefficient in the development of

$$\Sigma(\rho+1)(\rho+2)\ldots(\rho+n-1),$$

except the first and last, and the sum of these two, is divisible
by n when n is a prime number; and indeed we can actually express by aid of it the quotient of every intermediate coefficient
divided by n as the sum of separate integer terms free from the
sign of addition.

K, Woolwich Common,
  October 10, 1861.

Postscript. By an extension of the method of generating functions contained in the text above, it may easily be seen that the
number* of substitutions of n letters represented by the products
of r cyclical substitutions, where the number of letters of each
cycle leaves a given residue e in respect to a given *modulus* μ, may
be made to depend on the solution of the equation in differences

$$u_n - u_{n+\mu} = \frac{\rho}{n-e}\, u_{n-e}.$$

The case where $e=1$ is deserving of particular notice.

It may be shown by means of the above equation in differences, that the number of substitutions of n letters formed
by r cycles each of the form $\mu K+1$ (μ being constant), say
$\phi(n, r, \mu, 1)$, where $\dfrac{n-r}{\mu}$ is necessarily an integer, may be
found by taking in every possible way $\dfrac{n-r}{\mu}$ *distinct* groups of μ
consecutive terms of the series $1, 2, 3, \ldots (n-1)$; the sum of
the products of every such combination of groups is the value
required. For example, if

$$n=8, \quad r=3, \quad \mu=2,$$

---

* For this number, divided by $\Pi(n)$, is the coefficient of x^n in

$$\frac{1}{\Pi r}\left(\int_0^x \frac{dx\, x^{e-1}}{1-x}\right)^r, \text{ say } \frac{1}{\Pi r}(\phi x)^r,$$

and therefore of $x^n \rho^r$ in $e^{\rho \phi x}$, say $\psi(x, \rho)$, and therefore (since $\dfrac{d\psi}{dx} = \dfrac{x^{e-1}}{1-x^m}$
and ψ may be put under the form $\Sigma \dfrac{u_n}{n} x^n$) of ρ^r in $\dfrac{u_n}{n}$, where u_n is defined
as in the text.

$$\phi(8,3,2,1)=1.2.3.4.5.6+1.2.3.5.6.7+1.2.3.5.6.8$$
$$+1.2.3.6.7.8+2.3.4.5.6.7+2.3.4.6.7.8$$
$$+3.4.5.6.7.8.$$

And as a corollary, since it may easily be seen that $\phi(n, r, \mu, e)$ is always divisible by n when n is a prime and $\mu r + e < n$, it follows that the sum of all the possible products of (any given number) i distinct groups of a given number r of consecutive terms of the series $1, 2, 3, \ldots (n-1)$ will be divisible by n when n is a prime and $ir < n-1$*. When $r=1$, this theorem becomes identical with Wilson's, already referred to.

Finally, it may be noticed that the number of substitutions of n letters formed by *any* number of cycles, all of an *odd* order, will be the coefficient of x^n in $\left(\dfrac{1+x}{1-x}\right)^{\frac{1}{2}}$, *i. e.* $\left(1.3.5\ldots(n-1)\right)^2$ (the same as the number that can be formed with cycles all of an *even* order) when n is even, and $\left(1.3.5\ldots(n-2)\right)^2 n$ when n is odd †.

XLIX. *Notes on the Hexahedron inscribed in a Sphere.*
By CHARLES W. MERRIFIELD, *Esq.*‡

1. ITS six planes pass, in general, four by four through three points.

There is exception, as a singular case, where the intersections of two of the three pairs of opposite planes are parallel. In this case the intersection of the third pair is perpendicular to the parallels, and the inclination of the two planes equal, but opposite.

Let us consider the intersection of the sphere with any pair

* For instance, making $n=7$, $r=2$, $i=2$,
$$1.2.3.4+1.2.4.5+1.2.5.6+2.3.4.5+2.3.5.6+3.4.5.6=784$$
and is *divisible by* 7.

† By taking $\mu=2$ in the general theorem, it is an easy inference that if we write
$$(\tan^{-1}x)^r = x^r - \frac{A_2 x^{r+2}}{(r+1)(r+2)} + \frac{A_4 x^{r+4}}{(r+1)(r+2)(r+3)(r+4)} \mp \&c.,$$
A_{2i} will be the sum of all the products of $2i$ integers comprised between 1 and $r+2i-1$ that can be formed with combinations of i distinct pairs of consecutive integers; thus (*e. g.*) the coefficient of x^{2m} in $(\tan^{-1}x)^2$ ought to be
$$\frac{2}{m}\left(1+\frac{1}{3}+\frac{1}{5}+\cdots+\frac{1}{2m-1}\right),$$
which may be easily verified.

‡ Communicated by the Author.

of opposite planes separately. A cone can in general be drawn through the two circles of intersection, and the truncated portion of any four-sided pyramid inscribed in this cone will give an inscribed hexahedron in the sphere. Moreover, if a quadrangle be inscribed in one of the circles, as ABCD, and a plane be drawn through AB cutting the other circle in *ab*, and then two more planes be drawn through the points *a*AD and *b*BC cutting the second circle in *cd* respectively, it will be seen that the quadrangle CD*cd* will not be plane unless—

(α) The plane AB*ab*, and hence the three others, pass through the vertex of the cone, or

(β) Either pair of lines AB and CD or AC and BD be parallel to one another and perpendicular to the common diametral plane of the cone and sphere.

(α) is the general case, and the hexahedron is a frustum of a quadrangular pyramid. By considering in like manner the other two pairs of planes, it will be seen that the hexahedron is the common frustum of three four-sided pyramids.

(β) is the singular case of a four-sided prism, the two sections being equally inclined, but in opposite directions. Note: that this prism will not in general be inscribed in a right cylinder.

2. In the general case it obviously follows that the hexahedron has six diagonal planes, passing two by two through the three vertices of the pyramids.

3. The four diagonal lines of the hexahedron intersect in a point. This follows, in the general case, from their lying two by two on the six diagonal planes, and it is easily seen in the singular case.

4. Hence in both cases there are six diagonal planes, all intersecting in a point.

5. This point is the spherical pole to the plane passing through the three vertices. In the singular case the polar plane passes through the parallel lines of intersection. This property may be deduced from the similar one of plane quadrilaterals inscribed in a circle.

6. Since the sections of the cone by the planes ABCD, *abcd* are subcontrary, the corresponding angles A*a*, B*b*, C*c*, &c. are equal each to each. Hence the sum of the opposite angles BAD and *bcd* is equal to two right angles, and so of similarly opposite pairs. This is not, in general, true of the singular case.

7. Since six planes intersect two by two in fifteen lines, every hexahedron must have, associated with it, three external lines of intersection.

(a) If these three associated lines lie in one plane, they must intersect in three points, and then the hexahedron will have six diagonal planes and four diagonal lines all intersecting in one point.   This species may be considered as formed by the planes of two tetrahedra having a common base.

(*b*) If these three lines intersect in two points, but do not lie in the same plane, there will in general be four diagonal planes only, and the four linear diagonals will intersect in four points not in the same plane.

(*c*) If only two of the three lines intersect, there will in general be only two diagonal planes, upon each of which one pair only of diagonal lines will intersect.

(*d*) Lastly, if none of the three lines meet, there will in general be no diagonal planes, and the four diagonal lines will not meet.

Each of these four species may, however, have singular cases.

8. In species (*a*) all parallel or subcontrary planes, which divide or cleave the hexahedron into two other hexahedra, have similar quadrilaterals traced upon them.   In the other three species this is only true, in general, of selected planes.   In the singular case of the inscribed tetrahedron, the quadrilaterals obtained by parallel cleavage are not always similar.

9. The hexahedron inscribed in the sphere belongs in general, as has been seen, to species (*a*).   Its two tetrahedra have the corresponding plane angles at their vertices supplementary each to those of the other.   The singular case is a very restricted singular form of species (*c*), or it may be looked upon as an indeterminate form, arising out of singularity, in species (*a*); *i. e.* that when two of the three associated lines are parallel, the third may leave their plane, under certain conditions of symmetry.

30 Scarsdale Villas, Kensington, W.,
    October 18, 1861.

L. *Notices respecting New Books.*

An Elementary Treatise on the Theory of Equations, with a Collection of Examples.   By I. TODHUNTER, M.A.

MR. TODHUNTER'S merits, as a writer of some of our best elementary treatises on mathematics, are now so well established as to render it quite unnecessary to dwell upon the manner in which this, his last task, has been performed.   It will suffice, therefore, to inform the mathematical student that a thoroughly trustworthy, complete, and yet not too elaborate, treatise on the Theory of Equations is now within his reach; that, as far as the elementary character of the work would permit, the treatment of the subject has been brought up to the level of the science of our day; and that, in some branches of the subject, the more elaborate researches of modern authors have been carefully examined, their suitable portions judiciously selected, and now for the first time collected.

The three chapters on Determinants will be particularly acceptable; for, except in larger treatises especially devoted to the subject, the student will nowhere find the first principles of this beautiful and powerful method so clearly and satisfactorily explained. The only suggestion that occurs to us with respect to these chapters is that they might be transferred with advantage to future editions of the author's Treatise on Algebra; for experiment has long since convinced us that the method of determinants may be introduced with great profit even in schools, and as soon as simple equations involving two or more unknown quantities are studied. We have found that pupils of average intelligence rapidly acquire a knowledge of the more elementary properties of determinants, and that they invariably regard the method as a welcome augmentation of their computing power. More important than this, however, is the fact that, as a mental discipline, the study of the properties in question is certainly not inferior to that of any other branch of algebra.

In heartily recommending the work, we will merely add that it is enriched by a collection of well-chosen examples.

LI. *Proceedings of Learned Societies.*

ROYAL SOCIETY.

[Continued from p. 324.]

November 22, 1860.—Major-General Sabine, R.A., Treasurer and Vice-President, in the Chair.

THE following communications were read:—

" Researches on the Phosphorus-Bases." No. X.—Metamorphoses of Bromide of Bromethylated Triethylphosphonium. By A. W. Hofmann, LL.D., F.R.S. Received July 24, 1860.

Among the several products of transformation into which the bromide of bromethyl-triethylphosphonium is converted when submitted to the action of reagents, the substances formed by its union with bodies similar to ammonia, have hitherto almost exclusively occupied my attention. I have, however, of late examined a variety of other changes of this body, which deserve to be noticed.

When heated, the bromide begins to evolve hydrobromic acid at a temperature of about 200°, which continues for a considerable length of time. The product of this reaction is evidently the bromide of vinyl-triethylphosphonium,

$$[(C_2H_4Br)(C_2H_5)_3P]Br = HBr + [(C_2H_3)(C_2H_5)_3P]Br.$$

It is, however, difficult to obtain the substance pure by this process, since the temperature at which the last portion of hydrobromic acid is eliminated closely approximates the degree of heat at which the vinyl-body is entirely destroyed; and since the latter compound may be obtained with the greatest facility by other processes*, I have not followed up any further this direction of the inquiry.

---

* The hydrated di-oxide of ethylene-hexethyl-diphosphonium, when submitted to distillation, undergoes decomposition; two different phases are to be distinguished in this metamorphosis. At about 200° the base begins to disengage the

I have already mentioned, in a previous note, the deportment of the bromethylated bromide with oxide of silver; the whole of the bromine is eliminated in the form of bromide of silver, a new base being formed.

According to circumstances, this base may be the vinyl-compound previously mentioned, or another body differing from the latter by containing the elements of one molecule of water in addition. This substance, which is always formed when the reaction takes place in moderately dilute solutions, is the oxide of a phosphonium, with three molecules of ethyl substituted for three equivalents of hydrogen, the fourth equivalent of hydrogen being replaced by an oxygenated radical $C_2 H_5 O$, arising from the radical $C_2 H_4 Br$ by the insertion of HO in the place of Br

$$[(C_2 \overset{.}{H}_4 Br) (C_2 H_5), P] Br + 2\overset{Ag}{\underset{H}{}} \bigg\} O = 2AgBr + \overset{[(C_2 H_4 HO) (C_2 H_5), P]}{\underset{H}{}} \bigg\} O$$

I have fixed the nature of this compound by the analysis of the iodide, of the platinum-salt and of the gold-salt, and, moreover, by the study of several remarkable transformations which it undergoes when submitted to the action of reagents.

It appeared of some interest to ascertain whether the *oxethylated* might be reconverted into the *bromethylated* base. The chloride of the former is energetically attacked by pentabromide of phosphorus; oxybromide of phosphorus and hydrobromic acid are abundantly evolved, and the residue of the reaction contains the chloride of bromethylated triethylphosphonium.

$$[(C_2 H_5 O) (C_2 H_5), P] Cl + PBr_5 = HBr + POBr_3 + [(C_2 H_4 Br) (C_2 H_5), P] O.$$

Thus it is seen that the molecular group $C_2 H_5 O$, which we assume as hydrogen-replacing in this salt, suffers under the influence of pentabromide of phosphorus, alterations identical with those which it is known to undergo under similar circumstances, when conceived as a constituent of alcohol.

If we consider the facility with which the bromethylated triethyl-phosphonium is converted into the oxethylated compound, by the action of oxide of silver, and the simple re-formation of the first-mentioned body by means of pentabromide of phosphorus, a great variety of new experiments suggest themselves. As yet but little

vapour of triethylphosphine, the residuary solution retaining hydrated oxide of vinyl-triethylphosphonium,

$$[(C_2 H_4)''(C_2 H_5)_4 P]'' \underset{H_2}{\bigg\}} O_2 = (C_2 H_5)_3 P + H_2 O + \overset{[(C_2 H_3) (C_2 H_5), P]}{\underset{H}{}} \bigg\} O,$$

the latter yielding at a higher temperature the oxide of triethylphosphine together with ethylene,

$$[(C_2 H_3) (C_2 H_5)_3 P] \underset{H}{\bigg\}} O = C_2 H_4 + (C_2 H_5)_3 PO.$$

The vinyl-compound is even more readily obtained by the action of silver-salts, such as acetate of silver, at the temperature of 100°, on the bromethylated bromide.

$$_2 H_4 Br)(C_2 H_5), P] Br + 2 \left[\overset{(C_2 H_3 O)}{\underset{Ag}{}} \bigg\} O \right] = 2Ag Br + \overset{(C_2 H_3 O)}{\underset{[(C_2 H_3) (C_2 H_5), P]}{}} \bigg\} O + \overset{(C_2 H_3 O)}{\underset{H}{}}$$

progress has been made in this direction; one of the reactions, however, which I have studied deserves even now to be mentioned.

The salts of bromethylated and oxethylated triethylphosphonium may be regarded as tetrethyl-phosphonium-salts, in which an equivalent of hydrogen in one of the ethyl-molecules is replaced by bromine and by the molecular group HO respectively.

Bromide of tetrethylphosphonium $[(C_2H_4H\ )(C_2H_5)_3 P]$ Br,

Bromide of bromethylated tri-⎫
  phosphonium ....⎭ $[(C_2H_4Br)(C_2H_5)_3 P]$ Br ;

Bromide of oxethylated triethyl-⎫
  phosphonium ....⎭ $[(C_2H_4HO)(C_2H_5)_3 P]$ Br ;

and the question arose, whether the bromethylated compound might not be converted, simply by reduction, into a salt of tetrethylphosphonium. This transformation may, indeed, be effected without difficulty. On acidulating the solution of the bromethylated bromide with sulphuric acid and digesting the mixture with granulated zinc, the latent bromine is eliminated as hydrobromic acid, its place being at the same time filled by 1 equiv. of hydrogen,

$$[(C_2H_4Br)(C_2H_5)_3 P] Br + 2H = HBr + [(C_2H_5)_4 P] Br.$$

It was chiefly the facility with which a tetrethyl-phosphonium-compound may be obtained from the brominated bromide, that induced me to designate the hydrogen-replacing molecules C_2H_4Br, and C_2H_5O, which we meet with in the compounds above described, as *bromethyle* and *oxethyle.* It remained to be ascertained whether these compounds might actually be formed by means of direct substitution-products of ethyle-compounds. It was with the view of deciding this question that I have examined the deportment of triethylphosphine with the monochlorinated chloride and the monobrominated bromide of ethyle.

The former of these substances has been long known, having been investigated by Regnault many years ago; the latter had not been hitherto obtained. I have prepared it by submitting bromide of ethyle to the action of dry bromine under pressure[*] It is a heavy aromatic liquid boiling at 110°.

The chlorinated chloride and the brominated bromide of ethyle, although essentially different in their physical properties from dichloride and dibromide of ethylene, with which they are isomeric, nevertheless resemble the ethylene-compounds in their deportment with triethylphosphine.

In both cases the final product of the reaction is a salt of hexethylated ethylene-diphosphonium. I have identified these salts with those obtained by means of dichloride and dibromide of ethylene, both by a careful examination of their physical properties, and by the analysis of the characteristic iodide and of the platinum-salt. I have not been able to trace in the first of these reactions a salt of chlorethylated triethylphosphonium; but I have established by experiment that in the reaction between triethylphosphine and brominated bromide of

---

[*] In addition to the monobrominated bromide of ethyle, (C_2H_4Br) Br, there is also formed in this reaction the dibrominated bromide, $(C_4H_3Br_2)$ Br.

ethyle, the formation of bromethyl-triethylphosphonium invariably precedes the production of the diphosphonium-compound.

"Researches on the Phosphorus-Bases."—No. XI. Experiments in the Methyle- and in the Methylene-Series. By A. W. Hofmann, LL.D., F.R.S. Received July 24, 1860.

In former notes I have repeatedly called attention to the transformation of the bromide of bromethylated triethylphosphonium under the influence of bases. In continuing the study of these reactions, I was led to the discovery of a very large number of new compounds, the more important ones of which are briefly mentioned in this abstract.

Hybrids of Ethylene-diphosphonium.

Action of Trimethylphosphine upon Bromide of Bromethyl-triethylphosphonium.

These two bodies act upon each other with the greatest energy, and moreover exactly in the manner indicated by theory. The resulting compound was of course examined only so far as was necessary to establish the character of the reaction.

The dibromide of the hybrid diphosphonium is more soluble than the hexethylated compound formerly described, which in other respects it resembles. Oxide of silver eliminates the extremely caustic base

$$C_{11} H_{40} P_2 O_2 = \left.\begin{matrix} [(C_2 H_4)''(CH_3)_3(C_2 H_5)_3 P_2]'' \\ H_2 \end{matrix}\right\} O_2,$$

which yields with hydrochloric acid and dichloride of platinum a pale-yellow platinum-salt,

$$C_{11} H_{36} P_2 Pt_2 Cl_6 = \left[(C_2 H_4)'' \begin{matrix} (CH_3)_3 P \\ (C_2 H_5)_3 P \end{matrix} \right]'' Cl_2, 2 Pt Cl_2,$$

separating in scales from boiling water.

The salts of the hybrid diphosphonium crystallize like those of the hexethylated diphosphonium, but, so far as they have been examined, are somewhat more soluble. This remark applies especially to the iodide.

It seemed worth while to try whether the bromide of bromethylated triethylphosphonium was capable of fixing a molecule of phosphoretted hydrogen. It was found, however, that the two bodies do not act upon one another. Phosphoretted hydrogen gas, passed through the alcoholic solution of the bromide, either cold or boiling, did not seem to affect it in any way.

Action of Trimethylphosphine on Dibromide of Ethylene.

This reaction exhibits a repetition of all the phenomena observed in that which takes place between the dibromide and triethylphosphine. The process is completed sooner, if possible, than in the ethyle-series. The lower boiling-point and the overpowering odour of trimethylphosphine render it advisable to mix the materials with considerable quantities of alcohol or ether; and on account of the extreme oxidability of the phosphorus-compound, it is best to operate in vessels filled with carbonic acid and subsequently sealed before the blowpipe. After digestion for a short time at 100°, the mixture of the two liquids solidifies to a hard, dazzling, white, crystalline mass containing the two bromides,

$$C_3 H_{13} P Br_2 = [(C_2 H_4 Br)(C H_3)_3 P] Br,$$

$$C_8 H_{22} P_2 Br_2 = \left[(C_2 H_4)'' \frac{(C H_3)_3 P}{(C H_3)_3 P} \right]'' Br_2,$$

one or the other predominating according to the proportions in which the two bodies were allowed to act upon one another.

It was not difficult to establish the nature of these two compounds by numbers.

The solution of the saline mass .in absolute alcohol, deposits, on cooling, beautiful prismatic crystals, consisting of the bromide of bromethyl-trimethylphosphonium almost chemically pure, while the diphosphonium-bromide remains in solution. The nature of the monophosphonium-compound was fixed by a bromine determination in the bromide, and by the analysis of a platinum-salt beautifully crystallized in needles containing

$$C_6 H_{13} Br P PtCl_3 = [(C_2 H_4 Br)(C H_3)_3 P] Cl, PtCl_2.$$

Treatment of this platinum-salt with sulphuretted hydrogen yielded an extremely soluble and deliquescent chloride, which was not analysed, but submitted to the action of oxide of silver, when it furnished the oxide of the corresponding oxethylated compound

$$C_5 H_{15} PO_2 = \left. \genfrac{}{}{0pt}{}{[(C_2 H_4 O)(C H_3)_3 P]}{H} \right\} O.$$

The caustic liquid was converted by hydrochloric acid into the easily soluble chloride corresponding to the oxide; and this chloride, when treated with dichloride of platinum, deposited the platinum-salt of the oxethylated trimethylphosphonium in well-formed octahedra extremely soluble in water, containing

$$C_5 H_{14} P O PtCl_3 = [(C_2 H_4 O)(C H_3)_3 P]Cl, PtCl_2.$$

Salts of Hexmethylated Ethylene-diphosphonium.

Dibromide.—The preparation of this salt has already been mentioned. It is extremely soluble in water, and even in absolute alcohol, insoluble in ether. *In vacuo* over sulphuric acid it solidifies into a mass of acicular crystals, which are exceedingly deliquescent.

The dibromide, treated with oxide of silver, yields the corresponding dioxide,

$$C_8 H_{24} P_2 O_2 = \left. \genfrac{}{}{0pt}{}{[(C_2 H_4)''(C H_3)_4 P_2]''}{H_2} \right\} O_2,$$

which forms with acids a series of salts resembling the corresponding ethyle-compounds. Of these I have prepared only the

Di-iodide, which crystallizes in difficultly soluble needles of the composition

$$C_8 H_{22} P_2 I_2 = \left[(C_2 H_4)'' \frac{(C H_3)_3 P}{(C H_3)_3 P} \right]'' I_2,$$

surpassing in beauty the corresponding ethyle-compound; and the

Platinum-salt.—This is an apparently amorphous precipitate, which is nearly insoluble in water, dissolves with extreme slowness in boiling hydrochloric acid, and separates therefrom on cooling in golden-yellow laminæ, very much like those of the platinum-salt of the hybrid ethylene-trimethyl-triethyl-diphosphonium. It consists of—

$$\dot{C}_9 H_{22} P_2 Pt_2 Cl_6 = \left[(C_2 H_4)'' \frac{(C H_5)_3 P}{(C H_5)_3 P} \right]'' Cl_2, 2 PtCl_2.$$

METHYLENE GROUP.
Action of Triethylphosphine on Di-iodide of Methylene.

Triethylphosphine and di-iodide of methylene act so powerfully on one another, that it is necessary to moderate the reaction by the presence of a considerable quantity of ether. The reaction is very soon completed, even when the mixture is largely diluted, especially if it be heated to 100° in sealed tubes. The saline residue left after the evaporation of the ether is immediately seen to be a mixture of several compounds, one of which—a sparingly soluble iodide crystallizing in long needles—at once arrests attention.

From analogy we might expect to find in the saline mixture the compounds

$$[(C H_2 I)(C_2 H_5)_3 P] I,$$
or
$$[(C H_2)''(C_2 H_5)_6 P_2]'' I_2.$$

Experiment has, however, established the presence of the first only.

The difficultly soluble crystals just mentioned are easily purified, being readily soluble in water, sparingly in alcohol, insoluble in ether. Their solution in boiling alcohol yields splendid needles frequently an inch long, and possessing extraordinary lustre. Analysis prove this beautiful salt to be the first of the above-mentioned compounds.

The new iodide behaves with nitrate of silver like the bromide of bromethylated triethylphosphonium; half the iodine is eliminated in the form of iodide of silver. It differs, however, from the bromide in its deportment with oxide of silver which, after removal of the accessible iodine, leaves the latent iodine untouched, even after protracted ebullition. A powerfully alkaline solution is thus obtained containing the base

$$C_7 H_{18} I P O = \left. \begin{array}{c} [(C H_2 I)(C_2 H_5)_3 P] \\ H \end{array} \right\} O.$$

The crystals of the iodide were transformed into the chloride by means of chloride of silver, and the solution was precipitated by dichloride of platinum. The precipitate is very sparingly soluble in cold water, but may be recrystallized from a considerable quantity of boiling water. As the liquid cools, splendid needle-shaped crystals are deposited containing

$$C_7 H_{17} I P PtCl_3 = [(C H_2 I)(C_2 H_5)_3 P] Cl, PtCl_2.$$

The sparingly soluble iodide is present in proportionally small quantity only among the products of the action of di-iodide of methylene on triethylphosphine. I have in vain endeavoured to detect among these products anything of the nature of a diphosphonium-compound. On treating the mother-liquor of the sparingly soluble iodide with chloride of silver, and the dilute filtered solution with dichloride of platinum, a few needles of the iodated platinum-salt are still deposited; but after considerable evaporation the solution yields crystals, all of which exhibit an octahedral habitus. I was equally unsuccessful in a particular experiment, in which I subjected di-iodide of methylene to the action of a large excess of triethyl-phosphine; and a similar report must be made of the attempt to produce the desired body by treating the ready prepared iodide with triethylphosphine, according to the equation

$$[(C H_2 I)(C_2 H_5)_3 P]I + (C_2 H_5)_3 P = [(C_2 H_2)''(C_2 H_5)_6 P_2]''I_2.$$

The examination of the mother-liquor of the sparingly soluble iodide is a difficult and thankless proceeding; nevertheless, by a sufficient number of iodine- and platinum-determinations, it may be shown to be a mixture of four different compounds. The mother-liquor is thus found to contain, together with the hydriodate of the phosphorus-base, two crystallizable iodides differing in solubility, and to be separated from one another only by a great number of crystallizations. The more soluble salt is the iodide of oxymethylated triethyl-phosphonium, corresponding to the iodomethylated compound; the less soluble salt is the iodide of methyl-triethylphosphonium. The last mother-liquors contain considerable quantities of oxide of triethylphosphine.

Iodide of Oxymethyl-triethylphosphonium.

This salt is extremely soluble both in water and in alcohol, even in absolute alcohol, and crystallizes only after the alcohol has been completely evaporated. The crystals, resembling the frosty efflorescences on a window-pane, contain

$$C_7 H_{18} O P I = [(C H_3 O)(C_2 H_5)_3 P] I.$$

The iodide, treated with oxide of silver, is converted into the corresponding caustic oxide, which, when mixed with hydrochloric acid and dichloride of platinum, yields a rather easily soluble platinum-salt of an octahedral habitus.

Iodide of Methyl-triethylphosphonium.

The nature of the less soluble iodide was determined by an iodine-determination, and by the analysis of the platinum-salt. The iodide dissolves in water and in alcohol, but is insoluble in ether. By adding ether to the alcoholic solution, tolerable crystals are obtained. This compound is most conveniently purified by precipitating the alcoholic mother-liquor, after freeing it by crystallization as far as possible from the iodomethylated iodide, with a quantity of ether insufficient to precipitate the whole, so that the greater part of the iodides may remain in solution.

· The iodide thus prepared contains

$$C_7 H_{16} P I = [(C H_3) (C_2 \dot{H}_5), P]I.$$

For further verification of this formula the crystals were deiodized with silver-oxide, and the caustic liquid thus obtained was saturated with hydrochloric acid and precipitated by dichloride of platinum. The platinum-salt, which crystallizes in beautiful octahedra, was found to contain

$$C_7 H_{16} P PtCl_3 = [(C H_3) (C_2 H_5), P]Cl, PtCl_4.$$

The two iodides are accompanied by a considerable quantity of oxide of triethylphosphine, which immediately separates in oily drops on treating the last mother-liquor with potash. Its presence was likewise unmistakeably recognized by the preparation of the platinum-salt. If the last mother-liquor of the iodine-compounds be deiodized and mixed with hydrochloric acid and dichloride of platinum, a quantity of octahedral salts separates in the first place, which are removed by sufficient concentration; the remaining liquid, when mixed with alcohol and ether, yields a crystalline precipitate, which separates from alcohol by spontaneous evaporation in the beautiful large hexagonal tables consisting of the platinum-salt of the oxychloride of triethylphosphine, which has been more fully described in one of the previous notes on these researches.

The formation of the four compounds contained in the mother-liquor of the sparingly soluble iodide is illustrated by the following equations:—

$$C_2 H_5)_3 P] + CH_3 I_2 + H_2 O = [(C_2 H_5)_3 H P]I + [(CH_3 O)(C_2 H_5)_3 P] I$$
$$3C_2 H_5)_3 P] + CH_3 I_2 + H_2 O = [(C_2 H_5)_3 H P]I + [(CH_3)(C_2 H_5)_3 P]I + (C_2 H_5)_3 I$$

"Researches on the Phosphorus-Bases."—No. XII. Relations between the Monoatomic and the Polyatomic Bases. By A. W. Hofmann, LL.D., F.R.S. Received August 17, 1860.

In recording my experiments on the derivatives of triethylphosphine, I have had more than one opportunity of alluding to the energy and precision which characterize the reactions of this compound. The usefulness of triethylphosphine as an agent of research has more particularly manifested itself in the study of the polyatomic bases, the examination of which, in continuation of former inquiries, was naturally suggested by the beautiful researches on the polyatomic alcohols published during the last few years. In the commencement these studies were almost exclusively performed with reference to derivatives of ammonia; but the results obtained in the examination of triethylphosphine have, in a great measure, changed the track originally pursued, and of late I have generally preferred to solve the problems which I had proposed to myself, by the aid of the phosphorus-bases.

The light which the study of these compounds throws upon the nature of the polyatomic bases generally, will be fully appreciated by a retrospective glance at the deportment of triethylphosphine under the influence of dibromide of ethylene, and a comparison of the products formed in this reaction with the results suggested by theory.

A simple consideration shows that the action of diatomic bromides upon bases must give rise to the formation of several classes of compounds. Let us examine by way of illustration the products which may be expected to be formed in the reaction between ammonia and dibromide of ethylene.

The diatomic bromide being capable of fixing two molecules of ammonia, we have in the first place four diatomic bromides of the formulæ

$$[(C_2H_4)'' \ H_6 N_2]'' Br_2$$
$$[(C_2H_4)_2'' H_4 N_2]'' Br_2$$
$$[(C_2H_4)_3'' H_2 N_2]'' Br_2$$
$$[(C_2H_4)_4'' \quad N_2]'' Br_2.$$

These are, however, by no means the only salts which, in accordance with our present conception of diatomic compounds, may be formed in this reaction. Taking into consideration the general deportment of dibromide of ethylene, there could be no doubt that, under certain conditions, this body would act with ammonia as a monoatomic compound, giving rise to another series of bodies, in which the hydrogen would be more or less replaced by the monoatomic molecule $C_2 H_4 Br$, viz.

$$[(C_2 H_4 Br) \ H_3 N] Br$$
$$[(C_2 H_4 Br)_2 H_2 N] Br$$
$$[(C_2 H_4 Br)_3 H \ N] Br$$
$$[(C_2 H_4 Br)_4 \quad N] Br.$$

Further, if the reaction took place in the presence of water, it was to be expected that the latent bromine of these salts, wholly or partially eliminated in the form of hydrobromic acid, would be replaced by the molecular residue of water, and thus, independently of any mixed compounds containing simultaneously bromine and oxygen, a series of salts might be looked for, in which a molecule $C_2 H_4 (HO) = C_2 H_3 O$ would enter monoatomically,

$$[(C_2 H_3 O) \ H_3 N] Br$$
$$[(C_2 H_3 O)_2 H_2 N] Br$$
$$[(C_2 H_3 O)_3 H \ N] Br$$
$$[(C_2 H_3 O)_4 \quad N] Br.$$

Lastly, remembering the tendency exhibited by ethylene-compounds to resolve themselves in the presence of alkalies into vinyl-products, it appeared not improbable that a fourth series of bodies would likewise be formed,

$$[(C_2 H_3) \ H_3 N] Br$$
$$[(C_2 H_3)_2 H_2 N] Br$$
$$[(C_2 H_3)_3 H \ N] Br$$
$$[(C_2 H_3)_4 \quad N] Br.$$

In the experiments on the action of dibromide of ethylene upon ammonia, which I have already partly published, and which, in a more connected form, I hope soon to lay before the Royal Society, I have not, indeed, met with the whole of these compounds ; but in the place of the deficient members of the groups new products have made their appearance, whose formation in the present state of our knowledge could scarcely have been predicted, and thus the problem of disentangling the difficulties of this reaction becomes a task of very considerable difficulty. Nor did the action of dibromide of ethylene upon ethylamine, diethylamine, and triethylamine, which I subsequently studied, afford a sufficiently simple expression of the transformations suggested by theory. The difficulties disappeared at once when the experiment was repeated in the phosphorus-series. In the reaction with dibromide of ethylene, the sharply-defined characters of triethylphosphine exhibited themselves with welcome distinctness, and in consequence more especially of the absence of unreplaced hydrogen—whereby the formation of a large number of compounds of subordinate theoretical interest was excluded—the general character of the reaction, the recognition of which was the object of the inquiry, became at once perceptible.

I have shown that the action of dibromide of ethylene upon triethylphosphine gives rise to the formation of four different compounds, viz.

$$
\begin{array}{l}
[(C_2H_4)''\ (C_2H_5)_4\,P_2]''Br_2 \\
[(C_2H_4Br)(C_2H_5)_3\,P\,]\ Br \\
[(C_2H_3O)\ (C_2H_5)_3\,P\,]\ Br \\
[(C_2H_3)\ \ \ (C_2H_5)_3\,P\,]\cdot Br,
\end{array}
$$

each of which represents one of the *four* groups of compounds, which under favourable circumstances may arise from the mutual reaction between ammonia and dibromide of ethylene, the production of a greater number of terms being impossible on account of the ternary substitution of triethylphosphine.

Whilst going on with the researches on the phosphorus-bases which I have taken the liberty of submitting to the Royal Society, in notes sketched as I advanced, I have not altogether lost sight of the experiments in the nitrogen-series, which had originally suggested these inquiries. Numerous nitrogenated bases, both monoatomic and diatomic, with which I have become acquainted during this investigation, must be reserved for a future communication. I may here only remark, that these substances, although differing in several points, nevertheless imitate in their general deportment so closely the corresponding terms of the phosphorus-series, that the picture which I have endeavoured to delineate of the phosphorus-compounds, illustrates in a great measure the history of the nitrogen-bodies.

In conclusion, a few words about the further development of which the experiments on the polyatomic bases appear to be capable, and about the direction in which I propose to pursue the track which they have opened.

Conceived in its simplest form, the transition from the series of

monoatomic to that of diatomic bases, may be referred to the introduction of a monochlorinated or a monobrominated alcohol-radical
into the type ammonia, the chlorine and bromine thus inserted
furnishing the point of attack for a second molecule of ammonia.

If in bromide of ethylammonium—to pass from the phosphorus-
series to the more generally interesting nitrogen-series—we replace
1 equiv. of the hydrogen in ethyle by bromine, we arrive at bromide of bromethylammonium, which fixing a second equivalent of
ammonia, is converted into the dibromide of ethylene-diammonium,
the latent bromine becoming accessible to silver-salts.

$$[(C_2H_4Br)H_3N]Br+H_3N=[(C_2H_4)''H_8N_2]''Br_2.$$

The further elaboration of this reaction indicates two different
methods for the construction of the polyatomic bases of a higher
order. In the first place, the number of ammonia-molecules, to be
incorporated in the new system, may be increased by the gradually
advancing bromination of the radical. By the further bromination
of ethyle in bromide of bromethylammonium and the action of
ammonia on the bodies thus produced, the following salts may be
generated :—

$$[(C_2H_3Br_2)H_3N]Br+2H_3N=[(C_2H_3)'''H_9N_3]'''Br_3$$
$$[(C_2H_2Br_3)H_3N]Br+3H_3N=[(C_2H_2)''''H_{12}N_4]''''Br_4$$
$$[(C_2HBr_4)H_3N]Br+4H_3N=[(C_2H)'''''H_{15}N_5]'''''Br_5$$
$$[(C_2Br_5)H_3N]Br+5H_3N=[(C_2)''''''H_{18}N_6]''''''Br_6.$$

Again, the fixation of the ammonia-molecules may be attempted,
not by the progressive bromination of the ethyle, but by the accumulation of monobrominated ethyle-molecules in the ammonium-nucleus.
The bromide of di-bromethylammonium, when submitted to the
action of ammonia, would thus yield the tribromide of a triammo-
nium ; the bromide of tri-bromethylammonium, the tetrabromide of
a tetrammonium ; and lastly, the bromide of tetrabromethylammo-
nium, the pentabromide of a pentammonium.

$$[(C_2H_4Br)_2H_3N]Br+2H_3N=[(C_2H_4)_2''H_8N_3]'''Br_3$$
$$[(C_2H_4Br)_3H N]Br+3H_3N=[(C_2H_4)_3''H_{10}N_4]''''Br_4$$
$$[(C_2H_4Br)_4 N]Br+4H_3N=[(C_2H_4)_4''H_{12}N_5]'''''Br_5.$$

As yet the bromination of the alcohol-bases presents some difficulty ; appropriately selected reactions, however, will doubtless furnish the several brominated bases. They may probably be obtained
by indirect processes, similar to those by which years ago I succeeded
in preparing the chlorinated and brominated derivatives of phenylamine ; or these bodies may be generated by the action of pentachloride or pentabromide of phosphorus upon the oxethylated bases,
a process, which, to judge from the few experiments recorded in one
of the preceding sketches, promises a rich harvest of results.

I have but a faint hope that I may be able to trace these new
paths in the numerous directions which open in a variety at once
tempting and perplexing. Inexorable experiment follows but slowly
the flight of light-winged theory. The commencement is never-

theless made, and even now the triammonium- and tetrammonium-compounds begin to unfold themselves in unexpected variety. One of the most remarkable compounds belonging to the triammonium-group is *diethylene triamine,*

$$C_4 H_{13} N_3 = \left.\begin{matrix} (C_2 H_4)''_2 \\ H_5 \end{matrix}\right\} N_3.$$

This base, *the first triacid triammonia,* forms splendid salts of the formula
$$[(C_2 H_4)''_2 H_5 N_3]''' Cl_3,$$
which will be the subject of a special communication.

December 6.—Major-General Sabine, Treasurer and Vice-President, in the Chair.

The following communication was read :—
"On the Gyroscope." By Arthur Hill Curtis, Esq.

The object of this paper is to deduce on strict mechanical principles all the known properties of the gyroscope. The only assumption made is that the velocity of rotation impressed on the instrument is very great compared with that which the attached weight would produce on it if acting alone for an instant in a direction perpendicular to the axis. The theorems which the author establishes are the following :—

THEOREM I.—The curves described by the extremity of the axis of the gyroscope are a system of spherical cycloids generated by the motion of a point on the spherical radius of a circle, which, constantly remaining on the same sphere, rolls without sliding on the circumference of another fixed circle situated on the same sphere. These cycloids may be either ordinary, curtate, or prolate—including the case when the system degenerates into a circle, in which case the generating point becomes the centre of the rolling circle. Their species depends on the direction of the initial velocity communicated to the axis, the direction in which the instrument is set rotating, and the position of the attached weight ; when, for instance, no initial velocity is communicated to the axis, the cycloids will be ordinary *at first*, and would continue so if the gyroscope were a perfect instrument for illustrating the motion of a body round a fixed point ; but the inertia of the rings on which it is mounted, and of the attached weight, as well as the resistance of the air, after a short period has elapsed, has the effect of imparting to the axis a certain velocity which modifies the curves described by it, and at last causes the motion of the axis to become for a time sensibly one of uniform progression ; it then becomes oscillatory again, the amplitudes of the oscillations being smaller than before.

THEOREM II.—If the outer ring be fixed in any position so as to restrict the axis of the gyroscope to a fixed plane, the motion of the axis, when a weight is attached as above, is the same whether the instrument be set rotating or not. It is proved that the angular motion of the axis is determined by an equation of the same form as that of a circular pendulum, which does not involve the angular velocity of rotation impressed on the gyroscope.

THEOREM III.—If the gyroscope be set rotating rapidly, and its axis of figure be constrained, as in Theorem II., to move very freely in a plane fixed with regard to the horizon, the axis will tend to take the position of the projection on the given plane of the line drawn through the centre of gravity of the gyroscope, parallel to the axis of the earth, in such a way that the earth and the gyroscope may turn in the same direction; while, if the axis be perfectly free, it will move exactly in the same way as the axis of a telescope directed constantly towards the same fixed star, their initial positions being supposed parallel, as established experimentally by M. Léon Foucault (Comptes Rendus, September, 1852).

To prove this theorem, the angular velocity of the earth round its axis is resolved into an equal and codirectional motion of rotation round the line through the centre of gravity of the gyroscope parallel to the earth's axis, and a motion of translation, the direction of which is constantly changing, common to all parts of the earth. Of these motions the latter is communicated to the gyroscope by the friction of its base, and does not modify its position with regard to the horizon. The first alone requires to be considered. In order to estimate its effect, a rotation equal to it and round the same axis, but in an opposite direction, must be supposed to be communicated both to the earth and the gyroscope. This does not affect their relative motion, and simplifies the problem, as it enables us to consider the earth at rest. The relative motion of the gyroscope may therefore be found by adding to the three components, round its principal axis, of its instantaneous angular velocity of rotation, as found from its equations of absolute motion, the components of this introduced velocity of rotation, the moment of resistance of the given plane being taken into account in forming the equations of motion, and its intensity supposed such as to counteract that part of the total angular velocity of the axis which is perpendicular to the given plane. The equation which determines the motion of the axis is shown to be identical with that of a circular pendulum, and the motion consequently one of oscillation, the mean position of the axis being that in which it approaches, as close as the conditions of the question permit, to the line drawn through its centre of gravity parallel to the earth's axis, and in which it rotates in a direction similar to that of the earth's rotation. Similar reasoning establishes the second part of the theorem, which is theoretically true whether the gyroscope be set rotating or not. This result is, however, in practice modified by the effects of friction; but when a rapid rotatory motion has been impressed on the gyroscope, it acquires a stability which enables it to overcome to a great extent these effects.

December 13.—Major-General Sabine, Treasurer and Vice-President, in the Chair.

The following communication was read:—

"On the Surface-condensation of Steam." By J. P. Joule, LL.D., F.R.S.

In the author's experiments steam was passed into a tube, to the outside of which a stream of water was applied, by passing it along the concentric space between the steam-tube and a wider tube in which the steam-tube was placed. The steam-tube was connected at its lower end with a receiver to hold the condensed water. A mercury gauge indicated the pressure within the apparatus. The principal object of the author was to ascertain the conductivity of the tube under varied circumstances, by applying the formula suggested by Professor Thomson,

$$C = \frac{w}{a} \log \frac{V}{v},$$

where a is the area of the tube in square feet, w the quantity of water in pounds transmitted per hour, V and v the differences of temperature between the inside of the steam-tube, and the refrigerating water at its entrance and at its exit. The following are some of the author's most important conclusions.

1. The pressure in the vacuous space is sensibly the same in all parts.

2. It is a matter of indifference in which direction the refrigerating water flows in reference to the direction of the steam and condensed water.

3. The temperature of the vacuous space is sensibly equal in all its parts.

4. The resistance to conductivity must be attributed almost entirely to the film of water in immediate contact with the inside and outside surfaces of the tube, and is little influenced by the kind of metal of which the tube is composed, or by its thickness up to the limits of that of ordinary tubes.

5. The conductivity increases up to a limit as the rapidity of the stream of water is augmented.

6. By the use of a spiral of wire to give a rotary motion of the water in the concentric space, the conductivity is increased for the same head of water.

The author, in conclusion, gives an account of experiments with atmospheric air as the refrigerating agent; the conductivity is very small in this case, and will probably prevent air being employed for the condensation of steam except in very peculiar circumstances.

December 20.—Major-General Sabine, Treas. and V.P., in the Chair.

The following communication was read:—

" Preliminary Notice of Researches into the Chemical Constitution of Narcotine and of its Products of Decomposition." By A. Matthiessen, Esq., and George C. Foster, Esq.

I. *Composition of Narcotine.*

The announcement made by Wertheim[*] and Hinterberger[†] of the probable existence of various kinds of narcotine, rendered it necessary to commence the present investigation by a series of analyses of our material, in order to ascertain which variety of narcotine we were dealing with.

The narcotine employed was obtained from Mr. Morson, to whom

[*] Chem. Gaz., 1850, p. 141.   [†] Ibid., 1851, p. 309.

we are greatly indebted for the scrupulous care bestowed on its preparation and purification. He stated that it was extracted from the residues which had accumulated during the preparation of very large quantities of morphine and codeine, from opium of various qualities and from various sources. If, therefore distinct varieties of narcotine exist, there was reason to expect that our narcotine would prove to be a mixture of several of them. The results of all our analyses, however, agree with the formula $C^{22} H^{23} NO^{7}$, as shown by the following Table, which gives the highest, lowest, and mean results obtained:—

	Calculated.		Found.		
			Maxima.	Minima.	Mean.
C^{22}	264	63·92	64·00	63·42	63·79
H^{23}..........	23	5·57	6·05	5·69	5·81
N	14	3·39	3·40	3·26	3·32
O^{7}	112	27·12	27·53	26·72	27·08
$\overline{C^{22} H^{23} NO^{7}}$	413	100·00			

The formula which has been generally admitted since the publication of Wöhler's[*] and Blyth's[†] researches on narcotine, namely, $C^{23} H^{24} NO^{7}$, requires the following per-centages:—

Carbon .... 64·61
Hydrogen .... 5·85
Nitrogen .... 3·30
Oxygen .... 26·24

We may here remark that the recorded analyses of narcotine and its salts, with the exception of one by Dr. Hofmann, published by Blyth, agree at least as well with the former as with the latter formula; moreover, during the course of experiments made with several pounds of narcotine, we have observed nothing, either in the behaviour of this base itself, or in the nature or proportions of its products of decomposition, to indicate that it was variable in composition. Further data are, however, needed for the final decision of this question, and we shall accordingly feel very much indebted to any chemist who has a specimen of narcotine of well-ascertained origin, or which he believes to have a different composition from that given above, if he will kindly spare us a sufficient quantity for analysis.

II. *Composition of Cotarnine.*

The combustion of cotarnine with oxide of copper and oxygen, as well as the determination of the proportion of platinum in its chloroplatinate, leads us to adopt the formula $C^{12} H^{13} NO^{3}$ for this base. The formula usually adopted contains one more atom of carbon; but, independently of our analytical results, the supposition that cotarnine contains only twelve atoms of carbon is supported by the simple manner in which the action of oxidizing substances on narcotine can then be expressed, namely, by the equation

$$C^{22} H^{23} NO^{7} + O = C^{10} H^{10} O^{5} + C^{12} H^{13} NO^{3},$$

Narcotine.   Opianic acid.   Cotarnine.

[*] Ann. Chem. Pharm. vol. l. p. 1.
[†] Phil Mag. S. 3. vol. xxv. p. 363.

ahd, as will be shown hereafter, by the manner in which cotarnine is decomposed by dilute nitric acid.

III. *Decompositions of Opianic Acid.*

Opianic acid is ' readily decomposed when heated with strong hydriodic acid; no iodine is set free, but iodide of methyle is formed in considerable quantity at the same time as a non-volatile substance, very easily altered by heat and exposure to air, especially if in contact with alkali, the precise nature of which we have not yet been able to ascertain.

When opianic acid is heated with an excess of a very strong solution of potash, it splits up into meconine and hemipinic acid. These substances were found by experiment to be formed in proportions corresponding to the equation

$$2C^{10}H^{1}{}^{\prime}O^{6} = C^{10}H^{10}O^{4} + C^{10}H^{10}O^{6}.$$

Opianic acid.   Meconine.   Hemipinic acid.

The meconine thus produced has all the characters which have been ascribed by previous observers to meconine obtained by other processes; its identity was further established by analysis, and by the preparation of chloro- and nitro-meconine, the former of which was analysed. The hemipinic acid was also found to be identical with that obtained irectly from narcotine: the acid and its silver-salt were analysed.

Having thus found a method by which meconine and hemipinic acid can be produced with certainty and in large quantities, we intend to make an extended investigation of them and of opianic acid, in the hope of discovering the nature of the relationship of these three bodies to each other and to narcotine. The principal results which we have hitherto obtained in this direction are as follows.

Action of Hydriodic Acid on Meconine.—Meconine is decomposed by hydriodic acid like opianic acid, giving iodide of methyle and an easily alterable substance, the nature of which has not been determined.

Action of Hydriodic Acid on Hemipinic Acid.—Hemipinic acid, heated with concentrated hydriodic acid to within a few degrees of the boiling-point of the latter substance, is decomposed into iodide of methyle, carbonic acid, and an acid of the formula $C^{7}H^{6}O^{4}$. It was found by direct experiment that two atoms of iodide of methyle are formed from each atom of hemipinic acid, so that the following equation probably represents the reaction:—

$$C^{10}H^{10}O^{6}+2HI=2CH^{3}I+CO^{2}+C^{7}H^{6}O^{4}$$

Hemipinic acid.   New acid.

The new acid is moderately soluble in cold water, and very soluble in boiling water, alcohol, and ether; its solution has a strongly acid reaction with test-paper. It separates from hot water in small needle-shaped crystals containing $14\cdot80$ per cent. water of crystallization, which they lose at 100° (the formula $C^{7}H^{6}O^{4}+1\frac{1}{4}H^{2}O$ corresponds to $14\cdot92$ per cent. water); at a higher temperature the acid melts and sublimes without apparent alteration.

Dried at 100°, it gave the following results on analysis:—

	Calculated.		Found. (mean)
C⁷	84	54·55	54·38
H⁶	6	3·89	3·91
O⁴	64	41·56	41·71
	154	100·00	100·00

When the dry acid is heated in the air to a little above 100°, it slowly oxidizes and becomes brown; the same change takes place more rapidly when a solution of it, especially if neutral or alkaline, is evaporated. A solution of the acid immediately reduces ammonio-nitrate of silver, even in the cold; with sulphate of copper and a slight excess of potash it gives a yellowish-green solution, from which suboxide of copper is precipitated on warming. The free acid, or its ammonia-salt, gives a very intense blue coloration with perchloride of iron. The colour thus produced is changed to blood-red (exactly resembling the red produced by the sulphocyanates) by ammonia, and is destroyed by strong acids, being restored by dilution with water, or by neutralization by an alkali: like the colouring matter obtained by Anderson by the action of sulphuric acid on opianic acid, it is entirely removed from solution by alumina.

We have not yet obtained any of the salts of the new acid in a state fit for analysis, and prefer not to propose a name for it until its relationship to other bodies has been more thoroughly examined; its formula, however, assigns to it a place in the following series—

C⁷ H⁶ O .... Oil of bitter almonds.
C⁷ H⁶ O² .... Benzoic acid.
C⁷ H⁶ O³ .... Salicylic acid.
C⁷ H⁶ O⁴ .... New acid.
C⁷ H⁶ O⁵ .... Gallic acid.
C⁷ H⁶ O⁶ .... Tannoxylic acid (?).

It is remarkable that salicylic and gallic acids both give colorations with perchloride of iron much resembling that produced by the acid C⁷ H⁶ O⁴.

IV. *Action of dilute Nitric Acid on Cotarnine.*

By gently heating cotarnine with very dilute nitric acid, we have obtained nitrate of methylamine and a new acid, *cotarnic acid*, but have not hitherto found out the conditions necessary for the certain production of the latter substance.

Cotarnic acid dissolves easily in water, giving a solution which reacts strongly acid with litmus-paper; it dissolves only sparingly in alcohol, and is precipitated from its alcoholic solution by ether. Heated with an excess of sodium, it gives no trace of cyanide, and therefore contains no nitrogen. With perchloride of iron it gives no coloration; with acetate of lead it gives a white precipitate insoluble in excess of acetate; with nitrate of silver it gives a precipitate which is very slightly soluble in hot water. The silver-salt, crystallized from water, was found to contain $C^{11} H^{10} Ag^3 O^5$; on analysis it gave the following results:—

	Calculated.		Found (mean).
C^{11}	132	30·14	29·67
H^{10}	10	2·27	2·17
Ag^2	216	49·32	49·24
O^5	80	18·27	18·92
$C^{11}H^{10}Ag^2O^5$	438	100·00	100·00

The formation of cotarnic acid is therefore represented by the equation

$$C^{13}H^{13}NO^3 + 2H^2O + NHO^3 = C^{11}H^{13}O^5 + N(NCH^6)O^3$$

Cotarnine.   Cotarnic acid.   Nitrate of methylamine.

It is possible that the substance obtained by Anderson by the action of nitric acid on narcotine (Chem. Soc. Quart. Journ. vol. v. p. 265; Gerhardt, 'Traité,' vol. iv. p. 80), and supposed by him to be hydrate of meconine (*Opianyle*, Anderson), may have been cotarnic acid, with the composition of which Anderson's analyses closely agree, as shown by the following comparison:—

	Calculated.		Anderson.	
C^{11}	132	58·93	58·83	58·84
H^{12}	12	5·36	5·17	5·42
O^5	80	35·71	36·00	35·74
	224	100·00	100·00	100·00

If cotarnic acid be represented by the formula

$$\left.\begin{array}{c}(C^{11}H^{10}O^3)'' \\ H^2\end{array}\right\}.O^2,$$

cotarnine becomes methyl-cotarnimide—

$$\left.\begin{array}{c}(C^{11}H^{10}O^2)'' \\ C\;\;H^3\end{array}\right\}N;$$

if, however, we retain the formula $C^{13}H^{13}NO^3$ for cotarnine, no simple relation is apparent between it and cotarnic acid.

V. *Conclusion.*

In the absence of more definite knowledge of the constitution of meconine and opianic and hemipinic acids, it is obviously useless to try to assign a rational formula to narcotine. According to the formulæ which we have adopted for narcotine and cotarnine, narcotine contains the elements of cotarnine and meconine:—

$$C^{23}H^{23}NO^7 = C^{13}H^{13}NO^3 + C^{10}H^{10}O^4.$$

Narcotine.   Cotarnine.   Meconine.

It will be seen that these formulæ are the same as those of the methyl-narcotine and methyl-cotarnine of Hinterberger and Wertheim. The ground upon which Wertheim admitted the existence of ethyl- and propyl-narcotine was the formation of volatile bases containing C^3H^7N and C^3H^9N by the distillation of narcotine with potash. An experiment which we have made goes some way towards explaining the formation of these bases without assuming the existence of more than one variety of narcotine. Having so frequently observed the formation of methyle-compounds from the derivatives of narcotine, we tried the direct action of hydriodic acid on this base, expecting to obtain iodide of methyle. By distilling 20 grms. of narcotine with concentrated hydriodic acid, 19 grms. of pure iodide of methyle were obtained, a quantity which corresponds, as nearly as could be expected, with three atoms of iodide of methyle for one atom of narcotine[*],

$$(C^{22} H^{23} NO^7 : 3CH^3 I :: 413 : 436 \text{ or } 20 : 21\cdot1).$$

Narcotine therefore contains three atoms of methyle so combined as to be easily separable[†]; and it is very probable that when it is distilled with potash, according to the conditions of the experiment, sometimes nearly pure ammonia is evolved, while, at other times, methylamine, $CH^5 N$, dimethylamine, $C^2 H^7 N$, or trimethylamine, $C^3 H^9 N$, predominates.

We wish not to close without acknowledging our obligation to Dr. M. Holsmann for very valuable assistance rendered to us at the commencement of our investigation.

GEOLOGICAL SOCIETY.

[Continued from p. 326.]

June 19, 1861.—Leonard Horner, Esq., President, in the Chair.

The following communications were read:—

1. "On the Lines of Deepest Water around the British Isles." By the Rev. R. Everest, F.G.S.

By drawing on a chart a line traversing the deepest soundings along the English Channel and the Eastern Coast of England and Scotland, continuing it along the 100-fathom-line on the Atlantic side of Scotland and Ireland, and connecting with it the line of deepest soundings along St. George's Channel, an unequal-sided hexagonal figure is described around the British Isles, and a pentagonal figure around Ireland. A hexagonal polygon may be similarly defined around the Isle of Arran. These lines were described in detail by the author, who pointed out that they limited areas similar to the polygonal form that stony or earthy bodies take in shrinking, either in the process of cooling or in drying. The relations of the 100-fathom-line to the promontories, the inlets, and general contour of the coast were dwelt upon; and the bearings that certain lines drawn across the British Isles from the projecting angles of the polygon appear to have on the strike and other conditions of the strata were described. After some remarks on the probable effect that shrinkage of the earth's crust must have on the ejection of molten rock, the author observed that, in his opinion, the action of shrinking is the only one we know of that will afford any solution of the phenomena treated of in this paper, namely, long lines of depression accompanied by long lines of elevation, often, as in the case of the British Isles, Spain and Portugal, and elsewhere, belong-

---

[*] It is possible that narcotine will prove to be an economical, as it is certainly the most convenient, source of iodide of methyle.

[†] Gerhardt (Traité, iv. 64) had previously observed the production of a volatile substance, which he supposed to be nitrate of ethyle or of methyle, by the action of nitric acid on narcotine.

ing to parts of huge polygons broken up into small ones, as if the surface of the earth had once formed part of a basaltic causeway.

Several charts, plans, and drawings were provided by the author in illustration of the paper.

2. "On the Ludlow Bone-bed and its Crustacean Remains." By J. Harley, M.B.

Of the two bone-beds occurring near Ludlow, the lower one (seen in Ludford Lane and on the north-east slopes of Whitcliff) is that which has supplied the author with the materials for this paper. Besides spines, teeth, and shagreen-like remains of fish, the author finds in the Ludlow Bone-bed three kinds of minute organisms : 1st, conical bodies, the same as the "Conodonts" of Pander; 2ndly, bodies somewhat like the crown of a molar tooth; 3rdly, oblong plates. All these bodies possess the same chemical composition and microscopical structure—which is decidedly *Crustacean.* With *Pterygotus* they do not appear to have any relationship, unless some are the stomach-teeth : nor do they show any alliance with Trilobites ; but with *Ceratiocaris* they have a great resemblance as to structural characters, and some of them were probably the minute secondary spines of the tail of that Phyllopod. The plate-like forms might have belonged to Squilloid or Limuloid Crustaceans. To facilitate the recognition of these bodies, Mr. Harley places them all in one provisional genus with the name of *Astacoderma.* A letter from Dr. Volborth to the author was also read in confirmation of Mr. Harley's opinion that these bodies are identical with Dr. Pander's "Conodonts." Numerous original drawings illustrated the paper.

3. "On the Old Red Sandstone of Forfarshire." By James Powrie, Esq., F.G.S.

The author described the series of stratified rocks belonging to the Old Red Sandstone, upwards of 3000 feet in thickness, stretching southward from the Grampians to the coast of Fifeshire. 1st. Dark-red grits (with cornstones and flagstones) equivalent to the English "Tilestones." 2ndly. Thick conglomerates and the Arbroath paving-flags : *Pterygotus anglicus, Stylonurus, Parka decipiens, Cephalaspis, Diplacanthus gracilis,* and other fossils belong to this part of the series. 3rdly. Thick-bedded red sandstone (with cornstone) : *Cephalaspis* and *Pteraspis.* 4thly. Soft deep-red sandstones. 5thly. Spotted marls and shales : these are the uppermost, and may be the equivalent of the Holoptychian beds of Clashbinnie. The author showed that between the Grampians and the trappean hills of Bunnichen and Bunbarrow the series forms a great syncline; and between these hills and the sea the older beds are twice again brought to the surface; and he believes that the marls and sandstones at Whiteness are not unconformable, as Sir C. Lyell has represented them in his published section.

4. The Secretary gave a brief account of the discovery of an exposure of sandstone strata with two bands of clay full of calcareous nodules containing plentiful remains of *Coccosteus, Glyptolepis,* and other fishes belonging to the Old Red Sandstone, in a burn about 2½ miles from the Manse at Edderton, Ross-shire, on the south side

of Durnoch Firth. This information was contained in a letter from the Rev. J. M. Joass, of Edderton, communicated by Sir R. I. Murchison, V.P.G.S.

5. " On the Outburst of a Volcano near Edd, on the African coast of the Red Sea." By Capt. L. R. Playfair, R.N.

At Edd, lat. 13° 57′ N., long. 41° 4′ E., about half-way between Massouah and the Straits of Bab-el-Mandel, earthquake-shocks occurred on the night of the 7th of May or the morning of the 8th, during about an hour. At sunrise fine dust fell, at first white, afterwards red; the day was pitch-dark; and the dust was nearly knee-deep. On the 9th the fall of dust abated; and at night fire and smoke were seen issuing from Jebel Dubbeh, a mountain about a day's journey inland; and sounds like the firing of cannon were heard. At Perim these sounds were heard at about 2 A.M. on the 8th, and at long intervals up to the 10th or 11th. The dust was also met with at sea; and along the entire coast of Yemen the dust fell for several days. Several shocks were felt on the 8th at Mokha and Hodaida.

6. " Notice of the occurrence of an Earthquake on the 20th of March, 1861, in Mendoza, Argentine Confederation, South America." By C. Murray, Esq.

At about ¼ to 9 o'clock, the first shock, preceded by a thunder-clap, destroyed the city of Mendoza, killing (it is said) two-thirds of its 16,000 inhabitants. Altogether there were eighty-five shocks in ten days. The land-wave appears to have come from the south-east. Several towns S.E. of Buenos Ayres felt slight shocks. No earth-quake took place in Chile; but travellers crossing the Upsallata Pass of the Cordilleras met with a shower of ashes; the pass was obstructed by broken rocks; and chasms opened on all sides. At Buenos Ayres, 323 leagues from Mendoza, and elsewhere, it was ob-served in watch-makers' shops that the pendulums moving N. and S. were accelerated; those moving E. and W. were not affected.

7. " On the Increase of Land on the Coromandel Coast." By J. W. Dykes, Esq. In a Letter to Sir C. Lyell, F.G.S.

In the districts of the Kistna and Godavery, the land presents a parallel series of ridges and hollows near the coast, not in relation to the rivers but to the coast-line. These may have been formed by sedimentary deposits similar to what are now taking place on the Coromandel coast. By the strong currents alternately running N. and S., according to the monsoons, lines of sediment parallel with the coast are formed; and by the occasional interference of winds and tides dams are thrown across the hollows, and the latter soon become filled up. These parallel bands of coast-land become, in time, upheaved, and more or less affected by atmospheric agencies.

LII. *Intelligence and Miscellaneous Articles.*

ON A NEWLY DISCOVERED ACTION OF LIGHT.

BY M. NIÈPCE DE ST. VICTOR.

WHEN the freshly broken part of an opake porcelain plate was exposed to a strong sun for two or three hours, and then placed on choride of silver paper, after twenty-four hours' contact the silver

·was found to be reduced in the part corresponding to that which had been exposed to light, but there was no reduction in that part which had been preserved from light. Certain fine specimens of porcelain acquire this activity more easily.

A steel plate polished at one part, and roughened at another by the action of aquafortis, and well cleaned by alcohol, was exposed to the sun for two or three hours under the following conditions—half the polished and unpolished plate under an opake screen, and the other half under a white glass. The plate was then covered by a paper prepared with albuminized chloride of silver. After twenty-four hours' contact, an impression was formed on the unpolished part which had been exposed to the light, but none on the polished part, nor on the unpolished part under the screen. A roughened glass plate carefully cleaned gave similar results.

These experiments show that it is not necessary for the reduction of silver salts that there be a chemical action, as when a metallic salt is insolated with an organic matter. M. Arnaudon has repeated some of these experiments with different gases, and has obtained the same results as with air.

I may here recall a previous observation, that the insolated earth exhibits traces of this action to a depth of a metre, the thickness varying, of course, with the nature of the soil and the degree of insolation. The following experiment supports this view :—In a tin tube lined with pasteboard impregnated with tartaric acid, and insolated so as strongly to reduce silver salts, I placed in the middle of the tube, but not in contact, a small bladder containing a weak solution of starch ; after forty-eight hours this starch feebly reduced Barreswil's liquor, while other starch placed in the same conditions, excepting the insolation, produced no effect on the liquor.

The following experiments were made with a view of trying whether light could magnetize a steel bar, as has frequently been stated. Avoiding all sources of error, a knitting-needle suspended by a hair was entirely unattracted by another needle insolated for a very long time in a beam of light concentrated by a strong lens, whether the light was white or had traversed a violet glass.

I then enclosed a needle in a paper impregnated with nitrate of uranium, or tartaric acid, and insolated ; I also suspended a needle horizontally in tubes containing insolated pasteboard ; and the results were always negative, as also was the case with experiments made wit very feebly magnetized needles in the hope of demagnetizing them.

In conclusion, this persistent *activity* imparted by light to porous bodies cannot be the same as phosphorescence ; for, from Becquerel's experiments, it would not continue so long : it is probable that, as Foucault believes, it is a radiation invisible to our eyes, and which does not traverse glass.—*Comptes Rendus*, July 1, 1861.

ON TERRESTRIAL REFRACTION. BY M. BABINET.

A ray of light which traverses the layers of the atmosphere horizontally, is deflected from its rectilinear path towards the earth by a quantity which, in the mean, is a fifteenth of the terrestrial arc extending from the point where the ray enters to the point at which it arrives. Thus for a horizontal path of 1852 metres, which is equal to a minute of an arc on the terrestrial globe, the deflection or refraction of the ray would be $\frac{1}{15}$ of a minute, or 4″.

There are three things to be considered in this question;—

1. The trajectory of the ray is a circle.

2. There is a constant ratio *n* between the quantity by which the ray is inflected, and the terrestrial arc comprised between the point of entrance of the ray taken to be horizontal, and its point of arrival. Then let *s* be the angle at the earth's centre comprised between these two points, and *r* the refraction, we have

$$\frac{r}{s} = n = R(m-1)\frac{B}{0^m\cdot76}\frac{1}{(1+at)^a}\left\{\frac{1}{0^m\cdot76\,d} - \frac{a}{M}\right\}.$$

Here R is the mean radius of the earth; B is its atmospheric pressure reduced to zero; a is the coefficient of expansion $\frac{1}{1000}$ of the air for 1° C; d is the density of mercury as compared with air taken at zero; and (what is new and important) M is the height in metres corresponding to a diminution of one degree on the Centigrade scale.

This formula, expressed numerically, becomes

$$n = \frac{B}{0^m\cdot76}\frac{1}{(1+at)^a}\left\{0\cdot2345 - \frac{6^m\cdot867}{M}\right\}.$$

Several remarkable conclusions may be deduced from this, relating to the physical constitution of the atmosphere.

3. If the ray does not travel horizontally, but is inclined to the horizon at an angle *i*, the atmospheric refraction diminishes in the ratio of cos *i* to unity; but then the path of the ray being greater than its horizontal projection in the ratio of unity to cos *i*, a compensation is established; and calling *s* the angle at the centre of the earth comprised between the signal and the observer, we have, as before,

$$\frac{r}{s} = n = \frac{B}{0^m\cdot76}\frac{1}{(1+at)^a}\left\{0\cdot2345 - \frac{6^m\cdot867}{M}\right\}.$$

There would be no refraction, and the ray would travel in a right line, if $0\cdot2345 - \frac{6^m\cdot867}{M} = 0$, which gives $M = 29^m\cdot3$. Thus if the temperature of the air sank 1° for $29^m\cdot3$, there would be no refraction. On the other hand, taking $B = 0^m\cdot76$ and $t = 0$, we have

$$n = 0\cdot2345 - \frac{6^m\cdot867}{M};$$

taking this quantity as equal in the mean to $\frac{1}{15}$, or $0\cdot0667$, we have about 41 metres for M: all these quantities are much less than 200 metres, which is the height necessary to be traversed to have a diminution of one degree of temperature in the higher layers of the atmosphere.

The coefficient *n* varies from $0\cdot500$ to $0\cdot000$: it can even become negative, which corresponds to the case of mirage whenever M is less than $29^m\cdot3$. We shall see afterwards the great influence which the number M exerts on the stability of the atmosphere; but the formula which gives the value of n shows that in the vicinity of the soil the temperature decreases far more rapidly than aërostatic ascents would seem to indicate.

In a second note, M. Babinet gives a complete development of the above formula; and in a third note he gives a complete formula for refraction.

He supposed the heights h taken above the horizon of the observer, and not from a point of the surface corresponding to dh. Taking, as is necessary, the height h of a point of the trajectory of the ray on the vertical passing by this point, it follows that for horizontal refraction the atmospheric path is very limited, and that therefore the expression for refraction could never be a formula which becomes infinite for $z = 90°$. Supposing always a decrease of 1° C. for M metres, the complete differential formula is

$$dr = dh \frac{\text{R} \sin z}{\sqrt{h^2 + 2\text{R}h + \text{R}^2 \cos^2 z}} \left(1 + at - \frac{ah}{\text{M}}\right)^{\frac{\text{M}}{0.76\text{D}a} - 2} (m-1) \frac{\text{B}}{0.76}$$
$$\left(\frac{1}{0.76\text{D}} - \frac{a}{\text{M}}\right),$$

the integral being taken from $h = 0$ to $h = \dfrac{\text{M}(1 + at)}{a}$. It will probably be necessary to suppose that M varies with the height h, and to replace M by $\text{M} + kh$, k being so determined that, for instance, with a height of 7000 metres, 200 metres correspond to an increase of 1°. Thus $\text{M} + 7000k = 220$, and $k = \dfrac{220 - \text{M}}{7000}$.—*Comptes Rendus*, September 2 and 9, and October 7, 1861.

ON THE MAXIMUM DENSITY OF SEA-WATER.
BY M. V. NEUMANN.

Von Neumann, in an inaugural dissertation (Munich, 1861), has published a new determination of the maximum density of sea-water. Like Kopp and other physicists, who have made this determination for pure water, he measured the volume at different temperatures in a glass vessel analogous in construction to a thermometer, the coefficient of expansion of the glass being carefully determined. This method is well adapted for liquids whose freezing-point is above the point of greatest density. The sea-water used was obtained from Trieste, Genoa, and Heligoland, and was previously well mixed. Its freezing-point was found to be −2°·6 C., and its specific gravity at 0° C. 1·0281; its point of greatest density was −4°·7364 C.

This number is more than that obtained by Despretz (−3°·67 C. for sea-water of 1·0273 sp. gr.) and Erman (−3°·75 C.), but would probably agree with that of Marcet (−5°·25 C.) and Horner (−5°·56 C.), if a correction for the expansion of glass were introduced.—Poggendorff's *Annalen*, August 1861.

THE

LONDON, EDINBURGH and DUBLIN

PHILOSOPHICAL MAGAZINE

AND

JOURNAL OF SCIENCE.

[FOURTH SERIES.]

DECEMBER 1861.

LIII. *Explanation of a Projection by Balance of Errors for Maps applying to a very large extent of the Earth's Surface; and Comparison of this projection with other projections. By* G. B. AIRY, *Esq., Astronomer Royal*[*].

1. IN the projection for maps, whose principles I am about to explain, any point of the earth's surface (as Greenwich, Paris, &c.) may be adopted as the Centre of Reference, to be represented by the Central Point of the Map. But the projection which I propose, and those with which I shall compare it, are all subjected to the following conditions: that the azimuth of any other point on the earth, as viewed from the Centre of Reference, shall be the same as the azimuth of the corresponding point of the map as viewed from the central point of the map; and that equal great-circle distances of other points on the earth from the Centre of Reference, in all directions, shall be represented by equal radial distances from the central point of the map. These conditions include the Stereographic Projection, Sir H. James's Projection, and others; but they exclude Mercator's Projection, and the projections proposed by Sir John Herschel.

2. In projections like these, in which the relation of the surface represented to the surface representing it is the same in all directions from the central point, it is unnecessary for us to embarrass ourselves with considerations of the place of the pole and the forms of the curves representing arcs of meridian and parallels. It will be sufficient to consider what will be the radii of circles on the map which shall represent circles on the earth, of different radii (as measured by arcs of great circle), but having the centre of reference as their common centre; and when values are found for the radii of these circles on the map, means will

[*] Communicated by the Author.

easily be found for exhibiting, with their assistance, the points
of intersection of meridians and parallels which are referred to
the pole of the earth.

8. The two errors, to one or both of which all projections are
liable, are, Change of Area, and Distortion, as applying to small
portions of the earth's surface. On the one hand, a projection
may be invented (to which I shall give the name of "Projection
with Unchanged Areas") in which there is no Change of Area,
but excessive Distortion, for parts far from the centre; on the
other hand, the Stereographic Projection has no Distortion, but
has great Change of Area for distant parts. Between these lie
the projections which have usually been adopted by geographers,
with the tacit purpose of greatly reducing the error of one kind
by the admission of a small error of the other kind, but without
any distinctly-expressed principle (so far as I know) for their
guidance in the details of the projection.

4. My object in this paper is to exhibit a distinct mathematical process for determining the magnitudes of these errors, so
that the result of their combination shall be most advantageous.
This principle I call "The Balance of Errors." It is founded
upon the following assumptions and inferences:—

First. The Change of Area being represented by

$$\frac{\text{projected area}}{\text{original area}} - 1,$$

and the Distortion being represented by

$$\frac{\text{ratio of projected sides}}{\text{ratio of original sides}} - 1 = \frac{\text{projected length} \times \text{original breadth}}{\text{projected breadth} \times \text{original length}} - 1$$

(where the length of the rectangle is in the direction of the
great circle connecting the rectangle's centre with the Centre of
Reference, and the breadth is transverse to that great circle),
these two errors, when of equal magnitude, may be considered
as equal evils.

Second. As the annoyance produced by a negative value of
either of these formulæ is as great as that produced by a positive
value, we must use some even power of the formulæ to represent
the real amount of the evil of each. I shall take the squares.

Third. The total evil in the projection of any small part may
properly be represented by the sum of these squares.

Fourth. The total evil on the entire Map may therefore be
properly represented by the summation through the whole Map
(respect being had to the magnitude of every small area) of the
sum of these squares for every small area.

Fifth. The process for determining the most advantageous
projection will therefore consist in determining the laws ex-

pressing the " radii of map-circles " in article 2 in terms of the " great-circle radii on the earth," which will make the total evil, represented as has just been stated, the smallest possible.

5. Now let l and b be the length and breadth of a small rectangle on the earth's surface, and suppose that the length and breadth of the corresponding rectangle on the map are $l + \delta l$ and $b + \delta b$, and neglect powers of δl and δb above the first. (Although this does not apply with algebraic correctness to very great change of area and distortion, yet it will be found by the result that the theoretical failure introduces no practical inconvenience.) Then the Change of Area

$$= \frac{\text{projected area}}{\text{original area}} - 1 = \frac{(l + \delta l) \cdot (b + \delta b)}{lb} - 1 = \frac{\delta l}{l} + \frac{\delta b}{b}.$$

And the Distortion

$$= \frac{\text{projected length}}{\text{projected breadth}} \times \frac{\text{original breadth}}{\text{original length}} - 1 = \frac{l + \delta l}{b + \delta b} \times \frac{b}{l} - 1$$

$$= \frac{\delta l}{l} - \frac{\delta b}{b}.$$

The sum of their squares, or $\left(\frac{\delta l}{l} + \frac{\delta b}{b} \right)^2 + \left(\frac{\delta l}{l} - \frac{\delta b}{b} \right)^2$, is

$$2 \left(\frac{\delta l}{l} \right)^2 + 2 \left(\frac{\delta b}{b} \right)^2.$$

And therefore we may use $\left(\frac{\delta l}{l} \right)^2 + \left(\frac{\delta b}{b} \right)^2$ as the measure of the evil for each small rectangle.

6. Let θ be the length, expressed in terms of radius, of the arc of great circle on the earth connecting the centre of the small rectangle with the Centre of Reference; r the corresponding distance on the map, expressed in terms of the same radius, of the projection of the map's centre of the small rectangle from the centre of the map; the object of the whole investigation is to express r in terms of θ. Let the length of a small rectangle on the earth be $\delta \theta$, the corresponding length on the map δr. Also let ϕ be the minute angle of azimuth under which, in both cases, the breadth of the rectangle is seen from the Centre of Reference or the centre of the map. Then we have

$$l \dot= \delta \theta, \quad l + \delta l \dot= \delta r, \quad \delta l \dot= \delta r - \delta \theta;$$

$$b = \phi \cdot \sin \theta, \quad b + \delta b = \phi \cdot r, \quad \delta b = \phi \cdot (r - \sin \theta);$$

$$\left(\frac{\delta l}{l} \right)^2 + \left(\frac{\delta b}{b} \right)^2 = \left(\frac{\delta r}{\delta \theta} - 1 \right)^2 + \left(\frac{r}{\sin \theta} - 1 \right)^2.$$

This quantity expresses the evil on each small rectangle. The product of the evil by the extent of surface which it affects, omitting the general multiplier ϕ, is

$$\left\{ \left(\frac{\delta r}{\delta \theta} - 1 \right)^2 + \left(\frac{r}{\sin \theta} - 1 \right)^2 \right\} \times \sin \theta \cdot \delta \theta.$$

Consequently the summation of the partial evils for the whole map is represented by

$$\int d\theta \cdot \left\{ \left(\frac{dr}{d\theta} - 1 \right)^2 + \left(\frac{r}{\sin \theta} - 1 \right)^2 \right\} \times \sin \theta.$$

Or if $r - \theta = y$, and if we put p for $\frac{dy}{d\theta}$, the expression is

$$\int d\theta \cdot \left\{ p^2 \sin \theta + \frac{(y + \overline{\theta - \sin \theta})^2}{\sin \theta} \right\};$$

and this integral, through the surface to which the map applies, is to be minimum.

7. This is a case of the Calculus of Variations. The function V to be integrated exhibits values for the differential coefficients

$$M = \frac{dV}{d\theta}, \quad N = \frac{dV}{dy}, \quad P = \frac{dV}{dp}.$$

The equation of solution is $N - \frac{d(P)}{d\theta} = 0$. Now in consequence of the existence of a value for M, we cannot adopt the facilities of solution which present themselves when $M = 0$, and we must therefore take the equation $N - \frac{d(P)}{d\theta} = 0$ without modification.

Here

$$N = \frac{2(y + \overline{\theta - \sin \theta})}{\sin \theta}, \quad P = 2p \sin \theta, \quad \frac{d(P)}{d\theta} = 2q \sin \theta + 2p \cos \theta;$$

and the equation $N - \frac{d(P)}{d\theta}$ becomes

$$\frac{y + \overline{\theta - \sin \theta}}{\sin \theta} - \sin \theta \cdot \frac{d^2 y}{d\theta^2} - \cos \theta \cdot \frac{dy}{d\theta} = 0;$$

or

$$\sin^2 \theta \cdot \frac{d^2 y}{d\theta^2} + \sin \theta \cdot \cos \theta \cdot \frac{dy}{d\theta} - y = \theta - \sin \theta.$$

8. For $\theta - \sin \theta$ put the more general symbol Θ. To solve the equation, assume $z = \sin \theta \cdot \frac{dy}{d\theta} + y$. Then, by actual differentiation and substitution;

$$\sin\theta\cdot\frac{dz}{d\theta}-z=\sin^2\theta\cdot\frac{d^2y}{d\theta^2}+\sin\theta\cdot\cos\theta\cdot\frac{dy}{d\theta}-y,$$

or

$$\sin\theta\cdot\frac{dz}{d\theta}-z=\Theta.$$

This equation is integrable when multiplied by $\dfrac{1}{\sin^2\frac{\theta}{2}}$; the solution gives

$$z=\frac{1}{2}\tan\frac{\theta}{2}\cdot\int d\theta\,.\,\frac{\Theta}{\sin^2\frac{\theta}{2}}$$

Therefore

$$\sin\theta\cdot\frac{dy}{d\theta}+y=\frac{1}{2}\tan\frac{\theta}{2}\cdot\int d\theta\,.\,\frac{\Theta}{\sin^2\frac{\theta}{2}}.$$

This equation is integrable when multiplied by $\dfrac{1}{\cos^2\frac{\theta}{2}}$; the solution gives

$$y=\frac{1}{4}\cdot\cot\frac{\theta}{2}\cdot\int d\theta\,.\,\frac{\sin\frac{\theta}{2}}{\cos^3\frac{\theta}{2}}\cdot\int d\theta\,.\,\frac{\Theta}{\sin^2\frac{\theta}{2}}.$$

If $\dfrac{\theta}{2}=\psi$, the solution may be put in the form

$$y=\cot\psi\cdot\int d\psi\,.\,\frac{\sin\psi}{\cos^3\psi}\cdot\int d\psi\,.\,\frac{\Theta}{\sin^2\psi}\,;$$

or

$$y=\frac{1}{2}\cdot\frac{1}{\sin\psi\cdot\cos\psi}\cdot\int d\psi\,.\,\frac{\Theta}{\sin\psi}-\frac{1}{2}\cot\psi\cdot\int d\psi\,.\,\frac{\Theta}{\sin^2\psi\cdot\cos^2\psi}\,;$$

or

$$y=\frac{1}{2}\tan\psi\cdot\int d\psi\,.\,\frac{\Theta}{\sin^2\psi}-\frac{1}{2}\cot\psi\cdot\int d\psi\,.\,\frac{\Theta}{\cos^2\psi}.$$

9. If we substitute for Θ our value $\theta-\sin\theta$, we obtain

$$y=-\theta-2\cot\frac{\theta}{2}\cdot\log\cos\frac{\theta}{2}+C'\cdot\cot\frac{\theta}{2}+C''\cdot\frac{1}{\sin\theta}$$

(in which we may vary the arbitrary terms, remarking that

$$\frac{1}{\sin\theta}=\frac{1}{2}\tan\frac{\theta}{2}+\frac{1}{2}\cot\frac{\theta}{2}\Big)\,;$$

and if we determine the arbitrary constants so that, when $\theta=0$,

y shall $=0$ and $\frac{dy}{d\theta}$ shall $=0$ (that is, so that the central parts of the map shall correspond exactly with the region about the Centre of Reference),

$$y = \tan\frac{\theta}{2} - \theta - 2\cot\frac{\theta}{2} \cdot \log\cos\frac{\theta}{2},$$

and

$$r = \theta + y = \tan\frac{\theta}{2} + 2\cot\frac{\theta}{2} \cdot \log\sec\frac{\theta}{2},$$

in which the logarithm is the Napierian or hyperbolic logarithm.

This equation entirely defines the nature of the Projection by Balance of Errors. The numerical values of r, for a series of values of θ, will shortly be given in a tabular form.

10. In order to obtain a numerical estimate of the two errors of a Projection, we must make use of the formulæ,

$$\frac{\text{projected area}}{\text{original area}} = \frac{r}{\sin\theta} \cdot \frac{dr}{d\theta};$$

$$\frac{\text{projected breadth}}{\text{projected length}} \times \frac{\text{original length}}{\text{original breadth}} = \frac{r}{\sin\theta} \cdot \frac{1}{\dfrac{dr}{d\theta}};$$

and for all the projections which we desire to compare, we must express r and $\frac{dr}{d\theta}$ in terms of θ, and must substitute in these formulæ.

11. The Projections which I shall compare are the following (in the formulæ, ψ is put for $\frac{\theta}{2}$) :—

(1) The Projection with Equal Radial Degrees. In this,

$$r = \theta, \quad \frac{p.a.}{o.a.} = \frac{\theta}{\sin\theta}, \quad \frac{p.b.}{p.l.} \times \frac{o.l.}{o.b.} = \frac{\theta}{\sin\theta}.$$

(2) The Projection with Unchanged Areas. In this,

$$r = 2\sin\psi, \quad \frac{p.a.}{o.a.} = 1, \quad \frac{p.b.}{p.l.} \times \frac{o.l.}{o.b.} = \sec^2\psi.$$

(3) The Stereographic Projection. In this,

$$r = 2\tan\psi, \quad \frac{p.a.}{o.a.} = \sec^4\psi, \quad \frac{p.b.}{p.l.} \times \frac{o.l.}{o.b.} = 1.$$

(4) Sir H. James's Projection. Here

$$r = \frac{5\sin\theta}{3+2\cos\theta}, \quad \frac{p.a.}{o.a.} = \frac{25(3\cos\theta+2)}{(3+2\cos\theta)^3}, \quad \frac{p.b.}{p.l.} \times \frac{o.l.}{o.b.} = \frac{3+2\cos\theta}{3\cos\theta+2}.$$

This projection fails when $\cos\theta = -\frac{3}{2}$, or $\theta = 131°\ 49'$.

(5) The Projection by Balance of Errors. Here

$$r = \tan \psi + 2 \cot \psi \cdot \log \sec \psi,$$

$$\frac{p.\,a.}{o.\,a.} = \left(\frac{1}{2} \cdot \sec^2 \psi + \csc^2 \psi \cdot \log \sec \psi \right)$$

$$\times \left(1 + \frac{1}{2} \sec^2 \psi - \csc^2 \psi \cdot \log \sec \psi \right),$$

$$\frac{p.\,b.}{p.\,l.} \times \frac{o.\,l.}{o.\,b.} = \frac{\sec^2 \psi + 2 \csc^2 \psi \cdot \log \sec \psi}{2 + \sec^2 \psi - 2 \csc^2 \psi \cdot \log \sec \psi}.$$

From these formulæ the numbers in the following Tables are computed :—

12. Table of Radial Distances from the Centre of the Map, for different Great-Circle Distances θ from the Centre of Reference.

θ.	Equal Radial Degrees.	Unchanged Areas.	Stereographic.	Sir H. James.	Balance of Errors.
5	0·08727	0·08724	0·08732	0·08729	0·08728
10	0·17453	0·17431	0·17498	0·17471	0·17465
15	0·26180	0·26105	0·26331	0·26240	0·26218
20	0·34907	0·34730	0·35265	0·35047	0·34997
25	0·43634	0·43288	0·44339	0·43907	0·43811
30	0·52360	0·51764	0·53590	0·52831	0·52672
35	0·61087	0·60141	0·63060	0·61830	0·61589
40	0·69814	0·68404	0·72794	0·70915	0·70577
45	0·78540	0·76537	0·82843	0·80094	0·79650
50	0·87267	0·84524	0·93262	0·89375	0·88825
55	0·95994	0·92350	1·04113	0·98761	0·98121
60	1·04720	1·00000	1·15470	1·08253	1·07563
65	1·13447	1·07460	1·27414	1·17849	1·17178
70	1·22174	1·14715	1·40042	1·27535	1·27000
75	1·30901	1·21752	1·53465	1·37299	1·37068
80	1·39628	1·28558	1·67820	1·47105	1·47434
85	1·48354	1·35118	1·83266	1·56915	1·58157
90	1·57080	1·41421	2·00000	1·66666	1·69215
95	1·65807	1·47455	2·18262	1·76375	1·81002
100	1·74534	1·53209	2·38351	1·85623	1·93342
105	1·83261	1·58671	2·60645	1·94558	2·06492
110	1·91988	1·63830	2·85630	2·02673	2·20659
115	2·00714	1·68678	3·13937	2·10303	2·36118
120	2·09440	1·73205	3·46410	2·16506	2·53243
125	2·18167	1·77402	3·84196	2·21052	2·72550
130	2·26894	1·81262	4·28901	2·23412	2·94776
135	2·35620	1·84776	4·82843	After this	3·20996
140	2·44347	1·87939	5·49495	the radius	3·52847
145	2·53074	1·90743	6·34319	diminishes.	3·92934
150	2·61801	1·93185	7·46410		4·44831
155	2·70528	1·95259	9·02142		5·18929
160	2·79255	1·96962	11·34256		6·28868

13. Table of Exaggeration, as shown by the Proportions of Projected Area to Original Area, for different Great-Circle Distances θ from the Centre of Reference.

θ.	Equal Radial Degrees.	Unchanged Areas.	Stereographic.	Sir H. James.	Balance of Errors.
5	1·00127	1·00000	1·00382	1·00229	1·00191
10	1·00508	1·	1·01537	1·00917	1·00767
15	1·01152	1·	1·03496	1·02073	1·01735
20	1·02060	1·	1·06315	1·03706	1·03127
25	1·03245	1·	1·10071	1·05835	1·04961
30	1·04544	1·	1·14875	1·08485	1·07278
35	1·06501	1·	1·20871	1·11674	1·10131
40	1·08610	1·	1·28250	1·15432	1·13585
45	1·11072	1·	1·37255	1·19789	1·17728
50	1·13919	1·	1·48217	1·24774	1·22668
55	1·17186	1·	1·61512	1·30412	1·28549
60	1·20920	1·	1·77778	1·36719	1·35543
65	1·25174	1·	1·97644	1·43692	1·43894
70	1·30014	1·	2·22097	1·51302	1·53909
75	1·35517	1·	2·52426	1·59470	1·65992
80	1·41780	1·	2·90391	1·68043	1·80697
85	1·48920	1·	3·38436	1·76759	1·98777
90	1·57080	1·	4·00000	1·85185	2·21269
95	1·66439	1·	4·80028	1·92641	2·49650
100	1·77225	1·	5·85774	1·98088	2·86051
105	1·89724	1·	7·28135	1·99969	3·33627
110	2·04307	1·	9·23921	1·96010	3·97176
115	2·21462	1·	11·99861	1·82952	4·84226
120	2·41840	1·	16·00000	1·56250	6·05133
125	2·66332	1·	21·99771	1·09761	7·86206
130	2·96188	1·	31·34779	0·35540	10·58547
135	3·33216	1·	46·62740	After this	14·89565
140	3·80135	1·	73·07911	the projec-	22·28290
145	4·41219	1·	122·30176	tion fails.	35·68061
150	5·23598	1·	254·44946		70·86947
155	6·40119	1·	455·67252		123·61888
160	8·16480	1·00000	1099·81373		290·08199

14. Table of Distortion, as shown by the Proportions of the Transverse Side to the Radial Side, in the projection of an area originally square, for different Great-Circle Distances θ from the Centre of Reference.

θ.	Equal Radial Degrees.	Unchanged Areas.	Stereographic.	Sir H. James.	Balance of Errors.
5	1·00127	1·00191	1·00000	1·00076	1·00084
10	1·00508	1·00765	1·	1·00307	1·00382
15	1·01152	1·01733	1·	1·00696	1·00861
20	1·02060	1·03109	1·	1·01252	1·01526
25	1·03245	1·04915	1·	1·01986	1·02386
30	1·04544	1·07180	1·	1·02914	1·03444
35	1·06501	1·09941	1·	1·04057	1·04693
40	1·08610	1·13247	1·00000	1·05443	1·06137

Table (*continued*).

θ.	Equal Radial Degrees.	Unchanged Areas.	Stereographic	Sir H. James.	Balance of Errors.
45	1·11073	1·17157	1·00000	1·07107	1·07775
50	1·13919	1·21744	1·	1·09094	1 09604
55	1·17186	1·27099	1·	1·11461	1·11617
60	1·20920	1·33333	1·	1·14286	1·13812
65	1·25174	1·40586	1·	1·17668	1·16171
70	1·30014	1·49029	1·	1·21744	1·18679
75	1·35517	1·58879	1·	1·26695	1·21311
80	1·41780	1·70409	1·	1·32780	1·24033
85	1·48920	1·83966	1·	1·40365	1·26801
90	1·57080	2·00000	1·	1·50000	1·29559
95	1·66439	2·19095	1·	1·62533	1·32235
100	1·77225	2·42028	1·	1·79351	1·34743
105	1·89724	2·69840	1·	2·02883	1·36980
110	2·04307	3·03961	1·	2·37793	1·38832
115	2·21462	3·46391	1·	2·94308	1·40171
120	2·41840	4·00000	1·	4·00000	1·43395
125	2·66332	4·69017	1·	6·63459	1·40808
130	2 96188	5·59891	1·	23 93204	1·39883
135	3 33216	6·82842	1·	After this	1·38347
140	3·80135	8 54863	1·	the projec-	1·35228
145	4·41219	11 05901	1·	tion fails.	1·31519
150	5·23598	15·95147	1·		1·26275
155	6·40119	21·34648	1·		1·21965
160	8·16480	33·16345	1·00000		1·16546

15. The last two Tables, which enable us to compare numerically, the Exaggeration and the Distortion in the different systems, will give us the means of comparing the systems generally. I shall make this comparison for the two values of θ, 115°, and 135°: the former, because it is nearly the extreme value of θ in Sir H. James's maps; the latter because, while on the one hand it may be sometimes desirable, on the other hand it is the largest that is likely ever to be wanted.

16. For $\theta = 115°$, the Exaggeration of the stereographic system is 12·0; this will perhaps be judged so large as to exclude it from further consideration. The merits of the others stand in the order,—

      1. Unchanged Areas.
      2. Sir H. James's.
      3. Equal Radial Degrees.
      4. Balance of Errors.

But when we consider the Distortion, the order is,—

      1. Balance of Errors.
      2. Equal Radial Degrees.
      3. Sir H. James's.
      4. Unchanged Areas.

The distortion in Sir H. James's is expressed by 2·9; and this, in my opinion, is a more serious inconvenience than the exaggeration in the Balance of Errors, namely 4·8.   On the whole, I think that, for this value of θ, the Balance of Errors is preferable to Sir H. James's.   The nearest in merit, I think, is that of Equal Radial Degrees; but the distortion of Balance of Errors 1·4 is preferable to that of Equal Radial Degrees 2·2, while the exaggeration 4·8 is not much more injurious than 2·2.   I prefer the Balance of Errors.

17. For $\theta=135°$, Sir H. James's is inapplicable.   The remaining systems stand in these orders :—

Exaggeration.		Distortion.	
1. Unchanged Areas	1·0	1. Balance of Errors	1·4
2. Equal Radial Degrees	3·3	2. Equal Radial Degrees	3·3
3. Balance of Errors	14·9	3. Unchanged Areas	6·8

In my opinion, the Balance of Errors is here the best. A square whose sides are 1, 1, is projected into a parallelogram whose sides are 3·3, 4·6; and this is better than the parallelogram in the Equal Radial Degrees whose sides are 1, 3·3. This is on the supposition that we desire to preserve an intelligible representation of every part of the earth depicted in the map. I will shortly state under what circumstances this opinion may perhaps be modified.

18. If we take for Centre of Reference the point defined by longitude $3^h\,30^m$ east of Greenwich, latitude 23° north, the circle whose radius is 135° will contain every continent and large island, including Australia and New Zealand, omitting only the South Pacific Ocean. If we take for Centre of Reference the point defined by longitude 1^h east of Greenwich, latitude 10° south, the circle whose radius is 135° will contain every continent and large island as before, omitting only the North Pacific Ocean. In such maps, the countries which are found on the borders of the map are sufficiently extensive and important to require to be exhibited without much distortion; and all constructions are equally troublesome. For these maps, therefore, I should use the system of Balance of Errors.

19. If, however, we take for Centre of Reference the point defined by longitude $16^h\,20^m$ east of Greenwich, latitude 77° north, the circle whose radius is 135° will contain the same continents and large islands as before, including also the small islands of the Pacific Ocean, omitting only the Antarctic Seas. Such a map is of extraordinary value, because it not only contains all the known lands, but may also exhibit all the sea-courses between the southern capes. But where, as in this case, the boundary is touched by little more than the headlands, distortion is less im-

portant, and the objections to the system of Equal Radial Degrees are much diminished. And, as the Centre of Reference is so near to the north pole, no serious discordance, probably no perceptible discordance, will be produced, if we describe the Parallels as circles whose centres are in the north pole and whose radii increase by equal degrees, and the meridians by straight radii from the north pole with equal angles between each radius and the proximate radius; and if we afterwards limit the map by a boundary-circle whose centre is at longitude 16^h 20^m east of Greenwich, latitude $77°$ north. This construction would be extremely easy.

20. Reverting now to the general theory, it appears that while the Stereographic Projection, in which $r = 2 \tan \dfrac{\theta}{2}$, possesses the very great merit of being free from distortion in its small elements, yet a more acceptable map is given by advancing in some measure towards the Projection by Equal Radial Degrees, in which $r = \theta = \infty \times \tan \dfrac{\theta}{\infty}$. It is evident that this may be done conveniently by using larger numbers instead of the 2 and 2 which occur in the stereographic formula. Thus we may conveniently use $r = 8 \tan \dfrac{\theta}{8}$, in which, Exaggeration $= \dfrac{8}{\sin \theta} \cdot \tan \dfrac{\theta}{8} \cdot \sec^2 \dfrac{\theta}{8}$,

Distortion $= \dfrac{8}{\sin \theta} \cdot \tan \dfrac{\theta}{8} \cdot \cos^2 \dfrac{\theta}{8}$. Or $r = 4 \tan \dfrac{\theta}{4}$, which gives,

$$\text{Exaggeration} = \frac{4}{\sin \theta} \tan \frac{\theta}{4} \cdot \sec^2 \frac{\theta}{4} = \sec \frac{\theta}{2} \cdot \sec^4 \frac{\theta}{4},$$

$$\text{Distortion} = \frac{4}{\sin \theta} \cdot \tan \frac{\theta}{4} \cdot \cos^2 \frac{\theta}{4} = \sec \frac{\theta}{2}.$$

In either of these the Exaggeration is diminished, and Distortion is introduced, but more in the second than in the first.

21. I will now allude to the process by which any of these Projections can be adapted to any Point of Reference whatever. The process is in fact a transfer from one system of projection to another system of projection, and is founded upon this theorem: that if in one projection we describe a series of Circles whose common centre is the Centre of the Map (corresponding to the Point of Reference) having radii equal to values of r corresponding on that projection to values of θ which increase by uniform quantities as $5°$ or $10°$, and if we draw from that centre Radial Lines at equal angles of azimuth; and if we do the same thing for another projection; then all the intersections of Meridians and Parallels referred to the pole of the earth will occupy on one projection the same places, in reference to the circles and radial lines above-mentioned, which they occupy on the other projection.

Thus, if we possess a map in which the meridians and parallels are drawn through the circles and radial lines on one projection, then for any other projection we have merely to draw radial lines, and to describe circles with the radii given by the Table of article 12, by the formulæ of article 20, or by equivalent statements, for that other projection; and we can at once lay down among the radii and circles of the second projection the intersections corresponding to those of the first projection as seen among *its* radii and circles.

· 22. There is one projection in which the Meridians and Parallels are described with comparative facility, because all are accurately circular arcs, namely the Stereographic. This projection, therefore, will be most proper for use as the standard projection, by means of which any others may be drawn. As a termination to this paper, I will here place the formulæ required for drawing a Stereographic Map with any Centre of Reference.

Let a be the linear radius of the circle which would include a hemisphere of the earth, β the radius of the proposed map, in degrees. Let the Centre of Reference be in north latitude α. (If in south latitude, it will only be necessary to invert the map.)

(1) The linear radius of the entire map will be $a \cdot \tan \dfrac{\beta}{2}$.

(2) Through the centre of the map a line must be drawn as polar axis. On this line will lie the centres of all the circles representing parallels of latitude.

(3) Let x be the north latitude of any parallel which is to be drawn (x being treated as an algebraically negative quantity for parallels in south latitude). One intersection of the circle representing this parallel, with the polar axis, will be north of the centre of the map by $a \cdot \tan \dfrac{x - \alpha}{2}$; the other intersection will be north of the centre by $a \cdot \cot \dfrac{x + \alpha}{2}$. The centre of the circle will be north of the central point by half the sum of these quantities, or by

$$\frac{a}{2} \cdot \cos \alpha \cdot \sec \frac{x - \alpha}{2} \cdot \operatorname{cosec} \frac{x + \alpha}{2}.$$

The radius of the circle will be half the difference of these quantities, or

$$\frac{a}{2} \cdot \cos x \cdot \sec \frac{x - \alpha}{2} \cdot \operatorname{cosec} \frac{x + \alpha}{2}.$$

The rules of algebraic signs are to be severely followed.

· (4) The north pole is north of the central point by

Fig. 6.

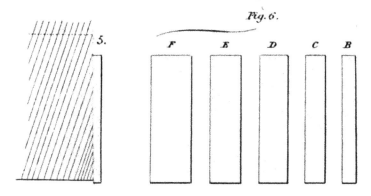

5.   F   E   D   C   B

8.

A

Fig. 11.

Fig. 12.

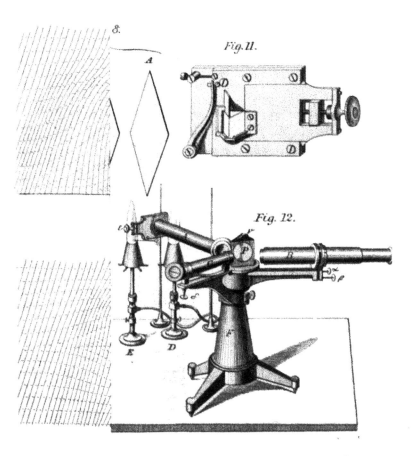

J. Basire sc.

$a \cdot \tan\left(45° - \frac{\alpha}{2}\right)$, and the south pole is south of the central point

by $a \cdot \tan\left(45° + \frac{\alpha}{2}\right)$, or $a \cdot \cotan\left(45° - \frac{\alpha}{2}\right)$.

(5) Bisect the line joining the poles (the point of bisection will be south of the central point by $a \cdot \tan \alpha$), and through the bisecting point draw an indefinite straight line at right angles to the polar axis. (This line represents the parallel for south latitude α.) On this transversal line will lie the centres of all the circles representing meridians.

(6) Let y be the angular measure of longitude east of the Point of Reference, of any meridian which is to be drawn. The eastern intersection, of the circle representing this meridian, with the transversal line, will be at the distance $a \cdot \mathrm{secant}\,\alpha \cdot \tan\frac{y}{2}$ from the polar axis; and the western intersection will be at the distance $a \cdot \mathrm{secant}\,\alpha \cdot \cotan\frac{y}{2}$. The western distance of the centre of the circle will be $a \cdot \mathrm{secant}\,\alpha \cdot \cotan y$, and the radius of the circle will be $a \cdot \mathrm{secant}\,\alpha \cdot \mathrm{cosecant}\,y$.

When, by means of these formulæ, the circles for Parallels and Meridians have been drawn, then concentric circles are to be described having the centre of the map for their centre, and with radii successively equal to $a \cdot \tan\frac{10°}{2}, a \cdot \tan\frac{20°}{2}$, &c., representing great-circle distances $10°, 20°$, &c., from the Point of Reference; and radii are to be drawn at equal azimuthal angles. Then the map may be used for laying down the intersections of meridians and parallels (as described in article 21) for any projection whatever, with the same Point of Reference.

Royal Observatory, Greenwich,
September 25, 1861.

LIV. *On the Deficiency of Rain in an elevated Rain-gauge, as caused by Wind.* By W. S. JEVONS, *B.A. of University College, London*[*].

[With a Plate.]

1. FROM the year 1767 many meteorologists have held the settled opinion that the larger part of the rain which falls upon the surface of the earth does not proceed from the clouds, as we should naturally suppose, but is derived from the lower strata of the atmosphere, within 200 or 300 feet of the

* Communicated by the Author, having been partly read at the meeting of the British Association at Manchester, September 1861.

surface. This paradox has been founded upon the fact that a rain-gauge, when placed at a moderate elevation in the atmosphere, is found to collect much less rain than if placed upon the ground. As the sudden increase of rain while it falls through the intervening air cannot be explained in accordance with the known laws of nature, many writers have spoken doubtfully of this subject, but have brought forward only scanty or palpably erroneous arguments to account for the experimental results.

2. I now hope to show that the observed differences of rainfall must be attributed to the influence of the wind upon our mode of experiment.

3. In observations with rain-gauges at different elevations, the higher gauges have been placed upon the roofs of houses, the summits of church-towers, or other erections which act as obstacles to the wind. It is obvious, too, that the rain-gauge is itself an obstacle, causing the wind to swerve aside, and to change the direction in which the rain-drops fall.

4. In order to determine the curves which the wind describes in meeting such obstacles, I have performed some small experiments. A vessel is formed of two oblong plates of glass, enclosing a layer of air about a quarter of an inch thick. One end of the vessel communicates through a pipe with a chimney or an aspirator, so that a regulated current of air may be drawn through it, to represent on a small scale a section of the wind moving over the surface of the earth. The curves described by the currents of air are shown very distinctly and beautifully by simply holding a piece of smoking brown paper in the draught of air which is about to enter the glass vessel. We may now place in the lower part of the current any small obstacle to represent a house or a rain-gauge placed in wind, and the curves described by the air will be depicted by the streams of smoke.

In trying such experiments, it is soon perceived that the curves are similar so long as the velocity of the current changes proportionally to the magnitude of the obstacle; and I am led to believe that the miniature experiment will indicate the course pursued by the actual wind meeting an obstacle, provided that the velocity of the wind and the magnitude of the obstacle bear somewhat the same proportion to each other as in the experiment. From such observations I have drawn the dotted lines in figs. 2 and 8, Pl. VI. They are intended to represent the course pursued by horizontal strata of air meeting an obstacle, such as a house (fig. 2), or a rain-gauge (fig. 3). Whatever may be the value of the experimental method, it cannot be denied that the air must move somewhat as shown in these figures.

5. A stream of air, then, meeting an obstacle leaps over it;

in so doing it is forced against the adjoining parallel stream of air, which must also diverge from the straight direction, and similarly impinge upon the next stream. But the increased pressure produced by the impact causes the streams of air to move more rapidly, and to diminish in thickness at the same time; and the disturbance of the streams of air will cease at the point where the total decrease of size of the streams is equal to the height of the obstacle. It is at least obvious that when a uniform wind meets an obstacle, some parts of the air must move more rapidly, just as a river moves most rapidly in the narrowest parts of its channel. It is quite in accordance, too, with our common experience, that an obstacle increases the velocity and force of the wind; thus the wind is always most fierce at the corner of a house, the end of a wall, or the summit of a hill.

6. We now have the whole explanation of the rain-observations in question. A drop of rain in falling is influenced at once by gravity and the motion of the air. It describes the diagonal of a rectangle, of which the perpendicular side represents the falling velocity of the drop, and the horizontal side the velocity communicated by the wind. In other words, we may say that the tangent of the angle of inclination (from the vertical direction) of the path of the falling drop varies nearly as the velocity of the wind.

Now conceive two equal drops of rain falling into a current of air at points where the velocity is not the same. They will not pursue parallel paths, but the one drop will either approach to, or recede from the other. The effect will be to increase or diminish the quantity of rain falling in the intermediate space.

To show clearly the nature of this effect, we may imagine the stream of air A B in Plate VI. fig. 1, to be suddenly contracted at C D to half its previous thickness, so that of course it must there commence to move with double velocity. At E F the stream dilates to its original size, and of course recovers its first velocity. The course of equidistant rain-drops falling into wind under such imaginary circumstances would be represented by the oblique black lines, and it is obvious that less rain would fall in the windward part of the contracted space than elsewhere.

7. To represent a real shower of rain falling upon an obstacle, we have only to conceive the drops of rain as falling through a great number of strata, all varying in velocity and thickness. I have thus conjecturally drawn the full lines in fig. 2 to represent the paths of the rain-drops in a shower falling through wind upon an obstacle such as a house, or tower which bears upon its summit an ordinary rain-gauge. In fig. 3, which is drawn upon a much different scale, the rain-gauge is the only obstacle, being

supposed fixed in mid-air. It is here, I venture to hope, rendered quite plain that *less rain will fall upon the summit of the obstacle than elsewhere, the surplus being carried forward to the lee side of the obstacle.* I entertain no doubt that we have in this process a sufficient explanation of the observed deficiency of rain in elevated places.

8. It is an evident corollary of this explanation, that no deficiency of rain would be observed did the measuring instrument cause no disturbance in the wind. But only a gauge of which the mouth is level with the ground fulfils this condition. Probably, indeed, the church-tower or house upon which a gauge is usually elevated occasions the chief part of the deficiency. Hence a gauge suspended in mid-air would collect more rain than if it were placed on a house. Yet a rain-gauge is itself an obstacle of some importance, and will cause a part of the rain to pass on unmeasured, as I have attempted to represent in fig. 3. The hollow of the funnel in this figure, it will be observed, is filled up with an *eddy* of wind.

9. In these drawings, I should observe, some little exaggeration must be excused: no notice, too, is taken of the motion of the wind in the third dimension of space, that is, round the obstacle instead of over it.

10. Thus having reason to suppose that the deficiency of rain at elevated points was due to the disturbance of the wind, I have examined all the observations and statements I could meet with bearing on the subject, and find my opinion, on the whole, strongly confirmed.

An intelligent observer, Mr. H. Boase of Penzance, after four months' experiment, remarks*, "Having observed that the difference between the first and the other gauges varied with more or less wind, its velocity has been registered from observation; but not having an accurate anemometer, we cannot yet offer any certain conclusion further than this, that the difference of the quantity of rain received in a gauge placed on the top of a building, and one at a level with the surface of the ground, is, for some reason or other, *proportional to the velocity of the wind.*"

11. Again, taking the measurements of rain† made by Luke Howard, and arranging them in the order of the ratio of the quantities in the lower and higher gauges, we find that we have also arranged them almost exactly in the order of the amount of accompanying wind, as indicated by the notes annexed. The results are as follows:—

---

* Annals of Philosophy (July 1822), new series, vol. iv. p. 18.
† Howard's 'Climate of London,' vol. ii. p. 158.

Ratio of rain in lower gauge to that in upper gauge.	Howard's remarks (in full).
3·00 . . .	Windy night; nimbus at sunset.
2·78 . . .	Stormy A.M.; wet P.M.
2·33 . . .	Cloudy; much wind; stormy night.
2·20 . . .	Much cloud with a fresh breeze.
2·00 . . .	Windy night.
1·75 . . .	Three currents in the air.
1·61 . . .	Showery day; cirrostratus evening.
1·60 . . .	Misty rain about midday; little wind veering from S.W. to E.
1·19 . . .	Cloudy; drizzling.
1·17 . . .	Rain by night.
1·17 . . .	(No remark.)
1·11 . . .	Showers chiefly by night.
1·10 . . .	Rain by night.
1·10 . . .	(No remark.)
1·08 . . .	Rain by night.
1·00 . . .	Clear A.M. with dew; nimbi; vane S.E. P.M., a heavy shower to S.; wind veered by S. to N.W.; then much cloud and rain.
1·00 . . .	Showers.

12. At the Greenwich Observatory, measurements of rain from three gauges placed at different heights have been daily recorded for about twenty years past. Examining the individual results, I was surprised to find great irregularity and want of accordance. Thus several hundredths of an inch of rain are often registered in the lowest gauge and *none* in the highest. Occasionally the middle gauge alone has caught any rain! The following will serve as a rather extreme specimen of these discordances:—

1844. Jan. and Feb.	30th. inch.	31st. inch.	2nd. inch.	6th. inch.	7th. inch.	9th. inch.
Highest gauge .	·01	·02	·08	·08	·21	·00
Middle „ .	·01	·00	·19	·17	·11	·33
Lowest „ .	·01	·04	·21	·18	·23	·16

These observations having been made by gentlemen of high ability and well-known scientific zeal, the discordances can only be attributed to the erroneous nature of the rain-gauge, and to the very unsuitable position and form of the Greenwich Observatory for rain observations: of course it is useless to look for any uniform law or ratio where such discrepancies may occur. The discordances, too, have no obvious relation with the force of the

wind, but might perhaps be explained by comparison with the
direction and force of the wind combined.   They serve me
here amply to establish the unsatisfactory nature of the best
rain measurements.

13. It is in this subject quite fallacious to appeal to *average
results*; for an appearance of uniformity and law will arise in the
long run, according to the doctrine of probabilities, however
irregular and various the causes which produce the difference.
A law of nature must appear in every case in which it acts alone,
reasonable error of observation being allowed for; but the discrepancies of individual rain observations at different altitudes
are such as can come under no law.   Even average or total
quantities for short periods are extremely discordant.   Prof.
Phillips's observations are stated in weekly totals[*]; but in the
week February 19 to 26, 1832, we find that the lowest gauge
received nearly six times as much rain as the upper one upon
the York Minster, while in the next succeeding week but one
the lower gauge contained only 1·22 times (or 1¼) as much
as the other.   The circumstances fully explain this difference,
"violent gales" having occurred in the former week and "perpendicular rain, without a trace of wind, in large drops" in the
latter.   This last statement will again be referred to (see par. 27).

14. Arago's results at the Paris Observatory, although pretty
uniform when stated in yearly averages, exhibit similar discordances in the separate months.   From a Table in the *Encyclopædia Metropolitana*, Art. Meteorology (p. 115), I extract the
results of the following three months, being the

Difference in Centimetres between results of Higher and Lower
Gauges at the Paris Observatory.

	1826.	1827.	1828.	1829.
March . . .	·285	1·207	·790	·174
May	·430	1·575	·210	·010
December . .	·810	1·220	·190	·030

Here in the same month, May, the difference varies from $\frac{1}{100}$ to
$\frac{157}{100}$ (centimetre)!

15. The deficiency of rain in an elevated rain-gauge varies greatly
according to the season of the year; and on an average the greatest
deficiency is found during the winter.   It is a phenomenon of a
wintry character, observes Prof. Phillips[†].   But of all the months
March generally shows the largest deficiency[‡]; and Prof. Phillips,

* Brit. Assoc. Report, 1833, Trans. Sections, p. 403.
† Brit. Assoc. Report, 1834, Trans. Sections, p. 562. See also Howard's
'Climate of London,' vol. i. p. 104; and Schouw, *Climat d'Italie*, p. 135:
"qu'elle est beaucoup plus forte en hiver qu'en été."
‡ See the observations of Dr. Heberden, Phil. Trans. vol. lix. (1769)

in discussing his observations*, adds the significant remark, "March very anomalous." Now March is in Europe the month in which strong, dry, north-east winds and equinoctial gales most occur, the very circumstances under which we should expect the results to be most erroneous.

16. I may lastly mention the observations of Dr. Buist, who having made four simultaneous measurements in the Island of Bombay, to determine the fall of rain at different heights below 200 feet, reported to the British Association, in 1852, that the results were entirely discordant. Although all proper precautions were taken†, "no satisfactory conclusion could be drawn, because the gauges at the several heights below and at 200 feet did not give uniform results,—sometimes the most elevated gauges having the greatest fall of rain, and at other times the lower. Nor did gauges at similar heights receive the same quantity of rain."

17. Although the effect of an obstacle upon the wind as causing a separation or approximation of the rain-drops, and a deficiency of rain in an elevated gauge, has now, I believe, for the first time been distinctly brought forward, several writers have made suggestions nearly to the same effect. Thus Howard speaks ‡ of strong winds as robbing the higher gauge. Dr. Trail says § of Prof. Phillips's observations, "These differences are too considerable to be attributed to anything but some imperfection in the instrument when much exposed to gales of wind; and it probably arises from eddies being formed round the rim of the funnel, which divert part of the water."

Again, H. Meikle writes in the 'Annals of Philosophy ‖,' "I can hardly pretend to give a complete solution of this well-known paradox, but am disposed to think it is in some way owing to the obstruction which the gauge itself offers to the wind. Perhaps the winds being made to rush with greater rapidity, and a little upward in beginning to pass over the mouth of the gauge, prevents the rain from falling into that part of it which is next the wind."

This almost coincides with my own explanation; but the remark is confined to the operation of the rain-gauge, which is usually an inconsiderable obstacle compared with the house or tower upon which the gauge is placed.

p. 359. Those by Bugge at Copenhagen, *Mém. de l'Acad. de Copenhagen*, nouv. sér. vol. v. p. 227; or in Schouw, *Climat d'Italie*, p. 131. In Arago's observations February is slightly more deficient than March.

* Brit. Assoc. Report, 1833, Trans. Sections, p. 408.
† Brit. Assoc. Report, 1852, Trans. Sections, p. 25.
‡ Climate of London, vol. i. p. 104.
§ Physical Geography (7th ed.), *Encyc. Brit.* reprint, p. 184.
‖ Vol. xiv. p. 312, for the year 1819.

Prof. Bache's "Note on the Effect of Deflected Currents of Air on the quantity of Rain collected by a Rain-gauge," communicated to the British Association in 1838, is to a different effect. It proves experimentally the immense differences which may occur between gauges placed at the different angles of a building, but does not show why a gauge on the top of an obstacle must on an average suffer a loss of rain. He found, however, that the gauges to the leeward received in general more rain than those to the windward, a fact fully in accordance with my theory.

18. It is hardly necessary to add that my explanation has no connexion with that of M. Flaugergues[*], who in an unfortunate moment mistook the sine for the radius of an angle, and argued that "less rain will fall into the horizontal opening of the rain-gauge when the rain is inclined than if it fell vertically, or in a direction less inclined." As long as the drops fall in parallel paths no such effect can be produced; it is the divergence of the rain-drops, owing to the varying velocity of the wind, which I assert to be the cause of the deficiency.

19. I will now approach the subject from an opposite point of view, and show *à priori* that the real increase of rain between the upper and lower gauges is not possible to any appreciable extent, according to the only physical explanation of the phenomenon which has ever been proposed. This theory was first suggested by Benjamin Franklin[†], who compared a drop of rain to a bottle of cold water condensing dew upon itself when brought into a warm room. That rain, even in our hottest days, he adds, comes from a very cold region, is obvious from its falling sometimes in the form of ice.

This explanation has been repeated and adopted by almost all who have expressed any belief in the phenomenon. But others have shown its utter inadequacy; and the single calculated example given by Sir J. Herschel, in his recent excellent 'Essay on Meteorology[‡],' may be adopted in our further discussions. "Admitting," he says, "a given weight of rain to arrive at 213 feet from the ground, with the temperature of the region at which it was formed unaltered, and supposing it to acquire in the remaining 213 feet the full temperature of the air (both of them extreme and, indeed, extravagant suppositions), admitting, too (though hardly less extravagant), the mean height of formation of the rain to be 12,000 feet, it would bring down with it a cold of $40°$ Fahr., which would condense (whether on the drops or in saturated air if diffused through it) only 40ths, or $\frac{1}{3\frac{1}{4}}$th

---

[*] Annals of Philosophy, vol. xiv. p. 114.
[†] See his letter to Dr. Thomas Percivall, dated London 1771, in the 'Memoirs of Thomas Percivall, M.D.,' Appendix B.
[‡] Page 104, as reprinted from the *Encyclopædia Britannica*, 8th ed.

$=0\cdot042$ of its weight, $=\frac{1}{17}$th of the quantity to be accounted for."

20. But, in reality, Sir J. Herschel's suppositions are far too favourable for the opinion which he opposes. In the first place, he makes no allowance for the heat derived from the gaseous air in addition to that received from the condensation of vapour. To estimate the amount of this, we may fairly make the assumption that has been found to give very exact results in the theory of the dry- and wet-bulb hygrometer. We may assume that the indefinitely thin film of air surrounding the drop of rain always takes the temperature of the drop, and yields up to it both the excess of its own sensible heat and the latent heat of the. condensed aqueous vapour (the sensible heat of the aqueous vapour may be neglected as very minute). Then according to the formulæ of M. Auguste,—

Let w = weight of a volume of air equal to that of the film at $0°$ (Cent.).

α = coefficient of dilatation of a gas per degree of temperature.

t = temperature of the air.

t' = temperature of the drop.

f = elastic force of aqueous vapour at temperature t, the air being supposed saturated.

f' = elastic force of aqueous vapour at temperature t'.

h = height of the barometer.

δ = specific gravity of aqueous vapour.

γ = specific heat of dry air.

λ = latent heat of aqueous vapour.

Then

$$w \cdot \frac{1}{1+\alpha \cdot t'} \cdot \frac{h-f'}{760} \cdot (t-t') \cdot \gamma$$

will nearly represent the sensible heat given out by the film of air in cooling from t to t', and

$$w \cdot \delta \cdot \frac{1}{1+\alpha t'} \cdot \frac{f-f'}{760} \cdot \lambda$$

will be the amount of latent heat given out by the vapour condensed. The ratio of these is

$$\frac{1}{\delta} \cdot \frac{h-f'}{f-f'} (t-t') \frac{\gamma}{\lambda},$$

which varies with the value of t'. In Sir J. Herschel's example, let us suppose the lowest 213 feet of air to have the temperature of $60°$ F. Then t' is at first $20°$ F., and the above formula (inserting for δ its value $\cdot6285$; for γ, $\cdot2669$; for λ, $640-t'$; for h, $\cdot760$; and for f and f' their values from the common tables of

elastic force of aqueous vapour) has the value 1·072; that is to say, rather more heat is at first received from the cooling of the air than from the condensation of vapour. When the drop has increased in temperature to 40° F. ($t' = 4°·44$ C.), the value will be ·822, or the condensed vapour yields the larger share of heat ; but even when the drop has the temperature 59°, the value has only diminished to ·624. Taking an average of these three determinations, we shall find that not more than 55 per cent. of the heat received by the drop will proceed from condensed vapour; consequently we must reduce Sir J. Herschel's first estimate almost to its half.

21. Again, considering that the temperature of the air increases uniformly from the elevation of 12,000 feet, at which Sir J. Herschel supposes the drop to be formed, down to its temperature at the surface, it is truly extravagant to suppose that a rain-drop should fall unaltered through 11,800 or 11,900 feet and then suddenly assume the full temperature of the air in the last 100 or 200 feet. A small drop falling very slowly will take the temperature of the air, or more strictly the temperature of evaporation, all the way down, and its degree of coldness on reaching the lowest stratum of air will be so slight as to produce no appreciable condensation even in perfectly moist air. On the other hand, a large drop falling so rapidly that it has no time to receive heat from the air, will indeed remain of a low temperature, but it will likewise have no time to receive heat from the lowest air. And drops of intermediate size, just in proportion as they fall more quickly and receive less heat from the upper strata of air, will be less able to receive heat from the lowest stratum.

22. Nor can it be argued that the rain-drop receives heat most freely in the lowest stratum of air because it there meets most vapour. For the humidity of the air invariably increases from the surface of the earth up to the first cloud, as was observed by Mr. Welch in each of his four balloon ascents. Even under the most rare or impossible hygrometric conditions the amount of condensation would be quite inappreciable. Under any usual or real conditions, it may be most confidently asserted that a falling drop of rain will either increase uniformly throughout its descent by an extremely minute quantity, or will, as is far more likely, evaporate and decrease by a small quantity. *Under no possible conditions will the increase within the last few hundred feet of descent be more than almost infinitesimal.*

23. It is of course perfectly well known and allowed that the temperature of rain is often much lower that that of the air at the surface. I have myself several times observed remarkably cold rain. So M. Boisgiraud* writes to the Paris Academy,

_____
* *Annales de Chimie et de Physique* (sér. 2) vol. xxxiii. p. 417.

that by experiment he has proved rain to be sufficiently cold to produce precipitation even when the air is far from being saturated.

In the Greenwich Meteorological Observations for 1843 (p. 123) it is stated that, in occasional observations on the temperature of the rain, "It has been always found that when the rain has been warm with respect to the temperature of the air at the time, no differences have existed in the quantities of rain collected at the different heights; but that when the temperature of the air has been higher than the temperature of the rain, a difference has always existed." It is quite surprising that these writers do not perceive that their experiments tell directly against their own conclusions, or at least tell nothing at all to the purpose. If it is the rain in the lower gauge, as is most likely, which was found to be cold, it simply proves that condensation of vapour *has not taken place*, otherwise the rain would have been warmed thereby. If, however, the rain in the higher gauge be found of a low temperature, it tells us nothing at all to the purpose, unless we likewise prove that the same rain, on reaching the lowest gauge, is of a much higher temperature. In short, we must have a change of temperature observed; and such an observation has never been recorded, so far as I am aware.

24. As a further objection to the condensation theory, it may be added that Arago, in stating it*, argues that the difference of the rain collected in the two gauges should be greater as the air is more moist, a consequence which he confesses is not at all conformable to experience. This remark is strikingly borne out by the fact already stated, that the apparent increase of rain between the higher and lower gauges is usually greatest during the month of March. Now this is the month of prevalent dry, cold, north-east winds and gales, the very circumstances under which the condensation theory is most utterly inadequate or inapplicable.

25. A single secondary argument in favour of the supposed increase of rain-drops remains to be disposed of. Arago has remarked† that the internal supernumerary fringes of a rainbow are never seen on the lower parts of the bow near the surface of the earth. Now the supernumerary bows were explained by Dr. Young on the theory of interference of undulations; and their appearance indicates that the drops of rain upon which the bows appear are of exactly uniform size. Not observing the

---

* *Annuaire du Bureau des Longitudes,* pour l'an 1824, p. 161.

† "Il faut donc que pendant leur descente verticale, les gouttes d'eau aient perdu les propriétés dont elles jouissaient d'abord; il faut qu'elles soient sorties des conditions d'interférence *efficaces*; il faut qu'elles aient beaucoup grossi."

supernumerary bows in the lower part, but only in the upper, Arago argues that the condition of efficacious interference of the drops must have been destroyed in descending into the lower part of the atmosphere. " Therefore," he concludes, " the drops of rain must have much increased in size*." Obviously this does not in the least follow; for the condition of efficacious interference is uniformity of size†; and uniform drops, condensing moisture upon themselves, or evaporating in the same circumstances, will remain uniform in size. The disappearance of the supernumerary bows near the surface no doubt arises from the more disturbed current of air there causing the drops to encounter each other and coalesce irregularly, so that some drops are produced two or three times as large as the others.

26. Distant showers of rain are often seen distinctly to evaporate, and sometimes entirely vanish during their fall; but I have never observed or heard of a shower being observed to increase in density visibly during its descent.

27. It is now only right to add that both Arago‡ and Prof. Phillips have recorded unequivocally that a deficiency of rain in the upper gauge occurs even during a perfect calm. We have already quoted one such observation by Prof. Phillips§; and two others are found in his second paper on this subject‖. Prof. Phillips, indeed, considers that falling rain itself produces a downward current of air, which, it is just conceivable, might, by flowing over the sides of the upper rain-gauge or its support, deflect the rain. Again, while a perfect calm prevails on the ground, a gentle wind is usually blowing at the top of a lofty tower. As my explanation of the deficiency of rain in an elevated gauge is certainly inapplicable in a calm, I confess that my hearers must choose for themselves between considering two distinguished scientific observers capable of mistake in the observation of wind and calm on the one hand, and overturning some of the best established facts of physical science on the other hand.

28. If the present explanation be accepted, all observations by rain-gauges elevated or exposed to wind must be rejected as fallacious and worse than useless. But it is improbable that the error in a gauge with its mouth not more than one or two feet above the ground is worth considering. Still I believe that during a heavy shower almost all gauges lose a little rain by splashing, and it is worthy of consideration whether more accu-

* *Annuaire du Bureau des Longitudes,* pour l'an 1836, p. 300.
† Herschel's 'Meteorology,' p. 219. It seems likely, however, that Arago argued upon some other view of the cause of this phenomenon, which has been much misunderstood.
‡ *Annuaire du Bureau des Longitudes,* pour l'an 1824, p. 160.
§ See above, paragraph 13.
‖ Report of the British Association, 1834, Trans. Sections, p. 561.

rate means of estimation should not be adopted in regular observatories. The most unexceptionable rain-gauge would consist of a sheet of metal, many feet square (for instance 10 feet), spread flat upon the ground in an open place, with a flat collecting vessel in the centre connected by a pipe with a sunken reservoir or recording apparatus. The edges of the collecting vessel should not be higher than an inch, so as to present no appreciable obstacle or hollow space to the wind. At the same time nothing would be lost by splashing, as the splashes within and without the vessel would be equal.

29. My conclusions, shortly stated, are:—

(1) An increase of the rainfall close to the earth's surface is incompatible with physical facts and laws.

(2) The individual observations on this subject are utterly discordant and devoid of law when separately examined, and the process of taking an average under such circumstances gives an apparent uniformity which is entirely fallacious.

(3) When daily measurements of rain, or even monthly totals, are examined with reference to the strength of the wind at the time, it becomes obvious that there is a connexion.

(4) Wind must move with increased velocity in passing over an obstacle. It follows demonstratively that rain-drops falling through such wind upon the windward part of the obstacle will be further apart, in horizontal distance, than where the wind is undisturbed and of ordinary velocity.

London, August 28, 1861.

LV. *On the Cubic Centres of a Line with respect to Three Lines and a Line.*—Second Note. *By* A. CAYLEY, *Esq.**

ON referring to my Note on this subject (Phil. Mag. vol. xx. pp. 418–423, 1860), it will be seen that the cubic centres of the line
$$\lambda x + \mu y + \nu z = 0$$
in relation to the lines $x = 0, y = 0, z = 0$, and the line $x + y + z = 0$, are determined by the equations
$$x : y : z = \frac{1}{\theta + \lambda} : \frac{1}{\theta + \mu} : \frac{1}{\theta + \nu},$$
where θ is a root of the cubic equation
$$\frac{1}{\theta + \lambda} + \frac{1}{\theta + \mu} + \frac{1}{\theta + \nu} - \frac{2}{\theta} = 0;$$
or as it may also be written,
$$\theta^3 - \theta(\mu\nu + \nu\lambda + \lambda\mu) - 2\lambda\mu\nu = 0.$$

* Communicated by the Author.

Two of the centres will coincide if the equation for θ has equal roots; and this will be the case if

$$\lambda^{-\frac{1}{2}}+\mu^{-\frac{1}{2}}+\nu^{-\frac{1}{2}}=0,$$

or, what is the same thing, if λ, μ, $\nu = a^{-3}$, b^{-3}, c^{-3}, where $a+b+c=0$. In fact, if $a+b+c=0$, then $a^3+b^3+c^3=3abc$, and the equation in θ becomes

$$\theta^3 - \frac{3\theta}{a^2b^3c^4} - \frac{2}{a^3b^3c^3}=0;$$

that is,

$$(abc\theta)^3-3(abc\theta)-2=0,$$

which is

$$(abc\theta+1)^2(abc\theta-2)=0.$$

So that the values of θ are $\dfrac{-1}{abc}$, $\dfrac{2}{abc}$. First, if $\theta=-\dfrac{1}{abc}$, then x, y, z will be the coordinates of the double centre. And we have

$$\theta+\lambda=\frac{1}{a^3}-\frac{1}{abc}=\frac{1}{2a^3bc}\,(2bc-2a^2)$$

$$=\frac{1}{2a^3bc}\,(-a^2-b^2-c^2);$$

or putting for shortness $\square=a^2+b^2+c^2$,

$$\theta+\lambda=-\frac{1}{2a^3bc}\,\square, \quad =-\frac{3}{abc}\cdot\frac{\square}{6a^2},$$

with similar values for $\theta+\mu$, $\theta+\nu$. But $\dfrac{1}{x}$, $\dfrac{1}{y}$, $\dfrac{1}{z}$ are proportional to $\theta+\lambda$, $\theta+\mu$, $\theta+\nu$; and we may therefore write

$$\frac{P}{x}=\frac{\square}{6a^2}, \quad \frac{P}{y}=\frac{\square}{6b^2}, \quad \frac{P}{z}=\frac{\square}{6c^2};$$

whence, in virtue of the equation $a+b+c=0$, we have for the locus of the double centre,

$$\sqrt{x}+\sqrt{y}+\sqrt{z}=0.$$

Or this locus is a conic touching the lines $x=0$, $y=0$, $z=0$ harmonically in respect to the line $x+y+z=0$, a result which was obtained somewhat differently in the paper above referred to.

Next, if $\theta=\dfrac{2}{abc}$, x, y, z will be the coordinates of the single centre. And we now have

$$\theta+\lambda=\frac{1}{a^3}+\frac{2}{abc}=\frac{1}{2a^3bc}\,(2bc-2a^2+6a^2)=\frac{1}{2a^3bc}\,(-\square+6a^2)$$

$$=-\frac{3}{abc}\frac{\square-6a^2}{6a^2},$$

with similar values for $\theta+\mu$, $\theta+\nu$. But $\dfrac{1}{x}$, $\dfrac{1}{y}$, $\dfrac{1}{z}$ are proportional to $\theta+\lambda$, $\theta+\mu$, $\theta+\nu$, and we may therefore write

$$\frac{P}{x}=\frac{\Box-6a^2}{6a^2}, \quad \frac{P}{y}=\frac{\Box-6b^2}{6b^2}, \quad \frac{P}{z}=\frac{\Box-6c^2}{6c^2},$$

from which equations, and the equation $a+b+c=0$, the quantities P, a, b, c have to be eliminated. I at first effected the elimination as follows: viz., writing the equations under the form

$$\frac{x}{x+P}=\frac{6a^2}{\Box}, \quad \frac{y}{y+P}=\frac{6b^2}{\Box}, \quad \frac{z}{z+P}=\frac{6c^2}{\Box},$$

we obtain

$$\frac{x}{x+P}+\frac{y}{y+P}+\frac{z}{z+P}=6,$$

$$\sqrt{\frac{x}{x+P}}+\sqrt{\frac{y}{y+P}}+\sqrt{\frac{z}{z+P}}=0,$$

which are easily transformed into

$$\frac{x}{x+P}+\frac{y}{y+P}+\frac{z}{z+P}=6,$$

$$\frac{yz}{(y+P)(z+P)}+\frac{zx}{(z+P)(x+P)}+\frac{xy}{(x+P)(y+P)}=9;$$

or, what is the same thing,

$$6(P+x)(P+y)(P+z)-x(P+y)(P+z)-y(P+z)(P+x)$$
$$-z(P+x)(P+y)=0,$$

$$9(P+x)(P+y)(P+z)-yz(P+x) \qquad -zx(P+y)$$
$$-xy(P+z)=0,$$

which give

$$6P^3+5P^2(x+y+z)+4P(x+y+z)+3xyz=0,$$
$$9P^3+9P^2(x+y+z)+8P(x+y+z)+6xyz=0.$$

Or, multiplying the first equation by 2, and subtracting the second,

$$3P+x+y+z=0;$$

and we thus obtain for the locus of the single centre the equation

$$\frac{x}{-2x+y+z}+\frac{y}{-2y+z+x}+\frac{z}{-2z+x+y}=2,$$

or, what is the same thing,

$$x^3+y^3+z^3-(yz^2+zx^2+xy^2+y^2z+z^2x+x^2y)+3xyz=0,$$

which may also be written,

$$-(-x+y+z)(x-y+z)(x+y-z)+xyz=0.$$

The same result may also be obtained as follows: viz., observing that $\square - 6a^2 = b^2 + c^2 - 5a^2 = -4a^2 - 2bc$, we have

$$\frac{x}{P} = \frac{-3a^2}{2a^2 + bc}, \quad \frac{y}{P} = \frac{-3b^2}{2b^3 + ca}, \quad \frac{z}{P} = \frac{-3c^2}{2c^2 + ab};$$

and then by means of the equation

$$\frac{a^2}{2a^2 + bc} + \frac{b^2}{2b^2 + ac} + \frac{c^2}{2c^2 + ab} - 1 = 0,$$

which is identically true in virtue of $a + b + c = 0$ (in fact, multiplying out, this gives

$$12a^2b^2c^2 + 4(b^3c^3 + c^3a^3 + a^3b^3) + abc(a^3 + b^3 + c^3)$$

$$-8a^2b^2c^2 - 4(b^3c^3 + c^3a^3 + a^3b^3) - 2abc(a^3 + b^3 + c^3) - a^2b^2c^2 = 0;$$

that is,

$$3a^2b^2c^2 - abc(a^3 + b^3 + c^3) = 0, \text{ or } abc(a^3 + b^3 + c^3 - 3abc) = 0,$$

where the second factor divides by $\overline{a + b + c}$), we find the above-mentioned equation,

$$x + y + z + 3P = 0.$$

We then have

$$\frac{-x + y + z}{P} = \frac{x + y + z}{P} - \frac{2x}{P} = -3 + \frac{6a^2}{2a^2 + bc} = -\frac{3bc}{2a^2 + bc};$$

that is,

$$\frac{-x + y + z}{P} = \frac{-3bc}{2a^2 + bc}, \quad \frac{x - y + z}{P} = \frac{-3ca}{2b^2 + c}, \quad \frac{x + y - z}{P} = \frac{-3ab}{2c^2 + ab}.$$

And forming the product of these functions, and that of the foregoing values of $\frac{x}{P}, \frac{y}{P}, \frac{z}{P}$, we find as before,

$$-(-x + y + z)(x - y + z)(x + y - z) + xyz = 0$$

for the equation of the locus of the single centre. The equation shows that the locus is a cubic curve which touches the lines $x = 0$, $y = 0$, $z = 0$ at the points where these lines are intersected by the lines $y - z = 0$, $z - x = 0$, $x - y = 0$ (that is, it touches the lines $x = 0$, $y = 0$, $z = 0$ harmonically in respect to the line $x + y + z = 0$), and besides meets the same lines $x = 0$, $y = 0$, $z = 0$ at the points in which they are respectively met by the line $x + y + z = 0$.

2 Stone Buildings, W.C.
September 25, 1861.

LVI. *On Earth-currents, and their Connexion with the Phenomena of Terrestrial Magnetism.* By the Rev. H. LLOYD, D.D., D.C.L.*

IN the year 1848 Mr. Barlow communicated to the Royal Society a paper "On the Spontaneous Electrical Currents observed in the Wires of the Electric Telegraph," in which he established the important fact, that a wire, whose extremities are connected with the earth at two distant points, is unceasingly traversed by electric currents, the intensity of which varies with the azimuth of the line joining the points of contact with the ground. The direction of these currents was proved to be the same at both extremities of the same wire, and was shown to depend on the relative positions of the earth-connexions, while it was wholly independent of the course followed by the wire itself. The currents cease altogether when either of the contacts with the earth is interrupted. From these facts Mr. Barlow concluded that "the currents are terrestrial, of which a portion is conveyed along the wire, and rendered visible by the multiplying action of the coil of the magnetometer."

Mr. Barlow further observed that, apart from sudden and occasional changes, the general direction of the needle of the galvanometer appeared to exhibit some regularity. He was thus led to institute a series of observations for fourteen days and nights, on two wires simultaneously, one from Derby to Rugby, and the other from Derby to Birmingham, the positions of the needles in both circuits being recorded every five minutes, day and night. From these observations he concluded—

"1. That the path described by the needle consisted of a regular diurnal motion, subject to disturbances of greater or less magnitude.

"2. That this motion is due to electric currents passing from the northern to the southern extremities of the telegraph wires, and returning in the opposite direction.

"3. That, exclusive of the irregular disturbances, the currents flowed in a southerly direction from about 8 or 9 A.M. until the evening, and in a northerly direction during the remainder of the twenty-four hours."

He was thus led to examine whether any relation subsisted between these movements and the daily changes of the horizontal magnetic needle. And having made for this purpose a series of simultaneous observations with a delicate declinometer, he came to the conclusion, that although generally the currents flow *southwards* during that part of the day in which the variation of

* Communicated by the Author, having been read at a meeting of the Royal Irish Academy, held November 11, 1861.

the horizontal needle is *westerly* (*i. e.* from 8 or 9 A.M. until the evening), and *northwards* when the variation is *easterly* (*i. e.* during the night and early part of the morning), " yet simulta. neous observations showed no similarity in the path described by the magnetic needle and the galvanometer."

An examination of Mr. Barlow's galvanometric observations led me, some time since, to an opposite conclusion ; and at the last meeting of the British Association I stated my conviction, founded on these observations, that the earth-currents, whose continuous flow Mr. Barlow has the merit of establishing, would eventually explain all the changes of terrestrial magnetism, both periodic and irregular. I now proceed to state the grounds of this conviction, and to show, from Mr. Barlow's observations, that the diurnal changes of the earth-currents correspond with those of the horizontal component of the earth's magnetic force *.

Let us suppose, then, that the forces which act upon the horizontal needle, and which cause it to deviate from its mean position, are due to electric currents traversing the upper strata of the earth in a horizontal direction ; and let ξ denote the intensity of the current in the magnetic meridian, *positive* when flowing *northwards*, and *vice versâ*; and η the intensity of the current perpendicular to the magnetic meridian, *positive* when flowing *eastward*, and *vice versâ*. Then the force of the current in any direction, making the angle ϵ with the magnetic meridian (measured to the east of north), is

$$\phi = \xi \cos \epsilon + \eta \sin \epsilon.$$

Now ξ is proportional to the force which deflects the freely suspended horizontal needle from its mean position, or to $X \Delta \psi$, X being the horizontal component of the earth's magnetic force, and $\Delta \psi$ the change of declination expressed in parts of radius. Similarly, η is proportional to the force which deflects from its mean position a magnet which is maintained (by torsion or

---

* The first proof of a correspondence between the magnetic variations, and the changes of the earth-currents, seems to be due to Dr. Lamont of Munich, in a letter dated July 29, 1861, which was read by the Astronomer Royal at the last meeting of the British Association. Dr. Lamont states that he has found " that electric currents, or (as they may be more properly termed) *electric waves*, varying in direction or intensity, are constantly passing at the surface of the earth, and that these waves correspond perfectly with the variations of terrestrial magnetism." The correspondence here referred to seems to relate to the *smaller* and more *rapid variations* of the terrestrial magnetic force. But in a letter to Prof. Heiss, dated September 1, Dr. Lamont expresses his conviction that the whole diurnal movements are due to these earth-currents. He adds, however, that he had hitherto been unable fully to verify this conclusion, owing to the continual changes produced in the collecting plates and in the wires by heat and moisture.

other means) in a position perpendicular to the magnetic meridian, and is measured, in terms of X, by the relative changes of the horizontal intensity taken negatively. Hence the force of the current in any given direction may be determined in terms of the same units.

Now $\qquad \epsilon = \alpha - \psi,$

in which α is the azimuth of the line connecting the two stations measured from the true meridian eastward, and ψ the magnetic declination measured in the same direction. The observations of Sir James Ross at Derby, give $\psi = -22° 25'$; and we have, for the line connecting Derby with Rugby,

$$\alpha = -18° 7', \quad \alpha - \psi = +9° 18';$$

and for the line joining Derby and Birmingham,

$$\alpha = +33° 27', \quad \alpha - \psi = +55° 52'.$$

The first column of the following Table contains the mean variations of the magnetic declination at the alternate hours for the month of May, as deduced from four years' observation of that element at the Dublin Magnetic Observatory; the second contains the corresponding values of the changes of the horizontal intensity, in ten thousandths of the whole intensity; and the third and fourth the calculated values of the deflecting forces in the line perpendicular to that connecting the earth-contacts at Derby and Rugby and at Derby and Birmingham respectively, and expressed in terms of the same units. These latter numbers are by hypothesis proportional to the intensities of the currents directed along the connecting wires.

TABLE I.—*Calculated* Values of the Intensity of the Currents traversing the Wires uniting Derby and Rugby, and Derby and Birmingham, respectively.

Hour.	$\Delta\psi$.	$\dfrac{\Delta X}{X}$.	Derby and Rugby.	Derby and Birmingham.
1 A.M.	1·8	0·4	5·1	2·6
3	2 5	− 1·6	7·6	5·5
5	3·9	− 3·7	11·9	9·5
7	5·2	− 8·4	16·2	15·4
9	2·1	−16·9	8·9	17·5
11	−4·1	−15·9	− 9·3	6·4
1 P.M.	−7·1	− 3·1	−19·8	− 9·0
3	−5·1	6·1	−15·7	−13·4
5	−1·8	14·2	− 7·6	−14·8
7	0·3	14 6	− 1·5	−11·6
9	1 0	9·0	1·3	− 5·9
11	1·3	5·2	2·9	− 2·2

The galvanometric observations instituted by Mr. Barlow on

these two lines were continued for fourteen consecutive days, commencing May 17, 1848. Of these days of observation, however, six are incomplete, viz. May 17, 19, 20, 23, 24, 30; and another day (May 27) appears from the Dublin observations to have been a day of considerable magnetic disturbance. Omitting these, as unsuited to furnish true mean results, the means of the remaining days are as follow. The *positive* numbers indicate currents proceeding *towards* Derby, and the *negative* currents in the contrary direction.

TABLE II.—Mean *observed* Values of the Intensity of the Currents traversing the Wires uniting Derby and Rugby, and Derby and Birmingham, respectively.

Hours.	Derby and Rugby.				Derby and Birmingham.			
	A.M.		P.M.		A.M.		P.M.	
1	−1·4	0·3	−5·0	−5·1	0·2	1·5	−9·1	−8·5
2	2·5		−5·5		2·9		−7·7	
3	1·6	1·7	−2·7	−3·3	0·9	1·3	−7·4	−7·4
4	1·1		−2·4		0·7		−7·2	
5	0·5	1·2	−1·8	−2·3	0·6	1·2	−3·6	−5·1
6	2·7		−3·2		2·8		−6·3	
7	3·1	3·0	−0·6	−1·1	3·9	4·1	−4·5	−4·7
8	3·1		−0·2		5·9		−3·4	
9	2·4	1·8	0·4	0·2	4·2	3·4	−0·8	−1·7
10	−0·9		0·1		−0·6		−1·7	
11	−4·3	−3·6	0·4	0·6	−7·2	−5·8	0·3	0·4
12	−5·1		1·7		−8·1		2·8	

It will be observed that the changes indicated by these numbers are very systematic. In the wire connecting Derby and Birmingham the current flows *southwards* from 10 A.M. to 10 P.M. inclusive, and *northwards* during the remaining hours. In the wire connecting Derby and Rugby the *southward* current lasts from 10 A.M. to 8 P.M. inclusive, and it is *northward* (with a single exception) during the remaining hours. There are, however, as might be expected in so short a series, some irregularities in the course of the changes. In order to lessen these, and at the same time to confine the results to such as are comparable with the preceding, I have given (in the alternate columns of the Table) the means corresponding to the alternate hours commencing at 1 A.M. computed by the formula

$$\tfrac{1}{4}(a + 2b + c).$$

The numbers so obtained are projected into curves in the annexed diagram, having been previously multiplied by constant coefficients in order to equalize the ranges with those of the computed results. The dotted lines in both cases are the corresponding projections of the calculated results. The agreement between these two sets of curves is probably as great as could be expected

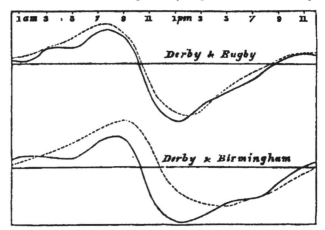

in the results of so short a series of observations; and we seem therefore entitled to conclude that the diurnal movements of the two horizontal magnetometers are accounted for by electric currents traversing the upper strata of the earth.

There is one point of difference to which it is important to draw attention. It will be seen that the *calculated* curves are for the most part *above the observed.* The reason of this will be evident upon a little consideration. The *zero* from which the calculated results are measured is the mean of the day; whereas that of the observed results is the *true zero,* corresponding to the absence of all current. Now the chief deflections of the galvanometer needle (as appears from the latter curves) are those in which the sun is above the horizon; and the *zero line* consequently divides the area of the diurnal curve unequally, being considerably nearer to the night observations than to those of the day. If the calculated curves be *displaced* by a corresponding amount, their agreement with the observed will be much closer.

The difference here noted is one of considerable theoretical importance. Magnetometric observations furnish merely *differential* results, the magnitude and the sign of which have reference solely to an *arbitrary zero.* We are accordingly ignorant even of the relative values of the effects, and are unable to compare them with their physical causes, whether real or supposed. In these respects the galvanometric observations have the advantage. In them, *positive* and *negative* are physically distinguished by the *direction* of the currents; and this, as well as the absence of all currents, is indicated by the instrument itself; the results therefore furnish the measures of the forces by which they are produced.

The next and most important step in this inquiry will be to assign the physical cause of these phenomena. The existence of electric currents traversing the earth's crust has hitherto been maintained as a *hypothesis*, on account of its supposed adequacy to explain the terrestrial magnetic changes. Now, however, their existence is proved not only to be a *fact*, but also a fact sufficient to explain the phenomena. It remains therefore only to ascertain their source; and it will be for those who deny that the sun operates by its *heat* in producing the phenomena of terrestrial magnetism, to assign to these currents a more probable origin.

P.S. While these pages were passing through the press, the writer received, by the kindness of Dr. Lamont, a copy of a further communication from him on the same subject, in a letter to Professor De la Rive, dated Oct. 10, 1861. In this letter, Dr. Lamont seems to recede from the view expressed in a former letter (see note, *suprà*), and expresses his belief that the diurnal variations of terrestrial magnetism cannot be explained by the direct action of electric currents propagated on the earth's surface; and he advances the hypothesis, that the regular portion of these variations is due to a peculiar influence of the sun, their irregular fluctuations alone being caused by the earth-currents. These conclusions seem to be irreconcilable with Mr. Barlow's observations, and are opposed to the inferences which I have drawn from them in the preceding pages.

The scientific public will therefore await with interest the detailed publication of Dr. Lamont's investigations on this important subject, in which, it may be hoped, he will give some clue to the explanation of this seeming discordance.

Trinity College, Dublin,
   Nov. 16, 1861.

LVII. *Considerations respecting the Original Formation of Aëro-lites.*—Part II. *By* W. HAIDINGER, *For. Mem. R.S.L. & E. and Director-General of the Geological Survey of Austria.*

[Concluded from p. 361.]

IF the phenomena attending the fall of meteorites upon our own earth offer serious difficulties, considerations concerning the condition of their previous existence is by far a more arduous task. It must not be forgotten that there are two cosmical or planetary bodies in question; the one a large one (our own globe), and a comparatively minute one (the meteorite). M. Leverrier, to whose talents and genius as an astronomer and mathematician we chiefly owe the discovery of the planet Neptune, felt himself authorized to pronounce, before the Paris Academy (October 1, 1860), a view or suspicion which he himself designates as "strange at the first aspect, but very possibly a reality*," viz. that in comparatively recent times new and small planets have been formed out of planetary matter existing at different distances around the sun, and possessing various degrees of density and volume†, but that their existence had remained unperceived till, during the last few years, the extraordinary amount of attention bestowed on the subject had at length been rewarded by a number of discoveries ‡.

The original formation and constitution of cosmical bodies have of late become the subject of the most diversified consideration. Some have tried to develope peculiarities previously more or less neglected; others (as my respected friend Prof. C. F. Naumann, in his classical 'Manual of Geology,' chapter on the Temperature of the Interior of the Globe, 2nd edit. 1857, vol. i. p. 86) have endeavoured to treat the question in a lucid and exhaustive synopsis, and to collect into a whole the opinions of men of the

* "Une idée, un soupçon, étrange peut-être au premier abord, mais qui peut très-bien être une réalité."—Moigno's *Cosmos*, 1860, vol. ix. p. 476.

† "L'espace autour du soleil est, on le sait, rempli de matière cosmique, et de matière cosmique de tous degrés de ténuité et de grosseur."—Ibid.

‡ As closely related to this portion of M. Haidinger's paper, the following extract from the 'Annual Register of Facts and Occurrences' for August 1861, may be here appropriately inserted:—"M. Leverrier, from the perturbations observed in the orbits of the planets Mercury, Venus, the Earth, and Mars, has still more recently come to the conclusion that there exists in our own system a considerable quantity of matter which has not hitherto been taken into account. In the *first* place, he supposes that there must exist within the orbit of Mercury, at about 0·17 of the Earth's distance from the Sun, a mass of matter nearly equal in weight to Mercury. As this mass of matter would probably have been observed before this, either in transit over the Sun's disc, or during total eclipses of the Sun, if it existed as one large planet, M. Leverrier supposes that it exists as a series of asteroids. *Secondly*, M. Leverrier sees reason to believe that there must be a mass of matter, equal to about one-tenth of the mass of the Earth, revolving around the Sun at very nearly the same distance as the Earth. This also he supposes split up into an immense number of asteroids [? meteorites]. *Thirdly*, M. Leverrier's researches have led him to the conclusion that the group of asteroids which revolve between Mars and Jupiter, sixty of which have already been seen and named, and had their elements determined, must have an aggregate mass equal to one-third of that of the Earth. He likewise thinks it is not unlikely that similar groups of asteroids exist between Jupiter and Saturn, Saturn and Uranus, and between Uranus and Neptune." See also *Cosmos* for June 1861, p. 639.—R. P. G.

highest authority, rather for the purpose of respectful study, than
to be made the subject of control or contradiction.   Proceeding
from simple correlations, I humbly venture to enunciate some
few considerations respecting the formation of meteorites, which,
eminently diversified as they are if taken individually, I must
yet consider, along with Sir David Brewster, Prof. Laurence
Smith, and other naturalists, to be fragments of a larger or more
voluminous body.

The formation of crystals requires a movement of molecules.
This is a general and most irrefragable theorem.   We see
crystals deposited from gaseous and liquid solutions, or wherever
the single molecules have acquired mobility under the influence
of high temperature, as in substances in a state of fusion.

Whenever solid bodies are undergoing metamorphic changes,
crystals form out of pulverulent, as well as out of relatively solid
substances, when they un ergo influences that make their intimate particles moveable. d We do not know that crystallization
can take place under any other circumstances, so long as the laws
of nature, as now known to us, remain in force.   We are entitled
therefore to conclude that these bodies, coming from cosmical
space into our atmosphere, took their point of departure from
matter either in a gaseous, liquid, or pulverulent condition.   The
real point of departure then is matter in the form of an *impal-
pable powder*, assumed to be the initial deposit of any substance
suspended in a gaseous or liquid solution.

Meteoric *stones*, almost pulverulent in their nature, with opake,
nearly earthy fracture (as those of Reichenbach's second family),
others whitish, without rounded particles, or dark-coloured (as
those of Bokkeveld), are connected, by a long series of intermediate forms, with the highly crystalline meteorites of Chassigny, Juvenas, Shalka, and the solid compact ones of Seres,
Tabor, Chantonnay, Segowlee, Parnallee, &c.   In the same way
a long series of structural transitions connect the non-crystalline
meteoric *irons* of the Cape of Good Hope and Hemalga with the
beautifully crystalline varieties of Agram, Elbogen, Lenarto,
Lockport, Red River, Nebraska, ending with the most perfect
type, that of Braunau.   The crystals of olivine contained in the
meteorites of Hainholz, Brahin, Atacama, and Krasnojarsk prove
the power of crystallization to have remained active during a long
period of time.

With our present knowledge of natural laws, these characteristically crystalline formations could not possibly have come
into existence except under the action of *high temperature* combined with *powerful pressure*; though we have to search in vain
for a *heated cosmical space*, as supposed by Poisson.

If we suppose within the glacial cold of space the existence of a pulverulent aggregate of all the substances found in meteorites, these could not be brought to crystallize gradually without some means or source by which heat could subsequently act upon them; and it may be questionable how far the *mutual pressure of masses,* or the attraction of a great whole on its isolated and still unconnected particles, may possibly suffice to produce such an effect.

I may here anticipate that a mere pulverulent aggregate having a rotatory movement in space must necessarily also acquire a spheroidal form dependent upon rotation, exactly like a liquid (according to Professor Plateau's experiments) not acted on by terrestrial attraction, and consequently in a state of free suspension.

A *septaria,* an object familiar to mineralogists and geologists, may serve to convey an idea of the effects of pressure acting from the circumference to the centre. Septariæ are spheroidal tuberiform bodies, occasionally slightly compressed in one direction (see fig 3), consisting of an external solid shell or crust of compact argillaceous sphærosiderite, filled up with the same substance, and intersected by numerous and somewhat imperfect veins of calcareous and magnesio-calcareous spar. Fig 3 is an autotype, taken from a specimen in the Imperial Museum of

Fig. 3.

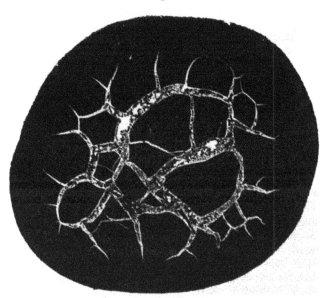

Vienna. The formation of such a septaria may be explained as follows:—within a stratum of clay, the particles richest in the carbonate of oxide of iron agglomerate or coalesce: the clay-

stratum, and with it the sphærosideritic agglomeration, under_
goes pressure, which, if sufficient, leaves in the interior a softer
portion, more impregnated with water than the external crust
from which that element has been squeezed more completely out.
The sphærosiderite is naturally inclined to assume throughout
the consistence of the external crust, which, like a vault or arch,
acts in every direction against further contraction. Contraction ensues, and the fissures produced in consequence are subsequently filled up with crystalline deposits of substances held in
solution by liquids penetrating, or *already contained* within, the
interstices. At first magnesian carbonate of lime, then calcareous spar (occasionally also iron pyrites) are separated and
deposited. Certainly there seems to exist a great analogy
between the process of formation of such septariæ and that
admissible as going on within a large pulverulent globe freely
suspended in space. There is indeed no external pressure, but
every stratum of ponderable matter exercises compression on the
whole.

The following figure (fig. 4) is taken from Professor C. Koppe's
"Physik und Meteorologie" (in Bädeker's collective publication,
Die gesammten Naturwissenschaften). A point A, attracted at
the surface by a material point B as a sum of many others,
undergoes also attraction from another point D in a similar
situation. The resulting line of direction falls between B and
D, and passes through the centre C, along the line C E.

No determinate direction could prevail in the centre itself, where the mass
of the sphere is uniformly distributed,
and there the action of gravitation would
completely cease (or be in equilibrium).
As each particle on the surface tends to
sink towards the centre, it finds an obstacle from another immediately subja-
cent, this from a third, &c., and this obstacle must be overcome or removed.
The particles, at first unconnected, join
or approximate more and more slowly;
pressure is beginning and increasing. As on the surface of our
globe, so we may suppose to have existed in meteorites, combinations of heterogeneous elements very different from each other
in their specific gravities. Among other substances found in
meteorites are oxygen, sulphur, phosphorus, carbon, chromium,
silicium, hydrogen, cobalt, nickel, iron, aluminium, magnesium,
calcium, potassium,—all of them extremely discrepant in density
and other physical qualities. It is doubtful whether these existed as elementary particles, or in chemical combinations. In

Fig. 4.

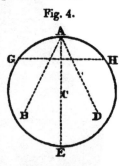

the present case, that first supposed may in reality have been precedent to the second mode or condition of existence.

Such a supposition may be considered more admissible than the views now more prevalent, that cosmical space possessed such an elevated temperature that the whole of matter existed in a gaseous state at the rate (as Vogt has calculated[*]) of only $\frac{15}{1,000,000}$ ths of a grain within the space of one cubic (German) mile. Such a supposition, however, lies far beyond us as regards experimental proof, even should we succeed, by connecting the past with the present, in producing correlations that might comparatively be considered "initial" ones. If the heavier metallic particles tend downwards along with others of less density which are pushed aside or even forced to ascend, while the whole surface pressing towards the centre is consequently continually diminishing in bulk, friction must unavoidably follow, and with it (as experience teaches) development of electricity and heat. We are, however, sufficiently acquainted with the phenomena attending the mutual combination of several among the above-named substances, as connected with combustion, oxidation, and chemical action in general, to enable us to pursue this part of our examination further.

On a former occasion (in "eine Leitform der Meteoriten," &c., Imp. Acad. Proceedings, vol. xl. p. 539) I mentioned an important communication from my respected colleague, Professor Schrötter, concerning the fact that substances whose mutual action under ordinary temperature goes on violently and with every appearance of intense combustion (as chlorine acting upon phosphorus, antimony, arsenic, or ammonia), when refrigerated to $-80°$ in a mixture of solid carbonic acid and ether (so that chlorine is liquified under ordinary barometric pressure), remain in a state of complete mutual indifference. Under these circumstances, a slight elevation of temperature, especially if care has not been taken to keep up a low temperature by rapid evaporation, may be the cause of dangerous explosions.

The same is the case with *alcohol* and chromic or chlorochromic acid, with *ammonia* and chloride of phosphorus, with iodine or bromine and phosphorus (see Professor Schrötter's *Die Chimie nach ihrem gegenwärtigen Zustande, &c.*, Vienna, 1847 vol. i. p. 129). Professor Dumas reported on this fact in the Paris Academy (*Comptes Rendus*, January 1845, No. 3, p. 193), remarking that he had not been able to observe a complete inactivity,—probably, as Professor Schrötter now objects, because, accelerated evaporation having not been duly provided for, the

[*] Nöggerath in "Geognosie und Geologie," in Bädeker's above-quoted 'Collective Publication.'

elevation of temperature in conducting the experiment took place too rapidly.

When chemical action has once commenced, a continuous increase of temperature easily takes place, till, beneath the uppermost dry and pulverulent surface still exposed to the intense cold of cosmical space, a crust or shell has been formed, within which the atoms of matter, following the influence of their own peculiar forces and properties, unite in chemical combinations, and individualize themselves into separate crystals whose elevated temperature (chemical action having ceased) effects more or less lithoid consistence.

The attempts to explain the central heat of the Earth by means of electrical and chemical action come near the views enounced by Sir Charles Lyell, Prof. De la Rive, &c. (see Naumann, *loc. cit.* p. 63), while the compressive action of the uppermost terrestrial strata, here taken for a point of departure, is quite adequate to the conditions required by an uninterrupted process of induction. The above-named mobility of particles once admitted, the frequent occurrence of globules in meteorites is no longer a matter of surprise. These globules, sometimes rather regularly rounded (as in some oolites), and in other cases angular or fragmentary (with edges occasionally rounded at the same time, however), are imbedded in an agglomeration of looser and frequently arenaceous particles, for which I have proposed the name of "meteoric tufa*." The surface of these globules is characteristically surrounded with particles of iron, as in the meteorites of Seres, Assam, Renazzo, Parnallee, and others. The meteorites presenting the aspect of crystalline rocks unmixed with native iron, as in those of Chassigny, Juvenas, Shalka, &c., stand far higher in the scale of development than even those most compact meteorites which include minute particles of metallic iron dispersed through an arenaceous, granular, or tufaceous aggregate of lithoid substance. The highest stage of development is exemplified by the pure and highly crystalline meteoric irons, partially resembling the contents of metalliferous veins (as in the Agram iron), and partly surrounded in all directions with smooth surfaces, a still unexplained circumstance even if superficial oxidation during their progress through the terrestrial atmosphere is taken into account. Instances of a vein-like disposition of metallic iron (as in the Macao meteoric stone), or of iron pyrites (in those of Pegu, Allahabad), as well as genuine planes of fissure (Stannern), rough (Allahabad), or specular (Ensisheim, Lixna, &c), exactly like those in our terrestrial rocks, are of no rare occurrence in many meteorites. The me-

* See Haidinger's paper, "Das von Herrn Dr. Auerbach entdeckte Meteoreisen von Tula," Imperial Academy, Meeting of November 29, 1860.

teoric iron of Tula, containing imbedded fragments of meteoric stone, discovered by Dr. Auerbach, proves beyond all doubt the occurrence of larger iron-masses in veins, and of their including fragments of the adjacent rocks*.

In his paper "On Meteorites in Meteorites" (Poggendorff's *Annalen*, 1860, vol. cxi. p. 353), Baron Reichenbach examines the mechanical composition of meteorites, paying particular attention to their rounded or angular particles, these last characterized as "fragments, broken and rolled pieces, and pebbles" (*loc. cit.* p. 384). Thirty-two meteorites (? stones), microscopically analysed, presented in their intimate or mechanical composition *five* distinct different substances, viz. sulphuret of iron (pyrites), native iron, oxidulated oxide of iron (magnetite), a grey, and a black substance†. Leaving aside some peculiarities in the terminology employed by Baron Reichenbach, as well as his criticism (p. 379) on the expression "secretion," stated to have been used by myself, while in fact I prefer the more neutral term "included substances," I could not give a better mode of considering in detail the structure of meteorites than has been rendered by Baron Reichenbach himself; and indeed the scientific world is obliged to him for it. There we have the character and nature of "meteoric tufa" pursued into their minutest details, indicating *successive* formation by the junction of the more intimate atoms of "cosmical dust,"—though, and this is the very foundation of either mode of consideration, this took place *not within the vaporous dust freely dispersed* through cosmical space, but within an *already* pre-existent and voluminous agglomeration, in which mutual attraction only became effective by producing real or absolute pressure. I really feel obliged to Baron Reichenbach for these statements, although undertaken with other intentions than to illustrate my own views on this matter.

The influence of solar heat has purposely been neglected in the preceding considerations, on account of the want of an atmosphere, in the strict sense of the term, in those spaces within which the formation of meteorites (in their *initial* condition or movement) may be admitted to take place. We know the temperature of planetary space to be far below that of the freezing-point, and we may assume an identical condition for the entire orbit of our globe, with a radius of 95,000,000 miles, as well as for the spaces beyond the orbit of Neptune (thirty times the distance of the Earth from the Sun); and even still further,

* See Haidinger's paper "On the Tula Meteoric Iron," *l. c.* note *ante.*

† These black and grey substances must refer to *stony particles.* I think Reichenbach's list might be extended so as to include magnetic pyrites (pyrrhotine), as well as a white substance.—R. P. G.

where probably more than one planet, and certainly comets, are pursuing their course, the solar distance of Neptune being itself only $\frac{1}{7000}$th of the interval between the Sun and the nearest fixed star*. During the period which the Earth takes to accomplish her annual revolution round the Sun, the latter, together with the whole solar system, has progressed (at the rate of about seven German miles a second) through a space of which the Earth's distance from the Sun is only the eleventh part†. Professor Koppe‡ says, "All circumstances agree in confirming the supposition that, for a period of 3300 years, the average temperature of Palestine has not undergone any notable change." During this period our globe has run in length some 36,300 times its own distance from the Sun—a course not to be achieved by light itself in less than 209 days, though this enormous distance is small indeed compared with the unlimited range of space itself!

Taking for granted that the weight of meteorites falling upon our earth's surface amounts yearly to 450,000 lbs. (Vienna weight), if not more§, and consequently to 450 millions. of pounds in a millennial period, Baron Reichenbach has brought under consideration the question, whether in the course of ages such an increase of ponderable matter would not be without notable influence on other as well as physical correlations connected with our globe in the solar system ‖. The length of such periods as are here taken into account is after all almost too enormous for our imagination to grasp : to accomplish the formation of a meteoric agglomeration equal in size to our globe would require 3000 trillions of years.

Another consideration, however, may here find appropriate notice. We may ask, if then our globe in the course of one solar revolution can thus admit of an increase of matter to the amount of 450,000 lbs., what would have ensued if it had followed a different path through space? Might not the increase have been nearly similar in amount in describing any orbit of equal length?

Mr. Greg's elaborate comparisons, indeed, prove meteoric falls to be less frequent at the time of perihelion than at the time of

---

* See Mädler's "Astronomie" in Bädeker's 'Collective Publication,' vol. iii. p. 595.

† See Mädler, *loc. cit.* p. 629. According to Arago and Herschel, the velocity of our sun in stellar space is only five English miles, or one German mile.—R. P. G.

‡ "Physik und Meteorologie," in Bädeker's 'Collective Publication,' vol. i. p. 169.

§ Probably too large an estimate by three-fourths. See Note at the end of this paper.—R. P. G.

‖ See Nöggerath's "Geologie und Geognosie," in Bädeker's 'Collective Publication,' p. 110.

aphelion. The Sun himself, however, as before shown, far from being stationary, is moving with considerable velocity through stellar space. While the Earth has received an increase of 450,000 lbs., it has completed its orbital movement round the Sun ($2r\pi$, r being the average distance of our globe from the Sun); and in the same space of time, by sharing in the progressive movement of the solar system as a whole, it has run through a space of $11r$. It may be sufficient to admit for the case in question, the approximative expression $r\sqrt{(121+4\pi_2)}$, and even $13r$ (instead of $12\cdot65r$). If we compare mutually the space (s) run through by the Earth, and the space (S) run through by the whole solar system in one year (even if we admit for s nearly double the diameter of Neptune's orbit—120 times the Earth's distance from the Sun,—and for the diameter of our globe itself, in round numbers say 2000 [German] miles, far exceeding it, its real value $=0\cdot0001r$), we obtain the following numbers:—

$$S : s = 120^2 \times 11 : 13 \times 0\cdot0001^2$$
$$= 14{,}400 \times 100{,}100{,}100 \times 11 : 13$$
$$= 1{,}440{,}000{,}000{,}000 \times 11 : 13$$
$$= 15{,}840{,}000{,}000 : 13$$
$$= 1{,}218{,}460{,}000{,}000 : 1.$$

The space-number of more than one billion, multiplied by 450,000 lbs. (the supposed yearly increase of our globe), gives in pounds the total weight or mass of meteoric matter existing and moving about in every direction within the space above assigned to our solar system. This sum, of over half a trillion of pounds, is, however, not very considerable when compared with the weight of our own globe, calculated to amount to $13\frac{1}{4}$ quadrillions of pounds[*]. If we suppose these 450,000 lbs. of meteorites to be united into one sphere, the diameter of this sphere would be to the diameter of our globe as 1 to $290\cdot8$.

The weight of the terrestrial globe is always to the total weight (450,000 lbs.) of meteorites moving in every direction within the space annually run through by our solar system as 24 millions are to unity. These are then the calculated results arising out of the above-mentioned supposition. A far greater proportion of solid matter distributed into small bodies would be obtained, if we were allowed to take into account the great number of meteors visible within our atmosphere in the shape of shooting-stars, and bolides that do not apparently deposit solid matter, and whose light is probably developed by compression of air, or, if not in every case by actual combustion, as supposed by Reichenbach, at least (as regards meteoric iron) after the manner of

[*] See Nöggerath's "Geologie und Geognosie," in Bädeker's 'Collective Publication,' p. 110.

"Callum's drops or globules." Professor T. H. Newton of Yale College, New Haven, Conn. U.S., says (*New York Tribune*, August 22, 1860), " it is calculated from perfectly reliable observation, that not less than 10 millions of meteors enter the atmosphere every day, and àre burnt up*."

This would then be 3650 millions per annum, which would again materially increase the total amount of meteoric matter contained within the above-mentioned space.   But is there not some probability that beyond our own system of fixed stars, all space is replete with such bodies, of which only a proportional *minimum*, and that transitorily, make up part of our own solar system ?   Not that all of them may be burned away or melted ; for the large 4 lb. stone of Segowlee, described by me, has its edges quite sharp and nearly unaltered, some of them being rounded only to a depth of not quite one-twelfth of an inch. Meteorites composed of less dense matter, while moving rapidly, may be frequently again repelled into space by the resistance of compressed air.   For this reason meteorites of earthy or carbonaceous-like consistence, as those of Bokkeveld, Alais, &c., are of particular importance, as well as rare.   Meteorites are far behind our terrestrial rocks with regard to diversity of mineralogical character.   The minerals composing granite, gneiss, micaschist, and others, representing the most solid basis of the terrestrial crust, are wanting in them ; and, to name a particularly important species, they are totally destitute of pure silica or quartz †.

* If, as I presume he does, Prof. Newton means that this represents the total number of meteors which enter our atmosphere daily, and that all these are, as a matter of course, consumed in it, I think he is mistaken, the number of meteors so consumed being in all probability limited to such only as burst or become dissipated in sparks.   These form but a small portion of those that apparently fly almost instantaneously through the upper strata of the air (at an average height of 65 miles, as recently proved by Prof. Secchi at Rome), and pass off again, perhaps tangentially, into interplanetary space.   Prof. Vaughan of Cincinnati, U.S., thinks it probable that the solid nucleus of an ordinary shooting-star is no bigger than a hailstone; and this is only analogous to what takes place with large aërolitic meteors.   There are cases of large and well-observed meteors, which after bursting, sometimes even with noise, into two or more parts at a height of 40 miles or so, have undoubtedly again passed into planetary space, —another proof also that the smaller shooting-stars may do the same.

It must not be forgotten in these rather speculative calculations, that the same groups of meteors may frequently repeat themselves, that is, return periodically without visible or material loss, and that, in fact, by far the greater number of meteors seen are doubtless those that are *periodical* and consequently belong to our own system.   Unless it can be otherwise proved, it would seem premature to suppose otherwise than that by far the larger portion of meteors and meteorites of all kinds belong to the solar system, and not to stellar space.—R. P. G.

† Since these lines were written, my highly respected friend Prof. G.

The subject of progressive changes in the world of meteorites, as hitherto exemplified by specimens within our knowledge, is thus for the present confined within narrow limits. Are these progressive changes of a character likely to terminate the proper existence of a celestial body by its definitive division into fragments? or is the possibility of such a "breaking up" justified by any precedent not opposed to the laws of nature as known to us? A few considerations made from this point of view may serve to supply a real *desideratum.*

I intend to sketch them here as briefly as possible, taking my point of departure from a *septaria,* whose constitution I have already explained. As in such a terrestrial concretion, so the outward crust of a cosmical body may become solid, presenting a stony appearance, under the centripetal pressure of gravitation, long before its interior has undergone a like degree of compression. I take our own globe as a point of comparison for data expressed in figures. Originally the particles of the solid terrestrial crust lying next each other may possibly have enjoyed a certain amount of mobility; this of course no longer exists. The maximum of pressure has its seat at a depth where the greater and more solid mass rests on the interior compressed by it in a descending direction. We are entitled to suppose this underlying mass is maintained by this very pressure in a state of incandescent fusion. Atmospheric pressure represented by the weight of a column of water 32 feet high, amounts to 1804·8 lbs. per square foot. A column of 10 feet average height of any substance whose specific gravity is $= 3·0$, acts nearly with the same degree of force. At the height of 1 German mile (24,000 feet) the pressure is $= 2400$ atmospheres; at 5 miles (25 miles English) (a measure generally adopted to express the solid terrestrial crust*) it would amount to no less than 12,000 atmospheres. A solid pressing on our globe with the weight of 1 lb. would in the Moon press only with $\frac{2}{13}$ lb., and if transported on to the Sun's surface with $28\frac{1}{4}$ lbs.† The pressure produced on

Rose of Berlin, has proved beyond doubt the occurrence of *quartz* in isolated crystals in the meteoric iron of Ziquipilco (Toluca).—W. H.

It may be here mentioned that Berzelius, Rammelsberg, and Dr. Laurence Smith have pointed out strongly and with much truth, the general resemblance that meteorites, in whole or part, not unfrequently bear to certain *volcanic* rocks. See Dr. Buchner's work, *Die Feuermeteore insbesondere Meteoriten,* &c., p. 175.—R. P. G.

* Important and more recent researches on the question of the thickness of the earth's crust, as conducted by Professor Hopkins and others in this country, may necessitate our raising M. Haidinger's estimate of only 25 miles of a solid crust to something like a *minimum* of 300 miles. This is more a question of degree, however, and does not materially affect M. Haidinger's line of argument.—R. P. G.

† Mädler, *loc. cit.* pp. 577 and 556.

our globe by a solid crust of 5 German miles thick, would require in the Moon a crust of $32\frac{1}{2}$ miles in thickness, and in the Sun a crust of only $\frac{3}{17}$ of a mile thick, or 4235 feet.

Original pressure takes place only so. long as a body is not completely solidified; from that moment perfect equilibrium is established within it; pressure, however, must again take place whenever change of temperature modifies the state of rigidity. A rigid body is always apt to conduct heat.   That kind of heat whose laws of increase we have to deal with while pursuing investigations on the central heat of the Earth, is *conducted* heat, transmitted from heat generated, or existing, at greater depths, as more immediately shown by volcanic eruptions.   In the regions where volcanic vents open in great numbers on the surface, "the smelted interior of our planet," as Humboldt emphatically says, "stands most in permanent communication with the atmosphere."   In our times this region is a zone between 75° W. and 125° E. long. of Paris, and 47° S. to 66° N. lat., running N.W. in the western portion of the Southern Ocean*.   It deserves consideration, that the whole continent of the Old World lies westward of this zone, separated from it towards the south-west only by the Indian Ocean, offering eastwards (as in the Southern Ocean itself) considerable "areas of subsidence†," and that towards the east of the Southern Ocean the American continent is again fringed with a series of. active volcanoes.   A remarkable connexion exists between these circumstances and the fact that the altitude of atmospheric strata at the same time rapidly decreases as we approach the antarctic pole,—just as though a mass of solid highlands had in those parts pressed on the interior of the globe at some early period of its existence, and the terrestrial crust had been broken, and its parts mutually dislocated into the general and more marked outlines now visible on its surface. Should the solidification of the crust proceed so far as to become stationary before the particles of primitive cosmical matter enclosed in it have completed their approximation, these might indeed commence a new and independent process of formation, giving rise to a second shell concentric with the first or external one, and enclosing another internal focus of volcanic activity, the primitive one having become meanwhile extinct.

If the sum of 65 miles, expressing the thickness of the Moon's crust taken double, is subtracted from her diameter $(=\frac{264}{1000}$ of the Earth's), and with a density of about 3·37‡,

---

* *Physical. und geognost. Erinnerungen. Reise der Novara um die Erde,* p. 20.

† See Darwin's 'Theory of the Formation of Coral Reefs.' Humboldt, *loc. cit.*

‡ Arago's 'Popular Astronomy,' translated by Hankel, vol. iv. p. 35.

`there would, at all events, remain an interior space of 403 miles, within which the formation of another such spherical shell might possibly proceed. Nevertheless it is not to be expected that further condensation out of the primitive molecular state should go on without some disturbance in a medium of such a temperature as prevails in planetary space. If contraction produces an actual internal vacuum, a *violent disruption* of the crust falls within the bounds of possibility. Admitting that a compensation of temperature by conduction or communication of heat to have already taken place, and supposing every solid shell to be hermetically sealed under a high temperature, an event quite opposite to the above-mentioned one might be expected with some degree of probability. Gases developed within this shell and brought to high tension, might indeed cause a violent explosion, exactly like that arising from ignited gunpowder enclosed within a hollow projectile.

What is the actual cause of the densities of the planetary bodies within our solar system being so different from each other? Does it merely arise from the natural correlation of the elements composing them, as in our globe, or from a progressive development in the earlier stages of their existence? The densities of these bodies are expressed by the following numbers:—Mercury, 6·71; Earth, 5·44; Mars, 5·15; Venus, 5·02; the Moon, 3·37; Sun, 1·37; Jupiter, 1·29; Neptune, 1·2; Uranus, 0·98; Saturn, 1·75.

Olbers is known to have first enounced the hypothesis that the minor planets Ceres and Pallas, discovered by Piazzi and himself, were probably mere fragments of a pre-existing and larger planet. After the discovery of Juno and Vesta, Lagrange* investigated the intensity of an explosive force sufficient to rend a planet into pieces, in order to permit a fragment of it to become a comet, or, to use a more accurate expression, move in an orbit similar to a comet. He found that an impulsion equal to the velocity of a cannon-ball multiplied by 12–15, that is 16,800–21,000 feet per second (the velocity of a cannon-ball being 1400 feet a second, and equal to that of a point at the equator in its diurnal rotation), would be sufficient to throw the fragments of a planet (the radius of its orbit being supposed to be equal to the distance of our globe from the Sun multiplied by 100) into progressive or retrogressive, elliptical or parabolical comet-orbits—the greater number of them even into hyperbolical ones, so that, after their first perihelion, they would disappear for ever from our system†.

* "Sur l'Origine des Comètes." Lu au Bureau des Longitudes, le 29 Janvier, 1812.—*Connaissance des Tems, &c.* pour l'an 1814, Avril 1812, p. 211.
† Baron Reichenbach has expressed an opinion that meteorites may

Certainly there is great difficulty in forming an idea where and how fragments of genuine solid rocks (as meteorites undoubtedly are) could be first violently broken from their parent repository and then hurled into distant solar systems; nevertheless their characteristic fragmentary form, together with the cosmical velocity of their course, leaves no room for any other solution. So daring a supposition, paying, however, due attention to Nature's laws as far as they are known to us, must, however, from time to time, provoke reiterated criticism.

I thought it desirable therefore to give here a short conspectus of such views concerning meteoric phenomena as have from time to time crossed my mind, or been the subject of distinct communications to the Academy, though at the same time I freely admit I may have been intruding into a region of natural science for the investigation of which I am but very imperfectly prepared. I must ask for some indulgence in this attempt to trace the outlines of views in some way different from current ones—the more so since they are intended to establish merely a kind of programme for more accurate investigations.

In an earlier period of development in human society, the "*nonum prematur in annum*" may have been more easily obeyed than it is in our times. Accelerated publication, however, has also its advantages, as contemporaneous investigators familiarized with the matter find in it a point of comparison for their own either analogous or contradictory views. For myself, some portion or other of my views have been more than once the subject of conversation and epistolary intercourse.

At least I hope I may have been successful in my endeavours to follow the strict rules of scientific induction for arriving at the result aimed at in this paper. During the whole course of these considerations I have made it my duty implicitly to obey the precept of our great master, Humboldt, that, "even within merely conjectural regions, uncontroled or arbitrary opinions, independent of induction, should never be allowed to prevail."

——————— —

Note.—Baron Reichenbach's estimate of an annual meteoric deposit on the surface of our globe, amounting to 450,000 (Vienna) pounds, is certainly considerably over the mark. In his paper on this

originally have been condensed from comet-dust; that this is quite contrary to M. Haidinger's opinion, I have good reason to believe. It may perhaps appear a little difficult to believe that, were any small planet or satellite of a planet to burst, some of the fragments would for ever be hurled beyond the influence of the sun,—though, as in the case possibly of the sixty asteroids, the original orbital conditions of the parent mass might become a good deal modified.—R. P. G.

subject (see Poggendorff's *Annalen*, vol. cv. p. 554 *et seq.*) he calcu-. lates there are 4500 meteoric falls per annum, averaging 100 lbs. per fall in weight. Assuming, as we perhaps may do, that he has not materially over-estimated the weight of each fall, he has certainly exaggerated their annual number. Supposing, in the first place, as I believe we may, that detonating meteors are equally aërolitic, whether stones are picked up or not, since most meteoric stones have resulted from a detonating meteor in the first instance, then for the last sixty years, over an area of 900,000 square miles, comprising the countries of Great Britain and Ireland, France, Germany (inclu- sive of Austria, Prussia, Hungary, &c.), and Italy, we find recorded (see my "Catalogue of Meteorites and Meteors" in the last volume of the British Association Report for the Oxford Meeting, p. 48) *sixty-nine* actual stone-falls, and *seventy-two* meteors accompanied with detonations from which no material *residuum* was obtained; say in all 144 cases of aërolitic phenomena. That is about $2\frac{1}{4}$ recorded instances per annum for an area of 900,000 square miles; and taking the superficial area of the whole globe at 197,000,000 square miles, we obtain rather over *five hundred* falls (511) as the number likely to be observed, were all the world covered by land and peopled in like manner by Europeans.

Now, what proportion this number would bear to those that abso- lutely do fall annually, but which are never noticed or not recorded in scientific works, it is not very easy to say; but from various reasons it may be fairly estimated at more than half of the entire number; Chladni and Humboldt have estimated the total number at 700.

There are several reasons for inducing us to increase the annual number of observed and recorded falls, viz. 500, to 800 or 900, as the *actual* number that fall, and not more. First, the fact that one-half the human race are supposed to be asleep or in their houses nearly twelve hours out of the twenty-four, must tend to limit considerably the number of observations; on the other hand, we are not without instances of stone-falls and other aërolitic phenomena, detonating meteors more especially, occurring during the night-time; while again, as I have shown in the tabulated results of my large Catalogue, p. 118, the greatest number of stone-falls seem to occur in the afternoon about 4 o'clock, not only as against falls taking place during the night, but as compared even with the corresponding hours in the forenoon, equally favourable as a time for such observations. Though stones have not frequently been picked up during the night-time, we may bear in mind that the night is a most favourable time for seeing large and brilliant aërolitic meteors, and that the darkness does not prevent us from hearing the violent detonations usually accompany- ing the explosion of an aërolitic meteor. Then, again, it is not unusual for an aërolitic meteor to pass overhead some hundreds of miles, and for the detonation to be heard over from twenty to forty miles square; and some persons would probably notice one or the other. Now as I have included as *aërolitic*, meteors from which no stones have been picked up, it will I think be admitted that to double the entire number of both classes actually recorded

in catalogues drawn from every available source is a reasonable estimate, especially when based on observations made in civilized and
densely peopled countries like England, France, and Germany. How
few persons, I may also add, are there who have ever in their life-
time either seen a meteoric stone fall, or even heard the always violent detonation of an exploding aërolitic meteor. There seems to
have been of the latter only three instances recorded as observed in
England during the last ten years, and no well-authenticated instance
of a meteoric-stone fall since 1835, and that a single one of about
2 lbs. in weight! To know whether a meteoric stone has fallen, it
is then not exactly necessary to calculate the proportions of waste or
forest ground, &c. that exists even in Europe, as Baron Reichenbach
argues, in order to arrive at the number of stones *not* picked up, if
we assume that aërolitic detonating meteors are seen and heard as a
rule over very large areas, and count as actual falls in our calculations. So striking indeed are the phenomena usually attending the
fall and appearance of aërolites and aërolitic meteors, that I much
question whether fully *two-thirds* of the real number would not certainly be recorded in the daily or scientific journals, say of England
and France. Instead, therefore, of placing the total weight of meteoric matter annually deposited on the earth as high as 450,000 lbs.,
as calculated by Baron Reichenbach, I am inclined to estimate it
at probably less than 100,000 lbs. This is, however, more a question of degree, and does not vitally affect the ulterior argument involved in the problem proposed by Baron Reichenbach and M. Haidinger—a problem not without importance and interest, though
somewhat speculative.—R.P.G.

The reader is requested to correct the following errata in the first portion of this paper:—

　Page 353, line 10 from top, *for* 24,000 *read* 24,000 × 7.
　—　— line 11 from top, *for* 124·4 *read* 1247.
　—　— line 14 from top, *for* 4080·32 *read* 40901·6.

LVIII. *A Sketch of M. Faye's "Examen d'un Mémoire de M.
Plana sur la force répulsive et le milieu résistant," with a few
Remarks thereon.* By HENRY S. BOASE, M.D., F.R.S. & G.S.*

THE concluding summary of M. Faye's memoir, published in
the 'Illustrated London News' for September 7 under the
head of "Scientific News," attracted my attention, and excited
a strong desire to see the memoir itself; for its statements concerning the duality and universality of the forces attraction and
repulsion, in celestial phenomena, seemed to indicate views very
similar to those advanced in my 'Philosophy of Nature.' Having

* Communicated by the Author.

since read this memoir in the August Number of the *Comptes Rendus* of the French Academy, I found that there was not the slightest foundation for the supposed similarity of opinions; but, its contents are so interesting, that I was induced to make copious extracts therefrom: and I am inclined to think that they will be generally acceptable; for even those most averse to speculations cannot refuse their attention to the opinions of such an illustrious man.

. M. Faye commences his memoir by briefly and clearly stating the points of the thesis which he proposes to prove, which are,— 1st, that the hypothesis of a resisting medium, as formulated by geometers, is unacceptable; 2nd, that if it be corrected so as to render it more rational, it becomes too indefinite for directing analysis; 3rd, that the theory of the repulsive force is the only one that can be scientifically constituted.

The hypothesis of a resisting medium, says M. Faye, applies very well as far as the motions of the periodical comets of three and seven years; it gives a clear and precise idea of the nature of their acceleration. According to M. Plana, this theory gives $29''\cdot5$ for the variation of the second comet's eccentricity, a result very near $34''\cdot6$ assigned by observation. Since M. Plana, M. Axel Möller advances a step further by introducing, after the suggestion of M. Valz, the variation of volume which a comet, supposing it compressible and not permeable to the surrounding medium, ought to experience when it penetrates into the gradually denser strata of this medium. The diminution of eccentricity is then equal to $32''$, that is, almost identical with the result of observation. But could we approach still nearer to $34''\cdot6$, the value of which is not indeed definitively fixed, I would still not the less persist in regarding these formulæ and calculations as purely empirical, inasmuch as it has not yet been proved that a ponderable medium, elastic or not, can exist around the sun without revolving round it.

. This idea of a heavy (or gravitating) and immoveable medium, says M. Faye, is no novelty; it may be traced back to the *materia cœlorum* of the ancients, which was supposed to fill the world after the manner of the extension of the atmosphere of a central body; and since the time of Newton, its existence has only been upheld for the purpose of conserving the conception of gravitation, as the one governing force of the universe. But this singular hypothesis ought to have vanished when Laplace made known the definite limits which mechanics imposes on the atmospheres of celestial bodies.

In vain we suppose, in order to evade the objection, that this medium is imponderable; for then we must have recourse to the æther of physicists. But in this case we must no longer attribute to the medium a density proportional to $\frac{1}{r^{y}}$, because in ceasing to gravitate towards the sun, the beds of this medium will cease to mutually press on one another in that direction; it therefore becomes requisite to attribute to the medium a constant density in space. Unfortunately, in this case, the beautiful agreement above alluded to between calculation and observation will disappear; for instead of $34''\cdot6$, the formula of M. Plana for this hypothesis will not give more than $14''$; and, moreover, there still remains to apply to this medium a velocity of translation equal and contrary to that which transports the masses of the solar system in indefinite space.

It is a long time ago, says M. Faye, since I first advanced this objection: no one has ever been able to answer it; and, notwithstanding, the hypothesis of an immoveable medium is persisted in. Can it be then that in itself the immobility or the movement of the medium is a thing of no consequence? This is what we are going to see.

Let us accept, then, continues M. Faye, the very different hypothesis of a medium revolving around the sun. It is then evident that such a medium can only resist the motion of comets in virtue of the excess of velocity. This excess, positive at perihelion, becomes negative at aphelion; consequently if the medium resists in the one case, it will push in the other. This circumstance alone is sufficient to show that the analysis of this new problem cannot be identified with that of the first. It indeed involves more than this; for what then will become of the law of density? No one can tell. When the medium was regarded as immoveable, it was admitted that its beds, in gravitating the one on the other and on the sun, would be mutually compressed in such a manner that the density would progressively increase towards the interior, according to the law approximately represented by $\frac{1}{r^{y}}$. But when the medium revolves, it ceases to gravitate, not only towards, but on the sun; its beds cease to press upon each other and the law of density becomes a perfectly indefinite problem.

M. Faye then gives some analyses for the periodical comet of seven years which bears his name; the result of which is $2' 40''$, instead of $34''\cdot6$ as given by observation. From this enormous discordance, he says, it must be concluded that a continuous ring of constant density is inadmissible. It is necessary, therefore, that the density of the medium should vary according to a peculiar law. Thus Encke's comet requires that this density should go on rapidly decreasing outside the orbit of Mercury, within which orbit the perihelion of this comet occurs. Again, the comet of seven years likewise requires, no less imperiously, that the density of such a ring should be well marked in the region of the orbit of Mars, rapidly decreasing in such a manner as to become imperceptible on approaching the orbit of Jupiter; for it is between these orbits that the motions of this remarkable comet are accomplished. These conditions can only be reconciled by adopting for the resisting medium a series of cosmical rings more or less resembling those of Saturn, but separated from one another by great intervals.

Such, in the opinion of M. Faye, is the only form under which the hypothesis of a resisting medium can be hereafter maintained. But he adds that nothing can be more indefinite than such a hypothesis; for the number of such rings, their respective limits, and the law of their interior density are completely arbitrary. It is impossible, for example, to extract from it any relation between δn, or $\delta \zeta$ and $\delta \phi$, on which, however, all the memoir of M. Plana proceeds.

In this manner, says M. Faye, the two first parts of his thesis have been justified: it has been shown that an immoveable medium is impossible; that a revolving medium is an indefinite hypothesis with which analysis can have no concern; and that a series of cosmical rings is so fanciful that such a hypothesis must be ranked with the transparent crystalline heavens and with the Cartesian vortices. And he lastly proceeds to the consideration of his theory of a repulsive force.

The several successive returns of the comet of three years have taught us that the duration of its revolution is constantly diminishing, whilst the other elements of its orbit remain unaltered. This is a most important fact; and Encke, the author of this great discovery, has concluded therefrom the existence of a force which is constantly opposed to the motion of the comet, and which therefore results in the comet's successive acceleration, without affecting its other elements, save the eccentricity. But this repulsive tangential force, is it real or apparent? If it be a real force, it may be asked what such a force can be which is able to contend in the heavens with gravitation, and thus to break the unity of astronomical science. Encke has declared for an apparent force, originating in the resistance of a medium; it was, it may be supposed, merely for the conservation of this threatened unity. It is doubtless an enlightened view, but arbitrary; for the unity of force is not, it is presumed, a scientific dogma. Let us then reserve our judgment concerning the nature of this repulsive force.

In studying, says M. Faye, the astonishing forms which comets present, their gigantic tails, the matter which they seem to dart towards the sun; but which soon returns to be confounded with the tail, &c., all the world will naturally say that such things come to pass because the sun exercises a repulsive action upon the atmosphere of the comets. Some will have it that this repulsive action is due to electricity, and others to magnetism, without once reflecting that these forces, so precise when applied to terrestrial phenomena, are vague and little understood in relation to celestial bodies. Others, again, have spoken of an apparent repulsion; it was the idea of Hooke and of Newton. Bessel, after a very profound study of certain phenomena, which he has, however, too much generalized, can only see in these cometary forms the effects of polar forces analogous to magnetism. But to form an opinion concerning the nature of such a force, one single order of facts is not sufficient; for it entails the necessity of guessing. We therefore again ask, what is this repulsive force?

Such was the state of the case, M. Faye says, when he took up the question:—on one side a repulsive tangential force, indicated by the motions of comets; on the other a radial repulsive force, indicated by their tails: on one hand, Encke with the ancient hypothesis of a resisting medium for the explanation of the former force; on the other, Bessel with his polar forces to account for the latter. A discussion, short but memorable, took place between these great astronomers. Bessel, who did not believe more than myself in a resisting medium, referred every thing to his radial forces; Encke, on the other hand, pointed out to him that this was impossible. So it would seem that the resisting medium adopted by the one is a physical impossibility; the play of forces, imagined by the other, in consideration of a single fact arbitrarily generalized, is still more inadmissible.

These two forces, real or apparent, are both repulsive; can they be reduced to a single force? If any celestial body exercise this singular action; it can be no other than the sun itself. But can it be imagined that a force emanating from the sun can act on any body in any other direction than that of the radius vector? Yes, certainly, the mechanician answers, if the body moves from right to left, and if the force is not instantaneously propagated like that of gravitation, but with an enormous velocity, indicated by the disproportion of the composing forces. So, then, the repulsive force exercised by the sun and endowed with a successive propagation, after the manner of luminous and calorific radiations, will furnish the two composing forces, the one radial and the other tangential, which are required to explain both the forms and the motions of comets. In studying the radial force under this point of view, it will be readily seen that this ought to be a force independent of the mass, and proportional to the extent of surface. The tangential force of this composition leads to precisely the same conclusions. The sun alone exercises it: it does not depend on the sun's mass, which is not operative in this case; it is probably only caused by the incandescence of the sun's surface; for this it is that distinguishes the sun from the planets, the vicinity of which does not affect the figure of the comets.

Such is an idea of the repulsive-force hypothesis, so far developed; a few steps more, and its astronomical formula may be attained. A repulsive force operating at all distances, but evidently becoming more feeble, and that rapidly as the distance increases; emanating from an incandescent polar surface; propagated with a velocity equal to that of radiant caloric; in the ratio of the surface and not of the mass; and pressing on the surfaces which it repels, in place of acting through all matter like gravitation:—Is there such a force in Nature?

If such a force exists, says M. Faye, it is probably the same repulsive force which is manifested in all material bodies under the names of dilatation, of expansion, of elasticity, &c. As in the case of the assumed astronomical force, the physical force which presides over these phenomena is due to heat; like it, its action is repulsive, does not extend through bodies, decreases rapidly with the distance, and relates to the surface and not to the mass; but there is a single difference between them: the physical, unlike the astronomical force, cannot act at a distance; so say many physicists; beyond molecular intervals it is imperceptible. But this, M. Faye, in common with Fresnel and others, regards only as an *à priori* opinion, not justified by experimental research: it ought to be proved that calorific repulsion, which acts from molecule to molecule in solids, liquids, and gases, that is to say, at intervals successively increasing at an enormous ratio, does not suddenly vanish at certain particular distances.

M. Faye says that for nearly three years he has worked on this subject; and in order to remind the Academy, he rapidly recounts the results up to the present time, viz. that his theory accounts for all the observed phenomena concerning the acceleration of cometary motions. Hereafter account must be taken of the variation of volume in the nucleus of comets, which doubtless will present some difficulties. He has also shown that, varied as the figures of comets are (which are complicated by the effects of perspective), they possess certain features in common—features which in their unity constitute in some measure a normal figure, which can be separated from accidental differences. And he thinks that the coexistence of several tails may be simply explained by the presence of substances having different specific gravities in the atmospheres of comets : the repulsive action of the sun, operating on them after the manner of the metallurgical washings of ores, would arrange them in trains more or less curved in the rear of the general motion according to the greater density of these substances. The repulsive force exercised by the sun on substances reduced to a great tenuity in cometary nebulosities, also explains the most general and important facts concerning the curvature of their tails, of their multiplicity, and of their form, not conical, as Arago supposed, but flat and fully displayed in the plane of their orbit, &c. The difficulty which so long attended the study of their heads and their atmospheres has disappeared as by enchantment;

· In conducting his experiments in verification of the repulsive force, M. Faye was guided by the memorable experiment of Cavendish on the mutual attraction between two solid bodies; in which the scientific world, he says, did not see so much the indispensable confirmation of the Newtonian theory, as an ingenious method of ascertaining the density of the earth. In experimenting, however, on the repulsive force, the question assumes a different aspect, as the earth *no longer* exercises this force in a sensible manner ; for at the present day the sun alone possesses this property, since it alone, in our little world, has preserved its primeval incandescence. Nor can we experiment on the solar action, because its repulsive force does not come within our reach ; it is dissipated on the superior beds of the atmosphere, in consequence of its incapacity of acting through all matter after the manner of gravitation. So we are reduced to study the feeble forces which can be produced by artificial means. In experimenting on the attraction between two bodies, the influence of the air and of its currents is a great obstacle to success ; and the same influence is still more obstructive in the case of the repulsive force, since one of the bodies employed requires to be in a state of incandescence. This is the reason why all the attempts have miscarried up to the present time. He then refers to the arrangements which he adopted for surmounting these difficulties, and announces a new series of experiments in hand, the incandescence being produced by the voltaic current in a vacuum rendered more perfect by chemical action; and he hopes to be able, not only to demonstrate repulsion at a distance, but also to measure it.

Lastly, M. Faye thus sums up the conclusions at which he has arrived by his labours. The celestial world does not obey one force alone, attraction, but a duality of forces, attraction and repulsion. The former depends solely on the mass, the latter on the surface and heat. The one is propagated instantaneously, the other successively. The one acts through all matter

without suffering any diminution, the other is intercepted even
by a simple screen. Both these forces are universal; for they
occur everywhere throughout the system of Nature.

Having at length finished the sketch of M. Faye's memoir, it
is now proposed to make a few remarks thereon: and these
remarks will principally apply to two points—an argument in
favour of the existence of a resisting medium, the æther of space,
and some objections to the proposed repulsive force.

But before proceeding, it may not be out of place to notice that
the appearance of speculations from such an illustrious man, and
in such a distinguished publication, is another notable sign of
the times, and proves that there is a growing desire in the
scientific world to emancipate itself from that thraldom to the
accumulation of facts which in the first instance is wise, if only
the means to an end, but which, if dogmatically adhered to as
intrinsically valuable in itself, can never enlarge the field of
science. It is like the folly of the miser who accumulates money,
but never applies it to its legitimate use. Facts we must have,
they are the raw material of science; but a good hypothesis is
of more value than a multitude of similar facts. To the same
effect M. Faye observes that his hypothesis has already proved
to be fruitful, and that it will in the hands of practical men
render much greater services; for to observe well, a good theory
is almost as necessary as a good telescope. Acting on this principle, how much Laplace has accomplished in France; and in
this country Newton has done more for science than any other
man by his theory of gravitation, which was not attained by
personal industry in accumulating facts, but by pondering on
known facts till utilized by generalization; for, as he says in
writing to the astronomer Flamsteed, "all the world knows that
I make no observations myself."

It would seem that the great difficulty in accepting the existence of a resisting medium depends on the discrepancy between
the calculated and the observed amount of the variation in the
eccentricity of Encke and Faye's comets: but are the data of
these calculations so perfectly reliable, that the failure in the
analysis can subvert the established opinion concerning the
nature of the æther of space? If it be ignored, what becomes of
the beautiful undulatory theory which explains the progression
of light and radiant heat? It is difficult to conceive that any
physical force, including gravitating force, can be transferred
from one part of the solar system to another without a successive
action and reaction on intermediate parts. In times past the
actio in distans was accepted because the phenomenon viewed
only superficially presented this appearance; and to revert to
such a doctrine, unless supported by indisputable facts, can be

no other than a retrograde movement in science. If any force can be said to act at a distance, it surely may be supposed to be that of electricity, for it has all the outward appearance of such an action; but Faraday's elaborate and searching inquiry into all the obscure conditions of the case has established beyond dispute that there is a communication of this force from molecule to molecule by the process of induction. M. Faye need not search after a perfect vacuum in which to perform experiments for proving the action of repulsive force at a distance; for the sodium he employs in removing the residue of oxygen after mechanical exhaustion will fill it with highly elastic vapour, the presence of which may be demonstrated by an electric discharge; and indeed the very electric incandescence will also fill the vessel with the vapour of the metallic poles, to say nothing of the æther of space, for this is the point in dispute; but surely if that æther be the means of transmitting radiant heat from the sun to all its planets, it must also be sufficient within an exhausted vessel. It does not, however, very clearly appear that M. Faye directly disputes luminous and calorific radiations; for he says that his radial repulsive force is exercised in the same manner, and that it is in all its properties identical with physical heat. If it be that which radiates from incandescent surfaces, it can be no other than radiant heat; it only remains to decide its mode of operation through space and other diathermanous media. Newton was decidedly opposed to the *actio in distans*; for in his third letter to Bentley he observes, " That one body may act upon another at a distance through a *vacuum* without the mediation of anything else, by and through which their action and force may be conveyed from one to another, is to me a great absurdity."

In the pursuit of science, the only legitimate method is to proceed from the known to the unknown. We know that in the case of a series of suspended ivory balls, the ball at one extremity will be put in motion by raising and letting fall the ball at the other extremity: here we have a visible and tangible medium through which force is transmitted, and by removal of which the extreme balls cannot affect each other. So in the electric telegraph, we know that force applied at one end of the conducting wire will produce motion at the other, but its transmission depends on the presence of the intervening medium. When, therefore, in experiments on radiant heat, we find that this force will affect bodies at a distance from its source, even through diathermanous screens, it is a legitimate inference that this must also take place through an appropriate medium, even though it be not tangible or visible. Such a medium has been acknowledged as adequate to the transmission of light by the vibrations of its atoms; and if so, it is capable of performing the same office for radiant heat, which always accompanies light in the sunbeam. Surely the admission of the universal presence of the æther of space as a medium for the transmission of physical forces, supported as it is by many facts that such forces are so propagated, is preferable to the bare assumption that, because celestial attraction and repulsion *appear* to act at a distance, therefore it is a fact that they do so act. It is clearly beginning at the wrong end: this question relates to the simple fact, the *an sit* of Aristotle; and in such a case no assumption is justifiable.

M. Faye's analyses relating to the acceleration of comets, seem to have rested on two data concerning a resisting medium: in the one case the resisting medium is regarded as gravitating but immoveable; in the other case as an imponderable (or immaterial) and revolving medium.

It does not appear in this memoir why such a distinction was adopted; and it is not easy to imagine the grounds for such conceptions concerning the æther of space. If it be a gravitating material fluid, it must necessarily partake of the common motion of the entire solar system; and if so, it remains to revise the data of the calculations, the resisting medium not being stationary, but spirally revolving around the sun with a velocity progressively increasing from the confines to the centre of the solar system. "When the medium revolves," says M. Faye, "it ceases to gravitate towards the sun, and its layers also cease to press each other according to the law of density." This statement is incomprehensible: atmospheres revolve with their respective central bodies, as æther with the sun, and yet their layers press on each other. If the medium possess any weight, however inappreciable, it cannot lose this by revolving; and if it be imponderable or immaterial, as the older physicists considered the æther to be, it could not revolve; for motion depends on a composition of forces essentially material or physical, that is, of attraction and repulsion, to the former of which the weight of bodies is due.

From these considerations it is evident that the notion concerning the nature of æther held by our neighbours is very different from that which we entertain. Grove thinks (and most of us agree with him) that æther is a highly elastic fluid having weight, though the amount of it is beyond the reach of determination, its excessive tenuity rendering it inappreciable. He regards this matter as the rarefied extensions of the atmospheres of celestial bodies. In my 'Philosophy of Nature' a different view is advanced; but we both agree in its being a material elastic fluid; and if so, it must follow the common law of fluids, and be more and more condensed in the vicinity of every celestial mass, and more particularly around the sun, which in mass transcends all the others. In calculating the amount of this density as progressively increasing towards the centre of the system, the intense heat of the sun need not be regarded as a modifying force; for its rays can in no way affect the æther, since it is perfectly diathermanous.

The grand objection to the æther as a resisting medium adequate to the explanation of cometary acceleration, is the statement of M. Faye, that each comet would require a distinct zone or ring in space, varying in density not regularly in the ratio of the sun's distance, but sometimes inversely, as in the case of some of Saturn's rings. Should this be established as a fact, some other disturbing influences must be sought for besides æther as a resisting medium; but still this would not annul the existence of æther, for it would ever remain a *datum* as a retarding influence *quantum valeat*.

In such a wide field for speculation, it is not difficult to imagine that such rings may exist in space as cosmical or nebular matter of extreme tenuity and yet really ponderous as compared with the æther in which they are situated. Such matter may be the residue of the solar zones, from which, according to Laplace's theory, each planet was evolved; or, a new form assumed by comets, which by the successive shortening of the major axis of their orbits, have been reduced to their mean distance from the sun, at which place the body is still very voluminous, and would be still more rarefied if reduced to the state of a ring revolving around the sun as in the case of Saturn.

Before concluding, a few words may be said concerning the operation of a repulsive force emanating from the sun, and which is said to be sufficient for the explanation of all the phenomena of comets, including the acceleration of the periodical comets: yet in truth we know but little concerning these phenomena, nothing concerning their rotation, although, as revolving bodies, they doubtless do rotate.

The radiant heat issuing from the incandescent surface of the sun is certainly the *vera causa* of many calorific phenomena; but that it can accelerate directly or indirectly the motion of a comet does not seem probable. A comet falling towards the sun with enormously increasing velocity, like the return of a rocket to the earth, will *pari passu* contract in volume, and consequently increase in density, in consequence of the increased influence of gravitation: when it has gained its perihelion, like a vibrating pendulum, it will be carried by its momentum from the sun with gradually diminishing velocity and increasing volume until it attains to its aphelion. If the comet, in its approach to the sun, passes through a resisting medium, especially if revolving in a direction in any way opposite to its motion, the friction must be tantamount to an increase of gravity, which, by reducing the orbit, must increase the centrifugal force, thereby accelerating its motion. This cause alone may not account for the entire alteration; but the subject is as yet in its infancy, and cannot be fairly condemned in the summary manner in which M. Faye has dealt with it. He gives us, it is true, another hypothesis in its place, but it does not seem to be in any way superior to that of Encke. We know the effect which radiant heat has on bodies when it impinges on them: if the surface be reflecting, accordto its degree the rays of heat are diverted; if it be absorbing, the heat assumes another phase and operates as an expansive force; if it be diathermanous, it passes freely through and renews its course on the opposite side of the body. In the case of comets, radiant heat can only be supposed to act in the two last-mentioned ways: in the last, in respect to a comet's motion, it must be perfectly inoperative; in the second, a certain amount of heat would be absorbed and enlarge the volume of the comet, thereby *pro tanto* counteracting the condensing effect of the sun's gravitating pressure.

M. Faye would seem to imply that the repulsive force of the sun is arrested by the surface of the comet, and thus is enabled to forcibly act on it as by impact; and agreeably to this notion he asserts, what is contrary to our daily experience, that "the repulsive force of the sun does not come within our reach; that it is dissipated on the superior beds of the atmosphere, in consequence of its incapacity of acting through all matter after the manner of gravitation." If it could be supposed that comets are enveloped by an impervious skin, as M. Valz suggested, and that this, moreover, had a good reflecting surface like some polished metals, then the impinging and reflecting of an enormous multitude of the sun's rays might by their aggregate force produce a sensible motion. But of what avail are suggestions of this kind: hypotheses arrived at inductively from facts may be tolerated, though they may prove to be invalid; but it is perfectly illogical to assume a fact, and then make it the subject of an argument.

It may be that M. Faye's repulsive force emanating from the sun may be well conceived, but imperfectly apprehended: it may prove to be akin to centrifugal force, which is the correlative of gravity, as set forth in the 'Philosophy of Nature.' But then such a mode of repulsion is quite different from that other mode of repulsion, which, as radiant heat, not only emanates from *incandescent*, but also from all *hot* bodies, and which is transmitted from body to body by the vibrations of ethereal atoms. Each mode of the physical forces can only be manifested by their phenomena, which are various kinds of motion : of these the principal motions seem to be radiations by undulations, and circuits by polarization. When the former relate to material molecules, the phenomena are best known as sound; when they relate to the medium æther, they become luminous and calorific phenomena : and so likewise the polar actions of molecules are known as electrical and magnetic phenomena ; those of æther, as the tangential currents due to gravitating and centrifugal forces. And should the same ratio of velocity as occurs between sound and light hold good between electricity and centrifugal force, and between magnetism and gravitation, the cosmical forces gravity and centrifugy must evidently be instantaneous in their operations, and not successive by undulations like radiant heat and actinism.

Claverhouse, near Dundee,
   October 1861.

LIX. *Observations on Lunar Radiation.*
By Professor TYNDALL, F.R.S.*

I HAD hoped, before the appearance of the present Number of this Magazine, to be able to prosecute the observations on Lunar Radiation referred to in my letter to Sir John Herschel to a definite issue; but I am so closely occupied with inquiries of another kind, that I must for the present content myself with recording the observations on which the remarks contained in the letter referred to were founded.

My place of observation was the roof of the Royal Institution in Albemarle Street, where I had a platform erected, sufficiently high to enable me to sweep a large portion of the heavens with my thermo-electric pile, without impediment from the chimney-pots. Wires were carried from the pile to an excellent galvanometer placed in the laboratory, the floor of which was about seventy-two feet below the platform.

On directing the axis of the pile towards the heavens, the chilling produced by radiation from its exposed face was so considerable†, and the consequent galvanometric deflection so great, that it was quite hopeless to operate on the needle in this position. To move it a single degree would have required many hundred times the quantity of heat or cold necessary to urge it through one of the lower degrees of the galvanometric scale; I therefore operated as follows :—

* Communicated by the Author.
† I intend to make this mixed action of our atmosphere and stellar space the subject of a special investigation. At midday also the refrigeration of the zenith is very great.

The galvanometer was a differential one; that is to say, two wires ran side by side round the astatic needle of the instrument. The ends of one of these wires were connected with the pile on the roof, the ends of the second wire were connected with a second pile, which was turned towards a vessel kept at a constant temperature by boiling water. The direction of the current caused by the heat below was opposed to that generated by the cold above; one of them in a great measure neutralized the other, and the needle was thus compelled to take up its place among the lower degrees of the scale.

I then ascended to the roof, fixed my pile at the proper angle, and directed it *off* the moon; I descended and observed the galvanometer; the needle oscillated between 10° and 20°, its mean position being therefore 15°.

I reascended and turned the pile *on* the moon; on descending I found the needle oscillating between 35° and 45°, the mean position being 40°.

The ascending and descending was repeated six times, and the following results were obtained:—

Mean deflection.	
Off the moon.	On the moon.
15	40
27	40
33	40

These numbers all show *cold*, the deflection being such as would be produced by the cooling of the face of the pile presented to the heavens; and the result is that *the chilling was in all cases greatest when the pile was directed towards the moon.*

The explanation given of this result in my letter to Sir John Herschel, I think, deals with a true cause. One hot body may, I think, be chilled by the presence of another in virtue of an action on the intervening medium. But whether the cause is *sufficient* may admit of question. It would not be sufficient if the height of our atmosphere were restricted to the limits which many assign to it. But if I understood the Astronomer Royal aright at Manchester, there is some reason for supposing the atmosphere to extend immeasurably beyond those limits. But then its extreme tenuity at great distances would probably be urged against the possibility of its producing any sensible effect. Tenuity in the abstract, however, hardly furnishes a sufficient argument. In a very few weeks I shall have occasion to show that the action of a stratum of vapour three feet thick, and possessing a tenuity which amounts only to a fraction of that assigned to our atmosphere at a height of eighty miles, is capable of accurate measurement. Nevertheless it would be a mere game of intellectual gymnastics to continue such speculations as these; for reflection on observations made before and since the publication of my letter to Sir John Herschel, leads me to conclude that in the atmosphere of London it is perfectly hopeless to obtain trustworthy results on this very delicate question.

For example, my place of observation was Albemarle Street, and my pile when turned on the moon looked nearly due south. The reflector of the instrument thus cleared in a great measure the buildings of Lambeth. I turned the instrument eastward, through a large arc, but in so doing came more over the mass of London. This *may* account for the diminished loss of heat. But even this, though apparently a natural one enough, I should hesitate to assign as the real cause of the result observed. Fresh experiments, under different conditions, will be required to decide the question.

I may add that I have furnished the pile with a conical reflector of polished tin of vast dimensions, hoping thereby to collect, not only the moon's luminous rays, but also her obscure rays, which even if they reached the earth, were effectually cut off by the polyzonal lens which Melloni used in his experiments on the moon. To protect the exposed face of the pile from currents of air, I have had the reflector furnished with screens of rock-salt. But these precautions led to no satisfactory result, the irregularities of the London atmosphere producing disturbances of the galvanometer far more than sufficient to mask the effect of the moon's rays.

LX. *On the Blue Band of the Lithium Spectrum.*
By Professor FRANKLAND, *F.R.S.*

Chemical Theatre, St. Bartholomew's Hospital, E.C.,
November 7, 1861.

MY DEAR TYNDALL,

ON throwing the spectrum of lithium upon the screen yesterday, I was surprised to see a magnificent blue band. At first I thought the chloride of lithium must be adulterated with strontium; but on testing it with Steinheil's apparatus, it yielded normal results without any trace of a blue band. I am just now reading the report of your Discourse in the 'Chemical News,' and I find that you have noticed the same thing. Whence does this blue line arise? Does it really belong to the lithium, or are the coke-points or ignited air guilty of its production? I find three blue bands with chloride of sodium, but they have not the definiteness and brilliancy of the lithium band. When lithium wire burns in air, it emits a splendid crimson light; plunge it into oxygen, and the light changes to bluish white.

This seems to indicate that a high temperature is necessary to bring out the blue ray.
Ever yours sincerely,
E. FRANKLAND.

P.S. I have just made some further experiments on the lithium spectrum, and they conclusively prove that the appearance of the blue line entirely depends upon temperature. The spectrum of chloride of lithium ignited in a Bunsen's-burner flame does not disclose the faintest trace of the blue line : replace the Bunsen's burner by a jet of hydrogen (the temperature of which is higher than that of the Bunsen's burner), and the blue line appears, faint, it is true, but sharp and quite unmistakeable; if oxygen be now slowly turned into the jet, the brilliancy of the blue line increases until the temperature of the flame rises high enough to fuse the platinum and thus put an end to the experiment.—E. F.

November 22, 1861.

[On the occasion referred to by Dr. Frankland, it was a general impression among the chemists present at the lecture that I had used the word lithium for strontium throughout the evening. This induced me to ask Dr. Miller to test my chloride of lithium, which he found quite pure. I afterwards showed the blue band, the splendour of which is unrivalled, to my class at the School of Mines. The coalpoints without the lithium show nothing of the kind; with the lithium the band always appears. Either therefore the substance itself is so altered by the exalted temperature that new periods of oscillation are possible to it, or the medium in which it vibrates is so changed in elasticity as to permit of the same thing. The observation appears to be one of considerable significance. I would also draw attention to the experiment by which the absorption of the yellow band by the sodium flame was effected on the same occasion, as one of the most striking class experiments in the whole range of optics. It is very easily performed, a band 18 inches long and $\frac{1}{4}$ of an inch wide being quite attainable within ordinary lecture-room limits. A salt flame 10 feet thick produced no such effect. Dr. Miller, I am informed, repeated this experiment with success before an evening meeting of the British Association at Manchester (see Phil. Mag. vol. xxii. p. 154).—J. T.]

LXI. *Proceedings of Learned Societies.*

ROYAL SOCIETY.

[Continued from p. 403.]

December 20, 1860.—Major-General Sabine, R.A., Treasurer and Vice-President in the Chair.

THE following communications were read :—
"Researches on the Arsenic-Bases." By A. W. Hofmann, LL.D., F.R.S.

In a previous note* I have shown the existence of a group of dia-

* Phil. Mag. for September, p. 245, " Researches on the Phosphorus-Bases. No. IX. Phospharsonium Compounds."

tomic bases, containing phosphorus and arsenic, which are formed by the action of monarsines on the bromethylated bromide, so frequently mentioned in my researches on the phosphorus-bases. The idea naturally suggested itself to examine the deportment of this salt under the influence of monostibines, with the view of producing the phospho-stibonium-compounds. The two bodies react upon one another, but only after protracted digestion or exposure to rather high temperatures. The product of the reaction is complex, yielding a comparatively small quantity of a difficultly soluble platinum-salt of diatomic appearance. I have repeatedly modified the circumstances and analysed the products in the form of platinum-salts; I omit to quote the detail of these experiments, since they have failed to disentangle the difficulties of the reaction.

Some experiments upon the deportment of dibromide of ethylene with triethylarsine were more successful. The reaction between these two bodies had been selected as a subject of inquiry by Mr. W. Valentin, to whom I am indebted for valuable assistance at the earlier stage of these researches. Circumstances have subsequently prevented Mr. Valentin from carrying out his plan, and I have therefore to take upon myself the responsibility for the following statements.

Action of Dibromide of Ethylene upon Triethylarsine.

MONARSONIUM SERIES.

The experience gathered during the examination of the phosphorus-bodies, enabled me to establish the nature of this reaction by a comparatively small number of platinum-determinations.

Bromide of Bromethyl-triethylarsonium.—To avoid as far as possible the formation of the second product, a mixture of triethylarsine with a very large excess of dibromide of ethylene was digested in sealed tubes at a temperature not exceeding 50° C. Notwithstanding the low temperature, the tubes invariably contained compressed gases; the product of the reaction was treated with water, which extracted a soluble bromide from the ethylene-compound unacted upon. On evaporation, a beautiful bromide was left, which being copiously soluble in boiling, and sparingly soluble in cold alcohol, could be readily recrystallized from absolute, and even from common alcohol. In water this substance is excessively soluble, and therefore scarcely crystallizable from an aqueous solution.

Analysis, as might have been expected, proved this salt to be the analogue of the bromethylated triethylphosphonium-salt. It contains

$$C_8 H_{19} As Br_2 = [(C_2 H_4 Br) (C_2 H_5)_3 As] Br*.$$

The bromide of bromethyl-triethylarsonium, the composition of which is sufficiently established by the analysis of the corresponding platinum-salt, can be obtained in beautiful crystals. Their form was determined by Quintino Sella; it corresponds exactly with that of the corresponding phosphorus-compound.

Platinum-salt.—The solution of the previous salt, converted by treatment with chloride of silver into the corresponding chloride,

* H=1; O=16; S=32; C=12.

yields with dichloride of platinum, splendid needles of a double salt, difficultly soluble in cold and even in boiling water, which contain

$$C_6 H_{19} Br As Pt Cl_4 = [(C_2 H_4 Br) (C_2 H_5)_3 As] Cl, Pt Cl_2.$$

Compounds of Vinyl-triethylarsonium.

The bromide of bromethyl-triethylarsonium, like the corresponding phosphorus-compound, loses its latent bromine under the influence of oxide of silver. If the solution of the bromide be precipitated by an excess of nitrate of silver, one half of the bromine separates as bromide of silver; the clear filtrate mixed with ammonia yields the second half of the bromine in the form of a dense precipitate. Nevertheless the reaction differs from that observed in the phosphorus-series. The bromide of the bromethylated phosphonium, as has been pointed out in a former part of the researches on the phosphorus-bases, is almost invariably converted into an oxethylated body, its transformation into a vinyl-compound being altogether exceptional. The bromide of the bromethylated arsonium, on the other hand, yields as a rule the vinyl-body of the series, the formation of an oxethylated compound taking place only under particular circumstances, in fact so rarely as to leave some doubt regarding the existence of this term of the series.

The bromide of bromethylated arsonium, treated with an excess of oxide of silver, yields a powerfully alkaline solution, the nature of which was determined by the analysis of the corresponding platinum-salt. Transformed into the chloride and precipitated with dichloride of platinum, this solution yielded beautiful rather soluble octahedra which were found to contain

$$C_8 H_{18} As Pt Cl_3 = [(C_2 H_3) (C_2 H_5)_3 As] Cl, Pt Cl_2.$$

The analysis of this salt shows that the transformation of the bromethylated compound ensues according to the following equation,

$$[(C_2 H_4 Br) (C_2 H_5)_3 As] Br + \left.\begin{matrix} Ag \\ Ag \end{matrix}\right\} O = \left.\begin{matrix} [(C_2 H_3) (C_2 H_5)_3 As] \\ H \end{matrix}\right\} O + 2 Ag Br.$$

The idea suggested itself that the vinyl-compound obtained in this reaction might be a secondary product resulting from the decomposition of an oxethylated compound of limited stability formed in the first instance,

$$\left.\begin{matrix} [(C_2 H_5 O) (C_2 H_5)_3 As] \\ H \end{matrix}\right\} O = \left.\begin{matrix} H \\ H \end{matrix}\right\} O + \left.\begin{matrix} [(C_2 H_3) (C_2 H_5)_3 As] \\ H \end{matrix}\right\} O.$$

It was with the view of avoiding this decomposition that in one of the operations the digestion was accomplished at the common temperature; the result, however, showed that even in this case the vinyl-compound was obtained.

Nevertheless the oxethylated body appears to exist: under circumstances which were not sufficiently well observed at the time, the action of oxide of silver upon bromide of bromethylated triethylarsonium yielded an octahedral platinum-salt, which on analysis furnished exactly the platinum-percentage of the oxethylated compound.

DIARSONIUM SERIES.
Dibromide of Ethylene-hexethyldiarsonium.

The bromide or chloride of the bromethylated arsonium-compound is but slowly acted upon by triethylarsine at 100° C. Two days' digestion at that temperature had produced but a slight impression ; at 150° the reaction is accomplished in two hours. The phenomena now to be recorded presented themselves in the succession repeatedly observed in the diphosphonium-series. The dibromide

$$C_{14} H_{34} As_2 Br_2 = (C_2 H_4)'' \begin{Bmatrix} (C_2 H_5)_3 As \\ (C_2 H_5)_3 As \end{Bmatrix}'' Br_2$$

yielded, when debromized, the powerful alkali

$$C_{14} H_{34} As_2 O_2 = \begin{Bmatrix} [(C_2 H_4)'' (C_2 H_5)_6 As_2]'' \\ H_2 \end{Bmatrix} O_2.$$

Treated with acids, this alkali produces a series of fine salts, amongst which the di-iodide deserves to be mentioned ; it equals in beauty the corresponding diphosphonium-compound.

I have fixed the composition of the series by the analysis of the platinum-salt and gold-salt.

Platinum-salt.—Pale-yellow crystalline precipitate, similar to the diphosphonium-compound, difficultly soluble in water, soluble in boiling concentrated hydrochloric acid, from which it crystallizes on cooling. It contains

$$C_{14} H_{34} As_2 Pt_2 Cl_6 = \left[(C_2 H_4)'' \begin{matrix} (C_2 H_5)_3 As \\ (C_2 H_5)_3 As \end{matrix} \right]'' Cl_2, 2 Pt Cl_2.$$

Gold-salt.—The dichloride obtained after separating the platinum in the previous analysis by sulphuretted hydrogen, was precipitated by trichloride of gold ; yellow slightly crystalline precipitate, soluble in hydrochloric acid, from which it crystallizes in golden-coloured plates. The formula of this salt is

$$C_{14} H_{34} As_2 Au_2 Cl_8 = \left[(C_2 H_4)'' \begin{matrix} (C_2 H_5)_3 As \\ (C_2 H_5)_3 As \end{matrix} \right]'' Cl_2, 2 Au Cl_3.$$

ARSAMMONIUM SERIES.

Bromide of bromethylated triethylarsonium, as might have been expected, is capable of fixing ammonia and monamines, giving rise to the formation of a group of compounds not less numerous than the bodies mentioned in the phosphorus-series. I have been satisfied to study the action of ammonia upon the bromide.

Dibromide of Ethylene-triethylarsammonium.

Reaction complete in two hours at 100°. The product contains the dibromide,

$$C_8 H_{22} As N Br_2 = \left[(C_2 H_4)'' \begin{matrix} (C_2 H_5)_3 As \\ H_3 N \end{matrix} \right] Br_2 ;$$

this salt is converted by oxide of silver into the stable caustic base

$$C_8 H_{24} As N O_2 = \begin{Bmatrix} [(C_2 H_4)'' (C_2 H_5)_3 H_3 As N]'' \\ H_2 \end{Bmatrix} O_2,$$

the composition of which was determined by the analysis of the platinum-salt and gold-salt.

Platinum-salt. — Needles, difficultly soluble in boiling water, soluble in concentrated hydrochloric acid, from which well-formed crystals are deposited, containing

$$C_6 H_{22} As N Pt_2 Cl_6 = \left[(C_2 H_4)'' \,^{(C_2 H_5)_2 As}_{H_2 N} \right]'' Cl_2, 2 Pt Cl_2.$$

Gold-salt. — Yellow compound precipitated from the dichloride obtained in the previous platinum-determination, on addition of trichloride of gold, soluble in hydrochloric acid, deposited from this solution in golden-yellow plates of the composition

$$C_6 H_{22} As N Au_2 Cl_8 = [(C_2 H_4)'' (C_2 H_5)_2 H_2 As N]'' Cl_2, 2 Au Cl_3.$$

I have also made a few experiments on the action of dibromide of ethylene upon triethylstibine. The reaction is slow, and requires long-continued digestion at temperatures higher than that of boiling water. The tubes invariably contained much gas; and the product of the reaction proved to be a complex mixture of several compounds, many of them secondary, which in no way invited me to a more minute examination of this process. I omit to quote the few platinum- and chlorine-determinations which were made, since they do not admit of a simple interpretation.

"Contributions towards the History of the Monamines."—No. IV. Separation of the Ethyle-Bases. By A. W. Hofmann, LL.D., F.R.S.

The preparation of the ethyle-bases by the action of ammonia upon iodide of ethyle, presents a difficulty which greatly interferes with the general application of this otherwise so convenient method. This difficulty consists in the simultaneous formation of all the four ethyle-bases. The equations

$$H_3 N + C_2 H_5 I = [(C_2 H_5) H_3 N] I*$$
$$(C_2 H_5) H_2 N + C_2 H_5 I = [(C_2 H_5)_2 H_2 N] I$$
$$(C_2 H_5)_2 H N + C_2 H_5 I = [(C_2 H_5)_3 H N] I$$
$$(C_2 H_5)_3 N + C_2 H_5 I = [(C_2 H_5)_4 N] I,$$

are an ideal representation of the four different phases through which ammonia passes during its transformation into iodide of tetrethyl-ammonium. In practice it is found impossible to carry out this transformation in the several steps indicated by these equations. The first substitution-product, generated as it is in the presence of the agent of substitution, is immediately acted upon again, the second product being formed, which in its turn may be converted into the third and even into the fourth compound. The following equations represent perhaps more correctly the final result of the several changes which are accomplished in the reaction of ammonia on iodide of ethyle.

$$H_3 N + C_2 H_5 I = [(C_2 H_5) H_3 N] I$$
$$2 H_3 N + 2 C_2 H_5 I = [(C_2 H_5)_2 H_2 N] I + [H_4 N] I$$
$$3 H_3 N + 3 C_2 H_5 I = [(C_2 H_5)_3 H N] I + 2 ([H_4 N] I).$$
$$4 H_3 N + 4 C_2 H_5 I = [(C_2 H_5)_4 N] I + 3 ([H_4 N] I).$$

* $H = 1; O = 16; C = 12, \&c.$

The mixture of iodides, when submitted to the action of potassa, yields ammonia, ethylamine, diethylamine, and triethylamine, the hydrate of tetrethylammonium, which is liberated, splitting into ethylene, triethylamine, and water. The separation of the three ethyle-ammonias presents unusual difficulties. The differences between their boiling-points being rather considerable,

Ethylamine, boiling-point .... 18°
Diethylamine, ,, ,, :..:...:.:.:. 57°·5
Triethylamine, ,, ,, ..:...:.:...:. 91°

it was thought that they might be readily separated by distillation. Experiments made with very large quantities showed, however, that even after ten fractional distillations the bases were far from being pure.

After many unsuccessful attempts, I have found a simple and elegant process by which the three ethyle-bases may be easily and perfectly separated. This process consists in submitting the anhydrous mixture of the three bases to the action of anhydrous oxalate of ethyle. By this treatment, ethylamine is converted into *diethyl-oxamide*, a beautifully crystalline body very difficultly soluble in water, diethylamine into *ethyl-oxamate of ethyle*, a liquid boiling at a very high temperature, whilst triethylamine is not affected by oxalic ether

By the action of oxalic ether upon ethylamine, two substances may be formed, viz. *ethyl-oxamate of ethyle* and *diethyl-oxamide*.

$$\left.\begin{array}{l}(C_2 O_2)'' \\ (C_2 H_5)_2\end{array}\right\} O_2 + \left.\begin{array}{l}C_4 H_5 \\ H \\ H\end{array}\right\} N = \left.\begin{array}{l}[(C_2 O_2)'' (C_4 H_5) H N] \\ (C_2 H_5)\end{array}\right\} \dot{O} + \left.\begin{array}{l}C_2 H_4 \\ H\end{array}\right\}$$

Oxalic Ether.　　Ethylamine.　　Ethyl-oxamate of ethyle.　　Alcohol.

$$\left.\begin{array}{l}(C_2 O_2)'' \\ (C_2 H_5)_2\end{array}\right\} O_2 + 2\left[\begin{array}{l}C_4 H_5 \\ H \\ H\end{array}\right\} N\right] = \left.\begin{array}{l}(C_2 O_2)'' \\ (C_2 H_5)_2 \\ H_2\end{array}\right\} N_2 + 2\left[\begin{array}{l}C_2 H_4 \\ H\end{array}\right\} O\right].$$

Oxalic Ether.　　Ethylamine.　　Diethyl-oxamide.　　Alcohol.

In practice it appears that the *second* of these compounds only is produced.

In the action of oxalate of ethyle upon diethylamine, two similar phases may be distinguished capable of producing respectively

Diethyl-oxamate of ethyle .... $\left.\begin{array}{l}[(C_2 O_2)'' (C_4 H_5)_2 N] \\ C_2 H_5\end{array}\right\} O$, and

Tetrethyl-oxamide .... $\left.\begin{array}{l}(C_2 O_2)'' \\ (C_4 H_5)_2 \\ (C_4 H_5)_2\end{array}\right\} N_2$.

In practice the *first* of these two compounds only is generated.

The action of oxalate of ethyle upon triethylamine might have involved the formation of the secondary oxalate of tetrethylammonium,

$$\left.\begin{array}{l}(C_2 O_2)'' \\ (C_2 H_5)_2\end{array}\right\} O_2 + 2\left[\begin{array}{l}C_2 H_5 \\ C_2 H_5 \\ C_2 H_5\end{array}\right\} N\right] = \left.[(C_2 H_5)_4 N]_2\right\} O_2;$$

under the circumstances under which I have worked, the two substances do not combine.

The product of the reaction of oxalate of ethyle upon the mixture of the ethyle-bases, when distilled in the water-bath, yields *triethylamine free from ethylamine and diethylamine.*

The residue in the retort solidifies on cooling into a fibrous mass of crystals of diethyloxamide, which are soaked with an oily liquid. They are drained from the oil and recrystallized from boiling water. Distilled with potassa, these crystals furnish *ethylamine free from diethylamine and triethylamine.*

The oily liquid is cooled to 0°, when a few more of the crystals are deposited; it is then submitted to distillation. The boiling-point rapidly rises to 260°. What distils at that temperature is pure diethyl-oxamate of ethyle, from which, by distillation with potassa, *diethylamine free from ethylamine and triethylamine* may be obtained.

January 10, 1861.—Major-General Sabine, R.A., Treasurer and Vice-President, in the Chair.

The following communication was read :—

"On the Lunar-diurnal Variation of the Magnetic Declination obtained from the Kew Photograms* in the years 1858, 1859, and 1860." By Major-General Edward Sabine, R.A., Treas and V.P.R.S.†

Having communicated to the Royal Society in a recent paper an analysis of the *disturbances* of the declination in the years 1858 and 1859, shown by the photograms of the Kew Observatory, I propose in the present paper to submit the results of the *lunar-diurnal variation* of the declination in the years 1858, 1859, and 1860, obtained from the same source. The directions of the declination magnet at the instant of the commencement of every solar-hour having been tabulated from the photograms, and the final normals for each month and hour computed, after the omission from the record of all the hourly directions which deviated 3′·3 from their final normals,—the *differences* were taken between each of the remaining hourly directions and the final normal of the same month and hour, and were entered afresh in *lunar* monthly tables, having the lunar days in successive horizontal lines, and the twenty-four lunar hours in vertical columns, each difference being placed under the lunar hour to which it most nearly approximated. The entries in these tables should consequently represent directly the lunar influence at the different lunar hours, subject only to minor disturb-

---

* The term Photogram is adopted in place of Photograph in conformity with modern usage.

† [Note added on February 8th, 1861.] When this communication was read to the Royal Society on January 10th, 1861, it contained the lunar-diurnal variation for the years 1858 and 1859 only: whilst it was passing through the press, the calculation of the lunar-diurnal variation for 1860 was completed, and the results in that year have been added.

ances; the effects of the solar-diurnal variation as well as of the larger disturbances having been eliminated. The differences were marked with a + sign when the north end of the magnet was east of its mean direction, and with the — sign when west of the same. The differences were then summed up, and hourly, monthly, and annual means taken by the non-commissioned officers of the Royal Artillery employed at Woolwich, under the superintendence of Mr. Magrath.

Having in the former paper exhibited the results of the disturbances at Kew in comparison with those at Hobarton, I propose to do the same with the lunar-diurnal variation treated of in this communication; believing that such comparisons are very conducive to a just appreciation of the systematic character and natural reality of the results, and instructive both by the agreements and disagreements which they exhibit. The lunar-diurnal variation at Hobarton has been obtained for the purpose of this comparison, by a similar process to that which has been described above, from observations at every solar hour during five years (Sundays excepted), from Oct. 1, 1843 to Sept. 30, 1848; omitting as disturbed such observations as deviated $2'·13$ from their respective final normals. The total number of hourly observations was 36,832; the disturbed observations 2606; and the number employed in the lunar-diurnal variation 34,226. As it has been customary to represent such periodical variations by formulæ of well-known character, the results at Kew and Hobarton are here represented by formulæ in which a, corresponding to x (the lunar time for which the lunar-diurnal variation is desired), is counted in hours and parts of an hour, multiplied by $15°$, from the epoch of the moon's upper culmination. The + sign corresponds (as before) to a deflection of the north end of the magnet to the east of its mean place, and the — sign to the west.

Kew $\quad \Delta x = +0''·64 - 2''·54 \sin(a+6°·2) - 9''·74 \sin(2a+59°·8)$.

Hobarton $\Delta x = -0''·1 + 1''·14 \sin(a+344°·7) + 6''·8 \sin(2a+43°·2)$.

In computing the lunar-diurnal variation by means of these formulæ, the coefficient of the term which includes the sine of twice the hour-angle is of principal importance: the subsequent terms are comparatively of little significance, and are therefore omitted on the present occasion. When all the terms are employed, the original observed values are reproduced.

Table I. exhibits, at Kew, in column 2 the lunar-diurnal variation as actually observed on the mean of the three years, and in column 3, the same computed by the formula. Column 4 is the lunar-diurnal variation at Hobarton on the mean of the five years as observed, and column 5 the same computed by the formula.

TABLE I.—Lunar-diurnal Variation at Kew and Hobarton.

Lunar Hours.	Kew.		Hobarton.		Lunar Hours.
	Observed.	Computed.	Observed.	Computed.	
Col. 1.	Col. 2.	Col. 3.	Col. 4.	Col. 5.	Col. 6.
0	− 6·0	− 8·0	+ 4·8	+ 4·3	0
1	−11·4	−10·0	+ 6·1	+ 6·4	1
2	− 8·6	− 9·3	+ 5·2	+ 6·8	2
3	− 5·0	− 6·2	+ 5·9	+ 5·4	3
4	− 3·2	− 1·7	+ 4·2	+ 2·7	4
5	+ 1·4	+ 3·0	0·0	− 0·6	5
6	+ 5·4	+ 6·5	− 4·9	− 3·7	6
7	+ 7·6	+ 8·0	− 6·1	− 5·5	7
8	+ 8·6	+ 7·0	− 4·9	− 5·6	8
9	+ 4·3	+ 3·9	− 3·3	− 4·1	9
10	+ 2·8	− 0·4	− 3·2	− 4·3	10
11	− 3·0	− 4·6	+ 3·6	+ 2·0	11
12	−10·6	− 7·5	+ 4·9	+ 4·9	12
13	−10·4	− 8·2	+ 6·6	+ 6·4	13
14	− 7·0	− 6·3	+ 5·9	+ 6·2	14
15	− 2·2	− 2·3	+ 4·1	+ 4·3	15
16	+ 4·8	+ 3·0	+ 1·4	+ 1·1	16
17	+10·4	+ 8·0	− 3·4	− 2·7	17
18	+13·2	+11·6	− 6·4	− 5·9	18
19	+12·6	+12·7	− 6·5	− 7·7	19
20	+ 7·2	+11·1	− 6·6	− 7·8	20
21	+ 6·2	+ 7·1	− 8·4	− 6·0	21
22	− 0·4	+ 1·7	− 1·9	− 2·9	22
23	− 1·4	− 4·6	+ 0·8	+ 0·9	23

The aspect of the lunar-diurnal variation at Kew and Hobarton presents features of great simplicity as well as accord. The form at both stations is a division of the 24 lunar hours into four equal or nearly equal portions, in which the magnet is attracted alternately to the east and to the west of its mean position, which is passed through four times in the progress of the magnet towards two extreme easterly and two extreme westerly deflections: the easterly extremes are about 12 hours apart, and the westerly the same. As far as our present experience goes, this appears to be the general form of the lunar-diurnal variation of the declination at all the stations at which it has been examined; it is also that of the corresponding variations of the Dip and Total force. At Hobarton, where the results are obtained from five years of observation, there is scarcely any difference deserving of notice between the amplitudes of the extremes on either side of the upper culmination and those on either side of the lower culmination. At Kew, where the results are obtained from only three years, the extreme deflections are not quite so symmetrical in amount, but they may become more so as additional years are brought into the account. The amplitude of the oscillation on a mean of the two alternations is 9″·74 at Kew and 6″·8 at Hobarton, a difference in correspondence with the difference in the opposite direction of the antagonistic retaining force of the earth's magnetism at the two stations, which is 3·7 at Kew and 4·5 at Hobarton. On inspecting the Table, we see that the lunar times when the moon's influence produces no deflection (or the times when the variation is zero), are four, and are nearly the same at Kew and at Hobarton, two of them being a little more than an hour before the moon's passage of the meridian, both at her upper and lower culminations, and the other two intermediate. So far the two stations are alike; but in regard to the direction towards which the magnet is deflected (if in conformity with general usage we speak in both hemispheres of the north end of the magnet, as is done in the Table), we see that the variation becomes west at Kew when it becomes east at Hobarton, and *vice versá*; the phases, while agreeing in hours at the two stations, having throughout opposite signs.

By extending the comparison of the lunar hours at which the lunar variation passes through its zero-points to other stations than Kew and Hobarton, we are made aware of differences which appear to deserve particular attention in theoretical respects. At Pekin, for example—which may be advantageously compared with Kew, being both in the same hemisphere, but Pekin some degrees nearer the equator—the variation is zero in the passage of the north end of the magnet from east to west at $20\frac{1}{2}$ lunar hours, or $2\frac{1}{2}$ hours earlier than the corresponding epoch at Kew. Again, at the Cape of Good Hope, situated in the same hemisphere with Hobarton, but some degrees nearer the equator, the variation is zero in the passage of the north end of the magnet from west to east also at $20\frac{1}{2}$ lunar hours, or $2\frac{1}{2}$ hours earlier than the corresponding epoch at Hobarton. Thus there is an accord of precisely the same kind between Pekin and the Cape of Good Hope that there is between Kew and Hobarton, whilst there is a difference between the two pairs of stations of $2\frac{1}{2}$ hours in the position of the moon relatively to the meridian at which she ceases to exercise a deflecting influence on the magnet. Again, at St. Helena, which is in the same (geographical) hemisphere as Hobarton and the Cape of Good Hope, but still nearer to the equator than either; the lunar influence is zero in the passage from west to east at $19\frac{1}{2}$ lunar hours, being one hour earlier than at the Cape, and $3\frac{1}{2}$ hours earlier than at Hobarton.

Where the whole range of the variation of which we have been treating is so small (not more than a few seconds of arc in each lunar day), it may be desirable to show by the accordance of the independent evidence obtained in single years, the degree of confidence which may be placed in the mean results of several years. This may be seen in the Table on the next page, which contains the separate results in each of the five successive years of observation at Hobarton, as well as their mean.

In this Table the principal features of the variation are seen to be substantially alike in each year. The individual results at the several hours in single years are of course somewhat less regular than in the mean of the five years: such small discrepancies are no doubt in great part due to the lesser disturbances which, being below the separating value of $2'\cdot13$, have been left in the body of the observations. They slightly disfigure the symmetry of the results in single years, but almost entirely disappear when the mean of several years is taken. In

.order to appreciate justly and fully the confidence to which the whole investigation is entitled, it must be borne in mind that every single entry in the Table (exclusive of course of the column which exhibits the mean of the five preceding columns) is derived from a wholly independent body of observations which belong to itself alone, and are not employed in the deduction of any of the other entries.

TABLE II. — Lunar-diurnal Variation at Hobarton in the several years from October 1843 to September 1848 ; omitting disturbed observations differing 2'·13 from their final normals.

Lunar Hours.	Years ending September 30th.					Means.	Lunar Hours.
	1844.	1845.	1846.	1847.	1848.		
0	+ 0·6	+ 7·8	+ 3·6	+ 2·4	+ 9·6	+ 4·8	0
1	+ 6·6	+ 9·0	+ 1·2	+ 0·6	+13·2	+ 6·1	1
2	+ 4·8	+. 5·4	+ 5·4	+ 4·2		+ 5·2	2
3	+ 9·6	+ 7·8	+ 7·8	+ 3·6	+ 0·6	+ 5·9	3
4	+ 4·8	+ 6·6	+ 6·0	+ 3·0	+ 0·6	+ 4·2	4
5	— 5·0	+ 2·4	+ 3·0	— 1·8	— 0·6	0·0	5
6	— 7·8	— 6·0	— 1·8	— 7·8	— 1·2	— 4·9	6
7	— 6·0	— 9·6	— 0·6	—10·8	— 3·6	— 6·1·	7
8	— 4·2	— 8·4	— 1·2	— 7·8	— 3·0	— 4·9	8
9	0·0	— 9·0	— 0·6	— 4·8	— 3·0	— 3·3	9
10	— 2·4	— 4·8	— 1·8	— 0·6	— 6·6	— 3·2	10
11	+ 8·0	+ 0·6	+ 3·0	+ 8·4	+ 3·0	+ 3·6	11
12	+ 7·2	+ 2·4	+ 4·8	+ 7·8	+ 2·4	+ 4·9	12
13	+12·0	+ 6·6	+ 6·6	+ 4·8	+ 3·0	+ 6·6	13
14	+ 8·0	+ 8·4	+ 7·2	+ 7·8	+ 3·0	+ 5·9	14
15	+ 7·8	+ 4·8	+ 4·2	+ 3·0	+ 0·6	+ 4·1	15
16	+ 1·8	+ 3·6	+ 0·6	+ 0·6	+ 0·6	+ 1·4	16
17	0·0	— 1·2	— 3·6	— 2·4	— 9·6	— 3·4	17
18	— 6·6	— 6·6	— 5·4	— 6·6	— 6·6	— 6·4	18
19	— 4·8	— 5·4	— 8·4	— 7·2	— 6·6	— 6·5	19
20	— 6·6	—10·2	—13·2	— 4·2	— 5·4	— 6·6	20
21	— 9·6	—12·6	—10·8	— 4·2	— 4·8	— 8·4	21
22	— 4·2	— 0·6	—10·8	+ 2·4	+ 3·6	— 1·9	22
23	— 2·4	0·0	— 2·4	+ 0·6	+ 8·4	+ 0·8	23

It may operate as an encouragement to those who have not yet subjected their observations to any process of examination or analysis, to perceive, by this example, how substantially satisfactory are the results which may be obtained from even a single year of hourly observations, after the larger disturbances and the solar-diurnal variation have been eliminated.

I have spoken in a recent paper of an unexceptionable test by which we may satisfy ourselves as to the confidence which may be reposed in a series of observations, whether obtained by the eye or tabulated from instrumental traces. Such a test is furnished when the entries at solar hours are rewritten according to the lunar hours to which they most nearly approximate, and when consequently their original order and relations are changed and are replaced by others which were wholly unforeseen, so that the observations must necessarily be free from the possibility of having been influenced by any mental bias. When we find the effects of a natural law, represented by such minute values as that of the lunar-diurnal variation, exhibited by the observations of a single year with the degree of symmetry shown in Table II., we may safely conclude that the observations themselves are worthy of the labour bestowed in eliciting their results. In this view the Hobarton observations prove themselves to have been not only a faithful, but also an extremely careful series, highly creditable to Captain Kay, R.N., and to the Naval Officers who with him and their Civil Assistant Mr. Jeffery, maintained for so many years the laborious and monotonous duty of hourly observation.

Table III. exhibits the separate results in each of the three years at Kew, as well as their mean.

TABLE III.—Lunar-diurnal Variation at Kew in the years 1858, 1859, and 1860 ; omitting disturbed observations differing 3′·3 from their final normals.

Lunar Hours.	Year ending December 31.			Means.	Lunar Hours.
	1858.	1859.	1860.		
0	− 6·0	+ 0·6	−12·6	− 6·0	0
1	−14·4	− 7·2	−12·6	−11·4	1
2	−10·8	− 9·6	− 5·4	− 8·6	2
3	− 7·8	− 4·2	− 3·0	− 5·0	3
4	− 3·0	− 4·2	− 2·4	− 3·2	4
5	+ 5·4	− 6·6	+ 5·4	+ 1·4	5
6	+ 2·0	+ 1·2	+ 3·0	+ 5·4	6
7	+ 9·0	+ 4·2	+ 9·6	+ 7·6	7
8	+19·6	+ 8·4	+ 7·8	+ 8·6	8
9	+ 7·2	+ 6·6	− 0·9	+ 4·3	9
10	+ 3·0	+ 7·2	− 1·8	+ 2·8	10
11	− 3·6	− 1·2	− 4·2	− 3·0	11
12	− 4·8	− 9·0	−18·0	−10·6	12
13	− 3·0	−13·2	−15·0	−10·4	13
14	− 3·0	− 8·4	− 9·6	− 7·0	14
15	− 7·2	− 3·6	+ 4·2	− 2·2	15
16	+ 3·0	+ 3·6	+ 7·8	+ 4·8	16
17	+ 7·8	+ 9·6	+13·8	+10·4	17
18	+ 7·8	+14·4	+17·4	+13·2	18
19	+ 4·8	+18·0	+15·0	+12·6	19
20	+ 3·0	+12·6	+ 6·0	+ 7·2	20
21	− 2·4	+18·6	+ 2·4	+ 6·2	21
22	− 7·8	+ 9·6	− 3·0	− 0·4	22
23	− 6·0	+ 5·4	− 3·6	− 1·4	23

In conclusion, it may be useful to call the attention of the Society, and of those Fellows in particular who interest themselves in tracing up the phenomena of nature to their physical causes, to the assemblage of facts which are now available for such inquiries, in a branch of magnetical science which may not inappropriately be called *celestial magnetism*. In the introductory discussion prefixed to the 2nd volume of the St. Helena Magnetical Observations, p. cxliv to cxlviii, the lunar-diurnal variation is given for each of the three magnetic elements, the Declination, the Dip, and the Intensity of the force, at the four stations of Toronto, St. Helena, the Cape of Good Hope and Hobarton, and for the Declination at two additional stations Kew and Pekin. The variations are given both in formulæ and in tables; the latter exhibiting the amount of the lunar influence at each of the 24 lunar hours, in the several magnetic elements at each station. These data are directly applicable to inquiries into the nature of the moon's magnetism; and into the mode by which the moon's magnetism acts either on the magnetism of the earth itself, or on the magnetic needle stationed at different points of the earth's surface, so as to produce a small but systematic and perfectly appreciable variation in each of the magnetic elements, having a double period in every lunar day.

The lunar-diurnal variation' of the Declination at Kew and Hobarton, as given in this communication, is slightly different from the figures in the 2nd St. Helena volume referred to, because the results at Kew are a mean of 3 years instead of 2, as in the St. Helena volume; and at Hobarton a lower standard has been taken for the disturbances, causing a larger number of the disturbed observations to be omitted in the calculation of the lunar-diurnal variation.

January 24.—Major-General Sabine, Treas. and V.P., in the Chair.

The following communications were read :—

"On the Calculus of Symbols, with Applications to the Theory of Differential Equations." By W. H. A. Russell, A.B.

"On the Properties of Liquid Carbonic Acid." By G. Gore, Esq.

In this communication the author has shown how a small quantity of liquid carbonic acid may be readily and safely prepared in glass tubes closed by stoppers of gutta percha, and be brought in a pure state into contact with any solid substance upon which it may be desired to ascertain its chemical or solvent action, or be submitted to the action of electricity by means of wires introduced through the stoppers. By immersing about fifty substances in the liquid acid for various periods of time, he has found that it is comparatively a chemically inert substance, and not deoxidized by any ordinary deoxidizing agent except the alkali-metals. Its solvent power is extremely limited; it dissolves camphor freely, iodine sparingly, and a few other bodies in small quantities; it does not dissolve oxygen-salts, and it does not redden solid extract of litmus; it penetrates gutta percha, dissolves out the dark-brown colouring matter, and leaves the gutta percha undissolved, and much more white. It also acts in a singular and somewhat similar manner upon india-rubber; the india-rubber whilst in the liquid acid exhibits no change, but immediately on being taken out it swells to at least six or eight times its original dimensions, and then slowly contracts to its original volume, evidently from expansion and liberation of absorbed carbonic acid; and it is found to be perfectly white throughout its substance. These effects upon gutta percha and india-rubber may prove useful for practical purposes.

The liquid acid is a strong insulator of electricity; sparks (from a Ruhmkorff's coil) which would pass readily through $\frac{9}{10}$nds of an inch of cold air, would with difficulty pass through about $\frac{1}{10}$th of an inch of the liquid acid.

In its general properties it is somewhat analogous to bisulphide of carbon, but it possesses much less solvent power over fatty substances.

LXII. *Intelligence and Miscellaneos Articles.*

LUNAR RADIATION.

To the Editors of the Philosophical Magazine and Journal.

GENTLEMEN,   Oxford and Cambridge Club, Nov. 19, 1861.

IN your November Number there is an account by Professor Tyndall of some observations with a thermo-electric pile, in the course of which it appeared that it "lost more heat when presented to the moon than when turned to any other portion of the heavens of the same altitude;" and there is a theoretical explanation of this fact as an indirect effect of the moon's heat, dispersing the "small quantity of precipitated vapour" which it appears was then floating in the atmosphere, and so facilitating radiation from the instrument.

Unless my memory is deceived, Sir John Herschel, in one of the earlier editions of his 'Astronomy,' described light clouds as, in like manner, dispersing *as they came between his telescope and the moon*; but in the edition of 1858, here at hand, I see the phrase is "the tendency to disappearance of clouds *under the full moon*," which may mean a very different thing, viz. a tendency to *clear skies when the moon is full.*

That the heat of the full moon may tend to clear the upper atmosphere, and so be the cause of cold below, may be true; but it does not appear to me that this can be the explanation of Professor Tyndall's fact, or of Sir J. Herschel's, if I state it correctly.

High in the air, in the region in which the moon is seen, there is cloud or vapour observed. The moon may have diminished, but it has not destroyed it *generally.* How then is that particular portion which happens to intervene between the observer's instrument and the moon more under her influence than any other equal portion? If a hundred observers were gazing at her at the same time within a few miles of each other, a hundred different portions of the haze would so intervene; and to suppose each of these dispersed, is to suppose the haze not to exist.

It is possible that a full examination of all the circumstances of Professor Tyndall's six experiments—the area embraced by his reflector, the probable height of the vapour in the air, the extent of the sweep he took with the instrument, &c.—might remove some of the difficulty I feel in admitting the explanation he proposes; and in the interest of exact science I venture to call his attention to the matter.   D. D. HEATH.

ON THE DIHEXAHEDRAL CRYSTALS OF SULPHATE OF POTASH, BY KARL RITTER VON HAUER[*].

The supposed dimorphism of the sulphate of potash, as K. von Hauer has proved, rests only on external appearance, as in reality this salt in a state of chemical purity constantly affects forms of the prismatic system, and when appearing in forms of the rhombohedral system invariably contains a certain quantity of anhydrous sulphate of soda. This bibasic salt is known to be produced at Glasgow, in the shape of hexagonal plate-like crystals, by the evaporation of a solution of kelp-ash. A mixture of both these sulphates (potash and soda), inspissated and left to crystallize, invariably gives no longer hexagonal plates, but exclusively dihexahedrons (double hexagonal

[*] Translated by Count Marschall.

pyramids), a form scarcely if ever met with among the crystals pro-·' duced by the above-described technical process. Analogous local actions are observed on natural minerals; so that in some cases an expert mineralogist may infer the place of origin of a mineral substance from its crystalline form only. On the other hand, conclusions as to the mode of formation of minerals founded on the results of laboratory experiments must be drawn with a certain degree of caution. In fact, the chemical forces, when acting on large quantities of substances, as in manufacturing processes, frequently produce results very different from those obtained by the chemist operating with comparatively small portions; and, still more, the results of natural operations, gigantic in quantity as in energy, and extending through immeasurable periods of time, may scarcely be comparable to mere laboratory investigations made with limited quantities in some few hours or days.

When immersed in solutions of other salts, the crystals of the bibasic sulphate in question show some curious phenomena. In a solution of sulphate of ammonia a hexagonal plate was gradually converted, by superposition on both of its larger planes, into a lengthened hexagonal prism, easily cleavable at any point in a direction perpendicular to its longitudinal axis. Thin plates of it taken from the newly added portion show the characteristic optical properties of the common prismatic sulphate of ammonia. This instance of episomorphism between a rhombohedral and a prismatic salt, or, in other words, of two substances belonging each to a different crystallographical system and nevertheless subject to the crystallographical laws of isomorphism, is highly interesting. The angular values of both (the rhombohedral and the prismatic combination) being, in this special case, very near each other, the existence of the fact here alluded to was to be decided by optical investigation. Trifling as the difference of the forms here in question may be, its existence is a fact not to be denied; and therefore such a formation as just described could not take place if the disposition of the molecules, by whose regular aggregation such crystals are formed, did not go on with mathematical exactitude. Observation shows deviations from the strict regularity of lines and angles to be of no rare occurrence in crystallogenetic processes; precise measurements of substances considered to be isomorphous have shown them not to be absolutely congruent; so that isomorphism, as far as it is concerned in this character, has only an approximate value. Two substances different in angular value, even when combined into one and the same crystal, cannot be considered as having totally lost their respective individuality. Their last constituent parts, representing the crystalline molecules of both salts (sulphates of potash and ammonia), are in juxtaposition to each other, as if they were but one homogeneous substance. Their superposition without preceding mixture is a proof that molecules of not absolute identity may be deposited on each other in the same way as analogous particles would be. Both these sulphates could be considered as absolutely isomorphous in the crystallographical sense, but for the optical phenomena characteristic of two distinct and mutually independent systems. Isomorphism, however, presupposes chemical analogy; now the potash being very prevalent in quantity in the rhombohedral bibasic salt, its chemical analogy with ammonia may be supposed to have been superseded by the comparatively small proportion of soda combined with it.   In the absence of isomorphism in the strictest sense, there are circumstances coming so near to it that the molecules of both substances still attract each other sufficiently to effect regular superposition.   The curved and disfigured planes of such crystals are at all events indicative of their origin under abnormal and, as it were, compulsory circumstances.—*Proceedings of the Vienna Imperial Institute,* April 16, 1861.

COMPARISON OF THE TEMPERATURE IN THE AIR AND OF THE
SOIL AT A DEPTH OF TWO METRES.   BY M. POURIAU.

From observations made during five consecutive years on the temperature of the soil at a depth of 2 metres compared with that of the air, it follows—

1. That the mean temperature in the air was $10°·21$, and in the soil $12°·79$.   Difference in favour of the soil $2·58$.

2. That the mean temperature of the soil in winter and autumn is higher than that of the air; that in summer it is about 2 degrees lower, and that in spring the mean temperatures are virtually equal.

3. That the mean of the extreme maximum temperatures in the air was $34°·5$, in the soil it was $19°·75$.   On the other hand, the mean of the extreme minima in air was $-12°·14$; in the soil this mean never sank below $+6°$.

4. While in air the mean of the total differences between the extreme maxima and extreme minima reached $46°·64$, in the soil this mean was only $13°·74$.

5. In 1860 the temperature of the air sank to $-20°$, in the soil the minimum was never less than $+5°·47$.

6. While in the air the maximum temperature usually occurs in July or August, and the minimum in December or January, the maximum temperature in the soil always corresponds to the end of August; the minimum always occurs at the end of February, or on the first days of March.

7. The changes of temperature in the soil at a depth of 2 metres may be thus stated:—

. While the mean temperature of the air usually begins to sink towards the end of July, in the soil the heat continues to accumulate in the superior layers under the influence of the intense solar radiation, and to extend to the lower layers, until the end of August. From this point the upper layers begin to lose more heat by radiation than they receive; the flow of heat changes its direction, it passes from the lower to the upper layers and becomes lost in the air; and this ascending motion, continuing until February, is more rapid as the external temperature is lower, that is, as the winter is longer and more severe.   Towards the middle of February or the beginning of March the upper layers begin to become heated under the influence of the solar rays, whose direction has become less oblique; the inferior layers give less and less heat to the upper ones; they begin, on the contrary, to receive some, and become then reheated, which continues until the end of August.—*Comptes Rendus,* October 7, 1861.

THE

LONDON, EDINBURGH AND DUBLIN

PHILOSOPHICAL MAGAZINE

AND

JOURNAL OF SCIENCE.

SUPPLEMENT to VOL. XXII. FOURTH SERIES.

LXIII. *Static and Dynamic Stability in the Secondary Systems.*
By DANIEL VAUGHAN, *Esq.**

SO small are the primary planets compared with their distances
from the sun, that they are regarded as material points in
the investigations of physical astronomy, and that no effects
arising from the unequal intensity of solar attraction on their
parts can vitiate in any sensible degree the results which analysis
gives for their movements. But in the systems of Jupiter and
Saturn many satellites are exposed to an enormous tidal force in
consequence of their proximity to their primaries; and the planetary theory requires some modification when applied to the revolutions of these minor worlds. My chief object at present is to
show that the unequal attraction of a primary occasions slow
secular changes in the orbits of its attendants, especially when
the presence of fluids on their surfaces brings tidal commotions
into play. But I deem it first necessary to prove what I have
assumed in my former communications, in regard to the physical
necessity for a synchronism of the orbital and rotatory motions
of these bodies, and for the small inclination of their equators
to the planes of their orbits.

On previous occasions I endeavoured to show that such an
arrangement would be the ultimate consequence of excessive
tides, when a satellite contained large bodies of fluid, or when,
from its close proximity to the primary, the solid matter of which
it may be composed were not possessed of sufficient cohesive
force to withstand the effects of the great disturbance. But the
same result would ultimately arise from slow secular changes
which must occur in every possible case. Let us suppose, for
instance, that the first satellite of Jupiter were composed entirely
of solid materials sufficiently strong to resist all crushing strains
to which they may be exposed, and that its form, in the absence
of all disturbing forces, were an exact sphere. Such a body,
turning on an axis perpendicular to the plane of its orbit and in

* Communicated by the Author.

a time different from that of its revolution, would have its equatorial gravity subject to a variation of about $1\frac{2}{3}$ per cent.; and the central pressure along the plane of the equator would undergo a periodical change of about 3000 pounds to the square inch. Now in consequence of the compressibility which belongs to every kind of solid matter, the satellite would be continually changed to an ellipsoid, the longest diameter always forming the same angle with the direction of the primary. If its component parts had a modulus of elasticity as great as that of iron, a difference of one-fifth of a mile may be expected between the major and mean axis; but were the mass of a more yielding character, or were it covered with fluid, its ever-changing form would deviate more considerably from a true sphere.

The effect which the attraction of the primary would exert on the rotation of an ellipsoidal satellite, the major axis of which had a constant inclination to the radius vector of its orbit, may be found by a method similar to that pursued for determining theoretically the amount of the precession of the equinoxes. Let A, B, and C be the major, mean, and minor semiaxes of the ellipsoid, the last being perpendicular to the plane of the orbit, and the first forming the angle ψ with the direction of the primary. Supposing the satellite homogeneous, the change in the velocity of rotation at the extremity of the major axis in a unit of time will be expressed by

$$\frac{3M}{2D^3}(A-B)\sin 2\psi, \quad \ldots \ldots \quad (1)$$

D being the distance of the primary, and M the measure of its attractive power. According to the theory of central forces, $\frac{M}{D^3} = \frac{4\pi^2}{T^2}$, π being put for 3·1416, and T for the time of revolution; the expression for the change in the equatorial movement thus becomes

$$\frac{6\pi^2(A-B)}{T^2}\sin 2\psi. \quad \ldots \ldots \quad (2)$$

Now, for a synchronism of the orbital and diurnal motions, the equator must have a velocity equal to $\frac{2\pi A}{T}$; and dividing this by the last expression, there results

$$T' = \frac{TA}{3\pi(A-B)\sin 2\psi}, \quad \ldots \ldots \quad (3)$$

T' denoting the time in which a satellite, having no primitive rotation, would acquire one sufficiently rapid for keeping the same point of its surface in perpetual conjunction with the primary.

In the case of a solid satellite composed of imperfectly elastic materials, it is necessary to take into consideration the slight change of density attending the constant alteration of form. Had this been done, the expressions (1) and (2) would be reduced to four-fifths of their value; while instead of formula (3) we should find

$$T = \frac{5TA}{12\pi(A-B)\sin 2\psi}. \quad \cdots \quad (4)$$

If the angle ψ were equal to 90 degrees, no change would be indicated in the rotation; but the angle could not have this magnitude except in the case of a solid satellite all parts of which were perfectly elastic, or in the case of one, consisting wholly or partially of fluid, which performed its tidal oscillations without friction.

By another investigation, which brevity compels me to omit, I have arrived at the same results in regard to the secular changes which the rotation of a secondary body must experience until it keeps pace with the orbital revolution. It will also readily appear that the ultimate effect of these changes is not affected by the inclination of the equator of the satellite to the plane of its orbit. But it will be necessary to show that the inclination is doomed to undergo a slow permanent diminution when the synchronism of the rotation and revolution is once established. For this purpose we may proceed in a manner similar to that employed in investigating the mutation of the earth's axis. Let I be the inclination of the equator of the satellite to the plane of its orbit, which for simplicity may be regarded as circular, and let L be the longitude of the satellite reckoned from the point of their intersection. Regarding the body as an ellipsoid, the tendency of the disturbing force to move the axis towards the plane of the orbit will be

$$\frac{3M}{2D^3}(A-C)\sin^2 I \sin 2L. \quad \cdots \quad (5)$$

If δI denote the change of inclination from this cause, then

$$\frac{d^2\delta I}{dt^2} = \frac{3M}{2D^3}\left(\frac{A-C}{A}\right)\sin^2 I \sin^2 L. \quad \cdot \quad (6)$$

On substituting nt for L, and regarding A and C as constant, the integration will give only periodical quantities; so that no permanent change would be indicated if the form were absolutely immutable. But supposing A—C to vary, either from the presence of large collections of fluid on the surface of the satellite, or from the necessary elasticity of its solid matter, the quantity

$\dfrac{A-C}{A}$ in the last equation must receive an increment, the principal term of which will be $C \cos(2L-2w)$, or

$$C(\cos 2L \cos 2w + \sin 2L \sin 2w), \quad . \quad . \quad . \quad (7)$$

in which w represents the increase of longitude during the interval between the times of high tides at any locality, and of the maximum intensity of tidal force. Denoting by N and N' the sine and cosine of w, which is constant, formula (6) becomes

$$\frac{d^2\delta I}{dt^2} = \frac{3M}{D^3}\left(\frac{A-C}{A}\sin^2 I \sin 2L + \frac{CN'}{2}\sin^2 I \sin 4L \right.$$
$$\left. - \frac{CN \sin^2 I \cos 4L}{2} + \frac{C \sin^2 IN}{2}\right). \quad . \quad . \quad . \quad (8)$$

If this equation be integrated, all the resulting terms of the second member will be periodical except the last, which will express the slow permanent diminution of I; but the term will disappear when w is exactly 90 degrees, as it should be if the oscillations on which the change of form depended were effected without any loss of force.

Although the influence of distant bodies in changing the plane of the orbit may prevent I from sinking to zero, yet we must recognize the tendency to the peculiar arrangement which reduces to the lowest scale the dynamic effects of the disturbing force on their surfaces of secondary planets. But though their times of rotation and the position of their axes may be adjusted for attaining this object, the eccentricity of the orbit would bring tidal action into existence; and any commotions which this might occasion in their seas must be attended with secular changes in the size and form of their orbits. This will appear evident when we consider that these tides could not reach their highest level on the parts of the satellite in conjunction with the primary, until some time after the disturbing force which produced them attained its greatest intensity; and the subordinate world would thus present a greater deviation from a true sphere, in passing from the lower to the higher apsis, than in returning to the former point. It would accordingly feel the restraint of the centripetal force more intensely when retiring from the primary than when approaching him; and its motion would be retarded during the former period to an extent slightly greater than that to which it is accelerated during the latter. We may therefore reasonably expect a secular alteration in its mean motion and the size of its orbit; but it may be advisable to show by analytical investigations, that such changes take place on a scale corresponding to the waste of tidal power.

Although this may be done without any hypothesis in regard

to form and density, yet we may more easily arrive at definite results by taking, as the most appropriate type of the figure of these bodies, the ellipsoid which a homogeneous fluid satellite must assume when its motions are adapted for keeping the same point of its surface always directed to the centre of the primary. Let A, B, and C represent the semiaxes; P, Q, and R the attractions at their extremities in the absence of all disturbing influences; and put $\frac{A^2-C^2}{A^2}=e^2$ and $\frac{B^2-C^2}{B^2}=e_l^2$.

By a course of investigation similar to that which I adopted for finding the attraction of a prolate spheroid in the Philosophical Magazine (vol. xx. p. 414) the following result may be obtained:—

$$P=\frac{2gk^2C^2}{A}\iint\frac{\cos^2\phi\sin\phi d\phi d\theta}{1-e^2\cos^2\phi-e_l^2\sin^2\phi\cos^2\theta}, \quad \cdot \quad (9)$$

in which g denotes the attractive force at the distance k of a small portion of the body, ϕ the angle formed with the axis A by any of the elementary pyramids extending from its extremity to the surface of the ellipsoid, and θ the angle which the projection of these pyramids on the plane of B and C forms with B. A double integration by series, rejecting the fourth and higher powers of e and e_l, gives

$$P=\frac{4\pi gk^2C^2}{3A}\left(1+\frac{3e^2}{5}+\frac{e_l^2}{5}\right). \quad \cdot \quad \cdot \quad \cdot \quad (10)$$

In like manner, by a slight modification of the process employed in the same article (page 415) for finding the attraction at the extremity of the minor axis of the prolate spheroid, we may obtain

$$Q=\frac{4\pi gk^2C^2}{3B}\left(1+\frac{e^2}{5}+\frac{3e_l^2}{5}\right), \quad \cdot \quad \cdot \quad \cdot \quad (11)$$

$$R=\frac{4\pi gk^2C}{3}\left(1+\frac{e^2}{5}+\frac{e_l}{5}\right). \quad \cdot \quad \cdot \quad \cdot \quad (12)$$

Let P′, Q′, and R′ represent the actual intensity of gravity at the extremity of each axis, taking into consideration the effects of centrifugal force and the disturbance of the primary, to which the axis A is always directed, while C is perpendicular to the plane of the orbit; for this condition is necessary for the equilibrium, as I have shown in the Philosophical Magazine for April 1861. Then

$$P'=\frac{4\pi gk^2C^2}{3A}\left(1+\frac{3e^2}{5}+\frac{e_l^2}{5}\right)-\frac{4\pi gk^2nh^3A}{D^3}, \quad \cdot \quad (13)$$

$$Q'=\frac{4\pi gk^2C^2}{3^2B}\left(1+\frac{e^2}{5}+\frac{3e_l^2}{5}\right), \quad \cdot \quad \cdot \quad \cdot \quad \cdot \quad \cdot \quad (14)$$

$$R'=\frac{4\pi gk^2C}{3}\left(1+\frac{e^2}{5}+\frac{e_l^2}{5}\right)+\frac{4\pi gk^2nh^3C}{3D^3}, \quad \cdot \quad \cdot \quad (15)$$

in which h denotes the radius of the primary supposed to be a sphere, D its distance, and n its density divided by that of the satellite. In the article just referred to, it has been shown that P', Q', and R' must be reciprocally proportional to A, B, and C ; and accordingly by equalling the values of AP', BQ', and CR' as deduced from the last equation, we obtain

$$\epsilon^2 = \frac{10nh^3}{D^3}, \quad \epsilon_i^2 = \frac{5nh^3}{D^3} ; \quad \cdots \cdots \quad (16)$$

whence

$$\epsilon_i^2 = \frac{\epsilon^2}{4} \text{ and } \frac{A-C}{A} = 4\left(\frac{B-C}{B}\right). \quad \cdots \quad (17)$$

If the satellite were not homogeneous, the ratio between the greatest and least ellipticities would vary between 4 and 6, the latter number expressing the ratio in the case in which the central matter alone is supposed to be endued with attractive power.

The extent to which the form of the satellite affects the intensity of the force which binds it to the primary, supposing this body to be a sphere, may be readily found by means of Ivory's theorem; and the application will be facilitated in the present case, in which the external point ranges with the axis A. The effect of the attractive force in moving the primary will be

$$\frac{4\pi k^2 g ABC}{3D^2}\left(1 + \frac{3}{5}\frac{A^2\epsilon^2}{D^2} - \frac{3}{10}\frac{A^2\epsilon_i^2}{D^2}\right). \quad \cdots \quad (18)$$

But the same amount of matter in a spherical form would attract the central orb with a force expressed by $\dfrac{4\pi k^2 g ABC}{3D^2}$; so that the excess of attractive power due to the ellipticity is

$$\frac{4\pi k^2 g ABC}{3D^4}\left(\frac{3}{5}A^2\epsilon^2 - \frac{3}{10}A^2\epsilon_i^2\right). \quad \cdots \quad (19)$$

Calling this F, and putting m for $\dfrac{4\pi k^2 g ABC}{3}$, and $\dfrac{\epsilon^2}{4}$ for ϵ_i^2, in accordance with formula (17),

$$F = \frac{21}{40}\frac{mA^2\epsilon^2}{D^4}. \quad \cdots \cdots \cdots \quad (20)$$

To show the effects of the change of form in consequence of the eccentricity of the orbit, which is to be regarded as deviating little from a circle, we must take the variation of the last formula. Then

$$\delta F = \frac{21}{20}\frac{mA\epsilon}{D^4}\left(A\delta\epsilon + \epsilon\delta A - \frac{2A\epsilon\delta D}{D}\right). \quad \cdots \quad (21)$$

But the volume of the ellipsoid is equal to $A^3\left(\dfrac{1-\epsilon^2}{\sqrt{1-\epsilon_i^2}}\right)$ or

$A^3 \left(1 - \dfrac{7}{8} e^2 \right)$ nearly, from which we obtain

$$A \delta e = \frac{12 \delta A}{7 e} \text{ nearly};$$

and formula (19) becomes

$$\delta F = \frac{9m A \delta A}{5 D^4} + \frac{21}{20} \frac{m A e^2 \delta A}{D^4} - \frac{21}{10} \frac{m A^2 e^2 \delta D}{D^5}, \quad (22)$$

δA is the change of level at the extremities of the major axis arising from the variation of the primary disturbance; and regarding these tides as conforming to dynamic principles, their maximum range must be proportional to the force producing them, multiplied by the square of its time of operation. As the force in these cases varies inversely as the fourth power of the distance, while the square of the time, according to Kepler's third law, is directly proportional to the cube of the same quantity, the maximum value of δA may be represented by $\dfrac{Se}{D}$. Now, if W be the angle which the satellite describes during the time the tidal force requires to produce its full effects, v being the true anomaly reckoned from the higher apsis, and e the relative eccentricity of the orbit, then

$$\delta A = - \frac{Se}{D} \cos (v - W). \quad \ldots \ldots \quad (23)$$

Substituting this value for δA, and for δD its approximate value $D_1 e \cos v$, formula (22) becomes

$$\delta F = - \frac{9m A Se \cos v - W}{5 D^5} - \frac{21}{10} \frac{m A^2 e^2 e D_1 \cos v}{D^5}, \quad (24)$$

the middle term of the second member being rejected as inconsiderable, and D_1 denoting the mean distance.

Formula (18) expresses the attractive force of the satellite on the primary supposed to be a sphere; F is the extent to which this force is augmented by the ellipticity of the satellite, and δF is the periodical change in the value of F in consequence of tidal fluctuations. The second term of the value of δF in equation (24) would be the same if the body were entirely solid; and accordingly it could not be expected to lead to non-periodical alterations in the orbit, but it has been retained to show that analysis leads to the same conclusion. Now to express the effect of these forces on the orbit which the satellite describes around the centre of the primary, the values of F and δF must be multiplied by $\dfrac{M+m}{m}$, M being the measure of the attractive energy of the central sphere. The disturbing force on the orbit thus becomes

$$\frac{M's}{r^4} - \frac{Mse\cos(v-W)}{r^5} - \frac{Ms''e\cos v}{r^5}, \quad . \quad . \quad (25)$$

M' being put for $(M+m)$, r for D, s for $\dfrac{21A^2\epsilon^2}{40}$, s' for $\dfrac{9AS}{5}$, and s'' for $\dfrac{21A^2\epsilon^2 D}{10}$. Accordingly, in the pro blem of the two bodies, the differential equation $\dfrac{d^2r}{dt^2} = \dfrac{a^2}{r^3} - \dfrac{M'}{r^2}$ becomes in the present case

$$\frac{d^2r}{dt^2} = \frac{a^2}{r^3} - \frac{M'}{r^2} - \frac{M's}{r^4} + \frac{M's'e}{r^5}\cos(v-w) + \frac{M's''e}{r^5}\cos v. \; . \; (26)$$

Multiplying by dr and integrating,

$$\frac{dr^2}{dt^2} = -b - \frac{a^2}{r^2} + \frac{2M}{r} + \frac{2}{3}\frac{M's}{r^3} + 2M's'e\int\frac{\cos(v-w)dr}{r^5}$$

$$+ 2M's''e\int\cos v\, dr. \quad . \quad . \quad . \quad . \quad . \quad . \quad . \quad (27)$$

In order to effect the integration of the last terms, which are extremely minute, we may substitute the elliptical values of r and dr in them. The last term integrated in this way gives only periodical quantities. But

$$\int\frac{\cos v - w}{r^5}dr = -\int\frac{e\sin v(1-e\cos v)^3}{p^4}(\cos v\cos W + \sin v\sin W)dv, \; (28)$$

p being the parameter of the orbit. Now the angle W being invariable, its sine and cosine may be expressed by the constant quantities H and H'; and the term becomes

$$-eH'\int\frac{\sin v\cos v(1-e\cos v)^3 dv}{p^4} - eH\int\frac{\sin^2 v(1-e\cos v)^3}{p^4}dv. \; (29)$$

It may be readily found that the first integral consists wholly of quantities multiplied by cosines of v and its multiples, and there-ore periodical, while the second is equivalent to

$$-\frac{eH}{2p^4}\int(1-\cos 2v)(1-e\cos v)^3 dv, \quad . \quad . \quad . \quad (30)$$

the secular part of which is found to be

$$-\frac{eHv}{2p^4}\left(1+\frac{e^2}{2}\right) \text{ or } -\frac{eHv}{2p^4} \text{ nearly.} \quad . \quad . \quad . \quad (31)$$

Equation (27) thus becomes

$$\frac{dr^2}{dt^2} = \frac{2M}{r} - \frac{a^2}{r^2} - b + \frac{2}{3}\frac{M's}{r^3} - \frac{M's'e^2Hv}{p^4} + \beta, \quad . \quad (32)$$

β denoting the periodical quantities arising from the disturbance. Since $\frac{2M's}{3r^3}$ is extremely minute, it may be replaced by $\frac{2M's}{3D_1{}^3}$, D_1 being the mean distance. The maximum and minimum values of r may then be found by a quadratic on making $\frac{dr^2}{dt^2} = 0$, and omitting the periodical terms denoted by β. From the coefficient of r in the resulting equation, it appears that

$$D_1 = \frac{M'}{b} - \frac{M^2}{b^3}\left(\frac{2D}{3D_1{}^3} + \frac{s'eHv}{p^4}\right) \quad . \quad . \quad . \quad (33)$$

$\frac{M'}{b}$ being the value of the mean distance in the absence of the disturbance, and s' being equal to $\frac{9AS}{5}$, the last expression becomes

$$D_1 = \text{Constant} - \frac{9AD^2Sev}{5p^4}\sin W; \quad . \quad . \quad . \quad (34)$$

and the secular diminution of D_1 during each revolution will be

$$\frac{5 \cdot 6548As\sin W}{D_1}\text{ nearly}, \quad . \quad . \quad . \quad . \quad (35)$$

in which s denotes the highest swell of the tides at the points of the satellite in conjunction and in opposition with the primary. The diminution which the disturbance occasions during the same period in the relative eccentricity or the eccentricity divided by D_1 will be

$$\frac{2 \cdot 8274As\sin W}{D_1{}^2}. \quad . \quad . \quad . \quad . \quad . \quad (36)$$

These results may also be obtained by investigating the variation of the elements of the orbit according to the method of Lagrange. If the tides could rise and fall on a satellite without any impediments from friction, W would become equal to 180 degrees, and there could be no permanent change in the ellipse which the body describes. It thus appears that the duration of the secondary planets is much dependent on the absence of tides from their surfaces; and perhaps the vast number of these attendants belonging to the remote planets may be indebted for their present existence to the intense cold, which keeps their oceans in a perpetually frozen condition.

Cincinnati, November 8, 1861.

LXIV. *Chemical Analysis by Spectrum-observations.*
By G. KIRCHHOFF *and* R. BUNSEN.
[With a Chromolithograph Plate.]
[Continued from p. 349.]

IV. *On Metallic Cæsium and some of its Compounds.*

a. *Metallic Cæsium.*

IF fused chloride of cæsium be placed in the circuit of a powerful zinc-carbon battery, exactly the same phenomena are noticed as when the chlorides of potassium or rubidium are thus treated.

The amalgam of cæsium is, however, not so easily formed from an aqueous solution of the chloride as is the rubidium-amalgam under similar circumstances. It can be obtained in a solid crystalline form only by the aid of a very powerful current. When thus prepared it is of a silver-white colour, exhibiting a granular structure. It undergoes oxidation on exposure to air much more rapidly than rubidium-amalgam, and quickly decomposes water. With a solution of chloride of potassium, it is found to be positively electric when compared with the amalgams of sodium, potassium, and rubidium; so that cæsium must be considered as the most electro-positive of all the known elementary bodies.

b. *Hydrated Oxide of Cæsium.*

The properties of fused chloride of cæsium, when acting as an electrolyte, show plainly that, like potassium, this metal forms a suboxide. We have not yet examined the compounds formed by cæsium with more than one atom of oxygen; the analogy of the metal with potassium would, however, render the existence of such compounds probable. The hydrated oxide, which is prepared in a similar manner to the corresponding rubidium compound, resembles the latter in all its properties. It contains one atom of water, which cannot be expelled by heat; it is in a high degree deliquescent, becomes strongly heated in contact with water, and is at least as powerful a caustic as potash or hydrated oxide of rubidium. It dissolves easily in alcohol, forming a syrupy liquid.

c. *Monocarbonate of Cæsium.*

Like the corresponding rubidium compound, this salt is most easily obtained by decomposing the boiling solution of the sulphate of cæsium with baryta-water, evaporating the caustic liquor to dryness with carbonate of ammonium, and separating any insoluble carbonate of barium by filtration. From the syrupy solution of the carbonate, the hydrated salt crystallizes in irregular masses, which soon deliquesce on exposure. The crystals, on heating, fuse in their water of crystallization, leaving a residue

of the anhydrous salt in the form of a sandy friable white mass,
which rapidly absorbs moisture from the air. At a red heat the
anhydrous salt melts; and it may be heated to whiteness, at which
temperature it begins to volatilize, without losing carbonic acid.
Placed on a platinum wire in the flame, it soon volatilizes completely. The aqueous solution of the salt possesses a strong
alkaline reaction and taste; when rubbed between the fingers, it
produces the peculiar soapy feeling characteristic of the alkalies,
and it acts as a cautery when it is allowed to remain for some time
in contact with the skin. Water containing $\frac{1}{10000}$th part of
the salt turns red litmus-paper distinctly blue. .

Monocarbonate of cæsium possesses a property which is re-
·markable in the alkaline carbonates, that, namely, of solubility
in absolute alcohol. 100 parts of alcohol dissolve, at 19° C.,
11·1, and at the boiling-point of the alcohol 20·1 parts of this
salt. The carbonate can be obtained in the form of small irregular crystals by quickly cooling the alcoholic solution. If the
cooling be carried on slowly to temperatures below 0° C., the
salt sometimes separates out in tabular crystals often 1 inch in
length, especially if some quantity of caustic oxide of cæsium be
present. 0·7921 grm. of the fused salt lost, on treatment with
dilute sulphuric acid, 0·1120 grm. carbonic acid. Hence the
salt contains—

		Calculated.	Found.
CsO . . . 131·35		85·65	85·86
CO^2 . . . 22·00		14·35	14·14
153·35		100·00	100·00

d. *Bicarbonate of Cæsium.*

A solution of monocarbonate of cæsium, exposed in an atmosphere of carbonic acid, passes into this salt in the course of a
few days. The solution, on standing in the air at the ordinary
temperature over sulphuric acid, deposits large but indistinctly
formed striated crystals, which are unalterable in the air, and
assume a prismatic form; they possess a feeble alkaline reaction;
their aqueous solution gives off carbonic acid on boiling, and in
outward properties they cannot be distinguished from the crystals of the corresponding rubidium salt. 0·8155 grm. of fused
monocarbonate of cæsium yielded 0·9761 grm. of bicarbonate
when exposed for some days in an atmosphere of carbonic acid,
and afterwards dried over sulphuric acid. Hence the composition
of the salt is—

		Calculated.	Found.
CsO . . . 131·35		71·25	71·56
2CO^2 . . . 44·00		23·87 }	28·44
HO. . . . 9·00		4·88 }	
184·35		100·00	100·00

e. *Nitrate of Cæsium.*

This salt contains no water of crystallization, it does not undergo alteration in the air, and may be obtained from its aqueous solution in the form of small shining crystals of a prismatic form, in which the faces of the prism are generally better defined than those at the summits. The crystals obtained by slow evaporation at 14° C. belong to the hexagonal system, and are isomorphous with nitrate of rubidium. The primary form is an obtuse hexagonal dodecahedron, with polar angles of 142° 56′, and basal angles of 78° 58′, corresponding to the following relation of the axes

$$1 : a = 1 : 0.71348.$$

The faces which could be observed (see Plate V. fig. 1) are as follows :—

$$P . \infty P . P2 . \infty P2 . OP . \tfrac{1}{4}P.$$
$$r \quad p \quad r_1 \quad p_1 \quad o \quad q.$$

	Calculated.	Found.
$p - p_1$	150 0	149 59
$p_1 - p$	150 0	149 58
$r - p_1*$		129 29
$r_1 - p_1$	125 30	125 28
$r - r_1$	161 28	161 41
$r - q$	172 14	172 0
$r_1 - o$	144 30	144 39

If the primary form be taken to be a hexagonal dodecahedron of the second order, the corresponding hexagonal dodecahedron of the first order yields as a hemihedral form a rhombohedron having polar angles of 106° 40′. Through this form, therefore, the isomorphism of the nitrates of cæsium and rubidium, and the potash and soda nitre, becomes apparent. We have—

Nitrate of cæsium	106 40
Nitrate of potassium	106 30
Nitrate of sodium	106 36

When crystallized quickly, the salt separates out in long needle-shaped prisms, longitudinally striated. It has the same saline bitter cooling taste as saltpetre—so much so that these salts cannot thus be distinguished from each other. On heating, the salt melts to a thin liquid at temperatures almost below the red heat; and when more strongly heated it evolves oxygen, and is converted first into nitrite, and afterwards, by absorption of moisture from the air, into caustic hydrate of cæsium, which

---

* This angle served as basis of calculation for the primary form.

attacks glass and platinum. In absolute alcohol the salt is very slightly soluble*.

Nitrate of cæsium is somewhat more difficultly soluble in water than the corresponding potassium compound; for whilst 100 parts of water at $+8°\cdot2$ C. dissolve $16\cdot1$ parts of the latter, $10\cdot58$ parts of nitrate of cæsium are dissolved under similar circumstances.

$3\cdot0567$ grms. of pure nitrate of cæsium gave $2\cdot8233$ grms. of the sulphate on decomposition and ignition with sulphuric acid. Hence the composition of the salt is—

	Calculated.	Found.
CsO ˙ . . . 131·35	70·87	70·80
NO⁵ . . . 54·00	29·18	29·20
185·35	100·00	100·00

f. *Bisulphate of Cæsium.*

Carbonate of cæsium is gradually heated with an excess of sulphuric acid until the temperature rises nearly to redness. The salt then consists of a transparent colourless liquid, which, on cooling, solidifies to a crystalline mass. Dissolved in water, the acid salt thus obtained crystallizes upon slow evaporation in the form of small short rhombic prisms, having rectangular terminations, and having the acute longitudinal edges equally bevelled. The crystals belong to the rhombic system. The relation of the horizontal axes is nearly

$$a : b = 1 : 1\cdot38.$$

The crystals obtained were badly formed, and their surfaces were not polished enough to enable us to make any accurate measurements with the reflecting goniometer. The relation of the principal axis to the horizontal axes could also not be obtained, as no faces were visible on the terminal edges of the prism. The crystals are represented by fig. 3, Plate V.

	Found.	Calculated.
$p-p$ on a . . .	107˚ 37'	108˚
$p-b$	126	

The salt has a strongly acid reaction and taste; it is, however, unalterable in the air. Heated gently, it melts quietly under a red heat; and when more strongly ignited, sulphuric anhydride escapes with effervescence, leaving a solid mass of neutral sulphate of cæsium, which melts at a temperature approaching a yellow heat.

---

* Saltpetre is by no means insoluble in alcohol, as Berzelius affirms. The slight solubility of the cæsium nitrate in alcohol cannot, therefore, be used, as one of us formerly proposed, as a distinctive reaction of these two salts.

g. *Neutral Sulphate of Cæsium.*

The aqueous solution of this salt possesses an insipid taste, but a bitter after-taste. It is far more soluble in water than the corresponding potassium-salt. 100 parts of water take up at −2° C. not less than 158·7 parts of sulphate of cæsium, whereas only 8·0 parts of sulphate of potassium are dissolved under similar circumstances. When the aqueous solution is allowed to evaporate slowly over sulphuric acid, small, irregularly formed, hard crystals are deposited, which generally are found to have the form of short flattened prisms, and often occur grouped together in irregular masses. The crystals are anhydrous, quite unaffected by exposure to air, and insoluble in alcohol. We have not succeeded in obtaining any individual crystals suitable for measurement.

The analysis of the salt was made by converting the carbonate into the sulphate. For this purpose 0·7921 grm. of fused carbonate of cæsium was treated with sulphuric acid, and yielded 0·8828 grm. of fused sulphate. Hence the composition of the salt is—

		Calculated.	Found.
CsO	. . . 131·35	76·66	76·85
SO^3	. . . 40·00	23·34	23·15
	171·35	100·00	100·00

With the sulphates of cobalt, nickel, magnesium, &c., sulphate of cæsium produces a series of beautifully crystallizing double salts, containing 6 atoms of water of crystallization, and isomorphous with the corresponding salts of potassium and rubidium. The following faces were observed in these crystals:—

$$OP \cdot \infty P \cdot + P \cdot [P \infty] \cdot + 2P \infty \cdot \infty P 2.$$

The sulphate of cæsium also forms with sulphate of aluminium a double salt containing 24 atoms of water, and crystallizing in the regular system, corresponding exactly to potassium and rubidium alums.

h. *Chloride of Cæsium.*

By neutralising the carbonate with hydrochloric acid and evaporating the solution, chloride of cæsium is obtained in the form of small anhydrous indistinct cubes. When quickly crystallized, the salt appears, like sal-ammoniac and chloride of potassium, as a mass of feathery crystals. Chloride of cæsium fuses at a low red heat, and volatilizes at a higher temperature much more easily than chloride of potassium, in the form of white vapours. The fused salt, on cooling, assumes the form of a white opake mass, which rapidly absorbs moisture from the air and deliquesces. When ignited for a long time in contact with air it becomes slightly basic.

According to the atomic-weight determinations, already described, 1·0124 grm. of chloride of cæsium, the solution of which was perfectly neutral, yielded 0·9133 grm. chloride of silver, and 0·0009 grm. of metallic silver from the filter-ash. This corresponds to the following numbers:—

						Calculated.	Found.
Cs					1·23·85	77·67	77·67
Cl					85·46	22·33	22·33
					158·81	100·00	100·00

i. *Double Chloride of Platinum and Cæsium.*

If, to an aqueous solution of chloride of cæsium, bichloride of platinum be added, a yellow precipitate is formed. The colour of this is somewhat lighter than that of the corresponding potassium-salt, because it is less soluble than the latter, and therefore is deposited in a finer state of division. The precipitate is anhydrous, and is composed of microscopic, honey-yellow, transparent regular octahedrons. 100 parts of water dissolve of this compound—

at 0 C.	0·021 part.	at 68 C.	0·234 part.
11	0·072 „	100	0·382 „
40	0·118 „		

These numbers are taken from the mean of a large number of careful determinations agreeing well amongst themselves.

As almost all the platinum which is found in commerce is very impure, and often possesses an atomic weight from 6 to 8 per cent. below the true value, we have previously purified the platinum which we used for the preparation of these as well as of the rubidium salts. This purification was effected by fusing the chloride of platinum and potassium in a platinum dish with a mixture of the carbonates of potassium and sodium, washing out the mass with water, and dissolving the residue in dilute aqua regia. When this operation had been repeated five times, it was found that the platinum attained an atomic weight varying but very slightly from 99·1.

The analysis of the double chloride was carried out as follows:—The salt was weighed out in a U-shaped tube of hard glass, after having been dried in a bath of fused chloride of zinc at a temperature of 160° to 170° C., and the tube, with the substance, bedded in magnesia and heated to dull redness, whilst a current of dry hydrogen was passed over the salt. The loss of weight thus obtained was determined, the chloride of cæsium separated by boiling with water from the insoluble platinum, both substances weighed, and the chlorine in the chloride of cæsium estimated with silver.

Experiment gave—

Chloride of platinum and cæsium employed.	8·6142 grms.
Loss on reduction with hydrogen	1·8725 „
Platinum separated	2·6188 „
Chloride of cæsium obtained	4·1544 „
Chloride of silver obtained	3·7506 „

Hence we obtain the following composition :—

			Calculated.	Found.
Bichloride of platinum	Pt .	99·10	30·14	30·25
	Cl² .	70·92	21·57	21·67
Chloride of cæsium .	Cs .	123·35	37·51	37·35
	Cl .	35·46	10·78	10·53
		328·83	100·00	99·80

It is interesting to compare the solubility of the double chlorides of platinum, rubidium, and cæsium with the potassium-platinum double salt. The solubility of the latter is seen from the following experiments, which were conducted with special care, the numbers being the mean of several well-agreeing determinations.

100 parts of water dissolve—

at 0 C.	0·724	chloride of platinum and potassium.	
6·8	0·873	„	„
13·8	0·927	„	„
46·5	1·776	„	„
71·0	3·018	„	„
100·0	5·199	„	„

By interpolation, the solubility of the cæsium, rubidium, and potassium-platinum chlorides is obtained for intervals of 10° C., and is found to be as follows :—

°C.	Potassium-salt.	Rubidium-salt.	Cæsium-salt.
0 C.	0·74	0·184	0·024
10	0·90	0·154	0·050
20	1·12	0·141	0·079
30	1·41	0·145	0·110
40	1·76	0·166	0·142
50	2·17	0·203	0·177
60	2·64	0·258	0·213
70	3·19	0·329	0·251
80	3·79	0·417	0·291
90	4·45	0·521	0·332
100	5·18	0·634	0·377

V. *Reactions of the Rubidium and Cæsium Compounds.*

Cæsium and rubidium are not precipitated either by sulphuretted hydrogen or by carbonate of ammonium. Hence both metals must be placed in the group containing magnesium, lithium, potassium, and sodium. They are distinguished from magnesium, lithium, and sodium by their reaction with bichloride of platinum, which precipitates them like potassium. Neither rubidium nor cæsium can be distinguished from potassium by any of the usual reagents. All three substances are precipitated by tartaric acid as white crystalline powders; by hydrofluosilicic acid as transparent opalescent jellies, and by perchloric acid as granular crystals; all three, when not combined with a fixed acid, are easily volatilized on the platinum wire, and they all three tinge the flame violet. The violet colour appears indeed of a bluer tint in the case of potassium, whilst the flame of rubidium is of a redder shade, and that of cæsium still more red. These slight differences can, however, only be perceived when the three flames are observed side by side, and when the salts undergoing volatilization are perfectly pure. In their reactions, then, with the common chemical tests, these new elements cannot be distinguished from potassium. The only method by means of which they can be recognized when they occur together is that of spectrum-analysis.

The spectra of rubidium and cæsium are highly characteristic, and are remarkable for their great beauty. In examining and measuring these spectra we have employed an improved form of apparatus, which in every respect is much to be preferred to that described in our first memoir. In addition to the advantages of being more manageable and producing more distinct and clearer images, it is so arranged that the spectra of two sources of light can be examined at the same time, and thus, with the greatest degree of precision, compared, both with one another and with the numbers on a divided scale.

The apparatus is represented by fig. 12, Plate VI. On the upper end of the cast-iron foot F a brass plate is screwed, carrying the flint-glass prism P, having a refracting angle of 60°. The tube A is also fastened to the brass plate; in the end of this tube, nearest the prism, is placed a lens, whilst the other end is closed by a plate in which a vertical slit has been made. Two arms are also fitted on to the cast-iron foot, so that they are moveable in a horizontal plane about the axis of the foot. One of these arms carries the telescope B, having a magnifying power of 8, whilst the other carries the tube C; a lens is placed in this tube at the end nearest to the prism, and at the other end is a scale which can be seen through the telescope by reflexion from the front surface of the prism. This scale is a photographic copy of a millimetre-scale, which has been produced in the camera, of about $\frac{1}{13}$ the original dimensions*. The scale is covered with tinfoil so that only the narrow strip upon which the divisions and the numbers are engraved can be seen.

The upper half only of the slit is left free, as is seen by reference to fig. 11, Plate VI.; the lower half is covered by a small equilateral glass prism, which sends by total reflexion the light of the lamp D, fig. 12, through the slit, whilst the rays from the lamp E pass freely through the upper and uncovered half. A small screen placed above the prism, prevents any of the light from D passing through the upper portion of the slit. By help of this arrangement the observer sees the spectra of the two sources of light immediately one under the other, and can easily determine whether the lines are coincident or not†.

We now proceed to describe the arrangement and mode of using the instrument.

The telescope B is first drawn out so far that a distant object is plainly seen, and screwed into the ring, in which it is held, care being taken to loosen the screws α and β beforehand. The tube A is then brought into its place, and the axis of B brought into one straight line with that of A. The slit is then drawn out until it is distinctly seen on looking through the telescope, and this latter is then fixed by moving the screws α and β, so that the middle of the slit is seen in about the middle of the field of view. After removing the small spring γ, the prism is next placed on the brass plate, and fastened in the position which is marked for it, and secured by screwing down the spring γ. If the axis of the tube A be now directed towards a bright surface, such as the flame of a candle, the spectrum of the flame is seen in the lower half of the field of the telescope on moving the latter through a certain angle round the axis of the foot F. When the telescope has been placed in position, the tube C is fastened on to the arm belonging to it; and this is turned through an angle round the axis of the foot such that, when a light is allowed to fall on the divided scale, the image of the scale is seen through the telescope B, reflected from the nearer face of the prism. This image is brought

---

* This millimetre-scale was drawn on a strip of glass covered with a thin coating of lampblack and wax dissolved in glycerine. The divisions and the numbers, which by transmitted light showed bright on a dark ground, were represented in the photograph dark on a light ground. It would be still better to employ, for the spectrum-apparatus, a scale in which the marks were light on a dark ground. Such scales are beautifully made by Salleron and Ferrier of Paris.

† This apparatus was made in the celebrated optical and astronomical atelier of C. A. Steinheil in Munich.

exactly into focus by altering the position of the scale in the tube C ; and by turning this tube on its axis, it is easy to make the line in which one side of. the divisions on the scale lie, parallel with the line dividing the two spectra, and by means of the screw δ to bring these two lines to coincide.

In order to bring the two sources of light, D and E, into position, two methods may be employed. One of these depends upon the existence of bright lines in the inner cone of the colourless gas-flame, which have been so carefully examined by Swan. If the lamp E be pushed past the slit, a point is easily found at which these lines become visible ; the lamp must then be pushed still further to the left, until these lines nearly or entirely disappear ; the right mantle of the flame is now before the slit, and into this the bead of substance under examination must be brought. In the same way the position of the source of light D may be ascertained.

The second method is as follows :—The telescope B is so placed that the brightest portion of the spectrum of the flame of a candle is seen in about the middle of the field of view; the flame is then placed before the ocular in the direction of the axis of the telescope, and the position before the slit determined in which the upper half of the slit appears to be the brightest; the lamp E is then placed so that the slit appears behind that portion of the flame from which the most light is given off after the introduction of the bead. In a similar way the position of the lamp D is determined by looking through the small prism and the lower half of the slit.

By means of the screw ε, the breadth of the slit can be regulated in accordance with the intensity of the light, and the degree of purity of spectrum which is required. To cut off foreign light, a black cloth, having a circular opening to admit the tube C, is thrown over the prism P and the tubes A and B. The illumination of the scale is best effected by means of a luminous gas-flame placed before it ; the light can, if necessary, be lessened by placing a silver-paper screen close before the scale. The degree of illumination suited to the spectrum under examination can then be easily found by placing this flame at different distances.

In order to obtain representations of the spectra of cæsium and rubidium corresponding to those of the other metals which we have given in our former paper, we have adopted the following course :—

We placed the tube C in such a position that a certain division of the scale, viz. No. 100, coincided with Fraunhofer's line " D " in the solar spectrum, and then observed the position of the dark solar lines A, B, C, D, E, F, G, H on the scale ;

these several readings we called A, B, C, &c. An interpolation scale was then calculated and drawn, in which each division corresponded to a division on the scale of the instrument, and in which the points corresponding to the observations A, B, C, &c. were placed at the same distances apart as the same lines on our first drawings of the spectrum. By help of this scale, curves of the new spectra were drawn, in which the ordinates express the degrees of luminosity at the various points on the scale, as judged of by the eye. The lithographer then made the designs represented in fig. 4, Plate V. from these curves*.

As in our first memoir, so here we have represented only those lines which, in respect to position, definition, and intensity, serve as the best means of recognition. We feel it necessary to repeat this statement, because it has not unfrequently happened that the presence of lines which are not represented in our drawings has been considered as indicative of the existence of new bodies.

We have likewise added a representation of the potassium spectrum to those of the new metals for the sake of comparison, so that the close analogy which the spectra of the new alkaline metals bear to the potassium-spectrum may be at once seen. All three possess spectra which are continuous in the centre, and decreasing at each end in luminosity. In the case of potassium this continuous portion is most intense, in that of rubidium less intense, and in the cæsium-spectrum the luminosity is least. In all three we observe the most intense and characteristic lines towards both the red and blue ends of the spectrum.

Amongst the rubidium lines, those splendid ones named Rbα and Rbβ are extremely brilliant, and hence are most suited for the recognition of the metal. Less brilliant, but still very characteristic, are the lines Rbδ and Rbγ. From their position they are in a high degree remarkable; as they both fall beyond Fraunhofer's line A; and the outer one of them lies in an ultra-red portion of the solar spectrum, which can only be rendered visible by some special arrangement. The other lines, which are found on the continuous part of the spectrum, cannot so well be used as a means of detection, because they only appear when the substance is very pure, and when the luminosity is very great. Nitrate of rubidium, and the chloride, chlorate, and perchlorate of rubidium, on account of their easy volatility, show

* The coincidence of this chromolithograph plate with that published in the former memoir is by no means complete, but this does not seriously interfere with the utility of either of the representations; for if the position of an observed line be found, by help of the scale above described, to be near to that of the line of any known substance, it is easy, by placing some of this substance in one of the flames, and in the other some of the body under examination, to see whether the lines are coincident or not.

these lines most distinctly. Sulphate of rubidium and similar salts also give very beautiful spectra. Even silicate and phosphate of rubidium yield spectra in which all the details are plainly seen.

The spectrum of cæsium is especially characterized by the two blue lines Csα and Csβ; these lines are situated close to the blue strontia line Srδ, and are remarkable for their wonderful brilliancy and sharp definition. The line Csδ, which cannot be so conveniently used, must also be mentioned. The yellow and green lines represented on the figure, which first appear·when the luminosity is great, cannot so well be employed for the purpose of detecting small quantities of the cæsium compounds; but they may be made use of with advantage as a test of the purity of the cæsium salt under examination. They appear much more distinctly than do the yellow and green lines in the potassium-spectrum, which, for this reason, we have not represented.

As regards *distinctness* of the reaction, the cæsium compounds resemble in every respect the corresponding rubidium salts: the chlorate, phosphate, and silicate gave the lines perfectly clearly. The *delicacy* of the reaction, however, in the case of the cæsium compounds is somewhat greater than in that of the corresponding compounds of rubidium. In a drop of water weighing 4 milligrammes, and containing only 0·0002 milligramme of chloride of rubidium, the lines Rbα and Rbβ can only just be distinguished; whilst 0·00005 milligramme of the chloride of cæsium can, under similar circumstances, easily be recognized by means of the lines Csα and Csβ.

If other members of the group of the alkaline metals occur together with cæsium and rubidium, the delicacy of the reaction is of course materially impaired, as is seen from the following experiments, in which the mixed chlorides contained in a drop of water weighing about 4 milligrammes, were brought into the flame on a platinum wire.

When 0·003 milligramme of chloride of cæsium was mixed with from 300 to 400 times its weight of the chlorides of potassium or sodium, it could be easily detected. Chloride of rubidium, on the other hand, could be detected with difficulty when the quantity of chloride of potassium or chloride of sodium amounted to 100 to 150 times its weight of the chloride of rubidium employed.

0·001 milligramme of chloride of cæsium was easily recognized when it was mixed with 1500 times its weight of chloride of lithium; whilst 0·001 milligramme of chloride of rubidium could not be recognized when the quantity of chloride of lithium added exceeded 600 times the weight of the rubidium salt.

———————

At the close of this memoir we cannot refrain from touching upon a question to which, on some future occasion, we must again·recur. Amongst the large number of those salts already examined by us, which, owing to their volatility in·the flame, render a spectrum-analysis possible, we have not found, in spite of the great variation in the elementary bodies combined with the metal, a single one which failed to produce the characteristic bright lines of the metal. Considering these numerous observations, made under the most widely differing circumstances, we might be led to suppose that in *all cases* the bright lines given out by a body occur quite independently of the other elements chemically combined with that body, and that therefore the relation of the elements, as regards the spectra of their vapours, is exactly the same, whether they are free or chemically combined. Yet this supposition is by no means founded on fact. We have repeatedly insisted that the bright lines in the spectrum of a luminous gas must coincide with the absorption lines which this gas produces in a continuous spectrum of a sufficient degree of luminosity. It is well known that the absorption lines of iodine vapour cannot be produced by hydriodic acid, and that, on the other hand, the absorption lines of nitrous acid are not visible in a mechanical mixture of nitrogen and oxygen. There is nothing to show why an influence of chemical combination· upon the absorption lines, similar to that here noticed at low temperatures, should not occur at a white heat. If, however, the state of chemical combination alters the absorption lines of a luminous gas, it *must* likewise alter the bright lines of its spectrum.

From these considerations one would conclude that in the case of the spectra of two different compounds of the same metal, different bright lines may appear; it is, however, possible that the salts which are volatilized in the flame cannot exist at the temperature of the flame, and are decomposed, so that it may be in reality the vapour of the free metal which produces the lines; and it would then appear quite possible that a chemical compound may produce bright lines differing from those produced by its constituent elements.

LXV. *On the* Rev. T. P. Kirkman's *Problem respecting certain Triadic Arrangements of Fifteen Symbols. By* W. S. B. Woolhouse, *F.R.A.S., F.I.A., F.S.S. &c.**

IN the 'Lady's and Gentleman's Diary' for the year 1844, I proposed the following mathematical prize question:—

* Communicated by the Author.

Determine the number of combinations that can be made out of s symbols, p symbols in each; with this limitation, that no combination of q symbols which may appear in any one of them shall be repeated in any other.

This question, which essentially involves a developed theory of partitions, is more difficult than would at first appear; and it has not yet received anything like an approach to a complete general investigation, although it has given rise to some able papers on cognate subjects by Professor Sylvester, Mr. Cayley, &c. in the Philosophical Magazine and other scientific journals. The Rev. T. P. Kirkman, who, like Professor Sylvester, has gone somewhat elaborately into the subject of partitions, and has brought considerable ingenuity to bear upon his researches, has made the largest contributions towards the solution of the problem referred to. His early investigations in this particular field of inquiry led him to construct the following curious triadic problem, which was proposed amongst the queries given in the 'Lady's and Gentleman's Diary' for 1850 :—

Fifteen young ladies in a school walk out three abreast for seven days in succession: it is required to arrange them daily so that no two shall walk twice abreast.

Two solutions, one of them by the talented proposer, were printed in the 'Diary' for 1851 ; but in these the results only were exhibited. Since that time the question has found its way into general society, and become somewhat noted as a fashionable puzzle, while, more scientifically considered, it has not failed to attract the attention of several eminent mathematicians.

Professor Sylvester, at the end of his paper "On a Four-valued Function," printed in the Philosophical Magazine for June last, page 520, has made some passing allusions to those mathematicians who, in common with himself, have contributed to the subject under consideration. On a recent perusal of this interesting paper I could not help noticing the summary character of these allusions, which first suggested to my mind the propriety of making the present communication with the view of pointing out the fact that the Rev. T. P. Kirkman originated this particular problem, and that it first appeared in the 'Diary.' I have at the same time been induced to give a systematic and comprehensive investigation of everything relating to it, in the 'Diary' for 1862, just published. In the investigation there given, it is shown that every solution to the problem must be contained in one or other of the three following systems :—

	1.	2.	3.	4.	5.	6.	7.
System No. 1.	$a_1 b_1 c_1$ $a_2 b_2 c_2$ $a_3 b_3 c_3$ $a_4 b_4 c_4$ $a_5 b_5 c_5$	$a_1 b_2 b_4$ $b_3 a_2 c_4$ $b_5 c_2 a_4$ $c_1 a_3 a_5$ $b_1 c_5 c_3$	$a_1 a_2 a_3$ $c_2 b_3 a_5$ $c_5 c_4 c_1$ $b_4 b_5 b_1$ $b_2 c_3 a_4$	$a_1 b_3 b_5$ $c_4 c_2 b_1$ $c_3 a_5 b_4$ $a_3 c_5 b_2$ $a_2 a_4 c_1$	$a_1 c_2 c_5$ $a_5 c_4 b_2$ $a_4 b_1 a_3$ $b_5 c_3 a_2$ $b_3 c_1 b_4$	$a_1 c_4 c_3$ $b_1 a_5 a_2$ $c_1 b_2 b_5$ $c_5 a_4 b_3$ $c_2 b_4 a_3$	$a_1 a_5 a_4$ $b_2 b_1 b_3$ $b_4 a_2 c_5$ $c_3 c_1 c_2$ $c_4 a_3 b_5$
System No. 2.	$b_1 a_1 c_1$ $a_2 c_2 b_2$ $a_3 b_3 c_3$ $b_4 c_4 a_4$ $c_5 a_5 b_5$	$b_1 a_2 a_5$ $c_2 a_3 b_4$ $b_2 b_5 c_1$ $c_3 a_4 b_3$ $a_1 c_4 c_3$	$b_1 c_2 c_4$ $a_3 b_2 c_5$ $b_4 c_3 a_5$ $a_1 b_3 b_5$ $a_2 a_4 c_1$	$b_1 a_3 a_4$ $b_2 b_4 a_1$ $c_5 c_1 c_4$ $a_2 b_5 c_3$ $c_2 b_3 a_5$	$b_1 b_2 b_3$ $b_4 c_5 a_2$ $a_1 a_5 a_4$ $c_2 c_3 c_1$ $a_3 b_5 c_4$	$b_1 b_4 b_5$ $c_5 a_1 c_2$ $a_2 c_4 b_3$ $a_3 c_1 a_5$ $b_2 c_3 a_4$	$b_1 c_5 c_3$ $a_1 a_2 a_3$ $c_2 a_4 b_5$ $b_2 a_5 c_4$ $b_4 c_1 b_3$
System No. 3.	$a_1 b_1 c_1$ $a_2 b_2 c_2$ $a_3 b_3 c_3$ $a_4 b_4 c_i$ $a_5 b_5 c_5$	$a_1 a_2 b_3$ $b_2 a_3 a_4$ $c_2 b_5 c_4$ $a_5 c_3 c_1$ $b_1 c_5 b_4$	$a_1 b_2 b_5$ $a_3 c_2 a_5$ $a_4 c_5 c_1$ $b_1 c_4 b_3$ $a_2 b_4 c_3$	$a_1 a_3 c_5$ $c_2 a_4 b_1$ $a_5 b_4 b_3$ $a_2 c_1 b_5$ $b_2 c_3 c_4$	$a_1 c_2 b_4$ $a_4 a_5 a_2$ $b_1 c_3 b_5$ $b_2 b_3 c_5$ $a_3 c_4 c_1$	$a_1 a_4 c_3$ $a_5 b_1 b_2$ $a_2 c_4 c_5$ $a_3 b_5 b_4$ $c_2 c_1 b_3$	$a_1 a_5 c_4$ $b\ a_2 a_3$ $b_2 c_1 b_4$ $c_2 c_5 c_3$ $a_4 b_3 b_5$

Here each of the seven combinations is derived from that which immediately precedes it by a fixed law of succession, which on continuation will circulate through exactly the same positions. Also the symbols occupying corresponding places, taken in order horizontally in each system, proceed according to two fixed cyclical series, each consisting of seven symbols, viz.—

$$\text{No. 1. } \begin{cases} b_1 & b_2 & a_2 & b_3 & c_2 & c_4 & a_5, \\ c_1 & b_4 & a_3 & b_5 & c_5 & c_3 & a_4; \end{cases}$$

$$\text{No. 2. } \begin{cases} a_1 & a_2 & c_2 & a_3 & b_2 & b_4 & c_5, \\ c_1 & a_5 & c_4 & a_4 & b_3 & b_5 & c_3; \end{cases}$$

$$\text{No. 3. } \begin{cases} b_1 & a_2 & b_2 & a_3 & c_2 & a_4 & a_5, \\ c_1 & b_3 & b_5 & c_5 & b_4 & c_3 & c_4. \end{cases}$$

The three systems are essentially independent of each other, and cannot be mutually transformed by symbolic substitution. The particular distribution of the primary combination of No. 2 has been modified, so as to exhibit the remarkable fact that the systems No. 1 and No. 2 admit of being made up of the same set of thirty-five triads, and so as to comprise the same triads in four of the seven combinations. Thus the 1st, 4th, 6th, and 7th combinations of No. 1 correspond to the 1st, 3rd, 2nd, and 5th combinations of No. 2. The thirty-five triads of No. 1 or No. 2 being collected, may be symmetrically arranged thus:—

$$a_1b_1c_1 + a_2b_2c_2 + a_3b_3c_3 + a_4b_4c_4 + a_5b_5c_5$$

$$+\,a_1\begin{vmatrix} a_2a_3+b_1 \\ a_4a_5 \\ b_2b_4 \\ b_3b_5 \\ c_2c_5 \\ c_3c_4 \end{vmatrix}\begin{vmatrix} b_2b_3+c_1 \\ b_4b_5 \\ c_2c_4 \\ c_3c_5 \\ a_2a_5 \\ a_3a_4 \end{vmatrix}\begin{vmatrix} c_2c_3+a_2 \\ c_4c_5 \\ a_2a_4 \\ a_3a_5 \\ b_2b_5 \\ b_3b_4 \end{vmatrix}\begin{vmatrix} b_3c_4+b_2 \\ b_4c_5 \\ b_5c_3 \end{vmatrix}\begin{vmatrix} c_3a_4+c_2 \\ c_5a_5 \\ c_5a_3 \end{vmatrix}\begin{vmatrix} a_3b_4 \\ a_4b_5 \\ a_5b_3 \end{vmatrix}$$

$$+\,a_5b_4c_3 + b_5c_4a_3 + c_5a_4b_3\,;$$

and it is a still more remarkable fact that from this same set of triads it is possible to construct no less than 1080 systematic arrangements according to the first and second systems, each of them fulfilling the conditions of the problem.

The triads contained in the system No. 3 are essentially unsymmetrical, and admit of only one systematic arrangement.

In the 'Diary' I have also briefly indicated a direct and effectual method of systematizing any given solution when presented under an irregular form, and of thereby ascertaining to which of the three systems it belongs. As an example of the application of this method, the numerical solution given in the 'Diary' for 1851 is here transcribed for comparison with the same after being so adjusted and arranged*.

Solution by Mr. Samuel Bills, of Hawton, near Newark-upon-Trent; Mr. Thomas Jones, Abbey Buildings, Chester; Mr. Thomas Wainman, Burley, near Leeds; and Mr. W. H. Levy, of Shalbourne, near Hungerford.

1st day.	2nd day.	3rd day.	4th day.	5th day.	6th day.	7th day.
1 2 3	1 4 5	1 6 7	1 8 9	1 10 11	1 12 13	1 14 15
4 8 12	2 8 10	2 12 14	2 13 15	2 4 6	2 5 7	2 9 11
5 11 14	3 12 15	3 8 11	3 5 6	3 13 14	3 9 10	3 4 7
6 9 15	6 11 13	4 9 13	4 10 14	5 9 12	4 11 15	5 8 13
7 10 13	7 9 14	5 10 15	7 11 12	7 8 15	6 8 14	6 10 12

The same duly arranged: System No. 1.

5 14 11	5 10 15	5 13 8	5 4 1	5 7 2	5 9 12	5 3 6
13 10 7	4 13 9	7 4 3	9 7 14	3 9 10	14 3 13	10 14 4
8 4 12	1 7 6	2 9 11	12 3 15	6 14 8	11 10 1	15 13 2
6 15 9	11 8 3	15 1 14	8 2 10	1 12 13	2 6 4	12 11 7
3 1 2	14 2 12	10 12 6	13 6 11	4 11 15	7 15 8	9 8 1

I have further to observe that in constructing the first system the seven resulting combinations will be essentially the same, though occurring in a different order, when the primary combi-

* If any correspondents should think it worth while to communicate other solutions, I shall willingly systematize them in like manner.

nation is successively put under the following twenty-four forms, which for greater simplicity are here represented by numerals:—

A.			B.			C.			D.			α.			β.			γ.			δ.		
1	2	3	1	2	3	1	2	3	1	2	3	1	3	2	1	3	2	1	3	2	1	3	2
4	5	6	7	8	9	11	12	10	15	13	14	4	6	5	7	9	8	11	10	12	15	14	13
7	8	9	4	5	6	5	6	4	6	4	5	7	9	8	4	6	5	5	4	6	6	5	4
10	11	12	13	14	15	8	9	7	12	10	11	13	15	14	10	12	11	14	13	15	9	8	7
13	14	15	10	11	12	14	15	13	9	7	8	10	12	11	13	15	14	8	7	9	12	11	10

A'.			B'.			C'.			D'.			α'.			β'.			γ'.			δ'.		
1	2	3	1	2	3	1	2	3	1	2	3	1	3	2	1	3	2	1	3	2	1	3	2
6	4	5	9	7	8	10	11	12	8	9	7	11	10	12	14	13	15	9	8	7	10	12	11
15	13	14	12	10	11	13	14	15	5	6	4	14	13	15	11	10	12	15	14	13	4	6	5
9	7	8	6	4	5	4	5	6	11	12	10	8	7	9	5	4	6	6	5	4	7	9	8
12	10	11	15	13	14	7	8	9	14	15	13	5	4	6	8	7	9	10	12	11	13	15	14

A''.			B''.			C''.			D''.			α''.			β''.			γ''.			δ''.		
1	2	3	1	2	3	1	2	3	1	2	3	1	3	2	1	3	2	1	3	2	1	3	2
5	6	4	8	9	7	12	10	11	13	14	15	6	5	4	9	8	7	10	12	11	14	13	15
11	12	10	14	15	13	9	7	8	10	11	12	15	14	13	12	11	10	13	15	14	8	7	9
14	15	13	11	12	10	15	13	14	7	8	9	12	11	10	15	14	13	7	9	8	11	10	12
8	9	7	5	6	4	6	4	5	4	5	6	9	8	7	6	5	4	4	6	5	5	4	6

In consequence of this flexibility in the disposition of the constituent triads of each combination, a solution obtained by a tentative process is most likely to belong to the first system. The seven combinations which result from the primaries A', A'' follow in the order of those from A when the latter are taken with strides of two and four respectively; and so of the others.

To determine the number of synthetic combinations of the fifteen symbols that can be formed out of a given set of thirty-five triads, suppose pqr to be a triad taken as one of a combination: it can be associated only with the sixteen of the remaining triads that do not contain p, q or r. Let $p'q'r'$, taken from these, be the second triad; then $p'q'r'$ can be associated only with the six of the sixteen triads that do not contain p', q' or r'. Again, let a third triad $p''q''r''$ be taken from these; then $p''q''r''$ can be associated only with the two of the six triads that do not contain p'', q'' or r''; and these last will be the fourth and fifth triads of the combination. The number of combinations that can thus be made, comprising the fifteen symbols, will therefore be $35 \times 16 \times 6 \times 2$; and as the results will comprehend every form of permutation of the five triads under each combination, the total number of such combinations that can be formed, without permutation, will be

$$\frac{35 \times 16 \times 6 \times 2}{5 \times 4 \times 3 \times 2} = 56.$$

These combinations are stated at length in the 'Diary.'

Before concluding, I may be permitted to take the opportunity of briefly stating another analogous problem that may possibly interest some of the numerous readers of the Journal:—

> Sixteen symbols may be arranged five times in the form of a square, so that every pair of symbols shall appear once both in a horizontal and a vertical line.

If not hereafter anticipated, I may take a future opportunity of communicating a discussion of this neat problem.

Alwyne Lodge, Canonbury,
November 8, 1861.

LXVI. *Chemical Notices from Foreign Journals.*
By E. ATKINSON, *Ph.D.*, *F.C.S.*

[Continued from p. 309.]

THERE are certain substances existing abundantly in nature, such as hydrogen, fluoride of silicon, and carbonic acid, which, without becoming fixed on the substances with which they come in contact, change them into mineral substances identical with those occurring in nature. These agents Deville* calls minera-lizing agents, and in a series of experiments has shown that hy-drochloric acid constitutes one. If sesquioxide of iron be heated to dull redness in a porcelain tube, and a rapid current of hydro-chloric acid passed through it, sesquichloride of iron is condensed on the cooler parts of the apparatus, and water escapes along with the excess of acid. But if the current is slow and regular, the sesquioxide is changed into crystals quite identical in form with those of specular iron ore; at the same time as much hydro-chloric acid escapes as enters the apparatus, not a trace of water being formed. If the temperature is very high, the crystals have the same form and the same angles as those of Elba iron ore; while if the temperature is lower, the crystals resemble volcanic specular iron ore.

Deville has also prepared† artificial cassiterite and rutile by the same method. Amorphous oxide of tin, obtained by the action of nitric acid on tin, is placed in a platinum tray which is heated in a porcelain tube to the fusing-point of copper, while a slow current of hydrochloric acid is transmitted through the tube. The oxide of tin remains behind in small but well-defined crystals, identical in form with those of the native mineral. When the current is rapid, a small quantity of bichloride of tin is formed, which is transported to the further end of the tube;

* *Comptes Rendus*, June 1861.   † Ibid. July 22, 1861.

by an ulterior reaction of aqueous vapour this is changed into crystals which are somewhat larger and more perfect.

Hydrochloric acid also exercises a very curious reaction on amorphous titanic acid, changing it into very small crystals which resemble anatase or rutile.   They are very lustrous, and have a bluish colour, like natural anatase.   This blue colour arises from a partial reduction of titanic acid; for when this body was treated in a reducing atmosphere by gaseous hydrochloric acid, small crystals were obtained of a deep indigo-blue colour, the analysis of which showed that they were a new oxide of titanium,

$$Ti^3 O^5 = TiO^2, Ti^2 O^3.$$

Artificial rutile may also readily be obtained by the following method devised by Deville in conjunction with Caron.   When a mixture of titanic acid and protoxide of tin is made, a titanate is formed at a red heat, which silica decomposes, forming a silicate and crystallized titanic acid.   The operation is effected by heating to redness in an earthen crucible titanic acid along with oxide of tin and a little sand.   A gangue is formed rich in tin, and in which are imbedded crystals of rutile 5 or 6 millims. in length. Their form is that of native rutile; they are pure and colourless if the materials are so, but they usually contain a trace of iron or manganese, which give them the colour of native rutile.

In the same manner Deville[*] has studied the formation of other minerals.

Magnetite.—Magnetic oxide of iron was prepared by heating protoxide of iron in a slow current of hydrochloric acid gas. Protochloride of iron was formed, but no water.   The crystals remaining in the tray were regular octahedra, and were found on analysis to have the formula $Fe^3 O^4$.   The protoxide of iron was prepared by Debray's method of heating sesquioxide of iron in a mixture of equal volumes of carbonic acid and carbonic oxide gases.

Periclase.—When calcined magnesia was heated in a slow current of hydrochloric acid gas, it was transformed without any loss into small crystals of periclase, which were sometimes white, sometimes greenish, and occasionally yellow from the presence of iron.   Their form is the regular octahedron, and when produced at a very high temperature they are of considerable size. Chloride of magnesium in vapour is also decomposed by water, forming transparent octahedra of periclase.

Martite.—When a mixture of calcined magnesia and sesquioxide of iron was heated in a slow regular current of hydrochloric acid, two distinct substances were produced; one of them was periclase, and the other consisted of lustrous octahedra. They were found on analysis to have the composition $Fe^2 O^3 MgO$, and are true spinelles.

Hausmannite.—By heating red oxide of manganese in hydrochloric acid gas, small dimetric octahedra of Hausmannite, $Mn^2 O^3 MnO$, were obtained.

Protoxide of Manganese.—This was prepared by reducing any oxide of manganese in hydrogen, and heating it in the apparatus with a little hydrogen and a few bubbles of hydrochloric acid gas succeeding each other at long intervals. The small quantity of the latter gas required is truly surprising, and it escapes from the apparatus unaltered. The crystals obtained have a remarkable lustre; their colour is emerald-green, and they appear to be highly refringent, but exercise no action on polarized light. Their form is that of the cube-octahedron.

The same chemist has made the following observations[*]. When fluoride of silicon is passed over oxide of zinc at a high temperature, a mixture is formed of silicate and fluoride of zinc, which dissolve each other. The latter being volatile, on being heated the silicate is left in hexagonal prisms large enough to be readily measured, by which, and by their analysis, they were identified with native Willemite, $3ZnO, SiO^3$.

Fluoride of zinc acting upon silica gave the same products; so that a small quantity of fluoride of silicon could mineralize an indefinite quantity of silica and oxide of zinc.

Daubrée had stated that he had obtained Willemite and zircon by the action of chloride of silicon upon oxide of zinc or of zirconia. Deville, who has repeated these experiments with care, has found that neither Willemite nor zircon is formed; in fact by passing chloride of zircon over Willemite this substance is destroyed. This result might be expected; for the chlorides of silicon, in acting upon mineral oxides, do so not only by their chlorine, but by the metalloid, which exerts a powerful reducing action; and as the metallic chlorides formed under the influence of the chloride of silicon never dissolve the silicates formed, there is no reason why they should crystallize. The reverse is the case with fluoride of silicon; it is to the solvent effects of this substance on the silicates that its mineralizing properties are due. A series of experiments on the action of chloride of silicon on various metallic oxides, and which led to negative results, proved the correctness of these views.

Schiel has communicated[†] a determination of the atomic weight of silicon.

[*] *Comptes Rendus,* June 1861.
[†] Liebig's *Annalen,* October 1861.

A small glass bulb containing chloride of silicon was placed in water containing some ammonia, and the bulb broken. After the decomposition, the liquid was still feebly alkaline; it was allowed to stand for some days, then heated to boiling, filtered, and the silicic acid well washed out. The chlorine was precipitated from the filtrate by adding solution of nitrate of silver acidulated with nitric acid, and the chloride of silver washed out with water containing nitric acid. In two determinations 0·6738 and 1·3092 grm. of chloride of silicon gave respectively 2·277 and 4·418 of chloride of silver, from which is obtained the average 28·01 as the equivalent of silicon. The same number also follows from the vapour-density of chloride of silicon (5·39) and of fluoride of silicon (3·57), inasmuch as only the formulæ $SiCl^4$ and SiF^4 correspond to the vapour-density. Hence silicon, like tin and titanium, is quadratomic.

M. Rosensthiehl[*] has investigated the action of anhydrous sulphuric acid upon common salt. To a quantity of the former substance, placed in a stoppered retort, a quantity of powdered fused chloride of sodium was added. As the anhydrous acid generally contained a trace of the hydrated acid, sufficient heat was disengaged by the reaction thereby set up to melt the anhydrous acid. When the first reaction was over, the mixture was distilled until the residue fused. The distillate was analysed, and was found to be monochlorinated sulphuric acid formed in virtue of the reaction,

$$NaCl + 2S^2O^6 = NaO\,2S\,O^3 + S^2O^5Cl$$

Anhydrous sulphuric acid.   Bisulphate of soda.   Monochlorinated sulphuric acid.

A determination of the vapour-density confirmed the accuracy of this formula.

The new body is an oily colourless liquid of the spec. grav. 1·762, boiling between 145° and 150°, fuming in the air rather less than anhydrous sulphuric acid.

Projected upon a crystal of chromate of potash, it forms immediately chloride of chromyle,

$$KO\,CrO^3 + S^2O^5Cl = KO\,S^2O^6 + CrO^2Cl.$$

It acts in the cold on anhydrous acetate of soda without carbonizing it, and forms chloride of acetyle. This proves that it is a good chlorinizing agent; and as it is easily prepared, it may sometimes replace chloride of phosphorus, over which it has the advantage of not disengaging noxious vapours.

Wilson has described[†] a method for determining the hardness

[*] *Comptes Rendus,* October 7, 1861.
[†] Liebig's *Annalen,* September 1861.

of water, which is·a modification of Clark's original method. He uses a solution of sulphate of lime prepared by dissolving 1 part CaO SO³, 2HO in 2543 parts of water. This corresponds to Clark's standard solution of 16 parts CaO CO² in 70,000 parts of water.

The solution of soap was prepared according to Faisst's method of dissolving 30 grms. of·soda oil-soap in alcohol of 56° F., and diluting this solution so that 32 cubic centims. were exactly enough to produce, when shaken with 100 cubic centims. of the normal gypsum solution, a froth which remained for five minutes. By adding 4 cubic centims. of a cold saturated solution of carbonate of soda the reaction is made more regular, inasmuch as it changes all lime compounds into the carbonate which remains dissolved.

The experiments were made in the following manner :—By adding to the normal gypsum solution of 16 degrees hardness corresponding quantities of distilled water, sixteen solutions were prepared of from 1 to 16 degrees hardness. Of these solutions 100 cubic centims. were placed in a stoppered glass cylinder of 400 cubic centims. capacity, with 4 cubic centims. of the cold saturated solution of carbonate of soda, ·and solution of soap added from a burette until, on agitation, a light froth was formed. The solution was then added very gradually with continual agitation, until after the addition of the last drop a froth was formed lasting five minutes. The experimental results showed that the use of every 2 cubic centims. of solution of soap corresponds to 1 degree of hardness.

In order to test the hardness of water, 100 cubic centims. are measured off, 4 cubic centims. of saturated solution of carbonate of soda added until a froth is formed which remains standing five minutes. The number of cubic centims. of soap-solution divided by 2, gives the corresponding degree of hardness.

This method is not applicable to waters of more than 16 degrees hardness. With water of 20 degrees hardness, a precipitate of carbonate of lime is formed on the addition of carbonate of soda. Such waters must be diluted to a proper extent by the addition of distilled water.

In order to ascertain whether salts of magnesia exhibit the same deportment, a solution of 1 part sulphate of magnesia (MgO SO³+7HO) was made in 1778 parts of water, which corresponds to a gypsum solution of 16 degrees hardness. By corresponding dilution with distilled water, solutions of 1, 4, 8, 12 degrees hardness were obtained, ·and these were estimated, after the addition of carbonate of soda, by means of soap-solution. The same results were obtained as with solutions of lime.

Wilson found, as Faisst had previously done, that mixtures of

LXVII. *On the Changes in the Induced Current by the employment of different Resistances.* By G. MAGNUS[*].

THE remarkably great power of hydrogen to conduct heat which I had observed, induced me to compare the electrical conducting power of this gas with that of other gases. I encountered difficulties, however, which at last compelled me to believe that under certain hitherto unobserved circumstances opposite currents are formed, and that by means of these the irregularities of the deflection of the magnetic needle which I noticed were brought about. It was therefore necessary to investigate the conditions under which these currents were originated.

Poggendorff has shown[†] that if in the wire that completes the circuit of an induction apparatus in which there is an "electric egg," and in which only currents of a given direction are circulating, a Leyden jar be introduced, both polar wires of the "electric egg" are covered with blue light. Also the magnetic needle of a galvanometer placed in the current, though previously undergoing deflection, is now no longer turned aside, from which he inferred that by the introduction of the Leyden jar opposite currents were established. Since then the presence of blue light at the two poles of the "electric egg" has almost always been looked upon as the sign of the existence of opposite currents, and especially so, seeing that Riess[‡] had even before this produced these appearances by opposite currents quickly following one another. It is certainly possible that blue light at both poles may not in all cases be an indubitable sign of the presence of opposite currents; still it is difficult to imagine that these appearances should have another cause. Nevertheless I shall not examine any further into this cause. To understand what follows, I emphatically remark that where the expression opposite currents is used, nothing more is signified than the presence of negative light at both polar wires. Dr. Paalzow[§], in a research which appeared a short time ago "On the different ways of discharging the Leyden Battery, and on the Direction of Principal and Secondary Currents," used a similar phenomenon as a test. He, however, employed the so-called Geissler tubes, and observed them between the poles of an active electromagnet. I have made use of short tubes, 75 to 150 millims. in length, and 5 to 15 millims. in diameter, which were sealed after the air that they contained had been rarefied to from 4

[*] Translated from the *Sitzungsberichte der Akademie der Wissenschaften zu Berlin*, June 6, 1861.

[†] Poggendorff's *Annalen*, vol. xciv. p. 328.
[‡] Ibid. vol. xci. p. 291.　　　　　[§] Ibid. vol. cxii. p. 567.

to 6 millims. pressure by means of the air-pump. Attached to their platinum wires and melted with them into the glass, are aluminium wires, the points of which are from 6 to 40 millims. apart. If wires are used made of platinum only, the tubes very soon become covered on their interior surface with a black coating which makes them almost opake. This is not the case when aluminium is employed; hence Geissler* has used this metal in the preparation of his tubes for a long time. Such tubes, when made use of for investigating direction, I shall call test-tubes.

For the experiments, induced currents only were used. For this purpose two induction-apparatus were at my disposal, made by Ruhmkorff of Paris,—a smaller and older one, the dimensions of which may be considered as known, and a larger one, finished only a few months ago, the induction-wire of which has a length of 40,000 metres, and (without taking into account the silk with which it is covered) a diameter of 0·13 of a millim.

For both instruments a battery of two of Bunsen's elements was employed, with which the large coil furnished in the open air a spark of from 3 to 4 centims. in length. If it were set in action by a large battery, a spark was obtained which had a length of 39 centims. The apparatus, however, could not be employed of such a strength for the experiments to be mentioned presently.

Besides the test-tube, another tube was used, in which were two platinum wires 1 millim. in thickness and rounded at the ends, which wires, by means of a stuffing-box, could be placed at any desired distance from one another. In order to rarefy the air in this tube, it was attached to the air-pump. It is therefore distinguished from the electric egg by its being narrower, as well as longer, and admitting of a greater removal of the wires. For the sake of distinction I shall call this tube the air-tube.

If this air-tube, together with the test-tube, be inserted in

* It has often been maintained that particles of platinum are thrown over from the negative to the positive wire. This seems to me to be without foundation in the case of induced currents; for if the discharge takes place through such a tube as has already been described, during a long time and always in the same direction, it becomes covered with a black coating only in that part of the tube where the negative wire is, whilst in the neighbourhood of the positive wire no deposit is perceived even after a much longer time. I believe, therefore, that the black coating originates thus:—the platinum of the negative wire is either volatilized or thrown away from it, but not exactly towards the positive wire. For if the tube contain aluminium wires which are so short that the negative light extends over a part of the platinum wire to which the aluminium is attached, then the black coating is only formed in the neighbourhood of the platinum which is entirely removed from the positive wire.

the circuit of the induced current of either of the two induction-apparatus, and if by a certain rarefaction and a certain separation of the ends of the two wires only single currents make their appearance, then opposite currents are always formed if the ends of the wires are separated to such an extent that the electricity no longer passes between them in a luminous line, but spreads itself out in a brush-like form. On a further separation of the wires, opposite currents in every case show themselves in the test-tube. Instead of moving the wires, the same result can be obtained by gradually increasing the density of the air in the tube; in this case also, as soon as the brush-like discharge commences in the tube, the currents begin to be in opposite directions.

I believe it must be inferred from this that an increase of resistance gives rise to the opposite currents; and I have therefore employed the resistance of liquid and solid conductors instead of that of air. For this purpose the air-tube was replaced by a glass tube 1 metre in length and 3 millims. in diameter, in which two platinum wires could be approached towards, or separated from one another at pleasure. If this tube be filled with a saline solution, even if it contained only 0·25 per cent. of sulphate of potash, and if the wires were 900 millims. apart, it was impossible to obtain opposite currents. When, however, the tube contained pure water, the result was like that obtained by employing the air-tube. By a certain removal of the wires, only single currents were produced, whereas by a further separation they were opposite. It is likewise possible to produce opposite currents by means of the resistance of metallic conductors; it only needs for this purpose (provided, in addition to the test-tube, no air- or water-tube is inserted) the spiral of the large induction-coil, 40,000 metres in length, as a means of resistance; for, on the production of currents by the small induction-apparatus, they then make their appearance with great distinctness.

Further, if the resistance is increased in other ways, negative light appears on both wires. If the sparks from the large induction-apparatus are allowed to traverse the air and then a test-tube is inserted into the conducting wire, as long as the spark traverses the air vigorously, negative light is seen at only one pole of the test-tube; but if the spark goes through the air with a hissing noise, negative light appears at both poles.

We likewise obtain, by the introduction of a thin plate of mica into the circuit, which latter, with the exception of the test-tube, consists of nothing but metallic conductors, negative light on both wires. The same effect is produced, as Poggendorff*

* Poggendorff's *Annalen*, vol. xciv. p. 326.

has already shown, when a Leyden jar is introduced into the circuit.

If, instead of introducing the test-tube into the induction wire, we attach it to one end of the same and conduct the other to the earth, we likewise obtain opposite currents; or, to explain myself more cautiously, negative light appears at both wires.

If we melt into a small tube, which contains very rarefied air, only one wire and then attach it to one end of the induction-coil whilst the other is in contact with the earth, we obtain, if the tube be freely suspended in air, a luminous appearance on the above wire, which appearance is always of negative light; the tube may be in connexion with either end of the induction-wire; or, supposing it to remain at the same end, the current may traverse the wire in either direction. The intensity of the light is increased if we approach the tube with a conductor from the outside.

In how peculiar a manner glass influences the discharge, is seen from the following observation. If the air-tube be introduced into the wire which completes the circuit, and if its wires be so far separated that only single currents circulate, whilst on their being further removed from each other opposite currents arise, the passage from one wire to the other ceases as soon as the tube, thus arranged, is grasped with the hand; whilst in the test-tube that also forms a part of the circuit, opposite currents are immediately perceived. At the same time we observe (if not always, it is frequently the case) that the electricity in the air-tube passes to the glass. This phenomenon likewise makes its appearance in the electric egg; it is necessary, however, as this instrument is much wider than the tubes, to make its whole circumference a conductor by means of a strip of tinfoil. When we remove either the hand or the tinfoil, some time usually elapses before the passage from one wire to another is re-established.

From what has been before cited it may be inferred, and experience conclusively confirms it, that if the distance of the polar wires in the tubes, filled with water or air, be so chosen that on the employment of the large induction-coil single currents are still formed, then on replacing the large apparatus by the small one, opposite currents make their appearance*. The resistance

---

* For both apparatus one and the same contact-breaker was always used, the one which Ruhmkorff makes for his large induction-apparatus. With it the breaking of the contact is effected by the separation of a platinum wire from amalgam, the separation being, on Neef's principle, brought about by a small special electro-magnet, whose magnetism is induced by a single Daniell's element.

The breaking of the contact by means of Neef's hammer, already mentioned, whereby a point separates from a plate, diminishes the strength of

of the air-tube is too great for the intensity of the current which this apparatus produces; consequently we also find that the discharge no longer takes place in the shape of a bright luminous line, but in a brush-like form.

Not only are opposite currents produced if the resistance be too great in proportion to the intensity of the current, but also if the same be too small in relation to the discharge.

If we choose such a separation of the polar wires in the air-tube that the employment of the small induction-apparatus produces single currents, and if we then exchange the above mentioned for the large coil, opposite currents are generated.

We can obtain a similar result with one and the same induction-apparatus. If we attach the two ends of the spiral of the large induction-apparatus to the test-tube, and likewise insert the air-tube, in which the air is rarefied as much as possible, we see, when the polar wires of the latter are brought sufficiently near together, that both are covered with intense negative light. Let these wires be separated from one another, and it will be found that the negative light on the positive wire will gradually decrease, and increase on the negative wire, until the positive wire is entirely devoid of light. Could the wires be separated sufficiently, we should again have opposite currents; but the tube is not long enough for this. This result is obtained, however, by gradually admitting air into the tube and thereby increasing the resistance.

The idea suggests itself that these opposite currents, formed under so trifling a resistance, may have their origin therein—that not only by the breaking, but also by the establishing of the circuit, a current may be induced. It is well known that Poggendorff* has shown that, if the ends of an induction-coil are connected by means of a metallic wire or a liquid which is a good conductor, induced currents are formed on completing as well as on

---

the induced current considerably more than the hammer used by Ruhmkorff for his small apparatus does, which hammer by its own weight completes the circuit. The latter has, however, an irregular action. I hoped by altering it in different ways to obtain a more regular movement, and for this purpose I employed two plates of osmium-iridium in order to avoid the adhesion, but I obtained no more favourable result. The contact-breaker made for the large induction-coil is in any case preferable. It likewise possesses the advantage, that we can accelerate its working at pleasure by shortening the pendulum attached to it. It is not, however, completely regular in its motion.

A short time ago Riess constructed an apparatus by means of which the breaking of the contact was brought about by clockwork—one of Mälzel's metronomes. It is possible that it will have a more regular action in consequence.

* Poggendorff's *Annalen*, vol. xciv. p. 309.

breaking the circuit, which circulate alternately to and fro. Since then, Gassiot* has called attention to the fact that we obtain a luminous appearance in tubes prepared by his method, on establishing the primary current, provided we use ten or more elements for the production of this current.

It was probable, therefore, that, if a test-tube were inserted which contained only a short stratum of very rarefied air, on establishing the principal current, generated in this case by two of Bunsen's elements, an induced current would be formed. This was found to be the case; for if the current was established by dipping the platinum wire of the contact-breaker only once with the hand into the amalgam, a luminous appearance was obtained in the test-tube, which, however, was considerably weaker than that produced by breaking the circuit. The opposite currents, observed with a very trifling resistance, depend therefore partly on the induced current which originates on completing the circuit. Still, I believe them to depend only *partly* thereon; for the current generated by once completing the circuit and then breaking it, also produces negative light at both polar wires. It might certainly be affirmed that we could produce no single breaking of the contact, that there is closing and breaking in succession; still there remains the remarkable fact, that by a single breaking of the contact the luminous appearance in the test-tube was always the same, whether the separation was produced quickly, by withdrawing a platinum point from the amalgam, or slowly, by disconnecting two copper surfaces like those in the rheotrope.

The following observation likewise shows the probability of the production of opposite currents in the induction-wire by a single breaking of the circuit.

It has just been mentioned that, if we choose such a separation of the polar wires in the air-tube that the employment of the small induction-apparatus produces single currents, and if we then replace this apparatus by the large coil, opposite currents are obtained. If, whilst the small induction-coil is in action, we first of all observe in the well-rarefied air-tube the negative wire, it appears to be covered to a considerable length with bluish light, whilst the positive, on the other hand, is entirely non-luminous. If we then employ the large induction-apparatus, a much smaller part of the negative wire is blue, whereas now a portion of the positive wire shines with this colour; the wires comport themselves just in this way if the circuit be broken once. It can scarcely be assumed that by such a single breaking of the contact a re-establishment of the circuit takes place which generates as strong a current as that which arises by the regular closing of the circuit. If, therefore, it is

* Phil. Mag. vol. xvi. p. 307.

not yet proved that opposite currents are produced by sufficiently weak resistance, still at the very least it is highly probable that such is the case.

Moreover, both Dr. Feddersen* and Dr. Paalzow have found
that when the Leyden battery is discharged, opposite currents
make their appearance if the resistance be trifling.

We may therefore regard as proved that induced currents are
only single with a certain amount of resistance. Let the resistance exceed a fixed limit, and they are opposite; let it likewise
sink below another certain limit, and they are also opposite.
These boundaries vary according to the intensity of the current.

On the Changes of Colour of Electric Light.

In the test-tubes that I have employed, the negative light,
which in rarefied atmospheric air is generally of an intense blue,
is almost white; and in like manner the light extending from
the positive pole to the dark intervening space was white, though
it is usually red. I have endeavoured to find out the cause of
these variations of colour.

When a newly-made tube of the prescribed kind is used, the
negative light is at first blue, and the space between the wires is
filled with red light; but soon afterwards both become brighter.
The space between the two wires becomes brown, and finally
white, and in the same way the negative light becomes entirely
whitish. When this change has once been effected, the colour
in the hermetically-sealed tube remains unchanged. If, however,
we use a tube which can be opened, and consequently can have
its air renewed, the negative light is at first blue and the intervening space red, but immediately afterwards both become white
again.

This change cannot depend on the union of oxygen with aluminium; for in nitrogen gas, which in this case would be
left, the colour of the electric light is very similar to what it
is in atmospheric air. The appearance is most like that of the
electric light in carbonic acid or hydrogen; but as neither of
these gases was present, it occurred to me that perhaps the aluminium, during its preparation, might have come in contact with
some foreign substance (some greasy matter for instance), and
that by this means the phenomenon was brought about. Two
aluminium wires, which were cut out of rolled plate, were consequently purified by scraping as much as possible, and without
being touched by the fingers were fused into the tube. Under
these circumstances the wire retained the light unchanged from
what it was at the first moment, that is, continually blue at the
negative wire, and red in the intervening space.

* Poggendorff's *Annalen*, vol. cxii. p. 452.

After the supposition, that the change in colour arose from the presence of a foreign substance, had thus been confirmed, I found that in so narrow a tube even the smallest quantity of fatty matter on the negative wire was sufficient to make the light white. Mere contact with the fingers is often sufficient for this purpose; and, indeed, not only by the employment of wires of aluminium, but likewise of copper, brass, platinum, and probably every other metal which is not volatile at the temperature in question, the same results are obtained. On the positive wire fat produces little or no effect; it may either be placed on its point, or at some distance from it.

Tallow, fatty oils, stearic acid, and wax behave all in a similar way. When we place some of either of the above on the negative wire, the greasy part appears red at first, whilst the rest of the wire shines with blue light. Immediately afterwards the spot is surrounded with a mantle of reddish light which gradually disappears. The blue light on the rest of the wire simultaneously turns white, and the red light between the two wires likewise changes to brown and white. It is probable that the fat was decomposed, but through the quantity being so small the decomposition could not be proved.

LXVIII. *On the Measurement of the Intensity of Electric Currents by means of a Tangent-galvanometer or a Multiplier.* By CH. V. ZENGER, *Professor of Physics, and Member of the Physical Institute*[*].

MR. G. JOHNSTONE STONEY published in this Journal (February 1858) a formula of correction for the length of the needle and for the derangement of its point of suspension. Mr. Stoney in his paper has mentioned the researches of MM. Gaugain and Bosscha, but he does not notice my researches on the corrections to be applied to tangent-galvanometers and to multipliers, although I published them three years ago[+], and although they were discussed in Liebig and Kopp's *Jahresbericht.* Mr. Johnstone Stoney examines two cases, viz. that of a common galvanometer, and that of Gaugain's arrangement; he found in the first case the formula of correction,

$$i = \mathrm{K} \tan \theta \left\{ 1 + \frac{15}{4} \lambda^2 \sin^2 \theta \right\},$$

K being a constant so long as the same needle is used, $\theta =$ the

* Communicated by the Author.
+ Proceedings of the Imperial Academy of Vienna, vol. xvii. April 19; and vol. xviii. November 2, 1855.

angle of deflection from the magnetic meridian, and $2\lambda =$ the length of the needle.

Though this equation is very similar to my formula of correction, yet both are not identical, and do not afford the same approximation to real intensity. As M. Gaugain's arrangement considerably lessens the sensibility of tangent-galvanometers, and as it cannot be applied to the construction of multipliers, it may perhaps be of use and interest to call to mind the formula of correction which I gave in 1855.

Conceive a circular or elliptical band of metal to be placed in the plane of the magnetic meridian, together with a magnetic needle, the centre of which coincides with the centre of the band; imagine now the action of an element A of an electric current to be p at the distance 1, and p' at a distance δ; then

$$p' = pf(\delta),$$

and the total action of the current

$$\left. \begin{array}{l} S = p' + p'' + p''' + \ldots + p_n = p\,\{f(\delta) + f(\delta') + f(\delta'') + \ldots f\delta_n\}, \\[2mm] S = pf(\delta)\left\{1 + \dfrac{f(\delta')}{f\delta} + \dfrac{f(\delta'')}{f(\delta)} + \ldots + \dfrac{f(\delta_n)}{f(\delta)}\right\} = pf(\delta)\Sigma f(\delta), \end{array} \right\} \quad (1)$$

$\Sigma f'(\delta)$ being a constant as long as the needle does not deviate from the magnetic meridian. The action of two different currents on the same galvanometer would be

$$S : S' = pf(\delta)\Sigma f'(\delta) : p'f(\delta)\Sigma f'(\delta) = p : p',$$

if the needle remained in the plane of the band.

The poles of the needle being deviated, the distances $\delta \ldots \delta_n$ become increased, and the magnetic action of the current decreases at the rate

$$\left. \begin{array}{l} S : S' = AN'^2 : AN^2, \\[2mm] S : S' = \dfrac{1 + 4\lambda(\delta + \lambda)}{\delta^2}\sin^2 \tfrac{1}{2}\theta : 1, \end{array} \right\} \quad \cdot \cdot \quad (2)$$

θ being the angle of deviation from the magnetic meridian, $2\lambda =$ the length of the needle, and $\delta =$ the distance AN. We find $AN + NO = AO$, or $\delta + \lambda = a$, $2a$ being the axis of the circular or elliptic band, and

$$S' = \frac{S}{1 + \dfrac{4a\lambda}{(a-\lambda)^2}\sin^2\tfrac{1}{2}\theta},$$

The constant $c = \dfrac{4a\lambda}{(a-\lambda)^2}$ rapidly increases when a nearly equals λ;

it is equal to 1 when $\frac{a}{\lambda} = 5\cdot828426$, that is to say, when the length of the needle is nearly one-sixth the length of the axis a. It then rapidly decreases; and as it has to be multiplied with $\sin^2\frac{1}{2}\theta$, it exerts but little influence on the measurements by means of a tangent-galvanometer, the needle of which has usually only one-seventh or one-eighth the length of the axis a. The needle of multipliers being very long, the influence of the term $c\sin^2\frac{1}{2}\theta$ becomes increased in such a manner as to prevent the possibility of measuring according to the law of tangents. For that reason I proposed to use multipliers consisting of an elliptic metal band or coil of copper wire with a single needle (made astatic by means of a magnetic bar approached to it with the same pole); the needle may then be taken of any length, and it may be made perfectly astatic if required. The needle being very long, c becomes so increased that it can be omitted in the equation

$$S : S' = 1 + c\sin^2\tfrac{1}{2}\theta : 1 + c\sin^2\tfrac{1}{2}\theta' = \frac{1}{c} + \sin^2\tfrac{1}{2}\theta : \frac{1}{c} + \sin^2\tfrac{1}{2}\theta'$$

$$S : S' = \sin^2\tfrac{1}{2}\theta : \sin^2\tfrac{1}{2}\theta'.$$

We find, in conformity with the law of tangents,

$$S' = H\tan\theta \text{ and } S = H\tan\theta(1 + c\sin^2\tfrac{1}{2}\theta),$$

$$S : S' = (1 + c\sin^2\tfrac{1}{2}\theta)\tan\theta : (1 + c\sin^2\tfrac{1}{2}\theta')\tan\theta'.$$

The needle being very long, we find with sufficient approximation,

$$S : S' = \sin^2\tfrac{1}{2}\theta\tan\theta : \sin^2\tfrac{1}{2}\theta'\tan\theta'.$$

To convey a clear idea of the accuracy to be attained by the formula of Mr. Johnstone Stoney and mine, I applied each of them to a series of observations made by means of a tangent-galvanometer, the needle of which had (with regard to the diameter of the circular metal band of 202·5 millims.) the enormous length of 190 millims. I observed the intensity of currents of a constant electromotive apparatus when 0·1, 0·2, and ultimately the whole length of the metal plates was immersed.

Immer-sion.	Deflections.		Average.	Immer-sion.	Deflections.		Average.
	North pole.	South pole.			North pole.	South pole.	
0·1	15·0	195·5	15 15	0·6	25·2	205·4	25 18
0·2	17·2	197·5	17 21	0·7	26·2	206·7	26 41
0·3	20·0	200·4	20 12	0·8	27·5	208·0	27 45
0·4	22·5	202·5	22 39	0·9	28·5	209·0	28 45
0·5	24·0	204·5	24 15	1·0	29·8	210·2	29 54

Intensity of the compared currents.		tan θ. tan θ'	Formulæ of correction.		Errors committed, in per cents.	
			Stoney.	Zenger.	Stoney.	Zenger.
					per cent.	per cent.
1 : 2	0·50000	0·86029	0·67236	0·58750	34·5	11·8
1 : 3	0·33333	0·72750	0·38854	0·38273	16·6	14·8
1 : 4	0·25000	0·65335	0·31520	0·27075	26·1	8·3
1 : 5	0·20000	0·60523	0·25120	0·21994	25·6	10·0
1 : 6	0·16667	0·57677	0·22136	0·19319	32·8	15·9
1 : 7	0·14286	0·54208	0·18846	0·16365	28·1	14·6
1 : 8	0·12500	0·51819	0·16765	0·14522	34·1	16·2
1 : 9	0·11111	0·49695	0·15074	0·13010	35·7	17·1
1 : 10	0·10000	0·47412	0·13394	0·11482	33·9	·14·8

Average .... 29·69 : 13·72

2·165 : 1

The errors arising from the use of Mr. Johnstone Stoney's formula considerably exceed those arising from the use of my own formula.   No doubt these numbers prove the advantages of this formula, and they suffice to show distinctly that, in conducting investigations in which accuracy is a point of importance, Mr. Johnstone Stoney's formula cannot be used.

Vienna, January 8, 1862.

LXIX. *On the Circularity of the Sun's Disc.* By G. B. AIRY, *Esq., Astronomer Royal*[*].

IT has been proposed lately to prepare an apparatus for the purpose of examining whether the Sun's disc is really circular, and in particular for ascertaining whether the diameters nearly perpendicular to the ecliptic are equal to those nearly parallel to the ecliptic.   I would not by any means discourage the trial of such apparatus; but I would unhesitatingly express my opinion that the result of the trial would be to show whether the apparatus is or is not trustworthy, and not to give any new information regarding the measure of the Sun's diameters in any degree comparable to that which we already possess.

Perhaps few persons except professional astronomers are aware of the enormous amount of evidence which already exists in reference to the values of the Sun's diameters, and of the way in which this evidence is growing every day in the ordinary routine of meridional observations. To make this fully understood, I will here explain what is prepared in the Nautical Almanac, what is observed at the Royal Observatory, how the observations are

[*] From the Monthly Notices of the Royal Astronomical Society, January 10, 1862.

reduced, and how the comparison of the reduced observations with the numbers of the Nautical Almanac bears upon the subject now before us.

For the calculations of the Nautical Almanac, an assumption is made as to the numerical value of the Sun's diameter as seen when the Earth is at its mean distance from the Sun. It matters not whether this assumed diameter is or is not correct, provided that it be used consistently in all the calculations of each year; and it matters not whether it be or be not changed from year to year, provided that each volume contain a statement of the assumed diameter which has actually been used in the calculations of that volume. Thus the assumed value of Sun's diameter, as seen at Earth's mean distance, in the Nautical Almanacs from 1836 to 1852 was 32′ 1″·80; that in the subsequent Nautical Almanacs is 32′ 3″·64.

With the diameter thus assumed, two sets of numbers are computed in the Nautical Almanac. One is the apparent diameter (or semidiameter) of the Sun at noon on every day; this is found by merely altering the assumed diameter in the inverse proportion of the Earth's varying distance from the Sun. The other is the duration of passage of the Sun's diameter across the meridian, or the measure of the sidereal time which elapses between the passage of the Sun's western limb and its eastern limb; this is found from the apparent diameter of the day, by introducing the consideration of the Sun's declination and of the Sun's motion in right ascension. And these numbers being prepared, it is evident that we have elements which correspond very closely with facts that may be observed, the elements being essentially based on the supposition that the Sun's disc is circular.

Corresponding to these two classes of computed elements, we have two classes of facts observed at the Royal Observatory and at other observatories. One is the zenith distance of the Sun's upper limb and that of the Sun's lower limb. When each of these is corrected separately for refraction and parallax, the true results of geocentric observation are obtained; and the difference between them gives the observed *vertical* diameter of the Sun on the day of observation. The other is the sidereal time shown by the transit-clock at the instant of transit of the Sun's western limb, and that at the transit of the Sun's eastern limb; the difference between these gives the observed duration of passage of the Sun's *horizontal* diameter across the meridian on the day of observation.

Now if we compare each of these numbers separately (namely, the observed vertical diameter and the observed duration of passage of horizontal diameter) with the corresponding numbers in the Nautical Almanac, and if we omit consideration of chance errors of observation, the effect of which may be supposed to be nearly eliminated in the mean of many observations, the following results ought to hold:—If the Sun's disc is really circular, and if the Nautical-Almanac assumed diameter at mean distance is correct, then the observed vertical diameter will agree with the Nautical-Almanac diameter for the day, and the observed duration of passage will agree with that of the Nautical Almanac. If the Sun's disc is really circular, but the assumed diameter incorrect, then neither of the compared measures will agree with the corresponding computation of the Nautical Almanac; each discordance (one of vertical diameter, the other of duration of passage of horizontal diameter) will indicate a numerical value of correction to be applied to the assumed diameter; but the two numerical values will absolutely agree. But if the Sun's disc is not really circular, then it is impossible that the comparison of observed vertical diameters on the one hand, and of observed durations of passage of horizontal diameters on the other hand, with elements computed on the supposition that the Sun is circular, can indicate the same correction to the assumed semi-diameter.

All that is necessary, therefore, for ascertaining whether the Sun's horizontal diameter and the Sun's vertical diameter are equal, is every day to compare the Sun's observed vertical diameter with the Nautical-Almanac diameter, and the observed duration of passage of Sun's horizontal diameter with the Nautical-Almanac duration, and to infer separately from these the correction to be made to the Nautical-Almanac assumed diameter. If the two results agree, the horizontal and vertical diameters are equal.

Now these comparisons are made every day in the routine of the Royal Observatory; and their results will be found in one of the late sections of each volume of the printed 'Greenwich Observations,' as well as in the more extensively distributed 'Results of the Greenwich Observations,' which contain that section; and the means of the numbers for each year are given in the Introduction to each volume. By extracting these numbers, the following Table is formed. I have thought it necessary to divide the Table into three parts, distinguished by the following circumstances:—From 1836 to 1850 the 4-inch telescope (I believe Dollond's) of the Mural Circle was used for the vertical diameters, and the 5-inch telescope (Dollond's) of the Transit for the horizontal passages; the diameter used in the computations of the Nautical Almanac was 32′ 1″·80. Through 1851 and 1852 the 8-inch telescope (Simms's) of the Transit Circle was used for both measures, the Nautical-Almanac assumed diameter being still 32′ 1″·80. From 1853 to 1860 the telescope of the Transit-Circle was used for both measures, but the Nautical Almanac assumed diameter was 32′ 3″·64.

Apparent Errors of the Duration of Passage of the Sun's Horizontal Diameter, and of the Sun's Vertical Diameter, as computed in the Nautical Almanac.

Year.	No. of obs. of horizontal diameter.	Mean value of N.-A. Ob.	No. of obs. of vertical diameter.	Mean value of N.-A. Ob.
		s		**″**
1836.	104	−0·17	116	−1·50
1837.	92	−0·18	122	−1·85
1838.	108	−0·14	115	−1·53
1839.	103	−0·14	114	−0·94
1840.	104	−0·18	112	−1·60
1841.	102	−0·14	109	−1·32
1842.	116	−0·14	121	−1·54
1843.	99	−0·14	107	−1·84
1844.	102	−0·18	117	−1·50
1845.	100	−0·17	113	−1·32
1846.	92	−0·10	101	−2·09
1847.	89	−0·05	89	−2·98
1848.	94	−0·06	102	−2·06
1849.	103	−0·11	101	−2·40
1850.	94	−0·11	86	−2·18
1851.	87	−0·08	106	−0·71
1852.	103	−0·12	112	−1·39
1853.	78	0·00	86	+0·58
1854.	109	+0·09	111	+1·29
1855.	84	+0·07	93	+0·65
1856.	104	+0·10	109	+1·17
1857.	113	+0·07	122	+0·99
1858.	126	+0·07	132	+0·92
1859.	109	+0·08	125	+0·99
1860.	72	+0·09	72	+1·64

If we take the sums of the numbers of observations and the means of the errors, and if we remark that the mean error of the horizontal diameter in arc may be obtained from the mean error of the duration of passage without sensible error by multiplying by 14, we obtain the following numbers :—

Period.	No. of obs.	Mean error of N.-A. duration of passage.	Mean error of N.-A. horizontal diameter.	No. of obs.	Mean error of N.-A. vertical diameter.
1836 to 1850.	1502	−0·134	−1·88	1625	−1·78
1851 and 1852.	190	−0·100	−1·40	218	−1·05
1853 to 1860.	795	+0·071	+0·99	851	+1·03

If we change the signs of these errors to form corrections, and apply them to the assumed diameter at mean distance of the Earth from the Sun (namely, 32′ 1″·80 to the end of 1852, and 32′ 3″·64 from the beginning of 1853), to produce a corrected diameter of the Sun at mean distance, we form the following Table :—

Instruments employed.	Period.	No. of obs.	Corrected horizontal diameter.	No. of obs.	Corrected vertical diameter.
Transit and Mural Circle }	1836 to 1850	1502	32′ 3″·68	1625	32′ 3″·58
Transit Circle {	{1851 and 1852}	{190}	{32 3·20}	{218}	{32 2·85}
	{1853 to 1860}	{795}	{32 2·65}	{851}	{32 2·61}
	Mean 1851 to 1860	985	32 2·76	1069	32 2·66

Thus the observations with both classes of instruments, in aggregate number 2487 for horizontal diameter, and 2694 for vertical diameter, agree in showing that the horizontal diameter exceeds the vertical diameter by only 0″·1, a quantity smaller than we can answer for in these or in any other methods of observation.

A consideration of the number and excellence of the observations fully supports the view which I have stated in introducing this subject,—that the only result which could be deduced from the trial of new apparatus would be to test the apparatus, but not to add to the certainty of the conclusion as to the equality of diameters.

The diameter adopted now in the Nautical Almanac was inferred from observations made with the Transit and Mural Circle, and therefore agrees very closely with that here deduced from the use of those instruments. That obtained with the Transit Circle is less by 0″·93.

Royal Observatory, Greenwich,
December 28, 1861.

LXX. *Proceedings of Learned Societies.*

ROYAL SOCIETY.

[Continued from p. 485.]

January 10, 1861.—Major-General Sabine, R.A., Treasurer and Vice-President, in the Chair.

THE following communications were read :—
"On the Equation for the Product of the Differences of all but one of the Roots of a given Equation." By Arthur Cayley, Esq., F.R.S.

"Description of a new Optical Instrument called the 'Stereotrope'." By William Thomas Shaw, Esq.

This instrument is an application of the principle of the stereoscope to that class of instruments variously termed thaumatropes, phantascopes, phenakistoscopes, &c., which depend for their results on "persistence of vision." In these instruments, as is well known, an object represented on a revolving disc, in the successive positions it assumes in performing a given evolution, is seen to execute the movement so delineated; in the stereotrope the effect of solidity is superadded, so that the object is perceived as if in motion and with an appearance of relief as in nature. The following is the manner in which I adapt to this purpose the refracting form of the stereoscope.

Having procured eight stereoscopic pictures of an object—of a steam-engine for example—in the successive positions it assumes in completing a revolution, I affix them, in the order in which they were taken, to an octagonal drum, which revolves on a horizontal axis beneath an ordinary lenticular stereoscope and brings them one after another into view. Immediately beneath the lenses, and with its axis situated half an inch from the plane of sight, is fixed a solid cylinder, 4 inches in diameter, capable of being moved freely on its axis. This cylinder, which is called the eye-cylinder, is pierced throughout its entire length (if we except a diaphragm in the centre inserted for obvious reasons) by two apertures, of such a shape, and so situated relatively to each other, that a transverse section of the cylinder shows them as cones, with their apices pointing in opposite directions, and with their axes parallel to, and distant half an inch from, the diameter of the cylinder. Attached to the axis of the eye-cylinder is a pulley, exactly one-fourth the size of a similar pulley affixed to the axis of the picture-drum, with which it is connected by means of an endless band. The eye-cylinder thus making four revolutions to one of the picture-drum, it is evident that the axes of its apertures will respectively coincide with the plane of sight four times in one complete revolution of the instrument, and that, consequently, vision will be permitted eight times, or once for each picture.

The cylinder is so placed that at the time of vision the *large* ends of the apertures are next the eyes, the effect of which is that when the *small* ends pass the eyes, the axes of the apertures, by reason of their eccentricity, do not coincide with the plane of sight, and vision is therefore impossible. If, however, the position of the cylinder be reversed end for end, vision will be possible only when the small ends are next the eyes, and the angle of the aperture will be found to subtend exactly the pencil of rays coming from a picture, which is so placed as to be bisected at right angles by the plane of sight. Hence it follows that, the former arrangement of the cylinder being reverted to, the observer looking along the upper side of the aperture will see a narrow strip extending along the top of the picture; then, moving the cylinder on and looking along the lower side of the aperture, he will see a similar strip at the bottom of the picture; consequently, in the intermediate positions of the aperture, the other parts of the picture will have been projected on the retinæ. The width of these strips is determined by that of the small ends of the apertures, which measure ·125 inch ; and the diameter of the large ends is 1·5 inch, the lenses being distant 9 inches from the pictures. The picture-drum being caused to revolve with the requisite rapidity, the observer will see the steam-engine constantly before him, its position remaining unchanged in respect of space, but its parts will appear to be in motion, and in solid relief, as in the veritable object. The stationary appearance of the pictures, notwithstanding the fact of their being in rapid motion, is brought about by causing their corresponding parts to be seen, respectively, *only* in the same part of space, and *that* for so short a time that while in view they make no sensible progression. As, however, there is an actual progression during the instant of vision, it is needful to take that fact into account—in order that it may be reduced as far as practicable—in regulating the diameter of the eye-cylinder, and of the apertures at their small ends ; and the following are the numerical data involved in the construction of an instrument with the relative proportions given above :—

The circumference of picture-drum $= 22\cdot5$ inches (A).

The circumference of eye-cylinder $= 12$ inches $\times 4$ revolutions $= 48$ inches (B).

The diameter of apertures at large ends $= 1\cdot5$ inch (C).

The diameter of apertures at small ends $= \cdot125$ inch (D).

While the large end is passing the eye, the picture under view progresses $\dfrac{1\cdot5 \ (C)}{48 \ (B)}$ of $22\cdot5$ (A), or ·703 inch.

This amount of progression (·703 in.), if perceived at one and the same instant, would be utterly destructive of all distinctness of definition ; but it is evident that the total movement brought under visual observation at any one moment is $\dfrac{\cdot125 \ (D)}{1\cdot5 \ (C)}$ of ·703 inch, or

·058 inch. This movement must necessarily occasion a corresponding slurring, so to speak, of the images on the retina ; and the fact of such slurring not affecting, to an appreciable extent, the distinctness of definition, seems to be referable to a faculty which the mind has of correcting or disregarding certain discrepant appearances or irregularities in the organ of vision ; as a further illustration of which I may cite the fact, mentioned by Mr. Warren De la Rue in his "Report on Celestial Photography," that the retinal image of a star is, at least under some atmospheric conditions, made up of " a great number of undulating points," which, however, the mind rightly interprets as the effect of the presence before the eye of a single minute object. That this corrective power is, as might be supposed, very limited, may be proved experimentally by this instrument ; for if the small ends be enlarged in only a slight degree, so as to increase this slurring on the retinæ, a very marked diminution in clearness of definition is the immediate result.

That form of the stereotrope, in which Professor Wheatstone's reflecting stereoscope is made use of, and which is better adapted for the exhibition of movements that are not only local but progressive in space, it is needless to describe here, because the principles it involves are essentially the same as those which are stated above.

Jan. 17.—"On the Homologies of the Eye and of its Parts in the Invertebrata." By J. Braxton Hicks, M.D. Lond., F.L.S.

Jan. 24.—"On the Calculus of Symbols, with Applications to the Theory of Differential Equations." By W. H. L. Russell, A.B.

January 31.—Major-General Sabine, R.A., Treasurer and Vice-President in the Chair.

The following communications were read:—

"On Systems of Linear Indeterminate Equations and Congruences." By H. J. Stephen Smith, Esq., M.A.

The present communication relates to the theory of the solutio in positive and negative integral numbers, of systems of linear indeterminate equations, having integral coefficients. In connexion with this theory, a solution is also given of certain problems relating to rectangular matrices, composed of integral numbers, which are of frequent use in the higher arithmetic. Of this kind are the two following:—

1. "Given (in integral numbers) the values of the determinants of any rectangular matrix of given dimensions, to find all the matrices, the constituents of which are integers, and the determinants of which have those given values.

2. "Given any rectangular matrix, the determinants of which have a given number D for their greatest common divisor, to find all the supplementary matrices, which, with the given matrix, form square matrices, of which the determinant is D."

A solution of particular, but still very important cases of these two problems, has been already given by M. Hermite. The method by which in this paper their general solution has been obtained, depends on an elementary, but apparently fertile principle in the theory of indeterminate linear systems; viz. that if *m* be the *index of indeterminateness* of such a system (*i. e.* the excess of the number of indeterminates above the number of really independent equations), it is always possible to assign a set of *m* solutions, such that the determinants of the matrix formed by them shall admit of no common divisor but unity.

Such a set of solutions is termed a *fundamental* set, and possesses the characteristic property, that every other solution of the system can be integrally expressed by means of the solutions contained in it. A set of *independent* solutions is one in which the determinants of the matrix have a finite common divisor, *i. e.* are not all zero. The theory of independent and fundamental sets of solutions in some respects resembles that of independent and fundamental systems of units in Lejeune Dirichlet's celebrated generalization of the solution of the Pellian equation.

By the aid of the same principle of fundamental sets, the follow-

ing criterion is obtained for the resolubility or irresolubility of indeterminate linear systems.

"A linear system is or is not resoluble in integral numbers, according as the greatest common divisor of the determinants of the
matrix of the system is or is not equal to the corresponding greatest
common divisor of its *augmented* matrix."

[The matrix of a linear system of equations is, of course, the
rectangular matrix formed by the coefficients of the indeterminates;
the *augmented* matrix is the matrix derived from that matrix, by
adding to it a vertical column composed of the absolute terms of the
equations.]

A system of linear congruences may, of course, be regarded as a
system of linear indeterminate equations of a particular form; and
the criterion for its resolubility or irresolubility is implicitly contained in that just given for any indeterminate system. But this
criterion may be expressed in a form in which its relation to the
modulus is very clearly seen.

Let

$$A_{i,1}x_1 + A_{i,2}x_2 + \ldots + A_{i,n}x_n \equiv A_{i,n+1}, \bmod M, \, i = 1, 2, 3, \ldots n$$

represent a system of congruences; let us denote by $\nabla_n, \nabla_{n-1}, \ldots$
∇_1, ∇_0, the greatest common divisors of the determinant, first minors,
&c., of the matrix of the system [so that, in fact, ∇_n is the determinant itself, ∇_1 the greatest common divisor of the coefficients
$A_{i,j}$, and $\nabla_0 = 1$]; by $D_n, D_{n-1}, \ldots D_1, D_0$ the corresponding
numbers for the *augmented* matrix; let also δ_i and d^i respectively
represent the greatest common divisors of M with $\dfrac{\nabla_i}{\nabla_{i-1}}$, and of M
with $\dfrac{D_i}{D_{i-1}}$; and put

$$m = d_n \times d_{n-1} \times \ldots \times d_1,$$
$$\mu = \delta_n \times \delta_{n-1} \times \ldots \times \delta_1, \ldots$$

Then the necessary and sufficient condition for the resolubility of
the system is

$$m = \mu;$$

and when this condition is satisfied, the number of solutions is precisely m.

The demonstration of this result (which seems to exhaust the
theory of these systems) is obtained by means of the following
theorem :—

"If $\|A\|$ represent any square matrix in integral numbers, ∇_n its
determinant, $\nabla_{n-1}, \nabla_{n-2}, \ldots \nabla_1, \nabla_0$ the greatest common divisors
of its successive orders of minors, it is always possible to assign two
unit-matrices $\|\alpha\|$ and $\|\beta\|$, of the same dimensions as $\|A\|$, and
satisfying the equation

$$\|A\| = \|\alpha\| \times \begin{Vmatrix} \dfrac{\nabla_n}{\nabla_{n-1}}, & 0, & 0, & \ldots & 0 \\[1.5ex] 0, & \dfrac{\nabla_{n-1}}{\nabla_{n-2}}, & 0, & \ldots & 0 \\[1.5ex] 0, & 0, & \dfrac{\nabla_{n-2}}{\nabla_{n-3}}, & \ldots & 0 \\[1ex] \cdot & \cdot & \cdot & \cdot & \cdot \\[0.5ex] 0, & 0, & \cdot & \cdot & \dfrac{\nabla_1}{\nabla_0} \end{Vmatrix} \times \|\beta\|."$$

The following result (among many which may be deduced from this transformation of a square matrix) admits of frequent applications:—

" If D be the greatest common divisor of the determinants of the matrix of any system of n independent linear equations; of the D^n sets of values (incongruous mod. D) that may be attributed to the absolute terms of the equations, the system is resoluble for D^{n-1}, and irresoluble for $D^{n-1}(D-1)$."

As an example of the use that may be made of this result, it is shown, in conclusion, that it supplies an immediate demonstration of a fundamental principle in the general theory of complex integral numbers, composed of the root of any irreducible equation, having its first coefficient unity, and all its coefficients integral; viz. that the number of incongruous residues, for any modulus, is always represented by the norm of the modulus. A demonstration of this principle has, however, already been given in the 'Quarterly Journal of Pure and Applied Mathematics,' in a paper signed *Lanavicensis*; to whom, therefore, the honour of priority in this inquiry is due.

"Contributions to the Physiology of the Liver—Influence of Alkalies." By Frederick W. Pavy, M.D,

Feb. 7.—The Bakerian Lecture.—On the Absorption and Radiation of Heat by Gases and Vapours, and on the Physical Connexion of Radiation, Absorption, and Conduction. By Professor Tyndall, F.R.S. (This paper was printed in full in the September and October Numbers of this Magazine.)

Feb. 14.—"On Magnetic Storms and Earth-Currents." By Charles V. Walker, Esq., F.R.S., F.R.A.S.

February 21.—Major-General Sabine, R.A., Treasurer and Vice-President, in the Chair.

The following communications were read:—

"On Terephthalic Acid and its derivatives." By Warren De la Rue, Ph.D., F.R.S. &c., and Hugo Müller, Ph.D., F.C.S.

Whilst pursuing our investigation of Burmese naphtha, an abstract of which we have already communicated to the Society, we noticed, among the products of the action of nitric acid on certain liquid hydrocarbons contained in Rangoon tar, an acid of peculiar properties. A very lengthened investigation of this acid and its derivatives we are about bringing to a close; but as the drawing up of this account will necessarily occupy a considerable time, we have thought it desirable to send a short abstract of the chief results we have obtained, with the view of its appearing in the ' Proceedings ' of the Society.

M. Caillot, about fifteen years ago, obtained a peculiar acid among the products of the action of dilute nitric acid on oil of turpentine, to which he gave the name of Terephthalic acid, on account of its generation from oil of turpentine and its isomerism with phthalic acid.   M. Caillot's account of his new acid was so brief and incomplete, that, although we recognized many points of resemblance between it and the acid we had obtained from Burmese naphtha, we were compelled to repeat his experiments on oil of turpentine before we could fix with certainty the identity of the two products.   In the course of these experiments, in which that identity was fully established, we noticed some interesting features in the compounds of the acid and the derivatives we discovered; more especially the relation of terephthalic acid to the well-known aromatic series,—a relation precisely analogous to that which succinic acid bears to the fatty acids.   The close relation which exists between terephthalic acid and benzoic acid is most strikingly manifested in the great number of derivatives which are obtained from the former; indeed, nearly all of the most characteristic benzoyl-compounds have their analogues amongst the derivatives of terephthalic acid.   Terephthalic acid being a bibasic acid, maintains its character throughout its various transformations, and it is this fact which claims particular interest.

Terephthalic acid, as well as its derivatives, forms the first term of a new series of well-characterized bodies, and may, as such, be considered the prototype of a great number of compounds still unknown.

Without dwelling at present on the tedious process by which terephthalic acid is produced, we may mention that it is obtainable from various sources.   We have, for instance, found that it is invariably formed, in a relatively small proportion, when toluylic acid is prepared from cymole; it is also formed when cymole is treated with fuming nitric acid for the purpose of preparing nitrotoluylic acid.   It is important to mention, that whether the cymole be prepared from oil of cumin or from camphor, the result is the same.

Subsequently, we found that insolinic acid, which was described some years ago by Hofmann as a new acid of the formula $C^9 H^6 O^4$, is in reality terephthalic acid.   The formation of this acid from oil of cumin or cuminic aldehyde by the action of chromic acid on these substances, turned out to be the most ready method of preparing terephthalic acid; and the principal part of our experiments were made with terephthalic acid which had been obtained from oil of cumin by this process.

Terephthalic acid being isomeric with phthalic acid, has the formula $C^8 H^6 O^4$ (Carbon=12, Oxygen=16), as already known. When pure, it forms a white opake powder; but if thrown down from a boiling dilute alkaline solution, it may be obtained in a crystalline state. When collected on a filter, these crystals dry in paper-like masses of a silky lustre. Terephthalic acid is not perceptibly soluble in ether, chloroform, acetic acid, water, or the other usual solvents. Concentrated sulphuric acid dissolves it to a considerable extent, especially when warm, without the formation of sulpho-terephthalic acid, and the acid separates unchanged on the addition of water. On heating, terephthalic acid sublimes without previously fusing. The sublimate, which is indistinctly crystalline, has the same composition and properties as the original acid, and therefore, unlike other bibasic acids, terephthalic acid cannot be converted into an anhydride by merely heating it. Terephthalic acid exhibits a remarkable deportment with regard to its salts; for although bibasic, there appear to exist no double salts; and even acid salts are only prepared with the greatest difficulty.

The alkaline terephthalates are all very soluble in water, but are insoluble in alcohol. The potassium, sodium, and ammonium compounds can be obtained in well-crystallized forms. The calcium and barium salts are less soluble than the before-named, and may be obtained in small scaly crystals. The copper salt is a pale blue crystalline powder. The silver and the lead salt occur as curdy precipitates when obtained by double decomposition. The compounds of terephthalic acid with the alcohol radicals possess a particular interest, as they furnish the most direct proof of the bibasic nature of the acid. There exist neutral and acid compounds. The neutral ethers are obtained either by the action of chloride of terephthalyle on the alcohols, or by means of the iodide of the alcohol radicals and terephthalate of silver or of potassium.

The methyl-terephthalic ether, $C^s H^4 (CH^3)^2 O^4$, is the most characteristic compound, and consequently may be used to detect the existence of terephthalic acid in the presence of other acids. It forms beautiful flat prismatic crystals several inches long, which fuse at a temperature above 100° (Cent.), and sublime without decomposition. It is readily soluble in warm alcohol, and slightly soluble in cold alcohol.

The ethyl-terephthalic-ether forms long prismatic crystals resembling urea, and is readily soluble in cold alcohol.

The amyl-terephthalic-ether forms scaly crystals of pearly lustre, is readily soluble in alcohol, and fuses in the temperature of the hand.

Phenyl-terephthalic-ether, a white crystalline substance, fuses at above 100° C.

The acid compounds are generally formed in small quantities, along with the neutral ethers, by the action of the iodide of the alcohol radicals on terephthalate of silver. They are well-defined monobasic acids, and form crystallizable substances soluble in alcohol.

Nitro-terephthalic acid, $C^s H^3 (NO^2) O^4$. This acid is formed by acting with a mixture of nitric and fuming sulphuric acid on terephthalic acid. When crystallized from certain solvents, it forms well-developed prismatic crystals of a faint yellow colour. From water, it deposits in cauliflower-like aggregations.

Nitro-terephthalic acid is readily soluble in warm alcohol and in warm water, and possesses the bibasic character of the terephthalic acid in a much higher degree. It forms well-defined crystallizable acid and neutral salts. The ethers of this acid are likewise crystallizable. They differ, however, from the terephthalic acid ethers by their greater solubility in alcohol and their depressed fusing-point.

Chloride of terephthalyle ($C^8 H^4 O^2 Cl^2$) is obtained, together with oxychloride of phosphorus (hydrochloric acid being evolved), when terephthalic acid is acted upon with pentachloride of phosphorus at a temperature of 40° (Cent.). Chloride of terephthalyle is a solid and beautifully crystalline substance, without odour at the ordinary temperature, but evolving, when heated, a very pungent smell like that of chloride of benzoyle, which it resembles in all its reactions. With the alcohols it forms terephthalic ethers, with ammonia an amide, and with the organic bases compound amides. Terephthalylamide, $C^8 H^6 N^2 O^2$, can only be obtained by acting with chloride of terephthalyle on ammonia; it is a white amorphous substance insoluble in all solvents. Terephthalylamide, when treated with fuming nitric acid, yields nitro-terephthalic amide, $C^8 H^5 (NO^2) N^2 O^2$, which crystallizes in beautiful prisms.

Terephthalamide shows a remarkable resemblance to benzamide when treated with substances capable of abstracting the elements of water. It loses two equivalents of water ($H^2 O$), and is converted into terephthalylnitrile, $C^8 H^4 N^2$. This remarkable substance is best formed by the action of anhydrous phosphoric acid on terephthalamide. It distils over in form of a liquid, which solidifies in the neck of the retort.

Terephthalylnitrile is colourless and without odour, and forms beautiful prismatic crystals. It is insoluble in water, readily soluble in boiling alcohol, less soluble in cold alcohol, and insoluble in benzole. When boiled with caustic alkalies, it is gradually decomposed, ammonia is given off, and terephthalic acid is reproduced.

. It is obvious that terephthalylnitrile, like all similar substances, may be considered as a cyanogen compound, which in this instance would be the cyanide of the bibasic radical phenylene, $C^6 H^4$, which is not yet discovered. If we could succeed in obtaining phenylene, the artificial production of terephthalic acid or an isomeric would probably be attended with little difficulty.

By acting on nitro-terephthalic acid with reducing agents, it undergoes the same change as other nitro-compounds. The product of this reaction is the oxy-terephthalamic acid, or the analogue of the glycocoll of the formula $C^8 H^7 NO^4$. This new member of the glycocolls is a lemon-yellow substance, crystallizing in thin prismatic, and sometimes moss-like forms. It is very slightly soluble in cold water, alcohol, ether, and chloroform. Like other substances of this kind, terephthal-glycocoll combines with bases as well as with acids. The salts formed with the bases are crystalline; they are readily soluble in water and dilute alcohol, yielding colourless solutions of most remarkable fluorescent properties, which have been investigated by Professor Stokes.

The aqueous and alcoholic solution of the pure terephthalic glycocoll shows the same properties. The compounds with acids crystallize well, and if dissolved in a large quantity of water decompose. They do not possess the fluorescence when in their acid solution.

The ether-like compounds of oxy-terephthalamic acid are obtained by acting upon the corresponding ethers of the nitro-terephthalic acid with reducing agents. The methylic ether is a beautiful crystalline substance, readily soluble in warm alcohol, but much less soluble in any of the solvents than nitro-terephthalate of methyle. The ethylic ether crystallizes in large crystals with an appearance resembling those of nitrate of uranium. The solutions of this ether possess the fluorescent property in the highest degree. Oxy-terephthalamate of methyle and ethyle combine with acids and form well-defined salts. Oxy-terephthalamic acid, as well as its ether, are readily acted upon by nitrous acid, this reaction giving rise to a number of new derivatives, which vary in their nature according to the condition in which the reaction takes place.

M. Griess has lately made us acquainted with a new class of remarkable substances which are obtained by the action of nitrous acid on a certain class of nitrogenous bodies. The several derivatives he obtained by this reaction from oxy-benzamic acid have their representatives in the bibasic terephthalyle series, and are obtained with the utmost facility. On acting with nitrous acid upon an aqueous solution of the oxy-terephthalamic acid instead of an alcoholic solution, as is employed in Griess's reaction, this substance is readily decomposed, nitrogen is given off in large quantities, and there gradually separates a whitish substance which is oxy-terephthalic acid, $C^{\circ} II^{\circ} O^{\circ}$. This acid is a substance of great interest, and its preparation offering much less difficulty than the analogous oxy-acids of the aromatic series, it affords an opportunity of studying to a fuller extent the nature of this class of acids, especially as it may be expected that the history of this acid will throw some light on the law of polybasicity. Oxy-terephthalic acid forms beautiful crystalline salts, which are less soluble than the corresponding terephthalates. The neutral ethers are liquid.

The chloride of oxy-terephthalyle is likewise a liquid readily decomposed by water and alcohols.

"Notes on the Generative Organs, and on the Formation of the Egg in the Annulosa."—Part I. By John Lubbock, Esq., F.R.S.

February 28.—Major-General Sabine, R.A., Treasurer and Vice-President, in the Chair.

The following communications were read :—

"Tables of the Weights of the Human Body and the Internal Organs in the Sane and Insane of both Sexes at various Ages." By Robert Boyd, M.D., F.R.C.P.

"On the Electric Conducting Power of Copper and its Alloys." By A. Matthiessen, Ph.D.

The difference in the numerical results obtained by Prof. W. Thomson (Proceedings of Roy. Soc. 1860, x. p. 300), and those by Dr. Holzmann and myself (Phil. Trans. 1860), on the conducting power of copper and its alloys, made it somewhat necessary to reinvestigate the subject, in order to ascertain the cause of these differences. For this purpose Professor Thomson kindly placed at my disposal all his alloys; and in the following Table I will give the results of the analyses and redeterminations of the conducting power of his set. The wires were in some cases very faulty, so that I was obliged to draw them finer; others drew so badly, that the values obtained could not very well agree with those already published. After having measured their resistances, I sent them back to Prof. Thomson for redetermination. Table I. gives the results so obtained, taking the alloy containing 99·75 copper and ·25 silver=100; and Table II. the values found for some specimens of pure copper :—

TABLE I.

Composition according to Messrs. Johnson and Matthey.	Analyses of Alloy.	Specific Conductivity.		Values found by myself.
		Values found by Professor Thomson.		
		Published Values.	Redetermined Values.	
Copper 99·75.. Silver 0·25....	Silver 0·24 p. c. traces of iron Suboxide of copper	} 100 {	100·1 99·9	100·37 at 17°* 99·73 at 17°
Copper 99·87.. Silver 0·13....	Silver 0·13 p. c. traces of iron Suboxide of copper	} 100·7 {	95·8 95·8	95·44 at 17°·8 94·58 at 17°·8
Copper 99·75.. Lead 0·25	Lead 0·2 per cent. traces of iron Suboxide of copper	} 103·9 {	102·7 103·1	102·80 at 17° 102·62 at 17°·6
Copper 99·75.. Tin 0·25......	Tin 0·23 per cent. traces of iron Suboxide of copper	} 94·6 {	100·7 101·0	99·89 at 18° 98·27 at 16°·4
Copper 99·87.. Tin 0·13......	Tin 0·07 per cent. traces of iron. Suboxide of copper	} 96·0 {	97·7 98·5	97·79 at 18° 97·62 at 18°
Copper 99·2 .. Zinc 0·8......	Zinc, with traces of iron, 1·06 per cent.	} 90·2 {	91·3 88·5	94·71 at 15°·4 90·67 at 15°·6
Copper 98·6 .. Zinc 1·4......	Zinc, with traces of iron, 1·47 per cent.	} 74·7 {	81·1 80·1	81·15 at 16°·8 80·13 at 17°·7
Copper 98·2 .. Zinc 1·8......	Zinc, with traces of iron, 1·75 per cent.	} ·· {	77·9 78·5	77·8 at 16°·4 78·0 at 17°
Pure copper ..	Contained suboxide of copper	} 100	98·6	
Copper 99·87.. Lead 0·13	·· ·· ·· ··	} 104·7		

* Compared with a hard-drawn gold-silver wire of equal diameter and length, whose conducting power is equal at 0° C. to 100, these values would be 603·7 and 600·5. (See my paper " On an Alloy which may be used as a Standard of Electrical Resistance," Phil. Mag. Feb. 1861.)

· TABLE II.

Composition according to Messrs. Johnson and Matthey.	Analyses of Alloy.	Specific Conductivity.		Values found by myself.
		Values found by Professor Thomson.		
		Published Values.	Redetermined Values.	
Pure copper electrotype from Messrs. De la Rue	.. ..	.. ..	107 at 9°	107·2 at 10°
Ditto from Messrs. Elkington and Co. ..	⎫ All not fused.	.. ..	107·5 at 12°	105·9 at 10°·5
Ditto from Mr. Matthews	.. ..	.. ..	108·7 at 12°	106·9 at 14°
Ditto, my own..	⎭	.. ..	107·7 at 12°	108·1 at 10°

All the above wires were hard-drawn. On looking at the above, we find that *pure copper conducts better than any of the alloys.*

With regard to the analyses, the quantity of each specimen was so small that they could not be checked by repetition; they, however, approach very closely to the composition assigned to them by Messrs. Johnson and Matthey (with the exception of the suboxide). The traces of iron will be due to the draw-plates. I will now make a few remarks on the above results.

I. That copper containing 0·25 per cent. of silver conducts better than that with 0·13 per cent., may be explained by assuming that the first contains less suboxide than the second; for it is very possible that copper containing silver will not absorb suboxide so readily as the purer metal. It must also be borne in mind that the copper employed for making these alloys was in all probability simply electrotype copper (not fused), and that the suboxide therefore was absorbed during the process of fusing the two metals together. This assumption explains how it is that the alloys contain almost the same amount of impurity as was originally alloyed with the copper; for had the copper employed contained suboxide, we should have expected to have found greater differences in the cases of the tin, lead, and zinc alloys, as some portion of those metals would have been oxidized at the expense of part of the suboxide of copper, and escaped as oxide to the surface of the melted metal.

II. That copper containing 0·25 per cent. tin conducts better than that containing 0·13 per cent., may also be explained by assuming that they absorbed different amounts of suboxide during the process of fusion; for although tin, in presence of suboxide of copper, would be oxidized, yet copper retains the suboxide so tenaciously, that portions will always remain with the copper.

III. The fact that the conducting powers of the alloy of copper containing 0·25 per cent. lead approaches the nearest of those which I analysed to that of pure copper, is, in my opinion, a proof that the alloy is probably a mechanical mixture of copper, traces of lead, and enough suboxide to allow its being drawn into wire, and not a solution of lead in copper; otherwise a much lower conducting power ought to have been found; for, according to my own experiments, it requires twice as many volumes per cent. of lead as of tin to reduce (within certain limits) the conducting power of a metal (bismuth, silver, &c., and copper, for it belongs to the same class) to the same value: thus, to reduce the conducting power of silver to 67, it would require 0·9 volume per cent. of lead, or about 0·4 volume per cent. of tin; to reduce it to 47·6, it would require 1·4 volume per cent. of lead, or 0·7 volume per cent. of tin, &c. (Phil. Trans. 1860). Dr. Holzmann and myself repeatedly tried to draw pure copper alloyed with 0·25 of lead without success; the alloy was perfectly rotten, which also seems to indicate a mechanical mixture.

IV. It is curious that the zinc alloys contained no suboxide.

The reason, therefore, of the difference in our results is simply that Messrs. Johnson and Matthey did not use those precautions in fusing their copper and its alloys which are necessary to ensure good results; for had they taken those precautions to prevent the absorption of oxygen by their copper and its alloys which Dr. Holzmann and myself did, and which are fully described in our paper on the subject (Phil. Trans. 1860), the lead-copper alloys which they supplied to Prof. Thomson would not have been superior in conductive quality to the unalloyed electrotype copper; and he would have been led to the same conclusion as that which Dr. Holzmann and myself arrived at, namely, *that there are no alloys of copper which conduct better than pure copper.* Professor Thomson, in his paper, states that it is his opinion that the differences he observed in the conducting powers of his alloys must depend upon very small admixtures of probably non-metallic impurities. This conclusion is completely borne out by the above, as well as by the investigation carried out by myself in conjunction with Dr. Holzmann.

The results obtained by Prof. Thomson show the marked influence of traces of foreign metals on the conducting power of pure copper,—which is fully confirmed in our research on the same subject. Professor Thomson's best-conducting alloy has a much higher conducting power than those found by some experimenters for electrotype copper; but it must be remembered that in all probability the copper had been previously fused, and therefore contained suboxide of copper. The fact that electrotype copper may be drawn without having been previously fused is, I believe, generally not known; Professor Buff of Giessen first drew my attention to it, and stated that he always obtained high values for the conducting powers of electrotype copper when drawn without previous fusion. I can confirm this statement, having tested a great many specimens, and found the values in all cases nearly the same.

March 7.—Major-General Sabine, R.A., Treasurer and Vice-
President, in the Chair.

The following communications were read :—

"On Combustion in Rarefied Air." By Dr. Edward Frankland,
F.R.S.

In the autumn of 1859, whilst accompanying Dr. Tyndall to the
summit of Mont Blanc, I undertook at his request some experiments
on the effect of atmospheric pressure upon the amount of combustible
matter consumed by a common candle. I found that, taking the
average of five experiments, a stearine candle diminished in weight
9·4 grammes when burnt for an hour at Chamounix; whilst its ignition for the same length of time on the summit of Mont Blanc,
perfectly protected from currents of air, reduced its weight to the
extent of 9·2 grammes.

This close approximation to the former number under such a
widely different atmospheric pressure, goes far to prove that the
rate of combustion is entirely independent of the density of the atmosphere.

It is impossible to repeat these determinations in a satisfactory
manner with artificially rarefied atmospheres, owing to the heating
of the apparatus which surrounds the candle, and the consequent
guttering and unequal combustion of the latter; but an experiment
in which a sperm candle was burnt first in air under a pressure of
28·7 inches of mercury, and then in air at 9 inches pressure, other
conditions being as similar as possible in the two experiments, the
consumption of sperm was found to be,—

At pressure of 28·7 inches 7·85 grms. of sperm per hour,
　　　　 ,,　　　 9·0 ,,　 9·10　　　 ,,　　　 ,,

thus confirming, for higher degrees of rarefaction, the result previously obtained.

In burning the candles upon the summit of Mont Blanc, I was
much struck by the comparatively small amount of light which they
emitted. The lower and blue portion of the flame, which under
ordinary circumstances scarcely rises to within a quarter of an inch
of the apex of the wick, now extended to the height of $\frac{1}{5}$th of an inch
above the cotton, thus greatly reducing the size of the luminous
portion of the flame.

On returning to England, I repeated the experiments under circumstances which enabled me to ascertain, by photometrical measurements, the extent of this loss of illuminating effect in rarefied air.
The results prove that a great reduction in the illuminating power of
a candle ensues when the candle is transferred from air at the ordinary atmospheric pressure to rarefied air. It was, however, found that,
owing to the circumstances mentioned above, no satisfactory quantitative experiments could be made with candles in artificially rarefied
air, and recourse was therefore had to coal-gas, which, although also
liable to certain disturbing influences, yet yielded results, during an extensive series of experiments, exhibiting sufficient uniformity to render
them worthy of confidence. The gas was in all cases passed through a
governor to secure uniformity of pressure in the delivery tubes. A
single jet of gas was employed as the standard of comparison, and

this was fixed at one end of a Bunsen's photometer, whilst the flame to be submitted to various pressures, and which I will call the experimental flame, was placed at the other. The experimental flame was made to burn a uniform amount of gas, viz. 0·65 cubic foot per hour in all the experiments.

The products of combustion were completely removed, so that the experimental flame, which burnt with perfect steadiness, was always surrounded with pure air, the supply of which was, however, so regulated as to secure a maximum of illuminating effect in each observation.

In all the following series of experiments, the illuminating power given under each pressure is the average of twenty observations, which accord with each other very closely. In each series, the maximum illuminating effect, that is the light given by the experimental flame when burning under the full atmospheric pressure, is assumed to be 100. The following is a summary of the results :—

1st Series.

Pressure of air in inches of mercury.	Illuminating power of experimental flame.
29·9	100·
24·9	75·0
19·9	52·9
14·6	20·2
9·6	5·4
6·6	·9

2nd Series.

Pressure of air in inches of mercury.	Illuminating power of experimental flame.
30·2	100·
28·2	91·4
26·2	80·6
24·2	73·0
22·2	61·4
20·2	47·8
18·2	37·4
16·2	29·4
14·2	19·8
12·2	12·5
10·2	3·6

These numbers indicate that even the natural oscillations of atmospheric pressure must produce a considerable variation in the amount of light emitted by gas-flames, and it was therefore important to determine, by a special series of observations, this variation in luminosity within, or nearly within, the usual fluctuations of the barometrical column. In order to attain greater delicacy in the pressure readings in these experiments, a water-gauge was used, but its indications are translated into inches of mercury in the following tabulated results, each of which represents, as before, the average of twenty observations.

3rd Series.

Pr. of air in in. of mercury.	Illum. power of exp. flame.
30·2	100·
29·2	95·0
28·2	89·7
27·2	84·4

It is thus evident that the combustion of an amount of gas which would give a light equal to 100 candles when the barometer stands at 31 inches, would give a light equal to only 84·4 candles if the barometer fell to 28 inches.

An inspection of all the above results shows that the rarefaction of air, from atmospheric pressure downwards, produces a uniformly diminishing illuminating power until the pressure is reduced to about 14 inches of mercury, below which the diminution of light proceeds at a less rapid rate. The above determinations give approximately 5·1 per cent. as the mean reduction of light for each diminution of 1 inch of mercurial pressure down to 14 inches. The following Table exhibits the actually observed light, compared with that calculated from this constant.

1st Series.

Pressure.	Illuminating power.	
	Observed.	Calculated.
29·9	100·	100·
24·9	75·0	74·5
19·9	52·9	49·0
14·6	20·2	22·0
9·6	5·4	− 3·5
6·6	·9	−18·8

2nd Series.

30·2	100·	100·
28·2	91·4	89·8
26·2	80·6	79·6
24·2	73·0	69·4
22·2	61·4	59·2
20·2	47·8	49·0
18·2	37·4	38·8
16·2	29·4	28·6
14·2	19·8	18·4
12·2	12·5	8·2
10·2	3·6	− 2·0

29·2

89·7

I am now extending this inquiry to pressures exceeding that of the atmosphere, and hope soon to lay before the Society the detailed results of the whole series, together with some observations on the causes of this variation of luminosity.

"On the Porism of the In-and-circumscribed Polygon." By Arthur Cayley, Esq., F.R.S.

"On a New Auxiliary Equation in the Theory of Equations of the Fifth Order." By Arthur Cayley, Esq., F.R.S.

LXXI. *Intelligence and Miscellaneous Articles.*

NOTE ON THE FREEZING OF SALINE SOLUTIONS.
BY M. RUDORFF.

WATER which contains saline substances in solution, freezes, as is well known, at a considerably lower temperature than pure water, and in passing into the solid state carries with it only a very small proportion of the dissolved salts. M. Rudorff proposed to investigate as large a number as possible of saline solutions, in reference to these two phenomena, and to ascertain if they were not subject to definite laws.

By experiments with solutions of common salt, of sulphate of copper, and of bichromate of potash, he first ascertained that the proportion of salt in the ice formed is always far less than that in the mother-liquor. The lamellar structure of the ice led him even to think that the salt which it contained was merely mechanically mixed, and that the ice, properly so called, was quite pure.

He then proceeded to a determination of the exact temperatures at which ice forms in different solutions. He prepared, with each of the salts, a series of solutions containing respectively 1, 2, 3 .... parts of salt to 100 of water. He placed each of them successively in a mixture of snow and salt, and stirred it with a thermometer which indicated twentieths of a degree Centigrade, until it solidified. Under these conditions, the freezing was sudden, simultaneous throughout all the mass, and accompanied by a marked increase of temperature. After this first determination (which was not entirely accurate, in consequence of the change in the composition of the liquor resulting from the instantaneous formation of a large quantity of ice) M. Rudorff made a second experiment, in which, without stirring the solution, he cooled it to about half a degree below the temperature at which the liquid froze in the preceding experiment. He then projected a few flakes of snow into the liquid, and observed that congelation took place accompanied by a slight increase of temperature, and he took the last indication of the thermometer for the freezing-point of the solution investigated. The formation of ice then continued for a long time without change of temperature; and it was only after a considerable quantity of ice had been formed that cooling commenced. If the solution was then taken out of the freezing-mixture and placed in a room at a temperature of 12°, the thermometer rose to the temperature of the freezing-point, and remained almost constant until the whole of the ice was melted.

The salts suitable for these experiments ought to be very soluble at low temperatures, and to exert a marked influence on the freezing-point; they are hence few in number, and in fact are almost confined to the alkaline chlorides and nitrates. In most cases, M. Rudorff found that the lowering of the freezing-point below zero was proportional to the quantity of *anhydrous* salt dissolved in 100 parts of water. The following numbers give the lowering produced by the addition of 1 part of salt to 100 parts of water :—

Sal-ammoniac	0·653
Common salt.............	0·600
Chloride of potassium	0·443
Nitrate of ammonia	0·384
Nitrate of soda...........	0·370
Carbonate of potash	0·317
Nitrate of lime	0·277
Nitrate of potash	0·267

For chloride of calcium, the lowering of temperature appeared to be proportional to the quantity of salt crystallized with six equivalents of water, and not to the quantity of anhydrous salt which the solution contains. Designating by M the quantity of anhydrous salt, by M_1 the quantity of hydrated salt dissolved in 100 parts of water, by T the lowering of temperature, the experiments gave the following results :—

M.	M_1.	T.	$\frac{T}{M}$	$\frac{T}{M_1}$.
2	4·02	0·90	0·450	0·224
4	8·21	1·85	0·462	0·225
8	17·20	3·90	0·487	0·225
14	31·89	7·40	0·528	0·232
18	43·05	10·00	0·555	0·232

We thus see that the value of $\frac{T}{M}$ increases somewhat rapidly with M, while that of $\frac{T}{M_1}$ is pretty constant.

At a temperature of −10°, crystals separate from a concentrated solution of common salt, which contain four equivalents of water for one of salt, and are rapidly destroyed by increase of temperature. This circumstance explains a remarkable result following from M. Rudorff's experiments. So long as the lowering of the freezing-point of a solution of common salt does not exceed 9°, it is proportional to the quantity of anhydrous salt dissolved. When it exceeds 9°, it becomes proportional to the quantity of salt crystallized with four equivalents of water. This will be seen from the

following Table, where the letters have the same meaning as in the
Table relative to chloride of calcium :—

M.	M₁.	T.	$\frac{T}{M}$	$\frac{T}{M_1}$
1	......	0·6	0·600	
2	......	1·2	id.	
4	......	2·4	id.	
6	......	3·6	id.	
8	......	4·8	id.	
10	......	6·0	id.	
12	......	7·2	id.	
14	......	8·4	id.	
15	27·04	9·2	0·613	0·340
16	29·06	9·9	0·619	0·341
17	31·07	10·6	0·623	0·341
18	33·17	11·4	0·633	0·343
19	35·29	12·1	0·637	0·342
20	37·38	12·8	0·640	0·342

Lastly, the lowering of the freezing-point in the case of chloride
of barium is proportional to the quantity of salt crystallized with
two equivalents of water. The coefficient of proportionality is
0·192.—*Proceedings of the Academy of Sciences at Berlin*, April 1861.

***

EXPERIMENTS ON SOME AMALGAMS.   BY J. P. JOULE, LL.D.

The weakness of the affinity which holds the constituents of amalgams in combination seemed to the author to offer the means of
studying the relationship between chemical and mechanical force.
His inquiries were extended to several amalgams, and gave results
of which the following is a summary.

Amalgam of iron was formed by precipitating iron on mercury
electrolytically. The solid amalgam containing the largest quantity
of mercury appeared to be a binary compound. Iron does not appear
to lose any of its magnetic virtue in consequence of its combination
with mercury. Its amalgamation has the effect of making it negative with respect to iron in the electro-chemical series. The affinity
between mercury and iron is so feeble that the amalgam is speedily
decomposed when left undisturbed, and almost immediately when
agitated. The application of a pressure of fifty tons to the square
inch drives out so much mercury as to leave only 30 per cent. of it
in the resulting button.

Amalgam of Copper.—By precipitating copper on mercury electrolytically, a mass of crystals is gradually formed. After a certain
time the crystals begin to get fringed with pink, indicating uncombined copper. In this state the amalgam is found to be nearly a
binary compound. On applying strong pressure to an amalgam containing excess of mercury, the latter is driven off, leaving a hard
mass composed of equivalents of the metals. If, however, the pres-

sure be continued for a long time, the resulting amalgam contains more than one equivalent of copper, indicating a partial decomposition.

The author gave an account of his experiments with amalgams of silver, platinum, lead, zinc, and tin. In the case of the latter amalgam, long-continued pressure drives off nearly the whole of the mercury, indicating in a striking manner the efficacy of mechanical means to overcome feeble chemical affinities.—*From the Proceedings of the Literary and Philosophical Society of Manchester*, No. 8. Session 1861–62.

PRELIMINARY NOTE ON THE PRODUCTION OF VIBRATIONS AND MUSICAL SOUNDS BY ELECTROLYSIS. BY GEORGE GORE, ESQ.

If a large quantity of electricity is made to pass through a suitable good conducting electrolyte into a small surface of pure mercury, and especially if the mercurial surface is in the form of a narrow strip about $\frac{1}{8}$th of an inch wide, strong vibrations occur; and symmetrical crispations of singular beauty, accompanied by definite sounds, are produced at the mutual surfaces of the liquid metal and electrolyte.

In my experiments the crispations and sounds were readily produced by taking a circular pool of mercury from 1 to 3 inches in diameter, surrounded by a ring of mercury about $\frac{1}{8}$th or $\frac{1}{10}$th of an inch wide, both being contained in a circular vessel of glass or gutta percha, covering the liquid metal to a depth of about $\frac{1}{2}$ an inch with a rather strong aqueous solution of cyanide of potassium, connecting the pool of mercury by a platinum wire with the positive pole of a battery capable of forcing a rather large quantity of electricity through the liquid, and connecting the ring of mercury with the negative platinum wire. The ring of mercury immediately became covered with crispations or elevated sharp ridges about $\frac{1}{8}$th of an inch asunder, all radiating towards the centre of the vessel, and a definite or musical sound was produced capable of being heard, on some occasions, at a distance of about 40 or 50 feet. The vibrations and sounds ceased after a short time, but were always reproduced by reversing the direction of the electric current for a short time, and then restoring it to its original direction. The loudness of the sound depends greatly upon the power of the battery; if the battery was too strong the sounds did not occur. The battery I have used consists of 10 pairs of Smee's elements, each silver plate containing about 90 square inches of immersed or acting surface; and I have used with equal success six Grove's batteries, arranged either as 2 or 3 pairs, each platinum plate being 6 inches long and 4 inches wide. If the cyanide solution was too strong, the sounds were altogether prevented.

Being occupied in investigating the conditions and relations of this phenomenon with the intention of submitting a complete account of the results to the notice of the Royal Society, I refrain from stating further particulars on the present occasion.—*From the Proceedings of the Royal Society*, April 11, 1861.

INDEX TO VOL. XXII.

END OF THE TWENTY-SECOND VOLUME.

PRINTED BY TAYLOR AND FRANCIS,
RED LION COURT, FLEET STREET.

ALERE　　FLAMMAM.

Lightning Source UK Ltd.
Milton Keynes UK
UKHW051454020119
334537UK00024B/81/P